D12 $D_x \cos u = -\sin u \cdot D_x u$

D13 $D_x \tan u = \sec^2 u \cdot D_x u$

D14 $D_x \cot u = -\csc^2 u \cdot D_x u$

D15 $D_x \sec u = \sec u \tan u \cdot D_x u$

D16 $D_x \csc u = -\csc u \cot u \cdot D_x u$

D17 $D_x \sin^{-1} u = \dfrac{1}{\sqrt{1-u^2}} \cdot D_x u \quad (-1 < u(x) < 1)$

D18 $D_x \cos^{-1} u = \dfrac{-1}{\sqrt{1-u^2}} \cdot D_x u \quad (-1 < u(x) < 1)$

D19 $D_x \tan^{-1} u = \dfrac{1}{1+u^2} \cdot D_x u$

D20 $D_x \sec^{-1} u = \dfrac{1}{|u|\sqrt{u^2-1}} D_x u, \ |u| > 1$

D21 $D_x \sinh u = \cosh u \, D_x u$

D22 $D_x \cosh u = \sinh u \, D_x u$

D23 $D_x \tanh u = \text{sech}^2 u \, D_x u$

D24 $D_x \coth u = -\text{csch}^2 u \, D_x u$

D25 $D_x \text{sech} \, u = -\text{sech} \, u \tanh u \, D_x u$

D26 $D_x \text{csch} \, u = -\text{csch} \, u \coth u \, D_x u$

D27 $D_x \sinh^{-1} u = \dfrac{1}{\sqrt{1+u^2}} D_x u$

D28 $D_x \cosh^{-1} u = \dfrac{1}{\sqrt{u^2-1}} D_x u, \ u > 1$

D29 $D_x \tanh^{-1} u = \dfrac{1}{1-u^2} D_x u, \ |u| < 1$

D30 $D_x \coth^{-1} u = \dfrac{1}{1-u^2} D_x u, \ |u| > 1$

D31 $D_x \text{sech}^{-1} u = \dfrac{-1}{u\sqrt{1-u^2}} D_x u \quad 0 < u < 1$

D32 $D_x \text{csch}^{-1} u = \dfrac{-1}{u\sqrt{1+u^2}} D_x u, \ u \neq 0$

Calculus
and Analytic Geometry

James E. Shockley

Virginia Polytechnic Institute and State University

SAUNDERS COLLEGE PUBLISHING

Philadelphia New York Chicago
San Francisco Montreal Toronto
London Sydney Tokyo Mexico City
Rio de Janeiro Madrid

Address orders to:
383 Madison Avenue
New York, NY 10017

Address editorial correspondence to:
West Washington Square
Philadelphia, PA 19105

This book was set in Times Roman by Progressive Typographers.
The editors were James Porterfield, Jay Freedman, Sally Kusch and Janet Wright.
The art & design director was Richard L. Moore.
The text design was done by Arlene Putterman.
The cover design was done by Richard L. Moore.
The artwork was drawn by ANCO/BOSTON.
The production manager was Tom O'Connor.
This book was printed by Von Hoffman.

Cover credit: Photograph of the dome of the First Bank of the United States, Philadelphia, PA. ©1981 by William T. Moore, Jr.

CALCULUS and Analytic Geometry   ISBN 0-03-018886-5

©1982 by CBS College Publishing. All rights reserved. Printed in the United States of America.
Library of Congress catalog card number 81-53075.

234   032   98765432

CBS COLLEGE PUBLISHING
Saunders College Publishing
Holt, Rinehart and Winston
The Dryden Press

Preface

This text is designed for a three-semester calculus course for students in engineering, the physical sciences, and the mathematical sciences. I have worked to develop the topics as simply and clearly as possible, and have organized them into sections that are designed to be covered in one class period. The book contains an abundance of exercises — some designed to reinforce the immediate concepts, others to help the student to synthesize concepts and to develop latent powers of mathematical reasoning. Figures illustrate the concepts wherever appropriate. Examples, illustrations, and applications immediately follow the discussion of each topic.

I have also worked to make the text as useful as possible for the student — both in the initial learning period and in the equally important reviewing period. Subheadings and marginal notes help to outline the section, and to locate and relate various concepts. The figures are larger than those in many texts, and are located near their references. Explanatory captions aid the reviewing process — many concepts can be understood directly from the figures. Important results are stated as theorems, not buried in the text.

Specific points of interest to potential users are discussed in the following paragraphs.

Organization The text follows one of the standard organizations for the calculus sequence. The main feature of this plan is that integration is covered as quickly as possible after a fairly thorough treatment of the derivative. Considering the flexibility of topics (discussed below), this book should be adaptable to most of the courses currently being taught.

Chapters 2 through 17 cover the standard subject matter of the three-semester sequence. Pertinent remedial work on algebra, analytic geometry, and trigonometry is provided in Chapter 1 and Appendix B. Chapters 18 (vector analysis) and 19 (differential equations) offer a choice of topics to round out longer courses.

Flexibility The text has been made as flexible as possible to meet the different needs of the departments served by the course. Engineering schools, in particular, frequently want special topics to be taught out of sequence. The following partial list shows how some of the common requests can be met, provided the homework problems are assigned with care.

Derivatives of the trigonometric functions (Chapter 9) can be introduced after the Chain Rule in Chapter 3, and integrals of these functions after Chapter 5. Analytic geometry (Chapter 7) can be introduced at any point after Chapter 4 and before Chapter 8. Topics in differential equations (from Chapter 19) can be integrated into the text at appropriate places after Chapter 9. Infinite series (Chapter 12) can be taught at any point after Chapter 10 and before Section 19-8. Improper integrals (from Chapter 10) can be taught after Chapter 6.

The Instructor's Manual has an organizational chart to help plan courses that differ markedly from the table of contents.

Rigor vs. Intuition Almost every sentence in an introductory calculus book involves a compromise between rigor and intuition. I have attempted to develop an intuitive understanding of the topics while presenting a fairly rigorous treatment of the main results.

The book contains a moderate number of formal arguments stated as theorems. I felt that it was better to provide a proof that could be omitted by the instructor than to omit one he might want to include. This approach has two obvious advantages: it helps the student to develop an awareness of formal mathematical statements; and (at a more primitive level) it helps him to recognize important results. I have followed the adage that the level of rigor should be appropriate to the level of student. Thus, many of the proofs would seem to be lacking in rigor to a senior mathematics major.

Exercises and Examples The exercises form the heart of any mathematics book. I have included three main types.

1. Routine exercises are designed to reinforce the techniques developed in the particular section. These are grouped into sets of similar problems, roughly graded by level of difficulty or complexity. Enough of these problems are provided to allow the homework assignments to be varied for several years without duplication.

2. More challenging problems require the student to synthesize several concepts or to break a problem into component problems. A typical problem of this type in Chapter 1 asks the student to find the equation of the perpendicular bisector of a line segment.

3. Theoretical problems ask the student to prove a result similar to a theorem or to complete the proof of a theorem. These problems are intended to provide immediate reference points for useful results that are not discussed in the text proper — for example, the formula for the distance from a point to a line. I would suggest that these problems be used for extra credit and not be assigned routinely to the average class.

More difficult and/or theoretical problems are separated from the simpler ones by a horizontal line.

Answers are provided for approximately half of the exercises. Sections in which the final answers give no hint about the method of solution (e.g., approximate integration) have answers for most of the problems. At the other extreme, sections in which the final answers completely describe the method of solution (e.g., simple graphing problems) have relatively few answers. Exercises with answers in the back of the book are marked by solid dots (●). The answers and solutions to the other problems can be found in the Instructor's Manual.

A few exercises are designed for students with scientific calculators or programmable calculators (e.g., Newton's method, calculation of partial sums of series). These exercises are marked with "c".

The exercises were worked independently by three people, including the author, and the answers were cross-checked for accuracy.

All of the techniques needed for the exercises are illustrated in the examples.

Applications The applications are designed to amplify the mathematical development, not to distract the student's attention from it. All of the elementary applications in physics, engineering, and mathematics (distance, velocity, work, volume, arc length, radioactive decay, harmonic motion, etc.) are included. In addition, a few nonstandard applications (business problems, learning curves, probability density functions, ballistic problems with air resistance) are included to indicate the scope of the calculus. Except for the most basic applications, they can be omitted without loss of continuity.

Other Features The text is richly illustrated. Figures are provided wherever appropriate, most with explanatory captions so that they can be reviewed by the student indepen-

Preface

dently of the text. For clarity, the figures are in two colors and are larger than those in many texts.

Notes in the margin are used to flag important concepts and results. These notes should help the serious student to outline the text by making the organization of the sections apparent.

Brief historical remarks about the concepts are occasionally included (see pages 78–79, for example), in the hope that some students will reach an awareness of mathematics as a developing discipline.

Where appropriate, special calculator problems are included. Many of these have two parts: one for students with scientific calculators, the other for students with programmable calculators. Such problems are especially useful for approximation techniques.

Accuracy Every effort has been made to get the text as free of errors as possible. Before publication, the exercises were worked independently by three people. After the first printing, the exercises and examples were reworked independently by two teams at different universities. The first, at the University of Mississippi, included Professors Eldon Miller, Glenn Hopkins, Mary Jane Causey, and Richard Tucker. The other, at the University of California, Santa Barbara, was a team of students under the direction of Professor Ken Millett. The discrepancies were checked by the author, and all errors were corrected before the second printing.

During the same period the author and developmental editor independently compared the printed text with the page proofs and (we hope) corrected all errors that remained.

Our aim is to make the text, the exercises, and the answers 100 percent error-free. If any errors remain, please notify us as soon as possible so that we can correct them in a future printing.

Acknowledgments I want to thank all of the reviewers and editors who worked with me during the five years the text was being developed.

The handwritten original draft was read by three of my friends, Leon Rutland of Virginia Tech, James Roche, and Evelyn Rubin, who made a large number of constructive suggestions concerning both content and style. The next several drafts were read by the following reviewers, who helped hammer it into semifinal form: Wesley Alexander, Elon College; Paul Baum, Brown University; Lowell Beinecke, Purdue University; George Benke, Georgetown University; Charles Benner, University of Houston; Lee Cook, University of Alabama; Dick Dahlke, University of Michigan; Duane Deal, Ball State University; Thomas Green, Contra Costa College; Mark Hale, Jr., University of Florida; Douglas Hall, Michigan State University; James Hurley, University of Connecticut; Melvin Lax, California State University; James Lewis, University of New Mexico; Lyle Mason, Phillips University; John Mathews, California State University; Ed Matzdorff, California State University; David Meredith, San Francisco State University; Eldon Miller, University of Mississippi; Maurice Monahan, South Dakota State University; John Montgomery, University of Rhode Island; Ronald Morash, University of Michigan; John Oman, University of Wisconsin; Theodore Palmer, University of Oregon; Donald Parker, University of Cincinnati; H. D. Perry, Texas A & M University; Nancy Poxon, California State University; David Rearick, University of Colorado; Lester Riggs, San Diego State University; David Rothchild, Herbert Lehman College (CUNY); John Russo, Indiana University; Philip Schaefer, University of Tennessee; Karen Schroeder, Bentley College; Steven Serbin, University of Tennessee; Donald Sherbert, University of Illinois; Charles Sherrill III, University of Colorado; Jerry Silver, University of Cincinnati; William Smith, University of North Carolina; Joel Stemple, Queens College (CUNY); John Thomas, New Mexico State University; Richard Tondra, Iowa State University; Jay Walton, Texas A & M

University; Gerald White, Western Illinois University; Sam Whitney, Diablo Valley College; and Jay Yellen, SUNY at Fredonia. A detailed review by Philip Schaefer helped to determine the final form.

The exercises were worked independently by Jay Humphrey, an undergraduate student at Virginia Tech, and by James Roche.

The staff at Saunders College Publishing has offered excellent support. I particularly want to thank developmental editor Jay Freedman and project editor Sally Kusch, who helped to make the work as painless as possible during the final production.

A special debt of gratitude is owed to James Roche of Mount St. Mary's College. Jim read the preliminary version of the manuscript, furnished the first draft of the exercises, worked all of the exercises in the final manuscript, and coauthored the Instructor's Manual. Without his help the book would probably not have been completed.

JAMES E. SHOCKLEY

Contents Overview

Chapter 1.	PRELIMINARY TOPICS	1
Chapter 2.	LIMITS AND DERIVATIVES	65
Chapter 3.	DERIVATIVES OF ALGEBRAIC FUNCTIONS	141
Chapter 4.	APPLICATIONS OF THE DERIVATIVE	192
Chapter 5.	THE INTEGRAL	257
Chapter 6.	APPLICATIONS OF THE INTEGRAL	306
Chapter 7.	TOPICS IN PLANE ANALYTICAL GEOMETRY	381
Chapter 8.	LOGARITHMIC AND EXPONENTIAL FUNCTIONS	445
Chapter 9.	TRIGONOMETRIC AND HYPERBOLIC FUNCTIONS	492
Chapter 10.	L'HÔPITAL'S RULE. IMPROPER INTEGRALS	547
Chapter 11.	TECHNIQUES OF INTEGRATION	581
Chapter 12.	INFINITE SERIES	616
Chapter 13.	PARAMETRIC EQUATIONS AND POLAR COORDINATES	686
Chapter 14.	VECTORS AND SOLID ANALYTIC GEOMETRY	734
Chapter 15.	THE CALCULUS OF VECTOR-VALUED FUNCTIONS	810
Chapter 16.	PARTIAL DIFFERENTIATION	857
Chapter 17.	MULTIPLE INTEGRATION	930
Chapter 18.	INTRODUCTION TO VECTOR ANALYSIS	1012
Chapter 19.	INTRODUCTION TO DIFFERENTIAL EQUATIONS	1079
Appendix A.	MATHEMATICAL INDUCTION	A.1
Appendix B.	THE TRIGONOMETRIC FUNCTIONS	A.6
Tables		A.22

Contents

1 **Preliminary Topics**   1

    1-1   Order   1
    1-2   Inequalities in One Variable   7
    1-3   Inequalities Involving Absolute Values   13
    1-4   Distance Formula. Midpoint Formula. Circles   19
    1-5   Lines in the xy-Plane   28
    1-6   Functions and Their Graphs   39
    1-7   Composition of Functions   47
    1-8   Inverse Functions   53
    Review Problems   61

2 **Limits and Derivatives**   65

    2-1   Introduction to Limits and Tangent Lines   65
    2-2   The Derivative   73
    2-3   Rates of Change   81
    2-4   Limits   88
    2-5   Some Properties of Limits   96
    2-6   Continuous Functions   104
    2-7   Properties of Continuous Functions   113
    2-8   Infinite Limits and Limits at Infinity. Asymptotes   120
    Review Problems   132
    Appendix: Proof of the Limit Theorems   135

3 **Derivatives of Algebraic Functions**   141

    3-1   Basic Derivative Formulas   141
    3-2   The Product, Quotient and General Power Rules   145
    3-3   Application to Approximation Theory. Differentials   150
    3-4   The Chain Rule   158
    3-5   Implicit Differentiation. Derivatives of Algebraic Functions   165
    3-6   Related Rates   171
    3-7   Vertical Tangent Lines   176
    3-8   Higher-order Derivatives   178
    3-9   Derivatives of Inverse Functions. Root Functions   182
    Review Problems   186
    Appendix: Continuity and Differentiability of Inverse Functions   188

Contents

4 Applications of the Derivative   192

    4-1 Introduction to Maximum and Minimum Problems   192
    4-2 Mean Value Theorem   199
    4-3 First Derivative Test for Local Extrema   205
    4-4 Some Applied Problems   210
    4-5 Concavity. Application to Curve Sketching   218
    4-6 The Second Derivative Test   225
    4-7* Application to Business Analysis   228
    4-8 Newton's Method   234
    4-9 Antiderivatives   242
    4-10 Application. Introduction to Differential Equations   249
    Review Problems   254

5 The Integral   257

    5-1 The Area Problem. Sigma Notation   258
    5-2 The Definite Integral   266
    5-3 Some Properties of Integrals   274
    5-4 Fundamental Theorem of Calculus   282
    5-5 The Indefinite Integral; the Substitution Principle   288
    5-6 Numerical Integration   294
    Review Problems   304

6 Applications of the Integral   306

    6-1 Area   306
    6-2 Volume I: Slices and Disks   315
    6-3 Volume II: The Shell Method   325
    6-4 Arc Length   333
    6-5 Work   343
    6-6 Fluid Pressure   348
    6-7 Moments I: Finite Mass Systems   356
    6-8 Moments II: Centroids of Regions   363
    6-9* Moments III: Applications. Theorems of Pappus   371
    Review Problems   378

7 Topics in Plane Analytic Geometry   381

    7-1 Curve Sketching, Asymptotes, Symmetry and Projections   381
    7-2 The Parabola. Translation of Axes   395
    7-3 The Ellipse   405
    7-4 The Hyperbola   414
    7-5* Eccentricity. Sections of a Cone   424
    7-6 Rotation of Axes   434
    Review Problems   442

8 Logarithmic and Exponential Functions   445

    8-1 The Natural Logarithm   445
    8-2 The Number *e*. The Natural Exponential Function   454
    8-3 Integration Formulas. General Exponential and Logarithmic Functions   462

*Sections marked with asterisks may be omitted without loss of continuity.

8-4　The Differential Equation $dy/dx = ky$　471
8-5　Laws of Growth　475
8-6　Laws of Motion　482
Review Problems　489

9　Trigonometric and Hyperbolic Functions　492

9-1　Derivatives and Integrals of the Sine and Cosine Functions　492
9-2　Additional Derivative and Integral Formulas　500
9-3*　Application. Simple Harmonic Motion　507
9-4　Inverse Trigonometric Functions　515
9-5　Integrals Obtained from Inverse Trigonometric Functions　526
9-6　Hyperbolic Functions　529
9-7　Inverse Hyperbolic Functions　534
9-8*　Application. The Hanging Cable　541
Review Problems　544

10　L'Hôpital's Rule. Improper Integrals　547

10-1　L'Hôpital's Rule. I: The Indeterminate Forms 0/0 and ∞/∞　547
10-2　L'Hôpital's Rule. II: Other Indeterminate Forms　553
10-3　Improper Integrals. I: Integrals of Discontinuous Functions　557
10-4　Improper Integrals. II: Integrals Defined over Infinite Intervals　563
10-5*　Application to Statistics. Average Value. Probability Density Function　569
10-6*　The Normal Probability Density Function　575
Review Problems　578

11　Techniques of Integration　581

11-1　Integration by Parts　581
11-2　Trigonometric Integrals　586
11-3　Trigonometric Substitutions　594
11-4　Partial Fractions　598
11-5　Quadratic Expressions　604
11-6　Miscellaneous Substitutions　607
11-7　Integral Tables　611
Review Problems　614

12　Infinite Series　616

12-1　Sequences and Series　616
12-2　Some Properties of Infinite Series. Positive-Term Series　622
12-3　The Integral Test. p-Series　630
12-4　The Comparison and Ratio Tests　638
12-5　Alternating Series　645
12-6　Absolute and Conditional Convergence　649
12-7　Power Series　657
12-8　Taylor Series　664
12-9　Taylor's Formula. Accuracy of Approximations　674
Review Problems　683

Contents

13 Parametric Equations and Polar Coordinates   686

- 13-1   Parametric Equations   686
- 13-2   Derivatives   694
- 13-3   Polar Coordinates   702
- 13-4   Graphs of Polar Equations   709
- 13-5   Circles, Lines, and Conic Sections   718
- 13-6   Area and Arc Length   724
- *Review Problems*   *731*

14 Vectors and Solid Analytic Geometry   734

- 14-1   Vectors in the xy-Plane   734
- 14-2   Three-Dimensional Space   742
- 14-3   Vectors in Three-Dimensional Space. Elementary Applications   749
- 14-4   The Inner Product   757
- 14-5   Applications to Geometry   763
- 14-6   Equations of Lines and Planes   768
- 14-7   The Vector Product   776
- 14-8   Some Problems Involving Vectors   781
- 14-9   Cylinders and Space Curves   788
- 14-10 Level Curves. Quadric Surfaces   793
- *Review Problems*   *803*
- *Appendix: Second-Order and Third-Order Determinants*   *805*

15 The Calculus of Vector-Valued Functions   810

- 15-1   Limits, Continuity, Derivatives, and Integrals   810
- 15-2   Velocity and Acceleration Vectors. Arc Length   818
- 15-3   Velocity and Acceleration Problems   825
- 15-4   Curvature. I: Plane Curves   833
- 15-5   Curvature. II: Unit Tangent and Normal Vectors. Space Curves   841
- 15-6   Tangential and Normal Components of Acceleration   848
- *Review Problems*   *854*

16 Partial Differentiation   857

- 16-1   Limits and Continuity   857
- 16-2   Partial Derivatives   867
- 16-3   Higher-Order Partial Derivatives   871
- 16-4   Increments and Approximations   875
- 16-5   The Chain Rule   882
- 16-6   The Directional Derivative   889
- 16-7   Properties of the Gradient   896
- 16-8   Normal Vectors and Tangent Planes   902
- 16-9   Local Maxima and Minima   907
- 16-10 Maxima and Minima   916
- 16-11 Extremal Problems with Constraints. Lagrange Multipliers   921
- *Review Problems*   *926*

17 Multiple Integration   930

- 17-1   The Double Integral   930
- 17-2   The Iterated Integral   938

17-3 Volume and Mass 948
17-4 Moments and Centroids 953
17-5 Double Integrals in Polar Coordinates 962
17-6 Surface Area 970
17-7 Triple Integration 976
17-8 Moments and Centroids 983
17-9 The Cylindrical and Spherical Coordinate Systems 990
17-10 Triple Integrals in Cylindrical and Spherical Coordinates 997
Review Problems *1008*

18 Introduction to Vector Analysis 1012

18-1 Vector Fields 1012
18-2 Line Integrals. I: Definitions and Properties 1017
18-3 Line Integrals. II: Independence of Path 1025
18-4 Applications 1033
18-5 Green's Theorem 1040
18-6 Surface Integrals 1048
18-7 Divergence and Curl 1055
18-8 The Divergence Theorem 1059
18-9 Stokes' Theorem 1068
Review Problems *1075*

19 Introduction to Differential Equations 1079

19-1 Review of Basic Concepts. Separation of Variables 1079
19-2 Exact Differential Equations 1086
19-3 First-Order Linear Differential Equations 1091
19-4 Homogeneous Second-Order Linear Differential Equations with Constant Coefficients. I: Real Solutions of the Characteristic Equation 1097
19-5 Homogeneous Second-Order Linear Differential Equations with Constant Coefficients. II: Imaginary Solutions of the Characteristic Equation 1103
19-6 Nonhomogeneous Second-Order Linear Differential Equations with Constant Coefficients 1108
19-7 Variation of Parameters 1114
19-8 Series Solutions 1117
19-9 Approximate Solutions 1121
Review Problems *1124*

Appendix A. MATHEMATICAL INDUCTION A.1

Appendix B. THE TRIGONOMETRIC FUNCTIONS A.6

Tables A.22

Answers A.43

Index I.1

1 Preliminary Topics

1-1 Order

Throughout this book, we assume that you are familiar with the elementary properties of sets (unions, intersections, and so on) and with the basic properties of the real numbers. In particular, you should be familiar with the representation of real numbers as points on the number line (Fig. 1–1), with the representation of ordered pairs of real numbers as points in the xy-plane (see Fig. 1–8), and with the concept of order (less than, greater than).

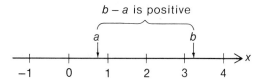

Figure 1–1 The number line: $a < b$ provided $b - a$ is positive.

In this section we show how the properties of order can be used to solve certain types of inequalities.

If a and b are real numbers we say that *a is less than b,* written

Order $\qquad a < b$

and that *b is greater than a,* written

$\qquad b > a$

provided that $b - a$ is a positive number. Using the standard orientation of the number line, this is equivalent to stating that $a < b$ provided the point that represents a on the number line is to the left of the point that represents b. (See Fig. 1–1.)

PROPERTIES OF ORDER

Most of the following properties can be proved from the definition of order by using the properties of positive and negative numbers. In particular, we need to recall that the sum and the product of positive numbers are positive, the product of two negative numbers is positive, and the product of a positive and a negative number is negative.

Order Properties

> **O1** (*The Trichotomy Property*) If a and b are real numbers, then
>
> $\qquad a < b \qquad$ or $\qquad a = b \qquad$ or $\qquad a > b$.
>
> **O2** (*The Transitive Property*) If $a < b$ and $b < c$, then $a < c$.
>
> **O3** If $a < b$ and c is a real number, then
>
> $\qquad a + c < b + c$.
>
> (*An inequality is preserved if a real number is added to both sides.*)
>
> **O4** If $a < b$ and c is positive, then $ac < bc$.
> (*An inequality is preserved if both sides are multiplied by a positive real number.*)
>
> **O5** If $a < b$ and c is negative, then $ac > bc$.
> (*An inequality is reversed if both sides are multiplied by a negative real number.*)

Properties O4 and O5 can be applied to division as well as multiplication, since dividing by c is equivalent to multiplying by $1/c$.

> **O4′** If $a < b$ and $c > 0$, then $\dfrac{a}{c} < \dfrac{b}{c}$.
>
> **O5′** If $a < b$ and $c < 0$, then $\dfrac{a}{c} > \dfrac{b}{c}$.

Example 1 establishes the transitive property. The basis of the argument is that the sum of two positive numbers is positive. Other proofs may require the fact that the product of two positive numbers is positive or the fact that the product of a positive number and a negative number is negative.

EXAMPLE 1 Let $a < b$ and $b < c$. Prove that $a < c$.

Solution Since $a < b$ and $b < c$, then

$\qquad b - a$ is positive

and

$\qquad c - b$ is positive.

Then

$\qquad c - a = (c - b) + (b - a)$

is positive because it is the sum of two positive numbers. Thus $a < c$. □

1-1 Order

LESS THAN OR EQUAL

We say that *a is less than or equal to b* ($a \leq b$) provided that $a = b$ or $a < b$. Under the same conditions, *b is greater than or equal to a* ($b \geq a$).

For example,

$$5 \leq 7 \quad \text{since } 5 < 7$$

and

$$-2 \leq -2 \quad \text{since } -2 = -2.$$

INTERVALS

Intervals

Certain subsets of the real numbers, called *intervals,* are important in the study of the calculus. A *finite interval* consists of all real numbers x between two fixed numbers a and b, where $a < b$. The numbers a and b, called the *endpoints* of the interval, may or may not belong to the interval. (See Figs. 1-2a-d.)

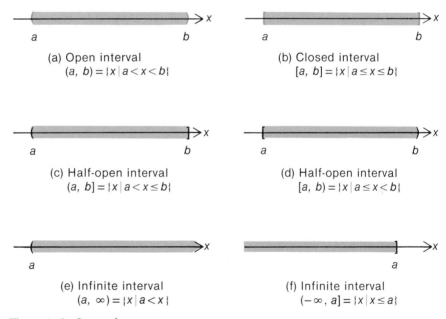

(a) Open interval
$(a, b) = \{x \mid a < x < b\}$

(b) Closed interval
$[a, b] = \{x \mid a \leq x \leq b\}$

(c) Half-open interval
$(a, b] = \{x \mid a < x \leq b\}$

(d) Half-open interval
$[a, b) = \{x \mid a \leq x < b\}$

(e) Infinite interval
$(a, \infty) = \{x \mid a < x\}$

(f) Infinite interval
$(-\infty, a] = \{x \mid x \leq a\}$

Figure 1-2 Intervals.

Open Interval

A finite interval with endpoints a and b is said to be *open* if its endpoints are not part of the interval. This open interval is denoted by the symbol (a, b). (See Fig. 1-2a).

Closed Interval

A finite interval is said to be *closed* if it includes both of its endpoints a and b. This closed interval is denoted by the symbol $[a, b]$. (See Fig. 1-2b.)

Half-Open Interval	A finite interval is said to be *half-open* (or *half-closed*) if it includes one endpoint but not the other. As in the open and closed intervals, a square bracket indicates that an endpoint is included, and a parenthesis that it is not. (See Figs. 1–2c, d.)

The four types of finite intervals are listed below:*

$$(a, b) = \{x \,|\, a < x < b\} \quad \text{(open interval)}$$
$$[a, b] = \{x \,|\, a \leq x \leq b\} \quad \text{(closed interval)}$$
$$\left.\begin{array}{l}[a, b) = \{x \,|\, a \leq x < b\} \\ (a, b] = \{x \,|\, a < x \leq b\}\end{array}\right\} \quad \text{(half-open intervals)}$$

Infinite Intervals	We also consider *infinite intervals*, that is, intervals that are not finite in length. As with finite intervals, a square bracket indicates that an endpoint is included in the interval, and a parenthesis indicates that it is not included. The five types of infinite intervals are listed below. (See Figs. 1–2e, f.)†

$$(a, \infty) = \{x \,|\, a < x\}$$
$$[a, \infty) = \{x \,|\, a \leq x\}$$
$$(-\infty, a) = \{x \,|\, x < a\}$$
$$(-\infty, a] = \{x \,|\, x \leq a\}$$
$$(-\infty, \infty) = \{x \,|\, x \text{ is real}\}.$$

INEQUALITIES

Inequality	An *inequality in x* is a statement, such as

$$3x - 2 > 5x + 4$$

that gives an order relation between two expressions involving x.

Solution of Inequality	By a *solution* of an inequality, we mean any number that, when substituted for the variable x, makes the inequality a true statement. The set of all solutions is called the *solution set*. For example, $x = -5$ is a solution of the above inequality because

$$3(-5) - 2 = -17 > 5(-5) + 4 = -21,$$

but $x = 0$ is not a solution because $3(0) - 2 < 5(0) + 4$. To *solve* an inequality, we must find its solution set.

Equivalent Equalities	Two inequalities are said to be *equivalent* if they have the same solution set. Most methods for solving an inequality are based on finding equivalent inequalities of

Set-building Notation	* The first statement within braces is read "The set of all x such that $a < x < b$." The *set-building notation*

$$\{x \,|\, \text{Statement about } x\}$$

is used to indicate the set of all x for which the statement is true.

† The symbol ∞ is read "infinity." This symbol does not represent a real number. It is used in the expression (a, ∞) to indicate that the interval extends forever in the positive direction.

Exercises 1–1

simpler form. The method for solving the inequality

$$3x - 2 > 5x + 4$$

in Example 2 is similar to the method for solving the linear equation

$$3x - 2 = 5x + 4.$$

EXAMPLE 2   *Solve the inequality* $3x - 2 > 5x + 4$.

Solution   We reduce the inequality to the form $ax > b$ and divide by the coefficient of x:

$$3x - 2 > 5x + 4$$
$$-2x > 6 \qquad \text{(Subtract } 5x, \text{ add 2 to both sides.)}$$
$$x < -3 \qquad \text{(Divide by } -2, \text{ use O5).}$$

Since each inequality is equivalent to the preceding one, the solution set of $3x - 2 > 5x + 4$ is

$$\{x \mid x < -3\} = (-\infty, -3). \quad \square$$

Exercises 1–1

1–8. Replace the commas by $<$, $=$, or $>$, so that the resulting statement will be true.

- •1 $1, -5$
- 2 $\frac{2}{7}, \frac{1}{3}$
- •3 $(\frac{1}{2} - \frac{1}{3}), \frac{1}{6}$
- 4 $\frac{1}{9}, 0.11$
- •5 $\pi, 3.14159$
- 6 $\sqrt{2}, 1.414$
- •7 $3^{-1}, 3^{-2}$
- 8 $16^{-1/2}, 16^{-1/4}$

9–16. These expressions define intervals. In exercises 9–12, express the intervals by the standard notation for intervals. In exercises 13–16, describe the intervals by inequalities. Which intervals are open? closed? Which are finite? infinite?

- •9 $\{x \mid 3 < x < 5\}$
- 10 $\{x \mid -1 \leq x < 7\}$
- •11 $\{x \mid -1000 < x < 83,000.1\}$
- 12 $\{x \mid 41 \leq x \leq 42\}$
- •13 $[0.0012, 0.0013)$
- 14 $(21.3, 25.4)$
- •15 $(-\infty, -2]$
- 16 $[3, \infty)$

17–24. Solve the inequalities. Sketch the solution sets on the number line.

- •17 $7x - 5 \geq 3x + 11$
- 18 $4x - 3 < 2x + 5$
- •19 $2 + 7x < 3x - 10$
- 20 $5x < 17x - 24$

NOTE: A dot placed before an exercise number indicates that an answer is given for this exercise in the Answer section at the back of the book.

- 21 $\dfrac{2x-3}{3} \leq 5$
- 22 $\dfrac{x}{6} - 1 > \dfrac{x}{3} + 4$
- 23 $\dfrac{x}{12} + 1 \geq \dfrac{x}{8} + 2$
- 24 $\dfrac{7-3x}{2} \geq 11$

25–30. By a solution to a pair of inequalities, we mean a number that solves them both. For example, the complete solution of the pair of inequalities $x > 2$ and $2x \leq 5$ is $\{x \mid 2 < x \leq \tfrac{5}{2}\}$. Solve these pairs of inequalities. Sketch the solution sets on the number line.

- 25 $x > 0$ and $5x - 3 > 2x - 9$
- 26 $x - 2 < 3$ and $x + 1 > 2$
- 27 $2x - 4 < 0$ and $3x + 1 < 6x - 2$
- 28 $2x - 1 > 5$ and $\dfrac{x-1}{3} \geq 4$
- 29 $\dfrac{1+x}{2} \leq 3$ and $x \geq 6$
- 30 $x > 0$ and $2 + 7x < 3x - 10$

31–38. Solve the inequalities. Sketch the solution sets on the number line.
[Hint for 31: $5 > 2 - 3x > -4$ means that $5 > 2 - 3x$ and $2 - 3x > -4$.]

- 31 $5 > 2 - 3x > -4$
- 32 $-1 < \dfrac{3-7x}{4} < 6$
- 33 $-6 \leq 2x - 4 \leq 2$
- 34 $10 < 3x - 2 < -5$
- 35 $7 < 5x - 3 < 12$
- 36 $-5 < \dfrac{3x-1}{2} < 1$
- 37 $7 \leq \dfrac{2x-3}{3} \leq 5$
- 38 $-5 < 2x + 5 < 3$

39–41. Use the definition of *order* to prove these properties. (*Hint:* Modify the argument of Example 1.)

39 O3: If $a < b$, then $a + c < b + c$.

40 O4: If $a < b$ and $c > 0$, then $ac < bc$.

41 O5: If $a < b$ and $c < 0$, then $ac > bc$.

42–49. Use the order properties O1–O5 to prove these propositions.

42 $a > b$ if and only if $-a < -b$.

43 If $0 < a < b$, then $a^2 < b^2$.

44 If $0 < a < b$ or $a < b < 0$, then $1/a > 1/b$.

45 If a is a real number, then $a^2 \geq 0$. If $a \neq 0$, then $a^2 > 0$.

46 If $a > 1$, then $a^2 > a$. If $0 < a < 1$, then $a^2 < a$.

47 If a and b are positive, then $a^2 + b^2 \leq (a+b)^2$.

48 $a^2 + b^2 \geq 2ab$.

49 If $a < b$, then $a < \dfrac{a+b}{2} < b$.

1-2 Inequalities in One Variable

Inequalities of the type

Linear Inequality

$$ax < b, \quad ax \leq b, \quad ax > b, \quad \text{or} \quad ax \geq b$$

are called *linear inequalities in one variable*. They can be solved by the method illustrated in Example 2 of Section 1–1. More complicated types of inequalities often can be solved by reducing them to cases that involve linear inequalities.

EXAMPLE 1 Solve $(x - 1)(x - 2) > 0$.

Solution For the product to be positive, both factors must be positive or both negative. We consider these two cases separately. Each case is solved by constructing a sequence of equivalent inequalities.

Case 1 $x - 1$ and $x - 2$ are both positive. Since

$$x - 1 > 0 \quad \text{and} \quad x - 2 > 0$$

then

$$x > 1 \quad \text{and} \quad x > 2 \quad \text{(by O3)}.$$

The solution set for this case is

$$S_1 = \{x \mid x > 1 \text{ and } x > 2\} = \{x \mid x > 2\} = (2, \infty).$$

Case 2 $x - 1$ and $x - 2$ are both negative. Since

$$x - 1 < 0 \quad \text{and} \quad x - 2 < 0$$

then

$$x < 1 \quad \text{and} \quad x < 2 \quad \text{(by O3)}.$$

The solution set for this case is

$$S_2 = \{x \mid x < 1 \text{ and } x < 2\} = \{x \mid x < 1\} = (-\infty, 1).$$

Since either Case 1 or Case 2 can hold, the solution set of the inequality is the union of the solution sets for the two cases:*

$$S = S_1 \cup S_2 = (2, \infty) \cup (-\infty, 1)$$
$$= \{x \mid x < 1 \text{ or } x > 2\}.$$

Figure 1–3 illustrates the two cases. □

A simple way of summarizing the work in Example 1 is by a table. Observe that the factors $x - 1$ and $x - 2$ change sign at $x = 1$ and $x = 2$, respectively. If we consider

Union Intersection

* The *union* of sets A and B, denoted by $A \cup B$, is

$$A \cup B = \{x \mid x \in A \text{ or } x \in B\}.$$

The *intersection* of A and B, denoted by $A \cap B$, is

$$A \cap B = \{x \mid x \in A \text{ and } x \in B\}.$$

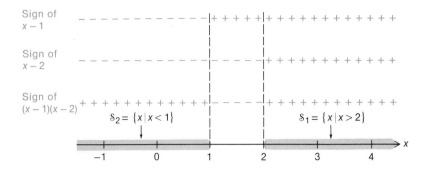

Figure 1-3 Example 1. The solution set of the inequality $(x - 1)(x - 2) > 0$.

values of x in the intervals $(-\infty, 1)$, $(1, 2)$ and $(2, \infty)$, we can determine the sign of $x - 1$, $x - 2$, and $(x - 1)(x - 2)$. The accompanying table summarizes the information.

	$(x - 1)$	$(x - 2)$	$(x - 1)(x - 2)$
$x < 1$	−	−	+
$1 < x < 2$	+	−	−
$x > 2$	+	+	+

Thus

$$(x - 1)(x - 2) > 0 \quad \text{if } x < 1 \text{ or } x > 2,$$
$$(x - 1)(x - 2) < 0 \quad \text{if } 1 < x < 2.$$

QUADRATIC INEQUALITIES

Inequalities of the form

$$ax^2 + bx + c > 0 \quad \text{or} \quad ax^2 + bx + c < 0$$

Quadratic Inequality

where a, b, c are real numbers, $a \neq 0$, are called *quadratic inequalities*. They can be solved by the method of Example 1 provided the quadratic polynomial

$$f(x) = ax^2 + bx + c$$

can be factored into real linear factors.

The most efficient method of factoring $f(x)$, except by inspection, is by the quadratic formula. The roots of $ax^2 + bx + c = 0$ are

$$r_1 = \frac{-b - \sqrt{b^2 - 4ac}}{2a} \quad \text{and} \quad r_2 = \frac{-b + \sqrt{b^2 - 4ac}}{2a}.$$

1-2 Inequalities in One Variable

It can be shown that $f(x)$ can be factored as

$$f(x) = ax^2 + bx + c$$
$$= a(x - r_1)(x - r_2).$$

If r_1 and r_2 are real, then the terms $x - r_1$ and $x - r_2$ change sign at r_1 and r_2. We can then proceed as in Example 1 or use a table.

If r_1 and r_2 are imaginary, then it can be shown that $f(x) = ax^2 + bx + c$ never changes sign. In this case we have either $f(x) > 0$ for all x or $f(x) < 0$ for all x.

EXAMPLE 2 Solve the inequality $x^2 + 4x < 1$.

Solution We rewrite the inequality as

$$x^2 + 4x - 1 < 0.$$

The roots of $x^2 + 4x - 1 = 0$ are

$$r_1 = \frac{-4 - \sqrt{16 + 4}}{2} = -2 - \sqrt{5} \quad \text{and} \quad r_2 = \frac{-4 + \sqrt{16 + 4}}{2} = -2 + \sqrt{5}.$$

Therefore $x^2 + 4x - 1 < 0$ can be factored as

$$[x - (-2 - \sqrt{5})][x - (-2 + \sqrt{5})] < 0,$$
$$(x + 2 + \sqrt{5})(x + 2 - \sqrt{5}) < 0.$$

For the product to be negative, one factor must be negative and the other positive. Thus we normally would consider two cases:

Case 1 $x + 2 - \sqrt{5} < 0$ and $x + 2 + \sqrt{5} > 0$.
Case 2 $x + 2 - \sqrt{5} > 0$ and $x + 2 + \sqrt{5} < 0$.

Observe that if x is any real number, however, we must have

$$x + 2 - \sqrt{5} < x + 2 + \sqrt{5},$$

and Case 2 cannot hold. Thus, we need to consider only Case 1:

$$x + 2 - \sqrt{5} < 0 \quad \text{and} \quad x + 2 + \sqrt{5} > 0.$$

Equivalently, we have $x < -2 + \sqrt{5}$ and $x > -2 - \sqrt{5}$ so that the solution set of the inequality is

$$S = \{x \mid -2 - \sqrt{5} < x < -2 + \sqrt{5}\}.$$

(See Fig. 1–4.) □

EXAMPLE 3 Solve the inequality $x^2 + 1 > 0$.

Solution Factor $x^2 + 1 > 0$ as

$$(x + i)(x - i) > 0.$$

Since the factors are imaginary, then either $x^2 + 1 > 0$ for all x or $x^2 + 1 < 0$ for all x. Since $0^2 + 1 > 0$ then $x^2 + 1 > 0$ for all x. The solution set is

$$S = \{x \mid x \text{ is real}\}. \quad \square$$

10 *Preliminary Topics*

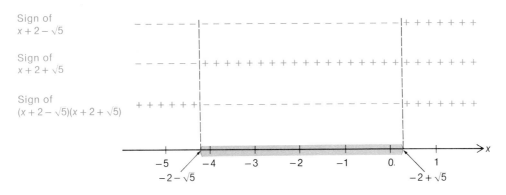

(a)

	$x + 2 + \sqrt{5}$	$x + 2 - \sqrt{5}$	$x^2 + 4x - 1$
$x < -2 - \sqrt{5}$	−	−	+
$-2 - \sqrt{5} < x < -2 + \sqrt{5}$	+	−	−
$x > -2 + \sqrt{5}$	+	+	+

(b)

Figure 1–4 Example 2. The solution set of $x^2 + 4x < 1$ is
$$S = \{x \mid -2 - \sqrt{5} < x < -2 + \sqrt{5}\}.$$

INEQUALITIES INVOLVING FRACTIONS

Inequalities involving fractions can usually be solved by considering cases as in the following example.

EXAMPLE 4 Solve the inequality $\dfrac{2x + 1}{x - 2} < 3$.

Solution We first rewrite the inequality as an inequality involving a single fraction:

$$\frac{2x + 1}{x - 2} < 3$$

$$\frac{2x + 1}{x - 2} - 3 < 0$$

$$\frac{2x + 1}{x - 2} - \frac{3(x - 2)}{x - 2} < 0$$

1-2 Inequalities in One Variable

$$\frac{2x + 1 - 3(x - 2)}{x - 2} < 0$$

$$\frac{-x + 7}{x - 2} < 0 \qquad \text{(combining terms)}$$

$$\frac{x - 7}{x - 2} > 0 \qquad \text{(by O5)}.$$

A fraction can be positive only if its numerator and denominator are both positive or both negative. This leads us to consider two cases:

Case 1 $x - 7 > 0$ and $x - 2 > 0$.

Then

$$x > 7 \qquad \text{and} \qquad x > 2.$$

The solution set for this case is

$$S_1 = \{x \mid x > 7\} = (7, \infty).$$

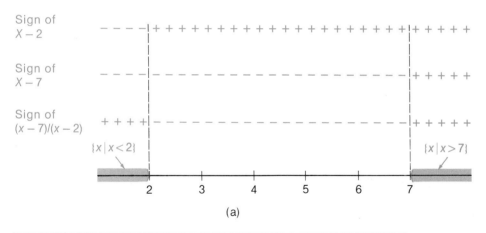

(a)

	$x - 2$	$x - 7$	$\dfrac{(x-7)}{(x-2)}$
$x < 2$	−	−	+
$2 < x < 7$	+	−	−
$x > 7$	+	+	+

(b)

Figure 1-5 Example 4. The solution set of $\dfrac{2x+1}{x-2} < 3$. (The inequality is equivalent to $\dfrac{x-7}{x-2} > 0$.)

Case 2 $x - 7 < 0$ and $x - 2 < 0$.

Then

$$x < 7 \quad \text{and} \quad x < 2.$$

The solution set for this case is

$$S_2 = \{x \mid x < 2\} = (-\infty, 2).$$

Since either Case 1 or Case 2 can hold, the solution set for the original inequality is

$$S = S_1 \cup S_2 = \{x \mid x < 2 \text{ or } x > 7\} = (-\infty, 2) \cup (7, \infty).$$

(See Fig. 1–5.) ☐

EXAMPLE 5   Solve $\dfrac{3x + 4}{x - 5} > 3$.

Solution   We rewrite the inequality as

$$\frac{3x + 4}{x - 5} - 3 > 0$$

$$\frac{3x + 4 - 3(x - 5)}{x - 5} > 0$$

$$\frac{19}{x - 5} > 0.$$

This inequality can hold only if $x - 5 > 0$. The solution set is

$$S = \{x \mid x > 5\} = (5, \infty). \quad \square$$

Exercises 1–2

Solve the following inequalities. Express the solution sets as intervals or unions of intervals.

- 1 $(2x + 3)(x - 2) < 0$
- 2 $(x + 2)(x - 5) > 0$
- 3 $x^3(x - 1) \leq 0$
- 4 $x^2 + x - 2 > 0$
- 5 $x^2 + 4x - 5 < 0$
- 6 $2x^2 - x - 3 \geq 0$
- 7 $x^2 + 3x - 4 > 0$
- 8 $x^2 - 5x - 6 \leq 0$
- 9 $2x^2 + 1 \leq 0$
- 10 $x^2 + 3 > 0$
- 11 $x^3 - 9x > 0$
- 12 $x^4 - 3x^2 - 4 < 0$
- 13 $x^2 - 4 < 0$
- 14 $x^2 + 4x < 0$

1–3 Inequalities Involving Absolute Values

- 15 $\dfrac{3}{x+5} > 2$
- 16 $\dfrac{4}{x} \le 2$
- 17 $\dfrac{x}{x-2} < 3$
- 18 $\dfrac{2x+3}{x-5} > 3$
- 19 $\dfrac{x+2}{x-1} \le 4$
- 20 $\dfrac{2}{t-2} \ge \dfrac{4}{t}$
- 21 $\dfrac{3t+2}{2t-8} \le 0$
- 22 $\dfrac{1}{3y-1} \le \dfrac{2}{3y+1}$
- 23 $\dfrac{3}{z^3} < \dfrac{1}{z^2}$
- 24 $\dfrac{3z^2+5z-2}{z+4} < 0$
- 25 $\dfrac{2}{3x-2} > \dfrac{3}{x+1}$
- 26 $\dfrac{x^2+1}{x-2} > 0$

1–3 Inequalities Involving Absolute Values

The *absolute value* of the real number a, denoted by $|a|$, is defined by the rule

Absolute Value

$$|a| = a \quad \text{if } a \ge 0$$

and

$$|a| = -a \quad \text{if } a < 0.$$

Geometrically, the absolute value of a equals the distance from the corresponding point on the number line to the origin. Thus,

$$|4| = 4 \text{ since 4 is four units from the origin,}$$
$$|-3| = 3 \text{ since } -3 \text{ is three units from the origin,}$$
$$|0| = 0 \text{ since 0 is at the origin,}$$

and so on. (See Fig. 1–6.) Note particularly that

$$|a| = |-a|$$

for every real number a.

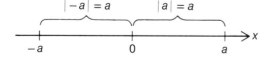

Figure 1–6 Absolute value. The absolute value of a is equal to the distance from the origin to the point represented by a.

$$|a| = \begin{cases} a & \text{if } a \ge 0 \\ -a & \text{if } a < 0. \end{cases}$$

The most basic of the properties concerning absolute value are given below:

Properties of Absolute Value

A1 $|ab| = |a| \cdot |b|$.
The absolute value of a product equals the product of the absolute values.

A2 $\left|\dfrac{a}{b}\right| = \dfrac{|a|}{|b|}$ provided $b \neq 0$
The absolute value of a quotient equals the quotient of the absolute values.

The Triangle Inequality

A3   (*The Triangle Inequality*) $|a + b| \leq |a| + |b|$
The absolute value of a sum is less than or equal to the sum of the absolute values.

The definition can be used to establish most of the properties of absolute value. To prove that a property holds, we consider two cases (x nonnegative and x negative) for each number x that is involved.

EXAMPLE 1   Prove A1: $|ab| = |a| \cdot |b|$.

Solution

We consider four cases (two each according to whether a and b are nonnegative or negative).

Case 1 $a \geq 0$ and $b \geq 0$. Then $|a| = a$, and $|b| = b$. Furthermore, $ab \geq 0$, so

$$|ab| = ab = |a| \cdot |b|.$$

Case 2 $a \geq 0$ and $b < 0$. Then $|a| = a$, and $|b| = -b$. In this case $ab \leq 0$, so

$$|ab| = -(ab) = a(-b) = |a| \cdot |b|.$$

Case 3 $a < 0$ and $b \geq 0$. The proof is similar to the one for Case 2.

Case 4 $a < 0$ and $b < 0$. Then $|a| = -a$, and $|b| = -b$. In this case $ab > 0$, so

$$|ab| = ab = (-a)(-b) = |a| \cdot |b|.$$

Since the result $|ab| = |a| \cdot |b|$ holds in every possible case, the theorem is proved. □

Two additional properties are worth noting:

A4 $\begin{cases} x^2 = a^2 & \text{if and only if } |x| = |a|. \\ x^2 < a^2 & \text{if and only if } |x| < |a|. \\ x^2 > a^2 & \text{if and only if } |x| > |a|. \end{cases}$

A5 $\sqrt{x^2} = |x|$ for any real number x.*

---

* The radical symbol $\sqrt{\ }$ denotes the nonnegative *square root*. Thus, $\sqrt{4} = 2$ even though $+2$ and -2 are both square roots of 4.

1-3 Inequalities Involving Absolute Values

EXAMPLE 2 Prove the first part of property A4.

Solution We must establish two results:

(1) If $x^2 = a^2$, then $|x| = |a|$.

and

(2) If $|x| = |a|$, then $x^2 = a^2$.

Part 1 Suppose that $x^2 = a^2$. Then

$$x^2 - a^2 = 0$$
$$(x - a)(x + a) = 0.$$

Therefore

$$x = \pm a = (\pm 1)a.$$

Then

$$|x| = |\pm 1| \cdot |a| = |a| \quad \text{(by A1)}.$$

Part 2 Suppose that $|x| = |a|$. Then

$$x^2 = |x|^2 = |a|^2 = a^2. \quad \square$$

To solve inequalities involving absolute values, we usually consider several cases.

EXAMPLE 3 Solve the inequality $2|x| > 3x - 10$.

Solution If we could somehow drop the absolute value signs, the inequality could be solved by the methods of Section 1-1. Observe that we could remove these signs if we knew whether x was nonnegative or negative. Thus, we consider two cases according to whether $x \geq 0$ or $x < 0$.

Case 1 If $x \geq 0$, then $|x| = x$. The inequality can be modified as follows:

$$
\begin{array}{ll}
2|x| > 3x - 10 & \text{and } x \geq 0. \\
2x > 3x - 10 & \text{and } x \geq 0. \\
-x > -10 & \text{and } x \geq 0 \quad \text{(by O3)}. \\
x < 10 & \text{and } x \geq 0 \quad \text{(by O5)}.
\end{array}
$$

Since the last inequalities are equivalent to the initial ones in this chain, the solution set for this case is

$$S_1 = \{x \mid 0 \leq x < 10\}.$$

Case 2 If $x < 0$, then $|x| = -x$. The inequality can be modified as follows:

$$
\begin{array}{lll}
2|x| > 3x - 10 & \text{and } x < 0 & \\
-2x > 3x - 10 & \text{and } x < 0 & \text{(since } |x| = -x\text{)} \\
-5x > -10 & \text{and } x < 0 & \text{(by O3)} \\
x < 2 & \text{and } x < 0 & \text{(by O5)}.
\end{array}
$$

The solution set for this case is

$$S_2 = \{x \mid x < 0\}.$$

The solution set for the inequality is

$$S = S_1 \cup S_2 = \{x \mid 0 \leq x < 10\} \cup \{x \mid x < 0\}$$
$$= \{x \mid x < 10\}. \quad \square$$

Some additional properties of absolute value are worth noting. In particular, if $|x| < a$, where a is a positive number, then the distance on the number line between x and the origin is less than a, so that x is between $-a$ and a. (See Fig. 1–7(a).) Since this argument can be reversed, it follows that

$$|x| < a \quad \text{if and only if } -a < x < a.$$

Similarly, if $a > 0$, then

$$|x| \geq a \quad \text{if and only if } x \leq -a \text{ or } x \geq a.$$

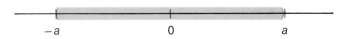

(a) The inequality $|x| < a$ is equivalent to the inequality $-a < x < a$.

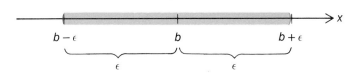

(b) The inequality $|x - b| < \epsilon$ is equivalent to the inequality $b - \epsilon < x < b + \epsilon$.

Figure 1–7

These remarks establish the following properties.

A6   Let $a > 0$. Then

$$|x| < a \quad \text{if and only if } -a < x < a$$

and $|x| \geq a \quad \text{if and only if } x \leq -a \text{ or } x \geq a.$

In Chapter 2 we work with inequalities of the form

$$|x - b| < \epsilon$$

1-3 Inequalities Involving Absolute Values

where ϵ (epilson) is a small number. We will need the following result:

A7 Let $\epsilon > 0$. Then

$|x - b| < \epsilon$ if and only if $b - \epsilon < x < b + \epsilon$.

This result also has a geometrical interpretation. The number x is a solution of $|x - b| < \epsilon$ if and only if the distance from x to b on the number line is less than ϵ. This holds if and only if x is between $b - \epsilon$ and $b + \epsilon$. (See Fig. 1-7b.)

EXAMPLE 4 Prove A7 by algebraic methods.

Solution Case 1 $x - b \geq 0$. Then $|x - b| = x - b$.
The inequality becomes

$$\begin{array}{lll} |x - b| < \epsilon & \text{and} & x - b \geq 0 \\ x - b < \epsilon & \text{and} & x - b \geq 0 \\ x < b + \epsilon & \text{and} & x \geq b. \end{array}$$

The solution set for this case is

$$S_1 = \{x \mid b \leq x < b + \epsilon\}.$$

Case 2 $x - b < 0$. Then $|x - b| = -(x - b) = -x + b$.

The inequality becomes

$$\begin{array}{lll} |x - b| < \epsilon & \text{and} & x - b < 0 \\ -x + b < \epsilon & \text{and} & x - b < 0 \\ b - \epsilon < x & \text{and} & x < b. \end{array}$$

The solution set for this case is

$$S_2 = \{x \mid b - \epsilon < x < b\}.$$

The solution set for the inequality is

$$S = S_1 \cup S_2 = \{x \mid b - \epsilon < x < b + \epsilon\}. \quad \square$$

EXAMPLE 5 Solve $|3x + 5| < 7$.

Solution We rewrite the inequality as

$$|(3x) - (-5)| < 7$$

and apply A7 (using $3x$ rather than x). The inequality is equivalent to

$$\begin{array}{c} -5 - 7 < 3x < 7 - 5 \\ -12 < 3x < 2 \\ -4 < x < \tfrac{2}{3}. \end{array}$$

The solution set is

$$S = \{x \mid -4 < x < \tfrac{2}{3}\}. \quad \square$$

EXAMPLE 6 Solve $(5x - 4)^2 < 13$.

Solution By A4 this inequality is equivalent to the inequality
$$|5x - 4| < \sqrt{13}.$$
Applying A7 (with $5x$ rather than x), we find that
$$4 - \sqrt{13} < 5x < 4 + \sqrt{13}$$
$$\frac{4 - \sqrt{13}}{5} < x < \frac{4 + \sqrt{13}}{5}. \quad \square$$

Exercises 1–3

1–22. Solve the inequalities.

- **1** $|2x + 3| \leq 5$
- **2** $|x + 3| \geq 7$
- **3** $|x - 1| \geq 8$
- **4** $|3x - 5| < -2$
- **5** $|2x - 5| > 0$
- **6** $\left|\dfrac{5 - 3x}{2}\right| < 1$
- **7** $\left|\dfrac{x}{3} - 4\right| > 5$
- **8** $|3 - 4x| > 5$
- **9** $|3 - x| \leq 5$
- **10** $|x^{-2} - 5| < 4$
- **11** $(x - 3)^2 < 4$
- **12** $(x + 4)^2 \leq 1$
- **13** $(2x + 5)^2 > 9$
- **14** $(3x - 2)^{-2} \geq 16$
- **15** $|x - 3| < \epsilon$
- **16** $|x + 5| < \epsilon$
- **17** $3|x| > x + 8$
- **18** $|x| > 2x + 9$
- **19** $2x + 1 < |x| - 2$
- **20** $3x - 5 < -4|x| + 9$
- **21** $2|x| - 1 > 3|x| - 3$
- **22** $|x + 5| > |x| - 5$

23–24. Use the Triangle Inequality to establish the given results.

23 $|x - 1| \geq |x| - 1$ (*Hint:* Consider $|(x - 1) + 1|$)

24 $|a - b| \geq |a| - |b|$

25–29. Prove the results.

25 The Triangle Inequality.

26 A5: $\sqrt{x^2} = |x|$.

27 A4: $x^2 < a^2$ if and only if $|x| < |a|$.

28 Let $b > 0$. Then $|a| < b$ if and only if $-b < a < b$.

29 Let $b \geq 0$. Then $|a| \geq b$ if and only if $a \geq b$ or $a \leq -b$.

30 The Triangle Inequality is $|a + b| \leq |a| + |b|$. Another useful inequality can be proved from the Triangle Inequality:

$$\big||a| - |b|\big| \leq |a + b|$$

Second Triangle Inequality

This new inequality is called the *Second Triangle Inequality*. It is usually combined with the Triangle Inequality to give

$$\big||a| - |b|\big| \leq |a + b| \leq |a| + |b|.$$

Prove the Second Triangle Inequality. (*Hint:* Show that $|a| - |b| \leq |a + b|$ and that $|b| - |a| \leq |a + b|$. For the first inequality, consider $|a + b - b|$.)

1-4 Distance Formula. Midpoint Formula. Circles

Points in the *xy*-plane can be represented by *coordinates,* ordered pairs of real numbers, as shown in Figure 1–8. The point $P(a, b)$ is located at the intersection of the vertical line through a on the *x*-axis and the horizontal line through b on the *y*-axis. The numbers a and b are called the *x*- and *y-coordinates*, respectively, of $P(a, b)$.

The coordinate axes divide the *xy*-plane into four *quadrants*. The quadrant that consists of points with both coordinates positive is called Quadrant I. We pass in order through Quadrants I, II, III, and IV as we move counterclockwise around the origin. (See Fig. 1–8.)

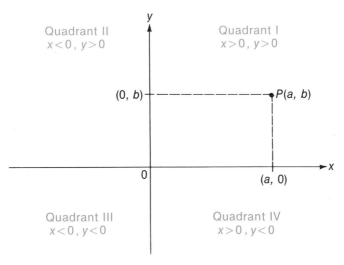

Figure 1–8 The coordinate plane.

DISTANCE FORMULA

If $P(a, b)$ is a point in the xy-plane, then $|a|$ is the distance from P to the y-axis and $|b|$ is the distance to the x-axis. More generally, if $P_1(x_1, y_1)$ and $P_2(x_2, y_2)$ are any two points, then

$$|x_2 - x_1| = \text{horizontal distance between } P_1 \text{ and } P_2,$$

and

$$|y_2 - y_1| = \text{vertical distance between } P_1 \text{ and } P_2.$$

(See Fig. 1–9.) We shall use this observation in conjunction with the Pythagorean Theorem to establish a formula for the distance between two points in the xy-plane.*

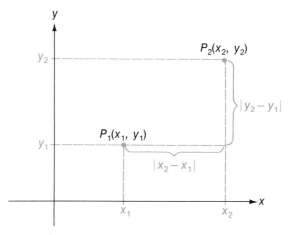

Figure 1–9 $|x_2 - x_1|$ = the horizontal distance between $P_1(x_1, y_1)$ and $P_2(x_2, y_2)$.
$|y_2 - y_1|$ = the vertical distance between $P_1(x_1, y_1)$ and $P_2(x_2, y_2)$.

In the work that follows, we denote the distance between points P_1 and P_2 by the symbol $|P_1P_2|$.

THEOREM 1–1   Let $P_1(x_1, y_1)$ and $P_2(x_2, y_2)$ be points in the xy-plane. The distance between P_1 and P_2 is

Distance Formula
$$|P_1P_2| = \sqrt{(x_2 - x_1)^2 + (y_2 - y_1)^2}.$$

Proof   Let Q be the point (x_2, y_1). (See Fig. 1–10.) Observe that P_1 and Q are on a line paral-

Pythagorean Theorem

* Pythagorean Theorem: If a and b are the lengths of the sides of a right triangle and c is the length of the hypotenuse, then $a^2 + b^2 = c^2$.

1-4 Distance Formula. Midpoint Formula. Circles

lel to the x-axis, and P_2 and Q are on a line parallel to the y-axis. Thus,

$$|P_1Q| = |x_2 - x_1| \quad \text{and} \quad |P_2Q| = |y_2 - y_1|.$$

Since the points P_1, Q, P_2 are the vertices of a right triangle with sides of length $|P_1Q|$, $|P_2Q|$, and hypotenuse $|P_1P_2|$, then by the Pythagorean Theorem,

$$\begin{aligned} |P_1P_2|^2 &= |P_1Q|^2 + |P_2Q|^2 \\ &= |x_2 - x_1|^2 + |y_2 - y_1|^2 \\ &= (x_2 - x_1)^2 + (y_2 - y_1)^2. \end{aligned}$$

If we take the positive square root of both sides of this last equation, we obtain

$$|P_1P_2| = \sqrt{(x_2 - x_1)^2 + (y_2 - y_1)^2}. \blacksquare$$

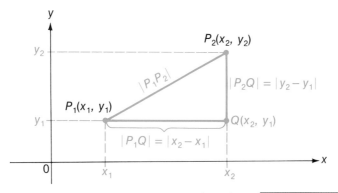

Figure 1-10 The distance formula. $|P_1P_2| = \sqrt{(x_2 - x_1)^2 + (y_2 - y_1)^2}$.

EXAMPLE 1   Calculate the distance between each pair of points: (a) $P_1(3, -1)$, $P_2(-1, 2)$; (b) $A(-4, 1)$, $B(3, 7)$.

Solution (a)
$$\begin{aligned} |P_1P_2| &= \sqrt{(x_2 - x_1)^2 + (y_2 - y_1)^2} \\ &= \sqrt{[-1 - 3]^2 + [2 - (-1)]^2} = \sqrt{16 + 9} = \sqrt{25} = 5. \end{aligned}$$

(b) $|AB| = \sqrt{[3 - (-4)]^2 + [7 - 1]^2} = \sqrt{49 + 36} = \sqrt{85} \approx 9.2195.$* □

MIDPOINT FORMULA

It is easy to find the midpoint of the line segment that connects the two points $P_1(x_1, y_1)$ and $P_2(x_2, y_2)$ in the xy-plane. The x-coordinate of the midpoint is the average of the x-coordinates; the y-coordinate of the midpoint is the average of the y-coordinates of P_1 and P_2. (See Fig. 1-11.) This result is stated formally in Theorem 1-2. The proof is outlined in Exercise 39.

* The symbol $\approx$ indicates that two quantities are *approximately equal*.

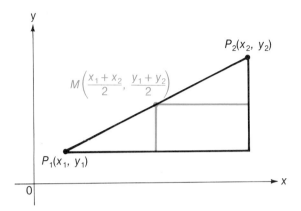

Figure 1-11 The midpoint formula. $M\left(\dfrac{x_1 + x_2}{2}, \dfrac{y_1 + y_2}{2}\right)$ is the midpoint of the line segment that connects $P_1(x_1, y_1)$ and $P_2(x_2, y_2)$.

THEOREM 1-2 (Midpoint Formula) Let $P_1(x_1, y_1)$ and $P_2(x_2, y_2)$ be points in the xy-plane. The point

Midpoint Formula
$$M\left(\frac{x_1 + x_2}{2}, \frac{y_1 + y_2}{2}\right)$$

is the midpoint of the line segment that connects P_1 and P_2.

For example, the midpoint between $P_1(3, -1)$ and $P_2(5, 7)$ is

$$M\left(\frac{3 + 5}{2}, \frac{-1 + 7}{2}\right) = M(4, 3).$$

GRAPHS OF EQUATIONS IN TWO VARIABLES

Graph of Equation

The set of all points $P(x, y)$ with coordinates that satisfy a given equation in x and y is called the *graph* of the equation. The graph of the equation $x^2 + y = 1$, for example, is

$$\{P(x, y) \mid x^2 + y = 1\}.$$

(See Fig. 1-12.) Thus, the point $(1, 0)$ is on the graph since $1^2 + 0^2 = 1$, while $(3, 1)$ is not on it since $3^2 + 1^2 \neq 1$.

To establish that a given geometrical figure is the graph of a stated equation, we must establish two facts:

(1) Every point that satisfies the equation is on the figure.
(2) Every point on the figure satisfies the equation.

1-4 Distance Formula. Midpoint Formula. Circles

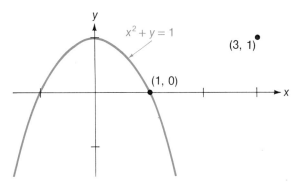

Figure 1-12 The graph of the equation $x^2 + y = 1$. The point $(1, 0)$ is on the graph, the point $(3, 1)$ is not.

THE CIRCLE

A circle is defined to be the set of all points $P(x, y)$ at a fixed distance (called the *radius*) from a fixed point C (called the *center*). It is a simple matter to use the distance formula to determine the equation of the circle with center $C(a, b)$ and radius $r > 0$.

THEOREM 1-3 The equation of the circle with center $C(a, b)$ and radius $r > 0$ is

Equation of Circle
$$(x - a)^2 + (y - b)^2 = r^2.$$

Proof We must show that a point $P(x, y)$ is on the circle if and only if it satisfies the equation

$$(x - a)^2 + (y - b)^2 = r^2.$$

Part 1 Suppose that $P(x, y)$ is on the circle. (See Fig. 1-13.) It follows from the definition of a circle that

$$|PC| = r$$
$$\sqrt{(x - a)^2 + (y - b)^2} = r \quad \text{(by the distance formula)}$$
$$(x - a)^2 + (y - b)^2 = r^2 \quad \text{(squaring both sides)}.$$

Thus, the coordinates of P satisfy the equation

$$(x - a)^2 + (y - b)^2 = r^2.$$

Part 2 Suppose that the coordinates of $P(x, y)$ satisfy the equation. Then

$$(x - a)^2 + (y - b)^2 = r^2.$$

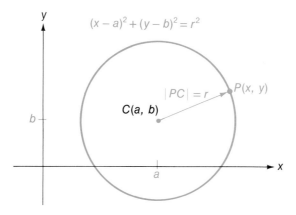

Figure 1–13 The circle $(x - a)^2 + (y - b)^2 = r^2$.

On taking positive square roots of both sides (and recalling that $r > 0$), we obtain
$$\sqrt{(x - a)^2 + (y - b)^2} = |r| = r.$$
Thus
$$|PC| = r$$
so that $P(x, y)$ is on the circle.

It follows from Parts 1 and 2 that the circle is the graph of the equation
$$(x - a)^2 + (y - b)^2 = r^2. \blacksquare$$

EXAMPLE 2   The graph of
$$(x + 1)^2 + (y - 2)^2 = 4$$
is the circle with center $C(-1, 2)$ and radius $r = 2$. (See Fig. 1–14.) □

GENERAL FORM OF THE EQUATION OF A CIRCLE

If we expand the squares and collect like terms, the equation $(x - a)^2 + (y - b)^2 = r^2$ can be rewritten in the form

General Form
$$x^2 + y^2 + Ax + By + C = 0.$$

This is known as the *general form* of the equation of a circle. For example, the equation of the circle in Example 2 can be rewritten as
$$x^2 + y^2 + 2x - 4y + 1 = 0.$$

1–4 Distance Formula. Midpoint Formula. Circles

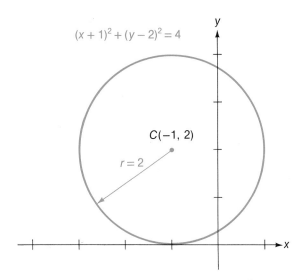

Figure 1–14 Example 2. The circle $(x + 1)^2 + (y - 2)^2 = 4$.

On the other hand, if we start with an equation

$$x^2 + y^2 + Ax + By + C = 0,$$

we can complete the square on the x-terms and on the y-terms and rewrite the equation as

$$(x - a)^2 + (y - b)^2 = c.$$

Three cases now arise:

(1) If $c > 0$, the graph is the circle with center (a, b) and radius $r = \sqrt{c}$.
(2) If $c = 0$, the graph consists of the single point (a, b).
(3) If $c < 0$, there is no graph. (The left-hand side of the equation is nonnegative for any choice of real x and y, and the right-hand side is negative.)

In the only case of real interest, the graph is a circle. In the other two cases we sometimes say, to be consistent, that the graph is a *degenerate circle*.

These remarks are illustrated in Example 3.

EXAMPLE 3 Determine the nature of the graphs of the following equations:
(a) $x^2 + y^2 - 4x + 6y = 3$
(b) $x^2 + y^2 + 2x + 2y + 4 = 0$
(c) $x^2 + y^2 - 10x - 8y + 41 = 0.$

Solution (a) $x^2 + y^2 - 4x + 6y = 3.$

We group the x terms and the y terms and complete the square on each set of terms:

$$(x^2 - 4x \quad) + (y^2 + 6y \quad) = 3$$
$$(x^2 - 4x + 4) + (y^2 + 6y + 9) = 3 + 4 + 9 = 16$$
$$(x - 2)^2 + (y + 3)^2 \quad = 4^2.$$

The graph is the circle with center $C(2, -3)$ and radius $r = 4$.

(b) $x^2 + y^2 + 2x + 2y + 4 = 0$.
We group the terms and complete the square:

$$(x^2 + 2x \quad) + (y^2 + 2y \quad) = -4$$
$$(x^2 + 2x + 1) + (y^2 + 2y + 1) = -4 + 1 + 1$$
$$(x + 1)^2 + (y + 1)^2 \quad = -2.$$

This equation has no solution with real x and y. Thus, there is no graph.

(c) $x^2 + y^2 - 10x - 8y + 41 = 0$.
We proceed as with the other examples:

$$(x^2 - 10x \quad) + (y^2 - 8y \quad) = -41$$
$$(x^2 - 10x + 25) + (y^2 - 8y + 16) = -41 + 25 + 16$$
$$(x - 5)^2 + (y - 4)^2 = 0.$$

This equation can be satisfied (for real values of x and y) only if $x = 5$ and $y = 4$. The graph is the single point $(5, 4)$. □

Exercises 1-4

1-4. Find the distance between the given pairs of points.

- **1** $(3, 2), (6, 6)$
- **2** $(5, 0), (0, 12)$
- **3** $(-2, 9), (5, 9)$
- **4** $(1, 2), (-1, -1)$

5-12. The given sets of points are vertices of triangles. Which of the triangles are right triangles? which isosceles?

- **5** $(2, 3), (-4, -3), (6, -1)$
- **6** $(2, 1), (5, -5), (8, 1)$
- **7** $(4, 4), (2, 1), (-1, 3)$
- **8** $(2, 1), (5, 0), (4, 7)$
- **9** $(0, -2), (4, -4), (5, 8)$
- **10** $(-3, -2), (4, -5), (5, 7)$
- **11** $(-3, 4), (-4, -7), (2, -1)$
- **12** $(1, 3), (5, 7), (5, -1)$

13-16. Find the midpoint of the line segment joining each pair of points.

- **13** $(-4, 4), (10, 6)$
- **14** $(-4, -1), (2, 3)$
- **15** $(2, 3), (-6, -7)$
- **16** $(4, 4), (10, 4)$

Exercises 1–4

●17. The ends of the base of an isosceles triangle are $(-1, 2)$ and $(7, 2)$. Its altitude is 10 units. What are the coordinates of the third vertex? (Two solutions are possible.)

●18 Two vertices of a square are $(-3, 2)$ and $(-3, 6)$. Find three pairs of other possible vertices.

19–26. Find the general equation of the circle that satisfies the stated conditions.

●19 Center $(1, 2)$, radius 4.

20 Center $(-3, 5)$, radius 1.

●21 Center $(4, -2)$, radius 3.

22 Center $(2, -4)$, passing through $(5, 8)$.

●23 Center $(3, -5)$, tangent to the y-axis.

24 Center $(-1, 4)$, tangent to the x-axis.

●25 The points $(-3, 4)$ and $(5, 2)$ are the endpoints of a diameter.

26 Center in Quadrant IV, radius 1, tangent to both coordinate axes.

27–36. Determine the nature of the graphs of the equations. If a graph is a circle, find its center and radius. If it is a single point or the empty set, state that fact.

●27 $x^2 + y^2 + 2x - 4y + 1 = 0$ 28 $x^2 + y^2 - 6x - 7 = 0$

●29 $x^2 + y^2 + 6x + 8y = 0$ 30 $x^2 + y^2 - 2x + 2y + 3 = 0$

●31 $x^2 + y^2 + 10x - 10y + 50 = 0$ 32 $x^2 + y^2 + 8x + 10y - 8 = 0$

●33 $4x^2 + 4y^2 - 16x + 4y + 1 = 0$ 34 $x^2 + y^2 + 6x - 2y + 10 = 0$

●35 $x^2 + y^2 + 4x - 6y + 15 = 0$ 36 $9x^2 + 9y^2 + 18x - 36y - 4 = 0$

●37 Determine whether the following points are inside, outside, or on the circle $x^2 + y^2 - 2x - 6y - 7 = 0$.
   (a) $(0, 0)$; (b) $(5, 2)$; (c) $(-\frac{1}{2}, -1)$; (d) $(-3, 4)$

●38 Find a simple equation satisfied by the set of all points $P(x, y)$ such that the distance from P to $A(3, -1)$ is twice the distance from P to $B(6, 5)$. Sketch the graph of the equation.

---

39 Prove the midpoint formula. This can be done in several ways. One method uses the triangles shown in Figure 1–11. (*Hint:* The triangles are similar.)

40 Let $x^2 + y^2 + Ax + By + C = 0$ be the equation of a circle. Let $P_0(x_0, y_0)$ be a point.
   (a) Prove that P_0 is inside the circle if
$$x_0^2 + y_0^2 + Ax_0 + By_0 + C < 0.$$

(b) Prove that P_0 is outside the circle if

$$x_0^2 + y_0^2 + Ax_0 + By_0 + C > 0.$$

(*Hint:* First work Exercise 37.)

41 Prove that the point $P(x, y)$ is equidistant from $A(0, 4)$ and the x-axis if and only if $x^2 - 8y + 16 = 0$.

•**42** Prove that the point $P(x, y)$ is equidistant from $A(5, 2)$ and $B(7, -3)$ if and only if $4x - 10y = 29$.

1–5 Lines in the xy-Plane

In this section we show that equation of a line in the xy-plane can be expressed in terms of the line's *slope* (the measure of its steepness) and the coordinates of any fixed point on the line. The work can be expedited by using *increments*.

Increment
An *increment* is a change measured along a line parallel to an axis. If we change from $P_1(x_1, y_1)$ to $P_2(x_2, y_2)$, the increments of change are Δx (read "delta x") and Δy, where

$$\Delta x = x_2 - x_1 \quad \text{(the change in the horizontal direction)},$$
$$\Delta y = y_2 - y_1 \quad \text{(the change in the vertical direction)}.$$

(See Fig. 1–15.)

If we change, for example, from $P_1(2, -3)$ to $P_2(1.4, -2.7)$, the increments of change are

$$\Delta x = \text{horizontal change} = x_2 - x_1 = 1.4 - 2 = -0.6,$$
$$\Delta y = \text{vertical change} = y_2 - y_1 = -2.7 - (-3) = 0.3.$$

In calculating increments of change, we subtract the coordinates of the original point from the coordinates of the new point. If we move from $P_2(x_2, y_2)$ to $P_1(x_1, y_1)$, for example, the increments are

$$\Delta x = x_1 - x_2 \quad \text{and} \quad \Delta y = y_1 - y_2.$$

EXAMPLE 1 Find the point $P(x, y)$ obtained by a change from $Q(2.5, 3.4)$ by increments of

$$\Delta x = 1.2 \quad \text{and} \quad \Delta y = -1.3.$$

Solution $\Delta x = x - 2.5 = 1.2$ and $\Delta y = y - 3.4 = -1.3$. Therefore,

$$x = 3.7 \quad \text{and} \quad y = 2.1.$$

The new point is $P(3.7, 2.1)$. (See Fig. 1–16.) □

1–5 Lines in the xy-Plane

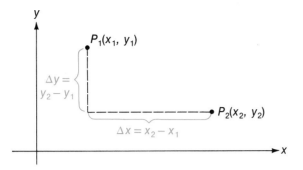

Figure 1–15 The increments Δx and Δy.
$\Delta x = x_2 - x_1 =$ the horizontal change from P_1 to P_2;
$\Delta y = y_2 - y_1 =$ the vertical change from P_1 to P_2.

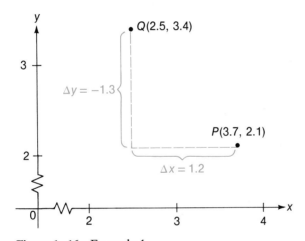

Figure 1–16 Example 1.

SLOPES OF LINES

Let $\mathscr{L}$ be a nonvertical line containing the distinct points $P_1(x_1, y_1)$ and $P_2(x_2, y_2)$. (See Fig. 1–17.) Let

$$\Delta x = x_2 - x_1, \qquad \Delta y = y_2 - y_1,$$

be the increments of change from P_1 to P_2. The ratio

Slope
$$m = \frac{\Delta y}{\Delta x} = \frac{y_2 - y_1}{x_2 - x_1}$$

is called the *slope* of line $\mathscr{L}$.

The quantity $\Delta y = y_2 - y_1$ is sometimes called the *rise* in the line from P_1 to P_2. Similarly, $\Delta x = x_2 - x_1$ is the *run*. (See Fig. 1–17.) The slope can be given as the "surveyor's formula"

$$m = \frac{\text{rise}}{\text{run}} = \frac{y_2 - y_1}{x_2 - x_1}.$$

EXAMPLE 2   The line through $P_1(3, 2)$ and $P_2(5, 0)$ has slope

$$m = \frac{\Delta y}{\Delta x} = \frac{y_2 - y_1}{x_2 - x_1} = \frac{0 - 2}{5 - 3} = \frac{-2}{2} = -1.$$

(See Fig. 1–18.) ☐

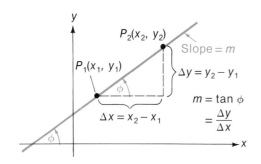

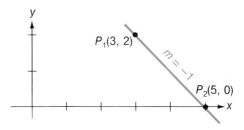

Figure 1–17   The slope of a line.

ϕ = the angle of inclination

m = the slope = $\tan \phi = \dfrac{y_2 - y_1}{x_2 - x_1} = \dfrac{\Delta y}{\Delta x}$

Figure 1–18   Example 2.

GEOMETRICAL INTERPRETATION OF THE SLOPE OF A LINE

Angle of Inclination

If the line $\mathscr{L}$ is not horizontal, then it intersects the *x*-axis. The smallest positive angle ϕ formed from the positive direction of the *x*-axis to the line $\mathscr{L}$ is called the *angle of inclination of $\mathscr{L}$*. As we see in Fig. 1–17,

$$\tan \phi = \frac{\Delta y}{\Delta x} = m,$$

so that *the slope equals the tangent of the angle of inclination.*

If the line $\mathscr{L}$ is horizontal, we *define* its angle of inclination to be zero. Then

$$m = \frac{\Delta y}{\Delta x} = \frac{y_2 - y_1}{x_2 - x_1} = \frac{0}{x_2 - x_1} = 0 = \tan 0$$

so that also in this case $\tan \phi = m$.

If the line $\mathscr{L}$ is vertical, then $\phi = 90°$. In this case $\Delta x = 0$, so that neither $\tan \phi$ nor

1–5 Lines in the xy-Plane

$\Delta y / \Delta x$ is defined. Thus a vertical line has no slope. (This is quite different from saying that a given line has zero slope.)

It follows from the above discussion that the slope of a nonvertical line is completely determined by its angle of inclination. Thus, we get the same value for the slope m no matter which points $P_1(x_1, y_1)$ and $P_2(x_2, y_2)$ are used to compute the value. Observe also that

$$m = \frac{y_2 - y_1}{x_2 - x_1} = \frac{y_1 - y_2}{x_1 - x_2}.$$

Thus, in applying the formula, we can label either point P_1 provided the other is P_2.

PROPERTIES OF SLOPES

Let ϕ be the angle of inclination of a line $\mathscr{L}$. Recall that $\tan \phi > 0$ if $0 < \phi < 90°$ and $\tan \phi < 0$ if $90° < \phi < 180°$. It follows that $\mathscr{L}$ has positive slope if it rises to the right and negative slope if it descends to the right (Figs. 1–19a, b). Horizontal lines have zero slope and vertical lines have no slope (Fig. 1–19c). The larger the absolute value

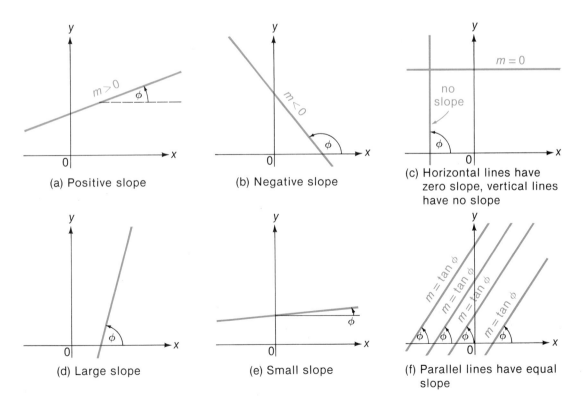

Figure 1–19 Properties of slopes.

of the slope, the nearer the line to the vertical; the smaller the absolute value of the slope, the nearer the line to the horizontal (Figs. 1–19d, e).

Two lines are parallel if and only if they have the same angle of inclination. It follows that two nonvertical lines are parallel if and only if they have the same slope. (See Fig. 1–19f.)

SLOPES OF PERPENDICULAR LINES

Let $\mathcal{L}_1$ and $\mathcal{L}_2$ be nonvertical, perpendicular lines. We shall show that their slopes are negative reciprocals, that is,

$$m_2 = -\frac{1}{m_1}.$$

To establish this fact, we work with their angles of inclination ϕ_1 and ϕ_2. For the sake of argument, we assume that $0 < \phi_1 < \phi_2$, as in Figure 1–20, so that

$$\phi_2 = \phi_1 + 90°.$$

Before proceeding further, we recall two formulas from trigonometry (Formulas T8 and T9 in Appendix B):

$$\sin(\alpha + \beta) = \sin\alpha \cos\beta + \cos\alpha \sin\beta,$$
$$\cos(\alpha + \beta) = \cos\alpha \cos\beta - \sin\alpha \sin\beta.$$

If we take $\alpha = \phi_1$ and $\beta = 90°$, we find that

$$\sin \phi_2 = \sin(\phi_1 + 90°) = \cos \phi_1$$

and

$$\cos \phi_2 = \cos(\phi_1 + 90°) = -\sin \phi_1.$$

Then

$$m_2 = \tan \phi_2 = \frac{\sin \phi_2}{\cos \phi_2} = \frac{\cos \phi_1}{-\sin \phi_1}$$
$$= -\cot \phi_1 = -\frac{1}{\tan \phi_1} = -\frac{1}{m_1}.$$

If $0 < \phi_2 < \phi_1$, we work through the same steps, changing all the 1 subscripts to 2 and all the 2 subscripts to 1.

EXAMPLE 3 The line $\mathcal{L}_1$ through $A(3, 4)$ and $B(4, 1)$ has slope

$$m_1 = \frac{1-4}{4-3} = -3.$$

1–5 Lines in the xy-Plane

The line $\mathscr{L}_2$ through $B(4, 1)$ and $C(-2, -1)$ has slope

$$m_2 = \frac{-1-1}{-2-4} = \frac{-2}{-6} = \frac{1}{3}.$$

Since $m_2 = -1/m_1$, then $\mathscr{L}_1$ and $\mathscr{L}_2$ are perpendicular (Fig. 1–21). □

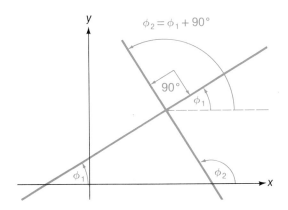

Figure 1–20 Perpendicular lines. The slope of one line is the negative reciprocal of the slope of the other, $m_2 = -\dfrac{1}{m_1}$.

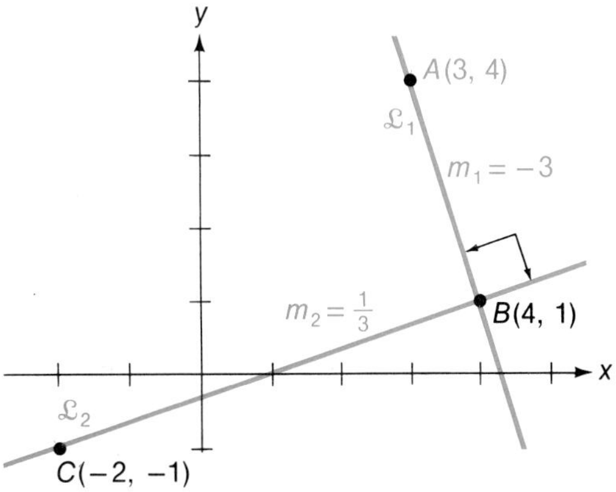

Figure 1–21 Example 3.

EQUATIONS OF LINES

A nonvertical line is completely determined by its slope and any single point that it contains. We shall establish that the equation of a nonvertical line can be expressed in terms of the slope and a point.

THEOREM 1–4 The equation of the line $\mathscr{L}$ through the point $P_0(x_0, y_0)$ with slope m is

$$y - y_0 = m(x - x_0).$$

Proof *Part 1* Let $P(x, y)$ be a point on the line. (See Fig. 1–22.) If $x \neq x_0$, then $P_0(x_0, y_0)$ and $P(x, y)$ can be used to calculate the slope m:

$$m = \frac{\Delta y}{\Delta x} = \frac{y - y_0}{x - x_0},$$

so that

$$y - y_0 = m(x - x_0).$$

Observe that this equation also is satisfied by the coordinates of $P_0(x_0, y_0)$. Thus, any point on the line satisfies the equation.

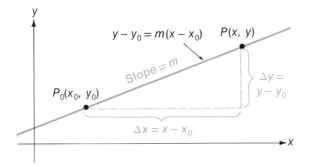

Figure 1-22 Theorem 1-4. The point-slope form of the equation of a line: $y - y_0 = m(x - x_0)$

Part 2 Let $P(x, y)$ be a point that satisfies the equation

$$y - y_0 = m(x - x_0).$$

If $x = x_0$, then $y - y_0 = m \cdot 0 = 0$, so that $y = y_0$. Thus P is the point (x_0, y_0), which is known to be on the line.

If $x \neq x_0$, then

$$m = \frac{y - y_0}{x - x_0}.$$

Thus, m is the slope of the line through $P(x, y)$ and $P_0(x_0, y_0)$. Since $\mathscr{L}$ is the only line through $P_0(x_0, y_0)$ with slope m, then $P(x, y)$ must be on the line $\mathscr{L}$.

Since every point on the line satisfies the equation and every point that satisfies the equation is on the line, then $\mathscr{L}$ is the graph of the equation

$$y - y_0 = m(x - x_0). \blacksquare$$

The equation

Point-Slope Form
$$y - y_0 = m(x - x_0)$$

is called the *point-slope form* of the equation of a line. It can be used to determine the equation when we are given a point and the slope.

EXAMPLE 4   The equation of the line through $P_0(3, -1)$ with slope $m = -\frac{1}{7}$ is

$$y - (-1) = -\tfrac{1}{7}(x - 3),$$

or rewritten more simply,

$$x + 7y = -4. \square$$

1-5 Lines in the xy-Plane

In the special case for which the point on the line is the y-intercept $(0, b)$ (see Fig. 1-23), the equation of the line is

$$y - b = m(x - 0)$$

which can be rewritten

Slope-Intercept Form
$$y = mx + b.$$

This last equation is called the *slope-intercept form* of the equation of a line. It can be used to find the slope and a point on the line from the equation.

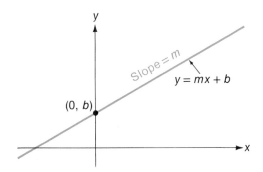

Figure 1-23 The slope-intercept form of the equation of a line.

EXAMPLE 5   Find the slope and a point on the line $2x + 3y = 6$.

Solution   We solve the equation for y: $y = -\frac{2}{3}x + 2$. The slope is $m = -\frac{2}{3}$; one point on the line is the y intercept $(0, 2)$. □

GENERAL FORM

In Examples 4 and 5, we encountered equations of form

General Form
$$ax + by = c.$$

This equation is called the *general form* of the equation of a line. If a and b are not both zero, the graph of this equation is a line. In particular:

(1) If $b \neq 0$, the equation can be solved for y,

$$y = -\frac{a}{b}x + \frac{c}{b}.$$

The graph is the line with slope $m = -a/b$ and y-intercept $(0, c/b)$.

(2) If $b = 0$ and $a \neq 0$, the equation can be rewritten

$$x = \frac{c}{a}.$$

Vertical Line

The graph of this equation is

$$\left\{ P(x, y) \,\big|\, x = \frac{c}{a} \right\}.$$

It is the vertical line through $(c/a, 0)$. (See Fig. 1–24.)

EXAMPLE 6   Find the equation of the perpendicular bisector of the line segment connecting the points $A(-2, 1)$ and $B(4, 3)$.

Solution   The slope of the line segment connecting A and B is

$$\text{Slope of line segment } AB = \frac{3 - 1}{4 - (-2)} = \frac{2}{6} = \frac{1}{3}.$$

The slope of the perpendicular bisector of this line segment is the negative reciprocal, $m = -3$. Let M be the midpoint of the line segment. By the midpoint formula, M is the point

$$M \left(\frac{-2 + 4}{2}, \frac{1 + 3}{2} \right) = M(1, 2).$$

Since the perpendicular bisector of the line segment AB passes through M, its equation is

$$y - 2 = (-3)(x - 1)$$
$$3x + y = 5.$$

(See Fig. 1–25.) □

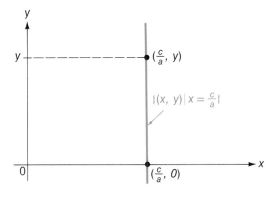

Figure 1–24 The graph of $ax = c$ is the vertical line through $(c/a, 0)$.

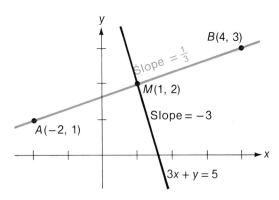

Figure 1–25 Example 6.

Exercises 1–5

1. Find the increments Δx and Δy that measure the change from P to Q:
   (a) $P(-1, 3)$, $Q(-0.95, 3.02)$
   (b) $P(2, 7)$, $Q(1.95, 6.87)$
   (c) $P(-3, 2)$, $Q(-3.98, 2.1)$
   (d) $P(-5, -3)$, $Q(-4.95, -3.01)$

2. Find the point $P(x, y)$ obtained by a change from Q by the increments Δx and Δy.
   (a) $Q(1, 3)$, $\Delta x = 0.01$, $\Delta y = -0.02$
   (b) $Q(0, -4)$, $\Delta x = -0.1$, $\Delta y = 0.002$
   (c) $Q(-1, 3)$, $\Delta x = -0.02$, $\Delta y = 0.03$
   (d) $Q(-1.25, 2.3)$, $\Delta x = 0.03$, $\Delta y = -0.04$

3–8. Find the slope of the line through each pair of points.

3. $(3, 1)$, $(5, 4)$

4. $(8, -2)$, $(-1, -2)$

5. $(2, -3)$, $(-5, 3)$

6. $(4, -3)$, $(4, 5)$

7. $(5, 7)$, $(2, 1)$

8. $(-3, 4)$, $(2, -8)$

9. The points $(-4, 1)$, $(0, -2)$, $(6, 6)$ and $(2, 9)$ are the vertices of a quadrilateral. Show that it is a parallelogram.

10. Use slopes to prove that the triangle with vertices $(3, -4)$, $(9, -4)$ and $(6, 2)$ is isosceles.

11–16. Determine whether or not the given points are collinear by (1) using distance and (2) using slopes.

11. $(-3, 2)$, $(4, 4)$, $(8, 5)$

12. $(-1, 0)$, $(11, 3)$, $(-9, -2)$

13. $(6, 2)$, $(1, 1)$, $(-4, 0)$

14. $(1, -3)$, $(-3, -11)$, $(3, 1)$

15. $(2, 7)$, $(4, 3)$, $(5, 1)$

16. $(-2, 1)$, $(2, 3)$, $(1, 2)$

17–26. Find the equation of the line that satisfies the given conditions.

17. Through $(3, -1)$ with slope $m = 2$.

18. Through $(-4, 2)$ with slope $m = -5$.

19. Vertical with x-intercept -1.

20. Horizontal with y-intercept 4.

21. With slope $m = -2$ and y-intercept 3.

22. With slope $m = 4$, x-intercept -5.

23. With x-intercept 2, y-intercept 5.

24. With x-intercept -5, y-intercept 1.

25. With angle of inclination $0°$, y-intercept 4.

26. With angle of inclination $135°$, y-intercept -2.

27–30. Find the slope and y-intercept of each of the lines.

•27 $3x + 8y - 24 = 0$ 28 $4x - 5y = 15$

•29 $4x - 3y + 6 = 0$ 30 $3 + 2y - 10x = 0$

31–34. Find the equations of the lines through the given point that are (a) parallel to and (b) perpendicular to the given line.

•31 $y = 2x - 3$, origin   32 $y = -3x + 5$, (6, 7)

•33 $4x + 6y = 3$, $(-2, -1)$ 34 $3x + y = 6$, (5, 1)

35–38. Find the points of intersection of the line and the circle.

•35 $x = 3$, $x^2 + y^2 = 25$ 36 $x + 2y = 10$, $x^2 + y^2 = 25$

•37 $x - y = 5$, $x^2 + y^2 - 4y = 1$ 38 $3x - 4y = 15$, $x^2 + y^2 = 10x$

•39 Show that the lines $2x - y = 5$, $x + 3y = 6$, $2x - y = -9$, and $x + 3y = -1$ form a parallelogram. Find its vertices.

•40 Find the equation of the line that is tangent to the circle $x^2 + y^2 - 8x + 6y + 8 = 0$ at the point (3, 1). (*Hint:* A line is tangent to a circle at a point P on the circle provided it passes through P and is perpendicular to the radius from the center to P.)

•41 A circle is tangent to the line $4x - 3y + 10 = 0$ at the point $(-1, 2)$ and to the line $3x + 4y - 30 = 0$ at (6, 3). Find the equation of the circle.

---

42 (*Intercept Form of the Equation of a Line*) A line intersects the x- and y-axes at the nonzero points a and b. Show that its equation is

Intercept Form

$$\frac{x}{a} + \frac{y}{b} = 1.$$

•43 The ends of the base of an isosceles triangle are $A(-2, 5)$ and $B(5, 4)$. The altitude is $5\sqrt{2}/2$. Find the coordinates of the third vertex. (Two solutions are possible.)

•44 Let A and B be the points $A(3, -1)$ and $B(10, 0)$. Find a number x such that $P(x, -4)$ is the vertex of a right angle APB.

•45 (a) A perpendicular is dropped from the point (1, 4) to the line $y - 2x + 8 = 0$. Find the coordinates of the foot of the perpendicular. What is the distance from the point to the line?
   (b) Work Part (a) for the point (4, 1) and the line $3x - 4y + 12 = 0$.

46 (*Three-Point Circle*) The perpendicular bisector of the line segment connecting any two points on a circle must pass through the center of the circle. Draw a figure that illustrates how any three points on a circle can be used to determine the center and radius of that circle. Explain how this problem also can be solved by considering a system of three equations in three unknowns.

47–50. Use Exercise 46 to find the equation of the circle determined by the set of three points.

1-6 Functions and Their Graphs

•47 (2, 3), (3, 2), (−4, 3) 48 (3, −5), (−1, −5), (−2, −4)

•49 (2, 8), (5, −1), (6, 0) 50 (0, 1), (1, 0), (4, 3)

51 (*Distance from a Point to a Line*) Generalize the method of Exercise 45. Let $ax + by + c = 0$ be the equation of a line; let $P_0(x_0, y_0)$ be a point. Prove that the distance from P_0 to the line is

Distance from Point to Line
$$d = \frac{|ax_0 + by_0 + c|}{\sqrt{a^2 + b^2}}.$$

•**52** Find the perpendicular distance between the parallel lines $2x + 3y = 7$ and $2x + 3y = 11$.

53 (*Angle of Intersection of Two Lines*) Two lines $\mathscr{L}_1$ and $\mathscr{L}_2$ with slopes m_1 and m_2 intersect at a point. Show that the smallest angle θ between the two lines satisfies the equation

Angle of Intersection
$$\tan \theta = \left| \frac{m_2 - m_1}{1 + m_2 m_1} \right| \qquad \text{provided } \theta \neq 90°.$$

•**54** Two lines with slopes m_1 and m_2 intersect at an angle $\theta = 45°$. Find m_2 if $m_1 = \frac{1}{2}$. (*Hint:* Use Exercise 53.)

1-6 Functions and Their Graphs

Many quantities depend on other quantities for their values. The number of seedlings required to reforest a region, for example, depends on the area of the region; the cost of launching a rocket depends on the weight of the rocket; and the area of a circle depends on the radius.

In mathematics we are particularly interested in relations such as the one between the area and the radius of a circle—relations in which one quantity (say y) is completely determined by another quantity (x). This type of relation defines y as a *function* of x. Observe that a function involves the following concepts: (1) the set $\mathscr{D}$ of allowable values of x; (2) a law of correspondence that can be used to calculate y from x; and (3) the set of values of y that can be calculated from the values of x. The formal definition follows.

DEFINITION

Function

A *function* from a set $\mathscr{D}$ to a set $\mathscr{S}$ is a rule, or law of correspondence, that assigns a unique element of $\mathscr{S}$ to each element of $\mathscr{D}$.*

* The following, more precise definition of the word *function* is favored by many mathematicians. A *function* is a collection of ordered pairs (x, y) such that no two distinct ordered pairs have the same first coordinate. In other words, if (x, y_1) and (x, y_2) are in the set, then y_1 must equal y_2. The set of all first coordinates of ordered pairs in the function is the *domain;* the set of all second coordinates is the *range*.

Domain
Range

The set $\mathcal{D}$ is called the domain of the function. If $x \in \mathcal{D}$, the element $y \in \mathcal{S}$ that the function assigns to x is called the *image* of x. The set $\mathcal{R}$ of all images of elements of $\mathcal{D}$ is called the *range* of the function. Observe that the range $\mathcal{R}$ is a subset of the set $\mathcal{S}$.

EXAMPLE 1 Let f be the function that assigns to each positive number x the number that is the area of a circle with radius x. The domain of f is

$$\mathcal{D} = \{x \mid x > 0\}.$$

The range is

$$\mathcal{R} = \{y \mid y > 0\}.$$

The law of correspondence is

$$y = \pi x^2. \quad \square$$

In most cases, if we know the law of correspondence and the domain, we can determine the range. Observe that the sets that compose the domain and the range may overlap. They may even be equal.

Remark In this textbook, we are concerned almost exclusively with *real* functions. These are functions in which both the domain and the range are sets of real numbers.

$f(x)$ NOTATION

Numerous notations have been developed to represent functions. The most useful in many mathematical settings is the $f(x)$ notation. If f represents the function and x is an element in the domain of f, then $f(x)$, read "*f of x*," denotes the number assigned to x by the function.

In Example 1, the function f assigns the number $y = \pi x^2$ to each $x > 0$. Thus,

$$f(x) = \pi x^2 \qquad \text{for each } x > 0.$$

In particular,

$$f(1) = \pi \cdot 1^2 = \pi$$
$$f(2) = \pi \cdot 2^2 = 4\pi$$
$$f(t) = \pi t^2 \text{ if } t > 0,$$

while $f(-1)$ and $f(0)$ are not defined. (Recall that the domain of f is the set of positive real numbers.)

Other symbols besides f can be used to represent functions. Mostly we shall use the letters f, g, h, F, G, and H. Certain special functions have been given common names, such as "ln" or "sin."

Functions can be represented schematically in many ways. Diagrams such as Figure 1–26a indicate that to each element x of $\mathcal{D}$ there corresponds a unique element $f(x)$ of $\mathcal{R}$. Such diagrams have the drawback, however, of suggesting that the domain and the range must be disjoint sets; they are not necessarily so.

A function can be illustrated by a mythical machine. (See Fig. 1–26b.) When a

1-6 Functions and Their Graphs

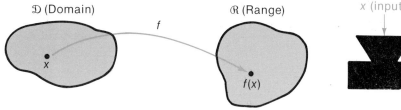
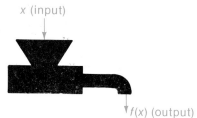

(a) Schematic representation. The function f assigns an element f(x) in the range ℛ to each element x in the domain 𝒟.

(b) A function represented by a machine. Each value of x causes the machine to generate a value f(x).

Figure 1-26 Two representations of a function.

number x is fed into the machine, the corresponding number $f(x)$ comes out. This approach is exemplified by the modern scientific calculator. If we enter a number such as 1.253 and push the **LN X** button, we immediately get the number 0.2255407 as output. The machine automatically calculates the value $f(1.253) = 0.2255407$ for the function $f(x) = $ **LN X**. It is not necessary for the user to understand the meaning of the symbol "ln x." He needs to know only that for certain numbers x the values of $\ln x$ can be calculated by pushing the **LN X** button. If your machine computes to 10 decimal places instead of 7, you get $f(1.253) = 0.2255406759$. Thus, the value you get depends on the machine you are using. There is, of course, a theoretical value for $f(1.253) = \ln(1.253)$, which we explain in Section 8-1. In most cases the numbers obtained from the calculators are only approximations of the exact value.

UNSPECIFIED DOMAINS

Rule for Unspecified Domains A function is not properly defined until we know both its domain and its law of correspondence. If only a law of correspondence is given, then by convention, the domain consists of all real numbers x that have real images. In other words, the domain of f consists of all real numbers for which $f(x)$ is defined and real.

EXAMPLE 2 Determine the domains of the functions f, g, and h defined as follows:
(a) $f(x) = 2x + 5$.
(b) $g(x) = \dfrac{3x - 1}{x + 2}$.
(c) $h(x) = \sqrt{x^2 - 1}$.

Solution (a) $f(x) = 2x + 5$. The domain consists of all real numbers x for which $2x + 5$ is real. Since this condition holds for any real x, then

$$\mathcal{D}_f = \{x \mid x \text{ is real}\}.$$

(b) $g(x) = (3x - 1)/(x + 2)$. Observe that $g(x)$ is defined and is real for any real number x except $x = -2$. (Division by zero is not allowed.) Thus,

$$\mathscr{D}_g = \{x \mid x \neq -2\}.$$

(c) $h(x) = \sqrt{x^2 - 1}$. For $\sqrt{x^2 - 1}$ to be real, we must have $x^2 - 1 \geq 0$. Then

$$x^2 \geq 1^2$$
$$|x| \geq |1| = 1 \quad \text{(by A4)}$$
$$x \leq -1 \text{ or } x \geq 1 \quad \text{(by A6)}.$$

The domain is

$$\mathscr{D}_h = \{x \mid x \leq -1 \text{ or } x \geq 1\}. \ \square$$

Equations in x and y are sometimes used to define functions. Not every such equation defines a function. If an equation defines y as a function of x, then for every x in the domain of the function, there must exist exactly one corresponding value of y.

EXAMPLE 3 **(a)** The equation $xy = 2x - 1$ defines y as a function of x:

$$y = f(x) = \frac{2x - 1}{x} = 2 - \frac{1}{x}.$$

The domain of the function is $\{x \mid x \neq 0\}$.

(b) The equation $y^2 = x^2 + 1$ does not define y as a function of x. For each real number x, there are two corresponding values of y for which the equation is satisfied:

$$y = \sqrt{x^2 + 1} \quad \text{and} \quad y = -\sqrt{x^2 + 1}. \ \square$$

Remark It is awkward to write out expressions such as "the function f defined by the law of correspondence $f(x) = 3x^2 - 2x + 7$." Mathematicians frequently shorten the expression and write simply "the function $f(x) = 3x^2 - 2x + 7$." We shall adopt this abuse of the language in the remainder of the book. You should recognize, however, that it is just that—an abuse of the language. Technically, the function is f and the law of correspondence of $f(x) = 3x^2 - 2x + 7$.

TYPES OF FUNCTIONS

Real functions are of four basic types:

(1) Polynomials A function of form

$$f(x) = a_n x^n + a_{n-1} x^{n-1} + \cdots + a_1 x + a_0,$$

where $a_0, a_1, \ldots, a_n$ are numbers and n is a nonnegative integer, is called a *polynomial*. For example, the functions

$$f(x) = \sqrt{2} x^2 - 1, \quad g(x) = 3x + 2, \quad \text{and} \quad h(x) = 0$$

are polynomials.

1–6 Functions and Their Graphs

(2) **Rational Functions** A function such as

$$F(x) = \frac{3x^2 - 2x + 5}{x^3 - 5x^2 + 7},$$

which is the quotient of two polynomials, is called a *rational function*.

(3) **Algebraic Functions** A function such as

$$G(x) = \frac{3\sqrt{x} - 5}{2x + 7} - \sqrt[3]{\frac{x^2 - 4\sqrt{x}}{3 - x}},$$

which can be built up from polynomials by taking a finite number of sums, differences, products, quotients, powers, and roots, is called an *algebraic function*.

(4) **Transcendental Functions** A function that is not an algebraic function is said to be *transcendental*. In the latter part of this book we consider several transcendental functions, such as $f(x) = \ln x$, $g(x) = e^x$, and $h(x) = \sin x$.

Observe that every polynomial is a rational function and that every rational function is algebraic. Every real function is either algebraic or transcendental.

To illustrate the function concept further, we give several examples of relations that may or may not define functions.

EXAMPLE 4 In parts (a), (b), (c) we use the set S, which consists of all positive integers between 1 and 10 (inclusive):

$$S = \{1, 2, 3, \ldots, 10\}.$$

(a) For each $x \in S$, let $f(x)$ be the largest element of S.
This is a function. Observe that $f(x) = 10$ for every $x \in S$.
(b) For each $x \in S$, let $g(x)$ be the largest element of S that is smaller than x.
This is not a function. It is defined for all $x > 1$ (for example, $g(7) = 6$) but not for $x = 1$.
(c) For each $x \in S$, let $h(x)$ be an element of S that is greater than or equal to x.
This is not a function. The statement does not assign a unique number $h(x)$ to each x. (For example, $h(6)$ could refer to 7, 8, 9, or 10.)
(d) For each real number x, let $j(x)$ be the smallest real number that is greater than x.
This is not a function. There is no smallest real number that is greater than x.
(e) For each real number x, let $F(x)$ be the largest integer that is not greater than x.
This is a function. For example, $F(2.5) = 2$, $F(7) = 7$, and $F(-2.8) = -3$.
(This function, known as the *greatest-integer function*, will be discussed in more detail at the end of the section.) □

GRAPHS OF FUNCTIONS

The *graph* of a function f is the set of all points (x, y) such that $y = f(x)$. For example, if

$$f(x) = \frac{3x - 1}{x + 2}$$

then $f(-1) = -4$, so that the point $(-1, -4)$ is on the graph of f. Observe that the graph of the function f is identical to the graph of the equation $y = f(x)$.

Most of the functions we shall consider have graphs that consist of one or more smooth curves. The graph of a typical function is pictured in Figure 1–27a.

Not every curve is the graph of a function. Recall that a function f has the property that for every x_0 in its domain there exists a unique number y_0 in its range such that $y_0 = f(x_0)$. Thus, there can be only one point (x_0, y_0) on the graph for each x_0 in the domain. It follows that a vertical line can intersect the graph of a function in at most one point. (See Fig. 1–27a.)

The curve in Figure 1–27b is not the graph of a function. The vertical line $x = x_0$ intersects the graph at two distinct points.

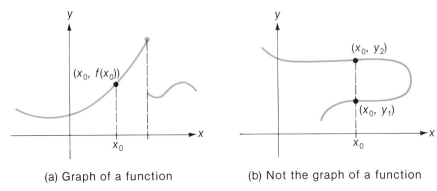

(a) Graph of a function   (b) Not the graph of a function

Figure 1–27 If a figure is the graph of a function, then each vertical line must intersect it in at most one point.

THE GREATEST-INTEGER FUNCTION

The greatest-integer function was considered briefly in Example 4(e). Since this function will be useful later, we discuss it in more detail here. It is standard to use the symbol $[x]$ to indicate the greatest-integer function. Thus,

Greatest Integer Function

$[x]$ = the largest integer that is not greater than x.

Observe that

$$[x] = -1 \quad \text{if } -1 \leq x < 0,$$
$$[x] = 0 \quad \text{if } 0 \leq x < 1,$$
$$[x] = 1 \quad \text{if } 1 \leq x < 2,$$
$$[x] = 2 \quad \text{if } 2 \leq x < 3,$$

and so on.

The graph of $f(x) = [x]$ is pictured in Figure 1–28. It consists of straight-line seg-

Exercises 1-6

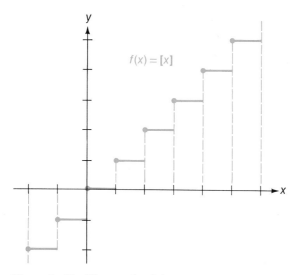

Figure 1-28 The graph of the greatest-integer function $f(x) = [x]$.

ments one unit long. The solid dots at the end of the line segments indicate the functional values of the integers. For example, $[2] = 2$, not 1. This function jumps in value at the integers.

Exercises 1-6

1-16. Find the domains of the functions defined by these laws of correspondence. (*Note:* $[x]$ is the greatest integer that is not greater than x).

- **1** $f(x) = \sqrt{1 - x}$
- **2** $f(x) = \dfrac{1}{x}$
- **3** $h(y) = \sqrt{y - 4}$
- **4** $f(t) = \sqrt{2t + 3}$
- **5** $g(z) = \sqrt[3]{z^2 - 4}$
- **6** $g(x) = \dfrac{1}{\sqrt{4 - x^2}}$
- **7** $g(x) = \dfrac{1}{9 - x^2}$
- **8** $h(x) = \dfrac{1}{\sqrt{x}(x - 2)}$
- **9** $g(z) = \sqrt[3]{z^2 - 1}$
- **10** $h(x) = \sqrt{\dfrac{x + 1}{x - 1}}$
- **11** $f(x) = \dfrac{1}{[x]}$
- **12** $g(x) = \dfrac{2x - 3}{x^2 - x}$

• 13 $f(x) = \dfrac{1}{\sqrt{x - 5}\sqrt{7 - x}}$

14 $f(y) = \dfrac{y}{\sqrt{(y - 1)(y + 2)}}$

• 15 $g(x) = \dfrac{[x]}{x + |x|}$

16 $f(x) = \dfrac{1}{x^2 + 1}$

17–22. Calculate **(a)** $f(1)$, **(b)** $f(2)$, **(c)** $f(a + h)$, and **(d)** $f(a) + f(h)$ for the given functions.

• 17 $f(x) = x^2 + 2x - 3$

18 $f(x) = x^3 + 3x^2 - 4x - 2$

• 19 $f(x) = \dfrac{1 + x}{1 - x}$

20 $f(z) = 2 - z - 3z^2$

• 21 $f(y) = y^2 + 2y - 2$

22 $f(t) = \dfrac{t^2 - 4}{t - 2}$

23 Which of the following propositions must be true for every real function f defined on the entire x-axis? If a proposition is false, find an example in which it is false. (Hint: First calculate the functional values for a "simple" function such as $f(x) = x^2 + 1$ or $f(x) = 2x - 3$. If the stated relation is true for this function, then see whether you can explain why it is true in general.)

(a) $f(3 - 2) = f(3) - f(2)$
(b) $f(3 - 2) = f(1)$
(c) $f(\sqrt{2}) = \sqrt{f(2)}$
(d) $f(2)f(3) = f(6)$
(e) $f\left(\dfrac{2}{3}\right) = \dfrac{f(2)}{f(3)}$
(f) $f(x^2) = [f(x)]^2$

24 Let c be a constant. Define f by $f(x) = \left(\dfrac{x - c}{x - c}\right)x$. Is $f(x) = x$ for all x? Explain.

25–28. Which of the equations define y as a function of x? Which define x as a function of y? Find the domains of all such functions.

• 25 $xy = x - 1$

26 $x^2 + y = 2x - xy$

• 27 $y^2 = x - 1$

28 $y^2 - 4xy + 4x^2 = 0$

29–34. Let f be a function. We shall use the expression

$$\dfrac{f(x) - f(a)}{x - a} \qquad (x \ne a)$$

in much of our subsequent work. Simplify this expression for the functions defined by these laws of correspondence.

• 29 $f(x) = c$

30 $f(x) = 3x$

• 31 $f(x) = x^2$

32 $f(x) = x$

• 33 $f(x) = \dfrac{1}{x}$

34 $f(x) = \sqrt{x}$

35 Let S be the set of positive integers between 1 and 10 (inclusive). For each $x \in S$, let $h(x)$ be the largest element of S that is less than or equal to every element of S that is greater than or equal to x. Does this statement define a function? If so, can its law of correspondence be stated more simply?

1–7 Composition of Functions

36–51. Plot several points and sketch the graph.

36. $f(x) = 4x + 3$
37. $f(x) = 4 - x^2$
38. $f(x) = -(4 + x^2)$
39. $f(x) = -3$
40. $f(x) = \dfrac{1}{x-4}$
41. $f(x) = |x - 4|$
42. $f(x) = \dfrac{x^2 - 4}{x - 2}$
43. $f(x) = [x]$
44. $f(x) = x - [x]$
45. $f(x) = x + [x]$
46. $f(x) = |x|$
47. $f(x) = |x| + x$
48. $f(x) = \dfrac{x}{|x|}$
49. $f(x) = \begin{cases} x + 1 & \text{if } x < 0 \\ x^2 & \text{if } x \geq 0 \end{cases}$
50. $f(x) = \begin{cases} -1 & \text{if } x < -1 \\ x & \text{if } -1 \leq x \leq 1 \\ 1 & \text{if } x > 1 \end{cases}$
51. $f(x) = \begin{cases} 0 & \text{if } x \text{ is not an integer} \\ 1 & \text{if } x \text{ is an integer} \end{cases}$

52. (a) Let a and b be real numbers. Show that

$$\frac{a + b}{2} + \frac{|a - b|}{2}$$

is equal to the larger of a and b. We indicate this fact by writing

$$\frac{a + b}{2} + \frac{|a - b|}{2} = \max\{a, b\}.$$

(b) Find a similar formula for $\min\{a, b\}$.

• (c) Let f and g be real functions of x. For each x in the common domain of f and g, let

$$h(x) = \max\{f(x), g(x)\}.$$

Obtain an algebraic formula for $h(x)$.

(d) Let $f(x) = x^2$, $g(x) = x + 1$. Sketch the graph of $h(x) = \max\{f(x), g(x)\}$.

1–7 Composition of Functions

There are several ways by which we can construct new functions from given functions. If f and g are given, for example, we can calculate $f(x) + g(x)$ for each x in the common part of the domain of f and g. This relation defines a new function, denoted

Sum of Functions

by $f + g$, called the *sum* of f and g. To be more precise:

(1) The domain of $f + g$ is the intersection of the domains of f and g.
(2) The law of correspondence for $f + g$ is given by

$$(f + g)(x) = f(x) + g(x).$$

For example, if

$$f(x) = x^2 + 3 \quad \text{and} \quad g(x) = 2x - 7,$$

then the law of correspondence is

$$(f + g)(x) = (x^2 + 3) + (2x - 7) = x^2 + 2x - 4.$$

The *difference, product,* and *quotient* of the functions f and g can be defined analogously. In more abstract courses, functions are used much as numbers are used in the elementary courses. The sum, difference, product, and quotient of functions are used to develop an *algebra of functions* that has many of the properties of elementary algebra.

EQUALITY

Although we shall not develop the algebra of functions, we do need one basic property.

DEFINITION Two functions f and g are *equal* provided their domains are equal and they have the same law of correspondence.

EXAMPLE 1 Let $f(x) = \dfrac{x^2 - 9}{x - 3}$ and $g(x) = x + 3$. The fact that

$$f(x) = \frac{x^2 - 9}{x - 3}$$

$$= \frac{(x - 3)(x + 3)}{x - 3} = x + 3 = g(x) \quad \text{provided } x \neq 3$$

does not mean that the functions f and g are equal. The domain of f consists of all real numbers different from 3, and the domain of g consists of all real numbers. The one point excluded from the domain of f makes the functions f and g unequal even though they have the same law of correspondence for all real numbers different from 3. □

COMPOSITION

In some cases, we need to calculate the value of a function of a function of x. If you have a scientific calculator, you may need, for example, to calculate $\sin(\sqrt{2})$.* This

---

* Here the argument of the sine function is measured in *radians*. The number $\sqrt{2}$ represents $\sqrt{2}$ radians, not $\sqrt{2}$ degrees.

1-7 Composition of Functions

can be accomplished in two steps, by pushing the "square root" key and the "sin" key in the proper order:

Step 1 Calculate $\sqrt{2} \approx 1.414213562$.
Step 2 Calculate $\sin(\sqrt{2}) \approx \sin(1.414213562)$
≈ 0.987765946.

If we let $f(u) = \sin u$ and $g(x) = \sqrt{x}$, then

$$f(g(x)) = \sin(\sqrt{x}),$$
and $\quad f(g(2)) = \sin(\sqrt{2}) \approx 0.987765946.$

Composition of Functions

The more general problem is determining the law of correspondence for $f(g(x))$ for given functions f and g. Letting $u = g(x)$ reduces the problem to calculating $f(u)$ where $u = g(x)$. The function defined by $f(u) = f(g(x))$ is called the *composition of f with g*.

Composition of functions is represented schematically in Figure 1–29. Observe that the composition of f with g involves two steps for each value of x.

(1) Calculate $g(x)$.
(2) Calculate $f(g(x))$.

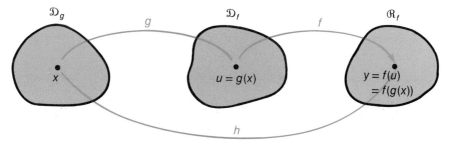

Figure 1–29 Schematic representation of composition of functions: $h(x) = f(g(x))$. For simplicity, in this diagram we consider the case where $\mathcal{R}_g$ (the range of g) is a subset of $\mathcal{D}_f$ (the domain of f). This is not always the case. In general, the domain of the composite function h is $\{x \mid x \in \mathcal{D}_g \text{ and } g(x) \in \mathcal{D}_f\}$.

To carry out these steps in the proper order, the following requirements must be met:

(1) x must be in the domain of g. (Otherwise $g(x)$ would not be defined.)
(2) $u = g(x)$ must be in the domain of f. (Otherwise $f(g(x))$ would not be defined.)

For this reason it is standard to *define the domain of the composition of f with g* to be

Domain of Composite Function

$$\{x \mid x \in \mathcal{D}_g \text{ and } g(x) \in \mathcal{D}_f\}.$$

The composition of f and g is denoted by the symbol

$$f \circ g.$$

That is

$$(f \circ g)(x) = f(g(x))$$

for all x in the domain of g for which $g(x)$ is in the domain of f.

The formal definition of composition follows.

DEFINITION   Let f and g be real functions. The function $f \circ g$, called the *composition of f with g*, is defined as follows:

$f \circ g$

(1) The domain of $f \circ g$ is

$$\mathcal{D}_{f \circ g} = \{x \mid x \in \mathcal{D}_g \text{ and } g(x) \in \mathcal{D}_f\}.$$

(2) The law of correspondence for $f \circ g$ is given by $(f \circ g)(x) = f(g(x))$.

EXAMPLE 2   Let $f(x) = x^2 - 1$ and $g(x) = \sqrt{x - 2}$.
(a) Determine the law of correspondence for $f \circ g$.
(b) Determine the domain of $f \circ g$.

Solution   (a) Let $u = g(x) = \sqrt{x - 2}$. Then

$$(f \circ g)(x) = f(g(x)) = f(u) = u^2 - 1$$
$$= (\sqrt{x - 2})^2 - 1 = x - 2 - 1 = x - 3.$$

(b) The domain of f is $\{x \mid x \text{ is real}\}$, and the domain of g is $\{x \mid x \geq 2\}$. The domain of the composite function $f \circ g$ is

$$\mathcal{D}_{f \circ g} = \{x \mid x \in \mathcal{D}_g \text{ and } g(x) \in \mathcal{D}_f\}$$
$$= \{x \mid x \geq 2 \text{ and } \sqrt{x - 2} \text{ is real}\}$$
$$= \{x \mid x \geq 2\}.$$

Note that if the law of correspondence $(f \circ g)(x) = x - 3$ were considered without regard to the way that it is obtained, we would come to the erroneous conclusion that the domain is the set of all real numbers.

The graph of this composite function is the ray shown in Figure 1–30a. □

EXAMPLE 3   (See Example 2.) Let $f(x) = x^2 - 1$ and $g(x) = \sqrt{x - 2}$.
(a) Determine the law of correspondence for $g \circ f$.
(b) Determine the domain of $g \circ f$.

Solution   (a) $(g \circ f)(x) = g(f(x)) = g(x^2 - 1)$
$$= \sqrt{(x^2 - 1) - 2}$$
$$= \sqrt{x^2 - 3}.$$
(b) $\mathcal{D}_{g \circ f} = \{x \mid x \in \mathcal{D}_f \text{ and } f(x) \in \mathcal{D}_g\}$
$$= \{x \mid x \text{ is real and } x^2 - 1 \geq 2\}$$
$$= \{x \mid x^2 \geq 3\} = \{x \mid x \leq -\sqrt{3} \text{ or } x \geq \sqrt{3}\} \quad \text{(by A6)}.$$

The graph of $g \circ f$ is shown in Figure 1–30b. □

1-7 Composition of Functions

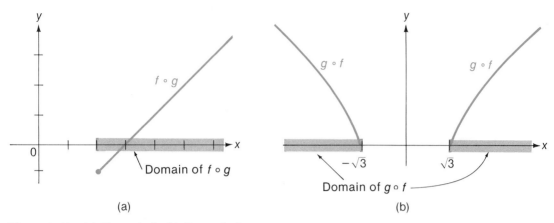

Figure 1-30 (a) Example 2; (b) Example 3.

Composition and Substitution

The principal use that we shall make of composition is in writing complicated functions in terms of simpler ones. Composition is involved whenever we simplify a function by a substitution.

EXAMPLE 4 Simplify the function

$$h(x) = \frac{3\left(\frac{2-x}{x^2+1}\right)^2 - 7\left(\frac{2-x}{x^2+1}\right) + 4}{2\left(\frac{2-x}{x^2+1}\right)^2 + 5}$$

by a substitution $u = g(x)$. Write h as the composition of two simpler functions.

Solution The obvious substitution is

$$u = g(x) = \frac{2-x}{x^2+1}.$$

Then

$$h(x) = f(u) = \frac{3u^2 - 7u + 4}{2u^2 + 5}.$$

Because $h(x) = f(u)$ where $u = g(x)$, then

$$h(x) = f(g(x)). \;\square$$

EXAMPLE 5 Let $g(x) = [x]$ (the greatest-integer function) and $f(x) = x^2$. Sketch the graph of $f(g(x))$ for $-2 \le x \le 2$.

Solution If $-2 \le x < -1$, then $g(x) = -2$ and $f(g(x)) = f(-2) = 4$.
If $-1 \le x < 0$, then $g(x) = -1$ and $f(g(x)) = f(-1) = 1$.

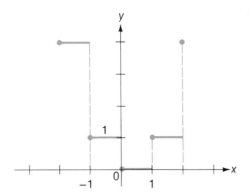

Figure 1–31 Example 5.

If $0 \le x < 1$, then $g(x) = 0$ and $f(g(x)) = f(0) = 0$.
If $1 \le x < 2$, then $g(x) = 1$ and $f(g(x)) = f(1) = 1$.
If $x = 2$, then $g(x) = 2$ and $f(g(x)) = f(2) = 4$.
The graph is shown in Figure 1–31. □

Exercises 1–7

c1 Students with scientific calculators can calculate the values of composite functions by pushing two or more buttons in succession. Set the calculator in the RADIAN mode. Let $f(x) = \cos(x)$, $g(x) = \sin(x)$, $h(x) = \tan(x)$, $j(x) = e^x$.* Calculate the following functional values.
  (a) $f(g(j(0.5)))$
  (b) $f(f(f(1)))$
  (c) $h(j(h(0)))$
  (d) $j(f(j(0)))$

2 Explain how to find the domain of the composite functions $f \circ g$ and $g \circ f$.

3–12. Find $(f \circ g)(x)$ and $(g \circ f)(x)$ and determine the domains.

• **3** $f(x) = x^2 + 1$, $g(x) = 2x + 7$ **4** $f(x) = \sqrt{x + 4}$, $g(x) = 3x^2 - 1$

• **5** $f(x) = \dfrac{1}{3x - 1}$, $g(x) = \dfrac{1}{x^2}$ **6** $f(x) = x^3$, $g(x) = x - 1$

• **7** $f(x) = 2x - 6$, $g(x) = \tfrac{1}{2}x + 3$ **8** $f(x) = 3x + 9$, $g(x) = \tfrac{1}{3}x - 3$

• **9** $f(x) = ax + b$, $g(x) = \dfrac{1}{a}x - \dfrac{b}{a}$, $a \ne 0$

10 $f(x) = \sqrt{2x + 1}$, $g(x) = \tfrac{1}{2}x^2 - \tfrac{1}{2}$

NOTE: **c** indicates calculator problem.
* Some calculators may denote the function e^x by **EXP (X)** or **INV LN (X)**

●11 $f(x) = 7$, $g(x) = 4$

12 $f(x) = x^3 + 1$, $g(x) = \sqrt[3]{x^2 + 1}$

●13 Let $f(x) = 3x + 1$. Calculate $f(x^2)$, $f(f(x))$ and $[f(x)]^2$.

14 Let $f(x) = x^2 - 1$. Calculate $f(x^2)$, $f(1/x)$, and $f(f(1/x))$.

●15 If $f(g(x)) = x^2 - 1$ and $g(x) = x$, what is $f(x)$?

16 If $f(g(x)) = x^2 + 2x - 1$ and $g(x) = x^2 + 2x - 1$, what is $f(x)$?

●17 If $f(g(x)) = x^2 - 2x + 1$ and $f(x) = x^2$, what is $g(x)$?

18 If $f(g(x)) = x^2$ and $f(x) = x - 1$, what is $g(x)$?

19–22. Simplify $h(x)$ by a substitution $u = g(x)$. Write h as the composition of two simpler functions.

●19 $h(x) = 3(x + 1)^2 - 2(x + 1) - 5$

20 $h(x) = \sqrt{5x - 3} + \dfrac{1}{(5x - 3)^2}$

●21 $h(x) = \sqrt{\dfrac{x - 1}{x + 1}} - \dfrac{x + 1}{x - 1}$

22 $h(x) = \dfrac{x^2 - 2x}{(x - 1)^2}$

1–8 Inverse Functions

Certain pairs of functions, f and g, have the property that each reverses the effect of the other. For example, if

$$f(x) = 2x - 4 \quad \text{and} \quad g(x) = 2 + \frac{x}{2},$$

then

$$f(g(x)) = f\left(2 + \frac{x}{2}\right) - 4 = x$$

and

$$g(f(x)) = g(2x - 4) = 2 + \tfrac{1}{2}(2x - 4) = x.$$

Thus, if we apply either f or g to a number x—that is, calculate the functional value—and then apply the other function, we always get the original number x. Using other symbols, we get

$$(f \circ g)(x) = x \quad \text{for every } x \text{ in the domain of } g$$

and

$$(g \circ f)(x) = x \quad \text{for every } x \text{ in the domain of } f.$$

DEFINITION

Inverse Functions

Two functions f and g are said to be *inverses* of each other provided

$(f \circ g)(x) = x$ for each x in the domain of g

and

$(g \circ f)(x) = x$ for each x in the domain of f.

This means, in other words, that

$$f(x) = y \text{ if and only if } g(y) = x$$

for every x in the domain of f and every y in the domain of g. If f assigns the value y to x, then g reverses the procedure and assigns x to y.

The definition is illustrated in Figure 1-32. Observe that the range of f must be the domain of g and the range of g must be the domain of f.

EXAMPLE 1   Let $f(x) = x^3$ and $g(x) = \sqrt[3]{x}$. If x is a real number, then

$$(f \circ g)(x) = f(g(x)) = f(\sqrt[3]{x}) = (\sqrt[3]{x})^3 = x$$

and

$$(g \circ f)(x) = g(f(x)) = g(x^3) = \sqrt[3]{x^3} = x.$$

Thus, f and g are inverses. □

EXAMPLE 2   (*Requires a scientific calculator set in the radian mode instead of the degree mode**)
Let $f(x) = \cos x$ and $g(x) = \cos^{-1} x$, where the domain of f is the set of numbers $0 \leq$

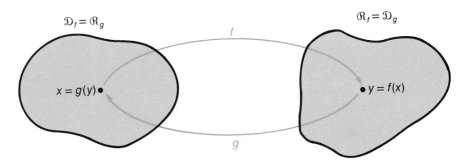

Figure 1-32 Inverse functions. Each function reverses the effect of the other. If $f(2) = 7$, then $g(7) = 2$. If $g(-1) = 3$, then $f(3) = -1$.

---

* On some calculators g may be labeled **INV COS** or **ARC COS**. The three names $\cos^{-1}$, inv cos, and arc cos are standard names for the same function—the one that reverses the effect of the cosine function.

1–8 Inverse Functions

$x \leq \pi \approx 3.14159$ and the domain of g is the set of all x such that $-1 \leq x \leq 1$. We illustrate in two numerical examples that $(g \circ f)(x) = x$.

If $x = 0.5732$, then

$$f(x) = \cos(0.5732) = 0.840169845$$

and

$$g(f(x)) = \cos^{-1}(0.840169845) = 0.5732 = x.$$

If $x = 1.036748$, then

$$f(x) = \cos(1.036748) = 0.5090221141$$

and

$$g(f(x)) = \cos^{-1}(0.5090221141) = 1.036748 = x.$$

These calculations, which look complicated on paper, are accomplished by simply pressing the keys **COS** and **COS**$^{-1}$ in succession. □

NOTATION

Let f and g be inverse functions. Let I be the *identity function* defined on the real numbers by $I(x) = x$. Then

$$(f \circ g)(x) = x = I(x) \quad \text{for each } x \text{ in the domain of } g$$

and

$$(g \circ f)(x) = x = I(x) \quad \text{for each } x \text{ in the domain of } f,$$

so that these composite functions coincide with the function I on their domains.

It can be shown that in composition the identity function I has properties like those of the number 1 relative to multiplication. In particular,

$$I \circ f = f \quad \text{and} \quad f \circ I = f$$

for every real function f. Similarly, the inverse functions f and g have properties like those of reciprocals. For this reason, and others, mathematicians frequently denote g, the inverse of f, by f^{-1}.

f^{-1}

For example, if $f(x) = x^3$, then

$$f^{-1}(x) = g(x) = x^{1/3},$$

and

$$g^{-1}(x) = f(x) = x^3.$$

Remark The symbol f^{-1} denotes the inverse of the function f, not the reciprocal of f. This notation helps to explain the symbols $\cos^{-1}$, $\sin^{-1}$, $\tan^{-1}$, and so on, for the inverse trigonometric functions.

ONE-TO-ONE FUNCTIONS

Only very special functions have inverses. It is easy to see that if f has an inverse g, then different elements x_1 and x_2 of the domain of f must have different functional values $f(x_1)$ and $f(x_2)$. This follows because the inverse function g must take $f(x_1)$ back to x_1 and $f(x_2)$ back to x_2. This would be impossible if $f(x_1)$ and $f(x_2)$ were equal when x_1 and x_2 were unequal.

DEFINITION

One-to-One Function

A function f is said to be *one-to-one* provided different elements x_1 and x_2 of its domain always have different functional values. That is, f is one-to-one provided

$$f(x_1) \neq f(x_2) \quad \text{whenever } x_1 \neq x_2.$$

The remarks preceding the definition establish that a function with an inverse must be one-to-one. It can be shown that conversely, if f is one-to-one, f has an inverse. The functions with inverses are just those that are one-to-one.

It is simple to tell whether a function is one-to-one by examining its graph. If f is not one-to-one, there must be two unequal numbers x_1 and x_2 such that $f(x_1) = f(x_2)$. Thus, a horizontal line $y = f(x_1)$ must intersect the graph in at least two points. (See Fig. 1–33a.) On the other hand, if f is a one-to-one function and $x_1 \neq x_2$, then $f(x_1)$ and $f(x_2)$ must be unequal. Thus, no horizontal line can intersect the graph at more than one point. (See Fig. 1–33b.)

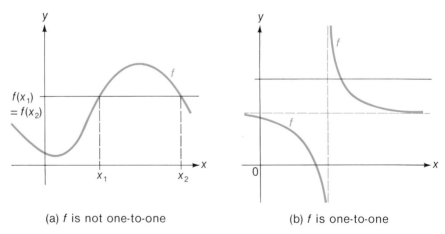

(a) f is not one-to-one   (b) f is one-to-one

Figure 1–33 A function f is one-to-one provided no horizontal line intersects the graph of f at more than one point.

It may not be convenient to examine the graph of a function to decide whether it is one-to-one. Observe that f is one-to-one if and only if the equation

$$f(x_1) = f(x_2)$$

1-8 Inverse Functions

implies that $x_1 = x_2$. In many cases this criterion gives us the simplest way to check whether a function is one-to-one.

EXAMPLE 3 Show that the function $f(x) = \dfrac{2x + 3}{5x - 7}$ is one-to-one.

Solution Suppose that $f(x_1) = f(x_2)$. Then

$$\frac{2x_1 + 3}{5x_1 - 7} = \frac{2x_2 + 3}{5x_2 - 7}$$
$$(2x_1 + 3)(5x_2 - 7) = (5x_1 - 7)(2x_2 + 3)$$
$$10x_1x_2 - 14x_1 + 15x_2 - 21 = 10x_1x_2 + 15x_1 - 14x_2 - 21$$
$$-29x_1 = -29x_2$$
$$x_1 = x_2.$$

Since $x_1 = x_2$ whenever $f(x_1) = f(x_2)$, then f is one-to-one. □

CALCULATION OF INVERSE FUNCTIONS

Let f and g be inverse functions. If f has a simple defining relation (law of correspondence), it may be possible to find the defining relation for g by solving the equation $y = f(x)$ for x as a function of y. Since $x = g(y)$ whenever $y = f(x)$, this expression gives us the functional relation for $g(y)$.

EXAMPLE 4 Let $f(x) = (2x + 3)/(5x - 7)$. (See Example 3.) Let g be the inverse function. Find the law of correspondence for $g(x)$.

Solution Let $y = f(x) = \dfrac{2x + 3}{5x - 7}$. We solve this equation for x, obtaining

$$(5x - 7)y = 2x + 3$$
$$5xy - 2x = 7y + 3$$
$$x = \frac{7y + 3}{5y - 2}.$$

Therefore,

$$g(y) = \frac{7y + 3}{5y - 2}.$$

Recall that the names of the variables are not important. If we use x for the symbol for the independent variable, then

$$g(x) = \frac{7x + 3}{5x - 2}.$$

Observe, as a check, that

$$(f \circ g)(x) = f(g(x)) = f\left(\frac{7x+3}{5x-2}\right)$$

$$= \frac{2\left(\frac{7x+3}{5x-2}\right) + 3}{5\left(\frac{7x+3}{5x-2}\right) - 7} = x.$$

Similarly, $(g \circ f)(x) = x$. □

EXAMPLE 5 Let $f(x) = x^3 + 2$, $-1 \leq x \leq 0$. (a) Show that f is one-to-one. (b) Find the law of correspondence and the domain of the inverse function.

Solution (a) Suppose that $f(x_1) = f(x_2)$, $-1 \leq x_1 \leq 0$, $-1 \leq x_2 \leq 0$. Then

$$x_1^3 + 2 = x_2^3 + 2$$
$$x_1^3 - x_2^3 = 0$$
$$(x_1 - x_2)(x_1^2 + x_1 x_2 + x_2^2) = 0.$$

Since $x_1^2 + x_1 x_2 + x_2^2 > 0$ unless $x_1 = x_2 = 0$, then we must have

$$x_1 - x_2 = 0$$
$$x_1 = x_2.$$

Therefore, f is one-to-one.

(b) Let $y = f(x)$. Let g be the inverse of f. We can determine g by solving the equation $y = f(x)$ for x. Then

$$y = x^3 + 2$$
$$x^3 = y - 2$$
$$x = \sqrt[3]{y-2}$$
$$g(y) = \sqrt[3]{y-2}.$$

Since x varies from -1 to 0, then y varies from 1 to 2. The domain of g is $\{y \mid 1 \leq y \leq 2\}$. □

GRAPHS OF INVERSE FUNCTIONS

A simple relation exists between the graph of f and the graph of its inverse function g. We shall show that the graph of g can be obtained from the graph of f by reflecting the graph of f across the 45°-line $y = x$ (Fig. 1–34).

Recall that the graph of f is

$$\{(x, y) \mid y = f(x)\}.$$

It is convenient to use symbols other than x and y in the following argument (since we think of x as a point on the x-axis and y as a point on the y-axis). Observe that the symbols x and y in the above statement act only as placeholders to describe how the

1-8 Inverse Functions

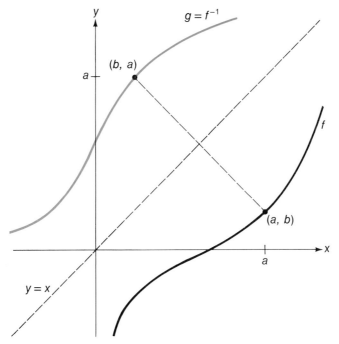

Figure 1-34 Graphs of inverse functions. The graph of f^{-1} is the reflection of the graph of f in the line $y = x$. The point (a, b) is on the graph of f if, and only if, the point (b, a) is on the graph of f^{-1}.

ordered pairs are formed. We could just as well describe the graph of f as

$$\{(a, b) \,|\, b = f(a)\} \quad \text{or} \quad \{(\alpha, \beta) \,|\, \beta = f(\alpha)\}.$$

To establish the relation between the graphs of f and g, recall that $b = f(a)$ if and only if $a = g(b)$. Thus, the graph of the inverse function g is

$$\begin{aligned}
\text{Graph of } g &= \{(x, y) \,|\, y = g(x)\} \\
&= \{(b, a) \,|\, a = g(b)\} \\
&= \{(b, a) \,|\, b = f(a)\}.
\end{aligned}$$

The graph of f, on the other hand, is

$$\text{Graph of } f = \{(a, b) \,|\, b = f(a)\}.$$

Thus, (a, b) is on the graph of f if and only if (b, a) is on the graph of g.

Graph of Inverse Functions

It is easy to see that the line $y = x$ is the perpendicular bisector of the line segment connecting the points (a, b) and (b, a). (See Exercise 22.) Thus, the point (b, a) on the graph of g can be obtained by reflecting the point (a, b) on the graph of f over the 45°-line $y = x$. (See Fig. 1-34.) It follows that the entire graph of g can be obtained in this way.

Exercises 1–8

1–4. Prove that f and g are inverses of each other.

1 $f(x) = 2x - 1$; $g(x) = \dfrac{x+1}{2}$

2 $f(x) = \sqrt{x}$, $x \geq 0$; $g(x) = x^2$, $x \geq 0$

3 $f(x) = \sqrt{4x - 1}$; $x \geq \dfrac{1}{4}$; $g(x) = \dfrac{(x^2 + 1)}{4}$, $x \geq 0$

4 $f(x) = x^3 - 1$; $g(x) = \sqrt[3]{x + 1}$

5–8. Decide whether these functions are one-to-one on the given domains.

• **5** $f(x) = x^3$, $\mathcal{D} = \{x \mid x \text{ is real}\}$ **6** $g(x) = 2x - 1$, $\mathcal{D} = \{x \mid x \text{ is real}\}$

• **7** $f(x) = x^2$, $\mathcal{D} = \{x \mid -1 < x < 0\}$ **8** $h(x) = x^2$, $\mathcal{D} = \{x \mid -1 \leq x \leq 2\}$

9 (a) Sketch the graph of a one-to-one function f with the properties that $f(2) = -1$, $f(5) = 7$, $f(6) = 8$.
   (b) Sketch the graph of the inverse of f.
   (c) Calculate $f^{-1}(-1)$ and $f^{-1}(7)$.

c10 Students with scientific calculators should verify for several values of x that the given pairs of functions are inverses of each other. (The domains of the functions f and g are given.)

(a) $f(x) = \textbf{SIN (X)}$, $-\dfrac{\pi}{2} \leq x \leq \dfrac{\pi}{2}$; $g(x) = \textbf{SIN}^{-1}\textbf{ (X)}$,* $-1 \leq x \leq 1$

(b) $f(x) = \textbf{LN (X)}$, $x > 0$; $g(x) = e^x$, x real†

11–20. (a) Show that f is one-to-one. (b) Find the inverse of f and state its domain.

• **11** $f(x) = 4x + 9$ **12** $f(x) = ax + b$, $a \neq 0$

• **13** $f(x) = 3x^2 - 1$, $x \geq 0$ **14** $f(x) = 3x^2 - 6x + 3$, $x \leq 1$

• **15** $f(x) = 1/x$ **16** $f(x) = \sqrt{4 - x^2}$, $0 \leq x \leq 2$

• **17** $f(x) = 8x^3 - 1$ **18** $f(x) = x$

• **19** $f(x) = 3\sqrt[3]{x} + 2$ **20** $f(x) = 6 - x^3$

21 (a) Let f be the inverse of g. Explain why g must be the inverse of f.
   (b) Show that the result of (a) implies that $(f^{-1})^{-1} = f$.

22 Show that the line $y = x$ is the perpendicular bisector of the line segment connecting $P(a, b)$ and $Q(b, a)$ provided $a \neq b$. (*Hint:* See Example 6 of Section 1–5.)

NOTE: **c** indicates calculator problem.
* Some calculators may give this function as **ARC SIN (X)** or **INV SIN (X)**. The calculator should be set in the RADIAN mode.
† Some calculators may give **LN (X)** as **LOG(X)**, and e^x as **EXP (X)** or **INV LN (X)**.

Review Problems

1 Explain the relation $a < b$ in terms of positive numbers and in terms of position on the number line.

2 What is meant by an "open" interval? a "closed" interval? Is $\{x \mid 2 \leq x < 3\}$ open, closed, or neither?

3 Prove order property O5: If $a < b$ and $c < 0$, then $ac > bc$.

4 State

    (a) The distance formula   (b) The midpoint formula

    (c) The Triangle Inequality   (d) The Second Triangle Inequality

    (e) The point-slope form of the equation of a line

    (f) The slope-intercept form of the equation of a line

    (g) The formula for the circle with center (a, b) and radius $r > 0$

5–12. Solve the inequalities.

- **5** $(2x - 7)(5x + 3) > 0$
- **6** $x^2 - 4 < 0$
- **7** $\dfrac{4x - 3}{x - 2} > 1$
- **8** $\dfrac{2x + 3}{x - 1} > 2$
- **9** $|3x - 5| < 4$
- **10** $|x^2 - 4| < 5$
- **11** $2x + |x| < 3$
- **12** $|x| < x - 1$

13 Derive the distance formula.

14 Derive the equation of the circle with center (a, b) and radius $r > 0$.

15–20. Sketch the graph if one exists.

- **15** $x^2 + y^2 - 4x - 4y + 4 = 0$
- **16** $x^2 + y^2 - 6x + 8y + 25 = 0$
- **17** $x^2 + y^2 - 5x + 4y + 13 = 0$
- **18** $3x + 2y = 7$
- **19** $f(x) = [x] + 1$
- **20** $f(x) = [2x - 1]$

- **21** Find the equation of the circle with $(-1, 2)$ and $(7, -6)$ at the ends of a diameter.

- **22** Is the point $(2.32, 3.79)$ inside, outside, or on the circle $x^2 + y^2 - 2x - 4y = 0$?

•23 A circle is tangent to both coordinate axes and passes through the point (1, 2). Find the center and the radius.

•24 Find the equations of the lines with slope 2 that are tangent to the circle $x^2 + y^2 + 4x - 8y = 25$.

•25 A circle is tangent to the line $3x - 4y + 20 = 0$ at the point (0, 5) and to the line $y = -4$. Find the two possible choices for the center. Illustrate with a sketch.

•26 Find the equations of the two lines through (7, 1) that are tangent to the circle $x^2 + y^2 = 25$. Illustrate with a sketch.

27 (a) What is meant by the slope of a line? by its angle of inclination? Explain how these two concepts are related.

(b) Explain why a nonvertical line is completely determined by its slope and a point on the line.

(c) Derive the point-slope form of the equation of the line with slope m that passes through (x_0, y_0).

28 Lines $\mathscr{L}_1$, $\mathscr{L}_2$, $\mathscr{L}_3$, $\mathscr{L}_4$, and $\mathscr{L}_5$ are shown in Figure 1–35. Which of the lines have positive slope? negative slope? no slope? Does $\mathscr{L}_1$ or $\mathscr{L}_2$ have the larger slope? What relation holds between the slopes of $\mathscr{L}_1$ and $\mathscr{L}_3$? between the slopes of $\mathscr{L}_3$ and $\mathscr{L}_4$?

•29 Find the equation of the line

(a) through (5, −2) with slope $m = -\tfrac{1}{3}$;

(b) through (2, 1) and (4, −7).

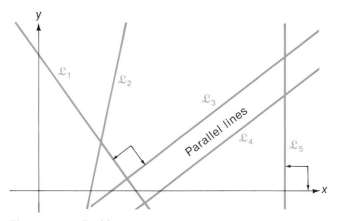

Figure 1–35 Problem 28.

Review Problems

- **30** Find the equation of the line through the point of intersection of the lines $2x - 3y = 9$ and $x - 4y + 8 = 0$ that is perpendicular to the line $2x - 3y = 9$.

- **31** Let $f(x) = x^2 + 1$. Calculate

    (a) $f(2x)$

    (b) $f\left(\dfrac{2}{x}\right)$

    (c) $f(f(2))$

    (d) $\dfrac{f(x + a) - f(x)}{a}$

32 Define the terms:

   (a) function

   (b) domain of a function

   (c) range of a function

   (d) polynomial

   (e) rational function

   (f) algebraic function

   (g) transcendental function

   (h) composition of functions

33 (a) What is the general rule when no domain is specified for a function?

- (b) Find the domain of $f(x) = \dfrac{1}{\sqrt{x^2 - 3}}$.

- (c) Find the domain of $g(x) = \sqrt{x^2 - 3}$.

- **34** Let $f(x) = \sqrt{x - 2}$ and $g(x) = \sqrt{16 - x^2}$.

    (a) Find the laws of correspondence for $f \circ g$ and $g \circ f$.

    (b) Find the domains of $f \circ g$ and $g \circ f$.

35 (a) What does it mean if f and g are inverses of each other?

   (b) What does it mean if f is one-to-one?

   (c) What relation holds between the concepts in (a) and (b)?

- **36** Let $f(x) = \sqrt{2x - 1}$, $x > \tfrac{1}{2}$. Find the law of correspondence and the domain for $g = f^{-1}$. Sketch the graphs of f and g on the same set of axes.

37 Invent situations in which the given functions can be used.

   (a) $f(x) = \begin{cases} 365 \text{ if } x \text{ is an integer not divisible by 4, } 1900 < x < 2100 \\ 366 \text{ if } x \text{ is an integer divisible by 4, } 1900 < x < 2100. \end{cases}$

   (b) $g(x) = \pi x^2$, $x \neq 0$

   - (c) $h(x) = x^3 + x^2$, $x > 0$

38 Which of the graphs in Figure 1–36 are graphs of functions? Which are not? Explain.

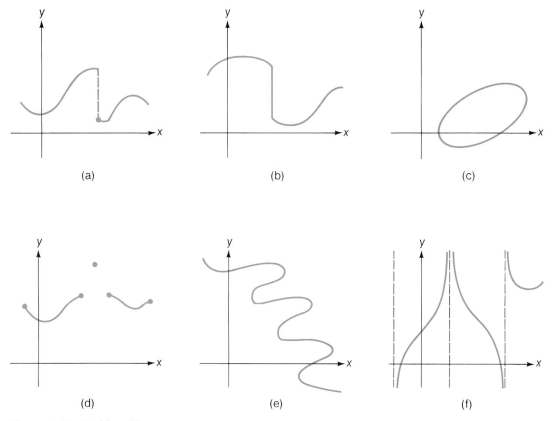

Figure 1-36 Problem 38.

39-40. Find the domain. Sketch the graph of f wherever it is defined on the interval $-5 \leq x \leq 5$.

•**39** $f(x) = \dfrac{1}{[x] + 1}$

40 $f(x) = [2x - 3] + 1$

•**41** Find the equation of the perpendicular bisector of the line segment connecting $P_1(x_1, y_1)$ and $P_2(x_2, y_2)$ where $x_1 \neq x_2$ and $y_1 \neq y_2$.

•**42** Find the minimum distance between the circles $x^2 + y^2 - 2x + 4y + 1 = 0$ and $x^2 + y^2 + 6x - 2y + 9 = 0$.

•**43** Find the minimum distance from the line $3x + 2y = 17$ to the circle $4x^2 + 4y^2 - 16x + 8y + 11 = 0$.

2 Limits and Derivatives

2-1 Introduction to Limits and Tangent Lines

A physicist needs to measure the electrical resistance of a wire at absolute zero (0 K), a temperature that cannot be obtained in the laboratory. He measures the resistance at a large number of temperatures close to 0 K and observes that these numbers seem to approach a particular number L as the temperature drops. He infers that L is the value of the resistance at 0 K.

Note that he never measures the resistance at 0 K. Instead he considers how it behaves near 0 K. A more reasonable conclusion is that *L is the limit of the resistance function as the temperature approaches* 0 K; this means that the functional values are close to L (closer than to any other number) when the temperature is sufficiently close to 0. If $r(t)$ is the resistance at temperature t, we write

$$\lim_{t \to 0} r(t) = L.$$

The concept of the limit of a function marks the main dividing line between the calculus and more elementary mathematics. Limits are basic to the development of the calculus and to many of its applications to engineering and science. Because the formal definition involves some subtle points, we first give a purely descriptive introduction to limits. After considering a few applications to the calculus, we shall return to the limit concept, giving a more rigorous treatment.

Intuitive Discussion of Limit

Roughly speaking, we say that the number L is the *limit of $f(x)$ as x approaches a* provided the functional values $f(x)$ are close to L (closer than to any other number) whenever x is close to a. If this is the case, we write

$$\lim_{x \to a} f(x) = L.$$

If x is close to 3, for example, then $2x$ is close to 6 and $2x + 5$ is close to 11. Thus,

$$\lim_{x \to 3} (2x + 5) = 11.$$

To understand the limit concept better, consider that if x is close to 3, then

$$x = 3 + \Delta x$$

where the increment Δx can be either positive or negative. Then

$$2x + 5 = 2(3 + \Delta x) + 5 = 11 + 2\Delta x.$$

If x is close to 3, then Δx is close to zero, so that

$$2x + 5 = 11 + 2\Delta x$$

is close to 11. Several values of $f(x)$ corresponding to values of x close to 3 are shown in Table 2–1.

Table 2–1

Δx	$x = 3 + \Delta x$	$f(x) = 2x + 5 = 11 + 2\Delta x$
-0.1	2.9	10.8
-0.01	2.99	10.98
-0.000001	2.999999	10.999998
. . . .	. . . .	. . . .
0.1	3.1	11.2
0.01	3.01	11.02
0.000001	3.000001	11.000002

It is convenient to write $x \to a$ to indicate that x approaches a. Similarly, we write

$$f(x) \to L \quad \text{as } x \to a$$

to indicate that $f(x)$ approaches L as x approaches a.

EXAMPLE 1 Let $f(x) = \dfrac{x^2 - 4}{3x - 6}$. Calculate $\lim_{x \to 2} f(x)$.

Solution We first calculate several values of $f(x)$ for x close to 2:

$$\begin{aligned} x &= 1.9, & f(x) &= 1.3 \\ x &= 1.99, & f(x) &= 1.33 \\ x &= 2.1, & f(x) &\approx 1.36666667 \\ x &= 2.01, & f(x) &\approx 1.33666667 \end{aligned}$$

It appears from these values that the limit might be $L = 1\frac{1}{3}$ as x approaches 2.

Observe that the function f is not defined at $x = 2$. If we substitute 2 for x, we get the meaningless expression $(4 - 4)/(6 - 6)$. In calculating the limit, however, we are concerned only with the values of $f(x)$ when x is close to 2, not with the value of f at 2. Observe that if $x \neq 2$, we can factor the numerator and denominator and simplify, obtaining

$$\frac{x^2 - 4}{3x - 6} = \frac{(x - 2)(x + 2)}{3(x - 2)} = \frac{x + 2}{3} \quad \text{(provided } x \neq 2\text{)}.$$

Since $f(x)$ and $(x + 2)/3$ have the same values except at $x = 2$, it follows that

$$\lim_{x \to 2} \frac{x^2 - 4}{3x - 6} = \lim_{x \to 2} \frac{x + 2}{3} = \frac{4}{3}. \quad \square$$

2-1 Introduction to Limits and Tangent Lines

EXAMPLE 2   Calculate the following limits:

(a) $\lim_{x \to 4} (3x^2 + 7x - 3)$;

(b) $\lim_{x \to 0} \dfrac{2x - 4}{x + 3}$;

(c) $\lim_{x \to 5} \dfrac{x^2 - 25}{x^2 - 7x + 10}$.

Solution   (a) Let $x \to 4$. Then $3x^2 \to 48$ and $7x \to 28$. Therefore,

$$3x^2 + 7x - 3 \to 48 + 28 - 3 = 73.$$

(b) Let $x \to 0$. Then $2x - 4 \to -4$ and $x + 3 \to 3$. Therefore,

$$\frac{2x - 4}{x + 3} \to \frac{-4}{3} = -\frac{4}{3}.$$

(c) We factor the numerator and denominator and simplify before calculating the limit:

$$\frac{x^2 - 25}{x^2 - 7x + 10} = \frac{(x - 5)(x + 5)}{(x - 2)(x - 5)} = \frac{x + 5}{x - 2} \to \frac{10}{3}$$

as $x \to 5$. □

Remark   Many of the limits of interest in the calculus are similar to the one in Example 2(c). They involve fractions in which both numerator and denominator approach zero as x approaches a. If the numerator and denominator are both polynomials, we may be able to factor out $x - a$ and then calculate the limit as in Example 2(c).

If the numerator and denominator are not both polynomials, it may be more difficult to calculate the limit of the fraction. We can show later that, for example,

$$\lim_{x \to 0} \frac{\sin 2x \cos 4x}{\tan 3x} = \frac{2}{3}.$$

(Such limits are considered in Chapter 9.)

Remark   The value of the limit (if any) of $f(x)$ as x approaches a may be completely independent of the value of the function at $x = a$. The function may not be defined at a, it may have a functional value $f(a)$ different from the limit, or the value $f(a)$ may equal the limit of $f(x)$ as x approaches a. As we shall see, many limits can be calculated by substituting $x = a$ in the law of correspondence for f. In general, however, limits cannot be calculated by substitution. We must find the value that $f(x)$ approaches as x approaches a.

Figure 2–1 shows two functions that have a limit L as x approaches a. In (a), the functional value $f(a)$ also equals L. In (b), the functional value $f(a)$ is different from the value of L. Observe, however, that $f(x) \to L$ as $x \to a$.

Figure 2–2 shows two functions that do not have limits as $x \to a$. In (a), the functional values approach L as x approaches a *from the left*, but approach M as x approaches a *from the right*. There is no well-defined limit as x approaches a. In (b), the functional values get infinitely large as x approaches a.

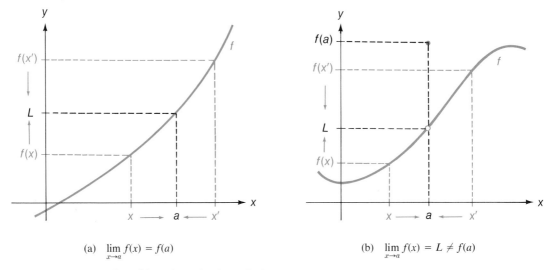

(a) $\lim_{x \to a} f(x) = f(a)$

(b) $\lim_{x \to a} f(x) = L \neq f(a)$

Figure 2–1 Examples of functions that have limits as x approaches a. The limit depends on the behavior of the function near a, but not on the value at a.

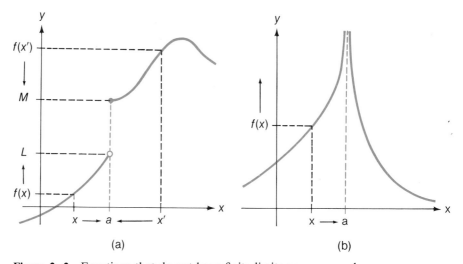

Figure 2–2 Functions that do not have finite limits as x approaches a.

TANGENT LINES

Wherever possible we shall interpret our work in geometrical terms. As we shall see later, many of the applications of the calculus can be interpreted in terms of lines tangent to curves.

2-1 Introduction to Limits and Tangent Lines

For our immediate purposes, we assume that you have worked with lines tangent to certain curves such as circles. In general, the line tangent to a curve at a point on the curve is the line that best approximates the curve in the vicinity of the point. In most cases, the tangent line appears to touch the curve at the point of contact, as in Figure 2–3. (We postpone the formal definition of a tangent line until later in this section.)

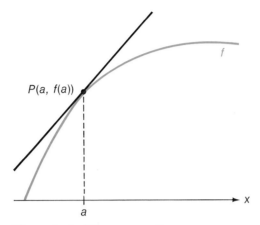

Figure 2–3 The tangent line to the graph of f at the point $P(a, f(a))$.

We now develop the formula for the slope of a nonvertical tangent line. You should refer to Figure 2–4 at each step of the derivation, locating the points and calculating the functional values involved.

Let $P(a, f(a))$ be the *point of tangency*, that is, the point at which the line is tangent to the graph of f. Let Q be a different point on the graph. (See Fig. 2–4a.) Let Δx be the horizontal increment of change from P to Q. Then Q has x-coordinate $a + \Delta x$ and y-coordinate $f(a + \Delta x)$. Thus, Q is the point $Q(a + \Delta x, f(a + \Delta x))$.

Now consider the secant line through the two points P and Q. This is not the tangent line but does approximate the tangent line if Q is sufficiently close to P. Thus, the slope of the secant line through P and Q should approximately equal the slope of the tangent line at P if Q is close to P. This is illustrated in Figure 2–4b for several different secant lines.

To calculate the slope of the secant line, we observe that the increment of change in the y direction from P to Q is

$$\Delta y = f(a + \Delta x) - f(a).$$

Thus, the slope of the secant line is

$$\frac{\Delta y}{\Delta x} = \frac{\text{change in } y}{\text{change in } x} = \frac{f(a + \Delta x) - f(a)}{\Delta x}.$$

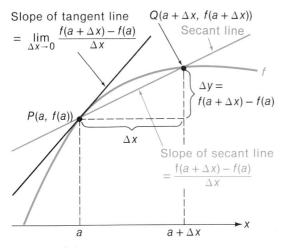
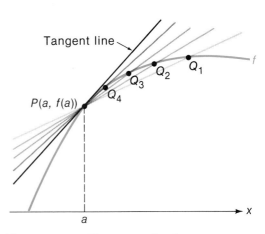

Figure 2-4(a)
Slope of secant line $= \dfrac{f(a + \Delta x) - f(a)}{\Delta x}$.

Slope of tangent line $= \lim\limits_{\Delta x \to 0} \dfrac{f(a + \Delta x) - f(a)}{\Delta x}$.

Figure 2-4(b) The tangent line is the limiting position of the secant lines.

As we noted above, this slope should approximate the slope of the tangent line if Q is close to P (that is, if Δx is close to zero). In the limiting case we should have equality:

$$m = \text{slope of nonvertical tangent line at } (a, f(a)) = \lim_{\Delta x \to 0} \frac{\Delta y}{\Delta x}$$

$$= \lim_{\Delta x \to 0} \frac{f(a + \Delta x) - f(a)}{\Delta x}.$$

Tangent Line In the preceding discussion we relied on your intuition about tangent lines. We are now ready to define the concept of a nonvertical tangent line. (Vertical tangent lines are considered in Section 3-7.)

DEFINITION Let a be in the domain of the function f. If the expression

$$\frac{f(a + \Delta x) - f(a)}{\Delta x}$$

has a finite limit m as $\Delta x \to 0$, then the line through $P(a, f(a))$ with slope equal to m is called the *tangent line to the graph of f at the point $P(a, f(a))$*.

EXAMPLE 3 Find the slope and the equation of the line tangent to the graph of $f(x) = x^2$ at the point $P(1, 1)$.

2-1 Introduction to Limits and Tangent Lines

Solution We take $a = 1$ in the above expression for the slope:

$$m = \lim_{\Delta x \to 0} \frac{f(1 + \Delta x) - f(1)}{\Delta x} = \lim_{\Delta x \to 0} \frac{(1 + \Delta x)^2 - 1^2}{\Delta x}$$

$$= \lim_{\Delta x \to 0} \frac{1 + 2\Delta x + \Delta x^2 - 1}{\Delta x} = \lim_{\Delta x \to 0} (2 + \Delta x) = 2.$$

The slope of the tangent line is $m = 2$.

The equation of the tangent line can be found from the point-slope form of the equation of a line using $m = 2$, $x_0 = 1$, $y_0 = 1$:

$$y - 1 = 2(x - 1)$$
$$2x - y = 1$$

(See Fig. 2–5.) □

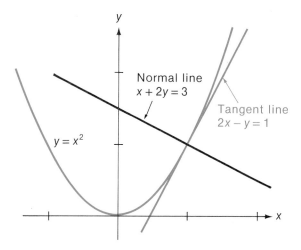

Figure 2–5 Example 3.

NORMAL LINES

The line perpendicular to the tangent line at its point of tangency is called a *normal line* relative to the curve. (See Fig. 2–5.) If $m \neq 0$ is the slope of the tangent line, then $-1/m$ is the slope of the normal line. If $m = 0$, the normal line is vertical. If the tangent line is vertical (no slope), then the normal line is horizontal.

In Example 3, we saw that the slope of the line tangent to the graph of $y = x^2$ at the point $(1, 1)$ is $m = 2$. The slope of the line normal to the graph at $(1, 1)$ is $-\frac{1}{2}$. The equation of this normal line is

$$y - 1 = -\tfrac{1}{2}(x - 1)$$
$$x + 2y = 3.$$

(See Fig. 2–5.)

Exercises 2-1

1-20. Determine the limits if they exist.

- 1 $\lim\limits_{x \to 2} (x^2 + 3x - 5)$
- 2 $\lim\limits_{x \to 2} \dfrac{x - 4}{x^2 + 3x + 1}$
- 3 $\lim\limits_{x \to 1} \dfrac{x^2 + x - 2}{x - 1}$
- 4 $\lim\limits_{x \to 2} \dfrac{x^2 - x - 2}{x - 2}$
- 5 $\lim\limits_{x \to 2} \dfrac{x - 2}{x^3 - 8}$
- 6 $\lim\limits_{x \to 0} \dfrac{x^3 - 4x}{x^2 + 4x}$
- 7 $\lim\limits_{x \to 9} \dfrac{\sqrt{x} - 3}{x - 9}$
- 8 $\lim\limits_{x \to -1} \dfrac{x^3 + 1}{x + 1}$
- 9 $\lim\limits_{x \to 2} \dfrac{(1/x) - (1/2)}{x - 2}$
- 10 $\lim\limits_{x \to 1} \dfrac{x^2 - 1}{x^4 - 1}$
- 11 $\lim\limits_{t \to -1} \dfrac{t - 1}{t^2 - 1}$
- 12 $\lim\limits_{t \to 2} \dfrac{t^3 - 8}{t^2 - 4}$
- 13 $\lim\limits_{h \to 0} \dfrac{(1 + h)^2 - 1}{h}$
- 14 $\lim\limits_{h \to 0} \dfrac{(a + h)^3 - a^3}{h}$
- 15 $\lim\limits_{y \to 0} y^0$
- 16 $\lim\limits_{z \to 0} \dfrac{|z|}{z}$
- 17 $\lim\limits_{z \to -1} \dfrac{1 - |z|}{1 + z}$
- 18 $\lim\limits_{x \to -1} \dfrac{x}{|x|}$
- 19 $\lim\limits_{t \to 1} [t]$
- 20 $\lim\limits_{t \to 1/2} [t]$

21-22. Work through the derivation in the text for the slope of the tangent line using the specific examples. In each case, draw a figure similar to Figure 2-4a, locating the points $P(a, f(a))$ and $Q(a + \Delta x, f(a + \Delta x))$. Label all parts of the figure as in Figure 2-4.

- 21 $f(x) = x^2$, $a = 2$
- 22 $f(x) = 1 - x^2$, $a = 1$

23-28. Find the slopes and the equations of the tangent and normal lines to each of the graphs at the indicated point.

- 23 $f(x) = x^2$, $P(-2, 4)$
- 24 $f(x) = x^2 - 3$, $P(2, 1)$
- 25 $f(x) = 2x^2 + 1$, $P(2, 9)$
- 26 $g(x) = (x + 1)^2$, $P(1, 4)$
- 27 $F(x) = x^2 - 3x + 1$, $A(-1, 5)$
- 28 $G(x) = 1 - 3x^2$, $B(1, -2)$

29-32. If you have a scientific calculator, use it to convince yourself that these results are correct. (Use the RADIAN mode for trigonometric functions.)

- c29 $\lim\limits_{x \to 0} \dfrac{\sin 2x \cos 3x}{\tan 4x} = \dfrac{1}{2}$
- c30 $\lim\limits_{x \to 0} \dfrac{\tan x}{x} = 1$

2–2 The Derivative

c31 $f(x) = \dfrac{1}{x^2}$ has no limit as $x \to 0$.

c32 $f(x) = \dfrac{2x}{(x-2)^2}$ has no limit as $x \to 2$.

2–2 The Derivative

The slope of the line tangent to the graph of a function f at the point $(a, f(a))$ is given by the limit

$$\lim_{\Delta x \to 0} \frac{f(a + \Delta x) - f(a)}{\Delta x}$$

provided the limit exists. This same type of limit is involved in the solution of many other types of problems. (Several will be discussed in Section 2–3.) It is so important in mathematics that it has been given a special name—the *derivative of f at a*, denoted by $f'(a)$.

DEFINITION Let f be a function of x; let a be in the domain of f. The *derivative of f at a*, denoted by $f'(a)$, is defined to be

Derivative
$$f'(a) = \lim_{\Delta x \to 0} \frac{f(a + \Delta x) - f(a)}{\Delta x}$$

provided the limit exists.

Differentiable Function
  The function f is said to be *differentiable at a* if $f'(a)$ exists. It is *differentiable on the set* S if it is differentiable at each point of S.

EXAMPLE 1 Let $f(x) = x^2 - 3x + 1$. Calculate $f'(2)$, $f'(a)$, and $f'(x)$.

Solution (a) By the above formula

$$f'(2) = \lim_{\Delta x \to 0} \frac{f(2 + \Delta x) - f(2)}{\Delta x}$$

$$= \lim_{\Delta x \to 0} \frac{[(2 + \Delta x)^2 - 3(2 + \Delta x) + 1] - [2^2 - 3 \cdot 2 + 1]}{\Delta x}$$

$$= \lim_{\Delta x \to 0} \frac{4\Delta x + (\Delta x)^2 - 3\Delta x}{\Delta x} = \lim_{\Delta x \to 0} (1 + \Delta x) = 1.$$

(b) If a is any real number, then

$$f'(a) = \lim_{\Delta x \to 0} \frac{f(a + \Delta x) - f(a)}{\Delta x}$$

$$= \lim_{\Delta x \to 0} \frac{[(a + \Delta x)^2 - 3(a + \Delta x) + 1] - [a^2 - 3a + 1]}{\Delta x}$$

$$= \lim_{\Delta x \to 0} \frac{(2a - 3)\Delta x + (\Delta x)^2}{\Delta x}$$

$$= \lim_{\Delta x \to 0} [(2a - 3) + \Delta x] = 2a - 3.$$

(c) Since $f'(a) = 2a - 3$ for every a, then

$$f'(x) = 2x - 3$$

for every x. □

Observe, as in Example 1(c), that the derivative defines a new (or *derived*) function of x. This derived function is defined by

$$f'(x) = \lim_{\Delta x \to 0} \frac{f(x + \Delta x) - f(x)}{\Delta x}$$

for each value of x for which the limit exists. In calculating the limit that defines $f'(x)$, we hold x fixed and let Δx vary.

If a different independent variable is used, then the derivative is calculated with respect to it. For example, if

$$f(t) = t^2 - 3t + 1$$

then

$$f'(t) = \lim_{\Delta t \to 0} \frac{f(t + \Delta t) - f(t)}{\Delta t} = 2t - 3.$$

If

$$f(z) = z^2 - 3z + 1$$

then

$$f'(z) = \lim_{\Delta z \to 0} \frac{f(z + \Delta z) - f(z)}{\Delta z} = 2z - 3,$$

and so on.

EXAMPLE 2 Calculate the derivatives of the following functions:
(a) $f(x) = C$ (a constant); (b) $g(t) = t$; (c) $h(z) = z^2$.

Solution (a) $f(x) = C$. Then

$$f'(x) = \lim_{\Delta x \to 0} \frac{f(x + \Delta x) - f(x)}{\Delta x} = \lim_{\Delta x \to 0} \frac{C - C}{\Delta x}$$

$$= \lim_{\Delta x \to 0} 0 = 0.$$

Thus, the derivative of a constant function is zero at each value of x.

2–2 The Derivative

(b) $g(t) = t$. Then

$$g'(t) = \lim_{\Delta t \to 0} \frac{g(t + \Delta t) - g(t)}{\Delta t} = \lim_{\Delta t \to 0} \frac{(t + \Delta t) - t}{\Delta t}$$

$$= \lim_{\Delta t \to 0} \frac{\Delta t}{\Delta t} = \lim_{\Delta t \to 0} 1 = 1.$$

(c) $h(z) = z^2$. Then

$$h'(z) = \lim_{\Delta z \to 0} \frac{h(z + \Delta z) - h(z)}{\Delta z} = \lim_{\Delta z \to 0} \frac{(z + \Delta z)^2 - z^2}{\Delta z}$$

$$= \lim_{\Delta z \to 0} \frac{z^2 + 2z\,\Delta z + \Delta z^2 - z^2}{\Delta z} = \lim_{\Delta z \to 0} \frac{2z\,\Delta z + \Delta z^2}{\Delta z}$$

$$= \lim_{\Delta z \to 0} (2z + \Delta z) = 2z. \quad \square$$

NOTATION

The derivative is basic to the study of the calculus. As we shall see, it is the key to solving diverse problems from science, mathematics, and economics.

Because of the wide range of applications, numerous symbols have been invented for the derivative. If $y = f(x)$ is the original function, then the derivative can be denoted by any of the symbols

Symbols for the Derivative

$$y', \quad f'(x), \quad \frac{dy}{dx}, \quad \frac{d(f(x))}{dx}, \quad D_x y, \quad D_x f(x), \quad \text{and} \quad Dy.$$

These symbols all represent the same thing—the derivative of f with respect to x:

$$f'(x) = \lim_{\Delta x \to 0} \frac{f(x + \Delta x) - f(x)}{\Delta x}.$$

Each of the above notations has been preferred by some principal group of mathematicians or scientists and has been found to be the most natural notation for solving an important class of problems. As a consequence, the multiplicity of notations has continued over three centuries.

We shall use mostly the notations $f'(x)$, y', $D_x y$, and dy/dx. Consider, as an illustration of the different notations, the result of Example 1. We established that if

$$y = f(x) = x^2 - 3x + 1,$$

then

$$f'(x) = 2x - 3.$$

If we use the other notations we can write

$$y' = 2x - 3,$$

$$\frac{dy}{dx} = \frac{d}{dx}(x^2 - 3x + 1) = 2x - 3,$$

and
$$D_x y = D_x(x^2 - 3x + 1) = 2x - 3.$$

Remark The notation dy/dx for the derivative is a consequence of the Δx notation for an increment. If we let
$$y = f(x + \Delta x) - f(x)$$
be the increment in f that corresponds to a change of Δx in x, then the derivative is
$$\frac{dy}{dx} = \lim_{\Delta x \to 0} \frac{f(x + \Delta x) - f(x)}{x} = \lim_{\Delta x \to 0} \frac{\Delta y}{\Delta x}.$$

The early developers of the calculus thought that Δy somehow changed into dy and Δx into dx when the limit was calculated.

EXAMPLE 3 Let $f(x) = \sqrt{x}$.
(a) Calculate the derivative dy/dx.
(b) Find the equation of the line tangent to the graph of f at the point (4, 2).

Solution (a)
$$\frac{dy}{dx} = f'(x) = \frac{d}{dx}(\sqrt{x})$$
$$= \lim_{\Delta x \to 0} \frac{f(x + \Delta x) - f(x)}{\Delta x}$$
$$= \lim_{\Delta x \to 0} \frac{\sqrt{x + \Delta x} - \sqrt{x}}{\Delta x}.$$

Observe that if we can remove the radicals from the numerator, the resulting expression can be simplified. To accomplish this, we multiply numerator and denominator by $\sqrt{x + \Delta x} + \sqrt{x}$:

$$\frac{dy}{dx} = \lim_{\Delta x \to 0} \frac{\sqrt{x + \Delta x} - \sqrt{x}}{\Delta x} \cdot \frac{\sqrt{x + \Delta x} + \sqrt{x}}{\sqrt{x + \Delta x} + \sqrt{x}}$$
$$= \lim_{\Delta x \to 0} \frac{x + \Delta x - x}{\Delta x(\sqrt{x + \Delta x} + \sqrt{x})}$$
$$= \lim_{\Delta x \to 0} \frac{\Delta x}{\Delta x(\sqrt{x + \Delta x} + \sqrt{x})} = \lim_{\Delta x \to 0} \frac{1}{\sqrt{x + \Delta x} + \sqrt{x}}$$
$$= \frac{1}{\sqrt{x} + \sqrt{x}} = \frac{1}{2\sqrt{x}} = f'(x).$$

(b) Recall that the value of the derivative $f'(4)$ equals the slope of the tangent line at $x = 4$. Thus, the slope of the tangent line at (4, 2) is
$$f'(4) = \frac{1}{2\sqrt{4}} = \frac{1}{4}.$$

2-2 The Derivative

The equation of the tangent line is

$$y - 2 = \tfrac{1}{4}(x - 4)$$
$$x - 4y + 4 = 0.$$

(See Fig. 2-6.) ☐

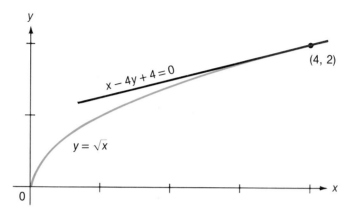

Figure 2-6 Example 3.

ALTERNATIVE FORM OF THE DIFFERENCE QUOTIENT

There is an alternative form for writing the limit

$$\lim_{\Delta x \to 0} \frac{f(a + \Delta x) - f(a)}{\Delta x}$$

that defines the derivative at $x = a$. We make the substitution

$$x = a + \Delta x,$$

so that $\Delta x = x - a$. Observe that $x \to a$ if and only if $\Delta x \to 0$. Therefore,

$$f'(a) = \lim_{\Delta x \to 0} \frac{f(a + \Delta x) - f(a)}{\Delta x} = \lim_{x \to a} \frac{f(x) - f(a)}{x - a}$$

$$\boxed{f'(a) = \lim_{x \to a} \frac{f(x) - f(a)}{x - a}.}$$

The expressions

Difference Quotient

$$\frac{f(a + \Delta x) - f(a)}{\Delta x} \quad \text{and} \quad \frac{f(x) - f(a)}{x - a}$$

are called *difference quotients*. In certain cases, it is much simpler to use the alternative form of the difference quotient than the original form.

EXAMPLE 4   Let $y = f(x) = 1/x$. Show that $f'(a) = -1/a^2$ provided $a \neq 0$.

Solution We use the alternative form of the difference quotient:

$$f'(a) = \lim_{x \to a} \frac{f(x) - f(a)}{x - a} = \lim_{x \to a} \frac{\dfrac{1}{x} - \dfrac{1}{a}}{x - a}$$

$$= \lim_{x \to a} \frac{\dfrac{a - x}{xa}}{x - a}$$

$$= \lim_{x \to a} \frac{a - x}{xa} \cdot \frac{1}{x - a} = \lim_{x \to a} \frac{-1}{xa} = -\frac{1}{a^2}. \quad \square$$

NEWTON AND LEIBNIZ

The concept of the derivative marks the beginning of the calculus. This branch of mathematics was invented independently by two of the intellectual giants of the seventeenth century—Isaac Newton (1642–1727) and Gottfried Wilhelm Leibniz (1646–1716).

Newton was a student at Cambridge when that university was closed because of the Black Plague in 1665. He returned to the family farm and spent the next two years in semi-isolation, contemplating basic problems in physics and mathematics. During this period, he formulated the law of universal gravitation, discovered many of the basic principles of optics, generalized the binomial theorem, and developed the basic principles of the calculus.

On his return to Cambridge, he published his work on optics, which was bitterly attacked by several of the leading physicists of the day. The resulting controversy upset Newton so much that he vowed never to publish again. Finally he relented somewhat and, in 1687, 20 years after making his discoveries in physics and astronomy, he published his monumental work, the *Principia mathematica*. This book also contained some of Newton's work on the calculus, which he used as a tool for theoretical physics.

The publication of the *Principia* established Newton as the foremost scientist of his day. He became a prominent public figure, was knighted by the king and elected president of the Royal Society, and was given the post of director of the mint (which left him no time to work on science). Toward the end of his life he published his other discoveries, including his more complete work on the calculus.

Leibniz, a many-sided genius, was one of the outstanding intellectuals of the seventeenth century. Although most of his life was spent in the German diplomatic service, he made important contributions to mathematics, philosophy, logic, theology, and linguistics.

Leibniz was almost ignorant of advanced mathematics until 1672, when he began a series of studies under the direction of Huygens, the Dutch scientist. He soon realized that several special results from various parts of mathematics could be unified into one important theory—and thereby invented the calculus.

Leibniz realized the importance of his discovery and spent the next few years perfecting it. He was especially concerned about proper notation.

Exercises 2-2

Isaac Newton (David Eugene Smith Collection, Rare Book and Manuscript Library, Columbia University).

Gottfried Wilhelm Leibniz (David Eugene Smith Collection, Rare Book and Manuscript Library, Columbia University).

His work in logic had made him aware that symbols should be chosen that appeal to the intuition and that suggest true properties. Realizing that the derivative has many properties in common with fractions, he invented the dy/dx notation. Thinking of the antiderivative as an infinite sum of infinitely small quantities led him to the symbol $\int$, an elongated S for the word *sum*.

These results were published between 1682 and 1692 and quickly began to revolutionize mathematics. After this work began to appear, Newton's friends pressed Newton to publish his work on the calculus; hence *Principia* was published.

A few years later, a serious controversy arose between English mathematicians, who believed that Leibniz had stolen his results from Newton, and continental mathematicians, who believed that Newton had stolen from Leibniz. The result was a rift in the mathematical community that lasted over a century. The continental mathematicians adopted the superior notation devised by Leibniz and used it as a principal tool in developing the calculus. The English mathematicians adopted the very awkward notation used by Newton and found that it was difficult to formulate even the simplest results and even more difficult to prove them. Not until the early nineteenth century did the English mathematicians adopt the Leibniz notation and move back into the mainstream of mathematics.

Exercises 2-2

1–14. **(a)** Calculate the derivatives of the functions. For 1–7 use the formula

$$f'(x) = \lim_{\Delta x \to 0} \frac{f(x + \Delta x) - f(x)}{\Delta x};$$

for 8–14 use

$$f'(a) = \lim_{x \to a} \frac{f(x) - f(a)}{x - a}.$$

(b) Find the equations of the tangent and normal lines to the graph of the functions at the given points.

- 1 $f(x) = 3x + 5$; (0, 5)
- 2 $f(x) = 6 - 4x$; (2, -2)
- 3 $f(x) = (x - 2)(x + 1)$; (1, -2)
- 4 $g(x) = 3x^2 + 2x - 1$; (1, 4)
- 5 $g(x) = \dfrac{x^4 - 2x^2}{x^2}$; (2, 2)
- 6 $h(x) = x^3 + 2x$; (1, 3)
- 7 $h(x) = 3x^3 - 4x - 5$; (-1, -4)
- 8 $f(x) = \sqrt{2x - 1}$; (5, 3)
- 9 $f(x) = \sqrt{x + 1}$; (3, 2)
- 10 $g(x) = \dfrac{1}{x - 1}$; (2, 1)
- 11 $f(t) = \sqrt{3 - t}$; (2, 1)
- 12 $g(u) = \dfrac{u}{u - 1}$; (2, 2)
- 13 $h(z) = \dfrac{1}{z^2}$; (2, $\tfrac{1}{4}$)
- 14 $f(v) = \dfrac{2}{\sqrt{v + 1}}$; (3, 1)

15 Let $f(x) = ax^2 + bx + c$, where a, b, c are constants. Show that $f'(x) = 2ax + b$. (This is the first of many derivative formulas that we encounter.)

16 Use the formula in Exercise 15 to show that:
  (a) If $f(x) = x^2$, then $f'(x) = 2x$.
  (b) If $f(x) = x$, then $f'(x) = 1$.
  (c) If $f(x) = 1$, then $f'(x) = 0$.

17–26. Use the formula in Exercise 15 to calculate the derivatives.

- 17 $f(x) = x^2 - x + 1$
- 18 $f(x) = 2x^2 - x - 5$
- 19 $g(x) = 2x^2 - 1$
- 20 $g(x) = 3x - 1$
- 21 $h(x) = \tfrac{1}{2}x^2 - x + 20$
- 22 $F(t) = 3t^2 + 1$
- 23 $G(t) = 3t + 4t^2$
- 24 $g(u) = 5$
- 25 $f(v) = 114 - 7v^2$
- 26 $H(z) = 5z^2 + 350z - 201$

27–34. For each of the functions find the points, if any, at which the slopes of the tangent lines equal zero.

- 27 $f(x) = 3x^2 + 3x - 1$
- 28 $g(x) = x^3 + 2x$
- 29 $h(x) = 2x^2 - x - 5$
- 30 $j(x) = \tfrac{1}{2}x^2 - x - 4$
- 31 $f(t) = 3t^3 - 4t + 2$
- 32 $g(v) = v^2 - v - 2$
- 33 $F(u) = \sqrt{u + 1}$
- 34 $G(w) = \dfrac{1}{w^2}$

35 (a) Show that $D_x(x^3) = 3x^2$.
   (b) Show that $D_x(x^4) = 4x^3$.
   (c) Use Example 2 or Exercise 15 to calculate $D_x(x^0)$, $D_x(x)$, and $D_x(x^2)$. Compare these results with (a) and (b). What general formula would you propose?

•36 Find all the tangent lines to the graph of $f(x) = x^2 + 2$ that pass through $(-2, 5)$.

37 (a) Sketch the graph of $y = x^2$, showing the tangent line at the point (a, a^2). The graph should show that this tangent line is under the curve except at the point of tangency.
   (b) Show that the tangent line in (a) has the equation $y = 2ax - a^2$.
   (c) Use Example 2 or Exercise 15 to calculate $D_x(x^0)$, $D_x(x)$, and $D_x(x^2)$. Compare these results with (a) and (b). What general formula would you propose?

2–3 Rates of Change

Consider a particle moving along a straight line. (The particle could represent a car traveling along a straight stretch of road, a raindrop falling straight down, or a rocket fired straight up into the air.) To measure distances along the line, we superimpose a coordinate axis on it, choosing a convenient point for the origin and a convenient unit of measure.

The position s of the particle on the axis at time t can be expressed as a function of t,

$$s = s(t).$$

Distance Function   It is customary to call s a *distance function*, although *position function* would be a more accurate designation.

We are interested in describing how the particle moves along the line. We should particularly like to know the speed and the direction in which it is moving at any instant. The following process can be used to define the *velocity* at any time t_0.

Over the period from t_0 to $t_0 + \Delta t$, the particle moves from position $s(t_0)$ to position $s(t_0 + \Delta t)$. Thus, the change in position is

$$s(t_0 + \Delta t) - s(t_0).$$

Average Velocity   The ratio of the change in position to the change in time is called the *average velocity* over the period:

$$v_{av} = \frac{\text{change in position}}{\text{change in time}} = \frac{s(t_0 + \Delta t) - s(t_0)}{\Delta t}.$$

The average velocity may be positive, negative, or zero. For example, if distance is measured in miles and the particle moves from position $s = +3$ to position $s = -5$ in

2 hours ($\Delta t = 2$), then the average velocity over the two-hour period is

$$v_{av} = \frac{(-5) - 3}{2} = \frac{-8}{2} = -4 \text{ miles per hour.}$$

If it then moves back to the original position $s = 3$ in three more hours, then the average velocity over the second period is

$$v_{av} = \frac{3 - (-5)}{3} = \frac{8}{3} \text{ miles per hour.}$$

The average velocity of the particle over the entire five-hour period, however, is

$$v_{av} = \frac{\text{change in position}}{\text{change in time}} = \frac{0}{5} = 0 \text{ miles per hour.}$$

The *(instantaneous) velocity* at time $t = t_0$, denoted by $v(t_0)$, is defined to be the limit of the average velocity $[s(t_0 + \Delta t) - s(t_0)]/\Delta t$ as the time period Δt approaches zero. Thus,

$$v(t_0) = \text{velocity at } t_0 = \lim_{\Delta t \to 0} \frac{s(t_0 + \Delta t) - s(t_0)}{\Delta t}.$$

Observe that the expression

$$\frac{s(t_0 + \Delta t) - s(t_0)}{\Delta t}$$

is the difference quotient for the derivative of s at time t_0. Therefore,

$$v(t_0) = \lim_{\Delta t \to 0} \frac{s(t_0 + \Delta t) - s(t_0)}{\Delta t} = s'(t_0).$$

In general the velocity at time t is

Velocity

$$v(t) = \text{velocity at time } t = s'(t)$$

provided the derivative exists.

Units of Measurement

Velocity is measured in units based on the units for distance and time. For example, if distance is measured in kilometers and time in hours, then velocity is measured in *kilometers per hour (km/hr)*. If distance is measured in feet and time in seconds, then velocity is measured in *feet per second (ft/sec)*.

EXAMPLE 1 A ball is rolled up an inclined plane and then allowed to roll freely down the plane. The position of the ball at time t is given by the distance function

$$s(t) = 3t^2 - 12t + 1$$

where distance is measured in feet from a fixed point on the plane and t is measured in seconds. Here the coordinate axis is oriented so that the positive direction is down the plane.

2-3 Rates of Change

The velocity at time t is

$$v(t) = s'(t) = 3 \cdot 2t - 12 = 6t - 12 \text{ ft/sec.}$$

(In calculating the derivative, we used the formula from Exercise 15, Section 2–2.) Thus, at time $t = 3$, the velocity is

$$v(3) = 6 \cdot 3 - 12 = 6 \text{ ft/sec.}$$

The positive velocity indicates that the ball is moving downward (in the positive direction). At time $t = 0$, on the other hand, the velocity is

$$v(0) = 6 \cdot 0 - 12 = -12 \text{ ft/sec,}$$

the negative velocity indicating that the ball is moving upward (in the negative direction). □

The absolute value of the velocity is called the *speed* of the particle. The ball in Example 1 has a speed of $|6| = 6$ ft/sec at time $t = 3$ and a speed of $|-12| = 12$ ft/sec at time $t = 0$.

Rate of Change

Velocity is an example of a *rate of change*. It measures the rate at which the position function changes with time. If the velocity has a constant value of 12 ft/min, then every minute the distance function increases by 12 ft. Every derivative can be interpreted as a rate of change. If $y = f(x)$ where f is differentiable, then

$$y' = f'(x)$$

measures the *rate of change of y with respect to x*.

ACCELERATION

Since the velocity $v(t)$ is a function of time, we can calculate its rate of change, which also is a function of time. This function is called the *acceleration* of the particle, denoted by $a(t)$. That is,

$$a(t) = \text{acceleration of particle at time } t$$
$$= \text{rate of change of velocity at time } t$$
$$= \lim_{\Delta t \to 0} \frac{v(t + \Delta t) - v(t)}{\Delta t} = v'(t)$$

Acceleration

$$a(t) = \text{acceleration at time } t = v'(t).$$

Units of Measurement

The acceleration is measured in feet per second per second (ft/sec²), miles per hour per hour (mi/hr²), and so on. If the acceleration is 20 mi/hr², for example, then the velocity increases 20 mi/hr over each 1-hr period.

EXAMPLE 2 A stone is dropped from a bridge 256 ft above the ground. It can be shown that its approximate distance above the earth at time t is given by the distance function

$$s(t) = -16t^2 + 256,$$

where s is measured in feet and t in seconds.

(a) Find the velocity and acceleration at time t.
(b) Find the velocity and speed at time $t = 2$.
(c) Find the speed at the instant the stone strikes the ground.

Solution (a) Since $s(t) = -16t^2 + 256$

then $\quad v(t) = s'(t) = -32t$

and $\quad a(t) = v'(t) = -32.$

(We used the formula from Exercise 15, Section 2–2, for calculating these derivatives.)

Since distance is measured in the positive direction above the earth, then the negative velocity indicates the stone is traveling in the opposite direction, namely, towards the earth. Observe that the acceleration is constant at -32 ft/sec^2. Thus, every second the velocity changes by 32 ft/sec.

(b) At time $t = 2$, the velocity is

$$v(2) = -32 \cdot 2 = -64 \text{ ft/sec}$$

and the speed is

$$|v(2)| = |-64| = 64 \text{ ft/sec}.$$

(c) The stone strikes the ground at the instant at which $s = 0$. To find this time, we set $s(t) = 0$ and solve for t:

$$s(t) = -16t^2 + 256 = 0$$
$$16t^2 = 256$$
$$t^2 = 16$$
$$t = \pm 4.$$

Since we are concerned only with nonnegative values of t, then the stone must strike the earth at the end of 4 sec. At this instant, its velocity is

$$v(4) = -32 \cdot 4 = -128 \text{ ft/sec}.$$

Its speed is

$$|v(4)| = |-128| = 128 \text{ ft/sec}.$$

If we convert the speed to more familiar units, we find that the stone is falling at a speed of approximately 87 mi/hr when it strikes the ground. $\square$

FALLING BODY PROBLEMS

Objects that have been thrown straight up into the air or that fall straight down obey a simple law of motion. As in Example 2, we measure the position $s(t)$ from ground level, with the positive direction upward.

Let v_0 be the velocity and s_0 the position at time $t = 0$. We shall show in Section 4–10 that the position after t sec is

Formula for Falling Objects

$$s(t) = -16t^2 + v_0 t + s_0.$$

2–3 Rates of Change

This formula holds from the instant the object is released until it strikes the ground or is deflected in some way.*

EXAMPLE 3   A stone is thrown upward from a height of 96 ft above the ground with an initial velocity of 80 ft/sec.
(a) Calculate the velocity and acceleration at time t.
(b) For which values of t are the formulas in (a) valid?
(c) What is the maximum height of the stone above the ground?

Solution   (a) The distance above the ground at time t is

$$s(t) = -16t^2 + v_0 t + s_0$$
$$= -16t^2 + 80t + 96.$$

The velocity function is the derivative of the distance function:

$$v(t) = s'(t) = -32t + 80 \text{ ft/sec}.$$

The acceleration is the derivative of the velocity:

$$a(t) = v'(t) = -32 \text{ ft/sec}^2.$$

(b) The stone strikes the earth when $s(t) = 0$. Setting $s = 0$ and solving for t, we get

$$s(t) = -16t^2 + 80t + 96 = 0$$
$$-16(t + 1)(t - 6) = 0$$
$$t = -1 \text{ or } t = 6.$$

Since we are concerned only with nonnegative values of t, we can regard the root $t = -1$ as extraneous for our problem. Thus, the stone strikes the ground at time $t = 6$ sec. Since the formulas for $s(t)$, $v(t)$, and $a(t)$ hold from the instant of release until the instant of impact, they are valid for $0 \leq t \leq 6$.

(c) Recall that $v(t) = -32t + 80$.

Thus,

$$v(t) > 0 \quad \text{if } 0 \leq t < 2.5$$

and

$$v(t) < 0 \quad \text{if } 2.5 < t \leq 6.$$

This means that the stone is rising for the first 2.5 sec and falling for the last 3.5 sec. Its maximum height above the ground is obtained at time $t = 2.5$ sec. This height is

$$s(2.5) = -16(2.5)^2 + 80(2.5) + 96 = 196 \text{ ft.} \quad \square$$

---

* This formula is obtained by making several simplifying assumptions. In actual practice, it is a good approximation to the actual distance formula if the object does not go too high and has little air resistance. A more exact formula is derived in Section 8–6.

INTERPRETATIONS OF THE DERIVATIVE

We have discussed several interpretations of the derivative:

(1) The derivative $f'(x_0)$ is the slope of the line tangent to the graph of f at the point $(x_0, f(x_0))$.
(2) If $s(t)$ is the directed distance from a fixed point on a line to a particle that moves along the line, then $v(t_0) = s'(t_0)$ is the velocity of the particle at time $t = t_0$.
(3) If $v(t)$ is the velocity of a particle moving on a line, then $a(t_0) = v'(t_0)$ is the acceleration of the particle at time $t = t_0$.

RATES OF CHANGE

Even when $f(x)$ does not measure distance or velocity, the quotient

$$\frac{\Delta y}{\Delta x} = \frac{f(x_0 + \Delta x) - f(x_0)}{\Delta x}$$

measures the ratio of the change in f to the change in x. As we noted earlier in this section, the derivative

$$\frac{dy}{dx} = \lim_{\Delta x \to 0} \frac{\Delta y}{\Delta x} = \lim_{\Delta x \to 0} \frac{f(x_0 + \Delta x) - f(x_0)}{\Delta x} = f'(x_0)$$

measures the *rate of change of f with respect to x* at $x = x_0$.

An additional example may help to clarify this concept.

EXAMPLE 4 A lamp casts a light on a wall with an intensity of illumination that depends on the distance of the light from the wall. If the proper units are used, the formula for the intensity of illumination is

$$I = \frac{500}{x^2}$$

where x is the distance of the lamp from the wall measured in feet.

If the lamp is moved towards the wall at the rate of 4 ft/min, starting from a point 20 ft from the wall, the illumination can be considered a function of time (t) as well as of distance (x). It makes sense to ask about the rate of change of I with respect to time as well as the rate of change with respect to distance.

Recall that the formula for I in terms of x is

$$I = f(x) = \frac{500}{x^2}.$$

Since $x = 20 - 4t$ (where t is measured in minutes), then

$$I = g(t) = \frac{500}{(20 - 4t)^2} = \frac{125}{4(5 - t)^2}.$$

If we set up the difference quotients and simplify, we can calculate that

$$\frac{dI}{dx} = f'(x) = \frac{-1000}{x^3} \quad \text{and} \quad \frac{dI}{dt} = g'(t) = \frac{125}{2(5-t)^3}.$$

When the lamp is x ft from the wall, the rate of change of the intensity of illumination is $-1000/x^3$ units per foot. At time t the rate of change of intensity with time is

$$\frac{125}{2(5-t)^3} \text{ units/min.} \quad \square$$

Exercises 2–3

1–8. The distance function for a particle moving along a straight line is given with s measured in meters and t in seconds.
   (a) Calculate the average velocity over the indicated interval of time.
   (b) Find the velocity, speed, and acceleration functions and evaluate them at $t = 1$.
   (c) At what time, if any, does the velocity equal zero?
   (d) Find a value of t such that the velocity $v(t)$ equals the average velocity over the indicated period of time.

- 1 $s(t) = 3t + 1$, $[0, 1]$
- 2 $s(t) = t^2 - 8t$, $[1, 3]$
- 3 $s(t) = 3t^2 - 12t + 10$, $[0, 2]$
- 4 $s(t) = 2t^2 - 4t - 5$, $[1, 4]$
- 5 $s(t) = t^2 + 4t - 6$, $[1, 2]$
- 6 $s(t) = -t^2 + 6t + 1$, $[-1, 2]$
- 7 $s(t) = t + \dfrac{1}{t}$, $[1, 2]$
- 8 $s(t) = t^3 + 1$, $[1, 3]$

9–12. An arrow is shot straight up from the ground with an initial velocity of v_0 ft/sec. Find the distance and velocity functions $s(t)$ and $v(t)$. When does the arrow reach its maximum height? Assume that the archer was lying on the ground when he shot the arrow. How much time does he have to move to avoid getting hit by the arrow on its way down?

- 9 $v_0 = 96$
- 10 $v_0 = 112$
- 11 $v_0 = 48$
- 12 $v_0 = 144$

13–16. A firecracker, set to explode in 2 sec, is dropped from an initial height of s_0 ft. Find the distance and velocity after t sec. Will the firecracker explode in the air, when it hits the ground, or after it hits the ground?

- 13 $s_0 = 100$ ft
- 14 $s_0 = 64$ ft
- 15 $s_0 = 36$ ft
- 16 $s_0 = 114$ ft

17–20. A ball is thrown upward from a distance of s_0 ft above the ground with an initial velocity of v_0 ft/sec. For the given conditions, find the maximum height of the ball above the ground, the length of time it is in the air, and its velocity when it strikes the ground.

- •17 $s_0 = 384$, $v_0 = 160$
- 18 $s_0 = 48$, $v_0 = 32$
- •19 $s_0 = 80$, $v_0 = 64$
- 20 $s_0 = 64$, $v_0 = 48$

•21   A car is traveling down a straight road at 60 mph when the driver applies the brakes, causing a constant deceleration. From the instant the brakes are applied until the car stops, the traveled distance is given by $s(t) = -4t^2 + 88t$, where t is measured in seconds and s in feet. How far does it travel before it stops?

22   Let f be a function of x. Let Δf be the change in f that results from a change in x from $x_0 - 1$ to x_0. Show that $\Delta f \approx f'(x_0)$. Illustrate this result graphically. (*Hint:* Consider the expression $[f(x_0 + \Delta x) - f(x_0)]/\Delta x$ for $\Delta x = -1$.)

2-4   Limits

Thus far we have relied on your intuition about limits. In calculating derivatives, we have assumed certain properties of limits that should seem reasonable, but we have not proved (or formally stated) these properties. Now we look more closely at the limit concept. Subsequently, we shall consider a general theorem that establishes most of the properties of limits needed for our work with the calculus.

The early developers of the calculus had vague ideas about limits. They thought of limits in terms of motion—*$f(x)$ moves toward the limit L as x moves toward a*. As time progressed, logical difficulties seemed to arise with the concept. Finally, in the early part of the nineteenth century, the French mathematician A. L. Cauchy (1789–1857) formulated the first modern definition of the limit concept. Cauchy's definition was still couched in terms of motion, but Karl Weierstrass (German, 1815–1897) later realized that it could be reworded to avoid motion altogether. Weierstrass's version of the limit definition was completely static—it was independent of the concept of motion.

The Cauchy-Weierstrass definition bore little resemblance to the earlier work on limits. It was more closely related to the concept of a tolerance range than to motion, and it did not appeal to the intuition. The value of the definition became apparent over the next few years, however, when Cauchy and other mathematicians put the calculus on a solid, rigorous basis for the first time.

As noted above, the modern definition is closely related to the idea of a tolerance range. If a is a number on the x-axis and δ (delta) is a small positive number, then the

Tolerance Range δ *tolerance range about* a consists of all numbers between $a - \delta$ and $a + \delta$. Thus,

$$\delta \text{ tolerance range about } a = \{x \mid a - \delta < x < a + \delta\}.$$

Similarly, if L is a number on the y-axis and ϵ (epsilon) is a small positive number, then the numbers between $L - \epsilon$ and $L + \epsilon$ form the ϵ *tolerance range about* L.

2–4 Limits

Thus,

$$\epsilon \text{ tolerance range about } L = \{y \mid L - \epsilon < y < L + \epsilon\}.$$

(See Fig. 2–7.)

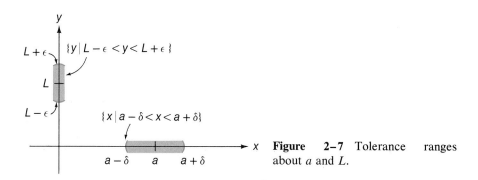

Figure 2–7 Tolerance ranges about a and L.

The Cauchy-Weierstrass definition is based on the idea that if L is the limit as x approaches a, then $f(x)$ must be "close" to L whenever x is "close" to a. The word *close*, however, is very imprecise. We use tolerance ranges to make it precise. According to the definition, $\lim_{x \to a} f(x) = L$ provided that the following holds. Given any ϵ tolerance range about L, there exists a corresponding δ tolerance range about

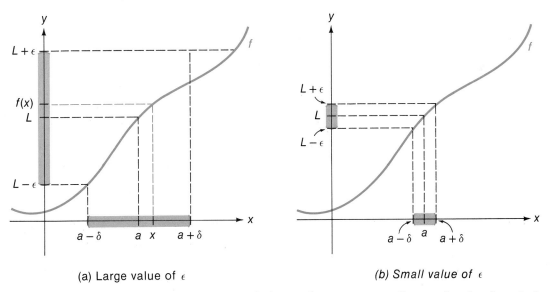

(a) Large value of ϵ (b) Small value of ϵ

Figure 2–8 $\lim_{x \to a} f(x) = L$. For every $\epsilon > 0$ there exists a corresponding number $\delta > 0$ such that $|f(x) - L| < \epsilon$ whenever $0 < |x - a| < \delta$.

a such that if x is a number different from a in the δ tolerance range, then the functional value $f(x)$ must be within the ϵ tolerance range about L.

This concept is illustrated in Figure 2-8. In (a), the value of ϵ is large. There exists a δ tolerance range about a such that if x is any number different from a in the δ range, then $f(x)$ is within the original ϵ range about L. A similar result holds in (b) for a small value of ϵ.

To get a more workable statement, observe that the ϵ tolerance range about L consists of the numbers y satisfying

$$|y - L| < \epsilon.$$

Similarly, the numbers different from a in the δ tolerance range about a are just those numbers satisfying

$$0 < |x - a| < \delta.$$

We are now ready to make the formal definition.

DEFINITION OF LIMIT Let f be a function of x. We say that

$$\lim_{x \to a} f(x) = L$$

provided that for every positive number ϵ there exists a positive number δ (which usually depends on ϵ) such that

$$|f(x) - L| < \epsilon \quad \text{whenever} \quad 0 < |x - a| < \delta.$$

In applying the definition, we first choose $\epsilon > 0$, then find $\delta > 0$ such that $|f(x) - L| < \epsilon$ whenever $0 < |x - a| < \delta$. Since such a δ must exist for every $\epsilon > 0$ (for the limit to exist), we usually treat ϵ as a variable rather than work with specific values. The corresponding δ can usually be expressed in terms of ϵ.

EXAMPLE 1 Use the definition of limit to prove that $\lim_{x \to 3} (2x + 5) = 11$.

Solution Let $\epsilon > 0$. We must find $\delta > 0$ such that

$$|f(x) - L| = |(2x + 5) - 11| < \epsilon \quad \text{whenever} \quad 0 < |x - 3| < \delta.$$

Before trying to find δ, we simplify the expression for $|f(x) - L| < \epsilon$:

$$|(2x + 5) - 11| < \epsilon$$
$$|2x - 6| < \epsilon$$
$$2|x - 3| < \epsilon$$
$$|x - 3| < \frac{\epsilon}{2}.$$

Observe that this last inequality is equivalent to the first one. Thus, we can take δ to be $\epsilon/2$ or any smaller positive number.

2-4 Limits

We now reverse the chain of calculations, starting with a positive number δ that is no greater than $\epsilon/2$. If x satisfies the inequality

$$0 < |x - 3| < \delta,$$

then

$$|x - 3| < \frac{\epsilon}{2}.$$

It follows from the equivalence of the chain of inequalities above that

$$|(2x + 5) - 11| < \epsilon.$$

Since $|(2x + 5) - 11| < \epsilon$ whenever $0 < |x - 3| < \delta$, then

$$\lim_{x \to 3} (2x + 5) = 11. \quad \square$$

The procedure used in Example 1 works only for limits of linear functions. If a function is nonlinear, considerable ingenuity may be needed to determine the value of δ to use for a given ϵ.

Example 2 illustrates the difficulty in establishing a limit for one type of nonlinear function.

EXAMPLE 2 Show that

$$\lim_{x \to 5} \left(\frac{1}{x}\right) = \frac{1}{5}.$$

Solution Let $\epsilon > 0$. We must find $\delta > 0$ such that

$$\left|\frac{1}{x} - \frac{1}{5}\right| < \epsilon \quad \text{whenever } 0 < |x - 5| < \delta.$$

First, observe that the following inequalities are equivalent provided $x \neq 0$:

$$\left|\frac{1}{x} - \frac{1}{5}\right| < \epsilon$$

$$\left|\frac{5 - x}{5x}\right| < \epsilon$$

$$|5 - x| < \epsilon \cdot |5x|.$$

By analogy to the procedure of Example 1, this expression provides a maximum value of δ. However, it is not useful in this form because of the factor $|5x|$ on the right, which would cause the value of δ to vary with x. We find a single maximum value for δ by recalling that we are interested in only those values of x that are "close" to 5. In particular, we may restrict x to values between 4 and 6. If that is done, then

$$20 < |5x| < 30$$

and

$$20\epsilon < |5x|\epsilon < 30\epsilon.$$

We now use this information to select δ. Let δ be a positive number that is less than 1 and less than 20ε. (The "20" comes from the inequality in the preceding paragraph.) Let $0 < |x - 5| < δ$. Since $δ < 1$, then $|x - 5| < 1$, so that x is between 4 and 6 and $|5x|$ is between 20 and 30. Since we also required that $δ < 20ε$, then

$$|x - 5| < 20ε < |5x| \cdot ε$$

$$\frac{|x - 5|}{|5x|} < ε$$

$$\left|\frac{1}{x} - \frac{1}{5}\right| = \left|\frac{5 - x}{5x}\right| = \frac{|5 - x|}{|5x|} < ε.$$

It follows from the definition that

$$\lim_{x \to 5} \left(\frac{1}{x}\right) = \frac{1}{5}. \quad \square$$

Remark   The definition of $\lim_{x \to a} f(x) = L$ specifically excludes the number $x = a$ from consideration. Given $ε > 0$, we must find $δ > 0$ such that if x is within the δ tolerance range about a *and is not equal to a*, then $|f(x) - L| < ε$. We do not care about the value of $f(a)$ at all. The function f may not be defined at $x = a$, or it may have a value $f(a)$ different from the value of the limit as x approaches a. (See Fig. 2–9.) Of course, the value of $f(a)$ may be within the $ε$ range about L. This possibility is not excluded.

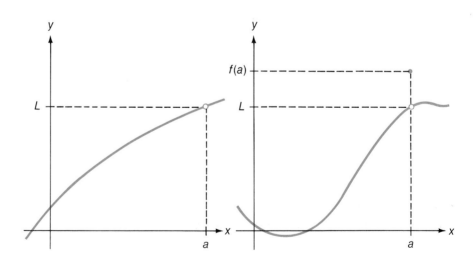

(a) $\lim_{x \to a} f(x) = L$, although f is not defined at a.

(b) $\lim_{x \to a} f(x) = L$, although $f(a) \neq L$.

Figure 2–9.

2–4 Limits

The definition of limit is primarily used in theoretical arguments, not in actual calculations of limits. In the Appendix to this chapter, we use the definition to establish some general properties that in turn can be used to calculate most of the limits needed in this book. (These properties are discussed in Section 2–5.)

Although the definition is not generally used for calculating limits, it is sometimes used to show that certain functions do not have limits at specific points. The key to this use is found in the following fact. For $\lim_{x \to a} f(x)$ to equal L, then for *every* $\epsilon > 0$ there must exist a corresponding number $\delta > 0$ such that $|f(x) - L| < \epsilon$ whenever $0 < |x - a| < \delta$. If a single ϵ can be found for which no such corresponding δ exists, then L is not the limit as x approaches a.

EXAMPLE 3 Show that the function pictured in Figure 2–10a has no limit as $x \to a$.

Solution The values of $f(x)$ are close to K if x is close to a and less than a, while the values are close to L if x is close to a and greater than a.

To show that f has no limit as $x \to a$, we first show that K is not the limit. Choose a positive value of ϵ small enough so that $K + \epsilon < L$. (See Fig. 2–10b.) Let δ be a small positive number. If x is between a and $a + \delta$, then

$$f(x) > L > K + \epsilon$$

so that

$$|f(x) - K| = f(x) - K > \epsilon.$$

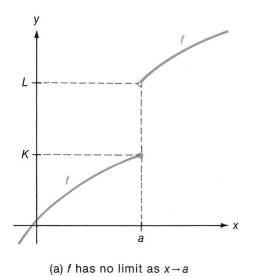

(a) f has no limit as $x \to a$

(b) (Example 3) K is not the limit as $x \to a$

Figure 2–10 One-sided limits: $\lim_{x \to a^-} f(x) = K$, and $\lim_{x \to a^+} f(x) = L$. The function f has no limit as x approaches a because the one-sided limits are unequal.

Thus, it is not possible to make $|f(x) - K| < \epsilon$ for all x satisfying $0 < |x - a| < \delta$. Since the conditions of the definition cannot be met, then K is not the limit as $x \to a$.

A similar argument can be used to show that L is not the limit as $x \to a$. A slight modification can be devised to show that no number besides K or L could be the limit as $x \to a$. Thus, $f(x)$ has no limit as $x \to a$. (See Exercise 16.) □

A function such as the one pictured in Figure 2–10 is said to be *discontinuous* at the point at which its graph is broken. (This subject will be discussed in more detail in Sections 2–6 and 2–7.)

ONE-SIDED LIMITS

Consider again the function pictured in Figure 2–10. This function has *one-sided limits* as $x \to a$. If x approaches a from the left, then $f(x) \to K$; and if x approaches a from the right, then $f(x) \to L$. We indicate this by writing

$$\lim_{x \to a^-} f(x) = K \quad (K \text{ is the limit as } x \text{ approaches } a \text{ from the left}),$$

$$\lim_{x \to a^+} f(x) = L \quad (L \text{ is the limit as } x \text{ approaches } a \text{ from the right}).$$

These one-sided limits are defined similarly to the limit concept. The essential difference is that in defining $\lim_{x \to a^+}$, for example, we consider only the values of x that are within a distance δ of a and are to the right of a.

DEFINITION

One-Sided Limits

We say that $\lim_{x \to a^+} f(x) = L$ provided that the following condition exists. If $\epsilon > 0$, there exists $\delta > 0$ such that

$$|f(x) - L| < \epsilon \quad \text{whenever } a < x < a + \delta.$$

We say that $\lim_{x \to a^-} f(x) = L$ provided that the following condition exists. If $\epsilon > 0$, there exists $\delta > 0$ such that

$$|f(x) - L| < \epsilon \quad \text{whenever } a - \delta < x < a.$$

For example (see Fig. 1–28), if $f(x) = [x]$, the greatest integer function, then

$$\lim_{x \to 1^-} f(x) = \lim_{x \to 1^-} [x] = \lim_{x \to 1^-} 0 = 0$$

and

$$\lim_{x \to 1^+} f(x) = \lim_{x \to 1^+} [x] = \lim_{x \to 1^+} 1 = 1.$$

Remark

Equality of One-Sided Limits

Observe that $\lim_{x \to a} f(x) = L$ if and only if both one-sided limits exist and are equal to L. Thus, for example, if $f(x) = [x]$, then f has no limit as $x \to 1$ since the left- and right-hand limits are not equal.

Exercises 2–4

1–4. Find $\delta > 0$ such that $|f(x) - L| < 0.01$ when $0 < |x - a| < \delta$.

- **1** $f(x) = 3x + 2$, $a = 1$, $L = 5$
- **2** $f(x) = -2x + 7$, $a = 2$, $L = 3$
- **3** $f(x) = x^2$, $a = 2$, $L = 4$
- **4** $f(x) = x^2 - 1$, $a = 3$, $L = 8$

5–10. Use the definition to establish the limits.

- **5** $\lim_{x \to 4} \tfrac{1}{4}x = 1$
- **6** $\lim_{x \to 3} 4 = 4$
- **7** $\lim_{x \to 3} (4x + 2) = 14$
- **8** $\lim_{x \to -1} (3x + 2) = -1$
- **9** $\lim_{x \to 2} (5 - x) = 3$
- **10** $\lim_{x \to -1} (2 - 3x) = 5$

11–15. Draw a graph and then determine the limit.

- **11** Let $f(x) = |x|/x$. Calculate $\lim_{x \to 0^+} f(x)$ and $\lim_{x \to 0^-} f(x)$. Does f have a limit as $x \to 0$? (*Hint:* Simplify the expression $|x|/x$ before calculating the one-sided limits.)

- **12** Let $f(x) = [x^2 + 1]$. (The brackets indicate the greatest integer function.) Calculate $\lim_{x \to 2^+} f(x)$ and $\lim_{x \to 2^-} f(x)$. Does f have a limit as $x \to 2$?

- **13** Let $f(x) = \begin{cases} \dfrac{x^2 - 9}{x + 3} & \text{if } x \neq -3 \\ -8 & \text{if } x = -3. \end{cases}$

  Calculate $\lim_{x \to -3^+} f(x)$ and $\lim_{x \to -3^-} f(x)$. Does f have a limit as $x \to -3$?

- **14** Let $f(x) = \begin{cases} x & \text{if } x < 0 \\ x^2 + 1 & \text{if } x > 0. \end{cases}$

  Calculate $\lim_{x \to 0^+} f(x)$ and $\lim_{x \to 0^-} f(x)$. Does f have a limit as $x \to 0$?

- **15** Let $f(x) = [x]$ (the greatest integer function) and let n be an integer. Show that f has no limit as $x \to n$.

16 Modify the argument of Example 3 to prove the following results:
  (a) L is not the limit as x approaches a.
  (b) No number except K or L could be the limit as x approaches a.
(*Hint:* If M is a number different from K and L, there exists $\epsilon > 0$ such that neither K nor L is in the ϵ range about M. Consider several cases, according to whether $M < K$, $K < M < L$, or $M > L$.)

17–18. Use your ingenuity! See whether you can prove the limits from the definition. (Example 2 may give you some ideas about Exercise 17. Start Exercise 18 by simplifying the inequality $|x^2 - 16| = |x - 4| \cdot |x + 4| < \epsilon$.)

17 $\lim_{x \to 7} \dfrac{2}{x} = \dfrac{2}{7}$

18 $\lim_{x \to 4} x^2 = 16$

2-5 Some Properties of Limits

The definition of limit is rarely used for the calculation of the value of a limit of a specific function. One glance at the amount of work involved in the examples and problems of the preceding section will explain this! Instead, the definition is used to prove a small number of basic properties of limits; these properties are, in turn, used to evaluate specific limits. The most important of these properties are summarized in the following theorem.

THEOREM 2-1	(*Limit Theorem*) Let f and g be functions of x that have limits L and M, respectively, as x approaches a. Then
Limit of Sum	**L1** $\quad \lim_{x \to a} [f(x) \pm g(x)] = L \pm M = \lim_{x \to a} f(x) \pm \lim_{x \to a} g(x).$ *The limit of a sum (or a difference) is the sum (or difference) of the limits.*
Limit of Product	**L2** $\quad \lim_{x \to a} [f(x) g(x)] = LM = \lim_{x \to a} f(x) \lim_{x \to a} g(x).$ *The limit of a product is the product of the limits.*
Limit of Quotient	**L3** $\quad \lim_{x \to a} \dfrac{f(x)}{g(x)} = \dfrac{L}{M} = \dfrac{\lim_{x \to a} f(x)}{\lim_{x \to a} g(x)} \quad$ provided $M \neq 0.$ *The limit of a quotient is the quotient of the limits provided the limit of the denominator is not zero.*

The proof of Theorem 2-1 is postponed until the Appendix at the end of this chapter.

The following two results have been used several times in the earlier sections.

THEOREM 2-2 $\lim_{x \to a} C = C$	**L4** $\quad$ If $f(x) = C$ for all x and a is a real number, then $\lim_{x \to a} f(x) = C.$ *The limit of a constant function is equal to the value of the constant.*
$\lim_{x \to a} x = a$	**L5** $\quad$ If $g(x) = x$ for all x and a is a real number, then $\lim_{x \to a} g(x) = a.$

Proof of L5 $\quad$ Let $\epsilon > 0$. We must find $\delta > 0$ such that $|g(x) - a| < \epsilon$ whenever $0 < |x - a| < \delta$. We can accomplish this by choosing $L = a$ and $\delta = \epsilon$ (see Fig. 2-11). If

$$0 < |x - a| < \delta,$$

then

$$|g(x) - a| = |x - a| < \delta = \epsilon.$$

2-5 Some Properties of Limits

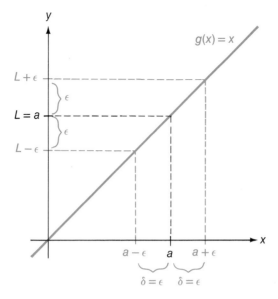

Figure 2–11 $\lim_{x \to a} x = a$.

Thus,
$$\lim_{x \to a} g(x) = \lim_{x \to a} x = a. \blacksquare$$

The proof of L4 is left for you to work. (See Exercise 24.)

Properties L1 and L2 are stated for only two functions. It should be apparent that these results can be extended to cover any finite number of functions. The proof depends on mathematical induction, which is discussed in Appendix A.

EXAMPLE 1 Use the Limit Theorem to justify the steps in proving the following limits:

(a) $\lim_{x \to 2} (3x^2 - 5) = 7$ \qquad (b) $\lim_{x \to -3} \dfrac{2x + 1}{3x + 7} = \dfrac{5}{2}$

Solution (a) $\lim_{x \to 2} (3x^2 - 5) = \lim_{x \to 2} 3x^2 - \lim_{x \to 2} 5$ \qquad (by L1)

$\qquad\qquad\qquad = \lim_{x \to 2} 3 \lim_{x \to 2} x \lim_{x \to 2} x - \lim_{x \to 2} 5$ \qquad (by L2)

$\qquad\qquad\qquad = (3)(2)(2) - 5 = 7$ \qquad (by L4, L5).

(b) $\lim_{x \to -3} \dfrac{2x + 1}{3x + 7} = \dfrac{\lim_{x \to -3} (2x + 1)}{\lim_{x \to -3} (3x + 7)}$ \qquad (by L3).

$\qquad\qquad\qquad = \dfrac{\lim_{x \to -3} 2 \lim_{x \to -3} x + \lim_{x \to -3} 1}{\lim_{x \to -3} 3 \lim_{x \to -3} x + \lim_{x \to -3} 7}$ \qquad (by L1, L2)

$\qquad\qquad\qquad = \dfrac{2(-3) + 1}{3(-3) + 7} = \dfrac{5}{2}$ \qquad (by L4, L5). $\square$

Remark Observe that each of the limits in Example 1 could have been calculated by substitution, that is, in both cases,

$$\lim_{x \to a} f(x) = f(a).$$

An argument like the one used in Example 1(a) can be used to show that every polynomial function has this property. An argument like the one in Example 1(b) can be used to establish the same result for a rational function. (See Exercises 20, 21, 30.)

ROOT FUNCTIONS

We shall establish in the Appendix to this chapter that

Limit of a Root Function

L6 $\quad \lim_{x \to a} \sqrt[n]{x} = \sqrt[n]{a} \quad$ provided $a > 0$ if n is even.

The restriction that $a > 0$ when n is even ensures that we deal with real roots, not imaginary ones.

EXAMPLE 2 $\quad \lim_{x \to 4} \dfrac{3x - 4}{\sqrt{x}} = \dfrac{\lim_{x \to 4} (3x - 4)}{\lim_{x \to 4} \sqrt{x}} \quad$ (by L3)

$$= \dfrac{\lim_{x \to 4} 3 \lim_{x \to 4} x - \lim_{x \to 4} 4}{\sqrt{4}} \quad \text{(by L1, L2, L6)}$$

$$= \dfrac{(3)(4) - 4}{2} = 4 \quad \text{(by L4, L5).} \ \square$$

It is apparent that a number of shortcuts can be taken in Examples 1 and 2. It seems to be true, for example, that if $\lim_{x \to a} f(x)$ exists and C is a constant, then

Limit of a Constant Times a Function

L7 $\quad \lim_{x \to a} C f(x) = C \lim_{x \to a} f(x).$

This result, which can be proved easily from L2 and L4, is left for you to do (Exercise 25). Observe that its use simplifies several of the steps in the above examples. A second useful result is that

Limit of a Power Function

$$\lim_{x \to a} x^n = a^n$$

for each positive integer n. This can be established from L2 and L5 after expressing x^n as $x \cdot x \cdot x \cdots x$ (n factors). (Also see Exercise 29.) This result can be combined

2-5 Some Properties of Limits

with L7 to establish that

$$\lim_{x \to a} C x^n = C a^n.$$

ONE-SIDED LIMITS

Properties L1–L7 all hold for one-sided limits. We can use this fact to calculate one-sided limits of functions defined by different laws of correspondence over different parts of their domains.

EXAMPLE 3 Let $f(x)$ be defined as follows:

$$f(x) = \begin{cases} 2x - 1 & \text{if } x < 1 \\ 3 & \text{if } x = 1 \\ 3 - x & \text{if } x > 1. \end{cases}$$

(See Fig. 2–12.) Then

$$\lim_{x \to 1^-} f(x) = \lim_{x \to 1^-} (2x - 1) = 2 \lim_{x \to 1^-} x - \lim_{x \to 1^-} 1 \quad \text{(by L1, L2, L7)}$$
$$= (2)(1) - 1 = 1. \quad \text{(by L4, L5)}$$
$$\lim_{x \to 1^+} f(x) = \lim_{x \to 1^+} (3 - x) = \lim_{x \to 1^+} 3 - \lim_{x \to 1^+} x \quad \text{(by L1)}$$
$$= 3 - 1 = 2 \quad \text{(by L4, L5).}$$

Observe that neither of the one-sided limits equals the value of the function at $x = 1$. □

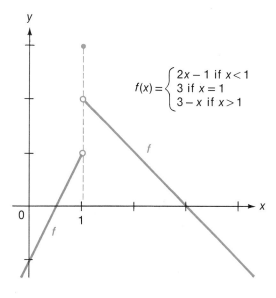

Figure 2–12 Example 3.
$\lim_{x \to 1^-} f(x) = 1$, $\lim_{x \to 1^+} f(x) = 2$.

FUNCTIONS BOUNDED BY OTHER FUNCTIONS

Recall from trigonometry that $-1 \leq \cos x \leq 1$ for all x, so that

$$-x^2 \leq x^2 \cos x \leq x^2$$

for all x. Thus, $x^2 \cos x$ is between $-x^2$ and x^2.

More generally, a function f may have its functional values between two other functions g and h; that is,

$$g(x) \leq f(x) \leq h(x)$$

for all x in which we are interested.

Suppose now that $f(x)$ is between $g(x)$ and $h(x)$ for all x on some open interval I that contains a as an interior point. Suppose also that $g(x)$ and $h(x)$ have the limit L as x approaches a. (See Fig. 2–13.) Since $f(x)$ is "caught between" $g(x)$ and $h(x)$, it seems that $f(x)$ should also have the limit L as $x \to a$. This is the gist of the following theorem, commonly known as the Sandwich Theorem.

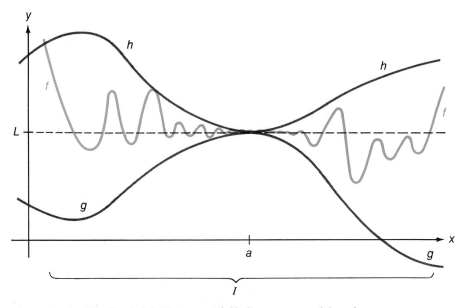

Figure 2–13 The Sandwich Theorem. If f is between g and h and $\lim_{x \to a} g(x) = \lim_{x \to a} h(x) = L$, then $\lim_{x \to a} f(x) = L$.

2–5 Some Properties of Limits

THEOREM 2–3 (*Sandwich Theorem*) Let I be an open interval that contains the number a. Let $g(x) \le f(x) \le h(x)$ for all x on I different from a. If $\lim_{x \to a} g(x) = \lim_{x \to a} h(x) = L$, then

$$\lim_{x \to a} f(x) = L.$$

(The proof of Theorem 2–3 is postponed until the Appendix.)

EXAMPLE 4 Use the Sandwich Theorem to calculate

$$\lim_{x \to 0} x^2 \cos x.$$

Solution As noted above, $-x^2 \le x^2 \cos x \le x^2$ for every x. Since

$$\lim_{x \to 0} (-x^2) = \lim_{x \to 0} x^2 = 0,$$

then it follows from the Sandwich Theorem that

$$\lim_{x \to 0} x^2 \cos x = 0. \ \square$$

POSITIVE AND NEGATIVE LIMITS

We need one further result about limits. If the limit of a function is positive as x approaches a, it seems reasonable that the function should be positive when x is close to a. Similarly, if the limit is negative, the function should be negative when x is close to a. This result is established in Theorem 2–4.

THEOREM 2–4 (a) If $\lim_{x \to a} g(x) = L > 0$, there exists a $\delta < 0$ such that if $0 < |x - a| < \delta$, then $g(x) > L/2 > 0$.

(b) If $\lim_{x \to a} g(x) = L < 0$, there exists a $\delta > 0$ such that if $0 < |x - a| < \delta$, then $g(x) < L/2 < 0$.

Proof of (a) Since $\lim_{x \to a} g(x) = L$, then for every $\epsilon > 0$ there exists $\delta > 0$ such that

$$|g(x) - L| < \epsilon \quad \text{whenever } 0 < |x - a| < \delta.$$

If we choose the specific value $\epsilon = L/2$ (see Fig. 2–14), it follows that

$$-\frac{L}{2} < g(x) - L < \frac{L}{2} \quad \text{whenever } 0 < |x - a| < \delta.$$

Then

$$L - \frac{L}{2} < g(x) < L + \frac{L}{2} \quad \text{whenever } 0 < |x - a| < \delta.$$

Thus, if $0 < |x - a| < \delta$, then

$$g(x) > \frac{L}{2} > 0.$$

The proof of (b) is similar. (See Exercise 28.) ■

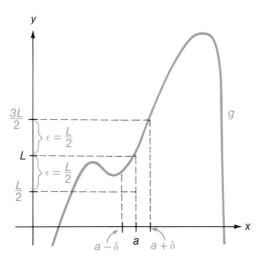

Figure 2–14 Theorem 2-4. If g has a positive limit as $x \to a$, then there exists a positive number δ such that $g(x) > 0$ whenever $0 < |x - a| < \delta$.

Exercises 2–5

1–16. Use properties L1–L7 to calculate each limit. State which of the properties is used at each step. (Boldface brackets indicate the greatest integer function.)

- 1 $\lim\limits_{x \to -1} (x^3 - 3x^2 + x - 1)$

2 $\lim\limits_{x \to 4} \dfrac{x^2 + \sqrt{x}}{x - 1}$

- 3 $\lim\limits_{x \to 2} \dfrac{x^2 + 2x - 5}{x^2 - x + 1}$

4 $\lim\limits_{x \to 3} \sqrt{3}$

- 5 $\lim\limits_{x \to 2} \dfrac{x^2 - 4}{x^3 - 8}$

6 $\lim\limits_{x \to 1} \dfrac{x^3 - 1}{x - 1}$

- 7 $\lim\limits_{x \to 3} \dfrac{1/x - 1/3}{x - 3}$

8 $\lim\limits_{x \to 9} \dfrac{x - 9}{\sqrt{x} - 3}$

- 9 $\lim\limits_{x \to -1} \dfrac{x^4 - 1}{x^2 + 3x + 2}$

10 $\lim\limits_{x \to -2} \dfrac{x^2 + x - 2}{x^4 - 16}$

- 11 $\lim\limits_{x \to -1^+} \dfrac{x^2 + 2x + 1}{x^4 - 2x^2 + 1}$

12 $\lim\limits_{x \to -8^-} \dfrac{x + 8}{\sqrt[3]{x} + 2}$

- 13 $\lim\limits_{x \to -1^-} \dfrac{2x - [x]}{3x + 1}$

14 $\lim\limits_{x \to -1^+} \dfrac{2x - [x]}{3x + 1}$

- 15 $\lim\limits_{x \to 1^+} \dfrac{[x + 1] - x}{3 - x}$

16 $\lim\limits_{x \to 1^-} \dfrac{[x + 1] - x}{3 - x}$

Exercises 2–5

17–19. If the conditions are not contradictory, then sketch graphs of functions that have the properties given. If the conditions are contradictory, explain why.

•17 $\lim_{x \to 1^-} f(x) = 2$, $\lim_{x \to 1^+} f(x) = -1$, $f(1) = 3$.

18 $\lim_{x \to 0^+} f(x) = 4$, $\lim_{x \to 0^-} f(x) = 2$, f is a polynomial function.

•19 $\lim_{x \to 1^+} f(x) = 5$, $\lim_{x \to 1^-} f(x) = 5$, f has no limit as $x \to 1$.

Limit of a Polynomial

20 Let $P(x)$ be the polynomial function

$$P(x) = c_n x^n + c_{n-1} x^{n-1} + \cdots + c_1 x + c_0.$$

Apply L1, L2, and L4 as many times as necessary to show that

$$\lim_{x \to a} P(x) = P(a)$$

for any real number a. (Also see Exercise 30.)

Limit of a Rational Function

21 Let $P(x)$ and $Q(x)$ be polynomial functions. Use Exercise 20 and L3 to show that

$$\lim_{x \to a} \frac{P(x)}{Q(x)} = \frac{P(a)}{Q(a)} \quad \text{provided } Q(a) \neq 0.$$

22 It can be proved that

$$1 - \frac{x^2}{2} \leq \cos x \leq 1 - \frac{x^2}{2} + \frac{x^4}{24} \quad \text{for all } x.$$

Use the Sandwich Theorem to show that $\lim_{x \to 0} (\cos x - 1)/x = 0$.

23 It can be proved that

$$x - \frac{x^3}{6} < \sin x < x + \frac{x^3}{6} \quad \text{if } 0 < x < 1.$$

Use the Sandwich theorem to show that $\lim_{x \to 0^+} ((\sin x)/x) = 1$.

24 Use the definition to prove L4: If $f(x) = C$ for all x, then $\lim_{x \to a} f(x) = C$ for any real number a.

25 Use L2 and L4 to prove L7: Let $g(x) = Cf(x)$. If $\lim_{x \to a} f(x) = L$, then $\lim_{x \to a} g(x) = CL$.

•26 (a) Suppose that $\lim_{x \to a} f(x)$ exists. Prove that

$$\lim_{x \to a} [f(x)]^2 = \left[\lim_{x \to a} f(x)\right]^2.$$

(b) Suppose that for some function f it is true that

$$\lim_{x \to a} [f(x)]^2 = L^2.$$

Show by example that it is not necessarily true that f has a limit as $x \to a$.

Limit of Absolute Value of a Function

27 Use the definition and the Second Triangle Inequality (see Exercise 30 of Section 1–3) to prove that:
(a) $\lim_{x \to a} |x| = |a|$.
(b) If $\lim_{x \to a} f(x) = L$, then $\lim_{x \to a} |f(x)| = |L|$.

28 Complete the proof of Theorem 2–4. Let $\lim_{x \to a} g(x) = L < 0$. Prove that there exists $\delta > 0$ such that $g(x) < L/2 < 0$ whenever $0 < |x - a| < \delta$.

29–30. Rigorous proofs of a number of propositions in mathematics require mathematical induction. (See Appendix A at the end of the book.)

29 Use L2 and mathematical induction to prove that $\lim_{x \to a} x^n = a^n$ for any positive integer n.

30 Use L1–L5 and mathematical induction to prove that if $P(x)$ is a polynomial function, then $\lim_{x \to a} P(x) = P(a)$. (*Hint:* Run induction on the degree of the polynomial. Show that a polynomial of degree $k + 1$ can be expressed in terms of polynomials of degree k and degree 1.)

2–6 Continuous Functions

Functions that arise in the solution of problems in the physical and social sciences are almost always "well behaved." The graph of such a function is like a string that, no matter how wildly it waves up and down, always has each point immediately adjacent to those on both sides. Functions which arise in other disciplines, however, such as the postage price function, may exhibit sudden jumps at certain values of the independent variable. We now seek a way to put these intuitive ideas into the language of mathematics so that we can discuss them precisely.

The functions in Example 1 of Section 2–5 have the property that

$$\lim_{x \to a} f(x) = f(a),$$

that is, the *limit of f(x) as x approaches a equals the value of f at a*. Such functions are of basic importance in the calculus.

DEFINITION

Continuity

The function f is said to be *continuous* at $x = a$ provided

$$\lim_{x \to a} f(x) = f(a).$$

For example, if $f(x) = \sqrt[3]{x}$, then (by L6)

$$\lim_{x \to a} f(x) = \lim_{x \to a} \sqrt[3]{x} = \sqrt[3]{a} = f(a).$$

Thus, f is continuous at each real number a.

2–6 Continuous Functions

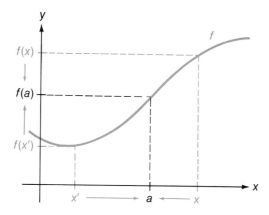

Figure 2–15 A continuous function. $\lim_{x \to a} f(x) = f(a)$.

Observe that three conditions must be met for a function f to be continuous at $x = a$. (See Fig. 2–15.)

(1) f must be defined at $x = a$.
(2) $f(x)$ must have a limit as x approaches a.
(3) $\lim_{x \to a} f(x)$ must equal the value of f at a.

EXAMPLE 1 Let $f(x) = x^2$. Show that f is continuous at each real number a.

Solution Since

$$\lim_{x \to a} f(x) = \lim_{x \to a} x^2 = \lim_{x \to a} x \lim_{x \to a} x \quad \text{(by L2)}$$
$$= a^2 \quad \text{(by L5)}$$

then f is continuous at a. □

Discontinuous Function   A function f is said to be *discontinuous* at a if it is not continuous there. This can happen in one of several ways:

(1) The function may not be defined at a (see Fig. 2–16a).
(2) The function may not have a limit as x approaches a (see Fig. 2–16b, c).
(3) The limit may exist but be different from the value $f(a)$ (see Fig. 2–16d).

Removable Discontinuity   Certain common types of discontinuity are illustrated in Figure 2–16. A function f has a *removable discontinuity* at a if it is possible to define or redefine f at this one value and make it continuous there. In other words, f has a removable discontinuity

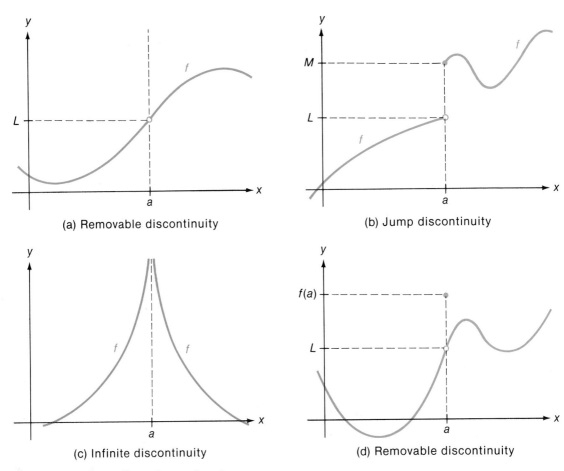

Figure 2–16 Some discontinuous functions.

at $x = a$ if $f(x)$ has a limit as $x \to a$ but either f is not defined at a or $\lim_{x \to a} f(x) \neq f(a)$ (see Fig. 2–16a, d).

Jump Discontinuity

A function f has a *jump discontinuity* at $x = a$ provided $f(x)$ has unequal one-sided limits as $x \to a$ (see Fig. 2–16b).

Infinite Discontinuity

A function f has an *infinite discontinuity* at a provided $|f(x)|$ gets larger and larger without bound as x approaches a from either the left or the right (Fig. 2–16c). (This subject is examined in more detail in Section 2–8.)

It is easy to use the limit properties to establish the following results about continuous functions.

THEOREM 2–5   Let f and g be continuous at $x = a$. Then:
(a) *If $s(x) = f(x) \pm g(x)$, then s is continuous at $x = a$. The sum (or difference) of two continuous functions is continuous.*

2-6 Continuous Functions

(b) If $p(x) = f(x)g(x)$, then p is continuous at $x = a$. *The product of two continuous functions is continuous.*

(c) If $q(x) = f(x)/g(x)$, then q is continuous at $x = a$ provided $g(a) \neq 0$. *The quotient of two continuous functions is continuous provided the denominator is not zero.*

Proof of (c)   If $g(a) \neq 0$, then

$$\lim_{x \to a} q(x) = \lim_{x \to a} \frac{f(x)}{g(x)} = \frac{\lim_{x \to a} f(x)}{\lim_{x \to a} g(x)} = \frac{f(a)}{g(a)} = q(a) \quad \text{(by L3)}.$$

The proofs of (a) and (b) are similar. ∎

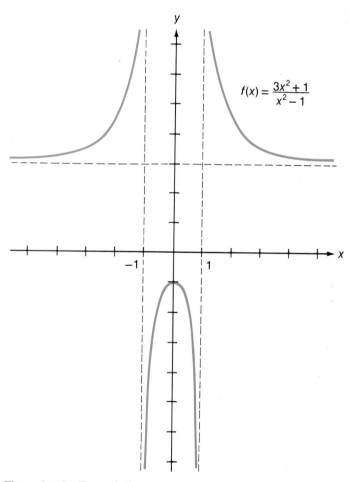

Figure 2–17   Example 2.

It follows from Theorem 2-5 and the properties of limits that:

(1) Every polynomial function is continuous everywhere.
(2) Every rational function is continuous everywhere except where the denominator is zero.

EXAMPLE 2 Let $h(x) = \dfrac{3x^2 + 1}{x^2 - 1}$. Discuss the continuity of h.

Solution Since h is a rational function, it is continuous everywhere except where the denominator $x^2 - 1$ is zero. Thus, h is continuous everywhere except at $x = \pm 1$, where it is discontinuous. (See Fig. 2-17.) □

CONTINUITY ON AN OPEN INTERVAL

In most of our work we deal with functions that are continuous at each point of an open interval. Such a function is said to be *continuous on the interval*. It can be shown that its graph is unbroken on the interval.

EXAMPLE 3 The greatest integer function is discontinuous at the integers but continuous on any interval that does not contain an integer. (See Fig. 2-18.) □

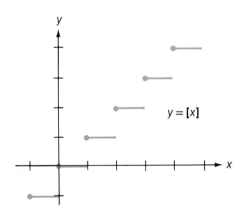

Figure 2-18 The greatest integer function: $f(x) = [x]$. The function is discontinuous at the integers but is continuous everywhere else.

CONTINUITY AT AN ENDPOINT OF THE DOMAIN

We now make two important exceptions to our definition of continuity at a point. If f is defined on an interval $[a, b]$ but not defined for $x < a$, then, we say, f is continuous at $x = a$ provided

$$\lim_{x \to a^+} f(x) = f(a).$$

2.6 Continuous Functions

Similarly, if f is defined on $[a, b]$ but not defined for $x > b$, then f is continuous at $x = b$ provided

$$\lim_{x \to b^-} f(x) = f(b).$$

EXAMPLE 4 If $f(x) = \sqrt{x}$, then $\lim_{x \to 0^+} f(x) = 0 = f(0)$. Therefore, f is continuous at $x = 0$ even though the limit is not defined as x approaches 0 from the left. (See Fig. 2–19.) □

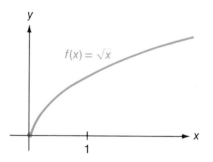

Figure 2–19 Examples 4 and 5. The function $f(x) = \sqrt{x}$ is continuous at $x = 0$ (an endpoint of its domain) because $\lim_{x \to 0^+} f(x) = 0 = f(0)$.

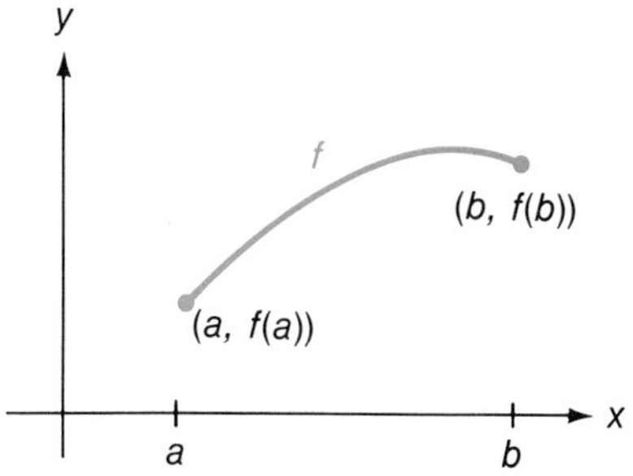

Figure 2–20 Continuity on the closed interval $[a, b]$.

CONTINUITY ON A CLOSED INTERVAL

Frequently it is desirable to restrict our attention to a function f on a closed interval $[a, b]$. (For all practical purposes this means that we consider the interval $[a, b]$ as if it were the domain of f.) The following definition extends the concepts of the preceding paragraphs. (See Fig. 2–20.)

DEFINITION We say that the function f is *continuous on the closed interval* $[a, b]$ provided:

(1) f is continuous at each point of the open interval (a, b).
(2) $\lim_{x \to a^+} f(x) = f(a)$.
(3) $\lim_{x \to b^-} f(x) = f(b)$.

EXAMPLE 5 The function $f(x) = \sqrt{x}$ is continuous at every x greater than 0 and has the property that

$$\lim_{x \to 0^+} f(x) = f(0).$$

(See Example 4.) In particular this implies that f is continuous on the closed interval [0, 1]. (See Fig. 2–19.) ∎

Remark If a function is continuous on an interval (open, closed, or neither), then its graph is unbroken on that interval. In most cases this means that the graph can be drawn over the interval without lifting the pencil from the paper. (See Fig. 2–20.)

DIFFERENTIABILITY AND CONTINUITY

Most of the functions we consider in this book have derivatives almost everywhere. Such functions are continuous almost everywhere as well. We shall show that a function g must be continuous at $x = a$ if it has a derivative there. This result can be extended to a comparable result for closed intervals.

THEOREM 2–6 If g is differentiable at $x = a$, then g is continuous at $x = a$.

Proof The function g must be defined at $x = a$ in order to have the derivative exist there. Furthermore, the limit $\lim_{x \to a} [g(x) - g(a)]/(x - a)$ must exist. We write the identity

$$g(x) = g(a) + \frac{g(x) - g(a)}{x - a}(x - a)$$

and take the limit as x approaches a:

$$\lim_{x \to a} g(x) = \lim_{x \to a} g(a) + \lim_{x \to a} \frac{g(x) - g(a)}{x - a} \lim_{x \to a} (x - a).$$
$$= g(a) + g'(a) \cdot 0 = g(a). \blacksquare$$

As we see in the following example, the converse of Theorem 2–6 is false. That a function is continuous at a point is not enough to make it differentiable there.

EXAMPLE 6 Let $f(x) = |x|$. It is easy to show that f is continuous everywhere. (See Exercise 29 and Figure 2–21.) On the other hand, if we attempt to use the difference quotient to calculate the derivative at $x = 0$, we find that

$$\lim_{x \to 0^+} \frac{f(x) - f(0)}{x - 0} = \lim_{x \to 0^+} \frac{|x| - 0}{x - 0} = \lim_{x \to 0^+} \frac{x}{x} = 1$$

and

$$\lim_{x \to 0^-} \frac{f(x) - f(0)}{x - 0} = \lim_{x \to 0^+} \frac{|x| - 0}{x - 0} = \lim_{x \to 0^-} \frac{-x}{x} = -1$$

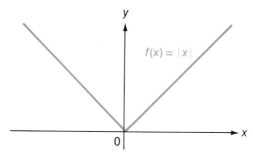

Figure 2–21 Example 6. The function $f(x) = |x|$ is continuous at $x = 0$ but is not differentiable there.

so that the difference quotient

$$\frac{f(x) - f(0)}{x - 0}$$

has no limit as $x \to 0$. It follows that f is continuous at $x = 0$ but not differentiable there. ☐

Exercises 2–6

1–18. Determine all points at which the functions are discontinuous. Identify each discontinuity by type (removable, jump, infinite). If a function has a removable discontinuity, explain how it can be redefined at the point to make it continuous there. Boldface brackets indicate the greatest integer function.

- 1 $f(x) = 3x^2 - 2x + 5$

2 $f(x) = \dfrac{|x|}{x}$

- 3 $f(x) = \dfrac{x - 2}{x - 4}$

4 $f(x) = \dfrac{x^2 - 4}{x - 2}$

- 5 $f(x) = x - [x]$

6 $f(x) = [x + 1] - [x]$

- 7 $f(x) = x + |x|$

8 $f(x) = \dfrac{1}{\sqrt{x}}$

- 9 $f(x) = \dfrac{x}{x^2 + 9}$

10 $f(x) = \dfrac{x + 1}{x^2 - 4}$

- 11 $f(x) = \dfrac{1}{x - \sqrt{x}}$

12 $f(x) = \dfrac{x + 1}{x^2 - 4x + 3}$

- 13 $f(x) = \dfrac{\sqrt{x - 1}}{x^2 - 3x + 2}$

14 $f(x) = \dfrac{x^2 + 2x + 1}{\sqrt[3]{x + 1}}$

•15 $f(x) = \begin{cases} \dfrac{x^2 - 1}{x + 1} & \text{if } x \neq -1 \\ 0 & \text{if } x = -1 \end{cases}$

16 $f(x) = \begin{cases} \dfrac{x^2 - 4}{x - 2} & \text{if } x \neq 2 \\ 4 & \text{if } x = 2 \end{cases}$

•17 $f(x) = \begin{cases} \dfrac{x^3 + 8}{x + 2} & \text{if } x \neq -2 \\ 6 & \text{if } x = -2 \end{cases}$

18 $f(x) = \begin{cases} \dfrac{|x|}{x} & \text{if } x \neq 0 \\ 0 & \text{if } x = 0 \end{cases}$

19–22. If the conditions are not contradictory, then sketch the graph of a function that satisfies them. If the conditions are contradictory, explain why.

•19 $\lim_{x \to 1^+} f(x) = \lim_{x \to 1^-} f(x) = L$, f is discontinuous at $x = 1$.

20 f is continuous on $[0, 1]$, $\lim_{x \to 0^-} f(x) \neq f(0)$.

•21 f is continuous at $x = 2$, but f has no derivative at $x = 2$.

22 $f'(1) = 2$, $f(1) = 3$, $\lim_{x \to 1} f(x) = 4$.

23–28. Are the functions continuous on the given closed intervals? Explain.

•23 $f(x) = 1/x$, $[0, 1]$

24 $f(x) = \sqrt[3]{x}$, $[-1, 1]$

•25 $f(x) = |x| - \sqrt{x}$, $[0, 1]$

26 $f(x) = 1/\sqrt{x}$, $[0, 1]$

•27 $f(x) = \begin{cases} [x] & \text{if } x \leq 1 \\ x^2 & \text{if } x > 1 \end{cases}$, $[1, 2]$

28 $f(x) = \begin{cases} 0 & \text{if } x < 1 \\ x & \text{if } 1 \leq x < 3 \\ [x] & \text{if } x \geq 3 \end{cases}$, $[1, 3]$

29 Prove that $f(x) = |x|$ is continuous everywhere. (See Exercise 27, Section 2–5.)

30 Prove Theorem 2–5(a).

31 Prove Theorem 2–5(b).

32 Let $f(x)$ be a rational function. Prove that f is continuous at each point of its domain. (See Exercise 21, Section 2–5.)

The criterion given in Exercise 33 is useful for some work with continuity.

33 Prove that f is continuous at $x = a$ if and only if for each $\epsilon > 0$ there exists a $\delta > 0$ such that

$$|f(x) - f(a)| < \epsilon \quad \text{wherever } |x - a| < \delta.$$

(*Hint:* Before trying to prove this result, carefully compare this statement with the definition of limit.)

34 Use Exercise 33 to prove that the square root function $f(x) = \sqrt{x}$ is continuous at each positive number a.

2-7 Properties of Continuous Functions

The continuous functions constitute one of the most important classes of functions studied in mathematics. All continuous functions possess several basic properties. The first property concerns extreme values.

MAXIMA AND MINIMA

Maximum   We say that *M is the maximum value of f on an interval* provided

(1) $f(x) \leq M$ for every x on the interval; and
(2) there is at least one number x_1 on the interval where $f(x_1) = M$.

Minimum   The *minimum value m of f on an interval* is defined similarly. We must have (1) $f(x) \geq m$ for every x on the interval; and (2) $f(x_2) = m$ for at least one number x_2 on the interval.

Extremum   A maximum or minimum is called an *extremum*, or an *extreme value*, of the function. These concepts are illustrated in Figures 2–22 and 2–23.

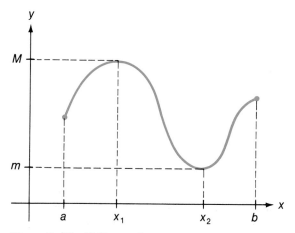

Figure 2–22 If f is continuous on the closed interval $[a, b]$, then f has a maximum M and a minimum m on the interval.

EXAMPLE 1   Let $f(x) = \sqrt{4 - x^2} + 1$. (The graph is the semicircle shown in Figure 2–23.) The maximum of f on $[-2, 2]$ is

$$M = 3$$

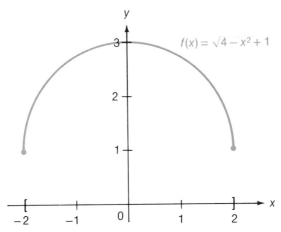

Figure 2–23 Example 1. $M = 3$ is the maximum and $m = 1$ the minimum of $f(x) = \sqrt{4 - x^2} + 1$ on $[-2, 2]$.

which is taken at $x = 0$. The minimum is

$$m = 1$$

which is taken at $x = \pm 2$. □

Not every function has an extreme value on a given interval. As we shall see, we cannot guarantee the existence of an extremum unless we know that the function is continuous and the interval is closed. Example 2 involves functions that fail to have maximum values.

EXAMPLE 2 (a) The function $f(x) = 1/x$ has a minimum value of 1, but no maximum value on the half-open interval $(0, 1]$. No matter how large a number M is chosen, we can find a number x_0 close to zero such that $1/x_0 > M$. (See Fig. 2–24a.) Observe that f is continuous on the interval but that the interval is not closed.
(b) Define f on $[0, 3]$ by

$$f(x) = \begin{cases} 2x & \text{if } 0 \leq x < 1 \\ 2 - x & \text{if } 1 \leq x \leq 3. \end{cases}$$

(See Fig. 2–24.) Then f has no maximum on $[0, 3]$. The values of f can be made as close to 2 as we wish, but every value of f is less than 2. The interval is closed, but f is not continuous. □

2-7 Properties of Continuous Functions

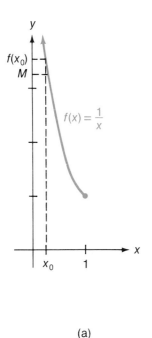

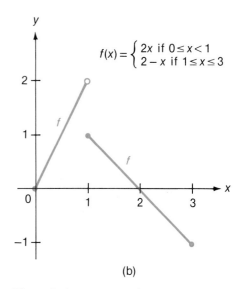

(a)

(b)

Figure 2–24(a) Example 2(a). The function $f(x) = 1/x$ has a minimum, but no maximum, on the half-open interval $(0, 1]$. For every number M there exists a number x_0 (which depends on M) such that $f(x_0) > M$.

Figure 2–24(b) Example 2(b). The function f has a minimum, but no maximum on the closed interval $[0, 3]$.

Theorem 2–7 states the basic result concerning extreme values of continuous functions on closed intervals. (The proof is beyond the scope of this book.)

THEOREM 2–7 Let f be continuous on the closed interval $[a, b]$. Then f has a maximum value M and a minimum value m on the interval.

In Chapter 3 we shall make an intensive study of extrema of functions. Theorem 2–7 will be basic to our work. It guarantees that a function f has a maximum and a minimum on every closed interval on which it is continuous. In particular, it follows from Theorems 2–6 and 2–7 that if f is differentiable at each point of a closed interval, then it has extreme values on that interval.

THE INTERMEDIATE VALUE THEOREM

Theorem 2–8 establishes that a continuous function can have no breaks in its graph. Before stating the theorem, we consider an example of a function that has a break in its graph. The function pictured in Figure 2–25 has a jump discontinuity at $x = c$. A number k between $f(a)$ and $f(b)$ is indicated on the y-axis. Observe that the horizontal line through this particular point does not intersect the graph. Theorem 2–8 shows that this situation cannot exist if f is continuous on $[a, b]$.

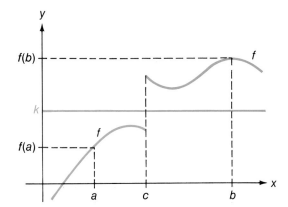

Figure 2–25 The graph of f has a jump discontinuity at $x = c$. The horizontal line $y = k$ does not intersect the graph.

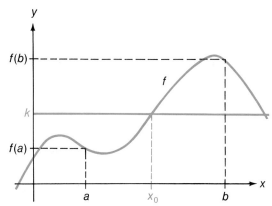

Figure 2–26 The Intermediate Value Theorem (Theorem 2–8). The function f is continuous on the closed interval $[a, b]$. If k is any number between $f(a)$ and $f(b)$, then the graph of f must intersect the horizontal line $y = k$ somewhere between a and b.

THEOREM 2–8 (*Intermediate Value Theorem*) Let f be continuous on $[a, b]$. Let k be any number between $f(a)$ and $f(b)$. There must exist at least one number x_0 between a and b such that $f(x_0) = k$.

[In other words, the horizontal line through the point k on the y-axis must intersect the graph of f at some point between $(a, f(a))$ and $(b, f(b))$. (See Fig. 2–26.)]

The Intermediate Value Theorem is primarily used as a theoretical tool in proving other results. As an example, we shall prove that the number 7 has a square root. The argument can be modified slightly to prove that any positive real number a has an nth root. (See Exercise 27.)

EXAMPLE 3 Prove that there exists a real number r between 2.64 and 2.65 such that $r^2 = 7$.

2-7 Properties of Continuous Functions

Solution Let $f(x) = x^2$. Since f is a polynomial, then it is continuous everywhere. In particular, f is continuous on the closed interval $[2.64, 2.65]$. Observe that

$$f(2.64) = (2.64)^2 = 6.9696 < 7$$

and

$$f(2.65) = (2.65)^2 = 7.0225 > 7.$$

The number $k = 7$ is between $f(2.64)$ and $f(2.65)$. By the Intermediate Value Theorem, there exists a number r between 2.64 and 2.65 such that

$$f(r) = r^2 = 7.$$

(See Fig. 2–27.) □

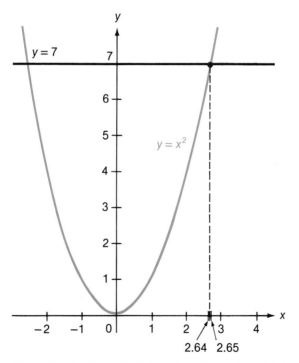

Figure 2–27 Example 3. The function $f(x) = x^2$ intersects the line $y = 7$ at some point between $x = 2.64$ and 2.65.

LIMITS OF COMPOSITE FUNCTIONS

Our final result enables us to fill an important gap in the limit formulas. Suppose, for example, that we need to calculate

$$\lim_{x \to 4} \sqrt[3]{x^2 + 1}.$$

Since $x^2 + 1 \to 17$ as $x \to 4$, it seems reasonable that the limit of $\sqrt[3]{x^2 + 1}$ should be $\sqrt[3]{17}$. Our final property will enable us to show that this is indeed the case.

The key to the above limit is that $h(x) = \sqrt[3]{x^2 + 1}$ can be written as the composition of two simple functions:

$$h(x) = f(g(x)),$$

where $f(u) = \sqrt[3]{u}$ and $g(x) = x^2 + 1$. We shall show that in general, if f is continuous at L and g has the limit L as $x \to a$, then

$$\lim_{x \to a} f(g(x)) = f\left(\lim_{x \to a} g(x)\right) = f(L).$$

In our example above,

$$\lim_{x \to 4} \sqrt[3]{x^2 + 1} = \sqrt[3]{\lim_{x \to 4} (x^2 + 1)} = \sqrt[3]{17}.$$

THEOREM 2–9 Let $\lim_{x \to a} g(x) = L$ and let f be continuous at L. Let $h(x) = f(g(x))$. Then

$$\lim_{x \to a} h(x) = \lim_{x \to a} f(g(x)) = f(L) = f\left(\lim_{x \to a} g(x)\right).$$

Proof We make the substitution $u = g(x)$. Since f is continuous at $u = L$, then

$$\lim_{u \to L} f(u) = f(L).$$

Let $\epsilon > 0$. There exists $\delta_0 > 0$ such that

$$|f(u) - f(L)| < \epsilon \quad \text{whenever } |u - L| < \delta_0.$$

(See Exercise 33 of Section 2–6.)

Since δ_0 is a positive number and $\lim_{x \to a} g(x) = L$, there exists $\delta > 0$ such that $|g(x) - L| < \delta_0$ whenever $0 < |x - a| < \delta$.

Thus, if $0 < |x - a| < \delta$, then $|u - L| = |g(x) - L| < \delta_0$ so that

$$|h(x) - f(L)| = |f(g(x)) - f(L)|$$
$$= |f(u) - f(L)| < \epsilon.$$

Therefore,

$$\lim_{x \to a} h(x) = f(L). \blacksquare$$

Exercises 2–7

This property is usually written in the following form:

> **L8** $\lim_{x \to a} f(g(x)) = f\left(\lim_{x \to a} g(x)\right)$
>
> provided f is continuous at $L = \lim_{x \to a} g(x)$.

Theorem 2–9 can be extended to cover one-sided limits. The use of the extended theorem is illustrated in Example 4.

EXAMPLE 4   Calculate $\lim_{x \to 2^+} \sqrt{3x - 6}$.

Solution   As $x \to 2^+$, the function $g(x) = 3x - 6 \to 0$ through positive values. Recall that the function $f(u) = \sqrt{u}$ is continuous for all $u \geq 0$. Then

$$\lim_{x \to 2^+} f(g(x)) = \lim_{x \to 2^+} \sqrt{3x - 6} = \sqrt{\lim_{x \to 2^+} (3x - 6)} = \sqrt{0} = 0. \quad \square$$

Exercises 2–7

1–6.   Use properties L1–L8 to calculate each limit. State which properties are used at each main step.

- **1** $\lim_{x \to 3} \sqrt{2x^3 - 5}$
- **2** $\lim_{x \to 1} \sqrt[3]{\dfrac{x^2 + 5x - 1}{x + 4}}$
- **3** $\lim_{x \to 1} \left(x + \dfrac{1}{\sqrt{x}}\right)^3$
- **4** $\lim_{x \to 2} \sqrt{\dfrac{x^2 - 4}{x - 2}}$
- **5** $\lim_{x \to 3} \sqrt[3]{\dfrac{x^3 - 27}{x - 3}}$
- **6** $\lim_{x \to 2} \sqrt[4]{11x^3 - 4x + 1}$

7–12.   Determine all points of discontinuity for the given functions.

- **7** $f(x) = \sqrt[3]{x + 1}$
- **8** $f(x) = \left(x + \dfrac{1}{\sqrt{x}}\right)^3$
- **9** $f(x) = \sqrt{\dfrac{x^2 - 4}{x^2 + 2}}$
- **10** $f(x) = \sqrt{9 - x^2}$
- **11** $f(x) = \sqrt{x^2 - 1}$
- **12** $f(x) = \sqrt{x^2 + 1}$

13–20.   Use graphs to find the maximum and minimum values of the function on the indicated interval. (Boldface brackets indicate the greatest integer function.)

- **13** $f(x) = |x|$ on $[-1, 1]$
- **14** $f(x) = x^2$ on $(0, 1)$
- **15** $f(x) = x^2$ on $(-1, 1)$
- **16** $f(x) = [x]$ on $[-1, 1]$
- **17** $f(x) = 2x - 1$ on $(0, 2]$
- **18** $f(x) = \begin{cases} x + 2 & \text{if } x < 0 \\ x - 1 & \text{if } x \geq 0 \end{cases}$ on $[-1, 2]$

•19 $f(x) = \begin{cases} [x] - x & \text{if } x < 0 \\ x^2 & \text{if } x \geq 0 \end{cases}$ on $[-1, 1]$ **20** $f(x) = \begin{cases} x - [x] & \text{if } x < 0 \\ \frac{1}{2} & \text{if } x = 0 \\ x^2 & \text{if } x > 0 \end{cases}$ on $[-1, 1]$

21–24. Use the Intermediate Value Theorem to prove that the graphs of the functions cross the x-axis on the given intervals.

•21 $f(x) = 2x^3 - x^2 - 3x$ on $[1, 2]$

22   $f(x) = 3x^3 + x^2 - 12x - 4$ on $[-1, 0]$

•23 $f(x) = 3x^4 - 14x^3 + 11x^2 + 16x - 12$ on $[0, 1]$

24   $f(x) = 4x^5 + 7x^4 - 2x^3 - 2$ on $[0, 1]$

25   Prove that the number 11 has a square root between 3.31 and 3.32. (*Hint:* See Example 3.)

•26   The velocity of an automobile is known to be a continuous function of time. An automobile is traveling at 20 mph at 3:00 P.M. Twenty minutes later it is traveling at 80 mph. Explain how we know that it must have traveled at 60 mph at some time between 3:00 and 3:20 P.M.

27   (a)   Let n be a positive integer greater than 1; let a be a real number greater than 1. Prove that a has a positive nth root. In other words, prove that there exists a positive real number r such that $r^n = a$. (*Hint:* Modify the argument in Example 3. Prove that such a number exists between 1 and a.)
    (b)   Use (a) to establish a similar result for the case $0 < a < 1$.

28   Let f be a continuous function, defined for all x, that has no real zeros. That is, there exists no real number r such that $f(r) = 0$. Prove that either $f(x) > 0$ for all x or $f(x) < 0$ for all x. [*Hint:* Suppose that f is neither positive nor negative for all x. Then there exist a and b such that $f(a) < 0$ and $f(b) > 0$. Use the Intermediate Value Theorem to show that f must have a real zero.]

29   Use Exercise 28 to prove the following theorem. Let $f(x) = ax^2 + bx + c$, where a, b, c are real numbers, $a \neq 0$. If $b^2 - 4ac < 0$, then either $f(x) > 0$ for all x or $f(x) < 0$ for all x. (*Hint:* First use the quadratic formula to find the zeros of $f(x)$.)

2–8 Infinite Limits and Limits at Infinity. Asymptotes

The limit concept can be extended in several ways. Consider, as an example of the first extension, the behavior of the function

$$f(x) = \frac{1}{x^2}$$

when x is close to 0. If x is a small number, then x^2 is a small positive number and $f(x) = 1/x^2$ is a large positive number. We can, in fact, make $1/x^2$ as large as we want

2.8 Infinite Limits and Limits at Infinity. Asymptotes

by choosing x sufficiently close to zero. (See Fig. 2–29a.) In this case we say that

$$\lim_{x \to 0} f(x) = \lim_{x \to 0} \frac{1}{x^2} = \infty.$$

This does not mean that we are considering a new number called infinity. The symbol indicates that $f(x)$ can be made as large as we wish by requiring x to be close enough to c. The following definition defines this concept more precisely.

DEFINITION

Infinite Limit

Let f be defined on some open interval about c except possibly at c. We say that

$$\lim_{x \to c} f(x) = \infty$$

provided that the following condition exists. For each $M > 0$ there exists $\delta > 0$ (δ depends on M) such that

$$f(x) > M \quad \text{whenever} \quad 0 < |x - c| < \delta.$$

(See Fig. 2–28a.)

A similar definition holds if $f(x)$ has negative values that get large in absolute value as $x \to c$. In that case we write

$$\lim_{x \to c} f(x) = -\infty.$$

(See Fig. 2–28b.)

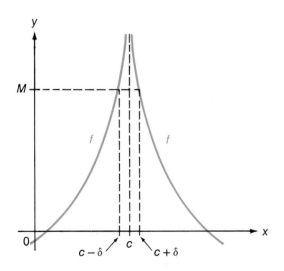
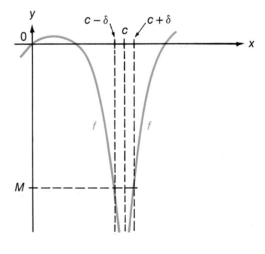

(a) $\lim_{x \to c} f(x) = \infty$. For every $M > 0$ there exists $\delta > 0$ such that $f(x) > M$ whenever $0 < |x - c| < \delta$.

(b) $\lim_{x \to c} f(x) = -\infty$. For every $M < 0$ there exists $\delta > 0$ such that $f(x) < M$ whenever $0 < |x - c| < \delta$.

Figure 2–28 Infinite limits as $x \to c$.

These concepts can be extended to cover one-sided limits. (See Exercises 26 and 29.)

Observe that if a function f has any of these infinite limits as $x \to c$, then x must be discontinuous at c. Thus, $f(x)$ has no limit as $x \to c$ in the sense of the original definition in Section 2-4.

EXAMPLE 1   (a) For the function $f(x) = 1/x^2$ (see first paragraph and also Fig. 2-29a), we have seen that

$$\lim_{x \to 0} f(x) = \infty.$$

(b) If we let x approach 1, the function

$$f(x) = \frac{-1}{(x-1)^2}$$

has negative values that get large in absolute value without bound. (See Fig. 2-29b.) Thus,

$$\lim_{x \to 1} \frac{-1}{(x-1)^2} = -\infty.$$

(c) If x approaches 0 through positive values, then $f(x) = 1/x$ gets large without bound. If x approaches 0 through negative values, then this same function has negative values that get large in absolute value without bound. (See Fig. 2-29c.) Thus,

$$\lim_{x \to 0^+} \frac{1}{x} = \infty \quad \text{and} \quad \lim_{x \to 0^-} \frac{1}{x} = -\infty. \quad \square$$

Many of the functions with infinite limits as $x \to c$ are of form $f(x)/g(x)$, where $f(x) \to L$ (a finite limit) and $g(x) \to 0$ as $x \to c$. The following simple rules can help us to evaluate most of these limits.

Two Special Rules for Infinite Limits

Rule 1   If the numerator $f(x)$ approaches a positive limit L and the denominator $g(x)$ approaches zero through positive values as $x \to c$, then

$$\frac{f(x)}{g(x)} \to \infty \quad \text{as } x \to c.$$

Rule 2   If the numerator $f(x)$ approaches a positive limit L and the denominator $g(x)$ approaches zero through negative values as $x \to c$, then

$$\frac{f(x)}{g(x)} \to -\infty \quad \text{as } x \to c.$$

The situation is reversed if $f(x)$ has a negative limit.

2.8 Infinite Limits and Limits at Infinity. Asymptotes

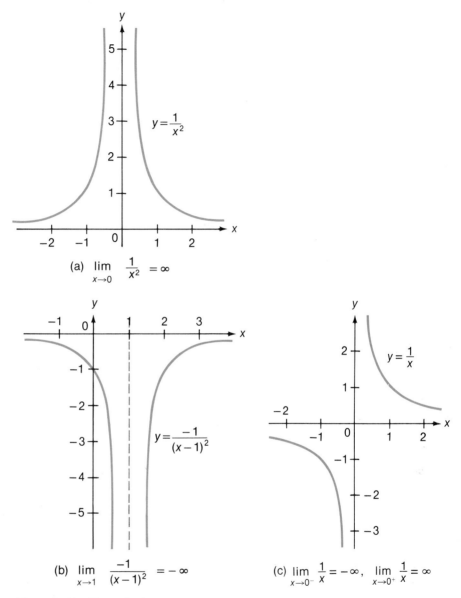

Figure 2-29 Example 1.

These rules can be extended to cover one-sided limits. Example 2 shows their use.

EXAMPLE 2   Calculate $\lim_{x \to 1^-} \frac{x+3}{x-1}$ and $\lim_{x \to 1^+} \frac{x+3}{x-1}$.

Solution   The numerator, $x + 3$, approaches 4 as x approaches 1 from either side.

(a) If $x \to 1^-$, then the denominator, $x - 1$, approaches zero through negative values. Thus,

$$\lim_{x \to 1^-} \frac{x + 3}{x - 1} = -\infty \quad \text{(by Rule 2)}.$$

(b) If $x \to 1^+$, then the denominator, $x - 1$, approaches zero through positive values. Thus,

$$\lim_{x \to 1^+} \frac{x + 3}{x - 1} = \infty \quad \text{(by Rule 1)}.$$

(See Fig. 2–30.) ☐

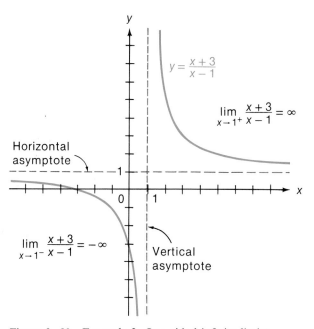

Figure 2–30 Example 2. One-sided infinite limits.

The following rules, which should be clear after some reflection, can be used to calculate some common limits of sums.

Special Rules for Infinite Limits of Sums

Rule 3 If $f(x) \to \infty$ and $g(x) \to \infty$ as $x \to c$, then

$$f(x) + g(x) \to \infty \quad \text{as } x \to c.$$

Rule 4 If $f(x) \to \infty$ and $g(x) \to L$ (a constant) as $x \to c$, then

$$f(x) + g(x) \to \infty \quad \text{as } x \to c.$$

Similar rules hold if $f(x) \to -\infty$ and $g(x) \to L$ or $g(x) \to -\infty$.

2.8 Infinite Limits and Limits at Infinity. Asymptotes

The only problem that can arise is when $f(x) \to \infty$ and $g(x) \to -\infty$ as $x \to c$. In that case no conclusion can be drawn without further information about the functions f and g. The sum may have no limit, may have a finite limit, or may have a limit of $+\infty$ or $-\infty$. (See Exercise 32.)

VERTICAL ASYMPTOTES

Observe in Figure 2–29a that the graph of

$$f(x) = \frac{1}{x^2}$$

gets closer and closer to the y-axis as x gets close to 0. In this case, we say that the y-axis is a vertical asymptote to the graph.

Vertical Asymptote   More generally we say that the graph of f has a *vertical asymptote* at $x = a$ provided

$$\lim_{x \to a^+} f(x) = \infty \text{ (or } -\infty\text{)} \quad \text{or} \quad \lim_{x \to a^-} f(x) = \infty \text{ (or } -\infty\text{)}.$$

(See Fig. 2–29.)

Observe that a vertical asymptote (if one exists at all) can exist only at a point of discontinuity. For example, the function $f(x) = (x + 3)/(x - 1)$, considered in Example 2, has $x = 1$ as its only point of discontinuity. This is the only point at which the graph could have a vertical asymptote. Since f has infinite limits as x approaches a from both sides, then the graph has a vertical asymptote at $x = 1$.

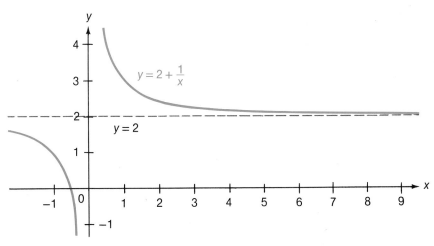

Figure 2–31 $\lim_{x \to \infty} \left(2 + \frac{1}{x}\right) = 2.$

LIMITS AT INFINITY

The function

$$f(x) = 2 + \frac{1}{x}$$

gets closer and closer to the number $L = 2$, as x increases through positive values. (See Fig. 2–31.) If we choose any small positive number ϵ, we can cause $f(x)$ to be between $L - \epsilon$ and $L + \epsilon$ by choosing a sufficiently large value of x. Under these circumstances we say that

$$\lim_{x \to \infty} f(x) = \lim_{x \to \infty} \left(2 + \frac{1}{x}\right) = 2.$$

The formal definition generalizes this concept.

DEFINITION

Limit at Infinity

Let f be defined for all x greater than some fixed number a. We say that

$$\lim_{x \to \infty} f(x) = L$$

provided that the following condition exists. For each $\epsilon > 0$ there exists a number N (which depends on ϵ) such that

$$|f(x) - L| < \epsilon \quad \text{whenever } x > N.$$

(See Fig. 2–32.)

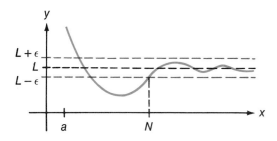
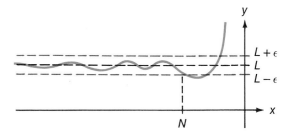

(a) $\lim_{x \to \infty} f(x) = L$: for every $\epsilon > 0$ there exists $N > 0$ such that $|f(x) - L| < \epsilon$ whenever $x > N$

(b) $\lim_{x \to -\infty} f(x) = L$: for every $\epsilon > 0$ there *exists* $N < 0$ such that $|f(x) - L| < \epsilon$ whenever $x < N$.

Figure 2–32 Limits at infinity.

The main results we need are listed in Theorems 2–10 and 2–11. Theorem 2–10 is a consequence of the fact that x^n gets large without bound as x gets large without bound. Theorem 2–11 is an extension of the limit theorem. The proofs are omitted.

THEOREM 2–10   If $n > 0$, then

$$\lim_{x \to \infty} \frac{C}{x^n} = 0.$$

2.8 Infinite Limits and Limits at Infinity. Asymptotes

THEOREM 2–11 (*Limit Theorem*) Let $\lim_{x \to \infty} f(x) = L$ and $\lim_{x \to \infty} g(x) = M$, where L and M are real numbers. Then

Limit Theorem for Limits at Infinity

L1 $\quad \lim_{x \to \infty} [f(x) \pm g(x)] = L \pm M = \lim_{x \to \infty} f(x) \pm \lim_{x \to \infty} g(x).$

L2 $\quad \lim_{x \to \infty} [f(x) g(x)] = LM = \lim_{x \to \infty} f(x) \lim_{x \to \infty} g(x).$

L3 $\quad \lim_{x \to \infty} \dfrac{f(x)}{g(x)} = \dfrac{L}{M} = \dfrac{\lim_{x \to \infty} f(x)}{\lim_{x \to \infty} g(x)}$ provided $M \neq 0.$

L4 $\quad \lim_{x \to \infty} C = C.$

L8 $\quad \lim_{x \to \infty} f(g(x)) = f(M) = f\left(\lim_{x \to \infty} g(x) \right)$ provided f is continuous at $M = \lim_{x \to \infty} g(x).$

The definition and properties discussed above can all be extended to cover limits of form

$$\lim_{x \to -\infty} f(x) = L.$$

In this case we want the functional values $f(x)$ to approach L as x gets numerically large without bound through negative values. (See Fig. 2–32b.)

We often need to modify a function somehow before calculating a limit. In particular, it is usually best, in calculating the limit as $x \to \pm\infty$ of a rational function (quotient of two polynomials), to divide numerator and denominator by the highest power of x in the denominator. As will be seen in parts (a) and (b) of Example 3, this strategy usually allows us to calculate the limit without difficulty.

EXAMPLE 3 Use the limit properties to evaluate the following limits:

(a) $\lim_{x \to \infty} \dfrac{3x^2 - 5x + 7}{2x^4 + 3x + 5}.$

(b) $\lim_{x \to \infty} \dfrac{7x^2 + 2x - 1}{5x^2 + x + 2}.$

(c) $\lim_{x \to \infty} \sqrt{3 + \dfrac{1}{x}}.$

Solution (a) We divide both numerator and denominator by x^4 (the largest power of x in the denominator) and then apply the limit theorem:

$$\lim_{x \to \infty} \frac{3x^2 - 5x + 7}{2x^4 + 3x + 5} = \lim_{x \to \infty} \frac{(3x^2 - 5x + 7)/x^4}{(2x^4 + 3x + 5)/x^4}$$

$$= \lim_{x \to \infty} \frac{\dfrac{3}{x^2} - \dfrac{5}{x^3} + \dfrac{7}{x^4}}{2 + \dfrac{3}{x^3} + \dfrac{5}{x^4}}$$

$$= \frac{0 - 0 + 0}{2 + 0 + 0} = 0$$

(by L3, L1, and Theorem 2–10).

(b) We divide numerator and denominator by x^2, the largest power of x in the denominator:

$$\lim_{x \to \infty} \frac{7x^2 + 2x - 1}{5x^2 + x + 2} = \lim_{x \to \infty} \frac{(7x^2 + 2x - 1)/x^2}{(5x^2 + x + 2)/x^2}$$

$$= \lim_{x \to \infty} \frac{7 + \dfrac{2}{x} - \dfrac{1}{x^2}}{5 + \dfrac{1}{x} + \dfrac{2}{x^2}}$$

$$= \frac{7 + 0 + 0}{5 + 0 + 0} = \frac{7}{5}$$

(by L3, L1, and Theorem 2-10).

(c) $\lim\limits_{x \to \infty} \sqrt{3 + \dfrac{1}{x}} = \sqrt{\lim\limits_{x \to \infty} \left(3 + \dfrac{1}{x}\right)} = \sqrt{3 + 0} = \sqrt{3}$

(by L8, L1, and Theorem 2-10). □

HORIZONTAL ASYMPTOTES

Observe in Figure 2-30 that

$$\lim_{x \to \infty} f(x) = \lim_{x \to \infty} \frac{x + 3}{x - 1} = 1.$$

Geometrically this means that the graph of f gets close to the line $y = 1$ as x gets large without bound. In this case, we say that the line $y = 1$ is a *horizontal asymptote* to the graph of f.

Horizontal Asymptote   More generally, we say that the line $y = L$ is a *horizontal asymptote* to the graph of the function f provided that either

$$\lim_{x \to \infty} f(x) = L \quad \text{or} \quad \lim_{x \to -\infty} f(x) = L.$$

(See Fig. 2-33.)

For example, the function

$$f(x) = \frac{7x^2 + 2x - 1}{5x^2 + x + 2}$$

of Example 3 has the property that $f(x) \to \frac{7}{5}$ as $x \to \infty$. Thus, the line $y = \frac{7}{5}$ is a horizontal asymptote to the graph.

INFINITE LIMITS AT INFINITY

There is one other type of infinite limit. If $f(x)$ gets large without bound as x gets large without bound we write

$$\lim_{x \to \infty} f(x) = \infty.$$

2.8 Infinite Limits and Limits at Infinity. Asymptotes

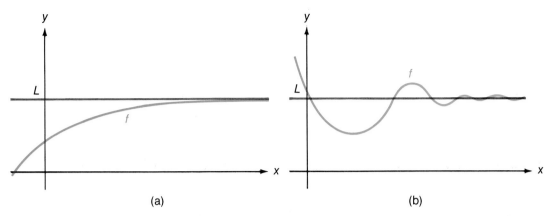

Figure 2–33 Horizontal asymptotes. The line $y = L$ is a horizontal asymptote as $x \to \infty$ if $\lim_{x \to \infty} f(x) = L$.

For example,

$$\lim_{x \to \infty} \sqrt{x} = \infty \quad \text{and} \quad \lim_{x \to \infty} (x^2 - 2) = \infty.$$

The formal definition follows:

DEFINITION

Infinite Limit at Infinity

Let $f(x)$ be defined for all x greater than some fixed number a. We say that

$$\lim_{x \to \infty} f(x) = \infty$$

provided that for each $M > 0$ there exists $N > 0$ (where N depends on M) such that

$$f(x) > M \quad \text{whenever } x > N.$$

(See Fig. 2–34.)

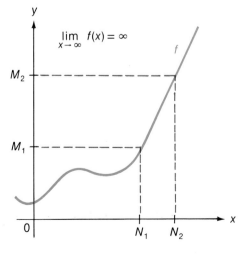

Figure 2–34 $\lim_{x \to \infty} f(x) = \infty$. For every $M > 0$ there exists $N > 0$ such that $f(x) > M$ whenever $x > N$.

Similar definitions hold for limits of the form

$$\lim_{x \to -\infty} f(x) = \infty, \quad \lim_{x \to \infty} f(x) = -\infty, \quad \text{and} \quad \lim_{x \to -\infty} f(x) = -\infty.$$

The meanings of these expressions should be intuitively clear on reflection. (Exercises 28, 30, 31.)

The properties of functions that become infinite are analogous to the properties of limits such as $\lim_{x \to a} f(x) = \infty$, which we studied at the beginning of this section. For example, if $f(x) \to \infty$ and $g(x) \to L$ as $x \to \infty$, then

$$f(x) + g(x) \to \infty.$$

If $f(x) \to \infty$ and $g(x) \to \infty$, then

$$f(x) + g(x) \to \infty.$$

If $f(x) \to \infty$ and $g(x) \to -\infty$, then nothing can be said about the limit of $f(x) + g(x)$ without further study of the particular functions. Some ingenuity is usually required to calculate such a limit.

EXAMPLE 4   Calculate (a) $\lim_{x \to \infty} (\sqrt{x+1} - \sqrt{x})$; (b) $\lim_{x \to \infty} \dfrac{2-x}{\sqrt{x+1}}$.

Solution   (a) The limit is of form $f(x) - g(x)$, where f and g get infinitely large. Since our formulas do not hold in this case, we must modify the form of the expression. To do this, we multiply numerator and denominator by $\sqrt{x+1} + \sqrt{x}$, thereby removing the radicals from the numerator.

$$\lim_{x \to \infty} (\sqrt{x+1} - \sqrt{x}) = \lim_{x \to \infty} \left[(\sqrt{x+1} - \sqrt{x}) \frac{\sqrt{x+1} + \sqrt{x}}{\sqrt{x+1} + \sqrt{x}} \right]$$

$$= \lim_{x \to \infty} \frac{x+1-x}{\sqrt{x+1} + \sqrt{x}}$$

$$= \lim_{x \to \infty} \frac{1}{\sqrt{x+1} + \sqrt{x}} = 0.$$

(b) We rewrite the function as

$$\frac{2-x}{\sqrt{x+1}} = \frac{3-(x+1)}{\sqrt{x+1}} = \frac{3}{\sqrt{x+1}} - \sqrt{x+1}.$$

Then

$$\lim_{x \to \infty} \frac{2-x}{\sqrt{x+1}} = \lim_{x \to \infty} \left(\frac{3}{\sqrt{x+1}} - \sqrt{x+1} \right) = -\infty. \quad \square$$

Remark   You should not jump to the conclusion that every function has some type of limit (possibly infinite) as $x \to \infty$. The function pictured in Figure 2–35, for example, has

Exercises 2-8

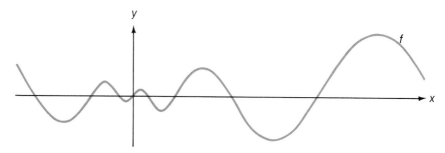

Figure 2-35 The function f has no limit as $x \to \infty$.

no limit. The functional values oscillate between positive and negative values, never approaching a limit as $x \to \infty$.

Exercises 2-8

1-8. **(a)** Calculate the limits if they exist.
(b) Find the vertical asymptotes, if any, for the graphs.

•1 $\lim\limits_{x \to 1^+} \dfrac{x^2 + 1}{x^2 - 1}$

2 $\lim\limits_{x \to 1^-} \dfrac{x^2 + 1}{x^2 - 1}$

•3 $\lim\limits_{x \to 2^+} \dfrac{x}{(x - 2)^3}$

4 $\lim\limits_{x \to 2^-} \dfrac{2x}{(x - 2)^2}$

•5 $\lim\limits_{x \to -2^+} \dfrac{x^2 + 2x - 1}{x + 2}$

6 $\lim\limits_{x \to -2^-} \dfrac{x^2 + 2x - 1}{x + 2}$

•7 $\lim\limits_{x \to 1^+} \dfrac{x^2}{x^2 + x - 2}$

8 $\lim\limits_{x \to 1} \dfrac{x^2 - 1}{x^2 + x - 2}$

9-20. **(a)** Use Theorems 2-10 and 2-11 to evaluate the limits. (In most cases it will be necessary to modify the form of the function.)
(b) Find the horizontal asymptotes, if any, for the graphs.

• 9 $\lim\limits_{x \to -\infty} \dfrac{1}{|x|}$

10 $\lim\limits_{x \to \infty} \dfrac{2}{x} - \dfrac{1}{x^2}$

•11 $\lim\limits_{x \to \infty} \dfrac{2 - x^2}{3 + 5x}$

12 $\lim\limits_{x \to \infty} \dfrac{x^3 - 2x + 1}{2x^2 + x - 3}$

•13 $\lim\limits_{x \to -\infty} \dfrac{4x^3 + x - 2}{3x^3 + x^2 - 2x}$

14 $\lim\limits_{x \to -\infty} \dfrac{2x + 3}{\sqrt{1 + 4x^2}}$

•15 $\lim\limits_{x \to \infty} \dfrac{2 - x + x^2}{4 - x^3}$

16 $\lim\limits_{x \to -\infty} \dfrac{2x^2 - 3}{4x^3 + 2x - 5}$

•17 $\lim_{x \to -\infty} \sqrt{\dfrac{1-4x}{4-x}}$ 18 $\lim_{x \to -\infty} \dfrac{\sqrt{1-2x}}{3x+5}$

•19 $\lim_{x \to \infty} [\sqrt{2+x^2} - x]$ 20 $\lim_{x \to \infty} [x - \sqrt{x^2-x}]$

21–25. If the conditions are not contradictory, then draw figures that illustrate them. If the conditions are contradictory, explain why.

•21 $\lim_{x \to 1^+} f(x) = \infty$, $\lim_{x \to 1^-} f(x) = 2$, $f(1) = 3$.

22 $\lim_{x \to 0} f(x) = -\infty$, $f(0) = 1$.

•23 $\lim_{x \to 0} f(x) = \infty$, f is continuous everywhere.

24 $\lim_{x \to \infty} f(x) = \infty$, f is continuous everywhere.

•25 $\lim_{x \to \infty} f(x) = \infty$, $\lim_{x \to \infty} g(x) = \infty$, $\lim_{x \to \infty} [f(x) - g(x)] = 1$.

26–31. State a formal definition of each of the following concepts and illustrate it with a sketch similar to Figure 2–28, 2–30, or 2–32.

26 $\lim_{x \to c^+} f(x) = \infty$ •27 $\lim_{x \to c} f(x) = -\infty$

28 $\lim_{x \to \infty} f(x) = -\infty$ •29 $\lim_{x \to c^-} f(x) = -\infty$

30 $\lim_{x \to -\infty} f(x) = L$ •31 $\lim_{x \to -\infty} f(x) = \infty$

32 This problem shows that no conclusion can be reached about the behavior of $f(x) + g(x)$ if $f(x) \to \infty$ and $g(x) \to -\infty$ unless additional information is given. Show in each part that $f(x) \to \infty$ and $g(x) \to -\infty$ as $x \to 0$.
  (a) Let $f(x) = 1/x^2$, $g(x) = (x^2 - 1)/x^2$. Show that $f(x) + g(x) \to 1$ as $x \to 0$.
  (b) Let $f(x) = 1/x^2$, $g(x) = -1/|x|$. Show that $f(x) + g(x) \to \infty$ as $x \to 0$. (*Hint:* Write $f(x) = 1/|x|^2$ and factor $f(x) + g(x)$.)
  (c) Construct an example like (b) in which $f(x) \to \infty$, $g(x) \to -\infty$, and $f(x) + g(x) \to -\infty$ as $x \to 0$.

33 (a) Let $f(x) \to \infty$ and $g(x) \to L$ as $x \to c$. Prove that $f(x) + g(x) \to \infty$ as $x \to c$.
  (b) State and prove a similar property if the same conditions hold when $x \to \infty$.

34 Prove Property L4 of Theorem 2–11.

35 Prove Theorem 2–10. (*Hint:* Let $N = \sqrt[n]{|C|/\epsilon}$.)

Review Problems

1 (a) Define the expression "$\lim_{x \to a} f(x) = L$."

  (b) Use the definition to prove that $\lim_{x \to 2} (3x - 5) = 1$.

2 (a) Define the expressions "$\lim_{x \to a^-} f(x) = L$" and "$\lim_{x \to a^+} f(x) = M$." (Give proper definitions involving ϵ and δ.) Illustrate with a sketch.

Review Problems

(b) Suppose that $\lim_{x \to a^-} f(x) = L$ and $\lim_{x \to a^+} f(x) = M$. Under what conditions will f have a limit as x approaches a? Under what conditions will f be continuous at $x = a$?

(c) Is it possible for a function f to have

$$\lim_{x \to 0^-} f(x) = \lim_{x \to 0^+} f(x)$$

and for f to be discontinuous at $x = 0$? Explain.

3. (a) Define the expressions "$\lim_{x \to a^-} f(x) = -\infty$" and "$\lim_{x \to a^+} f(x) = -\infty$." Illustrate with a sketch.

   (b) Define the expressions "$\lim_{x \to \infty} f(x) = L$" and "$\lim_{x \to a^+} f(x) = \infty$." Illustrate with a sketch.

4. State the basic limit properties (L1–L8).

5. Prove L4.

6. Prove L5.

7–12. (a) Calculate the limits (possibly infinite) if they exist.
   (b) Find all horizontal asymptotes for the graphs.
   (c) Find all vertical asymptotes.

• 7. $\displaystyle\lim_{x \to 2} \frac{x^2 - 4x + 4}{x^2 + x - 6}$

8. $\displaystyle\lim_{x \to 0} \frac{(x + 5)^2 - x^2 - 25}{x}$

• 9. $\displaystyle\lim_{x \to 2^-} \frac{x - 3}{x - 2}$

• 10. $\displaystyle\lim_{x \to 1^-} \frac{\sqrt{(x - 1)^2}}{x - 1}$

• 11. $\displaystyle\lim_{x \to \infty} \frac{x^2 + 2x + 3}{x^3 - x}$

12. $\displaystyle\lim_{x \to \infty} \frac{x^4 + x - 1}{x^3 + x}$

13–16. Find the equation of the line tangent to the graph at the indicated point.

• 13. $f(x) = 2x^2 - 1$, $x = 1$

14. $f(x) = \sqrt{3x + 2}$, $x = 0$

• 15. $f(x) = \dfrac{x - 1}{x + 1}$, $x = 1$

16. $f(x) = \sqrt{x + 1}$, $x = 3$

17. (a) Define the expression "the derivative of f at $x = a$."
    (b) State three different applications of the derivative.

• 18. Use the definition of the derivative to calculate:

   (a) $D_x \left(\dfrac{1}{\sqrt{x - 4}} \right)$

   (b) $D_x \left(\dfrac{x - 1}{x + 1} \right)$

•19 A ball is thrown upward into the air from a bridge 112 ft above the ground with an initial velocity of 96 ft/sec.

   (a) Find the maximum height of the ball above the ground.

   (b) When will it strike the ground? With what speed?

20 (a) Define the expression "f is continuous at a."

   (b) Define the expression "f is continuous on the closed interval $[a, b]$."

21 (a) State the properties of continuous functions that correspond to the limit properties L1–L3.

   (b) Use the limit properties to prove the properties in (a).

22 Sketch graphs of functions that satisfy the stated conditions (provided that the conditions are not contradictory):

   (a) f has a jump discontinuity at $x = a$.

   (b) f has a removable discontinuity at $x = a$.

   (c) f has an infinite discontinuity at $x = a$.

   (d) f is continuous at $x = a$ but not differentiable at $x = a$.

   (e) f is differentiable at $x = a$ but has no limit as x approaches a from the right.

23 Discuss the continuity of the following types of functions:

   (a) Polynomials   (b) Rational functions

   (c) Differentiable functions   (d) Quotients of continuous functions

   (e) Compositions of continuous functions

24 Let f be defined on the interval I. What conditions must be put on f and I to ensure that f has a maximum and a minimum on I.

25 State the Intermediate Value Theorem.

26 State the Sandwich Theorem.

27 Use the Intermediate Value Theorem to show that $f(x) = x^3 - 12$ has a zero between 2 and 3.

28 (a) Show that $x^4 \leq x^2$ if $-1 \leq x \leq 1$.

• (b) Suppose that $x^4 \leq f(x) \leq x^2$ for $-1 \leq x \leq 1$. Calculate $\lim_{x \to 0} f(x)$.

Appendix: Proofs of the Limit Theorems

This Appendix contains proper statements and proofs of the limit theorems. In particular, we prove properties L1, L2, L3, L6, and the Sandwich Theorem, which are not proved in the main text.

For convenience we list the various parts of the Limit Theorem (Theorem 2–1) as separate theorems.

THEOREM L1 Let $\lim_{x \to a} f(x) = L$ and $\lim_{x \to a} g(x) = M$. Let $h(x) = f(x) + g(x)$. Then

$$\lim_{x \to a} h(x) = L + M.$$

Proof Let $\epsilon > 0$. We must find $\delta > 0$ such that

$$|h(x) - [L + M]| < \epsilon \quad \text{whenever } 0 < |x - a| < \delta.$$

Before trying to find δ, we work with the expression $|h(x) - [L + M]|$:

$$\begin{aligned} |h(x) - [L + M]| &= |[f(x) + g(x)] - [L + M]| \\ &= |[f(x) - L] + [g(x) - M]| \\ &\leq |f(x) - L| + |g(x) - M| \quad \text{(by the Triangle Inequality)}. \end{aligned}$$

Observe that if we make each of these last terms less than $\epsilon/2$, then $|h(x) - (L + M)|$ is necessarily less than ϵ. This gives a clue to the following argument.

Since $\epsilon > 0$, then $\epsilon/2 > 0$. Since $\lim_{x \to a} f(x) = L$, there exists $\delta_1 > 0$ such that $|f(x) - L| < \epsilon/2$ whenever $0 < |x - a| < \delta_1$. Similarly, there exists $\delta_2 > 0$ such that $|g(x) - M| < \epsilon/2$ whenever $0 < |x - a| < \delta_2$.

Let δ equal the smaller of δ_1 and δ_2. If $0 < |x - a| < \delta$, then $0 < |x - a| < \delta_1$ and $0 < |x - a| < \delta_2$, so that

$$|h(x) - (L + M)| \leq |f(x) - L| + |g(x) - M| < \frac{\epsilon}{2} + \frac{\epsilon}{2} = \epsilon.$$

It follows from the definition of limit that $\lim_{x \to a} h(x) = L + M$. ∎

The following lemma is used in the proof of L2.

LEMMA 1 If $\lim_{x \to a} f(x) = L$, there exists $\delta > 0$ such that

$$|f(x)| < |L| + 1 \quad \text{whenever } 0 < |x - a| < \delta.$$

Proof Since $\lim_{x \to a} f(x) = L$ then, by Exercise 27 of Section 2–5,

$$\lim_{x \to a} |f(x)| = |L|.$$

We now apply the definition of limit with $\epsilon = 1$. There exists $\delta > 0$ such that if $0 < |x - a| < \delta$ then

$$\Big||f(x)| - |L|\Big| \leq |f(x) - L| < \epsilon = 1 \quad \text{(by Exercise 30 of Section 1–3).}$$

This inequality is equivalent to

$$|L| - 1 < |f(x)| < |L| + 1.$$

Thus, $|f(x)| < |L| + 1$ whenever $0 < |x - a| < \delta$. ∎

THEOREM L2 Let $\lim_{x \to a} f(x) = L$ and $\lim_{x \to a} g(x) = M$. Let $h(x) = f(x)g(x)$. Then

$$\lim_{x \to a} h(x) = LM.$$

Proof In this argument we must establish an inequality of form $|h(x) - LM| < \epsilon$. Before considering the technical details of the proof, we work with the expression $|h(x) - LM|$. We rewrite this expression by subtracting and adding $f(x) M$. This will enable us to isolate the terms $|g(x) - M|$ and $|f(x) - L|$.

$$\begin{aligned}
|h(x) - LM| &= |f(x)g(x) - LM| \\
&= |f(x)g(x) - f(x)M + f(x)M - LM| \\
&= |f(x)[g(x) - M] + [f(x) - L]M| \\
&\leq |f(x)| \cdot |g(x) - M| + |f(x) - L| \cdot |M|
\end{aligned}$$

(by the Triangle Inequality and A1).

Let $\epsilon > 0$. By Lemma 1 there exists $\delta_1 > 0$ such that

$$|f(x)| < |L| + 1 \quad \text{if } 0 < |x - a| < \delta_1.$$

Next, observe that the number $\epsilon/2(|M| + 1) > 0$, so there exists $\delta_2 > 0$ such that

$$|f(x) - L| < \frac{\epsilon}{2(|M| + 1)} \quad \text{whenever } 0 < |x - a| < \delta_2.$$

Similarly, there exists $\delta_3 > 0$ such that

$$|g(x) - M| < \frac{\epsilon}{2(|L| + 1)} \quad \text{whenever } 0 < |x - a| < \delta_3.$$

Let δ be the smallest of δ_1, δ_2, and δ_3. If $0 < |x - a| < \delta$, then all the above relations hold. By our previous work,

$$\begin{aligned}
|h(x) - LM| &\leq |f(x)| \, |g(x) - M| + |f(x) - L| \, |M| \\
&< (|L| + 1) \frac{\epsilon}{2(|L| + 1)} + \frac{\epsilon}{2(|M| + 1)} |M| \\
&= \frac{\epsilon}{2} + \frac{\epsilon}{2} \left(\frac{|M|}{|M| + 1} \right) \\
&< \frac{\epsilon}{2} + \frac{\epsilon}{2} = \epsilon.
\end{aligned}$$

Since $|h(x) - LM| < \epsilon$ whenever $0 < |x - a| < \delta$, then

$$\lim_{x \to a} h(x) = LM. \quad \blacksquare$$

Appendix: Proofs of the Limit Theorems

Lemma 2 establishes a special case of L3. It can be used with L2 to prove formula L3.

LEMMA 2 *(Special Case of L3)* Let $\lim_{x \to a} g(x) = M \neq 0$; let $h(x) = \dfrac{1}{g(x)}$. Then

$$\lim_{x \to a} h(x) = \lim_{x \to a} \frac{1}{g(x)} = \frac{1}{M}.$$

Proof In this argument we must establish an inequality of form

$$\left| h(x) - \frac{1}{M} \right| < \epsilon.$$

Observe that

$$\left| h(x) - \frac{1}{M} \right| = \left| \frac{1}{g(x)} - \frac{1}{M} \right| = \left| \frac{M - g(x)}{g(x)M} \right| = \frac{1}{|g(x)|} \frac{1}{|M|} |M - g(x)|.$$

Since $\lim_{x \to a} g(x) = M \neq 0$ then, by Exercise 27 of Section 2–5,

$$\lim_{x \to a} |g(x)| = |M| > 0.$$

It now follows from Theorem 2–4 that there exists $\delta_1 > 0$ such that

$$|g(x)| > \frac{|M|}{2} > 0 \quad \text{whenever } 0 < |x - a| < \delta_1.$$

In particular, if x is in this range, then $g(x) \neq 0$, which means that $h(x)$ is defined. Observe that this last inequality can be rewritten as

$$\frac{1}{|g(x)|} < \frac{2}{|M|} \quad \text{if } 0 < |x - a| < \delta_1.$$

This fact will enable us to replace a variable term in the expansion of $|h(x) - 1/M|$ with a constant.

Now let $\epsilon > 0$. Observe that the number $\epsilon |M|^2/2$ also is positive. Thus, there exists $\delta_2 > 0$ such that

$$|M - g(x)| = |g(x) - M| < \frac{\epsilon |M|^2}{2} \quad \text{whenever } 0 < |x - a| < \delta_2.$$

Let δ be the smaller of δ_1 and δ_2. If $0 < |x - a| < \delta$, then all of the above relations hold. Therefore,

$$\left| h(x) - \frac{1}{M} \right| = \frac{1}{|g(x)|} \frac{1}{|M|} |M - g(x)|$$

$$< \frac{2}{|M|} \frac{1}{|M|} \frac{\epsilon |M|^2}{2} = \epsilon.$$

Thus, $\lim_{x \to a} h(x) = 1/M$. ∎

THEOREM L3 Let $\lim_{x \to a} f(x) = L$ and $\lim_{x \to a} g(x) = M \neq 0$. Let $h(x) = f(x)/g(x)$. Then

$$\lim_{x \to a} h(x) = \frac{L}{M}.$$

The proof is left for you to work (Exercise 1).

THEOREM L6 Let $f(x) = \sqrt[n]{x}$; let a be a real number.
(a) If n is an odd integer, then $\lim_{x \to a} f(x) = \sqrt[n]{a}$.
(b) If n is an even integer with $a > 0$, then $\lim_{x \to a} f(x) = \sqrt[n]{a}$.

Proof We prove the theorem only for $a > 0$. The proofs of the other cases are left for you (Exercises 2 and 3).

Let $r = \sqrt[n]{a}$. Because of the restriction $a > 0$, we must have $r > 0$. Since we are interested only in values of x near a, we restrict ourselves to values in the range

$$\frac{a}{2} < x < \frac{3a}{2}.$$

In other words, x must be in the range defined by

$$|x - a| < \frac{a}{2}.$$

If x is in this range, then $x > 0$, so that

$$f(x) = \sqrt[n]{x} > 0.$$

Let ϵ be a positive number, and let δ be the smaller of the two numbers $a/2$ and $\epsilon \, r^{n-1}$. For the remainder of the proof, we restrict ourselves to values of x in the range $0 < |x - a| < \delta$, so that $f(x) > 0$ and $|x - a| < \epsilon \, r^{n-1}$.

We now factor $x - a$ as*

$$x - a = (\sqrt[n]{x})^n - (\sqrt[n]{a})^n = f(x)^n - r^n$$
$$= [f(x) - r][f(x)^{n-1} + f(x)^{n-2} r + \cdots + f(x) r^{n-2} + r^{n-1}],$$

and take absolute values:

$$|x - a| = |f(x) - r| \, |f(x)^{n-1} + f(x)^{n-2} r + \cdots + f(x) r^{n-2} + r^{n-1}|.$$

Consider the last expression on the right. By our restrictions, we know that $f(x) > 0$ and $r > 0$. Thus, each term is positive, which implies that

$$|f(x)^{n-1} + f(x)^{n-2} r + \cdots + f(x) \, r^{n-2} + r^{n-1}|$$
$$= f(x)^{n-1} + f(x)^{n-2} r + \cdots + f(x) \, r^{n-2} + r^{n-1}$$
$$> 0 + 0 + \cdots + 0 + r^{n-1} = r^{n-1}.$$

---

* The expression $f(x)^n$ represents $[f(x)]^n$.

Appendix: Proofs of the Limit Theorems

Therefore

$$|x - a| = |f(x) - r||f(x)^{n-1} + f(x)^{n-2}r + \cdots + f(x)r^{n-2} + r^{n-1}|$$
$$> |f(x) - r| r^{n-1}.$$

It follows from this last inequality and the fact that $|x - a| < \epsilon r^{n-1}$ that

$$|f(x) - r| r^{n-1} < |x - a| < \epsilon r^{n-1}.$$

Therefore

$$|f(x) - r| < \epsilon \quad \text{if } 0 < |x - a| < \delta,$$

so that

$$\lim_{x \to a} f(x) = r = \sqrt[n]{a}. \blacksquare$$

The proof is similar if n is odd and $a < 0$. A simple direct proof can be constructed if n is odd and $a = 0$ or if n is even, $a = 0$ and $x \to 0^+$.

PROOF OF THE SANDWICH THEOREM

Our final result is the proof of the Sandwich Theorem.

THEOREM 2–3 Let I be an open interval that contains the number a. Let $g(x) \leq f(x) \leq h(x)$ for all x on I different from a. If

$$\lim_{x \to a} g(x) = \lim_{x \to a} h(x) = L,$$

then

$$\lim_{x \to a} f(x) = L.$$

Proof There exists $\delta_1 > 0$ such that if $0 < |x - a| < \delta_1$, then x is in the interval I. Let $\epsilon > 0$. Since $\lim_{x \to a} g(x) = L$, there exists $\delta_2 > 0$ such that

$$L - \epsilon < g(x) < L + \epsilon$$

whenever $0 < |x - a| < \delta_2$.

Similarly, there exists $\delta_3 > 0$ such that

$$L - \epsilon < h(x) < L + \epsilon$$

whenever $0 < |x - a| < \delta_3$.

Let δ be the smallest of the numbers δ_1, δ_2 and δ_3. If $0 < |x - a| < \delta$, then all the above conditions hold, so that

$$L - \epsilon < g(x) \leq f(x) \leq h(x) < L + \epsilon.$$

Since $L - \epsilon < f(x) < L + \epsilon$ whenever $0 < |x - a| < \delta$, then $\lim_{x \to a} f(x) = L$. $\blacksquare$

Exercises

1. Use L2 and Lemma 2 to prove Theorem L3: If $\lim_{x \to a} f(x) = L$ and $\lim_{x \to a} g(x) = M \neq 0$, then
$$\lim_{x \to a} \frac{f(x)}{g(x)} = \frac{L}{M}.$$

2. Prove Theorem L6 for the special case in which n is odd and $a < 0$.

3. Prove Theorem L6 for the special case in which n is odd and $a = 0$.

4. Let n be an even positive integer. Modify the argument in Exercise 3 to prove that $\lim_{x \to 0^+} \sqrt[n]{x} = 0$.

5. Work through the proof of L1 using the specific example
$$f(x) = x + 1, \qquad g(x) = 2x, \qquad a = 1.$$

6. Work through the proof of L2 using the specific example
$$f(x) = x + 1, \qquad g(x) = 2x, \qquad a = 1.$$

7. Work through the proof of Lemma 2 using the specific example
$$g(x) = 2x, \, a = 1.$$

3 Derivatives of Algebraic Functions

3–1 Basic Derivative Formulas

Many of the problems considered in the remainder of this book require the calculation of derivatives at some point in their solution. As we have seen, it is cumbersome and time-consuming to use the definition for the calculation. Several basic formulas and techniques exist for calculating derivatives, and they reduce the labor for most calculations to a minimum.

You should acquire the habit of learning the derivative formulas as soon as they are introduced. Some of the later applications require a detailed knowledge of these formulas. If you are confused about them or remember them improperly, you will be at a distinct disadvantage in doing that work.

Two of the simplest formulas were established in Section 2–2 (Example 2):

> **D1** $D_x C = 0$.
> **D2** $D_x x = 1$.

We also established that $D_x(x^2) = 2x$ and $D_x(x^3) = 3x^2$. We could continue in the same fashion, establishing that

$$D_x(x^4) = 4x^3, \quad D_x(x^5) = 5x^4, \quad \ldots$$

Instead we shall establish the following *Power Rule* (to be proved in Theorem 3–1):

Power Rule   **D3** $D_x(x^n) = nx^{n-1}$ if n is a positive integer.

Using D3, we can calculate, for example, that

$$D_x(x^{75}) = 75x^{74},$$

$$\frac{d}{dx}(x^{2173}) = 2173x^{2172}$$

and

$$\frac{d}{dx}(x^6) = 6x^5.$$

THEOREM 3–1 (*The Power Rule*) Let $f(x) = x^n$, where n is a positive integer. Then

$$f'(x) = nx^{n-1}.$$

Proof Let a be a real number. We shall show that $f'(a) = na^{n-1}$. By definition

$$f'(a) = \lim_{x \to a} \frac{f(x) - f(a)}{x - a} = \lim_{x \to a} \frac{x^n - a^n}{x - a}.$$

It is established in elementary algebra that the numerator can be factored as

$$x^n - a^n = (x - a)(x^{n-1} + x^{n-2}a + x^{n-3}a^2 + \cdots + xa^{n-2} + a^{n-1}).$$

Then

$$f'(a) = \lim_{x \to a} \frac{x^n - a^n}{x - a}$$

$$= \lim_{x \to a} \underbrace{(x^{n-1} + x^{n-2}a + x^{n-3}a^2 + \cdots + xa^{n-2} + a^{n-1})}_{n \text{ terms}}$$

$$= \underbrace{a^{n-1} + a^{n-1} + \cdots + a^{n-1}}_{n \text{ terms}} = na^{n-1}.$$

Since $f'(a) = na^{n-1}$ for each real number a, then

$$f'(x) = nx^{n-1} \quad \text{for each real number } x. \blacksquare$$

The formulas in the following theorems greatly extend the use of formulas D1–D3.

THEOREM 3–2 (*D4*) Let f and g be differentiable on an interval I. Let $h(x) = f(x) \pm g(x)$. Then

Derivative of a Sum

$$\boxed{\text{D4} \quad h'(x) = f'(x) \pm g'(x)}$$

for each $x \in I$. *The derivative of a sum (or difference) is the sum (or difference) of the derivatives.*

Proof If $h(x) = f(x) + g(x)$, then

$$h'(x) = \lim_{\Delta x \to 0} \frac{h(x + \Delta x) - h(x)}{\Delta x}$$

$$= \lim_{\Delta x \to 0} \frac{[f(x + \Delta x) + g(x + \Delta x)] - [f(x) + g(x)]}{\Delta x}$$

3-1 Basic Derivative Formulas

$$= \lim_{\Delta x \to 0} \frac{[f(x + \Delta x) - f(x)] + [g(x + \Delta x) - g(x)]}{\Delta x}$$

$$= \lim_{\Delta x \to 0} \frac{f(x + \Delta x) - f(x)}{\Delta x} + \lim_{\Delta x \to 0} \frac{g(x + \Delta x) - g(x)}{\Delta x} \quad \text{(by L1)}$$

$$= f'(x) + g'(x).$$

The proof is similar if $h(x) = f(x) - g(x)$. ∎

THEOREM 3-3 (D5) Let f be differentiable on an interval I. Let $g(x) = Cf(x)$. Then

Derivative of Constant Times a Function

$$\boxed{\text{D5} \quad g'(x) = Cf'(x)}$$

for each $x \in I$. The derivative of a constant times a differentiable function equals the same constant times the derivative of the function.

The proof is left for the reader (Exercise 31.)

EXAMPLE 1 (a) $D_x(5x^2 - 3) = D_x(5x^2) - D_x(3)$ (by D4)
$= 5D_x(x^2) - 0 = 5 \cdot 2x = 10x$ (by D3, D1).

(b) $\dfrac{d}{dx}(14x^{63} - 8x^2 + 4x - 1)$

$= \dfrac{d}{dx}(14x^{63}) - \dfrac{d}{dx}(8x^2) + \dfrac{d}{dx}(4x) - \dfrac{d}{dx}(1)$ (by D4)

$= 14\dfrac{d}{dx}(x^{63}) - 8\dfrac{d}{dx}(x^2) + 4\dfrac{d}{dx}(x) - 0$ (by D5, D1)

$= 14 \cdot 63x^{62} - 8 \cdot 2x + 4 \cdot 1 = 882x^{62} - 16x + 4.$ □

POLYNOMIAL FUNCTIONS

The derivative of any polynomial function can be calculated by using formulas D1–D5 as in Example 1(b). In general, if

Derivative of a Polynomial

$$f(x) = a_n x^n + a_{n-1} x^{n-1} + \cdots + a_2 x^2 + a_1 x + a_0,$$

then

$$f'(x) = na_n x^{n-1} + (n - 1)a_{n-1} x^{n-2} + \cdots + 2a_2 x + a_1.$$

EXAMPLE 2 $\dfrac{d}{dx}(2x^7 - 5x^6 + 13x^5 + 4x^3 - x + 2)$

$= 2 \cdot 7x^6 - 5 \cdot 6x^5 + 13 \cdot 5x^4 + 4 \cdot 3x^2 - 1$
$= 14x^6 - 30x^5 + 65x^4 + 12x^2 - 1.$ □

Exercises 3-1

1-16. Use formulas D1-D5 to differentiate the functions.

- 1 $f(x) = \sqrt{2}$
- 2 $f(x) = 1 - 2x$
- 3 $f(z) = 3z + 5$
- 4 $g(x) = 4x^2 + 3x$
- 5 $h(t) = 3t^2 - 2t + 1$
- 6 $g(r) = 8r^6 - 4r^3 + 3r^2 - 1$
- 7 $f(x) = ax^2 + bx + c$
- 8 $h(y) = y^2/a - y/b + c$
- 9 $g(z) = -4z^3 + z - z^2 + 1$
- 10 $f(x) = -5x^8 + 4x^7 - 4x^3 + 5x - 10$
- 11 $h(t) = (3t - 2)^2$
- 12 $f(y) = (y + 1)(y^2 - 1)$
- 13 $f(x) = (2x)^3$
- 14 $h(v) = v(v^2 + 1)$
- 15 $f(t) = \dfrac{t^3 + 3t^2}{t^2}$
- 16 $g(z) = \dfrac{z^3 - 8}{z - 2}$

17-20. Find equations of the lines tangent and normal to the graph of f at the point P.

- 17 $f(x) = 3x^2 - 4x + 2$, $P(1, 1)$
- 18 $f(x) = 3x^5 + 4x^3 - 2x^2 + x - 1$, $P(1, 5)$
- 19 $f(x) = 4x^3 - 5x + 1$, $P(2, 23)$
- 20 $f(x) = x^4 + 3x^3 - 2x^2 + 5x - 3$, $P(-1, -12)$

21-26. Determine all points at which the line tangent to the graph of f is horizontal, that is, at which $f'(x) = 0$.

- 21 $f(x) = 2x^2 - 3x + 1$
- 22 $f(x) = 4x^2 + 8x - 3$
- 23 $f(x) = 2x^3 - 9x^2 - 24x + 4$
- 24 $f(x) = 3x^4 - 20x^3 + 36x^2 - 5$
- 25 $f(x) = 2x^3 - 3x^2 - 12x + 2$
- 26 $f(x) = \dfrac{x^3}{3} - 4x - 1$

- 27 Determine the values a, b, and c that make the graph of $y = ax^2 + bx + c$ pass through the point $(-1, 4)$ and be tangent to the line $y = 3x - 1$ at $(1, 2)$.

- 28 (a) Find the rate of change of the area of a circle with respect to the radius.
     (b) Find the rate of change of the volume of a sphere with respect to the radius.

- 29 Find the points at which the lines tangent to the graph of $f(x) = x^3 - 2x^2 + 3$ are perpendicular to the line $x + 7y = 5$.

- 30 A particle fired straight up from the ground reaches a maximum height of 400 ft. Find the initial velocity.

31 Prove Theorem 3-3: If f is a differentiable function of x and $g(x) = Cf(x)$, then $g'(x) = Cf'(x)$.

3-2 The Product, Quotient, and General Power Rules

32 Prove D4 for the case in which $h(x) = f(x) - g(x)$.

33 Let f_1, f_2, f_3 be differentiable functions. Let $g(x) = C_1 f_1(x) + C_2 f_2(x) + C_3 f_3(x)$, where C_1, C_2, C_3 are constants. Apply D4 twice to show that

$$g'(x) = C_1 f_1'(x) + C_2 f_2'(x) + C_3 f_3'(x).$$

What result holds for a similar sum involving n functions $f_1, f_2, \ldots, f_n$?

3-2 The Product, Quotient, and General Power Rules

PRODUCT RULE

Many beginning students conjecture that the derivative of a product should be the product of the derivatives. This conjecture is false, as you can see by calculating the derivative of almost any product. For example, if $f(x) = 2x$ and $g(x) = x^2$, the derivative of the product is $D_x(2x^3) = 6x^2$, and the product of the derivatives is $D_x(2x) \cdot D_x(x^2) = 2 \cdot 2x = 4x$. The correct rule, which will be proved in Theorem 3-4, is

Product Rule

D6 $\quad D_x(f(x) \cdot g(x)) = f(x)g'(x) + g(x)f'(x).$

The derivative of the product of two differentiable functions equals the first function times the derivative of the second plus the second function times the derivative of the first.

For example, if $f(x) = 2x$ and $g(x) = x^2$, then

$$D_x(f(x) \cdot g(x)) = f(x)g'(x) + g(x)f'(x) = 2x \cdot 2x + x^2 \cdot 2 = 6x^2.$$

The formal statement of the product rule is given in the following theorem.

THEOREM 3-4 (D6) Let f and g be differentiable at $x = a$. Let $h(x) = f(x)g(x)$. Then

$$h'(a) = f(a)g'(a) + g(a)f'(a).$$

Proof Before beginning the actual proof, we note that by Theorem 2-6, the function g must be continuous at $x = a$. Thus,

$$\lim_{x \to a} g(x) = g(a),$$

a fact that we need in the proof.

To establish the formula, we must calculate the limit of the difference quotient

$$h'(a) = \lim_{x \to a} \frac{h(x) - h(a)}{x - a} = \lim_{x \to a} \frac{f(x)g(x) - f(a)g(a)}{x - a}.$$

Before taking the limit, we reorganize the terms. To do this, we first add zero, written in the form $-f(a)g(x) + f(a)g(x)$, to the numerator. This is done to get the necessary terms for the difference quotients for $f'(a)$ and $g'(a)$.

Then

$$h'(a) = \lim_{x \to a} \frac{f(x)g(x) + 0 - f(a)g(a)}{x - a}$$

$$= \lim_{x \to a} \frac{f(x)g(x) + [-f(a)g(x) + f(a)g(x)] - f(a)g(a)}{x - a}$$

$$= \lim_{x \to a} \left[\frac{f(x) - f(a)}{x - a} g(x) + f(a) \frac{g(x) - g(a)}{x - a} \right]$$

$$= \left[\lim_{x \to a} \frac{f(x) - f(a)}{x - a} \cdot \lim_{x \to a} g(x) \right] + \left[\lim_{x \to a} f(a) \cdot \lim_{x \to a} \frac{g(x) - g(a)}{x - a} \right] \quad \text{(by L1, L2)}$$

$$= f'(a)g(a) + f(a)g'(a). \quad \blacksquare$$

QUOTIENT RULE

The formula for the quotient of two differentiable functions f and g is given by the formula

Quotient Rule **D7** $\quad D_x \left(\dfrac{f(x)}{g(x)} \right) = \dfrac{g(x) f'(x) - f(x) g'(x)}{[g(x)]^2} \quad$ provided $g(x) \neq 0$.

The derivative of the quotient of two differentiable functions equals the denominator times the derivative of the numerator minus the numerator times the derivative of the denominator, all divided by the square of the denominator. This formula holds for each value of x for which the denominator is not zero.

Theorem 3–5 is the formal statement of the Quotient Rule.

THEOREM 3–5   (*D7*) Let f and g be differentiable at $x = a$. Let $h(x) = f(x)/g(x)$. If $g(a) \neq 0$, then h is differentiable at $x = a$ and

$$h'(a) = \frac{g(a)f'(a) - f(a)g'(a)}{[g(a)]^2}$$

Proof   As in the proof of Theorem 3–4, we first note that

$$\lim_{x \to a} g(x) = g(a).$$

This proof requires the calculation of the difference quotient for the function h. Before calculating the limit of the difference quotient, we substitute $h(x) = f(x)/g(x)$ and rearrange the terms. At one step we add zero, written as $-f(a)g(a) + f(a)g(a)$, into the numerator. This is done to provide the necessary terms for the difference quotients for $f'(a)$ and $g'(a)$. The difference quotient for h is

$$\frac{h(x) - h(a)}{x - a} = \frac{f(x)/g(x) - f(a)/g(a)}{x - a} = \frac{f(x)g(a) - f(a)g(x)}{(x - a)g(x)g(a)}$$

$$= \frac{f(x)g(a) + [-f(a)g(a) + f(a)g(a)] - f(a)g(x)}{(x - a)g(x)g(a)}$$

3-2 The Product, Quotient, and General Power Rules

$$= \frac{[f(x) - f(a)]\, g(a) - f(a)[g(x) - g(a)]}{(x - a)g(x)g(a)}$$

$$= \frac{f(x) - f(a)}{x - a} \frac{g(a)}{g(x)g(a)} - \frac{f(a)}{g(x)g(a)} \frac{g(x) - g(a)}{x - a}.$$

We now take the limit as $x \to a$:

$$h'(x) = \lim_{x \to a} \frac{h(x) - h(a)}{x - a}$$

$$= \lim_{x \to a} \frac{f(x) - f(a)}{x - a} \lim_{x \to a} \frac{g(a)}{g(x)g(a)} - \lim_{x \to a} \frac{f(a)}{g(x)g(a)} \lim_{x \to a} \frac{g(x) - g(a)}{x - a}$$

$$= f'(a) \frac{g(a)}{g(a)^2} - \frac{f(a)}{g(a)^2} g'(a) = \frac{f'(a)g(a) - f(a)g'(a)}{g(a)^2}. \blacksquare$$

If we let $u = f(x)$ and $v = g(x)$, the product and quotient rules can be written in more compact form:

Product and Quotient Rules

D6 $D_x(uv) = u\, D_x v + v\, D_x u$

D7 $D_x\left(\dfrac{u}{v}\right) = \dfrac{v\, D_x u - u\, D_x v}{v^2}$ provided $v \neq 0$.

EXAMPLE 1

$$\frac{d}{dx}\left(\frac{4x + 1}{2x^2}\right)$$

$$= \frac{d}{dx}\left(\frac{u}{v}\right)$$

$$= \frac{v\left(\dfrac{du}{dx}\right) - u\left(\dfrac{dv}{dx}\right)}{v^2}$$

$$= \frac{2x^2 \dfrac{d}{dx}(4x + 1) - (4x + 1)\dfrac{d}{dx}(2x^2)}{(2x^2)^2}$$

$$= \frac{2x^2 \cdot 4 - (4x + 1) \cdot 4x}{4x^4} = -\frac{2x + 1}{x^3}. \; \square$$

GENERAL POWER RULE

Let u be a differentiable function of x, say $u = f(x)$. The special cases of the product rule D6 obtained by choosing $v = u$, $v = u^2$, $v = u^3$, and so on, are interesting:

$v = u$: $D_x(u^2) = D_x(u \cdot u) = u \cdot D_x u + u \cdot D_x u = 2u \cdot D_x u.$

$v = u^2$: $D_x(u^3) = D_x(u \cdot u^2) = u \cdot D_x(u^2) + u^2 \cdot D_x u$
$\qquad\qquad = u \cdot 2u \cdot D_x u + u^2 \cdot D_x u = 3u^2 \cdot D_x u.$

$v = u^3$: $\quad D_x(u^4) = D_x(u \cdot u^3) = u \cdot D_x(u^3) + u^3 \cdot D_x u$
$$= u \cdot 3u^2 \cdot D_x u + u^3 \cdot D_x u = 4u^3 \cdot D_x u.$$

If we continue in this fashion (this remark disguises a proof by mathematical induction—see Exercise 33), we get the *General Power Rule*.

$$D_x(u^n) = nu^{n-1} D_x u \quad \text{for any positive integer } n.$$

EXAMPLE 2   To calculate $D_x(2x^2 + 1)^5$, we let $u = 2x^2 + 1$. By the General Power Rule
$$D_x(2x^2 + 1)^5 = 5(2x^2 + 1)^4 \cdot D_x(2x^2 + 1)$$
$$= 5(2x^2 + 1)^4 \cdot 4x = 20x(2x^2 + 1)^4. \quad \square$$

It is relatively easy to extend the general power rule to cover negative exponents provided $u(x) \neq 0$. We simply write u^{-k} as $1/u^k$, apply the Quotient Rule D7, and simplify. The details are left for you to work out (Exercise 34).

It is trivial that the rule

$$D_x(u^n) = nu^{n-1} \cdot D_x u, \quad u(x) \neq 0,$$

holds when $n = 0$, for both sides of the equation reduce to zero. Since the rule holds when n is positive, negative, or zero, it holds for every integer n. Because this formula is a direct extension of Formula D3 of Section 3–1, we give it the same reference number:

General Power Rule   **D3** $D_x(u^n) = nu^{n-1} \cdot D_x u \quad$ if n is an integer and $u(x) \neq 0$ if $n \leq 0$.

EXAMPLE 3   (a) $D_x(x^{-3}) = -3x^{-4} D_x x = \dfrac{-3}{x^4}.$

(b) $D_x \left(\dfrac{1}{x^5} \right) = D_x(x^{-5}) = -5x^{-6} = \dfrac{-5}{x^6}.$

(c) $D_x \left(\dfrac{3}{2x + 1} \right) = 3D_x(2x + 1)^{-1} = -3(2x + 1)^{-2}(2) = \dfrac{-6}{(2x + 1)^2}. \quad \square$

Remark 1   Do not make the error of calculating nu^{n-1} for the derivative $D_x(u^n)$, forgetting the other factor $D_x u$. This practice is disastrous. The entire expression $nu^{n-1} D_x u$ must be calculated for $D_x(u^n)$.

The formal statement of the general Power Rule is given in Theorem 3–6.

THEOREM 3–6   (D3) Let f be differentiable at $x = a$; let n be a positive integer. Let $g(x) = [f(x)]^n$. Then
$$g'(a) = n[f(a)]^{n-1} f'(a).$$

This result also holds if n is a negative integer or zero, provided $f(a) \neq 0$.

Remark 2   We shall show in Section 3–5 that the Power Rule D3 can be extended to cover rational exponents. Thus, eventually we shall use it to calculate derivatives of functions such as $f(x) = \sqrt{3x^2 - 1}$, $g(x) = (3x - 5)^{5/3}$, and so on.

Exercises 3–2

Remark 3 Most derivatives can be calculated by several different methods. For example, if

$$f(x) = \frac{2x+1}{x-2}$$

we could calculate $f'(x)$ by the Quotient Rule D7, or we could write

$$f(x) = (2x+1)(x-2)^{-1}$$

and use the Product and Power rules. You should examine each derivative problem for possible approaches and try to use the most efficient method for the calculation.

EXAMPLE 4 Calculate the derivative of

$$f(x) = \frac{2x+1}{x-2}$$

by (a) the Quotient Rule and (b) the Product Rule.

Solution (a) $f'(x) = \dfrac{(x-2) \cdot 2 - (2x+1) \cdot 1}{(x-2)^2} = -\dfrac{5}{(x-2)^2}$ (by D7)

(b) $f'(x) = (2x+1)(x-2)^{-1} = (2x+1)(-1)(x-2)^{-2} + (x-2)^{-1} \cdot 2$ (by D6)

$= -\dfrac{2x+1}{(x-2)^2} + \dfrac{2}{x-2} = -\dfrac{5}{(x-2)^2}$. □

Exercises 3–2

1–18. Calculate the derivative. Try to use the most efficient method.

- **1** $f(x) = 3x + \dfrac{1}{x}$
- **2** $g(r) = (3r)^{-2}$
- **3** $h(z) = \dfrac{1}{z^3} + \dfrac{2}{z^2} - \dfrac{4}{z} + 1$
- **4** $f(t) = 3 + 2t^{-1} - 3t^{-2} + t^{-3} - 6t^{-5}$
- **5** $f(x) = (x^2 - 3x)^{45}$
- **6** $f(y) = (2y^3 + 4y^2 - 1)^3$
- **7** $q(x) = (x^2 - 3x + 1)(x^3 - 2x)$
- **8** $f(r) = (r^2 - 1)^3(2r + 3)^4$
- **9** $f(t) = (t^2 - 2t + 1)^5(t^2 - 1)^3$
- **10** $h(x) = (3x^2 - x + 1)^2(4x^3 + 2x^2 - 5)^4$
- **11** $f(y) = (2y^2 - 3y + 1)^{-20}$
- **12** $h(r) = \dfrac{r^2 - 2r + 3}{r^3 - 1}$
- **13** $f(t) = \dfrac{t-2}{t^3 + 2t}$
- **14** $h(x) = \dfrac{x^2 + 2}{x^2 - 2}$
- **15** $f(z) = \dfrac{z^3}{1 - z^2}$
- **16** $q(y) = \dfrac{1}{(y^3 + 2y - 1)^3}$
- **17** $f(x) = \left(\dfrac{x+2}{x-1}\right)^{-3}$
- **18** $f(r) = \left(\dfrac{r^2 + 2}{r^3 - 2r + 1}\right)^5$

19–22. Calculate $f'(x)$ by using (a) the Quotient Rule and (b) the Product Rule.

•19 $f(x) = \dfrac{x^4 + 3x^2 - 2}{x^3}$

20 $f(x) = \dfrac{x - 3}{x^2 + 2x - 1}$

•21 $f(x) = \dfrac{x^2}{x + 1}$

22 $f(x) = \dfrac{1 - 2x}{4 - x^2}$

23–24. Find the equations of the tangent and normal lines at the given point.

•23 $f(x) = 5x^2(x^2 - 2)^3$; $P(1, -5)$

24 $f(x) = \dfrac{x^2 + 2}{x^2 - 1}$; $P(2, 2)$

25–30. Find all points at which the tangent line to the graph of f is horizontal.

•25 $f(x) = (x^2 - 3x)^5$

26 $f(x) = (2x^2 - 3x + 1)^{10}$

•27 $f(x) = (x^2 - x - 12)^2$

28 $f(x) = \dfrac{x^2 + 3}{x - 1}$

•29 $f(x) = \dfrac{x^3}{x^2 - 3}$

30 $f(x) = (x - 3)^6(x + 3)^2$

•31 Find all points at which the line $y = -16x + 12$ is tangent to the graph of $y = 1/x^2$.

32 (a) Let u, v, and w be differentiable functions of x. Use D6 to show that
$$D_x(uvw) = uvw' + uv'w + u'vw.$$

• (b) Use the result of (a) to calculate
$$Dx[(5x^2 - 1)^4(4x + 2)^3(7x^2 + 1)^5].$$

33 Use D6 and mathematical induction to prove that $D_x(u^n) = nu^{n-1} D_x u$ if u is a differentiable function of x and n is a positive integer. (See Appendix A at the end of the book.)

34 Prove the Power Rule D3 for negative exponents. (*Hint:* If n is a negative integer, write $n = -k$ so that $u^n = u^{-k} = 1/u^k$.)

35 Many students assume that if a power of a function is differentiable, then the original function must be differentiable. The following example shows that this assumption is not correct.

Let $f(x) = |x|$.
  (a) Show that f does not have a derivative at $x = 0$.
  (b) Let $g(x) = [f(x)]^2$. Show that g has a derivative everywhere.

3–3 Application to Approximation Theory. Differentials

Here we take a temporary break from derivative formulas and consider an important application of the derivative—the approximation of small changes of functions.

3-3 Application to Approximation Theory. Differentials

Increment of y

Let $y = f(x)$ be differentiable at x_0. For each increment Δx, we define the corresponding increment Δy by

$$\Delta y = f(x_0 + \Delta x) - f(x_0).$$

Then Δy measures the vertical change between the points $(x_0, f(x_0))$ and $(x_0 + \Delta x, f(x_0 + \Delta x))$ on the graph of f. (See Fig. 3-1.) The increment Δy is equal to the change in f when x changes from x_0 to $x_0 + \Delta x$.

For example, if $f(x) = \pi x^2$, $x_0 = 3$, and $\Delta x = -0.001$, then

$$\Delta y = f(x_0 + \Delta x) - f(x_0) = \pi(2.999)^2 - \pi(3)^2$$
$$\approx -0.0188464.$$

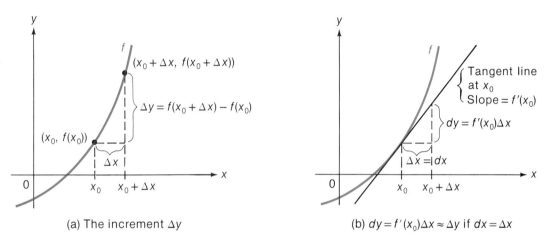

(a) The increment Δy
(b) $dy = f'(x_0)\Delta x \approx \Delta y$ if $dx = \Delta x$

Figure 3-1 Approximation of a small change in a function by the change along the tangent line: $\Delta y = f(x_0 + \Delta x) - f(x_0) \approx f'(x_0) \Delta x$.

Often the calculation of the exact value of Δy is messy. We shall show how derivatives can be used to approximate Δy with a minimum of effort. First, however, we show that Δy can be expressed in terms of $f'(x_0)$, Δx, and an "error" term.

Recall that

$$f'(x_0) = \lim_{\Delta x \to 0} \frac{f(x_0 + \Delta x) - f(x_0)}{\Delta x} = \lim_{\Delta x \to 0} \frac{\Delta y}{\Delta x}.$$

Thus, the number $f'(x_0)$ is approximately equal to $\Delta y/\Delta x$ if Δx is small. Since $f'(x_0)$ is not exactly equal to $\Delta y/\Delta x$, there is a small error (which depends on Δx) whenever $f'(x_0)$ is used as an approximation to $\Delta y/\Delta x$. Let ϵ be this error; that is, let

$$\epsilon = \frac{\Delta y}{\Delta x} - f'(x_0).$$

(Observe that ϵ is a function of Δx, not a constant.)

Thus far, ϵ is defined only when $\Delta x \neq 0$. Observe, however, that

$$\lim_{\Delta x \to 0} \epsilon = \lim_{\Delta x \to 0} \left[\frac{\Delta y}{\Delta x} - f'(x_0) \right] = f'(x_0) - f'(x_0) = 0.$$

Consequently, if we define ϵ to be zero when $\Delta x = 0$, the function ϵ will be continuous at $\Delta x = 0$. It is customary to make this definition so that

Error Term ϵ

$$\epsilon = \begin{cases} \dfrac{\Delta y}{\Delta x} - f'(x_0) & \text{if } \Delta x \neq 0 \\ 0 & \text{if } \Delta x = 0. \end{cases}$$

We can rewrite the defining equation for ϵ as

$$\Delta y = f'(x_0) \, \Delta x + \epsilon \, \Delta x \qquad \text{where } \epsilon \to 0 \text{ as } \Delta x \to 0.$$

Observe that this relation holds when $\Delta x = 0$ as well as when $\Delta x \neq 0$, for in this case Δy and ϵ both equal zero.

EXAMPLE 1   Let $f(x) = 2x^2 + 3x - 1$. Calculate the exact value of ϵ in the formula

$$\Delta y = f'(x_0) \, \Delta x + \epsilon \, \Delta x.$$

Solution   We work with a fixed value $x = x_0$, letting Δx vary. Then

$$\begin{aligned} \Delta y &= f(x_0 + \Delta x) - f(x_0) \\ &= [2(x_0 + \Delta x)^2 + 3(x_0 + \Delta x) - 1] - [2x_0^2 + 3x_0 - 1] \\ &= 4x_0 \, \Delta x + 2(\Delta x)^2 + 3 \, \Delta x. \end{aligned}$$

The quantity $f'(x_0) \, \Delta x$ equals

$$f'(x_0) \, \Delta x = (4x_0 + 3) \, \Delta x = 4x_0 \, \Delta x + 3 \, \Delta x.$$

Since $\Delta y = f'(x_0) \, \Delta x + \epsilon \, \Delta x$, then

$$4x_0 \, \Delta x + 2(\Delta x)^2 + 3 \, \Delta x = 4x_0 \, \Delta x + 3 \, \Delta x + \epsilon \, \Delta x$$

so that

$$\begin{aligned} 2(\Delta x)^2 &= \epsilon \, \Delta x \\ \epsilon &= 2 \, \Delta x. \end{aligned}$$

Observe that

$$\lim_{\Delta x \to 0} \epsilon = \lim_{\Delta x \to 0} 2 \, \Delta x = 0$$

as predicted by the formula. □

THE APPROXIMATION FORMULA

The exact value of ϵ can be calculated as in Example 1. Our main use for the formula, however, results from ignoring ϵ, and using $f'(x_0) \, \Delta x$ as a convenient approximation

3-3 Application to Approximation Theory. Differentials

to Δy. Observe that it makes sense to ignore the error term, because if Δx is close to zero, then the product $\epsilon \, \Delta x$ is extremely small. Thus,

Approximation Formula

$$\Delta y \approx f'(x_0) \, \Delta x \quad \text{if } \Delta x \text{ is small.}$$

This formula usually gives a very good approximation to Δy if Δx is close to zero.

To interpret this formula geometrically, recall that $f'(x_0)$ is the slope of the tangent line at $(x_0, f(x_0))$. Thus, $f'(x_0) \, \Delta x$ is the vertical change along the tangent line that corresponds to a horizontal change of Δx units from $(x_0, f(x_0))$. (See Fig. 3-1b.)

The approximation formula shows that the change in the function f when x changes from x_0 to $x_0 + \Delta x$ approximately equals the corresponding change along the tangent line to the graph of f at $(x_0, f(x_0))$. (See Fig. 3-1b.)

EXAMPLE 2 Use the approximation formula $\Delta y \approx f'(x_0) \, \Delta x$ to approximate the change in $f(x) = \pi x^2$ when x changes from 3 to 2.999.

Solution Let $x_0 = 3$, $\Delta x = -0.001$. The change in $y = f(x)$ corresponding to the change in x from x_0 to $x_0 + \Delta x$ is

$$\Delta y = f(x_0 + \Delta x) - f(x_0) \approx f'(x_0) \, \Delta x = 2\pi x_0 \, \Delta x$$
$$= 2\pi(3)(-0.001)$$
$$\approx -0.0188496.$$

This approximation agrees with the more accurate one obtained at the beginning of this section when both are rounded off to five decimal places. □

EXAMPLE 3 A flat circular metal plate has a radius $r = 4$ cm. When the plate is heated, the radius increases to 4.002 cm. Use the approximation formula to estimate (a) the change in the circumference and (b) the value of the new circumference.

Solution (a) The circumference is $C = f(r) = 2\pi r$. We need to approximate the value of ΔC (the change in C) when r changes from 4 to 4.002. Let $r_0 = 4$, $\Delta r = 0.002$. Then

$$\Delta C \approx f'(r_0) \, \Delta r$$
$$\approx 2\pi (0.002) = 0.004\pi \approx 0.01257.$$

(b) The new circumference equals

$$\underset{\substack{\text{old value of}\\\text{circumference}}}{C} + \underset{\substack{\text{change in}\\\text{circumference}}}{\Delta C} \approx 2\pi \cdot 4 + 0.004\pi \approx 8.004\pi$$

$$\approx 25.14531. \quad \square$$

EXAMPLE 4 A square metal plate has sides of length $x = x_0$. When the plate is heated, the sides increase to length $x = x_0 + \Delta x$. Draw a figure that pictures (a) ΔA, the exact change in the area; and (b) the approximate value ΔA obtained by the formula.

Solution The area is given by the formula $A = f(x) = x^2$. The exact change in the area is

$$\Delta A = f(x_0 + \Delta x) - f(x_0) = (x_0 + \Delta x)^2 - x_0^2 = 2x_0 \, \Delta x + \Delta x^2.$$

This quantity is represented by the area of the two shaded rectangles and the small square in Figure 3-2.

The approximate value of ΔA obtained by the formula is

$$\Delta A \approx f'(x_0) \Delta x = 2x_0 \Delta x.$$

This quantity is represented by the area of the two shaded rectangles in Figure 3-2.

The error term, $\epsilon \Delta x = (\Delta x)^2$, is represented by the area of the small shaded square in Figure 3-2. Observe that this term goes to zero very rapidly when $\Delta x \to 0$. □

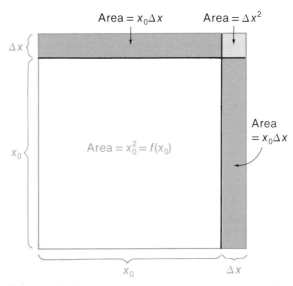

Figure 3-2 Example 4. Area of shaded region $= f(x_0 + \Delta x) - f(x_0) = \Delta A$.
Area of the two dark rectangles $= 2x_0 \cdot \Delta x = f'(x_0) \cdot \Delta x$.
Area of small square = error if $f'(x_0) \cdot \Delta x$ is used to approximate ΔA.

ALTERNATE FORM OF THE APPROXIMATION FORMULA

Recalling that

$$\Delta y = f(x_0 + \Delta x) - f(x_0)$$

we can write the approximation formula as

$$\Delta y = f(x_0 + \Delta x) - f(x_0) \approx f'(x_0) \Delta x.$$

Approximation Formula

$$f(x_0 + \Delta x) \approx f(x_0) + f'(x_0) \Delta x \qquad \text{when } \Delta x \text{ is small.}$$

This alternate version of the formula gives the value of f at the new point $x_0 + \Delta x$ as the value at the old point x_0 *plus* the approximate change in the function.

3-3 Application to Approximation Theory. Differentials

EXAMPLE 5 Let $f(x) = x^3$. Use the alternate form of the approximation formula to calculate the approximate value of $f(2.003) = (2.003)^3$.

Solution Let $x_0 = 2$, $\Delta x = 0.003$. Then

$$f(2.003) \approx f(2) + f'(2)\,\Delta x$$
$$(2.003)^3 \approx 2^3 + 3 \cdot 2^2(0.003)$$
$$= 8 + 12(0.003) = 8.036. \quad \square$$

THE DIFFERENTIAL

By the Liebniz notation, we can write the derivative of $y = f(x)$ as dy/dx. The early developers of the calculus treated the symbols dx and dy as if they were variables, freely multiplying and dividing by them. When Cauchy put the limit concept on a solid basis, he realized that the old concepts relating dx and dy were invalid. Because the manipulation of these symbols had proved of practical value, he redefined the concept to validate the manipulations.

The approximation formula

$$\Delta y \approx f'(x)\,\Delta x$$

is a good place to start. We shall define the differential dy so as to make it equal the right-hand side of this formula. (See Fig. 3-1b.)

More precisely, we let dx (called the *differential of* x) be an independent variable. Let $y = f(x)$ be a differentiable function of x. We define the *differential* dy by

Differential
$$dy = f'(x)\,dx.$$

EXAMPLE 6 Let $y = f(x) = 3x^2 - 2$. Then

$$dy = f'(x)\,dx = 6x\,dx.$$

If we evaluate dy when $x = 2$ and $dx = -1$, we get

$$dy = 6 \cdot 2(-1) = -12. \quad \square$$

Note that other symbols can be used for the variables. For example, if $y = 7t^3$ and $x = 5u^2$, then

$$dy = 21t^2\,dt \quad \text{and} \quad dx = 10u\,du.$$

Note also that since Δx and dx are both independent variables, we can use the approximation formula with $\Delta x = dx$, obtaining

$$\Delta y \approx f'(x)\,dx$$
$$\Delta y \approx dy.$$

EXAMPLE 7   Use differentials to work Example 2.

Solution   In differential notation, the approximation formula is

$$\Delta y \approx dy = f'(x)\, dx.$$

In this case we have $\Delta y \approx 2\pi x\, dx$. When $x = 3$ and $dx = -0.001$, we get

$$\Delta y \approx 2\pi \cdot 3 \cdot (-0.001) \approx -0.0188496. \ \square$$

The derivative $f'(x)$ can now be treated as if it were the quotient of two differentials. Since $dy = f'(x)\, dx$, then

Derivative as a Quotient

$$f'(x) = \frac{dy}{dx} = \frac{\text{differential of } y}{\text{differential of } x} \qquad \text{provided } dx \neq 0.$$

Thus, the Leibniz notation can be explained in terms of differentials.

DIFFERENTIAL FORMULAS

Each derivative formula has a corresponding differential formula. We derive, as an example, the formula for the differential of a product.

EXAMPLE 8   Let $u = f(x)$ and $v = g(x)$ be differentiable functions of x. Show that $d(uv) = u\, dv + v\, du$.

Solution   The differentials of u and v are

$$du = f'(x)\, dx \qquad \text{and} \qquad dv = g'(x)\, dx.$$

Therefore

$$\begin{aligned} d(uv) &= D_x(f(x)g(x))\, dx \\ &= [f(x)g'(x) + g(x)f'(x)]\, dx \\ &= f(x)g'(x)\, dx + g(x)f'(x)\, dx \\ &= u\, dv + v\, du. \ \square \end{aligned}$$

Exercises 3–3

1–4.   Calculate the exact value of ϵ in the formula $\Delta y = f'(x_0)\, \Delta x + \epsilon\, \Delta x$.

• 1 $f(x) = 4x + 5$ 2 $f(x) = 3 - 5x$

• 3 $f(x) = 2 - 2x - x^2$ 4 $f(x) = x^3$

5–10.   Use the approximation formula $\Delta y \approx f'(x_0)\, \Delta x$ to approximate the change in $y = f(x)$ for the given change in x.

• 5 $f(x) = x^2 - 2x + 1$, x changes from 2 to 1.99.

6   $f(x) = 3x^3 - 4x^2 + x - 5$, x changes from 1 to 1.02.

• 7 $f(x) = \dfrac{2}{x^2} - 3x$, x changes from -1 to -1.001.

Exercises 3-3

8 $f(x) = \dfrac{1}{1-x}$, x changes from 0 to 0.001.

•9 $f(x) = (3x^2 - 4x)^3$, x changes from 2 to 2.002.

10 $f(x) = \dfrac{x^2}{x+1}$, x changes from 1 to 1.004.

11–14. Use the approximation formula $f(x_0 + \Delta x) \approx f(x_0) + f'(x_0) \Delta x$ to approximate the given quantities.

•11 $\dfrac{1}{(0.98)^2}$ 　　　　　　　　　　**12** $\dfrac{1}{(1.95)^3}$

•13 $(3.002)^4$ 　　　　　　　　　　**14** $(1.01)^4 - \dfrac{1}{(1.01)^2}$

15–22. Calculate the differential dy.

•15 $y = x^2$ 　　　　　　　　　　**16** $y = x^{10} + x$

•17 $y = x^2 + 3x + 1$ 　　　　　**18** $y = (3x^2 - 5x + 2)^{-4}$

•19 $y = s + \dfrac{1}{s}$ 　　　　　　**20** $y = (t^2 + 3t)^5$

•21 $y = \dfrac{x}{x-1}$ 　　　　　　　**22** $y = \dfrac{x^2}{x+1}$

23–26. Evaluate dy for the given values of x and dx.

•23 $y = x - \dfrac{1}{x}$, $x = 2$, $dx = 3$ 　　**24** $y = 2x^2 - 3x + 1$, $x = 5$, $dx = \tfrac{1}{2}$

•25 $y = \dfrac{1}{x^2}$, $x = 2$, $dx = -0.5$ 　　**26** $y = x(x^2 + 1)$, $x = 2$, $dx = -2$

•27 Calculate the exact value of $(2.01)^3$ and the value obtained by use of the approximation formula with $x_0 = 2$, $\Delta x = 0.01$. What is the error if the approximation formula is used?

•28 The side of a square is measured as 10 in., with a possible error of not more than 0.01 in. Use differentials to estimate the maximum possible error for an area computed to be 100 in². (*Hint:* Let the "true" length be $x_0 + dx$, the measured length x_0.)

•29 The radius of a flat circular plate is 3 in. When the plate is heated, the radius increases to 3.02 in. Find the approximate change in the area of one side of the plate.

•30 The gas in a spherical balloon is heated, causing the diameter to increase from 12 to 12.1 ft. What is the approximate change in the surface area of the balloon? ($S = 4\pi r^2$.)

31–32. Let u and v be differentiable functions of x. Prove the differential formulas.

31 $d\left(\dfrac{u}{v}\right) = \dfrac{v\,du - u\,dv}{v^2}, \quad v \neq 0$

32 $d(u^n) = nu^{n-1}\,du$

•33 *Relative Change.* The ratio

$$\frac{\Delta y}{y} = \frac{f(x_0 + \Delta x) - f(x_0)}{f(x_0)}$$

is called the *relative change* in $y = f(x)$ corresponding to a change in x from x_0 to $x_0 + \Delta x$. In many applications it is more useful to know the relative change than the actual change. (For example, an increase in profit of $5000 in a year may be good for a small company but poor for a large one. The relative increase, however, furnishes a meaningful standard for judging the company.)

(a) Use the approximation formula to estimate the relative change in the area of a circle if the radius increases from r to $r + \Delta r$.

(b) Hold Δr fixed. Does the relative change in the area of a circle increase or decrease when the radius r increases?

3–4 The Chain Rule

You should reread Section 1–7 (Composition of Functions) before starting this section.

The most powerful of the derivative formulas is concerned with the derivative of a composite function. We have already developed one formula for a special type of composite function. If $y = u^n$ where u is a differentiable function of x, then by the general Power Rule D3,

$$\frac{d}{dx}(u^n) = nu^{n-1}\frac{du}{dx}.$$

Since $y = u^n$ and $dy/du = nu^{n-1}$, we can write this formula as

$$\frac{dy}{dx} = \frac{dy}{du}\frac{du}{dx}.$$

It can be shown that this last formula holds for an arbitrary composite function. If y is a differentiable function of u where u is a differentiable function of x, then y is a differentiable (composite) function of x. We shall show in Theorem 3–7 that the derivative of y with respect to x is given by the formula

Chain Rule | **D8** | $\dfrac{dy}{dx} = \dfrac{dy}{du}\dfrac{du}{dx}.$

3-4 The Chain Rule

Formula D8 is known as the *Chain Rule* since it allows us to calculate the derivative of a composite function by a chain of two steps:

Step 1 Calculate dy/du and du/dx.
Step 2 Multiply dy/du by du/dx to get dy/dx.

EXAMPLE 1   Let $y = 5u^6 - 3u^5 + 4u^2 + 5$, where $u = (3-x)/x^2$. Then

$$\frac{dy}{du} = 30u^5 - 15u^4 + 8u$$

and

$$\frac{du}{dx} = \frac{d}{dx}\left(\frac{3-x}{x^2}\right) = \frac{x^2(-1) - (3-x)(2x)}{x^4} \quad \text{(by D7)}$$

$$= \frac{-x^2 - 6x + 2x^2}{x^4} = \frac{x-6}{x^3}.$$

Thus,

$$\frac{dy}{dx} = \frac{dy}{du}\frac{du}{dx} = (30u^5 - 15u^4 + 8u)\frac{x-6}{x^3}$$

$$= \left[30\left(\frac{3-x}{x^2}\right)^5 - 15\left(\frac{3-x}{x^2}\right)^4 + 8\left(\frac{3-x}{x^2}\right)\right]\frac{x-6}{x^3}.$$

This answer is the same as the one obtained by first expressing y as a function of x and then calculating the derivative. □

We must be more careful in our formal statement of the Chain Rule than we indicated above. Let $y = f(u)$ where $u = g(x)$. Suppose that g is differentiable at x_0 and f is differentiable at $u_0 = g(x_0)$ (the value of u that corresponds to the value $x = x_0$). Let h be the composite function defined by

$$h(x) = f(g(x)).$$

We shall prove that

$$h'(x_0) = f'(u_0)g'(x_0) = f'(g(x_0))g'(x_0).$$

The key step in the proof of the Chain Rule requires the approximation formula

$$f(u_0 + \Delta u) - f(u_0) = f'(u_0)\Delta u + \epsilon \Delta u$$

where $\epsilon \to 0$ as $\Delta u \to 0$ and $\epsilon = 0$ if $\Delta u = 0$. You are advised to review the use of this formula in Section 3-3.

THEOREM 3-7   *(The Chain Rule)* Let h be the composite function defined by

$$h(x) = f(g(x))$$

where g is differentiable at x_0 and f is differentiable at $u_0 = g(x_0)$. Then h is differentiable at x_0 and

$$h'(x_0) = f'(u_0)g'(x_0) = f'(g(x_0))\,g'(x_0).$$

Proof Let $u = g(x)$. For each increment $\Delta x \neq 0$, let

$$\Delta u = g(x_0 + \Delta x) - g(x_0) \quad \text{(the change in } u\text{)}.$$

Then

$$g(x_0 + \Delta x) = g(x_0) + \Delta u = u_0 + \Delta u.$$

Observe that $\Delta u \to 0$ as $\Delta x \to 0$ (by Theorem 2-6) and that

$$\lim_{\Delta x \to 0} \frac{\Delta u}{\Delta x} = \lim_{\Delta x \to 0} \frac{g(x_0 + \Delta x) - g(x_0)}{\Delta x} = g'(x_0).$$

We now begin the calculation of $h'(x_0)$:

$$h'(x_0) = \lim_{\Delta x \to 0} \frac{h(x_0 + \Delta x) - h(x_0)}{\Delta x} = \lim_{\Delta x \to 0} \frac{f(g(x_0 + \Delta x)) - f(g(x_0))}{\Delta x}$$

$$= \lim_{\Delta x \to 0} \frac{f(u_0 + \Delta u) - f(u_0)}{\Delta x}.$$

At this point we use the approximation formula

$$f(u_0 + \Delta u) - f(u_0) = f'(u_0) \Delta u + \epsilon \Delta u$$

where $\epsilon \to 0$ as $\Delta u \to 0$. Since $\Delta u \to 0$ as $\Delta x \to 0$, we also must have $\epsilon \to 0$ as $\Delta x \to 0$. Thus,

$$h'(x_0) = \lim_{\Delta x \to 0} \frac{f(u_0 + \Delta u) - f(u_0)}{\Delta x} = \lim_{\Delta x \to 0} \frac{f'(u_0) \Delta u + \epsilon \Delta u}{\Delta x}$$

$$= \lim_{\Delta x \to 0} \left[f'(u_0) \frac{\Delta u}{\Delta x} + \epsilon \frac{\Delta u}{\Delta x} \right]$$

$$= f'(u_0) \cdot \lim_{\Delta x \to 0} \frac{\Delta u}{\Delta x} + \lim_{\Delta x \to 0} \epsilon \cdot \lim_{\Delta x \to 0} \frac{\Delta u}{\Delta x}$$

$$= f'(u_0) g'(x_0) + 0 \cdot g'(x_0) = f'(u_0) g'(x_0). \quad \blacksquare$$

EXAMPLE 2   Use the Chain Rule to calculate the derivative y' for

$$y = \left(\frac{5}{x^2}\right)^2 - 3\left(\frac{5}{x^2}\right) + 4 - 7\left(\frac{x^2}{5}\right).$$

Solution   The function can be simplified by the substitution $u = 5/x^2$, yielding

$$y = u^2 - 3u + 4 - \frac{7}{u}$$

$$= u^2 - 3u + 4 - 7u^{-1}.$$

3-4 The Chain Rule

By the Chain Rule,

$$y' = \frac{dy}{dx} = \frac{dy}{du}\frac{du}{dx}$$

$$= [2u - 3 + 0 - 7(-1)u^{-2}] \cdot (-10x^{-3})$$

$$= \left(2u - 3 + \frac{7}{u^2}\right)\left(\frac{-10}{x^3}\right)$$

$$= \left(\frac{10}{x^2} - 3 + \frac{7x^4}{25}\right)\left(\frac{-10}{x^3}\right) = -\frac{100}{x^5} + \frac{30}{x^3} - \frac{14x}{5}. \quad \square$$

$$u = 5x^{-2}$$

$$\frac{du}{dx} = -10x^{-3}$$

Substitutions and the Chain Rule

The Chain Rule is involved whenever we calculate the derivative of a function after making a substitution, as in Example 2. If we change the function $f(x)$ to the function $h(u)$ by a substitution $u = g(x)$, then

$$f'(x) = h'(u)g'(x) = h'(g(x))g'(x)$$

provided h and g are differentiable. If $y_0 = g(x_0)$, then

$$f'(x_0) = h'(u_0)g'(x_0).$$

EXAMPLE 3 Let $f(x) = 2(x^2 + 4)^8 + 5(x^2 + 4) + 2$. Use the Chain Rule to calculate $f'(1)$.

Solution Make the substitution $u = g(x) = x^2 + 4$. Then

$$f(x) = 2u^8 + 5u + 2 = h(u)$$

and

$$f'(x) = h'(u)g'(x)$$
$$= (16u^7 + 5) \cdot 2x.$$

Observe that $u = 5$ when $x = 1$, so that

$$f'(1) = (16 \cdot 5^7 + 5) \cdot 2 = 32 \cdot 5^7 + 10$$
$$= 2,500,010. \quad \square$$

EXAMPLE 4 Let g be a differentiable function. Use the Chain Rule to get a formula for the derivative of $g(2x^3 + 5)$ with respect to x.

Solution Let $u = 2x^3 + 5$. Then

$$g(2x^3 + 5) = g(u).$$

It follows from the Chain Rule that

$$D_x(g(2x^3 + 5)) = D_x(g(u)) = D_u(g(u)) \cdot D_x u$$
$$= g'(u)(6x^2)$$
$$= 6x^2 \cdot g'(2x^3 + 5).$$

This expression cannot be simplified further unless we know the function g. $\square$

As we shall see later, the Chain Rule is involved in four main types of problems:

Applications of the Chain Rule

(1) The calculation of derivatives of complicated functions (as in Examples 2, 3, and 4).

(2) The calculation of antiderivatives. Much of our work (after Chapter 6) will be concerned with finding and using *antiderivatives*— functions that have given functions for their derivatives. Many of the problems require one or more substitutions.

(3) The calculation of derivatives of implicit functions—functions defined by equations relating x and y. The equation $3x^2 + 5xy + 7y^3 = 8$, for example, defines y implicitly as a function of x. (In the Section 3-5 we consider techniques for calculating the derivatives of such functions.)

(4) The generalization of derivative formulas. We shall subsequently derive numerous formulas (Chapters 8 and 9). Almost all of them can be generalized to cover a much broader range of functions by means of the Chain Rule.

Remark The Leibniz notation makes the Chain Rule

$$\frac{dy}{dx} = \frac{dy}{du}\frac{du}{dx}$$

appear to be the consequence of laws of fractions. It is not. Leibniz chose this notation because it makes formulas such as the Chain Rule easy to remember. In other words, the notation was chosen because of the properties. The properties cannot be established from the notation.

Differentials can sometimes by used to simplify the Chain Rule. Suppose that $y = f(u)$ where $u = g(x)$, both functions being differentiable. There are two different ways by which we can calculate the differential dy in terms of x and dx:

(1) Calculate dy in terms of u and du; then calculate du in terms of x and dx and substitute the value in the first expression.

Differentials and the Chain Rule

(2) Write y as the composite function $y = h(x) = f(g(x))$ and calculate dy by use of the Chain Rule. We shall show that the two approaches yield the same result.

(1) If $y = f(u)$ where $u = g(x)$, then

$$dy = f'(u)\, du \text{ and } du = g'(x)\, dx.$$

Thus

$$dy = f'(u)\, du = f'(g(x))g'(x)\, dx.$$

(2) If $y = h(x) = f(g(x))$, then

$$dy = h'(x)dx = [f'(g(x))g'(x)]\, dx \quad \text{(by the Chain Rule)}$$
$$= f'(g(x))g'(x)\, dx.$$

Thus, the two methods give the same final result.

Exercises 3–4

Observe that the first of the above methods is much simpler than the second, since the Chain Rule becomes an automatic consequence of the differential formulas. Example 5 illustrates how this method can be used to calculate differentials and derivatives through several changes of variables.

EXAMPLE 5 Let $y = 2u^3$ where $u = \sqrt{3 + v^2}$ and $v = 1/x$. (a) Calculate dy in terms of dx. (b) Use these differentials to calculate the derivative dy/dx.

Solution (a) Observe that

$$dy = 6u^2\, du, \qquad du = \frac{v\, dv}{\sqrt{3 + v^2}}, \qquad \text{and} \qquad dv = -\frac{dx}{x^2}.$$

Then

$$dy = 6u^2\, du = 6u^2 \cdot \frac{v}{\sqrt{3 + v^2}}\, dv = 6u^2 \frac{v}{\sqrt{3 + v^2}} \left(-\frac{dx}{x^2}\right).$$

(b) Dividing by dx, we get

$$\frac{dy}{dx} = \frac{-6u^2 v}{x^2 \sqrt{3 + v^2}}.$$

If we wish, we can now substitute the expressions for u and v in the final answer and write dy/dx as a function of x. □

Exercises 3–4

1–10. Use the Chain Rule to calculate dy/dx.

•1 $y = u^2 - 2u + 1$, $u = (x^2 + 4)^2$

2 $y = (z^2 + 2z + 1)^3$, $z = 3x^2$

•3 $y = v^2 - \dfrac{2}{v}$, $v = x^2 - \dfrac{1}{x}$

4 $y = \dfrac{1}{u^2 - 1}$, $u = \dfrac{x^2}{x^2 + 3}$

•5 $y = v^2 - v$, $v = 3x^2 - 6x + 1$

6 $y = z^2 - z^{-2}$, $z = (1 - x^2)^3$

•7 $y = \dfrac{u^2 + 1}{u + 1}$, $u = \dfrac{1}{x} - \dfrac{x^2}{2}$

8 $y = \dfrac{2v}{v + 1}$, $v = x - \dfrac{1}{1 - x}$

•9 $y = (3u^2 - 1)^3$, $u = \dfrac{x + 1}{x - 1}$

10 $y = \dfrac{z + 1}{z - 1}$, $z = (x^2 + 1)^4$

11–14. Let f be a differentiable function. Calculate dy/dx. Leave your answer in terms of f and x. (*Hint:* First make a convenient substitution, as in Example 4.)

•11 $y = f(x^2 - 4)$

12 $y = f(x^3 - 2x + 1)$

•13 $y = f\left(\dfrac{x^2}{x^2 - 2}\right)$

14 $y = f((3 - x^2)^5)$

15–18. Suppose that ZIG and ZAG are functions of x and that $D_x(\text{ZIG}(x)) = -1/\text{ZAG}(x)$. Use the Chain Rule to calculate $f'(x)$. (*Hint:* Make a convenient substitution.)

• 15 $f(x) = \text{ZIG}(x^3 + x - 1)$

16 $f(x) = \text{ZIG}\left(\dfrac{x+2}{x-1}\right)$

• 17 $f(x) = \text{ZIG}(2x + 3)$

18 $f(x) = \text{ZIG}((x^2 - 5)^4)$

19–22. Make a convenient substitution $u = g(x)$ to simplify each of the functions; then calculate the derivative $f'(x)$ by the Chain Rule.

• 19 $f(x) = \left(\dfrac{x^2 - 1}{x}\right)^4 - 3\left(\dfrac{x^2 - 1}{x}\right)^2 + \left(\dfrac{x^2 - 1}{x}\right)$

20 $f(x) = \left(\dfrac{4x - 3}{x^3 - 1}\right)^2 + 2\left(\dfrac{4x - 3}{x^3 - 1}\right) + 2$

• 21 $f(x) = 2 \cdot \dfrac{3}{x^3} - 4 + \dfrac{5}{(3/x^3)^2}$

22 $f(x) = \dfrac{\left[3\left(\dfrac{1}{x^5}\right)^3 - 1\right]^2}{2\left(\dfrac{1}{x^5}\right)^4 + 5}$

23–26. Suppose that y is a function of x and that x and y are functions of u. Observe that if $dx/du \neq 0$, then the Chain Rule $\dfrac{dy}{du} = \dfrac{dy}{dx} \cdot \dfrac{dx}{du}$ can be written

$$\dfrac{dy}{dx} = \dfrac{dy/du}{dx/du}.$$

Use this fact to calculate dy/dx.

• 23 $y = u^3 + u,\ x = 3u - 5$

24 $x = u^3 - 3,\ y = u^5 + 4u - 1$

• 25 $y = u^3,\ x = u^4$

26 $x = \dfrac{u + 1}{u - 1},\ y = u^2$

27–28. Use differentials to calculate dy in terms of dx. Then calculate dy/dx.

• 27 $y = 5u^2 + 7u - 2,\ u = 4v - \dfrac{3}{v} + \dfrac{2}{v^2},\ v = \dfrac{2 - x}{2 + x}$

28 $y = 4w^2 - 5w,\ w = 3t^2 + 2t - 1,\ t = 5v + \dfrac{7}{v^2},\ v = 4x^2 - 3$

• 29 Let f, g, h be differentiable functions. State a generalization of the Chain Rule that can be used to calculate dy/dx if $y = f(w),\ w = g(s),\ s = h(x)$.

3-5 Implicit Differentiation. Derivatives of Algebraic Functions

Explicit and Implicit Functions

Most of the functions encountered thus far have been of the form $y = f(x)$. Such an equation is said to define y *explicitly* as a function of x. On the other hand, an equation relating x and y, such as

$$xy - 3 + 7x = 0,$$

is said to define y *implicitly* as a function of x provided it can be solved for y as a function of x. Since the preceding equation can be solved for y to give

$$y = \frac{3 - 7x}{x},$$

the original equation defines y implicitly as a function of x. In other cases it may be difficult or impossible to write an explicit expression for y; for instance, only the implicit expression is available for

$$x^5 + y^5 = 3xy^4.$$

Another example will warn you of a potential pitfall. Consider the equation

$$x^2 + y^2 = 4.$$

If we solve this equation for y, we get

$$y = \pm\sqrt{4 - x^2},$$

which is not a function of x. The equation does, however, define two functions of x:*

$$y = f_1(x) = \sqrt{4 - x^2} \quad \text{and} \quad y = f_2(x) = -\sqrt{4 - x^2}.$$

The graph of the first function is the upper semicircle and the graph of the second is the lower semicircle pictured in Figure 3-3a.

The following method, called *implicit differentiation*, enables us to calculate the derivative of a function defined implicitly by an equation. The process consists of three steps:

Steps in Implicit Differentiation

1. Write the equation that defines y as an implicit function of x.
2. Differentiate both sides of the equation, treating y as a differentiable function of x, and using the Chain Rule wherever functions of y appear.
3. Solve the resulting equation for $y' = dy/dx$.

---

* Actually, the equation $x^2 + y^2 = 4$ defines an infinite number of functions of x. The two that we consider are the only ones that are continuous for $-2 \le x \le 2$.

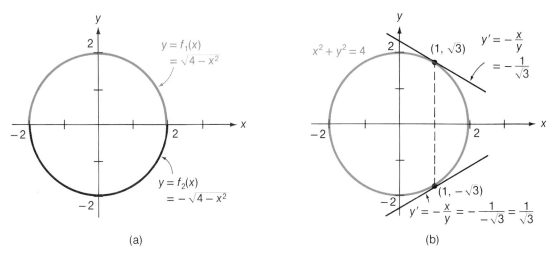

Figure 3-3 The equation $x^2 + y^2 = 4$ defines two functions of x.

EXAMPLE 1 Let $xy - 3 + 7x = 0$. Use implicit differentiation to calculate $y' = dy/dx$.

Solution We differentiate both sides of the defining equation, treating y as a function of x. The steps are:

1. Write down the equation:

   $xy - 3 + 7x = 0.$

2. Differentiate both sides of the equation with respect to x:

   $$\frac{d}{dx}(xy - 3 + 7x) = \frac{d}{dx}(0)$$

   $$\frac{d}{dx}(xy) - \frac{d}{dx}(3) + \frac{d}{dx}(7x) = \frac{d}{dx}(0) \quad \text{(by D4)}$$

   $$x\frac{dy}{dx} + y\frac{dx}{dx} - 0 + 7 = 0 \quad \text{(by D6, D1, D2)}$$

   $$xy' + y + 7 = 0.$$

3. Solve for y':

   $$y' = \frac{-y - 7}{x}$$

This problem can also be solved by first solving the equation for y and then differentiating. We obtain

$$y = \frac{3 - 7x}{x} = 3x^{-1} - 7;$$

$$y' = \frac{d}{dx}(3x^{-1} - 7) = 3(-1)x^{-2} = \frac{-3}{x^2}.$$

3–5 Implicit Differentiation. Derivatives of Algebraic Functions

To reconcile this answer with the one obtained in Step 3, we substitute the value of y into the expression for y':

$$y' = \frac{-y - 7}{x} = \frac{-[(3 - 7x)/x] - 7}{x} = \frac{-3}{x^2}$$

Thus, the two methods give the same answer. □

EXAMPLE 2 Let $x^2 + y^2 = 4$. Use implicit differentiation to calculate $y' = dy/dx$.

Solution
$$x^2 + y^2 = 4$$

$$\frac{d}{dx}(x^2 + y^2) = \frac{d}{dx}(4)$$

$$2x + 2y \cdot \frac{dy}{dx} = 0 \quad \text{(by the Chain Rule or the general Power Rule)}$$

$$y' = \frac{dy}{dx} = -\frac{x}{y}. \quad \square$$

Derivative Evaluated at a Point

How do we interpret the result obtained in Example 2? We know that the equation $x^2 + y^2 = 4$ defines y as two continuous functions of x. Which one of them have we differentiated? Actually, we have differentiated both. To evaluate the derivative $y' = -x/y$ at a point, we must know both x and y. Thus, we can use the result at any point (x, y) with $y \neq 0$ on either of the graphs

$$y = f_1(x) = \sqrt{4 - x^2} \quad \text{or} \quad y = f_2(x) = -\sqrt{4 - x^2}.$$

For example, at the point $(1, \sqrt{3})$ on the upper semicircle we have

$$y' = -\frac{x}{y} = -\frac{1}{\sqrt{3}}$$

while at the point $(1, -\sqrt{3})$ on the lower semicircle we have

$$y' = -\frac{1}{-\sqrt{3}} = \frac{1}{\sqrt{3}}.$$

(See Fig. 3–3b.)

In most cases, the graph of an equation consists of several branches. When we calculate dy/dx by implicit differentiation, we need to know which branch of the graph the particular point is on. Thus, we must generally know both coordinates of a point to get the value of the derivative there.

POWER RULE FOR FRACTIONAL EXPONENTS

We now make an important extension of the Power Rule D3. We shall show that

$$D_x(u^k) = ku^{k-1}D_xu$$

when k is a rational number, provided that u is a differentiable function of x and that u, u^k, and u^{k-1} are real functions.

We start with the function

$$y = x^{m/n}$$

where m and n are integers, $n > 0$, and $x^{m/n}$ is real. If we raise both sides of this equation to the power n, we get

$$y^n = x^m.$$

We now assume that y is a differentiable function of x and differentiate implicitly, obtaining

$$D_x(y^n) = D_x(x^m)$$
$$ny^{n-1}D_xy = mx^{m-1}.$$

Thus,

$$D_xy = \frac{m}{n}\frac{x^{m-1}}{y^{n-1}} = \frac{m}{n}\frac{x^{m-1}}{(x^{m/n})^{n-1}} = \frac{m}{n}\frac{x^{m-1}}{x^{m-m/n}}$$

$$= \frac{m}{n}x^{m-1-m+m/n} = \frac{m}{n}x^{m/n-1}.$$

Since $y = x^{m/n}$, then

$$D_x(x^{m/n}) = D_xy = \frac{m}{n}x^{m/n-1}.$$

Therefore, the formula

$$D_x(x^k) = kx^{k-1}$$

holds when $k = m/n$.

There is one gap in this argument. To use implicit differentiation, we must know that $y = x^{m/n}$ is differentiable. Although we do not now have the machinery to prove this fact, we shall establish in Theorem 3–11 that under suitable restrictions that make the function

$$f(x) = x^{1/n}$$

and its derivative real, it is differentiable. Thus, its mth power also is differentiable.

Having established that $D_x(x^n) = nx^{n-1}$ when n is a rational number, it is simple to use the Chain Rule and get the more general formula

$$D_x(u^n) = D_u(u^n) D_xy = nu^{n-1} D_xu \quad (n \text{ rational}).$$

This formula extends the General Power Rule D3

General Power Rule

D3 $\quad D_x(u^n) = nu^{n-1} D_xu \quad (n \text{ rational})$

provided u is differentiable and u, u^n, and u^{n-1} are real functions.

3–5 Implicit Differentiation. Derivatives of Algebraic Functions

EXAMPLE 3 Calculate $D_x(\sqrt{x^2 - 5})$.

Solution Let $u = x^2 - 5$. Then

$$D_x(\sqrt{x^2 - 5}) = D_x(u^{1/2}) = \frac{1}{2} u^{-1/2} D_x u \quad \text{(by D3)}$$

$$= \frac{1}{2}(x^2 - 5)^{-1/2} \cdot 2x = \frac{x}{\sqrt{x^2 - 5}}.$$

This formula holds for all x for which $x^2 - 5 > 0$, that is, for $x > \sqrt{5}$ and $x < -\sqrt{5}$. □

DERIVATIVES OF ALGEBRAIC FUNCTIONS

Any algebraic function can be differentiated by successively applying the general Power Rule D3 along with the other derivative formulas. It is only necessary to study the problem anew at each step to determine which formula to use.

EXAMPLE 4 Differentiate

$$f(x) = \sqrt{3x^7 - 2 + \frac{1}{\sqrt{x^2 - 1}}}.$$

Solution The basic form of the function is $f(x) = u^{1/2}$, where

$$u = 3x^7 - 2 + \frac{1}{\sqrt{x^2 - 1}}.$$

Thus, the first step is to use the General Power Rule, obtaining

$$f'(x) = \frac{1}{2} u^{-1/2} \frac{du}{dx}$$

where

$$\frac{du}{dx} = \frac{d}{dx}\left(3x^7 - 2 + \frac{1}{\sqrt{x^2 - 1}}\right)$$

$$= 21x^6 - 0 + \frac{d}{dx}\left(\frac{1}{\sqrt{x^2 - 1}}\right).$$

The final derivative can be calculated by the general Power Rule D3:

$$\frac{d}{dx}\left(\frac{1}{\sqrt{x^2 - 1}}\right) = \frac{d}{dx}(x^2 - 1)^{-1/2} = -\frac{1}{2}(x^2 - 1)^{-3/2}(2x) = \frac{-x}{(x^2 - 1)^{3/2}}.$$

The derivative of the original function is

$$\frac{1}{2} u^{-1/2}\left[21x^6 - \frac{x}{(x^2 - 1)^{3/2}}\right] = \frac{1}{2}\left(3x^7 - 2 + \frac{1}{\sqrt{x^2 - 1}}\right)^{-1/2}\left[21x^6 - \frac{x}{(x^2 - 1)^{3/2}}\right]. \quad □$$

Exercises 3-5

1–6. Find at least one function defined implicitly by each of the equations. What is the domain of the function?

- 1. $x + y - xy + 2 = 0$
- 2. $y^2 + 2x^2y - 2x^2 - 1 = 0$
- 3. $y^2 = 2xy + 3x^2$
- 4. $y^2 - x - 1 = 0$
- 5. $x^2y + 1 = x^3 + 4y + 2x$
- 6. $4y^2 = 1 + x^2y^2$

7–16. Assume that the given equation defines y as a differentiable function of x. Use implicit differentiation to calculate dy/dx.

- 7. $xy - x = 3$
- 8. $4x^2 + 9y^2 = 36$
- 9. $x^2 - xy + y^2 = 3$
- 10. $x^{-3} + y^{-1} = 4$
- 11. $x^2 - 16y^2 = 16$
- 12. $y + xy - x^2 = 3$
- 13. $x^{2/3} + y^{2/3} = 16$
- 14. $\sqrt{x} + \sqrt{y} = 1$
- 15. $xy^3 + 3x^2 + 1 = y^3 + 3x + x^3$
- 16. $x^2 + \sqrt{xy} + 2y = 1$

17–26. Use the general Power Rule and the other derivative formulas to calculate the derivatives of the functions.

- 17. $f(x) = 2^{1/6} + x^{3/2} + 3x^{4/3}$
- 18. $g(z) = \dfrac{1}{\sqrt{z}}$
- 19. $f(r) = 3r^{1/3} + 2r^{1/2} - 2r^{-1/2} + 3$
- 20. $h(w) = \left(w^2 - \dfrac{1}{w}\right)^{1/3}$
- 21. $f(t) = (3t^2 - 5t + 2)^{1/3}$
- 22. $h(x) = \sqrt{x^3 - 2x^2 + x}$
- 23. $h(r) = \sqrt{\dfrac{r+1}{r-1}}$
- 24. $g(t) = \dfrac{1}{(t^2 + 1)^{1/3}}$
- 25. $f(z) = \dfrac{\sqrt{16 - z^2}}{z}$
- 26. $f(x) = \sqrt{x^2 + 1}\,\sqrt[3]{x - 1}$

27–32. Find an equation of the tangent line to the graph of $y = f(x)$ at the indicated point. Use implicit differentiation.

- 27. $x^2 + y^2 = 25$, $P(3, 4)$
- 28. $x^2 + xy + y^2 = 7$, $P(2, 1)$
- 29. $x^2y - xy^2 + y - 8 = 0$, $P(-1, 2)$
- 30. $xy - 15 = 0$, $P(3, 5)$
- 31. $4x^3 - 2y^3 + x - 3 = 0$, $P(1, 1)$
- 32. $x^2 - 2xy + y^2 = x$, $P(4, 2)$

33–38. Find all points at which the tangent line to the graph of f is horizontal.

- 33. $f(x) = \sqrt{x} + \dfrac{1}{\sqrt{x}} - 2$
- 34. $f(x) = \left(\dfrac{x-1}{x+1}\right)^{3/2}$
- 35. $f(x) = \sqrt{x^3 + 6x^2 - 15x + 8}$
- 36. $f(x) = \sqrt{2x - 1}\,\sqrt{2x + 1}$
- 37. $f(x) = x^{1/3} + x^{-1/3}$
- 38. $f(x) = x\sqrt{1 - x^2}$

39–44. Use the alternative form of the approximation formula of Section 3–3 to approximate the numbers.

- 39 $\sqrt{99}$
- 40 $\sqrt[3]{65}$
- 41 $\sqrt{63.98}$
- 42 $1/\sqrt[3]{8.01}$
- 43 $\sqrt[3]{27.03}$
- 44 $(3.996)^{-1/2}$

45 (a) Use the identity $|x| = \sqrt{x^2}$ and the Chain Rule to show that $D_x|x| = x/|x|$ if $x \neq 0$.

  (b) Generalize the above formula, by the Chain Rule, to a formula for $D_x|u|$, where u is a differentiable function of x.

3–6 Related Rates

Rates of change furnish a meaningful application of implicit differentiation. Recall that if x changes with time, then dx/dt is the rate of change of x with respect to time. Suppose that another variable y is related to x by an equation. For example, x and y might satisfy the equation

$$x^2 + y^2 = 100.$$

We can calculate dy/dt in terms of dx/dt, x, and y by differentiating both sides of the equation with respect to t. In this case, we get

$$\frac{d}{dt}(x^2 + y^2) = \frac{d}{dt}(100)$$

$$2x\frac{dx}{dt} + 2y\frac{dy}{dt} = 0 \quad \text{(by the Chain Rule)}$$

$$\frac{dy}{dt} = -\frac{x}{y}\frac{dx}{dt}.$$

If x, y and dx/dt are known, we can calculate dy/dt.

The general steps in solving a related rate problem are as follows:

Steps in Solution of Related Rate Problems

1. Write down the equation that relates x and y.
2. Differentiate the equation implicitly with respect to time, obtaining an equation in x, y, dx/dt, and dy/dt.
3. Solve this last equation for dy/dt, evaluating dx/dt, x, and y at the required time.

EXAMPLE 1 A 10-ft ladder is propped against a wall. The foot of the ladder is pulled away from the wall at the rate of 1 ft/sec. How fast is the top of the ladder sliding down the wall at the instant when the foot of the ladder is 8 ft from the wall?

Solution Let x be the distance from the foot of the ladder to the wall and y the distance from the floor to the top of the ladder. (See Fig. 3–4.) We are given that $dx/dt = 1$ ft/sec. We want to find dy/dt at the instant when $x = 8$.

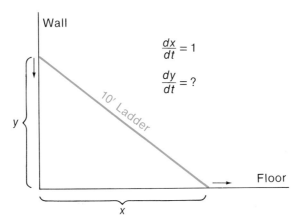

Figure 3–4 Example 1.

Step 1 Write down an equation involving x and y. By the Pythagorean Theorem

$$x^2 + y^2 = 10^2.$$

Step 2 Differentiate this equation with respect to t, treating x and y as functions of t.

$$\frac{d}{dt}(x^2 + y^2) = \frac{d}{dt}(10^2)$$

$$2x \cdot \frac{dx}{dt} + 2y \cdot \frac{dy}{dt} = 0 \quad \text{(by the Chain Rule)}.$$

Step 3 Solve for dy/dt and evaluate this quantity for the time when $x = 8$.

$$\frac{dy}{dt} = -\frac{2x}{2y}\frac{dx}{dt} = -\frac{x}{y}\frac{dx}{dt}.$$

At the instant at which $x = 8$, we must have $y = 6$ (by the Pythagorean Theorem). We also know that $dx/dt = 1$. Thus,

$$\frac{dy}{dt} = -\frac{x}{y}\frac{dx}{dt} = -\frac{8}{6} \cdot 1 = -\frac{4}{3} \text{ ft/sec}.$$

The ladder is sliding down the wall at the rate of $\frac{4}{3}$ ft/sec at the instant at which $x = 8$. □

3-6 Related Rates

Remark
Positive and Negative Derivatives

That dy/dt is negative in Example 1 indicates that y decreases as t increases. (The top of the ladder is sliding down the wall.) A positive derivative generally indicates a function that increases in value; and a negative derivative, one that decreases. (This subject is discussed in more detail in Section 4-2.)

EXAMPLE 2   A spherical balloon is inflated at the rate of 10 cubic inches (in.³) per second. At what rate is the radius increasing at the instant at which the radius is 3 in.?

Solution   The equation relating the volume and the radius of a sphere is

$$V = \tfrac{4}{3}\pi r^3.$$

We are given that $dV/dt = 10$. We want to find dr/dt at the instant at which $r = 3$. We differentiate the above equation with respect to time and solve for dr/dt.

$$\frac{d}{dt}(V) = \frac{d}{dt}\left(\frac{4}{3}\pi r^3\right)$$

$$\frac{dV}{dt} = \frac{4}{3}\pi \cdot 3r^2 \frac{dr}{dt} \quad \text{(by the Chain Rule)}$$

$$\frac{dV}{dt} = 4\pi r^2 \frac{dr}{dt}$$

$$\frac{dr}{dt} = \frac{1}{4\pi r^2}\frac{dV}{dt}.$$

When $r = 3$,

$$\frac{dr}{dt} = \frac{1}{4\pi \cdot 3^2} \cdot 10 = \frac{10}{36\pi} \approx 0.088.$$

At the instant at which $r = 3$, the radius is increasing at the rate of approximately 0.088 in./sec. □

EXAMPLE 3   Two airplanes in flight cross above a town at 1:00 P.M. One plane travels east at 400 mph; the other, north at 300 mph. At what rate does the distance between the planes change at 3:00 P.M.?

Solution   Let A be the point at which the paths cross. Let x be the distance from the first plane to A and y the distance from the second plane to A at time t. Let z be the distance between the two planes. (See Fig. 3-5.) By the Pythagorean Theorem, $z^2 = x^2 + y^2$. If we differentiate this equation with respect to time, we get

$$\frac{d}{dt}(z^2) = \frac{d}{dt}(x^2 + y^2)$$

$$2z\frac{dz}{dt} = 2x\frac{dx}{dt} + 2y\frac{dy}{dt}$$

$$\frac{dz}{dt} = \frac{x(dx/dt) + y(dy/dt)}{z}.$$

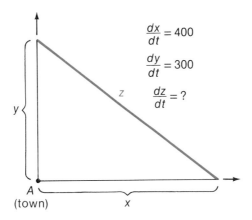

Figure 3-5 Example 3.

At 3:00 P.M., $x = 800$ and $y = 600$, so that

$$z = \sqrt{x^2 + y^2} = \sqrt{800^2 + 600^2}$$
$$= \sqrt{1{,}000{,}000} = 1000.$$

Thus,

$$\frac{dz}{dt} = \frac{x(dx/dt) + y(dy/dt)}{z}$$
$$= \frac{800 \cdot 400 + 600 \cdot 300}{1000} = \frac{500{,}000}{1000} = 500.$$

The planes are traveling apart at the rate of 500 mph at 3:00 P.M. □

Certain related rate problems can be solved by the direct use of the Chain Rule. This is the case when one variable depends on a second, which in turn depends on a third.

EXAMPLE 4 An airplane takes off from an airport at sea level and climbs at the constant rate of 5 m/sec. The outside air temperature T varies with altitude h according to the law

$$T = 15 - 0.0065h$$

where T is measured in degrees Celsius and h in meters above sea level. Find the rate of change of T with respect to time.

Solution Since $T = 15 - 0.0065h$ and $h = 5t$, then

$$\frac{dT}{dt} = \frac{dT}{dh}\frac{dh}{dt} = (-0.0065) \cdot 5 = -0.0325.$$

The outside air temperature decreases at the rate of 0.0325 °C per second. □

Exercises 3–6

1–4. Let x and y be differentiable functions of t. Calculate dy/dt at the instant when $x = 1$, given that $dx/dt = -2$ for all t.

• 1 $x + y = 2\sqrt{x}$

2 $x + y^2 = 5$

• 3 $x - xy = y + 2$

4 $x^2 + xy + y^2 = 2$

• 5 A 25-ft ladder leans against a wall with the foot of the ladder 7 ft from the base of the wall. The foot of the ladder is then pulled away from the wall at the rate of 1 ft/sec. How fast is the top of the ladder sliding down the wall 8 sec later?

• 6 Sand pouring from a pipe forms a conical pile with height equal to one-fourth the diameter of the base. If the sand pours out at the rate of 16 ft³/sec, how fast is the height increasing at the instant at which the pile is 4 ft high?

• 7 A man on a dock is pulling in a rope that is fastened to the bow of a small boat at the rate of 4 ft/sec. The man's hands are 12 ft higher than the point where the rope is attached to the boat. How fast is the boat approaching the dock when there is still 20 ft of rope out?

• 8 A spherical balloon is inflated at the rate of 12 ft³/min. At what rate is the surface area increasing at the instant at which the radius is 8 in.?

• 9 A boy is flying a kite at a constant height of 120 ft. The wind is causing the kite to move horizontally at the rate of 26 ft/sec. Find the rate at which the boy is paying out the string when there is 130 ft of string out. (Assume there is no sag in the string.)

• 10 A weather balloon is 100 ft above the ground and rising vertically at the rate of 4 ft/sec. A man passes beneath it, jogging along a straight road at the rate of 10 ft/sec. How fast is the distance between them changing 5 sec later?

• 11 A tank is shaped like a right circular cylinder with radius $r = 4$ ft. Water is pumped into the tank at the rate of 2 ft³/min. How fast does the surface rise?

• 12 Ship A leaves port at 2:00 P.M. and sails west at 8 mph. Ship B leaves the same port at 3:00 P.M. and sails south at 9 mph. At what rate are the ships moving apart at 5:00 P.M.?

• 13 A 6-ft-tall man walks at the rate of 4 ft/sec towards a lamp that is 18 ft above the ground. At what rate is the length of his shadow changing when he is 10 ft from the lamp post?

• 14 The weight of a person in the earth's gravitational field varies inversely with the square of his distance from the center of the earth. An astronaut weighed 200 lb at the surface of the earth. He is now in a rocket traveling away from the earth at the rate of 8 mi/sec. Find the rate at which his weight is changing with time (a) when the rocket is 1000 miles above the earth; (b) when it is 100 miles above the earth. (Assume that the radius of the earth is 4000 mi.)

• 15 Charles' law for an ideal gas states that $PV = kT$, where P is the pressure, V the volume, T the temperature, and k a constant of proportionality. The temperature of a gas is kept at 103°. Suppose that the pressure increases at the rate of 5 lb per in.² per min. At a certain instant the pressure is 20 lb/in.² and the volume is 22 in.³. At what rate is the volume changing at this instant?

●16 A piece of sculpture in the garden of a museum is a hemisphere of radius $r = 2$ ft, placed with its flat side down on a marble slab. An ice storm has coated the curved part of the hemisphere with a layer of ice 2 in. thick. Warm air is being blown on the sculpture, causing the ice to melt at the rate of 0.1 in./hr. Find the rate at which the volume of the ice changes with time at the instant at which it is 1 in. thick.

3-7 Vertical Tangent Lines

Thus far we have considered only tangent lines that are nonvertical. We now extend our work to vertical tangent lines. To illustrate the procedure consider the function $f(x) = x^{2/3}$. (See Fig. 3-6.) The derivative is

$$f'(x) = \frac{2}{3x^{1/3}}, \qquad x \neq 0.$$

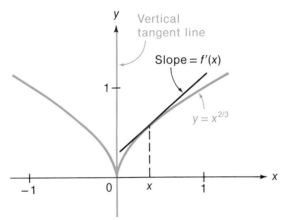

Figure 3-6 Example 1. $f(x) = x^{2/3}$ and $\lim_{x \to 0} |f'(x)| = \infty$. The graph of f has a vertical tangent line at $x = 0$.

Observe that if x is allowed to approach zero, then the slope of the tangent line gets larger and larger in absolute value and the tangent line becomes more and more inclined toward the vertical. In this situation, it seems natural that the graph should have a vertical tangent line at $x = 0$. This leads us to the following definition.

DEFINITION

Vertical Tangent Line

Let f be continuous on an open interval that contains c. Let $f'(x)$ exist at each point of the interval different from c. We say that the vertical line $x = c$ is *tangent to the graph of* f at $x = c$ provided that

$$\lim_{x \to c} |f'(x)| = \infty.$$

3-7 Vertical Tangent Lines

EXAMPLE 1 Let $f(x) = x^{2/3}$. Then

$$f'(x) = \frac{2}{3x^{1/3}}, \qquad x \neq 0.$$

The function f is continuous everywhere and is differentiable provided $x \neq 0$. Observe that

$$\lim_{x \to 0} |f'(x)| = \lim_{x \to 0} \left| \frac{2}{3x^{1/3}} \right| = \infty.$$

(See Fig. 3–6.) Thus, the graph of f has a vertical tangent line at $x = 0$. Observe in Figure 3–6 that a vertical tangent line may have properties different from the nonvertical tangent lines considered earlier. □

Vertical Tangent Line at Endpoint of the Domain

On occasion we relax the restrictions in the above definition. Suppose that f is continuous on the closed interval $[a, b]$ but not defined when $x < a$. (See Fig. 3–7.) In that case, we say that *the graph of f has a vertical tangent line at $x = a$* provided

$$\lim_{x \to a^+} |f'(x)| = \infty.$$

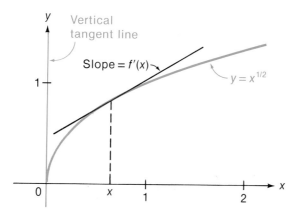

Figure 3–7 Example 2.

A similar rule holds if f is continuous on $[a, b]$ but not defined when $x > b$. In that case, we say that *the graph of f has a vertical tangent line at $x = b$* provided

$$\lim_{x \to b^-} |f'(x)| = \infty.$$

EXAMPLE 2 Let $f(x) = \sqrt{x}$. Then f is continuous on $[0, 1]$ but not defined (as a real function) when $x < 0$. The derivative is

$$f'(x) = \frac{1}{2\sqrt{x}}, \qquad x > 0.$$

Therefore

$$\lim_{x \to 0^+} |f'(x)| = \lim_{x \to 0^+} \left|\frac{1}{2\sqrt{x}}\right| = \infty.$$

The graph of f has a vertical tangent line at $x = 0$. (See Fig. 3–7.) □

Possible Locations for Vertical Tangent Lines

Observe that a vertical tangent line (if one exists at all) can occur only at a point at which f is continuous and the derivative f' is discontinuous. This observation can usually help us to find the possible locations for vertical tangent lines. For example, the function $f(x) = \sqrt{x}$, considered in Example 2, is continuous for all $x \geq 0$ and has a continuous derivative for all $x > 0$. Thus, $x = 0$ is the only possible location for a vertical tangent line.

Exercises 3–7

1–6. Find all vertical tangent lines (if any) to the graphs of these functions.

- 1 $f(x) = \sqrt{9x^2 - 36}$
- 2 $f(x) = \sqrt{144 - 36x^2}$
- 3 $f(x) = -\sqrt{x + 2}$
- 4 $f(x) = \sqrt{2x - 1}$
- 5 $f(x) = x^{1/3} + 1$
- 6 $f(x) = 5(x - 1)^{2/5} + 2$

7–10. The equations define one or more functions of x. Use implicit differentiation to find all vertical tangent lines.

- 7 $x^2 + y^2 = 100$
- 8 $4x^2 - 9y^2 = 36$
- 9 $x^{1/3} + y^{1/3} = 2$
- 10 $(x - 1)^{1/2} + (y - 2)^{1/2} = 1$

11 Explain why $f'(x)$ must be discontinuous at any point where the graph of f has a vertical tangent line.

3–8 Higher-Order Derivatives

Let $y = f(x)$. If f', the derivative of f, is itself a differentiable function, we can calculate its derivative, obtaining a new derived function called the *second derivative* of f, denoted by y'' or f''. Frequently the process may be repeated, yielding a third derivative, fourth derivative, and so on.

These higher-order derivatives are defined by

Higher Order Derivatives

$$y' = f'(x) = \frac{d}{dx}(f(x)) = \frac{dy}{dx} \quad \text{(first derivative)},$$

$$y'' = f''(x) = \frac{d}{dx}(f'(x)) = \frac{d}{dx}(y') \quad \text{(second derivative)},$$

3-8 Higher-Order Derivatives

$$y''' = f'''(x) = \frac{d}{dx}(f''(x)) = \frac{d}{dx}(y'') \quad \text{(third derivative)}.$$

EXAMPLE 1   If $y = f(x) = 3x^4 - 2x^3 + 4x^2 - 8x + 7$, then

$$\begin{aligned} y' &= f'(x) = 12x^3 - 6x^2 + 8x - 8 & \text{(first derivative)}, \\ y'' &= f''(x) = 36x^2 - 12x + 8 & \text{(second derivative)}, \\ y''' &= f'''(x) = 72x - 12 & \text{(third derivative)}, \\ y^{\text{iv}} &= f^{\text{iv}}(x) = 72 & \text{(fourth derivative)}, \\ y^{\text{v}} &= f^{\text{v}}(x) = 0 & \text{(fifth derivative)}. \end{aligned}$$

Note that the derivatives of order 5 or greater all equal zero. □

NOTATIONS

The second derivative is the derivative of the first derivative. Thus, it is

$$y'' = \frac{d}{dx}\left(\frac{dy}{dx}\right) = D_x(D_x y).$$

The similarity of these expressions to products has led mathematicians to use the symbols

$$\frac{d^2y}{dx^2} \quad \text{and} \quad D_x^2 y$$

for the second derivative. These are formal symbols and do not represent "squares of derivatives." Similarly the symbols

$$\frac{d^3y}{dx^3} \quad \text{and} \quad D_x^3 y$$

are used for third derivatives, and so on.

The symbols $f^{(n)}(x)$ and $y^{(n)}$ also are used for the nth derivative—especially when n is large. For example,

$$y^{(1)} = f^{(1)}(x) \text{ is the first derivative},$$
$$y^{(2)} = f^{(2)}(x) \text{ is the second derivative},$$

and

$$y^{(3)} = f^{(3)}(x) \text{ is the third derivative}.$$

EXAMPLE 2   If $y = x^{-1}$, then

$$y^{(1)} = y' = D_x y = \frac{dy}{dx} = \frac{d}{dx}(x^{-1}) = -x^{-2},$$

$$y^{(2)} = y'' = D_x^2 y = \frac{d^2y}{dx^2} = \frac{d}{dx}(-x^{-2}) = 2x^{-3},$$

$$y^{(3)} = y''' = D_x^3 y = \frac{d^3y}{dx^3} = \frac{d}{dx}(2x^{-3}) = -6x^{-4},$$

and

$$y^{(4)} = y^{iv} = D_x^4 y = \frac{d^4y}{dx^4} = \frac{d}{dx}(-6x^{-4}) = 24x^{-5}. \ \square$$

VELOCITY AND ACCELERATION

If $s(t)$ is a position function for straight-line motion, then

$$v(t) = \text{velocity at time } t = s'(t)$$

and

$$a(t) = \text{acceleration at time } t = v'(t).$$

Thus, the acceleration is the second derivative of the position function:

$$a(t) = v'(t) = s''(t).$$

EXAMPLE 3 If $s(t) = 3t^2 + \dfrac{1}{t+1}$, then

$$v(t) = s'(t) = D_t(3t^2 + (t+1)^{-1})$$

$$= 6t - (t+1)^{-2} = 6t - \frac{1}{(t+1)^2}$$

and

$$a(t) = s''(t) = D_t(6t - (t+1)^{-2})$$
$$= 6 + 2(t+1)^{-3}$$
$$= 6 + \frac{2}{(t+1)^3}. \ \square$$

HIGHER DERIVATIVES OF FUNCTIONS DEFINED IMPLICITLY

If we exercise care, we can calculate higher-order derivatives of functions defined implicitly. In calculating y'', we must remember to treat y and y' as functions of x. Similarly, in calculating y''', we treat y, y', and y'' as functions of x.

EXAMPLE 4 Calculate y' and y'' for $x^2 + y^2 = 1$.

Solution (a) $\quad D_x(x^2 + y^2) = D_x(1)$
$\qquad\qquad 2x + 2y \cdot y' = 0$

$$y' = -\frac{x}{y}.$$

(b)
$$y'' = D_x(y') = D_x\left(-\frac{x}{y}\right)$$
$$= -\frac{y\,D_x x - x\,D_x y}{y^2} = -\frac{y - xy'}{y^2} \quad \text{(by D7)}.$$

Recall from (a) that $y' = -x/y$. Thus,
$$y'' = -\frac{y - x(-x/y)}{y^2}$$
$$= -\frac{y^2 + x^2}{y^3}$$
$$= -\frac{1}{y^3} \quad \text{(since } x^2 + y^2 = 1\text{)}. \quad \square$$

Exercises 3–8

1–8. Calculate the first, second, and third derivatives of the functions.

- **1** $f(x) = 3x^2 - 4x + 1$
- **2** $f(t) = 2t^6 + 6t^2$
- **3** $g(y) = 9 - 4y^{-1} + 2y^{-2}$
- **4** $F(x) = x^{2/3} + 2x^{1/4}$
- **5** $g(y) = y^2 - \dfrac{1}{y}$
- **6** $h(x) = \sqrt{2x + 3}$
- **7** $f(x) = \sqrt[3]{2 - 3x}$
- **8** $f(t) = \left(t + \dfrac{1}{t}\right)^2$

9–14. Assume that y is a function of x. Use implicit differentiation to calculate y' and y''.

- **9** $x^2 + y^2 = 4$
- **10** $x^2 y - 2 = 0$
- **11** $\sqrt{x} + \sqrt{y} = 1$
- **12** $x^3 + y^3 = 1$
- **13** $x^3 - 3xy + y^3 = 1$
- **14** $x^2 y^2 + xy = 1$

15–22. **(a)** Let $s(t)$ be a position function for straight-line motion. Calculate $s'(t)$, the velocity function, and $s''(t)$, the acceleration function.
(b) Find all values of t for which the velocity is zero. Calculate the acceleration for these values.

- **15** $s(t) = 0.5gt^2 - t$
- **16** $s(t) = 3t^3 - 2t^2$
- **17** $s(t) = \sqrt{t} + 1/\sqrt{t}$
- **18** $s(t) = t^3 - 27t + 1$
- **19** $s(t) = t + \dfrac{4}{t},\ t < 0$
- **20** $s(t) = t^2 + \dfrac{16}{t},\ t > 0$
- **21** $s(t) = 1 + t - t^2 - t^3$
- **22** $s(t) = t + \dfrac{1}{2t^2}$.

23 Let $f(x) = 1/(1 - x)$.
   (a) Find the first four derivatives of f.
   •(b) Examine these derivatives to determine the general formula for the nth derivative.

•**24** Work Exercise 23 for $f(x) = \sqrt{x}$.

3–9 Derivatives of Inverse Functions. Root Functions

Properties of Inverse Functions

You should reread Section 1–8 (Inverse Functions) before starting this work.

Recall that functions f and g are inverses provided:

(1) The domain of f is the range of g, and the domain of g is the range of f,
(2) The functions "reverse" the effect of each other. That is, $y = f(x)$ if and only if $x = g(y)$. (See Fig. 1–32.)

The symbol f^{-1} is used for the inverse of f. It follows from (2) that

$$f(f^{-1}(x)) = x \quad \text{for } x \in \mathcal{D}_{f^{-1}}$$

and

$$f^{-1}(f(x)) = x \quad \text{for } x \in \mathcal{D}_f.$$

Recall also that f has an inverse if and only if it is *one-to-one*—that is, different elements of the domain of f always have different images. This is equivalent to stating that f has an inverse if and only if the equation $f(x_1) = f(x_2)$ has only the solution $x_1 = x_2$.

Finally, recall that the graph of f^{-1} is the reflection of the graph of f across the line $y = x$. (See Fig. 1–34.)

In the calculus we are concerned mainly with continuous functions. We shall see that continuous one-to-one functions are closely related to increasing and decreasing functions.

INCREASING AND DECREASING FUNCTIONS

Increasing and Decreasing Functions

The function f is said to *increase* on the interval I provided $f(x_1) < f(x_2)$ whenever x_1, x_2 are on I with $x_1 < x_2$. In other words, a function increases if its graph rises as we progress from left to right. (See Fig. 3–8a.)

Similarly, the function f *decreases* on the interval I provided $f(x_1) > f(x_2)$ whenever x_1, x_2 are on I with $x_1 < x_2$. In this case the graph "sinks" as we progress from left to right. (See Fig. 3–8b.)

It is easy to see that increasing and decreasing functions are one-to-one and thus have inverses. For example, let f be increasing on the interval I. Let $x_1, x_2 \in I$. If $x_1 \neq x_2$, then either $x_1 < x_2$, which implies that $f(x_1) < f(x_2)$, or $x_1 > x_2$, which implies that $f(x_1) > f(x_2)$. In either case, $f(x_1) \neq f(x_2)$, so that f is one-to-one.

Theorem 3–8 furnishes a converse to this result provided that the function is con-

3–9 Derivatives of Inverse Functions. Root Functions

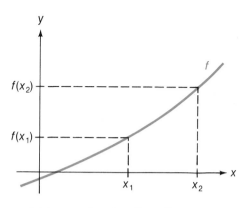
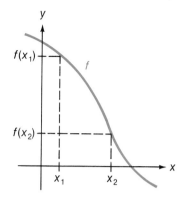

(a) Increasing function: if $x_1 < x_2$ then $f(x_1) < f(x_2)$

(b) Decreasing function: if $x_1 < x_2$ then $f(x_1) > f(x_2)$

Figure 3–8.

tinuous. That is, a continuous, one-to-one function must either increase or decrease. The proof, which depends on the Intermediate Value Theorem (Theorem 2–8), is left for the reader. (See Exercise 12.)

THEOREM 3–8 Let f be continuous and one-to-one on an interval I. Then f is either an increasing function or a decreasing function on I.

CONTINUITY AND DIFFERENTIABILITY

Our work with graphs shows that the graph of f^{-1} can be obtained by reflecting the graph of f across the line $y = x$. Consequently, if the graph of f has nice properties, we should expect the graph of f^{-1} to have corresponding nice properties. In particular, if the graph of f is unbroken, then the graph of f^{-1} should be unbroken. If the graph of f has a tangent line at the point $P(a, b)$, then the graph of f^{-1} should have a tangent line at the corresponding symmetric point $Q(b, a)$. These casual observations are stated formally in Theorems 3–9 and 3–10. The proofs can be found in the Appendix to this chapter.

THEOREM 3–9 Let f be continuous and one-to-one on the closed interval $[c, d]$. Then f^{-1} is continuous on the closed interval with endpoints $f(c)$ and $f(d)$.

A similar result holds on open intervals and on infinite intervals.

THEOREM 3–10 Let f be continuous and one-to-one on a closed interval that has a for an interior point. Let $f(a) = b$. Let $g = f^{-1}$ be the inverse of f. If f is differentiable at a and $f'(a) \neq 0$, then g is differentiable at b and

$$g'(b) = \frac{1}{f'(a)}.$$

Theorem 3–10 gives much more information than the mere fact that $g = f^{-1}$ is differentiable at b. The derivative of g at b is the reciprocal of the derivative of f at the corresponding point $(a, f(a))$. (See Fig. 3–9.)

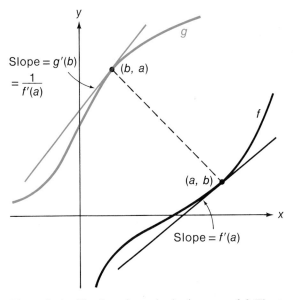

Figure 3–9 The function g is the inverse of f. The tangent line to the graph of g at (b, a) has slope equal to the reciprocal of the slope of the tangent line to the graph of f at the corresponding point (a, b).

DERIVATIVES OF ROOT FUNCTIONS

We can now fill a gap in our earlier work. We established in Section 3–5 that if the function $F(x) = x^{m/n}$ is differentiable, then its derivative is $F'(x) = (m/n)x^{m/n-1}$. We are now in a position to show that this function is differentiable, provided

(1) n is odd and $x \neq 0$, or
(2) n is even and $x > 0$.

Observe that it is enough to establish the result for the function $f(x) = x^{1/n}$, because $F(x) = x^{m/n} = [f(x)]^m$. Thus, if f is differentiable, it follows from D3 that F is also differentiable.

$f(x) = x^{1/n}$ is differentiable if $x > 0$.

Let $f(x) = x^{1/n}$ for $x > 0$. Let $g(x) = x^n$ for $x > 0$. Then f and g are inverse functions. Since g is differentiable at each point of its domain, it follows from Theorem 3–10 that f also is differentiable at each point of its domain.

A similar result holds if $x < 0$ provided n is odd. Thus, we have the following theorem.

THEOREM 3–11 Let $f(x) = x^{1/n}$. Then f is differentiable at $x = x_0$ provided

(1) n is odd and $x_0 \neq 0$, or
(2) n is even and $x_0 > 0$.

Exercises 3–9

1–2. Use the methods of Section 1–8 to show that f is one-to-one and to determine the law of correspondence for f^{-1}.

• 1 $f(x) = 3x + 1$ 2 $f(x) = x^5$

3 (a) Sketch the graph of a continuous, one-to-one function f such that $f(2) = -1$, $f(5) = 7$, $f'(2) = 3$.
   (b) Sketch the graph of the inverse function.
• (c) Let $g = f^{-1}$. Calculate $g(-1)$ and $g'(-1)$.

4 (a) Sketch the graph of a continuous, one-to-one function f such that $f(-1) = 4$, $f'(-1) = -2$, $f(2) = 0$, $f'(2) = 0$.
   (b) Sketch the graph of the inverse function.
   (c) Let $g = f^{-1}$. Calculate $g(4)$ and $g'(4)$. What about $g'(0)$?

5 (a) Sketch the graph of $f(x) = x^2$, $-3 \leq x \leq 3$.
   (b) Where is the function increasing? decreasing?
   (c) Is f one-to-one? Explain.

6 (a) Sketch the graph of a one-to-one function f defined on a closed interval that is neither an increasing function nor a decreasing function.
   (b) Explain why the function in (a) must be discontinuous.
   (c) Sketch the graph of f^{-1}.

7 Let f be a one-to-one function. Suppose that f^{-1} is not continuous. Can f be continuous? Explain.

8 Let f be a decreasing function defined on the closed interval $[a, b]$. Modify the argument that precedes Theorem 3–8 and prove that f is one-to-one.

• 9 The line $y = 7x - 10$ is tangent to the graph of a one-to-one function f at the point $(2, 4)$. What is the equation of the line tangent to the graph of the inverse function at the point $(4, 2)$?

10 Let f be a continuous, one-to-one function that is defined for every real number x. Must f^{-1} be defined for every real number x? If so, then explain. If not, then draw a graph that illustrates that fact.

11 Let f and g be one-to-one functions that are defined for every real number x.
   (a) Show that $f \circ g$ is one-to-one. (Hint: Suppose that $(f \circ g)(x_1) = f(g(x_1)) = (f \circ g)(x_2)$. Show that $x_1 = x_2$.)
   (b) Show that $(f \circ g)^{-1} = g^{-1} \circ f^{-1}$. (Hint: Let $y = (f \circ g)(x)$. Show that $(g^{-1} \circ f^{-1})(y) = x$. It may be convenient to let $u = g(x)$.]

12 Prove that if f is a continuous one-to-one function on the closed interval $[a, b]$, then f must be either an increasing function or a decreasing function on the interval. (Hint: If $f(a) < f(b)$, we must show that f is increasing on the interval. That is, if $a \leq$

$x_1 < x_2 \leq b$, then $f(x_1) < f(x_2)$. First show that $f(x_1)$ must be between $f(a)$ and $f(b)$. Then, by applying the same method to the interval $[x_1, b]$, show that $f(x_2)$ is between $f(x_1)$ and $f(b)$. To show that $f(x_1)$ is between $f(a)$ and $f(b)$, suppose that it is not. If $f(x_1)$ is greater than or equal to $f(b)$, apply the Intermediate Value Theorem to $[a, x_1]$ with $K = f(b)$. Show that there exists c, $a < c < x_1$, such that $f(c) = f(b)$—which is impossible if f is one-to-one. Hence, $f(x_1)$ cannot be greater than $f(b)$. Get a similar contradiction by assuming that $f(x_1) < f(a)$.)

Review Problems

1. State the definition of the derivative of the function f at $x = a$.

2. Give proper statements of the following derivative formulas: (a) the Product Rule; (b) the Quotient Rule; (c) the General Power Rule. (For what exponents does the General Power Rule hold?)

3–8. Calculate the derivative of y with respect to x.

- 3 $y = x^{-3} + 3x^{-2} + 5x^{-1} + 2$

- 4 $y = (x^3 - 3x^2 + 2x - 1)(x^5 + 2x^3 - x + 2)$

- 5 $y = \dfrac{3x^2 - 2}{x + 1}$

- 6 $y = \sqrt{u^2 - u}$ where $u = 3x^2 - 6x + 1$

- 7 $x^2y + xy^2 + 3x^2 - y^3 = 4$

- 8 $y = \sqrt{x + \sqrt{2x + \sqrt{3x}}}$

- 9 Find the points on the graph of
$$f(x) = \frac{\sqrt{x^2 - 1}}{x^2 + 1}$$
where the tangent line is (a) horizontal; (b) vertical.

- 10 Find the point on the graph of $y = (x + 2)^3/x^2$ at which the line $y = -4x + 24$ is tangent to the graph.

- 11 Use the approximation formula to approximate the change in
$$f(x) = [(x + 2)/(x - 1)]^2$$
when x changes from 2 to 1.98.

- 12 (a) Use the approximation formula to approximate $\sqrt{8.5}$ and $\sqrt[3]{8.5}$.

Review Problems

(b) Explain why it is best to use different values of x_0 and Δx in the two parts of (a).

13 Let $y = f(x)$ where f is a differentiable function. Define the differentials dx and dy.

•14 Calculate the differential dy for the function in (a) Exercise 3; (b) Exercise 5.

•15 The hypotenuse of a right isosceles triangle is measured as 4 in. with a possible error of no more than 0.02 in. Use the approximation formula to calculate the maximum possible error if the measured length is used to calculate the area.

•16 Make a convenient substitution $u = g(x)$ to simplify the following function f, then calculate its derivative $f'(x)$ by the Chain Rule:

$$f(x) = 7\left(\frac{5-x}{x^2}\right)^3 + \sqrt{\frac{5-x}{x^2}} - \frac{x^2}{5-x}.$$

•17 Find the equation of the line tangent to the graph of $x^{2/3} + y^{2/3} = 5$ at the point $P(1, 8)$.

•18 Use implicit differentiation to calculate y' and y'' for $x^3 + y^3 = 1$. Simplify the answers as much as possible.

•19 Find all vertical tangent lines to the graphs of the functions in (a) Problem 17; (b) Problem 18.

•20 A brick wall is located 60 ft from a 12-ft street light. A 6-ft-tall man is walking from the light directly towards the wall at the rate of 4 ft/sec. How fast is the man's shadow moving up the wall at the instant when he is 40 ft from the light?

•21 The ends of a water trough 10 ft long are isosceles triangles, with sides 3 ft and base (at the top of the triangle) 2 ft. Water is poured into the trough at the rate of 5 ft³/min. Find the rate at which the water level is rising at the instant when the water is 0.5 ft deep.

•22 A point P moves along the graph of the equation $y = x^2 - 3x + 4$ in such a way that the x-coordinate increases at the constant rate of 3 units/sec. Find the rate of change of the y-coordinate at the instant P passes through $(3, 4)$.

•23 Let $f(x) = \sqrt{2x - 1}$, $x > \frac{1}{2}$. Find the inverse function g. What is the domain of g? Sketch the graphs of f and g using the same set of axes.

24 A certain one-to-one function f has the property that $f'(x) = 1/x$ for all $x > 0$. Let g be the inverse function. Show that $g'(x) = g(x)$. (*Hint:* Let $y = g(x)$, so that $f(y) = x$. Use implicit differentiation.)

25 Let $u = f(x)$ and $v = g(x)$ where f and g are differentiable functions. Prove the differential formula $d(u + v) = du + dv$.

Appendix: Continuity and Differentiability of Inverse Functions

In this section we prove Theorems 3–9 and 3–10.

The proof of Theorem 3–9 requires the Intermediate Value Theorem (Theorem 2–8), which you should review at this time. It also requires the following proposition, (which is the converse of the condition for an increasing function): *Let f be an increasing function. If $f(x_1) < f(x_2)$, then $x_1 < x_2$.* (See Exercise 1.)

You should study Figure 3–10 carefully as you work through the proof of Theorem 3–9, locating the various points on the figure as they are defined.

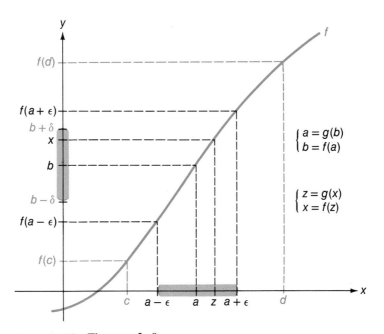

Figure 3–10 Theorem 3–9.

THEOREM 3–9 Let f be a one-to-one function that is continuous on the interval $[c, d]$. Let g be the inverse of f. Then g is continuous on the corresponding interval with endpoints $f(c)$ and $f(d)$.

Proof Recall from Theorem 3–8 that a continuous one-to-one function must be either an increasing or a decreasing function. We prove the theorem for the case in which f is an increasing function. The proof for the other case is similar. (See Exercise 3.)

Let b be in the domain of g. If $g(b) = a$, then $f(a) = b$. We consider the case in which $c < a < d$.

Appendix: Continuity and Differentiability of Inverse Functions

Let $\epsilon > 0$. We must find $\delta > 0$ such that $|g(x) - g(b)| < \epsilon$ whenever $|x - b| < \delta$. (See Exercise 33 of Section 2–6.)

Since $\epsilon > 0$, then $a - \epsilon < a < a + \epsilon$. Since f is an increasing function, then

$$f(a - \epsilon) < f(a) < f(a + \epsilon)$$

so that

$$f(a - \epsilon) < b < f(a + \epsilon).$$

We now choose a number $\delta > 0$ such that

$$f(a - \epsilon) < b - \delta < b < b + \delta < f(a + \epsilon).$$

Let x be any number such that $|x - b| < \delta$. Then

$$b - \delta < x < b + \delta$$

so that

$$f(a - \epsilon) < x < f(a + \epsilon).$$

It now follows from the Intermediate Value Theorem that there exists a number z between $a - \epsilon$ and $a + \epsilon$ such that $f(z) = x$. Since $a - \epsilon < z < a + \epsilon$, then $|z - a| < \epsilon$. Since $f(z) = x$, then $g(x) = z$. Thus, if $|x - b| < \delta$, then

$$|g(x) - g(b)| = |z - a| < \epsilon.$$

The proof is similar if a is an endpoint of the interval. In that case, we must work with one-sided limits. ∎

The proof of Theorem 3–10 is now comparatively easy.

THEOREM 3–10 Let f be a one-to-one function that is continuous on an interval containing a as an interior point. Let $f(a) = b$. Let g be the inverse of f on the interval. If f is differentiable at a and $f'(a) \neq 0$, then g is differentiable at b and

$$g'(b) = \frac{1}{f'(a)}.$$

Proof If we let $y = g(x)$, then $x = f(y)$. Similarly, since $f(a) = b$, then $g(b) = a$. Then

$$g'(b) = \lim_{x \to b} \frac{g(x) - g(b)}{x - b} = \lim_{x \to b} \frac{y - a}{f(y) - f(a)}$$

$$= \lim_{x \to b} \frac{1}{\frac{f(y) - f(a)}{y - a}}.$$

(See Exercise 4.) Since f is continuous, then by Theorem 3–8, g also is continuous. Therefore,

$$\lim_{x \to b} y = \lim_{x \to b} g(x) = g(b) = a,$$

so that $y \to a$ as $x \to b$. If we make this change in limits, we get

$$g'(b) = \lim_{x \to b} \frac{1}{\frac{f(y) - f(a)}{y - a}} = \lim_{y \to a} \frac{1}{\frac{f(y) - f(a)}{y - a}}$$

$$= \frac{1}{f'(a)}. \blacksquare$$

Remark The main result that we need from these theorems is that the inverse function g is differentiable at $b = f(a)$ provided the original function f is continuous on an interval that contains a, f is differentiable at a, and $f'(a) \neq 0$. As we shall see in the following argument, it is simple to calculate $g'(b)$ by implicit differentiation once we know that the derivative exists.

Given that $f'(a)$ and $g'(b)$ exist, we can show that

$$g'(b) = \frac{1}{f'(a)}$$

by an alternative argument. We let $y = g(x)$, so that $f(y) = x$. Using implicit differentiation gives

$$\frac{d}{dx}(f(y)) = \frac{d}{dx}(x)$$

$$f'(y)\frac{dy}{dx} = 1 \qquad \text{(by the Chain Rule)}$$

$$\frac{dy}{dx} = \frac{1}{f'(y)}.$$

Then

$$\frac{d}{dx}(g(x)) = \frac{1}{f'(y)} \qquad \text{(since } y = g(x))$$

$$g'(x) = \frac{1}{f'(y)}.$$

Since $y = a$ when $x = b$, then

$$g'(b) = \frac{1}{f'(a)}. \blacksquare$$

Exercises

1 Let f be an increasing function. Prove that if $f(x_1) < f(x_2)$, then $x_1 < x_2$. (*Remark:* This statement is not the same as the definition of an increasing function.)

2 Give the reason in the proof of Theorem 3–9 for the existence of the number $\delta >$

Exercises

0 satisfying the inequality

$$f(a - \epsilon) < b - \delta < b < b + \delta < f(a + \epsilon).$$

3 Prove Theorem 3–9 for the case in which f is a decreasing function.

4 It is assumed implicitly in the proof of Theorem 3–10 that if $x \neq b$, then $y \neq a$. Prove that this conclusion is indeed the case.

5 It is assumed implicitly in the proof of Theorem 3–9 that $c \leq a - \epsilon$ and $a + \epsilon \leq d$. What modifications must be made if either of these conditions is not met?

4 Applications of the Derivative

4–1 Introduction to Maximum and Minimum Problems

Maximum and Minimum

Many practical problems require us to find extreme values of functions. We want the maximum revenue, the minimum cost, the least amount of effort. The calculus often can be used to solve such problems.

Recall the following concepts from Section 2–7. We say that M is the *maximum* of the function f on the set S provided

(1) $f(x) \leq M$ for every $x \in S$, and
(2) $f(x_1) = M$ for at least one number $x_1 \in S$.

Similarly, m is the *minimum of f on S* provided

(1) $f(x) \geq m$ for every $x \in S$, and
(2) $f(x_2) = m$ for at least one number $x_2 \in S$.

(See Fig. 4–1.)

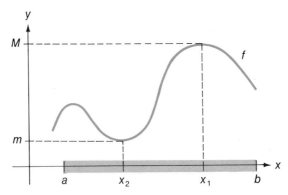

Figure 4–1 M is the maximum, m the minimum of f on the interval $[a, b]$.

4–1 Introduction to Maximum and Minimum Problems

Recall also from Theorem 2–7 that if I is a closed interval and f is continuous on I, then f must have a maximum and a minimum on I. This result does not necessarily hold if I is not a closed interval or if f is not continuous.

In the definitions of the terms *maximum* and *minimum* we did not restrict ourselves to closed intervals. Many of our applications will require us to find extreme values (if they exist) of functions defined on open intervals or infinite intervals. Thus, it is possible that these functions may fail to have the extreme values we want to find.

EXAMPLE 1   (a) The graph of $f(x) = x^3 - 6x^2 + 9x + 5$ is shown in Figure 4–2a. The maximum of f on $[0, 5]$ is 25, which is taken at $x = 5$. The minimum is 5, taken at $x = 0$ and $x = 3$. (Also see Example 3.)

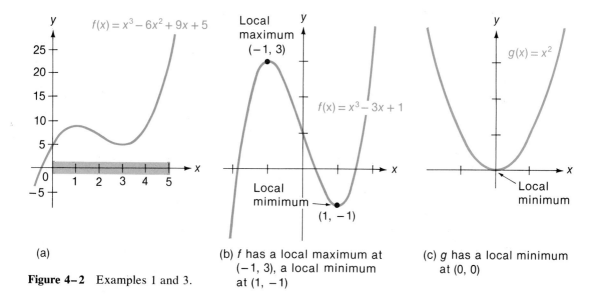

(a)

Figure 4–2   Examples 1 and 3.

(b) f has a local maximum at $(-1, 3)$, a local minimum at $(1, -1)$.

(c) g has a local minimum at $(0, 0)$.

(b) The graph of $f(x) = x^3 - 3x + 1$ is shown in Figure 4–2b. This function has no maximum and no minimum when x varies over the real numbers. The values increase without bound to the right of $x = 1$ (hence no maximum) and decrease without bound to the left of $x = -1$ (hence no minimum).

(c) The graph of $g(x) = x^2$ is shown in Figure 4–2c. This function has a minimum value of 0 at $x = 0$ but no maximum value. □

It is worth examining in more detail the function defined by $f(x) = x^3 - 3x + 1$ of Example 1(b). (See Fig. 4–2b.) Consider in particular the behavior of f in the vicinity of $x = 1$. Although $f(1) = -1$ is not the minimum value of f over the entire x-axis, it is the minimum when x is restricted to a small open interval about $x = 1$—the interval (0, 2), for example. We say in this case that f has a *local minimum* at $x = 1$. Similarly, f has a *local maximum* of 3 at $x = -1$. This is the maximum value of f on the interval $(-2, 0)$.

DEFINITION

Local Maximum and Minimum

The function f has a *local minimum* (or a *relative minimum*) at $x = c$ provided f is defined on an open interval I that contains c and $f(c)$ is the minimum value of f on I. (See Fig. 4–3a.)

Similarly, f has a *local maximum* (or a *relative maximum*) at $x = c$ provided f is defined on an open interval I that contains c and that $f(c)$ is the minimum value of f on I. (See Fig. 4–3a.)

The term *local extremum* refers to either a local maximum or a local minimum. (Much of the work of the next few sections is involved with finding local extrema.)

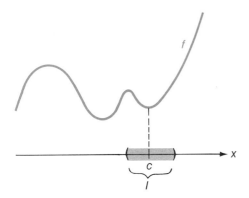

(a) f has a local minimum at $x = c$

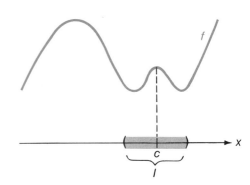

(b) f has a local maximum at $x = c$

Figure 4–3 Local extrema.

EXAMPLE 2 The graph in Figure 4–4 pictures a function with several local extrema. There are local maxima at x_1, x_3 and at each point of the closed interval $[x_6, x_7]$. There are local minima at x_2, x_5, and x_8 and at each point of the open interval (x_6, x_7). □

Possible Locations of Local Extrema

The lines drawn tangent to the graph in Figure 4–4 illustrate a very important point. *If a nonvertical tangent line exists at a local extremum, then it must be horizontal.* In other words, if f is differentiable at $x = c$ and has a local extremum there, then the derivative has the value zero. This result is established in Theorem 4–1.

THEOREM 4–1 **Let f have a local extremum at $x = c$. If f is differentiable at $x = c$, then $f'(c) = 0$.**

Proof We prove the theorem only for the case in which f has a local maximum at $x = c$. The other case is left for you to work out in Exercise 35.

4–1 Introduction to Maximum and Minimum Problems

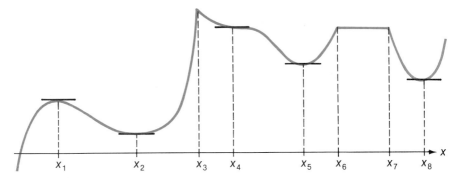

Figure 4–4 Example 2. A nonvertical tangent line at a local extremum must be horizontal.

If f has a local maximum at $x = c$, then there exists an open interval I containing c such that $f(x) \leq f(c)$ for every $x \in I$. Recall that

$$f'(c) = \lim_{x \to c} \frac{f(x) - f(c)}{x - c},$$

the limit being the same when x approaches c from the left or from the right. In calculating this limit, we are interested only in values of x that are "near" c. For the remainder of this argument, we restrict ourselves to values of x on the interval I.

Observe that if x is in I and is less than c, then

$$f(x) \leq f(c) \quad \text{and} \quad x < c$$

so that

$$\frac{f(x) - f(c)}{x - c} \geq 0.$$

(See Fig. 4–5a.) Since the difference quotient is nonnegative for all x near c and less than c, it cannot have a negative limit as x approaches c from the left. (See Theorem 2–4.) Thus,

$$f'(c) = \lim_{x \to c^-} \frac{f(x) - f(c)}{x - c} \geq 0.$$

Similarly, if x is in I and is greater than c, then

$$f(x) - f(c) \leq 0 \quad \text{and} \quad x - c > 0$$

so that

$$\frac{f(x) - f(c)}{x - c} \leq 0.$$

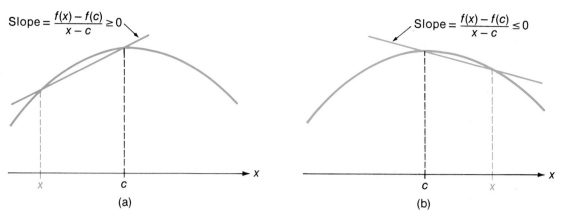

Figure 4–5 If f has a local maximum at $x = c$ and f is differentiable at $x = c$, then $f'(c) = 0$.

In this case

$$f'(c) = \lim_{x \to c^+} \frac{f(x) - f(c)}{x - c} \leq 0.$$

Since $f'(c)$ is both nonnegative and nonpositive, its value is zero. ∎

Remark Do not try to read too much into Theorem 4–1. We may have $f'(c) = 0$ even if f does not have a local extremum at $x = c$. For example, the function $f(x) = x^3$ has no local extremum at $x = 0$ even though $f'(x) = 3x^2$ has the value 0 at $x = 0$. (See Fig. 4–6a.)

Note also that the derivative may not exist at a local extremum. For example, the function $f(x) = x^{2/3}$ has a local minimum at $x = 0$, a point at which the derivative $f'(x) = 2/3x^{1/3}$ does not exist. (See Fig. 4–6b.)

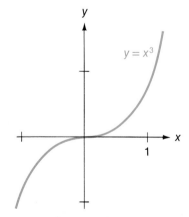
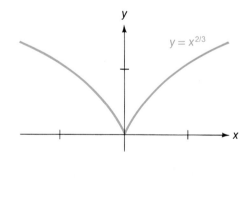

(a) $f(x) = x^3$ does not have a local extremum at the point where $f'(c) = 0$

(b) $f(x) = x^{2/3}$ has a local extremum at a point where the derivative is not defined

Figure 4–6.

4–1 Introduction to Maximum and Minimum Problems

STEPS FOR FINDING EXTREMA ON CLOSED INTERVALS

Theorem 4–1 is useful in finding extreme values of functions that are continuous over a closed interval. If f is such a function, then an extremum must occur at an interior point or at an endpoint. If it occurs at an interior point, then it is a local extremum and $f'(c)$ must be zero if it exists. These observations lead us to a procedure for finding extrema of functions that are continuous over closed intervals.

Let f be continuous on a closed interval $[a, b]$. To find the maximum (minimum) of f on the interval, we follow these steps:

Steps in Finding Extrema

(1) Calculate $f'(x)$.
(2) Find all x on the open interval (a, b) where $f'(x) = 0$.
(3) Find all x on the open interval (a, b) where $f'(x)$ does not exist.
(4) Calculate the values of f at the points found in (2) and (3) and at the endpoints a, b.
(5) The maximum (minimum) of f on $[a, b]$ is the largest (smallest) of the values calculated in (4).

Critical Points

The points at which a continuous function f has either a zero derivative or no derivative are called *critical points*. An extreme value of f on a closed interval must occur at a critical point or at one of the endpoints of the interval.

EXAMPLE 3 (See Example 1 (a).) Calculate the maximum and the minimum of $f(x) = x^3 - 6x^2 + 9x + 5$ on the closed interval $[0, 5]$. Find all local extrema on the interval.

Solution We follow the steps as outlined:

(1) $f'(x) = 3x^2 - 12x + 9 = 3(x - 1)(x - 3)$.
(2) $f'(x) = 0$ only in case $x = 1$ or $x = 3$. $\left.\right\}$ The critical points are
(3) $f'(x)$ exists everywhere on the interval. $\left.\right\}$ $x = 1$ and $x = 3$.
(4) We calculate the values of f at the critical points $x = 1$, $x = 3$ and at the endpoints $x = 0$, $x = 5$:

$f(0) = 5$ (value at left endpoint)
$\left.\begin{array}{l} f(1) = 9 \\ f(3) = 5 \end{array}\right\}$ (values at critical points)
$f(5) = 25$ (value at right endpoint).

(5) The minimum value occurs at the left endpoint and at $x = 3$. The maximum occurs at the right endpoint. The graph is shown in Figure 4–2a. Observe that there are local extrema at $x = 1$ and $x = 3$. □

EXAMPLE 4 Find all critical points for

$$f(x) = \frac{x}{x^2 + 2x + 10}.$$

Solution Since the denominator has no real zeros, then the function is continuous everywhere. The derivative is

$$f'(x) = \frac{(x^2 + 2x + 10)(1) - x(2x + 2)}{(x^2 + 2x + 10)^2} = \frac{-x^2 + 10}{(x^2 + 2x + 10)^2}.$$

The derivative exists everywhere and equals zero only at $x = \pm\sqrt{10}$, where the numerator is zero. The two critical points are $x = -\sqrt{10}$ and $x = \sqrt{10}$. □

Exercises 4–1

1–8. Find all critical points for f.

- 1 $f(x) = x^2 + x - 1$
- 2 $f(x) = \dfrac{\sqrt{x}}{x + 1}$
- 3 $f(x) = x^4 - 2x^2$
- 4 $f(x) = (3x^2 - 2x - 1)^4$
- 5 $f(x) = \sqrt[3]{x^2 + x - 2}$
- 6 $f(x) = 3x^4 + 10x^3 - 9x^2$
- 7 $f(x) = (x - 3)^6(x + 3)^2$
- 8 $f(x) = (x - 1)^3(x + 2)^4$

9–24. Find the maximum and minimum of f on the indicated interval.

- 9 $f(x) = 4x^2 + 8x;\ [-1, 2]$
- 10 $f(x) = x^2 - 2x - 2,\ [0, 3]$
- 11 $f(x) = 4x^3 + x^2 - 4x + 1;\ [-1, 2]$
- 12 $f(x) = 5 - 4x - 4x^2;\ [-1, 1]$
- 13 $f(x) = x^2 + 2x - 3;\ [0, 1]$
- 14 $f(x) = 2x^3 + 3x^2 - 12x - 5;\ [0, 3]$
- 15 $f(x) = x^4 - 2x^2;\ [0, 2]$
- 16 $f(x) = 2x^3 + 3x^2 - 36x;\ [0, 3]$
- 17 $f(x) = \dfrac{x^2}{x - 2};\ [3, 7]$
- 18 $f(x) = \dfrac{x - 1}{x^2 + 3};\ [-2, 2]$
- 19 $f(x) = x + \dfrac{1}{x};\ [\tfrac{1}{2}, 3]$
- 20 $f(x) = (x^2 - 1)^{2/3};\ [-1, 3]$
- 21 $f(x) = x^4 + 4x^3 + 4x^2 + 1;\ [-1, 1]$
- 22 $f(x) = x^{2/3};\ [-1, 8]$
- 23 $f(x) = 2x^5 + 5x^4 - 10x^3;\ [-1, 2]$
- 24 $f(x) = \dfrac{\sqrt{x}}{x^2 + 3};\ [0, 4]$

25–28. A particle moves along an axis with position $s(t)$ at time t. Find its maximum velocity on the indicated interval.

- 25 $s(t) = -\dfrac{t^3}{3} + 2t^2 + 5t + 1;\ [0, 4]$
- 26 $s(t) = 2t^3 + t^2 + 20t + 4;\ [-1, 1]$
- 27 $s(t) = \dfrac{t^4}{4} - \dfrac{27t^2}{2} + t + 1;\ [0, 4]$
- 28 $s(t) = t^2 + \dfrac{8}{t};\ [-4, -1]$

29–33. If the stated conditions are not contradictory, then sketch graphs of functions that satisfy them. If the conditions are contradictory, explain why.

●29 The function f is discontinuous at $x = 2$ and has a local minimum at $x = 2$, a local maximum at $x = 4$, and no other critical points.

30 The function f is continuous on the closed interval $[1, 5]$ and has its maximum value at $x = 1$, its minimum at $x = 5$, a local minimum at $x = 2$, a local maximum at $x = 4$, and no other critical points.

●31 The function f is continuous on the closed interval $[0, 6]$ and has its maximum value at $x = 3$, its minimum at $x = 6$, a local maximum at $x = 1$, and no other critical points.

32 The function f is differentiable on the closed interval $[0, 4]$ and has its maximum value at $x = 2$ and its minimum value at $x = 0$. The derivative $f'(x)$ equals 0 at $x = 0$ and 1 at $x = 2$.

●33 The function f is continuous everywhere and has a local maximum of 2 at $x = 0$, a local minimum of 4 at $x = 2$, and no other critical points.

34 Prove that a quadratic function has exactly one critical point.

35 Complete the proof of Theorem 4–1 by considering the case in which f has a local minimum at $x = c$.

36 This exercise refers to the definition of a local maximum. Suppose that $f(c)$ is the maximum value of f on an open interval I that contains c. Show that if J is any smaller open interval that contains c, then $f(c)$ is the maximum value of f on J. (In other words, the interval I is not unique. Any smaller interval will work just as well.)

4–2 Mean Value Theorem

Several of the applications of the derivative require us to determine properties of a function from information about its derivative. We might be asked, for example, questions about a position function s that has a given velocity function v. None of our work to this point has prepared us for this task.

The *Mean Value Theorem* is an important theorem that often can be used to determine properties of a function from properties of its derivative. A special application of this theorem is closely related to our work on local extrema.

ROLLE'S THEOREM

The following theorem, which is a special case of the Mean Value Theorem, is concerned with the behavior of a function g that is continuous on a closed interval $[a, b]$, differentiable on the corresponding open interval (a, b) and has the additional property that its values at the two endpoints a and b are equal. It is named after the seventeenth-century French mathematician Michel Rolle.

THEOREM 4–2   (*Rolle's Theorem*) Let g be continuous on the closed interval $[a, b]$ and differentiable on
Rolle's   the open interval (a, b). Suppose that $g(a) = g(b)$. Then there exists at least one number
Theorem c, with $a < c < b$, such that $g'(c) = 0$.

Proof   Since g is continuous on $[a, b]$, then g has a maximum and a minimum on $[a, b]$. Two cases arise: Either $g(x) = g(a)$ for all x on $[a, b]$ or g has an extremum different from $g(a)$.

Case 1   Suppose that $g(x) = g(a)$ on $[a, b]$. Then $g'(x) = 0$ for each x between a and b (Fig. 4–7a). We can choose any one of these numbers for c.

Case 2   Suppose that g has an extremum different from $g(a)$. Let this extremum be taken at c. Since $g(a) = g(b)$, then c must be between a and b; thus the extremum occurs at a critical point. (See Fig. 4–7b.) Therefore, $g'(c) = 0$, by Theorem 4–1.

In either Case 1 or Case 2 there exists a point c, with $a < c < b$, where $g'(c) = 0$. ∎

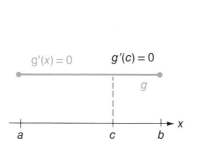
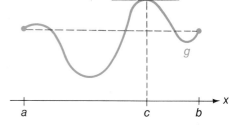

(a) *Case 1:* $g(x) = g(a)$ for $a \leq x \leq b$; then $g'(c) = 0$ for any c between a and b

(b) *Case 2:* g is not constant on $[a, b]$; then $g'(c) = 0$ at a local extremum

Figure 4–7   Rolle's theorem.

THE MEAN VALUE THEOREM

The Mean Value Theorem is concerned with a function f that is known to be continuous on $[a, b]$ and differentiable on (a, b). (We relax the restriction in Rolle's theorem that the functional values are equal at the endpoints a and b.) The Mean Value Theorem asserts that there is a point c between a and b at which the tangent line is parallel to the secant line connecting $(a, f(a))$ and $(b, f(b))$. Since the slope of the secant line is

$$m = \frac{f(b) - f(a)}{b - a},$$

this means that

$$\frac{f(b) - f(a)}{b - a} = f'(c)$$

for at least one number c between a and b. (See Fig. 4–8a.)

To prove the Mean Value Theorem, we construct an auxiliary function g that satisfies Rolle's theorem.

4-2 Mean Value Theorem

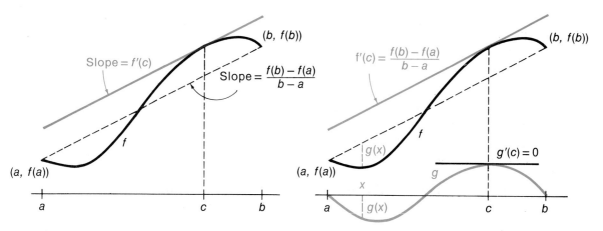

Figure 4-8 The Mean Value Theorem.

THEOREM 4-3

Mean Value Theorem

(*Mean Value Theorem*) **Let f be continuous on the closed interval $[a, b]$ and differentiable on the open interval (a, b). There exists at least one number c, where $a < c < b$, such that**

$$f'(c) = \frac{f(b) - f(a)}{b - a}.$$

Proof We define the auxiliary function g at each x in the domain of f by

$$g(x) = f(x) - \left[\frac{f(b) - f(a)}{b - a}(x - a) + f(a)\right]$$

Geometrically, g represents the directed vertical distance from the line connecting $(a, f(a))$ and $(b, f(b))$ to the graph of f. (See Fig. 4-8b.) Observe that g is continuous on $[a, b]$ and differentiable on (a, b) and has the property that

$$g(a) = g(b) = 0.$$

It follows from Rolle's theorem that there exists at least one number c, where $a < c < b$, such that $g'(c) = 0$. Since

$$g'(x) = f'(x) - \left[\frac{f(b) - f(a)}{b - a}(1 - 0) + 0\right]$$

then

$$g'(c) = f'(c) - \frac{f(b) - f(a)}{b - a} = 0.$$

It follows that

$$f'(c) = \frac{f(b) - f(a)}{b - a}. \blacksquare$$

Remark The statement

$$\frac{f(b) - f(a)}{b - a} = f'(c)$$

shows that the mean (average) rate of change of f over $[a, b]$ must equal the instantaneous rate of change at least once over the interval. For example, if a motorist travels 60 miles in one hour and his velocity is a continuous function of time, then he must have traveled at a speed of exactly 60 mph at one or more times during the hour.

INCREASING AND DECREASING FUNCTIONS

Recall that a function f *increases* on an interval I if

$$f(x_1) < f(x_2) \quad \text{whenever } x_1 < x_2 \text{ and } x_1, x_2 \in I$$

and *decreases* on I if

$$f(x_1) > f(x_2) \quad \text{whenever } x_1 < x_2 \text{ and } x_1, x_2 \in I.$$

(See Fig. 4–9.)

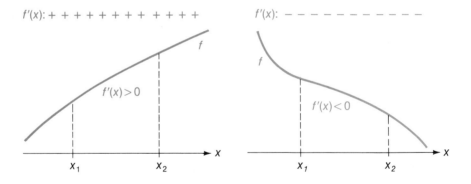

(a) Increasing function; if $x_1 < x_2$ then $f(x_1) < f(x_2)$

(b) Decreasing function; if $x_1 < x_2$ then $f(x_1) > f(x_2)$

Figure 4–9 Increasing and decreasing functions.

We shall use the Mean Value Theorem to show that a function with a positive derivative on an interval increases on the interval. Similarly, a function with a negative derivative on an interval decreases. We state the theorem for the special case in which the function is continuous on a closed interval $[a, b]$ and differentiable on the open interval (a, b). The theorem can be generalized to an arbitrary interval—open, closed, half-open, finite, or infinite.

4-2 Mean Value Theorem

THEOREM 4-4 Let f be continuous on $[a, b]$ and differentiable on (a, b). If $f'(x) > 0$ for each x on (a, b), then f is increasing on the interval $[a, b]$. If $f'(x) < 0$ for each x on (a, b), then f is decreasing on $[a, b]$.

Proof

Application to Increasing, Decreasing Functions

We consider only the case in which $f'(x) > 0$ on (a, b). The other case is left for you (Exercise 17).

To show that f increases on $[a, b]$, we must show that $f(x_1) < f(x_2)$ whenever $a \leq x_1 < x_2 \leq b$. We fix x_1 and x_2 in the range $a \leq x_1 < x_2 \leq b$ and apply the Mean Value Theorem to f on the interval $[x_1, x_2]$. It follows from Theorem 4-3 that for some number c, with $x_1 < c < x_2$, we have

$$\frac{f(x_2) - f(x_1)}{x_2 - x_1} = f'(c) > 0.$$

Since the denominator is positive, the numerator must be also. Thus,

$$f(x_2) > f(x_1).$$

Since x_1 and x_2 were arbitrary points on the interval, f increases on $[a, b]$. ∎

Theorem 4-4 is the basis for an important test for local extrema (discussed in Section 4-3).

OTHER APPLICATIONS OF THE MEAN VALUE THEOREM

It seems reasonable that only constant functions should have zero derivatives over intervals. Theorem 4-5 establishes this result.

THEOREM 4-5 Let f be continuous on the closed interval $[a, b]$ and differentiable on the open interval (a, b). Suppose that $f'(x) = 0$ for each x in (a, b). Then there exists a constant C such that

$$f(x) = C, \quad a \leq x \leq b.$$

Proof

A Function with Zero Derivative is Constant

Let x_0 be any fixed number in the interval $(a, b]$. Then f satisfies the conditions of the Mean Value Theorem on the interval $[a, x_0]$. Consequently, there exists a number c, with $a < c < x_0$, such that

$$\frac{f(x_0) - f(a)}{x_0 - a} = f'(c) = 0.$$

Thus, $f(x_0) = f(a)$ for each x_0 in the range $a < x_0 \leq b$. If we let $C = f(a)$, then

$$f(x) = C \quad \text{for each } x, \quad a \leq x \leq b. \blacksquare$$

Theorem 4-5 has an interesting corollary. Suppose that two functions have the same derivative at each point of an interval. It appears that their graphs should have identical shapes over the interval—that one graph should be a vertical translation of the other. Theorem 4-6 shows that this is indeed the case. The theorem is illustrated in Figure 4-10.

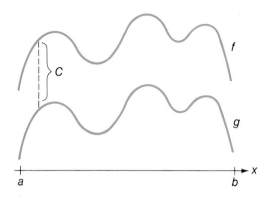

Figure 4-10 If f and g have the same derivative over an interval, then the functions differ by a constant (Theorem 4-6).

THEOREM 4-6

Functions with Same Derivative Differ by a Constant

Let f and g be continuous on $[a, b]$ and differentiable on (a, b). Suppose that

$$f'(x) = g'(x) \quad \text{for each } x, \quad a < x < b.$$

Then there exists a constant C such that

$$f(x) = g(x) + C, \quad a \leq x \leq b.$$

Proof Define the function h on the common domain of f and g by

$$h(x) = f(x) - g(x).$$

Then

$$h'(x) = f'(x) - g'(x) = 0 \quad \text{for } a < x < b.$$

By Theorem 4-5, there exists a constant C such that

$$h(x) = C \quad \text{if } a \leq x \leq b.$$

Thus,

$$f(x) - g(x) = C \quad \text{if } a \leq x \leq b,$$
$$f(x) = g(x) + C \quad \text{if } a \leq x \leq b. \blacksquare$$

Theorem 4-6 is the basic result needed for the theory of *antidifferentiation*. This theory involves finding all functions that have a given function as their derivative. Theorem 4-6 shows that any two antiderivatives of a given function must differ by a constant.

Exercises 4-2

1-10. Determine whether the Mean Value Theorem applies to the function on the given interval. If it applies, find an appropriate number c that satisfies the conclusion of the theorem. If it does not apply, explain why it does not. (Problem 7 involves the greatest integer function.)

4-3 First Derivative Test for Local Extrema

•1 $f(x) = x^2$ on $[0, 1]$

2 $f(x) = x^3 - x^2 + 2$ on $[-1, 1]$

•3 $f(x) = x + \dfrac{1}{x}$ on $[1, 2]$

4 $f(x) = x - \dfrac{1}{x}$ on $[-1, 2]$

•5 $f(x) = x^3 + 1$ on $[-2, 4]$

6 $f(x) = x^2 + 2x$ on $[-2, 0]$

•7 $f(x) = [x]$ on $[0, 1]$

8 $f(x) = x^{1/3}$ on $[0, 1]$

•9 $f(x) = x^{1/3}$ on $[-1, 1]$

10 $f(x) = \dfrac{x}{x + 1}$ on $[-2, 2]$

11 Verify in the proof of the Mean Value Theorem that $g(a) = g(b)$.

12 Let $f(x) = mx + n$, where m and n are constants. Prove that f satisfies the conditions of the Mean Value Theorem on every closed interval $[a, b]$ and that every number c between a and b satisfies the conclusion of the theorem.

13 Let $f'(x) = 2x + 3$. Use Theorem 4–6 to prove that $f(x) = x^2 + 3x + C$ for some constant C. (*Hint:* First show that $g(x) = x^2 + 3x$ has $f(x)$ for its derivative.)

14 Suppose that $f'(x) = x^2 - 2x + 1$ and that $f(1) = 4$. Use Theorem 4–6 to prove that $f(x) = x^3/3 - x^2 + x + 11/3$. (*Hint:* First show that $f(x) = x^3/3 - x^2 + x + C$; then use the fact that $f(1) = 4$ to determine C.)

•15 (a) A function f is known to be continuous on $[0, 2]$ and differentiable on the intervals $(0, 1)$ and $(1, 2)$. Must there exist a number c between 0 and 2 such that $f(2) = f(0) + 2f'(c)$? Explain.

(b) What if the function in (a) also is differentiable at $x = 1$?

16 An automobile travels 10 miles down a straight road at an average speed of 50 mph. Assume that the distance s is a differentiable function of time. Show that the automobile must have a velocity of exactly 50 mph at some point on the trip.

17 Complete the proof of Theorem 4–4. Let f be continuous on $[a, b]$ and differentiable on (a, b). Suppose that $f'(x) < 0$ for each x, in the interval (a, b). Prove that f decreases on $[a, b]$.

4-3 First Derivative Test for Local Extrema

We have examined maxima and minima of continuous functions defined over closed intervals (Section 4–1). In many problems we cannot restrict ourselves to such intervals. We may, for example, need to find the maximum of a function over the entire x-axis or the minimum over an open interval. We could formulate examples of such functions that do not have extreme values even though they may have local extrema. It is not hard to see, however, that if such a function does have an extreme value, then it occurs at a local extremum, or at an endpoint of an interval. (Recall the definition of a local extremum.)

In general it may be difficult to decide whether a given local extremum is an extreme value of the function. There is one special case, however, in which this can be established at once. This special case will enable us to solve many of our extremal problems.

THEOREM 4–7

Functions with a Single Critical Point

Let I be an interval (finite or infinite, open, half-open, or closed). Let f be continuous on I. Suppose that the only critical point for f on I occurs at the interior point $x = c$. If f has a local minimum (maximum) at c, then $f(c)$ is the minimum (maximum) of f on I.

The proof is omitted.

As we can see from Theorem 4–7, in many cases we need to know only whether a function has a local maximum (minimum) at a critical point in order to determine the maximum (minimum) of the function. An important test, the *first derivative test* can be used to test critical points for local extrema. A second test, the *three-point test*, is stated in Exercise 24.

Let f be continuous on an open interval (a, b) that contains the critical point c. Suppose that f increases on the interval (a, c) and decreases on the interval (c, b). (See Fig. 4–11a.) Then f must have a local maximum at c. Similarly, if f decreases on (a, c) and increases on (c, b), then f has a local minimum at c. (See Fig. 4–11b.) This leads us to the following test for local extrema.

THEOREM 4–8

First-Derivative Test

(*First Derivative Test*) Let f be continuous on an open interval (a, b) that contains the critical point c.
(1) If $f'(x) > 0$ for $a < x < c$ and $f'(x) < 0$ for $c < x < b$, then f has a local maximum at $x = c$ (See Fig. 4–11a.)
(2) If $f'(x) < 0$ for $a < x < c$ and $f'(x) > 0$ for $c < x < b$, then f has a local minimum at $x = c$. (See Fig. 4–11b.)
(3) If $f'(x) < 0$ (or $f'(x) > 0$) for all x different from c on the interval, then f has neither a local minimum nor a local maximum at $x = c$. (See Fig. 4–11c.)

EXAMPLE 1 Let $f(x) = x^2 - 4x$. Where is f increasing? decreasing?

Solution Since $f'(x) = 2x - 4 = 2(x - 2)$, then

$$f'(x) < 0 \quad \text{if } x < 2 \quad \text{(decreasing)}$$

and

$$f'(x) > 0 \quad \text{if } x > 2 \quad \text{(increasing)}.$$

It follows that f is a decreasing function on the infinite interval $(-\infty, 2)$ and an increasing function on $(2, \infty)$. (See Fig. 4–12a.) Thus f has a local minimum at $x = 2$. □

EXAMPLE 2 Let $f(x) = x^3 - 3x + 1$. Find all local extrema.

Solution $f'(x) = 3x^2 - 3 = 3(x - 1)(x + 1)$. Thus, $x = -1$ and $x = +1$ are the only critical

4-3 First Derivative Test for Local Extrema

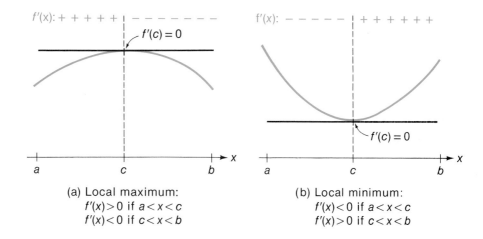

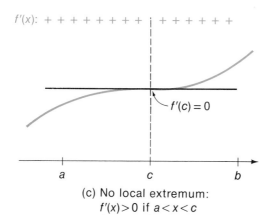

Figure 4-11 The first derivative test.

points. If we solve the inequalities $f'(x) > 0$ and $f'(x) < 0$ by the techniques of Section 1-2, we find that

$$
\begin{array}{lll}
f'(x) > 0 & \text{if } x < -1 & (f \text{ is increasing on } (-\infty, -1)), \\
f'(x) < 0 & \text{if } -1 < x < 1 & (f \text{ is decreasing on } (-1, 1)), \\
f'(x) > 0 & \text{if } x > 1 & (f \text{ is increasing on } (1, \infty)).
\end{array}
$$

It follows from the first-derivative test that f has a local maximum of $f(-1) = 3$ at $x = -1$ and a local minimum of $f(1) = -1$ at $x = +1$. (See Fig. 4-12b.) □

EXAMPLE 3   Let f be defined by the following rule:

$$f(x) = \begin{cases} x - 1 & \text{if } x \leq 4 \\ -2x + 11 & \text{if } x > 4 \end{cases}$$

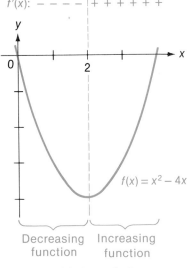
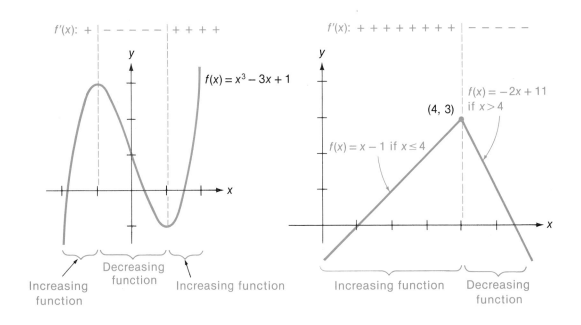

Figure 4-12

Exercises 4–3

(See Fig. 4–12c.) Find all local extrema.

Solution
$$f'(x) = +1 \quad \text{if } x < 4$$

and

$$f'(x) = -2 \quad \text{if } x > 4$$

while $f'(4)$ is not defined. Furthermore, since $\lim_{x \to 4} f(x) = 3 = f(4)$, then f is continuous at $x = 4$. Thus, the only critical point is $x = 4$, where the derivative does not exist. Since $f'(x)$ is positive to the left of $x = 4$ and negative to the right of $x = 4$, then f has a local maximum of 3 at $x = 4$. □

Exercises 4–3

1–10.   Use the first-derivative test to find all local extrema.

- **1** $f(x) = x^3 + 3x^2 + 1$
- **2** $f(x) = 1 - x^2$
- **3** $f(x) = x^4 - 2x^2$
- **4** $f(x) = x + \dfrac{1}{x}$
- **5** $f(x) = \dfrac{x^5}{5} - \dfrac{5x^3}{3} + 4x$
- **6** $f(x) = \dfrac{x^2}{x - 2}$
- **7** $f(x) = (x - 3)^6 (x + 3)^2$
- **8** $f(x) = x^{1/3} + x^{-1/3}$
- **9** $f(x) = x^2$ if $x < 1$, $f(x) = -x + 2$ if $x \geq 1$
- **10** $f(x) = x^2$ if $x \leq 1$, $f(x) = 1$ if $x > 1$

11–18.   The three-point test is described in Exercise 24. Use that test to find all local extrema of these functions.

- **11** $f(x) = x^2 - 4x + 1$
- **12** $f(x) = -x^2 + 6x - 7$
- **13** $f(x) = \dfrac{\sqrt[3]{x}}{x + 2}$
- **14** $f(x) = \dfrac{x - 1}{x^2 + 3}$
- **15** $f(x) = (x^2 - 1)^{2/3}$
- **16** $f(x) = (x + 1)^4$
- **17** $f(x) = \dfrac{x^2 + 2}{x^2 - 2}$
- **18** $f(x) = \dfrac{x^3}{x^2 - 3}$

- **19** Determine the constant c that makes the function $f(x) = x^2/2 + c/x$ have a local minimum at $x = -2$.

20 Determine the constants a, b, c, d that make the function $f(x) = ax^3 + bx^2 + cx + d$ have a local minimum of 0 at $x = 1$ and a local maximum of 4 at $x = -1$.

- **21** A dart is fired straight up into the air from a point 5 ft above the ground with an initial velocity of 160 ft/sec. Find the maximum height above the earth. (Neglect air resistance.)

22 The profit p of a corporation can be expressed as a function of output x by the formula

$$p = -\frac{x^3}{30} + \frac{x^2}{20} + 2x - 1$$

where x is the number of hundreds of units produced each week and p is measured in tens of thousands of dollars. Find the output that yields the maximum profit.

23 Show that the graph of $f(x) = ax^3 + bx^2 + cx + d$ has **(a)** no extremum if $b^2 \leq 3ac$; **(b)** a local maximum and a local minimum if $b^2 > 3ac$. (*Hint:* Use the quadratic formula to solve $f'(x) = 0$.)

24 (*The Three-Point Test*) Let f be continuous on the closed interval $[a, b]$. Suppose that c is the only critical point for f in the open interval (a, b).

Three-Point Test

(a) Prove that f has a local minimum at c if and only if $f(c)$ is less than both $f(a)$ and $f(b)$. (See Fig. 4–13a.)

(b) Prove that f has a local maximum at c if and only if $f(c)$ is greater than both $f(a)$ and $f(b)$. (See Fig. 4–13b.)

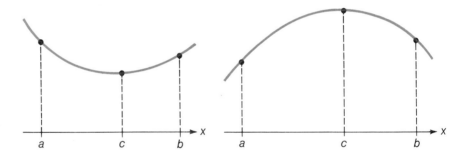

(a) Local minimum: f(c) is less than f(a) and f(b)

(b) Local maximum: f(c) is greater than f(a) and f(b)

Figure 4–13 The three-point test. The only critical point between a and b is at $x = c$.

4–4 Some Applied Problems

In this section we discuss several problems that can be solved by finding extreme values of functions on certain intervals.

EXAMPLE 1   A prominent government official has bought an estate near a small city. The telephone company has agreed to furnish additional communication lines to the estate, and wishes to minimize the installation cost. The location of the estate, as shown in Fig. 4–14, can be reached by traveling 10 km down a straight highway from point T in the city (the switching office) and then 6 km perpendicular to the highway. The com-

4-4 Some Applied Problems

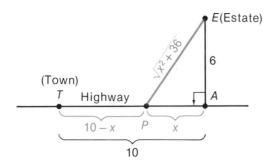

Figure 4-14 Example 1. The telephone line problem.

pany plans to extend new wires from T along the highway to a point P, using existing poles. From P, new lines and poles will be constructed across country to the estate (E). The cost is $500 per kilometer between T and P, and $1300 per kilometer between P and E. Where should P be located in order to minimize the total cost?

Solution Let A be the point on the highway closest to the estate. (See Fig. 4-14.) If x is the distance between A and P, then the distance between P and E is $\sqrt{x^2 + 36}$ (by the Pythagorean Theorem). Also, the distance between T and P is $10 - x$. The total cost (in hundreds of dollars) is

$$C(x) = 5(10 - x) + 13\sqrt{x^2 + 36}, \quad 0 \leq x \leq 10.$$

(If $x = 0$, the lines run from T to A to E; if $x = 10$, they run directly from T to E.)

To find the critical points for C, we calculate $C'(x)$, set it equal to zero, and solve for x:

$$C(x) = 5(10 - x) + 13(x^2 + 36)^{1/2}$$

$$C'(x) = -5 + \frac{13}{2}(x^2 + 36)^{-1/2}(2x) = -5 + \frac{13x}{\sqrt{x^2 + 36}}$$

If $C'(x) = 0$, then

$$13x = 5\sqrt{x^2 + 36}$$
$$169x^2 = 25(x^2 + 36)$$
$$144x^2 = 25 \cdot 36$$
$$12x = 5 \cdot 6 \quad \text{(since } x \geq 0\text{)}$$
$$x = 2.5$$

The only critical point is $x = 2.5$. Thus, the minimum value of $C(x)$ must occur at $x = 2.5$ or at one of the endpoints of the interval $[0, 10]$. Since $C(0) = 128$, $C(2.5) = 122$, and $C(10) \approx 151.6$, the minimum total cost is $12,200 for P located 2.5 km from A. □

STEPS IN THE SOLUTION OF APPLIED PROBLEMS

The following steps may help you to organize your work in solving problems like the one in Example 1.

212 *Applications of the Derivative*

Steps in Solution of
Applied Problems

(1) Try to draw a picture, diagram, or graph that illustrates the situation described in the problem statement. Label all lengths, angles, rates, and positions that are mentioned, using numbers if they are available or letters otherwise. Look for relationships that you know from elementary geometry (Pythagorean theorem, sum of angles in a triangle or circle, equal angles at intersection of straight lines). Identify the quantity to be maximized or minimized.

(2) Write an equation in which the left side consists of the quantity to be maximized or minimized, and the right side consists of other quantities found in the problem statement (some or all of which may be variables with unknown values). For each unknown on the right, analyze the situation to find a relation that will yield a known value. In addition to the geometric relations found in the sketch, consider (if appropriate) relationships involving rate, time, and distance. The goal is an equation that states the quantity to be maximized (or minimized) *as a function of a single variable*. Note any restrictions on the domain of this function; there is often a physical requirement that the independent variable be nonnegative or nonzero.

(3) As discussed in Section 4–1, find the interior critical points on the interval of definition of the function.

(4) Test each critical point and each endpoint to find the maximum (or minimum) on the interval.

(5) Stop for a moment to think about the answer you got. Is it reasonable in terms of the problem statement? Does it seem physically possible? If the function is simple or if you have a calculator at hand, test the function values a little to either side of the maximum (minimum) to see whether they are lower (higher) than the value at the critical point.

Keep in mind that *some extremal problems have no solution!* The function $f(x) = x^2$, for example, has no maximum value when x is allowed to range over the entire set of real numbers.

EXAMPLE 2   A 12-meter long Quonset hut has a semicircular cross-section with radius $r = 5$ meters. (See Fig. 4–15a.) The owner plans to build a rectangular room inside the building by constructing two vertical walls and a ceiling. What dimensions yield the maximum possible volume for the room?

Solution   A cross-section of the quonset hut is shown in Fig. 4–15, with a coordinate axis superimposed on the semicircle. In order to maximize the volume of the room, the walls must touch the roof of the hut. Furthermore, by the symmetry of the figure, the center of the semicircle should be midway between the two walls.

Now that we have visualized the situation, let us look for the function to be maximized. The volume of the room is the length of the quonset hut times the cross-sectional area of the room, or $V = 12(2XY) = 24XY$. However, this is still a function

4-4 Some Applied Problems

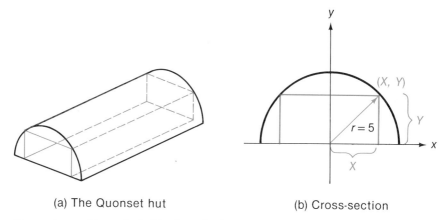

(a) The Quonset hut   (b) Cross-section

Figure 4–15 Example 2. The Inscribed Rectangle Problem.

of two variables, so we must look for another relation that expresses either X or Y in terms of the other. This relation is given by the equation of the semicircle:

$$Y = \sqrt{r^2 - X^2} = \sqrt{25 - X^2}, \qquad 0 \le X \le 5.$$

This allows us to write the volume function as

$$V = 24X\sqrt{25 - X^2}, \qquad 0 \le X \le 5$$

To maximize V, we calculate its derivative, set the result equal to zero, and solve for X:

$$\frac{dV}{dX} = 24X \cdot \frac{1}{2}(25 - X^2)^{-1/2}(-2X) + (25 - X^2)^{1/2}(24)$$

$$= \frac{24(-2X^2 + 25)}{(25 - X^2)^{1/2}}$$

Then $dV/dX = 0$ if and only if

$$-2X^2 + 25 = 0$$
$$X = \pm 5/\sqrt{2}$$

The only critical point between 0 and 5 is $X = 5/\sqrt{2}$. (The negative root of the equation, called an *extraneous root*, is outside the domain of V and is physically impossible.) If we apply the first derivative test, we see that V has a local maximum at this point. It now follows from Theorem 4–7 that the maximum possible volume occurs when $X = 5/\sqrt{2} \approx 3.54$ meters. The value of the maximum volume is

$$V_{max} = 24 \cdot \frac{5}{\sqrt{2}} \sqrt{25 - \left(\frac{5}{\sqrt{2}}\right)^2} = 300 \text{ m}^3. \quad \square$$

Example 2 is typical of a large number of practical problems in that it can be reduced to a problem in geometry. Since the length of the room is fixed at 12 m, the maximum volume occurs when the area of the cross-section is maximum. Consequently, the original problem can be reduced to the problem of finding the largest rectangle that can be inscribed in a semicircle. Most of the geometrical problems in the exercises of this section could be restated in terms of practical problems.

EXAMPLE 3   Find the cylinder of maximum volume that can be inscribed in a right circular cone of radius $r = 15$ and height $h = 20$.

Solution   Let x be the radius of a typical inscribed cylinder and y the height (Fig. 4–16a). The volume of the cylinder is $V = \pi x^2 y$.

Our first problem is to express y in terms of x. Once this is done we can express V in terms of x and use the derivative to find the maximum value of V.

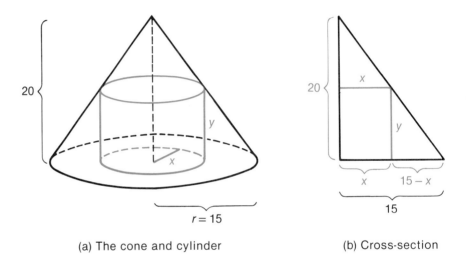

(a) The cone and cylinder   (b) Cross-section

Figure 4–16   Example 3.

To express y in terms of x, we consider the half of the cross section shown in Figure 4–16b. Since the triangles are similar, then the ratios of corresponding sides are equal. In particular, the ratio of height to base is

$$\frac{y}{15 - x} = \frac{20}{15} = \frac{4}{3}$$

so that

$$y = \frac{4}{3}(15 - x).$$

Therefore,

$$V = \pi x^2 y = \frac{4\pi}{3} x^2 (15 - x)$$

$$= 20\pi x^2 - \frac{4}{3}\pi x^3, \qquad 0 < x < 15.$$

4-4 Some Applied Problems

To find the critical points, we calculate dV/dx:

$$\frac{dV}{dx} = 40\pi x - 4\pi x^2 = 4\pi x(10 - x).$$

Since $dV/dx = 0$ only in case $x = 0$ or $x = 10$, the only critical point in the range $0 < x < 15$ is $x = 10$. It follows from the first-derivative test that V has a local maximum at $x = 10$. Thus, by Theorem 4–7, the maximum value of V is

$$V = 20\pi \cdot 10^2 - \frac{4}{3}\pi \cdot 10^3 = \frac{2000}{3}\pi \text{ cu units,}$$

which occurs when $x = 10$, $y = \frac{20}{3}$. □

The calculus can also be used in certain types of business problems.

EXAMPLE 4   The Macabre Book Company publishes an inexpensive line of murder mysteries. Each book costs the company $10,000 (fixed cost) plus $2 for every copy printed and distributed. The marketing division has found that 40,000 copies of a book can be sold at a wholesale price of $3.50 a book and that sales decrease by 100 copies for each cent the wholesale price is raised. (At a price of $3.51 only 39,900 copies would be sold. At $4 only 35,000 would be sold.)

Find the wholesale price that yields the maximum net profit to the company.

Solution   Let x = the wholesale price per copy (in dollars). Then $x - 3.50$ = the amount that x is greater than 3.50.

For each dollar that x is greater than $3.50, ten thousand fewer books will be sold. (One hundred books per penny is equivalent to 10,000 per dollar.) Thus,

Number of books that will be sold at $$x$ each
$= 40,000 - (x - 3.50)(10,000)$
$= 75,000 - 10,000x$ where $x \geq 3.50$

The gross revenue for the company is

$r(x)$ = (selling price) · (number sold)
$= x(75,000 - 10,000x)$
$= 75,000x - 10,000x^2.$

The total cost of producing these books is

$$c(x) = \underbrace{10,000}_{\substack{\text{fixed} \\ \text{cost}}} + \underbrace{2}_{\substack{\text{cost} \\ \text{per} \\ \text{copy}}} \cdot \underbrace{(75,000 - 10,000x)}_{\substack{\text{number of} \\ \text{copies}}} = 160,000 - 20,000x.$$

The net profit is

$p(x)$ = gross revenue − total cost
$= r(x) - c(x)$

$$= (75{,}000x - 10{,}000x^2) - (160{,}000 - 20{,}000x)$$
$$= -160{,}000 + 95{,}000x - 10{,}000x^2.$$

To find the selling price that maximizes the profit, we calculate $p'(x)$, set it equal to zero, and solve for x:

$$p'(x) = 95{,}000 - 20{,}000x.$$

Thus, $p'(x) = 0$ if and only if

$$x = \frac{95{,}000}{20{,}000} = 4.75.$$

It follows from the derivative test that $p(x)$ has a local maximum at $x = 4.75$. Thus, by Theorem 4–7, the maximum profit is obtained at a wholesale price of $4.75 a book. □

Exercises 4–4

- **1** Find the real number x that most exceeds its square.

2 Show that the rectangle with fixed perimeter p that has the largest area is a square.

- **3** Find the dimensions of the rectangle of maximum area that can be inscribed in a right triangle with sides of length 3, 4, and 5 in. with one corner at a vertex of the triangle.

- **4** Find the dimensions of the rectangle of maximum area that can be inscribed in a circle of radius 6 in.

- **5** Find the point on the graph of $y = \sqrt{x}$ that is closest to $(1, 0)$.

- **6** A closed right circular cylindrical can is to have a total surface area (side, top, and bottom) of 6π square feet (ft²). Find the ratio of the diameter to the height that yields the maximum volume.

- **7** An open-topped box with a square base is to be made of thin sheet metal. Determine the dimensions that require the minimum amount of sheet metal to get a total volume of 4 ft³.

- **8** A piece of thin sheet metal 18 inches square is to be made into an open-topped box by cutting squares out of the corners and turning up the sides. What is the maximum possible volume?

- **9** A right circular cone is to be inscribed in a sphere of radius 9 in. Find the dimensions that yield the maximum volume.

Exercises 4-4

- **10** A right circular cylinder is to be inscribed in a sphere of radius r.
  - (a) Find the dimensions that yield the maximum volume.
  - (b) Find the dimensions that yield the maximum lateral surface area.

- **11** A right circular cylinder is generated by revolving a rectangle of perimeter p about one of its edges. What dimensions of the rectangle yield the cylinder of (a) maximum volume; (b) maximum lateral surface area?

- **12** Find the length of the shortest ladder that can be propped against an 8-ft fence so as to reach to a house 1 ft beyond the fence.

- **13** A window, with total perimeter of 12 ft, is to be made in the shape of a rectangle surmounted by a semicircle. What dimensions will admit the most light?

- **14** The total area of a page of a book is to be 96 in.^{2}. The margins at the top and bottom of the page are to be 1.5 in. each, and those at the sides are to be 1 in. each. Find the dimensions of the page that yield the maximum printed area. What is this maximum printed area?

- **15** A long rectangular sheet of metal, 12 in. wide, is to be made into a rain gutter by turning up two sides at right angles to the sheet. How many inches should be turned up to give the gutter its maximum capacity?

- **16** A playing field is to be made in the shape of a rectangle with a semicircular region adjoined to each end. A 400-meter track is to form the perimeter of the field. What dimensions maximize the area of the rectangular part?

- **17** A farmer has 600 yd of fencing with which to enclose a rectangular field and then subdivide the field into two plots by constructing a fence parallel to a side. How should he fence in the field so as to get the greatest total area?

- **18** Two towns, Beeburg (B) and Ceeburg (C), located on the same side of a straight river, are to be supplied with water from a common pumping station at the river's edge. The towns B and C are located 2 and 3 miles, respectively, from the nearest points on the river, B' and C', which are 10 miles apart. (See Fig. 4–17.) Where should the pumping station be located to make the total length of pipeline a minimum?

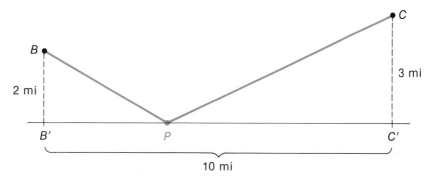

Figure 4–17 Problem 18.

●19 The strength of a rectangular beam is proportional to the product of its width and the square of its depth. Find the dimensions of the strongest wooden beam that can be cut from a circular cylindrical log of radius r.

●20 A man is in a boat at the south bank of a straight river 1 mile wide. He wants to go to a house on the north bank 5 miles east along the river by rowing in a straight line across the river to a point on the north bank, and then walking the rest of the way. He can row at the rate of 3 mph and walk at the rate of 5 mph. Find the shortest possible time in which he can complete the trip.

●21 The Beeburg Power Company needs to lay a cable from point A on one bank of an 800-ft wide, straight river to point B on the opposite bank 1600 ft downstream. It costs \$3 a foot to lay this cable on land and \$5 a foot to lay it in water. What route should be followed to minimize the total cost? What is the minimum possible cost?

●22 A company finds that if it produces x units a day, the cost is \$120 a day for overhead, \$1.95 a unit for material and labor, and $0.003x^2$ a day for maintenance and depreciation. How many units should be produced each day to minimize the cost per unit?

23 A triangle has vertices $(-a, 0)$, $(0, b)$, and $(c, 0)$, where a, b, c are positive. A rectangle is to be inscribed in the triangle with one side on the x-axis. Show that the maximum possible area of the rectangle is half the area of the triangle.

●24 The illumination at a point is directly proportional to the intensity of the light source and inversely proportional to the square of the distance from the light source. Two light sources of intensities I_1 and I_2 are located at the origin O and at the point $P(0, 6)$. Find the point between O and P at which the total illumination is a minimum.

4–5 Concavity. Application to Curve Sketching

Consider the graph of the function f pictured in Figure 4–18. Between a and b, the derivative is an increasing function. The graph bends upward away from each tangent line. Between b and c, the derivative is a decreasing function. The graph bends downward away from each tangent line. We say that the graph is *concave upward* between a and b and *concave downward* between b and c.

Concavity

More precisely, the graph of a function f is *concave upward on an interval I* provided its derivative f' is an increasing function on the interval. It is *concave downward on I* if f' is a decreasing function. This is illustrated in Figure 4–18. As we trace along the curve from a to b, the slope of the tangent line continually increases from negative to positive. Thus, f' increases from a negative value at a to a positive value at b. Similarly, f' decreases from a positive value at b to a negative value at c.

Recall from Theorem 4–4 that a function is known to be increasing over an interval if its derivative is positive there. Consequently, if the second derivative f'' is positive over an interval I, then f' must be increasing over I and f must be concave upward there. Similarly, if $f'' < 0$ over I, then f' decreases on I and f is concave downward there.

4–5 Concavity. Application to Curve Sketching

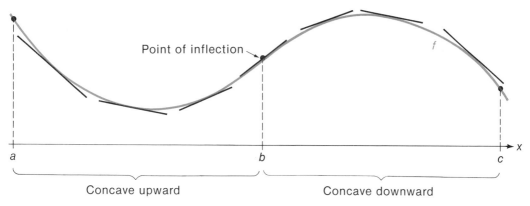

Figure 4–18 Concavity.

In summary:

(1) If $f''(x) > 0$ on an interval I, then the graph of f is concave upward on I.
(2) If $f''(x) < 0$ on an interval I, then the graph of f is concave downward on I.

POINT OF INFLECTION

Point of Inflection

A point on the graph of f at which f has a tangent line (possibly vertical) and which separates a portion of the graph that is concave upward from a portion that is concave downward is called a *point of inflection*. The graph in Figure 4–18, for example, has a point of inflection at $(b, f(b))$. The curve is smooth enough to have a tangent line there and is concave upward to the left of the point and concave downward to the right.

It can be proved that the second derivative f'' must equal zero at a point of inflection provided the second derivative exists there. Thus, the only possible locations for points of inflection are those points at which $f''(x) = 0$ or $f''(x)$ fails to exist.

EXAMPLE 1 Let $f(x) = x^3 - 6x^2 + 9x + 1$. Where is the graph of f concave upward? concave downward? Locate all points of inflection.

Solution
$$f(x) = x^3 - 6x^2 + 9x + 1.$$
$$f'(x) = 3x^2 - 12x + 9.$$
$$f''(x) = 6x - 12 = 6(x - 2).$$

Thus, $f''(x)$ is negative if $x < 2$ and positive if $x > 2$. The graph is concave downward if $x < 2$ and concave upward if $x > 2$. Since the concavity changes at $x = 2$, the point $(2, 3)$ is a point of inflection. See Fig. 4–19. (Also see Remark 2, p. 220.) □

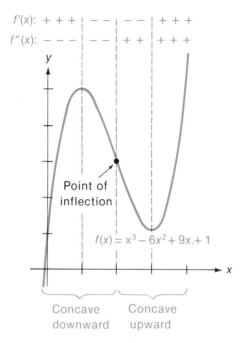

Figure 4-19 Example 1. The concavity of $f(x) = x^3 - 6x^2 + 9x + 1$.

Remark 1 That $f''(a) = 0$ is not enough to make $(a, f(a))$ a point of inflection. This can be illustrated with the function

$$f(x) = x^4.$$

The second derivative is $f''(x) = 12x^2$, which is positive if $x \neq 0$. It follows that the graph is concave upward everywhere even though $f''(0) = 0$.

Remark 2 The sign of the second derivative usually gives information about concavity on open intervals. In most cases, if a function is concave upward (downward) on an open interval, it has the same concavity on the corresponding closed interval provided it is defined at the endpoints. For example, the function of Example 1 is concave downward if $x \leq 2$, upward if $x \geq 2$.

APPLICATION TO CURVE SKETCHING

We can make quick and accurate sketches of many graphs by applying the information developed thus far. The steps are as follows:

Steps in Curve Sketching

(1) Locate all points of discontinuity.
(2) Find all horizontal and vertical asymptotes.
(3) Find all vertical tangent lines.
(4) Locate all critical points.
(5) Find all intervals on which f is increasing ($f'(x) > 0$) and all intervals on which f is decreasing ($f'(x) < 0$).

4–5 Concavity. Application to Curve Sketching

(6) Find all local maxima and minima.
(7) Find all intervals on which the graph is concave upward ($f''(x) > 0$) and all intervals on which it is concave downward ($f''(x) < 0$).
(8) Find all points of inflection.
(9) Plot a few reference points. Sketch the graph.

EXAMPLE 2 Sketch the graph of $f(x) = x^4 - 6x^2 - 8x + 15$.

Solution Since f is a polynomial function, there are no discontinuities. Furthermore, there are no vertical or horizontal asymptotes. We calculate the derivative:

$$f'(x) = 4x^3 - 12x - 8 = 4(x + 1)^2(x - 2).$$

The only critical points are $x = -1$ and $x = 2$. Observe that

$$\begin{aligned} f'(x) &< 0 \quad \text{if } x < 2 \text{ and } x \neq -1 \quad \text{(decreasing).} \\ f'(x) &> 0 \quad \text{if } x > 2 \quad \text{(increasing).} \end{aligned}$$

There is a local minimum at $x = 2$. The curve has a horizontal tangent line at $x = -1$ but no local extremum there. There is no vertical tangent line.

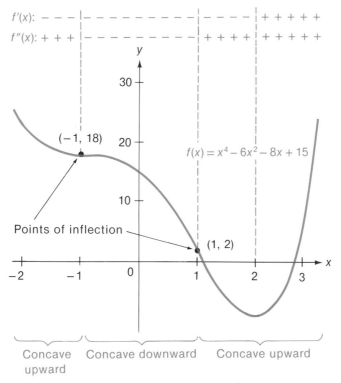

Figure 4–20 Example 2. The graph of $f(x) = x^4 - 6x^2 - 8x + 15$. (The vertical scale is contracted to save space.)

The second derivative is

$$f''(x) = 12x^2 - 12 = 12(x + 1)(x - 1).$$

Then

$$f''(x) = 0 \quad \text{only if } x = -1 \text{ or } x = +1.$$

Since

$$\begin{aligned} f''(x) > 0 & \quad \text{if } x < -1 & \text{(concave upward on } (-\infty, -1)), \\ f''(x) < 0 & \quad \text{if } -1 < x < 1 & \text{(concave downward on } (-1, 1)), \end{aligned}$$

and

$$f''(x) > 0 \text{ if } x > 1 \quad \text{(concave upward on } (1, \infty)),$$

then $(1, f(1))$ and $(-1, f(-1))$ are points of inflection.

If we now calculate the values of f at the points determined above and at a few additional reference points, we can sketch the curve. The graph is shown in Figure 4–20. □

APPLICATION TO POPULATION GROWTH

The graph of a function f is concave upward if f' is an increasing function and concave downward if f' is a decreasing function. A point of inflection occurs when f' stops increasing and starts to decrease, or vice versa. Figure 4–21 is the graph of an

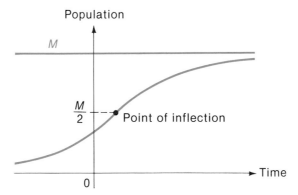

Figure 4–21 Graph of typical animal population. M is the maximum possible population.

Population Growth

animal population that increases according to a particular law of growth. The maximum possible population that can be supported is M. Observe that for this law of growth, f', the rate of increase of the population, is an increasing function (the curve is concave upward) until the population reaches $M/2$, half the maximum possible

4–5 Concavity. Application to Curve Sketching

population. After that point, the rate of increase is a decreasing function, the population almost stabilizing when it is close to M. By measuring the rates of increase of a population at various times, wildlife specialists can determine the value of $M/2$ (at which the concavity changes) and thus find M, the maximum possible population.

Suppose that a stable population is maintained by harvesting a fixed number of animals each year. Since the increase in population in year t is approximately equal to $f'(t)$, it follows that the maximum harvest is obtained when the population is stabilized at $M/2$, one half the theoretical maximum.

CONCAVITY AND TANGENT LINES

If we examine the graphs in Figures 4–18 and 4–19, we see that except at the points of tangency, the tangent lines are below the graph when the graph is concave upward on an interval and above the graph when the graph is concave downward. Furthermore, a tangent line at a point of inflection crosses the graph. We establish one of these results in Theorem 4–9. The other results can be established by similar arguments.

THEOREM 4–9

Relation Between Concavity and Tangent Line

Let $f''(x) > 0$ on the interval I. Let $a \in I$ and let $\mathcal{T}$ be the tangent line to the graph of f at the point $(a, f(a))$. If b is any point of I different from a, then the point $(b, f(b))$ on the graph of f is above the corresponding point $(b, f(a) + f'(a)(b - a))$ on the line $\mathcal{T}$. (See Fig. 4–22.)

Proof Since the equation of the tangent line is

$$y = f(a) + f'(a)(x - a),$$

we must show that

$$f(b) > f(a) + f'(a)(b - a).$$

(See Figure 4–22.) We apply the Mean Value Theorem twice—first to f, then to f'.

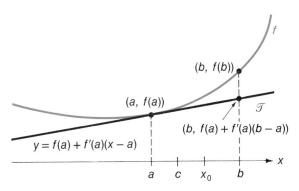

Figure 4–22 The graph of f is concave upward over the interval. The tangent line $\mathcal{T}$ is below the graph except at the point of tangency (Theorem 4–9).

Observe first that since $f''(x)$ exists at each point of I, then f and f' exist and are continuous at each point of I.

By the Mean Value Theorem, there exists x_0 between a and b such that

$$f(b) - f(a) = f'(x_0)(b - a).$$

Thus,

$$\begin{aligned} f(b) - f(a) - f'(a)(b - a) &= f'(x_0)(b - a) - f'(a)(b - a) \\ &= [f'(x_0) - f'(a)](b - a). \end{aligned}$$

We now apply the Mean Value Theorem to the function f' on the closed interval with endpoints a and x_0. There exists a number c between a and x_0 such that

$$f'(x_0) - f'(a) = f''(c)(x_0 - a).$$

Then

$$\begin{aligned} f(b) - f(a) - f'(a)(b - a) &= [f'(x_0) - f'(a)](b - a) \\ &= f''(c)(x_0 - a)(b - a). \end{aligned}$$

Observe that $f''(c) > 0$, and that $(x_0 - a)$ and $(b - a)$ are both positive or both negative. Thus,

$$f(b) - f(a) - f'(a)(b - a) = f''(c)(x_0 - a)(b - a) > 0$$

so that

$$f(b) > f(a) + f'(a)(b - a),$$

which was to be proved. ∎

Exercises 4–5

1–14 (a) Find the intervals where f is increasing and where f is decreasing. (b) Find the intervals where f is concave upward and where f is concave downward. (c) Find all local extrema and points of inflection. (d) Sketch the graph.

- **1** $f(x) = x^2 + x + 2$
- **2** $f(x) = x + \dfrac{1}{x}$

- **3** $f(x) = x^3 - 9x^2 + 24x - 10$
- **4** $f(x) = \dfrac{x}{x^2 + 1}$

- **5** $f(x) = x^3 - 3x + 5$
- **6** $f(x) = x^4 - 2x^2 + 1$

- **7** $f(x) = \dfrac{1}{x^2 + 3}$
- **8** $f(x) = x^{1/3}$

- **9** $f(x) = 1 + 10x - x^2$
- **10** $f(x) = x^3 - 6x^2 + 9x - 2$

4–6 The Second-Derivative Test

•11 $f(x) = \dfrac{x+1}{x}$

12 $f(x) = x^5 - 5x^4$

•13 $f(x) = x^2 + \dfrac{16}{x}$

14 $f(x) = x^4 + 4x^3 + 20$

15–17. If the conditions are not contradictory, then sketch the graph of a function that satisfies them. If the conditions are contradictory, explain why.

•15 f is continuous everywhere;

$f'(x) > 0$ if $x < 3$, $f'(x) < 0$ if $x > 3$;
$f''(x) > 0$ if $x < 1$ or $x > 4$; $f''(x) < 0$ if $1 < x < 4$.

16 f is continuous everywhere;

$f''(x) > 0$ if $x \neq 4$; $f(2) = f(4) = 5$.

•17 f is continuous everywhere;

$f'(x) < 0$ and $f''(x) < 0$ if $x < 3$;
$f'(x) > 0$ and $f''(x) > 0$ if $x > 3$.

18 Sketch the graphs of f, f', and f'' in different colors on the same axis system. What does the positive or negative nature of f' or f'' tell us about the graph of f?
  (a) $f(x) = x^3$
  (b) $f(x) = x^4$

4–6 The Second-Derivative Test

Another test for local extrema exists, which in many cases is the test most easily applied.

The proof of Theorem 4–10 depends on Theorem 2–4 (you should review this before going ahead).

THEOREM 4–10

Second Derivative Test

(*Second-Derivative Test*) Let f have a second derivative at the critical point c.
  (a) If $f''(c) > 0$, then f has a local minimum at c.
  (b) If $f''(c) < 0$, then f has a local maximum at c.
  (c) If $f''(c) = 0$, then a different test must be used.

Proof We consider only the case in which $f''(c) > 0$. The proof of (b) is left for you in Exercise 22. Part (c) is established in Example 3.

Since $f''(c)$ exists, then f and f' must exist and must be continuous at $x = c$. Since c is a critical point, then $f'(c)$ must be zero. Since $f''(c)$ is the derivative of f' at $x = c$, then

$$f''(c) = \lim_{x \to c} \dfrac{f'(x) - f'(c)}{x - c} > 0.$$

If we substitute $f'(c) = 0$, we get

$$\lim_{x \to c} \frac{f'(x)}{x - c} > 0.$$

We now apply Theorem 2–4 (using $g(x) = f'(x)/(x - c)$). By this theorem, there exists $\delta > 0$ such that if $0 < |x - c| < \delta$, then $(f'(x))/(x - c) > 0$.

Next, we restrict ourselves to values of x in the range $c - \delta < x < c$. For such x, the fraction $f'(x)/(x - c)$ is positive and the denominator, $x - c$, is negative. It follows that the numerator $f'(x)$ is negative. Thus, f is a decreasing function on the interval $(c - \delta, c)$ to the left of the critical point c.

Similarly, if $c < x < c + \delta$, then $f'(x)/(x - c) > 0$ and $x - c > 0$, so that $f'(x) > 0$. Thus, f is an increasing function to the right of the critical point.

Since f decreases to the left of c and increases to the right of c, then f has a local minimum at $x = c$ (Fig. 4–23a). ■

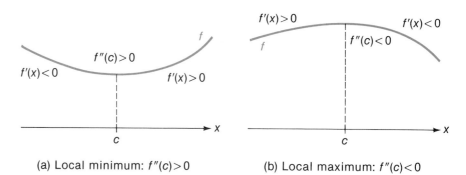

(a) Local minimum: $f''(c) > 0$ (b) Local maximum: $f''(c) < 0$

Figure 4–23 The second derivative test.

EXAMPLE 1   Show that $f(x) = x^2$ has a local minimum at $x = 0$.

Solution
$$f'(x) = 2x,$$
$$f''(x) = 2.$$

The only critical point is $x = 0$. Since $f''(0) = 2 > 0$, then f has a local minimum at $x = 0$. □

EXAMPLE 2   Find all local extrema of $f(x) = 2x^3 + 9x^2 - 24x + 5$.

Solution
$$f'(x) = 6x^2 + 18x - 24 = 6(x + 4)(x - 1),$$
$$f''(x) = 12x + 18.$$

The critical points are $x = -4$ and $x = 1$. At these points we have

$$f''(-4) = -30 < 0 \quad \text{(local maximum)}$$
$$f''(1) = 30 > 0 \quad \text{(local minimum)}.$$

The function has a local maximum at $x = -4$ and a local minimum at $x = 1$. □

Exercises 4–6 227

Example 3 shows that no conclusion can be reached from Theorem 4–10 if $f''(c) = 0$. The function may have a local maximum, a local minimum, or neither.

EXAMPLE 3 (a) Let $f(x) = x^3$. Then

$$f'(x) = 3x^2 \quad \text{and} \quad f''(x) = 6x.$$

The only critical point is $x = 0$. Observe that the function has no extremum and that $f''(0) = 0$. (See Fig. 4–24a.)

(b) Let $f(x) = x^4$. Then

$$f'(x) = 4x^3 \quad \text{and} \quad f''(x) = 12x^2.$$

The only critical point is $x = 0$. Observe that f has a local minimum at $x = 0$ and that $f''(0) = 0$. (See Fig. 4–24b.)

(c) Let $f(x) = -x^4$. Then

$$f'(x) = -4x^3 \quad \text{and} \quad f''(x) = -12x^2.$$

The only critical point is $x = 0$. Observe that f has a local maximum at $x = 0$ and that $f''(0) = 0$. (See Fig. 4–24c.) □

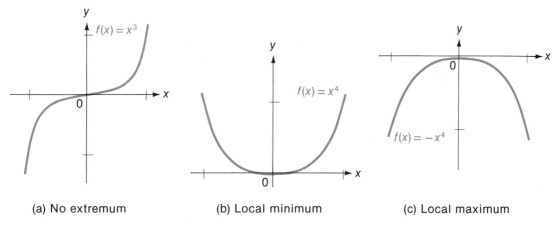

(a) No extremum (b) Local minimum (c) Local maximum

Figure 4–24 Example 3. The Second Derivative Test gives no information if $f'(c) = 0$ and $f''(c) = 0$.

Exercises 4–6

1–18. Find all local extrema and points of inflection. Use the second-derivative test whenever possible.

- 1 $f(x) = 5x^2 - 2x + 1$
- 2 $f(x) = 4x - x^2$
- 3 $f(x) = 5 + 16x - x^2$
- 4 $f(x) = 3x^5 - 20x^3$
- 5 $f(x) = x^3 - 3x^2 + 1$
- 6 $f(x) = 3x^2 - 2x^3$

- 7 $f(x) = x^5 - 5x^4$
- 8 $f(x) = x^{5/3} - 15x$
- 9 $f(x) = x^2 - \dfrac{x^3}{6}$
- 10 $f(x) = x + \dfrac{1}{x}$
- 11 $f(x) = \dfrac{x}{x^2 + 1}$
- 12 $f(x) = x^2 + \dfrac{16}{x}$
- 13 $f(x) = x^4 - 8x^2$
- 14 $f(x) = \sqrt{x} + \dfrac{1}{\sqrt{x}}$
- 15 $f(x) = (2x^2 - 3x + 1)^2$
- 16 $f(x) = \dfrac{x^3}{x^2 - 3}$
- 17 $f(x) = (x + 1)^4(2x - 4)^2$
- 18 $f(x) = (x - 3)^6(x + 3)^2$
- 19 A rectangular poster must contain 50 in.² of printed matter with margins of 4 in. at each of the top and bottom and 2 in. at each side. Find the dimensions that make the total area a minimum.

20 (a) Let $P(x, y)$ be a point on the graph of $y = x^2$. Use Exercise 51 of Section 1–5 to show that the distance from P to the line $4x - 3y = 12$ is

$$d(x) = \frac{|4x - 3x^2 - 12|}{5}.$$

(b) Use the result of (a) to show that the minimum distance between the graph of $y = x^2$ and the line $4x - 3y = 12$ is $d(\tfrac{2}{3}) = \tfrac{32}{15}$. (Exercise 45 of Section 3–5 may be useful.)

- 21 Painters, before painting an auditorium, must move a ladder down a 6-ft-wide corridor and turn at a right angle into a 5-ft-wide corridor. Find the length of the longest ladder that will make the turn. (*Remark.* This problem is equivalent to finding the *shortest* ladder that will reach from the far wall of one corridor to the far wall of the other while touching the corner formed by them.)

22 Complete the proof of Theorem 4–10. Let f have a second derivative at the critical point c. Show that if $f''(c) < 0$, then f has a local maximum at $x = c$.

4–7* Application to Business Analysis

The calculus originally was developed to solve problems in science and engineering. During the twentieth century, however, it has been used extensively to solve problems in business and the social sciences. This has been done by using functions that approximately describe real-life situations.

Almost all the functions used in economics are approximate. It is not possible to predict the exact profit that can be made by selling 20,000 sewing machines, for example. Instead, economists monitor the profit at different levels of production and then find a function that approximates these values. Thus, the mathematical analysis lacks the precision that scientists and engineers expect in their work.

4-7* Application to Business Analysis

DEMAND FUNCTIONS

Demand Function As a general rule, there is an inverse relation between the cost of a commodity and the demand for it. As the cost increases the demand decreases, and vice versa. This relation can be expressed as a *demand function*—an empirical function that expresses the *demand* for a commodity (the number of items that can be sold) as a function of its cost.

EXAMPLE 1 We have had one example of a demand function. The Macabre Book Company (of Example 4, Section 4–4) can sell 40,000 copies at a wholesale price of $3.50 a book, but sales decrease by 10,000 copies for each dollar the price is raised. If $x is the price per book, the demand function is

$$d(x) = 40{,}000 - (x - 3.50)(10{,}000), \qquad x \geq 3.50. \quad \square$$

EXAMPLE 2 The demand function of Thin Tomato Soup is

$$d(x) = 35 - \tfrac{5}{4}x$$

where x, the wholesale selling price per can, is measured in cents and $d(x)$, the number of cans that can be sold each week, is measured in hundreds of thousands of cans. What selling price yields the largest weekly revenue for the company?

Solution The revenue obtained from selling soup at x cents a can is

$$\begin{aligned} R(x) &= \text{(number of cans that can be sold)} \cdot \text{(price per can)} \\ &= 100{,}000\, d(x) \cdot x \text{ cents} \\ &= 100{,}000(35 - \tfrac{5}{4}x) \cdot x = 100{,}000(35x - \tfrac{5}{4}x^2) \text{ cents}. \end{aligned}$$

Then

$$R'(x) = 100{,}000(35 - \tfrac{5}{2}x)$$

so that $R'(x) = 0$ only in case $x = 14$. It follows from the second-derivative test that R has a local maximum at $x = 14$. Since this is the only critical point, the maximum revenue is

$$\begin{aligned} R(14) &= 100{,}000(35 \cdot 14 - \tfrac{5}{4} \cdot 14^2) \text{ cents} \\ &= 24{,}500{,}000 \text{ cents} = \$245{,}000. \quad \square \end{aligned}$$

MARGINAL ANALYSIS

Let $C(n)$ be the cost of producing and selling n items of a commodity. Then

$$C(n) - C(n-1)$$

"True" Marginal Cost is the cost of producing the nth item after the $n - 1$st item has been produced. This quantity is called the *true marginal cost* of the nth item.

If we apply the approximation formula from Section 3–3 to the cost function C, we have

$$C(x + \Delta x) - C(x) \approx C'(x)\, \Delta x.$$

In particular, if $x = n$ and $\Delta x = -1$, we have

$$C(n-1) - C(n) \approx C'(n) \cdot (-1)$$

so that

$$C(n) - C(n-1) \approx C'(n).$$

In other words, the "true" marginal cost of the nth item approximately equals the derivative of C at $x = n$.

In general the cost function $C(n)$ is only approximate. Thus, the derivative $C'(n)$ is probably as accurate an estimation of the cost of producing the nth item as is the "true" marginal cost $C(n) - C(n-1)$. For this reason economists usually define the *marginal cost of the nth item* to be $C'(n)$ rather than $C(n) - C(n-1)$. Thus, by convention,

Marginal Cost

$$\begin{aligned} C'(n) &= \text{marginal cost of } n\text{th item}, \\ &\approx \text{cost of producing } n\text{th item after } n - 1\text{st item has been produced} \\ &\approx C(n) - C(n-1). \end{aligned}$$

Similarly, if $R(n)$ and $P(n)$ are the *revenue* and *profit* from selling n items, then

Marginal Revenue

$$\begin{aligned} R'(n) &= \textit{marginal revenue} \text{ from } n\text{th item}, \\ &\approx \text{revenue from selling } n\text{th item after } n - 1\text{st item has been sold} \\ &\approx R(n) - R(n-1), \end{aligned}$$

and

Marginal Profit

$$\begin{aligned} P'(n) &= \textit{marginal profit} \text{ from } n\text{th item}, \\ &\approx \text{profit from selling } n\text{th item after } n - 1\text{st item has been sold} \\ &\approx P(n) - P(n-1). \end{aligned}$$

These functions are related by the equation

$$P(x) = R(x) - C(x)$$

(profit equals revenue minus cost). Taking derivatives, we get

$$P'(x) = R'(x) - C'(x)$$

(marginal profit equals marginal revenue minus marginal cost).

For many products, after a certain production level has been reached, cost increases for each successive item while revenue decreases. Consequently, if production is maintained at too high a level, the company will go bankrupt. The company should continue to increase production, however, so long as the marginal profit is positive. The company will make the maximum profit when marginal profit is zero. At this production level, the marginal cost equals the marginal revenue.

4-7* Application to Business Analysis

EXAMPLE 3 The cost and revenue functions for the ABC Block Company are

$$C(x) = 50 + 5x + \tfrac{1}{100}x^2$$

and

$$R(x) = 25 + 8x + \tfrac{1}{200}x^2 - \tfrac{1}{3000}x^3,$$

where x is the number of production runs per week.
(a) Find the marginal cost and marginal revenue when $x = 20$ and $x = 40$.
(b) Find the production level that yields the maximum possible profit.

Solution (a)
$$C'(x) = 5 + \frac{1}{100} \cdot 2x = 5 + \frac{x}{50}$$

$$R'(x) = 8 + \frac{1}{200} \cdot 2x - \frac{1}{3000} \cdot 3x^2 = 8 + \frac{x}{100} - \frac{x^2}{1000}.$$

When $x = 20$ and $x = 40$, we obtain

$$C'(20) = 5.4 \quad \text{and} \quad R'(20) = 7.8,$$
$$C'(40) = 5.8 \quad \text{and} \quad R'(40) = 6.8.$$

Observe that the marginal cost is greater and the marginal revenue is less when $x = 40$ than when $x = 20$. The marginal profit is still positive, however, so production at this higher level is more profitable than at $x = 20$.

(b) The profit function is

$$P(x) = R(x) - C(x)$$

$$= \left[25 + 8x + \frac{x^2}{200} - \frac{x^3}{3000}\right] - \left[50 + 5x + \frac{x^2}{100}\right]$$

$$= -25 + 3x - \frac{x^2}{200} - \frac{x^3}{3000}.$$

To find the maximum profit, we set $P'(x) = 0$, find the critical points, and test them for extrema:

$$P'(x) = 3 - \frac{x}{100} - \frac{x^2}{1000}.$$

Then $P'(x) = 0$ if and only if

$$3000 - 10x - x^2 = 0$$
$$(60 + x)(50 - x) = 0,$$
$$x = -60 \quad \text{or} \quad x = 50.$$

Since x represents production, the solution $x = -60$ is meaningless. Thus, the only critical point is $x = 50$.
The second derivative is

$$P''(x) = -\frac{1}{100} - \frac{x}{500}.$$

At $x = 50$, we get $P''(x) = P''(50) = -\tfrac{1}{100} - \tfrac{50}{100} < 0$. Thus, the profit function has a local maximum at $x = 50$. Since there is only one critical point, the maximum profit occurs at $x = 50$. Observe that at this number we have $C'(x) = R'(x)$, that is, marginal cost equals marginal revenue. □

OPTIMAL LOT SIZE

A company uses N units of a certain commodity each year. The consumption is distributed evenly over the year, the cost does not vary, the units may be ordered from a distributor and be received almost immediately, and there is no obsolescence. In short, there is no reason to stockpile the commodity and no reason to order only small quantities. The company wishes to determine the proper number of units to order in each shipment so as to keep the total cost of stocking the item to a minimum.

The total stock cost of the commodity for the year is the sum of three costs: (1) a fixed *unit cost* (price of each unit, shipping cost, and so on); (2) a *reorder cost* that is the same for each shipment, regardless of size; and (3) a *storage cost* (depreciation of warehouse facilities, and so on).

Let x be the number of units in each shipment; a, the unit cost; b, the reorder cost; and k, the cost of storing one unit for one year. Since there are N/x shipments a year, then the total cost of obtaining the N units is

$$(ax + b)\frac{N}{x}.$$

If we assume that each shipment arrives as soon as the inventory is zero, the average size of the inventory is $x/2$ and the storage cost for the year is $kx/2$. It follows that the total cost of stocking the commodity is

$$C(x) = (ax + b)\frac{N}{x} + \frac{kx}{2} = aN + \frac{bN}{x} + \frac{kx}{2}.$$

To find the value of x that minimizes C, we first calculate the derivative:

$$C'(x) = -\frac{bN}{x^2} + \frac{k}{2}.$$

Then $C'(x) = 0$ if and only if $x = \sqrt{2bN/k}$. (We can neglect the negative square root because x must be positive.) Since $C''(x)$ is positive at the only critical point, then C has a minimum at $x = \sqrt{2bN/k}$.

Optimal Lot Size Formula

The result $x = \sqrt{2bN/k}$ is called the *optimal lot size formula*. Observe that the optimal lot size is proportional to $\sqrt{N}$. Thus, if the scale of operation quadruples, the optimal lot size doubles. In that case, the company should double the size of its orders and cut the time between orders in half.

To illustrate the use of the formula, suppose that the company uses 1000 units a year, the unit cost is \$2, the reorder cost is \$5 per shipment and the yearly storage cost is \$0.25 per unit. Then the minimum cost of stocking the commodity occurs when the number of items in each shipment is

$$x_0 = \sqrt{\frac{2bN}{k}} = \sqrt{\frac{2 \cdot 5 \cdot 1000}{0.25}} = 200.$$

Exercises 4–7

1–4. Find the production level that yields the maximum profit in each of the given situations. In each case, x is the number of units, $C(x)$ is the cost of producing and selling x units, and $R(x)$ is the revenue from selling x units.

- 1 $C(x) = \dfrac{x^2}{2} - 50$, $R(x) = 6x + 7$

  2 $C(x) = \dfrac{4}{5}x^2 + 8x + 6$, $R(x) = -\dfrac{1}{30}x^3 + \dfrac{6}{5}x^2 + 10x - 4$

- 3 $C(x) = 9x^2 + 40x + 500$, $R(x) = -2x^3 + 30x^2 + 400x - 700$.

  4 $C(x) = \dfrac{x^2}{20} + 700x + 500$, $R(x) = 2700x - \dfrac{3x^2}{20}$, $2000 \le x \le 4000$

- 5 The cost function for a commodity is

  $$C(x) = x^4 - 36x^3 + 432x^2 + 500x + 1000, \quad 3 \le x \le 15,$$

  where x is the number of production runs.
  (a) Find the maximum marginal cost.
  (b) For what values of x is the marginal cost increasing? decreasing?

  6 The revenue function for a commodity is $R(x) = 240x + 57x^2 - x^3$.
  (a) Find the maximum marginal revenue.
  (b) Find the maximum possible revenue.

- 7 The demand function for a commodity is $d(x) = 20 - x/4$, where x is the wholesale selling price. What value of x yields the maximum revenue to the company?

  8 The demand function for a commodity is $d(x) = \sqrt{600 - x}$, where $100 \le x \le 300$. What value of x (the wholesale selling price) yields the maximum revenue to the company?

- 9 The Knight Chess Company finds that its fixed costs are $500, material and labor costs are $2 a unit, and the demand function is $d(x) = 450 - 30x$. What value of x yields the maximum profit for the company? How many units can be sold at this price?

  10 The Gasburner Car Leasing Company makes a profit of $750 a car if it leases no more than 500 cars. For each car over 500 that is leased, the profit per car decreases by $1. How many cars should the company lease to receive the maximum profit?

- 11 There are 70 apartments at the Hi-Rise Apartment complex. At $350 a month, all the apartments can be rented. For each $10-a-month increase in rent, one additional apartment will be unrented. What price per apartment yields the maximum revenue?

  12 The Beeburg World News has 1000 subscribers who pay $5 a month. For each 10-cent decrease in rates, the company can get 100 additional subscribers. What rate yields the maximum revenue?

●13 The R&S Production Company uses 100,000 magnets each year. The cost is $20 per thousand with a $10 reorder cost. The storage cost is $5 per thousand per year. How many magnets should be ordered in each shipment to keep the total cost of stocking the magnets at a minimum?

●14 The Electrical Circuit Company has found that a constant production level can be maintained if the relation between labor and machinery is given by

$$x^2 y^3 = 20^{10}$$

where x is the number of workers and y is the number of circuit-making machines. The total hourly cost of employing an average worker is $10 and the hourly cost of leasing and operating a machine is $15. How many workers and how many machines should be used to minimize the total production costs?

●15 Each month the Kil-A-Cycle Electronics Company can sell x shipments of 100 FM tuners at d dollars per tuner, where $d = (375 - 5x)/3$. The cost of producing x shipments is $5000 + 1500x$ dollars.

(a) How many tuners should the company produce each month to make the maximum profit? What should be the selling price of these tuners?

(b) The federal government plans to impose an emergency excise tax of $2 on each FM tuner (thus raising the selling price to $(d + 2)$ dollars, of which the company receives d dollars). How many tuners should the company now produce each month to make the maximum profit? What should be the new selling price? Who pays the tax—the company or the consumer? What other consequence does the tax have?

4-8 Newton's Method

A troublesome difficulty has plagued mathematicians and engineers from the time that mathematical analysis was first used to solve complicated problems. In many cases, the best that can be done is to reduce a problem to solving an equation that may be unsolvable by elementary algebra. In such a case, we search for numerical approximations to the exact roots.

For example, to find where the graph of $y = x^2 - 1$ intersects the graph of $y = \sin x$, we must solve the system of equations

$$\begin{cases} y = x^2 - 1 \\ y = \sin x. \end{cases}$$

on substituting one equation in the other, we have

$$x^2 - 1 = \sin x.$$

The solutions of this equation can be found graphically, but there is no algebraic method to calculate them exactly.

4–8 Newton's Method

Some other equations with roots that cannot be calculated exactly are

$$x^7 - 5x = 3,$$
$$2^x = x + 5,$$
$$\log(x^2 + 1) = x^2 - 1.$$

Unfortunately, the solution of a problem in which we have a vital interest may require us to solve such an equation. A practical method for approximating the solutions of such equations to any desired degree of accuracy is based on a method developed by Isaac Newton, one of the founders of the calculus.

Newton's method is laborious when the computations must be done by hand. Consequently, until recently, it was used primarily to calculate solutions of important equations but not as a daily tool. The electronic computer and the scientific calculator, however, have changed the situation. It is simple to program a computer or a programmable calculator for Newton's method. Thus, the method is now used to solve a wide variety of approximation problems.

Let f be a differentiable function of x. If the equation

$$f(x) = 0$$

has a solution $x = c$, then the graph of f crosses the x-axis at c. Let x_1 be our first approximation to c. (See Fig. 4–25a.) The line tangent to the graph at $(x_1, f(x_1))$ is an approximation to the graph in the vicinity of $x = x_1$. Thus, it also is an approximation to the graph in the vicinity of $x = c$.

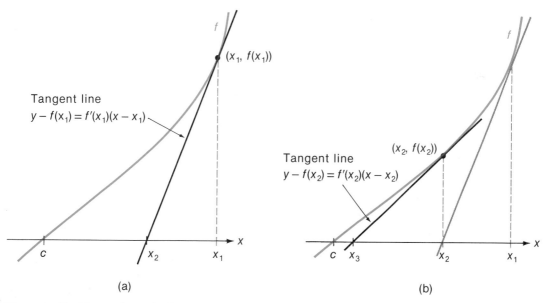

Figure 4–25 Newton's method.

The equation of the line tangent to the graph of f at $(x_1, f(x_1))$ is

$$y - f(x_1) = f'(x_1)(x - x_1).$$

If x_2 is the point at which this line crosses the x-axis, then

$$0 - f(x_1) = f'(x_1)(x_2 - x_1)$$

so that

$$x_2 = x_1 - \frac{f(x_1)}{f'(x_1)} \quad \text{provided } f'(x_1) \neq 0.$$

We now repeat the process, using x_2 instead of x_1, obtaining a point x_3 that we hope will be a better approximation to c than x_2 (see Fig. 4–25b):

$$x_3 = x_2 - \frac{f(x_2)}{f'(x_2)} \quad \text{provided } f'(x_2) \neq 0.$$

If we continue the process, we get the following sequence of (hopefully) better and better approximations of c:

Newton's Approximations

x_1 (First approximation—choose close to c)

$x_2 = x_1 - \dfrac{f(x_1)}{f'(x_1)}$ (Second approximation)

$x_3 = x_2 - \dfrac{f(x_2)}{f'(x_2)}$ (Third approximation)

$x_4 = x_3 - \dfrac{f(x_3)}{f'(x_3)}$ (Fourth approximation)

$x_5 = x_4 - \dfrac{f(x_4)}{f'(x_4)}$ (Fifth approximation)

. . . .

$x_{n+1} = x_n - \dfrac{f(x_n)}{f'(x_n)}$ (n + 1st approximation)

We shall see in Theorem 4–11 that Newton's method yields better and better approximations to the solution of $f(x) = 0$ provided the function f satisfies certain conditions on an interval that contains the solution. The functions in Examples 1 and 2 do satisfy these conditions.

EXAMPLE 1   Use Newton's method to approximate $\sqrt{7}$ correct to seven decimal places.

Solution   The number $c = \sqrt{7}$ is a solution of the equation $f(x) = 0$ where

$$f(x) = x^2 - 7.$$

The derivative is

$$f'(x) = 2x.$$

4–8 Newton's Method

Thus, the formula for the $n + 1$st term is

$$x_{n+1} = x_n - \frac{f(x_n)}{f'(x_n)} = x_n - \frac{x_n^2 - 7}{2x_n}.$$

Our first approximation, x_1, should be reasonably close to $\sqrt{7}$. Since $2 < \sqrt{7} < 3$, we choose $x_1 = 3$. (Any other number close to 2 or 3 would work about as well.) Then,

$$x_1 = 3$$
$$x_2 = x_1 - \frac{x_1^2 - 7}{2x_1} \approx 2.666666667$$
$$x_3 = x_2 - \frac{x_2^2 - 7}{2x_2} \approx 2.645833333$$
$$x_4 = x_3 - \frac{x_3^2 - 7}{2x_3} \approx 2.645751312$$
$$x_5 = x_4 - \frac{x_4^2 - 7}{2x_4} \approx 2.645751311$$

Observe that x_4 and x_5 both round off to 2.6457513. If we calculate $\sqrt{7}$ on a scientific calculator, we obtain

$$\sqrt{7} \approx 2.6457513.$$

Thus x_4 approximates $\sqrt{7}$ correct to seven decimal places. ☐

Newton's method can be used to find roots of equations such as

$$\tan x - x = 0$$

and

$$2^x - 5 = 0.$$

These equations, however, involve transcendental functions and at this point we have not considered the techniques for calculating their derivatives. Thus, for now we consider only algebraic functions. (Additional problems involving Newton's method will be considered later.)

EXAMPLE 2 Show that the polynomial $f(x) = x^3 - 3x^2 + 2x - 4$ has a zero between $x = 2$ and $x = 3$. Use Newton's method to approximate it.

Solution Since

$$f(2) = 2^3 - 3 \cdot 2^2 + 2 \cdot 2 - 4 = -4$$

and

$$f(3) = 3^3 - 3 \cdot 3^2 + 2 \cdot 3 - 4 = 2,$$

it follows from the Intermediate Value Theorem (Theorem 2–8) that there exists a number c between 2 and 3 such that $f(c) = 0$.

To apply Newton's method, we first calculate $f'(x)$:
$$f'(x) = 3x^2 - 6x + 2.$$
The general term in Newton's method is
$$x_{n+1} = x_n - \frac{f(x_n)}{f'(x_n)} = x_n - \frac{x_n^3 - 3x_n^2 + 2x_n - 4}{3x_n^2 - 6x_n + 2}.$$
As a starting point we choose $x_1 = 3$. Then

$x_1 = 3$

$x_2 = x_1 - \dfrac{f(x_1)}{f'(x_1)} \approx 2.818181818$

$x_3 = x_2 - \dfrac{f(x_2)}{f'(x_2)} \approx 2.796613026$

$x_4 = x_3 - \dfrac{f(x_3)}{f'(x_3)} \approx 2.796321956$

$x_5 = x_4 - \dfrac{f(x_4)}{f'(x_4)} \approx 2.796321903$

$x_6 = x_5 - \dfrac{f(x_5)}{f'(x_5)} \approx 2.796321903$

The solution of $x^3 - 3x^2 + 2x - 4 = 0$, correct to nine decimal places, is
$$c \approx 2.796321903. \quad \square$$

ACCURACY OF APPROXIMATION

Unfortunately, Newton's method does not work for every function. Consider as an example the curve in Figure 4-26. The shape of the graph is such that $x_1, x_2, x_3, \ldots$ are successively further away from the root c. This problem arises when the curve is not sufficiently well behaved in the vicinity of c. In this particular example, the approximation x_1 was chosen too far from the immediate vicinity of c.

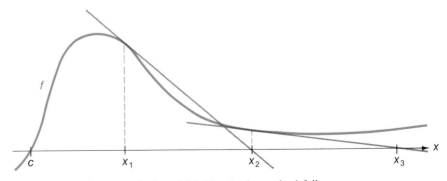

Figure 4-26 An example for which Newton's method fails.

4–8 Newton's Method

There are numerous conditions that can be used to ensure that Newton's method works. One of the simplest sets of conditions is given in Theorem 4–11. Observe that the theorem can be used to estimate the closeness of the nth approximation to the number c. The proof of the theorem, which depends on the Mean Value Theorem, is outlined in Exercise 15.

THEOREM 4–11

Accuracy of Newton's Approximations

(*Newton's Method*) Let c be a solution of the equation $f(x) = 0$. Let I be a closed interval that contains c as an interior point. Suppose that f has a nonzero first derivative and a continuous second derivative on I. Let x_1 be the first approximation to c and define x_2, $x_3, x_4, \ldots$ by Newton's method. That is,

$$x_{n+1} = x_n - \frac{f(x_n)}{f'(x_n)}.$$

Let

$$M = \text{maximum value of } |f''(x)| \text{ on } I,$$
$$m = \text{minimum value of } |f'(x)| \text{ on } I,$$

$$\epsilon = |x_1 - c|\frac{M}{m}.$$

If $\epsilon < 1$ and the numbers $x_1, x_2, \ldots$ are all in I, then the numbers $x_1, x_2, x_3, \ldots$ form a sequence of better and better approximations to c. To be more precise, these numbers satisfy the conditions

$$|x_2 - c| < \epsilon|x_1 - c|$$
$$|x_3 - c| < \epsilon^2|x_2 - c|$$
$$|x_4 - c| < \epsilon^4|x_3 - c|$$
$$|x_5 - c| < \epsilon^8|x_4 - c|$$
$$|x_6 - c| < \epsilon^{16}|x_5 - c|$$
$$\ldots\ldots\ldots\ldots\ldots$$

To illustrate the theorem, we return to the function $f(x) = x^2 - 7$ of Example 1. Since c, the solution of $f(x) = 0$, approximately equals 2.64, we let I be the closed interval $[2, 3]$. Then

$$M = \text{maximum of } |f''(x)| \text{ on } I = 2,$$
$$m = \text{minimum of } |f'(x)| \text{ on } I = 4.$$

We choose the first approximation to be

$$x_1 = 3.$$

Then

$$x_2 = x_1 - \frac{f(x_1)}{f'(x_1)} \approx 2.666666667.$$

Observe that x_1 and x_2 are both in I, as required in the hypothesis of the theorem. Also, $|x_1 - c| < 0.4$, so that

$$\epsilon = |x_1 - c| \cdot \frac{M}{m} < (0.4)\frac{2}{4} = 0.2.$$

Since $x_2, x_3, \ldots$ are all in I, then

$$|x_2 - c| < \epsilon|x_1 - c| < (0.2)(.4) = 0.08,$$
$$|x_3 - c| < \epsilon^2|x_2 - c| < (0.2)^2(.08) = 0.0032,$$
$$|x_4 - c| < \epsilon^4|x_3 - c| < (0.2)^4(.0032) = 0.00000512,$$

and

$$|x_5 - c| < \epsilon^8|x_4 - c| < (0.2)^8(0.00000512) \approx 0.000000000013.$$

These conditions ensure that x_4, as an approximation to c, is accurate to four decimal places, that x_5 is accurate to ten decimal places, and so on.

Exercises 4–8

1–4. Use Newton's method to approximate the given numbers:

- 1 $\sqrt{2}$
- 2 $\sqrt{5}$
- 3 $\sqrt[3]{2}$
- 4 $\sqrt[3]{9}$

5–6. Show that each of the polynomial equations has a solution between 1 and 2. Use Newton's method to approximate this number.

- 5 $x^4 + x - 3 = 0$
- 6 $x^4 - 2 = 0$

7–9. Use Newton's method to approximate these numbers.

- 7 The smallest positive solution of $x^3 + 5x - 3 = 0$.
- 8 The negative solution of $x^4 + x^3 - 1 = 0$.
- 9 The largest solution of $x^3 + 6x^2 + 9x + 2 = 0$.

10 Show that the equation $x^3 + 5x - 1 = 0$ has a solution c between 0 and 0.5. Let $x_1 = 0$.
- (a) Use Newton's method to approximate c to 4 decimal places.
- (b) Use Theorem 4–11 to show that the numbers $x_1, x_2, x_3, \ldots$ form a sequence of better and better approximations to c, the solution of the equation. How large must n be before we know from Theorem 4–11 that $|x_n - c| < 10^{-10}$? (*Hint*: Let $I = [0, 0.5]$. Show that $\epsilon < 0.3$.)

One situation in which Newton's method fails is illustrated in Figure 4–26. Another situation is illustrated by Exercise 11.

- 11 Let $f(x) = x^{1/3}$. Show that the solution of $f(x) = 0$ cannot be approximated by Newton's method. (The exact solution is $x = 0$. Show that if $x_1 \neq 0$, then x_1, x_2,

Exercises 4–8

$x_3, \ldots$ are successively further from 0.) Which of the hypotheses of Theorem 4–11 are not satisfied by this function?

Ancient Square Root Algorithm

12 (*Square Root Algorithm*) The following very efficient algorithm was used by the ancient Babylonians (about 1700 BC) to approximate the square root of a positive number a. Let x_1 be the first positive approximation. Let

$$x_2 = \frac{1}{2}\left(x_1 + \frac{a}{x_1}\right),$$

$$x_3 = \frac{1}{2}\left(x_2 + \frac{a}{x_2}\right),$$

and so on. The numbers $x_1, x_2, x_3, \ldots$ form a sequence of better and better approximations of $\sqrt{a}$.

(a) Use the Babylonian algorithm to approximate $\sqrt{2}$.

(b) Show that this algorithm can be obtained from Newton's method by choosing $f(x) = x^2 - a$.

(It is believed that Newton studied the Babylonian algorithm, discovered why it worked, and then generalized it, obtaining the method that bears his name.)

13 Let $a > 0$. We showed in Exercise 12 that Newton's method leads to the formula

$$x_{n+1} = \frac{1}{2}\left(x_n + \frac{a}{x_n}\right)$$

when we approximate $\sqrt{a}$.

(a) Set up a similar formula for $\sqrt[3]{a}$.

• (b) Generalize the formula in (a) to get a formula for $\sqrt[k]{a}$, where k is a positive integer.

14 (*Computer Program*) If you have access to a computer or programmable calculator, then write a program for Newton's method. The examples in the text can be used to test the program.

15 (*Newton's Method*) The following steps outline the proof of Theorem 4–11:

(1) Apply the Mean Value Theorem to show that there exists a number a between x_1 and c such that $f(x_1) - f(c) = f'(a)(x_1 - c)$. Since $f(c) = 0$, this implies that

$$x_1 - c = \frac{f(x_1)}{f'(a)}.$$

(2) Since $x_2 - x_1 = -f(x_1)/f'(x_1)$, then

$$x_2 - c = (x_1 - c) + (x_2 - x_1)$$

$$= \frac{f(x_1)}{f'(a)f'(x_1)}[f'(x_1) - f'(a)]$$

$$= \frac{1}{f'(x_1)}(x_1 - c)[f'(x_1) - f'(a)].$$

(3) Next apply the Mean Value Theorem to the function f'. There exists a number b between x_1 and a such that
$$f'(x_1) - f'(a) = f''(b)(x_1 - a).$$
Then
$$x_2 - c = \frac{f''(b)}{f'(x_1)}(x_1 - c)(x_1 - a).$$

(4) Taking absolute values, we get the fundamental inequality
$$|x_2 - c| = \frac{|f''(b)|}{|f'(x_1)|}|x_1 - c||x_1 - a| < \frac{M}{m}|x_1 - c|^2.$$

(5) The fundamental inequality implies that
$$|x_2 - c| < \epsilon|x_1 - c|.$$

(6) If we repeat the argument in Steps 1–4, we find that
$$|x_3 - c| < \frac{M}{m}|x_2 - c|^2.$$
Then
$$|x_3 - c| < \frac{M}{m} \cdot \epsilon|x_1 - c||x_2 - c| \quad \text{(by Step 5)}$$
$$\leq \epsilon^2 \cdot |x_2 - c|.$$

(7) Similarly,
$$|x_4 - c| < \frac{M}{m}|x_3 - c|^2$$
$$< \frac{M}{m} \cdot \epsilon^2|x_2 - c||x_3 - c|$$
$$< \frac{M}{m} \cdot \epsilon^2 \cdot \epsilon \cdot |x_1 - c||x_3 - c| \leq \epsilon^4|x_3 - c|.$$

(8) The above steps can be continued indefinitely. Show that each of the numbers $x_2, x_3, x_4, \ldots$ is successively closer to c.

4–9 Antiderivatives

In mathematics it is customary, whenever a new function or operation is studied, to ask whether another function or operation exists that "undoes" the effect of the first one. That is, does the first function have an inverse? (Take another look at Section 1–8 to refresh your understanding of inverses.)

Having defined the operation of differentiation of a function, therefore, we now investigate the possibility of an operation that can be performed on a derivative to

4-9 Antiderivatives

recover the original function. We have differentiated a function F to obtain another function, $F' = f$; we now look for an operation that, given f, returns the original function F. On the somewhat optimistic assumption that such an inverse operation exists, we give it the name antidifferentiation (and, by analogy, we call F the antiderivative of f). The formal definition follows.

DEFINITION

Antiderivative

The function F is said to be an *antiderivative* of the function f on an interval I provided that

$$F'(x) = f(x) \quad \text{for all } x \in I.$$

For example, since

$$D_x(3x^2 - 2x + 5) = 6x - 2 \quad \text{for all } x,$$

then the function

$$F(x) = 3x^2 - 2x + 5$$

is an antiderivative of the function

$$f(x) = 6x - 2$$

on every interval.

It is obvious that a function f can have many different antiderivatives. For example,

$$F_1(x) = 3x^2 - 2x + 1,$$
$$F_2(x) = 3x^2 - 2x + 2,$$

and

$$F_7(x) = 3x^2 - 2x + 7$$

are antiderivatives of $f(x) = 6x - 2$. It is easy to see that any function of form

$$F(x) = 3x^2 - 2x + C$$

is an antiderivative of $f(x) = 6x - 2$.

It follows from Theorem 4-6 that every antiderivative of a function f can be obtained from any given antiderivative F by adding an appropriate constant to F. The expression

$$F(x) + C$$

is called the *general antiderivative* of f.

We sometimes denote the general antiderivative of a function f by the symbol

$$D_x^{-1}(f(x)).$$

The superscript -1 emphasizes that antidifferentiation is the inverse operation to differentiation (represented by the symbol D_x). Thus, if F is an antiderivative of f, then

$$D_x^{-1}(f(x)) = F(x) + C.$$

EXAMPLE 1 (a) $D_x^{-1}(6x - 2) = 3x^2 - 2x + C$.

(b) Since $D_x(x^4 + x^2 + \sqrt{3}x) = 4x^3 + 2x + \sqrt{3}$, then
$$D_x^{-1}(4x^3 + 2x + \sqrt{3}) = x^4 + x^2 + \sqrt{3}x + C.$$

(c) Since $D_x(5x^{1/3} + 2x^{1/2} + 3) = \frac{5}{3}x^{-2/3} + x^{-1/2}$, then
$$D_x^{-1}\left(\frac{5}{3}x^{-2/3} + x^{-1/2}\right) = 5x^{1/3} + 2x^{1/2} + 3 + C_1,$$

where C_1 is an arbitrary constant. If we let $C = 3 + C_1$, then C also is an arbitrary constant, and we can write
$$D_x^{-1}\left(\frac{5}{3}x^{-2/3} + x^{-1/2}\right) = 5x^{1/3} + 2x^{1/2} + C. \ \square$$

The function $f = F'$ can be interpreted as giving the slope of the tangent line to F at every point along the graph of F. In this sense, f determines the shape of F, but it cannot determine the vertical distance between the graph of F and the x-axis. This is the reason for the inclusion of the constant C in the general antiderivative: the process of antidifferentiating f yields a family of curves, all of the same shape, with each displaced up or down by an amount determined by its particular value of C.

RULES FOR CALCULATING ANTIDERIVATIVES

The following rules simplify many of the calculations of antiderivatives. These rules, which correspond to the derivative formulas D4, D5, and D3, can be proved from those formulas.

Rule 1 Let $f(x)$ and $g(x)$ be functions that have antiderivatives. The sum of the antiderivatives of $f(x)$ and $g(x)$ is an antiderivative of $f(x) + g(x)$. Thus

Antiderivative of a Sum
$$D_x^{-1}(f(x) + g(x)) = D_x^{-1}(f(x)) + D_x^{-1}(g(x)) + C.$$

Rule 2 Let $f(x)$ have an antiderivative $F(x)$. Let k be a constant. The function $kF(x)$ is an antiderivative of $kf(x)$. Thus

Antiderivative of a Constant Times a Function
$$D_x^{-1}(kf(x)) = k\,D_x^{-1}(f(x)) + C.$$

Rule 3 The general antiderivative of kx^n is

Power Rule for Antiderivatives
$$D_x^{-1}(kx^n) = \frac{kx^{n+1}}{n+1} + C \qquad \text{provided } n \neq -1.$$

4-9 Antiderivatives

Remark 1 In the special case where $n = 0$, Rule 3 becomes

$$D_x^{-1}(k) = kx + C,$$

which enables us to calculate the antiderivative of a constant function.

Remark 2 The antiderivative of x^{-1} does exist but cannot be calculated by Rule 3. This antiderivative defines a function, the natural logarithm, which is discussed in Chapter 8.

Remark 3 In actual practice we do not usually add the constant until the final antiderivative has been calculated. Furthermore, we usually add a single arbitrary constant to the right-hand side instead of adding constants for each antiderivative. For example, we write

$$D_x^{-1}(2x + 3) = D_x^{-1}(2x) + D_x^{-1}(3)$$
$$= \frac{2x^2}{2} + 3x + C = x^2 + 3x + C$$

instead of

$$D_x^{-1}(2x + 3) = D_x^{-1}(2x) + D_x^{-1}(3) + C_1$$
$$= \left(\frac{2x^2}{2} + C_2\right) + (3x + C_3) + C_1$$
$$= x^2 + 3x + (C_1 + C_2 + C_3).$$

Observe that the final results are equivalent. If we let $C = C_1 + C_2 + C_3$, the two answers agree.

EXAMPLE 2 Calculate $D_x^{-1}\left(3\sqrt{x} + 2x^2 - \frac{1}{x^2} + 5\right)$.

Solution We first apply Rule 1, then Rule 3:

$$D_x^{-1}(3x^{1/2} + 2x^2 - x^{-2} + 5)$$
$$= D_x^{-1}(3x^{1/2}) + D_x^{-1}(2x^2) - D_x^{-1}(x^{-2}) + D_x^{-1}(5) \quad \text{(by Rule 1)}$$
$$= \frac{3x^{3/2}}{3/2} + \frac{2x^3}{3} - \frac{x^{-1}}{-1} + 5x + C \quad \text{(by Rule 3)}$$
$$= 2x^{3/2} + \frac{2x^3}{3} + \frac{1}{x} + 5x + C. \quad \square$$

In some cases we may need to find a particular antiderivative that satisfies a certain condition.

EXAMPLE 3 Find the particular antiderivative $F(x)$ of $f(x) = 3\sqrt{x} + 2x^2 - \frac{1}{x^2} + 5$ that satisfies the condition $F(1) = 4$.

Solution It follows from Example 2 that

$$F(x) = 2x^{3/2} + \frac{2}{3}x^3 + \frac{1}{x} + 5x + C$$

for some constant C. If $F(1) = 4$, we must have

$$4 = 2 \cdot 1^{3/2} + \frac{2}{3} \cdot 1^3 + \frac{1}{1} + 5 \cdot 1 + C$$

$$4 = 2 + \frac{2}{3} + 1 + 5 + C$$

$$C = -\frac{14}{3}$$

The particular antiderivative is

$$F(x) = 2x^{3/2} + \frac{2}{3}x^3 + \frac{1}{x} + 5x - \frac{14}{3}. \quad \square$$

EXAMPLE 4 Prove Rule 2 for antiderivatives.

Solution Let $F(x)$ be an antiderivative of $f(x)$ on the interval I. Then

$$F'(x) = f(x) \qquad \text{for } x \in I$$

and

$$D_x^{-1}(f(x)) = F(x) + C_1 \qquad \text{for } x \in I$$

where C_1 is an arbitrary constant.
Since

$$D_x(kF(x)) = kF'(x) = kf(x)$$

then $kF(x)$ is an antiderivative of $kf(x)$ on I. The general antiderivative is

$$D_x^{-1}(kf(x)) = kF(x) + C_2$$
$$= k[D_x^{-1}(f(x)) - C_1] + C_2$$
$$= kD_x^{-1}(f(x)) + C$$

where $C = C_2 - kC_1$. $\square$

EXTENDED POWER RULE

Rule 3, the Power Rule for antiderivatives, can be extended by the Chain Rule to calculate antiderivatives of expressions of form $u^n \, du/dx$. The extended power rule is

Extended Power Rule

Rule 3' $\quad D_x^{-1}\left(u^n \dfrac{du}{dx}\right) = \dfrac{u^{n+1}}{n+1} + C \qquad$ provided $n \neq -1$.

4–9 Antiderivatives

To apply Rule 3' to the problem of calculating

$$D_x^{-1}((5x^2 - 2x + 1)^5(10x - 2)),$$

for example, we must identify the function that we call u in the formula and make a substitution. If we let $u = 5x^2 - 2x + 1$, then $du/dx = 10x - 2$, and the expression becomes

$$D_x^{-1}((5x^2 - 2x + 1)^5(10x - 2)) = D_x^{-1}\left(u^5 \frac{du}{dx}\right)$$

$$= \frac{u^6}{6} + C = \frac{(5x^2 - 2x + 1)^6}{6} + C.$$

Remark In applying Rule 3', we must be certain that u, u^n, and du/dx are all real functions of x on the interval under consideration. It would not be proper to apply it to the function $u = -x^2 - 1$ with $n = \frac{1}{2}$, for example, because u^n would not be a real function of x.

In some cases it may be necessary to modify the form of an expression before we can calculate its antiderivative by Rule 3'. This is usually accomplished by first applying Rule 2 which, stripped of arbitrary constants, states that particular constants can be moved across the antiderivative sign:

$$D_x^{-1}(kf(x)) = kD_x^{-1}(f(x)).$$

If we divide both sides of this equation by k (assuming that $k \neq 0$), we get

$$D_x^{-1}(f(x)) = \frac{1}{k} D_x^{-1}(kf(x)).$$

Examples 5 and 6 illustrate how this process is used in particular problems.

EXAMPLE 5 Calculate $D_x^{-1}((3x^2 + 8)^{1/2} \cdot x)$.

Solution It would be natural to make the substitution $u = 3x^2 + 8$ and apply Rule 3'. Observe that $u^{1/2}$ is part of the expression that we wish to antidifferentiate but du/dx is not. Instead of du/dx, which is equal to $6x$, we have x. We shall change the function to the proper form, however, if we multiply it by 6. Then

$$D_x^{-1}((3x^2 + 8)^{1/2} \cdot x) = \frac{1}{6} D_x^{-1}((3x^2 + 8)^{1/2} \cdot 6x)$$

$$= \frac{1}{6} D_x^{-1}\left(u^{1/2} \frac{du}{dx}\right) \qquad \boxed{\begin{array}{l} u = 3x^2 + 8 \\[4pt] \dfrac{du}{dx} = 6x \end{array}}$$

$$= \frac{1}{6} \cdot \frac{u^{3/2}}{3/2} + C$$

$$= \frac{(3x^2 + 8)^{3/2}}{9} + C. \; \square$$

EXAMPLE 6   Calculate $D_x^{-1}((3x + 7)^{14})$.

Solution   Let $u = 3x + 7$. Then $du/dx = 3$. To get the expression in the proper form for Rule 3', we must multiply it by 3. Then

$$D_x^{-1}((3x + 7)^{14}) = \frac{1}{3} D_x^{-1}((3x + 7)^{14} \cdot 3)$$

$$= \frac{1}{3} D_x^{-1}\left(u^{14} \cdot \frac{du}{dx}\right)$$

$$= \frac{1}{3} \cdot \frac{u^{15}}{15} + C$$

$$= \frac{(3x + 7)^{15}}{45} + C. \quad \square$$

$\boxed{u = 3x + 7 \qquad \frac{du}{dx} = 3}$

Remark   Do not jump to the conclusion that every antiderivative can be calculated by the methods of this section. As we shall see, many antiderivatives cannot be calculated by elementary methods. Consider, as an example, the three functions

$$f(x) = x\sqrt{1 - x^2}, \quad g(x) = \sqrt{1 - x^2}, \quad \text{and} \quad h(x) = \sqrt{1 - x^3},$$

all of which have antiderivatives. The antiderivative of f can be calculated after making the substitution $u = 1 - x^2$. It is much more difficult to calculate the antiderivative of g—the inverse sine function from trigonometry is involved. It is still more difficult to calculate the antiderivative of h, since it cannot be expressed in terms of the standard functions studied in elementary calculus.

Exercises 4–9

1–12.   (a)   Find the general antiderivative $F(x)$ of each function $f(x)$.
(b)   Find the particular antiderivative $F(x)$ that satisfies the condition $F(1) = 2$.

- 1 $4x^3 - 3x + 2$
2   $x^3 - x^{-3}$
- 3 $x^2 + \sqrt{x} + 1$
4   $(2x - 1)^2$
- 5 $\dfrac{1}{x^2} - \dfrac{1}{\sqrt{x}}$
6   $\dfrac{x + 1}{\sqrt{x}}$
- 7 $(x + 1)(\sqrt{x} - 1)$
8   $(x + x^{-1})^2$
- 9 $\dfrac{x^2 + 2x - 1}{x^4}$
10   $5x^{3/2} - 2x^{-3/2}$
- 11 $x^{2/3} + x^{-2/3}$
12   $x^{-1/5} + x^{-6/5}$

13–26.   Substitute $u = g(x)$ and use Rule 3' to calculate the general antiderivative of each function.

4–10 Application. Introduction to Differential Equations

- 13 $(2x + 1)^{-3}$
- 14 $(x^2 + x + 1)^{-2}(2x + 1)$
- 15 $\dfrac{1}{\sqrt{1 - x}}$
- 16 $\left(\dfrac{x + 1}{2}\right)^4$
- 17 $x\sqrt{1 - x^2}$
- 18 $(1 - x)^3$
- 19 $(2x^4 + 4x^2 + 1)^{10}(x^3 + x)$
- 20 $(x^3 + 2x + 2)^{-5}(9x^2 + 6)$
- 21 $(2x + x^{-1})^2 \left(2 - \dfrac{1}{x^2}\right)$
- 22 $\sqrt{1 + 5x}$
- 23 $\dfrac{3x}{\sqrt{1 - x^2}}$
- 24 $x^2(1 + 2x^3)^{-2/3}$
- 25 $(1 + 3\sqrt{x})^3 \left(\dfrac{1}{\sqrt{x}}\right)$
- 26 $(x^2 + \sqrt[3]{x})^{1/2}(6x + x^{-2/3})$

Exercise 27 shows that only constants can be moved across the antiderivative sign.

27   Rule 2 states that $D_x^{-1}(kf(x)) = kD_x^{-1}(f(x)) + C$. Show by example that the corresponding rule may not hold if k is replaced by a nonconstant function of x. In other words, find functions $f(x)$ and $g(x)$ such that

$$D_x^{-1}(g(x)f(x)) \neq g(x)D_x^{-1}(f(x)) + C.$$

4–10 Application. Introduction to Differential Equations

In some applications, it is necessary to reconstruct a function from its derivative. It may be easier, for example, to measure the acceleration of a moving object than the velocity or the distance. The velocity function can then be determined as one of the antiderivatives of the acceleration function, and the distance function as one of the antiderivatives of the velocity function.

DISTANCE, VELOCITY, AND ACCELERATION

When a body falls, the force of gravity subjects it to a constant acceleration downward of 32 ft/sec².* If the air resistance is negligible, then this force is the only one affecting the body. Similar results hold for objects thrown straight up into the air.

Falling Body Problem
  Problem An object is thrown upward from a point s_0 ft above the earth with an initial velocity of v_0 ft/sec. Assume that the only force acting on the object is the acceleration due to gravity. Derive formulas for the velocity $v(t)$ and distance $s(t)$ at time t sec.

* Actually, the acceleration is not quite constant. It depends on the distance from the body to the center of the earth. If the body is not too high above the surface of the earth, acceleration varies very little.

Solution We measure distance as positive above the ground level. Since the acceleration due to gravity is in the opposite direction, it is negative:

$$a(t) = -32 \text{ ft/sec}^2.$$

The velocity is an antiderivative of the acceleration:

$$v(t) = D_t^{-1}(-32) = -32t + C_1.$$

To evaluate the constant C_1, recall that $v(0) = v_0$. Then

$$v(0) = v_0 = -32 \cdot 0 + C_1,$$

so that $C_1 = v_0$. The velocity function is

Velocity of Falling Body

$$v(t) = -32t + v_0.$$

The distance function is an antiderivative of the velocity:

$$s(t) = D_t^{-1}(-32t + v_0) = -16t^2 + v_0 t + C_2.$$

At time $t = 0$, the distance is s_0. Therefore

$$s(0) = s_0 = -16 \cdot 0^2 + v_0 \cdot 0 + C_2,$$

so that $C_2 = s_0$. The distance function is

Position of Falling Body

$$s(t) = -16t^2 + v_0 t + s_0.$$

These formulas hold from the instant the object is released until it strikes the ground or is deflected in some way.

The use of the above formulas was illustrated in Section 2–3. These formulas hold only for freely falling bodies, which are not affected by air resistance. Problems involving other types of straight-line motion must be worked by other methods. The following example illustrates another type of motion problem.

EXAMPLE 1 A man is driving along a straight road at 60 mph when the motor in his car stops, causing a constant deceleration. At the end of 20 sec he is traveling at 30 mph. How far will the car travel before it stops?

Solution Because of the small distances involved, we convert to feet per second. The initial velocity is 88 ft/sec. At the end of 20 sec, it is traveling 44 ft/sec.

Let a be the acceleration of the car after the motor stops. (Because the car is decelerating, a is negative.) If we measure time and distance from the point at which the motor stops running, we get the following formulas:

Velocity: $v(t) = D_t^{-1}(a) = at + C_1.$

4–10 Application. Introduction to Differential Equations

In determining the value of C_1, we recall that
$$v(0) = 88 = a \cdot 0 + C_1,$$
so that $C_1 = 88$. The formula for velocity is
$$v(t) = at + 88.$$

In determining the value of the constant a, recall that $v = 44$ when $t = 20$. Thus
$$v(20) = 44 = a \cdot 20 + 88$$
$$20a = -44$$
$$a = \frac{-44}{20} = \frac{-11}{5} \text{ ft/sec}^2.$$

When we substitute this value for a, we get the final form of the velocity formula,
$$v(t) = \frac{-11t}{5} + 88.$$

Distance: $s(t) = D_t^{-1}(v(t)) = D_t^{-1}\left(\frac{-11t}{5} + 88\right)$
$$= \frac{-11t^2}{10} + 88t + C_2.$$

In determining the value of C_2, we recall that $s(0) = 0$. Therefore,
$$s(0) = 0 = \frac{-11 \cdot 0^2}{10} + 88 \cdot 0 + C_2,$$
so that $C_2 = 0$. The distance function is
$$s(t) = \frac{-11t^2}{10} + 88t.$$

The car stops when $v(t) = 0$, that is, when
$$\frac{-11}{5}t + 88 = 0$$
$$t = \frac{5 \cdot 88}{11} = 40 \text{ sec.}$$

At the end of 40 sec, when the car comes to a complete stop, it has traveled
$$s(40) = \frac{-11 \cdot 40^2}{10} + 88 \cdot 40 = 1760 \text{ ft.} \quad \square$$

DIFFERENTIAL EQUATIONS

An equation, such as
$$s'(t) = -32t + v_0,$$
$$y' = 4x^2,$$

or
$$2y' + 3xy = 4x^2,$$

Differential Equation

which involves one or more of the derivatives of a function, is called a *differential equation*.

Solution

By a *solution* of a differential equation involving x, y, y', we mean any function $y = F(x)$, which when substituted in the differential equation makes it an identity. If the differential equation is of the simple form $y' = f(x)$, its solution can be obtained as an antiderivative of $f(x)$:

$$y = D_x^{-1}(f(x)) + C.$$

For example, if $y' = D_x y = 4x^2$, then

$$y = D_x^{-1}(4x^2) = \frac{4x^3}{3} + C.$$

Many applications of the calculus to science and engineering involve differential equations. Since it may be simpler to determine a relation that involves the derivative of a function than to work with the function itself, we can, if this situation arises, set up a differential equation that at least partially describes the function.

We usually are only interested in one of the infinite number of solutions of a differential equation. To find the particular solution of interest, we need additional information. This information is frequently given in the form of *initial conditions*—the values of the function and one or more of its derivatives at some special point.

For example, our earlier work on distance and velocity involved the differential equation $s''(t) = -32$, subject to the initial conditions $s'(0) = v_0$, $s(0) = s_0$. We found that the particular solution that satisfies the initial conditions is

$$s(t) = -16t^2 + v_0 t + s_0.$$

EXAMPLE 2 Find the particular solution $y = f(x)$ of the differential equation $dy/dx = 4x^2$ that satisfies the initial condition $f(1) = 7$.

Solution The solution of the differential equation is

$$y = D_x^{-1}(4x^2) = \frac{4x^3}{3} + C.$$

To satisfy the initial condition, we must have $y = 7$ when $x = 1$. If we substitute these values into the general form of the solution, we get

$$7 = \frac{4 \cdot 1^3}{3} + C$$

$$C = 7 - \frac{4}{3} = \frac{17}{3}.$$

The particular solution is

$$y = \frac{4x^3}{3} + \frac{17}{3}. \ \square$$

Exercises 4–10

1–6. Find the solution of the differential equation that satisfies the initial conditions.

- **1** $y' = 3x^2 - 6x + 1$, $y = 4$ when $x = 1$
- 2 $y' = x^3 + \dfrac{1}{x^2}$, $y = 3$ when $x = -1$
- **3** $f'(x) = \sqrt{x} - 3$, $f(4) = \tfrac{1}{3}$
- 4 $f'(x) = (x + 1)(x - 1)$, $f(2) = 4$
- **5** $f''(x) = 6x - 1$, $f'(1) = 4$, $f(1) = 8$
- 6 $f''(x) = \dfrac{1}{\sqrt{x}} + x$, $f'(1) = 14$, $f(1) = 4$

7–10. Find the distance function $s(t)$ that satisfies the given condition.

- **7** $v = t^2 + 2t$, $s(0) = 2$
- 8 $v = (2t - 6)^2$, $s(0) = 1$
- **9** $v = t^3 + t^{-1/2}$, $s(4) = 10$
- 10 $v = t\sqrt{1 - t^2}$, $s(0) = 2$

11–14. For each acceleration function $a(t)$, find the velocity function $v(t)$ and the distance function $s(t)$ satisfying the given conditions.

- **11** $a(t) = t - 1$, $v(4) = 5$, $s(0) = 3$
- 12 $a(t) = \dfrac{1}{(1 + t)^3}$, $v(0) = 1$, $s(1) = 2$
- **13** $a(t) = t^2 - 2$, $v(0) = -2$, $s(2) = -7$
- 14 $a(t) = t - 1/\sqrt{t}$, $v(4) = 0$, $s(0) = 5$

15 A projectile is fired vertically upward from the earth with an initial velocity of 96 ft/sec. Find its distance $s(t)$ above the earth at time t. Find its maximum height above the earth.

16 An automobile starts from rest. What constant acceleration will enable it to travel 112 ft in 4 sec? What is its speed at the end of 4 sec?

17 An automobile is traveling at 60 mph (88 ft/sec) when the driver applies the brakes, causing it to stop in 8 sec with constant deceleration. How far does the automobile travel before it stops?

18 An automobile starts from rest and with constant acceleration achieves a speed of 60 mph (88 ft/sec) in 11 sec. How far does it travel during the 11 sec?

19 A ball starts from rest and rolls down a 96-ft ramp with constant acceleration. At the end of 1 sec it is rolling with a speed of 12 ft/sec. What is its velocity when it reaches the end of the ramp?

20 The marginal cost of producing x units of a commodity is $30 - 0.02x$ dollars. Find the cost function if it costs $20 to produce one unit. What is the cost of producing 100 units?

•21 The rate of change of the marginal revenue function is $2 - 6x$, where x is the number of thousands of units produced. It is known that the revenue function satisfies the conditions $R(2) = 196$ and $R(3) = 300$. Find the revenue function.

•22 A ball starts from rest and rolls 125 ft down a ramp subject to a constant acceleration of 10 ft/sec². It then rolls onto another ramp, where it is subject to a constant acceleration of 20 ft/sec². Find the velocity of the ball at the instant when it has rolled 60 ft down the second ramp.

Review Problems

1 Define the following terms:
   (a) maximum of f on $[a, b]$ (b) local maximum
   (c) critical point   (d) point of inflection

2 State the following tests for local extrema:
   (a) The first derivative test.
   (b) The second derivative test.
   (c) The three point test.

3–5. Find the maximum and minimum of f on the indicated interval. Locate all critical points.

• 3 $f(x) = (x^2 - 1)^3$ on $[-2, 1]$

• 4 $f(x) = \sqrt[3]{x^3 + 1}$ on $[-1, 1]$

• 5 $f(x) = |x|$ on $[-1, 2]$

6 Explain why every number x that is not an integer is a critical point of the greatest integer function $f(x) = [x]$.

7–12. Find all local extrema and points of inflection.

• 7 $f(x) = 3x^2 + 2x + 2$ • 8 $f(x) = x^3 - 3x^2 + 2$

• 9 $f(x) = (x^2 - 1)^4$ •10 $f(x) = x(x - 1)^3$

•11 $f(x) = 3x^5 - 5x^4$ •12 $f(x) = x^{5/3} - 15x$

13–18. Determine where each of the functions in problems 7–12 is increasing, decreasing, concave upward, concave downward. Sketch the graph.

Review Problems

- **13** Problem 7
- **14** Problem 8
- **15** Problem 9
- **16** Problem 10
- **17** Problem 11
- **18** Problem 12

- **19** Find two numbers x and y such that $x + y = 10$ and $x^2 + y^2$, the sum of the squares, is a minimum.

- **20** An open-topped box with a square base has a total surface area of 108 ft². Find the dimensions that make the volume a maximum.

- **21** A piece of wire 40 ft long is to be cut into two pieces, one to be bent into a circle and the other into a square. Where should the wire be cut so as to make the combined area as small as possible?

- **22** A farmer has 500 ft of fencing to enclose a rectangular field. Find the dimensions of the field of largest area that can be enclosed, assuming that an existing fence can be used for one side of the field.

- **23** At 1:00 P.M. car A is 120 miles south of car B and is traveling north at the rate of 80 mph. Car B is traveling west at the rate of 40 mph. When will the distance between the two cars be a minimum?

- **24** Find the volume of the right circular cylinder of maximum surface area (side, top, and bottom) that can be inscribed in a right circular cone of radius 6 in. and height 18 in.

- **25** The demand function for a commodity is $d(x) = 750 + 60x - 2x^2$, where x is the selling price. Find the value of x that maximizes: **(a)** the marginal revenue; **(b)** the revenue.

26–27. Let $C(x)$ be the cost and $R(x)$ the revenue for the manufacture and sale of x units. Find the production level that yields the maximum possible profit.

- **26** $C(x) = \dfrac{1}{1500}x^3 + \dfrac{x^2}{40} + 20x + 2000$, $R(x) = -\dfrac{1}{3000}x^3 + \dfrac{x^2}{4} + 50x$

- **27** $C(x) = \dfrac{1}{20}x^2 + 700x + 500$, $R(x) = 2700x - \dfrac{3x^2}{20}$

- **28** A manufacturer sells shirts to department stores in lots of 125–200 shirts. The basic wholesale price per shirt is $4, but a discount of 2 cents a shirt is given for each shirt ordered above the minimum number of 125. For example, if 135 shirts are ordered, the discount is 20 cents per shirt, resulting in a price of $3.80 per shirt. If 175 shirts are ordered, the price is $3 per shirt. What size lot returns the greatest amount of money to the manufacturer?

●29 Show that the polynomial equation $x^3 - 5x = 1$ has a solution between 2 and 3. Use Newton's method to approximate it.

30 State the Mean Value Theorem. Draw a geometrical figure that illustrates it.

31–32. Use the Mean Value Theorem to prove the propositions.

●31 If f is continuous on $[-1, 2]$ and $f'(x) = 0$ for each x, $-1 < x < 2$, then $f(2) = f(-1)$.

32 If f is continuous on $[0, 1]$ and $f'(x) < 0$ for each x, where $0 < x < 1$, then $f(0) > f(1)$.

33 (a) What does it mean to say that $F(x)$ is an antiderivative of $f(x)$ on an open interval I?

34 Let $F(x)$ and $G(x)$ be antiderivatives of $f(x)$ on an open interval I. Explain why $F(x) - G(x)$ must be a constant. (Quote an appropriate theorem.)

35–38. Calculate general antiderivatives of the functions.

●35 $f(x) = x^2 + 3x - 1$

●36 $f(x) = \dfrac{x+1}{x^3}$.

●37 $f(x) = x^{-1/2} + x^{1/2} + 1$

●38 $f(x) = 5x^3 + x^{-2}$

39–40. Solve the differential equations subject to the condition that $y = 2$ when $x = 0$.

●39 $y' = 3x^2 - x + 1$

●40 $y' = \dfrac{1}{(x-1)^2}$

●41 A stone is thrown straight down from a bridge 200 ft above the water with an initial speed of 40 ft/sec. Find the speed of the stone when it strikes the water.

●42 The slope of the line tangent to the graph of a function f at the point $(x, f(x))$ is given by $m(x) = (2 - x)(3 - x^2)$. It is known that $f(1) = 5$. Find a formula for $f(x)$.

●43 A car is traveling down a straight road at 60 mph (88 ft/sec) when the driver suddenly applies the brakes, causing a constant deceleration that brings the car to a complete stop in 10 sec. How far does the car travel before it stops?

5 The Integral

In elementary geometry, you studied a number of formulas for the areas of plane regions—rectangles, triangles, circles, and so on. These formulas can be derived by a method of refining approximations. This method is so powerful that it can be applied to regions with fairly irregular boundaries—the region in Figure 5–1, for example. In this chapter we develop this method for the calculation of areas.* We will then prove an astonishing result: the area of a plane region can be calculated from the antiderivative of the bounding function. This link between an area and the slope of a tangent line, called the Fundamental Theorem of Calculus, is as powerful as it is unexpected. It will enable us to calculate not just areas but also such diverse quantities as volume and work (Chapter 6) and population growth (Chapter 8).

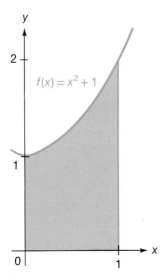

Figure 5–1 The area under the graph of $f(x) = x^2 + 1$, $0 \leq x \leq 1$.

---

* In the first part of this chapter we assume that the reader intuitively understands the concept of *area*. The term is defined in Section 6–1.

5-1 The Area Problem. Sigma Notation

How can we calculate the area of the shaded region pictured in Figure 5-1? Our approach will involve a limiting process using rectangles. We shall try to approximate the region with rectangles, measuring their areas. We can get better and better approximations by repeating our procedure, using more rectangles, with smaller bases than those used previously. If S_n is the sum of the areas of the rectangles when n rectangles are used and if S_n has a limiting value when n, the number of rectangles, increases without bound, then it seems reasonable that the area should equal this limiting value. In other words,

$$\text{Area} = \lim_{n \to \infty} S_n$$

provided a limiting value exists.

Figure 5-2 seems to bear out these remarks. It appears that S_8 (the sum of the areas of the eight rectangles shown in b) is a better approximation to the area of the region than S_4, and that S_{24} is even better.

These considerations lead us to a new type of limit:

$$\lim_{n \to \infty} S_n$$

(read "limit of S_n as n approaches infinity") where S_n is a function of the positive

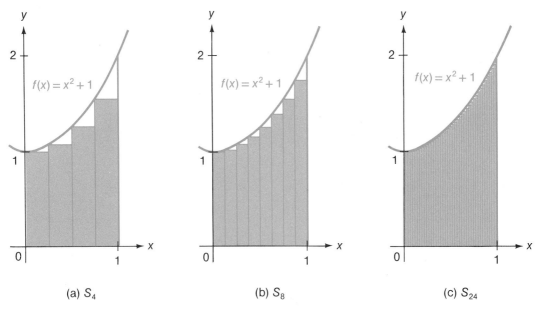

(a) S_4 (b) S_8 (c) S_{24}

Figure 5-2 Approximations of the area under a curve.

5-1 The Area Problem

integer n. At present we discuss the concept intuitively, postponing a more rigorous development to Section 5–2.

If S_n is defined for each positive integer n, then we write

$$\lim_{n \to \infty} S_n = L$$

provided the values of S_n approach L as n gets large without bound. For example, if

$$S_n = 2 - \frac{1}{n} + \frac{3}{n^2},$$

then n and n^2 both get large without bound as $n \to \infty$. Consequently, $1/n$ and $3/n^2$ both approach zero. It follows that

$$\lim_{n \to \infty} S_n = \lim_{n \to \infty} \left(2 - \frac{1}{n} + \frac{3}{n^2}\right) = 2 - 0 + 0 = 2$$

so that S_n approaches 2 as $n \to \infty$.

The limits we consider in this section can all be calculated by arguments like the one used for $S_n = 2 - 1/n + 3/n^2$.

We now return to the area problem and calculate the area of the region under the graph of $f(x) = x^2 + 1$, $0 \le x \le 1$. (See Fig. 5–1.)

Calculation of Area

Our main problem is to get a formula for S_n, the sum of the areas of the rectangles when n rectangles are inscribed in the figure. To do this, we partition the interval $[0, 1]$ into n equal subintervals and use these as the bases of the rectangles. Let Δx be the length of each subinterval. Then

$$\Delta x = \frac{1}{n}.$$

Let $x_0, x_1, x_2, \ldots, x_n$ be the endpoints of the subintervals. Then (see Fig. 5–3)

$x_0 = 0 =$ left-hand endpoint of first subinterval;
$x_1 = \Delta x =$ left-hand endpoint of second subinterval;
$x_2 = 2\,\Delta x =$ left-hand endpoint of third subinterval;
. . . .
$x_{n-1} = (n-1)\,\Delta x =$ left-hand endpoint of nth subinterval.

As indicated in Figure 5–3, we shall use the left-hand endpoints to determine the heights of the rectangles. Since the left-hand endpoint of the kth rectangle is $x_{k-1} = (k-1)\,\Delta x$, then the height of the kth rectangle is

$$f(x_{k-1}) = x_{k-1}^2 + 1 = [(k-1)\,\Delta x]^2 + 1$$
$$= (k-1)^2(\Delta x)^2 + 1 = \frac{(k-1)^2}{n^2} + 1,$$

and the area of the kth rectangle is

$$f(x_{k-1})\,\Delta x = \left[\frac{(k-1)^2}{n^2} + 1\right]\frac{1}{n} = \frac{(k-1)^2}{n^3} + \frac{1}{n}.$$

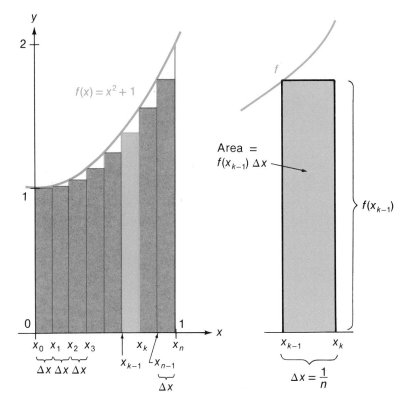

(a) $S_n = f(x_0)\,\Delta x + f(x_1)\,\Delta x + \cdots + f(x_{n-1})\,\Delta x$ (b) Area of kth rectangle $= f(x_{k-1})\,\Delta x$

Figure 5-3 Approximating sum for the area under a curve.

It follows that the sum of the areas of the n rectangles is

$$S_n = \underbrace{\left[\frac{(1-1)^2}{n^3} + \frac{1}{n}\right]}_{\substack{\text{area of first} \\ \text{rectangle} \\ (k=1)}} + \underbrace{\left[\frac{(2-1)^2}{n^3} + \frac{1}{n}\right]}_{\substack{\text{area of second} \\ \text{rectangle} \\ (k=2)}} + \underbrace{\left[\frac{(3-1)^2}{n^3} + \frac{1}{n}\right]}_{\substack{\text{area of third} \\ \text{rectangle} \\ (k=3)}}$$

$$+ \cdots + \underbrace{\left[\frac{(n-1)^2}{n^3} + \frac{1}{n}\right]}_{\substack{\text{area of } n\text{th} \\ \text{rectangle} \\ (k=n)}}$$

$$= \frac{0^2}{n^3} + \frac{1}{n} + \frac{1^2}{n^3} + \frac{1}{n} + \frac{2^2}{n^3} + \frac{1}{n} + \cdots + \frac{(n-1)^2}{n^3} + \frac{1}{n}$$

5-1 The Area Problem

$$= \frac{1}{n^3}[0^2 + 1^2 + 2^2 + \cdots + (n-1)^2] + \underbrace{\frac{1}{n} + \frac{1}{n} + \frac{1}{n} + \cdots + \frac{1}{n}}_{n \text{ terms}}.$$

$$= \frac{1}{n^3}[1^2 + 2^2 + \cdots + (n-1)^2] + n\frac{1}{n}.$$

The sum in this last expression is one of several that are considered in elementary algebra. It is shown that if m is a positive integer, then

Sums from Algebra

(1) $\quad 1 + 2 + 3 + \cdots + m = \dfrac{m(m+1)}{2}.$

(2) $\quad 1^2 + 2^2 + 3^2 + \cdots + m^2 = \dfrac{m(m+1)(2m+1)}{6}.$

(3) $\quad 1^3 + 2^3 + 3^3 + \cdots + m^3 = \dfrac{m^2(m+1)^2}{4}.$

If we choose $m = n - 1$ in the second formula, we get

$$1^2 + 2^2 + 3^2 + \cdots + (n-1)^2 = \frac{(n-1)n(2n-1)}{6} = \frac{2n^3 - 3n^2 + n}{6}.$$

If we substitute this value in the expression for S_n, we get

$$S_n = \frac{1}{n^3}\frac{2n^3 - 3n^2 + n}{6} + 1$$

$$= \frac{1}{3} - \frac{1}{2n} + \frac{1}{6n^2} + 1.$$

We now let $n \to \infty$, obtaining

$$\text{Area of region} = \lim_{n \to \infty} S_n = \lim_{n \to \infty}\left[\frac{1}{3} - \frac{1}{2n} + \frac{1}{6n^2} + 1\right]$$

$$= \frac{1}{3} - 0 + 0 + 1 = \frac{4}{3}.$$

The preceding example was analyzed in detail to acquaint you with certain basic concepts that will subsequently be developed. You may notice on reflection that this method raises an important problem: *Suppose we construct the "approximating" rectangles in several different ways. What assurance do we have that we should get the same value for the area by each construction?*

For example, we could construct the rectangles so that they cover the region instead of being contained within it (Figure 5-4a). We could use horizontal rectangles instead of vertical ones; that is, we could partition the y-axis instead of the x-axis (Figure 5-4b). We could use squares or rectangles stacked in some fashion, as in Figure 5-4c. We could even use geometrical figures different from rectangles—say

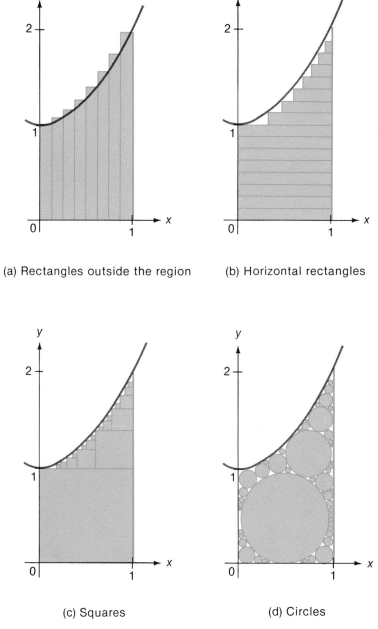

(a) Rectangles outside the region (b) Horizontal rectangles

(c) Squares (d) Circles

Figure 5-4 Alternative approaches to the area problem.

5-1 The Area Problem

circles, as in Figure 5–4d. It is not at all obvious that these approaches would lead to the same final value for the area of the region as what we have already determined. While it is possible to reconcile the approaches and show that we should get the same value for the area in each case, the proof is beyond the scope of this book.

As we shall see, the limiting process used for determining the area can be used to solve many other problems also. In the following sections we develop the basic theory for the more general concept of the *definite integral*. The results can then be interpreted in terms of *area, volume, profit, work,* or some other concept according to the problem at hand.

SUMMATION (SIGMA) NOTATION

We shall deal with complicated sums such as

$$S_n = \left[\frac{(1-1)^2}{n^3} + \frac{1}{n}\right] + \left[\frac{(2-1)^2}{n^3} + \frac{1}{n}\right] + \left[\frac{(3-1)^2}{n^3} + \frac{1}{n}\right] + \cdots$$
$$+ \left[\frac{(n-1)^2}{n^3} + \frac{1}{n}\right]$$

many times. Mathematicians have developed a convenient shorthand notation to represent such sums. We write, for example,

$$\sum_{k=1}^{m} k^2$$

(read "summation of k^2 from 1 to m") as a symbol for the sum $1^2 + 2^2 + \cdots + m^2$. Similarly,

Σ

$$\sum_{k=1}^{m} k^3 = 1^3 + 2^3 + \cdots + m^3.$$

In general, if f is a function defined on the integers $a, a + 1, a + 2, \ldots, b$, then

$$\sum_{k=a}^{b} f(k) = f(a) + f(a + 1) + f(a + 2) + \cdots + f(b)$$

The letter k, called a *dummy index,* does not actually occur in the final form of the sum. Thus, a different dummy index, such as i or j, could be used.

EXAMPLE 1

(a) $\sum_{k=1}^{3} (k^k + 5) = (1^1 + 5) + (2^2 + 5) + (3^3 + 5) = 47.$

(b) $\sum_{j=4}^{7} j = 4 + 5 + 6 + 7 = 22.$

(c) $\sum_{i=3}^{9} i^2 = 3^2 + 4^2 + 5^2 + 6^2 + 7^2 + 8^2 + 9^2 = 280.$ □

Using the sigma notation, we can write the formulas for the sums of the first three powers as:

Sums from Algebra

(1) $\displaystyle\sum_{k=1}^{m} k = \frac{m(m+1)}{2}.$

(2) $\displaystyle\sum_{k=1}^{m} k^2 = \frac{m(m+1)(2m+1)}{6}.$

(3) $\displaystyle\sum_{k=1}^{m} k^3 = \frac{m^2(m+1)^2}{4}.$

Two properties of sums are needed:

Properties of Sums

(4) $\displaystyle\sum_{k=1}^{m} cf(k) = c\sum_{k=1}^{m} f(k).$

(5) $\displaystyle\sum_{k=1}^{m} [f(k) \pm g(k)] = \sum_{k=1}^{m} f(k) \pm \sum_{k=1}^{m} g(k).$

Property 4 is the distributive property written in sigma notation; property 5 follows from the commutative and associative laws. These properties can be proved by expanding the sums.

We may occasionally find a sum involving a constant function. If we expand the sum, we see that

(6) $\displaystyle\sum_{k=1}^{m} c = \underbrace{c + c + \cdots + c}_{m \text{ terms}} = mc.$

EXAMPLE 2 Simplify $\displaystyle\sum_{k=1}^{m}(6k^2 + 2k - 5)$.

Solution

$$\sum_{k=1}^{m}(6k^2 + 2k - 5) = \sum_{k=1}^{m} 6k^2 + \sum_{k=1}^{m} 2k - \sum_{k=1}^{m} 5 \quad \text{(by (5))}$$

$$= 6\sum_{k=1}^{m} k^2 + 2\sum_{k=1}^{m} k - \sum_{k=1}^{m} 5 \quad \text{(by (4))}$$

$$= 6\frac{m(m+1)(2m+1)}{6} + 2\frac{m(m+1)}{2} - 5m$$

$$= 2m^3 + 4m^2 - 3m. \quad \square \qquad \text{(by (2), (1), and (6))}$$

Exercises 5–1

1–8. Use sigma notation to rewrite the sum more compactly.

- **1** $1 + 2 + 3 + 4 + 5$
- **2** $1^2 + 2^2 + 3^2 + 4^2 + \cdots + 8^2$
- **3** $1 \cdot 2^3 + 2 \cdot 3^4 + 3 \cdot 4^5 + 4 \cdot 5^6$
- **4** $\frac{3}{1} + \frac{4}{2} + \frac{5}{3} + \cdots + \frac{11}{9}$
- **5** $2^3 + 4^3 + \cdots + (2n)^3$
- **6** $5^2 + 6^2 + 7^2 + \cdots + (n+1)^2$
- **7** $m^3 + (m+1)^3 + (m+2)^3 + \cdots + n^3$
- **8** $\frac{1}{m} + \frac{1}{m+1} + \cdots + \frac{1}{n+1}$

9–18. Use formulas (1)–(6) to evaluate the sum.

- **9** $1 + \sum_{k=1}^{6} k^2$
- **10** $\sum_{k=1}^{6} (k^2 + 1)$
- **11** $\sum_{k=1}^{10} k(k+1)$
- **12** $\sum_{j=1}^{40} (-1)j$
- **13** $\sum_{i=4}^{50} 7$
- **14** $\sum_{t=3}^{20} 3$
- **15** $\sum_{k=1}^{n} (k^2 - k)$
- **16** $\sum_{i=1}^{n} (i^3 + i^2)$
- **17** $\sum_{j=5}^{n} (3j^2 + 2j)$
- **18** $\sum_{k=10}^{n} (k^3 - 2k^2 + 2k + 1)$

19–24. Evaluate the limit.

- **19** $\lim_{n \to \infty} \frac{1}{n^2} \sum_{j=1}^{n} j$
- **20** $\lim_{n \to \infty} \frac{2}{n^2} \sum_{j=3}^{n+1} j$
- **21** $\lim_{n \to \infty} \sum_{k=1}^{n} \frac{k^2}{n^3}$
- **22** $\lim_{n \to \infty} \sum_{k=1}^{n+1} \frac{k^2 + 1}{n^3}$
- **23** $\lim_{n \to \infty} \sum_{k=1}^{2n-1} \frac{2k}{n^2}$
- **24** $\lim_{n \to \infty} \sum_{k=1}^{n-1} \frac{(k-1)^2}{n^3}$

25–27. Use the method in the text to calculate the area of the region: (1) Partition the interval into n equal subintervals; (2) use either the right- or left-hand endpoint of each subinterval to construct an "approximating" rectangle; (3) calculate (in terms of n) the value of S_n, the sum of the areas of the n rectangles; (4) calculate the limit of S_n as $n \to \infty$.

Draw a figure like Figure 5–3, illustrating the approximating rectangles when $n = 8$.

- **25** The region under the graph of $f(x) = x + 1$ for $1 \leq x \leq 3$.
- **26** The region under the graph of $f(x) = x^2$ for $0 \leq x \leq 2$.
- **27** The region under the graph of $x - x^2$ for $0 \leq x \leq 1$.

28 Let m and n be integers, with $1 < m < n$. Prove that

$$\sum_{k=m}^{n} k = \frac{n(n+1)}{2} - \frac{m(m-1)}{2}.$$

(*Hint:* Use Summation Formula (1).)

---

29 Let c be a constant. Expand the sums to show that the following properties are correct.

(a) $\sum_{k=1}^{n} [f(k) + g(k)] = \sum_{k=1}^{n} f(k) + \sum_{k=1}^{n} g(k)$

(b) $\sum_{k=1}^{n} cf(k) = c \sum_{k=1}^{n} f(k)$

30–31. Show that the formula is correct.

30 $\sum_{k=0}^{n-1} (n-k)^2 = \sum_{k=1}^{n} k^2$

31 $\sum_{k=-n}^{n} (3k^2 + 1) = 2n^3 + 3n^2 + 3n + 1$

32 Use the formulas in the text to show that

$$1^3 + 2^3 + 3^3 + \cdots + m^3 = (1 + 2 + 3 + \cdots + m)^2.$$

33 Use the fact that $k^3 - (k-1)^3 = 3k^2 - 3k + 1$ to prove that

$$\sum_{k=1}^{m} k^2 = \frac{m(m+1)(2m+1)}{6}.$$

(*Hint:* Sum both sides of the equation from $k = 1$ to $k = m$ and simplify.)

5–2 The Definite Integral

In Section 5–1 we have just calculated the area of the region bounded by the graph of $f(x) = x^2 + 1$, the coordinate axes, and the vertical line $x = 1$. The process can be generalized. We now consider the equivalent calculation involving an arbitrary function f that is continuous over a closed interval [a, b]. If it happens that $f(x) \geq 0$ on [a, b], then the final result can be interpreted as the area under the graph of f.

Partition   We first need to develop some terminology. A *partition* $\mathscr{P}$ of an interval [a, b] is a set of points $\{x_0, x_1, x_2, \ldots, x_n\}$, where

$$a = x_0 < x_1 < x_2 < \cdots < x_n = b.$$

Each partition $\mathscr{P}$ splits the interval [a, b] into subintervals

$$[x_0, x_1], [x_1, x_2], \ldots, [x_{n-1}, x_n]$$

as in Figure 5–5.

5–2 The Definite Integral

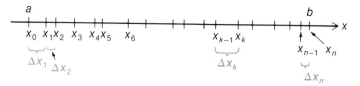

Figure 5–5 Partition of the interval $[a, b]$, with $\Delta x_k =$ length of the kth subinterval.

Let Δx_k be the length of the kth subinterval. Then

$\Delta x_1 = x_1 - x_0 =$ length of first subinterval;
$\Delta x_2 = x_2 - x_1 =$ length of second subinterval;
$\cdots\cdots\cdots\cdots\cdots\cdots\cdots\cdots\cdots\cdots$
$\Delta x_n = x_n - x_{n-1} =$ length of nth subinterval.

Norm of Partition

The *norm* of the partition $\mathcal{P}$, denoted by $\|\mathcal{P}\|$, equals the length of the largest subinterval. Thus,

$$\|\mathcal{P}\| = \text{largest of numbers } \Delta x_1, \Delta x_2, \ldots, \Delta x_n.$$

Let f be continuous on the closed interval $[a, b]$. Let $\mathcal{P}$ be the partition of $[a, b]$ defined by the numbers

$$a = x_0 < x_1 < x_2 < \cdots < x_n = b.$$

We choose numbers $\xi_1, \xi_2, \xi_3, \ldots, \xi_n$ on the n subintervals defined by the partition $\mathcal{P}$. That is,

$x_0 \leq \xi_1 \leq x_1$
$x_1 \leq \xi_2 \leq x_2$
$\cdots\cdots\cdots\cdots$
$x_{n-1} \leq \xi_n \leq x_n.$

We now form the sum

$$S_n = \sum_{k=1}^{n} f(\xi_k) \Delta x_k = f(\xi_1) \Delta x_1 + f(\xi_2) \Delta x_2 + \cdots + f(\xi_n) \Delta x_n.$$

The function f need not be nonnegative on $[a, b]$, but if it is, then the sum S_n equals the sum of areas of rectangles, as in Figure 5–6. Observe that the value of S_n depends on the partition $\mathcal{P}$ and on the numbers $\xi_1, \xi_2, \ldots, \xi_n$ on the subintervals, as well as on the function f and the interval $[a, b]$. In most cases if any of these numbers is changed, the value of S_n also changes.

We now consider all possible sums S_n that can be formed in the manner discussed above. In other words, we consider all possible partitions $\mathcal{P}$ of $[a, b]$. For each of

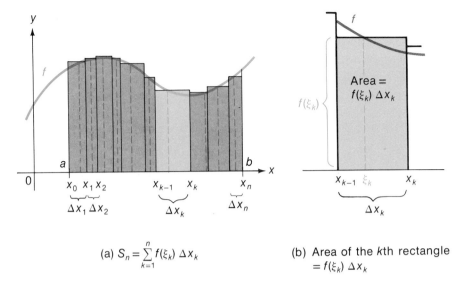

(a) $S_n = \sum_{k=1}^{n} f(\xi_k)\,\Delta x_k$

(b) Area of the kth rectangle $= f(\xi_k)\,\Delta x_k$

Figure 5–6 If f is nonnegative on $[a, b]$, then $S_n =$ the sum of the areas of the rectangles.

these partitions, we consider all possible choices of $\xi_1, \xi_2, \ldots, \xi_n$ on the subintervals and form the corresponding sums

$$S_n = \sum_{k=1}^{n} f(\xi_k)\,\Delta x_k.$$

It can be proved under our assumptions about f that this set of sums has a limit L in the sense that the sums S_n must be close to L whenever the norms of the corresponding partitions are close to zero.

This new limit is completely different from the types of limits that we have considered earlier (Chapter 2). It is denoted by the symbol

$$L = \lim_{\|\mathcal{P}\|\to 0} S_n = \lim_{\|\mathcal{P}\|\to 0} \sum_{k=1}^{n} f(\xi_k)\,\Delta x_k.$$

To a certain extent this notation is misleading, since it creates the impression that the norm of a single partition is tending towards zero. It is meant to indicate instead that S_n is close to L whenever $\mathcal{P}$ is a partition that has its norm close to zero. This limit is defined more precisely as follows.

DEFINITION

Limit of Sum Defined by Partitions

Let f be defined on $[a, b]$. Let $\mathcal{P}$ represent an arbitrary partition of $[a, b]$ into n subintervals; let Δx_k be the length and ξ_k a point on the kth subinterval. We say that

$$\lim_{\|\mathcal{P}\|\to 0} \sum_{k=1}^{n} f(\xi_k)\,\Delta x_k = L$$

5–2 The Definite Integral

provided that for each $\epsilon > 0$, there exists a $\delta > 0$ (δ depends on ϵ) such that

$$\left| \sum_{k=1}^{n} f(\xi_k) \Delta x_k - L \right| < \epsilon$$

whenever $\mathcal{P}$ is a partition that satisfies the condition $\|\mathcal{P}\| < \delta$.

In other words,

$$\lim_{\|\mathcal{P}\| \to 0} \sum_{k=1}^{n} f(\xi_k) \Delta x_k = L$$

if and only if we can make the value of the sum

$$S_n = \sum_{k=1}^{n} f(\xi_k) \Delta x_k$$

as close to L as we wish (between $L - \epsilon$ and $L + \epsilon$) by requiring that the length of each subinterval in a partition should be close to zero (less than δ). This holds regardless of the number we choose for ξ_k in the kth subinterval.

Let

$$L = \lim_{\|\mathcal{P}\| \to 0} S_n = \lim_{\|\mathcal{P}\| \to 0} \sum_{k=1}^{n} f(\xi_k) \Delta x_k.$$

Then L is completely independent of the partitions $\mathcal{P}$ and the numbers $\xi_1, \xi_2, \ldots, \xi_n$ used to define the individual sums S_n. The limit L depends only on the function f and the interval $[a, b]$. This limit L is called the *definite integral* of f over $[a, b]$, denoted by writing

Definite Integral

$$L = \int_a^b f(x)\, dx.$$

(See Fig. 5–7.)

Approximating (Riemann) Sum

The sum $S_n = \sum_{k=1}^{n} f(\xi_k) \Delta x_k$ is called an *approximating sum*, or a *Riemann sum*, for the integral $\int_a^b f(x)\, dx$. The numbers a and b are called the *limits of integration*.

As mentioned above, if f is continuous on $[a, b]$, then it can be proved that the definite integral exists. For reference purposes, we state this result as a theorem. The proof is omitted.

THEOREM 5–1

A Continuous Function is Integrable

Let f be continuous on the closed interval $[a, b]$. Then

$$\int_a^b f(x)\, dx$$

exists.

Remark 1 The entire symbol $\int_a^b f(x)\, dx$ is used for the definite integral. For the next few sections you should think of the symbol dx as a vestige of the Δx_k in the approximating sum S_n. We shall see (Section 5–5) that there is good reason to consider dx a differential.

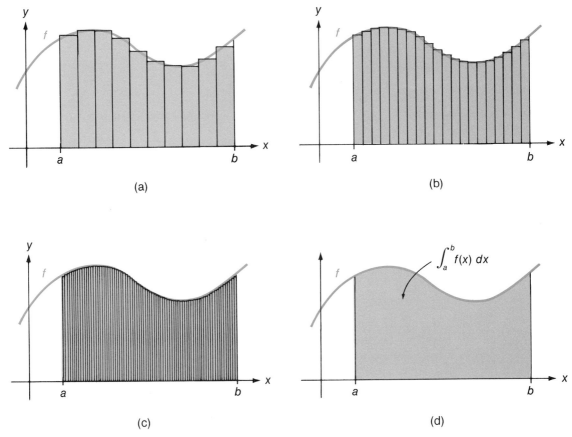

Figure 5–7 The sum $S_n = \sum_{k=1}^{n} f(\xi_k) \Delta x_k$ tends to a limit as $\|\mathcal{P}\| \to 0$ and $n \to \infty$.

Remark 2 In certain cases the definite integral $\int_a^b f(x)\,dx$ can be defined even when f is not continuous on $[a, b]$. (This extension of the theory is discussed in Sections 10–3 and 10–4.) The function f is said to be *integrable on* $[a, b]$ if $\int_a^b f(x)\,dx$ exists.

Remark 3

Special Partitions

If we know that f is continuous on $[a, b]$, then we have great freedom in choosing the partition into subintervals and the points $\xi_1, \xi_2, \ldots, \xi_n$ on the subintervals that are used to compute the approximating sums. We frequently choose the subintervals of equal length $\Delta x = (b - a)/n$ and let ξ_k be the right-hand or left-hand endpoint of the kth subinterval. This can be done, because any choice of subintervals and any choice of the ξ_k leads to the same value of the limit. The value of the limit is independent of the partitions and the numbers $\xi_1, \xi_2, \ldots, \xi_n$ on the subintervals.

The approximating sums have a particularly simple form if we choose all the subintervals of equal length,

$$\Delta x = \frac{b - a}{n},$$

5-2 The Definite Integral

and choose $\xi_k = x_k$, the right-hand endpoint of the kth subinterval. Then

$$x_0 = a,$$
$$x_1 = \xi_1 = a + \Delta x,$$
$$x_2 = \xi_2 = a + 2\Delta x,$$
$$\dots$$
$$x_n = \xi_n = a + n\Delta x.$$

The approximating sum S_n can then be written in the comparatively simple form

$$S_n = \sum_{k=1}^{n} f(\xi_k)\, \Delta x_k = \sum_{k=1}^{n} f(x_k)\, \Delta x.$$

Observe that in this case, $\|\mathcal{P}\| \to 0$ if and only if $n \to \infty$. Thus,

$$\int_a^b f(x)\, dx = \lim_{n\to\infty} S_n = \lim_{n\to\infty} \sum_{k=1}^{n} f(x_k)\, \Delta x.$$

EXAMPLE 1 Show that $\displaystyle\int_1^5 x\, dx = \frac{5^2 - 1^2}{2}$.

Solution Let $f(x) = x$. Partition the interval $[1, 5]$ into n equal subintervals of length $\Delta x = (5-1)/n = 4/n$ and let $\xi_k = x_k$ be the right-hand endpoint of the kth interval. Then

$$S_n = \sum_{k=1}^{n} f(x_k)\, \Delta x = \sum_{k=1}^{n} x_k\, \Delta x.$$

Recall that

$$x_0 = 1,$$
$$x_1 = 1 + \Delta x,$$
$$x_2 = 1 + 2\Delta x,$$

and so on. Then

$$S_n = \sum_{k=1}^{n} x_k\, \Delta x = \sum_{k=1}^{n} (1 + k\, \Delta x)\, \Delta x$$

$$= \sum_{k=1}^{n} \Delta x + \Delta x^2 \sum_{k=1}^{n} k = n\, \Delta x + \Delta x^2\, \frac{n(n+1)}{2}$$

$$= n\, \frac{4}{n} + \frac{16}{n^2}\, \frac{n(n+1)}{2} = 4 + 8\left(\frac{n^2 + n}{n^2}\right)$$

$$= 4 + 8\left(1 + \frac{1}{n}\right).$$

Thus

$$\int_1^5 x\, dx = \lim_{n\to\infty} S_n = \lim_{n\to\infty}\left[4 + 8\left(1 + \frac{1}{n}\right)\right]$$
$$= 4 + 8(1 + 0) = 12.$$

Since $(5^2 - 1^2)/2 = 12$, then

$$\int_1^5 x \, dx = \frac{5^2 - 1^2}{2}. \quad \square$$

The result of Example 1 can be generalized to show that

$$\int_a^b x \, dx = \frac{b^2 - a^2}{2}.$$

The details are left for you to work out. (See Exercise 6.)
A related integral formula is established in Example 2.

EXAMPLE 2 Show that $\int_a^b x^2 \, dx = \frac{b^3 - a^3}{3}$.

Solution We partition $[a, b]$ into n subintervals of length $\Delta x = (b - a)/n$ and let $\xi_k = x_k$ be the right-hand endpoint of the kth subinterval. The approximating sum is

$$S_n = \sum_{k=1}^n f(x_k) \, \Delta x = \sum_{k=1}^n x_k^2 \, \Delta x = \sum_{k=1}^n (a + k \, \Delta x)^2 \, \Delta x$$

$$= \sum_{k=1}^n (a^2 \, \Delta x + 2ak \, (\Delta x)^2 + k^2 \, (\Delta x)^3)$$

$$= \sum_{k=1}^n a^2 \, \Delta x + 2a \, (\Delta x)^2 \sum_{k=1}^n k + (\Delta x)^3 \sum_{k=1}^n k^2$$

$$= a^2 \, \Delta x \cdot n + 2a \, (\Delta x)^2 \, \frac{n(n+1)}{2} + (\Delta x)^3 \, \frac{n(n+1)(2n+1)}{6}$$

$$= a^2 \left(\frac{b-a}{n}\right) n + 2a \left(\frac{b-a}{n}\right)^2 \frac{n^2 + n}{2} + \left(\frac{b-a}{n}\right)^3 \frac{2n^3 + 3n^2 + n}{6}$$

$$= a^2(b - a) + a(b - a)^2 \left(1 + \frac{1}{n}\right) + \frac{(b-a)^3}{6}\left(2 + \frac{3}{n} + \frac{1}{n^2}\right).$$

Thus

$$\int_a^b x^2 \, dx = \lim_{n \to \infty} S_n$$

$$= a^2(b - a) + a(b - a)^2(1 + 0) + \frac{(b-a)^3}{6}(2 + 0 + 0),$$

which simplifies to

$$\int_a^b x^2 \, dx = \frac{b^3 - a^3}{3}. \quad \square$$

Remark 4 As we shall see later, the definite integral can be interpreted in several different ways. If $f(x) \geq 0$ over the interval $[a, b]$, then $\int_a^b f(x) \, dx$ equals the area under the graph of f.

Exercises 5–2

In other settings, the integral may equal total distance, volume of a solid, work done in moving an object from one point to another, pressure on a submerged plate, or mass of a figure cut from thin sheet metal. It will be useful, however, to think of the integral as area whenever possible. Many of our results can be explained easily if the integral $\int_a^b f(x)\,dx$ is thought of as the area under a curve, even though detailed proofs may be difficult and tedious.

Exercises 5–2

1–4. (a) Use the methods of this section to calculate the definite integral. Use the right-hand endpoint of each subinterval to determine the terms in a typical approximating sum S_n. (b) Sketch a region with area equal to the value of the integral.

•1 $\quad \int_{-2}^{0} x^2\,dx$

2 $\quad \int_{-1}^{2} x^2\,dx$

•3 $\quad \int_{1}^{3} (1 - 2x)\,dx$

4 $\quad \int_{5}^{7} (5 - 2x)\,dx$

5–7. Modify the argument of Example 2 to prove the formulas.

5 $\quad \int_a^b c\,dx = c(b - a)$

6 $\quad \int_a^b x\,dx = \dfrac{b^2 - a^2}{2}$

7 $\quad \int_a^b x^3\,dx = \dfrac{b^4 - a^4}{4}$

•8 It can be proved that $\int_a^b x^4\,dx = (b^5 - a^5)/5$. Compare this result with the results of Example 2 and Exercises 5, 6, and 7. What general result do you conjecture? Do you see any relation between $\int_0^b x^n\,dx$ and an antiderivative of the function x^n?

9–12. (a) Use the formulas in Example 2 and Exercises 5, 6, and 7 to evaluate the definite integral. (b) Sketch the graph of a region with area equal to the value of the integral.

•9 $\quad \int_3^6 x\,dx$

10 $\quad \int_1^5 \sqrt{2}\,dx$

•11 $\quad \int_{-1}^{\pi} x^2\,dx$

12 $\quad \int_1^3 x^3\,dx$

13–14. Calculate the integral by using the left-hand endpoint of each subinterval to determine the terms in the approximating sum.

13 Exercise 1   14 Exercise 2

15-16. Each of the integrals equals the area of a geometrical figure. Draw the figure and use a formula from geometry to calculate its area. Use this result to evaluate the original integral.

•15 $\int_1^4 4x\, dx$

•16 $\int_{-2}^2 \sqrt{4-x^2}\, dx$

17-18. Express the limits as definite integrals.

•17 $\lim_{n\to\infty} \frac{1}{n}\sum_{k=1}^{n} f\left(\frac{k}{n}\right) = \lim_{n\to\infty} \frac{1}{n}\left[f\left(\frac{1}{n}\right) + f\left(\frac{2}{n}\right) + \cdots + f\left(\frac{n}{n}\right)\right]$

•18 $\lim_{n\to\infty} \frac{2}{n}\sum_{k=1}^{n} f\left(1 + \frac{2k}{n}\right)$

5-3 Some Properties of Integrals

In the next section we prove an important theorem, the Fundamental Theorem of Calculus, that will enable us to calculate many definite integrals by means of antiderivatives. After we establish that theorem, we shall be able to calculate integrals such as

$$\int_1^2 (x^2 + 1)\, dx$$

in a few simple steps.

Several basic properties of integrals must be established before we can prove our main theorem. Most of these proofs are very technical and are omitted. We do, however, discuss the properties intuitively in terms of area and indicate the main ideas behind the proofs.

We shall find it convenient hereafter to have the definite integral defined when the upper limit of integration is less than the lower limit or equal to it. We define

$$\int_a^a f(x)\, dx = 0 \quad \text{provided } f \text{ is defined at } a$$

$$\int_b^a f(x)\, dx = -\int_a^b f(x)\, dx \quad \text{provided } f \text{ is continuous on } [a, b].$$

EXAMPLE 1 We have established (Section 5-2) that

$$\int_a^b x\, dx = \frac{b^2 - a^2}{2} \quad \text{and} \quad \int_a^b x^2\, dx = \frac{b^3 - a^3}{3}$$

5-3 Some Properties of Integrals

provided $b > a$. Then

(a) $\int_2^2 x \, dx = 0.$

(b) $\int_5^0 x \, dx = -\int_0^5 x \, dx = -\dfrac{5^2 - 0^2}{2} = -\dfrac{25}{2}.$

(c) $\int_3^1 x^2 \, dx = -\int_1^3 x^2 \, dx = -\dfrac{3^3 - 1^3}{3} = -\dfrac{26}{3}.$ □

ADDITION PROPERTY OF INTEGRALS

Let f be continuous and nonnegative on the interval $[a, c]$. Let b be a number on the interval $[a, c]$. Then the area under the graph from a to c equals the area from a to b added to the area from b to c. (See Fig. 5–8.) Using integrals, we have

$$\int_a^c f(x) \, dx = \int_a^b f(x) \, dx + \int_b^c f(x) \, dx.$$

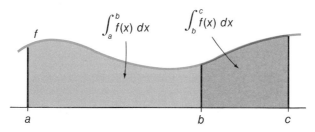

Figure 5–8 The addition property:
$\int_a^b f(x) \, dx + \int_b^c f(x) \, dx = \int_a^c f(x) \, dx.$

It can be shown that this property holds for any continuous function. The function f does not need to be nonnegative. The interpretation based on area, however, is valid only if f is nonnegative on $[a, c]$.

> **Property 1** Let f be continuous on $[a, c]$. If $a \leq b \leq c$, then
> $$\int_a^b f(x) \, dx + \int_b^c f(x) \, dx = \int_a^c f(x) \, dx.$$

The proof of this property is omitted. Essentially, the proof is based on the fact that each approximating sum for $\int_a^b f(x) \, dx$ can be added to an approximating sum for $\int_b^c f(x) \, dx$ to get an approximating sum for $\int_a^c f(x) \, dx$.

We can show that Property 1 holds for any order of a, b, and c. For example, if $a < c < b$, then

$$\int_a^b f(x)\, dx = \int_a^c f(x)\, dx + \int_c^b f(x)\, dx \quad \text{(by Property 1)}$$

$$= \int_a^c f(x)\, dx - \int_b^c f(x)\, dx$$

so that

$$\int_a^b f(x)\, dx + \int_b^c f(x)\, dx = \int_a^c f(x)\, dx.$$

Similar proofs can be devised for any other order of the numbers a, b, c. (See Exercise 16.) Thus, we get a more general property:

Addition Property

> **Property 1'** Let f be continuous on an interval that contains a, b, and c. Then
>
> $$\int_a^b f(x)\, dx + \int_b^c f(x)\, dx = \int_a^c f(x)\, dx.$$

COMPARISON PROPERTY

Let f and g be continuous on $[a, b]$ with $f(x) \leq g(x)$ for all x, $a \leq x \leq b$. If $f(x) \geq 0$ on $[a, b]$, it seems obvious that the area under the graph of f is less than or equal to the area under the graph of g for $a \leq x \leq b$. (See Fig. 5-9.) Using integrals, we obtain

$$\int_a^b f(x)\, dx \leq \int_a^b g(x)\, dx.$$

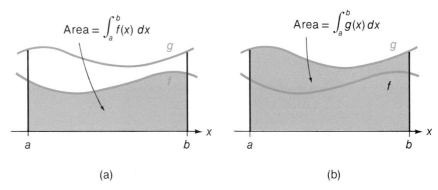

Figure 5-9 The comparison property. If $f(x) \leq g(x)$ for $a \leq x \leq b$, then $\int_a^b f(x)\, dx \leq \int_a^b g(x)\, dx$.

5–3 Some Properties of Integrals

It can be shown that this property holds for arbitrary continuous functions. The functions f and g do not need to be nonnegative on $[a, b]$. Thus, we get another property of integrals.

Comparison Property

Property 2 Let f and g be continuous on $[a, b]$. If $f(x) \leq g(x)$ for all x on $[a, b]$, then

$$\int_a^b f(x)\, dx \leq \int_a^b g(x)\, dx.$$

The basis of the proof is the fact that every partition $\mathcal{P}$ of $[a, b]$ leads to approximating sums for the two integrals with

$$\sum_{k=1}^n f(\xi_k)\, \Delta x_k \leq \sum_{k=1}^n g(\xi_k)\, \Delta x_k.$$

When we take the limit as $\|\mathcal{P}\| \to 0$, the same relation is preserved. Thus,

$$\int_a^b f(x)\, dx = \lim_{\|\mathcal{P}\| \to 0} \sum_{k=1}^n f(\xi_k)\, \Delta x_k \leq \lim_{\|\mathcal{P}\| \to 0} \sum_{k=1}^n g(\xi_k)\, \Delta x_k = \int_a^b g(x)\, dx.$$

The formal proof is omitted.

INTEGRALS WITH VARIABLE LIMITS OF INTEGRATION

Before we go on with our main results, we observe that the symbol x in the integral $\int_a^b f(x)\, dx$ is a dummy variable—it does not appear in the value of the integral, which depends only on the function f and the limits of integration, a and b. Thus, we could just as well use some other symbol, such as t, y, or α:

$$\int_a^b f(y)\, dy = \int_a^b f(t)\, dt = \int_a^b f(\alpha)\, d\alpha = \int_a^b f(x)\, dx.$$

For example, since $\int_a^b x^2\, dx = (b^3 - a^3)/3$, then

$$\int_2^4 t^2\, dt = \int_2^4 \alpha^2\, d\alpha = \frac{4^3 - 2^3}{3} = \frac{56}{3}.$$

Suppose now that f is continuous on an interval I that contains a. At each $x \in I$, we can define a new function $F(x)$ by the rule of correspondence

$$F(x) = \int_a^x f(t)\, dt.$$

In the special case where $f(x) \geq 0$ on I (see Fig. 5–10), it follows that

$$F(x) = \begin{cases} \text{the area under the graph of } f \text{ between } a \text{ and } x & \text{if } x > a \\ 0 & \text{if } x = a \\ \text{the negative of the area under the graph of } f \text{ between } x \text{ and } a \\ \quad \text{if } x < a. \end{cases}$$

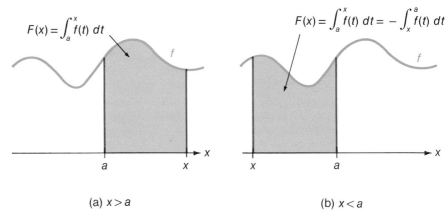

(a) $x > a$ (b) $x < a$

Figure 5–10 The indefinite integral $F(x) = \int_a^x f(t)\,dt$.

DEFINITION

Indefinite Integral

The function F defined by $F(x) = \int_a^x f(t)\,dt$ is called an *indefinite integral* (or an *integral function*) of the function f.

In defining the indefinite integral, we use the notation $\int_a^x f(t)\,dt$ instead of $\int_a^x f(x)\,dx$ to avoid confusion between x as a dummy variable and x as an upper limit of integration. Observe that the value of this integral depends on the upper limit x and not on the dummy variable t.

EXAMPLE 2 Let $f(x) = x$. Define

$$F(x) = \int_2^x f(t)\,dt = \int_2^x t\,dt.$$

Show that F is an antiderivative of f.

Solution Recall that

$$\int_a^b x\,dx = \int_a^b t\,dt = \frac{b^2 - a^2}{2}.$$

This formula holds for every choice of a and b. (See Exercise 13.) Thus,

$$F(x) = \int_2^x t\,dt = \frac{x^2 - 2^2}{2} = \frac{x^2}{2} - 2.$$

Then

$$F'(x) = D_x\left(\frac{x^2}{2} - 2\right) = \frac{2x}{2} - 0 = x = f(x)$$

for every x. □

Remark Example 2 gives a preview of the main theorem of this chapter, the Fundamental Theorem of Calculus. We shall prove (Section 5–4) that if f is continuous on an in-

5-3 Some Properties of Integrals

terval I that contains a and if F is the integral function $F(x) = \int_a^x f(t)\,dt$, then F is an antiderivative of f on the interval.

MEAN VALUE THEOREM FOR INTEGRALS

The final property of integrals that we consider here is a basic result in integration theory that is related to the average (mean) value of a continuous function f over an interval $[a, b]$. If $f(x) \geq 0$ on $[a, b]$, then this property has a simple geometrical interpretation in terms of area. (See Fig. 5–11.) It states that there is a number ξ between a and b such that the area of the rectangle with base $b - a$ and height $f(\xi)$ equals the area under the curve. Analytically, this result can be stated as

$$\int_a^b f(x)\,dx = f(\xi)(b - a)$$

for some ξ between a and b. The theorem does not give the value of ξ; it asserts that at least one such number must exist.

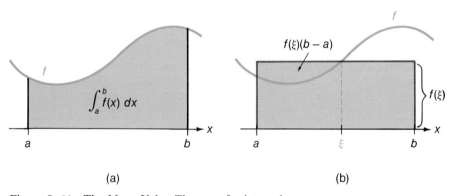

Figure 5–11 The Mean Value Theorem for integrals.

The general proof does not require that $f(x)$ be nonnegative on the interval. The proof uses Theorem 2–7, Property 2 of this section, Exercise 5 of Section 5–2, and the Intermediate Value Theorem (Theorem 2–8), which you should review before starting the proof.

THEOREM 5–2 *(Mean Value Theorem for Integrals)* Let f be continuous on an interval that contains a and b. There exists at least one number ξ between a and b such that

$$\int_a^b f(x)\,dx = f(\xi)(b - a).$$

Proof We prove the theorem for the special case in which $a < b$. The proof for the cases $a = b$ and $a > b$ are left for you to work out.

Let $a < b$. It follows from Theorem 2–7 that f has a maximum M and a minimum m on $[a, b]$. Then

$$m \leq f(x) \leq M \quad \text{for all } x \text{ on } [a, b].$$

It follows from Property 2 that

$$\int_a^b m \, dx \leq \int_a^b f(x) \, dx \leq \int_a^b M \, dx$$

$$m(b - a) \leq \int_a^b f(x) \, dx \leq M(b - a) \quad \text{(by Exercise 5, Section 5–2)}$$

$$m \leq \frac{1}{b - a} \int_a^b f(x) \, dx \leq M.$$

Let x_1 and x_2 be points at which f takes the values m and M, respectively. (See Fig. 5–12.) Then

$$f(x_1) = m \quad \text{and} \quad f(x_2) = M.$$

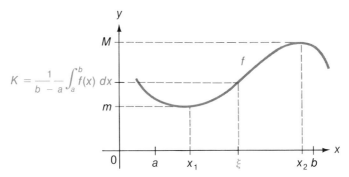

Figure 5–12 By the Intermediate Value Theorem, there exists ξ between x_1 and x_2 such that $f(\xi) = K$.

We now restrict ourselves to the closed interval with endpoints x_1 and x_2. Observe that f is continuous on this interval. Let

$$K = \frac{1}{b - a} \int_a^b f(x) \, dx.$$

Since

$$m \leq \frac{1}{b - a} \int_a^b f(x) \, dx \leq M,$$

then

$$f(x_1) \leq K \leq f(x_2).$$

Exercises 5–3

It follows from the Intermediate Value Theorem that there exists at least one number ξ between x_1 and x_2 such that

$$f(\xi) = K = \frac{1}{b-a} \int_a^b f(x)\, dx.$$

Thus,

$$\int_a^b f(x)\, dx = f(\xi)(b-a)$$

for at least one ξ between a and b. ∎

Exercises 5–3

The integrals in these exercises can be evaluated by the properties of this section and the formulas in Example 2 and Exercises 5, 6, and 7 of Section 5–2.

1–4. Calculate the integrals.

• 1. $\displaystyle\int_{-2}^{-4} x\, dx$

2. $\displaystyle\int_{7}^{-1} x^2\, dx$

• 3. $\displaystyle\int_{1}^{1} \frac{1}{x^2+1}\, dx$

4. $\displaystyle\int_{-5}^{-5} \frac{1}{x}\, dx$

5–8. Determine the integral function $F(x)$ defined by the indefinite integral. Calculate the derivative of F.

• 5. $\displaystyle\int_{-1}^{x} t^2\, dt$

6. $\displaystyle\int_{7}^{x} t^2\, dt$

• 7. $\displaystyle\int_{-20}^{x} t\, dt$

8. $\displaystyle\int_{4}^{x} \sqrt{2}\, dt$

9–12. Determine the function $F(x)$ defined by the integral.

• 9. $\displaystyle\int_{x}^{2x} t^2\, dt$

10. $\displaystyle\int_{3x^2}^{x^2} t\, dt$

• 11. $\displaystyle\int_{x}^{x+1} t^3\, dt$

12. $\displaystyle\int_{1}^{1+x} t^2\, dt$

• 13. Prove that $\displaystyle\int_a^b x\, dx = \frac{b^2 - a^2}{2}$ if $b = a$ or $b < a$.

• 14. Let $f(x) \leq x$ for $0 \leq x \leq 1$. Use Property 2 to show that $\int_{1/2}^{1} f(x)\, dx \leq \frac{3}{8}$.

• 15. Use Property 2 to show that $\frac{3}{4} \leq \int_{1}^{4} (1/x)\, dx \leq 3$. (Hint: Show that $\frac{1}{4} \leq 1/x \leq 1$ for $1 \leq x \leq 4$.)

16 Prove Property 1' for the given cases.
- (a) $b \leq a \leq c$
- (b) $b \leq c \leq a$
- (c) $c \leq a \leq b$
- (d) $c \leq b \leq a$

17–20. Find the specific number ξ guaranteed by Theorem 5–2 for the integral.

• **17** $\displaystyle\int_{-1}^{3} x \, dx$

18 $\displaystyle\int_{4}^{7} 5 \, dx$

• **19** $\displaystyle\int_{-2}^{4} x^2 \, dx$

• **20** $\displaystyle\int_{0}^{-3} x^2 \, dx$

• **21** Work through the proof of Theorem 5–2 with the example $f(x) = x^2$, $a = -1$, $b = 4$. Find the specific values of the numbers m, M, x_1, x_2, K, and ξ.

22 Analysts for the Big Bank Corporation describe the cash flow by two continuous functions:

$f(t)$ = amount received at time t;
$g(t)$ = amount paid out at time t.

(a) Suppose that $f(t) \geq g(t)$ for $t_0 \leq t \leq t_0 + \Delta t$. Explain why $[f(t_0) - g(t_0)] \Delta t$ is a reasonable approximation of the amount of cash accumulated over the period t_0 to $t_0 + \Delta t$.

(b) Use (a) to interpret the integral $\int_a^b [f(t) - g(t)] \, dt$. What does it mean if this integral is negative?

(c) Interpret the number

$$\frac{1}{b-a} \int_a^b [f(t) - g(t)] \, dt = f(\xi) - g(\xi)$$

obtained from the Mean Value Theorem for integrals.

5–4 Fundamental Theorem of Calculus

Thus far we have used the definition of the definite integral in calculating many definite integrals. This method is laborious, time-consuming, and inefficient—as evinced by the very few definite integrals that we have calculated. The Fundamental Theorem of Calculus will enable us to use antiderivatives to evaluate many definite integrals with comparative ease.

The result of the Fundamental Theorem is illustrated in Example 2 of Section 5–3. We established that if $f(x) = x$ and

$$F(x) = \int_2^x f(t) \, dt = \int_2^x t \, dt,$$

then $F'(x) = f(x)$ for every x. Newton and Leibniz discovered that this relation holds in general. This is the first part of Theorem 5–3.

5-4 Fundamental Theorem of Calculus

The second half of Theorem 5-3 is one of the most useful results in mathematics, for it enables us to calculate definite integrals by antiderivatives. As stated in the theorem, we can use any antiderivative of f to calculate $\int_a^b f(x)\,dx$. For simplicity, we usually choose an antiderivative that has the constant term C equal to zero.

THEOREM 5-3
Fundamental Theorem

(*Fundamental Theorem of Calculus*) Let f be continuous on the closed interval $[a, b]$.

Part 1 Define the function G on $[a, b]$ by

$$G(x) = \int_a^x f(t)\,dt, \quad a \leq x \leq b$$

Then G is an antiderivative of f on $[a, b]$.

Part 2 If F is any antiderivative of f on $[a, b]$, then

$$\int_a^b f(x)\,dx = F(b) - F(a).$$

Proof **Part 1** Let c be any number in the open interval (a, b). Then

$$G'(c) = \lim_{x \to c} \frac{G(x) - G(c)}{x - c} = \lim_{x \to c} \frac{\int_a^x f(t)\,dt - \int_a^c f(t)\,dt}{x - c}$$

$$= \lim_{x \to c} \frac{\int_a^c f(t)\,dt + \int_c^x f(t)\,dt - \int_a^c f(t)\,dt}{x - c} \quad \text{(by Property 1', Section 5-3)}$$

$$= \lim_{x \to c} \frac{\int_c^x f(t)\,dt}{x - c}.$$

It follows from the Mean Value Theorem for integrals that there exists a number ξ (which depends on x) between c and x such that

$$\int_c^x f(t)\,dt = f(\xi)(x - c).$$

Since ξ is between c and x, then $\xi \to c$ as $x \to c$. Thus

$$G'(c) = \lim_{x \to c} \frac{\int_c^x f(t)\,dt}{x - c} = \lim_{x \to c} \frac{f(\xi)(x - c)}{x - c}$$

$$= \lim_{\xi \to c} f(\xi) = f(c)$$

since f is continuous at c.

Since $G'(c) = f(c)$ for every c in the open interval (a, b), then

$$G'(x) = f(x) \quad \text{for } a < x < b.$$

A similar argument can be devised to show that

$$\lim_{x \to a^+} \frac{G(x) - G(a)}{x - a} = f(a) \quad \text{and} \quad \lim_{x \to b^-} \frac{G(x) - G(b)}{x - b} = f(b).$$

(See Exercise 25.) Therefore, G is an antiderivative of f on $[a, b]$.

Part 2 Let F be an antiderivative of f on $[a, b]$; let

$$G(x) = \int_a^x f(t)\, dt \quad \text{for each } x, \quad a \leq x \leq b.$$

Since F and G are antiderivatives of f on $[a, b]$, it follows from Theorem 4–6 that they differ by a constant C:

$$G(x) = F(x) + C \quad \text{for } a \leq x \leq b.$$

To evaluate C, we observe that

$$G(a) = \int_a^a f(t)\, dt = 0 = F(a) + C$$

so that $C = -F(a)$. Thus

$$\int_a^b f(x)\, dx = \int_a^b f(t)\, dt = G(b) = F(b) + C$$

$$= F(b) - F(a). \blacksquare$$

EXAMPLE 1 Calculate $\int_1^8 (2x - x^{-2/3})\, dx$.

Solution The general antiderivative of $f(x) = 2x - x^{-2/3}$ is

$$F(x) = \frac{2x^2}{2} - \frac{x^{1/3}}{1/3} + C = x^2 - 3x^{1/3} + C.$$

According to the Fundamental Theorem, we can use any antiderivative to calculate the integral. The antiderivative with constant terms equal to zero is

$$F(x) = x^2 - 3x^{1/3}.$$

Thus,

$$\int_1^8 (2x - x^{-2/3})\, dx = F(8) - F(1)$$

$$= [8^2 - 3 \cdot 8^{1/3}] - [1^2 - 3 \cdot 1^{1/3}]$$

$$= 64 - 6 - 1 + 3 = 60. \square$$

5-4 Fundamental Theorem of Calculus

NOTATION

$[F(x)]_a^b = F(b) - F(a)$

Expressions such as $F(b) - F(a)$ occur with such frequency in the calculus that we use a special notation for them. The symbol

$$[F(x)]_a^b$$

is used as a shorthand notation for the expression $F(b) - F(a)$.

For example,

$$[2x^2 - 1]_1^4 = [2 \cdot 4^2 - 1] - [2 \cdot 1^2 - 1] = 30$$

$$[\sqrt{x}]_2^3 = \sqrt{3} - \sqrt{2}$$

$$\int_a^b f(x) \, dx = [F(x)]_a^b = F(b) - F(a)$$

if F is an antiderivative of f on $[a, b]$.

EXAMPLE 2 (a) Since $F(x) = x^2/2$ is an antiderivative of $f(x) = x$, then

$$\int_0^2 x \, dx = \left[\frac{x^2}{2}\right]_0^2 = \frac{2^2}{2} - \frac{0^2}{2} = 2.$$

(b) Since $F(x) = x^3 - 3x^{4/3}/4 + 5x$ is an antiderivative of $f(x) = 3x^2 - x^{1/3} + 5$, then

$$\int_{-1}^1 (3x^2 - x^{1/3} + 5) \, dx = \left[x^3 - \frac{3x^{4/3}}{4} + 5x\right]_{-1}^1$$

$$= \left[1^3 - \frac{3 \cdot 1^{4/3}}{4} + 5 \cdot 1\right] - \left[(-1)^3 - \frac{3(-1)^{4/3}}{4} + 5(-1)\right]$$

$$= [1 - \tfrac{3}{4} + 5] - [-1 - \tfrac{3}{4} - 5] = 12.$$

(c) Since $F(x) = 2x^{3/2}/3$ is an antiderivative of $f(x) = x^{1/2}$, then

$$\int_0^4 \sqrt{x} \, dx = \int_0^4 x^{1/2} \, dx$$

$$= \frac{2}{3}\left[x^{3/2}\right]_0^4 = \frac{2}{3}\left[4^{3/2} - 0^{3/2}\right] = \frac{2}{3} \cdot 8 = \frac{16}{3}. \quad \square$$

Properties of antiderivatives can now be used to establish comparable properties of definite integrals. Example 3 shows how one such property can be proved.

EXAMPLE 3 Let f be continuous on $[a, b]$. Show that

$$\int_a^b k \cdot f(x) \, dx = k \int_a^b f(x) \, dx.$$

Solution Let $F(x)$ be an antiderivative of $f(x)$ on $[a, b]$ and let $G(x)$ be an antiderivative of $kf(x)$ on $[a, b]$. By Rule 2 of Section 4–9,

$$D_x^{-1}(k \cdot f(x)) = kD_x^{-1}(f(x)) + C$$

so that $G(x) = kF(x) + C$.

Integral of a Constant Times a Function

If we now apply Theorem 5–3, we get

$$\int_a^b kf(x)\, dx = G(b) - G(a)$$
$$= k[F(b) + C] - k[F(a) + C]$$
$$= k[F(b) - F(a)] = k\int_a^b f(x)\, dx. \quad \square$$

The Fundamental Theorem can also be applied to integrals with variable limits of integration. For example, since $F(t) = t^2$ is an antiderivative of $f(t) = 2t$, then

$$\int_1^x 2t\, dt = [t^2]_1^x = x^2 - 1.$$

If f is continuous at each point of $[a, b]$, it follows from the Fundamental Theorem that $\int_a^x f(t)\, dt$ is an antiderivative of f on $[a, b]$. Therefore

Derivative of Integral Function

$$D_x\left(\int_a^x f(t)\, dt\right) = f(x), \qquad a \le x \le b.$$

This result holds even though we may not have a convenient way of calculating the integral of f.

EXAMPLE 4 Calculate $D_x\left(\int_2^x \frac{1}{t}\, dt\right)$.

Solution Since $f(t) = 1/t$ is continuous for all $t > 0$, then

$$D_x\left(\int_2^x \frac{1}{t}\, dt\right) = f(x) = \frac{1}{x} \qquad \text{for } x > 0. \quad \square$$

EXAMPLE 5 Calculate $D_x\left(\int_1^{x^2}(3t^2 - 1)\, dt\right)$.

Solution If we let $u = x^2$, the derivative can be written

$$D_x\left(\int_1^u (3t^2 - 1)\, dt\right).$$

It follows from the Chain Rule that

$$D_x\left(\int_1^u (3t^2 - 1)\, dt\right) = D_u\left(\int_1^u (3t^2 - 1)\, dt\right) \cdot D_x u$$
$$= (3u^2 - 1)D_x u = (3x^4 - 1)2x = 6x^5 - 2x. \quad \square$$

Exercises 5–4

1–12. Use antiderivatives to calculate the definite integrals.

- 1. $\int_{2}^{3} (3x^2 + 2x + 1)\, dx$
- 2. $\int_{-1}^{2} (x^3 + 1)\, dx$
- 3. $\int_{4}^{9} (x^2 + \sqrt{x})\, dx$
- 4. $\int_{1}^{8} (x^{-2} - x^{-2/3})\, dx$
- 5. $\int_{1}^{\sqrt{2}} (x^3 + x^{-3})\, dx$
- 6. $\int_{2}^{3} \left(x + \dfrac{1}{x}\right)^2 dx$
- 7. $\int_{8}^{27} \sqrt[3]{t^2}\, dt$
- 8. $\int_{1}^{3} \dfrac{2}{3t^2}\, dt$
- 9. $\int_{1}^{2} \dfrac{t^2 + 2t - 1}{t^4}\, dt$
- 10. $\int_{1}^{4} \dfrac{t + 1}{\sqrt{t}}\, dt$
- 11. $\int_{-1}^{2} x(2x^2 + 3x)\, dx$
- 12. $\int_{1}^{4} (x - 4)(x - 1)\, dx$

13–14. Use antiderivatives to calculate the area of the region bounded by the graph of f, the x-axis, and the lines $x = a$ and $x = b$.

- 13. $f(x) = x + \sqrt{x}$, $a = 0$, $b = 1$
- 14. $f(x) = \dfrac{1}{x^2}$, $a = 1$, $b = 4$

15–22. Use Theorem 5–3 and the properties of definite integrals to calculate the derivative or integral.

- 15. $D_x \int_{1}^{x} \sqrt{1 + t^2}\, dt$
- 16. $D_x \int_{0}^{x} (t^2 + t + 1)\, dt$
- 17. $D_x \int_{x}^{2} \dfrac{1}{t^2 + 1}\, dt$
- 18. $D_x \left(3 \int_{x}^{-1} \dfrac{t}{\sqrt{t + 2}}\, dt\right)$
- 19. $D_x \int_{2}^{3} \dfrac{1}{\sqrt{t^2 + 1}}\, dt$
- 20. $D_x \int_{x^2}^{x} \dfrac{1}{t^2 + t + 1}\, dt$
- 21. $\int D_t \left(\dfrac{2}{\sqrt{1 - t^2}}\right) dt$
- 22. $\int_{2}^{3} D_t \left(\dfrac{1}{t^2 + 3}\right) dt$

- 23. Work through all the steps of Theorem 5–3 with the particular example $f(x) = x^2$, $a = 0$, $b = 1$. In Part 1 find the specific function $G(x)$.

- 24. Use the proof of Part 1 of Theorem 5–3 as a model to help you calculate the following limits.

  (a) $\lim\limits_{x \to c} \dfrac{\int_{c}^{x} \left(\dfrac{1}{t\sqrt{t^2 - 1}}\right) dt}{x - c}$

  (b) $\lim\limits_{h \to 0} \dfrac{\int_{x}^{x+h} \dfrac{1}{\sqrt{t^2 + 1}}\, dt}{h}$

•25 Complete the proof of Theorem 5-3 by proving the given properties.

(a) $\lim_{x \to a^+} \dfrac{G(x) - G(a)}{x - a} = f(a)$ (b) $\lim_{x \to b^-} \dfrac{G(x) - G(b)}{x - b} = f(b)$

5-5 The Indefinite Integral; the Substitution Principle

The Fundamental Theorem of Calculus changes the problem of calculating a definite integral, defined as the limit of a sum, into that of evaluating an antiderivative at two points, a and b. Because of this connection between the antiderivative and the definite integral, we use the symbol

$$\int f(x)\, dx \quad \text{(read ``integral of } f(x)\, dx\text{'')}$$

for any particular antiderivative of f. The antiderivative

$$\int f(x)\, dx + C$$

Indefinite Integral (Antiderivative)

is called the *general antiderivative* or the *indefinite integral* of f. The expression $f(x)\, dx$ is called the *integrand* of the integral.

The D_x^{-1} notation, which we have used until now for the antiderivative, was introduced to avoid confusion between the completely different concepts of the antiderivative and the definite integral. At this point we change to the conventional notation.

Our previous rules for antiderivatives can be used to calculate indefinite integrals; for example,

$$\int (2x - 3)\, dx = \dfrac{2x^2}{2} - 3x + C = x^2 - 3x + C$$

and

$$\int (5t^2 - t^{-2})\, dt = \dfrac{5t^3}{3} - \dfrac{t^{-1}}{-1} + C = \dfrac{5t^3}{3} + \dfrac{1}{t} + C.$$

Note that the definite integral of a continuous function can be written in integral notation as

$$\int_a^b f(x)\, dx = \left[\int f(x)\, dx \right]_a^b.$$

The first two of the antiderivative rules from Section 4-9 can be rewritten as

Basic Properties

I1 $\displaystyle\int (f(x) + g(x))\, dx = \int f(x)\, dx + \int g(x)\, dx + C$

I2 $\displaystyle\int kf(x)\, dx = k \int f(x)\, dx + C$

5–5 The Indefinite Integral; the Substitution Principle

Rule 3' is more interesting. Technically, it is

$$\int [g(x)]^n \, g'(x) \, dx = \frac{[g(x)]^{n+1}}{n+1} + C$$

provided $g(x)$ has a continuous first derivative and $n \neq -1$. This rule can be simplified considerably by making the formal substitution $u = g(x)$, $du = g'(x) \, dx$. The rule then becomes

Power Rule

I3 $\quad \int u^n \, du = \dfrac{u^{n+1}}{n+1} + C$

provided $u = g(x)$ has a continuous first derivative and $n \neq -1$.

It is a rather unusual experience to find an integral $\int f(x) \, dx$ that can be transformed directly into an integral $\int u^n \, du$ by a substitution $u = g(x)$. It is much more common to find that we need to multiply the integrand by a nonzero constant k before we can make the substitution. In that case, we first multiply the original integral by k/k, and then apply I2. Thus,

$$\int f(x) \, du = \frac{k}{k} \int f(x) \, dx = \frac{1}{k} \int f(x) \, k \, dx + C.$$

We have achieved the desired result by multiplying the integrand by k and, at the same time, multiplying the entire integral by $1/k$.

EXAMPLE 1 Calculate $\int x^2(2x^3 - 3)^2 \, dx$.

Solution Although a bit of trial-and-error experimentation may sometimes be necessary, it is frequently a good idea to let $u = g(x)$ represent the most complicated part of the integrand. In this case, it would be convenient to let

$$u = 2x^3 - 3$$

from which

$$du = 6x^2 \, dx$$

Comparing the product of u^2 and du to the original integrand, we find that we need to multiply the integrand by 6 in order to get the required product. According to I2, when we multiply the integrand by 6, we must multiply the whole integral by $\frac{1}{6}$ to maintain the value of the expression. Thus,

$$\int x^2(2x^3 - 3)^2 \, dx = \frac{1}{6} \int (2x^3 - 3)^2 \cdot 6x^2 \, dx = \frac{1}{6} \int u^2 \, du$$

$$= \frac{1}{6} \frac{u^3}{3} + C = \frac{(2x^3 - 3)^3}{18} + C. \ \square$$

Warning This strategy of substitution is useful *only* when the integrand differs from the ideal form $u^n \, du$ by a *constant* factor k. If any power of the independent variable x is missing, it is not permitted to multiply the integrand by x^n and the whole integral by $1/x^n$. All terms and factors involving x must be part of the original integrand.

Remark By this point you should be aware of the distinction between an indefinite integral and the definite integral. The general indefinite integral

$$\int f(x)\,dx + C$$

defines a class of functions—the set of all antiderivatives of f. The definite integral is a number defined by a certain limit process. The Fundamental Theorem of Calculus relates these two concepts.

THE SUBSTITUTION PRINCIPLE

The Power Rule I3 is a special case of the following general property, which depends on the Chain Rule.

THEOREM 5–4 (*The Substitution Principle*) Let f be a continuous function. Let g be a differentiable function. Suppose that the substitution $u = g(x)$, $du = g'(x)\,dx$ transforms the expression $f(x)\,dx$ into $h(u)\,du$, where h is continuous. Let H be an antiderivative of h. Then

Substitution Principle

$$\int f(x)\,dx = \int h(u)\,du + C = H(g(x)) + C.$$

Proof From the hypothesis, it follows that

$$f(x) = h(u)\frac{du}{dx}.$$

Also, since H is an antiderivative of h, then $H'(u) = h(u)$. By the Chain Rule,

$$D_x(H(g(x))) = D_x(H(u)) = D_u(H(u))\,D_x u$$
$$= h(u)\frac{du}{dx} = f(x),$$

so that $H(g(x))$ is an antiderivative of f. It follows from Theorem 5–3 that

$$\int f(x)\,dx = H(g(x)) + C = H(u) + C = \int h(u)\,du + C. \blacksquare$$

In applying the substitution principle, we substitute $u = g(x)$ and transform the original integral $\int f(x)\,dx$ into the integral $\int h(u)\,du$. We then integrate, obtaining $H(u) + C$, and substitute $g(x)$ for u, obtaining the final answer

$$\int f(x)\,dx = H(g(x)) + C.$$

Problems with Substitutions

You may experience two potential problems:

(1) The function h may not be continuous. In this event, the integral $\int h(u)\,du$ may not exist even though the original integral $\int f(x)\,dx$ does exist.

5-5 The Indefinite Integral; the Substitution Principle

(2) It would be natural to transform $f(x)\,dx$ into $h(u)\,du$ by a substitution for x as a function of u instead of u as a function of x. That is, we might substitute $x = k(u)$, $dx = k'(u)\,du$. This procedure may not yield valid results unless the function k is one-to-one. If k is one-to-one, then we can express u as a function of x, $u = k^{-1}(x)$, and we are back to the substitution principle.

DEFINITE INTEGRALS

The following corollary extends the substitution principle to definite integrals.

COROLLARY
Substitution with Definite Integrals

Let f be continuous and g differentiable on $[a, b]$. Suppose the substitution $u = g(x)$, $du = g'(x)\,dx$ changes $f(x)\,dx$ to $h(u)\,du$. Let $u = \alpha$ when $x = a$ and let $u = \beta$ when $x = b$. If h is continuous on $[\alpha, \beta]$, then

$$\int_a^b f(x)\,dx = \int_\alpha^\beta h(u)\,du.$$

Proof Observe that $g(a) = \alpha$ and that $g(b) = \beta$. Let H be an antiderivative of h on $[\alpha, \beta]$. By Theorem 5-4,

$$\int f(x)\,dx = H(g(x)) + C,$$

so that

$$\int_a^b f(x)\,dx = \left[H(g(x))\right]_a^b = \left[H(g(b)) - H(g(a))\right]$$

$$= H(\beta) - H(\alpha) = \int_\alpha^\beta h(u)\,du. \ \blacksquare$$

EXAMPLE 2 Calculate $\int_0^1 (2x - 3)^4 \,dx$.

Solution Make the substitution $u = 2x - 3$, $du = 2\,dx$. Then $u = -3$ when $x = 0$ and $u = -1$ when $x = 1$. Before making the substitution, we multiply the integrand by 2 to change dx into $du = 2\,dx$.

$$\int_0^1 (2x - 3)^4\,dx = \frac{1}{2}\int_0^1 (2x - 3)^4\, 2\,dx$$

$$= \frac{1}{2}\int_{-3}^{-1} u^4\,du \quad \text{(by the corollary)}$$

$$= \frac{1}{2}\left[\frac{u^5}{5}\right]_{-3}^{-1} = \frac{1}{2}\left[\frac{(-1)^5 - (-3)^5}{5}\right]$$

$$= \frac{1}{10}[-1 + 243] = \frac{242}{10} = \frac{121}{5}. \ \square$$

$u = 2x - 3$
$du = 2\,dx$
$x = 0 \Rightarrow u = -3$
$x = 1 \Rightarrow u = -1$

Integral formulas I1, I2, and I3 can be restated for definite integrals as follows:

I1 $\quad \int_a^b [f(x) + g(x)]\, dx = \int_a^b f(x)\, dx + \int_a^b g(x)\, dx.$

I2 $\quad \int_a^b kf(x)\, dx = k \int_a^b f(x)\, dx.$

I3 $\quad \int_a^b u^n\, du = \left[\dfrac{u^{n+1}}{n+1}\right]_a^b \quad$ provided $n \neq -1$.

EXAMPLE 3

$$\int_1^4 \left(3\sqrt{x} + \frac{12}{\sqrt{x}}\right) dx$$

$$= 3\int_1^4 \sqrt{x}\, dx + 12 \int_1^4 \frac{1}{\sqrt{x}}\, dx = 3\int_1^4 x^{1/2}\, dx + 12 \int_1^4 x^{-1/2}\, dx$$

$$= 3 \left[\frac{x^{3/2}}{3/2}\right]_1^4 + 12 \left[\frac{x^{1/2}}{1/2}\right]_1^4$$

$$= 3 \cdot \tfrac{2}{3} [x^{3/2}]_1^4 + 12 \cdot 2[x^{1/2}]_1^4$$

$$= 2[4^{3/2} - 1^{3/2}] + 24[4^{1/2} - 1^{1/2}]$$

$$= 2[8 - 1] + 24[2 - 1] = 38. \quad \square$$

Remark on Notation We originally introduced the symbol

$$\int_a^b f(x)\, dx$$

for the definite integral with the comment that the dx is a vestige of the Δx in the approximating sum

$$\sum_{k=1}^n f(x_k)\, \Delta x_k.$$

In this section we are treating the dx as if it is a differential.

To justify this change in approach, observe that if

$$F(x) = \int f(x)\, dx$$

is an antiderivative of f, then the differential of F is

$$dF = f(x)\, dx.$$

Thus, we can think of the integral $\int f(x)\, dx$ as an "antidifferential" of $f(x)\, dx$ as well as an antiderivative of f. If this is done, then the differential dx is a necessary part of the expression.

There is an advantage to this approach. Recall that the Chain Rule is automatic when we use differentials to calculate derivatives through several changes of vari-

Exercises 5–5

ables. (See Example 5 of Section 3–4.) The substitution principle can be thought of as a "reverse" chain rule. The use of differentials makes the change of variables automatic.

Exercises 5–5

1–12. Calculate the integrals.

- 1. $\int (4x^3 - 3x^2 + 2x + 1)\, dx$
- 2. $\int (3x^2 + x^{-2})\, dx$
- 3. $\int (2x - 1)(x - 2)\, dx$
- 4. $\int (x^2 - x^{-2})^2\, dx$
- 5. $\int 4(x^{1/2} + x^{-1/2})\, dx$
- 6. $\int \sqrt{x + 1}\, dx$
- 7. $\int \sqrt{5x - 1}\, dx$
- 8. $\int \sqrt{x^2 + 3} \cdot x \cdot dx$
- 9. $\int \dfrac{x}{\sqrt[3]{x^2 + 1}}\, dx$
- 10. $\int \sqrt[5]{32x^4}\, dx$
- 11. $\int \sqrt{(x^2 - 1)^2 + 4x^2}\, dx$
- 12. $\int \dfrac{(\sqrt{x + 1} - 5)^4}{\sqrt{x + 1}}\, dx$

13–26. Evaluate the definite integrals. If the substitution principle is used, then change the integral to a definite integral with new limits of integration.

- 13. $\int_{-1}^{1} (x + 1)\, dx$
- 14. $\int_{0}^{4} (x^{1/2} + 3x)\, dx$
- 15. $\int_{1}^{2} (3x - 1)^{-3}\, dx$
- 16. $\int_{0}^{2} \sqrt{1 + 4x}\, dx$
- 17. $\int_{-1}^{0} \dfrac{1}{\sqrt{1 - x}}\, dx$
- 18. $\int_{1}^{4} \dfrac{(1 - \sqrt{x})^3}{\sqrt{x}}\, dx$
- 19. $\int_{0}^{1} t\sqrt{1 - t^2}\, dt$
- 20. $\int_{0}^{2} \dfrac{z}{\sqrt{25 - 4z^2}}\, dz$
- 21. $\int_{0}^{1} \dfrac{x^2}{(x^3 - 2)^2}\, dx$
- 22. $\int_{1}^{2} \dfrac{6y}{\sqrt{5 - y^2}}\, dy$
- 23. $\int_{0}^{4} y\sqrt{y^2 + 9}\, dy$
- 24. $\int_{1}^{4} \dfrac{4x + 6}{\sqrt{x^2 + 3x - 3}}\, dx$
- 25. $\int_{0}^{3} \dfrac{(\sqrt{x + 1} - 2)^4}{\sqrt{x + 1}}\, dx$
- 26. $\int_{4}^{9} \dfrac{1}{\sqrt{t}(1 + \sqrt{t})^3}\, dt$

•27 Calculate the area of the region bounded by the graph of $f(x) = (\sqrt{x} + 1)^4/\sqrt{x}$, the x-axis, and the lines $x = 1$, $x = 4$.

•28 A particle moves along the x-axis with velocity at time t given by $v(t) = \sqrt{2t + 1}$. Use an integral to find the position function s if $s(0) = 4$.

29 Suppose that we transform the integral $\int_{-1}^{2} x^2 \, dx$ by substituting $u = 1/x$, $du = -dx/x^2$. Show that a direct substitution results in the new integral $\int_{-1}^{1/2} (-u^{-4}) \, du$. Is this a legal substitution? Explain.

30 Let $f(x) = 2x(\sqrt{x^2 + 1} + 1)$, $g(x) = \sqrt{x} + 1$. Let $\mathcal{R}$ be the region bounded by the graph of f, the x-axis, and the lines $x = 0$, $x = 3$. Let $\mathcal{S}$ be the region bounded by the graph of g, the x-axis, and the lines $x = 1$, $x = 10$. Use the corollary to Theorem 5–4 to explain why we know that the area of $\mathcal{R}$ equals the area of $\mathcal{S}$.

•31 The acceleration of a particle moving along the x-axis is given by $a(t) = t^2 - t$. Find the velocity and distance functions if it starts at the origin at time $t = 0$ with an initial velocity of 5 units per second.

•32 The slope of the tangent line to the graph of a function f at the point $(x, f(x))$ is given by $m(x) = 2 - 3/x^2$ provided $x \neq 0$.
  (a) Where is the graph concave upward? downward?
  (b) It is known that $f(1) = 2$. Is this enough information to determine $f(x)$ for all $x \neq 0$? Explain.

•33 The area of the region bounded by the graph of f, the x-axis, the y-axis, and the line $x = a$ equals $A = \sqrt[3]{a^2 + 4} - \sqrt[3]{4}$ for every $a > 0$. Find $f(x)$.

5–6 Numerical Integration

It may be difficult to find an antiderivative of a particular function f. We have not, for example, developed a formula for an antiderivative of

$$f(x) = \sqrt{x^3 + 1}.$$

Thus, we cannot apply the Fundamental Theorem to calculate the definite integral

$$\int_0^1 \sqrt{x^3 + 1} \, dx$$

even though the function f is integrable.

In this section we discuss three numerical methods that are especially useful for approximating the value of a definite integral that cannot be calculated exactly by an antiderivative.

APPROXIMATING SUMS

The simplest numerical method for approximating $\int_a^b f(x) \, dx$ is to use one of the approximating sums S_n. We partition the interval $[a, b]$ into n subintervals of equal

5–6 Numerical Integration

Method of Approximating Sums

length

$$\Delta x = \frac{b-a}{n}$$

and form the sum $\sum_{k=1}^{n} f(x_k) \Delta x$, where

$$x_0 = a, \quad x_1 = a + \Delta x, \quad x_2 = a + 2\Delta x, \quad \ldots, \quad x_n = a + n\Delta x = b.$$

Since

$$\int_a^b f(x)\,dx = \lim_{n \to \infty} \sum_{k=1}^{n} f(x_k)\,\Delta x$$

then

$$\int_a^b f(x)\,dx \approx \sum_{k=1}^{n} f(x_k)\,\Delta x$$

if n is large. If f is nonnegative on $[a, b]$, the approximating sum represents the sum of the areas of the rectangles shown in Figure 5–13.

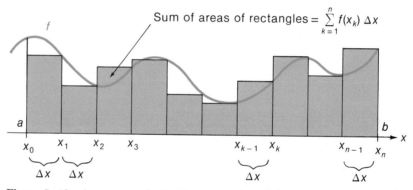

Figure 5–13 Area approximated by rectangles. $\int_a^b f(x)\,dx \approx \sum_{k=1}^{n} f(x_k)\,\Delta x.$

THE TRAPEZOIDAL RULE

A sizable error usually results if an approximating sum is used to approximate an integral. If $f(x) \geq 0$ on $[a, b]$, we can reduce much of this error by using trapezoids instead of rectangles in setting up the approximation. We partition the interval $[a, b]$ into n equal subintervals of length $\Delta x = (b-a)/n$ and let

$$x_0 = a, \quad x_1 = a + \Delta x, \quad x_2 = a + 2\Delta x, \quad \ldots, \quad x_n = a + n\Delta x = b.$$

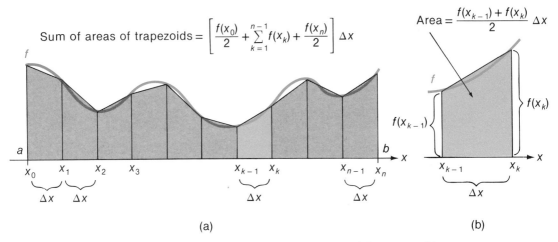

Figure 5-14 The Trapezoidal Rule involves area approximated by trapezoids: $\int_a^b f(x)\,dx \approx [f(x_0)/2 + \sum_{k=1}^{n-1} f(x_k) + f(x_n)/2]\,\Delta x$.

We now form the kth trapezoid over the interval $[x_{k-1}, x_k]$ by connecting the points $(x_{k-1}, f(x_{k-1}))$ and $(x_k, f(x_k))$ with a line segment. (See Fig. 5-14.)

Since the area of a trapezoid with bases B_1 and B_2 and height H is

$$\frac{B_1 + B_2}{2} \cdot H,$$

then the area of the kth trapezoid is

$$\frac{f(x_{k-1}) + f(x_k)}{2}\,\Delta x.$$

It follows that

$$\int_a^b f(x)\,dx \approx \underbrace{\frac{f(x_0) + f(x_1)}{2}\,\Delta x}_{\text{area of 1st trapezoid}} + \underbrace{\frac{f(x_1) + f(x_2)}{2}\,\Delta x}_{\text{area of 2nd trapezoid}} + \cdots$$

$$+ \underbrace{\frac{f(x_{n-1}) + f(x_n)}{2}\,\Delta x}_{\text{area of } n\text{th trapezoid}}$$

If we simplify this expression, we obtain the Trapezoidal Rule:

Trapezoidal Rule
$$\int_a^b f(x)\,dx \approx \left[\frac{f(x_0)}{2} + f(x_1) + f(x_2) + \cdots + f(x_{n-1}) + \frac{f(x_n)}{2}\right]\Delta x.$$

5-6 Numerical Integration

Observe that the formula for the Trapezoidal Rule differs from an approximating sum only in that $f(x_n)$ is replaced by $[f(x_0) + f(x_n)]/2$, the average of the values of f at the two endpoints. This formula also holds when f is negative on all or part of $[a, b]$. (See Exercise 9.)

EXAMPLE 1 Since the function $f(x) = 1/x$ is continuous for all $x > 0$, then $\int_2^3 (1/x)\, dx$ exists, even though we cannot now calculate it by a formula.
(a) Use the Trapezoidal Rule with $n = 10$ to approximate this integral.
(b) Use an approximating sum with $n = 10$ to approximate the integral.
(c) It can be shown that, correct to six decimal places,

$$\int_2^3 \frac{1}{x}\, dx \approx 0.405465.$$

Compare the results of (a) and (b) with this "true" approximation.

Solution (a) Since $n = 10$, then $\Delta x = 0.1$. The numbers in the partition of $[2, 3]$ are

$$x_0 = 2.0, \quad x_1 = 2.1, \quad x_2 = 2.2, \quad \ldots, \quad x_{10} = 3.0.$$

Using the Trapezoidal Rule, we obtain

$$\int_2^3 \frac{1}{x}\, dx \approx \left[\frac{1}{2} \cdot \frac{1}{2.0} + \frac{1}{2.1} + \frac{1}{2.2} + \frac{1}{2.3} + \frac{1}{2.4} + \frac{1}{2.5} + \frac{1}{2.6} \right.$$
$$\left. + \frac{1}{2.7} + \frac{1}{2.8} + \frac{1}{2.9} + \frac{1}{2} \cdot \frac{1}{3.0} \right] \cdot (0.1) \approx 0.405581.$$

(b) We use the same partition to set up the approximating sum

$$\sum_{k=1}^{n} f(x_k)\, \Delta x = \sum_{k=1}^{10} \frac{1}{x_k}\, \Delta x$$

obtaining

$$\int_2^3 \frac{1}{x}\, dx \approx \left[\frac{1}{2.1} + \frac{1}{2.2} + \frac{1}{2.3} + \frac{1}{2.4} + \frac{1}{2.5} + \frac{1}{2.6} + \frac{1}{2.7} \right.$$
$$\left. + \frac{1}{2.8} + \frac{1}{2.9} + \frac{1}{3.0} \right] \cdot (0.1) \approx 0.397247.$$

(c) The value obtained from the Trapezoidal Rule differs from the "true" value by 0.000116. The percentage error is

$$\frac{0.000116}{0.405465} \approx 0.029\%.$$

The value obtained from the approximating sum differs from the "true" value by 0.008218. The percentage error is

$$\frac{0.008218}{0.405465} \approx 2.03\%. \quad \square$$

SIMPSON'S RULE

The Trapezoidal Rule involves using line segments to approximate the graph of a function f. This means that the concavity of the graph of f is not considered. Another method of approximate integration partially utilizes the concavity. This method, known as *Simpson's Rule*, is based on using parabolic arcs to approximate the graph of f.

Let (x_0, y_0), (x_1, y_1), (x_2, y_2) be three noncollinear points on the graph of f, where $x_0 < x_1 < x_2$. There is a unique parabola

$$y = Ax^2 + Bx + C$$

that passes through these points. We shall use this parabola to approximate the graph of f on the interval $[x_0, x_2]$. (See Fig. 5–15.) Then

$$\int_{x_0}^{x_2} f(x)\,dx \approx \int_{x_0}^{x_2} (Ax^2 + Bx + C)\,dx.$$

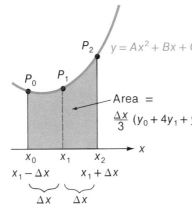

Figure 5–15 Area under a parabola.
$\int_{x_1-\Delta x}^{x_1+\Delta x}(Ax^2 + Bx + C)\,dx$
$= \dfrac{\Delta x}{3}(y_0 + 4y_1 + y_2)$

We shall see in the following lemma that the integral on the right can be evaluated by a simple formula provided x_0, x_1, and x_2 are equally spaced on the x-axis. Thus, we choose the points with x-coordinates

$$x_0 = x_1 - \Delta x, \qquad x_1, \qquad x_2 = x_1 + \Delta x$$

where Δx is a positive number.

LEMMA Let $P_0(x_1 - \Delta x, y_0)$, $P_1(x_1, y_1)$, and $P_2(x_1 + \Delta x, y_2)$ be noncollinear points. Let $y = Ax^2 + Bx + C$ be a parabola that passes through P_0, P_1, and P_2. Then

$$\int_{x_1-\Delta x}^{x_1+\Delta x} (Ax^2 + Bx + C)\,dx = \frac{\Delta x}{3}(y_0 + 4y_1 + y_2)$$

5-6 Numerical Integration

Proof We use the antiderivative of $Ax^2 + Bx + C$ to write

$$\int_{x_1-\Delta x}^{x_1+\Delta x} (Ax^2 + Bx + C) \, dx = \left[\frac{Ax^3}{3} + \frac{Bx^2}{2} + Cx\right]_{x_1-\Delta x}^{x_1+\Delta x}.$$

After simplification, this expression reduces to

$$\Delta x \left(2Ax_1^2 + \frac{2A \, \Delta x^2}{3} + 2Bx_1 + 2C\right).$$

We now show that $\Delta x(y_0 + 4y_1 + y_2)/3$ has the same value. Since the points P_0, P_1, and P_2 are on the graph of the parabola, their coordinates satisfy the equation $y = Ax^2 + Bx + C$. Therefore,

$$\begin{aligned}
y_0 &= A(x_1 - \Delta x)^2 + B(x_1 - \Delta x) + C \\
&= Ax_1^2 - 2Ax_1 \, \Delta x + A \, \Delta x^2 + Bx_1 - B \, \Delta x + C. \\
4y_1 &= 4Ax_1^2 \qquad\qquad\qquad\qquad\quad + 4Bx_1 \qquad\quad + 4C. \\
y_2 &= A(x_1 + \Delta x)^2 + B(x_1 + \Delta x) + C \\
&= Ax_1^2 + 2Ax_1 \, \Delta x + A \, \Delta x^2 + Bx_1 + B \, \Delta x + C.
\end{aligned}$$

If we add these equations and multiply by $\Delta x/3$, we get

$$\begin{aligned}
\frac{\Delta x}{3}(y_0 + 4y_1 + y_2) &= \frac{\Delta x}{3}(6Ax_1^2 + 2A \, \Delta x^2 + 6Bx_1 + 6C) \\
&= \Delta x \left(2Ax_1^2 + \frac{2A \, \Delta x^2}{3} + 2Bx_1 + 2C\right) \\
&= \int_{x_1-\Delta x}^{x_1+\Delta x} (Ax^2 + Bx + C) \, dx. \blacksquare
\end{aligned}$$

We are now ready to establish Simpson's Rule for the approximate integration of f over $[a, b]$.

We first partition the interval $[a, b]$ into an *even* number of subintervals of length Δx by the partition

$$a = x_0 < x_1 < x_2 < \cdots < x_n = b, \qquad n \text{ even}.$$

(See Fig. 5–16.) To each point x_k in the partition, there corresponds a point $P_k(x_k, y_k)$ on the graph of f, where $y_k = f(x_k)$. To each successive pair of subintervals, there corresponds a set of three points on the curve. The points $P_0(x_0, y_0)$, $P_1(x_1, y_1)$, $P_2(x_2, y_2)$ correspond to the first pair of subintervals, the points $P_2(x_2, y_2)$, $P_3(x_3, y_3)$, $P_4(x_4, y_4)$ to the second pair of subintervals, and so on. (See Fig. 5–16.) Through each set of three points, we now approximate the function by a parabola of form $y = Ax^2 + Bx + C$, as in the lemma. If Δx is small, the parabola approximates the graph of f, and the integral of $Ax^2 + Bx + C$ approximates the integral of f over the pair of subintervals. By the lemma,

$$\int_{x_0}^{x_2} f(x) \, dx \approx \frac{\Delta x}{3}[y_0 + 4y_1 + y_2]$$

$$\int_{x_2}^{x_4} f(x) \, dx \approx \frac{\Delta x}{3}[y_2 + 4y_3 + y_4]$$

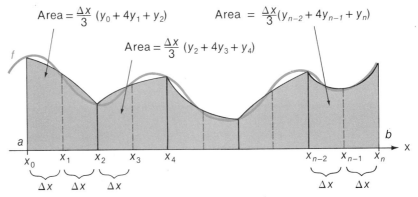

Figure 5-16 Simpson's rule: $\int_a^b f(x)\,dx \approx \dfrac{\Delta x}{3}[y_0 + 4y_1 + 2y_2 + 4y_3 + \cdots + 4y_{n-1} + y_n]$ (n even).

$$\int_{x_4}^{x_6} f(x)\,dx \approx \frac{\Delta x}{3}[y_4 + 4y_5 + y_6]$$

$$\cdots\cdots\cdots\cdots\cdots\cdots\cdots\cdots$$

$$\int_{x_{n-2}}^{x_n} f(x)\,dx \approx \frac{\Delta x}{3}[y_{n-2} + 4y_{n-1} + y_n].$$

It follows from Property 1 of Section 5–3 that

$$\int_a^b f(x)\,dx = \int_{x_0}^{x_n} f(x)\,dx$$

$$= \int_{x_0}^{x_2} f(x)\,dx + \int_{x_2}^{x_4} f(x)\,dx + \int_{x_4}^{x_6} f(x)\,dx + \cdots + \int_{x_{n-2}}^{x_n} f(x)\,dx$$

$$\approx \frac{\Delta x}{3}[y_0 + 4y_1 + y_2] + \frac{\Delta x}{3}[y_2 + 4y_3 + y_4] + \frac{\Delta x}{3}[y_4 + 4y_5 + y_6]$$

$$+ \cdots + \frac{\Delta x}{3}[y_{n-2} + 4y_{n-1} + y_n]$$

$$\approx \frac{\Delta x}{3}[y_0 + 4y_1 + 2y_2 + 4y_3 + 2y_4 + 4y_5 + \cdots + 2y_{n-2} + 4y_{n-1} + y_n]$$

Since $y_k = f(x_k)$ for each k, then

Simpson's Rule

$$\int_a^b f(x)\,dx \approx \frac{\Delta x}{3}[f(x_0) + 4f(x_1) + 2f(x_2) + 4f(x_3) + 2f(x_4) + \cdots + 2f(x_{n-2}) + 4f(x_{n-1}) + f(x_n)], \quad n \text{ even}.$$

This approximation formula is known as *Simpson's Rule*.

5-6 Numerical Integration

EXAMPLE 2 Use Simpson's Rule with $n = 10$ to approximate the integral

$$\int_2^3 \frac{1}{x}\,dx.$$

Solution As in Example 1, $\Delta x = 0.1$, $x_0 = 2.0$, $x_1 = 2.1$, $x_2 = 2.2$, $x_3 = 2.3$, and so on. Then

$$\int_2^3 \frac{1}{x}\,dx \approx \frac{0.1}{3}\left[\frac{1}{2.0} + 4\cdot\frac{1}{2.1} + 2\cdot\frac{1}{2.2} + 4\cdot\frac{1}{2.3} + 2\cdot\frac{1}{2.4} + 4\cdot\frac{1}{2.5}\right.$$
$$\left. + 2\cdot\frac{1}{2.6} + 4\cdot\frac{1}{2.7} + 2\cdot\frac{1}{2.8} + 4\cdot\frac{1}{2.9} + \frac{1}{3.0}\right]$$
$$\approx 0.405465,$$

which (see Example 1) is correct to six decimal places. □

ESTIMATES OF ERROR

We should like to have some way to estimate the error when we use the Trapezoidal Rule or Simpson's Rule. We are seldom fortunate enough to know the "true" value of the integral, as in Example 1.

The following results are proved in advanced calculus by an extension of the Mean Value Theorem. Observe that the function f must be at least twice differentiable on $[a, b]$ before either part of the theorem can be applied.

THEOREM 5-5

Accuracy of Approximations

(1) Let f be twice differentiable on $[a, b]$. Let M be a number such that $|f''(x)| \leq M$ for all x in $[a, b]$. The error obtained when the Trapezoidal Rule is used is no greater than

$$\frac{M(b-a)^3}{12n^2}.$$

(2) Let f be four times differentiable on $[a, b]$. Let N be a number such that $|f^{(4)}(x)| \leq N$ for all x in $[a, b]$. The error obtained when Simpson's Rule is used is no greater than

$$\frac{N(b-a)^5}{180n^4}.$$

EXAMPLE 3 Use Theorem 5-5 to get upper bounds on the errors when the Trapezoidal Rule and Simpson's Rule are used with $n = 10$ to approximate $\int_2^3 dx/x$.

Solution Since $f(x) = 1/x$, then

$$f'(x) = -\frac{1}{x^2}, \quad f''(x) = \frac{2}{x^3}, \quad f^{(3)}(x) = -\frac{6}{x^4}, \quad f^{(4)}(x) = \frac{24}{x^5}.$$

Then

$$|f''(x)| = \frac{2}{|x|^3} \leq \frac{2}{2^3} = \frac{1}{4}$$

and
$$|f^{(4)}(x)| = \frac{24}{|x|^5} \leq \frac{3}{4} \quad \text{when } 2 \leq x \leq 3.$$

Therefore, we choose
$$M = \tfrac{1}{4} \quad \text{and} \quad N = \tfrac{3}{4}.$$

The error when the Trapezoidal Rule is used is no greater than
$$\frac{M(b-a)^3}{12n^2} = \frac{\tfrac{1}{4}(3-2)^3}{12 \cdot 10^2} \approx 0.000208.$$

The error when Simpson's Rule is used is no greater than
$$\frac{N(b-a)^5}{180n^4} = \frac{\tfrac{3}{4}(3-2)^5}{180 \cdot 10^4} \approx 0.00000042. \quad \square$$

APPROXIMATE INTEGRATION ON THE COMPUTER

The methods of this section are especially easy to program for the electronic computer or the programmable calculator. Figure 5–17 contains a flowchart for the basic steps in the Trapezoidal Rule. A computer program can be constructed directly from this flowchart. (See Exercise 12.) A similar flowchart (and program) can be prepared for Simpson's Rule.

The symbol S is used in the flowchart for the temporary storage location of the "partial sums"—the numbers $f(a)/2$, $f(a)/2 + f(a + \Delta x)$, $f(a)/2 + f(a + \Delta x) + f(a + 2\Delta x)$, and so on. The final value obtained from the Trapezoidal Rule is stored in location V.

Exercises 5–6

1–6. (a) Use the Trapezoidal Rule with the given value of n to approximate the definite integral.
   (b) Use Simpson's Rule to approximate the integral.

•1 $\int_0^2 \dfrac{dx}{x+1}, \; n = 10$

•2 $\int_0^1 \dfrac{dx}{\sqrt{1+x^2}}, \; n = 4$

•3 $\int_0^2 \dfrac{dx}{1+x^2}, \; n = 8$

•4 $\int_1^5 \dfrac{dx}{x}, \; n = 8$

•5 $\int_1^2 \sqrt{1+x^2} \, dx, \; n = 10$

•6 $\int_0^2 \sqrt{1+x^4} \, dx, \; n = 8$

7–8. Use Theorem 5–5 to obtain upper bounds on the arrors if (a) the Trapezoidal Rule and (b) Simpson's Rule are used to approximate the integrals in Exercises 1 and 4.

Exercises 5–6

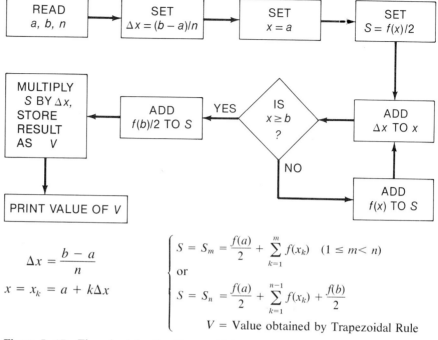

Figure 5–17 Flowchart for the Trapezoidal Rule.

$$\Delta x = \frac{b-a}{n}$$

$$x = x_k = a + k\Delta x$$

$$\begin{cases} S = S_m = \frac{f(a)}{2} + \sum_{k=1}^{m} f(x_k) & (1 \leq m < n) \\ \text{or} \\ S = S_n = \frac{f(a)}{2} + \sum_{k=1}^{n-1} f(x_k) + \frac{f(b)}{2} \end{cases}$$

$V = $ Value obtained by Trapezoidal Rule

• **7** Exercise 1   •**8** Exercise 4

9 Our discussion of the Trapezoidal Rule was based on areas of trapezoids. Show that the following procedure leads to the Trapezoidal Rule even if $f(x)$ is negative on part or all of $[a, b]$.
  (1) Partition $[a, b]$ into n equal subintervals of length $\Delta x = (b - a)/n$.
  (2) Let L be the approximating sum obtained by using the left-hand endpoints of the subintervals; let R be the approximating sum obtained by using the right-hand endpoints.
  (3) Take the average $(L + R)/2$.

•**10** Work through the steps of the flowchart in Figure 5–17 with the example $f(x) = x$, $a = 1$, $b = 3$, $n = 4$.

11 Prepare a flowchart like Figure 5–17 for Simpson's Rule

c12 (*For Students with Access to a Computer or Programmable Calculator*)
  (a) Prepare a program for the Trapezoidal Rule based on the flowchart in Figure 5–17. Run your program with the integral in Exercise 3.
  (b) Prepare a program for Simpson's Rule. Run your program with the integral in Exercise 3.

Review Problems

- **1** Write the following sum more compactly, using sigma notation.
$$\left(\frac{1}{2}\right)^2 + \left(\frac{2}{3}\right)^3 + \left(\frac{3}{4}\right)^4 + \cdots + \left(\frac{k-1}{k}\right)^k.$$

- **2** Evaluate the following sum. Then take the limit as $n \to \infty$:
$$\frac{1}{n^3} \sum_{k=1}^{n+1} (3k^2 - 2k + 4).$$

3. Let m and n be positive integers with $m < n$. Show that
$$\sum_{i=m}^{n} f(i) = \sum_{i=1}^{n} f(i) - \sum_{i=2}^{m} f(i-1).$$

- **4** Use the method of approximating sums to calculate:

   (a) $\displaystyle\int_1^3 x^2 \, dx$ (b) $\displaystyle\int_4^5 2x \, dx.$

5. Use the method of approximating sums to prove that if $a < b$, then

   (a) $\displaystyle\int_a^b k \, dx = k(b - a)$ (b) $\displaystyle\int_a^b x^2 \, dx = \frac{b^3 - a^3}{3}.$

6. (a) State three properties of definite integrals from Section 5–3.

   (b) State and prove the Mean Value Theorem for integrals. Illustrate the theorem with a figure.

7. State and prove both parts of the Fundamental Theorem of Calculus.

8. State the substitution principle (a) for indefinite integrals; (b) for definite integrals.

9–18. Use the substitution principle, if necessary, to calculate the integral.

- **9** $\displaystyle\int \frac{\sqrt{x} + 1}{x^2} \, dx$

- **10** $\displaystyle\int_1^2 \left(x - \frac{1}{x}\right)^2 dx$

- **11** $\displaystyle\int (\sqrt{x} + 1)^3 \, dx$

- **12** $\displaystyle\int_1^4 \frac{(\sqrt{x} + 1)^{13}}{\sqrt{x}} \, dx$

- **13** $\displaystyle\int \frac{x - 2}{(x^2 - 4x + 3)^7} \, dx$

- **14** $\displaystyle\int (8 + 7x^3)^{-2/3} \cdot x^2 \, dx$

- **15** $\displaystyle\int_2^{10} \frac{1}{\sqrt{5x - 1}} \, dx$

- **16** $\displaystyle\int_0^3 (3x - 1)^{4/3} \, dx$

Review Problems

●17 $\int (x^2 - 1)(x^2 + 1)\, dx$

●18 $\int (x^2 + 1)^{12}(x^2 - 1)^{12}x^3\, dx$

19–20. Use antiderivatives to calculate the area of the region enclosed by the graph of f and the x-axis.

●19 $f(x) = 4 - x^2$

●20 $f(x) = 2 + x - x^2$

●21 The slope of the tangent line to the graph of f at the point $(x, f(x))$ is given by $f'(x) = \sqrt{2x - 4}$ provided $x > 2$. Determine the function f if $(4, 2)$ is on the graph.

●22 (a) State the Trapezoidal Rule.

  (b) Use the Trapezoidal Rule with $n = 4$ to approximate the value of $\int_1^2 (dx/x)$.

  c(c) Use the Trapezoidal Rule with $n = 10$ to calculate the approximate value of

  $$\int_1^6 \frac{dx}{\sqrt{x^2 + 1}}.$$

●23 Work Problem 22 for Simpson's Rule rather than the Trapezoidal Rule.

●24 Calculate the derivative. (*Hint for* (b): Use the Chain Rule.)

  (a) $D_x \int_x^1 \frac{1}{t}\, dt$

  (b) $D_x \int_4^{x^2} \frac{1}{t}\, dt$

●25 A continuous function f has the property that $(x - 1)^{5/6} \le f(x) \le (x - 1)^{2/3}$ if $1 \le x \le 2$. Estimate the size of $\int_1^2 f(x)\, dx$.

6 Applications of the Integral

6–1 Area

The concept of the *area* of a region $\mathcal{R}$ in the xy-plane can properly be defined only in terms of a limiting process. The general outlines of the process can be easily understood. We construct a set of rectangles that approximates the region $\mathcal{R}$. (See Figs. 6–1a, b). If there are n such rectangles, we let A_n denote the sum of their areas. We then take the limit of A_n as $n \to \infty$ and the size of each rectangle (in some sense) shrinks towards zero. If the limit of A_n exists, then we define the *area* of $\mathcal{R}$ to be

$$\text{Area} = \lim_{n \to \infty} A_n.$$

This concept is illustrated in Figure 6–1. In (a) we have a set of rectangles that approximates the region. In (b) we have another set, each rectangle smaller than the rectangles in (a). As the sizes of the rectangles approach zero and n (the number of rectangles) approaches infinity, the number A_n (the sum of the areas of the rectangles) approaches a limit that equals the area of $\mathcal{R}$.

At this stage of our development, several technical problems prevent our making full use of the process described above. To define the concept of *area*, we shall use special rectangles based on a partition of an interval into subintervals. (The more general process will be considered in detail in Chapter 17.)

AREA OF A PLANE REGION

Consider a region $\mathcal{R}$ enclosed between the graphs of $y = f(x)$ and $y = g(x)$ for $a \leq x \leq b$, where f and g are continuous functions and $g(x) \leq f(x)$. (See Fig. 6–2.) To calculate the area of $\mathcal{R}$, we subdivide the interval $[a, b]$ into n subintervals by the partition $\mathcal{P}$ defined by

$$a = x_0 < x_1 < x_2 < \cdots < x_n = b.$$

6–1 Area

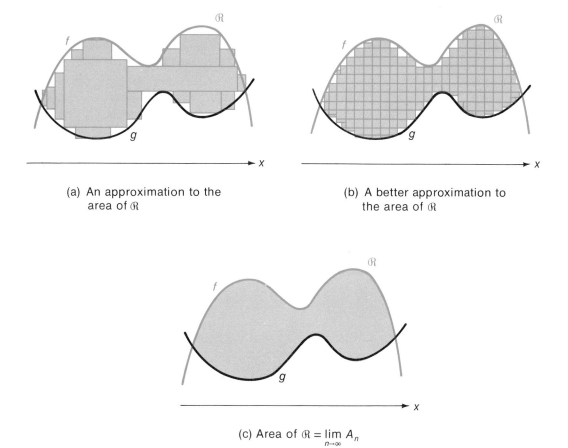

(a) An approximation to the area of $\mathcal{R}$

(b) A better approximation to the area of $\mathcal{R}$

(c) Area of $\mathcal{R} = \lim_{n \to \infty} A_n$

Figure 6–1 The general area problem: $A_n =$ sum of areas of n rectangles; Area of $\mathcal{R} = \lim_{n \to \infty} A_n$.

We choose points $\xi_1, \xi_2, \xi_3, \ldots, \xi_n$, with ξ_1 on the first subinterval, ξ_2 on the second, and so on. Next we construct rectangles over the n subintervals so that the kth rectangle has height $f(\xi_k) - g(\xi_k)$ and base $\Delta x_k = x_k - x_{k-1}$. The area of the kth rectangle is

$$[f(\xi_k) - g(\xi_k)] \Delta x_k.$$

Since the n rectangles approximate the region $\mathcal{R}$, then the sum of their areas should approximate the number that we commonly call the area of $\mathcal{R}$. Observe that the sum of the areas of the rectangles is

$$\sum_{k=1}^{n} [f(\xi_k) - g(\xi_k)] \Delta x_k,$$

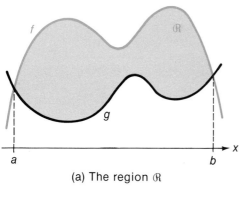

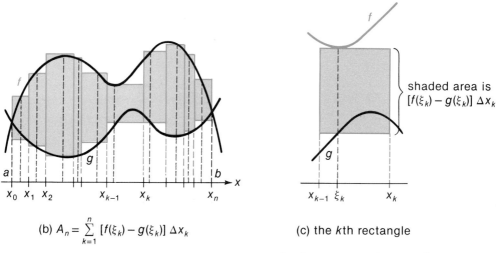

(a) The region $\mathcal{R}$

(b) $A_n = \sum_{k=1}^{n} [f(\xi_k) - g(\xi_k)] \Delta x_k$

(c) the kth rectangle

shaded area is $[f(\xi_k) - g(\xi_k)] \Delta x_k$

Figure 6–2 The area problem. Area of $\mathcal{R} = \lim_{\|\mathcal{P}\| \to 0} \sum_{k=1}^{n} [f(\xi_k) - g(\xi_k)] \Delta x_k = \int_a^b [f(x) - g(x)]\, dx$.

which approaches a limiting value as the norm of the partition approaches zero. This limiting value, which equals a definite integral, is independent of the partition $\mathcal{P}$ and the numbers $\xi_1, \xi_2, \ldots, \xi_n$. We define the area of the region $\mathcal{R}$ to be the definite integral defined by this limit.

DEFINITION

Area

Let f and g be continuous functions, with $f(x) \geq g(x)$ for $a \leq x \leq b$. Let $\mathcal{R}$ be the region bounded by the graphs of f and g and the vertical lines $x = a$ and $x = b$. Let the partition $\mathcal{P}$ and the numbers $\xi_1, \xi_2, \ldots, \xi_n$ be defined as in the above discussion. We define

$$\text{Area of } \mathcal{R} = \lim_{\|\mathcal{P}\| \to 0} \sum_{k=1}^{n} [f(\xi_k) - g(\xi_k)] \Delta x_k$$

6-1 Area

$$\text{Area} = \int_a^b [f(x) - g(x)]\, dx.$$

EXAMPLE 1   Calculate the area of the region bounded by the graphs of $y = x^2 - 1$ and $x - y + 1 = 0$.

Solution   The two graphs intersect at the points $(-1, 0)$ and $(2, 3)$. (These points can be found by solving the two equations simultaneously.) Observe that the line $y = x + 1$ is the upper boundary and the parabola $y = x^2 - 1$ is the lower boundary of the region and that x varies from -1 to 2. (See Fig. 6–3.)

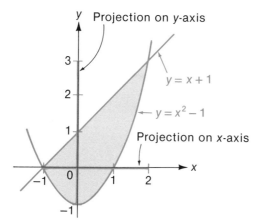

Figure 6–3   Example 1.

The area is

$$\text{Area} = \int_{-1}^{2} [(x + 1) - (x^2 - 1)]\, dx$$

$$= \int_{-1}^{2} (-x^2 + x + 2)\, dx$$

$$= \left[-\frac{x^3}{3} + \frac{x^2}{2} + 2x \right]_{-1}^{2} = \frac{9}{2}. \ \square$$

PROJECTIONS

Projection on an Axis

The *projection* of a region $\mathcal{R}$ on the x-axis is defined to be the set of all x such that the vertical line through x intersects the region $\mathcal{R}$. For example, the projection of the region in Figure 6–3 on the x-axis is the interval $[-1, 2]$. If we choose any point x, where $-1 \leq x \leq 2$, the vertical line through x intersects the region.

The limits of integration for the integral that defines the area

$$A = \int_a^b [f(x) - g(x)]\, dx$$

can be determined by projecting the region $\mathcal{R}$ on the x-axis.

The projection on the y-axis is defined similarly. For example, the region in Figure 6–3 has the projection $[-1, 3]$ on the y-axis.

REGIONS WITH RIGHT AND LEFT BOUNDARIES

Our previous integral formula for the area of $\mathcal{R}$ was obtained on the assumption that $\mathcal{R}$ has continuous upper and lower boundaries $y = f(x)$ and $y = g(x)$, respectively. A region such as the one in Figure 6–4 may have continuous right and left boundaries

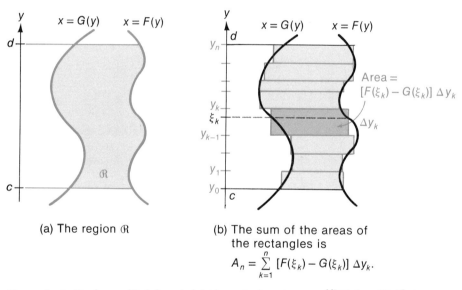

(a) The region $\mathcal{R}$

(b) The sum of the areas of the rectangles is
$$A_n = \sum_{k=1}^{n} [F(\xi_k) - G(\xi_k)]\, \Delta y_k.$$

Figure 6–4 Regions with left and right boundaries. Area $= \int_c^d [F(y) - G(y)]\, dy$.

$x = F(y)$ and $x = G(y)$, respectively, for $c \leq y \leq d$. In this case, we partition the interval $[c, d]$ on the y-axis into subintervals and use these subintervals to construct rectangles that extend across $\mathcal{R}$ as in Figure 6–4(b). A typical rectangle has area

$$[F(\xi_k) - G(\xi_k)]\, \Delta y_k$$

and the sum of the areas of the rectangles is

$$\sum_{k=1}^{n} [F(\xi_k) - G(\xi_k)]\, \Delta y_k.$$

6–1 Area

If we take the limit of this expression as $n \to \infty$ and $\|\mathcal{P}\| \to 0$, we get the integral for the area

Area of Region with Right and Left Boundaries

$$\text{Area of } \mathcal{R} = \int_c^d [F(y) - G(y)]\, dy.$$

Observe that $[c, d]$ is the projection of $\mathcal{R}$ on the y-axis.

ELEMENTS OF AREA

Element of Area

A simple device can help us to set up the integrals for many problems that involve area. An *element of area* of a region $\mathcal{R}$ is a thin rectangle that extends across the region from one boundary to the other.

Suppose, for example, that the upper and lower boundaries of $\mathcal{R}$ are the graphs of $y = f(x)$ and $y = g(x)$, respectively, where f and g are continuous, $a \leq x \leq b$. We choose a typical point x between a and b and draw a thin vertical rectangle across the region at x. (See Fig. 6–5a.) The height of this rectangle is $f(x) - g(x)$, and the width

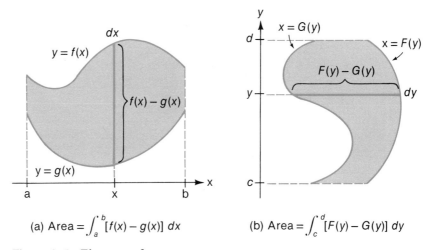

(a) Area = $\int_a^b [f(x) - g(x)]\, dx$ (b) Area = $\int_c^d [F(y) - G(y)]\, dy$

Figure 6–5 Elements of area.

is Δx. Recall that the differential dx can take on any value. If we let $\Delta x = dx$ be the base of this rectangle, then the area is

$$\text{Area of element} = [f(x) - g(x)]\, dx.$$

The area of the region is equal to the integral of this expression:

$$\text{Area of region} = \int_a^b [f(x) - g(x)]\, dx.$$

A similar result holds if the region has right and left boundaries $x = F(y)$ and $x = G(y)$, respectively. In that case, we taken our element of area between those boundaries, as in Figure 6–5b. The area of the element is

$$[F(y) - G(y)]\, dy$$

and the area of the region is

$$\text{Area of } \mathcal{R} = \int_c^d [F(y) - G(y)]\, dy.$$

Remark We shall use the element of area as a mnemonic device to help set up integrals and remember formulas. In the remainder of this chapter we shall derive general formulas by the use of approximating sums, then use elements of area to set up the integrals in specific problems.

EXAMPLE 2 Calculate the area of the region bounded by the graphs of $x = y^2$ and $x = 5 - (y - 1)^2$.

Solution The region is pictured in Figure 6–6. Observe that it would be awkward to set up an integral for the area by partitioning the region into vertical strips but it is easy to set up an integral after partitioning by horizontal strips.

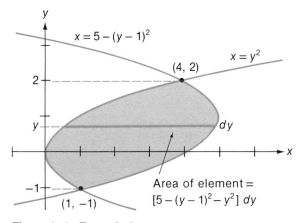

Figure 6–6 Example 2.

The points of intersection of the boundary curves are $(1, -1)$ and $(4, 2)$. These can be found by solving the system of equations

$$\begin{cases} x = y^2 \\ x = 5 - (y - 1)^2. \end{cases}$$

A typical horizontal element has area equal to

$$[(5 - (y - 1)^2) - y^2]\, dy = (4 + 2y - 2y^2)\, dy.$$

6-1 Area

The projection on the y-axis is $[-1, 2]$. Therefore the area of the region is

$$\text{Area} = \int_{-1}^{2} (4 + 2y - 2y^2) \, dy = \left[4y + y^2 - \frac{2y^3}{3} \right]_{-1}^{2} = 9. \quad \square$$

SPECIAL CASES

If the upper boundary of $\mathcal{R}$ is the graph of the continuous function f and the lower boundary is the x-axis, $a \leq x \leq b$, then (see Fig. 6–7) the area of $\mathcal{R}$ is

$$\text{Area} = \int_a^b [f(x) - 0] \, dx = \int_a^b f(x) \, dx,$$

the integral used for area in Chapter 5.

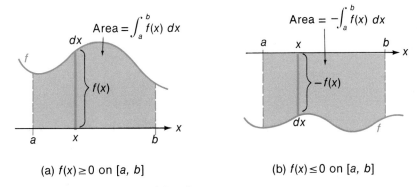

(a) $f(x) \geq 0$ on $[a, b]$ (b) $f(x) \leq 0$ on $[a, b]$

Figure 6–7 Areas of special regions.

Let f be negative on $[a, b]$. If the lower boundary of $\mathcal{R}$ is the graph of f and the upper boundary is the x-axis, then (see Fig. 6–7b) the area is

$$\text{Area} = \int_a^b [0 - f(x)] \, dx = -\int_a^b f(x) \, dx.$$

If the boundary graphs cross each other, then the integral must be split into two or more integrals before it can be evaluated. (See Property 1 of Section 5–3.) The process is illustrated in Example 3.

EXAMPLE 3 Calculate the area of the closed regions bounded by the x-axis and the graph of $y = x^3 - 6x^2 + 8x$.

Solution The regions have two parts, as shown in Figure 6–8.
Between 0 and 2 the upper boundary is defined by $y = x^3 - 6x^2 + 8x$ and the

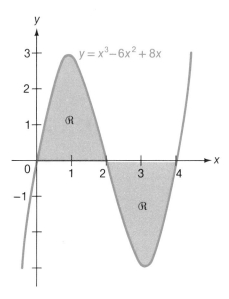

Figure 6-8 Example 3.

lower boundary by $y = 0$. Between 2 and 4 the upper boundary is defined by $y = 0$ and the lower boundary by $y = x^3 - 6x^2 + 8x$. The total area is

$$A = \int_0^2 [(x^3 - 6x^2 + 8x) - 0] \, dx + \int_2^4 [0 - (x^3 - 6x^2 + 8x)] \, dx$$

$$= \int_0^2 (x^3 - 6x^2 + 8x) \, dx - \int_2^4 (x^3 - 6x^2 + 8x) \, dx$$

$$= \left[\frac{x^4}{4} - 2x^3 + 4x^2 \right]_0^2 - \left[\frac{x^4}{4} - 2x^3 + 4x^2 \right]_2^4 = 4 - (-4) = 8. \quad \square$$

Exercises 6-1

1–12. Calculate the total area of the region bounded by the given curves and lines. Draw a figure that shows a typical element of area.

- **1** $y = 4x - x^2$ and x-axis
- **2** $y = x^2 - 4x + 3$ and x-axis
- **3** $y = x^3 - x^2 - 6x$ and $y = 0$
- **4** $y = x^3 - 3x^2 + 2x$ and x-axis
- **5** $x = y^2 - y^3$ and y-axis
- **6** $y = x^3$ and $y = 4x$
- **7** $x = 2y - y^2$ and $y - x = 2$
- **8** $y = x^2$ and $y = 8 - x^2$
- **9** $y = 1 - x^2$ and $y = x - 1$
- **10** $y = \sqrt{x}$, $y = -x$ and $y = x - 2$
- **11** $y^2 = -x$ and $y - x = 2$
- **12** $y = x$, $x + 3y - 4 = 0$, $y + 3x = 12$

- **13** The region bounded by the graph of $y = x^2$ and the line $y = 4$ is divided into two parts of equal area by the line $y = c$. Find c.

6–2 Volume I: Slices and Disks

14 Draw a figure that shows how the region in Exercise 11 can be decomposed into regions with upper and lower boundaries that are graphs of functions of x. Use this decomposition to calculate the area of the region.

15 Draw a figure like Figure 6–2b or 6–4b, which pictures an approximating sum for the area of the region. Write out a typical approximating sum based on a partition into n equal subintervals and take the limit as $n \to \infty$, obtaining an integral for the area.

•   (a) The region in Exercise 8   (b) The region in Exercise 11

16–17. Make a careful drawing of the regions bounded by the given curves and lines. Calculate the area.

•**16** $y = x^3 - 2x^2 - 3x$, $y = 5x$ •**17** $x = y^3 + 8$, $x = (y + 2)^2$

18 Suppose that, in integrating the function given in Example 3, you had forgotten to divide the integral into two parts (see Figure 6–8), and had simply calculated the definite integral from $x = 0$ to $x = 4$. What result would you have obtained? Do you see the advantage of making a rough sketch before trying to write a definite integral?

6–2 Volume I: Slices and Disks

The volumes of certain solids can be calculated by a method like the one used in calculating areas of plane regions. The general process is as follows.

Steps in Calculation of Volume

(1) Decompose the solid into small parts, each of which has a volume that can be approximated by an expression of form $f(\xi_k)\,\Delta x_k$. Then the total volume can be approximated by a sum of form

$$\sum_{k=1}^{n} f(\xi_k)\,\Delta x_k.$$

(2) Show that the approximation becomes better and better as $n \to \infty$ and each $\Delta x_k \to 0$. Thus,

$$\text{Volume} = \lim_{\|\mathcal{P}\| \to 0} \sum_{k=1}^{n} f(\xi_k)\,\Delta x_k.$$

(3) Express the above limit as a definite integral

$$\text{Volume} = \int_{a}^{b} f(x)\,dx.$$

(4) Evaluate the integral to determine the volume.

THREE-DIMENSIONAL SPACE

x-, y-, z- Axis System

In our work with volume, we consider the xy-plane to be part of three-dimensional space. We construct a third axis—the z-axis—perpendicular to both the x-axis and the y-axis. In this chapter we keep our usual orientation of the x- and y-axes and draw the z-axis as if it is pointing out of the book from the origin towards the reader. (See

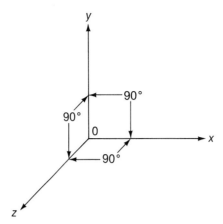

Figure 6–9 Coordinate axes in three-dimensional space. (A different axis system is used in the latter part of the book.)

Fig. 6–9.) (For certain technical reasons, we shall modify this arrangement of axes in the latter part of the book.)

VOLUMES BY SLICING

Let S be a solid in three-dimensional space that is contained between two planes perpendicular to the x-axis. Let these planes intersect the x-axis at $x = a$ and $x = b$. Slice the solid S by planes perpendicular to the x-axis, as in Figure 6–10. Let these

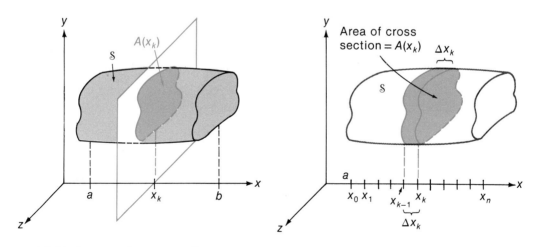

(a) $A(x_k)$ = area of cross section perpendicular to the x-axis

(b) Volume of kth slice $\approx A(x_k) \, \Delta x_k$

Figure 6–10 Volume of solid = $\lim_{\|\mathscr{P}\| \to 0} \sum_{k=1}^{n} A(x_k) \, \Delta x_k = \int_a^b A(x) \, dx$.

6-2 Volume I: Slices and Disks

planes intersect the x-axis at the points

$$a = x_0 < x_1 < x_2 < \cdots < x_n = b.$$

Note that these points form a partition of $[a, b]$.

Let $A(x_k)$ be the area of the cross section obtained by intersecting the solid S with the plane through x_k. Let $\Delta x_k = x_k - x_{k-1}$. Then

$$A(x_k) \, \Delta x_k$$

is an approximation to the volume of the slice between the planes at $x = x_{k-1}$ and $x = x_k$. (See Fig. 6–10b.) It follows that

$$\sum_{k=1}^{n} A(x_k) \, \Delta x_k$$

is an approximation to the volume of S.

If the function $A(x)$ is continuous on $[a, b]$, then we can get the exact value of the volume of S by taking the limit

$$\lim_{\|\mathcal{P}\| \to 0} \sum_{k=1}^{n} A(x_k) \, \Delta x_k.$$

Thus,

$$\text{Volume of } S = \lim_{\|\mathcal{P}\| \to 0} \sum_{k=1}^{n} A(x_k) \, \Delta x_k = \int_a^b A(x) \, dx.$$

EXAMPLE 1 A right pyramid 30 ft high has a square base measuring 40 ft on a side. Find its volume.

Solution For convenience in setting up the integral, we draw the pyramid on its side with the x-axis pointing through the vertex. (See Fig. 6–11a.) If we cut the pyramid at the point x by a plane perpendicular to the x-axis, then the cross section is a square. To calculate the length of one side of this square, we look at the cross section in Figure 6–11b. Since the small triangle is similar to the large one, then

$$\frac{y}{30 - x} = \frac{20}{30}$$

so that

$$y = \frac{20(30 - x)}{30} = \frac{2(30 - x)}{3}.$$

The length of the side of the square is

$$2y = \frac{4(30 - x)}{3}$$

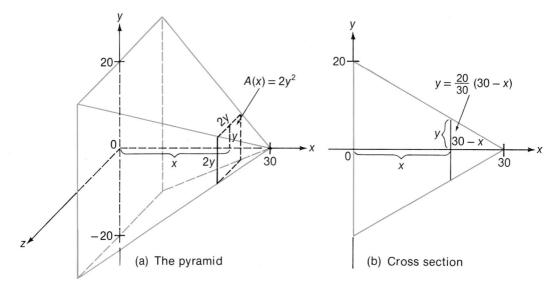

Figure 6–11 Example 1. Volume of a pyramid.

and the area of the square is

$$A(x) = (2y)^2 = \frac{16(30 - x)^2}{9}.$$

Since the projection of the pyramid on the x-axis is the interval $[0, 30]$, it follows that the volume is

$$\text{Volume} = \int_0^{30} A(x)\, dx = \int_0^{30} \frac{16(30 - x)^2}{9}\, dx$$

$$= \left[-\frac{16}{9} \cdot \frac{(30 - x)^3}{3} \right]_0^{30} = \frac{16}{9} \cdot \frac{30^3}{3} = 16{,}000 \text{ ft}^3. \ \square$$

SOLIDS OF REVOLUTION. DISK METHOD

The method of calculating volumes by slicing is easily applied to certain solids of revolution. The simplest situation involves a region bounded by the x-axis and the graph of a continuous function $y = f(x)$, $a \leq x \leq b$, that is revolved about the x-axis. (See Fig. 6–12.)

If we cut the solid of revolution at the point x by a plane perpendicular to the x-axis, the cross section is a circle of radius

$$R = |f(x)|.$$

(See Fig. 6–12c.) It follows that the area of the cross section is

$$A(x) = \pi R^2 = \pi |f(x)|^2 = \pi [f(x)]^2.$$

6–2 Volume I: Slices and Disks

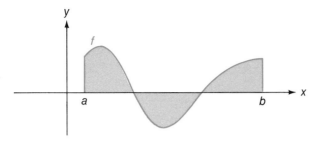

(a) the regions bounded by the graph of f and the x-axis

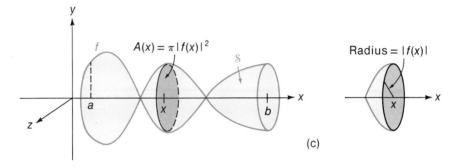

(b) The solid of revolution

(c) Area of cross section $A(x) = \pi |f(x)|^2$

Figure 6–12 Solid of revolution about the x-axis. Volume $= \pi \int_a^b [f(x)]^2\, dx$.

The volume of the solid of revolution is

$$\text{Volume} = \int_a^b A(x)\, dx = \int_a^b \pi [f(x)]^2\, dx.$$

Volume of Solid About x-Axis

$$\text{Volume} = \pi \int_a^b [f(x)]^2\, dx.$$

ELEMENTS OF VOLUME

The integral for volume is easily set up. A typical element of area, drawn across $\mathcal{R}$ perpendicular to the x-axis, has height $|f(x)|$ and thickness dx. When this element is revolved about the x-axis, it sweeps out a circular disk of volume $\pi (f(x))^2\, dx$, called an *element of volume*. The corresponding integral equals the volume of the solid of revolution. (See Fig. 6–13.)

As with area calculations, an element of volume is used as a mnemonic device to

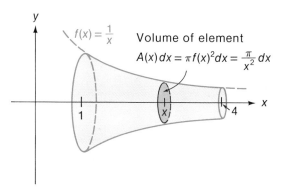

Figure 6-13 Example 2.

represent a typical term in an approximating sum. It is more convenient to write the term $\pi[f(x)]^2\, dx$, which leads directly to the integral, than the term $\pi[f(\xi_k)]^2\, \Delta x_k$, which leads to an approximating sum that in turn leads to the integral.

EXAMPLE 2 Let $\mathcal{R}$ be the region bounded above by the graph of $f(x) = 1/x$ and below by the x-axis, $1 \leq x \leq 4$. Calculate the volume of the solid obtained by revolving $\mathcal{R}$ about the x-axis.

Solution A typical element of volume has radius $f(x) = 1/x$, thickness dx, and volume

$$\pi[f(x)]^2\, dx = \frac{\pi\, dx}{x^2}.$$

The volume of the solid is

$$V = \pi \int_1^4 \frac{dx}{x^2} = \pi\left[-\frac{1}{x}\right]_1^4 = \pi\left[-\frac{1}{4} + 1\right] = \frac{3\pi}{4}.$$

(See Fig. 6-13.) □

REGIONS REVOLVED ABOUT THE y-AXIS

If the region $\mathcal{R}$ is bounded by the graph of $x = f(y)$ and the y-axis, $c \leq y \leq d$, and $\mathcal{R}$ is revolved about the y-axis, the volume of the solid of revolution is

Volume of
Solid About
y-Axis
$$V = \pi \int_c^d [f(y)]^2\, dy.$$

A typical element of volume has radius $|f(y)|$ and thickness dy. The volume of such a disk is $\pi[f(y)]^2\, dy$. (See Fig. 6-14.) When this expression is integrated, we get the volume of the solid of revolution.

The above method can be generalized to calculate the volume of a solid obtained by revolving a region about any horizontal or vertical line. In each case we draw a

6–2 Volume I: Slices and Disks

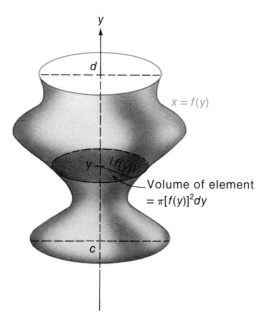

Figure 6–14 Solid of revolution about the y-axis. Volume $= \pi \int_c^d [f(y)]^2 \, dy$.

typical element of area, revolve it about the line, and calculate the resulting volume. The integral of this expression gives us the volume of the solid.

EXAMPLE 3 The closed region bounded by the graph of $x = y^2$ and the vertical line $x = 2$ is revolved about the line $x = 2$. Calculate the volume of the solid of revolution.

Solution The region is pictured in Figure 6–15a. A typical element has radius $2 - y^2$,

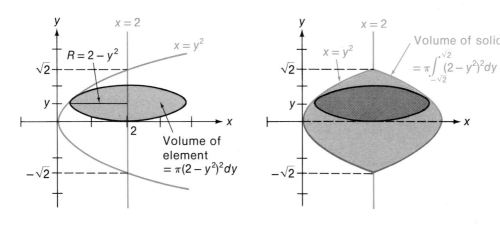

(a) Element of volume

(b) The solid of revolution

Figure 6–15 Example 3.

thickness dy, and volume $\pi(2 - y^2)^2\,dy$. The volume of the solid of revolution is

$$V = \pi \int_{-\sqrt{2}}^{\sqrt{2}} (2 - y^2)^2\,dy = \pi \int_{-\sqrt{2}}^{\sqrt{2}} (4 - 4y^2 + y^4)\,dy$$

$$= \pi \left[4y - \frac{4y^3}{3} + \frac{y^5}{5} \right]_{-\sqrt{2}}^{\sqrt{2}} = \frac{64\pi\sqrt{2}}{15}. \quad \square$$

REGIONS BOUNDED BY TWO CURVES

Let $\mathcal{R}$ be the region bounded by the graphs of the continuous functions $y = f(x)$ and $y = g(x)$, and the lines $x = a$, $x = b$, where $0 \leq g(x) \leq f(x)$. (See Fig. 6–16a.) The volume of the solid obtained by revolving $\mathcal{R}$ about the x-axis can be calculated in three steps.

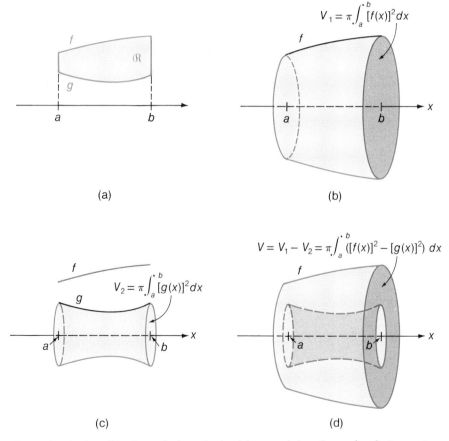

Figure 6–16 A solid of revolution obtained by revolving the region between two curves about the x-axis. Volume $= \pi \int_a^b [f(x)^2 - g(x)^2]\,dx$.

6–2 Volume I: Slices and Disks

Region Bounded by Two Curves Revolved About x-Axis

(1) Calculate the volume obtained by revolving the large region bounded by the graph of f and the x-axis about the x-axis (Fig. 6–16b). This volume is

$$V_1 = \pi \int_a^b [f(x)]^2 \, dx.$$

(2) Calculate the volume obtained by revolving the small region bounded by the graph of g and the x-axis about the x-axis (Fig. 6–16c.). This volume is

$$V_2 = \pi \int_a^b [g(x)]^2 \, dx.$$

(3) The volume of the solid obtained by revolving $\mathcal{R}$ about the x-axis is

$$V = V_1 - V_2 = \pi \int_a^b \left([f(x)]^2 - [g(x)]^2 \right) dx.$$

This calculation is equivalent to carving out a portion of the large solid and calculating the volume of the part that remains. (See Fig. 6–16d.)

EXAMPLE 4 Let $\mathcal{R}$ be the region bounded by the graphs of $f(x) = \sqrt{x}$ and $g(x) = x^2$, $0 \leq x \leq 1$. Calculate the volume of the solid obtained by revolving $\mathcal{R}$ about the x-axis. (See Fig. 6–17.)

Solution We follow the steps outlined above.

(1) The outer boundary is $f(x) = \sqrt{x}$. The volume obtained by revolving the region bounded by the graph of f and the x-axis about the x-axis is

$$V_1 = \pi \int_0^1 [f(x)]^2 \, dx = \pi \int_0^1 x \, dx$$

(see Fig. 6–17b).

(2) The inner boundary is $g(x) = x^2$. The volume obtained by revolving the region bounded by the graph of g and the x-axis about the x-axis is

$$V_2 = \pi \int_0^1 [g(x)]^2 \, dx = \pi \int_0^1 x^4 \, dx$$

(see Fig. 6–17c).

(3) The volume of the solid obtained by revolving $\mathcal{R}$ about the x-axis is

$$V = V_1 - V_2 = \pi \int_0^1 [f(x)^2 - g(x)^2] \, dx$$

$$= \pi \int_0^1 [x - x^4] \, dx = \pi \left[\frac{x^2}{2} - \frac{x^5}{5} \right]_0^1$$

$$= \pi \left[\frac{1}{2} - \frac{1}{5} \right] = \frac{3\pi}{10}$$

(see Fig. 6–17d). □

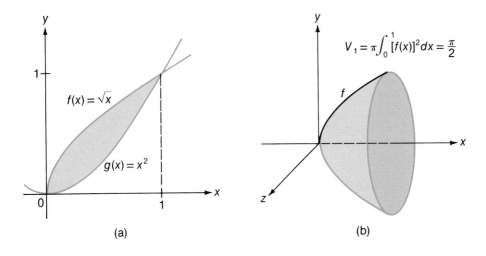

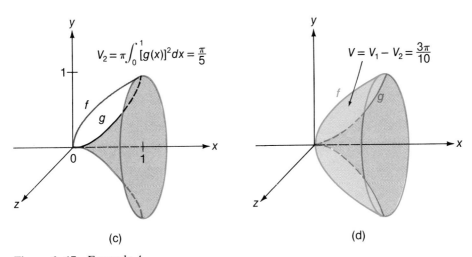

Figure 6–17 Example 4.

Exercises 6–2

1–8. Calculate the volume of the solid of revolution obtained by revolving the region $\mathscr{R}$, bounded by the given curves and lines, about the indicated line.

- **1** $y = \sqrt{9 - x^2}$ for $x \geq 0$, $x = 0$ and $y = 0$
  (a) About the x-axis
  (b) About the y-axis
- **2** $y = 4x - x^2$ and $y = 0$. About the x-axis.

- **3** $y = x^2$ and $y = 4$
  - (a) About the x-axis
  - (b) About the y-axis
  - (c) About the line $y = 4$

- **4** $y = \sqrt{x}$, $x = 4$ and $y = 0$
  - (a) About the x-axis
  - (b) About the y-axis
  - (c) About the line $x = 4$
  - (d) About the line $y = 2$

- **5** $y = x^3$, $y = 8$ and $x = 0$
  - (a) About the x-axis
  - (b) About the y-axis
  - (c) About the line $x = 2$
  - (d) About the line $y = 8$

- **6** $y = 1/x$, $x = 0$, $y = 1$ and $y = 3$. About the y-axis.

- **7** $y = x^2$ and $y = x$
  - (a) About the x-axis
  - (b) About the y-axis
  - (c) About the line $x = 3$
  - (d) About the line $y = 1$

- **8** $y = 3x - x^2$ and $y = x$. About the x-axis.

9 Show that the volume of a right circular cone with radius R and altitude H is $\pi R^2 H/3$. (*Hint:* Show how the cone can be obtained by revolving a triangular region about the x-axis.)

•10 Find the volume of the sphere obtained by revolving the circle $x^2 + y^2 = 9$ about the x-axis.

•11 The circle $x^2 + y^2 = 4$ is the base of a solid. Each cross section perpendicular to the x-axis is a square. Find the volume.

•12 A right pyramid has a rectangular base measuring 3 ft by 6 ft and a height of 9 ft. Express the volume as an integral. Calculate its value.

•13 The region bounded by the graphs of $y = x^2$ and $y = 4$ is the base of a solid. Each cross section perpendicular to the y-axis is an equilateral triangle. Find the volume.

•14 The region in the xy-plane bounded by the graphs of $x = y^2$ and $x = 4$ is the base of a solid. Each cross section parallel to the y-axis is a semicircle. Find the volume.

6–3 Volume II: The Shell Method

The disk method works well for calculating the volume of a solid of revolution when the region bounded by the graph of $y = f(x)$ and the x-axis is revolved about the x-axis. (See Fig. 6–18b.) The disk method may not work at all, however, if the same region is revolved about the y-axis. (See Fig. 6–18c.) The problem arises with the boundary curve $y = f(x)$. If we try to use disks to approximate the volume, we may have to break the disks into concentric rings when they cross the boundary curve. The resulting complications make the disk method impractical. The *shell method* is an alternative method for calculating the volumes of such solids.

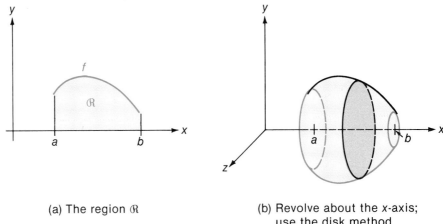
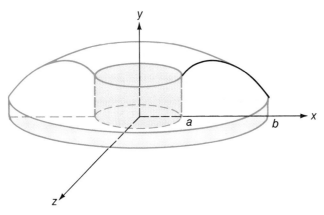

(a) The region $\mathcal{R}$

(b) Revolve about the x-axis; use the disk method

(c) Revolve about the y-axis; use the shell method

Figure 6–18 Solids of revolution.

SHELL METHOD

For simplicity, we derive the shell method for the special case in which a first quadrant region $\mathcal{R}$, bounded above by the graph of the continuous function f and below by the x-axis for $a \leq x \leq b$, is revolved about the y-axis. The procedure can easily be generalized to cover more complicated regions.

We begin by partitioning the interval $[a, b]$ into n subintervals by the partition $\mathcal{P}$ defined by

$$a = x_0 < x_1 < x_2 < \cdots < x_n = b.$$

6-3 Volume II: The Shell Method

We next construct rectangles to approximate the region $\mathcal{R}$. For convenience we use

$$\xi_k = \frac{x_{k-1} + x_k}{2},$$

the *midpoint* of the kth subinterval, to determine the height of the kth rectangle. (See Figs. 6–19a, b.) The height of the kth rectangle is $f(\xi_k)$, the thickness is Δx_k, and the area is $f(\xi_k)\,\Delta x_k$.

We now revolve these rectangles about the y-axis, obtaining a set of thin cylindrical shells that approximate the solid of revolution. (A typical shell is pictured in Fig. 6–19c.)

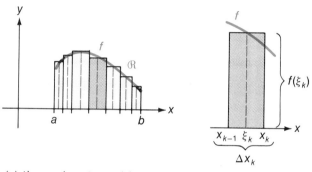

(a) the region $\mathcal{R}$ partitioned into rectangles

(b) The kth rectangle: ξ_k = midpoint of the interval

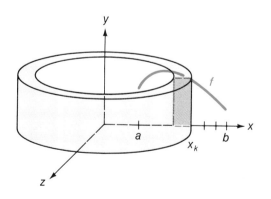

(c) Shell obtained by revolving the kth rectangle about the y-axis: volume of shell = $2\pi\,\xi_k f(\xi_k)\,\Delta x_k$.

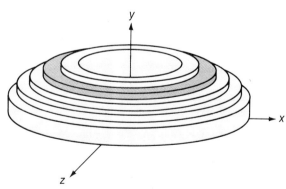

(d) Approximating shells: total volume of shells = $\sum_{k=1}^{n} 2\pi\,\xi_k f(\xi_k)\,\Delta x_k$.
(Two inner shells not shown.)

Figure 6–19 The shell method. Volume of solid of revolution = $\lim_{\|\mathcal{P}\|\to 0} \sum_{k=1}^{n} 2\pi\,\xi_k f(\xi_k)\,\Delta x_k = 2\pi\int_a^b x f(x)\,dx$.

The height of the kth shell is $f(\xi_k)$. The radius of the cylinder that forms the inside surface is x_{k-1} and the radius of the cylinder that forms the outside surface is x_k. The exact volume of this shell is

$$\begin{aligned} V_k &= \pi x_k^2 f(\xi_k) - \pi x_{k-1}^2 f(\xi_k) \\ &= \pi(x_k + x_{k-1})f(\xi_k)(x_k - x_{k-1}) \quad \text{(by factoring)} \\ &= 2\pi \left(\frac{x_k + x_{k-1}}{2}\right) f(\xi_k)(x_k - x_{k-1}) \\ &= 2\pi \xi_k f(\xi_k)\, \Delta x_k. \end{aligned}$$

The sum of the volumes of the n shells shown in Figure 6–19d is

$$\sum_{k=1}^{n} V_k = \sum_{k=1}^{n} 2\pi \xi_k f(\xi_k)\, \Delta x_k.$$

This sum is an approximation to the volume of the solid of revolution. Since f is continuous on $[a, b]$, then the approximation gets better and better as we increase the number of shells and make each of them thinner. Thus,

$$\text{Volume of solid} = \lim_{\|\mathcal{P}\| \to 0} \sum_{k=1}^{n} 2\pi \xi_k f(\xi_k)\, \Delta x_k$$

Basic Formula for Shell Method

$$\text{Volume} = 2\pi \int_a^b x f(x)\, dx.$$

EXAMPLE 1   Let $\mathcal{R}$ be the region bounded above by the graph of $f(x) = \sqrt{x}$ and bounded below by the x-axis, $0 \leq x \leq 1$. Calculate the volume of the solid obtained by revolving $\mathcal{R}$ about the y-axis.

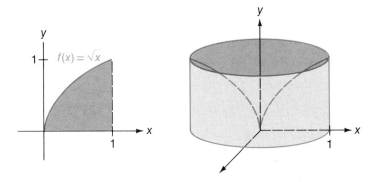

(a) The region   (b) The solid of revolution

Figure 6–20 Example 1.

6-3 Volume II: The Shell Method

Solution The volume is

$$V = 2\pi \int_0^1 xf(x)\, dx = 2\pi \int_0^1 x\sqrt{x}\, dx = 2\pi \int_0^1 x^{3/2}\, dx$$

$$= 2\pi \left[\frac{2x^{5/2}}{5}\right]_0^1 = \frac{4\pi}{5}.$$

(See Fig. 6–20.) □

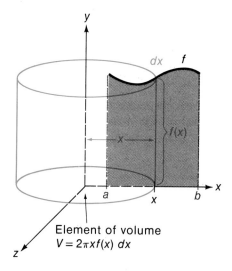

Element of volume
$V = 2\pi x f(x)\, dx$

Figure 6–21
Volume of element = $2\pi x f(x)\, dx$.
Volume of solid = $\int_a^b 2\pi x f(x)\, dx$.

Remark Elements of volume can be used to set up integrals for the shell method. The situation is illustrated in Figure 6–21. A typical element of area is drawn parallel to the y-axis. This element sweeps out a very thin cylindrical shell of radius x, height $f(x)$ and thickness dx. The volume of this shell approximately equals the inside surface area multiplied by the thickness dx. Thus

Elements of Volume to Set Up Integrals

Volume of typical shell ≈ $\underbrace{2\pi x \cdot f(x)}_{\text{surface area}} \cdot \underbrace{dx}_{\text{thickness}}$.

The corresponding integral is the volume of the solid:

$$\text{Volume} = 2\pi \int_a^b xf(x)\, dx.$$

The shell method also can be used with regions that have left and right boundaries and are revolved about the x-axis. It can be generalized to more complicated regions. The integrals usually can be set up by using elements of volume.

EXAMPLE 2 (See Example 4 of Section 6–2.) Let $\mathcal{R}$ be the region bounded above by the graph of $y = \sqrt{x}$ and below by the graph of $y = x^2$, $0 \le x \le 1$. Use the shell method to calculate the volume of the solid obtained by revolving $\mathcal{R}$ about the x-axis.

Solution The region is pictured in Figure 6–22a. To use the shell method, we set up a horizontal element of area and revolve it about the x-axis. (See Figs. 6–22a, b.) Observe that

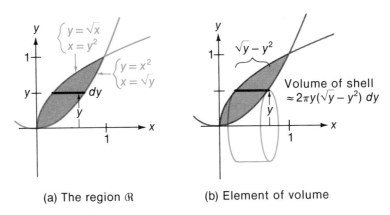

(a) The region $\mathcal{R}$

(b) Element of volume

Figure 6–22 Example 2.

the left-hand boundary is the graph of $x = y^2$ and the right-hand boundary is the graph of $x = \sqrt{y}$. A typical element of volume has the following properties:

$$\text{Radius} = y.$$
$$\text{Height} = \sqrt{y} - y^2.$$
$$\text{Thickness} = dy.$$
$$\text{Volume} \approx 2\pi \cdot \underbrace{y}_{\text{radius}} \cdot \underbrace{(\sqrt{y} - y^2)}_{\text{height}} \cdot \underbrace{dy}_{\text{thickness}}.$$

The volume of the solid of revolution is

$$V = 2\pi \int_0^1 y(\sqrt{y} - y^2)\, dy = 2\pi \int_0^1 (y^{3/2} - y^3)\, dy$$
$$= 2\pi \left[\frac{2}{5} y^{5/2} - \frac{y^4}{4} \right]_0^1 = 2\pi \left[\frac{2}{5} - \frac{1}{4} - 0 \right] = 2\pi \cdot \frac{3}{20}$$
$$= \frac{3\pi}{10}. \ \square$$

EXAMPLE 3 Let $\mathcal{R}$ be the closed region bounded above by the graph of $f(x) = x$ and below by the graph of $g(x) = x^2 - 2$. Calculate the volume of the solid obtained by revolving $\mathcal{R}$ about the line $x = 3$.

6-3 Volume II: The Shell Method

Solution A typical element of area is pictured in Figure 6-23. The shell swept out by this element has the following properties:

Radius $= 3 - x$.
Height $= f(x) - g(x) = x - (x^2 - 2) = x - x^2 + 2$.
Thickness $= dx$.

$$\text{Volume} \approx 2\pi \underbrace{(3 - x)}_{\text{radius}} \underbrace{(x - x^2 + 2)}_{\text{height}} \underbrace{dx}_{\text{thickness}}.$$

The limits of integration can be obtained by noting that if $(x, y) \in \mathcal{R}$, then $-1 \leq x \leq 2$. (See Fig. 6-23.)

The volume of the solid of revolution is

$$V = 2\pi \int_{-1}^{2} (3 - x)(x - x^2 + 2) \, dx$$

$$= 2\pi \int_{-1}^{2} (x^3 - 4x^2 + x + 6) \, dx$$

$$= 2\pi \left[\frac{x^4}{4} - \frac{4x^3}{3} + \frac{x^2}{2} + 6x \right]_{-1}^{2}$$

$$= 2\pi \left[\frac{45}{4} \right] = \frac{45\pi}{2}. \quad \square$$

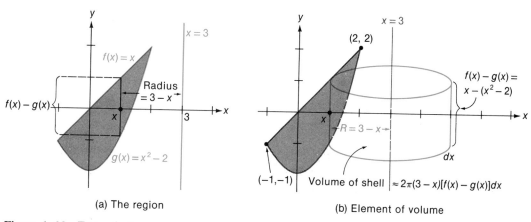

(a) The region

(b) Element of volume

Figure 6-23 Example 3.

Remark

Shell or Disk Method? A simple test can be used to decide when to use the shell method and when to use the disk method for the volume of a solid of revolution about a line $\mathcal{L}$. Draw a typical element of area across the region. If that element is parallel to the line $\mathcal{L}$, use the shell method. If it is perpendicular to $\mathcal{L}$, use the disk method.

In most cases, there is only one convenient way to draw the element of area. If, in a particular problem, it can be set up equally well either way, then we can use either the disk method or the shell method.

EXAMPLE 4 Let $\mathcal{R}$ be the closed region bounded by the graphs of $y = \sqrt[3]{x}$ and $y = x^2$. Calculate the volume of the solid obtained by revolving $\mathcal{R}$ about the line $y = -\frac{1}{2}$.

Solution The region is pictured in Figure 6–24a. Observe that we can use either the disk method or the shell method to calculate the volume. If we use the shell method, we must set up the element of area between the left and the right boundaries (the graphs of $x = y^3$ and $x = \sqrt{y}$).

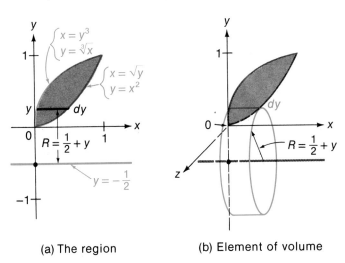

(a) The region

(b) Element of volume

Figure 6–24 Example 4.

A typical element of area and the corresponding shell are shown in Figure 6–24b. The properties of the shell are as follows:

Radius $= \frac{1}{2} + y$
Height $= \sqrt{y} - y^3$ (distance from left to right boundary).
Thickness $= dy$
Volume $\approx 2\pi(\frac{1}{2} + y)(\sqrt{y} - y^3)dy = \pi(1 + 2y)(\sqrt{y} - y^3)dy.$

The volume of the solid is

$$V = \pi \int_0^1 (1 + 2y)(\sqrt{y} - y^3)dy = \pi \int_0^1 (y^{1/2} + 2y^{3/2} - y^3 - 2y^4)\, dy$$

$$= \pi \left[\frac{2y^{3/2}}{3} + \frac{4y^{5/2}}{5} - \frac{y^4}{4} - \frac{2y^5}{5}\right]_0^1 = \frac{49\pi}{60}. \quad \square$$

Exercises 6–3

1–8. Use the shell method to calculate the volume of the solid of revolution obtained by revolving the region $\mathcal{R}$, bounded by the given curves and lines, about the indicated line.

- **1** $y = 4x - x^2$ and the x-axis
  - (a) About the y-axis
  - (b) About the line $x = 4$
- **2** $y = \dfrac{1}{x}$, $x = 0$, $y = 1$, and $y = 3$. About the x-axis.
- **3** $y = x^2 - 2x$ and $y = 0$. About the y-axis.
- **4** $y = x^2$ and $y = 4$. About the line $x = 2$.
- **5** $y = 2(x^2 + 1)$, $y = 0$, $x = 1$ and $x = 3$. About the line $x = -1$.
- **6** $y = x^2$ and $y = 8 - x^2$
  - (a) About the y-axis
  - (b) About the line $x = 4$.
- **7** $x = 2y - y^2$ and $x = 0$. About the x-axis.
- **8** $y = x$, $y = 2x$, and $x = 3$. About the y-axis.
- 9 Work Example 4 by the disk method.
- 10 Work Exercise 1 by the disk method.
- 11 Work Exercise 6 by the disk method.
- 12 Work Exercise 7 of Section 6–2 by the shell method.
- **13** A round hole of radius $r = \tfrac{1}{2}$ ft is bored through the center of a solid sphere of radius 2 ft. Find the total volume of the sawdust.
- **14** A round hole of radius 1 in. is bored through the center of a solid sphere of radius 4 in. Find the volume of the remaining part of the sphere.
- **15** Draw a figure similar to Figure 6–19d that pictures a typical approximating sum for the volume of the following solid. Write out a typical approximating sum based on a partition into n equal subintervals and take the limit as $n \to \infty$, obtaining an integral for the volume of the solid.
  - (a) The solid in Exercise 6(b)
  - (b) The solid in Exercise 14
- 16 (a) Sketch a region $\mathcal{R}$ in Quadrant I such that the volume of the solid of revolution obtained by revolving $\mathcal{R}$ about the y-axis is

$$V = \int_1^5 2\pi x \sqrt{x - 1}\, dx \qquad \text{(shell method)}.$$

  - (b) Write an equivalent integral for the volume of the solid by using the disk method. Evaluate this integral.

6–4 Arc Length

Smooth Graph

The graph of the function f is said to be *smooth* on the interval $[a, b]$ if f has a continuous first derivative f' on $[a, b]$. This use of the word *smooth* agrees with the usage in society. If a graph is smooth on $[a, b]$, then it has no sharp corners at points of the interval.

Process for Arc Length

We shall define the arc length of a smooth graph over the interval $[a, b]$ by a process that yields an integral. The process involves the following steps:

(1) Approximate the graph by many short line segments. (See Fig. 6-25a.)
(2) Approximate the length of the graph by the sum of the lengths of the line segments.
(3) Take the limit of the sum in (2) as the length of each line segment approaches zero and the number of line segments becomes infinite. This limit equals a definite integral. The *arc length* of the graph is defined to be the value of this integral.

Let the smooth curve be the graph of f. We begin the process by subdividing the interval $[a, b]$ into n subintervals by the partition $\mathscr{P}$ defined by

$$a = x_0 < x_1 < x_2 < \cdots < x_n = b.$$

Each point x_k in the partition corresponds to a point $P_k(x_k, f(x_k))$ on the curve. (See Fig. 6-25a.) By the distance formula, the length of the chord connecting $P_{k-1}(x_{k-1}, f(x_{k-1}))$ and $P_k(x_k, f(x_k))$ is

$$\Delta s_k = |P_{k-1}P_k| = \sqrt{[x_k - x_{k-1}]^2 + [f(x_k) - f(x_{k-1})]^2}.$$

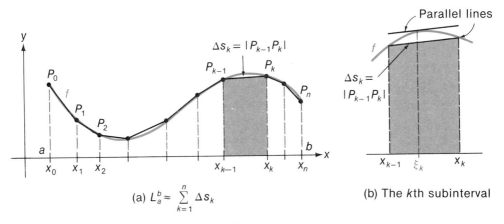

(a) $L_a^b \approx \sum_{k=1}^{n} \Delta s_k$ (b) The kth subinterval

Figure 6-25 Arc length. $L_a^b = \lim_{\|\mathscr{P}\| \to 0} \sum_{k=1}^{n} \Delta s_k$.

We now apply the Mean Value Theorem (Theorem 4-3) to the function f on the subinterval $[x_{k-1}, x_k]$. There exists a number ξ_k between x_{k-1} and x_k such that

$$\frac{f(x_k) - f(x_{k-1})}{x_k - x_{k-1}} = f'(\xi_k).$$

6-4 Arc Length

Then

$$f(x_k) - f(x_{k-1}) = f'(\xi_k)(x_k - x_{k-1}).$$

If we substitute this expression in the equation for Δs_k, we get

$$\Delta s_k = \sqrt{(x_k - x_{k-1})^2 + [f'(\xi_k)(x_k - x_{k-1})]^2}$$
$$= \sqrt{1 + [f'(\xi_k)]^2}\,(x_k - x_{k-1}) = \sqrt{1 + [f'(\xi_k)]^2}\,\Delta x_k.$$

The sum of the lengths of the chords is

$$\sum_{k=1}^{n} \Delta s_k = \sum_{k=1}^{n} \sqrt{1 + [f'(\xi_k)]^2}\,\Delta x_k.$$

Since f has a smooth graph, then f' is continuous on $[a, b]$. Thus, this sum has a limit as $\|\mathcal{P}\| \to 0$. This limit is defined to be the *arc length of f from* $(a, f(a))$ *to* $(b, f(b))$:

$$\text{Arc length} = \lim_{\|\mathcal{P}\| \to 0} \sum_{k=1}^{n} \sqrt{1 + [f'(\xi_k)]^2}\,\Delta x_k = \int_a^b \sqrt{1 + [f'(x)]^2}\,dx.$$

We denote the arc length from $(a, f(a))$ to $(b, f(b))$ by the symbol L_a^b:

Basic Arc Length Formula

$$L_a^b = \int_a^b \sqrt{1 + [f'(x)]^2}\,dx.$$

EXAMPLE 1 Let $f(x) = 2x^{3/2}$. Find the arc length from $(7, f(7))$ to $(11, f(11))$.

Solution The derivative is $f'(x) = 3\sqrt{x}$. Then

$$L_7^{11} = \int_7^{11} \sqrt{1 + [f'(x)]^2}\,dx = \int_7^{11} \sqrt{1 + 9x}\,dx.$$

To integrate this function, we make the substitution $u = 1 + 9x$, $du = 9\,dx$. Then $u = 64$ when $x = 7$ and $u = 100$ when $x = 11$. Therefore,

$$L_7^{11} = \frac{1}{9}\int_7^{11}(1 + 9x)^{1/2}(9\,dx) = \frac{1}{9}\int_{64}^{100} u^{1/2}\,du$$

$$= \frac{1}{9}\left[\frac{2u^{3/2}}{3}\right]_{64}^{100} = \frac{2}{27}[100^{3/2} - 64^{3/2}]$$

$$= \frac{2}{27}[1000 - 512] = \frac{2}{27} \cdot 488 = \frac{976}{27} \approx 36.15.\ \square$$

| $u = 1 + 9x$ |
| $du = 9\,dx$ |

| $x = 7 \Rightarrow u = 64$ |
| $x = 11 \Rightarrow u = 100$ |

If a graph is defined by a function $x = g(y)$, where g has a continuous first derivative for $c \le y \le d$, then the arc length from the points $(g(c), c)$ to $(g(d), d)$ is

Arc Length for $x = g(y)$

$$L_c^d = \int_c^d \sqrt{1 + [g'(y)]^2}\,dy.$$

EXAMPLE 2 If $x = g(y) = y^{2/3}$, then the arc length of the graph between the points at which $y = 1$ and $y = 8$ is

$$L_1^8 = \int_1^8 \sqrt{1 + [g'(y)]^2}\, dy = \int_1^8 \sqrt{1 + \left(\frac{2}{3} \cdot \frac{1}{y^{1/3}}\right)^2}\, dy$$

$$= \int_1^8 \sqrt{1 + \frac{4}{9y^{2/3}}}\, dy = \int_1^8 \frac{\sqrt{9y^{2/3} + 4}}{3y^{1/3}}\, dy$$

$$= \frac{1}{3}\int_1^8 \sqrt{9y^{2/3} + 4}\,(y^{-1/3}\,dy).$$

This integral can be evaluated by the substitution $u = 9y^{2/3} + 4$, $du = 6y^{-1/3}\,dy$. Observe that $u = 13$ when $y = 1$ and $u = 40$ when $y = 8$. Thus,

$$L_1^8 = \frac{1}{3}\int_1^8 \sqrt{9y^{2/3} + 4} \cdot (y^{-1/3}\,dy)$$

$$= \frac{1}{3} \cdot \frac{1}{6}\int_1^8 \sqrt{9y^{2/3} + 4} \cdot (6y^{-1/3}\,dy)$$

$$= \frac{1}{18}\int_{13}^{40} \sqrt{u}\,du = \frac{1}{18}\left[\frac{2u^{3/2}}{3}\right]_{13}^{40}$$

$$= \frac{2}{54}[40^{3/2} - 13^{3/2}]$$

$$= \frac{40^{3/2} - 13^{3/2}}{27} \approx 7.634.\ \square$$

| $u = 9y^{2/3} + 4$ |
| $du = 6y^{-1/3}\,dy$ |

| $y = 1 \Rightarrow u = 13$ |
| $y = 8 \Rightarrow u = 40$ |

Remark Many functions that we consider in this book have arc lengths which are almost impossible to calculate by the standard formulas. The graph of $y = x^3$, for example, has arc length

$$L_a^b = \int_a^b \sqrt{1 + 9x^4}\, dx,$$

which cannot be evaluated by the standard antiderivative formulas. Such integrals can be calculated by an approximation technique such as the Trapezoidal Rule or Simpson's Rule.

CURVES DEFINED PARAMETRICALLY

So far, we have considered graphs of equations and functions. There is an additional process for generating graphs. We can define x and y in terms of an auxiliary variable t by equations

Parametric Equations
$$x = f(t), \qquad y = g(t).$$

The variable t is called a *parameter*. The equations defining x and y are called *parametric equations*.

6–4 Arc Length

The graph of the parametric equations

$$x = f(t),\ y = g(t)$$

consists of all points $P(f(t), g(t))$ with t in the domain of f and g. For each number $t = t_0$ in the common domain of f and g, we calculate $x_0 = f(t_0)$, $y_0 = g(t_0)$, and locate the point (x_0, y_0) on the graph. (See Fig. 6–26.)

If f and g are continuous over an interval I (on the t-axis), then the corresponding portion of the graph is called a *curve*. A curve may cross itself many times or oscillate back and forth along an arc. A typical curve is shown in Figure 6–26. The arrows indicate how the curve is traced as t increases.

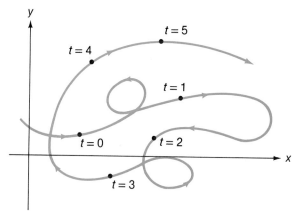

Figure 6–26 A curve defined parametrically. The arrows indicate how the curve can be traced out by a moving point. The locations of the moving point are shown for various values of t.

Smooth Curve

A curve is said to be *smooth* on the interval I if it is the graph of parametric equations $x = f(t)$, $y = g(t)$, where f and g have continuous first derivations on I which are not simultaneously zero except, possibly, at the end points of I.

This definition is an extension of the one for the graph of $y = F(x)$, because such a graph can be considered to be the graph of parametric equations. If we let

$$x = t,\ y = F(t),$$

the graph of the parametric equations is identical to the graph of

$$y = F(x).$$

(Parametric equations and curves are discussed in more detail in Section 13–1.)

We shall extend our work on arc length to include smooth curves defined parametrically. As we shall see, a technical problem prevents our making a complete derivation, but we can indicate the general lines of the argument.

Let the parametric equations $x = f(t)$, $y = g(t)$, $\alpha \leq t \leq \beta$, define a smooth curve

𝒞. We shall define the arc length of 𝒞 between the points at which $t = \alpha$ and $t = \beta$ to be

$$L_\alpha^\beta = \int_\alpha^\beta \sqrt{[f'(t)]^2 + [g'(t)]^2}\, dt.$$

To justify this definition, we subdivide the interval $[\alpha, \beta]$ (on the t axis) into subintervals by the partition $\mathscr{P}$ defined by

$$\alpha = t_0 < t_1 < t_2 < \cdots < t_n = \beta.$$

Each number t_k corresponds to a point $P_k(x_k, y_k)$, where $x_k = f(t_k)$ and $y_k = g(t_k)$ (Fig. 6–27a). The chord connecting P_{k-1} and P_k has length

$$\Delta s_k = |P_{k-1}P_k| = \sqrt{(x_k - x_{k-1})^2 + (y_k - y_{k-1})^2}$$

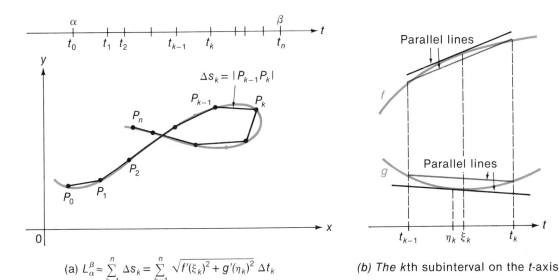

(a) $L_\alpha^\beta \approx \sum_{k=1}^{n} \Delta s_k = \sum_{k=1}^{n} \sqrt{f'(\xi_k)^2 + g'(\eta_k)^2}\, \Delta t_k$

(b) The kth subinterval on the t-axis

Figure 6–27 Arc length of curves defined parametrically.

and the sum of the lengths of these chords is

$$\sum_{k=1}^{n} \Delta s_k = \sum_{k=1}^{n} \sqrt{(x_k - x_{k-1})^2 + (y_k - y_{k-1})^2}.$$

If we apply the Mean Value Theorem to the functions f and g on the interval $[t_{k-1}, t_k]$, we find that there exist numbers ξ_k and η_k on the interval such that

$$\frac{f(t_k) - f(t_{k-1})}{t_k - t_{k-1}} = f'(\xi_k) \quad \text{and} \quad \frac{g(t_k) - g(t_{k-1})}{t_k - t_{k-1}} = g'(\eta_k).$$

6-4 Arc Length

(See Fig. 6-27b.) If we let $\Delta t_k = t_k - t_{k-1}$, then

$$f(t_k) - f(t_{k-1}) = f'(\xi_k) \Delta t_k \quad \text{and} \quad g(t_k) - g(t_{k-1}) = g'(\eta_k) \Delta t_k$$

so that

$$x_k - x_{k-1} = f'(\xi_k) \Delta t_k \quad \text{and} \quad y_k - y_{k-1} = g'(\eta_k) \Delta t_k.$$

We substitute these expressions in the sum of the Δs_k, obtaining

$$\sum_{k=1}^{n} \Delta s_k = \sum_{k=1}^{n} \sqrt{[f'(\xi_k) \Delta t_k]^2 + [g'(\eta_k) \Delta t_k]^2}$$

$$= \sum_{k=1}^{n} \sqrt{[f'(\xi_k)]^2 + [g'(\eta_k)]^2} \, \Delta t_k.$$

This last expression is slightly different from one of our approximating sums because generally ξ_k and η_k are different points in the kth subinterval. It can be shown, however, that as $\|\mathcal{P}\| \to 0$, this sum approaches the same limit as it would if ξ_k and η_k were equal for every k. Thus we define,

$$L_\alpha^\beta = \lim_{\|\mathcal{P}\| \to 0} \sum_{k=1}^{n} \Delta s_k = \lim_{\|\mathcal{P}\| \to 0} \sum_{k=1}^{n} \sqrt{[f'(\xi_k)]^2 + [g'(\eta_k)]^2} \, \Delta t_k$$

Arc Length for Parametric Equations

$$L_\alpha^\beta = \int_\alpha^\beta \sqrt{[f'(t)]^2 + [g'(t)]^2} \, dt.$$

EXAMPLE 3 Calculate the arc length of the curve defined parametrically by

$$x = f(t) = 2t^2, \, y = g(t) = (4t + 9)^{3/2}$$

for $0 \leq t \leq 4$.

Solution $f'(t) = 4t$ and $g'(t) = 6\sqrt{4t + 9}$. Then

$$L_0^4 = \int_0^4 \sqrt{[f'(t)]^2 + [g'(t)]^2} \, dt$$

$$= \int_0^4 \sqrt{16t^2 + 36(4t + 9)} \, dt = \int_0^4 \sqrt{4(4t^2 + 36t + 81)} \, dt$$

$$= 2 \int_0^4 \sqrt{(2t + 9)^2} \, dt = 2 \int_0^4 (2t + 9) \, dt$$

$$= 2 \left[t^2 + 9t \right]_0^4 = 2 \left[16 + 36 - 0 \right] = 104. \quad \square$$

THE ARC LENGTH FUNCTION

In our later work it will be convenient to consider the arc length to be a function of t. For each value of t, the value of this function is the arc length from some fixed point P_0 on the curve to the point defined by t. Let s denote the arc length function.

340   *Applications of the Integral*

To measure the arc length from the fixed point $P_0(f(t_0), g(t_0))$, we define

$s(t)$ = Arc Length from t_0 to t

$$s(t) = L_{t_0}{}^t = L_{t_0}{}^t = \int_{t_0}^{t} \sqrt{[f'(t)]^2 + [g'(t)]^2} \, dt.$$

Observe that

$$s(t) > 0 \quad \text{if } t > t_0$$

and

$$s(t) < 0 \quad \text{if } t < t_0.$$

Thus, in essence, the arc length function assigns a positive direction to the curve. The point $P(f(t), g(t))$ is in the positive direction from $P_0(f(t_0), g(t_0))$ if $s(t) > 0$ and in the negative direction if $s(t) < 0$. (See Fig. 6–28.)

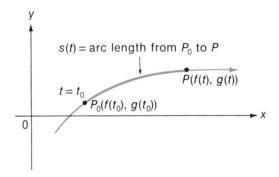

Figure 6–28   Arc length function.

Since $s(t)$ is an antiderivative of $\sqrt{[f'(t)]^2 + [g'(t)]^2}$, then

$$s'(t) = \sqrt{[f'(t)]^2 + [g'(t)]^2}$$

so that

$$[s'(t)]^2 = [f'(t)]^2 + [g'(t)]^2.$$

This formula is usually expressed in terms of differentials. If we multiply both sides of the equation by dt^2, we get the equivalent formulas

$$[s'(t) \, dt]^2 = [f'(t) \, dt]^2 + [g'(t) \, dt]^2$$

Differential Formula for $s(t)$

$$ds^2 = dx^2 + dy^2.$$

This last formula can be remembered from the Pythagorean Theorem. If we denote the horizontal side of a right triangle by dx, the vertical side by dy, and the hypotenuse by ds, then $ds^2 = dx^2 + dy^2$, as in our formula. (See Fig. 6–29.)

Exercises 6–4 341

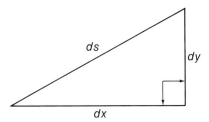

Figure 6–29 Differential arc length: $ds^2 = dx^2 + dy^2$.

The differential formula is the simplest arc length formula to remember. To use it to calculate the arc length from $P_1(f(t_1), g(t_1))$ to $P_2(f(t_2), g(t_2))$ we write

$$ds = \sqrt{dx^2 + dy^2}$$

and integrate this expression from $t = t_1$ to $t = t_2$:

$$s = \int_{t=t_1}^{t=t_2} ds = \int_{t=t_1}^{t=t_2} \sqrt{dx^2 + dy^2}.$$

To evaluate the integral, we must express dx and dy in terms of t and dt, simplify, and integrate.

EXAMPLE 4   Let $x = 8(t + 1)^{3/2}$, $y = 3t^2$. (a) Calculate the arc length from $t = 1$ to $t = 3$. (b) Calculate the arc length from $t = 3$ to $t = 1$.

Solution   (a) Since

$$dx = 12(t + 1)^{1/2}\, dt \quad \text{and} \quad dy = 6t\, dt,$$

then

$$ds = \sqrt{dx^2 + dy^2} = \sqrt{144(t + 1)\, dt^2 + 36t^2\, dt^2}.$$

The arc length is

$$L_1^3 = \int_{t=1}^{t=3} ds = \int_1^3 \sqrt{144(t + 1) + 36t^2}\, dt$$

$$= 6\int_1^3 \sqrt{t^2 + 4t + 4}\, dt = 6\int_1^3 \sqrt{(t + 2)^2}\, dt$$

$$= 6\int_1^3 (t + 2)\, dt = 6\left[\frac{t^2}{2} + 2t\right]_1^3 = 6\left[\frac{9}{2} + 6\right] - 6\left[\frac{1}{2} + 2\right] = 48.$$

(b) $$L_3^1 = \int_{t=3}^{t=1} ds = -\int_{t=1}^{t=3} ds = -48. \; \square$$

Exercises 6–4

1–6.   Set up, but do not evaluate, the integrals for the arc length.

•1 $f(x) = x^2$, $-1 \leq x \leq 2$

2 $f(x) = x - \dfrac{1}{x}$, $1 \leq x \leq 4$

- **3** $y = 2x^4 + x - 1$, $1 \leq x \leq 3$
- **4** $y = \sqrt{x}$, $1 \leq x \leq 4$
- **5** $x = t^2$, $y = 1 - \dfrac{1}{t}$, $1 \leq t \leq 9$
- **6** $x = t^2 - 2$, $y = t^4$, $-1 \leq t \leq 1$

7–20. Calculate the arc lengths of the given curves over the indicated intervals.

- **7** $y^3 = 27x^2$, $1 \leq x \leq 8$
- **8** $y^3 = 27x^2$, $-8 \leq x \leq -1$
- **9** $y = 2(x + 1)^{3/2}$, $-1 \leq x \leq 10$
- **10** $y = \dfrac{(x^2 + 2)^{3/2}}{3}$, $0 \leq x \leq 3$
- **11** $x^{2/3} + y^{2/3} = 4$, $1 \leq x \leq 8$, $y \geq 0$
- **12** $x = \dfrac{y^3}{3} + \dfrac{1}{4y}$, $2 \leq y \leq 4$
- **13** $x = \dfrac{2y^{3/2}}{3} - \dfrac{y^{1/2}}{2}$, $1 \leq y \leq 4$
- **14** $y = \dfrac{3x^{5/3}}{5} - \dfrac{3x^{1/3}}{4}$, $1 \leq x \leq 8$
- **15** $x = 2t^3$, $y = 3t^2$, $0 \leq t \leq 2$
- **16** $x = \dfrac{t^3}{3} - t$, $y = t^2$, $1 \leq t \leq 3$
- **17** $x = \dfrac{(2t + 1)^{3/2}}{3}$, $y = \dfrac{t^2}{2}$, $0 \leq t \leq 4$
- **18** $x = \dfrac{(2t + 1)^{3/2}}{3}$, $y = \dfrac{t^2}{2}$, $-4 \leq t \leq -2$
- **19** $x = \dfrac{(2t + 3)^{3/2}}{3}$, $y = \dfrac{t^2}{2} + t$, $0 \leq t \leq 3$
- **20** $x = \dfrac{(2t + 3)^{3/2}}{3}$, $y = \dfrac{t^2}{2} + t$, $-4 \leq t \leq 0$.

21–24. Draw a figure similar to Figure 6–25 or 6–27 that pictures a typical approximating sum for the arc lengths of the following curves. Write out a typical approximating sum based on the partition of the interval into n equal subintervals. Can you take the limit as $n \to \infty$, thereby obtaining an integral for the arc length without applying the Mean Value Theorem?

- **21** The curve in Exercise 7
- **22** The curve in Exercise 11
- **23** The curve in Exercise 15
- **24** The curve in Exercise 16

25–26. Use Simpson's Rule with the indicated value of n to approximate the arc length in the given exercises.

- **25** Exercise 1, $n = 6$
- **26** Exercise 4, $n = 6$

- **27** Let $s(t)$ be the arc length function for the graph of the parametric equations $x = t^2$, $y = t^3$, measured from the point at which $t = 1$.
  - (a) Express $s(t)$ as an integral and as an explicit function of t.
  - (b) Use the Fundamental Theorem of Calculus to calculate $s'(t)$.
  - (c) Use differentials to approximate the arc length when t varies from 1 to 1.2.

6–5 Work

When the application of a constant force F causes an object to move a distance d along a straight line, the *work* done, as defined by physicists, is

Work (Special Case)
$$W = \text{work} = Fd = \text{force} \cdot \text{distance}.$$

The work is measured in appropriate units such as *foot-pounds* (ft lb).

Suppose, for example, that we lift a 20-lb bucket of water a distance of 4 ft into the air. The force required to lift the bucket equals its weight, so that

$$W = Fd = 20 \cdot 4 = 80 \text{ ft lb}.$$

We can generalize the concept of work to cover the case in which an object is moved by a variable force. For example, if water leaks from a bucket through a hole in the bottom, then a different force is applied at each instant the bucket is being lifted. We shall also consider problems about water pumped to a new location, in which we must account for the various molecules of the water that are moved different distances.

Suppose that an object is moved from point a to point b, where $a < b$, along a line (axis) by having a variable force applied to it. Let $f(x)$ be the force that must be applied at point x to move the object. Suppose, further, that f is a continuous function of x.

To make a reasonable definition of work, we partition the interval $[a, b]$ into n subintervals by the partition $\mathcal{P}$, defined by

$$a = x_0 < x_1 < x_2 < \cdots < x_n = b.$$

If the kth subinterval, $[x_{k-1}, x_k]$ is small, then the force required to move the object will not vary much between x_{k-1} and x_k. Thus, the force required to move the object at each x between x_{k-1} and x_k is

$$f(x) \approx f(x_k).$$

In this case we would expect that the work done in moving it from x_{k-1} to x_k should approximately equal

$$\underbrace{f(x_k)}_{\text{approximate force}} \cdot \underbrace{(x_k - x_{k-1})}_{\text{distance}} = f(x_k)\, \Delta x_k.$$

The total work done in moving the object from a to b should approximately equal

$$\sum_{k=1}^{n} f(x_k)\, \Delta x_k.$$

It seems reasonable that the above sum should approximate the work better and better as $n \to \infty$ and each $\Delta x_k \to 0$. For this reason, we define the concept as follows:

DEFINITION Let $a < b$; let f be a continuous function. Let an object be moved from point a to point b along an axis by having a force $f(x)$ applied at each point x. The *work* done in moving the object from a to b is defined to be

Work
$$W = \lim_{\|\mathcal{P}\| \to 0} \sum_{k=1}^{n} f(x_k) \Delta x_k = \int_a^b f(x)\, dx.$$

EXAMPLE 1 Water leaks out of a bucket at a constant rate. The bucket is lifted from the floor at a constant speed and raised a distance of 4 ft into the air. At the instant at which it leaves the floor it weighs 20 lb. When it reaches the 4-ft level it weighs 18 lb. How much work was done in raising the bucket?

Solution Let x be the distance the bucket is above the floor at any given time. Since the force required to raise the bucket changes at a constant rate, then it is of form

$$f(x) = ax + b.$$

To evaluate a and b, recall that $f(x) = 20$ when $x = 0$ and $f(x) = 18$ when $x = 4$. Thus,

$$\begin{cases} 20 = a \cdot 0 + b \\ 18 = a \cdot 4 + b. \end{cases}$$

Solving these equations for a and b, we find that $a = -0.5$ and $b = 20$, so that $f(x) = -0.5x + 20$.

The work done in raising the bucket is

$$W = \int_0^4 f(x)\, dx = \int_0^4 (-0.5x + 20)\, dx = \left[\frac{(-0.5)x^2}{2} + 20x \right]_0^4$$
$$= \left[\frac{(-0.5)16}{2} + 80 \right] - \left[-0 + 0 \right] = 76 \text{ ft lb.} \quad \square$$

THE SPRING PROBLEM

Hooke's Law

According to Hooke's law, the force required to compress (or stretch) a spring is proportional to the amount by which it has already been compressed (or stretched) from its natural length. In other words, if the spring has been compressed a distance of x units from its natural length, then it requires a force

$$f(x) = cx$$

to be maintained in that position, where c is a constant that depends on the units of measurement and on the particular spring. A greater force will compress it still more. (See Fig. 6–30.) This law is the basis for many of the calculations used in designing mechanical devices.

6-5 Work

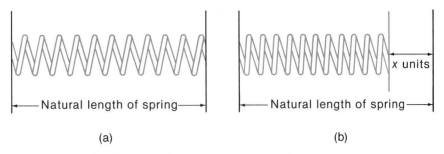

Figure 6-30 The force required to compress a spring is proportional to the amount it has been compressed from its natural length.

EXAMPLE 2   A certain spring can be compressed by 4 in. from its natural length of 12 in. when a force of 6 lb is applied. How much work is done in compressing the spring this distance?

Solution   If x is the amount by which the spring has been compressed at a given time, then the force required to maintain this compression is $f(x) = cx$, where c is a constant. To evaluate c we recall that $f(x) = 6$ when $x = 4$. Thus,

$$6 = c \cdot 4$$
$$c = 1.5.$$

The force is $f(x) = 1.5x$.

Compressing the spring can be considered as moving one end toward the other. Thus, the work done in compressing the spring by 4 in. is

$$W = \int_0^4 f(x)\, dx = \int_0^4 (1.5x)\, dx = \left[\frac{1.5x^2}{2}\right]_0^4$$
$$= (1.5)\tfrac{16}{2} - (1.5) \cdot 0 = 12 \text{ in. lb.} \ \square$$

PROBLEMS INVOLVING VARIABLE DISTANCES

When water is pumped from a container, some molecules must be moved further than others. The force required to move a molecule is constant, but the distance varies. In problems of this type, we decompose the quantity of water into very small quantities, each of which can be considered a unit, and calculate the work performed in moving each unit the necessary distance. This leads to an approximating sum for the total work. To change this approximating sum to an equivalent integral we let the number of small quantities increase without bound and calculate the total work as the value of the corresponding integral.

EXAMPLE 3   Water is stored in a cylindrical tank of radius 2 ft and height 6 ft. Calculate the work done in pumping all the water out of the tank to a point 4 ft above the top of the tank. (Water weighs 62.5 lb/ft^3.)

Applications of the Integral

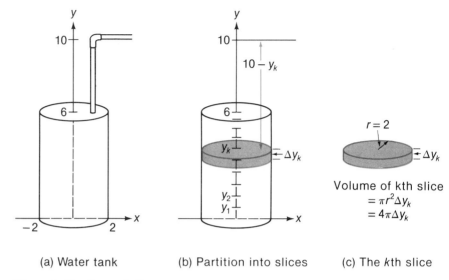

(a) Water tank   (b) Partition into slices   (c) The kth slice

Figure 6–31 Example 3.

Solution We construct an axis system with the origin at the bottom of the tank and the y-axis pointing upward. (See Fig. 6–31a.)

Since different particles of water must be moved different distances, we must decompose the water in the tank into small parts and calculate the distance each part must be moved.

We partition the interval $[0, 6]$ on the y-axis into n subintervals by the partition defined by

$$0 = y_0 < y_1 < y_2 < \cdots < y_n = 6.$$

We now imagine that the tank of water has been cut into n thin horizontal slices by horizontal planes through the points in the partition. The kth slice is shown in Figure 6–31c. This slice is a thin, flat, circular cylinder of radius 2 ft., thickness Δy_k, and volume

$$V_k = \pi \cdot 2^2 \, \Delta y_k = 4\pi \, \Delta y_k.$$

Since water weighs 62.5 lb/ft³, the force required to raise the kth thin cylinder of water is

$$(62.5)V_k = (62.5)4\pi \, \Delta y_k.$$

Observe that this thin cylinder of water must be raised a distance that is approximately

$$10 - y_k \text{ ft.}$$

Exercises 6–5

Thus, the work done in moving the kth slice of water to a point 4 ft above the top of the tank is

$$W_k \approx \underbrace{(62.5)4\pi \, \Delta y_k}_{\text{force}} \cdot \underbrace{(10 - y_k)}_{\substack{\text{approximate} \\ \text{distance}}}.$$

The total work done in moving all the water to a point 4 ft above the tank is

$$W = \sum_{k=1}^{n} W_k \approx \sum_{k=1}^{n} (62.5)4\pi \, \Delta y_k \, (10 - y_k).$$

This approximation should become better and better as $\|\mathscr{P}\| \to 0$. (As $\Delta y_k \to 0$, each slice becomes thinner and the distance $10 - y_k$ becomes more exact.) Thus,

$$W = \lim_{\|\mathscr{P}\| \to 0} \sum_{k=1}^{n} (62.5)4\pi(10 - y_k) \, \Delta y_k$$

$$= \int_0^6 (62.5)4\pi(10 - y) \, dy$$

$$= (62.5)4\pi \int_0^6 (10 - y) \, dy = (62.5)4\pi \left[10y - \frac{y^2}{2} \right]_0^6$$

$$= (62.5)4\pi[60 - 18] = 10{,}500\pi \approx 32986.7 \text{ ft lb}. \quad \square$$

Exercises 6–5

●1 A bag of sand weighs 50 lb when it is lifted from the floor by a windlass. The sand leaks out uniformly at such a rate that half the sand has been lost when the bag is 18 ft above the floor. The windlass raises the bag at the rate of 3 ft/min. Find the work done in lifting the bag this distance.

●2 A force of 6 pounds is needed to stretch a spring from its natural length of 8 in. to a length of 10 in. Find the work done in stretching the spring from its natural length to 12 in.

●3 A force of 10 lb is required to stretch a spring by 1 in. from its natural length of 6 in. Find the work done in stretching the spring from a length of 7 in. to a length of 8 in.

●4 A right circular cylindrical tank 16 ft in diameter and 20 ft deep is half-full of water. Find the work done in pumping the water over the top to empty the tank. (Water weighs 62.5 lb/ft^3.)

●5 A right circular conical tank with radius 10 ft and depth 15 ft is filled with water to within 5 ft of the top. Find the work done in pumping the water over the top to empty the tank. (Assume that the vertex of the cone is at the bottom.)

●6 A right circular conical tank is 4 ft across the top and 15 ft deep. At time $t = 0$, when pumping begins, it is filled with water. Find the work done in pumping out 4 ft of water to a point 10 ft above the top of the tank if it is pumped out at the rate of 1 ft^3/min. (The flat side is at the top of the tank.)

•7 A reservoir is in the shape of a hemisphere of radius 10 ft. with the flat side at the top of the tank. At time $t = 0$, it is filled with water. Then pumping begins at the rate of 2 ft³/min.
  (a) Find the work done in pumping all the water out over the top.
  (b) Find the work done in pumping all the water to a point 5 ft above the top of the tank.

•8 A uniformly dense rope that weighs 8 oz/ft has a leaky bucket attached to it. The rope is attached to a windlass above a vertical shaft. The bucket weighs 20 lb when 30 ft of rope is out and 10 lb when no rope is out. How much work is done in winding the rope on the windlass?

6–6 Fluid Pressure

If a flat plate is submerged horizontally in water, then the force exerted on the plate by the water equals the weight of the water directly above the plate. (See Fig. 6–32.) If the depth of the water is D ft and the area of the plate is A ft², then the force on the plate is

Force Due to Water Pressure

$$F = \underbrace{62.5}_{\substack{\text{weight of}\\\text{water}\\\text{per ft}^3}} \cdot \underbrace{D}_{\text{depth}} \cdot \underbrace{A}_{\substack{\text{area of}\\\text{plate}}} \text{ pounds}$$

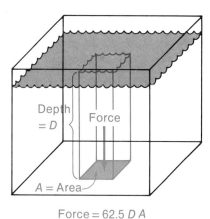

Figure 6–32 The force due to water pressure on one side of the submerged plate is equal to the weight of the water directly above the plate: $F = (62.5) \cdot D \cdot A$ pounds.

We now consider a more complicated problem—calculating the force on a plate that is submerged vertically in the water.

One of the remarkable facts about fluid pressure is that it is exerted equally in all directions. Thus, if a very small piece of the plate is submerged vertically, then the

6–6 Fluid Pressure

force on one side of the plate will be approximately the same as if that piece were submerged horizontally:

$$F \approx 62.5 \cdot D \cdot A$$

where D is the average depth and A is the area of the piece of plate. We get only an approximate value, because different parts of the plate are at different depths.

Our approach, when dealing with a large plate, will be to partition it into thin horizontal strips, calculate the force on each strip, add these forces, and finally, take the limit as the number of strips becomes infinite.

Let the plate be submerged vertically in the water. We construct an axis system that has the origin at some convenient point, with the x-axis horizontal, and the y-axis vertical. (See Fig. 6–33a.) Let the y-coordinate of the water level be L. For each y,

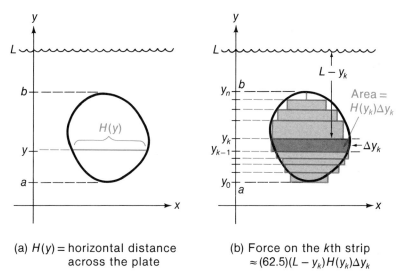

(a) $H(y) =$ horizontal distance across the plate

(b) Force on the kth strip $\approx (62.5)(L - y_k)H(y_k)\Delta y_k$

Figure 6–33 Force on one side of the plate $= \lim_{\|\mathcal{P}\| \to 0} \sum_{k=1}^{n} (62.5)(L - y_k)H(y_k) \Delta y_k$.

$a \leq y \leq b$, let $H(y)$ be the distance across the plate, measured perpendicularly to the y-axis. For the remainder of this discussion, we assume that H is a continuous function of y.

We divide the interval $[a, b]$ into n subintervals by the partition $\mathcal{P}$ defined by

$$a = y_0 < y_1 < y_2 < \cdots < y_n = b.$$

This partition corresponds to a subdivision of the plate into n horizontal strips. (See Fig. 6–33b.) The area of the kth strip is

$$A_k \approx H(y_k) \Delta y_k$$

350 Applications of the Integral

and its depth is approximately

$$L - y_k,$$

where L is the y-coordinate of the water level. (See Fig. 6–33.) The force on one side of the strip is

$$F_k \approx \underbrace{(62.5)}_{\substack{\text{wt of} \\ \text{water} \\ \text{per ft}^3}} \cdot \underbrace{(L - y_k)}_{\substack{\text{approx.} \\ \text{depth}}} \cdot \underbrace{H(y_k) \, \Delta y_k}_{\substack{\text{approx.} \\ \text{area of} \\ k\text{th strip}}},$$

The sum of the forces on the n strips is

$$\sum_{k=1}^{n} F_k \approx 62.5 \sum_{k=1}^{n} (L - y_k) H(y_k) \, \Delta y_k.$$

If we take the limit of this sum as $\|\mathcal{P}\| \to 0$, we get the exact value of the force on one side of the plate:

$$F = \lim_{\|\mathcal{P}\| \to 0} (62.5) \sum_{k=1}^{n} (L - y_k) H(y_k) \, \Delta y_k$$

Force Due to Water Pressure

$$F = 62.5 \int_{a}^{b} (L - y) H(y) \, dy.$$

EXAMPLE 1 A rectangular plate 2 ft high and 3 ft across is submerged vertically in water with the top of the plate 1 ft below the water level. Calculate the force due to the water pressure on one side of the plate.

Solution We set up our axis system with the bottom of the plate on the x-axis. (See Fig. 6–34.)

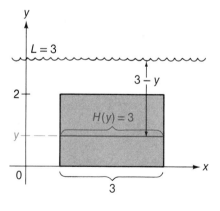

Figure 6–34 Example 1.

6-6 Fluid Pressure

The plate extends vertically from $y = 0$ to $y = 2$, and the water level is at the point $L = 3$.

For each y, $0 \leq y \leq 2$, the horizontal distance across the plate is

$$H(y) = 3.$$

It follows that the force on one side of the plate is

$$F = 62.5 \int_0^2 (L - y)H(y)\, dy = 62.5 \int_0^2 (3 - y) \cdot 3\, dy$$

$$= 3(62.5)\left[3y - \frac{y^2}{2}\right]_0^2 = 3(62.5)[6 - 2] - 0$$

$$= 750 \text{ lb.} \quad \square$$

Remark The integral for fluid pressure can be set up easily by using elements of area. A typical element has area

Elements of Area to Set up Integrals

$$H(y)\, dy$$

and is at a depth of $L - y$ below the water level. (See Fig. 6-35.) The force on this element is

$$62.5 \underbrace{(L - y)}_{\text{depth}} \underbrace{H(y)\, dy}_{\text{area}}.$$

The corresponding integral gives the force on one side of the plate.

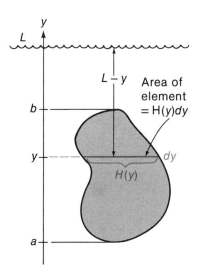

Figure 6-35 Force on element of area $= 62.5(L - y)H(y)\, dy$. Force on one side of plate $= 62.5\int_a^b (L - y)H(y)\, dy$.

EXAMPLE 2 A circular plate of radius 2 ft is submerged vertically in water with the center of the plate 6 ft below the water level. Calculate the force on one side of the plate due to the water pressure.

Solution For convenience, we set up the axis system with the origin at the center of the plate. (See Fig. 6–36.) Then the water level is at the point $L = 6$ on the y-axis, and the plate extends vertically from $y = -2$ to $y = 2$.

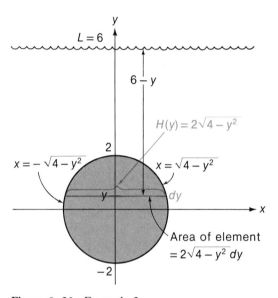

Figure 6–36 Example 2.

The equation of the plate is $x^2 + y^2 = 4$. If we solve this equation for x, we find that the equations of the right- and left-hand boundaries are

$$x = \sqrt{4 - y^2} \quad \text{and} \quad x = -\sqrt{4 - y^2},$$

respectively. For each value of y, the horizontal distance across the plate is the distance between these two boundaries:

$$H(y) = \sqrt{4 - y^2} - (-\sqrt{4 - y^2}) = 2\sqrt{4 - y^2}.$$

A typical element of area is shown in Figure 6–36. The area of this element is

$$H(y)\, dy = 2\sqrt{4 - y^2}\, dy$$

and its depth is $6 - y$. The force on this element equals

$$(62.5)(6 - y) \cdot 2\sqrt{4 - y^2}\, dy.$$

The total force on one side of the circular plate is

$$F = 62.5 \int_{-2}^{2} (6 - y) \cdot 2\sqrt{4 - y^2}\, dy.$$

6–6 Fluid Pressure

To evaluate this integral, we split it into two parts:

$$F = (62.5) \cdot 6 \cdot 2 \int_{-2}^{2} \sqrt{4 - y^2} \, dy + (62.5) \int_{-2}^{2} \sqrt{4 - y^2} \, (-2y \, dy).$$

The second integral can be calculated by the substitution $u = 4 - y^2$, $du = -2y \, du$. Its value is

$$(62.5) \left[\frac{2(4 - y^2)^{3/2}}{3} \right]_{-2}^{2} = 62.5[0 - 0] = 0.$$

The first integral cannot be evaluated by the formulas developed thus far. Observe, however, that $\int_{-2}^{2} \sqrt{4 - y^2} \, dy$ equals the area of a semicircle with radius $r = 2$. Thus,

$$\int_{-2}^{2} \sqrt{4 - y^2} \, dy = \frac{1}{2} \pi r^2 = \frac{1}{2} \pi \cdot 2^2 = 2\pi.$$

Then

$$F = (62.5) \cdot 6 \cdot 2 \int_{-2}^{2} \sqrt{4 - y^2} \, dy + (62.5) \int_{-2}^{2} \sqrt{4 - y^2} \, (-2y \, dy)$$
$$= (62.5)(12)(2\pi) + 0 = 1500\pi \approx 4712.4 \text{ lb.} \quad \square$$

EXAMPLE 3 Plans call for a rectangular flood gate in a dam to be 6 ft high and 3 ft wide. The gate is to be fastened in the dam by a pin in such a way that it will swing open when the water level is too high but remain closed when the water is at the proper level or lower. (See Figs. 6–37a, b.) Where should the pin be located so that the water pres-

(a) Floodgate is closed when the water is at the proper level

(b) Floodgate opens when the water is too high

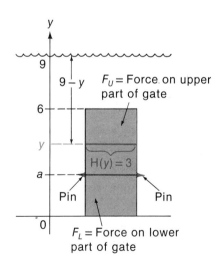

(c) Floodgate

Figure 6–37 Example 3.

sure on the part of the gate above the pin will equal the pressure on the part below the pin when the water is 3 ft above the top of the gate?

Solution We set up our axis system with the *x*-axis at the bottom of the gate. (See Fig. 6–37c.) Then the desired water level is at the point $L = 9$. Let a be the *y*-coordinate of the pin.

A typical element of area has area $3\,dy$ and depth $9 - y$. (See Fig. 6–37c.) The force on this element is

$$62.5 \underbrace{(9 - y)}_{\text{depth}} \cdot \underbrace{3\,dy}_{\text{area}}.$$

The force on the lower part of the gate is

$$F_L = 62.5 \int_0^a (9 - y)\,3\,dy = (62.5) \cdot 3 \left[9y - \frac{y^2}{2}\right]_0^a$$

$$= (62.5) \cdot 3 \left(9a - \frac{a^2}{2}\right).$$

Similarly, the force on the upper part of the gate is

$$F_U = (62.5) \int_a^6 (9 - y)\,3\,dy = (62.5) \cdot 3 \left[9y - \frac{y^2}{2}\right]_a^6$$

$$= (62.5) \cdot 3 \left(36 - 9a + \frac{a^2}{2}\right).$$

We want the value of a that makes $F_L = F_U$. If we set these forces equal and solve for a, we get

$$(62.5) \cdot 3 \left(9a - \frac{a^2}{2}\right) = (62.5) \cdot 3 \left(36 - 9a + \frac{a^2}{2}\right),$$

which simplifies to

$$a^2 - 18a + 36 = 0,$$
$$a = 9 \pm 3\sqrt{5}.$$

Observe that the solution $a = 9 + 3\sqrt{5}$ is impossible for our problem. (The pin would be above the gate.) Thus, we must have

$$a = 9 - 3\sqrt{5} \approx 2.29.$$

The pin should be located approximately 2.29 ft above the bottom of the gate.

You should verify that $F_L > F_U$ if the water level is less than 3 feet above the top of the gate and $F_L < F_U$ if it is more than 3 ft above the top of the gate. □

Exercises 6–6

• **1** A rectangular plate 2 ft high and 4 ft wide is submerged vertically in water with the top of the plate 2 ft below the water level. Find the force due to water pressure on one side of the plate.

Exercises 6–6

● **2** A dam has a submerged rectangular floodgate 4 ft wide and 2 ft high. Find the total force on the gate due to water pressure when the water level is 20 ft above the top of the gate.

●**3** A triangular plate with sides 5 ft, 5 ft, and 6 ft is submerged vertically in water with the 6-ft edge on the surface of the water. Find the force due to water pressure on one side of the plate.

● **4** The side of a dam is shaped like an inverted isosceles triangle 40 ft wide and 30 ft deep. Find what the total force due to water pressure on the dam is when the water level is at the top of the dam.

● **5** A circular gate in the vertical face of a dam is 4 ft in diameter. Find the total force on the gate due to water pressure when the water level is 3 ft above the top of the gate. (*Hint:* The integral can be written as the sum of two integrals. One can be calculated. The other is the area of a simple geometric figure.)

● **6** A horizontal cylindrical tank has a 12-ft diameter and is 30 ft long. Find the force on an end of the tank when it is half-full of water.

● **7** A 20-ft-long trough has a trapezoidal cross section. The cross section is 5 ft wide at the top, 3 ft wide at the bottom, and 3 ft deep. Find the force on one end of the trough when it is full of water.

8 A rectangular plate 4 ft long and 3 ft wide is submerged at the bottom of a 6-ft-deep pool with the 3-ft edge at the bottom of the pool and the 4-ft edge at a 30° angle to the bottom of the pool. (See Fig. 6–38 for a cross section.)

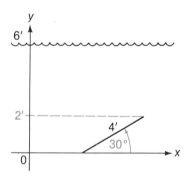

Figure 6–38 Problem 8.

(a) Modify the procedure in the text to get an approximating sum for the total force due to water pressure on one side of the plate.

(b) Show that the force equals $\int_0^2 6(62.5)(6 - y) \, dy$.

● **9** Work Exercise 8 for the same plate submerged at a 60° angle to the horizontal.

●**10** An 8 × 12 ft rectangular plate is submerged obliquely in water so that one 12-ft edge is at the surface of the water and the other is 6 ft below the surface. Find the force against one side of the plate.

•11 A rectangular water tank, 10 ft long, 10 ft wide, and 8 ft deep, was divided into two tanks by installing a vertical metal plate parallel to one side and 4 ft from that side. The smaller tank is full of oil that weighs 40 lb/ft^3 and the larger is half-full of water (62.5 lb/ft^3). Calculate the net force on the plate that separates the two tanks.

6–7 Moments I: Finite Mass Systems

Many formulas in physics are strictly true only when the total mass of an object is considered to be concentrated at a single point.* In computing the orbit of the earth about the sun, for example, we assume that the masses of the two bodies are concentrated at their centers, and we then apply the appropriate formulas. The point at which the mass of an object can be concentrated to make such a formula hold is called its *center of mass*. (The term will be defined precisely later.)

Most problems involving centers of mass are concerned with mass and distance. A simple illustration is furnished by a common seesaw. When children of masses m_1 and m_2 are placed at distances d_1 and d_2 on opposite sides of the fulcrum, then the seesaw balances, provided

$$d_1 m_1 = d_2 m_2.$$

If both children are now placed at the fulcrum, then the seesaw also balances. Thus, if $d_1 m_1 = d_2 m_2$ for the original mass system, then the mass can be concentrated at the fulcrum without changing the properties of balance. In this case the center of mass is at the fulcrum.

The illustration with the seesaw can be used to motivate several important concepts. First, we superimpose an axis system on the seesaw with the origin at the fulcrum. Next, we place weights of mass m_1 and m_2 at the points x_1 and x_2, respectively. Then the system will be in balance provided

$$x_1 m_1 + x_2 m_2 = 0.$$

The quantity

$$x_1 m_1 + x_2 m_2$$

is called the *moment of the mass system about the origin*, denoted by M_O:

Moment About Origin
$$M_O = x_1 m_1 + x_2 m_2.$$

---

* Mass is the theoretical property of matter that determines weight. It equals the ratio of the weight to the gravitational force: $m = w/g$. The mass remains unchanged when the gravitational attraction is changed. For example, an astronaut in orbit about the moon has the same mass as when he is on the earth, even though his weight is different.

6–7 Moments I: Finite Mass Systems

(a) $M_O = 0$. The seesaw balances

(b) $M_O > 0$. The seesaw tips downward on the positive side of the axis

Figure 6–39 The moment about the origin illustrated with a seesaw.

The seesaw is in equilibrium if $M_O = 0$. It tips downward on the positive side if $M_O > 0$, and downward on the negative side if $M_O < 0$. (See Fig. 6–39.) The greater the value of $|M_O|$, the greater the tendency of the seesaw to tip to one side.

These remarks hold for any number of weights. If k objects of masses m_1, $m_2, \ldots, m_k$ are placed at the points $x_1, x_2, \ldots, x_k$, respectively, then the *moment about the origin* is

Moment About Origin for k Weights

$$M_O = x_1 m_1 + x_2 m_2 + \cdots + x_k m_k.$$

The system is in equilibrium provided $M_O = 0$, tips downward to the positive side if $M_O > 0$, and so on. Observe that there is nothing special about the origin. We could just as well calculate the moment about any point $x = x_0$.

Suppose that the weight system described above does not balance at the origin. Suppose that instead it balances if the fulcrum is located at the point $\bar{x}$. In that case we could calculate the moment about $\bar{x}$:

$$M_{\bar{x}} = (x_1 - \bar{x})m_1 + (x_2 - \bar{x})m_2 + \cdots + (x_k - \bar{x})m_k.$$

Since the system balances at $\bar{x}$, then

$$M_{\bar{x}} = 0$$

so that

$$M_{\bar{x}} = x_1 m_1 + x_2 m_2 + \cdots + x_k m_k - \bar{x}(m_1 + m_2 + \cdots + m_k) = 0.$$

This last equation can be solved for $\bar{x}$, showing us where to locate the fulcrum in order to make the mass system balance:

Center of Mass

$$\bar{x} = \frac{x_1 m_1 + x_2 m_2 + \cdots + x_k m_k}{m_1 + m_2 + \cdots + m_k} = \frac{M_O}{M}$$

where M is the total mass.

The number $\bar{x} = M_O/M$ is called the *center of mass* of the system.

EXAMPLE 1 Weights of masses 3, 5, and 7 kg are located at the points $x_1 = -1$, $x_2 = 3$, $x_3 = 1$, respectively. Where should the fulcrum be located so that the weight system will balance?

Solution The moment about the origin is

$$M_O = x_1 m_1 + x_2 m_2 + x_3 m_3 = (-1) \cdot 3 + 3 \cdot 5 + 1 \cdot 7 = 19.$$

The total mass is

$$M = 3 + 5 + 7 = 15.$$

The center of mass is

$$\bar{x} = \frac{M_O}{M} = \frac{19}{15} \approx 1.267.$$

If the fulcrum is located at the point $\bar{x} = \frac{19}{15}$, the system will be in equilibrium. □

We now show that the center of mass $\bar{x}$ has the property described at the beginning of this section—if the total mass is concentrated at $\bar{x}$, the moment about the origin is unchanged.

Suppose that the masses $m_1, m_2, \ldots, m_k$ are located at the points $x_1, x_2, \ldots, x_k$. The moment about the origin is

$$M_O = x_1 m_1 + x_2 m_2 + \cdots + x_k m_k,$$

the total mass is

$$M = m_1 + m_2 + \cdots + m_k$$

and the center of mass is $\bar{x} = M_O/M$. If the total mass is concentrated at $\bar{x}$, then the new moment about the origin is

$$M'_O = (\text{distance}) \cdot (\text{mass}) = \bar{x} \cdot M$$

$$= \frac{M_O}{M} \cdot M = M_O.$$

Thus, the moment about the origin is unchanged if the mass is concentrated at the point $x = \bar{x}$.

FINITE MASS SYSTEMS IN THE PLANE

It is a simple matter to generalize the above concepts to mass systems in the xy-plane. Imagine the xy-plane to be a flat, horizontal rigid sheet in three-dimensional space. Weights of mass $m_1, m_2, \ldots, m_n$ are located at points $P_1(x_1, y_1)$, $P_2(x_2, y_2), \ldots, P_n(x_n, y_n)$, respectively. Imagine further that a sharp edge is placed under the y-axis as in Figure 6–40a. Under what conditions will the xy-plane tip to one side? Under what conditions will the plane balance on the sharp edge?

Obviously, the answers to these questions depend on the distances of the points

6-7 Moments I: Finite Mass Systems

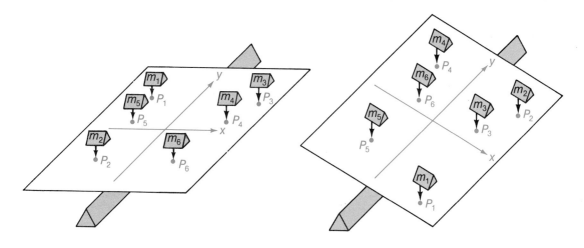

(a) $M_y = 0$. The plane balances

(b) $M_y > 0$. The plane tips downward to the side with positive x-coordinates

Figure 6-40 The moment about the y-axis illustrated with a rigid plane.

$P_1, P_2, \ldots P_n$ from the y-axis as well as on the size of the weights. We define the *moment about the y-axis* of the system to be

Moment About y-Axis
$$M_y = x_1 m_1 + x_2 m_2 + \cdots + x_n m_n$$

(the sum of the products of the directed distances from the y-axis and the masses of the weights). Just like the one-dimensional seesaw, the system will balance on the y-axis provided $M_y = 0$. It will tip to the side with positive x-coordinates if $M_y > 0$ (see Fig. 6-40b), and to the other side if $M_y < 0$.

In calculating the moment M_y, we multiply each mass by its directed distance from the y-axis. Thus, the sum involves several terms of form $x_k m_k$. (See Fig. 6-41a.) Observe that the y-coordinates of the points do not enter into the calculations.

The *moment about the x-axis* for the above system of weights is defined similarly:

Moment About x-Axis
$$M_x = \text{moment about x-axis} = y_1 m_1 + y_2 m_2 + \cdots + y_n m_n.$$

(See Fig. 6-41b.) In forming the sum, we multiply each mass by its directed distance from the x-axis.

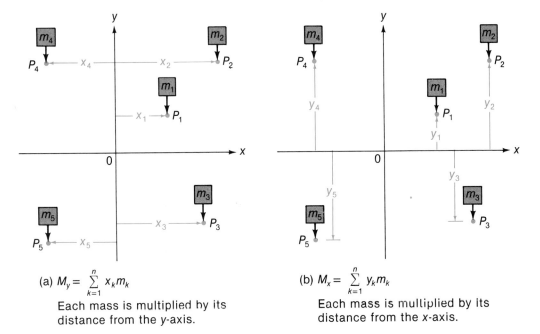

(a) $M_y = \sum_{k=1}^{n} x_k m_k$

Each mass is multiplied by its distance from the y-axis.

(b) $M_x = \sum_{k=1}^{n} y_k m_k$

Each mass is multiplied by its distance from the x-axis.

Figure 6–41 Moments about the y- and x-axes.

DEFINITION

Center of Mass

The point $(\bar{x}, \bar{y})$ at which the total mass can be concentrated without changing the moments M_x and M_y is called the *center of mass* of the weight system.

In calculating the coordinates of the center of mass, observe that if the entire mass is concentrated at the point $(\bar{x}, \bar{y})$, then the new moments about the y- and x-axes, respectively, are

$$\bar{x}M \quad \text{and} \quad \bar{y}M,$$

where $M = m_1 + m_2 + \cdots + m_n$ is the total mass. Since the new moments must equal the original moments, we must have

$$M_y = \bar{x}M \quad \text{and} \quad M_x = \bar{y}M$$

so that

$$\bar{x} = \frac{M_y}{M} \quad \text{and} \quad \bar{y} = \frac{M_x}{M}.$$

EXAMPLE 2 Weights of mass 2, 3, 5, and 3 grams (g) are placed at the points $P_1(3, 2)$, $P_2(4, -1)$, $P_3(-7, 1)$, and $P_4(-3, -2)$, respectively. (a) Calculate the moments M_x and M_y. (b) Find $(\bar{x}, \bar{y})$, the center of mass.

6-7 Moments I: Finite Mass Systems

Solution (a) $M_x = \sum_{k=1}^{4} y_k m_k = 2 \cdot 2 + (-1) \cdot 3 + 1 \cdot 5 + (-2) \cdot 3 = 0.$

$M_y = \sum_{k=1}^{4} x_k m_k = 3 \cdot 2 + 4 \cdot 3 + (-7) \cdot 5 + (-3) \cdot 3 = -26.$

(b) Since the total mass is

$$M = 2 + 3 + 5 + 3 = 13,$$

then

$$\bar{x} = \frac{M_y}{M} = \frac{-26}{13} = -2.$$
$$\bar{y} = \frac{M_x}{M} = \frac{0}{13} = 0.$$

The point $(-2, 0)$ is the center of mass. (See Fig. 6-42.) □

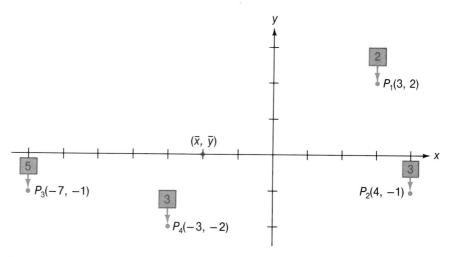

Figure 6-42 Example 2.

MOMENTS ABOUT ARBITRARY LINES

Directed Distance From a Line to a Point

It may seem that we have given undue emphasis to the x- and y-axes in our work. Obviously, we could have defined the moment of the weight system about any line $\mathscr{L}$. We merely need to assign a positive direction to points on one side of $\mathscr{L}$ and a negative direction to points on the other. If the line $\mathscr{L}$ is parallel to the y-axis, it is standard to let the points to the right of $\mathscr{L}$ be at a positive distance from $\mathscr{L}$ and the points to the left be at a negative distance. (See Fig. 6-43a.) If the line $\mathscr{L}$ does not parallel the y-axis, then the points above $\mathscr{L}$ are at positive distance from $\mathscr{L}$ and the points below are at a negative distance. (See Fig. 6-43b.)

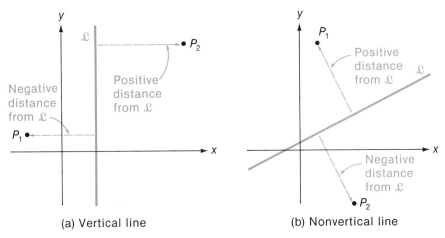

(a) Vertical line (b) Nonvertical line

Figure 6-43 Directed distances from a line $\mathscr{L}$.

If weights of mass $m_1, m_2, \ldots m_n$ are located at the points $P_1, P_2, \ldots, P_n$, respectively, and if these points are at directed distances $d_1, d_2, \ldots, d_n$, respectively, from the line $\mathscr{L}$, then the moment of the system about the line $\mathscr{L}$ is defined to be

Moment About Line $\mathscr{L}$
$$M_\mathscr{L} = d_1 m_1 + d_2 m_2 + \cdots + d_n m_n.$$

It can be shown that if the total mass of the system is concentrated at the center of mass $(\bar{x}, \bar{y})$, then the moment about *any* line $\mathscr{L}$ remains unchanged. Thus, we could have worked with any pair of nonparallel lines to find the center of mass. We chose to work with M_x and M_y because the calculations are relatively simple.

A second result is also of interest. If we take any line $\mathscr{L}$ that passes through the center of mass, then the moment of the system about the line $\mathscr{L}$ equals 0. The physical interpretation of this result is that the weight system will balance about any line through its center of mass.

These two results are discussed in more detail in Exercises 6 and 8.

Exercises 6-7

•1 Weights of mass 5, 2, 2, 1, 4 g are located at the points $x_1 = 1$, $x_2 = -2$, $x_3 = 5$, $x_4 = -4$, and $x_5 = 3$, respectively. Find the center of mass of the system.

2 Weights of mass 7, 3, 2, and 4 g are located at the points $x_1 = 6$, $x_2 = -2$, $x_3 = -4$ and $x_4 = -1$, respectively. Find the center of mass of the system.

•3 Weights of mass 8, 6, 3, and 5 g are located at the points $(-4, 4)$, $(0, 5)$, $(6, 4)$, and $(-3, -5)$, respectively. Find M_x, M_y, and the center of mass.

4 Weights of mass 6, 8, 6, 8, and 3 g are located at the points (0, 0), (5, 0), (0, −4), (−2, 0), and (0, 6), respectively. Find M_x, M_y, and the center of mass.

•5 Weights of mass 5, 3, 1, 4, 7, and 2 g are located at the points (1, 2), (−1, 0), (−4, 6), (5, 2), (1, 1), and (−1, 3), respectively. Find M_x, M_y, and the center of mass.

6 Let $\mathscr{L}$ be the line $x = a$. Let weights of mass $m_1, m_2, \ldots, m_n$ be located at the points $P_1(x_1, y_1), P_2(x_2, y_2), \ldots, P_n(x_n, y_n)$, respectively.
 (a) Show that $M_\mathscr{L}$, the moment of the system about the line $\mathscr{L}$, is unchanged if the total mass is concentrated at the center of mass $(\bar{x}, \bar{y})$.
 (b) Use (a) to prove that $(\bar{x}, \bar{y})$ is on $\mathscr{L}$ if and only if $M_\mathscr{L} = 0$.

Directed Distance

7 (*Directed Distances*) Directed distances are discussed briefly at the end of this section. (See Fig. 6–43.)
Let $\mathscr{L}$ be the line $ax + by + c = 0$, where $b > 0$.
 (a) Show that $P_0(x_0, y_0)$ is above the line $\mathscr{L}$ if and only if $ax_0 + by_0 + c > 0$, and below the line $\mathscr{L}$ if and only if $ax_0 + by_0 + c < 0$.
 (b) Use (a) and Exercise 51 of Section 1–5 to prove that if $b > 0$, then the directed distance from $\mathscr{L}$ to $P_0(x_0, y_0)$ is

$$d = \frac{ax_0 + by_0 + c}{\sqrt{a^2 + b^2}}.$$

8 Let $\mathscr{L}$ be the line $ax + by + c = 0$, where $b > 0$. Let weights of mass $m_1, m_2, \ldots, m_n$ be located at the points $P_1(x_1, y_1), P_2(x_2, y_2), \ldots, P_n(x_n, y_n)$, respectively.
• (a) Use Exercise 7 to find $M_\mathscr{L}$, the moment of the system about the line $\mathscr{L}$.
 (b) Let the center of mass of the weight system be $(\bar{x}, \bar{y})$, where $\bar{x} = M_y/\text{mass}$ and $\bar{y} = M_x/\text{mass}$. Show that the moment about $\mathscr{L}$ is unchanged if the total mass is concentrated at $(\bar{x}, \bar{y})$.
 (c) Use (b) to prove that $\mathscr{L}$ passes through $(\bar{x}, \bar{y})$ if and only if $M_\mathscr{L} = 0$.

6–8 Moments II: Centroids of Regions

The concepts developed in Section 6–7 can be extended to homogeneous slabs of matter.

Lamina

A *lamina* is a thin sheet of rigid material. In this chapter, we assume that each lamina has uniform density (mass per unit of surface area). Such a lamina is said to be homogeneous.

Suppose that a homogeneous lamina has been cut into a particular shape and placed over the *xy*-plane. We shall show how to find the balance point (center of mass).

Before going ahead, we note that the center of mass of a homogeneous rectangular lamina is at the *center* of the rectangle—the point of intersection of the diagonals. This lamina will balance on any line through its center. (See Fig. 6–44.)

364 *Applications of the Integral*

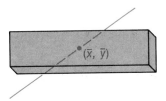

Figure 6-44 Rectangular lamina. The center of mass is at the center of the rectangle.

Let a homogeneous lamina of density ρ be bounded above by the graph of $y = f(x)$ and below by the graph of $y = g(x)$, $a \leq x \leq b$, where f and g are continuous functions. (See Fig. 6-45a.) We divide interval $[a, b]$ into n subintervals by the partition $\mathscr{P}$ defined by

$$a = x_0 < x_1 < x_2 < \cdots < x_n = b.$$

Let ξ_k be the midpoint of the kth subinterval. That is,

$$\xi_k = \frac{x_{k-1} + x_k}{2}.$$

We construct a rectangle over each subinterval in the partition, using the midpoints $\xi_1, \xi_2, \ldots, \xi_n$ to determine the heights. (See Figs. 6-45b, c.) The kth rectangle has the following properties:

$\quad$ *Height:* $f(\xi_k) - g(\xi_k)$
$\quad$ *Base:* Δx_k
$\quad$ *Area:* $[f(\xi_k) - g(\xi_k)] \Delta x_k$
$\quad$ *Mass:* $\rho[f(\xi_k) - g(\xi_k)] \Delta x_k$
$\quad$ *Center of mass:* $\left(\xi_k, \dfrac{f(\xi_k) + g(\xi_k)}{2}\right)$

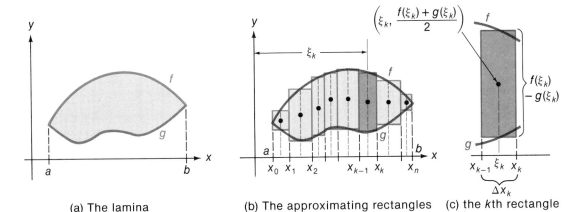

(a) The lamina $\quad$ (b) The approximating rectangles $\quad$ (c) the kth rectangle

Figure 6-45 Approximation of a homogeneous lamina by rectangles.

6–8 Moments II: Centroids of Regions

$$\left.\begin{array}{r}\text{Directed distance from } y\text{-axis}\\ \text{to center of mass}\end{array}\right\} \xi_k$$

$$\left.\begin{array}{r}\text{Directed distance from } x\text{-axis}\\ \text{to center of mass}\end{array}\right\} \frac{f(\xi_k) + g(\xi_k)}{2}.$$

We now assume that the moment of each rectangle can be calculated by concentrating its total mass at its center of mass. Then the moment about the y-axis of the kth rectangle is

$$\underbrace{\xi_k}_{\substack{\text{directed}\\ \text{distance}\\ \text{from } y\text{-axis}}} \cdot \underbrace{\rho[f(\xi_k) - g(\xi_k)] \Delta x_k}_{\text{mass}}.$$

The moment about the y-axis of the set of n rectangles is

$$\sum_{k=1}^{n} \xi_k \rho[f(\xi_k) - g(\xi_k)] \Delta x_k = \rho \sum_{k=1}^{n} \xi_k[f(\xi_k) - g(\xi_k)] \Delta x_k.$$

We define the *moment* of the original lamina about the y-axis to be the limit of the above sum as $\|\mathscr{P}\| \to 0$:

$$M_y = \lim_{\|\mathscr{P}\| \to 0} \rho \sum_{k=1}^{n} \xi_k[f(\xi_k) - g(\xi_k)] \Delta x_k$$

$$= \rho \int_a^b x[f(x) - g(x)] \, dx.$$

A similar argument can be used to motivate the definition of M_x, the moment of the lamina about the x-axis. Observe that if the mass of the kth rectangle is concentrated at its center of mass, then the moment of this rectangle about the x-axis is

$$\underbrace{\frac{f(\xi_k) + g(\xi_k)}{2}}_{\substack{\text{directed distance}\\ \text{from } x\text{-axis}}} \cdot \underbrace{\rho\,[f(\xi_k) - g(\xi_k)] \Delta x_k}_{\text{mass}}$$

$$= \frac{\rho}{2}[f(\xi_k)^2 - g(\xi_k)^2] \Delta x_k.$$

The moment about the x-axis of the set of n rectangles is

$$\sum_{k=1}^{n} \frac{\rho}{2}[f(\xi_k)^2 - g(\xi_k)^2] \Delta x_k.$$

We define the moment of the original lamina about the x-axis to be the limit of the above sum as $\|\mathscr{P}\| \to 0$:

$$M_x = \lim_{\|\mathscr{P}\| \to 0} \sum_{k=1}^{n} \frac{\rho}{2} [f(\xi_k)^2 - g(\xi_k)^2] \Delta x_k$$

$$= \frac{\rho}{2} \int_a^b [f(x)^2 - g(x)^2] \, dx.$$

Thus,

Moments About the Coordinate Axes

$$M_x = \frac{\rho}{2} \int_a^b [f(x)^2 - g(x)^2] \, dx$$

$$M_y = \rho \int_a^b x[f(x) - g(x)] \, dx.$$

The *center of mass* is the point $(\bar{x}, \bar{y})$ at which the mass can be concentrated without changing the moments M_x and M_y. Since the total mass of the lamina is

$$M = \rho \cdot area = \rho \int_a^b [f(x) - g(x)] \, dx,$$

then it follows as in the discrete case that

Center of Mass

$$\bar{x} = \frac{M_y}{M} \quad \text{and} \quad \bar{y} = \frac{M_x}{M}.$$

EXAMPLE 1 A lamina of density ρ has upper boundary $f(x) = \sqrt{x}$ and lower boundary $g(x) = x$, $0 \le x \le 1$,
(a) Calculate M_x and M_y.
(b) Find the center of mass.

Solution (a) The region is pictured in Figure 6–46. It follows from the formulas that

$$M_x = \frac{\rho}{2} \int_0^1 [f(x)^2 - g(x)^2] \, dx = \frac{\rho}{2} \int_0^1 (x - x^2) \, dx$$

$$= \frac{\rho}{2} \left[\frac{x^2}{2} - \frac{x^3}{3} \right]_0^1 = \frac{\rho}{12}.$$

$$M_y = \rho \int_0^1 x[f(x) - g(x)] \, dx = \rho \int_0^1 (x^{3/2} - x^2) \, dx$$

$$= \rho \left[\frac{2x^{5/2}}{5} - \frac{x^3}{3} \right]_0^1 = \frac{\rho}{15}.$$

6-8 Moments II: Centroids of Regions

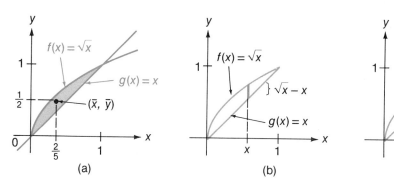

Figure 6-46 Examples 1 and 2.

(b) The mass of the lamina is

$$M = \rho \int_0^1 [f(x) - g(x)]\, dx = \rho \int_0^1 [x^{1/2} - x]\, dx$$

$$= \rho \left[\frac{2x^{3/2}}{3} - \frac{x^2}{2}\right]_0^1 = \frac{\rho}{6}.$$

Then

$$\bar{x} = \frac{M_y}{M} = \frac{\rho/15}{\rho/6} = \frac{2}{5}.$$

$$\bar{y} = \frac{M_x}{M} = \frac{\rho/12}{\rho/6} = \frac{1}{2}.$$

The center of mass is the point $(\frac{2}{5}, \frac{1}{2})$. □

Remark It is particularly easy to remember the formula

$$M_y = \rho \int_a^b x[f(x) - g(x)]\, dx$$

if we use elements of area. A typical element (set up vertically as in Fig. 6-47) has area equal to $[f(x) - g(x)]\, dx$ and mass equal to $\rho[f(x) - g(x)]\, dx$. Its directed distance from the y-axis is equal to x. The moment of the element about the y-axis is

$$\underbrace{x}_{\text{distance}} \cdot \underbrace{\rho\,[f(x) - g(x)]\, dx}_{\text{mass}}.$$

The corresponding integral gives the moment of the lamina about the y-axis.

The integral for M_x can be set up similarly by concentrating the mass of the element of area at its midpoint, the point $(x, [f(x) + g(x)]/2)$. Since the mass of the element is $\rho[f(x) - g(x)]\, dx$, the moment about the x-axis of the element is

$$\underbrace{\frac{f(x) + g(x)}{2}}_{\text{distance}} \cdot \underbrace{\rho[f(x) - g(x)]\, dx}_{\text{mass}} = \frac{\rho}{2}[f(x)^2 - g(x)^2]\, dx.$$

(See Fig. 6-47a.)

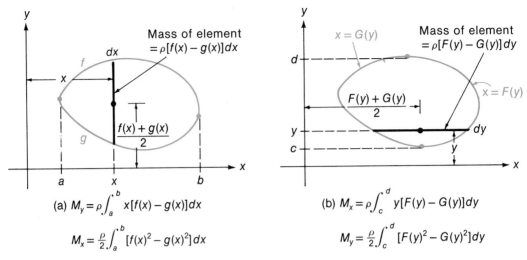

(a) $M_y = \rho \int_a^b x[f(x) - g(x)]dx$

$M_x = \frac{\rho}{2}\int_a^b [f(x)^2 - g(x)^2]dx$

(b) $M_x = \rho \int_c^d y[F(y) - G(y)]dy$

$M_y = \frac{\rho}{2}\int_c^d [F(y)^2 - G(y)^2]dy$

Figure 6-47 Elements of area used for moment problems.

If the lamina has continuous right- and left-hand boundaries, say $x = F(y)$ and $x = G(y)$, then we can set up the rectangles horizontally rather than vertically. In this case a typical element of area has mass $\rho[F(y) - G(y)]\,dy$ and directed distance from the x-axis equal to y. The moment of the element about the x-axis is

$$y \cdot \rho[F(y) - G(y)]\,dy.$$

Laminas with Right and Left Boundaries

The corresponding integral gives the moment of the lamina about the x-axis:

$$M_x = \rho \int_c^d y[F(y) - G(y)]\,dy,$$

where $[c, d]$ is the projection of the lamina on the y-axis. (See Fig. 6-47b.) Similarly, the moment of the lamina about the y-axis is

$$M_y = \frac{\rho}{2} \int_c^d [F(y)^2 - G(y)^2]\,dy.$$

EXAMPLE 2 Use the right- and left-hand boundaries to calculate M_x for the lamina of Example 1.

Solution The left-hand boundary is $G(y) = y^2$ and the right-hand boundary is $F(y) = y$. (See Fig. 6-46c.) A typical element has

$$\text{Mass} = \rho[F(y) - G(y)]\,dy = \rho[y - y^2]\,dy$$

and is y units from the x-axis. The moment of the element about the x-axis is

$$\rho y[y - y^2]\,dy.$$

6-8 Moments II: Centroids of Regions

The moment of the lamina is

$$M_x = \rho \int_0^1 y[y - y^2]\, dy = \rho \int_0^1 (y^2 - y^3)\, dy$$

$$= \rho \left[\frac{y^3}{3} - \frac{y^4}{4}\right]_0^1 = \frac{\rho}{12}. \quad \square$$

MOMENTS AND CENTROIDS OF PLANE REGIONS

Moments and centroids of a plane region are defined as if the region were covered by a lamina of density $\rho = 1$.

If the upper and lower boundaries of a region are the graphs of the continuous functions $y = f(x)$ and $y = g(x)$, for $a \le x \le b$, then the moments of the region about the coordinate axes are

Moments of a Plane Region

$$M_x = \tfrac{1}{2} \int_a^b [f(x)^2 - g(x)^2]\, dx,$$

$$M_y = \int_a^b x[f(x) - g(x)]\, dx.$$

If the right- and left-hand boundaries are the graphs of the continuous functions $x = F(y)$ and $x = G(y)$, for $c \le y \le d$, then

$$M_x = \int_c^d y[F(y) - G(y)]\, dy,$$

$$M_y = \tfrac{1}{2} \int_c^d [F(y)^2 - G(y)^2]\, dy.$$

The *centroid* of the region is defined as the point $(\bar{x}, \bar{y})$ that corresponds to the center of a mass of a lamina:

Centroid of a Plane Region

$$\bar{x} = \frac{M_y}{\text{area}}, \qquad \bar{y} = \frac{M_x}{\text{area}}.$$

EXAMPLE 3 Calculate the centroid of the region bounded by the graphs $y = x^2$, $x = 1$, and the x-axis.

Solution The region is pictured in Figure 6–48. The area and moments are

$$\text{Area} = \int_0^1 x^2\, dx = \left[\frac{x^3}{3}\right]_0^1 = \frac{1}{3}.$$

$$M_x = \int_0^1 y[1 - \sqrt{y}]\, dy = \int_0^1 [y - y^{3/2}]\, dy$$

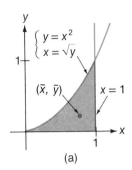
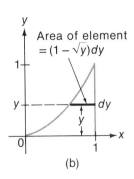
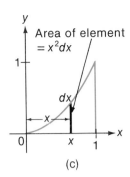

Figure 6–48 Example 3.

$$= \left[\frac{y^2}{2} - \frac{2y^{5/2}}{5}\right]_0^1 = \frac{1}{10}. \quad \text{(See Fig. 6–48b.)}$$

$$M_y = \int_0^1 x[x^2 - 0]\, dx = \int_0^1 x^3\, dx = \left[\frac{x^4}{4}\right]_0^1 = \frac{1}{4}.$$

(See Fig. 6–48c.) Then

$$\bar{x} = \frac{M_y}{\text{area}} = \frac{\frac{1}{4}}{\frac{1}{3}} = \frac{3}{4}.$$

$$\bar{y} = \frac{M_x}{\text{area}} = \frac{\frac{1}{10}}{\frac{1}{3}} = \frac{3}{10}.$$

The centroid is the point $(\frac{3}{4}, \frac{3}{10})$. □

Exercises 6–8

1–6. A lamina of density ρ is bounded by the graphs of the given curves and lines. Draw the lamina, showing a typical element of area. Calculate M_x and M_y. Find the center of mass.

- **1** $y = \sqrt{x}$, $x = 4$, and $y = 0$
- **2** $y = x^3$, $x = 2$, and $y = 0$
- **3** $y = 4x - x^2$ and the x-axis
- **4** $y = 1 - x^2$ and $y = x - 1$
- **5** $x = \sqrt{y}$, $x = -y$, and $y = 4$
- **6** $x = y^2$ and $x = 8 - y^2$

7–12. Sketch the graph of the region bounded by the given curves and lines, showing a typical element of area. Calculate M_x, M_y, and the coordinates of the centroid.

- **7** $y = x^3$ and $y = x^2$
- **8** $y = 4x - x^2$ and $y = x$
- **9** $x = 6 - y^2$ and $x = 3 - 2y$
- **10** $x = 4 - y^2$ and $x = y + 2$
- **11** $y = \sqrt{25 - x^2}$, $x + 6y + 5 = 0$, $x = 4y + 5$

6–9 Moments III: Applications. Theorems of Pappus

•12   Below by $y = x^2$, above by $y = 4x + 5$, $x + 2y = 10$

13   Set up approximating sums for M_x and M_y for the following regions. Convert the sums to integrals by taking the limit as $n \to \infty$.
(a)   The region in Exercise 7   (b)   The region in Exercise 9

14   (a)   Set up an approximating sum for the moment of the lamina in Exercise 2 about the line $3x + 4y + 5 = 0$.
•   (b)   Convert the sum in (a) to an integral by taking the limit as $n \to \infty$. Calculate the numerical value of the moment. (*Hint:* Set up the rectangles vertically after partitioning [0, 2]. Assume that the mass of each rectangle can be concentrated at its midpoint. Use the result of Exercise 7, Section 6–7.)

15   Let $\mathcal{R}_1$ and $\mathcal{R}_2$ be closed regions in the *xy*-plane that do not intersect except possibly at their boundaries. Let the areas and centroids of these regions be A_1, A_2, and $(\bar{x}_1, \bar{y}_1)$, $(\bar{x}_2, \bar{y}_2)$, respectively. Let the centroid of $\mathcal{R} = \mathcal{R}_1 \cup \mathcal{R}_2$ be $(\bar{x}, \bar{y})$. It can be proved that

Additive Properties of Centroids

$$\bar{x} = \frac{\bar{x}_1 A_1 + \bar{x}_2 A_2}{A_1 + A_2} \quad \text{and} \quad \bar{y} = \frac{\bar{y}_1 A_1 + \bar{y}_2 A_2}{A_1 + A_2}.$$

Prove this formula for the following special case: $\mathcal{R}$ is the region bounded by the graphs of the continuous functions f and g and the lines $x = a$, $x = c$, where $f(x) \geq g(x)$ for $a \leq x \leq c$. The number b is between a and c, $\mathcal{R}_1$ is the portion of $\mathcal{R}$ for $a \leq x \leq b$, $\mathcal{R}_2$ is the portion for $b \leq x \leq c$.

•16   Use the formula in Exercise 15 to calculate the centroid of $\mathcal{R}_1 \cup \mathcal{R}_2$, where $\mathcal{R}_1$ consists of the boundary and interior of the circle $x^2 + y^2 = 1$, and $\mathcal{R}_2$ consists of the boundary and interior of the square with vertices (1, 0), (1, 1), (2, 0), and (2, 1).

•17   Let $\mathcal{R}$ be the region bounded by the semicircle $y = \sqrt{4 - x^2}$ and the *x* axis. Let $\mathcal{R}_1$ be the interior of the square with vertices (0, 0), (1, 0), (0, 1), and (1, 1). Let $\mathcal{R}_2$ be the closed region obtained by removing $\mathcal{R}_1$ from $\mathcal{R}$. Use the formula in Exercise 15 to find the centroid of $\mathcal{R}_2$.

6–9* Moments III: Applications. Theorems of Pappus

Moments and centroids are involved in numerous theoretical and practical problems. We consider three simple applications.

FLUID PRESSURE

Fluid Pressure and Centroids

Let a flat plate be submerged vertically in water. We shall show that the force due to fluid pressure on one side of the plate equals

$$(62.5) \cdot D \cdot A$$

where D is the depth of the centroid and A is the area of the plate. (See Fig. 6–49a.)

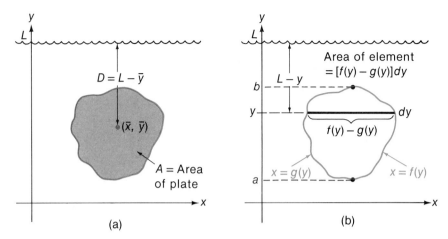

Figure 6-49 Force = $62.5(L - \bar{y}) \cdot A = (62.5)$(depth of centroid)(area).

Let $x = f(y)$ and $x = g(y)$ be the equations of the right- and left-hand boundaries of the plate, $a \leq y \leq b$, where f and g are continuous. (See Fig. 6-49b.) Let the water level be at the point L on the y-axis. A typical element has area

$$[f(y) - g(y)] \, dy$$

and is at a distance of $L - y$ units below the water level. The force on this element is

$$(62.5)(L - y)[f(y) - g(y)] \, dy.$$

It follows that the total force on one side of the plate is

$$F = 62.5 \int_a^b (L - y)[f(y) - g(y)] \, dy.$$

Observe that

$$\text{Area of plate} = \int_a^b [f(y) - g(y)] \, dy$$

and

$$M_x = \int_a^b y[f(y) - g(y)] \, dy.$$

Also recall that $M_x = \bar{y}A$, where A is the area of the plate. If we break the integral for F into two parts and apply these results, we get

$$F = 62.5 \int_a^b (L - y)[f(y) - g(y)] \, dy$$

$$= (62.5) \cdot L \cdot \int_a^b [f(y) - g(y)] \, dy - 62.5 \int_a^b y[f(y) - g(y)] \, dy$$

6–9 Moments III: Applications. Theorems of Pappus

$$= (62.5) \cdot L \cdot A - (62.5) \cdot M_x$$
$$= (62.5) \cdot L \cdot A - (62.5) \cdot \bar{y} A$$
$$= (62.5)(L - \bar{y}) \cdot A.$$

Since $L - \bar{y}$ is the depth of the centroid, then

$$F = (62.5) \cdot \text{(depth of centroid)} \cdot \text{(area of plate)}.$$

It can be proved that this result also holds when the plate is submerged horizontally, or at an angle to the vertical.

EXAMPLE 1 The area of one side of an irregular flat plate equals 17.3 ft². The plate is submerged so that its centroid is 19.2 ft below the surface of the water. The force on one side of the plate is

$$F = (62.5) \cdot \text{(depth of centroid)} \cdot \text{(area of plate)}$$
$$= (62.5)(19.2)(17.3) = 20{,}760 \text{ lb.} \ \square$$

VOLUMES OF SOLIDS OF REVOLUTION

Suppose that a region $\mathcal{R}$, bounded above by the graph of $y = f(x)$ and below by the graph of $y = g(x)$, $a \leq x \leq b$, where f and g are continuous functions and $a > 0$, is revolved about the y-axis. (See Figs. 6–50a, c.) If we apply the shell method, we find that the volume of the solid of revolution is

$$V = 2\pi \int_a^b x[f(x) - g(x)] \, dx.$$

Recall that $M_y = \int_a^b x[f(x) - g(x)] \, dx$ and that $\bar{x} = \dfrac{M_y}{\text{area}}$. Then

$$V = 2\pi \int_a^b x[f(x) - g(x)] \, dx = 2\pi M_y = 2\pi \bar{x} \cdot \text{area}.$$

When the region $\mathcal{R}$ is revolved about the y-axis, each point of $\mathcal{R}$ traces out a circle. The number $2\pi\bar{x}$ is the circumference of the circle traced out by the centroid. This number is called the *orbit of the centroid* (see Fig. 6–50b.):

$$\text{Orbit of centroid} = 2\pi\bar{x} = \text{circumference of circle traced}$$
$$\text{out by the centroid.}$$

Thus, the volume of the solid of revolution equals

Volume, Centroid and Area

$$V = \text{(orbit of the centroid)} \cdot \text{(area of the region)}$$

This result can be generalized to calculate the volume of any solid of revolution, provided the region $\mathcal{R}$ lies on one side of the line about which it is revolved and its

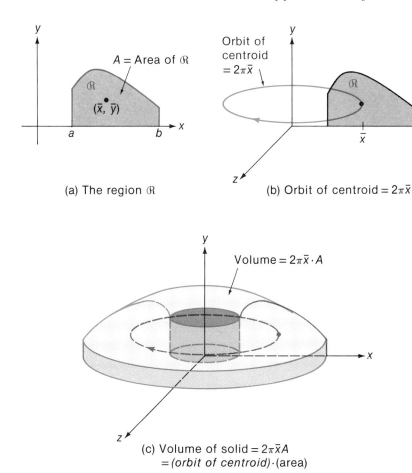

Figure 6-50 The First Theorem of Pappus.

boundary is not too complicated. This result is one of the *theorems of Pappus*, first proved by the ancient Greek mathematician Pappus (third century A.D.) without the use of the calculus.

FIRST THEOREM OF PAPPUS

Pappus' Theorem About Volume

Let $\mathcal{R}$ be a region in the xy-plane bounded by a curve $\mathcal{C}$. Let $\mathcal{L}$ be a line that lies completely on one side of $\mathcal{C}$. The volume of the solid of revolution obtained by revolving $\mathcal{R}$ about $\mathcal{L}$ is equal to the product of the area of $\mathcal{R}$ and the orbit of the centroid of $\mathcal{R}$ about $\mathcal{L}$.

EXAMPLE 2 A torus (doughnut-shaped solid) is formed by revolving the region bounded by the circle

$$(x - 1)^2 + (y - 2)^2 = 1$$

about the line $x + y + 1 = 0$. Find the volume of the torus.

6-9 Moments III: Applications. Theorems of Pappus

Solution The centroid of the circular region is the center of the circle—the point (1, 2). (See Fig. 6–51.) The distance from the centroid to the line $x + y + 1 = 0$ equals

$$d = \frac{1 + 2 + 1}{\sqrt{2}} = \sqrt{8}.$$

(See Exercise 51, Section 1–5.)
The orbit of the centroid is

$$\text{Orbit} = 2\pi d = 2\pi\sqrt{8}.$$

The area of the circle is

$$\text{Area} = \pi \cdot 1^2 = \pi.$$

It follows from Pappus' First Theorem that

$$\text{Volume of torus} = (\text{orbit of centroid}) \cdot (\text{area of region})$$
$$= 2\pi\sqrt{8} \cdot \pi = 2\pi^2\sqrt{8} \approx 55.83 \text{ cu units.} \quad \square$$

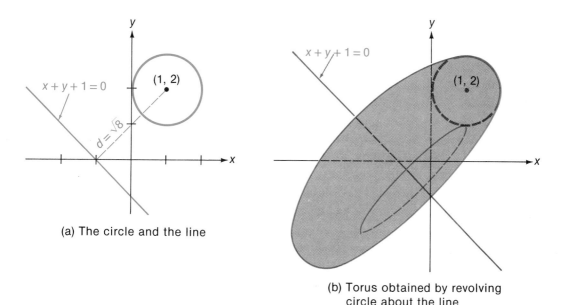

(a) The circle and the line

(b) Torus obtained by revolving circle about the line

Figure 6–51 Example 2. The torus.

Remark Pappus' First Theorem can be used to calculate the volume of almost any solid of revolution. All we need to know are the area of the original region and the location of its centroid. The approximate area of the region can be measured by a drafting instrument called a *planimeter*. The centroid can be located experimentally by cutting the region out of cardboard and hanging it up by a point on its edge. The centroid will always be directly under the point from which it is hung. If we do this twice, hanging

Applications of the Integral

the cardboard region from different points, we can locate the centroid as the intersection of two lines. Once the centroid and area have been found, we can calculate the volume of the solid of revolution.

SURFACE AREA

A second theorem of Pappus is concerned with the surface area of a solid of revolution. This theorem is analogous to the other theorem, but is involved with the *centroid of the boundary* of a region $\mathcal{R}$.

Centroid of a Boundary Curve

The centroid of the boundary of a region $\mathcal{R}$ can be described most easily by the following process. We bend a piece of thin homogeneous wire until it exactly fits the boundary of $\mathcal{R}$. If ρ is the mass of the wire per unit of length, then the wire will have a total mass of $L\rho$, where L is the length of the boundary of $\mathcal{R}$. Each particle of the wire has a certain moment with respect to an axis. The limit of the sum of the moments of all of the particles equals the moment of the boundary about the axis. The centroid $(\bar{\bar{x}}, \bar{\bar{y}})$ of the boundary is defined by

$$\bar{\bar{x}} = \frac{M_y}{mass}, \qquad \bar{\bar{y}} = \frac{M_x}{mass},$$

where M_x and M_y are the moments about the x- and the y-axis.

In most cases the centroid of a curve is not on the curve. The centroid of a circle, for example, is at the center of the circle, the centroid of the boundary of a rectangle is at the intersection of the diagonals. (See Figs. 6–52a, b.)

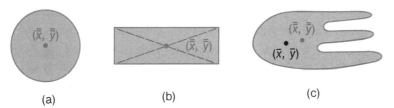

(a)   (b)   (c)

Figure 6–52 Centroids of boundaries: $(\bar{\bar{x}}, \bar{\bar{y}})$ is the centroid of the boundary, $(\bar{x}, \bar{y})$ is the centroid of the enclosed region. (a) The centroid of the boundary of a circle is at the center of the circle. It coincides with the centroid of the enclosed region. (b) The centroid of the boundary of a rectangle is at the center of the rectangle. It coincides with the centroid of the enclosed region. (c) In general the centroid of the boundary of a figure does not coincide with the centroid of the enclosed region.

Do not jump to the conclusion that the centroid of the boundary of a region always coincides with the centroid of the region. In general these points are distinct, as with the region in Figure 6–52c.

Pappus' Second Theorem has the following formulation.

SECOND THEOREM OF PAPPUS Let $\mathcal{R}$ be a region in the xy-plane bounded by a simple closed smooth curve $\mathcal{C}$. Let $\mathcal{L}$ be a line that lies completely on one side of $\mathcal{C}$. The surface area of the solid of revolution obtained by revolving $\mathcal{R}$ about the line $\mathcal{L}$ equals the product of the orbit of the centroid of $\mathcal{C}$ and the length of $\mathcal{C}$:

Pappus' Theorem About Surface Area

$$\text{Surface area} = (\text{orbit of centroid of boundary}) \cdot (\text{length of boundary})$$

EXAMPLE 3 Find the surface area of the torus in Example 2.

Solution The centroid of the boundary of the circular region is at the center $(1, 2)$. The orbit of the centroid is

$$\text{Orbit} = 2\pi\sqrt{8}.$$

The length of the boundary is

$$L = 2\pi.$$

Then

$$\text{Surface area} = (\text{orbit of centroid})(\text{length of boundary})$$
$$= (2\pi\sqrt{8}) \cdot (2\pi) = 4\sqrt{8}\pi^2 \text{ sq units.} \quad \square$$

Exercises 6-9

Use the methods of this section for all problems.

•1 A triangular plate with sides of length 5, 5, and 6 ft is submerged vertically in water with the 6-ft side at the surface. Find the centroid of the plate and use it to find the fluid pressure on one side.

•2 A circular floodgate in the vertical face of a dam is 4 ft in diameter. Calculate the fluid pressure on the gate when the water level is 3 ft above the top of the gate.

•3 A rectangular plate 8 ft by 12 ft is submerged in water at such an angle that one 12-ft edge is at the surface of the water and the other is 6 ft below the surface of the water. Find the force due to the water pressure on one side of the plate.

•4 Use Pappus' First Theorem to find the volume of a right circular cone of radius R and height H.

•5 A torus is formed by revolving the region bounded by the circle $x^2 + (y - 2)^2 = 4$ about the line $3y = 4x - 19$. Find its volume.

•6 Find the surface area of the torus in Exercise 5.

•7 A triangular plate, with sides 3, 4, and 5 units, is placed in the xy-plane in such a way that it does not intersect the x-axis nor the line $y = -x$ and its centroid is at the

point (2, 6). Find the volume of the solid generated by revolving the triangle about
(a) the x-axis;   (b) the line $y = -x$.

●8 The center of a circle of radius r is located R units from the line $\mathscr{L}$ where $R \geq r$. The region bounded by the circle is revolved about $\mathscr{L}$, generating a torus.
(a) Find the volume.   (b) Find the surface area.

Review Problems

1 (a) Define the term *area of a plane figure*. Why are we not allowed to use elements of area in the definition?

(b) What is meant by a *projection* of a region on an axis? How do we use projections to set up integrals for area?

2 Derive the formula for the area of the region bounded on the right by the graph of $x = f(y)$, on the left by the graph of $x = g(y)$, $c \leq y \leq d$, where f and g are continuous functions of y.

3–6. Calculate the areas of the regions bounded by the given curves and lines.

●3 $y = x^3$ and $y = x^2$

●4 $y = x^2$ and $x = y^2$

●5 $x = y^3 - 4y$ and $x = 4 - y^2$

●6 $y^2 = 4x$ and $y^2 + 4x - 8 = 0$

●7 A solid has the circle $x^2 + y^2 = 4$ as a base. The cross sections perpendicular to the x-axis are equilateral triangles. Find the volume.

●8 The base of a solid is the region bounded by the x-axis, the line $x = 9$, and the graph of $y = \sqrt{x}$. Each cross section perpendicular to the x-axis is an isosceles right triangle with the hypotenuse on the base of the solid. Find the volume.

9–15. Let $\mathscr{R}$ be the region bounded by the given curves and lines. Use either the disk method or the shell method to calculate the volume of the solid obtained by revolving $\mathscr{R}$ about the indicated line. Try to use the method that requires the least work.

●9 $y = \dfrac{2}{x+1}$, $y = 0$, $x = 1$, and $x = 3$, about the x-axis.

●10 $y = x^2$ and $y = 8 - x^2$, about line $y = -1$.

●11 Region in Exercise 10, about the y-axis.

●12 $y = \dfrac{1}{x}$, $x = 1$, $x = 4$, and $y = 0$, about the y-axis.

Review Problems

- **13** $y = x^2$, $x = 1$, $x = 3$, $y = 0$, about the line $x = 5$.

- **14** $y = x^3 + 1$, $x + y = 1$, $x = 1$, about the y-axis.

- **15** Region in Exercise 14, about the line $x = 2$.

16 (a) State the formula for the arc length of the graph of $y = f(x)$ between $x = a$ and $x = b$. Be sure to state the properties required for the function f.

  (b) Derive the formula in (a).

17 Define the concepts for graphs of parametric equations: (1) curve; (2) smooth curve.

18–21. Calculate the arc length.

- **18** $f(x) = 2x^{3/2}$, $0 \leq x \leq 7$
- **19** $y = (2x^{2/3} + 4)^{3/2}$, $0 \leq x \leq 1$
- **20** $x = -t^2/2$, $y = (2t + 9)^{3/2}$, $0 \leq t \leq 1$
- **21** $x = t^3$, $y = t^2$, $1 \leq t \leq 12$

- **22** A spring of natural length 8 in. requires a force of 2 lb to compress it by 1 in. Find the work done in compressing the spring from a length of 7 in. to a length of 4 in.

- **23** A right circular cylindrical tank 4 ft in diameter and 6 ft deep is full of water. Find the work done in pumping the water to a point 3 ft above the top of the tank.

- **24** A hemispherical punch bowl is 2 ft in diameter. How much work is done in sucking all the punch through a straw to a point 1 ft above the top of the bowl? (Assume that the punch weighs 62.5 lb/ft^3.)

- **25** A 12-ft-long trough has a vertical cross section that is an isosceles triangle with sides inclined 45° to the horizontal. Find the fluid pressure on one end of the trough when the water is 2 ft deep.

- **26** A rectangular board 1 ft wide and 3 ft long is immersed vertically in water with the 3-ft edge at the water level. Find the fluid pressure on one side of the board.

27 Define the moments M_x and M_y and the center of mass of the lamina of density ρ that has the graphs of $y = f(x)$ and $y = g(x)$, $a \leq x \leq b$, for its upper and lower boundaries, respectively.

28–31. Calculate the centroid of the region bounded by the curves and lines.

- **28** $y = \sqrt{4 - x^2}$, x-axis
- **29** $y = x^2$, $x - y + 2 = 0$
- **30** $x = \sqrt{4 - y}$, $x + 2y = 2$, y-axis

●31 $y = x + 1$, $y = 3x - 3$, $3x + y + 3 = 0$

●32 Weights of mass 3, 4, 5, and 12 g are located at the points $x = 1$, $x = 3$, $x = 6$, and $x = 8$, respectively, on a thin metal ruler that extends from $x = -1$ to $x = 9$. The ruler has a uniform density of 1 g per unit. Find the moment about the origin and the coordinates of the centroid.

33 State the two theorems of Pappus.

●34 The triangle bounded by the lines $x = 3$ $y = 2$, and $x + 2y = 9$ is revolved aboaut the y-axis, generating a solid of revolution. Use a theorem of Pappus to calculate the volume.

7 Topics in Plane Analytic Geometry

7–1 Curve Sketching, Asymptotes, Symmetry, and Projections

Analytic geometry is the study of relations between equations and their graphs. The graph of the equation

$$x^2 - 2x + y^2 + 6y - 12 = 0,$$

for example, is known to be a circle. It is possible to determine properties of this circle (location of center, size of radius) directly from the equation.

In the first part of this chapter we develop general methods for determining properties of graphs directly from their defining equations. Later we turn our attention to second-degree equations—equations of form

$$Ax^2 + Bxy + Cy^2 + Dx + Ey + F = 0.$$

We shall show that the graph of every second-degree equation is a parabola, an ellipse, or a hyperbola (possibly degenerate).

In most of this chapter we work with equations in x and y. Such equations can be written in the form

$$F(x, y) = 0$$

where F is a function of x and y. As a general rule, the graph of such an equation consists of one or more curves in the xy-plane. The derivatives dy/dx and dx/dy can be calculated at points on the graph by implicit differentiation.

We postpone a full discussion of curves until Section 13–1. We do, however, need one important property of curves.

A Portion of a Curve May Be the Graph of $y = f(x)$

Let $P_0(x_0, y_0)$ be a point on a curve $\mathscr{C}$ that is part of the graph of $F(x, y) = 0$. If the derivative dy/dx exists and is continuous in the vicinity of P_0, then there exists a differentiable function f such that the graph of f coincides with $\mathscr{C}$ in the vicinity of $P_0(x_0, y_0)$. (See Fig. 7–1.) Because of this fact, we can work with the graph of function $y = f(x)$ when it is more convenient to do so than to work with the graph of equation $F(x, y) = 0$.

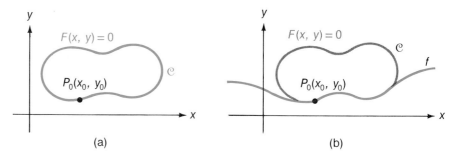

Figure 7–1 The curve $\mathscr{C}$ is the graph of the equation $F(x, y) = 0$. The derivative dy/dx is continuous on an interval containing $P_0(x_0, y_0)$. There exists a function f such that the graph of f coincides with $\mathscr{C}$ in the vicinity of P_0.

ASYMPTOTIC BEHAVIOR

Many functions behave erratically when x is small but very regularly when x is large. Two such functions are pictured in Figure 7–2. In (a), the functional values $f(x)$ ap-

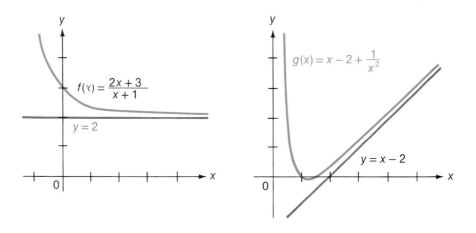

(a) Horizontal asymptote
$f(x) \to 2$ as $x \to \infty$

(b) Oblique asymptote
$g(x) - (x - 2) \to 0$ as $x \to \infty$

Figure 7–2 Horizontal and oblique asymptotes.

7-1 Curve Sketching, Asymptotes, Symmetry, and Projections

Asymptotic Equality

proach 2 as $x \to \infty$. In (b), the functional values $g(x)$ approach $x - 2$ as $x \to \infty$. We say that $f(x)$ *is asymptotically equal to* 2 as $x \to \infty$ and $g(x)$ *is asymptotically equal to* $x - 2$ as $x \to \infty$. The lines $y = 2$ and $y = x - 2$ are called *asymptotes* to the graphs of f and g respectively.

These concepts can be defined in terms of limits.

DEFINITION

Let f and g be functions of x that are defined for all x greater than some fixed number a. We say that $f(x)$ *is asymptotically equal to* $g(x)$ as $x \to \infty$ provided

$$\lim_{x \to \infty} [f(x) - g(x)] = 0.$$

In other words, $f(x)$ is asymptotically equal to $g(x)$ as $x \to \infty$ provided

$$f(x) = g(x) + d(x)$$

where $d(x) \to 0$ as $x \to \infty$. (See Fig. 7-3.)

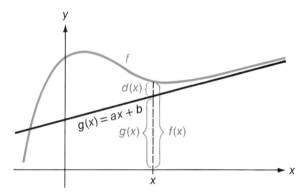

Figure 7-3 Asymptotic equality. The functions f and g are *asymptotically equal* provided the difference $d(x) = f(x) - g(x)$ approaches zero as a limit as $x \to \infty$.

In general, we want the function g to be simpler than the function f. For example, if

$$f(x) = x + 2 + \frac{3x}{x^2 - 1} \quad \text{and} \quad g(x) = x + 2$$

we would say that $f(x)$ is asymptotically equal to $g(x)$ as $x \to \infty$, but as a rule, we would not say that $g(x)$ is asymptotically equal to $f(x)$, although both relationships are technically correct.

We are particularly interested in the case where $g(x)$ is a linear function

$$g(x) = ax + b.$$

If that is the case, then the line $y = ax + b$ (the graph of g) is said to be an *asymptote* of the graph of f as $x \to \infty$. (See Fig. 7-3.)

Similar definitions and properties hold for asymptotes as $x \to -\infty$.

HORIZONTAL AND VERTICAL ASYMPTOTES

Horizontal Asymptotes

As we saw in Section 2–8, the line $y = L$ is a horizontal asymptote to the graph of f as $x \to \infty$ provided

$$\lim_{x \to \infty} f(x) = L.$$

A similar definition holds as $x \to -\infty$.

Vertical Asymptotes

Vertical asymptotes are defined in a slightly different manner. The line $x = a$ is a vertical asymptote of the graph of f if either (or both) of the one-sided limits

$$\lim_{x \to a^-} f(x) \quad \text{and} \quad \lim_{x \to a^+} f(x)$$

is infinite.

Recall that a vertical asymptote can exist only at a point of discontinuity of f. This, of course, does not mean that a graph always has a vertical asymptote at each point of discontinuity. The function $f(x) = (x^2 - 4)/(x - 2)$, for example, is discontinuous at $x = 2$ but does not have a vertical asymptote there, since $\lim_{x \to 2} f(x) = 4$.

EXAMPLE 1 Find all horizontal and vertical asymptotes to the graph of

$$f(x) = \frac{x - 2}{x - 1}.$$

Solution Since

$$\lim_{x \to \pm\infty} f(x) = \lim_{x \to \pm\infty} \frac{x - 2}{x - 1} = \lim_{x \to \pm\infty} \frac{1 - \frac{2}{x}}{1 - \frac{1}{x}} = 1,$$

then the line $y = 1$ is a horizontal asymptote as $x \to \infty$ or $x \to -\infty$.

The only point of discontinuity is $x = 1$. This is the only possible location for a vertical asymptote. The one-sided limits are

$$\lim_{x \to 1^+} \frac{x - 2}{x - 1} = -\infty \quad \text{and} \quad \lim_{x \to 1^-} \frac{x - 2}{x - 1} = \infty.$$

The line $x = 1$ is a vertical asymptote. (See Fig. 7–4.) □

OBLIQUE ASYMPTOTES

Oblique Asymptote

Let a and b be nonzero numbers. The line $y = ax + b$ is an *oblique asymptote* to the graph of f as $x \to \infty$ provided

$$f(x) = ax + b + d(x)$$

where $d(x) \to 0$ as $x \to \infty$. (See Figs. 7–3, 7–5.)

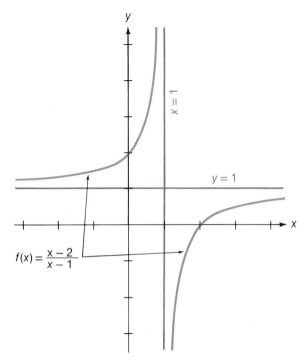

Figure 7–4 Example 1. $f(x) = (x - 2)/(x - 1)$.

EXAMPLE 2 Find an oblique asymptote to the graph of

$$f(x) = \frac{2x^2 + 3x}{x + 2}.$$

Solution We first use long division to divide $2x^2 + 3x$ by $x + 2$, obtaining

$$f(x) = \frac{2x^2 + 3x}{x + 2} = (2x - 1) + \frac{2}{x + 2}.$$

Then

$$\lim_{x \to \infty} [f(x) - (2x - 1)] = \lim_{x \to \infty} \frac{2}{x + 2} = 0$$

so the line $y = 2x - 1$ is an oblique asymptote to the graph of f. (See Fig. 7–5.) □

Special Case where an Oblique Asymptote Exists

Note that the procedure illustrated in Example 2 can be used to find oblique asymptotes for any rational function $g(x)/h(x)$ provided *the degree of $g(x)$ is one*

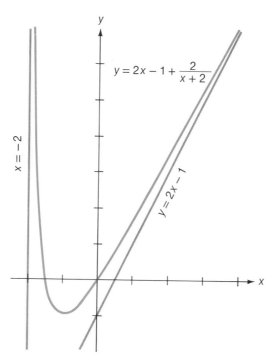

Figure 7–5 Example 2. $y = (2x^2 + 3x)/(x + 2)$.

greater than the degree of $h(x)$. We first divide $h(x)$ into $g(x)$ by long division. This gives us

$$\frac{g(x)}{h(x)} = ax + b + \frac{r(x)}{h(x)}$$

where $r(x)/h(x) \to 0$ as $x \to \infty$.

SYMMETRY

Symmetry A curve is said to be *symmetric about the line* $\mathscr{L}$ if for each point P on the curve there is a corresponding point Q on the curve such that $\mathscr{L}$ is the perpendicular bisector of the line segment PQ. (See Fig. 7–6.)

Symmetry can be very useful in sketching curves. If a graph is symmetric about both coordinate axes, for example, we need only to make a careful drawing of the part in Quadrant I. The part in Quadrant II can be obtained by using the symmetry about the y-axis; then the parts in Quadrants III and IV can be obtained by using the symmetry about the x-axis.

7–1 Curve Sketching, Asymptotes, Symmetry, and Projections

SYMMETRY ABOUT THE y-AXIS

If the graph of an equation is symmetric about the y-axis, then for each point $P(x, y)$ on the graph there must be a corresponding point $Q(-x, y)$ on the graph. (See Fig. 7–7.) Thus, $(-x, y)$ must satisfy the defining equation for the graph whenever (x, y) satisfies it. These remarks lead us to a test for symmetry.

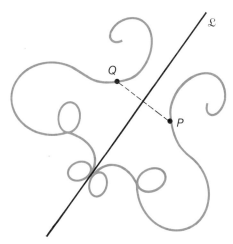

Figure 7–6 Symmetry about the line $\mathscr{L}$. The line $\mathscr{L}$ is the perpendicular bisector of the line segment PQ.

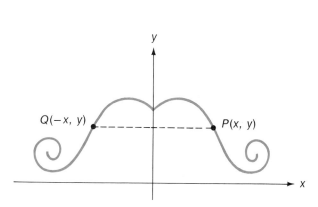

Figure 7–7 Symmetry about the y-axis. For each point $P(x, y)$ on the graph, there is a corresponding point $Q(-x, y)$ on the graph.

Test for Symmetry About y-Axis

> **Test for Symmetry about the y-Axis** Substitute $-x$ for x in the defining equation. If the new equation is equivalent to the original one, the graph is symmetric about the y-axis.*

EXAMPLE 3   The graph of

$$y = x^2 - 2 \quad \text{(original equation)}$$

is symmetric about the y-axis. If we substitute $-x$ for x, we get

$$y = (-x)^2 - 2 \quad \text{(new equation)}$$

which is equivalent to the original equation. ◻

Similar tests can be devised for symmetry about the x-axis and the line $y = x$. (See Fig. 7–8 and Exercises 51 and 52.) These tests are summarized in Table 7–1. Each

---

* Two equations are said to be equivalent if they have the same solution sets. For example, the equations $2x - 3y = 7$ and $-4x + 6y + 14 = 0$ are equivalent.

Table 7–1 Tests for Symmetry

Symmetry about . . .	Test
x-Axis	Substitute $-y$ for y.
y-Axis	Substitute $-x$ for x.
Origin	Substitute $-x$ for x and $-y$ for y.
Line $y = x$	Substitute x for y and y for x.

NOTE: A test is affirmative (indicating symmetry) if the equation obtained by the substitution is equivalent to the original equation. (See Figs. 7–7, 7–8, and 7–10.)

test is affirmative, indicating symmetry, if the new equation is equivalent to the original one. Otherwise, the test is negative, indicating a lack of symmetry.

EXAMPLE 4 Show that the graph of $y = 1/x$ is symmetric about the line $y = x$ but not about either axis.

Solution *Symmetry about the x-axis* If we substitute $-y$ for y, the new equation is

$$-y = \frac{1}{x} \quad \text{(new equation)}$$

which is not equivalent to the original equation. The graph is not symmetric about the x-axis.

Symmetry about the y-axis If we substitute $-x$ for x, the new equation is

$$y = \frac{1}{-x} \quad \text{(new equation)}$$

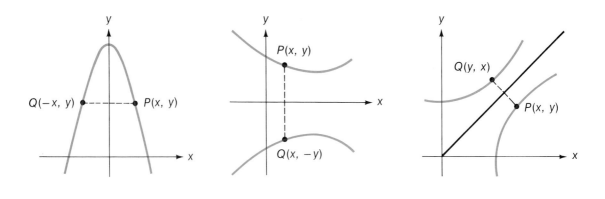

(a) Symmetry about the y-axis

(b) Symmetry about the x-axis

(c) Symmetry about the line $y = x$

Figure 7–8 Symmetry. (See Table 7–1)

7-1 Curve Sketching, Asymptotes, Symmetry, and Projections

which is not equivalent to the original equation. The graph is not symmetric about the y-axis.

Symmetry about the Line $y = x$ If we substitute y for x and x for y, the new equation is

$$x = \frac{1}{y} \quad \text{(new equation)}$$

which is equivalent to

$$y = \frac{1}{x} \quad \text{(original equation).}$$

Thus, the graph is symmetric about the line $y = x$ but not about either axis. (See Fig. 7–9.) □

SYMMETRY ABOUT THE ORIGIN

Symmetry About Origin

We consider one additional type of symmetry. If the point $Q(-x, -y)$ is on a curve whenever the point $P(x, y)$ is on it, then the curve is said to be *symmetric about the origin*. (See Figs. 7–9 and 7–10.)

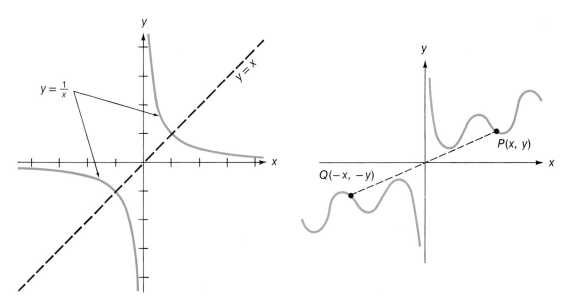

Figure 7–9 Examples 4 and 5. $y = 1/x$.

Figure 7–10 Symmetry about the origin.

To test an equation for symmetry about the origin, we substitute $-x$ for x and $-y$ for y. If the new equation is equivalent to the original one, then the test is affirmative, indicating symmetry about the origin. Otherwise it is negative, indicating no symmetry.

EXAMPLE 5   (*Continuation of Example 4*) Show that the graph of $y = 1/x$ is symmetric about the origin.

Solution   If we substitute $-x$ for x and $-y$ for y, the new equation is

$$-y = \frac{1}{-x} \quad \text{(new equation)}$$

which is equivalent to the original equation

$$y = \frac{1}{x}.$$

(See Fig. 7–9.) □

Remark   If a curve is symmetric about any two among the x-axis, the y-axis, and the origin, then it is symmetric about all three. (See Exercise 38.)

PROJECTIONS

The *projection of a curve on the x-axis* (or the *x-extent* of the curve) consists of all points on the x-axis that lie on vertical lines intersecting the curve. We can think of this projection as the shadow the curve casts on the x-axis. (See Figs. 7–11 and 7–12.) The *projection on the y-axis* (or the *y-extent*) is defined similarly.

If the x- and y-projections of a curve are the intervals $[a, b]$ and $[c, d]$ on the x- and

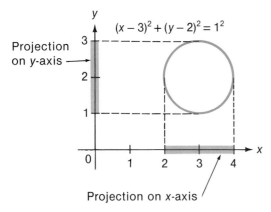

Figure 7–11 Projections on the coordinate axes.

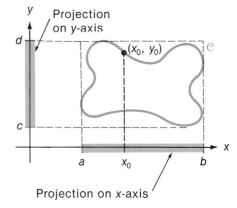

Figure 7–12 Projections on the coordinate axes. The number x_0 is in the x-projection if, and only if, the vertical line $x = x_0$ intersects the curve.

7-1 Curve Sketching, Asymptotes, Symmetry, and Projections

y-axes, respectively, then the curve is completely contained within the rectangle with sides formed by the projections. (See Fig. 7–12.)

If an equation $F(x, y) = 0$ is not too complicated, we can find its projections on the coordinate axes by the following process.

Determination of Projections

(1) *Solve the equation for y in terms of x.* The x-projection is the set of all real x for which the corresponding values of y are real.
(2) *Solve the equation for x in terms of y.* The y-projection is the set of all real y for which the corresponding values of x are real.

EXAMPLE 6 Find the projections of the graph of

$$x^2 + 4y^2 = 4$$

on the coordinate axes.

Solution *Projection on the x-axis* We solve the equation for y in terms of x:

$$4y^2 = 4 - x^2$$

$$y = \pm \frac{\sqrt{4 - x^2}}{2}.$$

Then y is real if and only if $4 - x^2 \geq 0$. Thus,

$$x\text{-projection} = \{x \,|\, 4 - x^2 \geq 0\} = \{x \,|\, x^2 \leq 4\}$$
$$= \{x \,|\, |x| \leq 2\} = \{x \,|\, -2 \leq x \leq 2\}.$$

Projection on the y-axis We solve the equation for x in terms of y:

$$x^2 = 4 - 4y^2$$

$$x = 2\sqrt{1 - y^2}.$$

Then,

$$y\text{-projection} = \{y \,|\, 1 - y^2 \geq 0\} = \{y \,|\, -1 \leq y \leq 1\}.$$

(See Fig. 7–13.) ☐

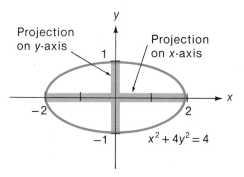

Figure 7–13 Example 6.

APPLICATION TO SKETCHING CURVES

The techniques developed thus far can be used with the results of Chapter 4 to make accurate sketches of most of the graphs in this book. The following steps outline the process.

Steps in Curve Sketching

(1) Find all points of discontinuity.
(2) Find all asymptotes—horizontal, vertical, and oblique.
(3) Test the equation for symmetry.
(4) Find the x- and y-projections.
(5) Find all points with horizontal or vertical tangent lines. Find all local extrema.
(6) Find where the graph is increasing, decreasing, concave upward, concave downward. Find all points of inflection.
(7) Plot a few reference points, including the local extrema and points of inflection. Sketch the tangent lines at these points. Sketch the curve, using the information obtained in the above steps.

EXAMPLE 7 Analyze the graph of $y = x + 1/x$. Sketch the graph.

Solution (1) *Points of Discontinuity.* The function is continuous everywhere except at $x = 0$. It is discontinuous there.
(2) *Asymptotes.* The line $y = x$ is an oblique asymptote; the y-axis is a vertical asymptote.
(3) *Symmetry.* The graph is symmetric about the origin but not about either axis nor about the line $y = x$.
(4) *Projections.* The projection on the x-axis is $\{x \mid x \neq 0\}$. To find the projection on the y-axis, we first solve for x in terms of y:

$$x^2 - xy + 1 = 0$$
$$x = \frac{y \pm \sqrt{y^2 - 4}}{2} \quad \text{(by the quadratic formula).}$$

The y-projection is $\{y \mid y^2 - 4 \geq 0\} = \{y \mid y \leq -2 \text{ or } y \geq 2\}$.

(5) *Local Extrema, Vertical Tangent Lines, and the Like.* The derivative is

$$y' = 1 - \frac{1}{x^2},$$

which is zero if and only if $x = \pm 1$. There is a local maximum at $(-1, -2)$ and a local minimum at $(1, 2)$.

There are no vertical tangent lines.

(6) *Increasing, Decreasing, and the Like.* The first derivative shows that the function is

increasing	if $x < -1$
decreasing	if $-1 < x < 0$ or $0 < x < 1$
increasing	if $x > 1$.

The second derivative shows that it is concave downward for $x < 0$ and concave upward for $x > 0$. There are no points of inflection.

(7) The graph is shown in Figure 7–14. ☐

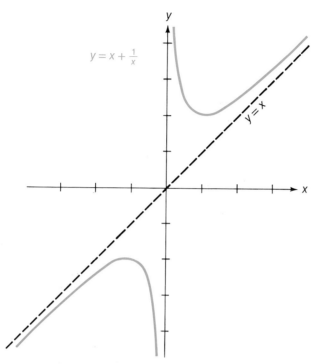

Figure 7–14 Example 7.

Exercises 7–1

1–10. Find all horizontal, vertical, and oblique asymptotes.

●1 $f(x) = \dfrac{x}{5 - x}$

2 $f(x) = \dfrac{1}{x - 4}$

●3 $f(x) = \dfrac{x + 3}{x^2 - 9}$

4 $f(x) = \dfrac{x^2 + 5x}{x^2 - 25}$

●5 $f(x) = \dfrac{x^2}{9 - x^2}$

6 $f(x) = \dfrac{x^2}{x^2 + x - 2}$

●7 $f(x) = \dfrac{x^2 + 5x}{x - 3}$

8 $f(x) = \dfrac{x^3 + 2x^2 + 2x}{x^2 + x - 2}$

•9 $f(x) = \dfrac{5x^3 + 7x^2 + 3x - 1}{x^2 - 3x + 2}$

10 $f(x) = \dfrac{4x^3 - 2x^2 + x - 4}{2x^2 + 3x + 1}$

11–12. Find the oblique asymptotes as $x \to \infty$ and $x \to -\infty$.

•11 $f(x) = \dfrac{3x^2 - 2x + 1}{|x + 1|}$

12 $f(x) = \dfrac{4x^2 - 2x + 5}{\sqrt{4x^2 - 4x + 1}}$

13–20. Find all vertical and horizontal tangent lines to the graphs of the equations. (*Hint for 20:* Use implicit differentiation.)

•13 $9x^2 - y^2 = 36$

14 $36x^2 + 4y^2 = 144$

•15 $x^2 + y^2 = 100$

16 $x^2 + 2x + y^2 - 6y = 26$

•17 $x^2 - 4x - 2y^2 + 4y = 14$

•18 $x^3 + y^3 - 8 = 0$

•19 $y^2 = x + 2$

•20 $x^2 + xy + y^2 - 12 = 0$

21–28. Determine whether the graphs of the equations are symmetric about the x-axis, the y-axis, the origin, the line $y = x$.

•21 $xy = 10$

22 $y = 4x^2$

•23 $x^2 + y^2 = 25$

24 $4x^2 - 9y^2 = 36$

•25 $x^4y^2 + y^2 = xy$

26 $x^3y - xy^3 + 2 = 0$

•27 $x^3 + y^3 - 1 = 0$

28 $x^2y - xy + xy^2 = 1$

29–36. Find the projections on the coordinate axes of the graphs of the equations.

•29 $y = \dfrac{1}{x - 4}$

30 $y = \dfrac{x}{5 - x}$

•31 $y - 1 = \dfrac{1}{x - 3}$

32 $y = 2x^2$

•33 $x^2 + 2x + y^2 - 6y = 26$

34 $y = -4x^2 + 24x - 35$

•35 $y^2 = x - 2$

36 $4y^2 - 25x^2 = 100$

37 A certain curve is symmetric about the y-axis and the origin. In Quadrant I the curve coincides with the graph of $y = x^2$ for $0 \le x \le 2$ and with the graph of $(x + 2)^2 + (y - 1)^2 = 25$ for $2 < x \le 3$. Sketch the curve.

38 Prove that if a curve is symmetric about any two of the following, then it is symmetric about all three: the x-axis, the y-axis, the origin. (*Hint:* Draw a figure that illustrates the situation.)

39–48. Analyze the equations in the exercises as noted for the properties discussed in this section.

39 Exercise 1

40 Exercise 2

41 Exercise 3

42 Exercise 4

43	Exercise 5	44	Exercise 6
45	Exercise 7	46	Exercise 8
47	Exercise 14	48	Exercise 18

49 Show that $f(x) = x^2 + (x - 1)/x^3$ is asymptotically equal to $g(x) = x^2$ as $x \to \pm\infty$. Draw a figure like Figure 7–3 that illustrates this fact.

50 Let $F(x, y) = (x - 1)^2 + (y + 2)^2 - 25$.
  (a) Find a continuous function f such that the graph of $y = f(x)$ coincides with the graph of $F(x, y) = 0$ in the vicinity of the point $(5, -5)$.
  (b) Does there exist such a function f in the vicinity of the point $(6, -2)$? Explain.

51 Prove the test for symmetry about the x-axis.

52 Prove the test for symmetry about the line $y = x$.

53 Devise a test for symmetry about the line $x = a$.

54 Devise a test for symmetry about the line $y = -x$.

55 Let f be a function of x. Show that the projection of the graph of $y = f(x)$ on the x-axis is the domain of f. What can be said about the projection on the y-axis?

7–2 The Parabola. Translation of Axes

Much of the remainder of this chapter is devoted to work with second-degree equations in x and y—equations of form

$$Ax^2 + Bxy + Cy^2 + Dx + Ey + F = 0$$

in which at least one of the three constants A, B, C is not zero. As we shall see later, there are only three basic types of graphs for such equations. Each graph is a parabola, an ellipse, a hyperbola, or a degenerate case of one of these curves.

The parabola, ellipse, and hyperbola occupy an interesting niche in the history of mathematics. They were studied extensively by the ancient Greeks as the curves of intersection of planes with cones—thus the name *conic sections*. (See Fig. 7–15.) For centuries afterward they were thought to be interesting textbook examples of curves defined by geometrical conditions. With the rise of science in the Western world it was discovered that all three conic sections have important scientific applications. (An automobile headlight reflector has a parabolic cross section; the planets move in elliptical orbits about the sun; electronic methods of range-finding are based on the hyperbola.)

We shall define the conic sections by geometrical conditions involving points and lines. Our approach is justified in Section 7–5, where we show that our definitions lead to the same curves as those studied by the ancient Greeks. We first examine the parabola.

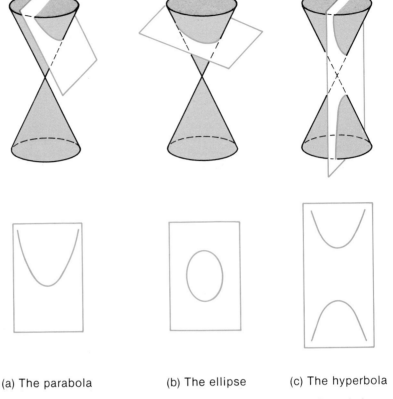

(a) The parabola (b) The ellipse (c) The hyperbola

Figure 7–15 The conic sections. The parabola, ellipse, and hyperbola are the curves of intersection of a cone with various planes.

DEFINITION

Definition of Parabola

Let $\mathcal{D}$ be a fixed line and F be a fixed point that is not on the line $\mathcal{D}$. The set of all points that are equidistant from F and $\mathcal{D}$ is called a *parabola*. (See Fig. 7–16.)

The point F is called the *focus* and the line $\mathcal{D}$ is called the *directrix* of the parabola. The point V on the parabola that is closest to the directrix is called the *vertex*.

Focus, Directrix, Vertex

We shall derive the equation of the parabola for the special case in which the focus is on the x-axis and the vertex is at the origin. If the focus is the point $F(p, 0)$, it follows that the directrix is the vertical line $x = -p$. (See Fig. 7–17.)

THEOREM 7–1   The equation of the parabola with focus $F(p, 0)$ and vertex $V(0, 0)$ is

Standard Equation

$$y^2 = 4px.$$

7–2 The Parabola. Translation of Axes

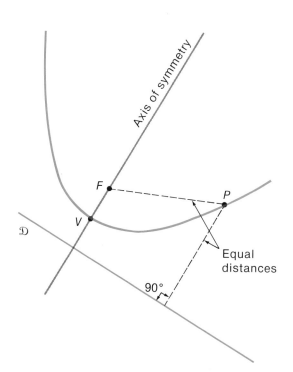

Fig 7–16 The parabola. Each point P on the parabola is equidistant from the focus F and the directrix $\mathscr{D}$.

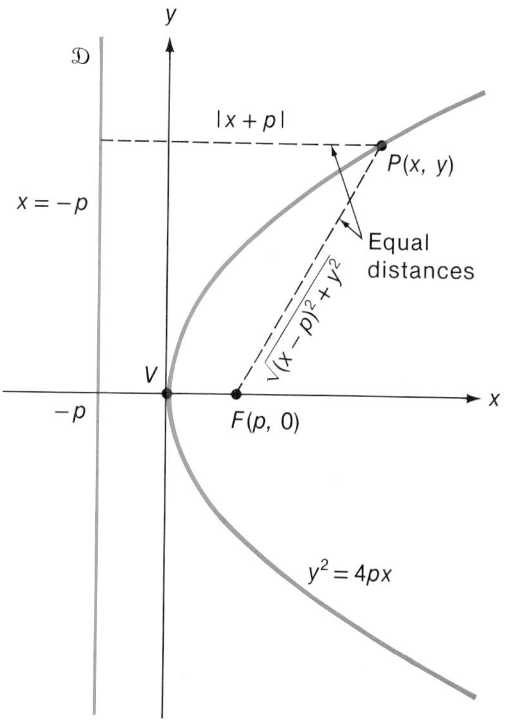

Figure 7–17 The parabola $y^2 = 4px$.

Proof The directrix is the line $x = -p$. If $P(x, y)$ is on the parabola, the distance from P to the focus $F(p, 0)$ is

$$|PF| = \sqrt{(x - p)^2 + (y - 0)^2},$$

while the distance from P to the directrix $\mathscr{D}$ (measured perpendicularly to the line) is

$$|x - (-p)| = |x + p|.$$

(See Fig. 7–17.) Since P is on the parabola, these distances are equal:

$$\sqrt{(x - p)^2 + y^2} = |x + p|.$$

If we square both sides and simplify, we obtain

$$x^2 - 2px + p^2 + y^2 = x^2 + 2px + p^2$$
$$y^2 = 4px.$$

On the other hand, if we start with a point that satisfies the equation, we can reverse all the above steps and show that the point is equidistant from the directrix and the focus. Thus, the parabola is the graph of the equation

$$y^2 = 4px.$$ ∎

The parabola $y^2 = 4px$ has the x-axis as an axis of symmetry. It opens to the right if $p > 0$, to the left if $p < 0$ (Figs. 7–18a, b). Observe that it crosses its axis of symmetry at the vertex.

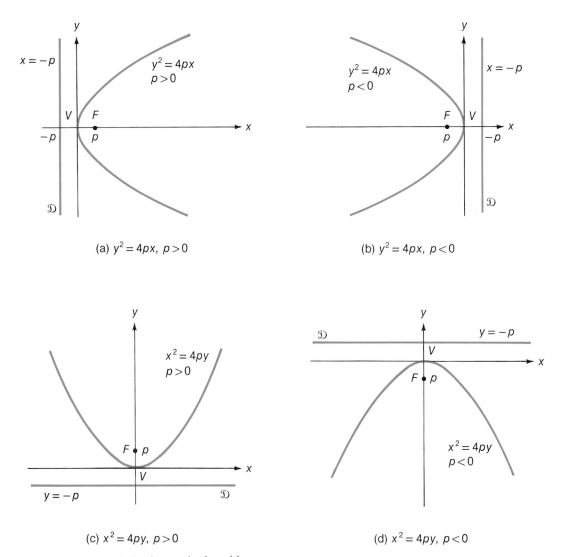

Figure 7–18 Parabolas in standard position.

7-2 The Parabola. Translation of Axes

THE PARABOLA $x^2 = 4py$

An argument similar to the proof of Theorem 7-1 can be used to show that the equation of the parabola with focus $F(0, p)$ on the y-axis and vertex at the origin is

Standard Equation with Focus on y-Axis

$$x^2 = 4py.$$

This parabola is symmetric about the y-axis. It opens upward if $p > 0$, downward if $p < 0$ (Figs. 7-18c, d).

The equation of a parabola is said to be in *standard form* if it is

$$y^2 = 4px \quad \text{or} \quad x^2 = 4py.$$

EXAMPLE 1 Locate the focus and directrix of the parabola

$$y^2 + 2x = 0.$$

Solution We rewrite the equation as

$$y^2 = -2x$$
$$y^2 = 4(-\tfrac{1}{2})x.$$

The equation is now of form $y^2 = 4px$ with $p = -\tfrac{1}{2}$. The parabola is symmetric about the x-axis and opens to the left. The focus is the point $F(-\tfrac{1}{2}, 0)$, the directrix the line $x = \tfrac{1}{2}$ (Fig. 7-19). □

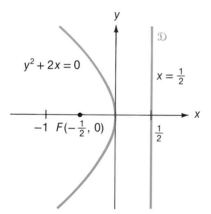

Figure 7-19 Example 1. $y^2 + 2x = 0$.

Remark

Unique Parabolic Shape

The geometrical conditions used to define the parabola show that there is essentially only one parabolic shape. In other words, if we photograph a parabola and project it onto a screen, we can make it coincide with any other parabola by choosing the proper degree of magnification. There are apparent differences in the shapes of the parabolas in the figures because only small portions of these curves are shown.

REFLECTION PROPERTY

Reflection Property   A remarkable reflection property of parabolas accounts for many of their applications. We can construct a parabolic reflector by revolving a parabola around its axis of symmetry and coating the inside of the resulting shell with silver. If a light source is put at the focus F, then the light rays are reflected along lines parallel to the axis of symmetry, causing a concentrated beam of light. (See Figs. 7–20 and 7–24.) The proof of this result is outlined in Exercise 29.

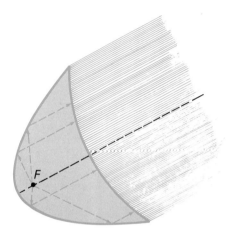

Figure 7–20 A parabolic reflector. A light source is placed at the focus of a parabolic reflector. The light rays are reflected out along lines parallel to the axis of symmetry.

This reflection property explains the use of parabolic reflectors in spotlights and automobile headlights. The same principle is used in reverse in both reflecting and radio telescopes. A modern optical telescope uses a parabolic mirror to collect light rays and concentrate them at the focus. A radio telescope uses a parabolic dish antenna to collect radio signals and concentrate them at the focus.

TRANSLATION OF AXES

The equation of a curve depends as much on its location in the xy-plane as on its shape. There is a simple technique to minimize the importance of the location and emphasize the properties of the curve. We construct a new set of axes, the $x'y'$-axes, and reduce the equation to a standard form with respect to the new axis system. Here x' and y' are new variables, not derivatives.

Let (a, b) be a point in the xy-plane. We construct a new axis system, the $x'y'$-axes, with origin at (a, b) and with the same scale and orientation as the xy-axes. (See Fig. 7–21.) Each point P now has two sets of coordinates. Relative to the xy-system it has coordinates $P(x, y)$. Relative to the $x'y'$-system it has coordinates $P(x', y')$.

7-2 The Parabola. Translation of Axes

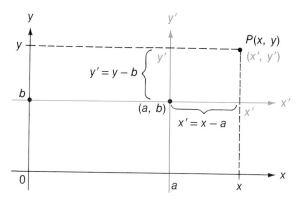

Figure 7-21 Translation of axes: $x' = x - a$, $y' = y - b$.

As we see in Figure 7-21, there is a simple relation between the coordinates in the two systems. In every case

Equations of Translation

$$x' = x - a \quad \text{and} \quad y' = y - b.$$

These equations can be used to convert equations in x and y to equations in x' and y' that have the same graph. The process is known as *translation of axes*.

Translation of axes enables us to graph equations after simplifying them by a substitution of form

$$x' = x - a, \quad y' = y - b.$$

The origin of the new system is at the point (a, b) in the original system.

EXAMPLE 2 Sketch the graph of

$$(x - 1)^2 = -4(y - 2).$$

Solution Locate the vertex and the focus.

If we make the substitution $x' = x - 1$, $y' = y - 2$, the equation becomes

$$x'^2 = -4y'.$$

The graph is a parabola that opens downward. The axis of symmetry is the y'-axis.

To sketch the graph, we first construct the $x'y'$-axis system with origin at the point $(1, 2)$. The parabola is then drawn in standard position relative to these axes. (See Fig. 7-22.)

With respect to the $x'y'$-system, the vertex is the point $(0, 0)$ and the focus the

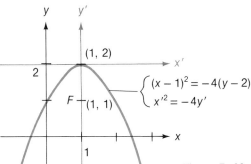

Figure 7-22 Example 2. $(x - 1)^2 = -4(y - 2)$.

point $(0, -1)$. These coordinates can be converted to xy-coordinates by use of the equations

$$x = x' + 1 \quad \text{and} \quad y = y' + 2.$$

With respect to the original xy-coordinate system, the vertex is the point $(1, 2)$, the focus is the point $(1, 1)$ and the directrix is the line $y = 3$. (See Fig. 7-22.) □

Reduction to Standard Form

Any equation of form

$$y = Ax^2 + Bx + C \quad \text{or} \quad x = Ay^2 + By + C \quad (A \neq 0)$$

can be reduced to a standard equation of a parabola by completing the square on the quadratic terms and then making a substitution

$$x' = x - a, \quad y' = y - b.$$

The technique is illustrated in Example 3.

EXAMPLE 3 Reduce the equation

$$3y = x^2 - 2x - 5$$

to a standard equation of a parabola by a translation of axes. Locate the vertex, focus, and directrix. Sketch the graph showing both sets of axes.

Solution We complete the square on the x-terms:

$$3y = x^2 - 2x + 1 - 6$$
$$3y + 6 = x^2 - 2x + 1$$
$$3(y + 2) = (x - 1)^2$$

If we make the substitutions $y' = y + 2$, $x' = x - 1$, the equation reduces to

$$x'^2 = 3y' = 4\left(\frac{3}{4}\right)y'$$

The new axis system has its origin at the point $(1, -2)$.

Exercises 7-2

	$x'y'$-Coordinate System	xy-Coordinate System
Vertex	$(0, 0)$	$(1, -2)$
Focus	$(0, \frac{3}{4})$	$(1, -\frac{5}{4})$
Directrix	Line $y' = -\frac{3}{4}$	Line $y = -\frac{11}{4}$

The parabola opens upward. Its vertex, focus, and directrix are given in both the xy-system and the $x'y'$-system in the accompanying table. The graph is shown in Figure 7–23. □

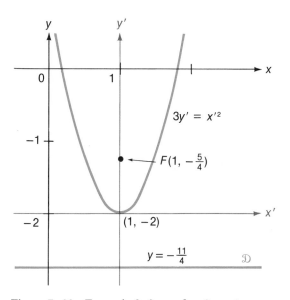

Figure 7–23 Example 3. $3y = x^2 - 2x - 5$.

Exercises 7–2

1–10. Find the equation of the parabola with directrix parallel to a coordinate axis that satisfies the given conditions.

- **1** $F(0, 1)$, $\mathcal{D}: y = -1$ **2** $F(2, 0)$, $V(1, 0)$
- **3** $V(0, 0)$, $\mathcal{D}: x = 4$ **4** $V(1, -1)$, $\mathcal{D}: y = 2$
- **5** $F(-2, -1)$, $\mathcal{D}: y = 3$ **6** $F(1, 1)$, $\mathcal{D}: x = -5$
- **7** $F(-1, 2)$, $V(2, 2)$ **8** $F(2, -3)$, $V(2, 1)$
- **9** $V(0, 0)$ with the point $(2, 1)$ on the parabola. (Two solutions.)
- **10** $V(-1, 2)$ with the point $(3, 6)$ on the parabola. (Two solutions.)

11–22. Simplify the equation by a translation of axes, if necessary, to get an equation in standard form for a parabola. Locate the vertex, focus, and directrix. Sketch the graph.

- 11 $x^2 + y = 0$
- 12 $x - 2y^2 = 0$
- 13 $(x + 1)^2 + y - 1 = 4$
- 14 $x + y^2 = 2$
- 15 $y + 3 = x^2 + 2x$
- 16 $y = 2x^2 + 12x + 15$
- 17 $x = y^2 + y$
- 18 $2x = -y^2 + 2y + 7$
- 19 $y = 3x^2 - 6x + 2$
- 20 $y = 6x^2 - 6x + 3$
- 21 $x = -4y^2 + 2y - 3$
- 22 $x = 4y^2 - 8y + 9$

23 Show that the line $x = -b/2a$ is the axis of symmetry for the parabola $y = ax^2 + bx + c$, $a \neq 0$.

24 Derive the equation for the parabola with focus $F(0, p)$ and directrix $y = -p$ for the special case where $p < 0$. Illustrate the derivation with a figure similar to Figure 7–17.

- 25 Derive the equation for the parabola with focus $F(8, 6)$ and directrix $4x + 3y = 0$. (Exercise 51 of Section 1–5 may be useful.)

- 26 Let a, b, c be real numbers, $a > 0$. Use the quadratic formula and a geometrical argument involving the parabola $y = ax^2 + bx + c$ to establish the following results:
  (a) If $b^2 - 4ac < 0$, then $ax^2 + bx + c > 0$ for all x.
  (b) If $b^2 - 4ac > 0$, then $ax^2 + bx + c < 0$ if
  $$\frac{-b - \sqrt{b^2 - 4ac}}{2a} < x < \frac{-b + \sqrt{b^2 - 4ac}}{2a},$$
  and $ax^2 + bx + c > 0$ if
  $$x < \frac{-b - \sqrt{b^2 - 4ac}}{2a} \quad \text{or} \quad x > \frac{-b + \sqrt{b^2 - 4ac}}{2a}.$$

- 27 Let $\mathcal{R}$ be the region bounded by the parabola $x^2 = 4py$ and the horizontal line through its focus.
  (a) Find the volume of the solid obtained by revolving $\mathcal{R}$ about the y-axis.
  (b) Find the centroid of $\mathcal{R}$.

- 28 A closed region $\mathcal{R}$ is formed by drawing a horizontal line through the focus of a parabola $y = ax^2$. Find the equation of the parabola if the centroid of $\mathcal{R}$ is $(0, -9)$.

Reflection Property

29 (*Reflection Property*) Imagine that the parabola $y^2 = 4px$ (where $p > 0$) is a mirror and that a light source has been placed at the focus $F(p, 0)$. (See Fig. 7–24.) The following facts are known from elementary physics:
  (1) A light ray that strikes the parabola at point $P_0(x_0, y_0)$ is reflected as if it had struck the line tangent to the parabola at P_0.
  (2) The angle of incidence α equals the angle of reflection β. (See Fig. 7–24a.)

7–3 The Ellipse

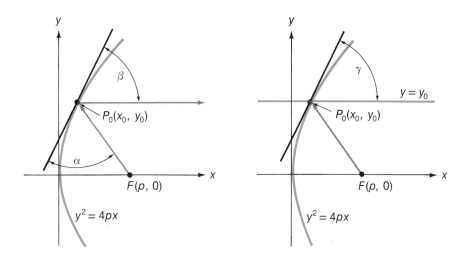

(a) The angle of incidence is equal to the angle of reflection

(b) γ is the angle between the line $y = y_0$ and the tangent line.

Figure 7–24 Exercise 29. The reflection property of the parabola.

The following steps show that the light rays from F are reflected from the parabola in lines parallel to the axis of symmetry (the x-axis).

(a) Use Exercise 53 of Section 1–5 to find $\tan \alpha$ which, by (2), equals $\tan \beta$.

(b) Let γ be the angle between the tangent line at P_0 and the line $y = y_0$. (See Fig. 7–24b.) Show that $\tan \beta = \tan \gamma$, so that $\beta = \gamma$.

Parabola by Paper Folding

30 Draw a parabola $x^2 = 4py$ with focus F and directrix $\mathcal{D}$. Fold the paper so that F is directly above the point $P(a, b)$ on $\mathcal{D}$. Prove that the line of the paper fold is tangent to the parabola. (*Hint:* The line of the paper fold is the perpendicular bisector of the line segment connecting P and F. Find the point at which the line of the paper fold intersects the parabola.) This result allows us to outline a parabola by a set of tangent lines.

7–3 The Ellipse

The ellipse is a curve that describes the path of planets around the sun. This fact about ellipses was discovered empirically by Johann Kepler (1571–1630) and proved theoretically by Isaac Newton in one of the first principal applications of the calculus. The ellipse is defined by the following conditions.

DEFINITION

Ellipse

Let F and F' be two fixed points. Let $2a$ be a positive number greater than the distance between F and F'. The set of all points P satisfying the condition

$$|FP| + |F'P| = 2a$$

is called an *ellipse*. (See Fig. 7–25.)

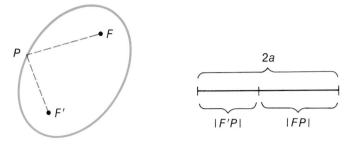

Figure 7-25 The ellipse. The sum of the distances from P to the foci F and F' is the constant $2a$.

The points F and F' are called the *foci* of the ellipse. The definition states that a point P is on the ellipse if and only if the sum of its distances from the two foci equals the constant $2a$.

We shall derive the equation of the ellipse for the special case in which the foci F and F' are on the x-axis with the origin midway between them.

THEOREM 7-2

Standard Equation

Let the foci F and F' be the points $F(c, 0)$ and $F'(-c, 0)$, where $c > 0$. Let $2a$ be a number greater than $2c$. The equation of the ellipse defined by the condition $|FP| + |F'P| = 2a$ is

$$\frac{x^2}{a^2} + \frac{y^2}{b^2} = 1$$

where $b^2 = a^2 - c^2$.

Proof If $P(x, y)$ is a point on the ellipse, then (see Fig. 7-26)

$$|FP| + |F'P| = 2a$$
$$\sqrt{(x - c)^2 + y^2} + \sqrt{(x + c)^2 + y^2} = 2a.$$

We rewrite this equation as

$$\sqrt{(x + c)^2 + y^2} = 2a - \sqrt{(x - c)^2 + y^2},$$

square both sides, and simplify, obtaining

$$a\sqrt{(x - c)^2 + y^2} = a^2 - cx.$$

We now square both sides of this new equation and simplify, obtaining

$$(a^2 - c^2)x^2 + a^2 y^2 = a^2(a^2 - c^2)$$

which reduces to

$$\frac{x^2}{a^2} + \frac{y^2}{a^2 - c^2} = 1.$$

7–3 The Ellipse

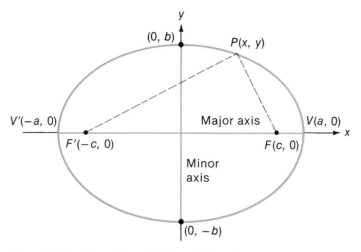

Figure 7–26 The ellipse $x^2/a^2 + y^2/b^2 = 1$.

All the steps in the above argument can be reversed. (You are invited to try it.) Thus, $P(x, y)$ is on the ellipse if and only if

$$\frac{x^2}{a^2} + \frac{y^2}{a^2 - c^2} = 1.$$

Recall that $a > c > 0$, so that $a^2 - c^2 > 0$. If we let $b = \sqrt{a^2 - c^2}$, then $a^2 - c^2 = b^2$, and the equation of the ellipse reduces to

$$\frac{x^2}{a^2} + \frac{y^2}{b^2} = 1 \quad \text{where } a > b > 0. \blacksquare$$

PROPERTIES OF THE ELLIPSE $x^2/a^2 + y^2/b^2 = 1$.

Properties of Ellipse

The following properties can be obtained by use of the methods developed in the first section of this chapter. (See Fig. 7–26.)

(1) The ellipse is symmetric about both axes and the origin.
(2) The projection on the x-axis is $\{x \mid -a \leq x \leq a\}$, the projection on the y-axis is $\{y \mid -b \leq y \leq b\}$. The x-intercepts are $V(a, 0)$, and $V'(-a, 0)$. The y-intercepts are $(0, b)$ and $(0, -b)$.
(3) The portion of the graph in Quadrant I is the graph of a decreasing function that is concave downward.
(4) The ellipse has horizontal tangent lines at its y-intercepts and vertical tangent lines at its x-intercepts.

Observe that we can sketch an ellipse easily once we locate its axes of symmetry and its intercepts with these axes.

TERMINOLOGY

Vertices, Major Axis, Minor Axis, Center

The points $V(a, 0)$ and $V'(-a, 0)$ are called the *vertices* of the ellipse $x^2/a^2 + y^2/b^2 = 1$. The line segment connecting V and V' is called the *major axis*. Similarly, the line segment connecting $(0, b)$ and $(0, -b)$ is called the *minor axis*. The point where these axes cross is called the *center*. Observe that the major axis is longer than the minor axis and that the foci are on the major axis. Both axes are portions of the axes of symmetry.

EXAMPLE 1 Find the major and minor axes, the vertices, the foci, and the ends of the minor axis of the ellipse

$$4x^2 + 9y^2 = 36.$$

Solution We rewrite the equation in the form

$$\frac{x^2}{3^2} + \frac{y^2}{2^2} = 1.$$

Then $a = 3$, $b = 2$. The major axis is on the x-axis, the minor axis is on the y-axis. The vertices are $V'(-3, 0)$ and $V(3, 0)$. The ends of the minor axis are $(0, -2)$ and $(0, 2)$.

To find the foci we recall that

$$c^2 = a^2 - b^2 = 9 - 4 = 5,$$

so that

$$c = \sqrt{5}.$$

The foci are $F(\sqrt{5}, 0)$ and $F'(-\sqrt{5}, 0)$. (See Fig. 7–27.) □

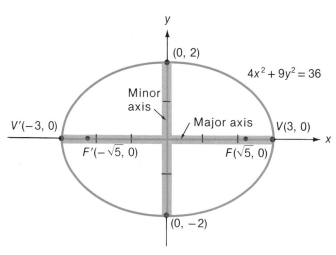

Figure 7–27 Example 1. $4x^2 + 9y^2 = 36$.

7–3 The Ellipse

THE ELLIPSE WITH FOCI ON THE y-AXIS

If the foci are the points $F(0, c)$ and $F'(0, -c)$ on the y-axis, an argument like the proof of Theorem 7–2 shows that the ellipse is the graph of

Standard Equation of Ellipse with Foci on y-axis

$$\frac{x^2}{b^2} + \frac{y^2}{a^2} = 1, \quad a > b > 0$$

where $a^2 = b^2 + c^2$. In this case the vertices are $V(0, a)$ and $V'(0, -a)$, the major axis is on the y-axis, the minor axis on the x-axis. (See Fig. 7–28.)

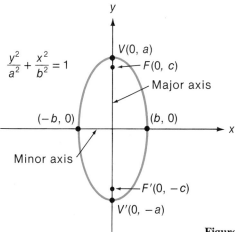

Figure 7–28 The ellipse $y^2/a^2 + x^2/b^2 = 1$.

There should be no danger of confusing the two forms. In each case $a > b > 0$ and $a^2 = b^2 + c^2$. To find the major axis, we simply locate the x- and y-intercepts. The intercepts furthest from the center are the vertices. They are on the major axis.

For example, the ellipse

$$\frac{x^2}{5^2} + \frac{y^2}{7^2} = 1$$

has x-intercepts $(\pm 5, 0)$ and y-intercepts $(0, \pm 7)$. The major axis is on the y-axis.

REDUCTION TO STANDARD FORM

The equation of an ellipse is said to be in *standard form* if it is

$$\frac{x^2}{a^2} + \frac{y^2}{b^2} = 1 \quad \text{or} \quad \frac{x^2}{b^2} + \frac{y^2}{a^2} = 1.$$

If the major axis of an ellipse is parallel to a coordinate axis, then the equation of the ellipse can be reduced to standard form by a translation of axes.

EXAMPLE 2   Sketch the graph of

$$25x^2 - 100x + 9y^2 - 18y = 116.$$

Solution   We complete the square on the terms involving x and on the terms involving y:

$$25(x^2 - 4x \quad) + 9(y^2 - 2y \quad) = 116$$
$$25(x^2 - 4x + 4) + 9(y^2 - 2y + 1) = 116 + 25 \cdot 4 + 9 \cdot 1$$
$$25(x - 2)^2 + 9(y - 1)^2 = 225$$
$$\frac{(x-2)^2}{3^2} + \frac{(y-1)^2}{5^2} = 1.$$

If we now make the translation of axes defined by $x' = x - 2$, $y' = y - 1$, the equation reduces to

$$\frac{x'^2}{3^2} + \frac{y'^2}{5^2} = 1.$$

	$x'y'$-Coordinate System	xy-Coordinate System
Major axis	y'-Axis	Vertical line $x = 2$
Minor axis	x'-Axis	Horizontal line $y = 1$
Center	$(0, 0)$	$(2, 1)$
Vertices	$V'(0, -5)$, $V(0, 5)$	$V'(2, -4)$, $V(2, 6)$
Foci	$F'(0, -4)$, $F(0, 4)$	$F'(2, -3)$, $F(2, 5)$
End of minor axis	$(-3, 0)$ $(3, 0)$	$(-1, 1)$, $(5, 1)$

The major axis is on the y'-axis. The origin of the new system is at the point (2, 1). The vertices, foci, and the like are given in the accompanying table. The graph is pictured in Figure 7–29. □

Reduction of Special Second Degree Equation to Standard Form

The graph of the second-degree equation

$$Ax^2 + Cy^2 + Dx + Ey + F = 0, \quad A \neq C,$$

where A and C are both positive or both negative, is an ellipse, a single point, or the empty set (there is no graph).

To show that this is the case, we complete the square on the x- and the y-terms and perform a translation of axes, reducing the equation to one of the forms

$$\frac{x'^2}{a^2} + \frac{y'^2}{b^2} = 1, \quad \frac{x'^2}{a^2} + \frac{y'^2}{b^2} = 0, \quad \text{or} \quad \frac{x'^2}{a^2} + \frac{y'^2}{b^2} = -1.$$

In the first case, the graph is an ellipse. In the second, it is a single point. In the third, there is no graph.

7-3 The Ellipse

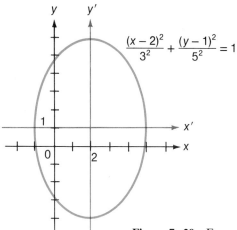

Figure 7-29 Example 2. $25x^2 - 100x + 9y^2 - 18y = 116$.

REFLECTION PROPERTY

Reflection Property

The ellipse has an interesting reflection property. If a light ray emanates from one focus F and strikes the side of the ellipse at a point P, it will be reflected to the other focus F'. (See Fig. 7-30a.) Furthermore, from the definition, it follows that all such rays travel the same total distance from F to P to F'.

This reflection property is the basis for the "whispering galleries" in some public buildings. A whispering gallery is a room with a ceiling that is a portion of an ellipsoid (a shell formed by revolving an ellipse about its major axis). One person stands with his head at focus F and whispers. All the sound waves that strike the ceiling are then reflected to the other focus F'. Since these waves all travel the same distance, they arrive at F' in phase. Thus, a person standing with his head at F' can hear every whisper from F even though the points may be many feet apart and the sounds cannot be heard between F and F'. (See Fig. 7-30b.)

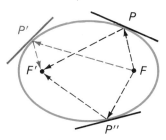

(a) Rays from one focus are reflected to the other focus

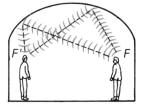

(b) Cross section of a whispering gallery

Figure 7-30 Reflection property of the ellipse.

Exercises 7-3

1-6. Find the equation of the ellipse that satisfies the given conditions.

- 1 Vertices $(\pm 10, 0)$, focus $(8, 0)$
- 2 Vertex $(0, -5)$, foci $(0, \pm 3)$
- 3 Vertices $(\pm 3, 5)$, length of minor axis $= 2$
- 4 Vertices $(1, \pm 6)$, passes through point $(4, 2)$
- 5 Foci $(2, \pm 2)$, length of minor axis $= 6$
- 6 Foci $(5, 1)$ and $(1, 1)$, length of major axis $= 10$

7-20. Simplify the equation by a translation of axes, if necessary, and reduce it to a standard equation of an ellipse. Find the major and minor axes, the center, the vertices, and the foci. Sketch the graph.

- 7 $\dfrac{x^2}{4} + \dfrac{y^2}{9} = 1$
- 8 $\dfrac{x^2}{25} + \dfrac{y^2}{4} = 1$
- 9 $x^2 + 9y^2 = 36$
- 10 $4x^2 + y^2 = 4$
- 11 $36x^2 + 4y^2 = 144$
- 12 $25x^2 + 4y^2 = 100$
- 13 $9x^2 - 18x + 4y^2 - 16y = 11$
- 14 $x^2 + 2x + 2y^2 - 4y = 13$
- 15 $x^2 - 4x + 3y^2 + 6y - 2 = 0$
- 16 $9x^2 + 36x + 5y^2 + 10y = 4$
- 17 $4x^2 - 8x + 3y^2 - 30y = 2$
- 18 $16x^2 + 64x + 25y^2 + 50y + 8 = 0$
- 19 $3x^2 + 6x + 5y^2 - 20y + 7 = 0$
- 20 $16x^2 - 32x + y^2 + 4y = 5$

- 21 The planet Earth moves in an elliptical orbit that has the sun as a focus. The maximum distance from the earth to the sun is 94.5 million miles and the minimum distance is 91.3 million miles (approximate). Set up a coordinate system that has the ellipse in standard position with the sun at $F(c, 0)$. Find the equation of the ellipse and sketch it.

- 22 Derive the equation of the ellipse in standard position with foci on the y-axis.

23 Prove that the equation of the line tangent to the graph of the ellipse $x^2/a^2 + y^2/b^2 = 1$ at the point (x_0, y_0) is the graph of $x_0 x/a^2 + y_0 y/b^2 = 1$.

24 A pool table has the shape of a perfect ellipse. Balls are placed at the foci F and F'. One ball is then struck with great force towards the side of the table. Use the reflection property to explain why (in theory, at least) the balls will continually strike each other at the foci.

25 A room is constructed by building curved vertical walls around an elliptical base. The inside walls are mirrored and a candle is placed at a focus. Describe the effect on an observer located (a) at the other focus; (b) at a point different from the other focus.

Exercises 7–3

Area of Ellipse

26 (*Area of an Ellipse*) Set up integrals for the area of the ellipse $b^2x^2 + a^2y^2 = a^2b^2$ and for the area of the circle $x^2 + y^2 = a^2$. Without evaluating the integrals, show that the area of the ellipse is b/a times the area of the circle. Use this result to show that the area of the ellipse is

$$\text{Area of ellipse} = ab\pi.$$

27 (*Reflection Property*) Prove the reflection property illustrated in Figure 7–30. (*Hint:* Modify the argument outlined in Exercise 29 of Section 7–2.)

28 (*Construction Methods*) Two of the methods for constructing ellipses are illustrated in Figures 7–31a, b. Explain why the resulting curves must be ellipses.
  (a) Fasten a string of length L to a board at points F and F', where the distance from F to F' is less than L. (See Fig. 7–31a.) Place the point of a pencil in

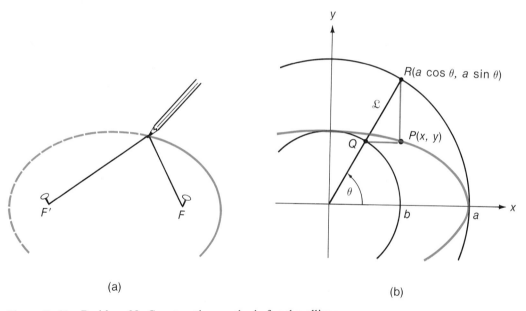

(a)   (b)

Figure 7–31 Problem 28. Construction methods for the ellipse.

the loop of the string, and trace it around the board, keeping the string taut. Show that the point of the pencil traces an ellipse.
  (b) Draw two concentric circles of radii a and b, where $a > b$, with centers at the origin. A ray $\mathcal{L}$ from the origin will strike the inner circle at a point Q and the outer circle at a point R. Let $P(x, y)$ be the point at which the horizontal line

through Q intersects the vertical line through R. Show that the set of all such points $P(x, y)$ is the ellipse

$$\frac{x^2}{a^2} + \frac{y^2}{b^2} + 1.$$

(*Hint:* Let θ be the angle from the positive x-axis to the ray $\mathscr{L}$. Express the coordinates of Q, R, and P in terms of θ. Use the Pythagorean identity $\cos^2 \theta + \sin^2 \theta = 1$.)

7–4 The Hyperbola

The last conic section is a *hyperbola*, a pair of curves that can be defined by geometrical conditions analogous to those used to define the ellipse.

DEFINITION

Hyperbola

Let F and F' be two fixed points. Let $2a$ be a positive real number that is less than the distance $|FF'|$. The set of all points P such that

$$|FP| - |F'P| = \pm 2a$$

is called a *hyperbola*.

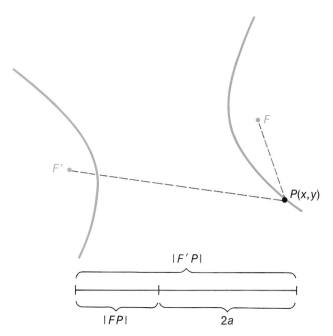

Figure 7–32 The hyperbola. A point P is on the hyperbola if, and only if, it is $2a$ units closer to one focus than the other.

7–4 The Hyperbola

The points F and F' are called the *foci* of the hyperbola. The definition states that P is on the hyperbola if and only if P is exactly $2a$ units closer to one focus than the other. (See Fig. 7–32.)

As we did with the ellipse, we derive the equation of a hyperbola for the special case where the foci F and F' are on the x-axis with the origin midway between them.

THEOREM 7–3   Let $F(c, 0)$ and $F'(-c, 0)$ be two fixed points on the x-axis, with $c > 0$. Let $2a$ be a positive number less than $2c$. The hyperbola defined as the set of all points $P(x, y)$ satisfying

$$|FP| - |F'P| = \pm 2a$$

has the equation

Equation in Standard Form

$$\frac{x^2}{a^2} - \frac{y^2}{b^2} = 1,$$

where $a^2 + b^2 = c^2$.

Proof   If $P(x, y)$ is on the hyperbola, then

$$|FP| - |F'P| = \pm 2a,$$

so that

$$\sqrt{(x - c)^2 + y^2} - \sqrt{(x + c)^2 + y^2} = \pm 2a.$$

(See Fig. 7–33.) If we follow the same basic steps as in the derivation of the equation

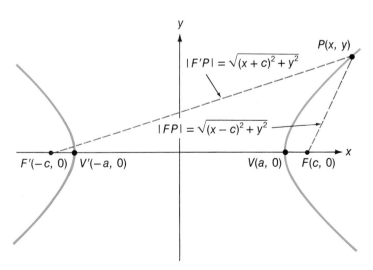

Figure 7–33   Theorem 7–3. The hyperbola $x^2/a^2 - y^2/b^2 = 1$.

of the ellipse, we find that

$$\frac{x^2}{a^2} - \frac{y^2}{c^2 - a^2} = 1.$$

Again, all the steps can be reversed, so the equation of the hyperbola is

$$\frac{x^2}{a^2} - \frac{y^2}{c^2 - a^2} = 1.$$

To simplify this equation, recall that $c > a > 0$, so that $c^2 - a^2 > 0$. If we let $b = \sqrt{c^2 - a^2}$, the equation reduces to

$$\frac{x^2}{a^2} - \frac{y^2}{b^2} = 1. \blacksquare$$

PROPERTIES OF THE HYPERBOLA

The hyperbola $x^2/a^2 - y^2/b^2 = 1$ is shown in Figure 7–34. The following properties can be obtained by the techniques discussed in Section 7–1.

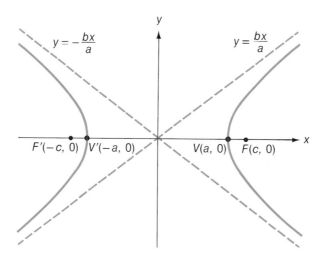

Figure 7–34 The hyperbola $x^2/a^2 - y^2/b^2 = 1$.

Properties of Hyperbola

(*1*) *Projections* The projections on the coordinate axes are

x-projection = $\{x \mid x \leq -a \text{ or } x \geq a\}$,
y-projection = $\{y \mid y \text{ is real}\}$.

The graph is isolated in two distinct parts of the xy-plane.

7-4 The Hyperbola

(2) *Symmetry* The hyperbola is symmetric about both axes and the origin. Thus, the entire graph can be obtained from the portion in Quadrant I by using the symmetry.

(3) *Properties Determined from Derivatives* The graph has vertical tangent lines at the points at which it crosses the x-axis. The portion of the curve in Quadrant I is the graph of an increasing function that is concave downward.

(4) *Asymptotes* The portion of the hyperbola in Quadrant I has the line $y = bx/a$ as an oblique asymptote. It follows from the symmetry that the line $y = bx/a$ is an oblique asymptote in Quadrants I and III, and the line $y = -bx/a$ is an oblique asymptote in quadrants II and IV.

To establish (4) we observe that the portion of the hyperbola in Quadrant I is the graph of

$$y = \frac{b\sqrt{x^2 - a^2}}{a}, \quad x \geq a.$$

Then

$$\lim_{x \to \infty} \left[\frac{b\sqrt{x^2 - a^2}}{a} - \frac{bx}{a} \right] = \frac{b}{a} \lim_{x \to \infty} \left[\sqrt{x^2 - a^2} - x \right]$$

$$= \frac{b}{a} \lim_{x \to \infty} \left[\frac{(\sqrt{x^2 - a^2} - x)(\sqrt{x^2 - a^2} + x)}{\sqrt{x^2 - a^2} + x} \right]$$

$$= \frac{b}{a} \lim_{x \to \infty} \frac{x^2 - a^2 - x^2}{\sqrt{x^2 - a^2} + x}$$

$$= \frac{b}{a} \lim_{x \to \infty} \frac{-a^2}{\sqrt{x^2 - a^2} + x} = 0.$$

TERMINOLOGY

Vertices, Transverse Axis, Conjugate Axis

The points $V(a, 0)$ and $V'(-a, 0)$ at which the hyperbola $x^2/a^2 - y^2/b^2 = 1$ crosses the x-axis are called the *vertices* of the hyperbola. The midpoint between the vertices—the point at which the two axes of symmetry cross—is the *center*. The line segment that connects the vertices is the *transverse axis*.

The line segment that connects the points $(0, b)$ and $(0, -b)$ is called the *conjugate axis* of the hyperbola. It should be clear that a hyperbola does not intersect its conjugate axis. (See Fig. 7–35.)

We can make a quick and accurate sketch of a hyperbola once we locate its vertices and asymptotes. Observe first that the vertices of the hyperbola

$$\frac{x^2}{a^2} - \frac{y^2}{b^2} = 1$$

are $V(a, 0)$ and $V'(-a, 0)$ on the x-axis.

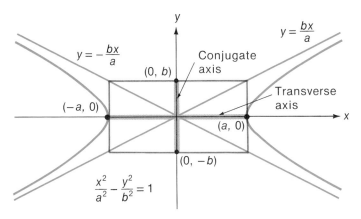

Figure 7-35 The asymptotes pass through the corners of the rectangle.

Construction of Asymptotes

The asymptotes can be drawn by the following method. Draw a rectangle with horizontal and vertical sides through the vertices ($\pm a$, 0) and the ends of the conjugate axis (0, $\pm b$). The asymptotes are the diagonal lines through the corners of this rectangle (Fig. 7-35).

THE HYPERBOLA $y^2/a^2 - x^2/b^2 = 1$.

By an argument similar to the proof of Theorem 7-3, we can show that the hyperbola with foci $F(0, c)$ and $F'(0, -c)$, where $c > 0$, has the equation

Standard Equation with Foci on y-Axis

$$\frac{y^2}{a^2} - \frac{x^2}{b^2} = 1$$

where $b = \sqrt{c^2 - a^2}$. (See Fig. 7-36.) The vertices are the points $V(0, a)$ and $V'(0, -a)$. The transverse axis is on the y-axis, the conjugate axis on the x-axis. The asymptotes are the lines $y = \pm ax/b$.

Remark There is no relation between the sizes of a and b in these formulas. This is in contrast to the situation for ellipses, where a is always the larger of the two numbers.

Observe, however, that the transverse axis can be determined from the equation

$$\frac{x^2}{a^2} - \frac{y^2}{b^2} = 1 \quad \text{or} \quad \frac{y^2}{a^2} - \frac{x^2}{b^2} = 1.$$

We simply check to see which variable has the positive coefficient.

EXAMPLE 1 Sketch the graphs of the hyperbolas $4x^2 - 9y^2 = 36$ and $4x^2 - 9y^2 = -36$.

7-4 The Hyperbola

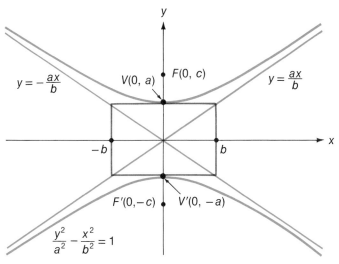

Figure 7-36 The hyperbola $y^2/a^2 - x^2/b^2 = 1$.

Solution The first equation can be written as

$$\frac{x^2}{3^2} - \frac{y^2}{2^2} = 1 \quad (a = 3, b = 2).$$

The transverse axis is on the x-axis; the vertices are $(\pm 3, 0)$; the asymptotes are the lines $y = \pm 2x/3$.

The second equation can be written as

$$\frac{y^2}{2^2} - \frac{x^2}{3^2} = 1 \quad (a = 2, b = 3).$$

The transverse axis is on the y-axis; the vertices are $(0, \pm 2)$; the asymptotes are $y = \pm 2x/3$.

The two hyperbolas are shown in Figure 7-37. □

REDUCTION TO STANDARD FORM

The equation of a hyperbola is said to be in *standard form* if it is

$$\frac{x^2}{a^2} - \frac{y^2}{b^2} = 1 \quad \text{or} \quad \frac{y^2}{a^2} - \frac{x^2}{b^2} = 1.$$

If the transverse axis of a hyperbola is parallel to a coordinate axis, then the equation of the hyperbola can be reduced to standard form by a translation of axes.

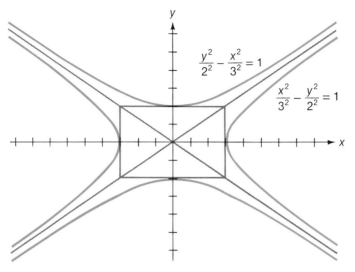

Figure 7–37 Example 1. The hyperbolas $4x^2 - 9y^2 = \pm 36$

EXAMPLE 2   Sketch the graph of

$$x^2 - 2x - 9y^2 - 36y = 26$$

showing the vertices, asymptotes, and foci.

Solution   We complete the square on the x-terms and on the y-terms:

$$(x^2 - 2x \quad) - 9(y^2 + 4y \quad) = 26$$
$$(x^2 - 2x + 1) - 9(y^2 + 4y + 4) = 26 + 1 - 9 \cdot 4$$
$$(x - 1)^2 - 9(y + 2)^2 = -9$$
$$-\frac{(x - 1)^2}{9} + \frac{(y + 2)^2}{1} = 1$$
$$\frac{y'^2}{1^2} - \frac{x'^2}{3^2} = 1$$

where $x' = x - 1$ and $y' = y + 2$.

This last equation is in standard form with respect to the $x'y'$-axis system.

The hyperbola opens along the y'-axis. Its vertices are the points at which $x' = 0$, $y' = \pm 1$. The asymptotes are the lines $y' = \pm x'/3$. To find the foci, we recall that $c^2 = a^2 + b^2 = 1 + 9 = 10$, so that $c = \sqrt{10}$. The foci are located at the points $(0, \pm\sqrt{10})$ in the $x'y'$-system. The conversion of these points and lines to xy-coordinates is given in the accompanying chart. The graph is shown in Figure 7–38. □

7–4 The Hyperbola

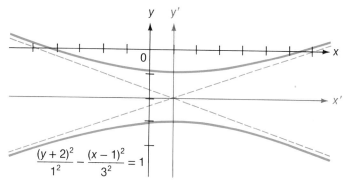

Figure 7–38 Example 2. $x^2 - 2x - 9y^2 - 36y = 26$.

	$x'y'$-Coordinate System	xy-Coordinate System
Transverse axis	y'-Axis	Vertical line $x = 1$
Conjugate axis	x'-Axis	Horizontal line $y = -2$
Center	$(0, 0)$	$(1, -2)$
Vertices	$V'(0, -1)$, $V(0, 1)$	$V'(1, -3)$, $V(1, -1)$
Foci	$F'(0, -\sqrt{10})$, $F(0, \sqrt{10})$	$F'(1, -2 - \sqrt{10})$, $F(1, -2 + \sqrt{10})$
Asymptotes	$y' = \pm\dfrac{x'}{3}$	$y + 2 = \pm\dfrac{x - 1}{3}$

Reduction of Second Degree Equation to Standard Form

The graph of the second-degree equation

$$Ax^2 + Cy^2 + Dx + Ey + F = 0$$

where A and C have opposite signs is either a hyperbola or a pair of intersecting lines.

To show that this is true, we complete the square on the x- and the y-terms and perform a translation of axes, reducing the equation to one of the forms

$$\frac{x'^2}{a^2} - \frac{y'^2}{b^2} = 1, \qquad \frac{x'^2}{a^2} - \frac{y'^2}{b^2} = -1, \quad \text{or} \quad \frac{x'^2}{a^2} - \frac{y'^2}{b^2} = 0.$$

In the first two cases, the graph is a hyperbola. In the third case, it is a pair of intersecting lines.

APPLICATION

The hyperbola is the basis for several methods of range-finding. Since all these methods involve similar principles, we discuss one that was used for artillery "spotting" by various armies about the time of World War I.

Because artillery shells travel for several miles, the explosions could not be seen from the command post. Even though the trajectories were worked out theoretically, the first few shells usually missed the target. It was necessary to determine where these shells exploded so that the aim could be corrected.

Observers were stationed with telephones at various points F, F', F'', and so on, as shown in Figure 7–39. Each observer telephoned a signal to the command post at the

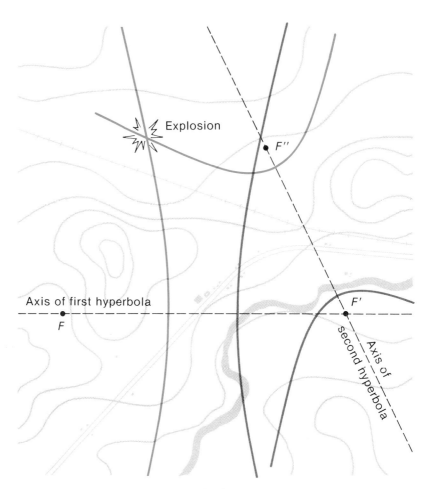

Figure 7–39 Application to range finding.

exact instant that he heard the explosion of the shell. The time lag between the signals was then used to determine how much closer one observer was to the explosion than another. (Sound travels at the approximate rate of 1100 ft/sec.)

Suppose, for example, that observation points F and F' were 4 miles apart and that

Exercises 7–4

P, the point of explosion, was 1 mile closer to F than to F'. This located P on a branch of a hyperbola (with foci F and F', $c = 2$ and $a = \frac{1}{2}$). A similar analysis using F' and F'' located P on a branch of a second hyperbola. The point at which the two curves intersect determined the location of the explosion.

Exercises 7–4

1–6. Find the equation of the hyperbola that satisfies the given conditions.

●1 Foci $(\pm 5, 0)$, vertex $(-3, 0)$

2 Focus $(0, 13)$, vertices $(0, \pm 12)$

●3 Vertex $(4, 0)$, asymptotes $2y = \pm x$

4 Foci $(2, 6)$ and $(2, -4)$, asymptote $4(y - 1) = 3(x - 2)$

●5 Vertices $(2, 2)$ and $(-6, 2)$, passes through the point $(6, 4)$

6 Foci $(1, 0)$ and $(1, 4)$, passes through the point $(13, 9)$.

7–20. Simplify each equation by a translation of axes, if necessary, and reduce it to a standard equation of a hyperbola. Locate the transverse and conjugate axes, the center, the vertices, and the foci. Sketch the graph.

●7 $\dfrac{x^2}{9} - \dfrac{y^2}{4} = 1$ 8 $\dfrac{x^2}{9} - \dfrac{y^2}{16} = -1$

●9 $x^2 - 9y^2 = 36$ 10 $9x^2 - y^2 = 9$

●11 $4x^2 - 36y^2 + 144 = 0$ 12 $25x^2 - 4y^2 + 100 = 0$

●13 $x^2 - 4x - 2y^2 + 4y = 14$ 14 $9x^2 - 18x - 4y^2 - 16y = 43$

●15 $9y^2 - 36y - x^2 + 12x = 36$ 16 $4y^2 + 40y - x^2 + 4x + 32 = 0$

●17 $-16x^2 - 32x + 3y^2 + 30y + 50 = 0$ 18 $-25x^2 + 100x + 4y^2 + 24y = 89$

●19 $9x^2 + 36x - 5y^2 + 10y + 6 = 0$ 20 $4x^2 - 8x - 10y^2 - 20y = 106$

21 Let $\mathscr{C}$ be the graph of the equation $(x - 2)(y - 3) = 1$. Perform a translation of axes to accomplish the following results.
    ●(a) $\mathscr{C}$ will have the x'- and y'-axes as asymptotes.
    ●(b) $\mathscr{C}$ will pass through the origin of the new system.
    (c) $\mathscr{C}$ will pass through the points $(0, 0)$ and $(-\frac{1}{2}, 1)$ in the new coordinate system.

●22 Find all hyperbolas that have the lines $y = \pm 2x$ for asymptotes.

23 (*Equilateral Hyperbolas*) A hyperbola with perpendicular asymptotes is called

Equilateral Hyperbola

an *equilateral hyperbola*. Find the equation of the equilateral hyperbola in standard position with vertex
- (a) $V(a, 0)$ on the x-axis;
- (b) $V(0, a)$ on the y-axis.

24 (a) Show that a hyperbola with transverse axis parallel to a coordinate axis must have an equation of form

$$\frac{(x-h)^2}{a^2} - \frac{(y-k)^2}{b^2} = 1 \quad \text{or} \quad \frac{(y-k)^2}{a^2} - \frac{(x-h)^2}{b^2} = 1.$$

(b) Show that the asymptotes of the hyperbolas in (a) can be found by setting the left-hand sides of the equations equal to zero and solving for y in terms of x.

•25 The hyperbola, in common with all other curves, can be described by several different sets of parametric equations. Show that each set of parametric equations defines part or all of a hyperbola. Sketch the graph showing the points that correspond to the particular values of t.
- (a) $x = -\sqrt{t^2 - 1}$, $y = t$, $t \geq 1$; points where $t = 1, 2$.
- (b) $x = 3 \sec \theta$, $y = 2 \tan \theta$, $-\pi/2 < \theta < \pi/2$; points where $t = -\pi/4, 0, \pi/4$. (Angles are measured in radians.)

26 Prove that the equation of the tangent line to the hyperbola $x^2/a^2 - y^2/b^2 = 1$ at the point (x_0, y_0) is

$$\frac{x_0 x}{a^2} - \frac{y_0 y}{b^2} = 1.$$

27 Derive the equation of the hyperbola in standard position with foci $F'(0, -c)$ and $F(0, c)$ on the y-axis.

The following exercise illustrates the principles of the LORAN system of navigation used by the United States Navy.

•28 A coordinate system in which units are measured in hundreds of miles is superimposed on a map. Radio stations are located at points $F(10, 0)$, $F'(-10, 0)$, and $F''(-10, -10)$. A ship is equipped with an electronic device that measures the time lags between signals sent simultaneously from the three radio stations. The navigational officer finds that the ship is 12 units closer to F' than to F and 6 units closer to F' than to F''. Locate the ship graphically.

7–5* Eccentricity. Sections of a Cone

If we hold the foci fixed at $F(c, 0)$ and $F'(-c, 0)$ and let the number a vary over the real numbers greater than c, the resulting ellipses vary from cigar-shaped to almost circular. (See Fig. 7–40.) In the first case, the ratio c/a is close to 1; in the second it is close to 0.

Eccentricity

The ratio c/a, called the *eccentricity* of an ellipse or hyperbola, determines the basic proportions of the conic. If we change a and c, but preserve the ratio c/a, we

7-5 Eccentricity. Sections of a Cone

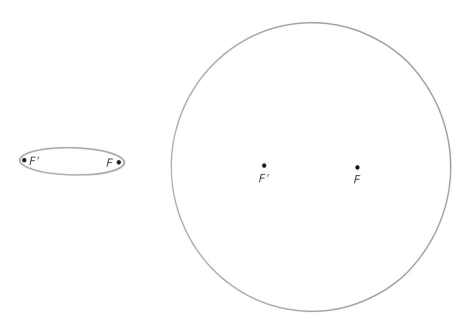

(a) c/a close to 1 (b) c/a close to zero

Figure 7-40. Ellipses with same foci but different eccentricity.

obtain a new conic that is *similar* to the original one in the sense that it either is identical to it in shape or can be enlarged or reduced so as to be identical.

We shall establish the result for ellipses in standard position with foci on the x-axis. The proof for the hyperbola is left for you to work out. (See Exercise 15.)

THEOREM 7-4 Let $x^2/a^2 + y^2/b^2 = 1$ and $x^2/a'^2 + y^2/b'^2 = 1$ be ellipses with foci $(\pm c, 0)$ and $(\pm c', 0)$, respectively. The ellipses are similar if and only if $c/a = c'/a'$.

Proof **Part 1** Suppose that $c/a = c'/a'$. If $a'/a = k > 0$, then $a' = ka$ and $c' = kc$.

Similar Conics Have the Same Eccentricity

Let the foci of the first ellipse be $F_1(-c, 0)$ and $F_2(c, 0)$, and those of the second ellipse be $F'_1(-c', 0)$ and $F'_2(c', 0)$. (See Fig. 7-41.) Let P and P' be the points $P(x, y)$ and $P'(kx, ky)$.

If the point $P(x, y)$ is on the first ellipse, then

$$|F_1P| + |F_2P| = 2a$$
$$\sqrt{(x + c)^2 + y^2} + \sqrt{(x - c)^2 + y^2} = 2a.$$

On multiplying this equation by k, we get

$$\sqrt{(kx + kc)^2 + (ky)^2} + \sqrt{(kx - kc)^2 + (ky)^2} = 2ka$$
$$\sqrt{(kx + c')^2 + (ky)^2} + \sqrt{(kx - c')^2 + (ky)^2} = 2a'$$
$$|F'_1P'| + |F'_2P'| = 2a',$$

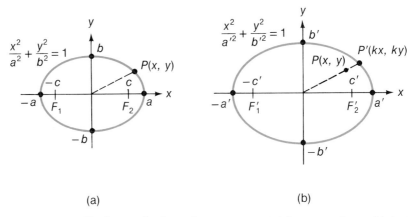

Figure 7-41 Similar conics have the same eccentricity: $e = c/a = c'/a'$.

so that $P'(kx, ky)$ is on the second ellipse. Thus, each point on the first ellipse corresponds to a point on the second ellipse that is on the same line through the origin and is k times as far from the origin. It follows that the two ellipses are similar.

Part 2 Suppose the ellipses are similar. Then the triangle with vertices $(0, 0)$, $(a, 0)$, and $(0, b)$ is similar to the triangle with vertices $(0, 0)$, $(a', 0)$, and $(0, b')$. It follows that

$$\frac{b'}{a'} = \frac{b}{a} \quad \text{so that} \quad b' = \frac{ba'}{a}.$$

Then

$$\frac{c'}{a'} = \frac{\sqrt{a'^2 - b'^2}}{a'} = \frac{\sqrt{a'^2 - b^2 a'^2/a^2}}{a'},$$

which reduces to c/a. ∎

Eccentricity of Parabola

The *eccentricity of a parabola* is defined to be

$$e = 1.$$

Since there is only one parabolic shape, then the shape of any conic is completely determined by its eccentricity.

Recall that $0 < c < a$ if the conic is an ellipse, and $0 < a < c$ if it is a hyperbola. Therefore, a conic is

Identification by Eccentricity

an *ellipse*	if $0 < e < 1$,
a *parabola*	if $e = 1$,
a *hyperbola*	if $e > 1$.

7-5 Eccentricity. Sections of a Cone

EXAMPLE 1 Calculate the eccentricity of the ellipse
$$9x^2 + 4y^2 = 36.$$

Solution Written in standard form the equation is
$$\frac{x^2}{2^2} + \frac{y^2}{3^2} = 1 \quad (a = 3, b = 2).$$

Then $c^2 = a^2 - b^2 = 9 - 4 = 5$, so that $c = \sqrt{5}$. The eccentricity is
$$e = \frac{c}{a} = \frac{\sqrt{5}}{3}. \quad \square$$

EXAMPLE 2 Find an equation and sketch the graph of a hyperbola with eccentricity $e = 2$.

Solution Since $e = c/a = 2$ then $c = 2a$. Since the conic is a hyperbola we have
$$c^2 = a^2 + b^2$$
$$4a^2 = a^2 + b^2$$
$$b = a\sqrt{3}.$$

If we draw the hyperbola in standard position with its foci on the y-axis, its equation is
$$\frac{y^2}{a^2} - \frac{x^2}{b^2} = 1,$$
$$\frac{y^2}{a^2} - \frac{x^2}{3a^2} = 1.$$

Any positive number can now be chosen for a. (The size of a determines the size of the hyperbola.) For example, if $a = 1$, the hyperbola has equation
$$3y^2 - x^2 = 3.$$

The graph of this hyperbola is shown in Figure 7–42. $\square$

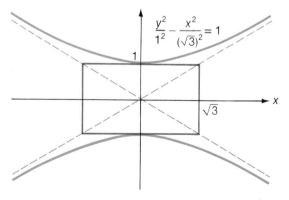

Figure 7–42 Example 2.

DIRECTRICES OF THE ELLIPSE AND HYPERBOLA

Directrices The foci of an ellipse or a hyperbola are located at points that are a distance of $c = ea$ units from the center. The lines perpendicular to the major axis (if ellipse) or transverse axis (if hyperbola) that are located at a distance of a/e units from the center are called the *directrices* of the conic. (See Fig. 7-43.)

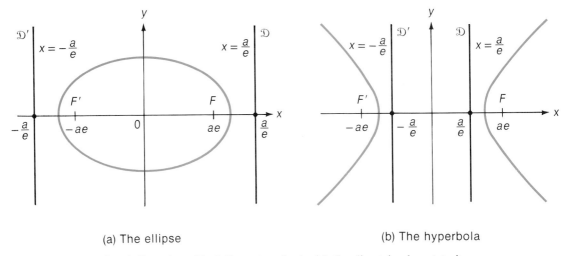

(a) The ellipse (b) The hyperbola

Figure 7-43 Foci and directrices. Each focus is paired with the directrix closest to it.

We normally pair the focus with the directrix that is on the same side of the center. Thus, if the foci are $F(ae, 0)$ and $F'(-ae, 0)$, the corresponding directrices are the vertical lines $x = a/e$ and $x = -a/e$. (See Fig. 7-43.)

Theorem 7-5 establishes that the points on an ellipse or a hyperbola satisfy the same basic focus-directrix relation as do the points on a parabola. In the proof we use the notation $|P\mathcal{D}|$ for the distance from the point P to the directrix $\mathcal{D}$.

THEOREM 7-5

Focus-Directrix Property of a Conic

Let F be a focus and $\mathcal{D}$ the corresponding directrix of an ellipse or hyperbola with eccentricity e. The points on the conic are exactly those points $P(x, y)$ that satisfy the geometrical condition

$$|PF| = e|P\mathcal{D}|.$$

Proof We set up an axis system so that the conic is in standard position with foci $(\pm ae, 0)$ on the x-axis and directrices $x = \pm a/e$. We work with the focus $F(ae, 0)$ and the corresponding directrix $x = a/e$. The proof is similar if the other focus and directrix are used.

For simplicity, we establish the result only for the ellipse. (The proof for the hyper-

7–5 Eccentricity. Sections of a Cone

bola is left for you to work.) The equation of the ellipse is

$$\frac{x^2}{a^2} + \frac{y^2}{b^2} = 1$$

where $b^2 = a^2 - c^2 = a^2 - a^2e^2 = a^2(1 - e^2)$. We first show that the points satisfying the geometrical condition

$$|PF| = e|P\mathcal{D}|$$

also satisfy this equation.

Let $P(x, y)$ satisfy the condition $|PF| = e|P\mathcal{D}|$. (See Fig. 7–44.) Since

$$|PF| = \sqrt{(x - ae)^2 + y^2} \quad \text{and} \quad |P\mathcal{D}| = \left|x - \frac{a}{e}\right|,$$

then

$$\sqrt{(x - ae)^2 + y^2} = e\left|x - \frac{a}{e}\right| = |xe - a|.$$

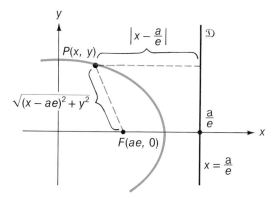

Figure 7–44.

Squaring both sides of this equation, we get

$$x^2 - 2aex + a^2e^2 + y^2 = x^2e^2 - 2aex + a^2$$
$$x^2(1 - e^2) + y^2 = a^2(1 - e^2)$$
$$\frac{x^2}{a^2} + \frac{y^2}{a^2(1 - e^2)} = 1.$$

Since $b^2 = a^2(1 - e^2)$, this equation reduces to

$$\frac{x^2}{a^2} + \frac{y^2}{b^2} = 1.$$

Thus, each point that satisfies the geometrical condition is on the ellipse.

Since all the above steps can be reversed, the ellipse is the set of all points that satisfy the geometrical condition

$$|PF| = e|P\mathscr{D}|. \blacksquare$$

Theorem 7–5 provides us with a common thread that connects the ellipse, parabola, and hyperbola. In each case, the conic is the set of all points satisfying the geometrical condition

$$|PF| = e|P\mathscr{D}|$$

where F is a focus, $\mathscr{D}$ the corresponding directrix, and e the eccentricity.

SECTIONS OF A CONE

Intersection of a Plane and a Cone

We now show that the parabola, ellipse, and hyperbola are the possible curves of intersection when a right circular cone is cut by a plane. (See Fig. 7–15.) In the following work, we allow degenerate cases of the conic sections. If the cutting plane passes through the vertex of the cone, for example, the intersection is either a point, a line, or a pair of intersecting lines; if it is perpendicular to the axis of the cone, the intersection is a circle or a point.

Let ϕ be the angle formed by the axis of the cone and its generating line. Let α be the angle formed by the axis of the cone and the cutting plane. (See Fig. 7–45.) We shall consider the case in which $0 < \alpha < 90°$ and the cutting plane does not intersect the cone at its vertex.

We inscribe a sphere in the cone between the vertex and the cutting plane, which is tangent to the cutting plane at a point F. (See Fig. 7–45a.) This sphere intersects the cone in a circle perpendicular to the axis of the cone. This circle defines a plane perpendicular to the axis of the cone that intersects the cutting plane at an angle $90° - \alpha$ (Fig. 7–45b).

Let $\mathscr{D}$ be the line of intersection of the plane of the circle and the cutting plane. We shall show that the curve of intersection of the cutting plane and the cone is a conic section with focus F and directrix $\mathscr{D}$.

Let P be a point on the curve of intersection of the cutting plane and the cone. Construct a line through P parallel to the axis of the cone. This line intersects the plane of the circle at a point Q.

Next construct a line from P to the vertex of the cone. This line intersects the circle at a point R.

Finally, construct a line from P to $\mathscr{D}$ perpendicular to $\mathscr{D}$. This line intersects $\mathscr{D}$ at a point S. Then $|P\mathscr{D}| = |PS|$.

Observe that line segments PR and PF emanate from the same point and are tangent to the sphere at their terminal points. Thus, they are equal in length:

$$|PR| = |PF|.$$

7–5 Eccentricity. Sections of a Cone

Observe also that triangles PQR and PQS are right triangles. (See Figs. 7–45b, c.) It follows that

$$|PQ| = |PR| \cos \phi$$

and

$$|PQ| = |PS| \cos \alpha.$$

(See Fig. 7–45c.) Therefore,

$$\frac{|PR|}{|PS|} = \frac{|PQ|/\cos \phi}{|PQ|/\cos \alpha} = \frac{\cos \alpha}{\cos \phi}$$

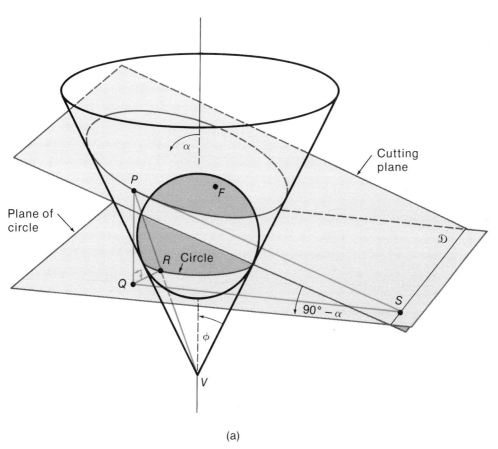

(a)

Figure 7–45 (a) The sphere inscribed in the cone.

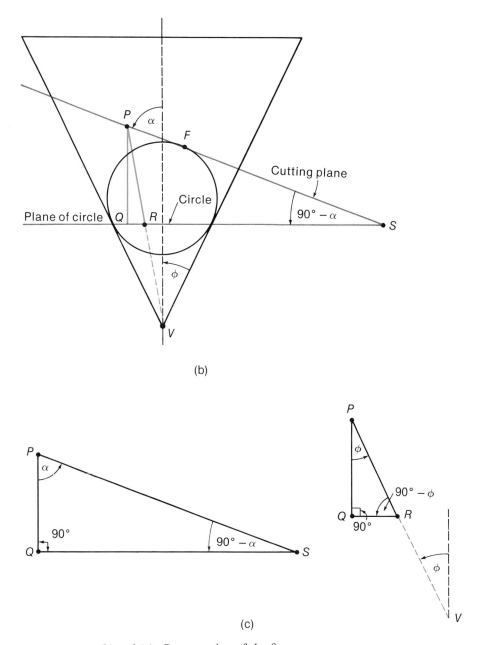

Figure 7–45 (b) and (c). Cross section of the figure.

so that

$$\frac{|PF|}{|P\mathcal{D}|} = \frac{|PR|}{|PS|} = \frac{\cos \alpha}{\cos \phi}.$$

If we now let $e = \cos \alpha / \cos \phi$, we get

$$|PF| = e|P\mathcal{D}|.$$

Since every point on the curve of intersection satisfies the condition

$$|PF| = e|P\mathcal{D}|,$$

then the curve is a conic section with focus F and direction $\mathcal{D}$. It is an ellipse if $e < 1$, a parabola if $e = 1$, a hyperbola if $e > 1$.

Since $e = \cos \alpha / \cos \phi$, we can express these conditions in terms of the angles α and ϕ: The curve is

an ellipse   if $\alpha > \phi$,
a parabola   if $\alpha = \phi$,
a hyperbola   if $\alpha < \phi$.

Exercises 7-5

1–6. Calculate the eccentricity and the equations of the directrices. Sketch the conic, showing all foci and directrices.

- **1** $144x^2 - 25y^2 = 3600$
- **2** $8x^2 + 9y^2 = 72$
- **3** $4x^2 - 8x - 10y^2 - 20y = 106$
- **4** $16x^2 + 64x + 25y^2 + 50y + 8 = 0$
- **5** $6x^2 - y - 3x + 5 = 0$
- **6** $4y^2 - x - 8y + 9 = 0$

7–10. Find an equation of a conic in standard position with the given eccentricity.

- **7** $e = 3$
- **8** $e = 1$
- **9** $e = \frac{1}{2}$
- **10** $e = \frac{1}{4}$

Conjugate Hyperbolas   **11** (*Conjugate Hyperbolas*) The pair of hyperbolas $x^2/a^2 - y^2/b^2 = \pm 1$ are said to be *conjugate*.

(a) Sketch the graph of this pair of conjugate hyperbolas. What relation holds between their asymptotes?
(b) Show that e_1 and e_2, the eccentricities of a pair of conjugate hyperbolas, satisfy the relation $1/e_1^2 + 1/e_2^2 = 1$.

12 What relation must hold between the positive constants a, b, c if the hyperbola $x^2/a^2 - y^2/b^2 = c^2$ has eccentricity $e = 2$?

13 On one set of axes, sketch ellipses with eccentricity $e_1 = \frac{1}{10}$, $e_2 = \frac{1}{2}$, and $e_3 = \frac{9}{10}$. How does the ellipse change when the eccentricity increases?

14 On one set of axes, sketch hyperbolas with eccentricity $e_1 = 1.1$, $e_2 = 2$, and $e_3 = 10$. How does the hyperbola change when the eccentricity increases?

15 (a) State a theorem comparable to Theorem 7–4 for hyperbolas.
    (b) Prove the theorem in (a).

16 (a) State a theorem comparable to Theorem 7–4 for parabolas.
    (b) Prove the theorem in (a).

7–6 Rotation of Axes

This section requires a knowledge of trigonometry. You may need to review Appendix B before studying it.

In Section 7–2 we showed how an equation could be changed by constructing new coordinate axes parallel to the original axes. We now consider what happens when new axes are constructed by rotating the original axes through an angle ϕ. (See Fig. 7–46.) The new axes—the x'- and y'-axes—have the same scale as the x- and y-axes. The y'-axis is obtained by a 90° counterclockwise rotation from the x'-axis.

Observe that each point P now has two sets of coordinates—(x, y) in the xy-system and (x', y') in the $x'y'$-system. (See Fig. 7–46b.) To determine the relations between these coordinates, we let r be the distance from the origin to P, and θ the angle from the positive x'-axis to the line segment OP.

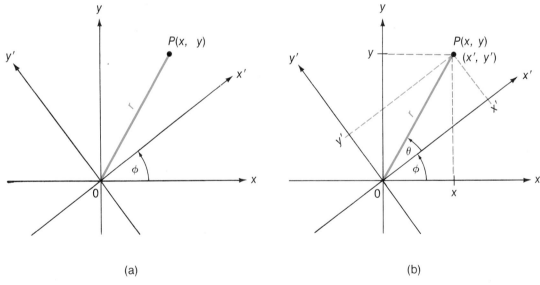

(a)   (b)

Figure 7–46 Rotation of axes.
$x = x' \cos \phi - y' \sin \phi$
$y = x' \sin \phi + y' \cos \phi$

7-6 Rotation of Axes

Using elementary trigonometry, we see that

$$x' = r\cos\theta, \qquad y' = r\sin\theta,$$

and

$$x = r\cos(\theta + \phi), \qquad y = r\sin(\theta + \phi).$$

If we now use trigonometric identities T8 and T9 from the Appendix on Trigonometry, we find that

$$\begin{aligned}x &= r(\cos\theta\cos\phi - \sin\theta\sin\phi)\\ &= (r\cos\theta)\cos\phi - (r\sin\theta)\sin\phi\\ &= x'\cos\phi - y'\sin\phi\end{aligned}$$

and

$$\begin{aligned}y &= r(\sin\theta\cos\phi + \cos\theta\sin\phi)\\ &= (r\sin\theta)\cos\phi + (r\cos\theta)\sin\phi\\ &= y'\cos\phi + x'\sin\phi.\end{aligned}$$

The equations relating the two sets of coordinates are

Equations of Rotation

$$\boxed{x = x'\cos\phi - y'\sin\phi \quad \text{and} \quad y = x'\sin\phi + y'\cos\phi.}$$

Since we could obtain the xy-axes by rotating the $x'y'$-axes through an angle of $-\phi$, then by the same argument,

$$x' = x\cos(-\phi) - y\sin(-\phi) = x\cos\phi + y\sin\phi$$

and

$$y' = x\sin(-\phi) + y\cos(-\phi) = -x\sin\phi + y\cos\phi.$$

Alternate Equations of Rotation

Thus, $\boxed{x' = x\cos\phi + y\sin\phi \quad \text{and} \quad y' = -x\sin\phi + y\cos\phi.}$

EXAMPLE 1 Find the coordinates of the point $(-3, 2)$ after a rotation of axes of $\phi = 30°$.

Solution The equations relating the coordinates are

$$x' = x\cos 30° + y\sin 30° = \frac{x\sqrt{3}}{2} + \frac{y}{2},$$

$$y' = -x\sin 30° + y\cos 30° = \frac{-x}{2} + \frac{y\sqrt{3}}{2}.$$

The new coordinates of the point are

$$x' = (-3)\frac{\sqrt{3}}{2} + 2\cdot\frac{1}{2} = \frac{-3\sqrt{3} + 2}{2}$$

and

$$y' = -(-3)\frac{1}{2} + 2 \cdot \frac{\sqrt{3}}{2} = \frac{3 + 2\sqrt{3}}{2}. \square$$

EXAMPLE 2 Find the new equation of $xy = 2$ after a 45° rotation of axes. Identify and sketch the graph.

Solution The equations relating the coordinates are

$$x = x' \cos 45° - y' \sin 45° = \frac{x' - y'}{\sqrt{2}};$$

$$y = x' \sin 45° + y' \cos 45° = \frac{x' + y'}{\sqrt{2}}.$$

On substituting these expressions in the equation $xy = 2$, we get

$$\frac{x' - y'}{\sqrt{2}} \cdot \frac{x' + y'}{\sqrt{2}} = 2$$

$$\frac{x'^2 - y'^2}{2} = 2$$

$$\frac{x'^2}{2^2} - \frac{y'^2}{2^2} = 1.$$

The graph is a hyperbola with asymptotes $y' = \pm x'$ (the original coordinate axes). See Figure 7–47. $\square$

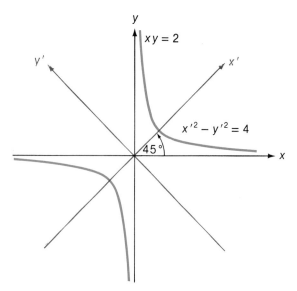

Figure 7–47 Example 2. $xy = 2$.

7–6 Rotation of Axes

THE GENERAL SECOND-DEGREE EQUATION

The equation

$$Ax^2 + Bxy + Cy^2 + Dx + Ey + F = 0$$

(A, B, C not all zero) is called the *general second-degree equation*. We now show that the "cross term" Bxy can be eliminated by an appropriate rotation of axis. The graph of the resulting equation is a conic section (possibly degenerate).

Before we find the proper rotation to eliminate the term Bxy, we consider the general problem of determining the new equation after a rotation of axis. If we perform a rotation through an angle ϕ, then

$$x = x' \cos \phi - y' \sin \phi \quad \text{and} \quad y = x' \sin \phi + y' \cos \phi.$$

If we substitute these expressions in the general second-degree equation, collect like terms, and simplify, we get a new second-degree equation of form

$$A'x'^2 + B'x'y' + C'y'^2 + D'x' + E'y' + F' = 0$$

where

$$A' = A \cos^2 \phi + B \cos \phi \sin \phi + C \sin^2 \phi,$$
$$B' = (-2A + 2C) \cos \phi \sin \phi + B(\cos^2 \phi - \sin^2 \phi),$$
$$C' = A \sin^2 \phi - B \cos \phi \sin \phi + C \cos^2 \phi,$$

and so on.

We need to find the proper value of ϕ so that $B' = 0$. If $B = 0$ in the original equation, there is no need for a rotation. If $B \neq 0$, we need

$$B' = -2(A - C) \cos \phi \sin \phi + B(\cos^2 \phi - \sin^2 \phi) = 0,$$
$$B (\cos^2 \phi - \sin^2 \phi) = (A - C) \cdot 2 \cos \phi \sin \phi$$
$$B \cos 2\phi = (A - C) \sin 2\phi \quad \text{(by T12, T13 of Appendix B)}$$
$$\frac{\cos 2\phi}{\sin 2\phi} = \frac{A - C}{B}$$
$$\cot 2\phi = \frac{A - C}{B} \quad \text{(by T2).}$$

Since the above steps can be reversed, it follows that the term $B'x'y'$ equals zero if and only if

Angle to Eliminate $x'y'$ Term

$$\cot 2\phi = \frac{A - C}{B}.$$

EXAMPLE 3 Simplify the equation

$$9x^2 - 24xy + 16y^2 - 400x - 300y + 2500 = 0$$

by a rotation of axes followed by a translation of axes.

Solution In this case $A = 9$, $B = -24$, $C = 16$. We need

$$\cot 2\phi = \frac{A - C}{B} = \frac{9 - 16}{-24} = \frac{7}{24}.$$

Observe that 2ϕ can be a first- or a third-quadrant angle. To be definite, we choose 2ϕ as a first-quadrant angle. Figure 7–48a pictures the angle 2ϕ. Observe that the corresponding reference angle has sides 7 and 24. By the Pythagorean Theorem, the hypotenuse is 25. Thus,

$$\cos 2\phi = \frac{7}{25} \quad \text{and} \quad \sin 2\phi = \frac{24}{25}.$$

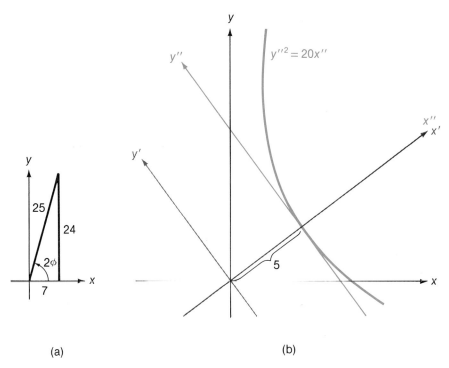

(a)   (b)

Figure 7–48 Example 3. $9x^2 - 24xy + 16y^2 - 400x - 300y + 2500 = 0$.

Using the half-angle formulas T16 and T17 from Appendix B, we get

$$\cos \phi = \sqrt{\frac{1 + \cos 2\phi}{2}} = \sqrt{\frac{1 + 7/25}{2}} = \sqrt{\frac{32}{50}} = \frac{4}{5}$$

and

$$\sin \phi = \sqrt{\frac{1 - \cos 2\phi}{2}} = \sqrt{\frac{1 - 7/25}{2}} = \sqrt{\frac{18}{50}} = \frac{3}{5}.$$

7–6 Rotation of Axes

The equations of rotation are

$$x = x' \cos \phi - y' \sin \phi = \frac{4x' - 3y'}{5};$$

$$y = x' \sin \phi + y' \cos \phi = \frac{3x' + 4y'}{5}.$$

If we substitute these expressions in the original equations and simplify, we get

$$\frac{625y'^2}{25} - 80(4x' - 3y') - 60(3x' + 4y') + 2500 = 0,$$

which reduces to

$$y'^2 = 20(x' - 5).$$

We now make the translation defined by

$$x'' = x' - 5, \quad y'' = y'.$$

The final form of the equation is

$$y''^2 = 20x''.$$

The graph of the equation, showing the three sets of axes, is in Figure 7–48b. □

DISCRIMINANT TEST

In many cases, a number can be assigned to a curve or an equation that will remain constant when the curve or equation is changed by some standard type of transformation. Such a number is called an *invariant* under the transformation.

Invariant

For example, the eccentricity is an invariant of a conic section under similarity. Two similar conics have the same eccentricity regardless of their size or orientation to an axis.

The following definition introduces an invariant of second-degree equations.

DEFINITION The quantity $B^2 - 4AC$ is called the *discriminant* of the second-degree equation

$$Ax^2 + Bxy + Cy^2 + Dx + Ey + F = 0.$$

It is left for you to show that the discriminant of a second-degree equation is an invariant under rotation and translation. (See Exercise 27.) In other words, if the equation

$$Ax^2 + Bxy + Cy^2 + Dx + Ey + F = 0$$

is transformed into

$$A'x'^2 + B'x'y' + C'y'^2 + D'x' + E'y' + F' = 0$$

by a rotation or translation of axes, then

$$B^2 - 4AC = B'^2 - 4A'C'.$$

This relation can be illustrated by the equation $xy - 2 = 0$, which is transformed into the equation

$$\frac{x'^2}{2} - \frac{y'^2}{2} - 2 = 0$$

by a 45°-rotation of axis. (See Example 2.) The discriminant of the original equation is

$$B^2 - 4AC = 1^2 - 4 \cdot 0 \cdot 0 = 1,$$

while the discriminant of the new equation is

$$B'^2 - 4A'C' = 0^2 - 4 \cdot \tfrac{1}{2}(-\tfrac{1}{2}) = 1.$$

Suppose now, that the original second-degree equation

$$Ax^2 + Bxy + Cy^2 + Dx + Ey + F = 0$$

has been reduced to the form

$$A'x'^2 + C'y'^2 + D'x' + E'y' + F' = 0$$

by a rotation of axes. The following cases occur.

Case 1 If A' or C' is zero, then the graph is a parabola (possibly degenerate). In this case, the discriminant is

$$B^2 - 4AC = B'^2 - 4A'C' = 0.$$

Case 2 If A' and C' are both positive or both negative, then the graph is an ellipse (possibly degenerate). In this case the discriminant is

$$B^2 - 4AC = B'^2 - 4A'C' < 0.$$

Case 3 If one of A', C' is positive and the other is negative, the graph is a hyperbola (possibly degenerate). In this case the discriminant is

$$B^2 - 4AC = B'^2 - 4A'C' > 0.$$

Since the three cases cover all possible situations, we have the following theorem.

THEOREM 7-6 (*The Discriminant Test*) Let

$$Ax^2 + Bxy + Cy^2 + Dx + Ey + F = 0$$

be a second-degree equation.

(1) If $B^2 - 4AC < 0$, the graph is an ellipse.
(2) If $B^2 - 4AC = 0$, the graph is a parabola.
(3) If $B^2 - 4AC > 0$, the graph is a hyperbola.

In each case the graph may be a degenerate case of the conic.

Exercises 7-6

EXAMPLE 4 Use the discriminant test to identify the graphs of the following equations:
(a) $4x^2 - 4xy + y^2 - 3x + 2y = 7$.
(b) $x^2 + 3xy + 3y^2 - 12 = 0$.
(c) $2x^2 - 9xy + 2y^2 = 87$.

Solution (a) $4x^2 - 4xy + y^2 - 3x + 2y = 7$. The discriminant is

$$B^2 - 4AC = (-4)^2 - 4 \cdot 4 \cdot 1 = 0.$$

The graph is a parabola (possibly degenerate).

(b) $x^2 + 3xy + 3y^2 - 12 = 0$. The discriminant is

$$B^2 - 4AC = 3^2 - 4 \cdot 1 \cdot 3 < 0.$$

The graph is an ellipse (possibly degenerate).

(c) $2x^2 - 9xy + 2y^2 = 87$. The discriminant is

$$B^2 - 4AC = (-9)^2 - 4 \cdot 2 \cdot 2 > 0.$$

The graph is a hyperbola (possible degenerate). $\square$

An interesting, if far-fetched, mnemonic device for the discriminant test is based on words similar to the names of the conic sections.

ellipse ↔ *ellipsis* (omission, something less)	<
parabola ↔ *parable* (analogy, sameness)	=
hyperbola ↔ *hyperbole* (exaggeration)	>

This same mnemonic device can be used to remember the relations for eccentricity.

Exercises 7-6

1–10. Reduce each of the equations to the standard form of a conic section by performing a rotation of axes followed by a translation of axes, if necessary. Sketch the graph showing all sets of axes.

- •1 $5x^2 + 6xy + 5y^2 = 32$
- •2 $29x^2 - 24xy + 36y^2 = 180$
- •3 $5x^2 - 26xy + 5y^2 = 72$
- •4 $21x^2 - 10\sqrt{3}xy + 31y^2 = 144$
- •5 $-x^2 + 6\sqrt{3}xy + 5y^2 = 16$
- •6 $25x^2 - 120xy + 144y^2 + 884x - 364y = 0$
- •7 $7x^2 + 6\sqrt{3}xy + 13y^2 - 16x - 16\sqrt{3}y = 0$
- •8 $12x^2 - 8\sqrt{3}xy + 4y^2 - x - \sqrt{3}y - 4 = 0$
- •9 $2x^2 + 4xy + 2y^2 - 3\sqrt{2}x - 5\sqrt{2}y + 2 = 0$
- •10 $3x^2 - 10xy + 3y^2 - 28\sqrt{2}x + 20\sqrt{2}y + 48 = 0$

•11–20. Use the discriminant test to identify the graphs of the equations in Exercises 1–10.

21–24. The equations have graphs that are degenerate conics. Classify the conics by the discriminant test. Sketch the graphs after a rotation of axes.

•21 $x^2 - 2xy + y^2 + 2x - 2y = 0$ 22 $x^2 - xy - 2y^2 = 0$

•23 $x^2 + y^2 - 2x + 4y + 4 = 0$ 24 $601x^2 + 360xy + 244y^2 + 400 = 0$

25 Show that the equation of a circle with center at the origin is unchanged (except for renaming the variables) after any rotation of axes.

26 (a)–(b) Perform a translation or rotation of axes that will accomplish the following result.
    (a) Make the line $3x + 4y = 0$ the x'-axis. (Two answers.)
    (b) Make the graph of $(4x - y)(x + 3y) = 17$ pass through the origin. (Infinite number of answers.)
    (c) Give simple geometrical arguments that explain why (a) has two solutions and (b) has an infinite number of solutions.

27 Prove that the discriminant of a second-degree equation is an invariant under the transformation.
    (a) Rotation of axes
    (b) Translation of axes.

Review Problems

1 (a) What is meant by symmetry about a line $\mathscr{L}$? about the origin?
   (b) State the tests for symmetry about
      (1) the x-axis;
      (2) the y-axis;
      (3) the origin;
      (4) the line $y = x$.

2 Prove that a curve that is symmetric about both coordinate axes is symmetric about the origin.

3–8. Find all horizontal, vertical, and oblique asymptotes, and all horizontal and vertical tangent lines. Find where the function is increasing and where it is decreasing. Check the concavity. Test for symmetry. Sketch the graph.

•3 $y = \dfrac{3x^2 - 4x + 7}{x - 2}$ •4 $y = \dfrac{x^2 - 4}{x^2 - 1}$

•5 $y = \dfrac{x}{x^2 - 9}$ •6 $y = \dfrac{x - 4}{x^2 - 1}$

Review Problems

• 7 $y = \dfrac{x^2}{x^2 - 9}$

• 8 $y = x\sqrt{x^2 - 8}$

9 State the formal definition of the parabola, ellipse, and hyperbola.

10–12. Use the definition of each of these conics to derive its equation in standard position with foci (focus) on the x-axis.

10 Parabola 11 Ellipse 12 Hyperbola

13–15. Find the equation of the conic that satisfies the given conditions.

• 13 Parabola with focus (3, 5) and directrix $y = 1$

• 14 Ellipse with center (2, −1), vertex (2, 4) and focus (2, −4)

• 15 Hyperbola with asymptotes $y - 1 = \pm 2(x - 2)$ and vertex (2, 5)

• 16 Use the definition to derive the equation of the ellipse with $a = 6$ and with foci (0, 0) and (6, 8).

17–24. Sketch the graph. If the graph is a conic, then locate the foci and vertices.

• 17 $x - 2y^2 = 8$

• 18 $4x^2 - 24x + y^2 + 2y + 21 = 0$

• 19 $x^2 + 8x - 16y^2 + 32y + 16 = 0$

• 20 $2x^2 + 4x + y = 0$

• 21 $x^2 + 10x + 9y^2 + 36y + 52 = 0$

• 22 $25x^2 - 150x - 4y^2 - 32y + 61 = 0$

• 23 $9x^2 - 4y^2 - 18x - 8y + 5 = 0$

• 24 $4x^2 + 5y^2 + 16x - 10y + 22 = 0$

25–32. Calculate the eccentricity and the equations of the directrices for the conics in the given exercises.

• 25 Exercise 17 • 26 Exercise 18

• 27 Exercise 19 • 28 Exercise 20

• 29 Exercise 21 • 30 Exercise 22

• 31 Exercise 23 • 32 Exericse 24

33–38. Simplify the equation by a rotation of axes. Sketch the graph showing both axis systems.

• 33 $12x^2 - 8\sqrt{3}xy + 4y^2 - x - \sqrt{3}y - 4 = 0$

• 34 $144x^2 - 120xy + 25y^2 + 247x - 286y = 169$

•35 $xy = -8$

•36 $x^2 - 2xy + y^2 + 8x + 8y = 16\sqrt{2}$

•37 $4x^2 + 24xy + 11y^2 + 20 = 0$

•38 $369x^2 - 384xy + 481y^2 = 5625$

39 State the discriminant test.

40–43. Use the discriminant test to classify the graphs of the equations in the given exercises.

•40 Exercise 33   •41 Exercise 34

•42 Exercise 35   •43 Exercise 38

44 What is meant by an invariant of a curve? Give two examples.

•45 A parabola with axis of symmetry parallel to the y-axis opens downward and passes through the points $(0, 0)$ and $(2, 6)$. Of all such parabolas, find the one such that the area of the region enclosed by the parabola and the x-axis is a minimum.

8 Logarithmic and Exponential Functions

8–1 The Natural Logarithm

Historically, the concept of an exponential function developed slowly. Positive integral exponents were used for centuries before it was realized that reciprocals and roots have similar properties.

Recall that positive integral exponents are defined by the relation

$$a^m = a \cdot a \cdots a \quad (m \text{ factors}).$$

Positive rational exponents are defined by

$$a^{p/q} = \sqrt[q]{a^p} = (\sqrt[q]{a})^p$$

provided p and q are positive integers. The zero exponent is defined by

$$a^0 = 1 \quad \text{provided } a \neq 0,$$

and negative exponents are defined by the law

$$a^{-m} = \frac{1}{a^m}$$

provided a^m is defined and $a \neq 0$.

For example,

$$2^3 = 2 \cdot 2 \cdot 2 = 8,$$

$$2^{-3/2} = \frac{1}{2^{3/2}} = \frac{1}{(\sqrt{2})^3},$$

and

$$2^0 = 1.$$

In the early seventeenth century, the Scottish mathematician John Napier (1550–1617) published his book on logarithms. It was soon realized that an equivalent system could be defined in terms of exponents. In the revised system, if a and b are positive numbers with $b \neq 1$, we define the *logarithm of a to the base b* to be the exponent to which b must be raised in order to have the power equal to a. In symbols,

$$\log_b a = L \quad \text{if and only if} \quad a = b^L.$$

(See Fig. 8–1.)

For example, since $2^3 = 8$, then

$$\log_2 8 = 3.$$

Similarly, since $b^0 = 1$, then

$$\log_b 1 = 0$$

for every positive number b different from 1.

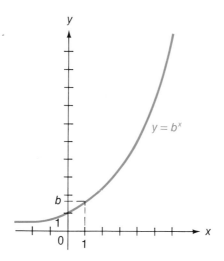
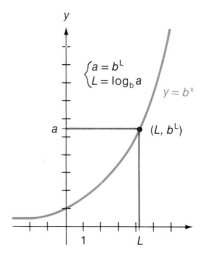

(a) The exponential function: $y = b^x$

(b) Logarithms: $\log_b a = L$ if, and only if, $a = b^L$

Figure 8–1.

As another example, if we are given that $\log_{10} 7 \approx 0.845098$, then

$$7 = 10^{\log_{10} 7} \approx 10^{0.845098}$$

To develop the logarithm concept along these lines, it is necessary to define b^x

8-1 The Natural Logarithm

when x is any real number. Thus, numbers such as $2^{\sqrt{3}}$ must be defined, as well as numbers such as $5^{1/2}$. (We discuss this matter in detail in Section 8-2.) Here we assume that it is possible to define b^x for every real number x and that the laws of exponents hold for these exponents as well as for rational exponents.

Three of the basic laws of exponents are:

(1) $\quad b^m \cdot b^n = b^{m+n}.$

(2) $\quad \dfrac{b^m}{b^n} = b^{m-n}.$

(3) $\quad (b^m)^k = b^{mk}.$

These laws can be used to prove equivalent laws of logarithms. If m, n, and b are positive numbers with $b \neq 1$, then:

Laws of Logarithms

(4) $\quad \log_b (mn) = \log_b m + \log_b n.$

(5) $\quad \log_b \left(\dfrac{m}{n}\right) = \log_b m - \log_b n.$

(6) $\quad \log_b (m^k) = k \cdot \log_b m.$

EXAMPLE 1 It can be proved that $\log_{10} 37 \approx 1.56820$ and $\log_{10} 82 \approx 1.91381$.
It follows from property (5) that

$$\log_{10}\left(\frac{82}{37}\right) = \log_{10} 82 - \log_{10} 37 \approx 1.91381 - 1.56820$$
$$\approx 0.34561.$$

If we look up this number in a table of common logarithms, we find that

$$\log_{10} 2.2162 \approx 0.34561.$$

It follows that

$$\frac{82}{37} \approx 2.2162. \quad \square$$

EXAMPLE 2 Use the definition of logarithm to prove that $\log_b (mn) = \log_b m + \log_b n.$

Solution Let $m = b^x$ and $n = b^y$. Then

$$x = \log_b m \quad \text{and} \quad y = \log_b n.$$

It follows that

$$mn = b^x \cdot b^y = b^{x+y}$$

so that

$$\log_b (mn) = x + y = \log_b m + \log_b n. \quad \square$$

Logarithms were developed originally as an aid in computation. With the invention of the calculus, however, it was discovered that logarithmic and exponential functions are involved in many of the applications of mathematics to science and engineering. Since World War II, the electronic computer and calculator have made logarithms virtually obsolete for computation. Their importance in mathematics is now based on their properties as functions.

In our development we will reverse the historical order. We first define a function (in terms of an integral) called the *natural logarithm function*, denoted by "ln x." Next we develop the properties of this function; and finally, we prove that it actually is a logarithm function—that there exists a number e such that

$$\ln a = c \quad \text{if and only if } a = e^c.$$

THE FUNCTION $y = \ln x$

Let $x > 0$. The *natural logarithm of x*, denoted by ln x, is defined by the integral

Natural Logarithm

$$\ln x = \int_1^x \frac{1}{t} \, dt, \quad x > 0.$$

Since the function $f(x) = 1/x$ is continuous for all $x > 0$, then this integral defines a new function of x. It follows from the Fundamental Theorem of Calculus (Theorem 5–3), that the derivative of the natural logarithm function is given by

$$D_x(\ln x) = D_x \left(\int_1^x \frac{1}{t} \, dt \right) = \frac{1}{x} \quad \text{provided } x > 0.$$

This formula can be generalized by the Chain Rule to

Basic Derivative Formula

D9 $\quad D_x(\ln u) = \frac{1}{u} \cdot D_x u \quad \text{provided } u > 0.$

EXAMPLE 3   (a) $\quad D_x(\ln (2x + 3)) = \frac{1}{2x + 3} \cdot D_x(2x + 3) = \frac{2}{2x + 3}.$

(b) $\quad D_x \left(\ln \left(\frac{1}{x^2 + 1} \right) \right) = \frac{1}{1/(x^2 + 1)} \cdot D_x \left(\frac{1}{x^2 + 1} \right)$

$$= (x^2 + 1)(-1)(x^2 + 1)^{-2} \cdot D_x(x^2 + 1)$$

$$= \frac{-2x}{x^2 + 1}. \quad \square$$

8-1 The Natural Logarithm

PROPERTIES OF THE NATURAL LOGARITHM FUNCTION

Several of the elementary properties can be established at once. Since

$$D_x(\ln x) = \frac{1}{x} > 0$$

and

$$D_x^2(\ln x) = D_x\left(\frac{1}{x}\right) = -\frac{1}{x^2} < 0$$

for $x > 0$, then the natural logarithm function is an increasing function that is concave downward. Observe also, that

$$\ln x = \int_1^x \frac{1}{t} dt < 0 \qquad \text{if } 0 < x < 1$$

$$\ln 1 = \int_1^1 \frac{1}{t} dt = 0$$

and

$$\ln x = \int_1^x \frac{1}{t} dt > 0 \qquad \text{if } x > 1.$$

(See Fig. 8–2.)

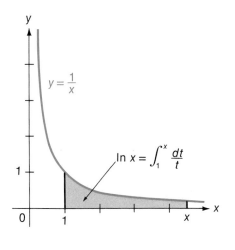

(a) $\ln x = \int_1^x \frac{dt}{t}$

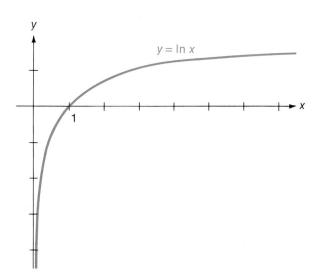

(b) Graph of the natural logarithm function

Figure 8–2 The natural logarithm function.

Logarithmic and Exponential Functions

We now establish that the properties of logarithms stated at the beginning of this section hold for the natural logarithm function.

THEOREM 8-1 Let a and b be positive numbers, k a rational number. Then:

Properties of Natural Logarithms

(7) $\quad \ln(ab) = \ln a + \ln b.$

(8) $\quad \ln\left(\dfrac{a}{b}\right) = \ln a - \ln b.$

(9) $\quad \ln(a^k) = k \ln a.$

Proof of (7) To prove the theorem, we shall work with functions of x. Let $f(x) = \ln ax$ and $g(x) = \ln x$, $x > 0$. Then

$$f'(x) = D_x(\ln ax) = \frac{1}{ax} \cdot a = \frac{1}{x} = g'(x), \qquad x > 0.$$

Since f and g have the same derivative, they differ by a constant C:

$$f(x) = g(x) + C,$$
$$\ln(ax) = \ln x + C.$$

To evaluate the constant, we substitute $x = 1$ in the equation, obtaining

$$\ln(a \cdot 1) = \ln 1 + C$$
$$\ln a = 0 + C$$
$$C = \ln a.$$

Then

$$\ln(ax) = \ln x + C = \ln x + \ln a.$$

If we evaluate this function at $x = b$, we get

$$\ln(ab) = \ln b + \ln a = \ln a + \ln b.$$

The proofs of properties (8) and (9) are left for you to work out. (See Exercises 33 and 34.) ∎

The three basic laws can be used to simplify many of the calculations involving logarithms. Examples 4 and 5 illustrate the procedure.

EXAMPLE 4 (a) $\quad \ln\left(\dfrac{x+1}{x+2}\right)^2 = 2\ln\left(\dfrac{x+1}{x+2}\right) \qquad$ (by (9))

$\qquad\qquad\qquad\qquad\qquad\quad = 2[\ln(x+1) - \ln(x+2)] \qquad$ (by (8)).
$\qquad\qquad\qquad\qquad\qquad\quad = 2\ln(x+1) - 2\ln(x+2).$

(b) $\quad \ln\left(\dfrac{1}{x}\right) = \ln 1 - \ln x = -\ln x \qquad$ (by (8)).

(c) $\quad \ln[(x^2+1)(x^3+3)] = \ln(x^2+1) + \ln(x^3+3) \qquad$ (by (7)). □

8–1 The Natural Logarithm

EXAMPLE 5 Use the properties of logarithms to simplify the expressions. Then calculate the derivatives.

(a) $D_x \left(\ln \left(\dfrac{1}{x^2 + 1} \right) \right)$. (See Example 3(b).)

(b) $D_x \left(\ln \left(\dfrac{x^2(x + 1)}{(x^3 + 1)^4} \right) \right)$.

Solution (a)
$$D_x \left(\ln \left(\dfrac{1}{x^2 + 1} \right) \right) = D_x(\ln (x^2 + 1)^{-1})$$
$$= D_x(-\ln (x^2 + 1))$$
$$= -\dfrac{1}{x^2 + 1} \cdot 2x = \dfrac{-2x}{x^2 + 1}.$$

(b) $D_x \left(\ln \left(\dfrac{x^2(x + 1)}{(x^3 + 1)^4} \right) \right) = D_x[\ln x^2 + \ln (x + 1) - \ln (x^3 + 1)^4]$

(by properties (7) and (8))

$$= D_x[2 \ln x + \ln (x + 1) - 4 \ln (x^3 + 1)]$$
$$= 2 \cdot \dfrac{1}{x} + \dfrac{1}{x + 1} - \dfrac{4}{x^3 + 1} \cdot 3x^2 = \dfrac{2}{x} + \dfrac{1}{x + 1} - \dfrac{12x^2}{x^3 + 1}. \quad \square$$

THE LIMITS $\lim_{x \to \infty} \ln x$ AND $\lim_{x \to 0^+} \ln x$

It appears in Figure 8–2 that

$$\lim_{x \to \infty} \ln x = \infty \quad \text{and} \quad \lim_{x \to 0^+} \ln x = -\infty.$$

To prove these limits, we need some information about particular values of the natural logarithm function. In particular, we need to know that

$$\ln 3 > 1.$$

This is established in Exercise 35 along with the related fact that $\ln 2 < 1$.

Actually, much better results can be established. If we use Simpson's Rule (see Section 5–6) with $n = 10$, we find that

$$\ln 2 = \int_1^2 \dfrac{1}{x} dx \approx 0.6932$$

and

$$\ln 3 = \int_1^3 \dfrac{1}{x} dx \approx 1.0987.$$

To establish that $\lim_{x \to \infty} \ln x = \infty$, we must show that $\ln x$ is greater than any given number n provided x is sufficiently large.

Let n be a positive number. Let $m = 3^n$. Recall that the natural logarithm function is an increasing function. Therefore, if $x > m = 3^n$, then

$$\ln x > \ln 3^n = n \ln 3.$$

Since $\ln 3 > 1$, then $\ln x > n$ if $x > m$. It follows that

$$\lim_{x \to \infty} \ln x = \infty.$$

It is now easy to establish the other limit. Observe that $1/x \to \infty$ as $x \to 0^+$. Since $\ln x = -\ln(1/x)$, then

$$\lim_{x \to 0^+} \ln x = \lim_{x \to 0^+} \left(-\ln \frac{1}{x}\right) = -\left[\lim_{x \to 0^+} \ln \left(\frac{1}{x}\right)\right]$$
$$= -\left[\lim_{(1/x) \to \infty} \ln \left(\frac{1}{x}\right)\right] = -\infty.$$

Exercises 8–1

1 Use the definition of logarithm to the base b to find the values.
  •(a) $\log_2 16$
  (b) $\log_3 \frac{1}{27}$
  •(c) $\log_{10} 0.00001$
  (d) $\log_5 125^6$.

2–5. Solve for y.

•2 $7x = \log_3 y$

3 $\log_{10} \dfrac{2}{y} = 5x + 1$

•4 $8^y = 13$

5 $4^{y+2} = 3^2$

6 It is known that $\log_5 3 \approx 0.68261$ and $\log_5 7 \approx 1.20906$. Use the properties of logarithms to calculate
  •(a) $\log_5 21$; (b) $\log_5 (\tfrac{7}{3})$; •(c) $\log_5 \sqrt[3]{7}$.

7 (a) Use the properties of exponents and logarithms to prove that

$$\log_b a = \frac{1}{\log_a b}.$$

(Hint: Let $x = \log_b a$, $y = \log_a b$. Show that $a^{xy} = a$.)
  •(b) Use (a) and Exercise 6 to calculate $\log_{\sqrt[3]{7}} 5$.

8 Use Table I at the end of this book and the properties of the natural logarithm function (or use a scientific calculator) to make an accurate graph of $y = \ln x$, $0.001 \le x \le 25$. Sketch the tangent lines, labeling them with their slopes, at the points where $x = 1$, 10, and 20.

9–12. Use the definition and the properties of the natural logarithm function to explain the equations. The numbers a, b, c are positive.

9 $\displaystyle\int_1^{12/7} \frac{dx}{x} = \int_7^{12} \frac{dx}{x}$

10 $\displaystyle\int_a^{a^2/b} \frac{dx}{x} = \int_b^{a} \frac{dx}{x}$

11 $\displaystyle\int_a^{ab} \frac{dx}{x} = \int_c^{cb} \frac{dx}{x}$

12 $\displaystyle\int_{a^3}^{b^3} \frac{dx}{x} = 3 \int_a^{b} \frac{dx}{x}$

Exercises 8–1

13–26. Calculate the derivatives of the functions with respect to x. Use the properties of natural logarithms to simplify the calculations as much as possible.

- 13 $\ln(x^2 - 1)$
- 14 $\ln(2x^2 - 1)^3$
- 15 $\ln \dfrac{1}{x}$
- 16 $\ln(x^x)$
- 17 $x^2 \ln x$
- 18 $[\ln(x^2)]^4$
- 19 $(\ln x)^3$
- 20 $\ln(x^3)$
- 21 $\ln\left(\dfrac{x+1}{x-1}\right)$
- 22 $\ln\left(\sqrt{\dfrac{5x^2+1}{\sqrt{x}}}\right)$
- 23 $\sqrt{\ln(x^2+1)}$
- 24 $\ln\sqrt{x^2+1}$
- 25 $[x + \ln(x^2)]^2$
- 26 $\ln(x + \sqrt{x^2+1})$

27–30. Use implicit differentiation to calculate y'.

- 27 $y + \ln(xy) = 2$
- 28 $y^2 - \ln(x/y) + x = 1$
- 29 $x \ln y + y \ln x = 1$
- 30 $\ln y = x \ln x + 1$

- 31   Use Simpson's Rule with $n = 10$ to approximate $\ln 5 = \displaystyle\int_1^5 \left(\dfrac{dx}{x}\right)$. Compare your answer with the value in Table I.

- 32   Let $f(x) = x - \ln x$. Find all local extrema and points of inflection. Sketch the graph.

---

33   Prove that $\ln(a/b) = \ln a - \ln b$.   34   Prove that $\ln(a^k) = k \ln a$.

35   In Section 8–2 we need to know that $\ln 2 < 1 < \ln 3$. The last part of this inequality was used in this section to prove that $\lim_{x \to \infty} \ln x = \infty$. These inequalities can be established by the following procedure. Partition $[1, 3]$ into eight equal subintervals of length $\Delta x = 0.25$.
   (a) Use rectangles of the type shown in Fig. 8–3a (p. 454) to prove that

$$\ln 2 = \int_1^2 \dfrac{1}{x}\, dx < \sum_{k=0}^{3} \dfrac{1}{x_k} \Delta x < 1.$$

   (b) Use the rectangles of the type shown in Fig. 8–3b to prove that

$$\ln 3 = \int_1^3 \dfrac{1}{x}\, dx > \sum_{k=1}^{8} \dfrac{1}{x_k} \Delta x > 1.$$

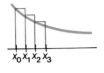

(a)

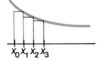

(b)   **Figure 8-3** Exercise 35.

8-2 The Number e. The Natural Exponential Function

You should review Section 1-8 on inverse functions before starting work on this section.

The natural logarithm function is continuous and steadily increasing. We have seen (Exercise 35, Section 8-1) that

$$\ln 2 < 1 \quad \text{and} \quad \ln 3 > 1.$$

It follows from the Intermediate Value Theorem (Theorem 2-8) that there exists a unique number e between 2 and 3 such that

The Number e | $\ln e = 1.$

(See Fig. 8-4.)

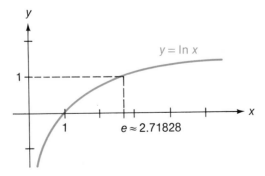

Figure 8-4 $\ln e = 1.$

8-2 The Number e. The Natural Exponential Function

The number e is *transcendental*, that is, it is not the root of an algebraic equation with rational coefficients. It can, however, be calculated to any desired number of decimal places. Correct to fifteen decimal places, for example,

$$e \approx 2.718281828459045.$$

Such approximations are useful in calculation but not particularly important for most theoretical work. The property we need for our development is that e is the number that has its natural logarithm equal to 1.

Our main goal in this section is to prove that the natural logarithm function is the inverse of an exponential function. We establish that this is indeed the case and that the number e is the base for the system. In symbols, we will show that

$$y = \ln x \quad \text{if and only if } x = e^y.$$

After we establish this result, we shall work with derivatives and integrals of the exponential function $f(x) = e^x$.

Our development is hampered because we have never properly defined the number a^x when x is an irrational number. For this reason we take a roundabout approach involving inverses of functions.

THE NATURAL EXPONENTIAL FUNCTION

Since the natural logarithm function is an increasing continuous function, then it has an inverse. This inverse function, denoted by the symbol "exp," is called the *natural exponential function*. That is,

Natural Exponential Function

$$y = \exp x \quad \text{if and only if } x = \ln y.$$

The graph of $y = \exp x$ is shown in Figure 8–5, along with the graph of the inverse function $y = \ln x$.

At this point, all we know about the function exp is that it is the inverse of the function ln. We shall establish that $\exp x = e^x$ for every x—a result that is not apparent from the definition. To establish this result, we need the following properties of the function exp, all of which can be proved from the equivalent properties of its inverse function ln. Most of these properties are illustrated in Figure 8–5. The proofs are left for you.

Properties of the Natural Exponential Function

(1) The function exp is defined and continuous for all real x. (Use Theorem 3–9.)
(2) $\exp x > 0$ for all real x.
(3) The function exp has a positive derivative everywhere. (Use Theorem 3–10.)
(4) $\exp 1 = e$.
(5) $\ln (\exp x) = x$ for every x.
(6) $\exp (\ln x) = x$ for every positive x.

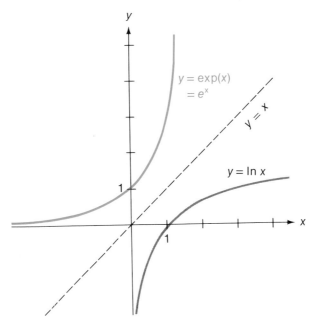

Figure 8-5 The natural exponential and natural logarithm functions.

Property (2) follows from the fact that ln x is defined only when $x > 0$; Property (4) from the fact that ln $e = 1$; Properties (5) and (6) from the definition of the natural exponential function.

THE DEFINITION OF a^x

Before going ahead, we show how a^x can be defined when x is an irrational number and $a > 0$. Our main consideration in defining a^x is that the resulting function $f(x) = a^x$ should be continuous for all x. That is, we want

$$\lim_{x \to k} a^x = a^k$$

for every k, rational or irrational. In essence, this means that we always want a^x to be "close" to a^k whenever x is "close" to k. Of course we also want the new definition to coincide with the original definition when a is rational.

It follows from Theorem 8-1 that if $a > 0$ and r is rational, then

$$\ln (a^r) = r \ln a$$

so that

$$\exp (r \ln a) = \exp (\ln a^r) = a^r$$

by Property (6). We use this relation to define a^x for any real number x.

8-2 The Number e. The Natural Exponential Function

DEFINITION Let $a > 0$, let x be a real number. Then

Definition of a^x
$$a^x = \exp(x \ln a).$$

Observe that a^x is now defined for irrational exponents as well as rational. Furthermore, this definition coincides with our previous definition when x is a rational number.

It is now easy to show that the function $f(x) = a^x$ is continuous.

EXAMPLE 1 Let $a > 0$; let $f(x) = a^x$. Show that f is continuous at every real number k.

Solution Since the function exp is continuous, then

$$\lim_{x \to k} f(x) = \lim_{x \to k} a^x = \lim_{x \to k} \exp(x \ln a)$$
$$= \exp(\lim_{x \to k} (x \ln a)) = \exp(k \ln a)$$
$$= a^k = f(k). \ \square$$

Example 2 extends Property (9) of Theorem 8–1 to cover all real exponents.

EXAMPLE 2 Let $a > 0$. Show that $\ln(a^x) = x \ln a$ for every real number x.

Solution $\ln a^x = \ln(\exp(x \ln a)) = x \ln a$ by Property (5). $\square$

THE FUNCTION e^x

Observe that if we take the particular value $a = e$ in the definition of a^x, we get

$$e^x = \exp(x \ln e) = \exp(x \cdot 1) = \exp x.$$

Thus,

$$\exp x = e^x \quad \text{for every real number } x.$$

In the remainder of this book we use e^x rather than $\exp x$ for the natural exponential function. The exp notation, however, is used in many advanced books, particularly for complicated expressions. It is easier to set $\exp(2^x + 1)$ into type than e^{2^x+1}.

The original definition of the natural exponential function can now be written as

Natural Exponents and Logarithms
$$y = e^x \quad \text{if and only if } x = \ln y.$$

Observe that this relation is equivalent to the one used to define logarithms (Section 8–1). Thus,

$$\ln y = \log_e y \quad \text{for every } y > 0.$$

PROPERTIES OF THE FUNCTION EXP $x = e^x$

For reference purposes, we restate the properties of the function $y = e^x$ that were stated above in the exp notation:

Properties of e^x

(1) The function $\exp(x) = e^x$ is defined and continuous for all real x.
(2) $e^x > 0$ for all x.
(3) $D_x(e^x) > 0$ for all x.
(4) $e^1 = e$.
(5) $\ln(e^x) = x$ for every x.
(6) $e^{\ln x} = x$ for every $x > 0$.

It is particularly easy to calculate the derivative $D_x(e^x)$.

THEOREM 8–2 $\quad D_x(e^x) = e^x.$

Proof Let $y = e^x$. Then $\ln y = x$. By implicit differentiation,

$$D_x(\ln y) = D_x(x)$$
$$\frac{1}{y} \cdot D_x y = 1$$
$$D_x y = y.$$

Since $y = e^x$, then

$$D_x(e^x) = e^x. \blacksquare$$

If we apply the Chain Rule, we get the more general formula

Basic Derivative Formula

D10 $\quad D_x(e^u) = e^u D_x u.$

EXAMPLE 3
(a) $D_x(e^{2x}) = e^{2x} \cdot D_x(2x) = 2e^{2x}.$
(b) $D_x(3e^{x^2}) = 3e^{x^2} \cdot D_x(x^2) = 6x\, e^{x^2}$
(c) $D_x(e^{1/x}) = e^{1/x} D_x\left(\frac{1}{x}\right) = -\frac{e^{1/x}}{x^2}. \square$

THE LIMIT $\lim_{x \to \infty} \left(1 + \frac{1}{x}\right)^x = e$

One of the most important limits in mathematics is the limit

$$\lim_{x \to \infty} \left(1 + \frac{1}{x}\right)^x = e.$$

This limit is established in Exercise 36. A simpler proof can be found in Example 3 of Section 10–2.

With the above limit, it is simple to show that

$$\lim_{x \to \infty} \left(1 + \frac{r}{x}\right)^x = e^r$$

and that $\lim_{x \to \infty} \left(1 + \frac{r}{x}\right)^{xt} = e^{rt}$. (See Exercises 29–30.)

These limits have an interesting interpretation in terms of compound interest. If an amount P_0 is loaned at an interest rate of r per year, compounded x times per year, then the value of the investment after t years is

$$P_0 \left(1 + \frac{r}{x}\right)^{xt}.$$

Continuous Compounding If we let $x \to \infty$, we obtain the value for *instantaneous* (or *continuous*) *compounding*. It follows that the value after t years of instantaneous compounding is

$$\lim_{x \to \infty} P_0 \left(1 + \frac{r}{x}\right)^{xt} = P_0 \, e^{rt}.$$

For example, suppose that $1000 is loaned at 6% interest for three years. If interest is compounded yearly, the value at the end of three years is

$$1000 \, (1 + 0.06)^3 \approx \$1191.02.$$

If it is compounded quarterly, the final value is

$$1000 \left(1 + \frac{0.06}{4}\right)^{12} \approx \$1195.62.$$

If it is compounded continuously, the final value is

$$1000 \, e^{(0.06)3} = 1000 \, e^{(0.18)} \approx \$1197.22.$$

Compound interest is considered in more detail in Section 8–5.

Exercises 8–2

1–4. Use the fact that $a = e^b$ if and only if $\ln a = b$ to solve the equations for y.

•1 $e^y = 3x$

2   $e^{y^2+2} = 5x^2 - 1$

•3 $\ln y = 2x + 4$

4   $\ln (y^2 + 1) = 4x^2 + 2$

5 Write the definition of each of the following numbers in symbols. Then calculate the value.
- •(a) $4^{\sqrt{2}}$
- (b) $10^{\ln 10}$
- •(c) $3^{\pi - \ln e}$
- (d) 5^e

6 Make accurate drawings of the graphs of $y = e^x$ ($-3 \leq x \leq 2.5$) and $y = \ln x$ ($0.1 \leq x \leq 10$) on the same axis system. (See Table II.)

7–18. Calculate the derivatives with respect to x.

- 7 $3e^{2x} + 1$
- 8 e^{x^2+x}
- 9 $(e^x + e^{-x})^5$
- 10 $(e^x + 1)^3$
- 11 $\dfrac{e^{2x} - 1}{e^{2x} + 1}$
- 12 $\dfrac{e^{2x-1}}{e^{2x+1}}$
- 13 $e^x \ln x$
- 14 $\dfrac{x^2}{e^x}$
- 15 $\ln \left(\dfrac{e^x + 1}{e^x - 1} \right)$
- 16 $\dfrac{\ln (e^x + 1)}{\ln (e^x - 1)}$
- 17 $\ln [(e^x + 2)^2]$
- 18 $[\ln (e^x + 2)]^2$

19–22. Find all local extrema and points of inflection. Sketch the graphs.

- 19 $y = \dfrac{e^x}{x}$
- 20 $y = e^{-x^2}$
- 21 $y = \dfrac{\ln x}{x}$
- 22 $y = (\ln x)^2$

23 Let $f(x) = Ce^{ax}$. Show that the rate of change of f with respect to x is proportional to the value of $f(x)$.

24–27. Find the values of m that make $y = e^{mx}$ a solution of the given differential equation. Show that if m_1 and m_2 are two such values, then $y = C_1 e^{m_1 x} + C_2 e^{m_2 x}$ is a solution. (*Hint:* Substitute $y = e^{mx}$ in the differential equation and solve for m.)

- 24 $y'' - 3y' + 2y = 0$
- 25 $4y'' + 5y' + y = 0$
- 26 $y'' - 7y' + 2y = 0$
- 27 $y'' + 4y' + 4y = 0$

28 (*Stirling's Formula for Factorials*) The number $n!$ is defined for the positive integer n by

$$n! = n(n - 1)(n - 2) \cdots 2 \cdot 1.$$

For example, $1! = 1$, $2! = 2 \cdot 1 = 2$, $5! = 5 \cdot 4 \cdot 3 \cdot 2 \cdot 1 = 120$. Stirling's formula is

Stirling's Formula

$$n! \approx \sqrt{2\pi n} \left(\dfrac{n}{e} \right)^n$$

if n is large.
- (a) Use Stirling's formula to approximate $50!$.
- c(b) Denote Stirling's approximation to $n!$ by $S(n)$. Calculate the ratio $S(n)/n!$ for $n = 10, 20, 30, 40, 50$, and 60. What tentative conclusion do you reach about the accuracy of Stirling's approximation?

Exercises 8–2

29–32. Use the fact that $\lim_{x\to\infty} (1 + 1/x)^x = e$ to establish the following limits.

29 $\lim_{x\to\infty} \left(1 + \dfrac{r}{x}\right)^x = e^r$

30 $\lim_{x\to\infty} \left(1 + \dfrac{r}{x}\right)^{xt} = e^{rt}$

31 $\lim_{x\to 0^+} (1 + x)^{1/x} = e$

32 $\lim_{x\to 0^-} (1 - x)^{1/x} = \dfrac{1}{e}$

---

33 Use the properties of the natural logarithm function and the fact that $\ln 2 < 1 < \ln 3$ to construct a rigorous proof that there exists one and only one number (which we call e) such that $\ln e = 1$. (*Hint:* See the discussion at the beginning of this section.)

34 Use the properties of the natural logarithm function and the fact that the function exp is the inverse of that function to prove the following properties of the function exp. (*Hint for (a)*: Calculate the natural logarithm of both sides of the equation. Use the fact that $\ln (\exp a) = a$.)

(a) $(\exp a)(\exp b) = \exp (a + b)$

(b) $\dfrac{\exp a}{\exp b} = \exp (a - b)$

(c) $(\exp a)^k = \exp (ak)$

Laws of Exponents

35 (*Laws of Exponents*) In algebra it is proved that the following laws of exponents hold when the exponents are rational numbers. Use the fact that $a^x = \exp (x \ln a)$ and Exercise 34 to prove that the following properties hold for any real exponents provided $a > 0$.

(a) $a^m \cdot a^n = a^{m+n}$

(b) $a^m / a^n = a^{m-n}$

(c) $(a^m)^k = a^{mk}$

36 Fill in the details of the following argument that

$$\lim_{x\to\infty} \left(1 + \dfrac{1}{x}\right)^x = e.$$

(1) Let $f(x) = \ln x$. Then

$$1 = f'(1) = \lim_{\Delta x\to 0} \dfrac{f(1 + \Delta x) - f(1)}{\Delta x} = \lim_{\Delta x\to 0} \ln (1 + \Delta x)^{1/\Delta x}.$$

(2) Since f is continuous, then

$$\lim_{\Delta x\to 0} [\ln (1 + \Delta x)^{1/\Delta x}] = \ln [\lim_{\Delta x\to 0} (1 + \Delta x)^{1/\Delta x}] = f'(1) = 1,$$

so that

$$\lim_{\Delta x\to 0} (1 + \Delta x)^{1/\Delta x} = e^1 = e.$$

(3) Replace Δx by $1/x$ and let $x \to \infty$.

8–3 Integration Formulas. General Exponential and Logarithmic Functions

For each derivative formula there is a corresponding integral formula. The formula corresponding to D9, for example, is

$$\int \frac{1}{u} \, du = \ln u + C, \, u > 0.$$

As it stands, this formula can be used only with positive functions; but a simple device enables us to extend it to negative functions also. Let

$$v = |x| = \begin{cases} x & \text{if } x \geq 0 \\ -x & \text{if } x < 0. \end{cases}$$

Then

$$D_x v = \begin{cases} 1 & \text{if } x > 0 \\ -1 & \text{if } x < 0. \end{cases}$$

It follows that

$$D_x(\ln |x|) = D_x(\ln v) = \frac{1}{v} D_x v = \begin{cases} \frac{1}{x}(1) & \text{if } x > 0 \\ \frac{1}{-x}(-1) & \text{if } x < 0 \end{cases}$$

so that

$$D_x(\ln |x|) = \frac{1}{x} \quad \text{if } x \neq 0.$$

This result can be extended by the Chain Rule to the following formula, also denoted by D9:

Extended Derivative Formula

D9 $\quad D_x(\ln |u|) = \dfrac{1}{u} D_x u \quad$ provided $u(x) \neq 0$.

For example,

$$D_x(\ln |2x^2 - 3x + 1|) = \frac{1}{2x^2 - 3x + 1} \cdot D_x(2x^2 - 3x + 1)$$

$$= \frac{4x - 3}{2x^2 - 3x + 1}.$$

This result holds at any x for which $2x^2 - 3x + 1 \neq 0$, that is, at any x except $x = 1$ and $x = \frac{1}{2}$.

8–3 Integration Formulas. General Exponential and Log Functions

The integration formula that corresponds to D9 is

Basic Integral Formula

I4 $\quad \int \dfrac{1}{u}\, du = \ln |u| + C \quad$ provided $u(x) \neq 0$.

In applying this formula, we must be certain that $u(x) \neq 0$ at any point of the interval that we are considering. Thus, we can use it to calculate

$$\int_a^b \dfrac{1}{x}\, dx$$

provided a and b are both positive or both negative, but not in other cases.

EXAMPLE 1
$$\int_{-4}^{-1} \dfrac{dx}{x} = \Big[\ln |x|\Big]_{-4}^{-1} = \ln |-1| - \ln |-4|$$
$$= \ln 1 - \ln 4 = -\ln 4. \quad \square$$

EXAMPLE 2 Calculate $\int \dfrac{x}{x^2 - 1}\, dx$.

Solution Let $u = x^2 - 1$, $du = 2x\, dx$. Then

$$\int \dfrac{x}{x^2-1}\, dx = \dfrac{1}{2} \int \dfrac{2x}{x^2-1}\, dx = \dfrac{1}{2} \int \dfrac{du}{u}$$
$$= \dfrac{1}{2} \ln |u| + C = \dfrac{1}{2} \ln |x^2 - 1| + C$$

$\boxed{u = x^2 - 1 \\ du = 2x\, dx}$

This formula holds over any interval that does not include $+1$ or -1. $\square$

EXAMPLE 3 Calculate $\int \dfrac{3x^2 + 4x + 1}{x - 1}\, dx$.

Solution We first simplify the function by dividing $x - 1$ into $3x^2 + 4x + 1$:

$$\dfrac{3x^2 + 4x + 1}{x - 1} = 3x + 7 + \dfrac{8}{x - 1}.$$

Then

$$\int \dfrac{3x^2 + 4x + 1}{x - 1}\, dx = \int \left(3x + 7 + \dfrac{8}{x - 1}\right) dx$$
$$= \int (3x + 7)\, dx + 8 \int \dfrac{dx}{x - 1}$$
$$= \dfrac{3x^2}{2} + 7x + 8 \ln |x - 1| + C. \quad \square$$

Logarithmic and Exponential Functions

Remark Any rational function of form

Integration of Certain Rational Functions

$$f(x) = \frac{a_n x^n + a_{n-1} x^{n-1} + \cdots + a_2 x^2 + a_1 x + a_0}{bx + c}$$

can be integrated by the technique illustrated in Example 3. We divide the denominator into the numerator by long division, simplify, and integrate.

THE EXPONENTIAL FUNCTION

Since

$$D_x(e^u) = e^u \cdot D_x u$$

then

Integral of Exponential Function

I5
$$\int e^u \, du = e^u + C.$$

EXAMPLE 4 Calculate $\int e^{2x} \, dx$.

Solution Let $u = 2x$, $du = 2 \, dx$. Then

$$\int e^{2x} \, dx = \frac{1}{2} \int e^{2x} \cdot 2 \, dx = \frac{1}{2} \int e^u \, du$$

$$= \frac{1}{2} e^u + C = \frac{1}{2} e^{2x} + C. \quad \square$$

$u = 2x$
$du = 2 \, dx$

EXAMPLE 5 Calculate $\int_1^4 \frac{e^{\sqrt{x}}}{\sqrt{x}} \, dx$.

Solution Let $u = \sqrt{x}$, $du = dx/2\sqrt{x}$. Observe that $u = 1$ when $x = 1$ and $u = 2$ when $x = 4$. Then

$$\int_1^4 \frac{e^{\sqrt{x}}}{\sqrt{x}} \, dx = 2 \int_1^4 e^{\sqrt{x}} \left(\frac{dx}{2\sqrt{x}} \right)$$

$$= 2 \int_1^2 e^u \, du = 2 \left[e^u \right]_1^2$$

$$= 2[e^2 - e]. \quad \square$$

$u = \sqrt{x}$
$du = \dfrac{dx}{2\sqrt{x}}$
$x = 1 \Rightarrow u = 1$
$x = 4 \Rightarrow u = 2$

EXAMPLE 6 Calculate the area of the region bounded by the graph of $y = 1/e^x$, the y-axis, and the line $y = \frac{1}{2}$.

8–3 Integration Formulas. General Exponential and Log Functions

Solution The region is shown in Figure 8–6. The graphs of $y = 1/e^x$ and $y = \frac{1}{2}$ intersect when

$$\frac{1}{e^x} = \frac{1}{2},$$
$$e^x = 2,$$
$$x = \ln 2.$$

The point of intersection is $(\ln 2, \frac{1}{2})$.

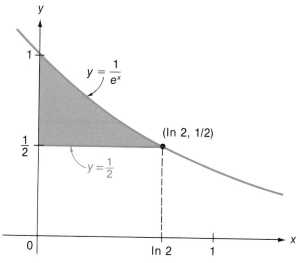

Figure 8–6 Example 6.

The interval $[0, \ln 2]$ is the projection of the region on the x-axis. The area is

$$A = \int_0^{\ln 2} \left(e^{-x} - \frac{1}{2} \right) dx = -\int_0^{\ln 2} e^{-x}(-dx) - \frac{1}{2}\left[x \right]_0^{\ln 2}$$

$$= -\left[e^{-x} \right]_0^{\ln 2} - \frac{1}{2}\left[x \right]_0^{\ln 2}$$

$$= -[e^{-\ln 2} - e^0] - \frac{1}{2}[\ln 2 - 0]$$

$$= -\left[\frac{1}{2} - 1\right] - \frac{1}{2}\ln 2 = \frac{1}{2}(1 - \ln 2) \;\square$$

For reference purposes we give one final integral. The proof of this formula is left for you. (See Exercise 41.)

Integral of Logarithmic Function

I6 $\displaystyle\int \ln u \, du = u \ln u - u + C$ provided $u(x) > 0$.

EXAMPLE 7

$$\int_1^e \ln x \, dx = \left[x \ln x - x \right]_1^e$$
$$= [e \ln e - e] - [1 \ln 1 - 1]$$
$$= e \cdot 1 - e - 1 \cdot 0 + 1 = 1. \; \square$$

GENERAL EXPONENTIAL AND LOGARITHMIC FUNCTIONS

Occasionally we may need to differentiate or integrate an exponential or a logarithmic function with base different from e. This can be accomplished by a simple substitution that essentially converts the function to base e.

The exponential function $f(x) = a^{u(x)}$, where $a > 0$, is easily converted to base e. Recall that

$$a = e^b \quad \text{if and only if } b = \ln a.$$

It follows that

$$a = e^b = e^{\ln a}$$

so that

$$a^{u(x)} = e^{u(x) \ln a}.$$

EXAMPLE 8 Calculate (a) $D_x(3^{2x})$; (b) $\int x \cdot 5^{x^2} \, dx$.

Solution (a) Since $3 = e^{\ln 3}$, then

$$D_x(3^{2x}) = D_x(e^{(\ln 3) \cdot 2x}) = e^{(\ln 3) \cdot 2x} D_x[(\ln 3) \cdot 2x]$$
$$= (e^{\ln 3})^{2x}[(\ln 3) \cdot 2] = 2 (\ln 3) \cdot 3^{2x}.$$

(b) Since $5 = e^{\ln 5}$, then

$$\int x \cdot 5^{x^2} \, dx = \int x \cdot e^{(\ln 5) \cdot x^2} \, dx.$$

Let $u = (\ln 5) x^2$, $du = 2(\ln 5) x \, dx$. Then

$$\int x \cdot 5^{x^2} \, dx = \int e^{(\ln 5) \cdot x^2} x \, dx = \frac{1}{2 \ln 5} \int e^{(\ln 5) x^2} (2(\ln 5)x \, dx)$$

$$= \frac{1}{2 \ln 5} \int e^u \, du = \frac{1}{2 \ln 5} e^u + C$$

$$= \frac{1}{2 \ln 5} e^{(\ln 5) x^2} + C = \frac{1}{2 \ln 5} 5^{x^2} + C. \; \square$$

$u = (\ln 5) \cdot x^2$

$du = 2(\ln 5) \cdot x \, dx$

8–3 Integration Formulas. General Exponential and Log Functions

The methods of Example 8 can be used to generalize formulas D10 and I5. The more general formulas are

Extended Derivative and Integral Formulas

$$D_x(a^u) = (\ln a)\, a^u\, D_x u$$

$$\int a^u\, du = \frac{1}{\ln a} a^u + C$$

where a is a positive constant.

Logarithmic functions require a different type of substitution. Suppose the logarithms are to the base b. It follows from the relation

$$a = b^c \quad \text{if and only if } c = \log_b a,$$

that

$$a = b^{\log_b a} \quad \text{if } a > 0.$$

Then

$$\ln a = \ln(b^{\log_b a}) = \log_b a \cdot \ln b$$

so that

Conversion Formula for Logarithms to Other Bases

$$\log_b a = \frac{\ln a}{\ln b}.$$

We use this substitution to convert logarithms from base b to natural logarithms.

EXAMPLE 9 Calculate (a) $D_x(\log_3(x^2 + 2))$; (b) $\int \log_2(3x + 1)\, dx$.

Solution (a) $D_x(\log_3(x^2 + 2)) = D_x\left(\dfrac{\ln(x^2 + 2)}{\ln 3}\right)$

$$= \frac{1}{\ln 3} \cdot \frac{1}{x^2 + 2} \cdot D_x(x^2 + 2)$$

$$= \frac{2x}{\ln 3 \cdot (x^2 + 2)}.$$

(b) $\displaystyle\int \log_2(3x + 1)\, dx = \int \frac{\ln(3x + 1)}{\ln 2}\, dx = \frac{1}{\ln 2}\int \ln(3x + 1)\, dx.$

Let $u = 3x + 1$, $du = 3\, dx$. Then

$$\int \log_2(3x + 1)\, dx = \frac{1}{3 \ln 2}\int \ln(3x + 1) \cdot 3\, dx$$

$$= \frac{1}{3 \ln 2}\int \ln u\, du$$

$\boxed{\begin{array}{l} u = 3x + 1 \\ du = 3\, dx \end{array}}$

$$= \frac{1}{3 \ln 2} \left[u \ln u - u \right] + C \quad \text{(by I6)}$$

$$= \frac{1}{3 \ln 2} \left[(3x + 1) \ln (3x + 1) - (3x + 1) \right] + C. \quad \square$$

LOGARITHMIC DIFFERENTIATION

Let f be a differentiable function that is not identically zero. It follows from D9 that if $f(x) \neq 0$, then

$$D_x(\ln |f(x)|) = \frac{f'(x)}{f(x)}$$

so that

$$f'(x) = f(x) \, D_x(\ln |f(x)|).$$

This relation is the basis for a method called *logarithmic differentiation*. We calculate the derivative of $\ln |f(x)|$ (or $\ln f(x)$ if $f(x) > 0$) and use it to calculate $f'(x)$.

Logarithmic differentiation is useful for functions that involve variables raised to variable powers (such as x^x) or functions that are complicated products and quotients. The following examples illustrate the method.

EXAMPLE 10   Calculate the derivative of $f(x) = x^x$ where $x > 0$.

Solution   We use logarithmic differentiation. Since

$$\ln f(x) = \ln (x^x) = x \ln x,$$

Example of Logarithmic Differentiation

then

$$D_x(\ln f(x)) = D_x(x \ln x)$$

$$\frac{1}{f(x)} \cdot f'(x) = x \cdot \frac{1}{x} + \ln x = 1 + \ln x.$$

Therefore,

$$f'(x) = f(x)(1 + \ln x) = x^x (1 + \ln x). \quad \square$$

EXAMPLE 11   Use logarithmic differentiation to calculate y' if

$$y = \frac{\sqrt[3]{3x^2 - 2}}{(x^3 - 1)(x^2 + 2)^4}.$$

Solution   We first calculate $\ln |y|$:

$$\ln |y| = \ln \left(\frac{\sqrt[3]{|3x^2 - 2|}}{|x^3 - 1| \cdot |x^2 + 2|^4} \right)$$

$$= \frac{1}{3} \ln |3x^2 - 2| - \ln |x^3 - 1| + 4 \ln |x^2 + 2|.$$

Exercises 8–3

Then

$$D_x(\ln |y|) = \frac{1}{3} \cdot \frac{1}{3x^2 - 2} \cdot 6x - \frac{1}{x^3 - 1} \cdot 3x^2 + 4 \cdot \frac{1}{x^2 + 2} \cdot 2x$$

$$\frac{1}{y} \cdot y' = \frac{2x}{3x^2 - 2} - \frac{3x^2}{x^3 - 1} + \frac{8x}{x^2 + 2}.$$

Therefore

$$y' = y \left[\frac{2x}{3x^2 - 2} - \frac{3x^2}{x^3 - 1} + \frac{8x}{x^2 + 2} \right]$$

$$= \frac{\sqrt[3]{3x^2 - 2}}{(x^3 - 1)(x^2 + 2)^4} \left[\frac{2x}{3x^2 - 2} - \frac{3x^2}{x^3 - 1} + \frac{8x}{x^2 + 2} \right]. \;\square$$

Exercises 8–3

1–22. Calculate the integrals.

- 1. $\displaystyle\int \frac{2}{x}\,dx$
- 2. $\displaystyle\int \frac{dx}{3x - 4}$
- 3. $\displaystyle\int \left(3 - \frac{4}{x}\right)^2 dx$
- 4. $\displaystyle\int \frac{4x^2 + 3x + 1}{x^2}\,dx$
- 5. $\displaystyle\int \frac{(\ln x)^3}{x}\,dx$
- 6. $\displaystyle\int [(\ln x)^2 + 2 \ln x + 5]\frac{dx}{x}$
- 7. $\displaystyle\int \frac{4x^2 + 3x + 1}{x - 1}\,dx$
- 8. $\displaystyle\int \frac{x^3 - x^2 - 2x + 4}{x + 2}\,dx$
- 9. $\displaystyle\int e^{3x}\,dx$
- 10. $\displaystyle\int e^{2x-4}\,dx$
- 11. $\displaystyle\int (x - 1)e^{x^2 - 2x + 2}\,dx$
- 12. $\displaystyle\int \frac{e^x}{e^x + 1}\,dx$
- 13. $\displaystyle\int \frac{e^x}{(e^x + 1)^2}\,dx$
- 14. $\displaystyle\int \frac{(e^x + 1)^2}{e^x}\,dx$
- 15. $\displaystyle\int \frac{e^{\sqrt{x}}}{\sqrt{x}}\,dx$
- 16. $\displaystyle\int \frac{e^{\sqrt{x}}}{\sqrt{x}\,(e^{\sqrt{x}} + 1)}\,dx$
- 17. $\displaystyle\int_0^1 \frac{1}{2 - x}\,dx$
- 18. $\displaystyle\int_0^1 \frac{x^3}{x^2 - 4}\,dx$
- 19. $\displaystyle\int_1^4 \frac{1}{\sqrt{x}\,(1 + \sqrt{x})}\,dx$
- 20. $\displaystyle\int_0^1 \frac{dx}{e^{3x}}$
- 21. $\displaystyle\int_1^2 \frac{e^x}{e^x + 1}\,dx$
- 22. $\displaystyle\int_0^{\ln 2} \frac{e^x - 1}{e^x}\,dx$

23–26. Calculate the derivatives with respect to x.

•23 5^x

24 3^{x^2}

•25 $\log_3 (x^2 + 1)$

26 $\log_{10} \left(\dfrac{2x + 3}{4x + 1} \right)$

27–30. Calculate the integrals.

•27 $\displaystyle\int 5^x \, dx$

28 $\displaystyle\int x \cdot 2^{x^2+1} \, dx$

•29 $\displaystyle\int \dfrac{4^{1/x}}{x^2} \, dx$

30 $\displaystyle\int \dfrac{3^{\ln x}}{x} \, dx.$

31–34. Use logarithmic differentiation to calculate the derivatives of these functions of x.

•31 $(1 + x)^x$ (for $x > -1$)

32 $x^{1/x}$ (for $x > 0$)

•33 $u = \dfrac{(5x^2 + 3x + 2)^{17}}{(4x + 3)^5}$

•34 $u = x^{x^2}$

35 Determine the following properties of the graph of $y = 2^{-x}$: (1) symmetry; (2) concavity; (3) asymptotes; (4) local extrema. Sketch the graph.

•36 Find the area of the region bounded by the graphs of $y = e^{2x}$, $y = e^x$, $x = \ln 5$.

•37 Let $\mathcal{R}$ be the region bounded by the graph of $f(x) = 1/\sqrt{x}$, the x-axis, and the lines $x = 1$, $x = 9$. Calculate the volume of the solid of revolution obtained by revolving $\mathcal{R}$ about
  (a) the x-axis; (b) the line $x = 9$.

•38 Find the centroid of the region $\mathcal{R}$ of Exercise 37.

•39 Find the centroid of the region bounded by the graph of $y = 1/x$, the vertical line $x = 3$, and the horizontal line $y = 2$.

•40 The base of a solid is the closed region in the xy-plane bounded by the graphs of $y = e^x$, $y = 0$, $x = 0$, and $x = 1$. Each cross section perpendicular to the x-axis is an equilateral triangle. Find the volume of the solid.

41 Prove Formula 16. (*Hint:* Show that $\ln x$ is the derivative of $x(\ln x - 1)$.)

42 An alternate proof that $D_x(\ln |u|) = (D_x u)/u$ can be obtained by differentiating the equation $2 \ln |u| = \ln u^2$ with respect to x. Show that this equation is an identity and use it to prove the derivative formula.

8–4 The Differential Equation $dy/dx = ky$

Differential equations were considered briefly in Section 4–10. Recall that a differential equation is an equation involving a function and its derivative, such as

$$y'' - 3y' + 2y = 4x,$$
$$y' = 3x,$$
$$y' = 2(y - 1).$$

Initial Conditions   Many problems lead to differential equations with *initial conditions*. These are statements that specify the values that the solution function and (possibly) some of its derivatives must have at a particular point in the domain. If the original problem involves distance, velocity, and acceleration, for example, then the initial conditions might be the velocity and position at time t_0.

Boundary Conditions   Other problems involve *boundary conditions*. These are statements that give values that the solution function must take at several different points in its domain. We might, for example, want the solution $f(x)$ to have the value 2 at $x = 1$, the value 7 at $x = 2$, and the value -5 at $x = 4$.

SEPARATION OF VARIABLES

A general method, known as *separation of variables*, can be used to solve certain types of differential equations. We write the derivative y' in the form dy/dx, which we regard as a quotient of differentials (Section 3–3). If possible, we then transform the differential equation by algebraic manipulation to the form

$$f(y)\, dy = g(x)\, dx.$$

If this last form can be obtained, then we integrate both sides of the equation to get an equation relating x and y. (This method is examined in more detail in Chapter 19.)

EXAMPLE 1   (a) Solve the differential equation $2xy' = 3$ by separation of variables.
(b) Find the solution that satisfies the initial condition $y = 2$ when $x = 1$.

Solution   (a) The differential equation can be rewritten as

$$2x \cdot \frac{dy}{dx} = 3$$

$$2\, dy = \frac{3\, dx}{x}.$$

Separation of Variables Example   To find y, we integrate both sides of this equation, obtaining

$$2\int dy = 3\int \frac{dx}{x} + C_1$$

$$2y = 3 \ln |x| + C_1$$

$$y = \frac{3}{2} \ln |x| + C$$

where $C = C_1/2$.

(b) To have $y = 2$ when $x = 1$, we must have

$$2 = \frac{3}{2} \ln |1| + C$$

$$2 = \frac{3}{2} \cdot 0 + C$$

$$C = 2.$$

The solution is

$$y = \frac{3}{2} \ln |x| + 2. \quad \square$$

THE DIFFERENTIAL EQUATION $dy/dt = ky$

The differential equation

$$\frac{dy}{dt} = ky$$

can be used to solve many applied problems. If the rate of change of a quantity is proportional to its size, then a function for the size at time t can be determined from the above differential equation.

Radioactive decay furnishes a good example. Over a fixed period, a certain proportion of a radioactive element will decay, changing into another element, and a certain proportion will remain unchanged. If $y = y(t)$ is the amount of the radioactive element at time t, then the rate of change of y with time is proportional to the size of y. Thus,

$$\frac{dy}{dt} = ky$$

where k is the constant of proportionality. For radioactive decay (where y is decreasing), we have $dy/dt < 0$, so that $k < 0$. Growth would be represented by $k > 0$.

EXAMPLE 2 Show that the solution of the differential equation

$$\frac{dy}{dt} = ky$$

is

Solution of $\frac{dy}{dt} = kt$

$$y = Ce^{kt}$$

where C is a constant.

Solution We separate the variables and integrate:

$$\frac{dy}{dt} = ky$$

8–4 The Differential Equation $dy/dx = ky$

$$\frac{dy}{y} = k\, dt$$

$$\int \frac{dy}{y} = k \int dt + C_1$$

$$\ln |y| = kt + C_1$$

where C_1 is the constant of integration.

This solution can be written in a more usable form by raising e to the power of the two sides of the equation:

$$e^{\ln |y|} = e^{kt+C_1} = e^{C_1}\, e^{kt}$$
$$|y| = e^{C_1}\, e^{kt}$$
$$y = (\pm e^{C_1})\, e^{kt}$$
$$y = Ce^{kt}$$

where $C = \pm e^{C_1}$. □

EXAMPLE 3 Radioactive carbon (carbon 14) has a *half-life* of 5568 years. At the end of that period, half of the original amount will remain.

In 1940 a laboratory collected 2 g of radioactive carbon.
(a) Derive the formula for the amount that will remain after t years.
(b) How much will remain in the year 2000?

Solution (a) If y is the amount at time t, then $dy/dt = ky$, where k is the constant of proportionality. Using the result of Example 2, we get

$$y = Ce^{kt}$$

where C is a constant.

Our problem now reduces to determining the unknown constants C and k. Let t be the elapsed time (in years) since 1940. Since $y = 2$ when $t = 0$, then

$$2 = Ce^{k \cdot 0} = C \cdot 1 = C$$
$$C = 2.$$

Thus, the solution of the differential equation has the form

$$y = 2e^{kt}.$$

To determine k, recall that y will equal 1 (one-half the original amount) when $t = 5568$. Thus,

$$1 = 2e^{k \cdot 5568}$$
$$\tfrac{1}{2} = e^{5568k}$$
$$5568k = \ln\left(\tfrac{1}{2}\right)$$
$$k = \frac{\ln\left(\tfrac{1}{2}\right)}{5568} \approx -0.000124$$

The particular solution is

$$y = Ce^{kt} = 2e^{\frac{\ln\left(\tfrac{1}{2}\right)}{5568} t}$$

This expression can be simplified somewhat if we recall that

$$e^{\ln(1/2)} = \tfrac{1}{2}.$$

The solution is

$$y = 2e^{\frac{\ln(\frac{1}{2})}{5568}t} = 2[e^{\ln(1/2)}]^{t/5568}$$
$$= 2(\tfrac{1}{2})^{t/5568}.$$

(b) In the year 2000 ($t = 60$), the amount remaining will be

$$y = 2(\tfrac{1}{2})^{60/5568} \approx 1.985 \text{ g.} \quad \square$$

Exercises 8–4

1–10. Use separation of variables to solve the differential equations for $y = f(x)$. (*Hint for 9 and 10:* Make the substitution $u = y'$.)

- •1 $\dfrac{dy}{dx} = \dfrac{x+1}{y+1}$
- 2 $\dfrac{dy}{dx} = \dfrac{y-1}{x+1}$
- •3 $y' = (x+1)(y-2)$
- 4 $y'(x+3) = 2xy$
- •5 $y' = \dfrac{2xy}{y^2+1}$
- 6 $y' = x - 4xy$
- •7 $y' = 4y$
- 8 $y' = -3(y-2)$
- •9 $y'' = 2y'$
- 10 $y'' = -4y'$

11–20. Solve the differential equations in Exercises 1–10 subject to the given initial or boundary conditions.

- •11 Exercise 1: $y = 2$ when $x = 0$
- 12 Exercise 2: $y = 5$ when $x = 1$
- •13 Exercise 3: $(0, -4)$ is on the graph
- 14 Exercise 4: $(1, 3e^2)$ is on the graph
- •15 Exercise 5: $f(4) = 1$
- 16 Exercise 6: $f(1) = 4$
- •17 Exercise 7: $f(0) = -4$
- 18 Exercise 8: $f(0) = 0$
- •19 Exercise 9: The graph passes through the points $(0, 1)$ and $(\ln 2, -5)$.
- •20 Exercise 10: The graph passes through $(0, 4)$ and has a tangent line with slope $m = \tfrac{1}{2}$ at this point.

21–24. The general solution of a differential equation of form $y^{(n)} = F(x)$ can be obtained by integration provided F is continuous. Find the solution $y = f(x)$ of the differential equation

$$y'' = \dfrac{4}{x^2} + 12x + 2$$

that satisfies the given initial or boundary conditions.

- •21 $f(1) = 2, f'(1) = 2$

22 $y = 0$ when $x = 1$, $y' = 1$ when $x = 1$

• 23 $f(1) = 2$, $f(2) = -1$

24 The points $(-1, 1)$ and $(-2, 3)$ are on the graph

• 25 Radium 226 has a half-life of 1620 years. At time $t = 0$ a mass of 200 g of radium is stored in a cylinder.
  (a) How much remains after t years?
  (b) In how many years will one-third of the original amount remain?

• 26 The decay rate of a radioactive element is such that there is a 40 percent loss over two years.
  (a) How much of the original amount remains after five years?
  (b) What is the half-life?

• 27 An analysis of an ancient campfire shows that one-tenth of the carbon 14 in the original ashes has decomposed. Use this information to date the campsite. (The half-life of carbon 14 is 5568 years.)

• 28 At the instant when a radioactive substance has a mass of 20 g, it is decaying at the rate of 0.06 g per week. Find the half-life.

• 29 The average rate of change of a function f from 0 to x is always equal to its rate of change at x. Find the function if $f(0) = 5$ and $f(1) = 2$.

Law of Cooling

30 (*Newton's Law of Cooling*) An object with an initial temperature $T = T_0$ is placed in a controlled environment with constant temperature K. According to Newton's law of cooling, the rate of change of the temperature T with respect to time t is proportional to the difference $T - K$.
  • (a) State Newton's law of cooling as a differential equation.
  (b) Show that the solution of the differential equation in (a) is

$$T = K + (T_0 - K)e^{mt}$$

where m is the constant of proportionality from the law of cooling.

• 31 An object with an intial temperature T_0 is placed in a controlled environment with temperature $K = 70°$. (See Exercise 30.) Find its temperature after 1 hour subject to the following conditions:
  (a) $T_0 = 90°$, temperature is $80°$ after $\frac{1}{2}$ hour;
  (b) $T_0 = 100°$, temperature is $75°$ after 40 minutes;
  (c) $T_0 = 40°$, temperature is $50°$ after 15 minutes;
  (d) $T_0 = 20°$, temperature is $30°$ after 5 minutes.

8–5 Laws of Growth

COMPOUND INTEREST

The limit

$$\lim_{x \to \infty} \left(1 + \frac{1}{x}\right)^x = e$$

was considered in Section 8–2. As we then saw, this limit can be interpreted in terms of compound interest. If an amount P_0 is invested at an interest rate of r per year and compounded n times per year, then the value of the investment after t years is

$$P_0 \left(1 + \frac{r}{n}\right)^{nt}.$$

If the interest is compounded continuously, then the value after t years is

$$\lim_{n \to \infty} P_0 \left(1 + \frac{r}{n}\right)^{nt} = P_0 e^{rt}.$$

These results on compound interest can be derived another way. If interest is compounded continuously at a rate of r per year, then at each instant, the rate of change of the value is proportional to the value, with r equal to the constant of proportionality. If $P = P(t)$ is the value of the investment at time t, then

$$\frac{dP}{dt} = rP.$$

If we solve this differential equation and use the fact that $P(0) = P_0$, we obtain the solution

$$P = P_0 e^{rt}.$$

EXAMPLE 1   A thousand dollars is put into a savings account that pays 5.5% interest, compounded continuously. What is the value of the account after one year? three years?

Solution   The value after one year ($t = 1$) is

$$Pe^{rt} = 1000 e^{0.055} \approx \$1056.54.$$

After three years ($t = 3$), the value is

$$Pe^{rt} = 1000 e^{(0.055) \cdot 3} \approx \$1179.39. \quad \square$$

EFFECTIVE RATES

A certain amount of confusion usually arises when interest is compounded continuously. The problem concerns the *actual rate* and the *effective annual rate*. The *effective annual rate* is the interest rate that, compounded once a year, would yield the same return in interest as does the actual rate compounded continuously.

For example, if $1 is invested at 5.5% interest per year, compounded continuously, then the value of the investment after one year is

$$V = 1 \cdot e^{(0.055) \cdot 1} \approx \$1.05654.$$

One dollar of this amount represents the original investment. The interest earned during the year is

$$\text{Interest} \approx \$0.05654.$$

8–5 Laws of Growth

Thus, the effective annual interest rate is

$$r \approx 5.654\%.$$

Similarly, there is an effective rate for continuous compounding that corresponds to any given annual rate.

EXAMPLE 2   A bank pays 6% interest per year, compounded annually. What is the effective rate for continuous compounding?

Solution   Let r be the effective rate for continuous compounding. The value of an investment of \$1 after one year is

$$1 \cdot e^{r \cdot 1} = 1 \cdot (1 + .06)$$
$$e^r = 1.06$$
$$r = \ln(1.06) \approx 0.05827.$$

The effective interest rate for continuous compounding is approximately 5.827%. This interest rate yields the same return as 6% interest, compounded annually. □

Numerous problems involving investments can be solved by using effective rates and continuous compounding.

EXAMPLE 3   Mr. Smith purchased \$100,000-worth of bank bonds that pay 7% interest after one year. He has owned the bonds for seven months and wishes to sell them now. (They cannot be redeemed for another five months.) What is their current value?

Solution   Assume that the interest is compounded continuously at an interest rate of r per year. After one year the value will be

$$V = 100{,}000 e^{r \cdot 1} = 107{,}000.$$

Then

$$e^r = 1.07$$
$$r = \ln(1.07) \approx 0.067659.$$

It follows that the value after t years is

$$v = 100{,}000 e^{rt} \approx 100{,}000 e^{0.067659 t}.$$

When $t = \frac{7}{12}$ (the present time),

$$V = 100{,}000 e^{(0.067659) \cdot 7/12} \approx \$104{,}025.70.\ \square$$

POPULATION GROWTH

In the study of population growth, it is assumed that the factors that influence growth do not change over a few generations. If this assumption holds, the result is a short-term situation for which the rate of change of the population is proportional to its size. For example, if a population increases by 5% one year and the factors influencing growth do not change, then it should increase by 5% the next year, and so on. Thus, the rate of change of the population is proportional to the size of the population.*

If $P = P(t)$ is the population at time t, then the assumption that the rate of change of the population is proportional to the size of the population can be expressed as the differential equation

$$\frac{dP}{dt} = kP$$

where k is the constant of proportionality. The solution is

Unrestricted Growth Law

$$P = Ce^{kt}$$

where C is a constant.

EXAMPLE 4 At time $t = 0$, 5 million bacteria were alive and well in George's lung. Three hours later the number had increased to 9 million. Assuming that the conditions for growth do not change over the next few hours, find the number of bacteria at the end of (a) 12 hr; (b) 36 hr.

Solution The population size at time t is

$$P(t) = Ce^{kt}$$

where C and k are constants.

To evaluate the constants, we use the facts that

$$P(0) = 5 \text{ and } P(3) = 9$$

where $P(t)$ is measured in millions of bacteria. Since $P(0) = 5$, then

$$5 = Ce^{k \cdot 0} = C.$$

Thus, the population function is of form

$$P(t) = 5e^{kt}.$$

Since $P(3) = 9$, then

$$9 = 5e^{k \cdot 3}$$

$$e^{3k} = \frac{9}{5}$$

$$3k = \ln \frac{9}{5}$$

$$k = \frac{1}{3} \ln \frac{9}{5} \approx 0.195929.$$

* In a strict sense, this assertion is not realistic. Over a very short period (say one-tenth of a second), the population probably will not change at all. Thus, the "instantaneous" rate of change, indicated by the derivative dP/dt, should be zero most of the time and nonzero at a few isolated moments. To make sense of the assumption, we must consider it as describing the rate of change at a mythical "average" time.

The population function is

$$P(t) = 5e^{t\ln(9/5)/3} \approx 5e^{0.195929t}.$$

Observe that the function can be written more compactly as

$$P(t) = 5[e^{\ln(9/5)}]^{t/3} = 5\left(\frac{9}{5}\right)^{t/3} = 5(1.8)^{t/3}.$$

(a) When $t = 12$,

$$P = P(12) = 5\left(\frac{9}{5}\right)^{12/3} = 5(1.8)^4 \approx 52.5 \text{ million.}$$

(b) When $t = 36$,

$$P = P(36) = 5(1.8)^{36/3} = 5(1.8)^{12} \approx 5784 \text{ million.} \quad \square$$

INHIBITED GROWTH RATES

It is obvious that the situation described in Example 4 cannot continue indefinitely. Eventually the growth rate must slow down. Either George will die or the defense system of his body will produce enough antibodies to kill the bacteria.

A similar situation must prevail in any population growth. The unrestricted growth described by the differential equation $dP/dt = kP$ cannot continue indefinitely. Eventually, if nothing else happens, the sheer size of the population makes the acquisition of food more difficult and thereby slows the rate of increase.

It makes sense to assume that the law of unrestricted growth can continue for a short period. Over a very long period, however, an inhibited growth rate is more realistic.

One law of inhibited growth is based on the assumption that the maximum possible size of the population that can be supported is a number M. In that case it seems reasonable to assume that the rate of change of the population is jointly proportional to P, which represents the potential to reproduce, and $M - P$, which represents the potential to expand. Thus, P is a solution of the differential equation

$$\frac{dP}{dt} = kP(M - P).$$

It is shown in Exercise 15 that this differential equation has the solution

Inhibited Growth Law
$$P = \frac{Me^{Mkt}}{e^{Mkt} + C}$$

where C is a constant. Additional work with this function is provided in Exercise 16.

Remark The solution of the differential equation $dP/dt = kP$, which leads to the law of unrestricted growth, is

$$P = Ce^{kt},$$

which is the same as the formula for instantaneous compounding of interest. It follows that there are equivalent problems involving effective rates and actual rates. Figure 8–7 shows how these restricted and unrestricted growth laws behave for the same initial populations and the same value of k.

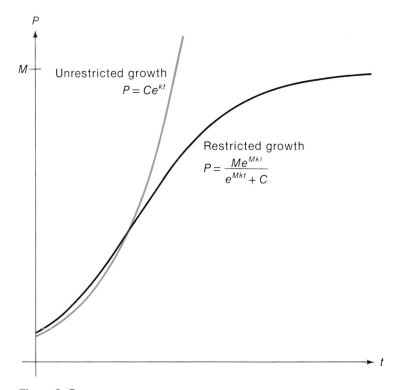

Figure 8–7.

Population rates are usually given as annual rates. When we say that a population increases by 5% annually, for example, we mean that it actually measures a 5% increase over any given one-year period. This does not mean that the number k equals 0.05 in the formulas

$$\frac{dP}{dt} = kP \quad \text{and} \quad P = ce^{kt}.$$

The number k, in an analogy with interest rates, is the population rate that must be

Exercises 8–5

compounded continuously to get an effective annual rate of 5% per year. Thus, k is slightly less than 0.05. The exact value can be calculated by the procedure used in Example 2.

Exercises 8–5

1–4. Find the effective rate for continuous compounding of interest for an annual rate r with interest compounded n times per year.

•1 $r = 6\%, n = 2$ 2 $r = 6\%, n = 12$

•3 $r = 8\%, n = 2$ 4 $r = 8\%, n = 4$

5–8. What is the effective annual rate of interest if the rate for continuous compounding is r per year?

• 5 $r = 5\%$ 6 $r = 6\%$

• 7 $r = 8\%$ 8 $r = 10\%$

• 9 A culture of bacteria contained 7 million bacteria at 3:00 P.M. and 14 million at 7:00 P.M. Assume that the conditions of growth do not change over the period under consideration.
   (a) How many were in the culture at 5:00 P.M.?
   (b) How many will there be at 10:00 P.M.?

•10 The population of a city increases at an annual rate of 5% per year. In how many years will the population double? triple?

•11 Five years ago a tract of land was purchased for $20,000. It is expected to be worth $60,000 in another 15 years. Assume that the value of the investment increases at a rate proportional to the value. Find its current value.

Depreciation •12 (*Depreciation*) Three years ago George bought a Gasdrinker Super Sport automobile for $9000. At the present time it is worth $6000. Assume that the rate at which the value changes is proportional to the value. What will it be worth in another seven years?

13 (*Depreciation*) Most household appliances have an expected useful life of ten years. At that time an appliance has a junk value approximately equal to one-tenth of its purchase price. Assume that the value changes at a rate proportional to the value.
   •(a) A refrigerator was purchased five years ago for $400. Find its current value.
   (b) Show that the value of an appliance is halved approximately every three years until it is ten years old.

•14 The population of the United States was approximately 151 million in 1950 and approximately 179 million in 1960. Use the formula for unrestricted growth to calculate the population expected in 1980. Compare this figure with the true population in 1980 as found in an almanac. What, if anything, does this comparison show?

15 *(Inhibited Growth Rate)* As discussed in the text, many populations increase according to the differential equation

$$\frac{dP}{dt} = kP(M - P)$$

where M is the maximum population that can be supported and k is a constant.

(a) Show that the solution of this differential equation is

$$P = \frac{Me^{Mkt}}{e^{Mkt} + C}$$

where C is a constant. [*Hint:* Separate the variables. Observe that

$$\frac{1}{P(M-P)} = \frac{1}{M}\left(\frac{1}{M-P} + \frac{1}{P}\right).]$$

(b) Show that the graph of $P(t)$ has a point of inflection at $([\ln C]/Mk, M/2)$. Now review the subsection "Application to Population Growth" near the end of Section 4–5.

•(c) Solve the differential equation in (a) subject to the following conditions: $M = 20$ million; $P = 5$ million when $t = 0$; $P = 8$ million when $t = 10$.

(d) Sketch the graph of the solution found in (c). (*Hint:* First find the horizontal asymptotes as $t \to \infty$ and $t \to -\infty$.)

16 *(Inhibited Growth Rate)* Three years ago a small lake was stocked with 1000 fish. At present the population has increased to 4000. The owners plan to maintain the population at a fixed number N by allowing a fixed number K of fish to be caught each year. They have been told by the state wildlife commission that the lake can support a maximum population of 10,000 fish.

•(a) Use the formula from Exercise 15 to determine what the population will be at time t provided that no fishing is allowed.

(b) Show that once fishing is allowed, the value of K will approximately equal dP/dt, evaluated at $P = N$.

(c) Show that the maximum value of K is reached when $N = 5000$. What is this maximum value of K?

8–6 Laws of Motion

Many of the applications of the calculus to science and engineering are consequences of *Newton's second law of motion:* If F is the force that causes a body of mass m to move in a straight line with acceleration a, then

Newton's Second Law of Motion

$$F = ma$$

(force equals mass times acceleration).

In Section 4–10 we considered the motion of a body in free fall in a vacuum near

8–6 Laws of Motion

the surface of the earth. We based our derivation on the fact that

$$\text{Acceleration} = \frac{dv}{dt} = -32 \text{ ft/sec}^2.$$

We now show how this relation can be derived.

Newton's Law of Gravitational Attraction

Newton's law of gravitational attraction states that the force of attraction between two bodies of masses m_1 and m_2 equals

$$\frac{Km_1m_2}{d^2}$$

where d is the distance between their centers of mass and K is a constant that depends on the units of measurement. If we use the centimeter-gram-second system of measurement, then

$$K \approx (6.66)10^{-8}.$$

(In this system, distance is measured in *centimeters*, mass in *grams*, and time in *seconds*. The unit of force is the *dyne*.) Since the mass of the earth is $M \approx (5.98)10^{27}$ grams and the radius is $R \approx (6.37)10^8$ cm, then an object of mass m, falling near the surface of the earth, is subject to a gravitational force

$$F = \frac{mMK}{R^2} \approx \frac{m(5.98)10^{27}(6.66)10^{-8}}{[(6.37)10^8]^2}$$

$$\approx 981m \text{ dynes}.$$

On the other hand, by Newton's second law of motion, $F = ma$, where a is the acceleration. Therefore,

$$F = ma \approx 981m$$

so that

$$a \approx 981 \text{ cm/sec}^2.$$

If we convert from centimeters to feet (recall that 1 in. $\approx$ 2.54 cm), we obtain

$$a \approx 32.17 \text{ ft/sec}^2.$$

It is customary to denote the "gravitational constant" by g and to write the force due to gravity as

$$F = mg$$

Gravitational Constant

where m is the mass. For convenience in our computations, we use the approximations

$$g = 981 \quad \text{in the centimeter-gram-second system}$$

and

$$g = 32 \quad \text{in the foot-pound-second system.}$$

MOTION SUBJECT TO AIR RESISTANCE

A realistic analysis of the motion of a falling body can be obtained by allowing for air resistance. Suppose, for example, that the air resistance is proportional to the velocity—a reasonable assumption for objects of certain shapes. In this case, the body is subject to two forces:

(1) A gravitational force equal to mg, which accelerates it toward the earth.
(2) A force kv, which acts in the direction opposite the velocity.

Thus, if $k > 0$, the resisting force is $-kv$.

If we orient upward the axis that we use to measure distance, then the total force acting on the object is

$$F = -kv - mg.$$

By Newton's second law of motion,

$$F = ma = -kv - mg$$

so that

$$a = -\frac{k}{m}v - g$$

Differential Equation for Falling Bodies Subject to a Particular Force Due to Air Resistance

$$\frac{dv}{dt} = -\frac{k}{m}v - g.$$

This differential equation can be solved by separation of variables, yielding the velocity function, which in turn can be integrated to get the distance (position) function.

In addition, several properties of the motion of a body in free fall can be obtained directly from the above differential equation. It can be shown, for example, that the velocity function has a limiting value when $t \to \infty$. This value is called the *terminal velocity*.

EXAMPLE 1   A body falls from rest from a point 2 kilometers (km) above the surface of the earth. It has a mass of 5 g and is subject to a resisting force numerically equal to kv, where $k = 0.5$.
(a) calculate $v(t)$.
(b) calculate $s(t)$.
(c) Find the terminal velocity.

Solution   (a) The velocity function satisfies the differential equation

Example of A Falling Body Subject to Air Resistance

$$\frac{dv}{dt} = -\frac{k}{m}v - g = -\frac{0.5}{5}v - 981$$
$$= -0.1(v + 9810).$$

8-6 Laws of Motion

Separating the variables and integrating, we get

$$\frac{dv}{v + 9810} = -0.1\,dt$$

$$\int \frac{dv}{v + 9810} = -0.1 \int dt + C_1$$

$$\ln|v + 9810| = -0.1t + C_1$$

$$v + 9810 = (\pm e^{C_1})e^{-0.1t} = Ce^{-0.1t}$$

where $C = \pm e^{C_1}$.

To evaluate the constant, we recall that $v = 0$ when $t = 0$. Therefore,

$$0 + 9810 = Ce^0 = C$$
$$C = 9810,$$

so that

$$v(t) = -9810 + 9810e^{-0.1t}.$$

Observe that the velocity function is negative for all $t > 0$.

(b) The distance function s is an antiderivative of the velocity function. Thus,

$$s(t) = \int v(t)\,dt = \int (-9810 + 9810e^{-0.1t})\,dt$$

$$= -9810t + \frac{9810}{-0.1}e^{-0.1t} + C_2$$

$$= -9810t - 98100e^{-0.1t} + C_2.$$

To evaluate the constant C_2, recall that $s = 2$ km $= 200{,}000$ cm when $t = 0$. Thus,

$$200{,}000 = (-9810)\cdot 0 - 98{,}100e^0 + C_2$$
$$C_2 = 200{,}000 + 98{,}100 = 298{,}100.$$

The distance function is

$$s(t) = -9810t - 98{,}100e^{-0.1t} + 298{,}100 \text{ cm.}$$

(c) The terminal velocity equals

$$\lim_{t \to \infty} v(t) = \lim_{t \to \infty}\left[-9810 + \frac{9810}{e^{0.1t}}\right]$$

$$= -9810 + 0 = -9810. \quad \square$$

WEIGHT AND MASS

The *weight* of an object is a measure of the force that gravity exerts on it. We get a different value, for example, when we weigh an object on the moon from what we get when we weigh it on earth. To get a measure that depends only on the intrinsic properties of the object, we divide its weight by the appropriate gravitational constant g, obtaining its *mass*.

Foot-Pound-Second System

In the *foot-pound-second* system, the *weight* (a measure of force) is in *pounds*. When we divide weight by $g = 32$, we get the mass,

$$\text{Mass} = \frac{\text{weight}}{32}$$

provided the object is at sea level. The unit of mass is the *slug*. For example, a man who weighs 160 lb at sea level has a mass of $\frac{160}{32} = 5$ slugs.

Metric System

In the *centimeter-gram-second* system, the weight (a measure of force) is in *dynes*, while the mass is in *grams*. To convert from weight to mass—a procedure that the following discussion shows we should almost never do—we should divide the weight at sea level by 981.

The weight-mass situation in the metric system is not so simple as it seems. The standard metric scales are calibrated to give the *mass* at sea level—not the weight, as in the foot-pound-second system. Thus, the two systems are basically incompatible. When we write the equation

$$1 \text{ kilogram} = 2.2046 \text{ pounds}$$

for example, we are comparing incompatible quantities (mass and force). A more correct equation would be

$$1 \text{ kilogram} = \frac{2.2046}{32} \text{ slugs,}$$

which compares mass with mass.

Conversion from One System to the Other

When we convert problems from one system to the other, all quantities must be converted, including the constants of proportionality. In the following example, which is stated in the foot-pound-second system, the weight (which must be converted to mass) is 4 oz, the gravitational constant is $g = 32$, the initial velocity is $v_0 = 128$ ft/sec, and the constant of proportionality is $k = 0.001$. If the same problem were to be restated in the centimeter-gram-second system, the mass would be 113.488 g, the gravitational constant $g = 981$, the initial velocity $v_0 = 3901.44$ cm/sec, and the constant of proportionality $k = (453.6)(32)(0.001) = 14.5152$. The numbers 453.6 and 32, which are used to change the constant of proportionality, are the mass of 1 pound and the gravitational constant, respectively.

EXAMPLE 2 A projectile that weighs 4 oz is fired straight up into the air with an initial velocity of 128 ft/sec. The retarding force due to air resistance equals 0.001 times the velocity.
(a) Calculate the velocity and position functions at time t.
(b) Find the maximum height above the earth.
(c) Find when the projectile strikes the earth.
(d) Find the impact velocity at the instant at which it strikes the earth.

Solution (a) The differential equation for the motion is

$$\frac{dv}{dt} = -\frac{k}{m}v - g.$$

8-6 Laws of Motion

Here $k = 0.001$ and $g = 32$. To convert the weight to mass, we divide the weight in pounds by 32:

$$m = mass = \frac{\frac{1}{4}}{32} = \frac{1}{128} \text{ slug.}$$

If we substitute these values in the differential equation, we get

$$\frac{dv}{dt} = \frac{-0.001}{\frac{1}{128}} v - 32$$
$$= -0.128v - 32 = -0.128(v + 250).$$

Separating the variables and integrating, we get

$$\frac{dv}{v + 250} = -0.128 \, dt$$

$$\int \frac{dv}{v + 250} = -0.128 \int dt$$

$$\ln |v + 250| = -0.128t + C_1$$

$$|v + 250| = e^{C_1} e^{-0.128t}$$

$$v + 250 = Ce^{-0.128t}$$

where $C = \pm e^{C_1}$.

To evaluate the constant C, we recall that $v = 128$ when $t = 0$. Thus,

$$128 + 250 = Ce^0 = C$$
$$C = 378.$$

The velocity function is

$$v(t) = -250 + 378e^{-0.128t}.$$

We get the distance function by integrating the velocity function:

$$s = \int (-250 + 378e^{-0.128t}) \, dt$$

$$= -250t + \frac{378}{-0.128} e^{-0.128t} + C_2$$

$$= -250t - 2953.125 e^{-0.128t} + C_2.$$

To evaluate the constant C_2, we recall that $s = 0$ when $t = 0$. Thus

$$0 = -250 \cdot 0 - 2953.125 e^0 + C_2$$
$$C_2 = 2953.125.$$

The distance function is

$$s(t) = -250t + 2953.125(1 - e^{-0.128t}).$$

(b) The maximum height above the earth is attained at the instant when $v = 0$. If we set $v = 0$ and solve for t, we get

$$0 = -250 + 378e^{-0.128t}$$

$$e^{-0.128t} = \frac{250}{378}$$

$$-0.128t = \ln \frac{250}{378} \approx -0.4134$$

$$t \approx \frac{-0.4134}{-0.128} \approx 3.23 \text{ sec.}$$

The maximum height is approximately equal to

$$s(3.23) = -250(3.23) + 2953.125(1 - e^{(-0.128)(3.23)})$$
$$\approx 192.5 \text{ ft.}$$

(c) To find the instant at which the projectile strikes the earth, we solve the equation $s(t) = 0$ for t. Thus, we must solve

$$-250t + 2953.125(1 - e^{-0.128t}) = 0$$

for t. This equation cannot be solved algebraically. If we apply Newton's method for the approximation of roots, however, we find that $t \approx 6.98$ sec when the projectile strikes the earth.

(d) The impact velocity is approximately equal to

$$v(6.98) = -250 + 378e^{(-0.128)(6.98)} \approx -95.3 \text{ ft/sec.} \quad \square$$

Exercises 8–6

1–4. An object moves in a straight line, subject only to a resisting force, which is proportional to its velocity. In other words, the *only* force acting on it is the resisting force, which equals $-kv$, where k is a positive constant. Use Newton's second law of motion to find the velocity and position functions v and s. The mass m, the value of k, the initial velocity v_0, and the initial distance s_0 are given.

•1 $k = 12$, $m = 4$ g, $v_0 = 150$ cm/sec, $s_0 = 0$

2 $k = 20$, $m = 5$ g, $v_0 = 60$ cm/sec, $s_0 = -10$ cm

•3 $k = 5$, $m = 5$ slugs, $v_0 = 200$ ft/sec, $s_0 = 0$

4 $k = 4$, $m = 8$ slugs, $v_0 = 75$ ft/sec, $s_0 = 80$ ft

5–8. An object in free fall is subject to gravity and to a resisting force that is proportional to its velocity. That is, the total force is $F = -kv - mg$, where m is the mass, g the gravitational constant, v the velocity, and k the constant of proportionality. At time $t = 0$, the velocity was v_0 and the distance above the ground was s_0. Find v and s as functions of t.

- 5 $m = \frac{1}{2}$ slug, $k = 0.1$, $v_0 = -30$ ft/sec, $s_0 = 1000$ ft
- 6 $m = 200$ g, $k = 3$, $v_0 = 0$ cm/sec, $s_0 = 20{,}000$ cm
- 7 weight $= 200$ lb, $k = 0.5$, $v_0 = 500$ ft/sec, $s_0 = 0$
- 8 $m = 10$ g, $k = 1$, $v_0 = 100$ m/sec, $s_0 = 0$.
- c●9 A 4-lb cannonball is fired straight up into the air from ground level with an initial velocity of 512 ft/sec. The retarding force due to air resistance equals 0.01 of the velocity.
   (a) Calculate the maximum height above the earth.
   (b) (*Requires Newton's method*) What is the velocity of the cannonball on its impact with the ground?

c●10 (*Requires Newton's method*) A skydiver with full gear weighs 192 lb. Assume that the resisting force is proportional to the velocity. In free fall, the constant of proportionality is $k = 1$. With her parachute open, it is $k = 4$. She jumps from a distance of 6000 ft, falls freely for 15 sec, opens the parachute, and floats to earth. Find her impact velocity on striking the earth.

Review Problems

1 Define:
   (a) The natural logarithm function ln.
   (b) The number e.
   (c) The natural exponential function exp.

2 Sketch the graphs of the functions ln and exp on one set of axes.

3–5. Prove the statements.

3 $D_x(\ln x) = \dfrac{1}{x}$, $x > 0$.

4 $D_x(\ln |x|) = \dfrac{1}{x}$, $x \neq 0$.

5 $\displaystyle\int \dfrac{dx}{x} = \ln |x| + C$, $x \neq 0$.

6 Prove that $\ln(ab) = \ln a + \ln b$ if $a > 0$ and $b > 0$. What is the corresponding formula if $a < 0$ and $b < 0$? if $a > 0$ and $b < 0$?

- 7 Use the properties of the natural logarithm function to explain why
$$\int_8^{27} \frac{dx}{x} = 3 \int_2^3 \frac{dx}{x}.$$

8–15. Calculate the derivatives with respect to x. Use the properties of the natural logarithm function to simplify the calculations wherever possible.

- **8** $\ln [(3x^2 - 2)^4(5x^3 + 7)^3]$
- **9** $\ln (x^{x^2})$
- **10** $(\ln x)(\ln 2x)$
- **11** $\ln (\ln x)$
- **12** xe^x
- **13** $e^{2 \ln x}$
- **14** $\ln (e^x + 1)$
- **15** $e^{\sqrt{x}}$

16–17. Use implicit differentiation to calculate y':

- **16** $e^{xy} + \ln y = 1$
- **17** $xe^y + x^2 - xy^2 = 0$

- **18** Use logarithmic differentiation to calculate y' for
$$y = \sqrt{\frac{x^3(x+1)}{x^2+3}}, \quad x > 0.$$

19–26. Calculate the integrals.

- **19** $\displaystyle\int \frac{dx}{e^{2x}}$
- **20** $\displaystyle\int \frac{dx}{x \ln x}$
- **21** $\displaystyle\int \frac{e^x + e^{-x}}{e^x - e^{-x}} dx$
- **22** $\displaystyle\int e^{x^2 + \ln x} dx$
- **23** $\displaystyle\int (3e^x + 2)^{20} e^x dx$
- **24** $\displaystyle\int (3e^x + 2)^2 dx$
- **25** $\displaystyle\int \frac{x^2 + 3x + 2}{x + 4} dx$
- **26** $\displaystyle\int \frac{x + 1}{x^2 + 2x + 5} dx$

- **27** The region bounded by the graph of $y = e^{-x}$, the x-axis, the y-axis, and the line $x = 1$ is revolved about the x-axis. Find the volume of the solid of revolution.

- **28** Derive the formula $D_x(e^x) = e^x$.

29–32. (a) Solve the differential equations by separation of variables.
        (b) Find the solution that satisfies the condition $y = 3$ when $x = 1$.

- **29** $y' = 2x + 1$
- **30** $y' = 2y + 1$
- **31** $y' = (x + 1)(y + 1)$
- **32** $y' = \dfrac{y}{x}$

Review Problems

●33 Let $p(x)$ be the atmospheric pressure at x meters above sea level, expressed in grams per square centimeter. It is known that $p(x)$ satisfies the differential equation

$$\frac{dp}{dx} = -0.00012p$$

and that the pressure at sea level is 10^3g/cm^2. Find the pressure at a height of 1000 m.

^{c}**●34** Phosphorus 32 has a half-life of 14.2 days. Ten grams are collected in a sample. How much remains after 30 days?

●35 The population of a city was 120,000 in 1960 and 200,000 in 1980. Assume that the rate of increase is proportional to the size of the population.

  (a) What was the population in 1970?

  (b) What is the expected population for the year 2020?

●36 A building purchased 10 years ago for $120,000 has a current value of $205,000. Assume that the rate of change of the value is proportional to the value at each instant. What will it be worth in another 7 years?

●37 The rate at which an appliance depreciates is proportional to its value. A stove, recently bought for $500, is expected to have a junk value of $25 in 15 years. Find its value when it is 6 years old.

●38 The rate of conversion of raw sugar to dextrose at time t is proportional to the amount of raw sugar present but not yet converted.

  (a) Express this relation by a differential equation.

  (b) At time $t = 0$, a solution contains 100 g of raw sugar and no dextrose. Fifteen minutes later 10 g has been converted. How much is converted over a two-hour period?

39 (a) Distinguish between mass and weight.

  (b) State Newton's second law of motion.

^{c}**40** A projectile with mass 5 slugs is fired upward from ground level with an initial velocity of 400 ft/sec. The projectile is subject to the force of gravity and to a resisting force equal to two-tenths of the velocity.

  (a) Set up a differential equation for the velocity.

  (b) Find the formulas for velocity and position at time t.

  ●(c) Find the masimum height above the earth.

9 Trigonometric and Hyperbolic Functions

9–1 Derivatives and Integrals of the Sine and Cosine Functions

Appendix B at the end of the book contains a brief discussion of the trigonometric functions and a list of important properties and identities. The reference numbers T1, T2, and so on, used in this chapter, refer to that Appendix. You should read Appendix B before proceeding further. In particular, it should be noted that all angles are measured in radians.

For centuries a primitive type of trigonometry was used to solve problems related to right triangles. In particular, surveying methods based on right triangles were developed by the ancient Babylonians and Egyptians and were used later by the Greeks, Romans, and Persians. During the Middle Ages, the Hindus refined the work of their predecessors and defined some of the trigonometric functions currently in use. Identities relating these functions were discovered empirically and later were proved by the European mathematicians.

Modern scientists and engineers do not spend much of their time solving right triangles. The important current uses of trigonometry are more closely connected with *periodic phenomena*—events that occur over and over with regularity. The trigonometric functions can be used to describe such events very accurately.

Periodic Phenomena

Consider a simple example of periodic behavior, the motion of a particle P on the edge of a spinning circle of radius r. Suppose that the circle spins counterclockwise with an angular velocity of ω revolutions per second. We shall show how to determine parametric equations for the x and y coordinates of P at time t.

To study the motion of the particle P, we superimpose a coordinate system over the circle so that the origin is at the center. (See Fig. 9–1a.) For convenience, we orient our axis system so that P is at the point $(r, 0)$ at time $t = 0$. At time t, the particle P is at a point (x, y), which is on the terminal side of an angle of ωt radians. It follows from elementary trigonometry that $P(x, y)$ is the point $P(r \cos \omega t, r \sin \omega t)$.

9–1 Derivatives and Integrals of the Sine and Cosine Functions

Therefore,

$$x = r \cos \omega t \quad \text{and} \quad y = r \sin \omega t.$$

These two functions define x and y parametrically in terms of the parameter t. The graphs of the functions, which are shown in Figure 9–1b, can be used to locate the point at any time t. Observe that both functions are periodic, with period $p = 2\pi/\omega$.

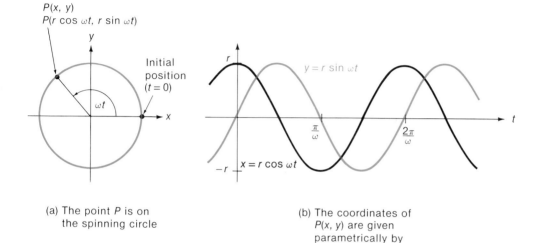

(a) The point P is on the spinning circle

(b) The coordinates of $P(x, y)$ are given parametrically by $x = r \cos \omega t$, $y = r \sin \omega t$

Figure 9–1 Periodic functions.

BASIC LIMITS

In applying the concepts of the calculus to the trigonometric functions, our first important task is to calculate the derivatives of the sine and cosine functions. To do this, we first establish the limits

Two Basic Limits

$$\lim_{x \to 0} \frac{\sin x}{x} = 1 \quad \text{and} \quad \lim_{x \to 0} \frac{1 - \cos x}{x} = 0,$$

which are needed in the calculation of the derivatives.

LEMMA 1 $\displaystyle\lim_{x \to 0} \frac{\sin x}{x} = 1.$

Proof We first consider the case in which x approaches 0 through positive values. Consequently, we restrict ourselves to values of x in the range $0 < x < \pi/2$. We construct angle x in standard position on the unit circle. Next we construct the two right triangles shown in Figure 9–2. It is obvious that

$$\text{Area of } \triangle OAB < \text{area of sector } OCB < \text{area of } \triangle OCD.$$

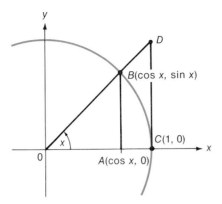

Figure 9–2 Lemma 1.

Observe that $|OC| = |OB| = 1$,

$$|OA| = \cos x \quad \text{and} \quad |AB| = \sin x.$$

Thus, the area of triangle OAB is $(\cos x \sin x)/2$.

In calculating the area of triangle OCD, observe that this triangle is similar to OAB, so that

$$\frac{|CD|}{|OC|} = \frac{|AB|}{|OA|} = \frac{\sin x}{\cos x}.$$

Since $|OC| = 1$, then $|CD| = \sin x/\cos x$ and

$$\text{Area of } \triangle OCD = \frac{|OC| \cdot |CD|}{2} = \frac{1 \cdot \sin x/\cos x}{2} = \frac{\sin x}{2 \cos x}.$$

Recall from Appendix B that the area of a sector of a circle with central angle α and radius r is $\alpha r^2/2$. It follows that

$$\text{Area of sector } OCB = \frac{x \cdot 1^2}{2} = \frac{x}{2}.$$

Since area of $\triangle OAB <$ area of sector $OCB <$ area of $\triangle OCD$, then

$$\frac{\cos x \sin x}{2} < \frac{x}{2} < \frac{\sin x}{2 \cos x}.$$

If we multiply these inequalities by $2/\sin x$ and take reciprocals, we obtain

$$\frac{1}{\cos x} > \frac{\sin x}{x} > \cos x.$$

Since $\cos x \to 1$ then $1/\cos x \to 1$ as $x \to 0^+$. It follows from the Sandwich Theorem (Theorem 2–3) that

$$\lim_{x \to 0^+} \frac{\sin x}{x} = 1.$$

We now consider the case in which x approaches 0 through negative values. Let

9-1 Derivatives and Integrals of the Sine and Cosine Functions

$u = -x$. Then

$$u \to 0^+ \quad \text{as } x \to 0^-.$$

It follows that

$$\lim_{x \to 0^-} \frac{\sin x}{x} = \lim_{x \to 0^-} \frac{\sin(-u)}{-u} = \lim_{u \to 0^+} \frac{-\sin u}{-u} \quad \text{(by T10)}$$

$$= \lim_{u \to 0^+} \frac{\sin u}{u} = 1.$$

Since $\lim_{x \to 0^+} (\sin x)/x = 1$ and $\lim_{x \to 0^-} (\sin x)/x = 1$, then

$$\lim_{x \to 0} \frac{\sin x}{x} = 1. \quad \blacksquare$$

LEMMA 2 $\quad \lim_{x \to 0} \dfrac{1 - \cos x}{x} = 0.$

Proof We first use trigonometric identity T5 and then apply Lemma 1:

$$\lim_{x \to 0} \frac{1 - \cos x}{x} = \lim_{x \to 0} \frac{1 - \cos x}{x} \cdot \frac{1 + \cos x}{1 + \cos x}$$

$$= \lim_{x \to 0} \frac{1 - \cos^2 x}{x(1 + \cos x)}$$

$$= \lim_{x \to 0} \frac{\sin^2 x}{x(1 + \cos x)} \quad \text{(by T5)}$$

$$= \lim_{x \to 0} \left(\frac{\sin x}{x} \cdot \frac{\sin x}{1 + \cos x} \right)$$

$$= \lim_{x \to 0} \frac{\sin x}{x} \cdot \lim_{x \to 0} \frac{\sin x}{1 + \cos x}$$

$$= 1 \cdot \frac{0}{1 + 1} = 0. \quad \blacksquare$$

Remark The arguments used to prove the lemmas depend on the limits

$$\lim_{x \to 0} \cos x = 1 \quad \text{and} \quad \lim_{x \to 0} \sin x = 0.$$

These limits are established in Exercise 43.

EXAMPLE 1 Show that $\lim_{x \to 0} \dfrac{\sin 3x}{x} = 3.$

Solution Let $t = 3x$. Since $t \to 0$ as $x \to 0$, then

$$\lim_{x \to 0} \frac{\sin 3x}{x} = \lim_{x \to 0} \frac{3 \cdot \sin 3x}{3x}$$

$$= 3 \lim_{t \to 0} \frac{\sin t}{t} = 3 \cdot 1 = 3. \quad \square$$

EXAMPLE 2 Show that $\lim_{x \to 0} \dfrac{\sin 2x}{\sin 5x} = \dfrac{2}{5}$.

Solution We first divide numerator and denominator by x:

$$= \lim_{x \to 0} \frac{\sin 2x}{\sin 5x} = \lim_{x \to 0} \frac{\dfrac{\sin 2x}{x}}{\dfrac{\sin 5x}{x}} = \lim_{x \to 0} \frac{\dfrac{2 \sin 2x}{2x}}{\dfrac{5 \sin 5x}{5x}}$$

$$= \frac{2}{5} \lim_{x \to 0} \frac{\dfrac{\sin 2x}{2x}}{\dfrac{\sin 5x}{5x}} = \frac{2}{5} \cdot \frac{1}{1} = \frac{2}{5}. \quad \square$$

DERIVATIVES OF THE SINE AND COSINE FUNCTIONS

THEOREM 9–1 $D_x(\sin x) = \cos x$.

Proof Let $f(x) = \sin x$. Then

$$D_x(\sin x) = f'(x) = \lim_{\Delta x \to 0} \frac{f(x + \Delta x) - f(x)}{\Delta x}$$

$$= \lim_{\Delta x \to 0} \frac{\sin(x + \Delta x) - \sin x}{\Delta x}$$

$$= \lim_{\Delta x \to 0} \frac{\sin x \cos \Delta x + \cos x \sin \Delta x - \sin x}{\Delta x} \quad \text{(by T8)}$$

$$= \lim_{\Delta x \to 0} \frac{\sin x \cdot (\cos \Delta x - 1) + \cos x \cdot \sin \Delta x}{\Delta x}$$

$$= \lim_{\Delta x \to 0} \sin x \cdot \lim_{\Delta x \to 0} \frac{\cos \Delta x - 1}{\Delta x} + \lim_{\Delta x \to 0} \cos x \cdot \lim_{\Delta x \to 0} \frac{\sin \Delta x}{\Delta x}$$

$$= \sin x \cdot 0 + \cos x \cdot 1 \quad \text{(by the lemmas)}$$

$$= \cos x. \quad \blacksquare$$

If we apply the Chain Rule, we get the more general formula

Derivative of Sine

D11 $D_x(\sin u) = \cos u \cdot D_x u$,

where u is a differentiable function of x.

EXAMPLE 3
$$D_x(\sin \sqrt{x}) = D_x(\sin x^{1/2})$$
$$= \cos x^{1/2} \cdot D_x(x^{1/2})$$
$$= \cos x^{1/2} \cdot \tfrac{1}{2} x^{-1/2}$$
$$= \frac{\cos \sqrt{x}}{2\sqrt{x}}. \quad \square$$

EXAMPLE 4
$$D_x(\sin(e^x)) = \cos(e^x) \cdot D_x(e^x)$$
$$= e^x \cos(e^x). \quad \square$$

9-1 Derivatives and Integrals of the Sine and Cosine Functions

The derivative of the cosine function can be calculated by an argument like the one for the sine function. A simpler argument, however, can be based on formulas T11:

$$\cos x = \sin\left(\frac{\pi}{2} - x\right) \quad \text{and} \quad \sin x = \cos\left(\frac{\pi}{2} - x\right).$$

EXAMPLE 5 Prove that $D_x(\cos x) = -\sin x$.

Solution
$$D_x(\cos x) = D_x\left(\sin\left(\frac{\pi}{2} - x\right)\right)$$
$$= \cos\left(\frac{\pi}{2} - x\right) \cdot D_x\left(\frac{\pi}{2} - x\right) = -\cos\left(\frac{\pi}{2} - x\right)$$
$$= -\sin x. \;\square$$

If we apply the Chain Rule to the result of Example 5, we get the formula

Derivative of Cosine

D12 $\quad D_x(\cos u) = -\sin u \cdot D_x u$

where u is a differentiable function of x.

INTEGRAL FORMULAS

The integration formulas corresponding to D11 and D12 are

Integrals of Sine and Cosine

I7 $\quad \displaystyle\int \cos u \, du = \sin u + C.$

I8 $\quad \displaystyle\int \sin u \, du = -\cos u + C.$

EXAMPLE 6 Calculate $\displaystyle\int \sin 2x \, dx$.

Solution
$$\int \sin 2x \, dx = \frac{1}{2}\int \sin 2x \cdot 2 \, dx \qquad \boxed{\begin{array}{l} u = 2x \\ du = 2\,dx \end{array}}$$
$$= \frac{1}{2}\int \sin u \, du$$
$$= -\frac{1}{2}\cos u + C = -\frac{1}{2}\cos 2x + C. \;\square$$

EXAMPLE 7 Calculate $\displaystyle\int \sin^3 x \cos x \, dx$.

Solution Let $u = \sin x$, $du = \cos x \, dx$. Then

$$\int \sin^3 x \cos x \, dx = \int u^3 \, du = \frac{u^4}{4} + C$$

$$= \frac{\sin^4 x}{4} + C. \quad \square$$

$u = \sin x$
$du = \cos x \, dx$

EXAMPLE 8 The graphs of $f(x) = \sin x$ and $g(x) = \cos x$ cross at $x = \pi/4$ and $x = 5\pi/4$. (See Fig. 9–3.) Calculate the area between these curves for $\pi/4 \leq x \leq 5\pi/4$.

Solution Since $\sin x \geq \cos x$ for $\pi/4 \leq x \leq 5\pi/4$, then

$$\text{Area} = \int_{\pi/4}^{5\pi/4} (\sin x - \cos x) \, dx = \left[-\cos x - \sin x \right]_{\pi/4}^{5\pi/4}$$

$$= \left[-\cos \frac{5\pi}{4} - \sin \frac{5\pi}{4} \right] - \left[-\cos \frac{\pi}{4} - \sin \frac{\pi}{4} \right]$$

$$= -\left(-\frac{1}{\sqrt{2}} \right) - \left(-\frac{1}{\sqrt{2}} \right) + \frac{1}{\sqrt{2}} + \frac{1}{\sqrt{2}} = \frac{4}{\sqrt{2}}. \quad \square$$

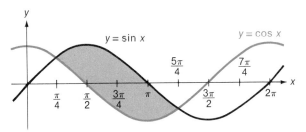

Figure 9–3 Example 8.

Exercises 9–1

1–8. Use the limits for $(\sin x)/x$ and $(\cos x - 1)/x$ to evaluate the limits. (*Hint* for 3: First apply identity T5.)

•1 $\lim\limits_{x \to 0} \dfrac{\sin 3x}{2x}$

2 $\lim\limits_{x \to 0} \dfrac{\sin^2 x}{x}$

•3 $\lim\limits_{x \to 0} \dfrac{1 - \cos x}{x^2}$

4 $\lim\limits_{x \to 0} \dfrac{x \csc x}{\sec x}$

•5 $\lim\limits_{t \to 0} \dfrac{2 \tan t}{3t}$

6 $\lim\limits_{t \to 0} \dfrac{\sin 2t}{4t^2 + t}$

•7 $\lim\limits_{\theta \to 0} \dfrac{1 - \cos 2\theta}{\theta}$

8 $\lim\limits_{\theta \to 0} \dfrac{4\theta^2}{1 - \cos^2 \theta}$

9–20. Calculate the derivatives of the functions.

• 9 $\cos 5x$

10 $x^2 \sin 2x$

•11 $\sin (x^2 + 5)$

12 $\cos (\sqrt{1 + x^3})$

•13 $\cos^3 x$

14 $\sin^2 3x$

Exercises 9-1

- 15 $\dfrac{\cos 2x}{\sin^2 x}$
- 16 $5(1 - \cos^2 x)$
- 17 $\sin 2x \cos 3x$
- 18 $\tan x$ (Use T1.)
- 19 $e^{\sin x^2}$
- 20 $\ln (\cos x)$

21–30. Calculate the integrals.

- 21 $\displaystyle\int \cos (2x - 1)\, dx$
- 22 $\displaystyle\int \cos \dfrac{x}{2}\, dx$
- 23 $\int (\sin 2x + \sin 4x)\, dx$
- 24 $\int x \sin 2x^2\, dx$
- 25 $\int \sin^2 3x \cos 3x\, dx$
- 26 $\int (1 - \cos 2x)^{5/2} \sin 2x\, dx$
- 27 $\displaystyle\int \dfrac{\sin \sqrt{x + 1}}{\sqrt{x + 1}}\, dx$
- 28 $\displaystyle\int \dfrac{\sin (\ln x)}{x}\, dx$
- 29 $\displaystyle\int_0^{\pi/2} \cos 3x\, dx$
- 30 $\displaystyle\int_{2\pi/3}^{\pi} \dfrac{\sin x}{\cos^3 x}\, dx$

31–36. Find the local extrema and the intervals where the functions are concave upward and concave downward. Sketch the graphs for $0 \le x \le 2\pi$.

- 31 $y = \sin x$
- 32 $y = \cos x$
- 33 $y = \tan x = \sin x / \cos x$
- 34 $y = \sec x = 1/\cos x$
- 35 $y = \cos x - \sin x$
- 36 $y = x - \sin x$

37 Show that the slope of the tangent line to the graph of $y = \sin x$ cannot be greater than 1 or less than -1.

38–39. Calculate the area of the region bounded by the graph of the function and the x-axis.

- 38 $y = \cos x,\ -\pi/2 \le x \le \pi/2$
- 39 $y = \sin 2x,\ 0 \le x \le \pi/2$

- 40 Let $\mathcal{R}$ be the region bounded by the graph of $y = \cos x$ and the x-axis, $-\pi/2 \le x \le \pi/2$. Find the volume of the solid obtained by revolving $\mathcal{R}$ about the x-axis. (*Hint:* Formula T14 can be rewritten as $\cos^2 \alpha = (1 + \cos 2\alpha)/2$.)

- 41 A light fixture is to be located directly above the center of a round poker table that is 6 ft in diameter. The intensity of the light varies directly with the cosine of the angle of incidence (the angle between the vertical and the ray of light) and inversely with the square of the distance. How high should the light be located above the table to provide the maximum illumination at the edge of the table?
  Is this solution feasible, or will the light interfere with the poker game?

42 Construct an argument like the proof of Theorem 9–1 to prove that $D_x(\cos x) = -\sin x$.

43 An argument that $\lim_{x \to 0} \sin x = 0$ and $\lim_{x \to 0} \cos x = 1$ can be based on Figure 9–4. Consider the case for which $x > 0$. Construct the point $P(\cos x, \sin x)$ at which the terminal side of angle x intersects the circle of radius 1 that has center at the origin. (See Fig. 9–4a.) Construct a circle of radius x with center at the point $(1, 0)$ (Fig.

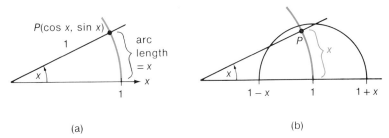

(a)   (b)

Figure 9-4 Exercise 43.

9–4b.). Explain why P must be interior to this circle. Use this fact to show that

$$1 - x < \cos x < 1 \quad \text{and} \quad 0 < \sin x < x.$$

Finally, apply the Sandwich Theorem (Theorem 2–3) as $x \to 0^+$.

44 (a) Show that the area of an n-sided regular polygon inscribed in a unit circle is

Area of Regular Polygon

$$A_n = n \cos \frac{\pi}{n} \sin \frac{\pi}{n} = \frac{n}{2} \sin \frac{2\pi}{n}.$$

(b) Evaluate this limit as $n \to \infty$. Is this result consistent with the limiting position of the polygons as $n \to \infty$?

45 A triangle is to be constructed with one side of length 20 and the opposite angle 30°. Show that the area is a maximum when the triangle is isosceles.

9–2 Additional Derivative and Integral Formulas

The derivatives of the remaining trigonometric functions can be calculated by expressing those functions in terms of sines and cosines and using the basic derivative formulas. The derivative formulas for the tangent, cotangent, secant, and cosecant functions are listed below. In each formula u represents a differentiable function of x.

Derivative Formulas

D13	$D_x(\tan u) = \sec^2 u \cdot D_x u.$
D14	$D_x(\cot u) = -\csc^2 u \cdot D_x u.$
D15	$D_x(\sec u) = \sec u \tan u \cdot D_x u.$
D16	$D_x(\csc u) = -\csc u \cot u \cdot D_x u.$

EXAMPLE 1   Prove formula D13.

Solution   Let u be a differentiable function of x. Then

$$D_x(\tan u) = D_x\left(\frac{\sin u}{\cos u}\right) \quad \text{(by T1)}$$

$$= \frac{\cos u \cdot D_x(\sin u) - \sin u \cdot D_x(\cos u)}{\cos^2 u} \quad \text{(by D7)}$$

$$= \frac{\cos u \cdot \cos u \cdot D_x u - \sin u \cdot (-\sin u \cdot D_x u)}{\cos^2 u}$$

9-2 Additional Derivative and Integral Formulas

$$= \frac{\cos^2 u + \sin^2 u}{\cos^2 u} \cdot D_x u = \frac{1}{\cos^2 u} \cdot D_x u \quad \text{(by T5)}$$

$$= \sec^2 u \cdot D_x u. \quad \square$$

EXAMPLE 2   Calculate $D_x(\sec^2 x \tan x)$.

Solution $D_x(\sec^2 x \tan x) = \sec^2 x \cdot D_x(\tan x) + \tan x \cdot D_x(\sec^2 x)$
$= \sec^2 x \cdot \sec^2 x + \tan x \cdot 2 \sec x \cdot \sec x \cdot \tan x \quad$ (by D13, D15)
$= \sec^4 x + 2 \sec^2 x \tan^2 x. \quad \square$

EXAMPLE 3   A rocket rising vertically is being tracked from a point 1 mile from the launch pad. At the instant when the rocket is 5 miles high, the angle of elevation is increasing at the rate of 0.1 degree per second. How fast is it climbing at this instant?

Solution   Let $y = y(t)$ be the height of the rocket and $\theta = \theta(t)$ the angle of elevation at time t. (See Fig. 9–5a.) Then dy/dt is the rate of change of the height and $d\theta/dt$ the rate of increase of the angle of elevation.
Since $\tan \theta = y/1 = y$, then

$$\frac{dy}{dt} = \frac{d}{dt}(\tan \theta)$$

$$= \sec^2 \theta \cdot \frac{d\theta}{dt}$$

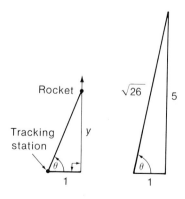

(a) (b)   **Figure 9–5**   Example 3.

At the instant when $y = 5$, we have

$$\cos \theta = \frac{1}{\sqrt{26}}$$

so that

$$\sec^2 \theta = 26$$

(See Fig. 9–5b.) At this instant, the angle of elevation is increasing at the rate of $1/10$ degree per second. Converting this value to radian measure (recall that $1° = \pi/180$ rad), we have

$$\frac{d\theta}{dt} = \frac{1}{10} \cdot \frac{\pi}{180} = \frac{\pi}{1800} \text{ rad/sec.}$$

Then

$$\frac{dy}{dt} = \sec^2 \theta \cdot \frac{d\theta}{dt} = \frac{26\pi}{1800} \text{ miles/sec}$$

$$\approx 0.04538 \text{ mile/sec} \approx 239.6 \text{ ft/sec.} \quad \square$$

EXAMPLE 4 Analyze and sketch the graph of $f(x) = \tan x$.

Solution *Continuity.* Since $f(x) = \tan x = (\sin x)/(\cos x)$, then the graph of f is differentiable (and therefore continuous) everywhere except at the points $x = \pm \pi/2, \pm 3\pi/2, \pm 5\pi/2, \ldots$, where the denominator is zero.

Symmetry. It follows from T10 that

$$f(-x) = \frac{\sin(-x)}{\cos(-x)} = \frac{-\sin x}{\cos x} = -\tan x = -f(x).$$

The graph is symmetric about the origin.

Periodicity. Since

$$f(x + \pi) = \frac{\sin(x + \pi)}{\cos(x + \pi)} = \frac{-\sin x}{-\cos x} = \tan x = f(x)$$

(by T8 and T9), then f is periodic. Since f is discontinuous only at $x = \pm \pi/2, \pm 3\pi/2, \ldots$, then no smaller number than π could be the period. Thus, the period is π.

It follows from the fact that $f(x + \pi) = f(x)$ that we need to sketch the graph for x only between $-\pi/2$ and $\pi/2$ (two points of discontinuity). The rest of the graph can be obtained by use of the periodicity. For the rest of this discussion we restrict ourselves to values of x in this range.

Vertical Asymptotes. The behavior of the function in the vicinity of $x = -\pi/2$, one of the two points of discontinuity, is as follows:

$$\lim_{x \to (-\pi/2)^+} f(x) = \lim_{x \to (-\pi/2)^+} \frac{\sin x}{\cos x} = -\infty;$$

$$\lim_{x \to (-\pi/2)^-} f(x) = \lim_{x \to (-\pi/2)^-} \frac{\sin x}{\cos x} = +\infty.$$

9-2 Additional Derivative and Integral Formulas

The graph has vertical asymptotes at $x = \pm \pi/2$.

Properties Determined from Derivatives. Since

$$f'(x) = \sec^2 x > 0 \quad \text{for } -\pi/2 < x < \pi/2,$$

then f is an increasing function on the open interval $(-\pi/2, \pi/2)$.
The second derivative is

$$\begin{aligned} f''(x) &= D_x(\sec^2 x) = 2 \sec x \cdot \sec x \cdot \tan x \\ &= 2 \sec^2 x \tan x. \end{aligned}$$

It follows that the graph is concave downward for $-\pi/2 < x < 0$ and concave upward for $0 < x < \pi/2$. There is a point of inflection at $x = 0$.

The graph is shown in Figure 9–6. Because of the periodicity, the branches of the graph are translates of the branch defined by $-\pi/2 < x < \pi/2$. □

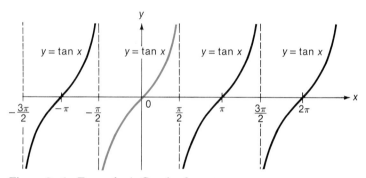

Figure 9–6 Example 4. Graph of $y = \tan x$.

INTEGRATION

The integral formulas corresponding to D13–D16 are listed below:

Integral Formulas

I9 $\quad \int \sec^2 u \, du = \tan u + C.$

I10 $\quad \int \csc^2 u \, du = -\cot u + C.$

I11 $\quad \int \sec u \tan u \, du = \sec u + C.$

I12 $\quad \int \csc u \cot u \, du = -\csc u + C.$

Observe that except for the sine and cosine function, we do not yet have integration formulas for the basic trigonometric functions. The missing formulas are given below. (See Examples 5 and 6 and Exercises 38, 39.)

I13 $\quad \int \tan u \, du = \ln |\sec u| + C = -\ln |\cos x| + C.$

I14 $\quad \int \cot u \, du = -\ln |\csc u| + C = \ln |\sin u| + C.$

I15 $\quad \int \sec u \, du = \ln |\sec u + \tan u| + C.$

I16 $\quad \int \csc u \, du = -\ln |\csc u + \cot u| + C.$

EXAMPLE 5   Derive formula I13.

Solution
$$\int \tan u \, du = \int \frac{\sin u}{\cos u} \, du.$$

Make the substitution $v = \cos u$, $dv = -\sin u \, du$. Then

$$\int \tan u \, du = -\int \frac{-\sin u \, du}{\cos u} = -\int \frac{dv}{v}$$

$\boxed{\begin{array}{l} v = \cos u \\ dv = -\sin u \, du \end{array}}$

$$= -\ln |v| + C = \ln \left(\frac{1}{|v|}\right) + C$$

$$= \ln \left(\frac{1}{|\cos u|}\right) + C = \ln |\sec u| + C. \ \square$$

EXAMPLE 6   Derive formula I15.

Solution   The two derivative formulas that involve the secant function are

$$D_x(\sec x) = \sec x \tan x$$

and

$$D_x(\tan x) = \sec^2 x.$$

Then

$$D_x(\sec x + \tan x) = \sec x \tan x + \sec^2 x$$
$$= \sec x \, (\sec x + \tan x),$$

so that

$$\sec x = \frac{D_x \, (\sec x + \tan x)}{\sec x + \tan x}.$$

If we make the substitution $u = \sec x + \tan x$ and integrate both sides of this equation, we obtain

$$\int \sec x \, dx = \int \frac{d(\sec x + \tan x)}{\sec x + \tan x}$$

$\boxed{\begin{array}{l} u = \sec x + \tan x \\ du = (\sec x + \tan x) \sec x \, dx \end{array}}$

$$= \int \frac{du}{u} = \ln |u| + C$$

$$= \ln |\sec x + \tan x| + C. \ \square$$

Exercises 9–2

EXAMPLE 7   Calculate $\int x \tan x^2 \, dx$.

Solution   We make the substitution $u = x^2$, $du = 2x \, dx$. Then

$$\int x \tan x^2 \, dx = \frac{1}{2} \int \tan x^2 (2x \, dx) = \frac{1}{2} \int \tan u \, du$$

$$= \frac{1}{2} \ln |\sec u| + C \quad \text{(by I13)}$$

$$= \frac{1}{2} \ln |\sec x^2| + C. \quad \square$$

$\boxed{\begin{array}{l} u = x^2 \\ du = 2x \, dx \end{array}}$

EXAMPLE 8   Let $\mathcal{R}$ be the region bounded by the graph of $y = \tan x$, the x-axis, and the line $x = \pi/4$. Find the volume of the solid obtained by revolving $\mathcal{R}$ about the x-axis.

Solution   We use the disk method. (See Fig. 9–7.) The volume is

$$V = \pi \int_0^{\pi/4} \tan^2 x \, dx.$$

While we have no integral formula for $\tan^2 x$, we do have one for $\sec^2 x$. Furthermore, it follows from T6 that $\tan^2 x = \sec^2 x - 1$. Therefore,

$$V = \pi \int_0^{\pi/4} (\sec^2 x - 1) \, dx = \pi \left[\tan x - x \right]_0^{\pi/4}$$

$$= \pi \left[\tan \frac{\pi}{4} - \frac{\pi}{4} \right] - \pi \left[\tan 0 - 0 \right]$$

$$= \pi \left[1 - \frac{\pi}{4} \right] - 0 = \pi \left[1 - \frac{\pi}{4} \right] \approx 0.6742. \quad \square$$

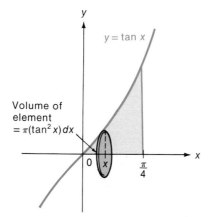

Figure 9–7   Example 8.

Exercises 9–2

1–12.   Calculate the derivative.

• 1 $\tan x^2$

2   $\sec 3x^2$

- 3 $\tan^2 x$
- 4 $\sec 2x \tan 2x$
- 5 $\tan^5 2x$
- 6 $\cot^2(3x^2 + 2x + 1)$
- 7 $\ln(\cot 4x)$
- 8 $\ln(\sec x + \tan x)$
- 9 $e^x \sec x$
- 10 $e^{\tan x}$
- 11 $\dfrac{1}{\sqrt{\tan x + 3}}$
- 12 $\sqrt[3]{1 + \csc^2 x}$

13–24. Calculate the integrals.

- 13 $\displaystyle\int \csc^2 2x\, dx$
- 14 $\displaystyle\int \sec 3x \tan 3x\, dx$
- 15 $\displaystyle\int x^2 \sec 2x^3\, dx$
- 16 $\displaystyle\int x \csc x^2\, dx$
- 17 $\displaystyle\int \tan 4x\, dx$
- 18 $\displaystyle\int \cot(3x - 1)\, dx$
- 19 $\displaystyle\int (1 + \tan x)^2\, dx$
- 20 $\displaystyle\int \dfrac{\sec^2 x\, dx}{\tan x}$
- 21 $\displaystyle\int \dfrac{\tan 2x\, dx}{\sec^3 2x}$
- 22 $\displaystyle\int \sqrt{\tan^3 x}\, \sec^2 x\, dx$
- 23 $\displaystyle\int_{\pi/12}^{\pi/6} \dfrac{\tan^2 2\theta + 1}{\tan 2\theta}\, d\theta$
- 24 $\displaystyle\int_{\pi/3}^{\pi/2} \cot \dfrac{x}{2} \csc \dfrac{x}{2}\, dx$

25–26. Find the period, all local extrema, points of inflection, and asymptotes. Sketch the graph.

- 25 $y = \sec x$
- 26 $y = \cot \dfrac{x}{2}$

- 27 Let $\mathcal{R}$ be the region bounded by the graph of $y = \sec x$ and the line $y = 2$, $-\pi/3 \le x \le \pi/3$. Find the area of $\mathcal{R}$.

- 28 Find the volume of the solid obtained by revolving the region in Exercise 27 about the x-axis.

- 29 A sheet of metal 24 in. wide and 20 ft long is to be formed into a drain gutter by bending up a 6-in. strip on each side of the sheet, making a cross section that is an open-topped trapezoid. Find the angle of bend that maximizes the capacity of the gutter.

- 30 A wall, in danger of collapsing, must be braced by a beam from the ground to the wall; this beam must pass over a lower wall that is 4 ft high and 6 ft from the first wall. What is the shortest beam that can be used?

- 31 A right circular cylinder is to be inscribed in a sphere of radius 4. Find the dimensions of the cylinder with the maximum possible lateral surface area.

●32 A plane 4 miles above the ground is flying directly away from an observer on the ground. At the instant at which the angle of elevation is 80° that angle is decreasing at the rate of 2.5° per second. Find the speed of the plane.

●33 A policeman, driving down a straight road at 30 mph, keeps his spotlight aimed at a man standing 60 ft back from the road. How fast is the angle made by the road, the spotlight, and the man changing at the instant when the angle is 30°?

●34 A light on the top of a police car revolves once each second projecting a spot of light on a wall 100 ft away. How fast does the spot travel at the instant when it is 100 ft from the point on the wall closest to the car?

35 (a) Sketch the graphs of $y = \sec x$ and $y = \ln |\sec x|$ for $-2\pi \le x \le 2\pi$.
●(b) Calculate the arc length of the graph of $y = \ln |\sec x|$ for $0 \le x \le \pi/3$.
(c) Can we use the integral formula for arc length to calculate the length of the graph of $y = \ln |\sec x|$ for $0 \le x \le \pi$? Explain.

36–39. Derive the formulas.

36 D14
37 D15
38 I14
39 I16

9–3* Application. Simple Harmonic Motion

The sine and cosine functions can be used to describe the motion of a vibrating weighted spring. To clarify the derivation, we make several simplifying assumptions. In particular, we assume that (1) the mass of the spring is negligible; (2) there is no resistance to the motion because of air resistance; and (3) there is no gravitational attraction to the mass. The resulting equation describes motion that continues forever without damping.

Because of our simplifying assumptions, you may think the results will be inaccurate. Actually, the opposite is true. The formula that we derive gives an excellent approximation to the true motion, provided we apply it for only a few oscillations.

THE OSCILLATING SPRING

Assume that a spring is suspended vertically and a weight is attached to the lower end of the spring, stretching it to an equilibrium position. (See Fig. 9–8a.) We stretch the spring downward a distance of x_0 units and release it. (See Fig. 9–8b.) The spring then contracts towards the equilibrium position but because of the force of the movement, overshoots this position and is compressed. It then expands and begins an oscillatory movement.

Let $x = x(t)$ be the displacement of the end of the spring at time t. We measure x positive if the end of the spring is below the equilibrium position, negative if it is above.

Recall from Section 6–5 that the force needed to maintain this displacement is proportional to the displacement. This same force, due to the potential energy of the

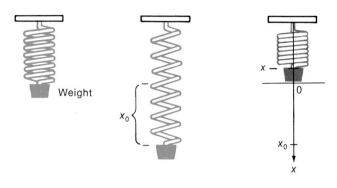

(a) Equilibrium position

(b) Initial position ($t = 0$)

(c) Displacement is measured on the vertical x-axis

Figure 9–8 The spring problem.

spring, acts to move it in the direction opposite the displacement. Thus, the force can be written as

$$F = -kx$$

where k is the positive constant of proportionality.

By Newton's second law of motion, the force equals the mass times the acceleration. Thus,

$$F = ma = m\frac{d^2x}{dt^2}$$

where m is the mass of the weight. Substituting these values into the first expression for F, we get

$$m\frac{d^2x}{dt^2} = F = -kx$$

so that

$$\frac{d^2x}{dt^2} + \frac{k}{m}x = 0.$$

Since k/m is positive, it is convenient to make the substitution

$$A^2 = k/m.$$

This reduces the differential equation to the form

$$\frac{d^2x}{dt^2} + A^2x = 0.$$

We first work with this differential equation and then return to the spring problem.

9–3* Application. Simple Harmonic Motion

THE DIFFERENTIAL EQUATION $\dfrac{d^2x}{dt^2} + A^2x = 0$

By direct verification we can show that if $x = C_1 \sin At$, then

$$\frac{d^2x}{dt^2} = -C_1 A^2 \sin At$$

so that

$$\frac{d^2x}{dt^2} + A^2x = -C_1 A^2 \sin At + A^2 C_1 \sin At = 0.$$

Thus $x = C_1 \sin At$ is a solution. Similarly, if $x = C_2 \cos At$, then x is a solution. We shall establish in Section 19–5 that the general solution is

$$x = C_1 \sin At + C_2 \cos At$$

where C_1 and C_2 are arbitrary constants that correspond to constants of integration. For reference purposes, this result is stated in Theorem 9–2.

THEOREM 9–2 The general solution of the differential equation

Solution of
$\dfrac{d^2x}{dt^2} + A^2x = 0$

$$\frac{d^2x}{dt^2} + A^2x = 0$$

is

$$x = C_1 \sin At + C_2 \cos At$$

where C_1 and C_2 are arbitrary constants.

We now show that the solution function

$$x = C_1 \sin At + C_2 \cos At$$

can be expressed in terms of the sine function alone. We let $C = \sqrt{C_1^2 + C_2^2}$ and write

$$x = C\left(\sin At \cdot \frac{C_1}{C} + \cos At \cdot \frac{C_2}{C}\right).$$

Since $(C_1/C)^2 + (C_2/C)^2 = 1$, there exists a number B such that

$$\cos B = \frac{C_1}{C} \quad \text{and} \quad \sin B = \frac{C_2}{C}.$$

Then

$$x = C(\sin At \cos B + \cos At \sin B).$$

If we apply T8, we get

$$x = C \sin (At + B)$$

where B and C are constants. (The graph is shown in Fig. 9–9.)

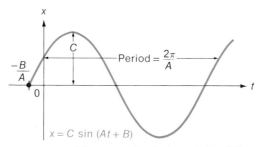

Figure 9-9 The general solution of the differential equation $d^2x/dt^2 + A^2x = 0$ is $x = C \sin (At + B)$.

THE SPRING PROBLEM (CONTINUED)

We now return to the spring problem. The differential equation that describes the motion is

$$\frac{d^2x}{dt^2} + A^2x = 0,$$

which has the general solution

$$x = C \sin (At + B),$$

where $A = \sqrt{k/m}$, k is the spring constant, m is the mass of the weight, and B and C are constants.

To evaluate the constants B and C, we use the initial conditions. Recall that the initial displacement is $x(0) = x_0$ and the initial velocity is $x'(0) = 0$. Then

$$x(0) = x_0 = C \sin (A \cdot 0 + B) = C \sin B$$

and

$$x'(0) = 0 = AC \cos (A \cdot 0 + B) = AC \cos B.$$

Note that A and C are nonzero, so that $\cos B = 0$. To get an initial value of B, we let $B = \pi/2$. Then

$$x_0 = C \sin \frac{\pi}{2} = C \cdot 1$$

so that

$$B = \frac{\pi}{2}, \quad C = x_0.$$

The equation for motion is

$$x = x_0 \sin \left(At + \frac{\pi}{2} \right) = x_0 \sin \left(\sqrt{\frac{k}{m}} t + \frac{\pi}{2} \right),$$

9–3* Application. Simple Harmonic Motion

which can be written as

$$x = x_0 \cos\left(\sqrt{\frac{k}{m}}\, t\right) \quad \text{(by T8)}.$$

(The graph is shown in Fig. 9–10.)

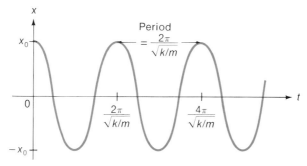

Figure 9–10 General solution of the spring problem with $x(0) = x_0$, $x'(0) = 0$.

Remark The choice of $B = \pi/2$ in the derivation seems arbitrary. Because of the periodicity of the trigonometric functions, however, we should have obtained the same final result with any other choice of B (provided $\cos B = 0$).

Motion described by the differential equation

$$\frac{d^2x}{dt^2} + A^2 x = 0$$

Simple Harmonic Motion is called *simple harmonic motion*. The motion of a weighted spring, subject to no external forces, is an example of simple harmonic motion. Some other examples are (1) a vibrating flexible shaft, (2) the projection of circular motion on a line, (3) the motion of a piston in an engine.

EXAMPLE 1 A suspended spring has a mass of 4 g attached to its free end. When stretched 3 in. from the equilibrium position and released, it oscillates with a period of 1 sec. Find the equation of the motion and calculate the spring constant k.

Solution The mass is $m = 4$ g and the initial displacement is $x_0 = 3$. The differential equation of motion is

$$\frac{d^2x}{dt^2} + A^2 x = 0.$$

The solution of the differential equation is

$$x = C \sin(At + B)$$

where $A = \sqrt{k/m}$, B, and C are constants.

As in the general derivation, we calculate that $B = \pi/2$ and $C = 3$. To find A, recall that the period is $2\pi/A$, which is equal to 1 sec. Thus

$$\text{Period} = \frac{2\pi}{A} = 1$$
$$A = 2\pi.$$

The equation of the motion is

$$x = 3 \sin(2\pi t + \pi/2) = 3 \cos(2\pi t).$$

To find the spring constant k, recall that $A = \sqrt{k/m}$, so that

$$k = mA^2 = 4(2\pi)^2 = 16\pi^2. \quad \square$$

Remark The theoretical springs discussed in this section oscillate forever. In actual practice, of course, an oscillating spring eventually slows to a stop. Surprisingly, this is not because of the gravitational attraction. The net effect of gravity is to change the amplitude and shift the displacement curve—essentially by establishing a new equilibrium position—not to damp the oscillations. (See Exercise 10.)

A very slight damping effect is due to air resistance. If the spring is immersed in a viscous fluid, this effect becomes much more pronounced. (See Exercise 14.)

The chief damping effect is from internal frictional forces in the spring. These forces cause some of the energy to be used for the generation of heat, thereby lessening the restoring force.

Example 2 is another demonstration of simple harmonic motion.

EXAMPLE 2 An object in simple harmonic motion has period $p = \frac{1}{4}$ sec. At time $t = 0$ (when we begin the observation), the object is at the equilibrium position $x = 0$. The maximum displacement (at time $t = \frac{1}{16}$ sec) is $x = 0.2$ cm.
(a) Find the equation of the motion.
(b) Calculate the velocity at time $t = 0$.

Solution (a) The equation of the motion is $x = C \sin(At + B)$, where $A > 0$.
First, observe that the period is

$$\text{Period} = \frac{2\pi}{A} = \frac{1}{4}$$

so that $A = 8\pi$.
To find B and C, we use the boundary conditions at times $t = 0$ and $t = \frac{1}{16}$.

$t = 0$: $\quad x = 0 = C \sin(A \cdot 0 + B)$
$\qquad\qquad C \sin B = 0$
$\qquad\qquad \sin B = 0$
$\qquad\qquad B = 0.$

$t = \frac{1}{16}$: $\quad x = 0.2 = C \sin\left(\frac{A}{16} + B\right)$

$\qquad\qquad 0.2 = C \sin\left(\frac{8\pi}{16} + 0\right)$

$$0.2 = C \sin \frac{\pi}{2} = C \cdot 1$$

$$C = 0.2.$$

The equation of the motion is $x = 0.2 \sin 8\pi t$.

(b) The velocity at time $t = 0$ is

$$\left(\frac{dx}{dt}\right)_{t=0} = ((0.2) 8\pi \cos 8\pi t)_{t=0}$$

$$= 1.6\pi \text{ cm/sec.} \quad \square$$

Exercises 9–3

1–3. An object is in simple harmonic motion. Find the equation of the motion from the information given.

•1 $x(0) = -4$, $x'(0) = 0$, period $= 2$

•2 $x(0) = 1$, $x'(0) = -1$, amplitude $= 2$

•3 Amplitude $= 1$, period $= 3$, $x(0) = -1/\sqrt{2}$

4 (a) Verify that $x = C_1 \cos At + C_2 \sin At + k/A^2$ is a solution of the differential equation

$$\frac{d^2x}{dt^2} + A^2 x = k.$$

It can be shown by the methods that we shall develop in Chapter 19 that this is the general solution of the differential equation.

(b) Show that this solution can be written in the form

$$x = C \sin (At + B) + k/A^2$$

where B and C are constants.

5–8. Use the result of Exercise 4 to find the particular solution of the differential equation

$$\frac{d^2x}{dt^2} + 4x = -5$$

that satisfies the given conditions.

•5 $x(0) = \frac{7}{4}$, $x'(0) = 0$ •6 $x(0) = -2$, $x'(\pi/6) = 0$

•7 $x(0) = -\frac{5}{4}$, $x(\pi/4) = -2$ •8 $x(0) = 0$, $x(\pi/2) = 0$

•9 The weight on the end of a spring has a mass of 2 kg. When the spring is stretched 1 m from the equilibrium position and released, it oscillates with a period of 3 sec. Assume that the simplifying assumptions mentioned in the text are satisfied. Find the equation of the motion and the value of the spring constant.

10 (*Effect of Gravity on the Spring Problem*) Let a weight of mass M be put at the end of a vertical spring. Assume that the mass of the spring is negligible. The spring is then stretched a distance of x_0 units from its natural length and released.

Gravity and the Spring Problem

(a) Use Newton's second law of motion to derive the differential equation of motion

$$m \frac{d^2x}{dt^2} = -kx + mg$$

where x measures the displacement from the natural length of the spring—not the displacement from the equilibrium position.

(b) Use Exercise 4 to solve this differential equation subject to the conditions $x(0) = x_0$, $x'(0) = 0$.

●(c) Sketch the graph of the solution in (b). What is the period? the amplitude? How does this differ from the solution obtained by ignoring the gravitational attraction?

●11 Let $f(x) = e^{-x} \cos x$, $g(x) = e^{-x} \sin x$. Calculate all local extrema and points of inflection for $0 \leq x \leq 3\pi$. Sketch the graphs. Use an exaggerated scale with 5 units on the y-axis for 1 unit on the x-axis.

12 Let A and B be positive constants with $B < A$. Verify that

$$f(t) = e^{-Bt} \cos \alpha t \quad \text{and} \quad g(t) = e^{-Bt} \sin \alpha t$$

are solutions of the differential equation

$$\frac{d^2x}{dt^2} + 2B \frac{dx}{dt} + A^2 x = 0,$$

where $\alpha = \sqrt{A^2 - B^2}$.

It can be shown by the methods of Chapter 19 that the general solution of this differential equation is

$$x = e^{-Bt} [C_1 \cos \alpha t + C_2 \sin \alpha t].$$

13 ●(a) Use the result of Exercise 12 to solve the differential equation

$$\frac{d^2x}{dt^2} + 6 \frac{dx}{dt} + 25x = 0$$

subject to the conditions $x(0) = 4$, $x'(0) = 0$.

c(b) Plot enough points on the graph of the solution in (a) to make a fairly accurate sketch for $-1 \leq t \leq 3$.

Damped Oscillations

14 (*Damped Oscillations*) Suppose that we immerse the weighted spring in a fluid that exerts a resisting force proportional to the velocity. Let all other factors remain as discussed in the text.

(a) Use Newton's second law of motion to derive the differential equation

$$m \frac{d^2x}{dt^2} = -kx - n \frac{dx}{dt},$$

where k is the spring constant, m is the mass, and n is the constant of proportionality for the resisting force.

Rewrite this differential equation in the form

$$\frac{d^2x}{dt^2} + 2B \frac{dx}{dt} + A^2 x = 0.$$

9–4 Inverse Trigonometric Functions

What are A and B?

(b) Use the result of Exercise 13 to find the equation of the motion for the special case in which $k = 100$, $m = 4$ and $n = 24$, subject to the initial conditions $x(0) = 4$, $x'(0) = 0$.

9–4 Inverse Trigonometric Functions

You should review Sections 1–8 and 3–9. (Inverse Functions) before starting work on this section.

If we examine the graphs of the trigonometric functions (see Appendix B), it is obvious that these functions are not one-to-one. For every number y in the range of a trigonometric function f, there exists an infinite number of x in the domain such that $f(x) = y$. Many applications, however, require us to invert these functions. To accomplish this, we must restrict their domains in such a way that the resulting functions are one-to-one and take all possible functional values of the original functions.

We now turn our attention to some particular trigonometric functions.

SIN^{-1} x

To define the inverse sine function, we first restrict the domain of the sine function to the interval $[-\pi/2, \pi/2]$. For all practical purposes, this means that we work with the function S, defined by

$$S(x) = \sin x, \quad -\pi/2 \leq x \leq \pi/2.$$

(See Fig. 9–11.) The *inverse sine function*, denoted by "sin^{-1}" or "arc sin" is defined to be the inverse of the function S. Since the range of S is the interval $[-1, 1]$,

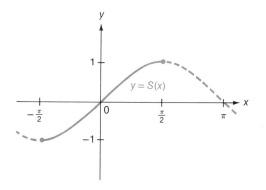

(a) Graph of $S(x) = \sin x$, $-\pi/2 \leq x \leq \pi/2$; the function S coincides with the sine function for $-\pi/2 \leq x \leq \pi/2$

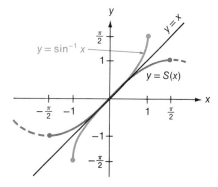

(b) Graph of $y = \sin^{-1} x$; the function sin^{-1} is the inverse of the function S

Figure 9–11 The inverse sine function.

then the domain of $\sin^{-1}$ is the interval $[-1, 1]$. The range of $\sin^{-1}$ is the interval $[-\pi/2, \pi/2]$. The formal definition follows.

DEFINITION

Inverse Sine

For each x, $-1 \leq x \leq 1$, we define

$$\sin^{-1} x = y \text{ if and only if } S(y) = x.$$

(This is equivalent to stating that $\sin^{-1} x = y$ if and only if $\sin y = x$, where $-\pi/2 \leq y \leq \pi/2$.)

Observe that the graph of $\sin^{-1}$ can be obtained by reflecting the graph of S across the 45° line $y = x$. (See Fig. 9–11b.)

All scientific calculators have a key for the inverse sine function. These keys are denoted by any of the symbols

SIN^{-1} X,   ARC SIN X,   or   INV SIN X.

EXAMPLE 1   (a) Calculate $\sin^{-1} 0.48818$.
(b) Calculate $\sin^{-1} (-\sqrt{2}/2)$.

Solution   (a) If we use the SIN^{-1} X key on a scientific calculator, we find that

$$\sin^{-1} 0.48818 \approx 0.51.$$

If no calculator is available, we can use Table III. First, we let

$$y = \sin^{-1} 0.48818.$$

Then

$$\sin y = 0.48818 \qquad \text{where } -\pi/2 \leq y \leq \pi/2.$$

Since $\sin y > 0$, it follows that y must be a first-quadrant angle. Thus, $0 < y \leq \pi/2$. By Table III, $\sin 0.51 \approx 0.48818$, so that

$$y = \sin^{-1} 0.48818 \approx 0.51.$$

(b) Let $y = \sin^{-1} (-\sqrt{2}/2)$. Then

$$\sin y = \frac{-\sqrt{2}}{2} \qquad \text{and} \qquad -\frac{\pi}{2} \leq y \leq \frac{\pi}{2}.$$

Since $\sin y < 0$, then y is a fourth-quadrant angle. Thus

$$\sin y = \frac{-\sqrt{2}}{2} \qquad \text{and} \qquad -\frac{\pi}{2} \leq y \leq 0.$$

The only angle y that satisfies those conditions is $y = -\pi/4$. Thus

$$y = \sin^{-1} \frac{-\sqrt{2}}{2} = -\frac{\pi}{4}. \quad \square$$

Scientific calculators are convenient when we calculate the values of $\sin^{-1} x$ for particular values of x. When we deal with more complicated problems, however, the type of analysis used in Example 1(b) is necessary.

EXAMPLE 2   Calculate (a) $\cos (\sin^{-1} (-12/13))$; (b) $\tan (\sin^{-1} (x/2))$.

9-4 Inverse Trigonometric Functions

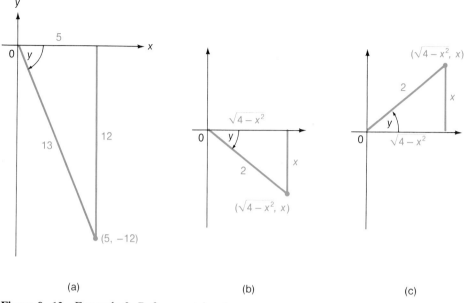

Figure 9-12 Example 2. Reference triangles.

Solution (a) Let $y = \sin^{-1}(-12/13)$. We need to calculate the cosine of y. The problem is simplified by drawing the angle y in standard position. (See Fig. 9-12a.) Observe that the reference triangle has sides of length 5, 12, and 13 and that the point $(5, -12)$ is on the terminal side of angle y. Therefore,

$$\cos\left(\sin^{-1}\frac{-12}{13}\right) = \cos y = \frac{5}{13}.$$

(b) Let $y = \sin^{-1}(x/2)$. Then

$$\sin y = \frac{x}{2} \quad \text{where} \quad -\frac{\pi}{2} \leq y \leq \frac{\pi}{2}.$$

Observe that

$$-\frac{\pi}{2} \leq y < 0 \quad \text{if } -2 \leq x < 0 \quad \text{(Figure 9-12b)},$$

while

$$0 \leq y \leq \frac{\pi}{2} \quad \text{if } 0 \leq x \leq 2 \quad \text{(Figure 9-12c)}.$$

In either case the reference triangle has sides of length 2, $|x|$, and $\sqrt{4-x^2}$ and the point $(\sqrt{4-x^2}, x)$ is on the terminal side of y. Therefore,

$$\tan\left(\sin^{-1}\frac{x}{2}\right) = \tan y = \frac{x}{\sqrt{4-x^2}}. \quad \square$$

DERIVATIVE OF THE INVERSE SINE FUNCTION

It is simple to calculate the derivative of the inverse sine function.

THEOREM 9–3 $D_x(\sin^{-1} x) = \dfrac{1}{\sqrt{1-x^2}}$ provided $-1 < x < 1$.

Proof It follows from Theorem 3–10 that the function $\sin^{-1}$ is differentiable at each point x where $-1 < x < 1$.

Let $y = \sin^{-1} x$, where $-1 < x < 1$. Then

$$\sin y = x,$$

where $-\pi/2 < y < \pi/2$. If we use implicit differentiation, we get

$$D_x(\sin y) = D_x(x)$$
$$\cos y \cdot D_x y = 1$$
$$D_x y = \frac{1}{\cos y}.$$

We now need to express $D_x y$ in terms of $\sin y$. By the Pythagorean Identity T5, $\cos^2 y = 1 - \sin^2 y$. Since $-\pi/2 < y < \pi/2$, then $\cos y > 0$. Therefore,

$$\cos y = \sqrt{1 - \sin^2 y} = \sqrt{1 - x^2}.$$

It follows that

$$D_x y = \frac{1}{\cos y} = \frac{1}{\sqrt{1 - \sin^2 y}} = \frac{1}{\sqrt{1 - x^2}}$$

so that

$$D_x(\sin^{-1} x) = \frac{1}{\sqrt{1 - x^2}}, \quad -1 < x < 1. \ \blacksquare$$

This formula can be generalized by use of the Chain Rule, yielding

Derivative of Inverse Sine

D17 $D_x(\sin^{-1} u) = \dfrac{1}{\sqrt{1 - u^2}} \cdot D_x u$ provided $-1 < u(x) < 1$,

where u is a differentiable function of x.

EXAMPLE 3 $\quad D_x(\sin^{-1}(2x^2)) = \dfrac{1}{\sqrt{1 - (2x^2)^2}} \cdot D_x(2x^2) = \dfrac{4x}{\sqrt{1 - 4x^4}}. \ \square$

Observe that even though the inverse sine function is transcendental, its derivative is an algebraic function. We shall show that this is true of all the inverse trigonometric functions.

9–4 Inverse Trigonometric Functions

$\cos^{-1} x$

To define the inverse cosine function, we restrict the domain of the cosine function to the interval $[0, \pi]$. Define the function C by

$$C(x) = \cos x, \quad 0 \leq x \leq \pi.$$

The *inverse cosine function* is defined to be the inverse of the function C.

DEFINITION
Inverse Cosine

For each x, $-1 \leq x \leq 1$, we define

$$\cos^{-1} x = y \quad \text{if and only if } C(y) = x.$$

(This is equivalent to stating that $\cos^{-1} x = y$ if and only if $\cos y = x$, where $0 \leq y \leq \pi$. See Figure 9–13.)

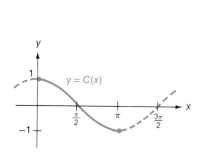
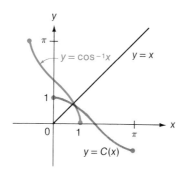

(a) Graph of $y = C(x)$; the function C coincides with the cosine function for $0 \leq x \leq \pi$

(b) Graph of $y = \cos^{-1} x$; the function $\cos^{-1}$ is the inverse of the function C

Figure 9–13 The inverse cosine function.

EXAMPLE 4 Calculate $\tan(\cos^{-1}(-4/5))$.

Solution Let $y = \cos^{-1}(-4/5)$. Then

$$\cos y = -\frac{4}{5}.$$

Since $0 \leq y \leq \pi$, then y must be a second-quadrant angle. The reference triangle is shown in Figure 9–14. This triangle has sides of length 3, 4, and 5 units. Therefore,

$$\tan y = \tan\left(\cos^{-1}\left(-\frac{4}{5}\right)\right) = -\frac{3}{4}. \quad \square$$

The derivative formula for the inverse cosine function is

Derivative of Inverse Cosine

D18 $\quad D_x(\cos^{-1} u) = \dfrac{-1}{\sqrt{1 - u^2}} D_x u, \quad -1 < u(x) < 1,$

where u is a differentiable function of x. The proof is left for you (Exercise 30).

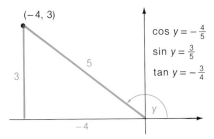

Figure 9–14 Example 4.

EXAMPLE 5 Show that $\cos^{-1} x + \sin^{-1} x = \pi/2$.

Solution **Method 1** Let $\theta = \cos^{-1} x$ and $\phi = \sin^{-1} x$. Then $0 \le \theta \le \pi$ and $-\frac{\pi}{2} \le \phi \le \frac{\pi}{2}$, so that

$$-\frac{\pi}{2} \le \theta + \phi \le \frac{3\pi}{2}.$$

Since $\cos \theta = \sin \phi = x$, then it follows from the Pythagorean Identity that

$$\sin \theta = \cos \phi = \sqrt{1 - x^2}.$$

Then

$$\sin(\theta + \phi) = \sin \theta \cos \phi + \cos \theta \sin \phi \quad \text{(by T8)}$$
$$= \sqrt{1 - x^2} \sqrt{1 - x^2} + x \cdot x = 1.$$

Since $\theta + \phi$ is between $-\pi/2$ and $3\pi/2$, then

$$\theta + \phi = \frac{\pi}{2}$$

$$\cos^{-1} x + \sin^{-1} x = \frac{\pi}{2}.$$

Method 2 Let $f(x) = \cos^{-1} x + \sin^{-1} x$, with $-1 < x < 1$. Then

$$f'(x) = \frac{-1}{\sqrt{1 - x^2}} + \frac{1}{\sqrt{1 - x^2}} = 0.$$

Since $f'(x) = 0$, there exists a constant C such that

$$f(x) = \cos^{-1} x + \sin^{-1} x = C.$$

To evaluate the constant, we choose the specific value $x = 0$. Then

$$f(0) = \cos^{-1} 0 + \sin^{-1} 0 = \frac{\pi}{2} + 0 = C.$$

Thus, $C = \pi/2$, so that

$$\cos^{-1} x + \sin^{-1} x = \frac{\pi}{2}. \quad \square$$

9-4 Inverse Trigonometric Functions

TAN⁻¹ x

To define the inverse tangent function, we restrict the domain of the tangent function to the open interval $(-\pi/2, \pi/2)$. This gives us one complete branch of the graph of the tangent function. See Figure 9-15.

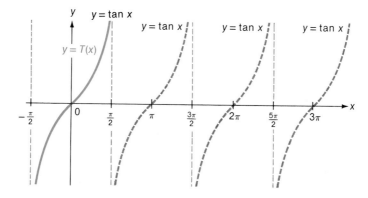

(a) Graph of $y = T(x)$; the function T coincides with the tangent function for $-\pi/2 < x < \pi/2$

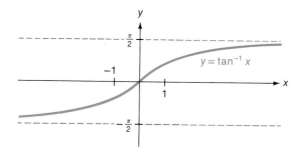

(b) Graph of $y = \tan^{-1} x$; the function $\tan^{-1}$ is the inverse of the function T

Figure 9-15 The inverse tangent function.

DEFINITION
Inverse Tangent

Let $T(x) = \tan x$, for $-\pi/2 < x < \pi/2$. The function $\tan^{-1}$ is defined to be the inverse of the function T, that is,

$$\tan^{-1} x = y \quad \text{if and only if } T(y) = x.$$

This statement is equivalent to the statement that $\tan^{-1} x = y$ if and only if $\tan y = x$, where $-\pi/2 < y < \pi/2$.

EXAMPLE 6 Show that $\sec(\tan^{-1} x) > 0$ for every x.

Solution Let $y = \tan^{-1} x$. Then $-\pi/2 < y < \pi/2$. Since y is a fourth-quadrant or first-quadrant angle, then $\sec y = 1/\cos y > 0$. □

The derivative formula for the inverse tangent function is stated in D19. The proof is left for you (see Exercise 31).

Derivative of Inverse Tangent

D19 $\quad D_x(\tan^{-1} u) = \dfrac{1}{1+u^2} \cdot D_x u$

where u is a differentiable function of x.

SEC^{-1} x

We consider one final inverse trigonometric function. The inverse secant is defined by restricting the domain of the secant function to the interval $[0, \pi]$, where $x \neq \pi/2$.* By adopting this convention, the restricted domain of the secant function will coincide as closely as possible to the restricted domain of the cosine function used to define $C(x)$.

DEFINITION Let $s(x) = \sec x$, $0 \leq x \leq \pi$, $x \neq \pi/2$. The function sec^{-1} is defined to be the inverse of the function s, that is

Inverse Secant $\qquad \sec^{-1} x = y \quad$ if and only if $s(y) = x$.

(This statement is equivalent to the statement that $\sec^{-1} x = y$ if and only if $\sec y = x$ where $0 \leq y \leq \pi$, $y \neq \pi/2$.)

The graphs of $\sec x$, $s(x)$, and $\sec^{-1} x$ are shown in Figures 9–16a, b. Observe that the domain of the function $\sec^{-1}$ is

$$\mathscr{D} = \{x \mid |x| \geq 1\}.$$

The formula for the derivative of the inverse secant function is stated in Theorem 9–4.

THEOREM 9–4 $\quad D_x(\sec^{-1} x) = \dfrac{1}{|x|\sqrt{x^2 - 1}} \quad$ provided $|x| > 1$.†

Proof It follows from Theorem 3–10 that the inverse secant function is differentiable at each point of its domain. Let $y = \sec^{-1} x$, so that $\sec y = x$ where $0 \leq y \leq \pi$, $y \neq \pi/2$. Then

---

* There is a fundamental disagreement among textbook writers about the inverse secant function. Some authors restrict the domain to the interval $[0, \pi]$, others to the union of the intervals $[-\pi, -\pi/2)$ and $[0, \pi/2)$. There are advantages and disadvantages to both approaches. Either method leads to a discontinuous function.

† Other books may give the formula as

$$D_x(\sec^{-1} x) = \dfrac{1}{x\sqrt{x^2 - 1}}$$

This formula is obtained by choosing the different restricted domain for the secant function that was mentioned in the preceding footnote.

9-4 Inverse Trigonometric Functions

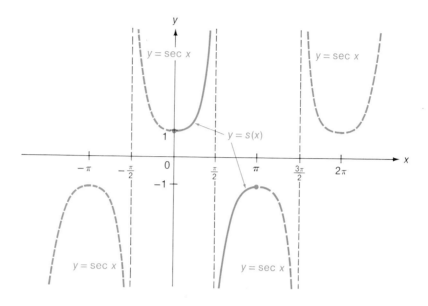

(a) Graph of $y = s(x)$; the function s coincides with the secant function for $0 \leq x \leq \pi$, $x \neq \pi/2$

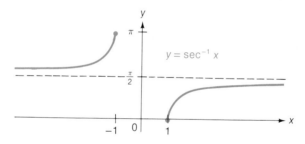

(b) Graph of $y = \sec^{-1} x$; the function $\sec^{-1}$ is the inverse of the function s

Figure 9-16 The inverse secant function.

$$D_x(\sec y) = D_x y$$
$$\sec y \tan y \cdot D_x y = 1$$
$$D_x y = \frac{1}{\sec y \tan y}.$$

Observe that $\tan^2 y = \sec^2 y - 1$ (by T6), so that

$$\tan y = \pm\sqrt{\sec^2 y - 1}.$$

We now consider two cases. If $x > 1$, then $0 < y < \pi/2$ and $\tan y = \sqrt{\sec^2 y - 1}$.

In this case

$$\sec y \tan y = \sec y \sqrt{\sec^2 y - 1} = x\sqrt{x^2 - 1} = |x|\sqrt{x^2 - 1}.$$

If $x < -1$, then $\pi/2 < y < \pi$ and $\tan y = -\sqrt{\sec^2 y - 1}$. In this case

$$\sec y \tan y = \sec y (-\sqrt{\sec^2 y - 1}) = -x\sqrt{x^2 - 1} = |x|\sqrt{x^2 - 1}.$$

In either case

$$\sec y \tan y = |x|\sqrt{x^2 - 1}.$$

Therefore

$$D_x(\sec^{-1} x) = D_x y = \frac{1}{\sec y \tan y} = \frac{1}{|x|\sqrt{x^2 - 1}}. \blacksquare$$

If we apply the Chain Rule, this formula generalizes to

Derivative of Inverse Secant

D20 $\quad D_x(\sec^{-1} u) = \dfrac{1}{|u|\sqrt{u^2 - 1}} D_x u, \quad$ provided $|u| > 1,$‡

where u is a differentiable function of x.

Remark We are more interested in the integral formulas that correspond to the derivative formulas of this section than in the derivative formulas themselves. Observe that we have not worked with the inverse cotangent or inverse cosecant function. These functions can be defined, but they are of little practical value to us because the corresponding integral formulas are almost identical to the ones obtained from the inverse tangent and inverse secant functions.

Exercises 9–4

1–10. Evaluate the numbers. In 1–3 use a calculator or Table III in the Appendix. In 4–10 find the exact value.

- **1** $\cos^{-1}(0.26750)$
- **2** $\tan^{-1}(-0.89121)$
- **3** $\sec^{-1}(-1.1395)$
- **4** $\tan^{-1}(\sqrt{3})$
- **5** $\sec(\tan^{-1}(-1))$
- **6** $\csc(\tan^{-1}(-3))$
- **7** $\cos\left(\sin^{-1}\left(\frac{1}{2}\right) + \tan^{-1}\left(\frac{1}{3}\right)\right)$
- **8** $\tan\left(\sin^{-1}\left(\frac{1}{3}\right) + \cos^{-1}\left(\frac{1}{-\sqrt{2}}\right)\right)$
- **9** $\tan(\cos^{-1}(2x))$
- **10** $\sec\left(\tan^{-1}\frac{2x}{3}\right)$

‡ See preceding note.

Exercises 9-4

11-20. Calculate the derivative with respect to x.

- 11 $\cos^{-1}(\sqrt{x})$
- 12 $\tan^{-1}(\cos x)$
- 13 $\sin^{-1} x + \cos^{-1} 2x$
- 14 $\sec^{-1}\left(\dfrac{x}{2}\right)$
- 15 $\ln(\cos^{-1} 4x)$
- 16 $\sec^{-1}(\ln 3x)$
- 17 $\dfrac{1}{\tan^{-1} x}$
- 18 $\sqrt{\sin^{-1} 2x}$
- 19 $(1 - \sin^{-1} 2x)^2$
- 20 $x (\sin^{-1} 2x)^2$.

21-22. Use implicit differentiation to calculate y'.

- 21 $x \cos^{-1} y + \sin^{-1} y = x$
- 22 $e^y + \tan^{-1} y = x^2 y$.

23   Use differentiation to prove that

$$\tan^{-1} x + \tan^{-1}\left(\frac{1}{x}\right) = \begin{cases} \dfrac{\pi}{2} & \text{if } x > 0, \\ -\dfrac{\pi}{2} & \text{if } x < 0. \end{cases}$$

24   Use differentiation to prove that

$$\sec^{-1} x + \sec^{-1}(-x) = \pi \quad \text{if } |x| \geq 1.$$

- 25   Determine the concavity and local extrema (if any) for the graph of

$$f(x) = \cos^{-1}\left(\frac{x-1}{2}\right).$$

Sketch the graph. How is this graph related to the graph of the function $g(x) = \cos^{-1} x$?

26   (a) Describe and sketch the graph of $y = \tan^{-1}(x + 1)$. How is this graph related to the graph of $y = \tan^{-1} x$?
      •(b) Find all points on the graph of $y = \tan^{-1}(x + 1)$ at which the tangent line has slope $m = 1$.

27   Use inverse trigonometric functions to work Exercise 32 of Section 9-2.

28   Use inverse trigonometric functions to work Problem 34 of Section 9-2.

- 29   A rectangular mural 5 ft high is painted on a wall with the bottom of the mural 6 ft above eye level. How far from the wall should you stand to get the best view of the mural?

30-31. Derive the derivative formulas.

30   D18   31   D19

9-5 Integrals Obtained from Inverse Trigonometric Functions

The integral formulas that correspond to the derivative formulas for the inverse trigonometric functions are used extensively. These formulas enable us to integrate a wide variety of algebraic functions that cannot be integrated by the formulas and techniques so far studied. The first two formulas are:

$$\text{I17'} \quad \int \frac{du}{\sqrt{1-u^2}} = \sin^{-1} u + C, \quad -1 < u < 1.$$

$$\text{I18'} \quad \int \frac{du}{1+u^2} = \tan^{-1} u + C.$$

The integral formula involving the inverse secant function is

$$\int \frac{du}{|u|\sqrt{u^2-1}} = \sec^{-1} u + C, \quad |u| > 1.$$

This formula can be rewritten in a more usable form if we consider the cases $u > 1$ and $u < 1$ separately.

If $u > 1$, then $|u| = u$. In this case, we get

$$\int \frac{du}{u\sqrt{u^2-1}} = \sec^{-1} u + C = \sec^{-1} |u| + C.$$

If $u < 1$, then $|u| = -u$. In this case, we get

$$\int \frac{du}{-u\sqrt{u^2-1}} = \sec^{-1} u + C_1$$

$$\int \frac{du}{u\sqrt{u^2-1}} = -\sec^{-1} u - C_1.$$

Recall from Exercise 24 of Section 9-4 that $\sec^{-1} x + \sec^{-1}(-x) = \pi$. It follows that

$$\int \frac{du}{u\sqrt{u^2-1}} = -\sec^{-1} u - C_1$$

$$= \sec^{-1}(-u) - \pi - C_1$$

$$= \sec^{-1} |u| + C$$

where $C = -\pi - C_1$.

In either case we have

$$\text{I19'} \quad \int \frac{du}{u\sqrt{u^2-1}} = \sec^{-1} |u| + C \quad \text{provided } |u| > 1.$$

Example 1 shows how these formulas can be generalized by the substitution principle.

9-5 Integrals Obtained from Inverse Trigonometric Functions

EXAMPLE 1 Show that

$$\int \frac{du}{a^2 + u^2} = \frac{1}{a} \tan^{-1}\left(\frac{u}{a}\right) + C.$$

Solution We rewrite the integral as

$$\int \frac{du}{a^2 + u^2} = \int \frac{du}{a^2(1 + (u/a)^2)} = \frac{1}{a}\int \frac{du/a}{1 + (u/a)^2}.$$

We now let $v = u/a$, $dv = du/a$. Then

$$\int \frac{du}{a^2 + u^2} = \frac{1}{a}\int \frac{dv}{1 + v^2}$$

$$= \frac{1}{a}\tan^{-1} v + C$$

$$= \frac{1}{a}\tan^{-1}\left(\frac{u}{a}\right) + C. \quad \square$$

$$\boxed{\begin{array}{c} v = \dfrac{u}{a} \\[4pt] dv = \dfrac{du}{a} \end{array}}$$

As indicated in Example 1, formulas I17′, I18′, and I19′ can be generalized to integrate expressions of form $du/\sqrt{a^2 - u^2}$, $du/(a^2 + u^2)$ and $du/u\sqrt{u^2 - a^2}$. The more general formulas are listed below. The constant a is positive.

Integral Formulas

I17 $\displaystyle\int \frac{du}{\sqrt{a^2 - u^2}} = \sin^{-1}\left(\frac{u}{a}\right) + C,\ |u| < a.$

I18 $\displaystyle\int \frac{du}{a^2 + u^2} = \frac{1}{a}\tan^{-1}\left(\frac{u}{a}\right) + C.$

I19 $\displaystyle\int \frac{du}{u\sqrt{u^2 - a^2}} = \frac{1}{a}\sec^{-1}\left|\frac{u}{a}\right| + C,\ |u| > a.$

EXAMPLE 2 Calculate $\displaystyle\int_{\sqrt{2}}^{2} \frac{dx}{x\sqrt{x^2 - 1}}.$

Solution

$$\int_{\sqrt{2}}^{2} \frac{dx}{x\sqrt{x^2 - 1}} = \left[\sec^{-1}|x|\right]_{\sqrt{2}}^{2} \quad \text{(by I19)}.$$

$$= \sec^{-1} 2 - \sec^{-1} \sqrt{2}.$$

Since $\sec \dfrac{\pi}{4} = \sqrt{2}$ and $\sec \dfrac{\pi}{3} = 2$, then

$$\sec^{-1} \sqrt{2} = \frac{\pi}{4} \quad \text{and} \quad \sec^{-1} 2 = \frac{\pi}{3}.$$

Therefore

$$\int_{\sqrt{2}}^{2} \frac{dx}{x\sqrt{x^2 - 1}} = \sec^{-1} 2 - \sec^{-1} \sqrt{2} = \frac{\pi}{3} - \frac{\pi}{4} = \frac{\pi}{12}. \quad \square$$

Trigonometric and Hyperbolic Functions

EXAMPLE 3 Calculate $\int \dfrac{e^x\,dx}{e^{2x}+4}$.

Solution Let $u = e^x$, $du = e^x\,dx$. Then

$$\int \frac{e^x\,dx}{e^{2x}+4} = \int \frac{du}{u^2+2^2} = \frac{1}{2}\tan^{-1}\left(\frac{u}{2}\right) + C$$

$$= \frac{1}{2}\tan^{-1}\left(\frac{e^x}{2}\right) + C. \;\square$$

$$\boxed{\begin{array}{l} u = e^x \\ du = e^x\,dx \end{array}}$$

EXAMPLE 4 Calculate the area of the region bounded by the x-axis and the graph of $y = 1/(x^2 + 1)$, $0 \le x \le \sqrt{3}$.

Solution The area is

$$A = \int_0^{\sqrt{3}} \frac{dx}{x^2+1} = \Big[\tan^{-1} x\Big]_0^{\sqrt{3}} = \tan^{-1}\sqrt{3} - \tan^{-1} 0.$$

Since $\tan(\pi/3) = \sqrt{3}$ and $\tan 0 = 0$, then

$$\tan^{-1}\sqrt{3} = \frac{\pi}{3} \quad \text{and} \quad \tan^{-1} 0 = 0.$$

Therefore,

$$A = \tan^{-1}\sqrt{3} - \tan^{-1} 0 = \frac{\pi}{3}. \;\square$$

Exercises 9–5

1–16. Evaluate the integrals.

- 1. $\displaystyle\int \frac{dx}{9+x^2}$
- 2. $\displaystyle\int \frac{dx}{1+4x^2}$
- 3. $\displaystyle\int \frac{2\,dx}{\sqrt{4-x^2}}$
- 4. $\displaystyle\int \frac{x+1}{\sqrt{1-16x^2}}\,dx$
- 5. $\displaystyle\int \frac{e^x\,dx}{\sqrt{1-e^{2x}}}$
- 6. $\displaystyle\int \frac{e^{2x}\,dx}{1+e^{2x}}$
- 7. $\displaystyle\int \frac{\sin x\,dx}{1+\cos^2 x}$
- 8. $\displaystyle\int \frac{\cos x\,dx}{\sqrt{4-\sin^2 x}}$
- 9. $\displaystyle\int \frac{dx}{x\sqrt{9x^2-4}}$
- 10. $\displaystyle\int \frac{2x\,dx}{\sqrt{1-9x^2}}$
- 11. $\displaystyle\int_0^2 \frac{dx}{x^2+4}$
- 12. $\displaystyle\int_2^4 \frac{dx}{x\sqrt{x^2-4}}$
- 13. $\displaystyle\int_0^{\sqrt{3}/2} \frac{dx}{\sqrt{1-x^2}}$
- 14. $\displaystyle\int_{-1}^1 \frac{x+1}{x^2+1}\,dx$
- 15. $\displaystyle\int_0^{1/\sqrt{2}} \frac{x\,dx}{\sqrt{1-x^4}}$
- 16. $\displaystyle\int_0^{\pi/2} \frac{\sin\theta\cos\theta\,d\theta}{1+\cos^4\theta}$

9-6 Hyperbolic Functions

●17 Let $\mathcal{R}$ be the region bounded by the x-axis and the graph of $f(x) = 1/\sqrt{x^2 + 1}$, $0 \le x \le 1$. Find the volume of the solid obtained by revolving $\mathcal{R}$ about
(a) the x-axis (b) the y-axis.

●18 Find the arc length of the graph of $f(x) = \sqrt{1 - x^2}$, $0 \le x \le \frac{1}{2}$.

9-6 Hyperbolic Functions

We showed in Section 9-3 that the functions $\sin ax$ and $\cos ax$ are solutions of the differential equation

$$\frac{d^2y}{dx^2} + a^2y = 0,$$

which defines harmonic motion. Similarly, it can be shown that the functions e^{ax} and e^{-ax} are solutions of the related differential equation

$$\frac{d^2y}{dx^2} - a^2y = 0,$$

which also is involved with certain types of motion.

The general solution of the differential equation $d^2y/dx^2 - a^2y = 0$ is $y = C_1 e^{ax} + C_2 e^{-ax}$, where C_1 and C_2 are arbitrary constants. Certain combinations of e^{ax} and e^{-ax}, obtained by choosing particular values of C_1 and C_2, appear with such frequency that they have been given special names—the *hyperbolic sine* and the *hyperbolic cosine*. The formal definitions follow.

DEFINITIONS

Hyperbolic Sine and Cosine

Let x be a real number. The *hyperbolic sine of x* ($\sinh x$) and *hyperbolic cosine of x* ($\cosh x$) are defined as follows:

$$\sinh x = \frac{e^x - e^{-x}}{2} \quad \text{and} \quad \cosh x = \frac{e^x + e^{-x}}{2}.$$

These functions have properties analogous to those of the sine and cosine functions. (See Example 1 and Exercises 20-27.) Furthermore, they define points on a hyperbola much as the sine and cosine functions define points on a circle—thus the names *hyperbolic sine* and *hyperbolic cosine*. (See Exercise 33.)

DERIVATIVE FORMULAS

It is particularly easy to calculate the derivatives of the hyperbolic sine and hyperbolic cosine. We express the functions in terms of e^x and e^{-x} and use formula D10. Thus

$$D_x(\sinh x) = D_x\left(\frac{e^x - e^{-x}}{2}\right)$$
$$= \frac{e^x - e^{-x}(-1)}{2}$$
$$= \frac{e^x + e^{-x}}{2} = \cosh x.$$

Similarly,

$$D_x(\cosh x) = \sinh x.$$

If we apply the Chain Rule, we obtain the more general formulas:

Derivatives of Hyperbolic Sine and Cosine

D21	$D_x(\sinh u) = \cosh u \cdot D_x u,$
D22	$D_x(\cosh u) = \sinh u \cdot D_x u,$

where u is a differentiable function of x.

GRAPHS OF THE HYPERBOLIC SINE AND COSINE

If we apply the techniques for sketching curves that we developed in Section 7–1, we see that the graph of the hyperbolic sine function has the following properties (See Fig. 9–17a.)

(1) The graph is a smooth curve that is defined for all x.
(2) It is symmetric about the origin and passes through the origin.
(3) It is an increasing function for all x.
(4) It is concave downward for $x < 0$ and concave upward for $x > 0$.

The graph of the hyperbolic cosine function has the following properties (see Fig. 9–17b):

(1) The graph is a smooth curve that is defined for all x.
(2) It is symmetric about the y-axis and passes through the point $(0, 1)$.

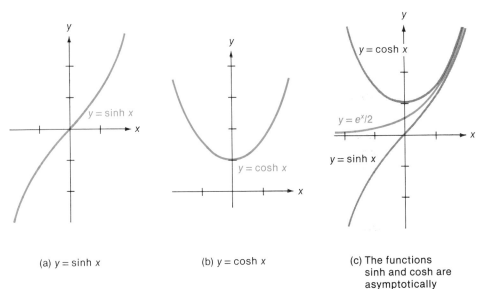

(a) $y = \sinh x$

(b) $y = \cosh x$

(c) The functions sinh and cosh are asymptotically equal to $e^x/2$ as $x \to \infty$

Figure 9–17 The hyperbolic sine and cosine functions.

9-6 Hyperbolic Functions

(3) It is a decreasing function for $x < 0$ and an increasing function for $x > 0$.

(4) It is concave upward for all x.

Since $e^{-x} \to 0$ as $x \to \infty$, both the hyperbolic sine and hyperbolic cosine functions are dominated by the term $e^x/2$ when x is large. That is,

$$\sinh x = \frac{e^x - e^{-x}}{2} \approx \frac{e^x}{2}$$

and

$$\cosh x = \frac{e^x + e^{-x}}{2} \approx \frac{e^x}{2}$$

when x is large. Thus, both functions are asymptotically equal to $e^x/2$ as $x \to \infty$. The graphs of the three functions are shown for comparison in Figure 9–17c.

IDENTITIES

Many identities similar to those for the trigonometric functions hold for the hyperbolic functions. These identities can be established by expressing the hyperbolic functions in terms of exponential functions.

EXAMPLE 1 Prove the identity: $\cosh^2 x - \sinh^2 x = 1$.

Solution

Basic Identity

$$\cosh^2 x - \sinh^2 x = \left(\frac{e^x + e^{-x}}{2}\right)^2 - \left(\frac{e^x - e^{-x}}{2}\right)^2$$

$$= \frac{(e^{2x} + 2 + e^{-2x}) - (e^{2x} - 2 + e^{-2x})}{4}$$

$$= \frac{4}{4} = 1. \ \square$$

The preceding identity is comparable to the Pythagorean Identity T5. Various other identities can be found in Exercises 20–27. These should be compared with the trigonometric identities in Appendix B.

Remark The hyperbolic identities are just different enough from the trigonometric identities to cause trouble. For this reason, you are advised not to memorize them. In most cases it is better to derive them as needed or look them up in tables.

OTHER HYPERBOLIC FUNCTIONS

The remaining hyperbolic functions are defined by rules analogous to those for the corresponding trigonometric functions.

Tanh $\qquad\qquad \tanh x = \dfrac{\sinh x}{\cosh x} \qquad\qquad$ (hyperbolic tangent).

Coth $\qquad\qquad \coth x = \dfrac{\cosh x}{\sinh x}, \quad x \neq 0 \qquad$ (hyperbolic cotangent).

Sech $\quad \operatorname{sech} x = \dfrac{1}{\cosh x} \quad$ (hyperbolic secant).

Csch $\quad \operatorname{csch} x = \dfrac{1}{\sinh x}, \quad x \neq 0 \quad$ (hyperbolic cosecant).

The derivative formulas for these functions are listed below. The proofs are left for the reader. (See Exercises 29–32.)

Derivative Formulas

D23 $\ D_x(\tanh u) = \operatorname{sech}^2 u \cdot D_x u,$
D24 $\ D_x(\coth u) = -\operatorname{csch}^2 u \cdot D_x u,$
D25 $\ D_x(\operatorname{sech} u) = -\operatorname{sech} u \tanh u \cdot D_x u,$
D26 $\ D_x(\operatorname{csch} u) = -\operatorname{csch} u \coth u \cdot D_x u,$

where u is a differentiable function of x.

There are, of course, corresponding integration formulas. Since we do not work extensively with hyperbolic function in this book, these formulas are omitted from the text. They are listed inside the front cover as formulas I20–I25. These formulas are analogous to the corresponding formulas for the trigonometric functions.

EXAMPLE 2 Calculate $\int \operatorname{sech} 2x \tanh 2x \, dx$.

Solution Let $u = 2x$, $du = 2\, dx$; then

$$\begin{aligned}
\int \operatorname{sech} 2x \tanh 2x \, dx &= \tfrac{1}{2} \int \operatorname{sech} 2x \tanh 2x \cdot 2 \, dx \\
&= \tfrac{1}{2} \int \operatorname{sech} u \tanh u \, du \\
&= -\tfrac{1}{2} \operatorname{sech} u + C \quad \text{(by I24)}. \\
&= -\tfrac{1}{2} \operatorname{sech} 2x + C
\end{aligned}$$

$u = 2x$
$du = 2\, dx$

EXAMPLE 3 Calculate $\displaystyle\int \tanh x \, dx$.

Solution
$$\int \tanh x \, dx = \int \dfrac{\sinh x}{\cosh x} \, dx.$$

Let $u = \cosh x$, $du = \sinh x \, dx$. Then

$$\int \tanh x \, dx = \int \dfrac{\sinh x \, dx}{\cosh x} = \int \dfrac{du}{u}$$

$u = \cosh x$
$du = \sinh x \, dx$

$$= \ln |u| + C = \ln (\cosh x) + C. \ \square$$

Exercises 9–6

1–8. Calculate the derivative with respect to x.

•1 $\sinh (x^4)$ \qquad 2 $\cosh^2 2x$

•3 $\tanh (2x^2 + 1)$ \qquad 4 $\coth \left(\dfrac{1}{x}\right)$

Exercises 9–6

- 5 ln (coth $2x$)
- 6 ln (coth x + csch x)
- 7 e^{2x} csch (x^2)
- 8 $\dfrac{4}{\sinh(3x+1)}$

9–16. Use formulas I20–I25, if applicable, to calculate the given integrals.

- 9 $\int \sinh(2x+1)\,dx$
- 10 $\int x \operatorname{sech}^2(4x^2+1)\,dx$
- 11 $\int \cosh^3 2x \sinh 2x\,dx$
- 12 $\int \coth x\,dx$
- 13 $\int \sinh x \operatorname{sech}^4 x\,dx$
- 14 $\int \operatorname{sech}^4 x\,dx$ (use Exercise 26)
- 15 $\displaystyle\int \dfrac{dx}{(e^x - e^{-x})^2}$
- 16 $\displaystyle\int \dfrac{e^x - e^{-x}}{(e^x + e^{-x})^2}\,dx$

17–19. Find all obvious symmetries, all local extrema, all points of inflection, the concavity, and all asymptotes for the graph of the function. Make an accurate sketch of the graph.

- 17 $y = \sinh x$
- 18 $y = \cosh x$
- 19 $y = \tanh x$

20–27. Prove the identity. Then compare it to a trigonometric identity in Appendix B.

20 $\sinh(x+y) = \sinh x \cosh y + \cosh x \sinh y$

21 $\cosh(x+y) = \cosh x \cosh y + \sinh x \sinh y$

22 $\sinh 2x = 2 \sinh x \cosh x$

23 $\cosh 2x = \cosh^2 x + \sinh^2 x$

24 $\tanh(x+y) = \dfrac{\tanh x + \tanh y}{1 + \tanh x \tanh y}$

25 $\tanh 2x = \dfrac{2 \tanh x}{1 + \tanh^2 x}$

26 $1 - \tanh^2 x = \operatorname{sech}^2 x$

27 $\sinh \dfrac{x}{2} = \pm \sqrt{\dfrac{\cosh x - 1}{2}}$

- 28 Verify that

$$y = C_1 e^{ax} + C_2 e^{-ax} \quad \text{and} \quad y = K_1 \cosh ax + K_2 \sinh ax$$

are solutions of the differential equation

$$\dfrac{d^2y}{dx^2} - a^2 y = 0.$$

Is it possible to get either of these solutions from the other? If so, what relation holds among the constants C_1, C_2, K_1, and K_2?

29–32. Derive the derivative formulas.

- 29 D23
- 30 D24
- 31 D25
- 32 D26

33 Let $x = \cosh t$, $y = \sinh t$. Show that the point (x, y) is on the graph of the hyperbola $x^2 - y^2 = 1$. Is the entire hyperbola traced out as t varies over the real numbers? If not, then which parts are traced out and which parts are not? This relation may help to explain the names *hyperbolic sine* and *hyperbolic cosine*.

9-7 Inverse Hyperbolic Functions

The inverse hyperbolic functions can be defined by methods like the ones used to define the inverse trigonometric functions. If necessary, we first restrict the domain of a particular function and then define its inverse. These functions can be used to integrate several types of algebraic functions that would otherwise be difficult to integrate. The formulas, which are listed later in this section, are very similar in form to those we derived from the inverse trigonometric functions.

Recall that the hyperbolic functions can be expressed in terms of exponential functions. As we shall see, the inverse hyperbolic functions can be expressed in terms of logarithmic functions—the inverses of the exponential functions.

SINH^{-1} x

Since the hyperbolic sine function is one-to-one, it has an inverse. We define

Inverse of Hyperbolic Sine

$$\sinh^{-1} x = y \quad \text{if and only if} \quad \sinh y = x.$$

(See Fig. 9-18b.) The domain and range of the function $\sinh^{-1}$ consist of all real numbers.

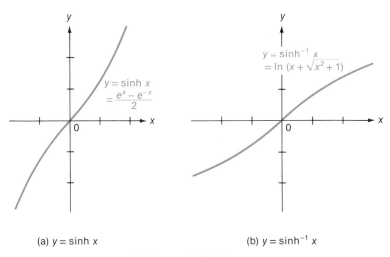

(a) $y = \sinh x$ (b) $y = \sinh^{-1} x$

Figure 9-18 Graphs of $\sinh x$ and $\sinh^{-1} x$.

Examples 1 and 2 show how the inverse hyperbolic sine function can be differentiated and how it can be expressed by the natural logarithmic function.

9–7 Inverse Hyperbolic Functions

EXAMPLE 1 Show that

$$\mathbf{D27} \quad D_x(\sinh^{-1} x) = \frac{1}{\sqrt{1 + x^2}}.$$

Solution It follows from Theorem 3–10 that the function $\sinh^{-1}$ is differentiable. Let $y = \sinh^{-1} x$. Then

Derivative of Inverse Hyperbolic Sine

$$\sinh y = x.$$

Using implicit differentiation, we get

$$D_x(\sinh y) = D_x x$$
$$\cosh y \cdot D_x y = 1$$

$$D_x y = \frac{1}{\cosh y} = \frac{1}{\sqrt{1 + \sinh^2 y}} = \frac{1}{\sqrt{1 + x^2}} \quad \text{(by Example 1, Section 9–6)}$$

$$D_x(\sinh^{-1} x) = \frac{1}{\sqrt{1 + x^2}}.$$

You should compare this formula with D17. ☐

The following example shows how the inverse hyperbolic sine can be expressed by logarithms.

EXAMPLE 2 Show that

Inverse Hyperbolic Sine Expressed as a Logarithm

$$\sinh^{-1} x = \ln (x + \sqrt{x^2 + 1}).$$

Solution Let $y = \sinh^{-1} x$. Then $x = \sinh y = (e^y - e^{-y})/2$. This equation can be rewritten as

$$e^y - \frac{1}{e^y} = 2x$$

$$e^{2y} - 2x e^y - 1 = 0,$$

which reduces to

$$z^2 - 2xz - 1 = 0,$$

where $z = e^y$. The solution of this quadratic equation is

$$z = \frac{2x \pm \sqrt{4x^2 + 4}}{2} = x \pm \sqrt{x^2 + 1}.$$

Since $z = e^y > 0$, we can eliminate the solution $z = x - \sqrt{x^2 + 1}$, which is always negative. Therefore,

$$e^y = z = x + \sqrt{x^2 + 1}$$
$$y = \ln (x + \sqrt{x^2 + 1}). \ \square$$

COSH⁻¹ x

The hyperbolic cosine function is not one-to-one. (See Fig. 9-19a.) To define the inverse hyperbolic cosine, we first restrict the domain of the hyperbolic cosine function to the numbers that are greater than or equal to zero. We define

Inverse Hyperbolic Cosine

$$y = \cosh^{-1} x \quad \text{if and only if} \quad x = \cosh y,$$

where $x \geq 1$ and $y \geq 0$.

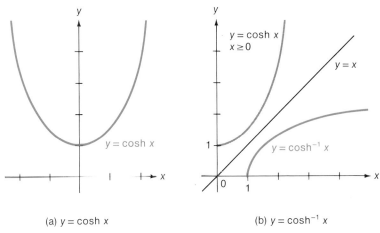

(a) $y = \cosh x$ (b) $y = \cosh^{-1} x$

Figure 9-19 Graphs of $\cosh x$ and $\cosh^{-1} x$.

The graph of the inverse hyperbolic cosine function is shown in Figure 9-19b. Observe that the domain consists of all $x \geq 1$ and the range of all $y \geq 0$.

Arguments like those in Examples 1 and 2 can be used to establish that

Derivative of Inverse Hyperbolic Cosine

D28 $\quad D_x(\cosh^{-1} x) = \dfrac{1}{\sqrt{x^2 - 1}}, \quad x > 1$

and

Logarithmic Form

$$\cosh^{-1} x = \ln(x + \sqrt{x^2 - 1}), \quad x \geq 1.$$

The proofs are left for you. (See Exercises 16 and 20.)

OTHER INVERSE HYPERBOLIC FUNCTIONS

Since the hyperbolic secant is the reciprocal of the hyperbolic cosine, its domain must also be restricted to values of $x \geq 0$ before defining its inverse function. The other hyperbolic functions are one-to-one. Their inverses can be defined without restricting the domains. The definitions follow.

9-7 Inverse Hyperbolic Functions

DEFINITIONS
(1) $y = \text{sech}^{-1} x$ if and only if $x = \text{sech } y$, where $0 < x \le 1$ and $y \ge 0$.
(2) $y = \tanh^{-1} x$ if and only if $x = \tanh y$.
(3) $y = \coth^{-1} x$ if and only if $x = \coth y$.
(4) $y = \text{csch}^{-1} x$ if and only if $x = \text{csch } y$.

The graphs of these functions are shown in Figure 9–20. All can be expressed as logarithmic functions. (See Exercises 16–19.) Their derivatives are listed below. The proofs, which parallel the comparable proofs for the inverse trigonometric functions, are left for you. (See Exercises 20–23.)

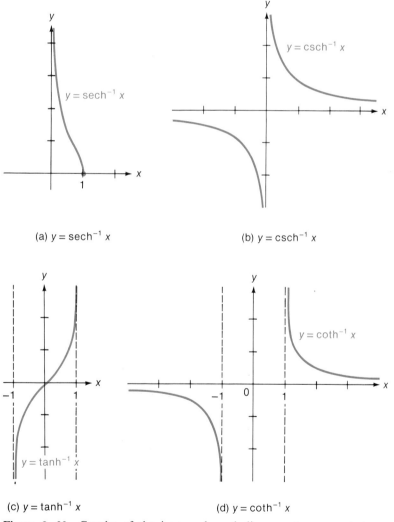

(a) $y = \text{sech}^{-1} x$ (b) $y = \text{csch}^{-1} x$

(c) $y = \tanh^{-1} x$ (d) $y = \coth^{-1} x$

Figure 9–20 Graphs of the inverse hyperbolic secant, cosecant, tangent, and cotangent functions.

Derivative Formulas

D29 $\quad D_x(\tanh^{-1} x) = \dfrac{1}{1-x^2}, \quad |x| < 1.$

D30 $\quad D_x(\coth^{-1} x) = \dfrac{1}{1-x^2}, \quad |x| > 1.$

D31 $\quad D_x(\operatorname{sech}^{-1} x) = \dfrac{-1}{x\sqrt{1-x^2}}, \quad 0 < x < 1.$

D32 $\quad D_x(\operatorname{csch}^{-1} x) = \dfrac{-1}{x\sqrt{1+x^2}}, \quad x \neq 0.$

INTEGRALS

The inverse hyperbolic functions are useful for integrating certain algebraic functions. The formulas corresponding to the derivative formulas of this section are as follows. The constant a is positive.

I26 $\quad \displaystyle\int \dfrac{du}{\sqrt{u^2 + a^2}} = \sinh^{-1}\left(\dfrac{u}{a}\right) + C.$

I27 $\quad \displaystyle\int \dfrac{du}{\sqrt{u^2 - a^2}} = \begin{cases} \cosh^{-1}\left(\dfrac{u}{a}\right) + C & \text{if } u > a \\ -\cosh^{-1}\left|\dfrac{u}{a}\right| + C & \text{if } u < -a. \end{cases}$

I28 $\quad \displaystyle\int \dfrac{du}{a^2 - u^2} = \begin{cases} \dfrac{1}{a}\tanh^{-1}\left(\dfrac{u}{a}\right) + C & \text{if } |u| < a. \\ \dfrac{1}{a}\coth^{-1}\left(\dfrac{u}{a}\right) + C & \text{if } |u| > a. \end{cases}$

I29 $\quad \displaystyle\int \dfrac{du}{u\sqrt{a^2 - u^2}} = -\dfrac{1}{a}\operatorname{sech}^{-1}\left|\dfrac{u}{a}\right| + C \qquad \text{provided } 0 < |u| < a.$

I30 $\quad \displaystyle\int \dfrac{du}{u\sqrt{a^2 + u^2}} = -\dfrac{1}{a}\operatorname{csch}^{-1}\left|\dfrac{u}{a}\right| + C.$

EXAMPLE 3 Prove formula I29.

Solution First, observe that Formula D31 implies that

$$\int \dfrac{dx}{x\sqrt{1-x^2}} = -\operatorname{sech}^{-1} x + C, \qquad \text{provided } 0 < x < 1.$$

Suppose that $0 < u < a$. If we make the substitution $u = ax$, $du = a\,dx$, the integral on the left-hand side of I29 reduces to

$$\int \dfrac{du}{u\sqrt{a^2 - u^2}} = \int \dfrac{a\,dx}{ax\sqrt{a^2 - a^2x^2}} = \dfrac{1}{a}\int \dfrac{dx}{x\sqrt{1-x^2}}$$

$$= -\frac{1}{a}\operatorname{sech}^{-1} x + C = -\frac{1}{a}\operatorname{sech}^{-1}\left(\frac{u}{a}\right) + C$$

$$= -\frac{1}{a}\operatorname{sech}^{-1}\left|\frac{u}{a}\right| + C.$$

Next, suppose that $-a < u < 0$. Let $v = |u| = -u$, $dv = -du$. Then

$$\int \frac{du}{u\sqrt{a^2 - u^2}} = \int \frac{-du}{-u\sqrt{a^2 - u^2}} = \int \frac{dv}{v\sqrt{a^2 - v^2}}$$

$$= -\frac{1}{a}\operatorname{sech}^{-1}\left|\frac{v}{a}\right| + C$$

$$= -\frac{1}{a}\operatorname{sech}^{-1}\left|\frac{u}{a}\right| + C. \quad \square$$

ALTERNATIVE FORM OF THE INTEGRAL FORMULAS

Formulas I26–I30 can be rewritten by expressing the inverse hyperbolic functions in terms of natural logarithms. The alternative forms are listed below.

Integral Formulas

I26 $\quad \int \dfrac{du}{\sqrt{u^2 + a^2}} = \ln(u + \sqrt{u^2 + a^2}) + C.$

I27 $\quad \int \dfrac{du}{\sqrt{u^2 - a^2}} = \ln|u + \sqrt{u^2 - a^2}| + C \quad$ if $|u| > a$.

I28 $\quad \int \dfrac{du}{a^2 - u^2} = \dfrac{1}{2a}\ln\left|\dfrac{u + a}{u - a}\right| + C \quad$ if $u \neq a$.

I29 $\quad \int \dfrac{du}{u\sqrt{a^2 - u^2}} = -\dfrac{1}{a}\ln\left|\dfrac{a + \sqrt{a^2 - u^2}}{u}\right| + C \quad$ if $0 < |u| < a$.

I30 $\quad \int \dfrac{du}{u\sqrt{a^2 + u^2}} = -\dfrac{1}{a}\ln\left|\dfrac{a + \sqrt{a^2 + u^2}}{u}\right| + C \quad$ if $u \neq 0$.

Exercises 9–7

1–8. Calculate the derivative with respect to x.

- **1** $\sinh^{-1}(2x^2)$
- **2** $\tanh^{-1}(4 - x)$
- **3** $e^{\cosh^{-1} x}$
- **4** $\ln(\sinh^{-1} 3x)$
- **5** $\sqrt{\operatorname{sech}^{-1} x}$
- **6** $\cosh^{-1}(\ln 3x^2)$
- **7** $x \coth^{-1}(\sqrt{x})$
- **8** $\sinh^{-1}(\tan x)$

9–14. Calculate the integral. Express the answer in inverse hyperbolic functions and in natural logarithms.

- **9** $\int \dfrac{dx}{1 - 9x^2}$
- **10** $\int \dfrac{dx}{6 - 5x^2}$

●11 $\int \dfrac{3\,dx}{\sqrt{4x^2 - 9}}$

12 $\int \dfrac{dx}{x\sqrt{9 - 4x^2}}$

●13 $\displaystyle\int_{\ln 3}^{\ln 4} \dfrac{e^x\,dx}{4 - e^{2x}}$

14 $\displaystyle\int_0^{\pi/2} \dfrac{\cos x\,dx}{\sqrt{1 + \sin^2 x}}$

●15 Find the arc length of the graph defined parametrically by
$$x = f(t) = \ln t, \qquad y = g(t) = \sec^{-1} t, \qquad \tfrac{5}{3} \le t \le \tfrac{13}{5}.$$

16–19. Prove the given relation.

16 $\cosh^{-1} x = \ln(x + \sqrt{x^2 - 1}),\ x \ge 1$

17 $\tanh^{-1} x = \dfrac{1}{2} \ln \dfrac{1 + x}{1 - x},\ |x| < 1$

18 $\coth^{-1} x = \dfrac{1}{2} \ln \dfrac{x + 1}{x - 1},\ |x| > 1.$

19 $\operatorname{sech}^{-1} x = \ln \left(\dfrac{1 + \sqrt{1 - x^2}}{x} \right),\ 0 < x \le 1$

20–23. Derive the derivative formulas.

20 D28

21 D29

22 D30

23 D31.

24–27. Prove the integral formulas.

24 I26

25 I27

26 I28

27 I30

Falling Bodies Subject to a Particular Type of Air Resistance

28 (*Falling Bodies Subject to Air Resistance*) Suppose that an object dropped from rest falls towards the earth subject to gravity and to a resisting force proportional to the square of the velocity. Modify the argument in Section 8–6 to establish the following results.

(a) The velocity at time t satisfies the differential equation
$$m \dfrac{dv}{dt} = -mg + kv^2$$
where k is the constant of proportionality.

(b) The solution of this differential equation is
$$v = -\sqrt{\dfrac{mg}{k}} \tanh\left(t\sqrt{\dfrac{kg}{m}}\right).$$

(c) The terminal velocity is $-\sqrt{mg/k}$.

(d) The height above the earth at time t is
$$S = -\dfrac{m}{k} \ln \cosh\left(t\sqrt{\dfrac{kg}{m}}\right) + s_0$$
where s_0 is the initial height.

9–8* Application. The Hanging Cable

Catenary   The graph of $y = a \cosh(x/a)$ is called a *catenary*. This curve has a superficial resemblance to a parabola but quite different properties. The catenary has a respectable place in classical physics. As we shall see, it is the shape assumed by a hanging flexible cable of uniform mass per unit of length. (We must assume that the cable is "ideal" and does not stretch when tension is applied to it, which is what happens with all real cables and ropes.)

We consider the case in which the cable is suspended from two points A and B. Let D be the lowest point of the cable between A and B. (See Fig. 9–21a.) We set up an

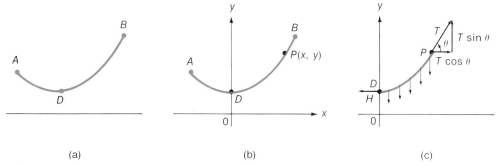

(a)   (b)   (c)

Figure 9–21   The hanging cable problem.

axis system with D on the y-axis, as in Figure 9–21b. Let $y = f(x)$ be the equation satisfied by the hanging cable. We shall derive a differential equation that is satisfied by the function f, then solve it to show that the graph of f is a catenary.

Let $P(x, y)$ be an arbitrary point on the cable. We imagine that the cable has been cut at D and P and attached to supports at these points. Then the cable will maintain its original shape between D and P. (See Fig. 9–21c.)

Observe that three forces act on the remaining piece of cable:

(1) A force due to gravitational attraction, which tends to pull the cable towards the earth. The numerical value of this force is mgs, where m is the mass per unit length, s is the arc length of the piece of cable, and g is the gravitational constant.

(2) A horizontal tension force at D represented by the symbol H in Figure 9–21c.

(3) A tangential tension force at P represented by the symbol T in Figure 9–21c. This force can be decomposed into horizontal and vertical component forces. If θ is the angle shown in Figure 9–21c, then
Horizontal component of $T = T \cos \theta$,
Vertical component of $T = T \sin \theta$.

It is proved in elementary physics that if a body is in equilibrium, then the sum of the component forces acting in any direction must be zero. (Otherwise, by Newton's second law of motion, there would be motion in that direction.) It follows that the sum of the horizontal forces acting on the cable must be zero and the sum of the vertical forces must be zero. Thus

$$T \cos \theta - H = 0 \quad \text{(horizontal forces)}$$

and

$$T \sin \theta - mgs = 0 \quad \text{(vertical forces)}.$$

It follows that

$$\tan \theta = \frac{T \sin \theta}{T \cos \theta} = \frac{mgs}{H}.$$

Observe, however, that θ is the angle of inclination of the tangent line at $P(x, y)$. Thus

$$f'(x) = \tan \theta = \frac{mgs}{H}.$$

At this point there are two ways by which we can go ahead. (1) We can express s as an integral

$$s = \int_0^x \sqrt{1 + [f'(t)]^2} \, dt$$

and substitute this expression in the above equation for $f'(x)$. Or (2) we can differentiate the equation for $f'(x)$ and then substitute $\sqrt{1 + [f'(x)]^2}$ for ds/dx. The first approach seems less promising than the second one because it leads to an equation containing both a derivative and an integral. Thus, we use the second approach. If we differentiate the equation

$$f'(x) = \frac{mgs}{H},$$

recalling that m, g, and H are constants, we get

$$f''(x) = \frac{mg}{H} \frac{ds}{dx} = \frac{mg}{H} \sqrt{1 + [f'(x)]^2},$$

so that $y = f(x)$ is a solution of the differential equation

$$y'' = \frac{mg}{H} \sqrt{1 + (y')^2}.$$

To solve this differential equation, we first rewrite it as

$$\frac{d}{dx}(y') = \frac{mg}{H} \sqrt{1 + (y')^2}$$

and substitute $z = y'$, getting

$$\frac{dz}{dx} = \frac{mg}{H} \sqrt{1 + z^2}.$$

$\boxed{z = y'}$

Next, we separate the variables, obtaining

$$\frac{dz}{\sqrt{1+z^2}} = \frac{mg}{H} dx$$

$$\int \frac{dz}{\sqrt{1+z^2}} = \frac{mg}{H} \int dx + C_1$$

$$\sinh^{-1} z = \frac{mg}{H} x + C_1$$

$$z = \sinh\left(\frac{mg}{H} x + C_1\right).$$

In evaluating the constant C_1, observe that the tangent line to the graph of f is horizontal at point D, so that

$$z = \frac{dy}{dx} = 0 \quad \text{when } x = 0.$$

Therefore,

$$0 = \sinh\left(\frac{mg}{H} \cdot 0 + C_1\right) = \sinh C_1,$$

so that $C_1 = 0$. Thus,

$$z = \frac{dy}{dx} = \sinh\left(\frac{mg}{H} x\right).$$

To find $y = f(x)$, we integrate this last expression, obtaining

$$f(x) = y = \int \sinh\left(\frac{mg}{H} x\right) dx + C$$

$$= \frac{H}{mg} \cosh\left(\frac{mg}{H} x\right) + C.$$

Since m, g, H, and C are constants, the hanging cable has the shape of a catenary:

$$y = a \cosh\left(\frac{x}{a}\right) + C \quad \text{where } a = \frac{H}{mg}.$$

Exercises 9–8

c1 Show that the catenary $y = \cosh x$ and the parabola

$$y = \frac{e^2 - 2e + 1}{2e} x^2 + 1$$

pass through the same points when $x = -1, 0,$ and 1. Make an accurate large-scale drawing, showing both graphs for $-2 \le x \le 2$. How do the graphs differ?

2   The hanging cable discussed in the text has the equation

$$y = a \cosh\left(\frac{x}{a}\right) + C, \quad \text{where } a = \frac{H}{mg}.$$

Let $P_1(x_1, y_1)$ and $P_2(x_2, y_2)$ be points on the cable with $x_1 < x_2$. Show that the length of the cable between P_1 and P_2 is

$$a \sinh\left(\frac{x_2}{a}\right) - a \sinh\left(\frac{x_1}{a}\right).$$

3   Let $y = f(x) = a \cosh(x/a)$ be the equation of a hanging cable. Let D be the point at which the cable crosses the positive y-axis. (See Fig. 9–21.) Let s be the length of the cable between D and $P(x, y)$:

$$s = \int_0^x \sqrt{1 + f'(t)^2}\, dt.$$

(Observe that s is negative if $x < 0$.)

Show that s can be used as a parameter for x and y. In particular, show that

$$x = a \sinh^{-1}\left(\frac{s}{a}\right) \quad \text{and} \quad y = \sqrt{s^2 + a^2}.$$

c
•4   A 64-ft cable hangs from points $A(-20, 40)$ and $B(20, 40)$. Find the point at which it crosses the y-axis. (*Hint:* Write $y = a \cosh(x/a) + C$. Use Exercise 2 to show that a is a solution of $x \sinh(20/x) = 32$. Use Newton's method to approximate a. Then use the fact that B is on the curve to evaluate C.)

Review Problems

1–14.   Calculate the derivative with respect to x.

- 1 $\dfrac{\cos x}{1 + \sin x}$
- 2 $e^{\sin 2x}$
- 3 $\sec x \tan x$
- 4 $\ln(\csc(x^2))$
- 5 $(\tan x + \sec x)^3$
- 6 $5^{\cot x}$
- 7 $x \cos^{-1}(x^2)$
- 8 $\sin^{-1}(\sqrt{x^2 - 1})$
- 9 $(\sinh 2x)^2$
- 10 $\cosh x \sinh x$
- 11 $e^x \tanh x$
- 12 $\text{sech}(4x - 1)$
- 13 $\cosh^{-1}(x^2)$
- 14 $x \sinh^{-1}(2x)$

15–16.   Use implicit differentiation to calculate y'.

- 15 $\sin(x + y) = x$
- 16 $xy = \tan y$

Review Problems

17–34. Evaluate the integrals.

●17 $\displaystyle\int_0^{\pi/2} e^{\cos x} \sin x \, dx$

●18 $\displaystyle\int_0^{\pi/2} \frac{\cos x \, dx}{1 + \sin x}$

●19 $\displaystyle\int \csc 3x \cot 3x \, dx$

●20 $\displaystyle\int_0^{\pi/4} \sec x \, dx$

●21 $\displaystyle\int (2 + \cot x)^2 \, dx$

●22 $\displaystyle\int \frac{e^{\tan 2x} \, dx}{\cos^2 2x}$

●23 $\displaystyle\int \frac{dx}{4 + 9x^2}$

●24 $\displaystyle\int \frac{2 \, dx}{\sqrt{4 - 9x^2}}$

●25 $\displaystyle\int_{-1/2}^{1/2} \frac{dx}{\sqrt{1 - x^2}}$

●26 $\displaystyle\int_{-3}^{3} \frac{dx}{x^2 + 9}$

●27 $\displaystyle\int \frac{x^2 \, dx}{x^6 + 25}$

●28 $\displaystyle\int \frac{x^3 \, dx}{\sqrt{4 - x^8}}$

●29 $\displaystyle\int \frac{\sinh \sqrt{x}}{\sqrt{x}} \, dx$

●30 $\displaystyle\int \sinh^4 x \cosh x \, dx$

●31 $\displaystyle\int x \tanh (x^2) \, dx$

●32 $\displaystyle\int \frac{dx}{4 - 9x^2}$

●33 $\displaystyle\int \frac{2 \, dx}{\sqrt{x^2 + 4}}$

●34 $\displaystyle\int \frac{dx}{\sqrt{3x^2 + 2}}$

35–36. Prove the limits. Use a geometrical argument for 35. Use 35 for 36.

35 $\displaystyle\lim_{x \to 0} \frac{\sin x}{x} = 1$

36 $\displaystyle\lim_{x \to 0} \frac{1 - \cos x}{x} = 0$

37–38. Use the limit in Problem 35 to evaluate the given limits.

●37 $\displaystyle\lim_{x \to 0} \frac{3 \sin^2 x}{x}$

●38 $\displaystyle\lim_{x \to 0} \frac{\tan 2x}{3x \cos x}$

39 Derive the formula for $D_x(\sin x)$.

40 (a) Define "cosh x," "sinh x," and "tanh x."

(b) Prove that $D_x(\cosh x) = \sinh x$ and that $D_x(\sinh x) = \cosh x$.

41–44. Determine the local extrema, concavity, and asymptotes. Sketch the graph.

●41 $f(x) = \cot x$

●42 $f(x) = \sin^{-1} x$

●43 $f(x) = \sinh x$

●44 $f(x) = \sinh^{-1} x$

45–48. Derive the formulas.

45 $D_x(\sin^{-1} x) = \dfrac{1}{\sqrt{1-x^2}}$

46 $D_x(\sinh^{-1} x) = \dfrac{1}{\sqrt{1+x^2}}$

47 $D_x(\tan^{-1} x) = \dfrac{1}{1+x^2}$

48 $D_x(\tanh^{-1} x) = \dfrac{1}{1-x^2}$

49 Use derivatives to prove that

$$\cos^{-1} x = \tan^{-1}\left(\dfrac{\sqrt{1-x^2}}{x}\right), \qquad 0 < x < 1.$$

50 Define *simple harmonic motion*. Give two examples of such motion.

51–52. Solve the differential equation $d^2x/dt^2 + 16x = 0$ subject to the stated conditions.

●51 $x(0) = 2,\ x'(0) = 1$

●52 $x(0) = 1,\ x\left(\dfrac{\pi}{6}\right) = 1$

●53 A vertical spring has a weight of mass 3 kg attached to its lower end. When the spring is stretched 0.5 m from the equilibrium position and released, it oscillates with a period of 3 sec. Find the equation of the motion.

54 Prove the identity $\cosh^2 x - \sinh^2 x = 1$.

55 Prove that $\sinh^{-1} x = \ln(x + \sqrt{x^2 + 1})$.

●56 Let $\mathscr{R}$ be the region bounded by the graph of $y = \sin x$ and the x-axis, $0 \leq x \leq \pi$.

(a) Calculate the area of $\mathscr{R}$.

(b) Calculate the volume of the solid obtained by revolving $\mathscr{R}$ about the x-axis. [*Hint:* Use T15.]

●57 A plane 10 miles above the surface of the earth is traveling at 600 mph in a line that is directly above a lake. An automatic camera in the plane is taking pictures of the lake. How fast must the camera be turning to keep the center of the lake in the center of the picture at the instant at which the angle of depression from the plane to the center of the lake is 60°?

10 L'Hôpital's Rule. Improper Integrals

10-1 L'Hôpital's Rule. I: The Indeterminate Forms 0/0 and ∞/∞

We used a geometrical argument in Section 9–1 to evaluate the limit

$$\lim_{x \to 0} \frac{\sin x}{x}$$

We now turn our attention to a powerful method that can be used to evaluate certain limits. Its use will enable us to evaluate the above limit in one step (See Example 1).

THE INDETERMINATE FORM 0/0

A limit of the type

Form 0/0
$$\lim_{x \to a} \frac{f(x)}{g(x)}$$

is said to have the *indeterminate form* 0/0 if $f(x)$ and $g(x)$ both approach zero as x approaches a.

We shall establish in Theorem 10–2 that under suitable conditions on f and g, limits of form 0/0 can be evaluated by the rule

$$\lim_{x \to a} \frac{f(x)}{g(x)} = \lim_{x \to a} \frac{f'(x)}{g'(x)}.$$

This result is known as *L'Hôpital's rule*.*

Suppose, for example, that $f(x) = \sin x$ and $g(x) = x$. We know that $f(x)/g(x)$ has

---

* L'Hôpital's Rule is named after the Marquis G. de l'Hôpital (1661–1704), who first published it, rather than after Jean Bernoulli (Swiss mathematician, 1667–1748), who discovered and proved it. The marquis had a contract that allowed him to publish Bernoulli's discoveries in return for partial support.

the indeterminate form $0/0$ as $x \to 0$ and that $f(x)/g(x) \to 1$ as $x \to 0$. Observe that

$$\frac{f'(x)}{g'(x)} = \frac{\cos x}{1} \to 1 \quad \text{as } x \to 0,$$

so that

$$\lim_{x \to 0} \frac{f(x)}{g(x)} = \lim_{x \to 0} \frac{f'(x)}{g'(x)} = 1.$$

Thus, L'Hôpital's rule holds for these particular functions.

We prove L'Hôpital's rule after the following theorem.

CAUCHY'S THEOREM

The proof of L'Hôpital's rule is based on the following generalization of the Mean Value Theorem to two functions f and g.

THEOREM 10-1

Cauchy's Theorem

(*Cauchy's Theorem*) Let f and g be continuous on the closed interval $[a, b]$ and differentiable on the open interval (a, b). If $g'(x) \neq 0$ for $a < x < b$, then there exists a number ξ between a and b such that

$$\frac{f(b) - f(a)}{g(b) - g(a)} = \frac{f'(\xi)}{g'(\xi)}.$$

Proof It follows from the Mean Value Theorem that $g(b) \neq g(a)$. (See Exercise 14a.)
Define the auxilliary function h by

$$h(x) = [f(x) - f(a)] - \frac{f(b) - f(a)}{g(b) - g(a)} [g(x) - g(a)].$$

The function h satisfies the conditions of Rolle's theorem. (See Exercise 14(b).) Thus, there exists a number ξ between a and b such that

$$h'(\xi) = 0.$$

If we calculate $h'(\xi)$, we find that

$$h'(\xi) = [f'(\xi) - 0] - \frac{f(b) - f(a)}{g(b) - g(a)} [g'(\xi) - 0] = 0$$

so that

$$\frac{f'(\xi)}{g'(\xi)} = \frac{f(b) - f(a)}{g(b) - g(a)}. \blacksquare$$

L'HÔPITAL'S RULE

We are now ready to prove our main result.

THEOREM 10-2

L'Hôpital's Rule

(*L'Hôpital's Rule*) Let f and g be differentiable at each point of an interval I except possibly at the interior point $x = a$. Suppose that g has a nonzero derivative at each point of I different from a, and that $\lim_{x \to a} f(x) = \lim_{x \to a} g(x) = 0$.

10–1 L'Hôpital's Rule. I: The Indeterminate Forms 0/0 and ∞/∞

If $\lim_{x \to a} \dfrac{f'(x)}{g'(x)} = L$, then

$$\lim_{x \to a} \frac{f(x)}{g(x)} = \lim_{x \to a} \frac{f'(x)}{g'(x)} = L.$$

This result also holds for one-sided limits and holds if L is replaced by ∞ or $-\infty$. It also holds (under slightly different hypotheses) if a is replaced by ∞ or $-\infty$. (See Exercise 16.)

Proof We prove L'Hôpital's rule for the case where

$$\lim_{x \to a} \frac{f'(x)}{g'(x)} = L$$

The generalization obtained by replacing a or L by ∞ or $-\infty$ are left for you to work out. (See Exercises 16 and 17.)

A minor complication exists because f and g may not be continuous at $x = a$. To get around this problem, we work with functions F and G defined by

$$F(x) = \begin{cases} f(x) & \text{if } x \neq a \\ 0 & \text{if } x = a \end{cases}, \quad G(x) = \begin{cases} g(x) & \text{if } x \neq a \\ 0 & \text{if } x = a \end{cases}.$$

Then F and G are continuous on the interval I and coincide with f and g, respectively, at each point of I different from $x = a$.

Let x be a number in I that is greater than a. Since the functions F and G satisfy the conditions of Cauchy's theorem on the interval $[a, x]$, there exists a number ξ between a and x such that

$$\frac{F(x) - F(a)}{G(x) - G(a)} = \frac{F'(\xi)}{G'(\xi)}.$$

It follows from the definition of F and G that

$$\frac{f(x) - 0}{g(x) - 0} = \frac{f'(\xi)}{g'(\xi)}$$

Since ξ is between a and x, then $\xi \to a^+$ as $x \to a^+$. Therefore,

$$\lim_{x \to a^+} \frac{f(x)}{g(x)} = \lim_{x \to a^+} \frac{f'(\xi)}{g'(\xi)} = \lim_{\xi \to a^+} \frac{f'(\xi)}{g'(\xi)} = L.$$

A similar argument shows that

$$\lim_{x \to a^-} \frac{f(x)}{g(x)} = L.$$

(See Exercise 15.)

Since $f(x)/g(x) \to L$ as x approaches a from either the right or the left, then

$$\lim_{x \to a} \frac{f(x)}{g(x)} = L = \lim_{x \to a} \frac{f'(x)}{g'(x)}.$$

L'Hôpital's Rule for One-Sided Limits The individual parts of the above argument establish L'Hôpital's rule for one-sided limits.

The proof that L'Hôpital's rule holds if a is replaced by $\pm\infty$ is left for you to do. (See Exercise 16.) ∎

Our first example involves the limit from Section 9–1.

EXAMPLE 1   Calculate $\lim\limits_{x\to 0}\dfrac{\sin x}{x}$.

Solution   The limit is of form 0/0. By L'Hôpital's rule

$$\lim_{x\to 0}\frac{\sin x}{x}=\lim_{x\to 0}\frac{\cos x}{1}=\frac{1}{1}=1.\ \square$$

EXAMPLE 2   Calculate $\lim\limits_{x\to 0}\dfrac{\tan 2x}{x^3}$.

Solution   By L'Hôpital's rule,

$$\lim_{x\to 0}\frac{\tan 2x}{x^3}=\lim_{x\to 0}\frac{2\sec^2 2x}{3x^2}.$$

Since the numerator, $2\sec^2 2x$, approaches 2 and the denominator, $3x^2$, approaches 0 through positive values as $x\to 0$, then

$$\lim_{x\to 0}\frac{\tan 2x}{x^3}=\lim_{x\to 0}\frac{2\sec^2 2x}{3x^2}=\infty.\ \square$$

To solve Example 3, L'Hôpital's rule must be applied twice.

EXAMPLE 3   Calculate $\lim\limits_{x\to 0}\dfrac{\cos 3x-1}{x^2}$.

Solution   By L'Hôpital's rule,

$$\lim_{x\to 0}\frac{\cos 3x-1}{x^2}=\lim_{x\to 0}\frac{-3\sin 3x}{2x}.$$

Observe that this limit also is of form 0/0. If we apply L'Hôpital's rule a second time, we get

$$\lim_{x\to 0}\frac{\cos 3x-1}{x^2}=\lim_{x\to 0}\frac{-3\sin 3x}{2x}=\lim_{x\to 0}\frac{-9\cos 3x}{2}=-\frac{9}{2}.\ \square$$

THE INDETERMINATE FORM ∞/∞

A limit of the type

$$\lim_{x\to a}\frac{f(x)}{g(x)}$$

is said to have the indeterminate form ∞/∞ if $f(x)\to\pm\infty$ and $g(x)\to\pm\infty$ as $x\to a$.

Theorem 10–3 extends L'Hôpital's rule to limits of this type. The proof is omitted.

10–1 L'Hôpital's Rule. I: The Indeterminate Forms 0/0 and ∞/∞

THEOREM 10–3

L'Hôpital's Rule for Form ∞/∞

(*L'Hopital's Rule for the Form* ∞/∞) Let f and g be differentiable at each point of an interval I except possibly at the interior point $x = a$. Suppose that g has a nonzero derivative at each point of I different from a and that $f(x)/g(x)$ has the indeterminate form ∞/∞ as $x \to a$. If

$$\lim_{x \to a} \frac{f'(x)}{g'(x)} = L,$$

then

$$\lim_{x \to a} \frac{f(x)}{g(x)} = \lim_{x \to a} \frac{f'(x)}{g'(x)} = L.$$

This result also holds for one-sided limits and it holds if L is replaced by ∞ or $-\infty$. It also holds (under slightly different hypotheses) if a is replaced by ∞ or $-\infty$. (If a is replaced by ∞, then f and g must have derivatives for all x greater than some number b. Furthermore, $g'(x) \neq 0$ if $x > b$.)

EXAMPLE 4 Calculate $\lim\limits_{x \to \infty} \dfrac{\sqrt{x}}{\ln x}$.

Solution The limit is of form ∞/∞. By L'Hôpital's rule,

$$\lim_{x \to \infty} \frac{\sqrt{x}}{\ln x} = \lim_{x \to \infty} \frac{\frac{1}{2\sqrt{x}}}{1/x} = \lim_{x \to \infty} \frac{x}{2\sqrt{x}} = \lim_{x \to \infty} \frac{\sqrt{x}}{2} = \infty. \quad \square$$

Remarks Several comments about the application of L'Hôpital's rule are in order.

(1) L'Hôpital's rule can be applied only to the indeterminate forms 0/0 and ∞/∞. If we try to apply L'Hôpital's rule several times (as in Example 3), we must be certain that we have one of these forms each time.

For example, we might incorrectly calculate that

$$\text{``}\lim_{x \to 0} \frac{\sin x}{x^3} = \lim_{x \to 0} \frac{\cos x}{3x^2} = \lim_{x \to 0} \frac{-\sin x}{6x} = \lim_{x \to 0} \frac{-\cos x}{6} = -\frac{1}{6}.\text{''}$$

The error occurs in the second step. Since $\cos x \to 1$ and $3x^2 \to 0$ as $x \to 0$, L'Hôpital's rule cannot be applied.

(2) It may be necessary to apply L'Hôpital's rule many times to calculate a certain limit. For example, to calculate $\lim_{x \to \infty} x^{1000}/e^x$, we apply L'Hôpital's rule 1000 times:

$$\lim_{x \to \infty} \frac{x^{1000}}{e^x} = \lim_{x \to \infty} \frac{1000 x^{999}}{e^x} = \lim_{x \to \infty} \frac{(1000)(999) x^{998}}{e^x} = \cdots = \lim_{x \to 0} \frac{1000!}{e^x} = 0.$$

(3) We should try to simplify the function obtained after each application of L'Hôpital's rule. The limit may then be obvious, or the next application of L'Hôpital's rule may be made easier.

Example 4 illustrates why simplification is important. If we do not simplify the expressions, then L'Hôpital's rule must be applied an infinite number of times,

$$\lim_{x\to\infty} \frac{\sqrt{x}}{\ln x} = \lim_{x\to\infty} \frac{x^{-1/2}/2}{x^{-1}} = \lim_{x\to\infty} \frac{-x^{-3/2}/4}{-x^{-2}} = \lim_{x\to\infty} \frac{3x^{-5/2}/8}{2x^{-3}} = \cdots,$$

and we never determine the limit.

Exercises 10-1

1-10. Use L'Hôpital's rule to evaluate the limits.

•1 $\lim\limits_{x\to 1^+} \dfrac{\ln x}{x - 1}$
2 $\lim\limits_{x\to\infty} \dfrac{\ln x}{\sqrt{x}}$

•3 $\lim\limits_{x\to\infty} \dfrac{e^x}{x^3}$
4 $\lim\limits_{x\to 0} \dfrac{\sin x}{3x - x^3}$

•5 $\lim\limits_{x\to 0} \dfrac{\sin x - x \cos x}{x^3}$
6 $\lim\limits_{x\to\infty} \dfrac{2x^3 - 2x + 3}{3x^3 + 5x^2 - 7}$

•7 $\lim\limits_{x\to 0^+} \dfrac{\ln(x + \sin x)}{\cot x}$
8 $\lim\limits_{x\to\infty} \dfrac{\cosh x}{x^4}$

•9 $\lim\limits_{x\to 0} \dfrac{e^x - \sin x - 1}{\cos x - 1}$
10 $\lim\limits_{x\to 0^-} \dfrac{\sin^2 x - 2\sin x}{x^4}$

11 Explain why L'Hôpital's rule cannot be used to evaluate

$$\lim_{x\to\infty} \frac{\tan x}{x}.$$

(Or can it?)

12 Use L'Hôpital's rule and mathematical induction to prove that if n is a positive integer, then (a) $\lim_{x\to\infty}(x^n/e^x) = 0$; (b) $\lim_{x\to\infty}(\ln x)^n/x = 0$.

13 Show that the Mean Value Theorem (Theorem 4-3) is the special case of Theorem 10-1 for $g(x) = x$.

14 (a) Prove that under the hypothesis of Theorem 10-1 $g(b) \neq g(a)$.
(b) Show that the function h, defined in the proof of Theorem 10-1, satisfies the hypothesis of Rolle's theorem.

Exercises 15, 16, and 17 complete the proof of Theorem 10-2 (L'Hôpital's rule for the form 0/0).

15 Modify the argument in the text to show that

$$\lim_{x\to a^-} \frac{f(x)}{g(x)} = \lim_{x\to a^-} \frac{f'(x)}{g'(x)} = L.$$

(*Hint:* This argument requires little more than reversing certain inequality signs.)

16 Let f and g be differentiable for all x greater than some fixed positive number b and

10-2 L'Hôpital's Rule. II: Other Indeterminate Forms

L'Hôpital's Rule as $x \to \infty$

$g'(x) \neq 0$ if $x > b$. Suppose that $\lim_{x \to \infty} f(x) = \lim_{x \to \infty} g(x) = 0$ and that $\lim_{x \to \infty} f'(x)/g'(x) = L$. Prove that

$$\lim_{x \to \infty} \frac{f(x)}{g(x)} = \lim_{x \to \infty} \frac{f'(x)}{g'(x)} = L.$$

(*Hint:* Let $t = 1/x$ for $x > b$. Define $F(t) = f(1/t)$ and $G(t) = g(1/t)$. Use the knowledge that

$$\lim_{x \to \infty} \frac{f(x)}{g(x)} = \lim_{t \to 0^+} \frac{F(t)}{G(t)} = \lim_{t \to 0^+} \frac{F'(t)}{G'(t)}.)$$

17 Modify the proof of Theorem 10-2 to show that if $\lim_{x \to a^+} (f'(x)/g'(x)) = \infty$, then $\lim_{x \to a^+} (f(x)/g(x)) = \infty$.

18 This problem will be used in Exercise 21 of Section 12-8. Let

$$f(x) = \begin{cases} e^{-1/x^2} & \text{if } x \neq 0 \\ 0 & \text{if } x = 0. \end{cases}$$

(a) Prove that f is continuous at $x = 0$.
(b) Use mathematical induction and L'Hôpital's rule to prove that $f^{(n)}(0) = 0$ for every positive integer n. (*Hint:* To establish that $f'(0) = 0$, show that the difference quotient for $f'(0)$ can be written as u/e^{u^2} where $u = 1/x$. To establish the general result, show that the nth derivative of e^{-1/x^2} can be written as

$$f^{(n)}(x) = \frac{P(1/x)}{e^{1/x^2}} \quad (\text{if } x \neq 0)$$

where P is a polynomial. Then make the substitution $u = 1/x$ before taking the limit of the difference quotient.)

10–2 L'Hôpital's Rule. II: Other Indeterminate Forms

Many indeterminate forms can be reduced to one of the forms $0/0$ or ∞/∞ by various techniques.

THE INDETERMINATE FORMS $0 \cdot \infty$ AND $\infty - \infty$

If $f(x) \to 0$ and $g(x) \to \pm \infty$, the limit of the product $f(x) \cdot g(x)$ is represented by the form $0 \cdot \infty$. This form can be changed to the form $0/0$ or ∞/∞ by rewriting the product $f(x)g(x)$ as

$$\frac{f(x)}{1/g(x)} \quad \text{or} \quad \frac{g(x)}{1/f(x)}.$$

EXAMPLE 1 Calculate $\lim_{x \to 0^+} x^2 \ln x$.

Solution The form is $0 \cdot \infty$. We rewrite the limit in the form ∞/∞ and use L'Hôpital's rule:

$$\lim_{x \to 0^+} x^2 \ln x = \lim_{x \to 0^+} \frac{\ln x}{1/x^2} = \lim_{x \to 0^+} \frac{1/x}{-2/x^3} = \lim_{x \to 0^+} \left(-\frac{x^2}{2}\right) = 0. \quad \square$$

If $f(x) \to \infty$ and $g(x) \to \infty$, the limit of the difference $f(x) - g(x)$ is represented by the form $\infty - \infty$. If it is possible to change this limit to one of the forms $0/0$ or ∞/∞ by algebraic manipulation, then L'Hôpital's rule can be applied.

EXAMPLE 2 Calculate $\lim\limits_{x \to 0^+} \left[\dfrac{1}{x} - \dfrac{1}{\ln(x+1)} \right]$.

Solution The limit is of form $\infty - \infty$. We write the function as one fraction with common denominator $x \ln(x+1)$:

$$\lim_{x \to 0^+} \left[\frac{1}{x} - \frac{1}{\ln(x+1)} \right] = \lim_{x \to 0^+} \frac{\ln(x+1) - x}{x \ln(x+1)}.$$

The limit is now of form $0/0$. We apply L'Hôpital's rule:

$$\lim_{x \to 0^+} \left[\frac{1}{x} - \frac{1}{\ln(x+1)} \right] = \lim_{x \to 0^+} \frac{\ln(x+1) - x}{x \ln(x+1)}$$

$$= \lim_{x \to 0^+} \frac{\dfrac{1}{(x+1)} - 1}{\dfrac{x}{(x+1)} + \ln(x+1)} = \lim_{x \to 0^+} \frac{-\dfrac{1}{(x+1)^2} - 0}{\dfrac{1}{(x+1)^2} + \dfrac{1}{(x+1)}}$$

(Applying L'Hôpital's Rule again.)

$$= \frac{-\dfrac{1}{1^2} - 0}{\dfrac{1}{1^2} + \dfrac{1}{1}} = -\frac{1}{2}. \quad \square$$

EXPONENTIAL FORMS 0^0, ∞^0, 1^∞

If $f(x) > 0$ and $g(x) \to 0$, then the limit of $f(x)^{g(x)}$ is represented by the form 0^0. The forms ∞^0 and 1^∞ have the meanings that are suggested naturally by the forms.

The three forms 0^0, ∞^0, and 1^∞ can be reduced to the form $0 \cdot \infty$ by logarithms. We then rewrite the limit in the form $0/0$ or ∞/∞ and apply L'Hôpital's rule.

The general procedure can be illustrated with limits of form 0^0. Suppose that $f(x) > 0$ for all x near a and that $f(x) \to 0$ and $g(x) \to 0$ as $x \to a$.

Procedure for Exponential Forms

(1) We define y by $y = f(x)^{g(x)}$. Then

$$\ln y = g(x) \ln f(x)$$

so that

$$\lim_{x \to a} [\ln y] = \lim_{x \to a} [g(x) \ln f(x)].$$

This last limit is of form $0 \cdot \infty$.

(2) We rewrite the limit

$$\lim_{x \to a} [g(x) \ln f(x)]$$

10–2 L'Hôpital's Rule. II: Other Indeterminate Forms

in the form $0/0$ or ∞/∞ and apply L'Hôpital's rule to calculate

$$\lim_{x \to a} [\ln y] = \lim_{x \to a} [g(x) \ln f(x)].$$

(3) Suppose that this limit is a real number L. Since the logarithm function is continuous, then (by L8)

$$\lim_{x \to a} [\ln y] = \ln [\lim_{x \to a} y] = L,$$

so that

$$\lim_{x \to a} y = e^L,$$

$$\lim_{x \to a} f(x)^{g(x)} = e^L.$$

(4) If the limit in Step 2 is ∞, then $\lim_{x \to a} [\ln y] = \infty$, so that

$$\lim_{x \to a} y = \lim_{x \to a} f(x)^{g(x)} = \infty.$$

(5) If the limit in Step 2 is $-\infty$, then $\lim_{x \to a} [\ln y] = -\infty$, so that

$$\lim_{x \to a} y = \lim_{x \to a} f(x)^{g(x)} = 0.$$

EXAMPLE 3 Calculate $\lim_{x \to \infty} \left(1 + \dfrac{1}{x}\right)^x$.

Solution The limit is of form 1^∞. Let $y = (1 + 1/x)^x$. Then

$$\ln y = x \ln \left(1 + \frac{1}{x}\right) = \frac{\ln \left(1 + \dfrac{1}{x}\right)}{1/x}$$

By L'Hôpital's rule,

$$\lim_{x \to \infty} [\ln y] = \lim_{x \to \infty} \left[\frac{\ln \left(1 + \dfrac{1}{x}\right)}{\dfrac{1}{x}}\right] = \lim_{x \to \infty} \frac{\dfrac{1}{(1 + 1/x)}\left(-\dfrac{1}{x^2}\right)}{-\dfrac{1}{x^2}}$$

$$= \lim_{x \to \infty} \frac{1}{1 + 1/x} = 1.$$

Since $\ln [\lim_{x \to \infty} y] = \lim_{x \to \infty} [\ln y] = 1$, then

$$\lim_{x \to \infty} y = e^1 = e$$

$$\lim_{x \to \infty} \left(1 + \frac{1}{x}\right)^x = e. \;\square$$

EXAMPLE 4 Calculate $\lim_{x \to \infty} x^{2/x}$.

Solution The limit is of form ∞^0. Let $y = x^{2/x}$. Then

$$\ln y = \frac{2}{x} \ln x$$

and

$$\lim_{x \to \infty} [\ln y] = \lim_{x \to \infty} \frac{2 \ln x}{x} = \lim_{x \to \infty} \frac{2/x}{1} \quad \text{(by L'Hôpital's rule)}$$
$$= 0.$$

It follows that $\ln [\lim_{x \to \infty} y] = \lim_{x \to \infty} [\ln y] = 0$, so that

$$\lim_{x \to \infty} y = e^0 = 1$$
$$\lim_{x \to \infty} x^{2/x} = \quad 1. \ \square$$

EXAMPLE 5   Calculate $\lim_{x \to 0^+} (x + 1)^{1/(\cos x - 1)}$.

Solution   The limit is of form 1^∞. Let $y = (x + 1)^{1/(\cos x - 1)}$. Then

$$\ln y = \frac{\ln (x + 1)}{\cos x - 1}$$

and

$$\lim_{x \to 0^+} [\ln y] = \lim_{x \to 0^+} \frac{\ln (x + 1)}{\cos x - 1}.$$

This limit is of form 0/0. By L'Hôpital's rule,

$$\lim_{x \to 0^+} [\ln y] = \lim_{x \to 0^+} \frac{\ln (x + 1)}{\cos x - 1} = \lim_{x \to 0^+} \frac{1/(x + 1)}{- \sin x} = -\infty.$$

Since $\ln y \to -\infty$ as $x \to 0^+$, then $y \to 0$ as $x \to 0^+$. Therefore,

$$\lim_{x \to 0^+} (x + 1)^{1/(\cos x - 1)} = \lim_{x \to 0^+} [y] = 0. \ \square$$

Exercises 10-2

1–14.   Classify the indeterminate form. Use L'Hôpital's rule or some other method to evaluate the limit.

- 1 $\lim_{x \to 0^+} \sqrt{x} \ln x$
- 2 $\lim_{x \to 0^+} x \csc x$
- 3 $\lim_{x \to \infty} [x - \sqrt{x^2 + 4x + 3}]$
- 4 $\lim_{x \to \infty} [\ln (2e^x + 1) - \ln (e^x + 3)]$
- 5 $\lim_{x \to 0^+} x^x$
- 6 $\lim_{x \to 0^+} (\sec \sqrt{x})^{1/x}$
- 7 $\lim_{x \to 1^+} x^{1/(x-1)}$
- 8 $\lim_{x \to 1^-} x^{1/(x-1)^2}$
- 9 $\lim_{x \to \infty} \left(1 - \frac{2}{x}\right)^x$
- 10 $\lim_{x \to \infty} (5x + 2 \ln x)^{3/\ln x}$
- 11 $\lim_{x \to \infty} (\sinh x)^{1/x}$
- 12 $\lim_{x \to \infty} (\sinh x)^{1/x^2}$
- 13 $\lim_{x \to 0^+} x^{\sin x}$
- 14 $\lim_{x \to 0^+} (e^x - 1)^{1/x}$

10–3 Improper Integrals. I: Integrals of Discontinuous Functions

Thus far we have considered integrals of functions that are continuous over finite closed intervals. There are several natural ways to extend our results.

(1) We can consider integrals that have discontinuous integrands at one or more points of the interval of integration. We can, for example, work with integrals such as

$$\int_0^1 \frac{1}{\sqrt{x}}\, dx \quad \text{or} \quad \int_{-1}^1 \frac{1}{x^2}\, dx,$$

where the integrands are discontinuous at $x = 0$.

(2) We can consider integrals defined over infinite intervals. We can, for example, consider integrals such as

$$\int_1^\infty \frac{1}{x^2}\, dx, \quad \int_{-\infty}^0 \frac{x-1}{x^3-8}\, dx, \quad \text{or} \quad \int_{-\infty}^\infty \frac{x}{(x^2+1)^2}\, dx.$$

(3) Finally, we can combine the above types of integrals and consider integrals such as

$$\int_0^\infty \frac{1}{(x-1)^2}\, dx,$$

where the interval is infinite and the integrand is discontinuous at an intermediate point.

Integrals of these three types are called *improper integrals*.

THE INTEGRAL $\int_a^b f(x)\, dx$, f DISCONTINUOUS AT AN ENDPOINT

Let f be a function that is continuous everywhere on the closed interval $[a, b]$ except at the left-hand endpoint a. If t is any other point of the interval, the integral $\int_t^b f(x)\, dx$ is well defined. Since the value of this integral depends on t, then it defines a new function,

$$F(t) = \int_t^b f(x)\, dx.$$

If this function F has a finite limit L as t approaches a through values on the interval (a, b), then we define the value of integral $\int_a^b f(x)\, dx$ to be this limit. That is

$\int_a^b f(x)\, dx$.
f discontinuous at a

$$\int_a^b f(x)\, dx = \lim_{t \to a^+} \int_t^b f(x)\, dx = L$$

provided the limit exists. (See Fig. 10–1a.) In this case we say that the original integral, $\int_a^b f(x)\, dx$, *converges* to L. If no finite limit exists, then the original integral is said to *diverge*.

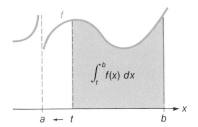
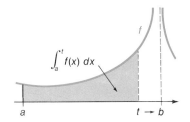

(a) *f* discontinuous at the left-hand endpoint:
$$\int_a^b f(x)\, dx = \lim_{t \to a^+} \int_t^b f(x)\, dx$$

(b) *f* discontinuous at the right-hand endpoint:
$$\int_a^b f(x)\, dx = \lim_{t \to b^-} \int_a^t f(x)\, dx$$

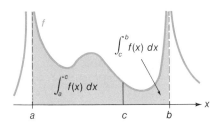

(c) *f* discontinuous at both endpoints:
$$\int_a^b f(x)\, dx = \int_a^c f(x)\, dx + \int_c^b f(x)\, dx$$
provided the integrals converge

Figure 10-1 Integrals of discontinuous functions.

EXAMPLE 1 Calculate $\int_0^1 \frac{1}{\sqrt{x}}\, dx$.

Solution If $t > 0$, then

$$\int_t^1 \frac{1}{\sqrt{x}}\, dx = \int_t^1 x^{-1/2}\, dx = \left[2x^{1/2}\right]_t^1 = 2 - 2\sqrt{t}.$$

Therefore,

$$\int_0^1 \frac{1}{\sqrt{x}}\, dx = \lim_{t \to 0^+} \int_t^1 \frac{1}{\sqrt{x}}\, dx = \lim_{t \to 0^+} [2 - 2\sqrt{t}] = 2.$$

The integral converges to the number 2. □

If f is continuous everywhere on $[a, b]$ except at b, we choose $t < b$, calculate $\int_a^t f(x)\, dx$, then let t approach b through values on the interval. If the limit of the integral exists and is equal to L, we define

10–3 Improper Integrals. I: Integrals of Discontinuous Functions

$\int_a^b f(x)\,dx$,
f discontinuous at b

$$\int_a^b f(x)\,dx = \lim_{t \to b^-} \int_a^t f(x)\,dx = L.$$

(See Fig. 10–1b.) In this case, we say that the original integral *converges to L*. If no finite limit exists, the integral *diverges*.

EXAMPLE 2   Calculate $\int_{-1}^{0} \dfrac{dx}{x^{2/3}}$.

Solution   If $t < 0$, then

$$\int_{-1}^{t} \frac{dx}{x^{2/3}} = \int_{-1}^{t} x^{-2/3}\,dx = \left[3x^{1/3}\right]_{-1}^{t} = 3t^{1/3} + 3.$$

Therefore,

$$\int_{-1}^{0} \frac{dx}{x^{2/3}} = \lim_{t \to 0^-} \int_{-1}^{t} \frac{dx}{x^{2/3}} = \lim_{t \to 0^-} [3t^{1/3} + 3] = 3 \cdot 0 + 3 = 3. \ \square$$

If f is continuous on the open interval (a, b) but discontinuous at both endpoints, we choose a convenient intermediate point c and define

$\int_a^b f(x)\,dx$,
f discontinuous at a and b

$$\int_a^b f(x)\,dx = \int_a^c f(x)\,dx + \int_c^b f(x)\,dx$$

provided both integrals on the right converge (see Fig. 10–1c.). If at least one integral diverges, then the original integral also diverges.

Remark   This definition makes it appear that the value of $\int_a^b f(x)\,dx$ depends on the particular intermediate point c. It does not. We get the same final result for any other choice of c (see Exercise 19).

FUNCTIONS DISCONTINUOUS AT INTERMEDIATE POINTS OF AN INTERVAL

If f is continuous on $[a, b]$ except at the points $c_1, c_2, \ldots, c_n$, where

$$a < c_1 < c_2 < \cdots < c_n < b,$$

we define

$\int_a^b f(x)\,dx$,
f discontinuous at intermediate points

$$\int_a^b f(x)\,dx = \int_a^{c_1} f(x)\,dx + \int_{c_1}^{c_2} f(x)\,dx + \cdots + \int_{c_n}^{b} f(x)\,dx$$

provided all the improper integrals on the right converge. (See Fig. 10–2.) If any one integral diverges, then the original integral also diverges.

EXAMPLE 3   Calculate $\int_{-1}^{8} \dfrac{dx}{\sqrt[3]{x}}$ if the integral converges.

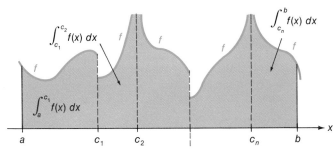

Figure 10-2 If f is continuous on $[a,b]$ except at $c_1, c_2, \ldots, c_n$, then $\int_a^b f(x)\,dx = \int_a^{c_1} f(x)\,dx + \int_{c_1}^{c_2} f(x)\,dx + \cdots + \int_{c_n}^b f(x)\,dx$ provided the integrals converge.

Solution The function $f(x) = 1/\sqrt[3]{x}$ is continuous everywhere except at $x = 0$. Therefore,

$$\int_{-1}^8 f(x)\,dx = \int_{-1}^0 f(x)\,dx + \int_0^8 f(x)\,dx$$

provided the integrals on the right converge.
We calculate

$$\int_{-1}^0 f(x)\,dx = \lim_{t \to 0^-} \int_{-1}^t x^{-1/3}\,dx = \lim_{t \to 0^-} \left[\frac{3x^{2/3}}{2}\right]_{-1}^t$$

$$= \lim_{t \to 0^-} \left[\frac{3t^{2/3}}{2} - \frac{3(-1)^{2/3}}{2}\right] = 0 - \frac{3}{2} = -\frac{3}{2}$$

and

$$\int_0^8 f(x)\,dx = \lim_{t \to 0^+} \int_t^8 x^{-1/3}\,dx = \lim_{t \to 0^+} \left[\frac{3x^{2/3}}{2}\right]_t^8$$

$$= \lim_{t \to 0^+} \left[\frac{3 \cdot 8^{2/3}}{2} - \frac{3t^{2/3}}{2}\right] = \frac{3 \cdot 4}{2} - 0 = \frac{12}{2}.$$

Therefore,

$$\int_{-1}^8 f(x)\,dx = \int_{-1}^0 f(x)\,dx + \int_0^8 f(x)\,dx = -\frac{3}{2} + \frac{12}{2} = \frac{9}{2}. \quad \square$$

EXAMPLE 4 Does the integral $\int_{-1}^1 \frac{dx}{x^3}$ converge or diverge?

Solution Since $f(x) = 1/x^3$ is discontinuous only at $x = 0$, then

$$\int_{-1}^1 \frac{dx}{x^3} = \int_{-1}^0 \frac{dx}{x^3} + \int_0^1 \frac{dx}{x^3}$$

provided both integrals on the right converge.

10-3 Improper Integrals. I: Integrals of Discontinuous Functions

We calculate

$$\int_{-1}^{0} \frac{dx}{x^3} = \lim_{t \to 0^-} \int_{-1}^{t} x^{-3} \, dx$$

$$= \lim_{t \to 0^-} \left[-\frac{1}{2x^2} \right]_{-1}^{t} = \lim_{t \to 0^-} \left[-\frac{1}{2t^2} + \frac{1}{2} \right] = -\infty.$$

Since at least one of the integrals diverges, then the integral $\int_{-1}^{1} dx/x^3$ diverges. (See Fig. 10-3.) □

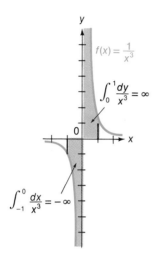

Figure 10-3 Example 4.

AREA

Concepts defined by integrals, such as area, volume, arc length, and moments, can be extended to improper integrals. The essential condition is that each integral must converge. For example, let f and g be continuous on $[a, b]$ except at a finite number of points. Let $f(x) \geq g(x)$ wherever f and g are defined on $[a, b]$. Let $\mathcal{R}$ be the region defined by the graphs of f and g and the lines $x = a$ and $x = b$. We say that the *area* of $\mathcal{R}$ is

Area
$$A = \int_a^b [f(x) - g(x)] \, dx$$

provided the integral converges.

Volume, arc length, and so on are defined analogously.

EXAMPLE 5 Let $f(x) = [x]$ (greatest integer function). Calculate the area of the region bounded by the graph of f, the x-axis and the vertical lines $x = 1$ and $x = 3$.

Solution Area $= \int_1^3 f(x) \, dx = \int_1^2 [x] \, dx + \int_2^3 [x] \, dx.$

We calculate

$$\int_1^2 [x]\,dx = \lim_{t \to 2^-} \int_1^t [x]\,dx = \lim_{t \to 2^-} \int_1^t 1\,dx = 1$$

$$\int_2^3 [x]\,dx = \lim_{t \to 3^-} \int_2^t [x]\,dx = \lim_{t \to 3^-} \int_2^t 2\,dx = 2.$$

Therefore, the area of the region is

$$A = \int_1^2 [x]\,dx + \int_2^3 [x]\,dx = 1 + 2 = 3.$$

(See Fig. 10–4.) □

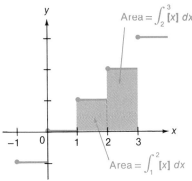

Figure 10–4 Example 5. $\int_1^3 [x]\,dx = \int_1^2 [x]\,dx + \int_2^3 [x]\,dx.$

Exercises 10–3

1–10. Calculate the value of the definite integral if it converges. If an integral diverges, determine whether it diverges to ∞ or $-\infty$ or neither. Boldface brackets indicate the greatest integer function.

•1 $\displaystyle\int_{-1}^0 \frac{1}{\sqrt{x+1}}\,dx$

2 $\displaystyle\int_{-1}^1 \frac{1}{\sqrt[3]{x}}\,dx$

•3 $\displaystyle\int_{-1}^0 \frac{1}{(x+1)^2}\,dx$

4 $\displaystyle\int_0^2 \frac{1}{(1-x)^2}\,dx$

•5 $\displaystyle\int_{-1}^4 2[x]\,dx$

•6 $\displaystyle\int_{-1}^3 [x^2]\,dx$

•7 $\displaystyle\int_{-2}^2 \frac{x}{\sqrt{4-x^2}}\,dx$

8 $\displaystyle\int_0^4 x(16-x^2)^{-3/2}\,dx$

•9 $\displaystyle\int_0^e (\ln x + 1)\,dx$

10 $\displaystyle\int_0^2 e^{x \ln x}(\ln x + 1)\,dx$

11–13. Calculate the area of the region bounded by the given curves and lines if it is defined.

• 11 $y = \dfrac{1}{x^2}$ and x-axis, $0 < x \le 1$

12   $y = \dfrac{1}{\sqrt{x}}$ and x-axis, $0 < x \le 1$

• 13 $y = (x - 8)^{-2/3}$, x-axis, y-axis, and line $x = 8$.

14   Revolve the region in Exercise 11 about the x-axis. Calculate the volume of the solid of revolution if it is defined.

• 15   Revolve the region in Exercise 13 about the x-axis. Calculate the volume of the solid of revolution if it is defined.

• 16   Find the arc length of the graph of $x^{2/3} + y^{2/3} = 4$, $0 \le x \le 8$, $y \ge 0$.

17   Show that $\int_0^1 dx/x^n$ converges if $n < 1$ and diverges to ∞ if $n > 1$.

Exercise 18 shows that the formula $\int_a^b f(x)\,dx = F(b) - F(a)$ does not necessarily hold for improper integrals.

18   Let $f(x) = \begin{cases} \dfrac{1}{x^{1/2}} & \text{if } x > 0 \\ \dfrac{1}{x^{2/3}} & \text{if } x < 0. \end{cases}$ Let $F(x) = \begin{cases} 2\sqrt{x} + 1 & \text{if } x > 0 \\ 3\sqrt[3]{x} - 2 & \text{if } x < 0. \end{cases}$

(a) Show that F is an antiderivative of f everywhere except at $x = 0$ (where the functions are discontinuous).
(b) Show that $\int_{-1}^{1} f(x)\,dx \ne F(1) - F(-1)$.

19   This exercise shows that the value of $\int_a^b f(x)\,dx$ does not depend on the particular intermediate point that is used to split up the integral. Let f be continuous on the open interval (a, b) but discontinuous at the endpoints a and b. Let c_1 and c_2 be intermediate points with $a < c_1 < c_2 < b$. Suppose that $\int_a^{c_1} f(x)\,dx$ and $\int_{c_1}^{b} f(x)$ have finite values. Prove that

$$\int_a^{c_1} f(x)\,dx + \int_{c_1}^{b} f(x)\,dx = \int_a^{c_2} f(x)\,dx + \int_{c_2}^{b} f(x)\,dx.$$

10–4 Improper Integrals. II: Integrals Defined over Infinite Intervals

Let f be continuous for all x greater than or equal to some fixed number a. We define

$\int_a^\infty f(x)\,dx$

$$\int_a^\infty f(x)\,dx = \lim_{t \to \infty} \int_a^t f(x)\,dx$$

provided the limit on the right exists. (See Fig. 10–5.)

If $\lim_{t \to \infty} \int_a^t f(x)\,dx$ has a finite limit L, we say that $\int_a^\infty f(x)\,dx$ *converges* to L. Otherwise, the integral is said to *diverge*.

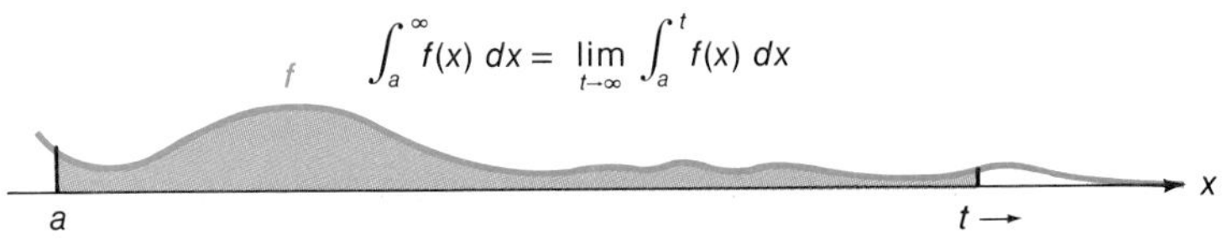

$$\int_a^\infty f(x)\,dx = \lim_{t\to\infty} \int_a^t f(x)\,dx$$

Figure 10–5 $\int_a^\infty f(x)\,dx = \lim_{t\to\infty} \int_a^t f(x)\,dx.$

EXAMPLE 1 Calculate (a) $\int_1^\infty \frac{1}{x^2}\,dx;$ (b) $\int_1^\infty \frac{1}{\sqrt{x}}\,dx.$

Solution (a)
$$\int_1^\infty \frac{1}{x^2}\,dx = \lim_{t\to\infty} \int_1^t x^{-2}\,dx = \lim_{t\to\infty}\left[-\frac{1}{x}\right]_1^t$$
$$= \lim_{t\to\infty}\left[-\frac{1}{t}+1\right] = 0 + 1 = 1.$$

The integral converges to $L = 1$.

(b)
$$\int_1^\infty \frac{1}{\sqrt{x}}\,dx = \lim_{t\to\infty} \int_1^t x^{-1/2}\,dx = \lim_{t\to\infty}\left[2\sqrt{x}\right]_1^t$$
$$= \lim_{t\to\infty}[2\sqrt{t} - 2] = \infty.$$

The integral diverges to ∞. □

EXAMPLE 2 Show that $\int_0^\infty \cos x\,dx$ diverges.

Solution
$$\int_0^t \cos x\,dx = \left[\sin x\right]_0^t = \sin t - \sin 0 = \sin t.$$

Since the values of the sine function oscillate between -1 and 1, then $\sin t$ has no limit as $t \to \infty$. Therefore, $\int_0^\infty \cos x\,dx$ diverges, but not to ∞ or to $-\infty$. □

The improper integral $\int_{-\infty}^b f(x)\,dx$ is defined in a similar manner. Let f be continuous for all $x \le b$. We define

$\int_{-\infty}^b f(x)\,dx$
$$\int_{-\infty}^b f(x)\,dx = \lim_{t\to-\infty} \int_t^b f(x)\,dx$$

provided the limit exists.

THE INTEGRAL $\int_{-\infty}^\infty f(x)\,dx$

Let f be continuous for all x. We choose a convenient point c and define

$\int_{-\infty}^\infty f(x)\,dx$
$$\int_{-\infty}^\infty f(x)\,dx = \int_{-\infty}^c f(x)\,dx + \int_c^\infty f(x)\,dx$$

provided both integrals on the right converge. (See Fig. 10–6.)

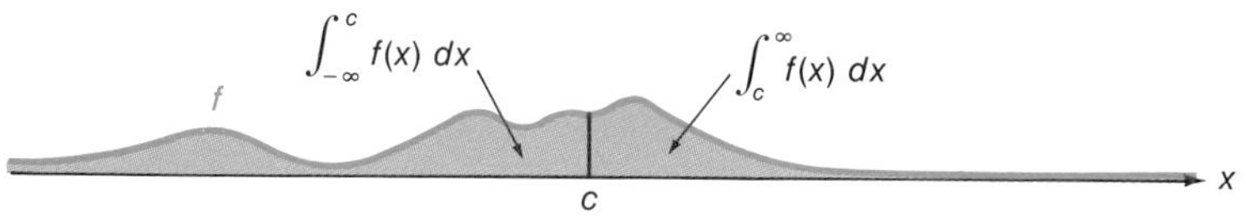

10–4 Improper Integrals. II: Integrals Defined over Infinite Intervals

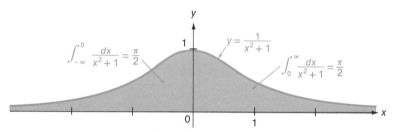

Figure 10–6 $\int_{-\infty}^{\infty} f(x)\,dx = \int_{-\infty}^{c} f(x)\,dx + \int_{c}^{\infty} f(x)\,dx$ provided the integrals on the right converge.

It can be shown that the final value of $\int_{-\infty}^{\infty} f(x)\,dx$, assuming it converges, is independent of the number c used in the definition. (See Exercise 19.)

EXAMPLE 3 Calculate $\displaystyle\int_{-\infty}^{\infty} \frac{dx}{x^2 + 1}$.

Solution

$$\int_{-\infty}^{\infty} \frac{dx}{x^2 + 1} = \int_{-\infty}^{0} \frac{dx}{x^2 + 1} + \int_{0}^{\infty} \frac{dx}{x^2 + 1}$$

provided the integrals on the right converge. (See Fig. 10–7.)

Figure 10–7 Example 3. $\displaystyle\int_{-\infty}^{\infty} \frac{dx}{x^2 + 1} = \pi$.

It follows from I18 that

$$\int_{0}^{\infty} \frac{dx}{x^2 + 1} = \lim_{t \to \infty} \int_{0}^{t} \frac{dx}{x^2 + 1} = \lim_{t \to \infty} \left[\tan^{-1} x\right]_{0}^{t}$$

$$= \lim_{t \to \infty} [\tan^{-1} t - \tan^{-1} 0] = \lim_{t \to \infty} \tan^{-1} t.$$

Recall from Chapter 9 that $\lim_{x \to \infty} \tan^{-1} x = \pi/2$. (See Fig. 9–15.) Therefore

$$\int_{0}^{\infty} \frac{dx}{x^2 + 1} = \lim_{t \to \infty} \tan^{-1} t = \frac{\pi}{2}.$$

Similarly

$$\int_{-\infty}^{0} \frac{dx}{x^2 + 1} = \lim_{t \to -\infty} [0 - \tan^{-1} t] = \frac{\pi}{2}.$$

Therefore

$$\int_{-\infty}^{\infty} \frac{dx}{x^2 + 1} = \frac{\pi}{2} + \frac{\pi}{2} = \pi. \quad \square$$

EXAMPLE 4 Show that $\int_{-\infty}^{\infty} 2x \, dx$ diverges.

Solution We choose $c = 0$ and calculate

$$\int_{-\infty}^{0} 2x \, dx = \lim_{t \to -\infty} \int_{t}^{0} 2x \, dx = \lim_{t \to -\infty} \left[x^2 \right]_{t}^{0}$$
$$= \lim_{t \to -\infty} [0^2 - t^2] = -\infty.$$

Since $\int_{-\infty}^{0} 2x \, dx$ diverges, then $\int_{-\infty}^{\infty} 2x \, dx$ also diverges. □

INTEGRALS OF DISCONTINUOUS FUNCTIONS DEFINED OVER INFINITE INTERVALS

Let f be continuous for $x > a$ but discontinuous at a. To define $\int_{a}^{\infty} f(x) \, dx$, we choose a convenient number $c > a$ and define

$\int_{a}^{\infty} f(x) \, dx,$
f discontinuous at a

$$\int_{a}^{\infty} f(x) \, dx = \int_{a}^{c} f(x) \, dx + \int_{c}^{\infty} f(x) \, dx$$

provided the integrals on the right converge.

EXAMPLE 5 Calculate $\int_{0}^{\infty} \frac{1}{x^2} \, dx$ if it converges.

Solution We use $c = 1$. Then

$$\int_{0}^{\infty} \frac{1}{x^2} \, dx = \int_{0}^{1} \frac{1}{x^2} \, dx + \int_{1}^{\infty} \frac{1}{x^2} \, dx$$

provided the integrals on the right converge. We calculate

$$\int_{0}^{1} \frac{1}{x^2} \, dx = \lim_{t \to 0^{+}} \int_{t}^{1} x^{-2} \, dx = \lim_{t \to 0^{+}} \left[-\frac{1}{x} \right]_{t}^{1}$$
$$= \lim_{t \to 0^{+}} \left[-1 + \frac{1}{t} \right] = \infty.$$

Therefore, the original integral diverges. □

The integral $\int_{-\infty}^{b} f(x) \, dx$, where f is continuous for all $x < b$ and discontinuous at b, is defined similarly:

$\int_{-\infty}^{b} f(x) \, dx,$
f discontinuous at b

$$\int_{-\infty}^{b} f(x) \, dx = \int_{-\infty}^{c} f(x) \, dx + \int_{c}^{b} f(x) \, dx,$$

where c is a convenient number less than b, provided both integrals on the right converge.

10-4 Improper Integrals. II: Integrals Defined over Infinite Integrals

If f is continuous for all $x > a$ except the numbers $c_1 < c_2 < \cdots < c_n$, we define

$\int_a^\infty f(x)\, dx$, f discontinuous at intermediate points

$$\int_a^\infty f(x)\, dx = \int_a^{c_1} f(x)\, dx + \int_{c_1}^{c_2} f(x)\, dx + \cdots + \int_{c_n}^\infty f(x)\, dx$$

provided all the integrals on the right converge.

A similar definition holds for $\int_{-\infty}^\infty f(x)\, dx$ if f is continuous everywhere except at $c_1 < c_2 < \cdots < c_n$. In this case,

$\int_{-\infty}^\infty f(x)\, dx$, f discontinuous at intermediate points

$$\int_{-\infty}^\infty f(x)\, dx = \int_{-\infty}^{c_1} f(x)\, dx + \int_{c_1}^{c_2} f(x)\, dx + \cdots + \int_{c_n}^\infty f(x)\, dx$$

provided all the integrals on the right converge.

AREA, VOLUME, ETC.

Area, volume, arc length, moments, and the like are defined for this type of improper integral analogously to the way they were defined in Section 10-3 for integrals of discontinuous functions. In each case all the integrals must converge. (See Fig. 10-8, for example.)

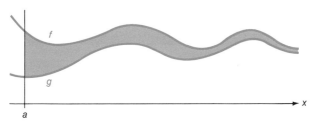

Figure 10-8 The area of an infinite region:
$A = \int_a^\infty [f(x) - g(x)]\, dx$ provided the integral converges.

EXAMPLE 6 Let $\mathcal{R}$ be the region bounded by the positive x-axis, the y-axis, and the graph of $f(x) = e^{-x}$, $x \geq 0$. Find the volume of the solid obtained by revolving $\mathcal{R}$ about the x-axis.

Solution We use the disk method. Provided the integral converges, the volume is

$$\pi \int_0^\infty e^{-2x}\, dx.$$

We calculate

$$V = \pi \int_0^\infty e^{-2x}\, dx = \pi \lim_{t \to \infty} \int_0^t e^{-2x}\, dx$$

$$= \pi \lim_{t \to \infty}\left[-\frac{1}{2}e^{-2x}\right]_0^t = \pi \lim_{t \to \infty}\left[-\frac{1}{2}e^{-2t} + \frac{1}{2}e^0\right]$$

$$= \pi \lim_{t \to \infty}\left[-\frac{1}{2e^{2t}} + \frac{1}{2}\right] = \pi\left[0 + \frac{1}{2}\right] = \frac{\pi}{2}. \quad \square$$

EXAMPLE 7 Let $\mathcal{R}$ be the region bounded by the positive x-axis, the line $x = 1$, and the graph of $f(x) = 1/x^4$, $x \geq 1$. Find the centroid of $\mathcal{R}$.

Solution The area and moments about the axes are defined by

$$A = \int_1^\infty \frac{dx}{x^4}, \quad M_x = \frac{1}{2}\int_1^\infty \left(\frac{1}{x^4}\right)^2 dx, \quad M_y = \int_1^\infty x \cdot \frac{1}{x^4} dx$$

provided the integrals converge. We calculate

$$A = \int_1^\infty \frac{dx}{x^4} = \lim_{t \to \infty} \int_1^t x^{-4}\, dx = \lim_{t \to \infty}\left[\frac{x^{-3}}{-3}\right]_1^t$$

$$= \lim_{t \to \infty}\left[-\frac{1}{3t^3} + \frac{1}{3}\right] = \frac{1}{3}.$$

$$M_x = \frac{1}{2}\int_1^\infty \frac{dx}{x^8} = \frac{1}{2}\lim_{t \to \infty}\int_1^t x^{-8}\,dx = \frac{1}{2}\lim_{t \to \infty}\left[\frac{x^{-7}}{-7}\right]_1^t$$

$$= \frac{1}{2}\lim_{t \to \infty}\left[-\frac{1}{7t^7} + \frac{1}{7}\right] = \frac{1}{2}\cdot\left[0 + \frac{1}{7}\right] = \frac{1}{14}.$$

$$M_y = \int_1^\infty x \cdot \frac{1}{x^4}\,dx = \int_1^\infty \frac{dx}{x^3} = \lim_{t \to \infty}\int_1^t x^{-3}\,dx$$

$$= \lim_{t \to \infty}\left[\frac{x^{-2}}{-2}\right]_1^t = \lim_{t \to \infty}\left[-\frac{1}{2t^2} + \frac{1}{2}\right] = \frac{1}{2}.$$

Since the area and moments are all defined, then the centroid is $(\bar{x}, \bar{y})$, where

$$\bar{x} = \frac{M_y}{A} = \frac{1/2}{1/3} = \frac{3}{2},$$

$$\bar{y} = \frac{M_x}{A} = \frac{1/14}{1/3} = \frac{3}{14}. \quad \square$$

Exercises 10-4

1–10. Calculate the value of the improper integral if it converges. If an integral diverges, try to determine whether it diverges to ∞, $-\infty$, or neither.

• 1 $\displaystyle\int_1^\infty \frac{1}{\sqrt[3]{x}}\,dx$

2 $\displaystyle\int_1^\infty \frac{1}{x^{4/3}}\,dx$

• 3 $\displaystyle\int_{-\infty}^{-1} \frac{1}{x^2}\,dx$

4 $\displaystyle\int_1^\infty \frac{1}{x\sqrt{x}}\,dx$

• 5 $\displaystyle\int_2^\infty \frac{1}{(x-1)^{2/3}}\,dx$

6 $\displaystyle\int_{-\infty}^0 \frac{x}{\sqrt{1+x^2}}\,dx$

10–5 Application to Statistics. Probability Density Functions

●7 $\quad \int_3^\infty \dfrac{dx}{x \ln^2 x}$

8 $\quad \int_4^\infty \left(1 - \dfrac{x+2}{\sqrt{x^2 + 4x + 1}}\right) dx$

●9 $\quad \int_e^\infty \left(\dfrac{1}{x-1} - \dfrac{1}{x+1}\right) dx$

10 $\quad \int_{-\infty}^\infty \dfrac{dx}{e^x + e^{-x}}$

11–14. Calculate the area, if it exists, of the region bounded by the graphs of the equations.

●11 $\quad f(x) = \dfrac{1}{x^3},\ g(x) = -\dfrac{1}{x^4},\ x \geq 1$

12 $\quad f(x) = x^{-3/2}$, x-axis, $x \geq 4$

●13 $\quad f(x) = (1-x)^{-1/2}$, x-axis, $x < 1$

14 $\quad f(x) = -\dfrac{1}{x^2}$, x-axis, $x \geq 2$

●15 Let $\mathcal{R}$ be the region bounded by $f(x) = 1/x^2$, the line $x = 1$, the x-axis for $x \geq 1$. Calculate the volume of the solid of revolution obtained by revolving $\mathcal{R}$ about the x-axis.

●16 Let $\mathcal{R}$ be the region in Exercise 11. Find the centroid of $\mathcal{R}$.

17 Show that $\int_1^\infty x^n\, dx$ converges if $n < -1$ and diverges if $n > -1$.

18 Let $\mathcal{R}$ be the region bounded by the graph of $y = 1/x$, the line $x = 1$, and the x-axis for $x \geq 1$. Let $\mathcal{S}$ be the solid obtained by revolving $\mathcal{R}$ about the x-axis.
  (a) Show that the area of $\mathcal{R}$ is not defined. Explain why no finite quantity of paint could ever cover $\mathcal{R}$.
  (b) Show that the volume of $\mathcal{S}$ is finite.
  (c) If we think of $\mathcal{S}$ as a hollow shell, then (b) tells us that it can be filled with a finite amount of paint, while (a) shows that no amount of paint is enough to paint a cross section. Can you explain this apparent paradox?

19 Let f be continuous for all x. Show that the value of $\int_{-\infty}^\infty f(x)\, dx$, if the integral converges, is independent of the number c used in the definition. In other words, show that if c_1 and c_2 are real numbers, then

$$\int_{-\infty}^{c_1} f(x)\, dx + \int_{c_1}^\infty f(x)\, dx = \int_{-\infty}^{c_2} f(x)\, dx + \int_{c_2}^\infty f(x)\, dx.$$

10–5* Application to Statistics. Average Value. Probability Density Functions

AVERAGE VALUE

The *mean* (or *average*) of the n numbers $a_1, a_2, \ldots, a_n$ is defined to be

$$\dfrac{a_1 + a_2 + \cdots + a_n}{n}.$$

We shall show how to extend this definition to calculate the average value of a continuous function f over a closed interval $[a, b]$.

We first partition the interval into n equal subintervals by the partition $\mathscr{P}$ defined by

$$a = x_0 < x_1 < x_2 < \cdots < x_n = b,$$

where

$$x_k = a + k\,\Delta x, \qquad \Delta x = \frac{b-a}{n}.$$

Observe that the mean value of f at the points $x_1, x_2, x_3, \ldots, x_n$ is

$$\frac{f(x_1) + f(x_2) + \cdots + f(x_n)}{n} = \frac{\sum_{k=1}^{n} f(x_k)}{(b-a)/\Delta x} = \frac{1}{b-a} \sum_{k=1}^{n} f(x_k)\,\Delta x.$$

The *mean (average) value of f on $[a, b]$* is defined to be the limit of the above sum as the norm of the partition approaches zero:

Average Value

$$\text{Average value} = \lim_{\|\mathscr{P}\|\to 0} \frac{1}{b-a} \sum_{k=1}^{n} f(x_k)\,\Delta x_k$$

$$= \frac{1}{b-a} \int_a^b f(x)\,dx.$$

This definition coincides with our intuitive ideas about average value and with the value obtained from the Mean Value Theorem for integrals. For example, if v is a velocity function, then

$$\int_a^b v(t)\,dt$$

is the total distance traveled from time a to time b and $b - a$ equals the elapsed time. Then

$$\text{Average value} = \frac{1}{b-a} \int_a^b v(t)\,dt = \frac{\text{total distance}}{\text{elapsed time}},$$

which is how we originally defined average velocity in Section 2-3.

EXAMPLE 1 Calculate the average value of the function $f(x) = x^{3/2}$, $1 \le x \le 4$.

Solution
$$\text{Average value} = \frac{1}{4-1} \int_1^4 x^{3/2}\,dx = \frac{1}{3}\left[\frac{2x^{5/2}}{5}\right]_1^4 = \frac{2}{15}[4^{5/2} - 1^{5/2}]$$

$$= \frac{2}{15}[32 - 1] = \frac{62}{15} \approx 4.133. \quad \square$$

This definition can be extended to average values of functions that are discontin-

10–5 Application to Statistics. Probability Density Functions

uous at a finite number of points on $[a, b]$. We define

$$\text{Average value of } f \text{ on } [a, b] = \frac{1}{b-a} \int_a^b f(t)\, dx$$

provided the improper integral converges.

EXAMPLE 2 The average value of $f(x) = 1/\sqrt{x}$ over the interval $[0, 16]$ is

$$\frac{1}{16-0} \int_0^{16} \frac{1}{\sqrt{x}}\, dx = \frac{1}{16} \lim_{t \to 0^+} \int_t^{16} x^{-1/2}\, dx = \frac{1}{16} \lim_{t \to 0^+} \left[2\sqrt{x} \right]_t^{16}$$

$$= \frac{1}{16} \lim_{t \to 0^+} [2\sqrt{16} - 2\sqrt{t}]$$

$$= \frac{1}{16} [8 - 0] = \frac{1}{2}. \quad \square$$

PROBABILITY DENSITY FUNCTIONS

Roughly speaking, the *probability* of an event is the likelihood that it will occur. For example, if we toss a fair coin, then *heads* or *tails* is equally likely. We say that the *probability of heads is* $\frac{1}{2}$ and the *probability of tails is* $\frac{1}{2}$, indicating that approximately half the tosses will produce heads and half tails. This type of result can be established by an elementary argument concerning events with a finite number of possible outcomes, all equally likely. The probability of a particular outcome is

Probability
$$p = \frac{\text{number of favorable outcomes}}{\text{number of possible outcomes}}.$$

(The word *favorable* refers to outcomes that yield the result in which we are interested. Thus a "favorable" outcome—such as a patient's dying of cancer—might be most unfavorable by any other criterion.)

In our coin-tossing example, there are two possible outcomes, equally likely. One is heads, the other is tails. Thus,

$$\text{Probability of heads} = \frac{\text{number of favorable outcomes}}{\text{number of possible outcomes}} = \frac{1}{2}.$$

Many probability problems involve an infinite number of possible outcomes. For example, if we throw a dart at a dartboard, the number of possible locations at which the dart could strike the board is infinite. Obviously, the probability that any given point will be struck is zero. We might ask, however, for the probability that the dart will strike the board within a certain distance of the center of the bullseye. Such problems are best handled by *probability density functions*.

DEFINITION A *probability density function* is an integrable function f, defined on the entire x-axis, such that

Probability Density Function

(1) $\quad f(x) \geq 0 \quad$ for all x,
(2) $\quad \int_{-\infty}^{\infty} f(x)\, dx = 1.$

If an event has possible outcomes that can be described by numerical values and we have chosen the correct probability density function, then the *probability* that the outcome will be in the range $a \leq x \leq b$ is

Probability Determined from Density Function

$$p = \int_a^b f(x)\, dx.$$

(See Fig. 10–9.)

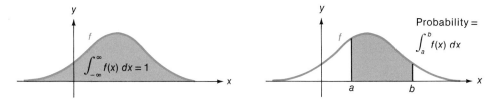

(a) Probability density function: $f(x) \geq 0$ for all x; $\int_{-\infty}^{\infty} f(x)\, dx = 1$

(b) The probability that an event has its value between a and b is $p = \int_a^b f(x)\, dx$

Figure 10–9 Probability density function.

The particular probability density function is determined by the type of problem under consideration. Much of the work in theoretical statistics involves finding probability density functions for particular types of problems.

In our work we give the probability density functions without derivation.

EXAMPLE 3 An event consists of randomly choosing a point on the closed interval [1, 4]. The probability density function is

$$f(x) = \begin{cases} \frac{1}{3} & \text{if } 1 \leq x \leq 4 \\ 0 & \text{if } x < 1 \text{ or } x > 4. \end{cases}$$

(See Fig. 10–10.)

(a) Verify that f is a probability density function.
(b) Calculate the probability that the point will be chosen between 3.0 and 3.2.

Solution (a) The definition of f shows that $f(x) \geq 0$ for all x. Then

$$\int_{-\infty}^{\infty} f(x)\, dx = \int_{-\infty}^{1} f(x)\, dx + \int_{1}^{4} f(x)\, dx + \int_{4}^{\infty} f(x)\, dx$$

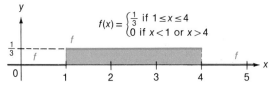

Figure 10–10 Example 3.

10-5 Application to Statistics. Probability Density Functions

$$= \int_{-\infty}^{1} 0 \cdot dx + \int_{1}^{4} \tfrac{1}{3} dx + \int_{4}^{\infty} 0 \cdot dx$$

$$= 0 + \frac{1}{3}\left[x\right]_{1}^{4} + 0 = \frac{1}{3}\left[4 - 1\right] = 1.$$

Thus, p is a probability density function.

(b) The probability that the selected point will be between 3.0 and 3.2 is

$$p = \int_{3.0}^{3.2} \tfrac{1}{3} dx = \tfrac{1}{3}\left[x\right]_{3.0}^{3.2} = \tfrac{1}{3} \cdot 0.2 = \tfrac{1}{15}. \ \square$$

The final example is typical of certain problems that can be solved by using probability density functions.

EXAMPLE 4 The Brokken China Company manufactures 1000 china plates a day. The probability density function for the proportion of plates that pass inspection is

$$f(x) = \begin{cases} 13 \cdot 14 x^{12}(1-x) & \text{if } 0 \leq x \leq 1 \\ 0 & \text{if } x < 0 \text{ or } x > 1. \end{cases}$$

(a) Verify that f is a probability density function.
(b) Find the probability that between 500 and 800 plates pass inspection on a given day.
(c) Find the probability that no more than 800 plates pass inspection on a given day.

Solution (a) By definition, $f(x) \geq 0$ for all x. To calculate $\int_{-\infty}^{\infty} f(x)\,dx$, we split the integral into three parts:

$$\int_{-\infty}^{\infty} f(x)\,dx = \int_{-\infty}^{0} f(x)\,dx + \int_{0}^{1} f(x)\,dx + \int_{1}^{\infty} f(x)\,dx$$

$$= \int_{-\infty}^{0} 0 \cdot dx + \int_{0}^{1} 13 \cdot 14(x^{12} - x^{13})\,dx + \int_{1}^{\infty} 0 \cdot dx$$

$$= 0 + 13 \cdot 14 \left[\frac{x^{13}}{13} - \frac{x^{14}}{14}\right]_{0}^{1} + 0 = 13 \cdot 14 \left[\frac{1}{13} - \frac{1}{14}\right] = 1.$$

Thus, f is a probability density function.

(b) If between 500 and 800 plates pass inspection on a given day, then x, the proportion of plates that pass, is between 0.5 and 0.8. The probability that x is within this range is

$$p = \int_{0.5}^{0.8} f(x)\,dx = \int_{0.5}^{0.8} 13 \cdot 14 x^{12}(1 - x)\,dx = 13 \cdot 14 \left[\frac{x^{13}}{13} - \frac{x^{14}}{14}\right]_{0.5}^{0.8}$$

$$= 13 \cdot 14 \left[\frac{14 x^{13} - 13 x^{14}}{13 \cdot 14}\right]_{0.5}^{0.8} = \left[14 x^{13} - 13 x^{14}\right]_{0.5}^{0.8}$$

$$= [14(0.8)^{13} - 13(0.8)^{14}] - [14(0.5)^{13} - 13(0.5)^{14}] \approx 0.197.$$

(c) The probability that no more than 800 plates pass inspection on a given day is the probability that x is between 0 and 0.8:

$$p = \int_{0}^{0.8} f(x)\,dx = 13 \cdot 14 \left[\frac{14 x^{13} - 13 x^{14}}{13 \cdot 14}\right]_{0}^{0.8}$$

$$= \left[14 x^{13} - 13 x^{14}\right]_{0}^{0.8} = 14(0.8)^{13} - 13(0.8)^{14} \approx 0.198. \ \square$$

Exercises 10–5

1–4. Calculate the average value of the function f over the given interval.

- 1 $f(x) = 4x - x^2$ on $[0, 4]$
- 2 $f(x) = \sqrt{x}$ on $[0, 9]$
- 3 $f(x) = x^3 - x^2$ on $[0, 2]$
- 4 $f(x) = x^2 - 4x + 3$ on $[0, 1]$
- 5 Find the average value of $f(x) = \sqrt{ax}$ from $x = 0$ to $x = 9a$.
- 6 A falling object satisfies the distance formula

$$s(t) = -16t^2 - 80t + 2400.$$

  (a) Find the average velocity from time $t = 0$ until the object strikes the earth.
  (b) Find the average value of the velocity from time $t = 0$ until the object strikes the earth. How does this answer compare with the one for (a)?

7–10. A function f is given.
  (a) Verify that f is a probability density function.
  (b) Find the probability that an event will have its numerical value in the given range.

- 7 $f(x) = \begin{cases} \frac{1}{2} & \text{if } 0 \leq x \leq 2, \\ 0 & \text{if } x < 0 \text{ or } x > 2, \end{cases}$ $0 \leq x \leq .5$

- 8 $f(x) = \begin{cases} 2x & \text{if } 0 \leq x \leq 1, \\ 0 & \text{if } x < 0 \text{ or } x > 1, \end{cases}$ $0.5 \leq x \leq 7$

- 9 $f(x) = \begin{cases} \dfrac{3(1 - x^2)}{4} & \text{if } -1 \leq x \leq 1, \\ 0 & \text{if } x < -1 \text{ or } x > 1, \end{cases}$ $-1 \leq x \leq -0.5$ or $0 \leq x \leq 0.5$

- 10 $f(x) = \begin{cases} 1 - x & \text{if } 0 \leq x \leq 1, \\ 1 + x & \text{if } -1 \leq x < 0, \\ 0 & \text{if } x < -1 \text{ or } x > 1, \end{cases}$ $-0.5 \leq x \leq 0.5.$

11 A probability density function f is of form

$$f(x) = \begin{cases} 0 & \text{if } x < 1, \\ \dfrac{A}{x^5} & \text{if } x \geq 1. \end{cases}$$

  (a) Find the constant A.
  (b) Find the probability that a number x is greater than or equal to 4.

- 12 Use the probability density function in Example 4.
  (a) Find the probability that at least 60 percent of the plates pass inspection in one day.
  (b) Find the probability that no more than 60 percent of the plates pass inspection in one day.

- 13 Professor Jones tries to arrive at his college between 9:20 and 9:30 each morning. Because of varying traffic conditions, the actual time of arrival is between 9:10 and 9:40 and is described by the probability density function (where t is the number

10-6 The Normal Probability Density Function

of minutes after 9:00):

$$f(t) = \begin{cases} 0 & \text{if } t < 10 \text{ or } t > 40, \\ \frac{1}{225}(t - 10) & \text{if } 10 \le t \le 25, \\ -\frac{1}{225}(t - 40) & \text{if } 25 \le t \le 40. \end{cases}$$

Find the probability that the professor arrives between 9:20 and 9:30 on a given morning.

10-6* The Normal Probability Density Function

When the length of an object is measured repeatedly, small random errors will occur. If the measurements are plotted along a number line, there will be some spreading of the data along the line, with most of the points clustered near some central value. This central value is in a sense the "true" length of the object. Similar clustering of measured values occurs in many experimental situations, such as measurement of the weight of many members of the same species, or the SAT scores of many students.

It is proved in courses on statistics that the probability density function that describes such outcomes is the *normal probability density function*

Normal Probability Density Function

$$f(x) = \frac{1}{\sqrt{2\pi}\,\sigma} e^{-(x-\mu)^2/2\sigma^2}, \qquad \sigma > 0,$$

Mean Standard Deviation

where the constants μ and σ are called the *mean* and the *standard deviation*, respectively. the mean is the central value around which the data points cluster, and the standard deviation is related to the extent to which the points spread out on either side of the mean. The graph of the normal probability density function is the familiar bell-shaped curve (Fig. 10-11a). It is symmetric about the line $x = \mu$; as σ increases, the curve becomes wider and flatter.

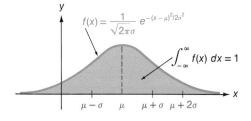

(a) The normal probability density function:
$f(x) = \frac{1}{\sqrt{2\pi}\,\sigma} e^{-(x-\mu)^2/2\sigma^2}$

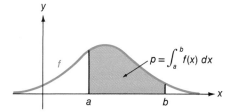

(b) The probability that an event has its value between a and b is
$$p = \int_a^b \frac{1}{\sqrt{2\pi}\,\sigma} e^{-(x-\mu)^2/2\sigma^2}\, dx$$

Figure 10-11 The normal probability density function.

The function f is always positive, since $e^u > 0$ for all u. It also satisfies the condition

$$\int_{-\infty}^{\infty} f(x)\, dx = 1,$$

although the proof of this fact is beyond our scope. Thus, the given formula meets both of the requirements to be called a probability density function.

If data are described by the normal probability density function ("normally distributed," in statisticians' jargon), with mean μ and standard deviation σ, then the probability that a given data item will be between a and b, where $a < b$, is

$$p = \int_a^b \frac{1}{\sqrt{2\pi}\sigma} e^{-(x-\mu)^2/2\sigma^2}\, dx$$

(see Fig. 10–11b). Since this integral is not one of the "elementary" integrals and cannot be calculated in terms of antiderivatives, then numerical methods (Section 5–6) have been used to prepare tables for p. A small part of such a table is reproduced in Table 10–1.

Table 10–1 Area under the Normal Probability Density Curve from μ to $\mu + k\sigma$

k	Area	k	Area	k	Area	k	Area
.00	.0000	1.00	.3413	2.00	.4773	3.00	.4987
.25	.0987	1.25	.3944	2.25	.4878	3.25	.4994
.50	.1915	1.50	.4332	2.50	.4938	3.50	.4998
.75	.2734	1.75	.4599	2.75	.4970	3.75	.4999

Each entry in Table 10–1 gives the area under the curve $f(x)$ bounded by the vertical lines $x = \mu$ and $x = \mu + k\sigma$. For instance, the area under the curve between μ and $\mu + 0.5\sigma$ is 0.1915. Because the graph of f is symmetric about $x = \mu$, the area under the curve from $\mu - k\sigma$ to μ must equal the area from μ to $\mu + k\sigma$. For example, the area from $\mu - 0.5\sigma$ to μ must also be 0.1915.

Each of the areas listed in the table corresponds to the value of p calculated for the appropriate limits of integration, and thus to the probability that a data item falls between μ and $\mu + k\sigma$. However, most applications in statistics inquire about the probability that a given data item will fall between $\mu + k_1\sigma$ and $\mu + k_2\sigma$. To get these values, we use the additivity property of integrals to write

$$p = \int_{\mu+k_1\sigma}^{\mu+k_2\sigma} f(x)\, dx = \int_\mu^{\mu+k_2\sigma} f(x)\, dx - \int_\mu^{\mu+k_1\sigma} f(x)\, dx.$$

It is worth noting that the area under the curve from $\mu - \sigma$ to $\mu + \sigma$ is $2(0.3413) = 0.6826$, and the area from $\mu - 2\sigma$ to $\mu + 2\sigma$ is $2(0.4773) = 0.9546$. Thus, any given data item has approximately a 68 percent chance of falling within one standard devia-

Exercises 10-6

EXAMPLE 1   Intelligence quotient (IQ) scores are distributed normally with mean $\mu = 100$ and standard deviation $\sigma = 15$. (The raw scores are adjusted so that the final scores have this distribution.) What is the probability that a person selected at random has an IQ in the range (a) between 100 and 115? (b) greater than 130?

Solution   (a) The probability that the score is between μ (the number 100) and $\mu + \sigma$ (the number 115) is

Probability = area between μ and $\mu + \sigma$
= .3413.

(b) The probability that a given score is greater than $\mu + 2\sigma$ (the number 130) is equal to the area under the curve from $\mu + 2\sigma$ to ∞.

Recall that the area under the curve from μ to ∞ is 0.5. (This is half the total area under the curve, which is 1.) From Table 10-1, the area from μ to $\mu + 2\sigma$ is .4773. Thus,

Probability = area from $\mu + 2\sigma$ to ∞
= (area from μ to ∞) − (area from μ to $\mu + 2\sigma$)
= .5000 − .4773 = .0227. □

The results of Example 1 can be interpreted in a slightly different way. The probability that a person has an IQ in a certain range is equal to the percentage of the population with IQ scores in that range. Thus, slightly more than one-third of the population have IQ scores between 100 and 115. Approximately 2 percent of the population have IQ scores greater than 130.

Exercises 10-6

1 The scores on the SAT mathematics aptitude test are distributed normally with mean $\mu = 500$ and standard deviation $\sigma = 100$. (Because of the very small percentage of scores below $\mu - 3\sigma$ and above $\mu + 3\sigma$, the scores are adjusted so that the minimum score is 200 and the maximum score 800.) What is the probability that a randomly selected student scores in the following range?
- •(a) $400 \leq x \leq 500$
- (b) $500 \leq x \leq 700$
- •(c) $450 \leq x \leq 550$
- (d) $750 \leq x \leq 800$

2 A high school student made the following score on the SAT mathematics aptitude test. What percentage of students had higher scores? lower scores? (See Exercise 1 for an explanation of SAT scores.)
- •(a) 450
- (b) 750

3 Weights of college women are distributed normally with mean $\mu = 127$ lb and standard deviation $\sigma = 17$ lb. What is the probability that a randomly chosen college woman will have her weight in the following range?
- •(a) Between 110 and 144 lb
- (b) Less than 93 lb
- •(c) Greater than 178 lb
- (d) Less than 161 lb

4 A commuter tries to arrive at his office at exactly 8:45 each morning. The probability that he will arrive at a given time is given by a normal density function with mean $\mu = 8{:}45$ and standard deviation $\sigma = 4$ min. What is the probability that he will arrive during the given period?
- •(a) Between 8:43 and 8:49
- (b) Between 8:46 and 8:50
- •(c) Before 8:40
- (d) After 9:00

5 Sketch the graph of the normal density function with the following mean and standard deviation.
- (a) $\mu = 50, \sigma = 2$
- (b) $\mu = 20, \sigma = 20$

6 It can be proved that $\lim_{b\to\infty} \int_0^b e^{-x^2/2}\, dx = \sqrt{\pi/2}$.
- (a) Use this result to show that if $\sigma > 0$, then

$$\int_{-\infty}^{\infty} \frac{1}{\sqrt{2\pi}\sigma} e^{-x^2/2\sigma^2}\, dx = 1.$$

- (b) Use the result of (a) to show that if $\sigma > 0$, then

$$\int_{-\infty}^{\infty} \frac{1}{\sqrt{2\pi}\sigma} e^{-(x-\mu)^2/2\sigma^2}\, dx = 1.$$

Standard Normal Probability Density Function

7 The function $f(x) = e^{-x^2/2}/\sqrt{2\pi}$, which is the normal probability density function with mean $\mu = 0$ and standard deviation $\sigma = 1$, is known as the *standard normal probability density function*.
- •(a) Determine the following properties of the graph of f: (1) local extrema; (2) concavity; (3) points of inflection; (4) asymptotes.
- (b) Sketch the graph of f.
- (c) Show that the graph of the normal density function with mean μ and standard deviation σ can be obtained from the graph of f by performing the following in order:

  (1) Change the horizontal scale by a factor of σ.
  (2) Change the vertical scale by a factor of $1/\sigma$.
  (3) Shift the graph horizontally a distance of μ units.

Review Problems

1 State Cauchy's theorem. Compare this theorem with the Mean Value Theorem.

2 State L'Hôpital's rule for the form $0/0$ and the form ∞/∞.

3–16. Use L'Hôpital's rule to evaluate the limit.

- • 3 $\lim_{x\to 0} \dfrac{\cos x - 1}{x^2}$
- • 4 $\lim_{x\to 0} \dfrac{\sinh x - \sin x}{x^3}$

- • 5 $\lim_{x\to 0} \dfrac{e^x - x - \cos x}{x^2}$
- • 6 $\lim_{x\to 1} \dfrac{\ln x}{x - 1}$

Review Problems

- **7** $\lim_{x \to 0} \dfrac{x^2 + \cos x - 1}{x^2 + \cos 2x - 1}$
- **8** $\lim_{x \to \infty} \dfrac{\sinh x}{x}$
- **9** $\lim_{x \to 0^+} \dfrac{\ln(\sin 2x)}{\ln x}$
- **10** $\lim_{x \to 0} (\cos x)^{1/x^2}$
- **11** $\lim_{x \to \infty} (e^x + 1)^{1/2x}$
- **12** $\lim_{x \to 0^+} x^{\sin x}$
- **13** $\lim_{x \to (\pi/2)^-} (\tan x)^{\cos x}$
- **14** $\lim_{x \to \infty} \left(1 - \dfrac{1}{x}\right)^x$
- **15** $\lim_{x \to 0} (\sin 2x + \cos 3x)^{\cot x}$
- **16** $\lim_{x \to \infty} (e^{3x} + x)^{1/x}$

17 Define the improper integral.

    (a) $\int_a^b f(x)\, dx$, f continuous on $(a, b]$, discontinuous at a

    (b) $\int_a^b f(x)\, dx$, f continuous on $[a, b]$ except at the interior poiont c

18 Define the improper integral.

    (a) $\int_a^\infty f(x)\, dx$, f continuous for $x \ge a$

    (b) $\int_a^\infty f(x)\, dx$, f continuous for $x > a$, discontinuous at a

    (c) $\int_{-\infty}^\infty f(x)\, dx$, f continuous for all x

19–26. Evaluate the improper integral if it converges. If it diverges, then determine, if poissible, whether it diverges to ∞, $-\infty$, or neither.

- **19** $\displaystyle\int_0^3 \dfrac{x\, dx}{(x^2 - 1)^{1/3}}$
- **20** $\displaystyle\int_0^4 \dfrac{2x}{(x^2 - 4)^2}\, dx$
- **21** $\displaystyle\int_1^\infty \dfrac{dx}{(2x + 1)^2}$
- **22** $\displaystyle\int_0^\infty \dfrac{dx}{(x + 3)^{1/2}}$
- **23** $\displaystyle\int_0^\infty \dfrac{x\, dx}{\sqrt{x^2 + 1}}$
- **24** $\displaystyle\int_0^\infty \dfrac{x^3}{(x^4 + 1)^3}\, dx$
- **25** $\displaystyle\int_0^1 \dfrac{dx}{x^n}$ $(n > 1)$
- **26** $\displaystyle\int_1^\infty \dfrac{dx}{x^n}$ $(n > 1)$

- **27** Find the centroid, if it exists, of the infinite region bounded by the curve and lines: $y = 1/x^4$, $x \ge 2$, the x-axis, the line $x = 2$.

- **28** Find the average value of the distance from the origin to the line segment connecting $(1, 0)$ to $(1, 1)$ by the following methods.

    (a) Consider the distance from $(0, 0)$ to $(1, y)$ to be a function of the angle α that is the inclination of the line through the two points.

(b) Consider the distance from (0, 0) to (1, y) to be a function of y. (*Hint:* Use I57 to evaluate the integral.)

29 Let
$$f(x) = \begin{cases} 0 & \text{if } x < 1, \\ \dfrac{1}{x^2} & \text{if } x \geq 1. \end{cases}$$

(a) Verify that f is a probability density function.

•(b) Calculate the probability that $-1 \leq x \leq 3$.

•(c) Calculate the probability that $x \geq 5$.

30 (a) What does it mean to say that data are distributed *normally?*

(b) Sketch the graph of the normal probability distribution function.

•31 Certain data are distributed normally with mean $\mu = 50$ and standard deviation $\sigma = 5$. Use Table 10–1 to find the probability that x is in the range

(a) $40 \leq x \leq 55$ (b) $x \geq 60$

11 Techniques of Integration

You will find that many integrals cannot be evaluated by methods we have studied so far. For example, the area of the region bounded by the graph of $y = \sqrt{1 - x^2}$, the x-axis, the y-axis, and the line $x = \frac{1}{2}$ is

$$A = \int_0^{1/2} \sqrt{1 - x^2}\, dx,$$

an integral that we cannot evaluate by the formulas from Chapters 1 to 10.

In this chapter we develop several other techniques to evaluate special types of integrals. Unfortunately, no one technique can be used for all of them. The best we can do is classify the integrals. Those of one type can be worked by one method; those of another type by another method. If an integral cannot be evaluated by any of these methods, we may have to use an approximation technique, such as Simpson's rule, to tabulate a set of values for the integral.

11–1 Integration by Parts

One of the most powerful techniques of integration is based on the formula for the differential of a product

D6 $\quad d(uv) = u\, dv + v\, du,$

where u and v are functions of x.

If we integrate both sides of D6 and recall that $\int dv = v + C$, we get

$$\int d(uv) = \int u\, dv + \int v\, du$$
$$uv + C = \int u\, dv + \int v\, du.$$

We now solve this last equation for $\int u\, dv$, obtaining the *integration by parts* formula

Integration by Parts Formula

I31 $\quad \int u\, dv = uv - \int v\, du + C.$

To apply the formula to calculate $\int f(x)g(x)\,dx$, we choose one of the factors, say $f(x)$, to be u and the remaining factor, $g(x)\,dx$, to be dv. It is generally best to try to choose the factors so that $v\,du$ can be integrated more easily than the original integral.

EXAMPLE 1 Calculate $\int xe^x\,dx$.

Solution We choose $u = x$, $dv = e^x\,dx$. Then $du = dx$ and $v = e^x$. (Actually, we should add an arbitrary constant to $v = e^x$, but it would cancel out later so we omit it. See Exercise 35. Then

$$\int x e^x\,dx = \int u\,dv = uv - \int v\,du + C$$

$$= x e^x - \int e^x\,dx + C$$

$$= x e^x - e^x + C. \;\square$$

$u = x$	$dv = e^x\,dx$
$du = dx$	$v = e^x$

As you might suspect after reflecting on Example 1, integration by parts is especially useful for integrals of form $\int x f(x)\,dx$, where $f(x)$ can be integrated twice. Thus, it is the most natural method for such integrals as

$\int x e^x\,dx$ and $\int x \sin x\,dx$.

Integration by parts also is a valuable tool for integrating functions that have fairly simple derivatives but complicated antiderivatives. To calculate $\int f(x)\,dx$, where f is such a function, we choose $u = f(x)$ and $dv = dx$. Then

$$\int f(x)\,dx = uv - \int v\,du + C$$

$$= x f(x) - \int x f'(x)\,dx + C.$$

In some cases (but not all) it is easier to calculate this last integral than it is the original one.

EXAMPLE 2 Calculate $\int \tan^{-1} x\,dx$.

Solution Let $u = \tan^{-1} x$, $dv = dx$. Then $du = dx/(1 + x^2)$ and $v = x$. The integral becomes

$$\int \tan^{-1} x\,dx = uv - \int v\,du + C$$

$$= x \tan^{-1} x - \int x \cdot \frac{1}{1 + x^2}\,dx + C$$

$$= x \tan^{-1} x - \frac{1}{2} \int \frac{2x}{1 + x^2}\,dx + C$$

$$= x \tan^{-1} x - \frac{1}{2} \ln(1 + x^2) + C. \;\square$$

$u = \tan^{-1} x$	$dv = dx$
$du = \dfrac{dx}{1 + x^2}$	$v = x$

11-1 Integration by Parts

REDUCTION FORMULAS

Formulas in which a given integral is expressed in terms of similar integrals of simpler form are called *reduction formulas*. Most of these formulas are derived by integration by parts.

EXAMPLE 3 Let n be a positive integer. Use integration by parts to derive the reduction formula

$$\int x^n e^x \, dx = x^n e^x - n \int x^{n-1} e^x \, dx + C.$$

Solution Let $u = x^n$, $dv = e^x \, dx$. Then $du = nx^{n-1} \, dx$, $v = e^x$:

$$\int x^n e^x \, dx = uv - \int v \, du + C$$

$$= x^n e^x - \int e^x (nx^{n-1} \, dx) + C$$

$$= x^n e^x - n \int x^{n-1} e^x \, dx + C. \quad \square$$

$u = x^n$
$du = nx^{n-1} \, dx$
$dv = e^x \, dx$
$v = e^x$

EXAMPLE 4 Use the reduction formula of Example 3 to calculate $\int x^2 e^x \, dx$ and $\int x^3 e^x \, dx$.

Solution **(a)** $n = 2$: $\int x^2 e^x \, dx = x^2 e^x - 2 \int x e^x \, dx + C_1$

$$= x^2 e^x - 2(xe^x - e^x + C_2) + C_1 \quad \text{(by Example 1)}$$
$$= x^2 e^x - 2xe^x + 2e^x + C.$$

(b) $n = 3$: $\int x^3 e^x \, dx = x^3 e^x - 3 \int x^2 e^x \, dx + C_1$

$$= x^3 e^x - 3(x^2 e^x - 2xe^x + 2e^x + C_2) + C_1 \quad \text{(by part a)}$$
$$= x^3 e^x - 3x^2 e^x + 6xe^x - 6e^x + C$$

where $C = -3C_2 + C_1$. $\square$

Some other reduction formulas are given in Exercises 23–26.

It may be necessary to apply integration by parts twice. Occasionally this leads to a situation in which the original integral appears on both sides of an equation. When that happens, we may be able to solve the resulting equation algebraically, getting the integral as the solution.

EXAMPLE 5 Calculate $\int e^x \sin x \, dx$.

Solution Let $u = \sin x$, $dv = e^x \, dx$. Then

$$\int e^x \sin x \, dx = uv - \int v \, du + C_1$$

$$= e^x \sin x - \int e^x \cos x \, dx + C_1.$$

$u = \sin x$
$du = \cos x \, dx$
$dv = e^x \, dx$
$v = e^x$

To evaluate the integral on the right we apply integration by parts again. Let $\alpha = \cos x$, $d\beta = e^x\, dx$. Then

$$\int e^x \cos x\, dx = \alpha\beta - \int \beta\, d\alpha + C_2$$

$$= e^x \cos x - \int (-e^x \sin x)\, dx + C_2$$

$$= e^x \cos x + \int e^x \sin x\, dx + C_2.$$

$\alpha = \cos x$
$d\alpha = -\sin x\, dx$
$d\beta = e^x\, dx$
$\beta = e^x$

We substitute this expression in the equation for $\int e^x \sin x\, dx$, obtaining

$$\int e^x \sin x\, dx = e^x \sin x - \left\{ e^x \cos x + \int e^x \sin x\, dx + C_2 \right\} + C_1$$

$$= e^x \sin x - e^x \cos x - \int e^x \sin x\, dx + C_1 - C_2.$$

We now solve this equation algebraically for $\int e^x \sin x\, dx$, obtaining

$$\int e^x \sin x\, dx = \frac{e^x (\sin x - \cos x)}{2} + C$$

where $C = (C_1 - C_2)/2$. □

Remark In our examples we have written a new constant of integration with each application of integration by parts. In actual practice the constant is not usually written until the last integral is calculated. In fact, the integration by parts formula is often written as

$$\int u\, dv = uv - \int v\, du$$

without the constant of integration. This form is technically incorrect and may lead to difficulties. No problems occur when the integral on the right is actually calculated and the constant added at that time. They arise when the integral on the right is combined with the one on the left (as in Example 5) and the resulting equation solved algebraically. If the constant is not written out, the final form may be incorrect. (See Exercise 36, for example.)

Constant of Integration

DEFINITE INTEGRALS

The formula for integration by parts can be used with definite integrals. In that case, the formula becomes

Integration by Parts Formula for Definite Integrals

$$\int_a^b u(x)\, d(v(x)) = \Big[u(x)v(x) \Big]_a^b - \int_a^b v(x)\, d(u(x))$$

provided $u(x)$ and $v(x)$ have continuous derivatives on the open interval (a, b).

Exercises 11–1

For example, it follows from our work in Example 1 that

$$\int_0^{\ln 2} xe^x \, dx = \left[xe^x\right]_0^{\ln 2} - \int_0^{\ln 2} e^x \, dx = \left[xe^x - e^x\right]_0^{\ln 2}$$

$$= [\ln 2 \, e^{\ln 2} - e^{\ln 2}] - [0 - 1]$$

$$= \ln 2 \cdot 2 - 2 + 1 = 2 \ln 2 - 1. \ \square$$

Exercises 11–1

1–18. Use integration by parts to calculate the integral.

- 1. $\int x \sec^2 x \, dx$
- 2. $\int x^2 \ln 2x \, dx$
- 3. $\int x \cos x \, dx$
- 4. $\int x \sin (2x - 1) \, dx$
- 5. $\int \ln (1 + x^2) \, dx$
- 6. $\int e^{-x} \cos 2x \, dx$
- 7. $\int x^3 \sin x^2 \, dx$
- 8. $\int x \tan^{-1} x \, dx$
- 9. $\int \sin x \ln (\cos x) \, dx$
- 10. $\int \cos (\ln x) \, dx$
- 11. $\int x^2 \sin x \, dx$
- 12. $\int x^3 \sqrt{x^2 + 1} \, dx$
- 13. $\int_1^e \ln x \, dx$
- 14. $\int_0^1 \sin^{-1} x \, dx$
- 15. $\int_0^1 x^2 e^{-x} \, dx$
- 16. $\int_0^1 x^2 \tan^{-1} x \, dx$
- 17. $\int_0^\pi e^{-2x} \sin 3x \, dx$
- 18. $\int_0^{\ln \pi} e^{2x} \cos (e^x) \, dx$

19–22. Use integration by parts, followed by L'Hôpital's rule, to evaluate the improper integral provided it converges.

- 19. $\int_1^\infty \frac{\ln x}{x^2} \, dx$
- 20. $\int_0^\infty \frac{x}{e^x} \, dx$
- 21. $\int_0^1 \sqrt{x} \ln x \, dx$
- 22. $\int_1^\infty \frac{e^x(x-1)}{x^2} \, dx$

Reduction Formulas

23–26. Use integration by parts to get the given reduction formula.

23. $\int x^m (\ln x)^n \, dx = \frac{x^{m+1} (\ln x)^n}{m + 1} - \frac{n}{m + 1} \int x^m (\ln x)^{n-1} \, dx + C;\ m, n \neq -1$

24 $\int x^n \cos x \, dx = x^n \sin x - n \int x^{n-1} \sin x \, dx + C$

25 $\int x^n \sin x \, dx = -x^n \cos x + n \int x^{n-1} \cos x \, dx + C$

26 $\int \sec^n x \, dx = \dfrac{\sec^{n-2} x \tan x}{n-1} + \dfrac{n-2}{n-1} \int \sec^{n-2} x \, dx + C; \ n \neq 1$

27–32. Use the reduction formulas from Example 3 and Exercises 23–26 to calculate the integrals.

- 27 $\int \sec^3 x \, dx$
- 28 $\int (\ln x)^2 \, dx$
- 29 $\int x^2 \sin x \, dx$
- 30 $\int x^3 \cos x \, dx$
- 31 $\int x^4 e^x \, dx$
- 32 $\int x^3 \ln x \, dx$

- 33 Let $\mathcal{R}$ be the region bounded by the x-axis, the graph of $y = \ln x$, and the vertical lines $x = 1$, $x = e$. Calculate the volume of the solid of revolution obtained by revolving $\mathcal{R}$ about (a) the x-axis; (b) the y-axis.

- 34 Find the centroid of the region bounded by the graphs of $f(x) = e^x$, $y = 0$, $x = 0$, $x = 2$.

35 Show that if we use $v = e^x + K$ in Example 1, where K is a constant, we get the same final solution.

36 Apply the formula for integration by parts to $\int dx/x$, letting $u = 1/x$ and $dv = dx$. Show that if the constant of integration is ignored, then the resulting equation implies that $0 = 1$.

11–2 Trigonometric Integrals

Many problems require us to integrate powers of trigonometric functions. As we shall see, certain combinations of powers can be integrated easily. The key to calculating our first type of integral is the observation that integrals of form

$$\int f(\sin x) \cos x \, dx$$

can be changed to the form

$$\int f(u) \, du$$

by the substitution $u = \sin x$. Similarly, integrals of form $\int f(\cos x) \sin x \, dx$ can be changed to the form $-\int f(u) \, du$ by the substitution $u = \cos x$.

11–2 Trigonometric Integrals

$\int \sin^m x \cos^n x \, dx$, m OR n ODD

Let m and n be real numbers, at least one of which is an odd positive integer. If n is the odd positive integer, we write $n = 2k + 1$ for some integer k. We rewrite the integral

$$\int \sin^m x \cos^{2k+1} x \, dx = \int \sin^m x \, (\cos^2 x)^k \cos x \, dx$$

Odd Power of the Sine or Cosine

and substitute $1 - \sin^2 x$ for $\cos^2 x$, obtaining

$$\int \sin^m x \, (1 - \sin^2 x)^k \cos x \, dx.$$

The integral is now of form $\int f(\sin x) \cos x \, dx$ and can be simplified by the substitution $u = \sin x$, $du = \cos x \, dx$.

A similar technique works if m is an odd positive integer. In that case, we keep one $\sin x$ and change the other terms to $\cos x$ to reduce the integral of one of form

$$\int f(\cos x) \sin x \, dx.$$

If both m and n are odd positive integers, then either substitution can be used.

EXAMPLE 1 Calculate $\int \sin^6 x \cos^5 x \, dx$.

Solution We rewrite the integral as

$$\int \sin^6 x \cos^4 x \cos x \, dx = \int \sin^6 x \, (\cos^2 x)^2 \cos x \, dx$$

$$= \int \sin^6 x \, (1 - \sin^2 x)^2 \cos x \, dx$$

$$= \int u^6 (1 - u^2)^2 \, du \qquad \boxed{\begin{array}{l} u = \sin x \\ du = \cos x \, dx \end{array}}$$

$$= \int u^6 (1 - 2u^2 + u^4) \, du$$

$$= \int (u^6 - 2u^8 + u^{10}) \, du$$

$$= \frac{u^7}{7} - \frac{2u^9}{9} + \frac{u^{11}}{11} + C$$

$$= \frac{\sin^7 x}{7} - 2\frac{\sin^9 x}{9} + \frac{\sin^{11} x}{11} + C. \ \square$$

EXAMPLE 2 Calculate $\int \sqrt{\cos x} \sin^3 x \, dx$.

Solution We rewrite the integral as

$$\int \sqrt{\cos x}\, \sin^3 x\, dx = \int (\cos x)^{1/2} \sin^2 x \cdot \sin x\, dx$$

$$= -\int (\cos x)^{1/2} (1 - \cos^2 x)(-\sin x\, dx)$$

and make the substitution $u = \cos x$, $du = -\sin x\, dx$. The integral becomes

$$-\int u^{1/2}(1 - u^2)\, du = -\int (u^{1/2} - u^{5/2})\, du$$

$$= -\frac{2u^{3/2}}{3} + \frac{2u^{7/2}}{7} + C$$

$$= -\frac{2(\cos x)^{3/2}}{3} + \frac{2(\cos x)^{7/2}}{7} + C.$$

This result is valid on any interval where $\cos x \geq 0$. ☐

$$\int \sin^m x\, \cos^n x\, dx, \quad m \text{ AND } n \text{ EVEN}$$

If m and n are even nonnegative integers, then the integral

$$\int \sin^m x\, \cos^n x\, dx$$

Even Powers of Sine and Cosine

can be simplified by using the double-angle formulas T14 and T15:

$$\sin^2 \alpha = \frac{1 - \cos 2\alpha}{2} \quad \text{and} \quad \cos^2 \alpha = \frac{1 + \cos 2\alpha}{2}.$$

EXAMPLE 3 Calculate $\int \sin^2 x \cos^2 x\, dx$.

Solution
$$\int \sin^2 x \cos^2 x\, dx = \int \frac{1 - \cos 2x}{2} \cdot \frac{1 + \cos 2x}{2}\, dx$$

$$= \frac{1}{4} \int (1 - \cos^2 2x)\, dx$$

$$= \frac{x}{4} - \frac{1}{4} \int \cos^2 2x\, dx.$$

We apply the double-angle formula again. The original integral equals

$$\int \sin^2 x \cos^2 x\, dx = \frac{x}{4} - \frac{1}{4} \int \frac{1 + \cos 4x}{2}\, dx$$

$$= \frac{x}{4} - \frac{x}{8} - \frac{1}{8} \int \cos 4x\, dx$$

$$= \frac{x}{8} - \frac{1}{8} \frac{\sin 4x}{4} + C$$

$$= \frac{x}{8} - \frac{\sin 4x}{32} + C. \ \square$$

11-2 Trigonometric Integrals

$\int \tan^m x \, \sec^n x \, dx$, n EVEN

Let m be a real number, and n a positive even integer. The integral

$$\int \tan^m x \, \sec^n x \, dx$$

Even Powers of the Secant

can be reduced to an integral of form $\int f(\tan x) \sec^2 x \, dx$ by the following technique. First, we write $n = 2k + 2$, where k is a nonnegative integer (recall that n is an even positive integer). Then

$$\int \tan^m x \, \sec^n x \, dx = \int \tan^m x \, (\sec^2 x)^k \sec^2 x \, dx.$$

Next, we make the substitution $\sec^2 x = 1 + \tan^2 x$, obtaining

$$\int \tan^m x \, (1 + \tan^2 x)^k \sec^2 x \, dx.$$

Finally, we make the substitution $u = \tan x$, $du = \sec^2 x \, dx$ to reduce this integral to

$$\int u^m (1 + u^2)^k \, du.$$

EXAMPLE 4

$$\int \tan^4 x \, \sec^4 x \, dx = \int \tan^4 x \, \sec^2 x \, \sec^2 x \, dx$$

$$= \int \tan^4 x \, (1 + \tan^2 x) \sec^2 x \, dx \quad \boxed{\begin{array}{l} u = \tan x \\ du = \sec^2 x \, dx \end{array}}$$

$$= \int u^4 (1 + u^2) \, du$$

$$= \int (u^4 + u^6) \, du = \frac{u^5}{5} + \frac{u^7}{7} + C$$

$$= \frac{\tan^5 x}{5} + \frac{\tan^7 x}{7} + C. \quad \square$$

Powers of the Tangent

A similar approach can be used in integrating powers of the tangent function. The integral

$$\int \tan^m x \, dx,$$

where m is an integer greater than one, can be reduced to a simpler integral by first factoring $\tan^m x$ as

$$\tan^m x = \tan^{m-2} x \, \tan^2 x$$

and then substituting $\tan^2 x = \sec^2 x - 1$. The resulting integral can be written as

$$\int \tan^m x \, dx = \int \tan^{m-2} x \, (\sec^2 x - 1) \, dx$$

$$= \int \tan^{m-2} x \, \sec^2 x \, dx - \int \tan^{m-2} x \, dx.$$

The first of these integrals can be evaluated after making the substitution $u = \tan x$, $du = \sec^2 x \, dx$. The second one has the same form as the original one. If $m - 2 \geq 2$, we repeat the process. If $m - 2 = 1$, we use the fact that

$$\int \tan x \, dx = \int \frac{\sin x}{\cos x} \, dx = \ln |\sec x| + C.$$

EXAMPLE 5
$$\int \tan^3 x \, dx = \int \tan x \tan^2 x \, dx$$
$$= \int \tan x \, (\sec^2 x - 1) \, dx$$
$$= \int \tan x \sec^2 x \, dx - \int \tan x \, dx$$
$$= \int u \, du - \int \tan x \, dx = \frac{u^2}{2} - \ln |\sec x| + C \qquad \boxed{\begin{array}{l} u = \tan x \\ du = \sec^2 x \, dx \end{array}}$$
$$= \frac{\tan^2 x}{2} - \ln |\sec x| + C. \quad \square$$

$$\int \tan^m x \sec^n x \, dx, \; m \text{ ODD}$$

Let m be an odd positive integer, n a real number. The integral

Odd Powers of
the Tangent
$$\int \tan^m x \sec^n x \, dx$$

can be reduced to one of form

$$\int f(\sec x) \sec x \tan x \, dx$$

by the following process.
We write $m = 2k + 1$. Then

$$\int \tan^m x \sec^n x \, dx = \int \tan^{2k+1} \sec^n x \, dx$$
$$= \int (\tan^2 x)^k \sec^{n-1} x \tan x \sec x \, dx$$
$$= \int (\sec^2 x - 1)^k \sec^{n-1} x \, (\sec x \tan x \, dx).$$

This integral is now of form

$$\int f(\sec x) \sec x \tan x \, dx,$$

which can be simplified by the substitution $u = \sec x$, $du = \sec x \tan x \, dx$.

EXAMPLE 6
$$\int \tan^5 x \sec^5 x \, dx = \int (\tan^2 x)^2 \sec^5 x \tan x \, dx$$

11-2 Trigonometric Integrals

$$= \int (\sec^2 x - 1)^2 \sec^4 x \, (\sec x \tan x \, dx)$$

$$= \int (u^2 - 1)^2 u^4 \, du \qquad \boxed{\begin{array}{l} u = \sec x \\ du = \sec x \tan x \, dx \end{array}}$$

$$= \int (u^8 - 2u^6 + u^4) \, du = \frac{u^9}{9} - \frac{2u^7}{7} + \frac{u^5}{5} + C$$

$$= \frac{\sec^9 x}{9} - \frac{2 \sec^7 x}{7} + \frac{\sec^5 x}{5} + C. \; \square$$

Remark 1 The integral

$$\int \tan^m x \sec^n x \, dx$$

where m is not an odd integer and n is not an even integer must be integrated by some technique different from the ones discussed in this section. Integration by parts will work in some examples.

EXAMPLE 7 Calculate $\int \sec^3 x \, dx$.

Solution Let $u = \sec x$, $dv = \sec^2 x \, dx$. Then $du = \sec x \tan x \, dx$ and $v = \tan x$. Using integration by parts, we obtain

$$\int \sec^3 x \, dx = uv - \int v \, du + C_1$$

$$= \sec x \tan x - \int \tan x \, (\sec x \tan x) \, dx + C_1 \qquad \boxed{\begin{array}{l} u = \sec x \\ du = \sec x \tan x \, dx \\ dv = \sec^2 x \, dx \\ v = \tan x \end{array}}$$

$$= \sec x \tan x - \int \tan^2 x \sec x \, dx + C_1$$

$$= \sec x \tan x - \int (\sec^2 x - 1) \sec x \, dx + C_1$$

$$= \sec x \tan x - \int \sec^3 x \, dx + \int \sec x \, dx + C_1$$

$$= \sec x \tan x - \int \sec^3 x \, dx + \ln |\sec x + \tan x| + C_1.$$

Therefore,

$$2 \int \sec^3 x \, dx = \sec x \tan x + \ln |\sec x + \tan x| + C_1$$

$$\int \sec^3 x \, dx = \frac{\sec x \tan x}{2} + \frac{\ln |\sec x + \tan x|}{2} + C$$

where $C = C_1/2$. $\square$

Remark 2 The integral

$$\int \cot^m x \csc^n x \, dx,$$

where m is odd or n is even, can be integrated by the same technique that would be used for the integral $\int \tan^m x \sec^n x \, dx$. Similar methods can be used to integrate powers of the hyperbolic functions.

Remark 3   Some trigonometric integrals different from the types discussed in this section can be calculated after all the functions have been expressed in terms of the sine and cosine functions. A substitution may be suggested by the new form, or integration by parts may be natural.

EXAMPLE 8   Calculate $\int \sin x \, (\tan x + \cot x) \, dx$.

Solution   We change all the trigonometric functions to sines and cosines and simplify:

$$\int \sin x \, (\tan x + \cot x) \, dx = \int \sin x \left(\frac{\sin x}{\cos x} + \frac{\cos x}{\sin x} \right) dx$$

$$= \int \sin x \left(\frac{\sin^2 x + \cos^2 x}{\cos x \sin x} \right) dx$$

$$= \int \frac{\sin x}{\cos x \sin x} \, dx \quad \text{(by T5)}$$

$$= \int \frac{dx}{\cos x} = \int \sec x \, dx$$

$$= \ln |\sec x + \tan x| + C. \quad \square$$

TRIGONOMETRIC PRODUCTS

One form is very useful in dealing with the integrals encountered in the branch of higher mathematics known as *Fourier analysis*. Integrals of form

$$\int \sin mx \cos nx \, dx,$$

where m and n are positive integers, can be calculated by means of the product formula T19:

$$\sin \alpha \cos \beta = \tfrac{1}{2} [\sin (\alpha + \beta) + \sin (\alpha - \beta)].$$

EXAMPLE 9   Calculate $\int \sin 5x \cos 2x \, dx$.

Solution

$$\int \sin 5x \cos 2x \, dx = \int \frac{1}{2} [\sin 7x + \sin 3x] \, dx$$

$$= \frac{1}{2} \left(-\frac{\cos 7x}{7} \right) + \frac{1}{2} \left(-\frac{\cos 3x}{3} \right) + C$$

Use of Product Formula

$$= -\frac{\cos 7x}{14} - \frac{\cos 3x}{6} + C. \quad \square$$

An integral of form

$$\int \sin mx \sin nx \, dx \quad \text{or} \quad \int \cos mx \cos nx \, dx$$

can be calculated likewise by the appropriate product formula.

Exercises 11–2

1–32. Evaluate the integral.

- 1. $\int \sin^3 2x \cos^3 2x \, dx$
- 2. $\int \sin^3 x \cos^6 x \, dx$
- 3. $\int \sin^5 x \, dx$
- 4. $\int \sqrt{\sin x} \cos^3 x \, dx$
- 5. $\int \dfrac{\cos^3 x}{\sin^2 x} \, dx$
- 6. $\int \dfrac{\cos^3 x}{\sin x} \, dx$
- 7. $\int \dfrac{dx}{\cos^2 x}$
- 8. $\int \dfrac{dx}{\sin^4 x}$
- 9. $\int \sin 3x \cot^2 3x \, dx$
- 10. $\int \tan^2 2x \sec^4 2x \, dx$
- 11. $\int \tan x \sec^5 x \, dx$
- 12. $\int \tan^5 x \sec x \, dx$
- 13. $\int \tan^3 x \sec^3 x \, dx$
- 14. $\int \tan x \sec^6 x \, dx$
- 15. $\int \tan^5 x \, dx$
- 16. $\int \cot^3 3x \, dx$
- 17. $\int \csc^3 x \, dx$
- 18. $\int \csc^4 3x \cot 3x \, dx$
- 19. $\int \dfrac{\tan 2x}{\sec 2x} \, dx$
- 20. $\int \sec^4 2x \, dx$
- 21. $\int \cot x \csc^3 x \, dx$
- 22. $\int (\tan x + \cot x)^2 \, dx$
- 23. $\int \sin 3x \sin 2x \, dx$
- 24. $\int \sin 4x \cos 5x \, dx$
- 25. $\int \cos x \cos 3x \, dx$
- 26. $\int \cos 2x \sin 3x \, dx$
- 27. $\int_0^{\pi/4} \tan^2 x \, dx$
- 28. $\int_0^{\pi/4} \dfrac{\sin^2 x}{\cos x} \, dx$
- 29. $\int_0^{\pi/4} \cos^4 x \sin^2 x \, dx$
- 30. $\int_0^{\pi/4} \cos^2 3x \, dx$
- 31. $\int_0^{\pi/3} \sin^4 2x \, dx$
- 32. $\int_{\pi/3}^{\pi/2} \cot^3 x \csc^3 x \, dx$.

33. Show that $\int \sin 2x \, dx = -(\cos 2x)/2 + C$ and that $\int \sin 2x \, dx = \sin^2 x + C$. Does this mean that $2 \sin^2 x = -\cos 2x$? Explain.

- 34. Find the centroid of the region bounded by the graph of $y = \sin x$ and the x-axis, $0 \le x \le \pi$.

- 35. Find the centroid of the region bounded by the graphs of $y = \sin x$ and $y = \cos x$, $\pi/4 \le x \le 5\pi/4$.

•36 Find the volume of the solid of revolution obtained by revolving the region in Exercise 35 about (a) the y-axis; (b) the line $y = -1$.

11-3 Trigonometric Substitutions

Integrals involving expressions of form $\sqrt{a^2 - x^2}$, $\sqrt{a^2 + x^2}$ or $\sqrt{x^2 - a^2}$ can often be simplified by trigonometric substitutions. In many cases, the resulting integrals can be calculated without difficulty.

EXAMPLE 1   Calculate $\frac{1}{2} \int \sqrt{4 - x^2}\, dx$.

Solution   The values of x must be between -2 and $+2$ to have the integrand defined and real. Observe that the substitution $x = 2 \sin \theta$ would simplify the expression $\sqrt{4 - x^2}$. To make this substitution, we define $\theta = \sin^{-1}(x/2)$. Then θ is between $-\pi/2$ and $\pi/2$, $x = 2 \sin \theta$ and $dx = 2 \cos \theta\, d\theta$. The integral becomes

$$\frac{1}{2} \int \sqrt{4 - x^2}\, dx = \frac{1}{2} \int \sqrt{4 - 4\sin^2 \theta}\; 2 \cos \theta\, d\theta$$

$$= 2 \int \sqrt{1 - \sin^2 \theta} \cos \theta\, d\theta$$

$$= 2 \int \sqrt{\cos^2 \theta} \cos \theta\, d\theta.$$

$\boxed{\begin{array}{l} \theta = \sin^{-1} \dfrac{x}{2} \\ x = 2 \sin \theta \\ dx = 2 \cos \theta\, d\theta \end{array}}$

Since θ is between $-\pi/2$ and $\pi/2$, then $\cos \theta > 0$, so that

$$\sqrt{\cos^2 \theta} = |\cos \theta| = \cos \theta.$$

The integral becomes

$$\frac{1}{2} \int \sqrt{4 - x^2}\, dx = 2 \int \cos \theta \cdot \cos \theta\, d\theta = 2 \int \frac{1 + \cos 2\theta}{2}\, d\theta \quad \text{(by T14)}$$

$$= \int (1 + \cos 2\theta)\, d\theta = \theta + \frac{\sin 2\theta}{2} + C.$$

The final answer should be expressed in x, the original variable. To do this recall that $\sin 2\theta = 2 \sin \theta \cos \theta$ and that $\sin \theta = x/2$. To evaluate $\cos \theta$, we draw a reference triangle for the angle θ. (See Fig. 11-1.) Observe from the figure that $\cos \theta = \sqrt{4 - x^2}/2$. Consequently,

$$\frac{1}{2} \int \sqrt{4 - x^2}\, dx = \theta + \frac{\sin 2\theta}{2} + C = \theta + \sin \theta \cos \theta + C$$

$$= \sin^{-1}\left(\frac{x}{2}\right) + \frac{x}{2} \cdot \frac{\sqrt{4 - x^2}}{2} + C$$

$$= \sin^{-1}\left(\frac{x}{2}\right) + \frac{x\sqrt{4 - x^2}}{4} + C. \quad \square$$

The substitution needed for Example 1 is $x = 2 \sin \theta$. To have θ in the proper range, we first define θ to be $\sin^{-1}(x/2)$. A similar procedure works for integrals involving other quadratic expressions. Table 11-1 gives the substitutions needed to simplify expressions of form $\sqrt{a^2 - x^2}$, $\sqrt{a^2 + x^2}$, and $\sqrt{x^2 - a^2}$. The reference triangles corresponding to these substitutions are illustrated in Figures 11-2a, b, c.

11-3 Trigonometric Substitutions

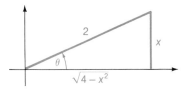

Figure 11-1 Example 1. $\theta = \sin^{-1}\left(\dfrac{x}{2}\right)$, $\sin\theta = \dfrac{x}{2}$, $\cos\theta = \dfrac{\sqrt{4-x^2}}{2}$.

These figures can be used to determine the other trigonometric functions of the angle θ as in Example 1.

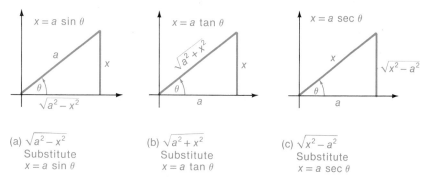

(a) $\sqrt{a^2 - x^2}$
Substitute
$x = a\sin\theta$

(b) $\sqrt{a^2 + x^2}$
Substitute
$x = a\tan\theta$

(c) $\sqrt{x^2 - a^2}$
Substitute
$x = a\sec\theta$

Figure 11-2 Trigonometric substitutions. See Table 11-1.

Substitutions for Particular Terms

Table 11-1 Trigonometric Substitutions

Expression to Be Simplified	Substitution	Equivalent Substitution
$\sqrt{a^2 - x^2}$	$x = a\sin\theta$	$\theta = \sin^{-1}(x/a)$
$\sqrt{a^2 + x^2}$	$x = a\tan\theta$	$\theta = \tan^{-1}(x/a)$
$\sqrt{x^2 - a^2}$	$x = a\sec\theta$	$\theta = \sec^{-1}(x/a)$

The trigonometric substitutions illustrated in Figure 11-2 are based on the Pythagorean Identities T5-T7. For example, in simplifying the expression $\sqrt{a^2 + x^2}$, where $a > 0$, we recall that $1 + \tan^2\theta = \sec^2\theta$, so that

$$a^2 + a^2\tan^2\theta = a^2\sec^2\theta$$
$$\sqrt{a^2 + a^2\tan^2\theta} = a\,|\sec\theta|.$$

If we now make the substitution $\theta = \tan^{-1}(x/a)$, then $x = a\tan\theta$ and

$$\sqrt{a^2 + x^2} = \sqrt{a^2 + a^2\tan^2\theta} = a\,|\sec\theta| = a\sec\theta.$$

(We can drop the absolute value sign because θ must be a Quadrant IV or a Quadrant I angle.)

Conversion Back to the Original Variable

If a problem involves an indefinite integral, the final answer should be converted back to the original variable. The appropriate triangle in Figure 11–2 can be used to make this change. If the original substitution, for example, is $\theta = \tan^{-1}(x/a)$, then (using Fig. 11–2b), we find that

$$\sin\theta = \frac{x}{\sqrt{a^2+x^2}}, \qquad \cos\theta = \frac{a}{\sqrt{x^2+a^2}}, \qquad \tan\theta = \frac{x}{a}$$

$$\csc\theta = \frac{\sqrt{a^2+x^2}}{x}, \qquad \sec\theta = \frac{\sqrt{x^2+a^2}}{a}, \qquad \cot\theta = \frac{a}{x}.$$

EXAMPLE 2 Calculate $\int \frac{dx}{(9+x^2)^{3/2}}$.

Solution Since the integral involves an expression of form $\sqrt{3^2+x^2}$, we let $\theta = \tan^{-1}(x/3)$. Then $x = 3\tan\theta$, $dx = 3\sec^2\theta\,d\theta$. The integral is

$$\int \frac{dx}{(9+x^2)^{3/2}} = \int \frac{3\sec^2\theta\,d\theta}{(9+9\tan^2\theta)^{3/2}} = \int \frac{3\sec^2\theta\,d\theta}{27(1+\tan^2\theta)^{3/2}}$$

$$= \frac{1}{9}\int \frac{\sec^2\theta\,d\theta}{(\sec^2\theta)^{3/2}} = \frac{1}{9}\int \frac{\sec^2\theta\,d\theta}{\sec^3\theta} = \frac{1}{9}\int \frac{d\theta}{\sec\theta}$$

$$= \frac{1}{9}\int \cos\theta\,d\theta = \frac{1}{9}\sin\theta + C.$$

$\boxed{\begin{array}{l}\theta = \tan^{-1}\dfrac{x}{3}\\ x = 3\tan\theta\\ dx = 3\sec^2\theta\,d\theta\end{array}}$

To express the answer in terms of x, we use Figure 11–2b, obtaining

$$\int \frac{dx}{(9+x^2)^{3/2}} = \frac{1}{9}\sin\theta + C = \frac{x}{9\sqrt{9+x^2}} + C. \quad \square$$

EXAMPLE 3 Calculate $\int \frac{\sqrt{4x^2-9}}{x}\,dx$, where $x \geq \frac{3}{2}$.

Solution To get the expression in the standard form used in Table 11–1, we first make the substitution $u = 2x$, $du = 2\,dx$:

$$\int \frac{\sqrt{4x^2-9}}{x}\,dx = \int \frac{\sqrt{(2x)^2-9}(2\,dx)}{2x} = \int \frac{\sqrt{u^2-9}\,du}{u}.$$

$\boxed{\begin{array}{l}u = 2x\\ du = 2\,dx\end{array}}$

Next, we make the substitution $\theta = \sec^{-1}(u/3)$, $u = 3\sec\theta$, $du = 3\sec\theta\tan\theta\,d\theta$. The integral becomes

$$\int \frac{\sqrt{u^2-9}}{u}\,du = \int \frac{\sqrt{9\sec^2\theta-9}}{3\sec\theta}\cdot 3\sec\theta\tan\theta\,d\theta$$

$\boxed{\begin{array}{l}u = 3\sec\theta\\ du = 3\sec\theta\tan\theta\,d\theta\end{array}}$

$$= \int \frac{3\sqrt{\sec^2\theta-1}\,(3\sec\theta\tan\theta\,d\theta)}{3\sec\theta}$$

$$= 3\int \frac{\sqrt{\tan^2\theta}\,\sec\theta\tan\theta}{\sec\theta}\,d\theta = 3\int \tan^2\theta\,d\theta \qquad (\text{since } x \geq \tfrac{3}{2})$$

$$= 3\int (\sec^2\theta - 1)\,d\theta = 3\tan\theta - 3\theta + C.$$

To express $\tan \theta$ in terms of u, we refer to Figure 11–2c. Then

$$\int \frac{\sqrt{4x^2 - 9}}{x} \, dx = \int \frac{\sqrt{u^2 - 9}}{u} \, du = 3 \tan \theta - 3\theta + C$$

$$= \frac{3\sqrt{u^2 - 9}}{3} + 3 \sec^{-1}\left(\frac{u}{3}\right) + C$$

$$= \sqrt{4x^2 - 9} + 3 \sec^{-1}\left(\frac{2x}{3}\right) + C. \quad \square$$

Remark Do not forget that we have several techniques of integration. When you are faced with a particular integral, make a preliminary classification. *Is there a substitution that would change the integral to a simpler form? Would integration by parts simplify the problem or make it more difficult? Does the integral contain expressions of form $\sqrt{a^2 - x^2}$, $\sqrt{a^2 + x^2}$ or $\sqrt{x^2 - a^2}$?* This classification becomes more important as additional techniques of integration are developed.

Exercises 11–3

1–28. Evaluate the integral.

- **1** $\displaystyle\int \frac{\sqrt{25 - x^2}}{x^2} \, dx$
- **2** $\displaystyle\int \frac{\sqrt{9 - x^2}}{x} \, dx$
- **3** $\displaystyle\int e^x \sqrt{9 - e^{2x}} \, dx$
- **4** $\displaystyle\int \frac{dx}{x\sqrt{9 + (\ln x)^2}}$
- **5** $\displaystyle\int \frac{x^2}{\sqrt{1 - x^2}} \, dx$
- **6** $\displaystyle\int \frac{dx}{x\sqrt{16 - x^2}}$
- **7** $\displaystyle\int \frac{dx}{x^2\sqrt{1 - 9x^2}}$
- **8** $\displaystyle\int \frac{dx}{x\sqrt{25 + x^2}}$
- **9** $\displaystyle\int \frac{dx}{(4 - 9x^2)^{3/2}}$
- **10** $\displaystyle\int \frac{x^2 \, dx}{(36 - x^2)^{3/2}}$
- **11** $\displaystyle\int (4 - x^2)^{3/2} \, dx$
- **12** $\displaystyle\int \frac{x^2 \, dx}{(25 - x^2)^2}$
- **13** $\displaystyle\int \frac{dx}{x(4 + x^2)^{3/2}}$
- **14** $\displaystyle\int \frac{dx}{x(16 + x^2)^2}$
- **15** $\displaystyle\int \frac{x^2 \, dx}{(1 + x^2)^3}$
- **16** $\displaystyle\int x^3(4 + x^2)^{3/2} \, dx$
- **17** $\displaystyle\int x^3\sqrt{9 + x^2} \, dx$
- **18** $\displaystyle\int x^3(4x^2 - 25)^{3/2} \, dx$
- **19** $\displaystyle\int \frac{e^x \, dx}{\sqrt{e^{2x} - 1}}$
- **20** $\displaystyle\int \frac{dx}{e^{2x}\sqrt{4e^{2x} - 1}}$
- **21** $\displaystyle\int x\sqrt{4 + x^4} \, dx$
- **22** $\displaystyle\int \frac{x^5 \, dx}{(16 + x^4)^{3/2}}$

- 23. $\int_0^2 \dfrac{x^2\,dx}{(16 - x^2)^{3/2}}$

- 24. $\int_1^2 \dfrac{x^3}{\sqrt{1 + 4x^2}}\,dx$

- 25. $\int_1^2 \dfrac{dx}{x^2\sqrt{1 + 9x^2}}$

- 26. $\int_0^3 \dfrac{x^2}{9 + x^2}\,dx$

- 27. $\int_5^{16} \dfrac{x^2\,dx}{\sqrt{x^2 + 144}}$

- 28. $\int_3^5 x^3\sqrt{x^2 - 9}\,dx$

29–32. Evaluate the improper integrals provided they converge.

- 29. $\int_0^2 \dfrac{dx}{x^2 - 4}$

- 30. $\int_4^5 \dfrac{x^2\,dx}{\sqrt{x^2 - 16}}$

- 31. $\int_1^\infty \dfrac{dx}{(25x^2 - 9)^{3/2}}$

- 32. $\int_0^\infty \dfrac{dx}{(1 + 16x^2)^2}$

- 33. Find the arc length of the graph of $y = x^2/2$, $0 \le x \le 1$.

- 34. Use trigonometric substitution to calculate the area of the ellipse $x^2/a^2 + y^2/b^2 = 1$.

- 35. Let $\mathcal{R}$ be the region in Quadrant I bounded by the x-axis, the y-axis, the line $x = 1$, and the semicircle $y = \sqrt{2 - x^2}$. Find the area of $\mathcal{R}$.

- 36. The region $\mathcal{R}$ in Exercise 35 is revolved about the x-axis. Find the volume of the resulting solid.

- 37. The base of a solid is the region bounded by the x-axis and the semicircle $y = \sqrt{1 - x^2}$. The cross section through the point $(x, 0)$ perpendicular to the x-axis is a rectangle with height $1/(|x| + 1)$. Calculate the volume.

11–4 Partial Fractions

When we add fractions, we add the numerators over a common denominator. For example,

$$\frac{1}{x} + \frac{2}{x + 1} + \frac{3}{x + 2} = \frac{(x + 1)(x + 2) + 2x(x + 2) + 3x(x + 1)}{x(x + 1)(x + 2)}$$

$$= \frac{6x^2 + 10x + 2}{x^3 + 3x^2 + 2x}.$$

The method of partial fractions enables us to reverse the above process and decompose a fraction into a sum of "partial fractions," such as

$$\frac{6x^2 + 10x + 2}{x^3 + 3x^2 + 2x} = \frac{1}{x} + \frac{2}{x + 1} + \frac{3}{x + 2}.$$

Observe that it is much easier to integrate this last expression than the first one:

$$\int \frac{6x^2 + 10x + 2}{x^3 + 3x^2 + 2x}\,dx = \int \left(\frac{1}{x} + \frac{2}{x + 1} + \frac{3}{x + 2}\right) dx$$

$$= \ln |x| + 2 \ln |x + 1| + 3 \ln |x + 2| + C.$$

11-4 Partial Fractions

In our development, we work with rational functions (quotients of polynomials) in which the numerator has lower degree than the denominator. If this condition is not met, we first divide the denominator into the numerator by long division and reduce the problem to an equivalent one in which the numerator has lower degree than the denominator.

For example, if our original function is

$$\frac{2x^3 + x^2 - x + 4}{x^2 - 1},$$

we divide $x^2 - 1$ into $2x^3 + x^2 - x + 4$, obtaining

$$\frac{2x^3 + x^2 - x + 4}{x^2 - 1} = 2x + 1 + \frac{x + 5}{x^2 - 1}.$$

To integrate the original expression, we have only to decompose $(x + 5)/(x^2 - 1)$ into partial fractions.

In the following example, we show that

$$\frac{x + 5}{x^2 - 1} = \frac{3}{x - 1} - \frac{2}{x + 1}.$$

It follows that

$$\int \frac{2x^3 + x^2 - x + 4}{x^2 - 1} \, dx = \int \left(2x + 1 + \frac{x + 5}{x^2 - 1} \right) dx$$

$$= \int \left(2x + 1 + \frac{3}{x - 1} - \frac{2}{x + 1} \right) dx$$

$$= x^2 + x + 3 \ln |x - 1| - 2 \ln |x + 1| + C.$$

Example 1 shows the basic procedure for decomposing a fraction into partial fractions.

EXAMPLE 1   Decompose $\frac{x + 5}{x^2 - 1}$ into partial fractions.

Solution

$$\frac{x + 5}{x^2 - 1} = \frac{x + 5}{(x - 1)(x + 1)}.$$

Assume that the fraction can be decomposed into a sum of two fractions of the type

$$\frac{A}{x - 1} + \frac{B}{x + 1}$$

where A and B are constants. Adding these fractions over a common denominator, we obtain

$$\frac{x + 5}{(x - 1)(x + 1)} = \frac{A}{x - 1} + \frac{B}{x + 1}$$

$$= \frac{A(x + 1) + B(x - 1)}{(x - 1)(x + 1)}$$

$$= \frac{(A + B)x + (A - B)}{(x - 1)(x + 1)}.$$

Since the two equal fractions have the same denominator, their numerators must also be equal. Equating the numerators, we get

$$x + 5 = (A + B)x + (A - B).$$

Since the polynomial on the left is identically equal to the one on the right, the corresponding coefficients must be equal. Therefore,

$$A + B = 1 \quad \text{(coefficient of } x\text{)}$$

and

$$A - B = 5 \quad \text{(constant term)}.$$

It follows that A, B is the solution of the system of equations

$$\begin{cases} A + B = 1 \\ A - B = 5, \end{cases}$$

so that $A = 3$ and $B = -2$.

The partial fraction decomposition is

$$\frac{x + 5}{(x - 1)(x + 1)} = \frac{A}{x - 1} + \frac{B}{x + 1} = \frac{3}{x - 1} - \frac{2}{x + 1}. \quad \square$$

IRREDUCIBLE POLYNOMIALS

Before we consider the general partial fraction decomposition of a rational function we must develop a few background topics.

Reducible and Irreducible Polynomials

A real polynomial $G(x)$ of positive degree is said to be *reducible* if it is the product of two or more real polynomials, each of positive degree. It is *irreducible* otherwise.

For example, the polynomial $2x^2 - 1$ is reducible. It can be written

$$2x^2 - 1 = (\sqrt{2}x - 1)(\sqrt{2}x + 1).$$

The polynomial $2x^2 + 1$, on the other hand, is irreducible. It cannot be factored into first-degree polynomials with real coefficients.

It is proved in algebra that all irreducible real polynomials are of degree 1 or 2. If a real polynomial has degree greater than 2, then it has irreducible real factors of degree 1 or 2. It follows that every real polynomial $G(x)$ of degree greater than 2 can (theoretically, at least) be written as a product of irreducible linear and quadratic real polynomials. If we collect like terms, we write

$$G(x) = [g_1(x)]^{n_1}[g_2(x)]^{n_2} \cdots [g_k(x)]^{n_k}$$

where $g_1(x), g_2(x), \ldots, g_k(x)$ are distinct irreducible polynomials.*

---

* By distinct, we mean that no two of these polynomials are equivalent.

11-4 Partial Fractions

GENERAL PROCEDURE FOR PARTIAL FRACTIONS

Let $F(x)$ and $G(x)$ be real polynomials with $F(x)$ of lower degree than $G(x)$. Factor $G(x)$ into a product of powers of irreducible linear and quadratic polynomials:

$$G(x) = [g_1(x)]^{n_1}[g_2(x)]^{n_2} \cdots [g_k(x)]^{n_k}.$$

We consider the partial fraction decomposition of $F(x)/G(x)$.

Family of Terms   It is proved in algebra that each power of an irreducible polynomial in the denominator defines a *family of terms* in the partial fraction decomposition. Each term in a given family has the same type of numerator (constant or linear) and a denominator that is a power of the original irreducible factor.

Linear Factors   A term $(ax + b)^n$ in the denominator gives rise to a family of n terms, each with a constant numerator

Decomposition for Linear Factors

$$\frac{A_1}{ax + b} + \frac{A_2}{(ax + b)^2} + \cdots + \frac{A_n}{(ax + b)^n}$$

Each numerator has lower degree than the original linear factor $ax + b$. Observe that the first n powers of $ax + b$ are represented in the decomposition.

For example, we can write

$$\frac{3x^2 - 2x + 1}{(2x + 1)^4} = \frac{A_1}{2x + 1} + \frac{A_2}{(2x + 1)^2} + \frac{A_3}{(2x + 1)^3} + \frac{A_4}{(2x + 1)^4}$$

If we solve the resulting system of equations we find that

$$A_1 = 0, \; A_2 = \frac{3}{4}, \; A_3 = -\frac{5}{2}, \; A_4 = \frac{11}{4}$$

Quadratic Factors   A term $(ax^2 + bx + c)^n$ in the denominator gives rise to a family of n terms, each with a linear polynomial for its numerator

Decomposition for Quadratic Factors

$$\frac{A_1 x + B_1}{ax^2 + bx + c} + \frac{A_2 x + B_2}{(ax^2 + bx + c)^2} + \cdots + \frac{A_n x + B_n}{(ax^2 + bx + c)^n}$$

Again, each numerator is of lower degree then the original irreducible factor $ax^2 + bx + c$ and the first n powers of $ax^2 + bx + c$ are represented.

For example, the fraction $(5x^4 + 7x + 1)/(2x^2 + 3)^3$ can be decomposed into partial fractions of form

$$\frac{5x^4 + 7x + 1}{(2x^2 + 3)^3} = \frac{A_1 x + B_1}{2x^2 + 3} + \frac{A_2 x + B_2}{(2x^2 + 3)^2} + \frac{A_3 x + B_3}{(2x^2 + 3)^3}$$

where $A_1, A_2, \ldots, B_3$ are constants. The constants can be evaluated by solving the resulting system of six equations.

In most cases we must work with both types of partial fractions. For example, the fraction

$$\frac{3x^2 - 4x + 7}{(x + 1)^3(x^2 + 1)^2(x - 7)}$$

can be decomposed into partial fractions of form

$$\frac{A_1}{x+1} + \frac{A_2}{(x+1)^2} + \frac{A_3}{(x+1)^3} + \frac{B_1 x + C_1}{x^2+1} + \frac{B_2 x + C_2}{(x^2+1)^2} + \frac{A_4}{x-7}.$$

Once we know the form of the partial fraction decomposition, we can calculate the coefficients $A_1, A_2, A_3, \ldots$ by the procedures illustrated in the following examples.

EXAMPLE 2 Calculate $\displaystyle\int \frac{3x^2 + 2x + 3}{(x^2+1)^2} dx$.

Solution We first decompose the fraction into partial fractions—

$$\frac{3x^2 + 2x + 3}{(x^2+1)^2} = \frac{Ax + B}{x^2+1} + \frac{Cx + D}{(x^2+1)^2}.$$

If we add the fractions on the right and collect like terms, we get

$$\frac{3x^2 + 2x + 3}{(x^2+1)^2} = \frac{(Ax + B)(x^2+1) + Cx + D}{(x^2+1)^2}$$

$$= \frac{Ax^3 + Bx^2 + (A+C)x + (B+D)}{(x^2+1)^2}.$$

Equating the numerators, we get

$$3x^2 + 2x + 3 = Ax^3 + Bx^2 + (A+C)x + (B+D).$$

Equating the coefficients on the left- and right-hand sides of this equation, we get

$$\begin{cases} A = 0 & \text{(coefficient of } x^3) \\ B = 3 & \text{(coefficient of } x^2) \\ A + C = 2 & \text{(coefficient of } x) \\ B + D = 3 & \text{(constant term)} \end{cases}$$

This system of equations has the solution $A = 0, B = 3, C = 2$, and $D = 0$. The partial fraction decomposition is

$$\frac{3x^2 + 2x + 3}{(x^2+1)^2} = \frac{3}{x^2+1} + \frac{2x}{(x^2+1)^2}.$$

The integral is

$$\int \frac{3x^2 + 2x + 3}{(x^2+1)^2} dx = \int \frac{3\, dx}{x^2+1} + \int \frac{2x\, dx}{(x^2+1)^2}$$

$$= 3\int \frac{dx}{x^2+1} + \int (x^2+1)^{-2}(2x\, dx)$$

$$= 3\tan^{-1} x + \frac{(x^2+1)^{-1}}{-1} + C$$

$$= 3\tan^{-1} x - \frac{1}{x^2+1} + C. \quad \square$$

The technique for calculating the coefficients $A_1, A_2, \ldots$ illustrated in Examples 1 and 2 has one main drawback. It may produce a very large system of equations in

11-4 Partial Fractions

many unknowns. An alternative method that can be used to calculate some of these numbers—often with less labor—is shown in Example 3.

EXAMPLE 3 Calculate $\int \dfrac{11x^2 - 3x}{(x^2 - 1)(x + 3)}\, dx$.

Solution We first decompose the fraction into partial fractions and add the fractions over a common denominator. Since $x^2 - 1 = (x - 1)(x + 1)$, then

$$\frac{11x^2 - 3x}{(x^2 - 1)(x + 3)} = \frac{A}{x - 1} + \frac{B}{x + 1} + \frac{C}{x + 3}$$

$$= \frac{A(x + 1)(x + 3) + B(x - 1)(x + 3) + C(x - 1)(x + 1)}{(x - 1)(x + 1)(x + 3)}.$$

Equating the numerators, we get

$$11x^2 - 3x = A(x + 1)(x + 3) + B(x - 1)(x + 3) + C(x - 1)(x + 1).$$

Observe that this equation is an identity. It must hold for all values of x, including values for which the original fraction is not defined. We shall substitute certain numbers for x and work with the resulting equations. Specifically, we shall substitute the values $x = 1$, $x = -1$, and $x = -3$. These numbers make most of the products on the right-hand side reduce to zero, producing very simple equations.

$x = 1$:
$$11 \cdot 1^2 - 3 \cdot 1 = A(2)(4) + B \cdot 0 + C \cdot 0$$
$$11 - 3 = 8A.$$
$$A = 1$$

$x = -1$:
$$11(-1)^2 - 3(-1) = A \cdot 0 + B(-2)(2) + C \cdot 0$$
$$11 + 3 = -4B$$
$$B = -\frac{14}{4} = -\frac{7}{2}.$$

$x = -3$:
$$11(-3)^2 - 3(-3) = A \cdot 0 + B \cdot 0 + C(-4)(-2)$$
$$99 + 9 = 8C$$
$$108 = 8C$$
$$C = \frac{27}{2}.$$

The partial fraction decomposition is

$$\frac{11x^2 - 3x}{(x^2 - 1)(x + 3)} = \frac{1}{x - 1} - \frac{7/2}{x + 1} + \frac{27/2}{x + 3}.$$

Therefore,

$$\int \frac{11x^2 - 3x}{(x^2 - 1)(x + 3)}\, dx = \int \frac{1}{x - 1}\, dx - \int \frac{7/2}{x + 1}\, dx + \int \frac{27/2}{x + 3}\, dx$$

$$= \ln|x - 1| - \frac{7}{2}\ln|x + 1| + \frac{27}{2}\ln|x + 3| + C. \square$$

Exercises 11-4

1-18. Use partial fractions to evaluate the integrals.

- 1. $\int \dfrac{x-7}{x^2+x-2}\,dx$
- 2. $\int \dfrac{dx}{x^2-x}$
- 3. $\int \dfrac{2}{x^2-9}\,dx$
- 4. $\int \dfrac{2x-11}{2x^2-7x+3}\,dx$
- 5. $\int \dfrac{2x^3-x^2-14x+26}{x^2+x-6}\,dx$
- 6. $\int \dfrac{x^4-2x^2+4}{x^3-4x}\,dx$
- 7. $\int \dfrac{x^3+3x^2-2}{x^2-x}\,dx$
- 8. $\int \dfrac{4x^2+5x-3}{x^3-x}\,dx$
- 9. $\int \dfrac{x^2-2x+1}{x^3+x}\,dx$
- 10. $\int \dfrac{4x^2-3x+8}{(x+2)(x^2+1)}\,dx$
- 11. $\int \dfrac{4x^2+3x+4}{(x^2+1)^2}\,dx$
- 12. $\int \dfrac{2x^5+5x^3-3x^2-8x+12}{x^4+4x^2}\,dx$
- 13. $\int \dfrac{4x^2-18x-16}{x^3+2x^2-8x}\,dx$
- 14. $\int \dfrac{11x^2+20}{x^4-16}\,dx$
- 15. $\int \dfrac{x^3-2x^2+4x-2}{(x^2+4)(x^2+1)}\,dx$
- 16. $\int \dfrac{x^2+x+9}{(2x+1)(x^2+1)}\,dx$
- 17. $\int \dfrac{5x^3-4x^2+2x-3}{x^4+x^2}\,dx$
- 18. $\int \dfrac{x^6+2x^4-5x^2-8}{x^4+4x^2}\,dx.$
- 19. Find the arc length of the graph of $y = \ln x$, $1/e \leq x \leq e$.

11-5 Quadratic Expressions

A real quadratic polynomial $ax^2 + bx + c$ is irreducible if it cannot be factored into two real linear factors. It follows from the Factor Theorem of algebra that a quadratic polynomial is irreducible if and only if it does not have real zeros.* The zeros can be found by using the quadratic formula.

EXAMPLE 1 (a) The zeros of the quadratic polynomial

$$f(x) = 2x^2 + 5x + 7$$

are

$$x = \frac{-5 \pm \sqrt{5^2 - 4 \cdot 2 \cdot 7}}{4} = \frac{-5 \pm i\sqrt{31}}{4}.$$

---

* The Factor Theorem states that the number r is a zero of the polynomial $f(x)$ of degree k greater than 0 if and only if $f(x)$ can be factored as $f(x) = (x - r)g(x)$ where $g(x)$ is a polynomial of degree $k - 1$.

11-5 Quadratic Expressions

Since the zeros are not real, the polynomial is irreducible.

(b) The zeros of the quadratic polynomial

$$g(x) = 2x^2 - 3x - 6$$

are

$$x = \frac{3 \pm \sqrt{3^2 - 4 \cdot 2 \cdot (-6)}}{4} = \frac{3 \pm \sqrt{57}}{4}.$$

Since the zeros are real, the polynomial is reducible. It can be factored as

$$2x^2 - 3x - 6 = 2\left(x - \frac{3 + \sqrt{57}}{4}\right)\left(x - \frac{3 - \sqrt{57}}{4}\right)$$

$$= 2\left(\frac{4x - 3 - \sqrt{57}}{4}\right)\left(\frac{4x - 3 + \sqrt{57}}{4}\right). \square$$

We frequently obtain an irreducible quadratic polynomial as the denominator of a fraction when we use partial fractions. It is usually best to rewrite such polynomials in the form $u^2 + a^2$. This can be accomplished by first completing the square and then making an appropriate substitution. The technique is illustrated in Example 2.

EXAMPLE 2 Calculate $\int \frac{x + 5}{x^2 + 4x + 12} dx$.

Solution We first complete the square and write $x^2 + 4x + 12$ in the form $u^2 + a^2$:

$$x^2 + 4x + 12 = (x^2 + 4x + 4) + 8$$
$$= (x + 2)^2 + (\sqrt{8})^2$$
$$= u^2 + (\sqrt{8})^2 \quad \text{where } u = x + 2.$$

We make the substitution $u = x + 2$ in the original integral and break the integral into two integrals:

$$\int \frac{x + 5}{x^2 + 4x + 12} dx = \int \frac{u + 3}{u^2 + (\sqrt{8})^2} du \qquad \boxed{\begin{array}{c} u = x + 2 \\ du = dx \end{array}}$$

$$= \int \frac{u \, du}{u^2 + (\sqrt{8})^2} + 3 \int \frac{du}{u^2 + (\sqrt{8})^2}$$

$$= \frac{1}{2} \int \frac{2u \, du}{u^2 + 8} + 3 \int \frac{du}{u^2 + (\sqrt{8})^2}$$

$$= \frac{1}{2} \ln (u^2 + 8) + \frac{3}{\sqrt{8}} \tan^{-1} \left(\frac{u}{\sqrt{8}}\right) + C$$

$$= \frac{1}{2} \ln (x^2 + 4x + 12) + \frac{3}{\sqrt{8}} \tan^{-1} \left(\frac{x + 2}{\sqrt{8}}\right) + C. \square$$

This method can also be used when quadratic expressions (either reducible or irreducible) occur under radical signs. If the quadratic is irreducible, it is changed to the form $u^2 + a^2$ by completing the square. If it is reducible, it is changed to the form $u^2 - a^2$.

EXAMPLE 3 Calculate $\displaystyle\int \frac{x\,dx}{\sqrt{x^2 - 2x - 3}}$.

Solution We complete the square on the quadratic polynomial in the radical:

$$\int \frac{x\,dx}{\sqrt{x^2 - 2x - 3}} = \int \frac{x\,dx}{\sqrt{(x^2 - 2x + 1) - 4}} = \int \frac{x\,dx}{\sqrt{(x - 1)^2 - 2^2}}.$$

We now make the substitution $u = x - 1$, getting

$$\int \frac{x\,dx}{\sqrt{x^2 - 2x - 3}} = \int \frac{(u + 1)\,du}{\sqrt{u^2 - 2^2}} = \int \frac{u\,du}{\sqrt{u^2 - 2^2}} + \int \frac{du}{\sqrt{u^2 - 2^2}}.$$

The first integral can be calculated by making the substitution $v = u^2 - 2^2$, $dv = 2u\,du$. The second can be calculated by using the trigonometric substitution $\theta = \sec^{-1}(u/2)$, $u = 2\sec\theta$, $du = 2\sec\theta\tan\theta\,d\theta$.

If we make the substitutions indicated, the original integral becomes

$$\int \frac{x\,dx}{\sqrt{x^2 - 2x - 3}} = \frac{1}{2}\int \frac{2u\,du}{\sqrt{u^2 - 2^2}} + \int \frac{du}{\sqrt{u^2 - 2^2}}$$

$$= \frac{1}{2}\int \frac{dv}{v^{1/2}} + \int \sec\theta\,d\theta$$

$$= v^{1/2} + \ln|\sec\theta + \tan\theta| + C_1$$

$$= \sqrt{u^2 - 4} + \ln\left|\frac{u}{2} + \frac{\sqrt{u^2 - 4}}{2}\right| + C_1$$

$$= \sqrt{u^2 - 4} + \ln|u + \sqrt{u^2 - 4}| + C \quad \text{(where } C = C_1 - \ln|2|\text{)}$$

$$= \sqrt{x^2 - 2x - 3} + \ln|x - 1 + \sqrt{x^2 - 2x - 3}| + C. \quad \square$$

$v = u^2 - 2^2$
$dv = 2u\,du$

$\theta = \sec^{-1}\dfrac{u}{2}$

$u = 2\sec\theta$
$du = 2\sec\theta\tan\theta\,d\theta$

Remark Many powers of quadratic expressions can be integrated by a procedure like the one shown in Example 3. If the quadratic is irreducible, we write it in the form $u^2 + a^2$. Observe that integrals of form

$$\int \frac{u\,du}{(u^2 + a^2)^n}$$

can be integrated after making the substitution $v = u^2 + a^2$, $dv = 2u\,du$, while a trigonometric substitution is natural for integrals of form

$$\int \frac{du}{(u^2 + a^2)^n}.$$

Exercises 11–5

1–16. Calculate the integrals.

- 1 $\displaystyle\int \frac{2x\,dx}{x^2 - 2x + 5}$
- 2 $\displaystyle\int \frac{dx}{x^2 + 6x + 13}$
- 3 $\displaystyle\int \frac{dx}{\sqrt{7 - 4x - 2x^2}}$
- 4 $\displaystyle\int \frac{(2x + 3)\,dx}{\sqrt{21 - 4x - x^2}}$

- 5. $\int \sqrt{5 - 4x^2 - 8x}\, dx$
- 6. $\int \dfrac{x\, dx}{\sqrt{16 - 18x - 9x^2}}$
- 7. $\int \dfrac{x\, dx}{4x^2 - 8x + 13}$
- 8. $\int \dfrac{dx}{4x^2 - 24x + 27}$
- 9. $\int \dfrac{dx}{(5 + 4x - x^2)^{3/2}}$
- 10. $\int \dfrac{x\, dx}{(7 - 18x - 9x^2)^{3/2}}$
- 11. $\int (x + 3)^3 \sqrt{x^2 + 6x + 8}\, dx$
- 12. $\int \dfrac{dx}{(x^2 - 2x + 5)^2}$
- 13. $\int \dfrac{3x\, dx}{\sqrt{x^2 - 4x + 13}}$
- 14. $\int \dfrac{dx}{4x^2 - 12x + 5}$
- 15. $\int \dfrac{x\, dx}{x^4 - 6x^2 + 25}$
- 16. $\int \dfrac{(x + 3)\, dx}{x^2 + 4x - 5}$

17. Use the Factor Theorem to show that a real quadratic polynomial $f(x)$ is reducible if and only if it has real zeros (which may be equal).

11–6 Miscellaneous Substitutions

In this section we discuss two special types of substitution that simplify some integrals when other techniques fail.

Substitutions for Troublesome Terms

Many integrals contain troublesome factors, such as $\sqrt{2x + 3}$, which are multiplied by other terms. Such integrals frequently can be simplified by making substitutions that eliminate these factors. If we have a factor such as $\sqrt{2x + 3}$, we can let $u = \sqrt{2x + 3}$ and change the integral to a new one expressed in terms of the variable u.

EXAMPLE 1 Calculate $\int x\sqrt{2x + 3}\, dx$.

Solution The troublesome term is $\sqrt{2x + 3}$. Let $u = \sqrt{2x + 3}$. Then

$$x = \frac{u^2 - 3}{2} \quad \text{and} \quad dx = u\, du.$$

On substituting these values in the integral, we get

$$\int x\sqrt{2x + 3}\, dx = \int \frac{u^2 - 3}{2} \cdot u \cdot u\, du = \frac{1}{2}\int (u^4 - 3u^2)\, du$$

$$= \frac{u^5}{10} - \frac{u^3}{2} + C = \frac{(2x + 3)^{5/2}}{10} - \frac{(2x + 3)^{3/2}}{2} + C. \quad \square$$

EXAMPLE 2 Calculate $\int x^5 \sqrt[3]{x^2 + 5}\, dx$.

Solution The troublesome term is $\sqrt[3]{x^2 + 5}$. We let $u = \sqrt[3]{x^2 + 5}$. Then

$$x^2 = u^3 - 5 \quad \text{and} \quad 2x\, dx = 3u^2\, du.$$

On substituting these values into the integral, we get

$$\int x^5 \sqrt[3]{x^2 + 5}\, dx = \frac{1}{2}\int (x^2)^2 \sqrt[3]{x^2+5} \cdot 2x\, dx \qquad \boxed{\begin{array}{l} u = \sqrt[3]{x^2+5} \\ 3u^2\, du = 2x\, dx \end{array}}$$

$$= \frac{1}{2}\int (u^3 - 5)^2 \cdot u \cdot 3u^2\, du$$

$$= \frac{3}{2}\int (u^6 - 10u^3 + 25)u^3\, du$$

$$= \frac{3}{2}\int (u^9 - 10u^6 + 25u^3)\, du$$

$$= \frac{3}{2}\left[\frac{u^{10}}{10} - \frac{10u^7}{7} + \frac{25u^4}{4}\right] + C$$

$$= \frac{3}{2}\left[\frac{(x^2+5)^{10/3}}{10} - \frac{10(x^2+5)^{7/3}}{7} + \frac{25(x^2+5)^{4/3}}{4}\right] + C. \quad \square$$

EXAMPLE 3 Calculate $\int \sin(\sqrt{x})\, dx$.

Solution The troublesome term is $\sqrt{x}$. Make the substitution $w = \sqrt{x}$, $x = w^2$, $dx = 2w\, dw$. Then

$$\int \sin(\sqrt{x})\, dx = \int \sin w\, (2w\, dw). \qquad \boxed{\begin{array}{l} w = \sqrt{x},\ x = w^2 \\ dx = 2w\, dw \end{array}}$$

We now apply integration by parts, getting

$\boxed{\begin{array}{ll} u = w & dv = \sin w\, dw \\ du = dw & v = -\cos w \end{array}}$

$$2\left[-w\cos w + \int \cos w\, dw\right]$$

$$= -2w \cos w + 2 \sin w + C$$

$$= -2\sqrt{x} \cos(\sqrt{x}) + 2 \sin(\sqrt{x}) + C. \quad \square$$

Remark The substitution method illustrated above works well in some cases but not others. Example 2 is a good illustration. If we try the same method on the similar integral

$$\int x^4 \sqrt[3]{x^2 + 5}\, dx$$

we get

$$\frac{3}{2}\int (u^3 - 5)^{3/2}\, u^3\, du,$$

which is about as bad as the original integral. Thus, the method works for the integral $\int x^5 \sqrt[3]{x^2+5}\, dx$ but not the integral $\int x^4 \sqrt[3]{x^2+5}\, dx$.

RATIONAL FUNCTIONS OF sin x AND cos x

Rational functions of $\sin x$ and $\cos x$ can be transformed to rational functions of u by the substitution $x = 2\tan^{-1} u$, $u = \tan(x/2)$. To see why this substitution works, ob-

11–6 Miscellaneous Substitutions

serve first, that

$$\cos\left(\frac{x}{2}\right) = \frac{1}{\sec\left(\frac{x}{2}\right)} = \frac{1}{\sqrt{1 + \tan^2\left(\frac{x}{2}\right)}} = \frac{1}{\sqrt{1 + u^2}}$$

and

$$\sin\left(\frac{x}{2}\right) = \tan\left(\frac{x}{2}\right)\cos\left(\frac{x}{2}\right) = \frac{u}{\sqrt{1 + u^2}}.$$

Then

$$\sin x = 2\sin\left(\frac{x}{2}\right)\cos\left(\frac{x}{2}\right) = 2\frac{u}{\sqrt{1 + u^2}} \cdot \frac{1}{\sqrt{1 + u^2}} = \frac{2u}{1 + u^2}$$

and

$$\cos x = \cos^2\left(\frac{x}{2}\right) - \sin^2\left(\frac{x}{2}\right) = \frac{1}{1 + u^2} - \frac{u^2}{1 + u^2} = \frac{1 - u^2}{1 + u^2}.$$

Furthermore, since $x = 2\tan^{-1} u$, then

$$dx = \frac{2}{1 + u^2}\,du.$$

If we make these substitutions, the integral of a rational function of $\sin x$ and $\cos x$ will be transformed to the integral of a rational function of u.

We repeat the substitutions for reference:

Substitutions for Rational Functions of Sine and Cosine

$$u = \tan\left(\frac{x}{2}\right) \qquad \cos x = \frac{1 - u^2}{1 + u^2}$$

$$\sin x = \frac{2u}{1 + u^2} \qquad dx = \frac{2}{1 + u^2}\,du$$

EXAMPLE 4 Calculate $\int \frac{\cos x}{\sin x + \cos x}\,dx$.

Solution We make the substitution $u = \tan(x/2)$. The integral is transformed to

$$\int \frac{\cos x}{\sin x + \cos x}\,dx = \int \frac{\frac{1 - u^2}{1 + u^2}}{\frac{2u}{1 + u^2} + \frac{1 - u^2}{1 + u^2}} \cdot \frac{2}{1 + u^2}\,du$$

$$= 2\int \frac{1 - u^2}{(1 + u^2)(1 + 2u - u^2)}\,du.$$

If we use partial fractions we can write this integral as

$$2\int \left[\frac{\frac{1}{2}(1 - u)}{1 + u^2} + \frac{\frac{1}{2}(1 - u)}{1 + 2u - u^2}\right]du = \int \frac{1 - u}{1 + u^2}\,du + \int \frac{1 - u}{1 + 2u - u^2}\,du.$$

The first of these integrals is

$$\int \frac{1-u}{1+u^2} du = \int \frac{du}{1+u^2} - \frac{1}{2}\int \frac{2u}{1+u^2} du$$

$$= \tan^{-1} u - \frac{1}{2} \ln(1+u^2) + C_1.$$

The second can be integrated by the substitution $v = 1 + 2u - u^2$, $dv = (2 - 2u) du$:

$$\int \frac{1-u}{1+2u-u^2} du = \frac{1}{2}\int \frac{2-2u}{1+2u-u^2} du = \frac{1}{2}\int \frac{dv}{v}$$

$$= \frac{1}{2} \ln|1 + 2u - u^2| + C_2.$$

Thus,

$$\int \frac{\cos x}{\sin x + \cos x} dx = \tan^{-1} u - \frac{1}{2}\ln(1+u^2) + \frac{1}{2}\ln|1+2u-u^2| + C$$

$$= \frac{x}{2} - \frac{1}{2}\ln\left(1 + \tan^2\left(\frac{x}{2}\right)\right)$$

$$+ \frac{1}{2}\ln\left|1 + 2\tan\left(\frac{x}{2}\right) - \tan^2\left(\frac{x}{2}\right)\right| + C,$$

where $C = C_1 + C_2$. □

Exercises 11-6

1-18. Evaluate the integral. (*Hint for 9:* Let $x = u^6$.)

- 1 $\int \dfrac{dx}{x + \sqrt{x}}$
- 2 $\int \dfrac{dx}{1 + e^x}$
- 3 $\int \dfrac{dx}{x + 1 - \sqrt{x+1}}$
- 4 $\int \dfrac{(x-3)\,dx}{(2x-5)^{3/2}}$
- 5 $\int (x+2)\sqrt{3-x}\,dx$
- 6 $\int \dfrac{3x+1}{\sqrt{x+1}}\,dx$
- 7 $\int \dfrac{x^5\,dx}{\sqrt{x^3+2}}$
- 8 $\int x\sqrt[3]{1-x}\,dx$
- 9 $\int \dfrac{dx}{\sqrt{x} + \sqrt[3]{x}}$
- 10 $\int \dfrac{\sqrt{x}}{1 + \sqrt[4]{x}}\,dx$
- 11 $\int e^{\sqrt{x}}\,dx$
- 12 $\int \cos(\sqrt[3]{x})\,dx$
- 13 $\int \dfrac{dx}{1 + \cos x}$
- 14 $\int \dfrac{dx}{1 + \sin x}$

- 15 $\int \dfrac{\cos x \, dx}{1 + \cos x}$

- 16 $\int \sqrt{\dfrac{1 + \sin x}{1 + \cos x}} \, dx$

- 17 $\int \dfrac{dx}{\sin x + 2 \cos x - 1}$

- 18 $\int \dfrac{dx}{1 + \cos x - \sin x}$

19 What happens if we change the integral $\int \sqrt{\sin^3 x + 1} \, dx$ by a substitution that eliminates the "troublesome" term $\sin^3 x + 1$? Is it any easier to evaluate the resulting integral?

11–7 Integral Tables

As you will discover, most advanced mathematics students use tables of integrals. The standard mathematics reference books all contain tables of moderate length. For example, the Rinehart tables* list 494 integrals, while the CRC tables list 463.** While these are adequate for most elementary work, more extensive (and expensive) tables listing thousands of integrals are available for the specialist.

The usual table has integrals classified according to type—exponential forms, logarithmic forms, forms involving $\sqrt{a^2 - x^2}$, forms involving $\sqrt{ax + b}$, and so on. To avoid trivial duplication, only certain basic forms are given. Related integrals must be reduced to these forms by appropriate substitutions or other methods.

Table V at the end of this book contains a list of integrals selected from the Rinehart tables. Observe that the constant of integration is omitted. Natural logarithms are denoted by "ln" as in this book, rather than by "log" as is customary in most advanced books and tables.

The following examples illustrate the use of the tables. The reference numbers refer to Table V.

EXAMPLE 1 Evaluate $\int \sin^3 \left(2x - \dfrac{\pi}{2}\right) dx$.

Solution Formula I98 in Table V gives the "best fit" to our problem. To use it, we make the substitution $u = 2x - \pi/2$, $du = 2 \, dx$. Then

$$\int \sin^3 \left(2x - \dfrac{\pi}{2}\right) dx = \dfrac{1}{2} \int \sin^3 \left(2x - \dfrac{\pi}{2}\right) \cdot 2 \, dx \qquad \boxed{\begin{array}{l} u = 2x - \dfrac{\pi}{2} \\ du = 2 \, dx \end{array}}$$

$$= \dfrac{1}{2} \int \sin^3 u \, du = \dfrac{1}{2} \left[\dfrac{\cos^3 u}{3} - \cos u\right] + C \qquad \text{(by I98)}$$

$$= \dfrac{\cos^3 (2x - \pi/2)}{6} - \dfrac{\cos (2x - \pi/2)}{2} + C. \quad \square$$

* H. D. Larsen, *Rinehart Mathematical Tables, Formulas and Curves.* New York: Holt, Rinehart and Winston, 1953.

** C. D. Hodgman, *CRC Standard Mathematical Tables,* 12th ed. Cleveland: Chemical Rubber Publishing Company.

EXAMPLE 2   Evaluate $\int x^3 \sqrt{2x + 3}\, dx$.

Solution   We first use formula I46, then I45:

$$\int x^3 \sqrt{2x + 3}\, dx$$

$$= \frac{2}{2(2 \cdot 3 + 3)} \left[x^3 \sqrt{(2x + 3)^3} - 3 \cdot 3 \int x^2 \sqrt{2x + 3}\, dx \right] \quad \text{(by I46)}$$

$$= \frac{1}{9} x^3 \sqrt{(2x + 3)^3} - \frac{2(15 \cdot 2^2 \cdot x^2 - 12 \cdot 2 \cdot 3x + 8 \cdot 3^2)}{105 \cdot 2^3} \sqrt{(2x + 3)^3} + C \quad \text{(by I45)}$$

$$= \frac{1}{9} x^3 \sqrt{(2x + 3)^3} - \frac{5x^2 - 6x + 6}{35} \sqrt{(2x + 3)^3} + C$$

$$= \frac{35x^3 - 45x^2 + 54x - 54}{315} \sqrt{(2x + 3)^3} + C. \quad \square$$

NUMERICAL INTEGRATION

The functions $\sqrt[3]{\ln x}$, e^{x^2}, and $5^{\sin x}$ cannot be integrated by any of the standard techniques. These functions have antiderivatives that cannot be expressed in terms of the "elementary" functions studied in the calculus.

Unfortunately, many applied problems require us to integrate functions such as e^{x^2} or $\sqrt[3]{\ln x}$. As a rule, this can be accomplished by one of the approximation techniques such as Simpson's rule or the Trapezoidal Rule. In Chapter 12, we shall use infinite series to obtain another method.

What if a physics or engineering problem requires us to integrate e^{x^2} for x between 0 and 10? There are two ways by which we might proceed. First, we can define

$$F(x) = \int_0^x e^{t^2}\, dt$$

and choose the values of x for which $F(x)$ is to be calculated. For example, we might decide to calculate $F(x)$ for $x = 0, 0.01, 0.02, \ldots, 9.98, 9.99$, and 10.00. It is simple to program the electronic computer to use Simpson's rule to calculate the approximate values of F for these numbers and to print out the functional values in tabular form. We can then refer to the table whenever we need one of these values.

A more sophisticated method is to set up a program for solving the main problem (the one in which the solution involves $\int e^{x^2}\, dx$) and to have a subroutine for calculating $F(x)$ whenever it is needed. These values can then be used to calculate the complete solution of the main problem.

Exercises 11-7

1-40.   Use Table V to evaluate the following integrals, if possible. If a definite integral cannot be evaluated by Table V, then use Simpson's rule to approximate its

Exercises 11–7

value. (*Note:* It may be necessary to simplify an integral by a substitution before using the table.)

1. $\displaystyle\int \frac{e^x}{e^x - 1} \, dx$

2. $\displaystyle\int \frac{x-1}{(x^2 - 2x + 5)^2} \, dx$

3. $\displaystyle\int x \ln(x^2 + 1) \, dx$

4. $\displaystyle\int \frac{x \, dx}{\sqrt{4 - 9x^2}}$

5. $\displaystyle\int \frac{dx}{x \cot(\ln x)}$

6. $\displaystyle\int \frac{dx}{x[4 + (\ln x)^2]}$

7. $\displaystyle\int \frac{e^{\sin x} \, dx}{\sec x}$

8. $\displaystyle\int e^x \sec(e^x - 4) \, dx$

9. $\displaystyle\int \frac{\cos x \, dx}{\sqrt{1 - \sin^2 x}}$

10. $\displaystyle\int \frac{x \, dx}{\cos(x^2 - 1)}$

11. $\displaystyle\int \frac{e^x \, dx}{\sqrt{1 - e^{2x}}}$

12. $\displaystyle\int \frac{e^{2x} \, dx}{\sqrt{e^{2x} - 1}}$

13. $\displaystyle\int \frac{3x^2 \, dx}{\sqrt{4 - 9x^2}}$

14. $\displaystyle\int 5x \sqrt{2x - 3} \, dx$

15. $\displaystyle\int x^2 \sqrt{25 + 4x^2} \, dx$

16. $\displaystyle\int e^{\cos x} \sin x \cos x \, dx$

17. $\displaystyle\int \frac{\sqrt{4 + x^2}}{x} \, dx$

18. $\displaystyle\int \frac{\sqrt{9 - (\ln x)^2}}{x} \, dx$

19. $\displaystyle\int \sin^2(4x) \, dx$

20. $\displaystyle\int x^4 (\ln x)^2 \, dx$

21. $\displaystyle\int \frac{\sqrt{9 - x^2}}{x} \, dx$

22. $\displaystyle\int x^3 e^{-x} \, dx$

23. $\displaystyle\int e^x \cos^4(e^x) \, dx$

24. $\displaystyle\int x^2 \cos\left(\frac{x}{2}\right) dx$

•25. $\displaystyle\int_1^2 \frac{dx}{4x^2 + 1}$

•26. $\displaystyle\int_1^5 \frac{x^2 \, dx}{2x - 1}$

•27. $\displaystyle\int_0^{\pi/4} \frac{\sec^2 x \, dx}{\tan x + 1}$

•28. $\displaystyle\int_0^{\sqrt{3}/2} \frac{\tan^{-1} 2x \, dx}{1 + 4x^2}$

•29. $\displaystyle\int_0^1 e^{x^2} \, dx$

•30. $\displaystyle\int_0^{\pi} e^{\sin x} \, dx$

•31. $\displaystyle\int_1^e \sqrt{\ln x} \, dx$

•32. $\displaystyle\int_1^e \ln(\sqrt{x}) \, dx$

•33. $\displaystyle\int_0^1 \frac{dx}{3 + 2e^{4x}}$

•34. $\displaystyle\int_1^4 \frac{\sqrt{1 + 9x^2}}{x^2} \, dx$

•35. $\displaystyle\int_0^{\pi/2} 3x \sin 2x \, dx$

•36. $\displaystyle\int_1^e (\ln x)^3 \, dx$

•37. $\displaystyle\int_0^{\ln 2} \frac{e^{2x} \, dx}{(4e^x + 1)^3}$

•38. $\displaystyle\int_0^{\pi/2} \sin x \sqrt{9 + \cos^2 x} \, dx$

- **39** $\displaystyle\int_0^2 \frac{dx}{2x^2 - 5x - 3}$
- **40** $\displaystyle\int_0^2 \frac{x\,dx}{\sqrt{4 + x^4}}$

c41 (*For students with access to a computer*) Let $F(x) = \int_0^x e^{t^2/2}\,dt$. Use Simpson's rule to make a table of values for $F(x)$ for $x = 0, 0.01, 0.02, \ldots, 1.00$.

Review Problems

1–26. **(a)** Classify the integral according to the technique of integration most likely to be successful.
(b) Evaluate the integral.

- **1** $\displaystyle\int \cos^7 2x\,dx$
- **2** $\displaystyle\int \tan^2 x\,dx$
- **3** $\displaystyle\int \frac{(x+1)\,dx}{x^2 - 6x + 5}$
- **4** $\displaystyle\int \ln(2x)\,dx$
- **5** $\displaystyle\int \frac{x\,dx}{4x^2 + 1}$
- **6** $\displaystyle\int \frac{x^2\,dx}{9 - x^2}$
- **7** $\displaystyle\int \frac{5x^3 - 11x^2 + 5x - 1}{x^2(x-1)^2}\,dx$
- **8** $\displaystyle\int \frac{\sin x + 2}{2\cos x + 1}\,dx$
- **9** $\displaystyle\int x^3 e^{x^2}\,dx$
- **10** $\displaystyle\int \frac{\cot x\,dx}{\ln \sin x}$
- **11** $\displaystyle\int \frac{\sqrt{x+3}}{x-1}\,dx$
- **12** $\displaystyle\int \frac{(2x-3)\,dx}{x^2 - 4x + 8}$
- **13** $\displaystyle\int \frac{\cos 3x}{\sin^4 3x}\,dx$
- **14** $\displaystyle\int e^x \cos x\,dx$
- **15** $\displaystyle\int x^3 \sqrt{9 + x^2}\,dx$
- **16** $\displaystyle\int (x-1)\sqrt{3x+1}\,dx$
- **17** $\displaystyle\int \frac{3\sin x - 4}{3\cos x + 5}\,dx$
- **18** $\displaystyle\int \frac{4x+1}{x^4 - x^2}\,dx$
- **19** $\displaystyle\int \frac{dx}{2\sin x - 2\cos x - 1}$
- **20** $\displaystyle\int \tan x \sec^7 x\,dx$
- **21** $\displaystyle\int \cos 5x \sin 3x\,dx$
- **22** $\displaystyle\int \frac{1}{9 - 4x^2}\,dx$
- **23** $\displaystyle\int \ln\sqrt{x+1}\,dx$
- **24** $\displaystyle\int \sqrt{6x - x^2}\,dx$

•25 $\displaystyle\int \frac{3x^2 + x + 1}{x^3 + x}$

•26 $\displaystyle\int \frac{\tan^2 x}{\sec^2 x}\, dx$

27–28. Evaluate the improper integral if it converges.

•27 $\displaystyle\int_1^\infty \frac{\ln x}{x^4}\, dx$

•28 $\displaystyle\int_0^\infty \left(\frac{\pi}{2} - \frac{x}{1 + x^2} - \tan^{-1} x\right) dx$

29 Draw diagrams that illustrate the three basic substitutions used in trigonometric substitution.

30 Derive the formulas

$$\cos x = \frac{1 - u^2}{1 + u^2}, \qquad \sin x = \frac{2u}{1 + u^2}, \qquad dx = \frac{2\, du}{1 + u^2}$$

based on the substitution $u = \tan(x/2)$.

31–32. Find the centroid of the region bounded by the graphs of the equations.

•31 $y = \ln x$, $y = 0$, $x = e$

•32 $f(x) = e^{-x}$, the positive x-axis, the y-axis

33–34. Find the arc length of the graph.

•33 $y = x^2$, $0 \le x \le \dfrac{1}{2}$

•34 $f(x) = \ln \cos x$, $0 \le x \le \dfrac{\pi}{4}$

35 Derive the reduction formula

$$\int x^n e^{ax}\, dx = \frac{x^n e^{ax}}{a} - \frac{n}{a}\int x^{n-1} e^{ax}\, dx + C, \qquad a \ne 0.$$

36 (a) Derive the reduction formula

$$\int \sin^n x\, dx = -\frac{\sin^{n-1} x \cos x}{n} + \frac{n-1}{n}\int \sin^{n-2} x\, dx.$$

•(b) Let n be an odd positive integer. Use the formula in (a) to obtain a formula for

$$\int_0^{\pi/2} \sin^n x\, dx.$$

(*Hint:* First work out the formula for $n = 1$, 3, and 5.)

12 Infinite Series

12-1 Sequences and Series

Sequence

On several occasions we have worked with functions defined on the positive integers. Such a function is called a *sequence*. For example, if $f(n) = 2n + 1$ for every positive integer n, then f is a sequence. Consider, for a second example, Newton's Method for approximating the roots of $f(x) = 0$. We start with a first approximation x_1, then define

$$x_2 = x_1 - \frac{f(x_1)}{f'(x_1)}$$

$$x_3 = x_2 - \frac{f(x_2)}{f'(x_2)},$$

and so on. The numbers $x_1, x_2, x_3, \ldots$ constitute a sequence. For every positive integer n, there is a corresponding functional value x_n.

Subscript Notation

It is customary to use subscripts to denote the functional values of a sequence. For example, if

$$f(n) = \frac{2n-1}{n}$$

for every positive integer n, we usually write

$$a_n = \frac{2n-1}{n}$$

In particular,

$$a_1 = 1, \quad a_2 = \tfrac{3}{2}, \quad a_3 = \tfrac{5}{3}, \quad a_4 = \tfrac{7}{4}.$$

We can indicate the entire sequence by writing

$$\{a_n\} \quad \text{or} \quad \left\{\frac{2n-1}{n}\right\}.$$

The number a_n is called the *n*th *term* of the sequence $\{a_n\}$.

It should be obvious that the sequence $\{a_n\}$ defined by

$$a_n = \frac{2n-1}{n} = 2 - \frac{1}{n}$$

12–1 Sequences and Series

has a limiting value of 2 as n gets large without bound. We represent this fact by writing

$$\lim_{n \to \infty} a_n = \lim_{n \to \infty} \left(2 - \frac{1}{n}\right) = 2 \quad \text{or} \quad \{a_n\} \to 2.$$

The formal definition of such a limit follows.

DEFINITION

Limit of a Sequence

We say that the sequence $\{a_n\}$ *converges to L*, indicated by writing

$$\lim_{n \to \infty} a_n = L \quad \text{or} \quad \{a_n\} \to L$$

provided that for every $\epsilon > 0$ there exists a positive integer N (which depends on ϵ) such that

$$|a_n - L| < \epsilon \quad \text{whenever } n \geq N.$$

The definition can be restated in the form $\{a_n\} \to L$ if and only if for each $\epsilon > 0$ there is a positive integer N such that

$$L - \epsilon < a_n < L + \epsilon \quad \text{whenever } n \geq N.$$

In other words, all the terms in the sequence after the $N - 1$st term are between $L - \epsilon$ and $L + \epsilon$. (See Fig. 12–1.) This condition must hold for every positive number ϵ.

Figure 12–1 $\lim a_n = L$. If ϵ is any positive number, there exists a positive integer N such that a_N and all succeeding terms of the sequence are between $L - \epsilon$ and $L + \epsilon$.

Divergent Sequence

If a sequence $\{a_n\}$ does not converge to a limit L, then it is said to *diverge*. If its functional values increase without bound as $n \to \infty$, then it *diverges to infinity*. In that case we write

$$\lim_{n \to \infty} a_n = \infty \quad \text{or} \quad \{a_n\} \to \infty.$$

Similarly, if the functional values decrease without bound as $n \to \infty$, then we write

$$\lim_{n \to \infty} a_n = -\infty \quad \text{or} \quad \{a_n\} \to -\infty.$$

The formal definitions of these concepts are similar to the related concepts for functions of a continuous variable. (See Exercise 26.)

EXAMPLE 1 (a) The sequence defined by $a_n = (-1)^n$ diverges. The functional values alternate between $+1$ and -1 and never approach a limit.

(b) The sequence $\{(-n)^n\}$ diverges. The functional values alternate between positive and negative numbers, getting larger in absolute value as n increases.

(c) The sequence $\left\{4 - \frac{2}{n} + \frac{1}{n^2}\right\}$ converges to 4. □

The following theorem extends the Limit Theorem to sequences. The proof is omitted.

THEOREM 12–1
Limit Theorem for Sequences

(*The Limit Theorem*) Let $\{a_n\} \to L$ and $\{b_n\} \to M$, where L and M are real numbers. Then

L1 $\quad \lim\limits_{n \to \infty} (a_n \pm b_n) = L \pm M = (\lim\limits_{n \to \infty} a_n) \pm (\lim\limits_{n \to \infty} b_n).$

L2 $\quad \lim\limits_{n \to \infty} (a_n b_n) = LM = (\lim\limits_{n \to \infty} a_n) \cdot (\lim\limits_{n \to \infty} b_n).$

L3 $\quad \lim\limits_{n \to \infty} \dfrac{a_n}{b_n} = \dfrac{L}{M} = \dfrac{\lim\limits_{n \to \infty} a_n}{\lim\limits_{n \to \infty} b_n} \quad$ provided $M \neq 0.$

L6 $\quad \lim\limits_{n \to \infty} \sqrt[k]{a_n} = \sqrt[k]{L} = \sqrt[k]{\lim\limits_{n \to \infty} a_n} \quad$ provided $L > 0$ if k is even.

L7 $\quad \lim\limits_{n \to \infty} (ca_n) = cL = c \lim\limits_{n \to \infty} a_n \quad$ for each real number c.

Rules like those stated in Section 2–8 hold if $\{a_n\} \to \pm\infty$ or if $\{b_n\} \to \pm\infty$. For example, if $\{a_n\} \to -5$ and $\{b_n\} \to -\infty$, then

$$\{a_n + b_n\} \to -\infty, \qquad \{a_n b_n\} \to \infty, \qquad \text{and} \qquad \left\{\dfrac{a_n}{b_n}\right\} \to 0.$$

EXAMPLE 2

(a) $\lim\limits_{n \to \infty} \dfrac{5n^2 + 4n - 1}{2n^2 + 3n} = \lim\limits_{n \to \infty} \dfrac{5 + (4/n) - (1/n^2)}{2 + 3/n} = \dfrac{5}{2}.$

(b) $\lim\limits_{n \to \infty} (e^n - e^{-n}) = \lim\limits_{n \to \infty} \left(e^n - \dfrac{1}{e^n}\right) = \infty.$

(c) $\lim\limits_{n \to \infty} \dfrac{2n^2 - 3}{n} = \lim\limits_{n \to \infty} \left(2n - \dfrac{3}{n}\right) = \infty.\ \square$

Remark Sequences are used extensively for computing approximate solutions to problems. We have already mentioned how Newton's method leads to a sequence. Another example is furnished by the Trapezoidal Rule for approximate integration. Let f be continuous on $[a, b]$. If t_n is the approximation to $\int_a^b f(x)\, dx$ obtained by using the Trapezoidal Rule with n subintervals, then the sequence $\{t_1, t_2, t_3, \ldots\}$ converges to the value of the integral.

INFINITE SERIES

Infinite Series

Let $\{a_n\}$ be a sequence. The expression

$$\sum_{k=1}^{\infty} a_k = a_1 + a_2 + a_3 + \cdots + a_n + \cdots,$$

called an *infinite series*, is assigned a meaning by the following process.

We define the sequence $\{s_n\}$, called the *sequence of partial sums* of the series, by

Sequence of Partial Sums

$s_1 = a_1$
$s_2 = s_1 + a_2 = a_1 + a_2$

12–1 Sequences and Series

$$s_3 = s_2 + a_3 = a_1 + a_2 + a_3$$
$$s_4 = s_3 + a_4 = a_1 + a_2 + a_3 + a_4$$
$$\cdots\cdots\cdots\cdots\cdots\cdots\cdots$$
$$s_n = s_{n-1} + a_n = a_1 + a_2 + a_3 + \cdots + a_n.$$

Convergent Series   If the sequence of partial sums $\{s_n\}$ converges to a limit L, then L is, by definition, the *value* or *sum* of the series. In that case, we say that the series $\sum_{n=1}^{\infty} a_n$ *converges to the limit* L. We indicate this fact by writing

$$\sum_{n=1}^{\infty} a_n = L.$$

In other words,

$$\sum_{n=1}^{\infty} a_n = L$$

if and only if the sequence of partial sums $\{s_n\}$ converges to L as a limit.

Divergent Series   If the sequence of partial sums $\{s_n\}$ diverges, we say that the series $\sum_{n=1}^{\infty} a_n$ *diverges*. If the sequence $\{s_n\} \to \infty$, we say that the series $\sum_{n=1}^{\infty} a_n$ *diverges to* ∞. In that case we write

$$\sum_{n=1}^{\infty} a_n = \infty.$$

EXAMPLE 3   (*Decimal Fractions*) Decimal fractions, such as $0.333333\ldots$ and $0.24242424\ldots$, are defined as infinite series. These two fractions, for example, are defined by the series

Decimal Fractions

$$0.333333\ldots = \frac{3}{10} + \frac{3}{10^2} + \frac{3}{10^3} + \frac{3}{10^4} + \cdots,$$

$$0.24242424\ldots = \frac{2}{10} + \frac{4}{10^2} + \frac{2}{10^3} + \frac{4}{10^4} + \frac{2}{10^5} + \frac{4}{10^6} + \cdots.$$

The first of these series can be evaluated by the sequence of partial sums. Recall that a finite geometric series can be summed by the formula

$$a + ar + ar^2 + \cdots + ar^n = \frac{a(1 - r^{n+1})}{1 - r} \qquad \text{provided } r \neq 1.$$

Then

$$s_1 = \frac{3}{10}$$

$$s_2 = \frac{3}{10} + \frac{3}{10^2} = \frac{3}{10}\left(\frac{1 - (1/10)^2}{1 - (1/10)}\right)$$

$$s_3 = \frac{3}{10} + \frac{3}{10^2} + \frac{3}{10^3} = \frac{3}{10}\left(\frac{1 - (1/10)^3}{1 - (1/10)}\right)$$

$$s_4 = \frac{3}{10} + \frac{3}{10^2} + \frac{3}{10^3} + \frac{3}{10^4} = \frac{3}{10}\left(\frac{1 - (1/10)^4}{1 - (1/10)}\right)$$

and so on. Then

$$0.333333\cdots = \frac{3}{10} + \frac{3}{10^2} + \frac{3}{10^3} + \frac{3}{10^4} + \cdots$$

$$= \lim_{n\to\infty} s_n = \lim_{n\to\infty} \frac{3}{10}\left[\frac{1-(1/10)^n}{1-(1/10)}\right]$$

$$= \frac{3}{10}\left[\frac{1-0}{1-(1/10)}\right] = \frac{3}{10}\cdot\frac{1}{9/10} = \frac{1}{3}.$$

The second series can be evaluated by a similar process. (See Exercise 21.) □

Remark When we write $\Sigma_{n=1}^{\infty} a_n = L$, we are using the equality sign differently from how it is used in algebra. Technically, the series is not *equal* to L—its sequence of partial sums *converges* to L. Thus, when we write

$$0.333\cdots = \tfrac{1}{3},$$

we mean that the sequence of partial sums defined by

$$s_1 = 0.3, \qquad s_2 = 0.33, \qquad s_3 = 0.333, \qquad \text{and so on,}$$

converges to $\tfrac{1}{3}$ as a limit.

EXAMPLE 4   Find the sum of the series $\sum_{n=1}^{\infty} \frac{1}{2^n} = \frac{1}{2} + \frac{1}{2^2} + \frac{1}{2^3} + \cdots$.

Solution   The sequence of partial sums is

$$s_1 = \frac{1}{2}$$

$$s_2 = \frac{1}{2} + \frac{1}{4} = \frac{1}{2}\left[\frac{1-(1/2)^2}{1-1/2}\right] = 1 - \frac{1}{4}$$

$$s_3 = \frac{1}{2} + \frac{1}{4} + \frac{1}{8} = \frac{1}{2}\left[\frac{1-(1/2)^3}{1-1/2}\right] = 1 - \frac{1}{8}$$

and so on. In general,

$$s_n = \frac{1}{2} + \frac{1}{4} + \frac{1}{8} + \cdots + \frac{1}{2^n} = \frac{1}{2}\left[\frac{1-(1/2)^n}{1-(1/2)}\right] = 1 - \frac{1}{2^n}.$$

Then

$$\sum_{n=1}^{\infty} \frac{1}{2^n} = \lim_{n\to\infty} s_n = \lim_{n\to\infty}\left(1 - \frac{1}{2^n}\right) = 1 - 0 = 1. \quad \square$$

EXAMPLE 5   Show that the series $\Sigma_{n=1}^{\infty} n$ diverges to infinity.

Solution   The sequence of partial sums is

$$s_1 = 1$$
$$s_2 = 1 + 2 = 3$$
$$s_3 = 1 + 2 + 3 = 6$$

$$s_n = 1 + 2 + 3 + \cdots + n = \frac{n(n+1)}{2}$$

Then $\sum\limits_{n=1}^{\infty} n = \lim\limits_{n \to \infty} s_n = \lim\limits_{n \to \infty} \frac{n(n+1)}{2} = \infty.$ □

Exercises 12–1

1–6. Write out the first four terms of the sequence $\{a_n\}$ and calculate the limit if it exists.

•1 $a_n = 2 + \dfrac{1}{2^n}$ 　　　　　　　　2 $a_n = \dfrac{2n-3}{n+1}$

•3 $a_n = \dfrac{(-1)^n n^2}{n+1}$ 　　　　　　　4 $a_n = \dfrac{(-1)^n n}{n^2+1}$

•5 $a_n = e$ 　　　　　　　　　　　6 $a_n = (-1)^n$

7–10. Let $\{x_n\}$ be the sequence obtained by using Newton's method to approximate the solution of $f(x) = 0$. Calculate x_n for $n = 2, 3, 4,$ and 5. To what number does x_n converge?

•7 $f(x) = x^2 - 3, \; x_1 = 1$ 　　　　　8 $f(x) = x^2 - 5, \; x_1 = -2$

•9 $f(x) = \ln x - 1, \; x_1 = 2$ 　　　　10 $f(x) = e^x - 1, \; x_1 = 1$

11–12. Let $\{t_n\}$ be the sequence obtained by using the Trapezoidal Rule with n subintervals to approximate the integral. Let $\{s_n\}$ be the sequence obtained by using Simpson's rule with $2n$ subintervals.

　　(a) Calculate t_n for $n = 1, 2, 3,$ and 4.
　　(b) Calcualte s_n for $n = 1, 2, 3, 4$.
　　(c) To what number do these sequences converge?

•11 $\displaystyle\int_0^1 x^2 \, dx$ 　　　　　　　　•12 $\displaystyle\int_0^2 e^x \, dx$

Recursion Formulas

13–16. (*Recursion Formulas*) Many of the advanced problems solved on the computer involve sequences defined by *recursion formulas*—formulas that express the value of a_{n+1} in terms of the previously computed number a_n. The given sequences are defined recursively. Calculate the first five terms of each sequence. Can you decide whether these sequences converge?

•13 $a_1 = 1, \; a_{n+1} = a_n + \dfrac{1}{2^n}$ 　　　14 $a_1 = 3, \; a_{n+1} = -2a_n$

•15 $a_1 = 1, \; a_{n+1} = a_n + \dfrac{1}{n!}$ 　　　16 $a_1 = 3, \; a_{n+1} = a_n$

17–20. Write the first five terms of $\{s_n\}$, the sequence of partial sums of the given series. If possible, find a general formula for s_n? Does the series converge to a limit? If so, find it. (*Hint for 19 and 20:* Modify the argument in Example 4.)

- 17 $\sum_{n=1}^{\infty} 1^n$
- 18 $\sum_{n=1}^{\infty} (-1)^n$
- 19 $\sum_{n=1}^{\infty} \frac{1}{3^n}$
- 20 $\sum_{n=1}^{\infty} \left(-\frac{2}{5}\right)^n$

21–24. Modify the method of Example 3 to calculate the exact value of the decimal fractions. (*Hint for 21:*

$$0.24242424\cdots = \frac{24}{100} + \frac{24}{100^2} + \frac{24}{100^3} + \cdots .)$$

- 21 $0.24242424\cdots$
- 22 $1.37373737\cdots$
- 23 $0.555555\cdots$
- 24 $0.019019019019\cdots$

25 A student with an electronic calculator set in the radian mode starts with an arbitrary number x_1 and repeatedly presses the COS key, obtaining a sequence $x_1, x_2, x_3, \ldots$, where $x_{n+1} = \cos x_n$. Eventually he gets the single number 0.7390851332.

Assume that the sequence $\{x_n\}$ converges. Explain why it converges to the unique solution of the equation $x = \cos x$.

26 State formal definitions of the following limits.

- (a) $\lim_{n \to \infty} a_n = \infty$
- (b) $\lim_{n \to \infty} a_n = -\infty$

27 Use the definition of *limit* to prove that if $\{a_n\} \to L$, then $\{|a_n|\} \to |L|$.

28 Prove that the sequence $\{a_n\} \to 0$ if and only if the related sequence $\{|a_n|\} \to 0$. Does a similar result hold if one of these sequences converges to a nonzero number? (See Exercise 27 for a partial answer.)

12–2 Some Properties of Infinite Series. Positive-Term Series

Several theorems relate to convergence of series. The first theorem we consider states a specialized result that is primarily used in a negative way—to show that a given series does not converge.

THEOREM 12–2   If the series $\sum_{n=1}^{\infty} a_n$ converges to a number L, then the sequence $\{a_n\}$ converges to 0.

Proof   Let $\{s_n\}$ be the sequence of partial sums. Since the series converges to L, then $\{s_n\} \to L$.

Observe that the nth term in the sequence of partial sums is

$$s_n = a_1 + a_2 + \cdots + a_{n-1} + a_n = s_{n-1} + a_n.$$

It follows from Theorem 12–1 that

$$\lim_{n \to \infty} a_n = \lim_{n \to \infty} (s_n - s_{n-1}) = \lim_{n \to \infty} s_n - \lim_{n \to \infty} s_{n-1}$$
$$= L - L = 0. \blacksquare$$

12-2 Some Properties of Infinite Series. Positive-Term Series

COROLLARY

Test for Divergence of a Series

If the sequence $\{a_n\}$ does not converge or if it converges to a number different from 0, then the series $\sum_{n=1}^{\infty} a_n$ diverges.

EXAMPLE 1 Let $a_n = \dfrac{2n + 3}{3n + 1}$. Show that the series $\sum_{n=1}^{\infty} a_n$ diverges.

Solution
$$\lim_{n \to \infty} a_n = \lim_{n \to \infty} \frac{2n + 3}{3n + 1} = \lim_{n \to \infty} \frac{2 + (3/n)}{3 + (1/n)} = \frac{2}{3}.$$

Since $\lim_{n \to \infty} a_n \neq 0$, then the series

$$\sum_{n=1}^{\infty} a_n = \sum_{n=1}^{\infty} \frac{2n + 3}{3n + 1}$$

must diverge. □

Remark 1 We must be careful not to read more into Theorem 12-2 than actually is there. Nothing is said about what happens to the series $\Sigma\, a_n$ if the sequence $\{a_n\}$ converges to zero. As we shall see later, such a series may converge or may diverge.

Remark 2 Theorem 12-2 is typical of many of our results in that we get information about the convergence or divergence of a series $\Sigma\, a_n$ from the convergence or divergence of a related sequence. In this case, we find that the series must diverge if the sequence of terms $\{a_n\}$ does not converge to zero.

SOME GENERAL RESULTS ABOUT SERIES

Several computational results about series are stated in Theorem 12-3. The proofs, which involve using the partial sums, are omitted.

THEOREM 12-3 Let $\sum_{n=1}^{\infty} a_n = L$ and $\sum_{n=1}^{\infty} b_n = M$; let $\sum_{n=1}^{\infty} c_n$ be a divergent series.

Arithmetic Properties of Series

(a) The series $\sum_{n=1}^{\infty} (a_n + b_n)$ converges to $L + M$. That is,

$$\sum_{n=1}^{\infty} (a_n + b_n) = L + M = \sum_{n=1}^{\infty} a_n + \sum_{n=1}^{\infty} b_n.$$

If two convergent series are added term by term, the resulting series converges to the sum of their values.

(b) The series $\sum_{n=1}^{\infty} (a_n + c_n)$ diverges.

If a convergent series is added term by term to a divergent series, the resulting series diverges.

(c) Let k be a nonzero constant. Then

$$\sum_{n=1}^{\infty} ka_n = kL = k \sum_{n=1}^{\infty} a_n$$

and the series

$$\sum_{n=1}^{\infty} kc_n \text{ diverges.}$$

A nonzero constant multiple of a convergent series converges; a nonzero constant multiple of a divergent series diverges.

POSITIVE-TERM SERIES

A series

$$\sum_{n=1}^{\infty} a_n$$

Positive-term Series

in which each term a_n is positive is called a *positive-term series*. As we shall see, it is especially convenient to work with such series.

Observe that if $\Sigma \, a_n$ is a positive-term series, then its sequence of partial sums $\{s_n\}$ has the property that

$$s_1 < s_2 < s_3 < \cdots.$$

Monotone Sequence

Such a sequence is said to be *monotone-increasing*. Similarly, a sequence that satisfies the relation

$$s_1 \leq s_2 \leq s_3 \leq \cdots$$

is *monotone-nondecreasing;* one that satisfies

$$s_1 > s_2 > s_3 > \cdots$$

is *monotone-decreasing;* and so on. Observe that every monotone increasing sequence is monotone-nondecreasing.

We shall show in Theorem 12–4 that a monotone-nondecreasing sequence must either converge to a finite limit or diverge to infinity. Before we establish this result, we consider a basic property of the real number system—the property of completeness.

COMPLETENESS

Let S be a nonempty set of real numbers. The number b is called an *upper bound* for S if every element $s \in S$ is less than or equal to b. (See Fig. 12–2.)

An upper bound for a set is not unique—in fact, any larger number is also an upper bound. For example, if

$$S = \{x \,|\, x^2 < 7\},$$

then $\sqrt{7}$, 5, 20, and 187.43 are upper bounds for S.

12–2 Some Properties of Infinite Series. Positive-Term Series

Figure 12–2 The number b is an upper bound for set S. The least upper bound is u. If ϵ is any positive number, there exists an element of S that is greater than $u - \epsilon$.

DEFINITION The number u is called the *least upper bound* for S provided that:

(1) u is an upper bound for S.
(2) No number smaller than u is an upper bound for S.

For example, if $S = \{x \mid x^2 < 7\}$, then the least upper bound for S is $u = \sqrt{7}$. This number is an upper bound, and there is no smaller upper bound.

The following axiom of the real number system states that every set with an upper bound has a least upper bound.

COMPLETENESS AXIOM Let S be a nonempty set of real numbers that is bounded above. Then S has a least upper bound.

We remark in passing that similar concepts and results hold for *lower bounds*. If a set S has a lower bound, then it must have a *greatest lower bound*—a lower bound v such that no number greater than v is a lower bound.

These results on upper and lower bounds can be extended to cover sequences. We say that b is an *upper bound* for a sequence $\{a_n\}$ if $a_n \leq b$ for every b. The upper-bound l is the *least upper bound* of the sequence if no smaller number is an upper bound. It follows from the completeness axiom that every sequence that is bounded above has a least upper bound.

MONOTONE SEQUENCES

We are now ready to establish our basic result for monotone sequences.

THEOREM 12–4 Let $\{a_n\}$ be a monotone-nondecreasing sequence.
(a) If $\{a_n\}$ is bounded above, then $\{a_n\}$ converges to its least upper bound.
(b) If $\{a_n\}$ is not bounded above, then $\{a_n\} \to \infty$.

Proof (a) Let L be the least upper bound of the sequence $\{a_n\}$. Let $\epsilon > 0$. Since $L - \epsilon < L$, then $L - \epsilon$ is not an upper bound for $\{a_n\}$. Consequently, there must exist a term a_N of the sequence such that

$$a_N > L - \epsilon.$$

Since the sequence is nondecreasing, then

$$L - \epsilon < a_N \leq a_{N+1} \leq a_{N+2} \leq \cdots.$$

Figure 12-3 A monotone-nondecreasing sequence converges to its least upper bound.

(See Fig. 12-3.) Since L is an upper bound for the sequence, then each of these terms is less than or equal to L. Therefore,

$$L - \epsilon < a_N \leq a_{N+1} \leq a_{N+2} \leq \cdots \leq L < L + \epsilon,$$

so that

$$|a_n - L| < \epsilon \quad \text{if } n \geq N.$$

It follows from the definition of limit that

$$\lim_{n \to \infty} a_n = L.$$

The proof of part (b) is left for you to work. (See Exercise 23.) ∎

Theorem 12-5 applies Theorem 12-4 to positive-term series.

THEOREM 12-5

Basic Theorem for Positive-term Series

Let $\sum_{n=1}^{\infty} a_n$ be a positive-term series. This series converges if and only if its sequence $\{s_n\}$ of partial sums is bounded above. If L is the least upper bound of the sequence $\{s_n\}$, then

$$\sum_{n=1}^{\infty} a_n = L.$$

Proof (1) Suppose that the sequence of partial sums $\{s_n\}$ is bounded above. Let L be its least upper bound. It follows from Theorem 12-4(a) that $\lim_{n \to \infty} s_n = L$, so that

$$\sum_{n=1}^{\infty} a_n = \lim_{n \to \infty} s_n = L.$$

The series converges to L.

(2) If the sequence of partial sums is unbounded, then it follows from Theorem 12-4(b) that $\lim_{n \to \infty} s_n = \infty$. Therefore

$$\sum_{n=1}^{\infty} a_n = \lim_{n \to \infty} s_n = \infty.$$

The series diverges to infinity. ∎

HARMONIC SERIES

We show, as an example of how Theorem 12-5 can be applied, that the *harmonic series*

Harmonic Series

$$\sum_{n=1}^{\infty} \frac{1}{n} = 1 + \frac{1}{2} + \frac{1}{3} + \frac{1}{4} + \frac{1}{5} + \cdots$$

12-2 Some Properties of Infinite Series. Positive-Term Series

diverges. Observe that Theorem 12-2 gives no information in this case, since

$$\lim_{n \to \infty} \{a_n\} = \lim_{n \to \infty} \left\{\frac{1}{n}\right\} = 0.$$

EXAMPLE 2   Prove that the harmonic series

$$\sum_{n=1}^{\infty} \frac{1}{n} = 1 + \frac{1}{2} + \frac{1}{3} + \frac{1}{4} + \frac{1}{5} + \cdots$$

diverges.

Solution   Observe in Figure 12–4 that $1/n$, the nth term of the series, is greater than $\int_n^{n+1} dx/x$, the area under the graph of $f(x) = 1/x$ from $x = n$ to $x = n + 1$. It fol-

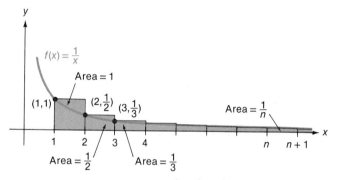

Figure 12–4   Example 2. $s_n = \frac{1}{1} + \frac{1}{2} + \frac{1}{3} + \cdots + \frac{1}{n} > \int_1^{n+1} \frac{dx}{x}$

lows that s_n, the nth term in the sequence of partial sums, satisfies the inequality

$$s_n = 1 + \frac{1}{2} + \frac{1}{3} + \frac{1}{4} + \cdots + \frac{1}{n}$$

$$> \int_1^2 \frac{dx}{x} + \int_2^3 \frac{dx}{x} + \int_3^4 \frac{dx}{x} + \cdots + \int_n^{n+1} \frac{dx}{x}$$

$$= \int_1^{n+1} \frac{dx}{x} = \left[\ln x\right]_1^{n+1} = \ln(n+1).$$

Since $\ln(n+1) \to \infty$ as $n \to \infty$, the sequence $\{s_n\}$ is unbounded. Therefore, by Theorem 12–5,

$$\sum_{n=1}^{\infty} \frac{1}{n} = \lim_{n \to \infty} s_n = \infty.$$

The harmonic series diverges to infinity. □

The technique used in Example 2 will be generalized in Section 12–3 to prove a powerful test for convergence—the integral test.

GEOMETRIC SERIES

Our development proceeds along two parallel lines. One line leads to the development of properties of series—tests for convergence and the like. The other line leads to practical applications and to the evaluation of specific series.

As part of this development, we need to construct a catalogue of series that are known to be either convergent or divergent. In Example 2, we considered the harmonic series (divergent). We now establish conditions for the convergence or divergence of the so-called *geometric series*—the series

Geometric Series

$$a + ar + ar^2 + ar^3 + \cdots$$

where $a \neq 0$.

THEOREM 12–6 The geometric series

$$a + ar + ar^2 + ar^3 + ar^4 + \cdots$$

converges to $a/(1 - r)$ if $|r| < 1$ and diverges if $|r| \geq 1$.

Proof Let $s_n = a + ar + ar^2 + \cdots + ar^{n-1}$ be the nth term in the sequence of partial sums. It is proved in elementary algebra that

$$s_n = a + ar + ar^2 + \cdots + ar^{n-1} = \frac{a(1 - r^n)}{1 - r}$$

if $r \neq 1$. (See Table VI at the end of this book.)

Suppose that $|r| < 1$. Then $r^n \to 0$ as $n \to \infty$. (See Exercise 24.) It follows that

$$\lim_{n \to \infty} s_n = \lim_{n \to \infty} \frac{a(1 - r^n)}{1 - r} = \frac{a(1 - 0)}{1 - r} = \frac{a}{1 - r}.$$

Therefore,

$$a + ar + ar^2 + \cdots = \lim_{n \to \infty} s_n = \frac{a}{1 - r}.$$

Suppose that $|r| \geq 1$. Then the nth term of the series is

$$a_n = ar^{n-1},$$

which in absolute value is greater than or equal to the first term. Therefore, the sequence of terms does not converge to zero. It follows from Theorem 12–2 that the series $a + ar + ar^2 + \cdots$ diverges. ∎

EXAMPLE 3 The series

$$5 - \frac{5}{3} + \frac{5}{3^2} - \frac{5}{3^3} + \cdots$$

is a geometric series with $r = -\frac{1}{3}$. Since $|r| < 1$ the series converges to

$$\frac{a}{1 - r} = \frac{5}{1 - (-1/3)} = \frac{5}{4/3} = \frac{15}{4}.$$

Consequently,

$$5 - \frac{5}{3} + \frac{5}{3^2} - \frac{5}{3^3} + \frac{5}{3^4} - \cdots = \frac{15}{4}. \quad \square$$

Exercises 12–2

1–6. Calculate the first four partial sums for the following series. Decide whether the series converges or diverges. If a series converges, then find its sum.

● 1 $\sum_{n=1}^{\infty} \frac{(-1)^n}{4^n}$

2 $\sum_{n=1}^{\infty} \frac{(-3)^n}{4^{n+1}}$

● 3 $\sum_{n=1}^{\infty} \frac{e^n}{2n+1}$

4 $\sum_{n=1}^{\infty} \frac{-2}{n}$

● 5 $\sum_{n=1}^{\infty} \frac{2n+1}{3n+2}$

6 $\sum_{n=1}^{\infty} \frac{n^2+n-1}{2n^2-3}$

7–10. Rewrite the given series using summation notation. Use Theorem 12–3 and the other results of this chapter to decide whether the series converge or diverge. If a series converges, find its limit.

● 7 $\left(1 - \frac{1}{3}\right) + \left(\frac{1}{2} + \frac{1}{3^2}\right) + \left(\frac{1}{2^2} - \frac{1}{3^3}\right) + \left(\frac{1}{2^3} + \frac{1}{3^4}\right) + \left(\frac{1}{2^4} - \frac{1}{3^5}\right) + \cdots$

8 $\left(1 + \frac{1}{2}\right) + \left(\frac{1}{2} + \frac{1}{2^2}\right) + \left(\frac{1}{3} + \frac{1}{2^3}\right) + \left(\frac{1}{4} + \frac{1}{2^4}\right) + \left(\frac{1}{5} + \frac{1}{2^5}\right) + \cdots$

● 9 $\frac{3}{1} + \frac{3}{2} + \frac{3}{3} + \frac{3}{4} + \frac{3}{5} + \frac{3}{6} + \cdots$

10 $\frac{5}{7^3} + \frac{5}{7^5} + \frac{5}{7^7} + \frac{5}{7^9} + \cdots$

● 11 A ball is dropped from a height of 8 ft. On each bounce it bounces two-thirds as far as on the preceding bounce. Find the total distance it travels before it stops.

12 It will be established in Section 12–8 that the series

$$1 + \frac{x}{1!} + \frac{x^2}{2!} + \frac{x^3}{3!} + \frac{x^4}{4!} + \frac{x^5}{5!} + \cdots$$

converges to e^x for every real number x.

(a) Show that

$$e = 1 + \frac{1}{1!} + \frac{1}{2!} + \frac{1}{3!} + \frac{1}{4!} + \frac{1}{5!} + \cdots$$

●(b) Calculate the partial sums s_1, s_2, s_3, s_4, s_5 for the series for e. How closely does s_5 approximate e?

●(c) Students with programmable calculators or with access to a computer should calculate the partial sum s_{20}. How closely does s_{20} approximate e?

13 It can be shown that the series

$$\frac{x}{1} - \frac{x^2}{2} + \frac{x^3}{3} - \frac{x^4}{4} + \frac{x^5}{5} - \frac{x^6}{6} + \cdots$$

converges to $\ln(x+1)$ if $-1 < x \le 1$.

(a) Show that

$$\ln 2 = 1 - \frac{1}{2} + \frac{1}{3} - \frac{1}{4} + \frac{1}{5} - \frac{1}{6} + \cdots.$$

(This series is called the *alternating harmonic series*.)

(b) Calculate the partial sums $s_1, s_2, \ldots, s_5$ for the series for $\ln 2$. How closely does s_5 approximate $\ln 2$?

(c) Students with programmable calculators should calculate s_{20} and s_{100}. How closely does s_{100} approximate $\ln 2$?

14 Let $\{a_n\}$ be a sequence that converges to a number L. Suppose that $M \le a_n \le N$ for every positive integer n. Prove that $M \le L \le N$.

15–18. Show that the sets are bounded above and below. Find the least upper bound and the greatest lower bound of each set.

•**15** $\{x \mid 0 < x \le 1\}$ **16** $\{x \mid 2 < |x| < 3\}$

•**17** $\left\{\frac{1}{2}, \frac{1}{2^2}, \frac{1}{2^3}, \frac{1}{2^4}, \ldots\right\}$ **18** $\{x \mid x \ge 0 \text{ and } x^2 < 2\}$

19 Show by examples that the least upper bound of a set is not necessarily in the set, although it may be.

---

20 Prove Theorem 12–3(a). (*Hint:* Let $s_n = a_1 + a_2 + \cdots + a_n$, $t_n = b_1 + b_2 + \cdots + b_n$, and $u_n = (a_1 + b_1) + (a_2 + b_2) + \cdots + (a_n + b_n)$. Show that $\{u_n\} \to L + M$.)

21 Prove Theorem 12–3(c). (*Hint:* Let $s_n = a_1 + a_2 + \cdots + a_n$, $t_n = ka_1 + ka_2 + \cdots + ka_n$. Show that $\{t_n\} \to kL$.

22 Prove Theorem 12–3(b). (*Hint:* Suppose that $\Sigma(a_n + c_n)$ converges. Use Exercises 20 and 21 to show that $\Sigma(a_n + c_n - a_n)$ also converges, contrary to the fact that Σc_n diverges.)

23 Prove Theorem 12–4(b). (*Hint:* Let N be a positive number. Explain why there must exist a positive integer n such that $s_n > N$. What about $s_{n+1}, s_{n+2}, \ldots$?)

24 (a) Prove that $\lim_{n \to \infty} |r|^n = 0$ if $0 \le |r| < 1$. (*Hint:* Show that $\lim_{n \to \infty} n \ln |r| = -\infty$.)
(b) Use (a) to prove that $\lim_{n \to \infty} r^n = 0$ if $0 \le |r| < 1$.

12–3 The Integral Test. *p*-Series

The problem of determining the convergence or divergence of a series is central in our work. As we shall see, most of the functions encountered in the calculus can be associated with infinite series. If such a series converges, then in most cases, it con-

12-3 The Integral Test. p-Series

verges to the original function and can be used to calculate the values of that function. If it does not converge, then it is worthless for most applications.

As an example of these remarks, the series

$$\frac{x}{1} - \frac{x^2}{2} + \frac{x^3}{3} - \frac{x^4}{4} + \frac{x^5}{5} - \cdots$$

converges to $\ln(x+1)$ for each value of x in the range $-1 < x \leq 1$. It diverges for all other values of x. Thus, we can use the series to compute decimal approximations such as

$$\ln(0.1) \approx \frac{(-0.9)}{1} - \frac{(-0.9)^2}{2} + \frac{(-0.9)^3}{3} - \frac{(-0.9)^4}{4}$$

and

$$\ln 2 \approx \frac{1}{1} - \frac{1}{2} + \frac{1}{3} - \frac{1}{4} + \frac{1}{5} - \frac{1}{6} + \frac{1}{7} - \frac{1}{8}$$

but not $\ln 3$.

Since determining convergence and divergence is so important, we give considerable attention to establishing tests for convergence in this section and the next.

THE INTEGRAL TEST

Our first test is specialized. It can be applied only to a series that meets stringent conditions. If these conditions are met, then the convergence or divergence of the series can be determined from the convergence or divergence of a related improper integral.

THEOREM 12–7

Integral Test

(*The Integral Test*) Let f be a positive-valued, continuous, and nonincreasing function for $x \geq 1$. Let

$$a_n = f(n) \qquad (n = 1, 2, 3, \ldots).$$

(1) Either the series $\sum_{n=1}^{\infty} a_n$ and the improper integral $\int_1^{\infty} f(x)\, dx$ both converge or both diverge to ∞.

(2) If the improper integral $\int_1^{\infty} f(x)\, dx$ converges to L, then the series $\sum_{n=1}^{\infty} a_n$ converges to a number between L and $a_1 + L$.

Proof **Part 1** Observe in Figure 12–5a that $a_n = f(n)$ equals the area of the nth rectangle. This number is greater than or equal to the area under the graph of f from $x = n$ to $x = n+1$. Thus,

$$a_n = f(n) \geq \int_n^{n+1} f(x)\, dx.$$

It follows that

$$a_1 + a_2 + \cdots + a_n \geq \int_1^2 f(x)\, dx + \int_2^3 f(x)\, dx + \cdots + \int_n^{n+1} f(x)\, dx$$

$$= \int_1^{n+1} f(x)\, dx.$$

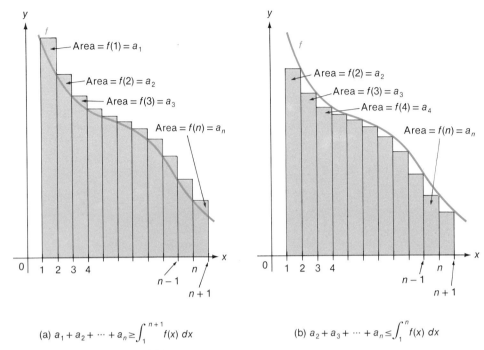

(a) $a_1 + a_2 + \cdots + a_n \geq \int_1^{n+1} f(x)\, dx$

(b) $a_2 + a_3 + \cdots + a_n \leq \int_1^n f(x)\, dx$

Figure 12-5 The Integral Test.

Similarly, we see in Figure 12-5b that if $n \geq 2$, then $a_n = f(n)$ is less than or equal to the area under the graph of f from $x = n - 1$ to $x = n$. It follows that

$$a_2 + a_3 + \cdots + a_n \leq \int_1^n f(x)\, dx,$$

so that

$$a_1 + a_2 + a_3 + \cdots + a_n \leq a_1 + \int_1^n f(x)\, dx.$$

If we combine these inequalities, we get

$$\int_1^{n+1} f(x)\, dx \leq a_1 + a_2 + \cdots + a_n \leq a_1 + \int_1^n f(x)\, dx.$$

Part 2 Suppose that the improper integral $\int_1^\infty f(x)\, dx$ converges to L, that is, that

$$\lim_{n \to \infty} \int_1^n f(x)\, dx = \int_1^\infty f(x)\, dx = L.$$

Let $s_n = a_1 + a_2 + \cdots + a_n$ be the nth term in the sequence of partial sums for the

12-3 The Integral Test. p-Series

series. It follows from Part 1 that

$$s_n = a_1 + a_2 + \cdots + a_n$$
$$\leq a_1 + \int_1^n f(x)\, dx$$
$$< a_1 + \int_1^\infty f(x)\, dx = a_1 + L.$$

Since the sequence of partial sums is bounded above by the number $a_1 + L$, it follows from Theorem 12-5 that the series Σa_n converges to a number less than or equal to $a_1 + L$.

Furthermore, from the inequality in Part 1, we know that

$$\int_1^{n+1} f(x)\, dx < s_n < a_1 + \int_1^n f(x)\, dx.$$

Since all these expressions have limits as $n \to \infty$, we get

$$\lim_{n\to\infty} \int_1^{n+1} f(x)\, dx \leq \lim_{n\to\infty} s_n \leq \lim_{n\to\infty} \left(a_1 + \int_1^n f(x)\, dx \right)$$

$$\int_1^\infty f(x)\, dx \leq \sum_{n=1}^\infty a_n \leq a_1 + \int_1^\infty f(x)\, dx$$

$$L \leq \sum_{n=1}^\infty a_n \leq a_1 + L.$$

Therefore, the series converges to a number between L and $L + a_1$.

Part 3 If $\int_1^\infty f(x) = \infty$, then $\int_1^n f(x)\, dx$ can be made arbitrarily large by choosing a sufficiently large value of n. Since

$$s_n = a_1 + a_2 + \cdots + a_n \geq \int_1^{n+1} f(x)\, dx \geq \int_1^n f(x)\, dx,$$

then s_n can also be made arbitrarily large. Consequently, the sequence of partial sums is unbounded. It follows from Theorem 12-5 that

$$\sum_{n=1}^\infty a_n = \lim_{n\to\infty} s_n = \infty. \quad \blacksquare$$

EXAMPLE 1 Use the integral test to show that the series

$$\sum_{n=1}^\infty \frac{1}{n^2} = \frac{1}{1^2} + \frac{1}{2^2} + \frac{1}{3^2} + \frac{1}{4^2} + \cdots$$

converges. Find bounds for the limit.

Solution Let $f(x) = 1/x^2$. Then f satisfies the conditions of the integral test. It follows from Theorem 12-7 that the series converges if and only if the related improper integral converges.

The integral is

$$\int_1^\infty \frac{dx}{x^2} = \lim_{n\to\infty} \int_1^n x^{-2}\, dx = \lim_{n\to\infty} \left[-\frac{1}{x}\right]_1^n = \lim_{n\to\infty} \left[-\frac{1}{n} + 1\right] = 1.$$

Thus the series converges.

It follows from Part 2 of the integral test that

$$\int_1^\infty \frac{1}{x^2}\, dx \le \sum_{n=1}^\infty \frac{1}{n^2} \le 1 + \int_1^\infty \frac{1}{x^2}\, dx.$$

Therefore

$$1 \le \sum_{n=1}^\infty \frac{1}{n^2} \le 2.$$

The series converges to a number between 1 and 2. □

BOUNDS FOR THE LIMIT OF A SERIES

It is obvious that the bound obtained in the integral test is generally very poor. (For example, see the bound obtained in Example 1.) A much better bound can be obtained by the following argument.

Suppose that the convergent series $\sum_{n=1}^\infty a_n$ and the function f satisfy the conditions of Theorem 12-7. If we ignore the first term of the series, then Theorem 12-7(b) implies that

$$\int_2^\infty f(x)\, dx \le \sum_{n=2}^\infty a_n \le a_2 + \int_2^\infty f(x)\, dx.$$

(This result is obtained by working with the series $\sum_{n=2}^\infty a_n$ rather than the original series.) If we add a_1 to this inequality, we get

$$a_1 + \int_2^\infty f(x)\, dx \le \sum_{n=1}^\infty a_n \le (a_1 + a_2) + \int_2^\infty f(x)\, dx.$$

If we delete the first two terms of the series and use a similar argument, we get

$$(a_1 + a_2) + \int_3^\infty f(x)\, dx \le \sum_{n=1}^\infty a_n \le (a_1 + a_2 + a_3) + \int_3^\infty f(x)\, dx.$$

If we continue in this manner, we find that

$$(a_1 + a_2 + \cdots + a_m) + \int_{m+1}^\infty f(x)\, dx \le \sum_{n=1}^\infty a_n$$

$$\le (a_1 + a_2 + \cdots + a_{m+1}) + \int_{m+1}^\infty f(x)\, dx$$

for every positive integer m. This proves the following theorem.

12-3 The Integral Test. p-Series

THEOREM 12-8

Improved Bounds from the Integral Test

Let the series $\sum_{n=1}^{\infty} a_n$ and the function f satisfy the conditions of the integral test. Let $s_m = a_1 + a_2 + \cdots + a_m$ be the mth partial sum. If the improper integral $\int_1^{\infty} f(x)\, dx$ converges, then

$$s_m + \int_{m+1}^{\infty} f(x)\, dx \leq \sum_{n=1}^{\infty} a_n \leq s_{m+1} + \int_{m+1}^{\infty} f(x)\, dx.$$

EXAMPLE 2 Use Theorem 12-8 with $m = 9$ to estimate the value of the series $\sum_{n=1}^{\infty} \frac{1}{n^2}$.

Solution We first calculate the partial sums s_9 and s_{10}.

$$s_9 = \frac{1}{1^2} + \frac{1}{2^2} + \cdots + \frac{1}{9^2} \approx 1.539767731.$$

$$s_{10} = \frac{1}{1^2} + \frac{1}{2^2} + \cdots + \frac{1}{9^2} + \frac{1}{10^2} \approx 1.549767731.$$

Also

$$\int_{10}^{\infty} \frac{1}{x^2}\, dx = \lim_{n \to \infty} \int_{10}^{n} x^{-2}\, dx = \lim_{n \to \infty} \left[-\frac{1}{x}\right]_{10}^{n}$$

$$= \lim_{n \to \infty} \left[-\frac{1}{n} + \frac{1}{10}\right] = 0.1.$$

It follows from Theorem 12-8 that

$$s_9 + \int_{10}^{\infty} \frac{dx}{x^2} \leq \sum_{n=1}^{\infty} \frac{1}{n^2} \leq s_{10} + \int_{10}^{\infty} \frac{dx}{x^2}$$

$$1.539767731 + 0.1 \leq \sum_{n=1}^{\infty} \frac{1}{n^2} \leq 1.549767731 + 0.1$$

$$1.639767731 \leq \sum_{n=1}^{\infty} \frac{1}{n^2} \leq 1.649767731. \quad \square$$

p-SERIES

p-Series

The series

$$\sum_{n=1}^{\infty} \frac{1}{n^p} = \frac{1}{1^p} + \frac{1}{2^p} + \frac{1}{3^p} + \cdots$$

where p is a positive real number, is called a *p-series*. the integral test can be used as in Example 1 to prove Theorem 12-9.

THEOREM 12-9 (*p-Series*) The *p*-series

$$\sum_{n=1}^{\infty} \frac{1}{n^p} = \frac{1}{1^p} + \frac{1}{2^p} + \frac{1}{3^p} + \cdots$$

converges if $p > 1$ and diverges to ∞ if $0 < p \leq 1$.

The details of the proof are left for you to do. (See Exercise 13.)

EXAMPLE 3   (a) The series

$$\sum_{n=1}^{\infty} \frac{1}{n\sqrt{n}} = \frac{1}{1\sqrt{1}} + \frac{1}{2\sqrt{2}} + \frac{1}{3\sqrt{3}} + \frac{1}{4\sqrt{4}} + \cdots$$

is a *p*-series with $p = \frac{3}{2}$. Since $p > 1$, it converges.

(b) The series

$$\sum_{n=1}^{\infty} \frac{1}{\sqrt{n}} = \frac{1}{\sqrt{1}} + \frac{1}{\sqrt{2}} + \frac{1}{\sqrt{3}} + \frac{1}{\sqrt{4}} + \cdots$$

is a *p*-series with $p = \frac{1}{2}$. Since $0 < p < 1$, it diverges to ∞.

(c) The harmonic series

$$\sum_{n=1}^{\infty} \frac{1}{n} = \frac{1}{1} + \frac{1}{2} + \frac{1}{3} + \frac{1}{4} + \cdots$$

is a *p*-series with $p = 1$. It diverges to ∞. □

Remark   The first few terms of a series have no influence on its convergence or divergence. Thus, the series

$$a_1 + a_2 + a_3 + \cdots + a_k + a_{k+1} + a_{k+2} + \cdots$$

will converge or diverge as the series

$$a_k + a_{k+1} + a_{k+2} + \cdots$$

converges or diverges. (See Exercises 14 and 15.) It should be clear, however, that the two series have different values unless $a_1 = a_2 = \cdots = a_{k-1} = 0$.

Effect of Discarding a Few Terms of a Series

If all the terms of a series except the first few satisfy the conditions of the integral test, then we can delete the first few terms and apply the test to the remaining terms. Consider, as an example, the series

$$\sum_{n=1}^{\infty} \frac{\ln n}{n} = \frac{\ln 1}{1} + \frac{\ln 2}{2} + \frac{\ln 3}{3} + \frac{\ln 4}{4} + \cdots.$$

The first term is zero and the second is less than the third. From the third term on, however, the terms are monotone-decreasing:

$$\frac{\ln 3}{3} > \frac{\ln 4}{4} > \frac{\ln 5}{5} > \cdots.$$

Therefore, the series

$$\frac{\ln 3}{3} + \frac{\ln 4}{4} + \frac{\ln 5}{5} + \cdots$$

satisfies the conditions of the integral test. If we apply the test to this series, we find that it diverges. (See Exercise 7.) Since it diverges, so does the original series.

Exercises 12–3

1–8. (a) Use the integral test to decide whether the series converge or diverge. (*Hint for 7 and 8:* See the Remark at the end of the section.)
  c(b) Use Theorem 12–8 with $m = 5$ to obtain bounds for the series that converge.
  c(c) (*For students with programmable calculators or access to a computer*) Work Part (b), using $m = 50$.

•1 $\sum_{n=1}^{\infty} \dfrac{n}{n^2 + 1}$
2 $\sum_{n=1}^{\infty} \dfrac{1}{n^2 + 1}$

•3 $\sum_{n=1}^{\infty} \dfrac{1}{(n + 1)(\ln (n + 1))^2}$
4 $\sum_{n=1}^{\infty} \dfrac{1}{(n + 2) \ln (n + 2)}$

•5 $\sum_{n=1}^{\infty} \dfrac{e^n}{e^{2n} + 1}$
6 $\sum_{n=1}^{\infty} \dfrac{\cosh n}{\sinh^2 n}$

•7 $\sum_{n=1}^{\infty} \dfrac{\ln n}{n}$
8 $\sum_{n=1}^{\infty} \dfrac{2n - 1}{n^2 - n - 5}$

9–12. Test the *p*-series for convergence. If a series converges, then use Theorem 12–8 with $m = 5$ to obtain bounds on its value.

• 9 $\sum_{n=1}^{\infty} \dfrac{1}{n^{2/3}}$
10 $\sum_{n=3}^{\infty} \dfrac{1}{\sqrt[4]{n}}$

•11 $\sum_{n=1}^{\infty} \dfrac{1}{n^3}$
12 $\sum_{n=3}^{\infty} \dfrac{1}{n^2 \sqrt{n}}$

13 Prove that the *p*-series converges if $p > 1$ and diverges to ∞ if $0 < p \le 1$. (*Hint:* Use the integral test.)

14 (a) Let $\sum_{n=1}^{\infty} a_n$ converge to L. Prove that the series $\sum_{n=2}^{\infty} a_n$ converges to $L - a_1$. (*Hint:* Use the sequences of partial sums of the two series.)
  (b) Use mathematical induction to prove that $\sum_{n=N}^{\infty} a_n$ converges to $L - (a_1 + a_2 + \cdots + a_{N-1})$ for each positive integer N.

15 (a) Let $\sum_{n=1}^{\infty} a_n$ diverge. Prove that the series $\sum_{n=2}^{\infty} a_n$ also diverges.
  (b) Use mathematical induction to prove that $\sum_{n=N}^{\infty} a_n$ diverges for each positive integer N.

12–4 The Comparison and Ratio Tests

Three useful tests for convergence of a positive-term series will be established in this section. The first test requires that we compare the terms of the given series with those of a series known to be convergent or to be divergent.

THE COMPARISON TEST

We say that the positive-term series Σb_n *dominates* the positive-term series Σa_n (or is *term by term greater than or equal to* Σa_n) provided

$$a_n \leq b_n$$

for every positive integer n. We indicate this relation by writing

Comparison of Series
$$\sum_{n=1}^{\infty} a_n \ll \sum_{n=1}^{\infty} b_n.$$

For example, since

$$\frac{1}{2^n + 3} \leq \frac{1}{2^n}$$

for every positive integer n, then

$$\sum_{n=1}^{\infty} \frac{1}{2^n + 3} \ll \sum_{n=1}^{\infty} \frac{1}{2^n}.$$

It is easy to show that a positive-term series Σa_n must converge if it is dominated by a convergent series. This is the first half of Theorem 12–10.

THEOREM 12–10 *(The Comparison Test)* Let Σa_n and Σb_n be positive term series with

$$\sum_{n=1}^{\infty} a_n \ll \sum_{n=1}^{\infty} b_n.$$

Comparison Test

(a) If the dominant series $\sum_{n=1}^{\infty} b_n$ converges to M, then the series $\sum_{n=1}^{\infty} a_n$ converges to a number L that is less than or equal to M.

(b) If the series $\sum_{n=1}^{\infty} a_n$ diverges, then the dominant series $\sum_{n=1}^{\infty} b_n$ also diverges.

Proof Let $\{s_n\}$ and $\{t_n\}$ be the sequences of partial sums for the series Σa_n and Σb_n, respectively. Then

$$s_n = a_1 + a_2 + \cdots + a_n \leq b_1 + b_2 + \cdots + b_n = t_n$$

for every positive integer n.

(a) If the series Σb_n converges to M, then by Theorem 12–5, M is the least upper bound of the sequence $\{t_n\}$. Since $s_n \leq t_n \leq M$ for every positive integer n, then the sequence $\{s_n\}$ is bounded above by M. It follows from Theorem 12–5 that the sequence $\{s_n\}$ converges to its least upper bound L where $L \leq M$. Therefore,

$$\sum_{n=1}^{\infty} a_n = \lim_{n \to \infty} s_n = L \leq M.$$

12-4 The Comparison and Ratio Tests

(b) If the series Σa_n diverges, then by Theorem 12-5 the sequence $\{s_n\}$ is unbounded. Since

$$t_n \geq s_n$$

for every positive integer n, then the sequence $\{t_n\}$ also is unbounded. Therefore, the series Σb_n diverges to ∞. ∎

As we shall see, the comparison test is one of the most useful convergence tests. Our previous work with p series and geometric series has given us a supply of known convergent and divergent series to use with the comparison test.

EXAMPLE 1 Show that the series

$$\sum_{n=1}^{\infty} \frac{2}{3^n + 1} = \frac{2}{3^1 + 1} + \frac{2}{3^2 + 1} + \frac{2}{3^3 + 1} + \frac{2}{3^4 + 1} + \cdots$$

converges to a number less than or equal to 1.

Solution Since $2/(3^n + 1) < 2/3^n$, then

$$\sum_{n=1}^{\infty} \frac{2}{3^n + 1} \ll \sum_{n=1}^{\infty} \frac{2}{3^n} = \frac{2}{3} + \frac{2}{3^2} + \frac{2}{3^3} + \cdots.$$

The dominant series is a geometric series (with $a = \tfrac{2}{3}$ and $r = \tfrac{1}{3}$) that converges to

$$\frac{a}{1-r} = \frac{2/3}{1 - 1/3} = 1.$$

It follows from the comparison test that the series $\Sigma\, 2/(3^n + 1)$ converges to a number that is no greater than 1. □

EXAMPLE 2 Show that the series $\displaystyle\sum_{n=1}^{\infty} \frac{\ln(n+2)}{n}$ diverges.

Solution Since $\dfrac{\ln(n+2)}{n} \geq \dfrac{1}{n}$ if $n \geq 1$, then

$$\sum_{n=1}^{\infty} \frac{\ln(n+2)}{n} \gg \sum_{n=1}^{\infty} \frac{1}{n}.$$

Since the given series dominates a divergent series, then it diverges. □

REMARKS ABOUT THE COMPARISON TEST

Several problems arise when we try to use the comparison test to decide whether a given positive-term series Σc_n converges. The main problem is finding another series to use for the comparison. To find such a series, we first need to make an intelligent guess about whether the original series converges or diverges. The following remarks may help to solve this problem.

(1) It may help to simplify the nth term of the given series as much as possible, discarding those parts that do not significantly affect its value when n is large. This

produces a new series, which usually has the same convergence properties as the given one and may possibly be used for comparison.

Let the given series be, for example,
$$\sum_{n=1}^{\infty} \frac{5\sqrt{n}+1}{3n^2-2}.$$

The nth term is
$$\frac{5\sqrt{n}+1}{3n^2-2} \approx \frac{5\sqrt{n}}{3n^2} = \frac{5}{3n^{3/2}}.$$

Thus, it would be natural to try to compare the given series with the series
$$\sum_{n=1}^{\infty} \frac{5}{3n^{3/2}} = \frac{5}{3} \sum_{n=1}^{\infty} \frac{1}{n^{3/2}}.$$

Since the new series is a multiple of a convergent p-series, we should expect the original series to converge.

Use of a Multiple of a Series for the Comparison Test

(2) It may be impossible to compare the given series directly with one that would be natural to use for the comparison test. We saw above, for example, that it would be natural to try to compare the series
$$\sum_{n=1}^{\infty} \frac{5\sqrt{n}+1}{3n^2-2} \quad \text{with} \quad \sum_{n=1}^{\infty} \frac{5}{3n^{3/2}}.$$

Unfortunately, the second series is term by term less than the given series. Thus, it cannot be used to show that the given series converges.

In a situation of this type, it may be possible to compare the given series with a multiple of the second series. In our example, we need to find a positive number k such that
$$\sum_{n=1}^{\infty} \frac{5\sqrt{n}+1}{3n^2-2} \ll k \sum_{n=1}^{\infty} \frac{5}{3n^{3/2}}.$$

It is easy to see that $k = 2$ will work in this problem. Since the series $2\Sigma 5/3n^{3/2}$ is convergent, then the original series
$$\sum_{n=1}^{\infty} \frac{5\sqrt{n}+1}{3n^2-2}$$
is convergent.

(3) As mentioned in the Remark at the end of Section 12–3, the first few terms of a series have no effect on its convergence or divergence. If the first few terms are radically different in form from the remaining terms, then we should temporarily discard them and try to use the comparison test on the series formed by the remaining terms.

For example, the nth term of the series
$$\sum_{n=1}^{\infty} \frac{\ln n}{n}$$

12–4 The Comparison and Ratio Tests

is $(\ln n)/n$. If $n = 1$, then $(\ln n)/n < 1/n$, but if $n \geq 3$, then $(\ln n)/n > 1/n$. If we discard the first two terms, we get the series

$$\frac{\ln 3}{3} + \frac{\ln 4}{4} + \frac{\ln 5}{5} + \frac{\ln 6}{6} + \cdots \gg \frac{1}{3} + \frac{1}{4} + \frac{1}{5} + \frac{1}{6} + \cdots.$$

Since the series

$$\frac{1}{3} + \frac{1}{4} + \frac{1}{5} + \frac{1}{6} + \cdots$$

is obtained by deleting the first two terms of the harmonic series, it diverges. It follows from the comparison test that the series

$$\frac{\ln 3}{3} + \frac{\ln 4}{4} + \frac{\ln 5}{5} + \frac{\ln 6}{6} + \cdots$$

diverges. Thus, the original series diverges.

THE SECOND COMPARISON TEST

Theorem 12–10 is one of several comparison tests. When it cannot be applied directly because of the difficulty mentioned in the second remark, another test may be possible. One candidate is given in the following theorem.

THEOREM 12–11 *(The Second Comparison Test)* Let Σa_n and Σb_n be positive-term series.

Second Comparison Test

(a) If

$$\lim_{n \to \infty} \frac{a_n}{b_n} = L$$

where L is a positive number, then either both series converge or both diverge.

(b) If the series Σb_n converges and $\lim_{n \to \infty} (a_n/b_n) = 0$, then the series Σa_n converges.

(c) If the series Σb_n diverges and $\lim_{n \to \infty} (a_n/b_n) = \infty$, then the series Σa_n diverges.

Proof of (a) Let ϵ be a positive number that is less than $L/2$. There exists an integer N such that

$$\left| \frac{a_n}{b_n} - L \right| < \epsilon < \frac{L}{2}$$

whenever $n \geq N$. It follows that

$$L - \frac{L}{2} < \frac{a_n}{b_n} < L + \frac{L}{2}$$

so that

$$\frac{Lb_n}{2} < a_n < \frac{3Lb_n}{2} \quad \text{whenever } n \geq N.$$

Suppose that the series $\sum_{n=1}^{\infty} b_n$ converges. Then the two series

$$\sum_{n=1}^{\infty} \frac{3Lb_n}{2} \quad \text{and} \quad \sum_{n=N}^{\infty} \frac{3Lb_n}{2}$$

also converge. Since

$$\sum_{n=N}^{\infty} a_n \ll \sum_{n=N}^{\infty} \frac{3Lb_n}{2}$$

then $\sum_{n=N}^{\infty} a_n$ converges, so that $\sum_{n=1}^{\infty} a_n$ converges.

If the series $\sum_{n=1}^{\infty} b_n$ diverges, then so do

$$\sum_{n=1}^{\infty} \frac{Lb_n}{2} \quad \text{and} \quad \sum_{n=N}^{\infty} \frac{Lb_n}{2}.$$

In this case we have

$$\sum_{n=N}^{\infty} \frac{Lb_n}{2} \ll \sum_{n=N}^{\infty} a_n$$

so that the series $\sum_{n=N}^{\infty} a_n$ and $\sum_{n=1}^{\infty} a_n$ both diverge. The proofs of (b) and (c) are omitted. ∎

EXAMPLE 3 Use Theorem 12–11 to show that the series

$$\sum_{n=1}^{\infty} \frac{5\sqrt{n} + 1}{3n^2 - 2}$$

converges.

Solution Since the nth term is

$$\frac{5\sqrt{n} + 1}{3n^2 - 2} \approx \frac{5}{3n^{3/2}}$$

it is natural to compare the original series with the series

$$\sum_{n=1}^{\infty} \frac{5}{3n^{3/2}}.$$

The limit of the sequence of ratios of the nth terms is

$$\lim_{n \to \infty} \frac{\frac{5\sqrt{n}+1}{3n^2-2}}{\frac{5}{3n^{3/2}}} = \lim_{n \to \infty} \frac{5n^{1/2}+1}{3n^2-2} \cdot \frac{3n^{3/2}}{5}$$
$$= \lim_{n \to \infty} \frac{15n^2 + 3n^{3/2}}{15n^2 - 10} = \lim_{n \to \infty} \frac{15 + (3/\sqrt{n})}{15 + (10/n^2)} = 1.$$

Since the p-series $\sum 5/3n^{3/2}$ converges, it follows from Theorem 12–11 that the given series converges. □

Remark This is the same series that was shown to be convergent in the second remark above. For the second comparison test, however, we do not have to find a multiplier k that

12–4 The Comparison and Ratio Tests

makes the comparison series dominant. It is helpful to eliminate guesswork whenever possible.

THE RATIO TEST

The following test is proved by using the comparison test. It is the simplest test to apply for many series.

THEOREM 12–12 *(The Ratio Test)* Let $\sum_{n=1}^{\infty} a_n$ be a positive-term series. Let

$$r_n = \frac{a_{n+1}}{a_n}$$

Ratio Test be the ratio of the $n + 1$st term to the nth term of the series. Suppose that the sequence $\{r_n\}$ converges to a number r.
(a) If $r < 1$, the original series converges.
(b) If $r > 1$, the original series diverges.
(c) If $r = 1$, a different test must be used.

Proof To prove this theorem, we construct a geometric series that can be compared with the original series.
(a) Suppose that

$$\lim_{n \to \infty} r_n = \lim_{n \to \infty} \frac{a_{n+1}}{a_n} = r < 1.$$

Let s be a number that is greater than r and less than 1. Let $\epsilon = s - r$. Since $r_n \to r$ as $n \to \infty$, there exists a positive integer N such that

$$r - \epsilon < r_n < r + \epsilon = s \qquad \text{if } n \geq N.$$

(See Fig. 12–6.)

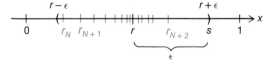

Figure 12–6.

In particular,

$$r_N = \frac{a_{N+1}}{a_N} < s, \qquad \text{so that } a_{N+1} < a_N s,$$

$$r_{N+1} = \frac{a_{N+2}}{a_{N+1}} < s, \qquad \text{so that } a_{N+2} < a_{N+1} s < a_N s^2,$$

and so on. In general, we find that

$$a_{N+k} < a_N s^k \qquad \text{for each positive integer } k.$$

Therefore,

$$a_N + a_{N+1} + a_{N+2} + a_{N+3} + \cdots \ll a_N + a_N s + a_N s^2 + a_N s^3 + \cdots.$$

Since $0 < s < 1$, the geometric series on the right converges. It follows from the comparison test that the series

$$a_N + a_{N+1} + a_{N+2} + \cdots$$

also converges. Since this series was obtained from the series $\sum_{n=1}^{\infty} a_n$ by deleting the first few terms, then that series also converges.

(b) The proof of (b) is similar to the proof of (a). We choose a number s in the range $1 < s < r$ and find a positive integer N such that

$$a_N + a_{N+1} + a_{N+2} + a_{N+3} + \cdots \gg a_N + a_N s + a_N s^2 + a_N s^3 + \cdots.$$

Since the geometric series on the right diverges, so does the one on the left. Thus, the original series must diverge. The details are left for you to work. (See Exercise 14.)

(c) The third part of the proof is by example. We need only to find a convergent series and a divergent series for which $r = 1$. (See Exercise 13.) ∎

EXAMPLE 4 Show that the series $\sum_{n=1}^{\infty} \dfrac{2^n}{n!}$ converges.

Solution The ratio of the $n + 1$st to the nth term is

$$r_n = \frac{a_{n+1}}{a_n} = \frac{2^{n+1}/(n+1)!}{2^n/n!} = \frac{2^{n+1}}{(n+1)!} \cdot \frac{n!}{2^n} = \frac{2}{n+1}.$$

Then

$$r = \lim_{n \to \infty} r_n = \lim_{n \to \infty} \frac{2}{n+1} = 0.$$

Since $r < 1$, the series converges. □

EXAMPLE 5 Does the series $\sum_{n=1}^{\infty} \dfrac{n^2}{2^n}$ converge or diverge?

Solution The ratio of the $n + 1$st term to the nth term is

$$r_n = \frac{a_{n+1}}{a_n} = \frac{(n+1)^2/2^{n+1}}{n^2/2^n} = \frac{(n+1)^2}{n^2} \cdot \frac{2^n}{2^{n+1}} = \frac{n^2 + 2n + 1}{n^2} \cdot \frac{1}{2}.$$

Then

$$r = \lim_{n \to \infty} r_n = \lim_{n \to \infty} \frac{n^2 + 2n + 1}{n^2} \cdot \frac{1}{2} = \frac{1}{2}.$$

The series converges. □

Remark The ratio test usually yields good information if the nth term of a series is an exponential function of n or a product involving $n!$ It is always inconclusive if the nth term is a rational function of n. (See Exercise 16.)

Exercises 12-4

1–12. Use one of the three tests of this section to decide whether the series converge or diverge.

- 1. $\sum_{n=1}^{\infty} \dfrac{n}{3^n}$

2. $\sum_{n=1}^{\infty} \dfrac{\sqrt{n}}{n+1}$

- 3. $\sum_{n=1}^{\infty} \dfrac{\sqrt{n}}{n^2+2}$

4. $\sum_{n=1}^{\infty} \dfrac{\sqrt{n}+1}{n^2-2}$

- 5. $\sum_{n=1}^{\infty} \dfrac{\cosh n}{\sqrt{n}}$

6. $\sum_{n=1}^{\infty} \dfrac{\tan^{-1} n}{n^2+1}$

- 7. $\sum_{n=1}^{\infty} \dfrac{2^{n+1}+1}{3^n-1}$

8. $\sum_{n=1}^{\infty} \dfrac{|\sin n|^n}{2^n-1}$

- 9. $\sum_{n=1}^{\infty} \dfrac{3^n}{(n+1)!}$

10. $\sum_{n=1}^{\infty} \dfrac{5^{2n}}{2^n \, n!}$

- 11. $\sum_{n=1}^{\infty} \dfrac{1}{2n-1}$

12. $\sum_{n=1}^{\infty} \dfrac{2^n}{(2n)!}$

13. Show that the ratio test is always inconclusive when applied to a p-series. Use this fact to find a convergent series and a divergent series with $r = 1$.

14. Work through the detailed proof of part (b) of the ratio test.

15. Let $a_1, a_2, a_3, \ldots$ be integers in the range $0 \le a_i \le 9$. The series

$$\frac{a_1}{10} + \frac{a_2}{10^2} + \frac{a_3}{10^3} + \frac{a_4}{10^4} + \cdots$$

Proper Decimal Fraction is called a *proper decimal fraction* (usually written as $0.a_1 a_2 a_3 a_4 \cdots$). Use the comparison test to prove that every proper decimal fraction converges to a number x in the range $0 \le x \le 1$. (*Hint:* Compare the series with the series for $0.99999 \cdots$.)

16. Show that the ratio test is inconclusive if the nth term of the series is a rational function of n. (*Hint:* Let

$$a_n = \frac{b_r n^r + b_{r-1} n^{r-1} + \cdots + b_1 n + b_0}{c_s n^s + c_{s-1} n^{s-1} + \cdots + c_1 n + c_0}$$

where b_r and c_s are nonzero. Show that $\lim_{n \to \infty} \dfrac{a_{n+1}}{a_n} = 1$.)

12-5 Alternating Series

Many series have some terms that are positive and others that are negative. We shall subsequently develop (Section 12-6) some general methods for working with such series—methods based on our tests for positive-term series. We now turn our attention to a special type of series—one with terms that alternate between positive and negative, such as the series

$$1 - \tfrac{1}{2} + \tfrac{1}{3} - \tfrac{1}{4} + \tfrac{1}{5} - \tfrac{1}{6} + \tfrac{1}{7} - \cdots.$$

Alternating Series Such series are called *alternating series*.

It is customary to write an alternating series in one of the two forms

$$a_1 - a_2 + a_3 - a_4 + \cdots = \sum_{n=1}^{\infty} (-1)^{n+1} a_n$$

or

$$-a_1 + a_2 - a_3 + a_4 - \cdots = \sum_{n=1}^{\infty} (-1)^n a_n$$

where $a_1, a_2, a_3, \ldots$ are positive numbers. Here we restrict ourselves to the first type of alternating series. Similar results hold for the second type.

A simple convergence test for alternating series is stated in Theorem 12–13.

THEOREM 12–13 (*Convergence Test for Alternating Series*) Let $a_1 - a_2 + a_3 - a_4 + \cdots$ be an alternating series with the properties

Convergence Test for Alternating Series

(1) $a_1 \geq a_2 \geq a_3 \geq a_4 \geq \cdots$

and

(2) $\lim_{n \to \infty} a_n = 0.$

Then the series converges to a limit L. Furthermore, the number L is between s_n and s_{n+1} for every positive integer n, where $\{s_n\}$ is the sequence of partial sums.

Proof We first work with $\{s_{2n}\}$, the sequence of partial sums with even subscripts. Observe that

$$s_2 = (a_1 - a_2),$$
$$s_4 = (a_1 - a_2) + (a_3 - a_4) \geq s_2,$$
$$s_6 = (a_1 - a_2) + (a_3 - a_4) + (a_5 - a_6) \geq s_4,$$

and so on. It follows that

$$s_2 \leq s_4 \leq s_6 \leq s_8 \leq \cdots$$

so that this sequence is monotone-nondecreasing. Observe also that

$$s_2 = a_1 - a_2 < a_1,$$
$$s_4 = a_1 - (a_2 - a_3) - a_4 < a_1,$$
$$s_6 = a_1 - (a_2 - a_3) - (a_4 - a_5) - a_6 < a_1,$$

and so on. Thus, the monotone-nondecreasing sequence $\{s_{2n}\}$ is bounded above by a_1. It follows from Theorem 12–4 that the sequence $\{s_{2n}\}$ converges to its least upper bound L.

We now consider s_{2n+1}, the sequence of partial sums with odd subscripts. Observe

12-5 Alternating Series

that

$$s_{2n+1} = (a_1 - a_2 + a_3 - a_4 + \cdots - a_{2n}) + a_{2n+1}$$
$$= s_{2n} + a_{2n+1}.$$

It follows that

$$\lim_{n \to \infty} s_{2n+1} = \lim_{n \to \infty} s_{2n} + \lim_{n \to \infty} a_{2n+1}$$
$$= L + 0 = L.$$

Since the sequence of partial sums with even subscripts and the sequence of partial sums with odd subscripts converge to the same limit L, then

$$\lim_{n \to \infty} s_n = L.$$

It follows that

$$a_1 - a_2 + a_3 - a_4 + \cdots = \lim_{n \to \infty} s_n = L.$$

If we calculate all the partial sums, we find that

$$s_2 \leq s_4 \leq s_6 \leq \cdots \leq s_5 \leq s_3 \leq s_1.$$

That is, all partial sums with even subscripts are less than or equal to all partial sums with odd subscripts. It follows that

$$s_2 \leq s_4 \leq s_6 \leq \cdots \leq L \leq \cdots \leq s_5 \leq s_3 \leq s_1.$$

(See Fig. 12–7.) Therefore, L is between s_n and s_{n+1} for every positive integer n. ∎

Figure 12–7.

EXAMPLE 1 (a) Use Theorem 12–13 to show that the alternating series

$$1 - \frac{1}{2} + \frac{1}{3} - \frac{1}{4} + \frac{1}{5} - \frac{1}{6} + \cdots$$

converges.

(b) Use s_{10} and s_{11}, the tenth and eleventh partial sums, to obtain bounds for the limit of the series.

Solution (a) The sequence of absolute values of the terms $\{a_n\}$ is defined by

$$a_n = \frac{1}{n}.$$

Since $1/n > 1/(n + 1)$ for every positive integer n, then $a_n > a_{n+1}$ ($n = 1, 2, 3, \ldots$). Furthermore

$$\lim_{n \to \infty} a_n = \lim_{n \to \infty} \frac{1}{n} = 0.$$

It follows from Theorem 12–13 that the alternating series

$$1 - \frac{1}{2} + \frac{1}{3} - \frac{1}{4} + \frac{1}{5} - \cdots$$

converges to a number L.

(b) The tenth and eleventh partial sums are

$$s_{10} = 1 - \tfrac{1}{2} + \tfrac{1}{3} - \cdots + \tfrac{1}{9} - \tfrac{1}{10} \approx 0.64563492$$
$$s_{11} = 1 - \tfrac{1}{2} + \tfrac{1}{3} - \cdots + \tfrac{1}{9} - \tfrac{1}{10} + \tfrac{1}{11} \approx 0.73654401.$$

It follows from Theorem 12–13 that

$$0.64563492 < \sum_{n=1}^{\infty} \frac{(-1)^{n+1}}{n} < 0.73654401. \ \square$$

Remark Theorem 12–13 cannot be used to determine divergence. If the conditions of the theorem are not satisfied, then some other test must be used.

EXAMPLE 2 Does the series $\sum_{n=1}^{\infty} (-1)^{n+1} \dfrac{3n+1}{4n+1}$ converge or diverge?

Solution The series is an alternating series with $a_1 > a_2 > a_3 > \cdots$. Observe, however, that

$$\lim_{n \to \infty} a_n = \lim_{n \to \infty} \frac{3n+1}{4n+1} = \frac{3}{4}.$$

Since the nth term does not approach zero as $n \to \infty$, it follows from Theorem 12–2 that the series diverges. $\square$

Remark The ratio test, comparison tests, integral test and alternating series test can be used to determine the convergence or divergence of most of the series of this chapter.

Exercises 7–14 are a miscellaneous collection of series to test your skill. You should first classify all these series before testing them for convergence. They should be classified according to the convergence tests that are likely to be successful and the tests that are likely to be unsuccessful. If a comparison test is to be used, find a possible series to use for the comparison.

Exercises 12–5

1–6. (a) Use Theorem 12–13 or some other test to decide whether the series converge. (*Hint for 6:* Write the series as a sum of two series. Use Theorem 12–13 for one of them.)

(b) If the series converges, then use Theorem 12–13 with $n = 5$ to estimate the value.

c(c) (*For Students with Computers or Programmable Calculators*) Work part (b) using $n = 50$.

•1 $\sum_{n=1}^{\infty} \dfrac{(-1)^{n+1}}{\sqrt{n}}$

2 $\sum_{n=1}^{\infty} \dfrac{(-1)^{n+1}}{\tan^{-1} n}$

•3 $\sum_{n=1}^{\infty} \frac{(-1)^n(3n+2)}{n^2+1}$

4 $\sum_{n=1}^{\infty} \frac{(-1)^{n+1} 2n}{n^2-10}$

•5 $\sum_{n=1}^{\infty} \frac{(-1)^n \ln n}{n}$

•6 $\sum_{n=1}^{\infty} \frac{2+(-1)^n n}{n^{3/2}}$

7–14. (*Review Series*) Use the tests developed in this chapter to decide whether the series converge.

•7 $\sum_{n=1}^{\infty} \frac{e^n}{e^{2n}+1}$

8 $\sum_{n=1}^{\infty} \frac{|\cos n|}{n^{3/2}}$

•9 $\sum_{n=1}^{\infty} \frac{e^n}{3^{n-1}}$

10 $\sum_{n=1}^{\infty} \frac{n^e+1}{e^n}$

•11 $\sum_{n=1}^{\infty} \frac{2n-3}{n^{3/2}+1}$

12 $\sum_{n=1}^{\infty} \frac{\sqrt{n}}{n^3-5}$

•13 $\sum_{n=1}^{\infty} \frac{\cosh n}{\sinh n}$

14 $\sum_{n=1}^{\infty} \frac{n \cos(n\pi)}{n^{3/2}+1}$

Accuracy of Approximation Using Alternating Series

15 (*Accuracy of Approximation*) Let $a_1 - a_2 + a_3 - a_4 + \cdots$ be an alternating series that converges to L. Use Theorem 12–13 to show that the error obtained by using the partial sum s_n to approximate L is less than a_{n+1}, the absolute value of the first term that is missing from s_n.

16 Find a value of n such that the partial sum s_n approximates the limit of the series to five decimal places. (*Hint:* We must find a value of n such that $|s_n - L| < 5/10^6$. Use the result of Exercise 15.)

•(a) $\sum_{n=1}^{\infty} \frac{(-1)^n}{\sqrt{n}}$

(b) $\sum_{n=1}^{\infty} \frac{(-1)^n}{n^4}$

17 Work through the details of Theorem 12–13 to show that

$$s_1 \geq s_3 \geq s_5 \geq \cdots .$$

12–6 Absolute and Conditional Convergence

Series were used for computations for over a century before their properties were examined rigorously. The first mathematicians to use series, in common with most beginning students of the calculus, assumed that series had properties like those of ordinary sums.

Consider, as an example, the problem of rearranging the terms of a series. The terms in a finite sum can be arranged in any order with the same result. For example,

$$1 - \tfrac{1}{2} + \tfrac{1}{3} - \tfrac{1}{4} + \tfrac{1}{5} = 1 + \tfrac{1}{3} + \tfrac{1}{5} - \tfrac{1}{2} - \tfrac{1}{4}.$$

It was natural to assume that the same property holds for infinite series. If the series Σa_n converges to L and the series Σb_n is obtained by rearranging the terms of this series, then it seemed clear that Σb_n should also converge to L.

In the nineteenth century, Cauchy startled the mathematical community by showing that such "obvious facts" might not be true. He classified convergent series into two types—*absolutely convergent* and *conditionally convergent*. He showed that the absolutely convergent series have properties similar to those of ordinary sums and the conditionally convergent ones do not.

After we examine the criteria that Cauchy used for his classification, we describe several properties of the two classes of series—mostly without proof. Finally, in a specific example, we give a rearrangement of a conditionally convergent series that converges to a different number than the original series.

Cauchy's discovery was at least partially responsible for the emphasis on rigor in the calculus. He showed that intuition cannot always be trusted—that "obvious facts" may not even be true.

ABSOLUTE AND CONDITIONAL CONVERGENCE

Absolute Convergence

A series Σa_n is said to *converge absolutely* if the related series

$$\sum_{n=1}^{\infty} |a_n|,$$

formed by summing the absolute values of the terms, converges.

EXAMPLE 1   The series

$$\sum_{n=1}^{\infty} \frac{(-1)^{n+1}}{n^2} = \frac{1}{1^2} - \frac{1}{2^2} + \frac{1}{3^2} - \frac{1}{4^2} + \cdots$$

is absolutely convergent. The related series of absolute value terms is the convergent p series

$$\sum_{n=1}^{\infty} \frac{1}{n^2} = \frac{1}{1^2} + \frac{1}{2^2} + \frac{1}{3^2} + \frac{1}{4^2} + \cdots. \quad \square$$

Conditional Convergence

A convergent series that is not absolutely convergent is said to be *conditionally convergent*.

EXAMPLE 2   The alternating harmonic series

$$\sum_{n=1}^{\infty} \frac{(-1)^{n+1}}{n} = 1 - \frac{1}{2} + \frac{1}{3} - \frac{1}{4} + \frac{1}{5} - \frac{1}{6} + \cdots$$

is conditionally convergent. Since it is an alternating series that meets the conditions of Theorem 12-13, it converges. The series of absolute-value terms is the harmonic series

$$\sum_{n=1}^{\infty} \left|\frac{(-1)^{n+1}}{n}\right| = \sum_{n=1}^{\infty} \frac{1}{n} = \frac{1}{1} + \frac{1}{2} + \frac{1}{3} + \frac{1}{4} + \frac{1}{5} + \cdots,$$

which diverges. $\square$

12–6 Absolute and Conditional Convergence

ABSOLUTE CONVERGENCE AND CONVERGENCE

It is not obvious that an absolutely convergent series must converge. Theorem 12–14 establishes that this is the case.

THEOREM 12–14
Basic Theorem for Absolute Convergence

If the series Σa_n converges absolutely, then it converges. To be more precise, if the series $\Sigma |a_n|$ converges to L, then the series Σa_n converges to a number M that is no greater than L.

Proof Let

$$s_n = a_1 + a_2 + \cdots + a_n$$

and

$$t_n = |a_1| + |a_2| + \cdots + |a_n|$$

be the nth partial sums of the two series.

If $\sum_{n=1}^{\infty} |a_n|$ converges to L, then $\{t_n\} \to L$ as $n \to \infty$. Since the sequence $\{t_n\}$ is monotone-nondecreasing, it follows from Theorem 12–4 that

$$t_n = |a_1| + |a_2| + \cdots + |a_n| \leq L$$

for every n.

It follows from the above remarks that

$$\begin{aligned} s_n + t_n &= (a_1 + a_2 + \cdots + a_n) + (|a_1| + |a_2| + \cdots + |a_n|) \\ &= (a_1 + |a_1|) + (a_2 + |a_2|) + \cdots + (a_n + |a_n|) \\ &\leq (|a_1| + |a_1|) + (|a_2| + |a_2|) + \cdots + (|a_n| + |a_n|) \leq 2L \end{aligned}$$

for every n. Observe also, that the sequence $\{s_n + t_n\}$ is monotone-nondecreasing. It follows from Theorem 12–4 that the sequence $\{s_n + t_n\}$ converges to a number N that is no greater than $2L$:

$$\lim_{n \to \infty} (s_n + t_n) = N \leq 2L.$$

Since $s_n = (s_n + t_n) - t_n$, then

$$\lim_{n \to \infty} s_n = \lim_{n \to \infty} (s_n + t_n) - \lim_{n \to \infty} t_n = N - L.$$

If we let $M = N - L$, it follows that

$$\sum_{n=1}^{\infty} a_n = \lim_{n \to \infty} s_n = M = N - L \leq 2L - L = L. \quad \blacksquare$$

EXAMPLE 3 Prove that the series

$$\frac{1}{1^{3/2}} + \frac{1}{2^{3/2}} - \frac{1}{3^{3/2}} + \frac{1}{4^{3/2}} + \frac{1}{5^{3/2}} - \frac{1}{6^{3/2}} + \cdots$$

converges.

Solution The related series of absolute value terms is

$$\frac{1}{1^{3/2}} + \frac{1}{2^{3/2}} + \frac{1}{3^{3/2}} + \frac{1}{4^{3/2}} + \frac{1}{5^{3/2}} + \frac{1}{6^{3/2}} + \cdots,$$

which is a convergent *p*-series. The given series is absolutely convergent, so it is convergent. □

THE RATIO TEST

It should be apparent that the convergence tests for positive-term series (Sections 12–3 and 12–4) can be used to develop tests for absolute convergence. It is necessary only to delete all the zero terms from a given series and then take the absolute values of the remaining terms. If the new series converges, then the original series is absolutely convergent and therefore convergent.

The most important of these modified tests is the ratio test, which we restate for absolute convergence.

THEOREM 12–15 (*The Ratio Test for Absolute Convergence*) Let $\sum_{n=1}^{\infty} a_n$ be a series with nonzero terms. Let

$$\lim_{n \to \infty} \frac{|a_{n+1}|}{|a_n|} = r.$$

(a) If $0 \leq r < 1$, the series is absolutely convergent.

(b) If $r > 1$, the series is divergent.

(c) If $r = 1$, a different test must be used. The series may be absolutely convergent, conditionally convergent, or divergent.

Proof (a) If $0 \leq r < 1$, then it follows from Theorem 12–12 that the series

$$|a_1| + |a_2| + \cdots + |a_n| + \cdots$$

is convergent. Thus, the given series is absolutely convergent.

(b) Let $r > 1$. Let ϵ be a positive number less than $r - 1$. There exists an integer N such that

$$\left| \left| \frac{a_{n+1}}{a_n} \right| - r \right| < \epsilon$$

if $n \geq N$. This implies that

$$1 < r - \epsilon < \left| \frac{a_{n+1}}{a_n} \right| < r + \epsilon$$

so that $|a_{n+1}| > |a_n|$. In particular, it follows that

$$|a_{N+1}| > |a_N|$$
$$|a_{N+2}| > |a_{N+1}| > |a_N|$$
$$\cdots \cdots \cdots \cdots \cdots$$
$$|a_{N+k}| > |a_N|.$$

12-6 Absolute and Conditional Convergence

Since $|a_{N+k}| > |a_N| > 0$ for every integer k, then the sequence $\{a_n\}$ cannot converge to zero. Thus the series $\sum_{n=1}^{\infty} a_n$ must diverge (by Theorem 12-2).

Part (c) is established by the example. (See Exercises 9, 10, 11.) ∎

EXAMPLE 4 Test the series

$$-\frac{1}{2^1} - \frac{2}{2^2} + \frac{3}{2^3} - \frac{4}{2^4} - \frac{5}{2^5} + \frac{6}{2^6} - \cdots$$

for convergence.

Solution Note that the series does not alternate. If a_n is the nth term, then $|a_n| = n/2^n$. It follows that

$$\frac{|a_{n+1}|}{|a_n|} = \frac{\dfrac{n+1}{2^{n+1}}}{\dfrac{n}{2^n}} = \frac{n+1}{n} \cdot \frac{2^n}{2^{n+1}} = \left(1 + \frac{1}{n}\right) \cdot \frac{1}{2}$$

so that

$$\lim_{n \to \infty} \frac{|a_{n+1}|}{|a_n|} = r = \frac{1}{2}.$$

The series is absolutely convergent. □

SUBSERIES

Subseries

Absolutely and conditionally convergent series have a number of "relatives" that may or may not share their properties. One family consists of subseries. A *subseries* of a series $\sum_1^{\infty} a_n$ is obtained by deleting one or more terms from the original series. For example, three of the many subseries of the alternating harmonic series

$$1 - \tfrac{1}{2} + \tfrac{1}{3} - \tfrac{1}{4} + \tfrac{1}{5} - \tfrac{1}{6} + \tfrac{1}{7} - \cdots$$

are

$$1 + \tfrac{1}{3} + \tfrac{1}{5} + \tfrac{1}{7} + \tfrac{1}{9} + \cdots \quad \text{(subseries of positive terms)}$$
$$-\tfrac{1}{2} - \tfrac{1}{4} - \tfrac{1}{6} - \tfrac{1}{8} - \tfrac{1}{10} - \cdots \quad \text{(subseries of negative terms)}$$

and

$$1 - \tfrac{1}{2} - \tfrac{1}{4} + \tfrac{1}{5} + \tfrac{1}{7} - \tfrac{1}{8} - \cdots \quad \text{(delete every third term)}.$$

It is not difficult to show that every infinite subseries of an absolutely convergent series must be absolutely convergent. On the other hand, a conditionally convergent series can have divergent subseries. This is illustrated in the above example of the alternating harmonic series. The subseries

$$1 + \tfrac{1}{3} + \tfrac{1}{5} + \tfrac{1}{7} + \tfrac{1}{9} + \cdots \quad \text{and} \quad -\tfrac{1}{2} - \tfrac{1}{4} - \tfrac{1}{6} - \tfrac{1}{8} - \cdots$$

are both divergent.

This illustrates the general situation. It can be proved that the subseries of positive terms of a conditionally convergent series must diverge, and so must the subseries of negative terms. This result is stated, for reference purposes, in Theorem 12–16. The proof is omitted.

THEOREM 12–16 (a) Every infinite subseries of an absolutely convergent series is absolutely convergent.
(b) The subseries that consists of all positive terms of a conditionally convergent series is divergent. The subseries of negative terms also is divergent.

REARRANGEMENTS OF SERIES

Another family consists of rearrangements of the parent series. A series $\Sigma_1^\infty b_n$ is said to be a *rearrangement* of the series $\Sigma_1^\infty a_n$ provided the two series have the same terms, possibly in a different order. For example, the two series

$$-\tfrac{1}{2} + 1 - \tfrac{1}{4} + \tfrac{1}{3} + \tfrac{1}{5} - \tfrac{1}{8} + \tfrac{1}{7} - \tfrac{1}{6} - \cdots$$

and

$$1 + \tfrac{1}{3} - \tfrac{1}{2} + \tfrac{1}{5} + \tfrac{1}{7} - \tfrac{1}{4} + \tfrac{1}{9} + \tfrac{1}{11} - \tfrac{1}{6} + \cdots$$

are rearrangements of the alternating harmonic series

$$1 - \tfrac{1}{2} + \tfrac{1}{3} - \tfrac{1}{4} + \tfrac{1}{5} - \tfrac{1}{6} + \tfrac{1}{7} - \tfrac{1}{8} + \cdots.$$

It can be proved that any rearrangement of an absolutely convergent series must be absolutely convergent and must converge to the same limit as the original series. This property should surprise no one. It is what we should intuitively expect of all series.

This property does not hold, however, for conditionally convergent series. In fact, the situation is more bizarre than we might imagine. If a series is conditionally convergent, then there must exist a rearrangement that diverges to ∞ and another rearrangement that diverges to $-\infty$. Furthermore, if k is any real number, then there exists a rearrangement of the original series that converges to k. These facts are summarized in Theorem 12–17. The proof is omitted.

THEOREM 12–17 (a) Any rearrangement of an absolutely convergent series is absolutely convergent and converges to the limit of the original series.
(b) Let Σa_n be a conditionally convergent series. Let k be a real number. There exists a rearrangement of this series that diverges to ∞, another rearrangement that diverges to $-\infty$, and another that converges to k.

For example, since the series

$$1 - \tfrac{1}{2} + \tfrac{1}{3} - \tfrac{1}{4} + \tfrac{1}{5} - \tfrac{1}{6} + \cdots$$

is conditionally convergent, then there exists a rearrangement that converges to $\sqrt{3}$, another that converges to $-\pi$, another that diverges to ∞, and so on.

A final example shows a rearrangement of a conditionally convergent series that converges to a number different from the original series. Before considering the example, we observe that a few zero terms can be inserted in a series without changing

12-6 Absolute and Conditional Convergence

its value. We can, in fact, insert a finite number of zeros between any two terms of a series and the new series will converge to the same value as the original one.

For example, the three series

$$1 + \tfrac{1}{4} + \tfrac{1}{8} + \tfrac{1}{16} + \tfrac{1}{32} + \cdots,$$
$$0 + 1 + 0 + \tfrac{1}{4} + 0 + \tfrac{1}{8} + 0 + \tfrac{1}{16} + 0 + \tfrac{1}{32} + 0 + \cdots,$$

and

$$1 + 0 + \tfrac{1}{4} + 0 + 0 + \tfrac{1}{8} + 0 + 0 + 0 + \tfrac{1}{16} + \cdots$$

converge to the same limit.

EXAMPLE 5 *(Rearrangements of Series)* It follows from the alternating-series test that the alternating harmonic series

$$1 - \frac{1}{2} + \frac{1}{3} - \frac{1}{4} + \frac{1}{5} - \frac{1}{6} + \frac{1}{7} - \cdots$$

is conditionally convergent. If we add several thousand terms of this series, we find that it converges to a number L, where

$$L \approx 0.693147.$$

(In fact, $L = \ln 2$.) We shall show that the rearrangement

$$1 + \frac{1}{3} - \frac{1}{2} + \frac{1}{5} + \frac{1}{7} - \frac{1}{4} + \frac{1}{9} + \frac{1}{11} - \frac{1}{6} + \cdots$$

converges to $3L/2 \approx 1.03972$.

Since $1 - \tfrac{1}{2} + \tfrac{1}{3} - \tfrac{1}{4} + \tfrac{1}{5} - \tfrac{1}{6} + \tfrac{1}{7} - \cdots = L$, then, by Theorem 12-3(c),

$$\frac{1}{2} - \frac{1}{4} + \frac{1}{6} - \frac{1}{8} + \frac{1}{10} - \frac{1}{12} + \frac{1}{14} - \cdots = \frac{L}{2}.$$

The limit of this series is not changed if we insert a zero term at the beginning and between each pair of terms. Thus,

$$0 + \frac{1}{2} + 0 - \frac{1}{4} + 0 + \frac{1}{6} + 0 - \frac{1}{8} + 0 + \frac{1}{10} + 0 - \cdots = \frac{L}{2}.$$

We now add this series to the original series term by term (see Theorem 12-3a):

$$\begin{array}{c} 1 - \dfrac{1}{2} + \dfrac{1}{3} - \dfrac{1}{4} + \dfrac{1}{5} - \dfrac{1}{6} + \dfrac{1}{7} - \dfrac{1}{8} + \dfrac{1}{9} - \dfrac{1}{10} + \dfrac{1}{11} - \cdots = L \\ 0 + \dfrac{1}{2} + 0 - \dfrac{1}{4} + 0 + \dfrac{1}{6} + 0 - \dfrac{1}{8} + 0 + \dfrac{1}{10} + \;\;0\;\; - \cdots = \dfrac{L}{2} \\ \hline 1 + 0 + \dfrac{1}{3} - \dfrac{2}{4} + \dfrac{1}{5} + 0 + \dfrac{1}{7} - \dfrac{2}{8} + \dfrac{1}{9} + 0 + \dfrac{1}{11} - \cdots = \dfrac{3L}{2}. \end{array}$$

If we delete the zero terms from this series and simplify the fractions, we get

$$1 + \frac{1}{3} - \frac{1}{2} + \frac{1}{5} + \frac{1}{7} - \frac{1}{4} + \frac{1}{9} + \frac{1}{11} - \cdots = \frac{3L}{2} \approx 1.03972.$$

Thus, this rearrangement of the alternating harmonic series converges to a value different from the original series. ∎

Exercises 12-6

1-8. Are the series conditionally convergent, absolutely convergent, or divergent?

•1 $\sum_{n=1}^{\infty} \dfrac{(-1)^n}{2n+1}$

2 $\sum_{n=1}^{\infty} \dfrac{(-1)^n \ln n}{n}$

•3 $\sum_{n=1}^{\infty} \dfrac{(-1)^{(n^2+n)/2}}{2n^{3/2}}$

4 $\sum_{n=1}^{\infty} \dfrac{(-1)^n n^2}{(n+1)\, 2^n}$

•5 $\sum_{n=1}^{\infty} \dfrac{(-1)^n\, n!}{2^n + 1}$

6 $\sum_{n=2}^{\infty} \dfrac{(-1)^n}{n(\ln n)^3}$

•7 $\sum_{n=1}^{\infty} \dfrac{\sin n}{n^2 - 2}$

8 $\sum_{n=1}^{\infty} \dfrac{(n!)^2}{(2n)!}$

9-11. (*Ratio Test*) Find a series Σa_n with $\lim_{n \to \infty} |a_{n+1}|/|a_n| = 1$ that has the stated property. (*Hint:* Review Exercise 13 of Section 12-4.)

9 Absolutely convergent

10 Conditionally convergent

11 Divergent

---

12 (a) Let $\{a_n\}$ be a sequence of nonzero numbers with the property that $\lim_{n \to \infty} |a_{n+1}|/|a_n| = L < 1$. Show that $\{a_n\} \to 0$ as $n \to \infty$. (*Hint:* Use Theorems 12-15, 12-2.)

(b) Let a and c be real numbers. Use (a) to show that

$$\lim_{n \to \infty} \dfrac{ca^n}{n!} = 0.$$

13 Let $L = 1 - \tfrac{1}{2} + \tfrac{1}{3} - \tfrac{1}{4} + \tfrac{1}{5} - \tfrac{1}{6} + \tfrac{1}{7} - \tfrac{1}{8} + \cdots$. Modify the argument in Example 5 to show that the rearrangement

$$\underline{1 - \tfrac{1}{2} + \tfrac{1}{3}} + \underline{\tfrac{1}{5} - \tfrac{1}{6} + \tfrac{1}{7}} - \tfrac{1}{4} + \underline{\tfrac{1}{9} - \tfrac{1}{10} + \tfrac{1}{11}} + \underline{\tfrac{1}{13} - \tfrac{1}{14} + \tfrac{1}{15}} - \tfrac{1}{8} + \underline{\tfrac{1}{17}} + \cdots$$

converges to $5L/4$. (*Hint:* Write $L/4$ in the form

$$\dfrac{L}{4} = 0 + 0 + 0 + \dfrac{1}{4} + 0 + 0 + 0 - \dfrac{1}{8} + 0 + 0 + 0 + \dfrac{1}{12} + 0 + \cdots.)$$

14 (*Insertion of Zero Terms*) Let the series Σb_n be obtained by adding a "few" zero terms to the series Σa_n. (No more than a finite number of zero terms can be added between every pair of terms of the original series.) Let $\{s_n\}$ and $\{t_n\}$ be the sequences of partial sums for Σa_n and Σb_n, respectively.

(a) Show that every partial sum s_n equals a partial sum t_k, and *vice versa*, provided b_1 is not an "added zero."

(b) Use the definition of limit to show that $\lim_{n \to \infty} s_n = L$ if and only if $\lim_{n \to \infty} t_n = L$. Interpret this result in terms of the series Σa_n and Σb_n.

12–7 Power Series

Power Series A series of form

$$\sum_{n=0}^{\infty} a_n x^n = a_0 + a_1 x + a_2 x^2 + \cdots + a_n x^n + \cdots$$

or

$$\sum_{n=0}^{\infty} a_n (x - c)^n = a_0 + a_1(x - c) + a_2(x - c)^2 + \cdots + a_n(x - c)^n + \cdots$$

where the a_k are real numbers, is called a *power series*.

For the first part of this section, we restrict ourselves to power series of the first type—series in powers of x instead of $x - c$.

If we substitute the real number x_0 for x in the power series $\Sigma a_n x^n$, we get a series of constants $\Sigma a_n x_0^n$. This series either converges to a unique real number or diverges. Thus, a power series defines a real function f at each point at which it converges. The function f is defined by

$$f(x) = \sum_{n=0}^{\infty} a_n x^n.$$

Its domain is the set of all x for which the series converges.

EXAMPLE 1 Determine the domain of f if

$$f(x) = \sum_{n=0}^{\infty} \frac{x^n}{3^n}$$

Solution Let $u_n = x^n/3^n$ be the nth term of the series. We apply the ratio test.

$$r = \lim_{n \to \infty} \frac{|u_{n+1}|}{|u_n|} = \lim_{n \to \infty} \frac{|x|^{n+1}/3^{n+1}}{|x|^n/3^n}$$

$$= \lim_{n \to \infty} \frac{|x|^{n+1}}{3^{n+1}} \cdot \frac{3^n}{|x|^n} = \lim_{n \to \infty} \frac{|x|}{3} = \frac{|x|}{3}.$$

Since the series converges if $r < 1$ and diverges if $r > 1$, then it converges if $-3 < x < 3$ and diverges if $x < -3$ or $x > 3$.

We now test the series at the endpoints of the interval $[-3, 3]$. At $x = -3$ the series reduces to

$$\sum_{n=0}^{\infty} (-1)^n = 1 - 1 + 1 - 1 + 1 - \cdots,$$

which diverges. At $x = 3$ it reduces to

$$\sum_{n=0}^{\infty} 1^n = 1 + 1 + 1 + \cdots,$$

which also diverges. Thus, the function is defined for $-3 < x < 3$ but no other values of x. □

We shall show that the result of Example 1 is typical of power series. If a power series $\Sigma a_n x^n$ converges for at least one nonzero real number but not all real numbers, then there is a real number R such that the series converges absolutely if $|x| < R$ and diverges if $|x| > R$.

Theorem 12–18 is basic to our work.

THEOREM 12–18 (a) If the power series $\Sigma_{n=0}^{\infty} a_n x^n$ converges for $x = x_1$ where x_1 is a nonzero real number, then it converges absolutely for all x such that $|x| < |x_1|$.

(b) If the series $\Sigma a_n x^n$ diverges for $x = x_2$, then it diverges for all x such that $|x| > |x_2|$.

Proof (a) If $\Sigma_{n=1}^{\infty} a_n x_1^n$ converges, then it follows from Theorem 12–2 that

$$\lim_{n \to \infty} a_n x_1^n = 0,$$

which implies that

$$\lim_{n \to \infty} |a_n| \cdot |x_1|^n = 0.$$

If we let $\epsilon = 1$ in the definition of the limit of a sequence, we find that there exists a positive integer N such that

$$0 \leq |a_n| \cdot |x_1|^n < 1 \quad \text{if } n \geq N.$$

Let x_0 be a number such that $|x_0| < |x_1|$. Let

$$r = \frac{|x_0|}{|x_1|} < 1.$$

Then

$$|a_n| \cdot |x_0|^n = |a_n| \cdot |x_1|^n \cdot \frac{|x_0|^n}{|x_1|^n} < 1 \cdot \left(\frac{|x_0|}{|x_1|}\right)^n = r^n \quad \text{if } n \geq N.$$

It follows that

$$|a_N| \cdot |x_0|^N + |a_{N+1}| \cdot |x_0|^{N+1} + |a_{N+2}| \cdot |x_0|^{N+2} + \cdots$$
$$\ll r^N + r^{N+1} + r^{N+2} + \cdots.$$

Since the second series is a convergent geometric series, then it follows from the comparison test that the series

$$|a_N| \cdot |x_0|^N + |a_{N+1}| \cdot |x_0|^{N+1} + |a_{N+2}| \cdot |x_0|^{N+1} + \cdots$$

converges. Since the convergence of a series is not affected by its first few terms, then the series

$$|a_0| + |a_1| \cdot |x_0| + |a_2| \cdot |x_0|^2 + \cdots + |a_N| \cdot |x_0|^N + |a_{N+1}| \cdot |x_0|^{N+1} + \cdots$$

also converges. Therefore, the series

$$a_0 + a_1 x_0 + a_2 x_0^2 + a_3 x_0^3 + \cdots$$

12-7 Power Series

converges absolutely. Since x_0 is an arbitrary number such that $|x_0| < |x_1|$, then the series

$$a_0 + a_1 x + a_2 x^2 + a_3 x^3 + \cdots$$

converges absolutely whenever $|x| < |x_1|$.

Proof **(b)** Suppose that the series $\Sigma a_n x_2{}^n$ diverges. Let x_0 be a number such that $|x_0| > |x_2|$. The series $\Sigma a_n x_0{}^n$ either converges or diverges. If the series converges, then it follows from part (a) that the series $\Sigma a_n x_2{}^n$ also converges—an impossibility. Consequently, the series $\Sigma a_n x_0{}^n$ must diverge. Since x_0 is an arbitrary number such that $|x_0| > |x_2|$, then the series

$$a_0 + a_1 x + a_2 x^2 + a_3 x^3 + \cdots$$

diverges if $|x| > |x_2|$. ∎

EXAMPLE 2 The series

$$1 + \frac{x}{2} + \frac{x^2}{3} + \frac{x^3}{4} + \frac{x^4}{5} + \cdots$$

converges if $x = -1$ and diverges if $x = 1$. It follows from Theorem 12–18 that this series converges absolutely if $-1 < x < 1$ and diverges if $x < -1$ or $x \geq 1$. □

We now prove the main result of this section.

THEOREM 12–19

Convergence Properties of a Power Series

Let $\Sigma a_n x^n$ be a power series. Exactly one of the following is true.

(1) The series converges for all x.
(2) The series converges for $x = 0$ and diverges for all other values of x.
(3) There exists a positive number R such that the series converges absolutely if $|x| < R$ and diverges if $|x| > R$.

Proof We must show that (3) holds if neither (1) nor (2) holds. Suppose that the series Σa_n does not satisfy conditions (1) or (2) in the statement of the theorem. Then there must exist a positive number x_1 such that $\Sigma a_n x_1{}^n$ converges and a positive number x_2 such that $\Sigma a_n x_2{}^n$ diverges. (See Exercise 12.) It follows from Theorem 12–18 that the series

$$a_0 + a_1 x + a_2 x^2 + \cdots$$

converges if $|x| < x_1$ and diverges if $|x| > x_2$.

Let $\mathcal{S}$ be the set of all positive numbers x for which the series converges. Then $\mathcal{S}$ is nonempty (since $x_1 \in \mathcal{S}$) and is bounded above by x_2. Thus, $\mathcal{S}$ has a least upper bound R. We shall show that R has the properties stated in the theorem.

Let x be a number in the range $-R < x < R$. Since R is the least upper bound of $\mathcal{S}$, there exists a positive number x_3 in $\mathcal{S}$ that is less than R and greater than $|x|$ (see Fig. 12–8a). Since $x_3 \in \mathcal{S}$, then the series $\Sigma a_n x_3{}^n$ converges. It follows from Theorem 12–18(b) that the series $\Sigma a_n x^n$ is absolutely convergent.

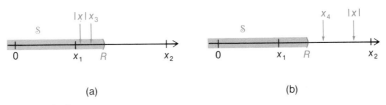

(a)   (b)

Figure 12–8.

Now let x be a number such that $|x| > R$. Let x_4 be a number between R and $|x|$. Since R is the least upper bound for S and $x_4 > R$, then $x_4 \notin S$. (See Fig. 12–8b.) Therefore, the series $\Sigma a_n x_4^n$ diverges. Since $|x| > |x_4|$, it follows from Theorem 12–18(b) that the series $\Sigma a_n x^n$ diverges.

We have proved that the series $\Sigma a_n x^n$ converges absolutely if $-R < x < R$ and diverges if $x < -R$ or $x > R$. As we see from the examples, the series may either converge or diverge at $x = \pm R$. ∎

It follows from Theorem 12–19 that a power series $\Sigma a_n x^n$ converges only at $x = 0$ *or* converges for all x *or* converges on one of the intervals $[-R, R]$, $(-R, R)$, $[-R, R)$, $(-R, R]$ for some real number $R > 0$. The set of points for which the series converges is called its *interval of convergence*.

Interval and Radius of Convergence

The number R is called the *radius of convergence*. (See Fig. 12–9.) In the event that the series converges for all x, we define the radius of convergence to be $R = \infty$. If it converges only for $x = 0$, we define the radius of convergence to be $R = 0$.

Series converges absolutely if $-R<x<R$

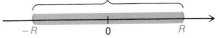

Figure 12–9 The interval of convergence. The power series $\Sigma a_n x^n$ converges absolutely if $|x| < R$ and diverges if $|x| > R$.

As we shall see, the radius of convergence is important in problems involving derivatives and integrals of series.

EXAMPLE 3   It follows from our work in Example 2 that the radius of convergence of the series

$$1 + \frac{x}{2} + \frac{x^2}{3} + \frac{x^4}{4} + \frac{x^5}{5} + \cdots$$

is $R = 1$. The interval of convergence is $[-1, 1)$. □

Results similar to Theorem 12–19 hold for series of the form

$$a_0 + a_1(x - c) + a_2(x - c)^2 + a_3(x - c)^3 + \cdots$$

If we let $y = x - c$, this series can be rewritten

$$a_0 + a_1 y + a_2 y^2 + a_3 y^3 + \cdots$$

12–7 Power Series

If R is the radius of convergence of this last series, then the original series will converge if $|x - c| < R$ and diverge if $|x - c| > R$. As with the other type of power series, the number R is called the *radius of convergence* of the series. These results are stated in the accompanying corollary.

COROLLARY TO THEOREM 12–19   Let

$$a_0 + a_1(x - c) + a_2(x - c)^2 + a_3(x - c)^3 + \cdots + a_n(x - c)^n + \cdots$$

be a power series. Exactly one of the following is true.
   (1) The series converges for all x.
   (2) It converges for $x = c$ and diverges for all other values of x.
   (3) There exists a positive number R such that the series converges absolutely if $|x - c| < R$ and diverges if $|x - c| > R$.

DERIVATIVES AND INTEGRALS

It can be proved that power series can be differentiated and integrated term by term. These new series have the same radii of convergence as the original series.

For reference purposes we state this result as a theorem. The proof is omitted.

THEOREM 12–20

Derivative and Integral of a Power Series

Let

$$f(x) = a_0 + a_1 x + a_2 x^2 + a_3 x^3 + a_4 x^4 + \cdots .$$

Then

$$f'(x) = a_1 + 2a_2 x + 3a_3 x^2 + 4a_4 x^3 + \cdots$$

and

$$\int f(x)\, dx = C + a_0 x + \frac{a_1 x^2}{2} + \frac{a_2 x^3}{3} + \frac{a_3 x^4}{4} + \cdots .$$

These new series have the same radii of convergence as the given power series. Similar results hold for series of the form

$$a_0 + a_1(x - c) + a_2(x - c)^2 + \cdots + a_n(x - c)^n + \cdots .$$

Theorem 12–20 is a powerful result. It can often be used to determine the power series for a particular function.

EXAMPLE 4   Show that

$$\tan^{-1} x = x - \frac{x^3}{3} + \frac{x^5}{5} - \frac{x^7}{7} + \cdots \quad \text{if} \quad -1 < x < 1.$$

Solution   Recall that

$$\tan^{-1} x = \int \frac{dx}{1 + x^2} + C.$$

Also note that the geometric series

$$1 - x^2 + x^4 - x^6 + \cdots$$

converges to $1/(1 + x^2)$ if $-1 < x < 1$. Then

$$\tan^{-1} x = \int \frac{dx}{1 + x^2} + C$$
$$= \int (1 - x^2 + x^4 - x^6 + \cdots) \, dx + C$$
$$= C + x - \frac{x^3}{3} + \frac{x^5}{5} - \frac{x^7}{7} + \cdots \quad \text{(by Theorem 12–20)}.$$

This series converges to $\tan^{-1} x$ if $-1 < x < 1$.

To evaluate the constant of integration, recall that $\tan^{-1} 0 = 0$. Thus

$$0 = C + 0 - \frac{0^3}{3} + \frac{0^5}{5} - \frac{0^7}{7} + \cdots$$

so that $C = 0$. The series is

$$\tan^{-1} x = x - \frac{x^3}{3} + \frac{x^5}{5} - \frac{x^7}{7} + \cdots, \quad \text{provided} \quad -1 < x < 1.$$

The results for integrals of power series are even more powerful than indicated in Theorem 12–20. If the power series for an integral converges at an endpoint of its interval of convergence, then the function defined by the series is continuous there.

For example, the series for $\tan^{-1} x$

$$x - \frac{x^3}{3} + \frac{x^5}{5} - \frac{x^7}{7} + \cdots$$

converges if $x \pm 1$. Thus,

$$\tan^{-1} x = x - \frac{x^3}{3} + \frac{x^5}{5} - \frac{x^7}{7} + \cdots, \quad -1 \leq x \leq 1.$$

Leibniz used this series to obtain a series representation for π. Note that

$$\frac{\pi}{4} = \tan^{-1} 1 = 1 - \frac{1^3}{3} + \frac{1^5}{5} - \frac{1^7}{7} + \cdots.$$

Thus

Series for π

$$\pi = 4(1 - \tfrac{1}{3} + \tfrac{1}{5} - \tfrac{1}{7} + \cdots).$$

This series for π converges very slowly. We would need to use thousands of terms to get a reasonable approximation to π. A better series based on the formula in Example 4 is developed in Exercise 11.

Exercises 12–7

1–6. Find all values of x for which the power series is convergent. Where is it absolutely convergent?

•1 $\displaystyle\sum_{n=1}^{\infty} \frac{2^{n+1} x^n}{(n+1)!}$

2 $\displaystyle\sum_{n=1}^{\infty} \frac{(-1)^n x^{2n+1}}{(2n+1)!}$

•3 $\displaystyle\sum_{n=1}^{\infty} \frac{2^{n+1} x^n}{n+1}$

4 $\displaystyle\sum_{n=1}^{\infty} \frac{(-1)^n (x-1)^{2n+1}}{2n+1}$

•5 $\displaystyle\sum_{n=1}^{\infty} \frac{(x+3)^n n!}{2n+1}$

6 $\displaystyle\sum_{n=1}^{\infty} \frac{n(x-2)^n}{2^n}$

•7 A power series $\Sigma a_n x^n$ is known to converge when $x = 1$ and to diverge when $x = -1.5$. What can be said about its radius of convergence? its interval of convergence?

8 A power series $\Sigma a_n (x - 2)^n$ is known to converge when $x = 4$ and diverge when $x = 0$. What can be said about its radius of convergence? its interval of convergence?

9 A function f is defined by the power series

$$f(x) = 1 - \frac{x}{2!} + \frac{x^2}{4!} - \frac{x^3}{6!} + \frac{x^4}{8!} - \cdots.$$

(a) Find the domain of f.
•(b) Find the equation of the tangent line to the graph of f at the point at which $x = 0$.
•(c) Calculate $\int f(x)\, dx$.

10 A function f is defined by the power series

$$f(x) = 1 + \frac{x}{1!} + \frac{x^2}{2!} + \frac{x^3}{3!} + \frac{x^4}{4!} + \cdots.$$

(a) Show that f is defined for all real x.
(b) Show that $f'(x) = f(x)$. Use this result, along with the value of $f(0)$, to show that $f(x) = e^x$.
c(c) Use (b) to show that

$$e = 1 + \frac{1}{1!} + \frac{1}{2!} + \frac{1}{3!} + \frac{1}{4!} + \frac{1}{5!} + \cdots.$$

Use the first six terms of this series to approximate e.
c(d) Find a series for $e^{1/2}$ comparable to the one in (c) for e. Use the first six terms to approximate $e^{1/2}$.

11 (a) Show that $\tan^{-1} \frac{2}{5} + \tan^{-1} \frac{3}{7} = \pi/4$. (Hint: Let $\alpha = \tan^{-1} \frac{2}{5}$. Use the formula

$$\tan(\alpha + \beta) = \frac{\tan \alpha + \tan \beta}{1 - \tan \alpha \tan \beta}.)$$

Approximation of π

(b) Show that

$$\frac{\pi}{4} = \left[\frac{2}{5} + \frac{3}{7}\right] - \frac{1}{3}\left[\left(\frac{2}{5}\right)^3 + \left(\frac{3}{7}\right)^3\right] + \frac{1}{5}\left[\left(\frac{2}{5}\right)^5 + \left(\frac{3}{7}\right)^5\right] - \cdots$$

$$= \sum_{n=1}^{\infty} \frac{(-1)^{n-1}}{2n-1}\left[\left(\frac{2}{5}\right)^{2n-1} + \left(\frac{3}{7}\right)^{2n-1}\right].$$

•**(c)** Use the approximating sum s_4 to approximate $\pi/4$. How good is the approximation?

12 Justify the statements made in the first paragraph of the proof of Theorem 12–19. In particular, explain why *positive* numbers x_1 and x_2 exist with the specified properties.

12–8 Taylor Series

A power series

$$\sum_{n=0}^{\infty} a_n(x-c)^n = a_0 + a_1(x-c) + a_2(x-c)^2 + a_3(x-c)^3 + \cdots$$

defines a function $f(x)$ on its interval of convergence. If we are given the function f, we can get the coefficients $a_0, a_1, a_2, \cdots$.

We shall find it convenient to make a slight change in our notation for partial sums when we work with power series. If the series is

$$a_0 + a_1(x-c) + a_2(x-c)^2 + a_3(x-c)^3 + \cdots,$$

then we define the *sequence of partial sums* (all of which are functions of x) by

Sequence of Partial Sums for a Power Series

$$s_0(x) = a_0$$
$$s_1(x) = a_0 + a_1(x-c)$$
$$s_2(x) = a_0 + a_1(x-c) + a_2(x-c)^2$$
$$\cdots\cdots\cdots\cdots\cdots\cdots\cdots$$
$$s_n(x) = a_0 + a_1(x-c) + a_2(x-c)^2 + \cdots + a_n(x-c)^n.$$
$$\cdots\cdots\cdots\cdots\cdots\cdots\cdots$$

Thus, $s_n(x)$ is a polynomial of degree no greater than n.

It follows from the definition of convergence of a series that

$$f(x) = a_0 + a_1(x-c) + a_2(x-c)^2 + \cdots$$

for a given value of x if and only if

$$\{s_n(x)\} \to f(x).$$

In this section we show how a series can be obtained for a given function. In Section 12–9 we consider the problem of deciding whether the series converges to this function or to some other function.

We illustrate the general method by finding a series representation for e^x.

12-8 Taylor Series

EXAMPLE 1 Show that if $f(x) = e^x$ can be represented by a series of form

$$a_0 + a_1 x + a_2 x^2 + a_3 x^3 + a_4 x^4 + \cdots,$$

Power Series for e^x

then the series is

$$1 + \frac{x}{1!} + \frac{x^2}{2!} + \frac{x^3}{3!} + \frac{x^4}{4!} + \cdots.$$

Solution Suppose that $f(x) = e^x$ can be represented by a series

$$f(x) = e^x = a_0 + a_1 x + a_2 x^2 + a_3 x^3 + \cdots + a_n x^n + \cdots.$$

Since

$$f(0) = e^0 = 1 = a_0 + a_1 \cdot 0 + a_2 \cdot 0^2 + \cdots,$$

then

$$a_0 = 1.$$

Since a power series can be differentiated term by term, then

$$f'(x) = e^x = a_1 + 2a_2 x + 3a_3 x^2 + 4a_4 x^3 + \cdots + na_n x^{n-1} + \cdots.$$

If we evaluate this series at $x = 0$, we find that

$$f'(0) = e^0 = 1 = a_1 + 2a_2 \cdot 0 + 3a_3 \cdot 0^2 + \cdots$$

so that

$$a_1 = 1.$$

We now repeat the process, calculating the second derivative with respect to x and evaluating the new series at $x = 0$:

$$f''(x) = e^x = 2a_2 + 3 \cdot 2a_3 x + 4 \cdot 3a_4 x^2 + \cdots + n(n-1)a_n x^{n-2} + \cdots$$
$$f''(0) = e^0 = 1 = 2a_2$$
$$a_2 = \tfrac{1}{2}.$$

We repeat the process again, getting

$$f'''(x) = e^x = 6a_3 + 4 \cdot 3 \cdot 2a_4 x + \cdots + n(n-1)(n-2)a_n x^{n-3} + \cdots$$
$$f'''(0) = e^0 = 1 = 6a_3$$
$$a_3 = \tfrac{1}{6}.$$

This process can be continued indefinitely, giving us

$$a_0 = 1, \quad a_1 = 1, \quad a_2 = \frac{1}{2} = \frac{1}{2!}, \quad a_3 = \frac{1}{6} = \frac{1}{3!}, \quad a_4 = \frac{1}{24} = \frac{1}{4!}, \cdots.$$

In general we find that

$$a_n = \frac{1}{n!} \quad (n = 0, 1, 2, 3, \cdots).$$

Thus, the series is

$$1 + \frac{x}{1!} + \frac{x^2}{2!} + \frac{x^3}{3!} + \frac{x^4}{4!} + \cdots + \frac{x^n}{n!} + \cdots. \;\square$$

Remark Example 1 does not establish that

$$e^x = 1 + \frac{x}{1!} + \frac{x^2}{2!} + \frac{x^3}{3!} + \frac{x^4}{4!} + \cdots.$$

It establishes that if e^x can be written as a power series, then that power series is the one that is stated. To establish that the power series converges to e^x, we would need to show that

$$\lim_{n \to \infty} s_n(x) = \lim_{n \to \infty} \left(1 + \frac{x}{1!} + \frac{x^2}{2!} + \cdots + \frac{x^n}{n!}\right) = e^x$$

for each real number x.

TAYLOR SERIES

The work of Example 1 can be generalized as follows. Let $f(x)$ have derivatives of all orders at $x = c$. If $f(x)$ can be represented by a series of form

$$f(x) = a_0 + a_1(x - c) + a_2(x - c)^2 + \cdots + a_n(x - c)^n + \cdots,$$

then

$$a_0 = f(c), \quad a_1 = \frac{f'(c)}{1!}, \quad a_2 = \frac{f''(c)}{2!}, \quad \ldots, \quad a_n = \frac{f^{(n)}(c)}{n!}, \quad \ldots$$

Thus, the series is

Taylor Series for f

$$f(c) + \frac{f'(c)}{1!}(x - c) + \frac{f''(c)}{2!}(x - c)^2 + \frac{f'''(c)}{3!}(x - c)^3 + \cdots$$
$$+ \frac{f^{(n)}(c)}{n!}(x - c)^n + \cdots.$$

Such a series is called a *Taylor series* (after Brook Taylor, English mathematician, 1685–1731).

It is usually simple to calculate the coefficients for a Taylor series. The hard work begins after we have the series. First we must calculate its interval of convergence, then decide whether the series converges to the original function on that interval. We content ourselves here with determining the series and the interval of convergence. Later we consider the problem of determining whether the series converges to the original function (Section 12–9).

MACLAURIN SERIES

If the number c equals zero, then the Taylor series has the particularly simple form

Maclaurin Series for f

$$f(0) + \frac{f'(0)}{1!}x + \frac{f''(0)}{2!}x^2 + \frac{f'''(0)}{3!}x^3 + \cdots + \frac{f^{(n)}(0)}{n!}x^n + \cdots.$$

12–8 Taylor Series

Such a series is called a *Maclaurin series* (after the Scottish mathematician Colin Maclaurin, 1698–1746).

EXAMPLE 2 In Example 1 we found the Maclaurin series for e^x to be

$$1 + \frac{x}{1!} + \frac{x^2}{2!} + \frac{x^3}{3!} + \frac{x^4}{4!} + \cdots + \frac{x^n}{n!} + \cdots.$$

Assume that this series converges to e^x on its interval of convergence.
(a) Determine the interval of convergence for this series.
(b) Use $s_8(x)$ to approximate $\sqrt{e}$.

Solution (a) We apply the ratio test:

$$\lim_{n \to \infty} \frac{|u_{n+1}|}{|u_n|} = \lim_{n \to \infty} \frac{|x|^{n+1}/(n+1)!}{|x|^n/n!}$$

$$= \lim_{n \to \infty} \frac{|x|^{n+1}}{|x|^n} \cdot \frac{n!}{(n+1)!}$$

$$= \lim_{n \to \infty} |x| \cdot \frac{1}{n+1} = 0.$$

It follows from the ratio test that the series converges for all real x. The interval of convergence is $(-\infty, \infty)$.

(b) If we assume that the series actually converges to e^x, then it follows that

$$e^x = 1 + \frac{x}{1!} + \frac{x^2}{2!} + \frac{x^3}{3!} + \cdots.$$

In particular

$$\sqrt{e} = e^{1/2} = 1 + \frac{1/2}{1!} + \frac{(1/2)^2}{2!} + \frac{(1/2)^3}{3!} + \cdots + \frac{(1/2)^n}{n!} + \cdots.$$

If we use $s_8(\tfrac{1}{2})$ to approximate $e^{1/2}$, then we have

$$\sqrt{e} = e^{1/2} \approx 1 + \frac{(1/2)}{1!} + \frac{(1/2)^2}{2!} + \cdots + \frac{(1/2)^8}{8!}.$$

The terms are calculated as follows:

$$1 = 1.0000000$$
$$(\tfrac{1}{2})/1! = 0.5000000$$
$$(\tfrac{1}{2})^2/2! = 0.1250000$$
$$(\tfrac{1}{2})^3/3! \approx 0.0208333$$
$$(\tfrac{1}{2})^4/4! \approx 0.0026042$$
$$(\tfrac{1}{2})^5/5! \approx 0.0002604$$
$$(\tfrac{1}{2})^6/6! \approx 0.0000217$$
$$(\tfrac{1}{2})^7/7! \approx 0.0000016$$
$$(\tfrac{1}{2})^8/8! \approx 0.0000001$$

Sum: $s_8(\tfrac{1}{2}) \approx 1.6487213$

Thus $\sqrt{e} \approx s_8(\tfrac{1}{2}) \approx 1.6487213$. □

EXAMPLE 3 (a) Find the Taylor series for ln x for powers of $x - 1$.
(b) Find the interval of convergence.
(c) Assume that the Taylor series converges to ln x on its interval of convergence. Use the series to calculate ln 3.

Solution (a) The Taylor series for $f(x) = \ln x$ is

$$f(x) = f(1) + \frac{f'(1)}{1!}(x-1) + \frac{f''(1)}{2!}(x-1)^2 + \frac{f'''(1)}{3!}(x-1)^3$$

$$+ \cdots + \frac{f^{(n)}(1)}{n!}(x-1)^n + \cdots.$$

The derivatives are

$$f(x) = \ln x$$

$$f'(x) = x^{-1} = \frac{1}{x}$$

$$f''(x) = -x^{-2} = -\frac{1}{x^2}$$

$$f'''(x) = 2x^{-3} = \frac{2}{x^3}$$

$$f^{(4)}(x) = -2 \cdot 3x^{-4} = -\frac{2 \cdot 3}{x^4}$$

$$f^{(5)}(x) = 2 \cdot 3 \cdot 4x^{-5} = \frac{2 \cdot 3 \cdot 4}{x^5}$$

. . . .

It appears from these terms that

$$f^{(n)}(x) = (-1)^{n+1}\frac{(n-1)!}{x^n} \quad (n \geq 1).$$

This result can be established by mathematical induction (see Exercise 24). Thus,

$$f^{(n)}(1) = (-1)^{n+1}(n-1)!$$

Since $f(1) = 0$, then the Taylor series for ln x is

$$0 + \frac{0!}{1!}(x-1) - \frac{1!}{2!}(x-1)^2 + \frac{2!}{3!}(x-1)^3 - \cdots$$

$$+ \frac{(-1)^{n+1}(n-1)!}{n!}(x-1)^n + \cdots$$

$$= (x-1) - \frac{(x-1)^2}{2} + \frac{(x-1)^3}{3} - \frac{(x-1)^4}{4} + \cdots$$

$$+ \frac{(-1)^{n+1}(x-1)^n}{n} + \cdots.$$

12–8 Taylor Series

(b) We apply the ratio test:

$$r = \lim_{n\to\infty} \frac{|u_{n+1}|}{|u_n|} = \lim_{n\to\infty} \frac{|x-1|^{n+1}/(n+1)}{|x-1|^n/n}$$

$$= \lim_{n\to\infty} \frac{|x-1|^{n+1}}{|x-1|^n} \cdot \frac{n}{n+1}$$

$$= \lim_{n\to\infty} |x-1| \cdot \frac{n}{n+1} = |x-1|.$$

The series converges absolutely if

$$r = |x-1| < 1$$

and diverges if

$$r = |x-1| > 1.$$

Thus, it converges absolutely if

$$0 < x < 2$$

and diverges if

$$x < 0 \text{ or } x > 2.$$

We now test the series at the endpoints of the interval of convergence. If $x = 0$, the series is

$$-1 - \frac{1}{2} - \frac{1}{3} - \frac{1}{4} - \frac{1}{5} - \cdots,$$

which diverges (a multiple of the harmonic series). If $x = 2$, the series is the alternating series

$$1 - \frac{1}{2} + \frac{1}{3} - \frac{1}{4} + \frac{1}{5} - \frac{1}{6} + \cdots,$$

which converges. Thus, the series converges if

$$0 < x \leq 2$$

and diverges if

$$x \leq 0 \quad \text{or} \quad x > 2.$$

(c) We cannot apply the series directly to calculate ln 3 because 3 is outside the interval of convergence. Recall, however that $\ln 3 = -\ln \frac{1}{3}$ and that $\ln \frac{1}{3}$ can be calculated from the series. Thus

$$\ln \frac{1}{3} = \left(\frac{1}{3} - 1\right) - \frac{(\frac{1}{3} - 1)^2}{2} + \frac{(\frac{1}{3} - 1)^3}{3} - \frac{(\frac{1}{3} - 1)^4}{4} + \cdots$$

$$= \left(-\frac{2}{3}\right) - \frac{(-\frac{2}{3})^2}{2} + \frac{(-\frac{2}{3})^3}{3} - \frac{(-\frac{2}{3})^4}{4} + \cdots$$

$$= -\frac{2}{3} - \frac{2^2}{3^2 \cdot 2} - \frac{2^3}{3^3 \cdot 3} - \frac{2^4}{3^4 \cdot 4} - \cdots$$

If we use $s_{10}(\frac{1}{3})$ to approximate $\ln \frac{1}{3}$, we get

$$\ln \frac{1}{3} \approx s_{10}\left(\frac{1}{3}\right) = -\frac{2}{3} - \frac{2^2}{3^2 \cdot 2} - \frac{2^3}{3^3 \cdot 3} - \cdots - \frac{2^{10}}{3^{10} \cdot 10}.$$

Then

$$\ln 3 = -\ln \frac{1}{3} \approx \frac{2}{3} + \frac{2^2}{3^2 \cdot 2} + \frac{2^3}{3^3 \cdot 3} + \cdots + \frac{2^{10}}{3^{10} \cdot 10}.$$

These terms are calculated as follows:

$$\begin{aligned}
2/3 &\approx 0.6666667 \\
2^2/(3^2 \cdot 2) &\approx 0.2222222 \\
2^3/(3^3 \cdot 3) &\approx 0.0987654 \\
2^4/(3^4 \cdot 4) &\approx 0.0493827 \\
2^5/(3^5 \cdot 5) &\approx 0.0263374 \\
2^6/(3^6 \cdot 6) &\approx 0.0146319 \\
2^7/(3^7 \cdot 7) &\approx 0.0083611 \\
2^8/(3^8 \cdot 8) &\approx 0.0048773 \\
2^9/(3^9 \cdot 9) &\approx 0.0028903 \\
2^{10}/(3^{10} \cdot 10) &\approx 0.0017342 \\
\hline
\ln 3 &\approx 1.0958692
\end{aligned}$$

Note that this answer agrees with the decimal approximation of $\ln 3$ to only two decimal places. If we want greater accuracy, we must use more terms of the series. ☐

Taylor series for several important functions are given below. We shall establish in Section 12–9 that the series converge to the functions on their intervals of convergence.

Taylor Series for Some Important Functions

$$e^x = 1 + \frac{x}{1!} + \frac{x^2}{2!} + \frac{x^3}{3!} + \frac{x^4}{4!} + \cdots \qquad \text{(all real } x\text{)}.$$

$$\sin x = x - \frac{x^3}{3!} + \frac{x^5}{5!} - \frac{x^7}{7!} + \frac{x^9}{9!} - \cdots \qquad \text{(all real } x\text{)}.$$

$$\cos x = 1 - \frac{x^2}{2!} + \frac{x^4}{4!} - \frac{x^6}{6!} + \frac{x^8}{8!} - \cdots \qquad \text{(all real } x\text{)}.$$

$$\ln x = (x - 1) - \frac{(x-1)^2}{2} + \frac{(x-1)^3}{3} - \frac{(x-1)^4}{4} + \cdots \qquad (0 < x \leq 2).$$

$$\sinh x = x + \frac{x^3}{3!} + \frac{x^5}{5!} + \frac{x^7}{7!} + \frac{x^9}{9!} + \cdots \qquad \text{(all real } x\text{)}.$$

$$\cosh x = 1 + \frac{x^2}{2!} + \frac{x^4}{4!} + \frac{x^6}{6!} + \frac{x^8}{8!} + \cdots \qquad \text{(all real } x\text{)}.$$

Remark The partial sums $\{s_n(x)\}$ form a sequence of polynomial approximations to $f(x)$. This is

12-8 Taylor Series

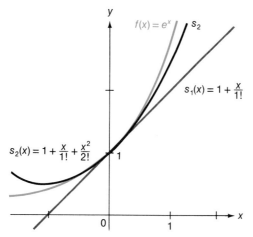

Figure 12-10 Two partial sums for the series for e^x: $s_1(x) = 1 + x$, $s_2(x) = 1 + x + x^2/2$.

illustrated in Figure 12-10 for the function $f(x) = e^x$. Observe that

$$s_1(x) = 1 + \frac{x}{1!}$$

is a very poor approximation to e^x except when x is very close to zero but

$$s_2(x) = 1 + \frac{x}{1!} + \frac{x^2}{2!}$$

is a fairly good approximation when x is between -1 and 1. The graph of

$$s_3(x) = 1 + \frac{x}{1!} + \frac{x^2}{2!} + \frac{x^3}{3!}$$

is not shown in the figure. That function approximates $f(x)$ so closely when $-1.5 \le x \le 2$ that the graphs seem to coincide.

LIMITS EVALUATED BY SERIES

Power series can be used instead of L'Hôpital's rule to evaluate certain limits.

EXAMPLE 4 Calculate $\lim\limits_{x \to 0} \dfrac{\sin x - x + (x^3/6)}{x^5}$.

Solution Since

$$\sin x = x - \frac{x^3}{3!} + \frac{x^5}{5!} - \frac{x^7}{7!} + \frac{x^9}{9!} - \cdots ,$$

Limits Evaluated by Series

then

$$\frac{\sin x - x + (x^3/6)}{x^5} = \frac{(x^5/5!) - (x^7/7!) + (x^9/9!) - \cdots}{x^5}$$

$$= \frac{1}{5!} - \frac{x^2}{7!} + \frac{x^4}{9!} - \cdots.$$

Therefore,

$$\lim_{x \to 0} \frac{\sin x - x + (x^3/6)}{x^5} = \lim_{x \to 0} \left(\frac{1}{5!} - \frac{x^2}{7!} + \frac{x^4}{9!} - \cdots \right) = \frac{1}{5!} = \frac{1}{120}. \quad \square$$

EXAMPLE 5 Calculate $\lim\limits_{x \to 0} \dfrac{e^x - 1}{x}$.

Solution Since

$$e^x = 1 + \frac{x}{1!} + \frac{x^2}{2!} + \frac{x^3}{3!} \cdots,$$

then

$$\frac{e^x - 1}{x} = \frac{(x/1!) + (x^2/2!) + (x^3/3!) + \cdots}{x} = \frac{1}{1!} + \frac{x}{2!} + \frac{x^2}{3!} + \cdots.$$

Therefore,

$$\lim_{x \to 0} \frac{e^x - 1}{x} = \lim_{x \to 0} \left(\frac{1}{1!} + \frac{x}{2!} + \frac{x^2}{3!} + \cdots \right) = \frac{1}{1!} = 1. \quad \square$$

Exercises 12-8

1–6. Assume that the series converge to the original functions on their intervals of convergence.

(a) Derive the Maclaurin series for the given functions. Use the ratio test to find the values of x for which the series converges.

(b) Use $s_4(\frac{1}{2})$ to approximate $f(\frac{1}{2})$. How good is this approximation?

c(c) (*For students with programmable calculators or computers*) Use $s_{20}(\frac{1}{2})$ to approximate $f(\frac{1}{2})$.

•1 $\sin x$ 2 $\cos x$

•3 $\sinh x$ 4 $\cosh x$

•5 $\ln(x + 1)$ 6 e^{2x}

7–10. Use series to calculate the given limits. Compare your answers with the answers obtained by using L'Hôpital's rule.

•7 $\lim\limits_{x \to 0} \dfrac{\sinh x - x}{x^3}$ 8 $\lim\limits_{x \to 0} \dfrac{e^{2x} - 1 - 2x - 2x^2}{x^3}$

Exercises 12–8

●9 $\lim_{x \to 0} \dfrac{\tan^{-1}(2x)}{x}$

10 $\lim_{x \to 0} \dfrac{\sin x}{3x - x^3}$

11–14. Express the integrals as power series.

●11 $\displaystyle\int e^{-x^2}\, dx$

12 $\displaystyle\int_0^1 \sin\left(\dfrac{x^3}{2}\right) dx$

●13 $\displaystyle\int \ln(x + \sqrt{x^2 + 1})\, dx$

14 $\displaystyle\int \cos^{-1} x\, dx$

(First three nonzero terms only)   (First three nonzero terms only)

15–16. Write the Maclaurin series for the function. Construct a graph similar to Figure 12–10, showing $f(x)$, $s_1(x)$, $s_2(x)$, and $s_3(x)$.

15 $f(x) = \sin x$, $-\pi \le x \le 2\pi$

16 $f(x) = \ln(x + 1)$, $-\tfrac{1}{2} \le x \le \tfrac{1}{2}$

17 Use Maclaurin's series for e^x to show that

$$e^{-x} = 1 - \dfrac{x}{1!} + \dfrac{x^2}{2!} - \dfrac{x^3}{3!} + \dfrac{x^4}{4!} - \cdots$$

and

$$e^{x^2} = 1 + \dfrac{x^2}{1!} + \dfrac{x^4}{2!} + \dfrac{x^6}{3!} + \dfrac{x^8}{4!} + \cdots.$$

18 Use the Maclaurin series for e^x and e^{-x} (Exercise 17) to show that

$$\sinh x = \dfrac{e^x - e^{-x}}{2} = \dfrac{x}{1!} + \dfrac{x^3}{3!} + \dfrac{x^5}{5!} + \dfrac{x^7}{7!} + \cdots$$

and

$$\cosh x = \dfrac{e^x + e^{-x}}{2} = 1 + \dfrac{x^2}{2!} + \dfrac{x^4}{4!} + \dfrac{x^6}{6!} + \cdots.$$

19 What goes wrong if we try to find a Maclaurin series for $f(x) = \ln x$? Could this Section 10–1.) You may assume that this result is known.

Exercises 20 and 21 show that a function f may have a Maclaurin series that converges to a function different from f.

20 Define the function f by

$$f(x) = \begin{cases} \dfrac{1}{e} & \text{if } x \le -1 \\ e^x & \text{if } -1 < x < 1 \\ e & \text{if } x \ge 1. \end{cases}$$

(a) Find the Maclaurin series for $f(x)$.
(b) Show that this series converges for all x.
(c) Show that when $x = 2$, the Maclaurin series does not converge to $f(2)$. (Hint: It may help to graph the function f and the function defined by the Maclaurin series for f.)

21 Define the function f by

$$f(x) = \begin{cases} e^{-1/x^2} & \text{if } x \neq 0 \\ 0 & \text{if } x = 0. \end{cases}$$

It can be proved that $f^{(n)}(0) = 0$ for every nonnegative integer n. (See Exercise 18 of Section 10–1.) You may assume that this result is known.

    (a) Construct the Maclaurin series for f. Show that this series converges for all x.

    (b) Show that the Maclaurin series for f converges to $f(x)$ only when $x = 0$.

    (c) Let n be a positive integer, let $x \neq 0$. What is $s_n(x)$?

22 Let $a_0 + a_1 x + a_2 x^2 + \cdots = 0$ for all x in the interval $[0, 1]$. Prove that $a_0 = a_1 = a_2 = \cdots = 0$. [*Hint:* Let $f(x) = a_0 + a_1 x + a_2 x^2 + \cdots$. Show that the radius of convergence is greater than 0. Evaluate $f(0), f'(0), f''(0)$, and so on.]

23 (*Euler's Identity*) Assume that Maclaurin's series for e^x, $\cos x$, and $\sin x$ also are valid when x is an imaginary number.

    ●**(a)** Calculate the series for $\sin(ix)$ and compare it with the series for $\sinh x$.

    ●**(b)** Calculate the series for $\cos(ix)$ and compare it with the series for $\cosh x$.

    (c) Assume that the algebraic rules for working with series also hold when imaginary numbers are involved. Show that

Euler's Identity

$$e^{ix} = \cos x + i \sin x.$$

This result is known as *Euler's identity*. It is fundamental in the study of complex valued functions.

    (d) Use Euler's identity to show that $e^{i\pi} = -1$, an equation that relates four of the most important constants in mathematics.

24 Use mathematical induction to show that if $f(x) = \ln x$, then

$$f^{(n)}(x) = \frac{(-1)^{n+1}(n-1)!}{x^n} \qquad (n = 1, 2, 3, \cdots).$$

12–9 Taylor's Formula. Accuracy of Approximations

We have established that any function with derivatives of all orders at $x = c$ has a Taylor series in powers of $x - c$. If the function is f, then the corresponding Taylor series is

$$f(c) + \frac{f'(c)}{1!}(x - c) + \frac{f''(c)}{2!}(x - c)^2 + \frac{f'''(c)}{3!}(x - c)^3 + \cdots.$$

Recall that except for the number $x = c$, the series may not converge to $f(x)$ even if it converges. To determine whether it converges to $f(x)$, we must consider the sequence of partial sums $\{s_n(x)\}$, where

$$s_n(x) = f(c) + \frac{f'(c)}{1!}(x - c) + \cdots + \frac{f^{(n)}(c)}{n!}(x - c)^n.$$

12-9 Taylor's Formula. Accuracy of Approximations

The series converges to $f(x)$ for a given value of x if and only if

$$\lim_{n\to\infty} s_n(x) = f(x).$$

We shall find it convenient to work with the sequence of *remainder terms* $\{r_n(x)\}$ defined by

Remainder Polynomial

$$r_n(x) = f(x) - s_n(x).$$

(The value of $r_n(x)$ is the error that results if $s_n(x)$ is used to approximate $f(x)$.) Since $f(x) = s_n(x) + r_n(x)$, then

$$\{s_n(x)\} \to f(x) \quad \text{if and only if} \quad \{r_n(x)\} \to 0.$$

Thus, to show that the series converges to $f(x)$, we must establish that the sequence of remainder terms converges to 0.

The following theorem frequently can be used to establish bounds on $r_n(x)$.

THEOREM 12-21

Taylor's Formula

(*Taylor's Formula*) Let f have derivatives of all orders on an interval that contains c and x. Let $r_n(x)$ be the nth remainder term obtained from the Taylor series for $f(x)$ in powers of $x - c$:

$$r_n(x) = f(x) - s_n(x)$$

$$= f(x) - \left[f(c) + \frac{f'(c)}{1!}(x-c) + \cdots + \frac{f^{(n)}(c)}{n!}(x-c)^n \right].$$

Then

$$r_n(x) = \int_c^x \frac{(x-t)^n}{n!} f^{(n+1)}(t)\, dt.$$

Proof The argument is more easily followed if we fix the value of x at a particular number $x = b$.

Observe first that by the Fundamental Theorem of Calculus,

$$\int_c^b f'(t)\, dt = \left[f(t) \right]_c^b = f(b) - f(c)$$

so that

(1) $$f(b) = f(c) + \int_c^b f'(t)\, dt.$$

We now calculate this same integral using integration by parts with $u = f'(t)$, $dv = dt$, $du = f''(t)dt$, $v = t - b$:

$$\int_c^b f'(t)\, dt = \left[f'(t)(t-b) \right]_c^b - \int_c^b (t-b) f''(t)\, dt$$

$$= 0 - f'(c)(c - b) - \int_c^b (t - b)f''(t)\,dt$$

$$= \frac{f'(c)}{1!}(b - c) + \int_c^b \frac{(b - t)f''(t)}{1!}\,dt.$$

If we substitute this expression in formula (1), we get

(2) $$f(b) = f(c) + \frac{f'(c)}{1!}(b - c) + \int_c^b \frac{(b - t)f''(t)}{1!}\,dt,$$

which shows that

$$r_1(b) = \int_c^b \frac{(b - t)f''(t)}{1!}\,dt.$$

To show that $r_2(b)$ has the form given in the statement of the theorem, we use integration by parts to evaluate

$$\int_c^b \frac{(b - t)}{1!} f''(t)\,dt.$$

Let $u = f''(t)$, $dv = (b - t)/1!$, $du = f'''(t)\,dt$, $v = -(b - t)^2/2!$. Then

$$\int_c^b \frac{b - t}{1!} f''(t)\,dt = \left[\frac{-f''(t)(b - t)^2}{2!}\right]_c^b - \int_c^b \frac{-(b - t)^2}{2!} \cdot f'''(t)\,dt$$

$$= 0 + \frac{f''(c)(b - c)^2}{2!} + \int_c^b \frac{(b - t)^2}{2!} f'''(t)\,dt.$$

On substituting in (2), we get

$$f(b) = f(c) + \frac{f'(c)}{1!}(b - c) + \frac{f''(c)}{2!}(b - c)^2 + \int_c^b \frac{(b - t)^2}{2!} f'''(t)\,dt$$

so that

$$r_2(b) = \int_c^b \frac{(b - t)^2}{2!} f'''(t)\,dt.$$

If we continue this process (this disguises a proof by mathematical induction), we find that

$$r_n(b) = \int_c^b \frac{(b - t)^n}{n!} f^{(n+1)}(t)\,dt \qquad (n = 1, 2, 3, \ldots). \blacksquare$$

EXAMPLE 1 Show that the Maclaurin series for the cosine function

$$1 - \frac{x^2}{2!} + \frac{x^4}{4!} - \frac{x^6}{6!} + \cdots$$

converges to $\cos x$ for every $x > 0$.

Solution We fix the value of $x > 0$. It follows from Theorem 12–21 that

$$\cos x = 1 - \frac{x^2}{2!} + \frac{x^4}{4!} - \frac{x^6}{6!} + \cdots + \frac{f^{(n)}(0)}{n!} x^n + r_n(x)$$

12-9 Taylor's Formula. Accuracy of Approximations

where

$$r_n(x) = \int_0^x \frac{(x-t)^n}{n!} f^{(n+1)}(t) \, dt.$$

Since $f(t) = \cos t$, then $f^{(n+1)}(t)$ is one of the functions $\cos t$, $-\cos t$, $\sin t$, or $-\sin t$. In every case $|f^{n+1}(t)| \le 1$.

It can be proved from the properties of integrals in Section 5-3 that if g is continuous on $[a, b]$, then

$$\left| \int_a^b g(x) \, dx \right| \le \int_a^b |g(x)| \, dx.$$

(See Exercise 14.) Then

$$|r_n(x)| = \left| \int_0^x \frac{(x-t)^n}{n!} f^{(n+1)}(t) \, dt \right| \le \int_0^x \frac{|x-t|^n}{n!} |f^{(n+1)}(t)| \, dt$$

$$\le \int_0^x \frac{|x-t|^n}{n!} \cdot 1 \cdot dt = \int_0^x \frac{(x-t)^n}{n!} \, dt$$

$$= -\int_0^x \frac{(x-t)^n}{n!} (-dt) = -\left[\frac{(x-t)^{n+1}}{(n+1)!} \right]_0^x = -0 + \frac{x^{n+1}}{(n+1)!}.$$

It follows from Exercise 12 of Section 12-6 that

$$\lim_{n \to \infty} \frac{x^{n+1}}{(n+1)!} = 0.$$

Since $0 \le |r_n(x)| \le x^{n+1}/(n+1)!$, then

$$\lim_{n \to \infty} |r_n(x)| = 0$$

and therefore

$$\lim_{n \to \infty} r_n(x) = 0.$$

Since the sequence of remainder terms converges to zero for any positive value of x, then

$$\{s_n(x)\} \to \cos x \qquad \text{if } x > 0.$$

Therefore,

$$\cos x = 1 - \frac{x^2}{2!} + \frac{x^4}{4!} - \frac{x^6}{6!} + \cdots \qquad \text{for } x > 0. \ \square$$

Remark The Maclaurin series for the cosine function converges to $\cos 0$ if $x = 0$. An argument like the one in Example 1 can be used to show that it converges to $\cos x$ for any $x < 0$. (See Exercise 9.) Therefore

$$\cos x = 1 - \frac{x^2}{2!} + \frac{x^4}{4!} - \cdots + \frac{(-1)^n x^{2n}}{(2n)!} + \cdots \qquad (x \text{ real}).$$

LAGRANGE'S FORMULA

Lagrange established an alternative way of writing the remainder $r_n(x)$. Lagrange's formula is more convenient to use than Taylor's formula with many of the functions in the calculus because it involves a derivative instead of an integral. The proof of Lagrange's formula, which is based on Theorem 12–21, is omitted.

THEOREM 12–22 *(Lagrange's Formula)* Let f have derivatives of all orders on an interval that contains c and x. Let $r_n(x)$ be the remainder term obtained from the Taylor series in powers of $(x - c)$. There exists a number ξ between c and x such that

$$r_n(x) = \frac{f^{(n+1)}(\xi)(x - c)^{n+1}}{(n + 1)!}.$$

EXAMPLE 2 Use Lagrange's formula to prove that for every x

$$e^x = 1 + \frac{x}{1!} + \frac{x^2}{2!} + \frac{x^3}{3!} + \cdots.$$

Solution We showed in Example 1 of Section 12–8 that the Maclaurin series for e^x is

$$1 + \frac{x}{1!} + \frac{x^2}{2!} + \frac{x^3}{3!} + \cdots.$$

We must now show that this series converges to e^x for every x. To do this, we must show that $\{r_n(x)\} \to 0$ for every fixed value of x.

By Lagrange's formula, if $x \ne 0$, there exists a number ξ between 0 and x such that

$$r_n(x) = \frac{f^{(n+1)}(\xi) \, x^{n+1}}{(n + 1)!} = \frac{e^\xi x^{n+1}}{(n + 1)!}.$$

Suppose that $x > 0$. Since $2 < e < 3$ and $0 < \xi < x$, then

$$|r_n(x)| = \left| \frac{e^\xi x^{n+1}}{(n + 1)!} \right| < \frac{3^x x^{n+1}}{(n + 1)!}.$$

It follows from Exercise 12 of Section 12–6 that

$$\lim_{n \to \infty} \frac{3^x x^{n+1}}{(n + 1)!} = 0.$$

Since $|r_n(x)| < 3^x x^{n+1}/(n + 1)!$ for every n, then

$$\lim_{n \to \infty} |r_n(x)| = 0.$$

Therefore,

$$\lim_{n \to \infty} r_n(x) = 0$$

and the series

$$1 + \frac{x}{1!} + \frac{x^2}{2!} + \frac{x^3}{3!} + \cdots$$

converges to e^x for every positive number x.

12-9 Taylor's Formula. Accuracy of Approximations

A similar argument holds if $x < 0$. In that case, we can show that

$$|r_n(x)| = \left|\frac{e^\xi x^{n+1}}{(n+1)!}\right| < \frac{|x|^{n+1}}{(n+1)!},$$

which implies that $\{r_n(x)\} \to 0$ as $n \to \infty$. (See Exercise 10.) □

ESTIMATION OF ERROR TERMS

One of the central problems in any work with power series is finding the number of terms that must be used to approximate a number to a desired degree of accuracy. As we shall see, Theorems 12-21 and 12-22 can help us solve this problem.

If we need, for example, to approximate $e^{1.3}$ with six-decimal-place accuracy from the Maclaurin series

$$e^x = 1 + \frac{x}{1!} + \frac{x^2}{2!} + \frac{x^3}{3!} + \frac{x^4}{4!} + \cdots,$$

then we must choose a positive integer n and use the approximation

$$e^{1.3} \approx s_n(1.3) = 1 + \frac{1.3}{1!} + \frac{(1.3)^2}{2!} + \frac{(1.3)^3}{3!} + \cdots + \frac{(1.3)^n}{n!}.$$

Our problem is to determine a value of n that will ensure that this approximation is correct to six decimal places. (We cannot calculate $s_n(1.3)$ for several different values of n and compare the answers. We must find n before carrying out the calculations.)

The specific problem mentioned above can be restated as follows:
Let $r_n(1.3) = e^{1.3} - s_n(1.3)$. Find a value of n that makes

$$|r_n(1.3)| < \frac{5}{10^7}.$$

Example 3 shows how Theorem 12-22 can be used to solve this problem.

EXAMPLE 3 Find a value of n such that

$$s_n(1.3) = 1 + \frac{1.3}{1!} + \frac{(1.3)^2}{2!} + \frac{(1.3)^3}{3!} + \cdots + \frac{(1.3)^n}{n!}$$

approximates $e^{1.3}$ with six-decimal-place accuracy. Check the result by computing $s_n(1.3)$ and $e^{1.3}$.

Solution We must find a positive integer n such that

$$|r_n(1.3)| < \frac{5}{10^7}.$$

It follows from Theorem 12-22 that there exists a number ξ between 0 and 1.3 such that

$$r_n(1.3) = \frac{e^\xi (1.3)^{n+1}}{(n+1)!}.$$

Since $0 < \xi < 1.3 < 2$ and $e < 3$, then

$$|r_n(1.3)| = \left|\frac{e^\xi (1.3)^{n+1}}{(n+1)!}\right| = \frac{e^\xi (1.3)^{n+1}}{(n+1)!}$$
$$< \frac{3^\xi (1.3)^{n+1}}{(n+1)!} < \frac{3^2 (1.3)^{n+1}}{(n+1)!} = \frac{9(1.3)^{n+1}}{(n+1)!}.$$

If we can find an integer n such that

$$\frac{9(1.3)^{n+1}}{(n+1)!} < \frac{5}{10^7},$$

it will follow that $|r_n| < 5/10^7$.

Table 12-1 contains the approximate values of $9(1.3)^{n+1}/(n+1)!$ for $n = 1, 2, 3, \cdots, 12$. Observe that this number is less than $5/10^7$ when $n \geq 11$. Thus, $s_{11}(1.3)$ approximates $e^{1.3}$ correct to six decimal places.

Table 12-1 Example 3

n	$\dfrac{9(1.3)^{n+1}}{(n+1)!}$	n	$\dfrac{9(1.3)^{n+1}}{(n+1)!}$	n	$\dfrac{9(1.3)^{n+1}}{(n+1)!}$
1	7.605	5	0.060	9	0.00003
2	3.296	6	0.011	10	0.000004
3	1.071	7	0.002	11	0.0000004
4	0.278	8	0.0003	12	0.00000004

To check, we calculate $s_{11}(1.3)$:

$$s_{11}(1.3) = 1 + \frac{1.3}{1!} + \frac{(1.3)^2}{2!} + \frac{(1.3)^3}{3!} + \cdots + \frac{(1.3)^{11}}{11!}$$
$$\approx 3.669296614.$$

This result agrees with the "true" value of $e^{1.3}$ (as calculated on a scientific calculator) to six decimal places. □

Remark The number

$$|r_n(1.3)| = |e^{1.3} - s_n(1.3)|$$

is the error that results if $s_n(1.3)$ is used to approximate $e^{1.3}$. Our work in Example 3 involves finding bounds for this error. It follows from the data in Table 12-1 that the error is no greater than 0.28 if $s_4(1.3)$ is used to approximate $e^{1.3}$, is no greater than 0.06 if $s_5(1.3)$ is used to approximate $e^{1.3}$, and so on.

Exercises 12-9

1-8. (a) Calculate the Taylor series for the function f in powers of $x - a$. Use the ratio test to find the radius of convergence.

(b) Use Lagrange's formula to show that the series converges to $f(x)$ when $x = 1.5$.

Exercises 12-9

(c) Use Lagrange's formula to find a positive integer n such that $s_n(1.5)$ approximates $f(1.5)$ correct to five decimal places.

- **1** $f(x) = \cos x$, $a = \dfrac{\pi}{3}$
- **2** $f(x) = \sin x$, $a = \dfrac{\pi}{2}$
- **3** $f(x) = \ln x$, $a = 1$
- **4** $f(x) = x^{1/2}$, $a = 1$
- **5** $f(x) = x^3 + 7x^2 - 5x + 2$, $a = 2$
- **6** $f(x) = 2x^4 - 1$, $a = 1$
- **7** $f(x) = \dfrac{1}{x}$, $a = 1$
- **8** $f(x) = e^x$, $a = 1$

9 Modify the argument in Example 1 to show that the series

$$1 - \frac{x^2}{2!} + \frac{x^4}{4!} - \frac{x^6}{6!} + \cdots$$

converges to $\cos x$ if $x < 0$.

10 Modify the argument in Example 2 to show that the series

$$1 + \frac{x}{1!} + \frac{x^2}{2!} + \frac{x^3}{3!} + \frac{x^4}{4!} + \cdots$$

converges to e^x if $x < 0$.

11 (*The Binomial Series*) The Binomial Theorem, studied in elementary algebra, is

$$(a + b)^n = a^n + \frac{n}{1!} a^{n-1} b + \frac{n(n-1)}{2!} a^{n-2} b^2 + \frac{n(n-1)(n-2)}{3!} a^{n-3} b^3$$

$$+ \cdots + \frac{n(n-1)(n-2) \cdots 1}{n!} \cdot b^n$$

provided n is a positive integer.

(a) Generalize the Binomial Theorem to cover the case for which the exponent n is not a positive integer. Show that the Maclaurin series for $(1 + x)^r$ is

$$1 + \frac{r}{1!} x + \frac{r(r-1)}{2!} x^2 + \frac{r(r-1)(r-2)}{3!} x^3 + \cdots$$

provided r is a real number and $(1 + x)^r$ is real. It can be proved that this series converges to $(1 + x)^r$ if $|x| < 1$. Thus

Binomial Series
$$\boxed{(1 + x)^r = 1 + \frac{r}{1!} x + \frac{r(r-1)}{2!} x^2 + \frac{r(r-1)(r-2)}{3!} x^3 + \cdots \quad (-1 < x < 1).}$$

(b) Use the ratio test to show that the series in (a) converges if $|x| < 1$.
- (c) Use the series in (a) to calculate $\sqrt[3]{1.5}$.

12 (a) Use the Binomial Series (Exercise 11) to show that

$$\sqrt{1 + 2x} = 1 + \frac{x}{1!} - \frac{x^2}{2!} + \frac{3x^3}{3!} - \frac{3 \cdot 5 x^4}{4!} + \frac{3 \cdot 5 \cdot 7 x^5}{5!} - \cdots.$$

What is the radius of convergence?

(b) Show that the general term of the series is

$$\frac{(-1)^{n-1}(2n-2)!x^n}{2^{n-1}(n-1)!n!}.$$

•(c) For what values of x does the polynomial

$$1 + x - \frac{x^2}{2}$$

approximate $\sqrt{1 + 2x}$ correct to four decimal places?

•13 (a) Use the Binomial Series (Exercise 11) to get the Maclaurin series for

$$f(x) = \frac{1}{\sqrt{1 - x^2}}.$$

What is the radius of convergence?

(b) Integrate the series in (a) term by term to get a series for $\sin^{-1} x$.

14 Let g be continuous on $[a, b]$. Prove that

$$\left| \int_a^b g(x)\, dx \right| \leq \int_a^b |g(x)|\, dx.$$

(*Hint:* Use Property 2 of Section 5–3 to show that $\int_a^b g(x)\, dx$ and $-\int_a^b g(x)\, dx$ are both less than or equal to $\int_a^b |g(x)|\, dx$.)

Review Problems

1. (a) Define the terms *sequence* and *series*.

   (b) What is meant by the "limit" of a sequence?

   (c) What is meant by saying that "a series Σa_n converges to L"?

2. Various sequences can be associated with a series Σa_n, including the sequence of terms $\{a_n\}$, the sequence of ratios of successive terms $\{a_{n+1}/a_n\}$, and so on. State three tests for convergence or divergence of a series based on associated sequences.

3. (a) Define the terms *monotone-increasing* and *monotone-nondecreasing* as applied to sequences.

   (b) Suppose that a monotone-increasing sequence is bounded above. What do we know about its convergence properties?

   (c) Answer (b) for a monotone-increasing sequence that is bounded below.

4. State the completeness axiom. Define all terms used in the definition.

5–8. Calculate the limits of the sequences that converge.

•5 $a_n = 2 + \dfrac{1}{2^n}$

•6 $a_n = 2 + (-1)^n$

Review Problems

- **7** $a_n = \dfrac{1 + 2 + 3 + \cdots + n}{n^2}$
- **8** $a_n = 1$

9 ●(a) Express the decimal fraction $0.0123123123\cdots$ as an infinite series. Evaluate the limit.

(b) Is every repeating decimal fraction equal to a rational number? Explain.

10 (a) Let $\{a_n\}$ be a positive-term sequence that converges to 0. Either explain why the series $\sum_{n=1}^{\infty} a_n$ converges or by example show that it may diverge.

(b) Let $\sum_{n=1}^{\infty} a_n$ converge. Either explain why $\{a_n\} \to 0$ or by example show that the sequence may not converge to 0.

11 Prove: If Σa_n and Σb_n converge to L and M, respectively, then $\Sigma(a_n + b_n)$ converges to $L + M$.

12 Suppose that Σa_n and Σb_n both diverge. Show by examples that $\Sigma(a_n + b_n)$ can either converge or diverge.

13–24. Test the series for convergence.

- **●13** $\displaystyle\sum_{n=1}^{\infty} \dfrac{3^n}{4^{n+2}}$
- **●14** $\displaystyle\sum_{n=1}^{\infty} \left(\dfrac{1}{3^n} + \dfrac{1}{n^3}\right)$
- **●15** $\displaystyle\sum_{n=1}^{\infty} \dfrac{\ln n}{n}$
- **●16** $\displaystyle\sum_{n=1}^{\infty} \dfrac{\cosh n}{\sinh n}$
- **●17** $\displaystyle\sum_{n=1}^{\infty} \dfrac{1}{n^2 + 4}$
- **●18** $\displaystyle\sum_{n=1}^{\infty} \dfrac{4}{\sqrt{n}}$
- **●19** $\displaystyle\sum_{n=1}^{\infty} \dfrac{\sqrt{n-1}}{n^2}$
- **●20** $\displaystyle\sum_{n=1}^{\infty} \dfrac{\cos n}{2^n}$
- **●21** $\displaystyle\sum_{n=1}^{\infty} \dfrac{2^n}{(2n)!}$
- **●22** $\displaystyle\sum_{n=1}^{\infty} \dfrac{(2n+1)5^n}{(n-1)!}$
- **●23** $\displaystyle\sum_{n=1}^{\infty} \dfrac{(-1)^{n+1}\sqrt{n}}{n+1}$
- **●24** $\displaystyle\sum_{n=1}^{\infty} \dfrac{(-1)^n \ln n}{n+2}$

25 (a) State the integral test.

(b) State both of the comparison tests.

(c) State the ratio test (both forms).

(d) State the alternating-series test.

26 Prove that the geometric series Σar^{n-1} converges if $|r| < 1$ and diverges if $|r| \geq 1$.

27 Use the integral test to prove that the p series $\Sigma \, 1/n^p$ converges if $p > 1$ and diverges if $0 < p \leq 1$.

28 **(a)** What is meant by absolute convergence? conditional convergence?

(b) Suppose a series converges absolutely. Must it converge?

(c) State two properties possessed by absolutely convergent series but not conditionally convergent series.

29–32. Find the interval of convergence. Test for convergence at the endpoints of the interval (if finite). Where does the series converge absolutely?

•**29** $\displaystyle\sum_{n=1}^{\infty} \frac{x^{n+1}}{n \cdot 3^n}$

•**30** $\displaystyle\sum_{n=1}^{\infty} \frac{(2x)^n}{(n-1)!}$

•**31** $\displaystyle\sum_{n=1}^{\infty} \frac{(x-2)^{2n+1}}{n+1}$

•**32** $\displaystyle\sum_{n=1}^{\infty} \frac{(2x+5)^n (-1)^{n+1}}{2n-1}$

33–36. Derive the Maclaurin series for the given functions. Find the intervals of convergence for the series.

33 e^x

34 $\sin x$

35 $\cos x$

•**36** $\ln(2x+1)$

37–38. Write Taylor's series for the functions in powers of $x - c$.

•**37** $f(x) = \ln x$, $c = 1$

•**38** $f(x) = \sin x$, $c = \dfrac{\pi}{4}$

39–40. Use the Maclaurin series for e^x to write the series for the function or number.

•**39** e^{x^2}

•**40** $\sqrt{e}$

41–44. Use power series to calculate the integrals.

•**41** $\displaystyle\int \cos(\sqrt{x})\, dx$

•**42** $\displaystyle\int \frac{dx}{1+x^4}$

•**43** $\displaystyle\int \frac{\cos x - 1}{x^2}\, dx$

•**44** $\displaystyle\int \frac{e^x}{x}\, dx$

45–48. Use series to evaluate the limits. Compare your answers with those obtained from L'Hôpital's rule.

•**45** $\displaystyle\lim_{x \to 0} \frac{\cos x - 1}{x^2}$

•**46** $\displaystyle\lim_{x \to 0} \frac{\sinh x - \sin x}{x^3}$

Review Problems

•47 $\lim_{x \to 0} \dfrac{e^x - x - \cos x}{x^2}$

•48 $\lim_{x \to 1} \dfrac{\ln x}{x - 1}$

49 State Taylor's formula and Lagrange's formula.

50 Prove that the Maclaurin series for e^x converges to e^x for every x.

•51 Find a positive integer n such that the partial sum $s_n(x)$ obtained from the Maclaurin series for e^x must approximate e^x correct to five decimal places if $-1 \le x \le 1$.

13 Parametric Equations and Polar Coordinates

13–1 Parametric Equations

Problems involving the motion of a particle can often be solved most effectively by using parametric equations. As discussed earlier (Section 6–4), these are equations of form

$$x = f(t), \qquad y = g(t),$$

in which x and y are expressed as functions of an auxiliary variable t, called a *parameter*. It is convenient to think that the parameter t represents time. In that case $P(f(t), y(t))$ represents the location of the particle at time t. (See Fig. 13–1.)

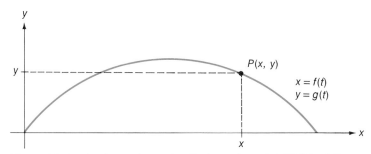

Figure 13–1 At time t the projectile is at the point $P(f(t), g(t))$.

Suppose, for example, that a projectile of mass m is fired from the surface of the earth at an angle of elevation α. We assume (as in Section 8–6) that the projectile is subject to two forces—a gravitational force mg, which acts downward, and a retarding force, due to air resistance, which is proportional to the velocity and acts in the direction opposite the direction of motion.

13–1 Parametric Equations

It is clear from the physical problem that y, the height of the projectile at time t, is a function of x, the horizontal distance from the point at which it was fired. (See Fig. 13–1.) We will later show that we can break the motion into horizontal and vertical components, which depend on time (t); that is, there exist functions f and g such that

$$x = f(t) \quad \text{and} \quad y = g(t).$$

This problem is discussed in detail in Section 15–3.

GRAPHS AND CURVES

Recall from Section 6–4 that the graph of parametric equations

$$x = f(t), \quad y = g(t)$$

is the set of all $P(x, y)$ such that $x = f(t)$, $y = g(t)$, where t is in the common part of the domains of f and g. That is, the graph is

$$\{P(f(t), g(t)) \mid t \in \mathcal{D}_f \cap \mathcal{D}_g\}.$$

Curve
Smooth Curve

If the common domain of f and g is an interval I on the t-axis on which f and g are continuous, then the graph is called a *curve*. (See Fig. 13-2a.) The curve is *smooth* provided f and g have continuous first derivatives on I which are not simultaneously 0 except possibly at the endpoints of I.

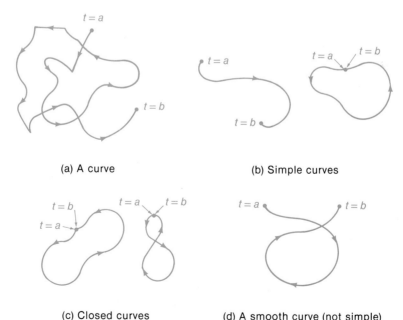

(a) A curve

(b) Simple curves

(c) Closed curves

(d) A smooth curve (not simple)

Figure 13–2 Curves.

Several curves defined over a closed interval $[a, b]$ are pictured in Figure 13–2. The points corresponding to the values $t = a$ and $t = b$ are labeled. The arrows indicate how the curves are traced out as t varies from a to b. Observe that a "curve" may have straight portions and may cross itself many times.

Simple Curve
Closed Curve

Two particular types of curves illustrated in Figures 13–2b and c are of interest in higher-level courses. A *simple curve* defined over $[a, b]$ is one that does not intersect itself except possibly at the points corresponding to the endpoints a and b. A *closed curve* defined over $[a, b]$ is one with the property that $f(a) = f(b)$ and $g(a) = g(b)$. (See Exercises 1–4.)

A good example of a curve is furnished by the path traced out by a bug as it crawls over a sheet of paper. The path might be like the curve in Figure 13–2a. At time t the bug is at a point $P(x, y)$, where x and y depend on t. The path may cross itself many times and even retrace itself. Thus, to determine the properties of the motion, one must know the location of the bug at each instant. It is not enough to know the final path; we must know how the path was traced out.

Thus far we have used time as our parameter. In other settings it may be useful to use other parameters. We may, for example, use an angle ϕ for the parameter, or the arc length s. In these cases we write

$$x = f(\phi), \qquad y = g(\phi)$$

or

$$x = f(s), \qquad y = g(s).$$

REDUCTION TO RECTANGULAR EQUATIONS

If at least one of the equations $x = f(t)$, $y = g(t)$ has a simple enough form, we can get an equation that relates x and y directly. When this is done, however, we lose all knowledge of how the curve is traced out with respect to time.

EXAMPLE 1   Let $x = \sqrt{t} - 2$, $y = t - 1$ for $t \geq 0$. Find an equation satisfied by x and y. Sketch the graph of the original parametric equations.

Solution   We solve the first equation for t and substitute in the second equation:

$$t = (x + 2)^2$$
$$y = t - 1 = (x + 2)^2 - 1$$
$$y + 1 = (x + 2)^2.$$

The graph of this last equation is a parabola that opens upward. The vertex is at the point $(-2, -1)$. (See Fig. 13–3a.)

It follows that every point on the graph of the parametric equations is on the parabola $y + 1 = (x + 2)^2$. We must now find the points on the parabola that satisfy the parametric equations.

Note that f and g are continuous functions of t. Since $t \geq 0$, then

$$x = \sqrt{t} - 2 \geq -2 \qquad \text{and} \qquad y = t - 1 \geq -1.$$

Observe also that x and y increase as t increases and that $x \to \infty$ and $y \to \infty$ as $t \to \infty$.

13-1 Parametric Equations

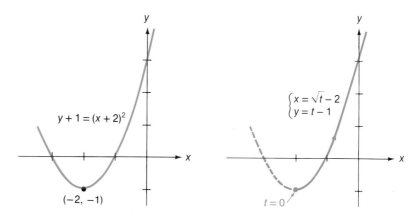

(a) The parabola $y + 1 = (x + 2)^2$

(b) Graph of the parametric equations $x = \sqrt{t} - 2$, $y = t - 1$

Figure 13-3 Example 1.

It follows that the graph of the parametric equations is the portion of the parabola that satisfies the condition $x \geq -2$. (See Fig. 13-3b.) ☐

EXAMPLE 2 Let $x = a \cos \theta$, $y = a \sin \theta$, where $a > 0$.
(a) Find an equation in rectangular coordinates satisfied by x and y.
(b) Sketch the graph of the parametric equations for $0 \leq \theta \leq 2\pi$.
(c) Sketch the graph for $-\pi/4 \leq \theta \leq \pi/2$.

Solution (a) By the Pythagorean Identity T5,

$$\left(\frac{x}{a}\right)^2 + \left(\frac{y}{a}\right)^2 = \cos^2 \theta + \sin^2 \theta = 1$$

so that

$$x^2 + y^2 = a^2.$$

The points on the graph of the parametric equation are on the circle with center at the origin and radius a.

(b) As θ increases from 0 to $\pi/2$, the values of x decrease from a to 0 and the values of y increase from 0 to a. Thus, the arc of the circle connecting $(a, 0)$ and $(0, a)$ is traced out counterclockwise as θ varies over this part of its domain. A similar analysis shows that the entire circle is traced out once as θ varies from 0 to 2π. (See Fig. 13-4a.)

(c) As θ varies from $-\pi/4$ to $\pi/2$, x increases from $-a/\sqrt{2}$ to a and then decreases to 0, while y increases from $-a/\sqrt{2}$ to a. The graph is the arc shown in Figure 13-4b. It is traced out counterclockwise. ☐

Remark Examples 1 and 2 show that the graph of a set of parametric equations may be only part of the graph of the rectangular equation obtained by eliminating the parameter. It

690 *Parametric Equations and Polar Coordinates*

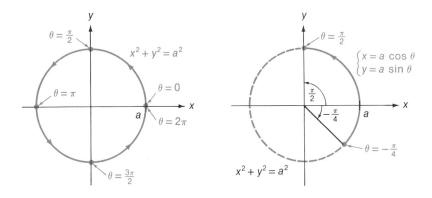

(a) Portion of the graph for $0 \leq \theta \leq 2\pi$

(b) Portion of the graph for $-\frac{\pi}{4} \leq \theta \leq \frac{\pi}{2}$

Figure 13-4 Example 2: $x = a \cos \theta$, $y = a \sin \theta$.

Relation Between Parametric and Rectangular Equations

may help to think of the parametric equations as describing motion around a race track—the graph of the rectangular equation being the track. Although the moving particle is restricted to the track, it does not necessarily move over all of it. In Example 2(c) the track is the circle $x^2 + y^2 = a^2$. The particle moves around three-eighths of the track as θ varies from $-\pi/4$ to $\pi/2$.

THE CYCLOID

The next example is a curve with equations that can be derived conveniently only in terms of a parameter.

Suppose that we place a light at a point P on a circle and then roll the circle along a line. The path traced out by the light is called a *cycloid*. (See Fig. 13-5.)

To derive parametric equations for the cycloid, we assume that the point P is initially located at the origin and that the circle is rolled along the positive x-axis. Let a be the radius of the circle. After the circle is rolled a certain distance, the situation is as pictured in Fig. 13-5b, with the point P located at (x, y). Let ϕ be the central angle through which the circle has rolled. We shall use ϕ as a parameter for x and y.

Observe that the circle has rolled a distance equal to the arc length $a\phi$. Thus, the center of the circle is the point $C(a\phi, a)$. Let $\theta = 3\pi/2 - \phi$. Then θ is an angle in standard position with the line segment CP as terminal side. (See Figs. 13-5b, c.)

We now set up a new coordinate axis system, the $x'y'$-system, with origin at C. (See Fig. 13-5c.) Relative to this system, P has coordinates

$$x' = a \cos \theta = a \cos (3\pi/2 - \phi) = -a \sin \phi,$$
$$y' = a \sin \theta = a \sin (3\pi/2 - \phi) = -a \cos \phi$$

(by trigonometric identities T8, T9). With respect to the original xy-system, P has coordinates (x, y), where

$$x = x' + a\phi = -a \sin \phi + a\phi = a(\phi - \sin \phi),$$
$$y = y' + a\ \ = -a \cos \phi + a = a(1 - \cos \phi).$$

13-1 Parametric Equations

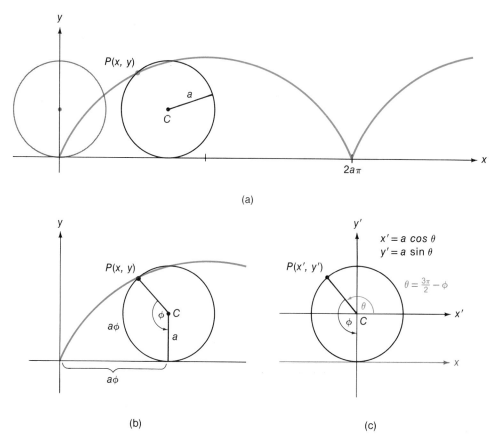

Figure 13–5 The cycloid. $x = a(\phi - \sin \phi)$, $y = a(1 - \cos \phi)$.

The parametric equations for the cycloid are

Standard Equations for the Cycloid

$$x = a(\phi - \sin \phi),$$
$$y = a(1 - \cos \phi).$$

Remark Two final observations about parametric equations are in order. First, it is possible to artificially parametricize any function. If the original function is defined by $y = g(x)$, we can let $x = f(t) = t$ and $y = g(t)$. We then have x and y expressed in terms of the artificial parameter t.

Different Parameters The second observation is that the parametric equations for a graph are not unique. In fact, any graph has an infinite number of different sets of parametric equations. As we shall see in Example 3, the parametric equations $x = \sqrt{t} - 2$, $y = t - 1$, where $t \geq 0$, have the same graph as the parametric equations $x = 2(\theta - 1)$, $y = 4\theta^2 - 1$, where $\theta \geq 0$.

EXAMPLE 3   Show that the graph of
$$x = 2(\theta - 1), \quad y = 4\theta^2 - 1, \quad \theta \geq 0$$
is the same as the graph of
$$x = \sqrt{t} - 2, \quad y = t - 1, \quad t \geq 0.$$

Solution   Let $t = 4\theta^2$, $\theta \geq 0$. Then $\sqrt{t} = 2\theta$, so that
$$x = \sqrt{t} - 2 = 2\theta - 2 = 2(\theta - 1),$$
$$y = t - 1 = 4\theta^2 - 1.$$

Since t and θ are both greater than or equal to zero, it follows that the two sets of parametric equations have the same graph. ☐

Exercises 13–1

1–4. Sketch the graphs of the given parametric equations $x = f(t)$, $y = g(t)$. Do not try to eliminate the parameter. Label several of the points with the corresponding values of t. Use arrows to indicate how the graphs are traced out. Identify all simple curves and closed curves. (*Hint:* It may help first to graph f and g as functions of t.)

- 1. $x = f(t) = \begin{cases} 0 & \text{if } 0 \leq t \leq 1 \\ t - 1 & \text{if } 1 \leq t \leq 2 \end{cases}$

    $y = g(t) = \begin{cases} t & \text{if } 0 \leq t \leq 1 \\ 1 & \text{if } 1 \leq t \leq 2 \end{cases}$

2. $x = f(t) = t$ if $-1 \leq t \leq 1$

    $y = g(t) = \begin{cases} t & \text{if } -1 \leq t \leq 0 \\ -t & \text{if } 0 < t \leq 1 \end{cases}$

3. $x = f(t) = \begin{cases} \cos t & \text{if } 0 \leq t \leq 2\pi \\ 1 & \text{if } 2\pi < t \leq 4\pi \end{cases}$

    $y = g(t) = \begin{cases} 2 & \text{if } 0 \leq t \leq 2\pi \\ 2 + \sin t & \text{if } 2\pi < t \leq 4\pi \end{cases}$

4. $x = f(t) = \begin{cases} t & \text{if } -1 \leq t \leq 1 \\ 2 - t & \text{if } 1 < t \leq 3 \end{cases}$

    $y = g(t) = \begin{cases} \sqrt{1 - t^2} & \text{if } -1 \leq t \leq 1 \\ 0 & \text{if } 1 < t \leq 3 \end{cases}$

5–16. Eliminate the parameters to get equations satisfied by x and y. Sketch the graphs of the parametric equations over the given domains. Label several points with the corresponding values of t. Use arrows to indicate how the graphs are traced out.

- 5. $x = -2t^4$, $y = t^2$, t is real
6. $x = t^2 - 1$, $y = t^4 + t^2$, t is real
- 7. $x = e^t$, $y = e^{-2t}$, t is real
8. $x = e^{3t}$, $y = e^t$, $t \geq 0$
- 9. $x = \sqrt{t} + 1$, $y = 3 + \sqrt{t}$, $t \geq 0$
10. $x = t^2 + 1$, $y = t^2 - 1$, t is real
- 11. $x = 3 \sin t$, $y = 2 \cos t$, $0 \leq t \leq 2\pi$
12. $x = 2 \sin t - 1$, $y = 3 \cos t + 2$, $0 \leq t \leq 2\pi$
- 13. $x = \cos 2t$, $y = \cos t$, $0 \leq t \leq 2\pi$
14. $x = 1 - \cos 2t$, $y = \sin^2 t$, $0 \leq t \leq 2\pi$

Exercises 13–1

- **15** $x = \cosh t,\ y = -\sinh t,\ t \leq 0$

16 $x = \ln t,\ y = t^2,\ t > 0$

17 (*Parametric Equations of a Line*) Let $P_1(x_1, y_1)$ and $P_2(x_2, y_2)$ be distinct points.
  (a) Show that

Parametric Equations of a Line
$$x = (x_2 - x_1)t + x_1,\quad y = (y_2 - y_1)t + y_1,$$

where t ranges are all real numbers, are parametric equations of the line through P_1 and P_2.
- (b) What restrictions must be placed on the parameter t to get the line segment from P_1 to P_2 as the graph of the parametric equations?

Arc Length as a Parameter
- **18** In many cases arc length is a suitable parameter. Find parametric equations for the catenary $y = \cosh x$, using arc length s as a parameter. Measure s from the point $(0, 1)$. Assume that s is negative if the arc length is measured from $(0, 1)$ to a point that has a negative x-coordinate and is positive if measured from $(0, 1)$ to a point that has a positive x-coordinate.

Hypocycloid
19 An *hypocycloid* is traced out by a point on a small circle that rolls around the inside of a fixed larger circle. (See Fig. 13–6a.) Let the radius of the fixed circle be a and that of the rolling circle be b. Let the fixed circle have its center at the origin. Suppose the initial position of the tracing point is $(a, 0)$. Show that parametric equations for the epicycloid are

$$x = (a - b)\cos\theta + b\cos\left(\frac{a-b}{b}\theta\right),$$

$$y = (a - b)\sin\theta - b\sin\left(\frac{a-b}{b}\theta\right),$$

where θ is the angle shown in Fig. 13–6a.

Epicycloid
20 A *epicycloid* is traced out by a point on a circle that rolls around the outside of a fixed circle. (See Fig. 13–6b.) Let the radius of the fixed circle be a and of the rolling circle b. Let the fixed circle have its center at the origin. Suppose the initial position of the tracing point is $(a, 0)$. Show that parametric equations for the hypocycloid are

$$x = (a + b)\cos\theta - b\cos\left(\frac{a+b}{b}\theta\right),$$

$$y = (a + b)\sin\theta - b\sin\left(\frac{a+b}{b}\theta\right).$$

- **21** A small reflector is attached to a 30-in.-diameter bicycle wheel 5 in. from the edge. At time $t = 0$ the wheel is positioned on the x-axis, touching the origin, with the reflector at the point $(0, 5)$. It then rolls along the positive x axis. Modify the argument for the cycloid. Find parametric equations for the curve traced out by the reflector. (One unit = 1 in.)

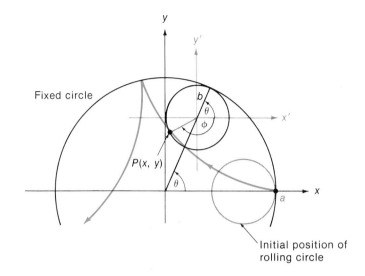

(a) The hypocycloid

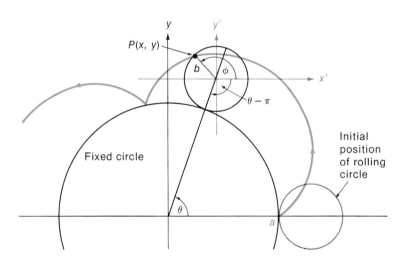

(b) The epicycloid

Figure 13–6 Exercises 19 and 20.

13–2 Derivatives

Many useful properties of graphs can be determined from a study of the derivatives dy/dx and d^2y/dx^2. Our problem is to determine these derivatives when x and y are defined parametrically. One complication arises. The graph of parametric equations $x = f(t)$, $y = g(t)$ may pass through the same point several times, having different

13-2 Derivatives

tangent lines for different values of t. This situation is illustrated in Figure 13–7. The curve passes through P twice, once when $t = t_1$ and again when $t = t_2$.

Notation $(dy/dx)_t$

Using the symbol $(dy/dx)_t$ to indicate the value of the derivative dy/dx at a specific value of t, we observe in Fig. 13–7 that, since the tangent lines have different slopes,

$$\left(\frac{dy}{dx}\right)_{t_1} \neq \left(\frac{dy}{dx}\right)_{t_2}$$

even though $P(f(t_1), g(t_1))$ is the same point as $P(f(t_2), g(t_2))$.

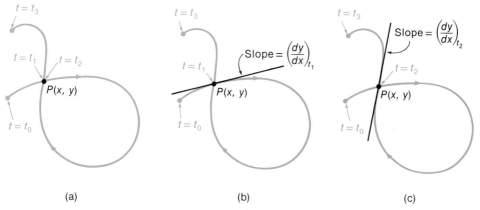

Figure 13–7 The value of the derivative dy/dx depends on the value of t which defines x and y.

Our task is to calculate $(dy/dx)_t$ when the graph of $x = f(t)$, $y = g(t)$ is a smooth curve. We shall show that under suitable conditions, on f and g,

$$\left(\frac{dy}{dx}\right)_t = \left(\frac{dy/dt}{dx/dt}\right)_t = \frac{g'(t)}{f'(t)}.$$

The exact result is found in Theorem 12–1. The following lemma simplifies the proof of that theorem considerably.

LEMMA   Let f have a continuous first derivative in the vicinity of $t = t_0$. If $f'(t_0) \neq 0$, there exists a number $\delta > 0$ such that f is a one-to-one function on the interval $(t_0 - \delta, t_0 + \delta)$.

Proof   Suppose that $f'(t_0) > 0$. By Theorem 2–4, there exists $\delta > 0$ such that

$$f'(t) > 0 \quad \text{for } t_0 - \delta < t < t_0 + \delta.$$

It follows that f is an increasing continuous function on the interval $(t_0 - \delta, t_0 + \delta)$. Thus, f is one-to-one on the interval.

The proof is similar if $f'(t_0) < 0$. ∎

Recall that the one-to-one functions are those functions that have inverses. We are now ready to prove our main result.

THEOREM 13–1 Let the graph of $x = f(t)$, $y = g(t)$ be a smooth curve. Let t_0 be a point in the common domain of f and g. Then

$$\left(\frac{dy}{dx}\right)_{t_0} = \frac{g'(t_0)}{f'(t_0)} = \left(\frac{dy/dt}{dx/dt}\right)_{t_0}$$

provided $f'(t_0) \neq 0$.

Proof Since the graph is a smooth curve, then f and g have continuous first derivatives. Since $f'(t_0) \neq 0$, it follows from the lemma that there exists an interval $(t_0 - \delta, t_0 + \delta)$ on which f is one-to-one.

We now restrict t to the interval $(t_0 - \delta, t_0 + \delta)$. Since f is one-to-one on this restricted domain, then it has an inverse function f^{-1} defined by

$$t = f^{-1}(x) \quad \text{if and only if } x = f(t).$$

On this restricted domain y is a function of x:

$$y = g(t) = g(f^{-1}(x)).$$

It follows from our work on inverse functions that y is a differentiable function of x. By the Chain Rule

$$\frac{dy}{dt} = \frac{dy}{dx} \cdot \frac{dx}{dt}.$$

If we divide this equation by $f'(t) = dx/dt$, which is known to be nonzero, we get

$$\frac{dy}{dx} = \frac{dy/dt}{dx/dt} = \frac{g'(t)}{f'(t)}.$$

In particular, when $t = t_0$,

$$\left(\frac{dy}{dx}\right)_{t_0} = \left(\frac{dy/dt}{dx/dt}\right)_{t_0} = \frac{g'(t_0)}{f'(t_0)}. \quad \blacksquare$$

Remark The derivative formula for parametric equations can be remembered most easily by using the Leibniz notation. In that notation, we have

dy/dx in Terms of a Parameter

$$\frac{dy}{dx} = \frac{dy/dt}{dx/dt}$$

which resembles the cancellation rule for fractions.

EXAMPLE 1 Let $x = t^2 - 4t$, $y = \sin \pi t$.
(a) Calculate dy/dx in terms of t;
(b) Find the equations of the tangent lines at the point $P(0, 0)$.

Solution (a) $\quad \dfrac{dy}{dt} = \pi \cos \pi t \quad$ and $\quad \dfrac{dx}{dt} = 2t - 4.$

13–2 Derivatives

Then

$$\left(\frac{dy}{dx}\right)_t = \frac{dy/dt}{dx/dt} = \frac{\pi \cos \pi t}{2t - 4}.$$

This formula holds provided $t \neq 2$.

(b) The curve passes through the point $P(0, 0)$ when $t = 0$ and when $t = 4$. (See Fig. 13–8.) When $t = 0$, the slope of the tangent line is

$$\left(\frac{dy}{dx}\right)_{t=0} = \left(\frac{\pi \cos \pi t}{2t - 4}\right)_{t=0} = \frac{\pi \cdot 1}{-4} = -\frac{\pi}{4}.$$

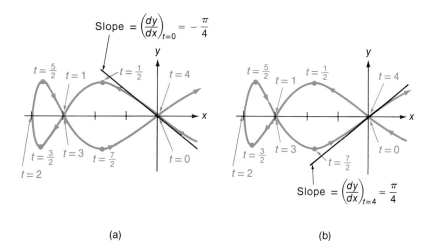

(a)   (b)

Figure 13–8 Example 1.

The equation of the tangent line is

$$y - 0 = -\frac{\pi}{4}(x - 0)$$

$$4y + \pi x = 0.$$

(See Fig. 13–8a.) When $t = 4$, the slope of the tangent line is

$$\left(\frac{dy}{dx}\right)_{t=4} = \left(\frac{\pi \cos \pi t}{2t - 4}\right)_{t=4} = \frac{\pi}{8 - 4} = \frac{\pi}{4}.$$

The equation of the tangent line is

$$y - 0 = \frac{\pi}{4}(x - 0)$$

$$4y - \pi x = 0. \quad \text{(See Fig. 13–8b.)} \quad \square$$

HIGHER DERIVATIVES

Higher derivatives can be calculated by the process described above. If $x = f(t)$ and $y = g(t)$, then

$$y' = \frac{dy/dt}{dx/dt}$$

also is a function of t. Then

$$y'' = \frac{d^2y}{dx^2} = \frac{d}{dx}(y').$$

It follows, from the same rule that we used to calculate dy/dt, that

Second Derivative in Terms of a Parameter

$$y'' = \frac{dy'/dt}{dx/dt}.$$

Similarly,

$$\frac{d^3y}{dx^3} = \frac{d}{dx}(y'') = \frac{dy''/dt}{dx/dt},$$

and so on.

EXAMPLE 2 Let $a > 0$. Show that the cycloid $x = a(\phi - \sin\phi)$, $y = a(1 - \cos\phi)$ is concave downward everywhere except at the points where $\phi = 0, \pm 2\pi, \pm 4\pi, \ldots$.

Solution

$$\frac{dy}{d\phi} = \frac{d}{d\phi}(a - a\cos\phi) = a\sin\phi$$

and

$$\frac{dx}{d\phi} = \frac{d}{d\phi}(a\phi - a\sin\phi) = a - a\cos\phi,$$

so that

$$\frac{dy}{dx} = y' = \frac{dy/d\phi}{dx/d\phi} = \frac{a\sin\phi}{a - a\cos\phi} = \frac{\sin\phi}{1 - \cos\phi}.$$

The derivative of y' with respect to ϕ is

$$\frac{dy'}{d\phi} = \frac{d}{d\phi}\left(\frac{\sin\phi}{1-\cos\phi}\right)$$

$$= \frac{(1 - \cos\phi)\cos\phi - \sin\phi(\sin\phi)}{(1 - \cos\phi)^2}$$

$$= \frac{\cos\phi - \cos^2\phi - \sin^2\phi}{(1 - \cos\phi)^2}$$

$$= \frac{\cos\phi - 1}{(1 - \cos\phi)^2} = \frac{1}{\cos\phi - 1}.$$

13-2 Derivatives

Finally,

$$y'' = \frac{dy'}{dx} = \frac{dy'/d\phi}{dx/d\phi} = \frac{1/(\cos\phi - 1)}{a - a\cos\phi} = \frac{-1}{a(1-\cos\phi)^2}.$$

Thus, $y'' < 0$ except at the points where $\phi = 0, \pm 2\pi, \pm 4\pi$, and so on. At these points y'' is not defined, because the denominator $a(1 - \cos\phi)^2$ equals zero. □

EXAMPLE 3 Find all local extrema for the graph of the parametric equations given in Example 1.

Solution The slope of the tangent line to the graph was found to be

$$\left(\frac{dy}{dx}\right)_t = \frac{\pi \cos \pi t}{2t - 4}, \quad t \neq 2$$

The graph has a vertical tangent line at $t = 2$, which corresponds to the point $x = -4$, $y = 0$. This is not an extremum (see Fig. 13-8).

We now set the derivative $(dy/dx)_t$ equal to zero and solve for t. Since we have $2t - 4 \neq 0$, this results in

$$\cos \pi t = 0$$

which is true for the values

$$t = n + \tfrac{1}{2}, \quad n = 0, \pm 1, \pm 2, \ldots$$

To decide which of these values yield maxima and which yield minima, calculate the second derivative, $(d^2y/dx^2)_t$. First calculate

$$\frac{dy'}{dt} = \frac{(2t - 4)(-\pi^2 \sin \pi t) - (\pi \cos \pi t)(2)}{(2t - 4)^2}$$

Then

$$\left(\frac{d^2y}{dx^2}\right)_t = \frac{(2t - 4)(-\pi^2 \sin \pi t) - 2\pi \cos \pi t}{(2t - 4)^3}$$

At each value of t that we consider, we have $\cos \pi t = 0$, so we can drop the second term in the numerator and cancel one factor $(2t - 4)$ from both numerator and denominator. This leaves

$$\left(\frac{d^2y}{dx^2}\right)_t = \frac{-\pi^2 \sin \pi t}{(2t - 4)^2}$$

The constant $-\pi^2/(2t - 4)^2$ is negative for all values of t, so the second derivative is negative when $\sin \pi t > 0$ and is positive when $\sin \pi t < 0$. Therefore, the graph has a local maximum when

$$t = 2n + \tfrac{1}{2}, \quad n = 0, \pm 1, \pm 2, \ldots$$

and has a local minimum when

$$t = 2n + \tfrac{3}{2}, \quad n = 0, \pm 1, \pm 2, \ldots$$

Some sample points are (see Fig. 13-8): local minimum at $(\tfrac{9}{4}, -1)$ when $t = -\tfrac{1}{2}$; local maximum at $(-\tfrac{7}{4}, 1)$ when $t = \tfrac{1}{2}$; local minimum at $(-\tfrac{15}{4}, -1)$ when $t = \tfrac{3}{2}$; and local maximum at $(-\tfrac{15}{4}, 1)$ when $t = \tfrac{5}{2}$. □

Parametric Equations and Polar Coordinates

CURVE SKETCHING

All our previous work on sketching curves can be applied to graphs of parametric equations $x = f(t)$, $y = g(t)$. We still need to determine local extrema, points of inflection, concavity, asymptotes, and so on. Our calculations now, however, will be in terms of the parameter t. We need to find the values of t that correspond to local extrema or points of inflection on the graph in the xy-plane. We need to find intervals on the t-axis for which the graph (in the xy-plane) is concave upward or downward.

Much information about the graph of the parametric equations $x = f(t)$, $y = g(t)$ can be determined from the graphs of f and g with respect to the t-axis. The type of information is illustrated in the following discussion of the graph in Figure 13–9.

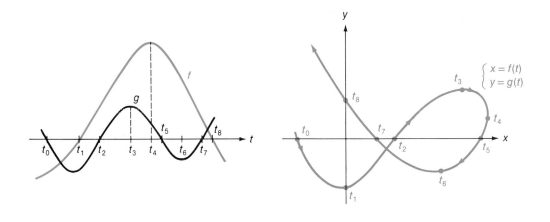

(a) Graphs of f and g

(b) Graph of the parametric equations

Figure 13–9.

Observe first, that the graph of the parametric equations crosses the x-axis whenever $g(t) = 0$. (This occurs at t_0, t_2, t_5, and t_7 in Figure 13–9a.) Similarly, it crosses the y-axis whenever $f(t) = 0$. It is below the x-axis when $g(t) < 0$, above the x-axis when $g(t) > 0$, to the left of the y-axis when $f(t) < 0$, and to the right of the y-axis when $f(t) > 0$.

Next, observe that x increases as t increases whenever $f'(t) > 0$ on an interval. (This occurs on the interval (t_0, t_4), for example.) Thus, the point that traces out the curve moves towards the right when $f'(t) > 0$. Similarly, it moves towards the left when $f'(t) < 0$. It moves upward if $g'(t) > 0$ and downward if $g'(t) < 0$.

To sketch the graph, we first plot the points at which the tracing point crosses the coordinate axes. Next we find the intervals on the t-axis that correspond to arcs traced out to the right, to the left, upward, and downward. Next, we plot on the graph the points at which the tracing point changes direction from right to left, from upward to downward, and so on. If we connect these points, in order, with a smooth curve, we should have a fairly good approximation of the graph of the parametric equations. (See Fig. 13–9b.)

Exercises 13–2

1–8. **(a)** Calculate dy/dx and d^2y/dx^2 for the parametric equations. At what points, if any, does either derivative fail to exist?

(b) Find the equation of the tangent line to the graph of the parametric equations at the indicated point.

- •1 $x = t - 3, y = t + 4; (-4, 3)$
- 2 $x = -2t^4, y = t^2; t = 1$
- •3 $x = \ln t, y = t^2, t > 0; (0, 1)$
- 4 $x = e^t, y = e^{-t}; (1, 1)$
- •5 $x = 3 \sin t, y = 2 \cos t; (0, -2)$
- 6 $x = \sec t, y = 2 \tan t; t = \pi/4$
- •7 $x = 2 - \sin t, y = \cos^2 t; t = \pi/6$
- 8 $x = \cos 2t, y = \cos t; (-\tfrac{1}{2}, \tfrac{1}{2})$

9–12. Find all local extrema for the graphs of the given parametric equations.

- • 9 $x = \sin t, y = \cos 2t$
- 10 $x = \cos t, y = \sin^2 t$
- •11 $x = t^2 - 1, y = t^4 - 8t^2$
- 12 $x = t - 1/t, y = t + 1/t$

13–14. Find all points where the tangent lines are either horizontal or vertical. Sketch the graphs, showing the horizontal and vertical tangent lines.

- •13 $x = t - \sin t, y = 1 + \cos t$
- 14 $x = t^2, y = t^3 - 12t$

•15 Find the equation of the tangent line to the graph of $x = \ln t, y = t^2 - 2t, t > 0$, that is

(a) parallel to the line $24x - y = 4$;

(b) perpendicular to the line $x + 12y = 5$.

16 The graphs of functions f and g are shown in Figure 13–10. Sketch the graph of the parametric equations $x = f(t), y = g(t)$ for the functions shown in

(a) Figure 13–10a;   (b) Figure 13–10b.

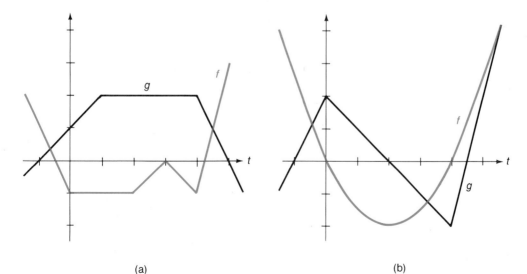

(a)   (b)

Figure 13–10 Exercise 16.

●17 (a) Find the slope of the tangent line to the cycloid $x = a(\phi - \sin \phi)$, $y = a(1 - \cos \phi)$ at the point at which $\phi = \phi_0$.
(b) Find the points where the cycloid has a vertical tangent line.

13–3 Polar Coordinates

Many different coordinate systems have been invented over the centuries to solve special types of problems. The original system Descartes used when he invented analytic geometry, for example, had oblique instead of perpendicular axes. A more current example is the log-log system engineers use, in which the coordinate axes are marked with logarithmic scales.

Two special coordinate systems—the *rectangular system* and the *polar system*—have continued in general use because of their many applications. These are most useful for graphing large classes of functions.

The rectangular coordinate system, which we have used thus far, is good for solving problems involving polynomial equations, distance, and straight-line motion. The polar coordinate system is best for problems involving circular motion and angular velocity.

THE POLAR COORDINATE SYSTEM

Polar Coordinates

Fix a point O (the *origin*) in the plane and draw a ray (the *polar axis*) that emanates from O. By convention the polar axis is drawn horizontally so that it coincides with the positive x axis. Each point P in the plane can be located by a pair of coordinates (r, θ), where r is the distance from O to P and θ is the angle, in standard position, from the polar axis to the line segment OP. (See Fig. 13–11a.) We indicate this assignment of coordinates by writing $P(r, \theta)$.

For example, the point (0, 1) (rectangular coordinates) has polar coordinates $(1, \pi/2)$. The origin has coordinates $O(0, \theta)$, where θ can be any angle. (See Fig. 13–11b.)

For each r and θ, there is a unique point $P(r, \theta)$. This uniqueness does not go the other way, however. Since many different angles have the same terminal side, a given point P has many different sets of coordinates. This is illustrated in Figure

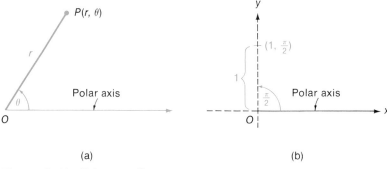

Figure 13–11 Polar coordinates.

13–3 Polar Coordinates

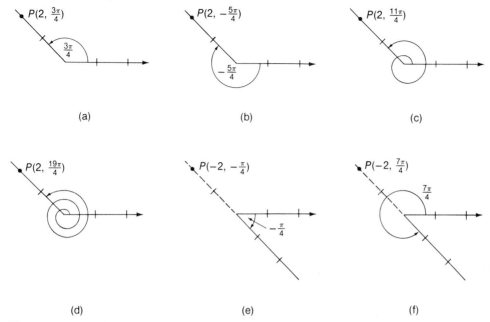

Figure 13–12 Different sets of coordinates for the point $P(2, 3\pi/4)$.

13–12 with the point $P(2, 3\pi/4)$. This same point has coordinates $(2, -5\pi/4)$, $(2, 11\pi/4)$, $(2, 19\pi/4)$, and so on.

We shall find it convenient to define $P(r, \theta)$ when r is negative as well as nonnegative. To do this, we extend the terminal side of angle θ across the origin and locate P at a distance $|r|$ from O on this extension.

For example, the point $P(2, 3\pi/4)$ also has coordinates $(-2, -\pi/4)$ and $(-2, 7\pi/4)$. (See Figs. 13–12e, f.)

GRAPHS

Graph of Polar Equation

The *graph* of a polar equation $F(r, \theta) = 0$ is the set of all points $P(r, \theta)$ with coordinates that satisfy the equation.

EXAMPLE 1 The graph of the polar equation

$$r = 2$$

consists of all points $P(r, \theta)$ such that $r = 2$—that is, all points that are two units from 0. The graph is a circle with center at O and radius 2. (See Fig. 13–13a.) Observe that this same circle also is the graph of the equation $r = -2$. □

EXAMPLE 2 The graph of the equation

$$\theta = \frac{\pi}{3}$$

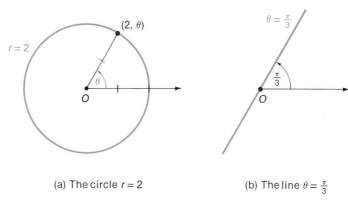

(a) The circle $r = 2$
(b) The line $\theta = \frac{\pi}{3}$

Figure 13–13 Examples 1 and 2.

consists of all points $P(r, \pi/3)$. Since r can be negative or zero as well as positive, the graph is the line through the origin with angle of inclination $\theta = \pi/3$. (See Fig. 13–13b.) ☐

If a polar equation is simple enough, we can sketch its graph by plotting numerous points and connecting them with a smooth curve. Unfortunately, this primitive method often fails when the equation is complicated. In the next section we develop more sophisticated methods of graphing. In the remainder of this section we show how to convert polar equations to equivalent equations in rectangular coordinates or to parametric equations. In some cases it is easier to graph the resulting equations in x and y than it is to graph the original ones.

RECTANGULAR AND POLAR COORDINATES

It is convenient to be able to use either rectangular or polar coordinates. To accomplish this, we superimpose the polar axis on the nonnegative x-axis with O at the origin of the xy-system. Each point P has two sets of coordinates: $P(x, y)$ and $P(r, \theta)$. (See Fig. 13–14.)

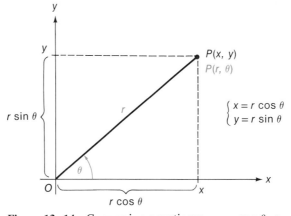

Figure 13–14 Conversion equations: $x = r \cos \theta$, $y = r \sin \theta$.

13-3 Polar Coordinates

To determine the relation between coordinates in the two systems, observe that the angle θ is in standard position relative to the xy-coordinate system, so that

$$\cos \theta = x/r \quad \text{and} \quad \sin \theta = y/r.$$

It follows that

Conversion Equations
$$x = r \cos \theta \quad \text{and} \quad y = r \sin \theta.$$

EXAMPLE 3 The point $P(2, \pi/3)$ (polar coordinates) has rectangular coordinates $P(1, \sqrt{3})$. This follows from the equations

$$x = r \cos \theta = 2 \cos \frac{\pi}{3} = 2 \cdot \frac{1}{2} = 1,$$

$$y = r \sin \theta = 2 \sin \frac{\pi}{3} = 2 \cdot \frac{\sqrt{3}}{2} = \sqrt{3}.$$

(See Fig. 13–15a.) □

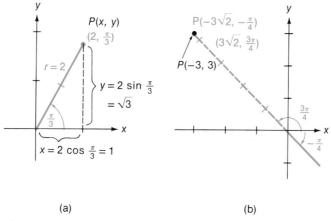

(a)   (b)

Figure 13–15 Examples 3 and 4.

To change from rectangular to polar coordinates, observe that

$$x^2 + y^2 = r^2 \cos^2 \theta + r^2 \sin^2 \theta = r^2$$

and

$$\tan \theta = \frac{r \sin \theta}{r \cos \theta} = \frac{y}{x} \quad \text{provided } x \neq 0.$$

Therefore,

$$r = \pm\sqrt{x^2 + y^2} \quad \text{and} \quad \tan \theta = \frac{y}{x} \quad \text{provided } x \neq 0.$$

The exact values of r and θ can be calculated by determining the quadrant in which the point is located. To get the polar coordinates, we must decide which angle to take for θ and whether r should be positive or negative. In particular, if we choose θ between $-\pi/2$ and $\pi/2$, then $\theta = \tan^{-1}(y/x)$, provided $x \neq 0$.

EXAMPLE 4 Let the point P have rectangular coordinates $P(-3, 3)$. If we choose θ between $-\pi/2$ and $\pi/2$, then

$$\theta = \tan^{-1}\left(\frac{y}{x}\right) = \tan^{-1}\left(\frac{3}{-3}\right) = \tan^{-1}(-1) = -\frac{\pi}{4}.$$

Since θ is a fourth quadrant angle and the point P is in Quadrant II, then r must be negative:

$$r = -\sqrt{x^2 + y^2} = -\sqrt{(-3)^2 + 3^2} = -3\sqrt{2}.$$

The point has polar coordinates $(-3\sqrt{2}, -\pi/4)$. (See Fig. 13-15b.)

On the other hand, if we choose θ to be the smallest positive angle from the x-axis to the line segment OP, then the coordinates of P are $(3\sqrt{2}, 3\pi/4)$. (See Fig. 13-15b.) □

The substitutions $x = r \cos \theta$, $y = r \sin \theta$, and $x^2 + y^2 = r^2$ can be used to convert some polar equations to rectangular equations. Example 5 illustrates the process.

EXAMPLE 5 Convert the polar equation $r = 2 \cos \theta$ to an equation in rectangular coordinates. What is the graph?

Solution It would be easy to make the substitution $r \cos \theta = x$ if we had $2r \cos \theta$ on the right-hand side of the equation instead of $2 \cos \theta$. To get the equation into this form, we first multiply both sides by r:

$$r = 2 \cos \theta$$
$$r^2 = 2r \cos \theta$$
$$x^2 + y^2 = 2x$$
$$x^2 - 2x + y^2 = 0$$
$$(x - 1)^2 + y^2 = 1^2.$$

Thus, each point on the graph of the polar equation is on the circle with center at $(1, 0)$ and radius 1. (See Fig. 13-16a.) Observe that the entire circle is traced out as θ varies from 0 to π and then is traced out again as θ varies from π to 2π. □

INTERSECTIONS OF POLAR CURVES

Some of our later work will require us to locate the points of intersection of two polar curves. A certain amount of care must be exercised since each point has an infinite number of polar coordinates. For example, a point (r, θ) may be on one curve be-

13–3 Polar Coordination

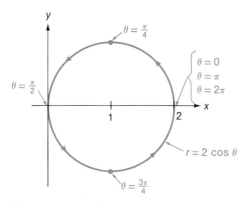

Figure 13–16a Example 5. Graph of $r = 2 \cos \theta$.

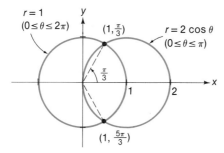

Figure 13–16b

cause (r, θ) satisfies its equation, and on a different curve because $(-r, \pi + \theta)$ satisfies its equation. If so, it may be impossible to find the point by solving the equations simultaneously as we do with rectangular coordinates. In some cases, it may be necessary to make accurate graphs of the functions and locate the points of intersection approximately from the graphs. The following example illustrates the difficulty.

EXAMPLE 6   The circles $r = 1$ (traced out as θ varies from 0 to 2π) and $r = 2 \cos \theta$ (traced out as θ varies from 0 to π) intersect at two points. (See Figure 13–16b.)

If we solve the equations simultaneously, however, using the fact that $0 \leq \theta \leq \pi$ for the circle $r = 2 \cos \theta$, we obtain

$$r = 2 \cos \theta = 1$$
$$\cos \theta = \tfrac{1}{2}$$
$$\theta = \frac{\pi}{3}$$

Thus, we find only the point $(1, \pi/3)$.

The other point is $(1, 5\pi/3) = (-1, 2\pi/3)$. It is on the graph of $r = 1$ when $\theta = 5\pi/3$ and on the graph of $r = 2 \cos \theta$ when $r = -1$ and $\theta = 2\pi/3$. □

POLAR FUNCTIONS AND PARAMETRIC EQUATIONS

A polar function $r = f(\theta)$ defines x and y in terms of the parameter θ:

$$x = r \cos \theta = f(\theta) \cos \theta;$$
$$y = r \sin \theta = f(\theta) \sin \theta.$$

Thus, all our previous work on parametric equations can be applied to polar functions. In particular, you should know that the graph of an equation obtained by converting a polar equation to rectangular coordinates may contain more points than the graph of the original polar equation.

Exercises 13–3

1 Polar coordinates of the points are given. Find two additional sets of polar coordinates for each point, one with $r > 0$ and the other with $r < 0$. Convert each set to rectangular coordinates.

- •(a) $(-1, -\pi/3)$
- •(c) $(-\sqrt{2}, \pi/4)$
- •(e) $(4, 7\pi/6)$
- (b) $(2, \pi/6)$
- (d) $(1, 0)$
- (f) $(\sqrt{3}, -5\pi/6)$

2 Change from rectangular to polar coordinates. Find at least two sets of polar coordinates for each point, one with $r > 0$ and one with $r < 0$.
- •(a) $(-2, 0)$
- •(c) $(-1, -\sqrt{3})$
- (b) $(4, 4)$
- (d) $(-\sqrt{2}, \sqrt{2})$

3–10. Convert the equation to a rectangular equation with the same graph.

- •3 $r \cos \theta = 2$
- •5 $\tan \theta = 1$
- •7 $r = -1$
- •9 $r = \sec \theta + \csc \theta$
- 4 $\sin \theta = -3/r$
- 6 $\theta = \pi/6$
- 8 $r = -2 \sin \theta$
- 10 $r(1 + 2 \sin \theta) = 2.$

11–16. Change the rectangular equation to a polar equation with the same graph.

- •11 $x = 2$
- •13 $x^2 + y^2 = 36$
- •15 $2x = y^2 - 1$
- 12 $y = -4$
- •14 $4x^2 + 3(y + 1)^2 = 12$
- 16 $4y = 4x^2 - 1$

17–18. Use the conversion equations $x = r \cos \theta$, $y = r \sin \theta$ to find parametric equations for x and y in terms of θ.

- •17 $r = \cos 2\theta$
- 18 $r = 1 + \sin \theta$

13-4 Graphs of Polar Equations

19–26. Sketch the graphs of the polar equations. (*Hint:* In some problems it is best to change to rectangular coordinates. In others, consider how r behaves as θ increases.)

- **19** $r = -3$
- **20** $\tan \theta = -1$
- **21** $r = 2 \sin \theta$
- **22** $r \sin \theta = 2$
- **23** $r = 3 \csc \theta$
- **24** $r = \sin\left(\frac{\pi}{2} - \theta\right)$
- **25** $r\theta = 1, \theta > 0$ (*spiral*)
- **26** $r = \theta, \theta \geq 0$ (*spiral*)

27–28. Find the points of intersection of the graphs.

- **27** $r = 2 \cos \theta, r = -1$
- **28** $r = 2 \cos 2\theta, r = 2$

13-4 Graphs of Polar Equations

In this section we first establish tests for general properties of polar equations and then consider some important graphs. We assume throughout that the polar system is superimposed on the xy-coordinate system.

SYMMETRY

The graph of a polar equation $F(r, \theta) = 0$ is symmetric about the x-axis if the point $Q(r, -\theta)$ is on the graph whenever $P(r, \theta)$ is on it. (See Fig. 13–17a.) This observation leads to the following test.

Test for Symmetry about the x-Axis Substitute $-\theta$ for θ in the defining equation. If the new equation is equivalent to the original equation, then the graph is symmetric about the x-axis.

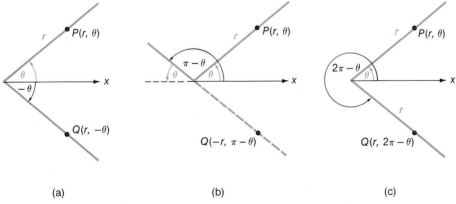

(a)   (b)   (c)

Figure 13–17 Symmetry About the x axis. Each diagram corresponds to a different test for symmetry.

EXAMPLE 1 Let $r = 1 + \cos \theta$. If we substitute $-\theta$ for θ, we get

$$r = 1 + \cos(-\theta) \quad \text{(new equation)}$$

which, since $\cos(-\theta) = \cos \theta$, is equivalent to the original equation. The graph is symmetric about the x-axis. □

Remark A word of warning is in order. The above test for symmetry about the x-axis has as its basis the fact that the symmetric point $Q(r, -\theta)$ must be on the graph whenever $P(r, \theta)$ is on it. Unfortunately, even if the point Q is always on the curve, the coordinates (r, θ) may not satisfy the equation. For example, we may need to write the coordinates as $(-r, \pi - \theta)$ or $(r, 2\pi - \theta)$ to satisfy the equation. (See Figs. 13–17b, c.) Each of these sets of coordinates leads to a different test for symmetry about the x-axis. If we use the coordinates $Q(-r, \pi - \theta)$, for example, we get the following test.

Second Test for Symmetry about the x-Axis Substitute $-r$ for r and $\pi - \theta$ for θ in the defining equation. If the new equation is equivalent to the original one, then the graph is symmetric about the x-axis.

EXAMPLE 2 Let $r = \sin 2\theta$. If we substitute $-\theta$ for θ, we get

$$r = \sin(-2\theta) \quad \text{(new equation)}$$
$$r = -\sin 2\theta$$

which is not equivalent to the original equation. Thus, the first test gives no information.
  If we substitute $-r$ for r and $\pi - \theta$ for θ, we get

$$-r = \sin(2\pi - 2\theta) \quad \text{(new equation)}$$
$$-r = \sin(-2\theta)$$
$$-r = -\sin 2\theta,$$

which is equivalent to the original equation.
  The graph is symmetric about the x-axis. □

Table 13–1 Tests for Symmetry

Symmetry about . . .	Test
x-axis	Substitute $-\theta$ for θ
y-axis	Substitute $\pi - \theta$ for θ
Origin	Substitute $-r$ for r

Tests for symmetry about the x-axis, the y-axis, and the origin are stated in Table 13–1 and illustrated in Figure 13–18. In each case, only one test is given, although an infinite number of other tests exist. (See Exercises 15 and 16.) A test is affirmative, indicating symmetry, if the new equation obtained by the substitution is equivalent to the original equation; however, a negative test does not indicate a lack of symmetry. (Two additional tests for symmetry about the x-axis are shown in Figure 13–17b and c.)

13-4 Graphs of Polar Equations

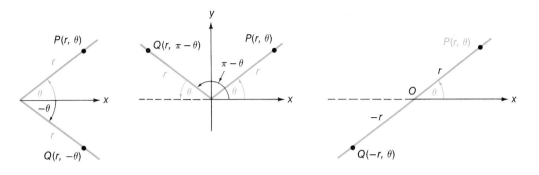

(a) Symmetry about x-axis
(b) Symmetry about y-axis
(c) Symmetry about origin

Figure 13–18 Tests for symmetry. Each diagram corresponds to a test for symmetry. (See Table 13–1.)

PROPERTIES DETERMINED FROM DERIVATIVES

Let $r = f(\theta)$ be a function of θ. If f is continuous, then its graph is traced out continuously as θ increases. If f has a continuous first derivative, then the graph is a smooth curve.

It may be instructive to think of the curve as traced out by a moving point on a rotating ray through the origin. As the ray rotates the point moves in and out along the ray—tracing out the graph.

Suppose now that $dr/d\theta = f'(\theta) > 0$. Then $r = f(\theta)$ is an increasing function of θ. If $f(\theta)$ is positive, then the tracing point moves away from the origin. (See Fig. 13–19a.) If $f(\theta)$ is negative, then the tracing point moves towards the origin in the opposite quadrant (Fig. 13–19b).

The situation is reversed if $dr/d\theta < 0$. If $f(\theta)$ is positive, then the tracing point moves towards the origin; if $f(\theta)$ is negative, it moves away from the origin in the opposite quadrant. (See Figs. 13–19c, d.)

Derivative properties based on the xy-coordinate system, such as slopes of tangent lines, concavity, local extrema, and so on, can be determined from the derivatives dy/dx and d^2y/dx^2, calculated in terms of the parameter θ.

Since
$$x = r \cos \theta = f(\theta) \cos \theta,$$
$$y = r \sin \theta = f(\theta) \sin \theta,$$

then

dy/dx in Polar Coordinates

$$y' = \frac{dy}{dx} = \frac{dy/d\theta}{dx/d\theta} = \frac{f(\theta) \cos \theta + f'(\theta) \sin \theta}{-f(\theta) \sin \theta + f'(\theta) \cos \theta}.$$

The second derivative d^2y/dx^2 can be calculated similarly. Since $y' = dy/dx$ is a

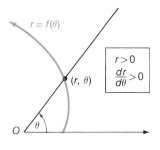
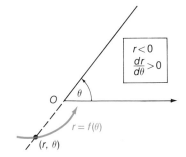

(a) $r>0$, $\frac{dr}{d\theta}>0$: the tracing point moves away from the origin as θ increases

(b) $r<0$, $\frac{dr}{d\theta}>0$: the tracing point moves towards the origin as θ increases

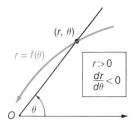
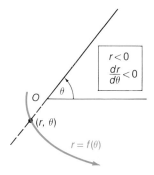

(c) $r>0$, $\frac{dr}{d\theta}<0$: the tracing point moves towards the origin as θ increases

(d) $r<0$, $\frac{dr}{d\theta}<0$: the tracing point moves away from the origin as θ increases

Figure 13–19 Properties of polar graphs determined by r and $dr/d\theta$.

function of θ, then the second derivative can be calculated as

$$\frac{d^2y}{dx^2} = \frac{d}{dx}(y') = \frac{d(y')/d\theta}{dx/d\theta}.$$

TANGENT LINES AT THE ORIGIN

It is particularly easy to find the tangent lines at the origin. Suppose that the graph of $r = f(\theta)$ passes through the origin when $\theta = \theta_0$. Then

$$r_0 = f(\theta_0) = 0.$$

The slope of the tangent line at (r_0, θ_0) is

$$\left(\frac{dy}{dx}\right)_{\theta=\theta_0} = \frac{f(\theta_0)\cos\theta_0 + f'(\theta_0)\sin\theta_0}{-f(\theta_0)\sin\theta_0 + f'(\theta_0)\cos\theta_0}$$

13–4 Graphs of Polar Equations

$$= \frac{f'(\theta_0) \sin \theta_0}{f'(\theta_0) \cos \theta_0}$$

$$= \tan \theta_0$$

provided $f'(\theta_0) \neq 0$. Since $(dy/dx)_{\theta=\theta_0} = \tan \theta_0$, then θ_0 is the angle of inclination of the tangent line. (See Fig. 13–20.)

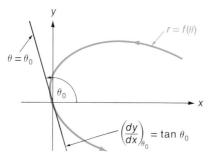

Figure 13–20 The tangent lines at the origin are found among the values of θ for which $f(\theta) = 0$.

It follows from the above remarks that the tangent lines at the origin can be found among the values of θ for which $r = f(\theta) = 0$. This result also holds if $\theta_0 = \pm \pi/2$, even though $\tan \theta_0$ is not defined. In that case the tangent line at the origin is vertical.

EXAMPLE 3 Let $r = 2 \cos \theta$. The tangent lines at the origin are found by setting $r = 0$ and solving for θ. If $2 \cos \theta = 0$, then

$$\cos \theta = 0$$
$$\theta = \pm \pi/2, \pm 3\pi/2, \pm 5\pi/2, \ldots$$

Thus, the y-axis is the only tangent line at the origin. (See Fig. 13–16.) □

Similar Graph We now turn our attention to some special polar curves. In each case, we give an equation that defines a particular type of curve. Curves of similar shape but different size or orientation can be obtained by replacing cosine functions with sine functions, by multiplying the function by a constant, and so on.

THE CARDIOID

The graph of the equation

$$r = 1 + \cos \theta$$

is called a *cardioid* (heart-shaped figure). (See Fig. 13–21.) This curve is traced out one time as θ varies from 0 to 2π.

Symmetry The curve is symmetric about the x-axis. (See Example 1.)

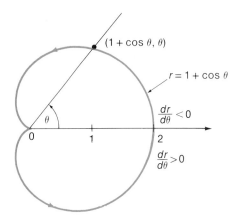

θ	0	$\frac{\pi}{4}$	$\frac{\pi}{2}$	$\frac{3\pi}{4}$	π	$\frac{5\pi}{4}$	$\frac{3\pi}{2}$	$\frac{7\pi}{4}$	2π
r	2	$1+\frac{\sqrt{2}}{2}$	1	$1-\frac{\sqrt{2}}{2}$	0	$1-\frac{\sqrt{2}}{2}$	1	$1+\frac{\sqrt{2}}{2}$	2

Figure 13–21 The cardioid $r = 1 + \cos\theta$.

Tangent Lines at the Origin If we set $r = 0$ and solve for θ, we get

$$1 + \cos\theta = 0$$
$$\cos\theta = -1$$
$$\theta = \pm\pi, \pm 3\pi, \pm 5\pi, \ldots.$$

The only possible tangent line at the origin is the line $\theta = \pi$.

The derivative of y with respect to x is given in terms of θ as

$$\frac{dy}{dx} = \frac{\sin^2\theta - \cos^2\theta - \cos\theta}{\sin\theta(1 + 2\cos\theta)}$$

which is not defined at $\theta = \pi$. Thus, there is no tangent line at the origin. The negative x-axis does, however, have some properties similar to those of a tangent line. If we calculate the limit of $(dy/dx)_\theta$ as $\theta \to \pi$ from either side, we find that

$$\lim_{\theta \to \pi^-}\left(\frac{dy}{dx}\right)_\theta = \lim_{\theta \to \pi^+}\left(\frac{dy}{dx}\right)_\theta = 0.$$

Thus, the negative x-axis is a limiting position of the tangent lines as $\theta \to \pi$.

Other Properties The function $f(\theta) = 1 + \cos\theta$ is nonnegative for all θ. It is a decreasing function for $0 < \theta < \pi$ and an increasing function for $\pi < \theta < 2\pi$.

The graph is shown in Figure 13–21.

THE LIMAÇON

The graph of the equation

$$r = 1 + 2\cos\theta$$

13-4 Graphs of Polar Equations

is an example of a *limaçon*. (See Fig. 13-22a.) It is traced out one time as θ varies from 0 to 2π. The following properties can be established. The details are left for you.

Symmetry The limaçon $r = 1 + 2 \cos \theta$ is symmetric about the x-axis.

Tangent Lines at the Origin The lines

$$\theta = \frac{2\pi}{3} \quad \text{and} \quad \theta = \frac{4\pi}{3}$$

are tangent to the graph at the origin.

Other Properties

(1) The function $f(\theta) = 1 + 2 \cos \theta$ decreases for $0 < \theta < \pi$ and increases for $\pi < \theta < 2\pi$.
(2) The function decreases from $r = 3$ to $r = 0$ as θ increases from 0 to $2\pi/3$, then decreases from 0 to -1 (tracing out the bottom half of the inner loop) as θ varies from $2\pi/3$ to π.
(3) The function increases from -1 to 0 (tracing out the top half of the inner loop) as θ varies from π to $4\pi/3$, then increases from 0 to 3 as θ varies from $4\pi/3$ to 2π.

The graph is pictured in Figure 13-22a.

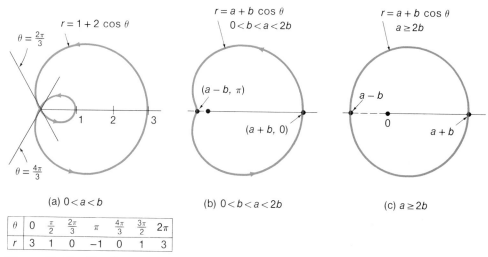

θ	0	$\frac{\pi}{2}$	$\frac{2\pi}{3}$	π	$\frac{4\pi}{3}$	$\frac{3\pi}{2}$	2π
r	3	1	0	-1	0	1	3

Figure 13-22 The limaçon $r = a + b \cos \theta$.

Remark The limaçon has several different basic shapes, as illustrated in Figure 13-22. The fundamental curve is defined by an equation of form

$$r = a + b \cos \theta$$

where a and b are positive constants, $a \neq b$. As indicated in the figure, the shape is determined by the relative sizes of a and b. There is one shape if $a < b$, another shape if $b < a < 2b$, and a third shape if $a \geq 2b$. (See Exercise 21.) The cardioid is the limiting shape for the limaçon as $a \to b$.

THE n-LEAVED ROSE

Equations of form

$$r = \cos n\theta,$$

where n is a positive integer greater than 1, form a set of "roses." If n is odd, the rose has n leaves, each of which is traced out twice as θ varies from 0 to 2π. If n is even, the rose has $2n$ leaves, each of which is traced out once as θ varies from 0 to 2π. (See Fig. 13–23.)

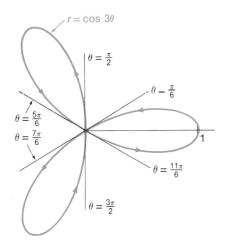

(a) Three-leaved Rose $r = \cos 3\theta$

θ	0	$\frac{\pi}{6}$	$\frac{\pi}{3}$	$\frac{\pi}{2}$	$\frac{2\pi}{3}$	$\frac{5\pi}{6}$	π
r	1	0	-1	0	1	0	-1

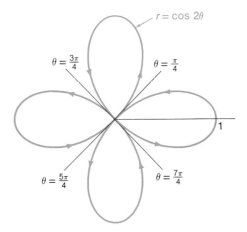

(b) Four-leaved Rose $r = \cos 2\theta$

θ	0	$\frac{\pi}{4}$	$\frac{\pi}{2}$	$\frac{3\pi}{4}$	π	$\frac{5\pi}{4}$	$\frac{3\pi}{2}$	$\frac{7\pi}{4}$	2π
r	1	0	-1	0	1	0	-1	0	1

Figure 13–23.

THE THREE-LEAVED ROSE

The three-leaved rose

$$r = \cos 3\theta,$$

pictured in Figure 13–23a, has the following properties.

Symmetry The graph is symmetric about the x-axis.

Tangent Lines at the Origin The lines

$$\theta = \frac{\pi}{6}, \ \frac{\pi}{2}, \ \frac{5\pi}{6}, \ \frac{7\pi}{6}, \ \frac{3\pi}{2}, \ \text{and} \ \frac{11\pi}{6}$$

Exercises 13–4

are tangent to the curve at the origin. Observe that some of these lines coincide with others.

Other Properties

(1) The function $r = f(\theta) = \cos 3\theta$ decreases from $r = 1$ to $r = 0$ as θ varies from 0 to $\pi/6$, tracing out part of a leaf in Quadrant I.
(2) The function decreases from 0 to -1 and then increases from -1 to 0 as θ varies from $\pi/6$ to $\pi/2$, tracing out a leaf in Quadrant III.
(3) Other leaves are traced out similarly as θ varies from $\pi/2$ to π.
(4) The entire graph is traced out a second time as θ varies from π to 2π.

The graph is pictured in Figure 13–23, along with the graph of the four-leaved rose $r = \cos 2\theta$.

Exercises 13–4

1–2. Sketch the graph of a polar curve $r = f(\theta)$ that satisfies the following conditions—provided the conditions are not contradictory.

1 r decreases from 1 to 0 as θ varies from 0 to $\pi/4$, decreases from 0 to -2 as θ varies from $\pi/4$ to $\pi/2$, increases from -2 to 0 as θ varies from $\pi/2$ to $3\pi/4$, decreases from 0 to -1 as θ varies from $3\pi/4$ to π.

2 $f(\theta) \begin{cases} >0 \text{ if } 0 < \theta < \pi/4 \\ <0 \text{ if } \pi/4 < \theta < 3\pi/4 \\ >0 \text{ if } 3\pi/4 < \theta < \pi \end{cases}$ $f'(\theta) \begin{cases} <0 \text{ if } 0 < \theta < \pi/2 \\ >0 \text{ if } \pi/2 < \theta < \pi \end{cases}$

3–14. Find all obvious symmetries and the tangent lines at the origin. Sketch the graph. If the graph is one of the standard curves discussed in this section, then identify it by name.

- **3** $r = 2(1 + \sin \theta)$
- **4** $r = 3(1 - \cos \theta)$
- **5** $r = 2 - \sin \theta$
- **6** $r = 1 - 2 \sin \theta$
- **7** $r = 3 - \cos \theta$
- **8** $r = 3 - 3 \sin \theta$
- **9** $r = \sin 3\theta$
- **10** $r = 2 \sin 2\theta$
- **11** $r^2 = 4 \sin 2\theta$ (lemniscate)
- **12** $r^2 = 9 \cos 2\theta$ (lemniscate)
- **13** $r = 2 \sin^2 \dfrac{\theta}{2}$
- **14** $r = -2 \cos^2 \dfrac{\theta}{2}$

Alternative Test for Symmetry About y-axis

15 (a) Let $F(r, \theta) = 0$ be a polar equation. Devise a test for symmetry about the y-axis having the basis that $P(r, \theta)$ and $Q(-r, -\theta)$ are symmetric points about the y-axis.
(b) Use the test in (a) and the one described in the text to test the equation $r^2 = \cos \theta$ for symmetry about the y-axis.

Symmetry About Origin

16 (a) Let $F(r, \theta) = 0$ be a polar equation. Devise a test for symmetry about the origin having the basis that $Q(r, \pi + \theta)$ and $P(r, \theta)$ are symmetric points about the origin.

(b) Use the test in (a) and the one described in the text to test the equation $r = \cos 2\theta$ for symmetry about the origin.

17–20. Find the points on the graph that are the furthest from the x-axis.

• 17 $r = 1 + \cos \theta$ 18 $r = 1 + 2\cos \theta$

• 19 $r = \cos 2\theta$ 20 $r = 2\sqrt{\cos 2\theta}$

21 (*The Limaçon* $r = a + b \cos \theta$) Let a and b be positive numbers, $a \neq b$. Interpret the results of (a), (b), and (c) in terms of Figure 13–22.

Properties of a Limaçon

(a) Prove that the limaçon $r = a + b \cos \theta$ has an inner loop if $a < b$ and no inner loop if $a > b$.

(b) Prove that there are four distinct values of θ, $0 \leq \theta < 2\pi$, where the limaçon $r = a + b \cos \theta$ has a vertical tangent line if $b < a < 2b$.

(c) Prove that there are two distinct values of θ, $0 \leq \theta < 2\pi$, where the limaçon $r = a + b \cos \theta$ has a vertical tangent line if $a \geq 2b$.

13–5 Circles, Lines, and Conic Sections

As a rule, curves with simple equations in one coordinate system have complicated equations in any system in which coordinates are assigned in a radically different manner. An example is furnished by the four-leaved rose $r = \cos 2\theta$. The equation for this rose in rectangular coordinates is

$$(x^2 + y^2)^3 = (x^2 - y^2)^2,$$

which is difficult to graph without changing back to polar coordinates.

Circles, lines, and conic sections have simple equations in rectangular coordinates and fairly complicated equations in polar coordinates unless the figures are located in special positions with respect to the axis system.

LINES

The general equation of the line $ax + by = c$ in polar coordinates is

General Equation of a Line

$$r(a \cos \theta + b \sin \theta) = c.$$

It is difficult to graph this without conversion to rectangular coordinates. (See Fig. 13–24.)

In the special case in which a line is parallel to a coordinate axis, the equation reduces to one of the forms

Vertical and Horizontal Lines

$$r \cos \theta = A \quad \text{or} \quad r \sin \theta = B.$$

(See Fig. 13–24b.)

13–5 Circles, Lines, and Conic Sections

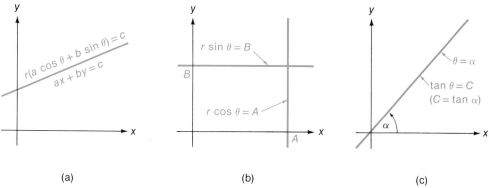

Figure 13–24 Equations of lines in polar coordinates.

If a line passes through the origin with angle of inclination α, then its polar equation can be written as

Lines Through the Origin

$$\theta = \alpha \quad \text{or} \quad \tan \theta = C$$

where $C = \tan \alpha$. (See Fig. 13–24c.)

CIRCLES

The circle $r = a$ has its center at the origin and radius $|a|$. (See Fig. 13–25a.)

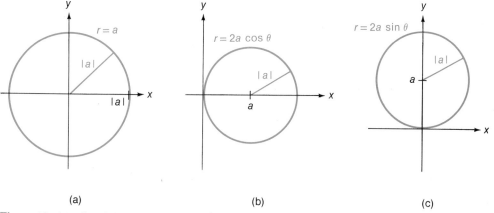

Figure 13–25 Special equations of circles in polar coordinates.

There are two other special circles of interest. The graph of

Circle with Center on x-Axis, Tangent to y-Axis

$$r = 2a \cos \theta$$

can be changed to rectangular coordinates without difficulty. We multiply the equation by r, getting

$$r^2 = 2ar\cos\theta$$
$$x^2 + y^2 = 2ax$$

which reduces to $(x - a)^2 + y^2 = a^2$. Thus the center is the point $(a, 0)$ on the x-axis (rectangular coordinates) and its radius is $|a|$. It is tangent to the y-axis at the origin. (See Fig. 13–25b.)

The graph of

Circle with Center on y-Axis, Tangent to x-Axis

$$r = 2a\sin\theta$$

is a circle with center at the point $(0, a)$ on the y-axis (rectangular coordinates) and radius $|a|$. It is tangent to the x-axis at the origin. (See Fig. 13–25c.)

CONIC SECTIONS

If a conic section has a focus at the origin and the corresponding directrix parallel to a coordinate axis, then its polar equation has a comparatively simple form.

THEOREM 13–2 Let $k > 0$. Let a conic section have a focus at the origin and the line $y = k$ as the corresponding directrix. The polar equation of the conic is

Equation of Conic with Focus at Origin, Directrix Parallel to x-Axis

$$r = \frac{ke}{1 + e\sin\theta}$$

where e is the eccentricity (not to be confused with the base of the natural logarithm system).

Proof We consider the case for which the conic is a parabola or ellipse. The other case is left for you.

Let $P(r, \theta)$ be a point on the conic. (See Fig. 13–26.) It follows from Theorem 7–5 that

$$|PF| = e|P\mathcal{D}|$$

where

$|PF|$ = distance from the focus F to P;
$|P\mathcal{D}|$ = distance from the directrix $\mathcal{D}$ to P.

Since the focus F is at the origin, then

$$|PF| = r.$$

Since the conic is a parabola or an ellipse, then it is below the directrix. Therefore,

$$|P\mathcal{D}| = k - y\text{-coordinate of } P = k - r\sin\theta.$$

13–5 Circles, Lines, and Conic Sections

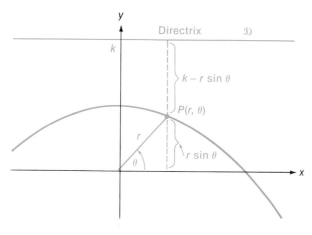

Figure 13–26.

Since $|PF| = e|P\mathcal{D}|$, then

$$r = e(k - r \sin \theta).$$

If we solve this last equation for r, we get

$$r = \frac{ke}{1 + e \sin \theta}. \blacksquare$$

Recall from Section 7–5 that a conic section is

Classification of a Conic by Eccentricity

an *ellipse* if $0 < e < 1$,
a *parabola* if $e = 1$,
a *hyperbola* if $e > 1$.

We can use this information and Theorem 13–2 to identify a conic by type and to make a rough sketch.

EXAMPLE 1 (a) Identify the graph of the polar equation

$$r(4 + 5 \sin \theta) = 9.$$

(b) Rewrite the equation in rectangular form.

Solution (a) The equation can be rewritten as

$$r = \frac{9}{4 + 5 \sin \theta} = \frac{\frac{9}{4}}{1 + \frac{5}{4} \sin \theta} = \frac{\frac{9}{5} \cdot \frac{5}{4}}{1 + \frac{5}{4} \sin \theta}.$$

The graph is a conic with eccentricity $e = \frac{5}{4}$. It has a focus at the origin. The corresponding directrix is the line $y = \frac{9}{5}$. Since $e > 1$, the conic is a hyperbola.

(b) To change the equation to rectangular form, we first rewrite it as

$$4r + 5r \sin \theta = 9$$
$$\pm 4\sqrt{x^2 + y^2} + 5y = 9$$
$$16(x^2 + y^2) = (9 - 5y)^2 = 81 - 90y + 25y^2.$$

If we now complete the square on the y terms, we can rewrite the equation in the form

$$\frac{(y-5)^2}{4^2} - \frac{x^2}{3^2} = 1. \quad \square$$

STANDARD EQUATIONS OF CONIC SECTIONS

Let $k > 0$. Results similar to Theorem 13–2 hold if the directrix is the line $y = -k$ or either of the lines $x = k$ or $x = -k$, where $k > 0$. (See Fig. 13–27.) The results are summarized below:

(1) The graph of

$$r = \frac{ek}{1 - e \sin \theta}$$

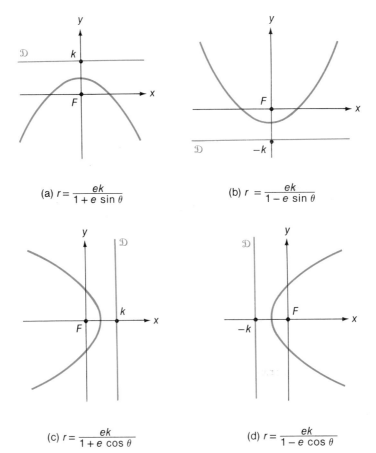

(a) $r = \dfrac{ek}{1 + e \sin \theta}$ (b) $r = \dfrac{ek}{1 - e \sin \theta}$

(c) $r = \dfrac{ek}{1 + e \cos \theta}$ (d) $r = \dfrac{ek}{1 - e \cos \theta}$

Figure 13–27 Standard equations of conic sections in polar coordinates.

is a conic section with focus at the origin. The line $y = -k$ is the corresponding directrix (Fig. 13–27b).

(2) The graph of

$$r = \frac{ek}{1 + e \cos \theta}$$

is a conic section with focus at the origin. The line $x = k$ is the corresponding directrix (Fig. 13–27c).

(3) The graph of

$$r = \frac{ek}{1 - e \cos \theta}$$

is a conic section with a focus at the origin. The line $x = -k$ is the corresponding directrix (Fig. 13–27d).

Exercises 13–5

1–6. Identify and sketch the graphs of the given lines and curves.

- **1** $r = 2 \sec \theta$
- **2** $4r \sin \theta = 3$
- **3** $4 \sin^2 \theta = 3$
- **4** $\sec^2 \theta = 2$
- **5** $r = -2 \cos \theta$
- **6** $r = -3 \sin \theta$

7–12. Identify and sketch the graphs of the conics. Convert each equation to an equivalent rectangular equation

- **7** $r = \dfrac{3}{2 + 2 \cos \theta}$
- **8** $r = \dfrac{5}{2 - 6 \cos \theta}$
- **9** $r = \dfrac{6}{3 + \cos \theta}$
- **10** $r = \dfrac{1}{2 + 6 \sin \theta}$
- **11** $r = \dfrac{4}{\sin \theta - 2}$
- **12** $r = \dfrac{3}{2 - 2 \sin \theta}.$

13–16. Write the equation in polar coordinates of the conic with a focus at the origin that satisfies the given conditions.

- **13**   Parabola, vertex $(1, -\pi/2)$.
- **14**   Ellipse, vertices $(4, 0)$ and $(6, \pi)$.
- **15**   Hyperbola, vertices $(1, 3\pi/2)$ and $(-5, \pi/2)$.
- **16**   Ellipse, end of minor axis $(2, \pm\pi/4)$.

17–19. The maximum and minimum distances of five planets from the sun are given in the table.
   (a) Calculate the eccentricity of the elliptical orbit of the planet.
   (b) Write the standard equation of the orbit in polar coordinates with the sun (focus) at the origin. (Hint: Use the definition of an ellipse to show that the maximum and minimum distances occur at the ends of the major axis.)

•**17** Mercury   **18** Venus   •**19** Earth

Planet	Distance from Sun (millions of miles)	
	Maximum	Minimum
Mercury	43.4	28.6
Venus	67.6	66.7
Earth	94.5	91.3
Mars	154.8	128.8
Jupiter	506.7	459.9

13–6 Area and Arc Length

It is not difficult to express as an integral the area of a region bounded by a polar curve. As we shall see, however, the integral formula is different from the corresponding formula in rectangular coordinates.

Let $\mathcal{R}$ be the region bounded by the graph of $r = f(\theta)$ and the rays $\theta = \alpha$, $\theta = \beta$, where f is continuous for $\alpha \leq \theta \leq \beta$. (See Fig. 13–28a.) Partition the angle from α to β into n subangles by the rays $\theta = \theta_k$, $k = 0, 1, 2, \ldots, n$, where

$$\alpha = \theta_0 < \theta_1 < \theta_2 < \cdots < \theta_n = \beta.$$

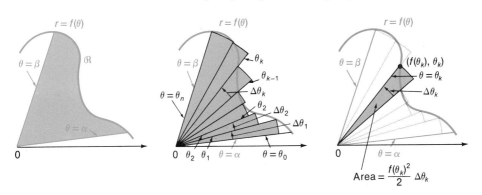

(a) The region $\mathcal{R}$

(b) The area of the region is approximated by sectors of circles

(c) The area of the kth sector is $A_k = \dfrac{f(\theta_k)^2}{2} \Delta\theta_k$

Figure 13–28   Area $= \dfrac{1}{2} \displaystyle\int_\alpha^\beta f(\theta)^2 d\theta.$

13-6 Area and Arc Length

(See Fig. 13–28b.) Let

$$\Delta\theta_k = \theta_k - \theta_{k-1}$$

be the kth subangle.

The area of the portion of $\mathcal{R}$ between the rays θ_{k-1} and θ_k is approximately equal to the area of the sector of a circle with central angle $\Delta\theta_k$ and radius $|f(\theta_k)|$. (See Fig. 13–28c.) It is known from trigonometry that the area of a sector of a circle with central angle α and radius r is $\alpha r^2/2$. (See Appendix B.) Thus, the area of the sector shown in Figure 13–28c is

$$\frac{1}{2}|f(\theta_k)|^2 \, \Delta\theta_k = \frac{1}{2}[f(\theta_k)]^2 \, \Delta\theta_k.$$

It follows that

$$\text{Area of } \mathcal{R} \approx \sum_{k=1}^{n} \frac{1}{2}[f(\theta_k)]^2 \, \Delta\theta_k.$$

Observe that this approximation to the area improves as each $\Delta\theta_k \to 0$. Thus,

$$\text{Area of } \mathcal{R} = \lim_{\Delta\theta_k \to 0} \sum_{k=1}^{n} \frac{1}{2}[f(\theta_k)]^2 \, \Delta\theta_k$$

Basic Area Formula for Polar Coordinates

$$\text{Area of } \mathcal{R} = \frac{1}{2}\int_{\alpha}^{\beta} [f(\theta)]^2 \, d\theta.$$

Remark When we apply the area formula we must determine the exact values of θ that are used to trace out the appropriate part of the graph. The limits of integration, α and β, must be chosen so that the graph is traced out one time as θ increases from α to β.

EXAMPLE 1 Calculate the area of the region inside the cardioid

$$r = f(\theta) = 1 + \cos\theta.$$

Solution The graph is traced out one time as θ varies from 0 to 2π. Therefore,

$$\text{Area} = \frac{1}{2}\int_0^{2\pi} (1 + \cos\theta)^2 \, d\theta = \frac{1}{2}\int_0^{2\pi} (1 + 2\cos\theta + \cos^2\theta) \, d\theta.$$

$$= \frac{1}{2}\int_0^{2\pi} \left(1 + 2\cos\theta + \frac{1 + \cos 2\theta}{2}\right) d\theta$$

$$= \frac{1}{2}\int_0^{2\pi} \left(\frac{3}{2} + 2\cos\theta + \frac{\cos 2\theta}{2}\right) d\theta$$

$$= \frac{1}{2}\left[\frac{3\theta}{2} + 2\sin\theta + \frac{\sin 2\theta}{4}\right]_0^{2\pi}$$

$$= \frac{1}{2}\left[\frac{3 \cdot 2\pi}{2} + 2\sin 2\pi + \frac{\sin 4\pi}{4}\right] - \frac{1}{2}\left[\frac{3 \cdot 0}{2} + 2\sin 0 + \frac{\sin 0}{4}\right]$$

$$= \frac{3\pi}{2}. \quad \square$$

REGIONS BETWEEN TWO CURVES

Let $0 \leq g(\theta) \leq f(\theta)$ for $\alpha \leq \theta \leq \beta$. (See Fig. 13–29a.) The area of the region $\mathcal{R}$ between the graphs of f and g, $\alpha \leq \theta \leq \beta$, can be found in three stages:

(1) The region $\mathcal{R}_1$ bounded by the graph of f and the rays $\theta = \alpha$ and $\theta = \beta$ has area

$$A_1 = \frac{1}{2} \int_\alpha^\beta [f(\theta)]^2 \, d\theta. \qquad \text{(See Fig. 13–29b.)}$$

(2) The region $\mathcal{R}_2$ bounded by the graph of g and the rays $\theta = \alpha$ and $\theta = \beta$ has area

$$A_2 = \frac{1}{2} \int_\alpha^\beta [g(\theta)]^2 \, d\theta. \qquad \text{(See Fig. 13–29c.)}$$

(3) The area of $\mathcal{R}$ is

$$\begin{aligned} A &= A_1 - A_2 \\ &= \frac{1}{2} \int_\alpha^\beta [f(\theta)]^2 \, d\theta - \frac{1}{2} \int_\alpha^\beta [g(\theta)]^2 \, d\theta \\ &= \frac{1}{2} \int_\alpha^\beta [f(\theta)^2 - g(\theta)^2] \, d\theta. \qquad \text{(See Fig. 13–29a.)} \end{aligned}$$

EXAMPLE 2 Calculate the area of the region inside the cardioid $r = f(\theta) = 1 + \cos \theta$ and outside the circle $r = g(\theta) = 1$.

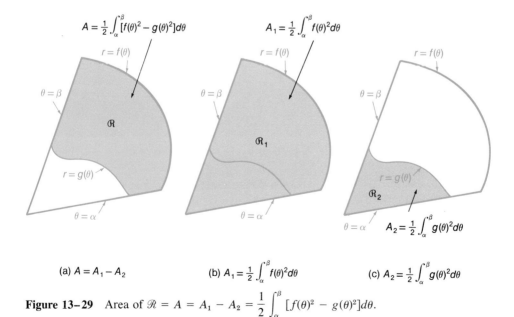

(a) $A = A_1 - A_2$ (b) $A_1 = \frac{1}{2} \int_\alpha^\beta f(\theta)^2 \, d\theta$ (c) $A_2 = \frac{1}{2} \int_\alpha^\beta g(\theta)^2 \, d\theta$

Figure 13–29 Area of $\mathcal{R} = A = A_1 - A_2 = \frac{1}{2} \int_\alpha^\beta [f(\theta)^2 - g(\theta)^2] \, d\theta$.

13-6 Area and Arc Length

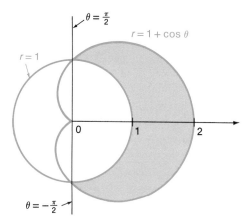

Figure 13-30 Example 2.

Solution The region is pictured in Figure 13-30. Observe that both boundary curves are traced out as θ varies from $-\pi/2$ to $\pi/2$. Therefore,

$$\text{Area} = \frac{1}{2} \int_\alpha^\beta [f(\theta)^2 - g(\theta)^2] \, d\theta$$

$$= \frac{1}{2} \int_{-\pi/2}^{\pi/2} [(1 + \cos \theta)^2 - 1^2] \, d\theta$$

$$= \frac{1}{2} \int_{-\pi/2}^{\pi/2} [2 \cos \theta + \cos^2 \theta] \, d\theta$$

$$= \frac{1}{2} \int_{-\pi/2}^{\pi/2} \left[2 \cos \theta + \frac{1 + \cos 2\theta}{2} \right] d\theta$$

$$= \frac{1}{2} \left[2 \sin \theta + \frac{\theta}{2} + \frac{\sin 2\theta}{4} \right]_{-\pi/2}^{\pi/2}$$

$$= \frac{1}{2} \left[2 \sin \frac{\pi}{2} + \frac{\pi}{4} + \frac{\sin \pi}{4} \right] - \frac{1}{2} \left[2 \sin \left(-\frac{\pi}{2}\right) - \frac{\pi}{4} + \frac{\sin (-\pi)}{4} \right]$$

$$= \frac{1}{2} \left[2 + \frac{\pi}{4} + 0 \right] - \frac{1}{2} \left[-2 - \frac{\pi}{4} + 0 \right]$$

$$= 2 + \frac{\pi}{4}. \quad \square$$

Remark It may be possible to use symmetry to simplify some of the integrals. For example, the region in Example 2 is symmetric about the *x*-axis. It follows that

$$\text{Area} = \tfrac{1}{2} \int_{-\pi/2}^{\pi/2} [(1 + \cos^2 \theta)^2 - 1] \, d\theta = 2 \cdot \tfrac{1}{2} \int_0^{\pi/2} [(1 + \cos^2 \theta)^2 - 1] \, d\theta.$$

If a region is much more complicated than the one shown in Figure 13-30, or if the graphs of *f* and *g* are traced out through different values of θ, then it may be necessary to break the region up into simpler subregions and calculate their areas separately.

EXAMPLE 3   Calculate the area of the region between the two loops of the limaçon

$$r = f(\theta) = 1 + 2\cos\theta.$$

Solution   The region is pictured in Figure 13–31a. Observe that the entire limaçon is traced out as θ varies from $-2\pi/3$ to $4\pi/3$. More precisely, the outer loop is traced out as θ varies from $-2\pi/3$ to $2\pi/3$ and the inner loop as θ varies from $2\pi/3$ to $4\pi/3$.

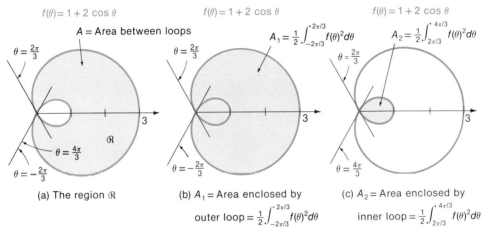

Figure 13–31   Example 3.

The total area enclosed by the outer loop (see Fig. 13–31b) is

$$\begin{aligned}
A_1 &= \frac{1}{2} \int_{-2\pi/3}^{2\pi/3} [f(\theta)]^2 \, d\theta \\
&= \frac{1}{2} \int_{-2\pi/3}^{2\pi/3} (1 + 4\cos\theta + 4\cos^2\theta) \, d\theta \\
&= \frac{1}{2} \int_{-2\pi/3}^{2\pi/3} \left(1 + 4\cos\theta + \frac{4(1 + \cos 2\theta)}{2}\right) d\theta \\
&= \frac{1}{2} \left[3\theta + 4\sin\theta + \sin 2\theta\right]_{-2\pi/3}^{2\pi/3} = 2\pi + \frac{3\sqrt{3}}{2}.
\end{aligned}$$

The total area inside the inner loop (see Fig. 13–31c) is

$$\begin{aligned}
A_2 &= \frac{1}{2} \int_{2\pi/3}^{4\pi/3} [f(\theta)]^2 \, d\theta = \frac{1}{2} \int_{2\pi/3}^{4\pi/3} (1 + 4\cos\theta + 4\cos^2\theta) \, d\theta \\
&= \frac{1}{2} \left[3\theta + 4\sin\theta + \sin 2\theta\right]_{2\pi/3}^{4\pi/3} = \pi - \frac{3\sqrt{3}}{2}.
\end{aligned}$$

The area of the region between the loops is

$$A = A_1 - A_2 = [2\pi + 3\sqrt{3}/2] - [\pi - 3\sqrt{3}/2] = \pi + 3\sqrt{3} \approx 8.3377. \quad \square$$

13-6 Area and Arc Length

ARC LENGTH

Recall from Section 6-4 that the arc length of a smooth curve defined parametrically by $x = F(t)$, $y = G(t)$, $a \le t \le b$, is

Arc Length of Curves Defined Parametrically

$$L_a^b = \int_a^b \sqrt{F'(t)^2 + G'(t)^2}\, dt.$$

Suppose now that $r = f(\theta)$ defines a smooth polar curve for $\alpha \le \theta \le \beta$. The equations

$$x = r \cos \theta = f(\theta) \cos \theta = F(\theta),$$
$$y = r \sin \theta = f(\theta) \sin \theta = G(\theta)$$

define x and y parametrically in terms of θ. Then

$$F'(\theta) = -f(\theta) \sin \theta + f'(\theta) \cos \theta,$$
$$G'(\theta) = f(\theta) \cos \theta + f'(\theta) \sin \theta,$$

so that

$$L_\alpha^\beta = \int_\alpha^\beta \sqrt{F'(\theta)^2 + G'(\theta)^2}\, d\theta$$
$$= \int_\alpha^\beta \sqrt{f(\theta)^2(\cos^2 \theta + \sin^2 \theta) + f'(\theta)^2(\cos^2 \theta + \sin^2 \theta)}\, d\theta$$

This formula reduces to

Basic Formula for Arc Length of a Polar Curve

$$L_\alpha^\beta = \int_\alpha^\beta \sqrt{f(\theta)^2 + f'(\theta)^2}\, d\theta$$

EXAMPLE 4 Find the total length of the boundary of the cardioid

$$r = f(\theta) = 1 + \cos \theta.$$

Solution The calculation of the integral is simplified by using the symmetry about the x-axis. Since the entire cardioid is traced out as θ varies from 0 to 2π, then half of it is traced out as θ varies from 0 to π. Thus,

$$\text{Arc length} = L_0^{2\pi} = 2L_0^\pi$$
$$= 2 \int_0^\pi \sqrt{f(\theta)^2 + f'(\theta)^2}\, d\theta$$
$$= 2 \int_0^\pi \sqrt{(1 + 2\cos \theta + \cos^2 \theta) + \sin^2 \theta}\, d\theta$$
$$= 2 \int_0^\pi \sqrt{2(1 + \cos \theta)}\, d\theta = 2 \int_0^\pi \sqrt{4 \cos^2 \frac{\theta}{2}}\, d\theta \quad \text{(by T17)}.$$

Since θ is between 0 and π, then $\theta/2$ is between 0 and $\pi/2$, so that $\cos(\theta/2) \geq 0$. Therefore

$$L_0^{2\pi} = 2L_0^\pi = 2\int_0^\pi 2\cos\frac{\theta}{2}\,d\theta = 8\left[\sin\frac{\theta}{2}\right]_0^\pi$$

$$= \left[8\sin\frac{\pi}{2} - 8\sin 0\right] = 8. \quad \square$$

Remark When we calculate either area or arc length, it is essential that we know how the curve is traced out as θ increases. Recall, for example, that the circle $r = 2\cos\theta$ is traced out once as θ varies from 0 to π and again as θ varies from π to 2π. If we use the incorrect area formula

$$\frac{1}{2}\int_0^{2\pi}(2\cos\theta)^2\,d\theta$$

we get 2π—twice the actual area. Observe that we calculate the complete area of this circle each time we trace it out.

A similar error occurs if we use the incorrect formula

$$\int_0^{2\pi}\sqrt{f(\theta)^2 + f'(\theta)^2}\,d\theta$$

to calculate the arc length of the circle. We get 4π—twice the actual arc length.

Exercises 13-6

1-8. Identify and sketch each curve. Calculate the area of the region each curve encloses.

- •1 $r = 2\sin\theta$
- 2 $r = 3 - \cos\theta$
- •3 $r = 2 - 2\sin\theta$
- 4 $r = 2 + \sin\theta$
- •5 $r^2 = 9\cos 2\theta$ (lemniscate)
- 6 $r^2 = 4\sin 2\theta$ (lemniscate)
- •7 $r = \cos 3\theta$
- 8 $r = \cos 2\theta$

9-14. Find the area of the region that is inside the graph of the first equation and outside the graph of the second one.

- • 9 $r = 2\cos 3\theta$, $r = 1$
- •10 $r = 1 + \sin\theta$, $r = 1$
- •11 $r = 3\sin\theta$, $r = 1 + \sin\theta$
- •12 $r^2 = 2\cos 2\theta$, $r = 1$
- •13 $r = 2\cos 2\theta$, $r = \sqrt{2}$
- •14 $r = 1 + \cos\theta$, $r = \cos\theta$

15-20. Find the lengths of the curves.

- •15 $r = e^{\theta/2}$, $1 \leq \theta \leq 2$
- 16 $r = \cos^2\left(\dfrac{\theta}{2}\right)$, $0 \leq \theta \leq 2\pi$
- •17 $r = \theta$, $0 \leq \theta \leq 2\pi$
- 18 $r = 2\sin\theta$

●19 $r = \sin^3\left(\dfrac{\theta}{3}\right),\ 0 \le \theta \le \dfrac{3\pi}{2}$ 20 $r = \sin^2\left(\dfrac{\theta}{2}\right),\ 0 \le \theta \le \pi$

●21 Find the area enclosed by the inner loop of the limaçon $r = \sqrt{3} - 2\sin\theta$.

●22 Find the total area enclosed by the outer loop of the limaçon $r = \sqrt{3} - 2\sin\theta$.

●23 Find the area of the region between the loops of the limaçon $r = \sqrt{3} - 2\sin\theta$. (See Exercises 21 and 22.)

24–25. Find the area of the region that is inside both of the graphs.

●24 $r = \sin\theta$ and $r = \sqrt{3}\cos\theta$ ●25 $r = 3\sin\theta$ and $r = 1 + \sin\theta$

Arc Length of a Cycloid

26 Show that the length of one arch of the cycloid $x = a(\phi - \sin\phi),\ y = a(1 - \cos\phi)$ is equal to four times the diameter of the rolling circle.

Area of a Cycloid

27 Let $\mathscr{R}$ be the region bounded by the x-axis and one arch of the cycloid in Exercise 26. Show that the area of $\mathscr{R}$ is three times the area of the rolling circle. (*Hint:* If we consider y as a function of x, then

$$A = \int_{x=0}^{x=2\pi a} y\, dx,$$

where a is the radius of the circle. Change the integral to one involving the parameter ϕ.)

Witch of Agnesi

28 (*The Witch of Agnesi*) Let $\mathscr{C}$ be the circle of radius a with center at $(0, a)$ on the y-axis. Let $\mathscr{L}$ be the line $y = 2a$. For each angle $\phi,\ 0 \le \phi \le \pi$, let $Q_\phi(a_\phi, b_\phi)$ be the point where the line through the origin with inclination ϕ intersects the line $\mathscr{L}$, and let $R_\phi(c_\phi, d_\phi)$ be the point different from the origin where it intersects the circle $\mathscr{C}$. Let P_ϕ be the point $P_\phi(a_\phi, d_\phi)$. [The x-coordinate of P_ϕ is the x-coordinate of Q_ϕ and the y-coordinate is the y-coordinate of R_ϕ.]

The curve traced out by P_ϕ as ϕ varies from 0 to π is called the *witch of Agnesi*.

(a) Sketch the witch, showing the circle $\mathscr{C}$ and the line $\mathscr{L}$.
(b) Show that parametric equations for the witch are

$$x = 2a\cot\phi,\qquad y = 2a\sin^2\phi.$$

(c) Show that the rectangular equation of the witch is

$$y = \dfrac{8a^3}{x^2 + 4a^2}.$$

●(d) Calculate the total area of the region bounded by the witch and the x-axis.

Review Problems

1–4. Eliminate the parameter to get an equation in rectangular coordinates satisfied by x and y. Sketch the graphs of the rectangular equation and the parametric equations.

●1 $x = \cos^2 t,\ y = \sin^2 t,\ 0 \le t \le 2\pi$

2 $x = \sqrt{t},\ y = t - 4,\ t \ge 1$

●3 $x = \sec t$, $y = 2 \tan t$, $-\pi/2 < t < \pi/2$

4 $x = e^t$, $y = e^{-2t} + 1$

5 Derive the equation of the cycloid shown in Figure 13–5.

6 Define the following concepts. Illustrate each with a sketch.

    (a) Curve   (b) Smooth curve

    (c) Simple curve   (d) Closed curve

7–8. Calculate dy/dx and d^2y/dx^2 in terms of the parameter.

●7 $x = t^2 + t$, $y = 7t^2 + t^3$ ●8 $r = 2 \cos \theta$

9–10. Find the equations of the tangent lines to the graphs in Problems 7 and 8 at the indicated points.

●9 Problem 7: $t = 2$ ●10 Problem 8: $\theta = \dfrac{\pi}{6}$

11–12. Find all points at which the tangent line is either horizontal or vertical.

●11 $x = t^2 + 1$, $y = t^2 - t$ ●12 $r = 2 + 2 \cos \theta$

●13 Let $x = f(t)$, $y = g(t)$ be parametric equations for a curve, where f and g are twice differentiable and $f'(t) \neq 0$. Express d^2y/dx^2 as a function of t.

●14 Find the equation of the tangent line to the graph of $x = 8t - 2$, $y = t^4 + 4$ that is parallel to the line $y - 4x + 2$.

15–18. (a) Convert the polar equation to a rectangular equation with the same graph.

        (b) Convert the polar equation to parametric equations $x = f(\theta)$, $y = g(\theta)$ with the same graph.

●15 $r = \sin \theta + \cos \theta$ ●16 $r = \sec \theta \tan \theta$

●17 $r^2 = 2 \csc 2\theta$ ●18 $r(1 - 3 \cos \theta) = 2$

19–22. Convert the rectangular equation to a polar equation with the same graph.

●19 $y = \sqrt{3}x$ ●20 $(x - 2)^2 + y^2 = 4$

●21 $(x^2 + y^2)^3 = 4x^2y^2$ ●22 $y^2 = x$

Review Problems

23–28. Test for symmetry. Find the tangent lines at the origin. Identify and sketch each graph.

- 23 $r = \sin 2\theta$
- 24 $r^2 = 4 \cos 2\theta$
- 25 $r = -4 \cos \theta$
- 26 $r = 4 \sin 3\theta$
- 27 $r = 1 - 2 \sin \theta$
- 28 $r = 1 + \cos \theta$

29–30. Identify and sketch each conic section. Locate the directrix that corresponds to the focus located at the origin.

- 29 $r = \dfrac{2}{1 - 3 \cos \theta}$
- 30 $r = \dfrac{6}{1 + \sin \theta}$

31–34. Calculate the area of the region.

- 31   Inside the graph of $r = 2 \sin \theta$ and outside the graph of $r = -1$.
- 32   Inside the graph of $r = 5 \cos \theta$ and outside the graph of $r = 2 + \cos \theta$.
- 33   Inside the graph of $r^2 = 4 \cos 2\theta$.
- 34   Between the loops of the limaçon $r = 1 - \sqrt{2} \sin \theta$.

35–38. Calculate the length of the curve.

- 35 $r = \cos \theta,\ 0 \leq \theta \leq \dfrac{\pi}{3}$
- 36 $r = 1 - \cos \theta$
- 37 $r = e^\theta,\ 0 \leq \theta \leq 2\pi$
- 38 $r = 2e^{-\theta},\ 0 \leq \theta \leq 2\pi$

39   Devise two tests for symmetry about the origin for the graph of $F(r, \theta) = 0$. Illustrate the tests with a sketch.

40   Derive the formula

$$\text{Area} = \tfrac{1}{2} \int_\alpha^\beta [f(\theta)]^2\, d\theta$$

for the area of the region bounded by the graph of the continuous polar function $r = f(\theta)$ and the rays $\theta = \alpha,\ \theta = \beta$.

41   Use the formula for the arc length of the graph of parametric equations $x = F(t)$, $y = G(t)$ to derive the corresponding formula for the arc length of the polar curve $r = f(\theta)$.

14 Vectors and Solid Analytic Geometry

14–1 Vectors in the xy-Plane

Many mathematical and physical concepts involve both a magnitude and a direction. To cause an object to move, for example, we must apply a force. This force has a magnitude and must be applied in a certain direction.

Mathematicians and physicists have traditionally used *vectors* to represent forces. Intuitively we can think of a vector as an arrow. The length of the arrow represents the magnitude of the force and the direction of the arrow the direction in which force is applied.

Every arrow drawn in the plane has an *initial point* (at the "tail" of the arrow) and a *terminal point* (at the "head" of the arrow). We can think of the arrow as describing the change in the x- and y-coordinates necessary to move from the initial point to the terminal point. (See Fig. 14–1.) This leads us to the following definition.

DEFINITION

Vector

A *two-dimensional vector* v is an ordered pair of real numbers a, b, written

$$v = \langle a, b \rangle,$$

which represents the change as we move from a point $P_0(x_0, y_0)$ to a point $P(x_0 + a, y_0 + b)$ in the xy-plane.

Components

The number a is called the *x-component* (or *horizontal component*) of the vector $\langle a, b \rangle$. The number b is the *y-component* (or *vertical component*).

Zero Vector

The *zero vector* is

$$\mathbf{0} = \langle 0, 0 \rangle.$$

If $v = \langle a, b \rangle$ is a nonzero vector, then v can be represented by an arrow from a point

14-1 Vectors in the xy-Plane

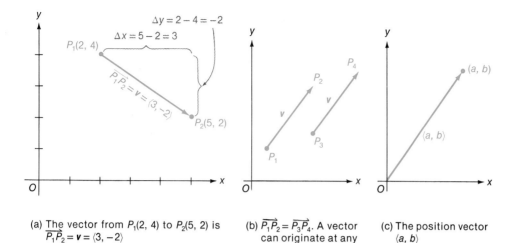

(a) The vector from $P_1(2, 4)$ to $P_2(5, 2)$ is $\overrightarrow{P_1P_2} = \mathbf{v} = \langle 3, -2 \rangle$

(b) $\overrightarrow{P_1P_2} = \overrightarrow{P_3P_4}$. A vector can originate at any point

(c) The position vector $\langle a, b \rangle$

Figure 14-1 Vectors in the xy-plane.

$P_1(x_1, y_1)$ to a point $P_2(x_2, y_2)$, where

$$x_2 = x_1 + a \quad \text{and} \quad y_2 = y_1 + a.$$

Geometric Vector This arrow is sometimes called a *geometric vector*. The geometric vector corresponding to the zero vector $\langle 0, 0 \rangle$ is represented by a single point.

In most cases the initial and terminal points of a vector are not specified. Thus, the vector

$$\mathbf{v} = \langle a, b \rangle$$

can represent the change when we move from any point $P_1(x_1, y_1)$ in the xy-plane to a corresponding point $P_2(x_1 + a, y_1 + b)$. (See Fig. 14–1b.).

Position Vector A geometric vector that originates at the origin (that is, the origin is its initial point) is called a *position vector*. If the vector is $\langle a, b \rangle$, then its initial point is $(0, 0)$ and its terminal point is (a, b). (See Fig. 14–1c.)

EXAMPLE 1 (a) The vector from $(0, 0)$ to $(-1, 2)$ is the position vector $\langle -1, 2 \rangle$.
(b) The vector from $P_1(5, 7)$ to $P_2(4, -1)$ is $\langle 4 - 5, -1 - 7 \rangle = \langle -1, -8 \rangle$.
(c) The vector from $(2, 4)$ to $(2, 4)$ is the zero vector $\langle 0, 0 \rangle$. □

NOTATIONS

Several notations are used for vectors. In textbooks it is customary to indicate them by boldface type. Thus, we write

$$\mathbf{v} = \langle a, b \rangle$$

for the vector $\langle a, b \rangle$.

In handwritten work we usually indicate vectors by arrows over letters. Thus, we write

$$\vec{v} = \langle a, b \rangle$$

for the vector v.

The vector from $P_1(x_1, y_1)$ to $P_2(x_2, y_2)$ is indicated by the symbol $\overrightarrow{P_1P_2}$. Thus,

$$\overrightarrow{P_1P_2} = \langle x_2 - x_1, y_2 - y_1 \rangle.$$

(See Fig. 14–1a.)

The zero vector $\langle 0, 0 \rangle$ is indicated by either of the symbols $\vec{0}$ or $\boldsymbol{0}$.

EQUALITY

Two vectors u and v are considered *equal* only in case they measure the same horizontal and vertical changes. Thus, they are equal if and only if their corresponding components are equal.

EXAMPLE 2   The vector from $P_1(2, 5)$ to $P_2(x_2, y_2)$ is $u = \langle 4, -1 \rangle$. What is the point P_2?

Solution   The vector from P_1 to P_2 is

$$\overrightarrow{P_1P_2} = \langle x_2 - 2, y_2 - 5 \rangle.$$

We are given that

$$\overrightarrow{P_1P_2} = \langle x_2 - 2, y_2 - 5 \rangle = u = \langle 4, -1 \rangle.$$

Since the corresponding components must be equal, then

$$x_2 - 2 = 4 \quad \text{and} \quad y_2 - 5 = -1,$$
$$x_2 = 6 \quad \text{and} \quad y_2 = 4.$$

The point P_2 is $P_2(6, 4)$. □

ADDITION

The *sum* of the vectors $u = \langle a, b \rangle$ and $v = \langle c, d \rangle$ is defined to be the vector obtained by adding the corresponding components:

$$u + v = \langle a + c, b + d \rangle.$$

For example, if $u = \langle 5, 7 \rangle$ and $v = \langle 3, -2 \rangle$, then

$$u + v = \langle 5 + 3, 7 - 2 \rangle = \langle 8, 5 \rangle.$$

There is a simple geometrical interpretation of addition. First, draw the vector $u = \langle a, b \rangle$ with its initial point at the origin and terminal point (tip of the arrow) at (a, b). Next, draw $v = \langle c, d \rangle$ with its initial point at (a, b). As seen in Figure 14–2, the terminal point of v is $(a + c, b + d)$. Thus, the vector $u + v$ originates at the origin and ends at the terminal point of v. Observe that the vectors u, v, and $u + v$ form the sides of a triangle.

The vector $u + v = \langle a + c, b + d \rangle$ is often called the *resultant vector* obtained by adding u and v.

14-1 Vectors in the xy-Plane

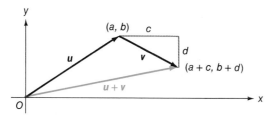

Figure 14-2 Vector addition. $\langle a, b \rangle + \langle c, d \rangle = \langle a + c, b + d \rangle$.

SCALAR MULTIPLICATION

When one is working with vectors, the real numbers are called *scalars*.* We now define the product of a scalar k and a vector v.

DEFINITION Let k be a scalar, $u = \langle a, b \rangle$ a vector. The *product* of k and u is defined to be the vector

$$ku = \langle ka, kb \rangle.$$

EXAMPLE 3 Let $u = \langle 3, -2 \rangle$ and $v = \langle -4, 3 \rangle$. Then
(a) $5u = 5\langle 3, -2 \rangle = \langle 15, -10 \rangle$;
(b) $-3v = -3\langle -4, 3 \rangle = \langle 12, -9 \rangle$;
(c) $-2u + 5v = \langle -6, 4 \rangle + \langle -20, 15 \rangle = \langle -26, 19 \rangle$. □

The product of a scalar and a vector has a convenient interpretation in terms of the arrows that represent vectors. We speak of the *length*, or the *magnitude*, of a vector, meaning the length of the arrow that represents it. That is, the magnitude of a vector equals the distance between its initial and terminal points.

If $k > 0$, then ku is a vector that points in the same direction as u and has length equal to k times the length of u. (See Fig. 14-3a.)

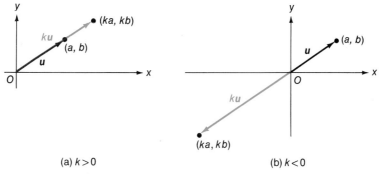

(a) $k > 0$ (b) $k < 0$

Figure 14-3 Scalar multiplication. $k\langle a, b \rangle = \langle ka, kb \rangle$.

---

* In a more general setting, vectors are defined in terms of components from some number field. The numbers from that field are called *scalars*.

If $k < 0$, than ku points in the direction opposite u and has length equal to $|k|$ times the length of u. (See Fig. 14–3b.) For example, the vector $3u$ points in the same direction as u and is three times as long as u. The vector $-2u$ points in the opposite direction and is twice as long as u.

For convenience, we write u/k or $\dfrac{u}{k}$ to indicate the vector

$$\frac{1}{k} u.$$

Thus, if $u = \langle -2, 1 \rangle$, then $u/3 = \langle -2/3, 1/3 \rangle$ and

$$\frac{u}{-5} = \left\langle \frac{2}{5}, -\frac{1}{5} \right\rangle.$$

MAGNITUDES OF VECTORS

As mentioned above, the *magnitude* (or *length*) of a vector equals the length of the arrow that represents it. We use the symbol $\|v\|$ to represent the magnitude of the vector $v = \langle a, b \rangle$. It follows from the Pythagorean Theorem that

Magnitude of a Vector

$$\|v\| = \|\langle a, b \rangle\| = \sqrt{a^2 + b^2}.$$

(See Fig. 14–4.)

For example, if $v = \langle 3, -4 \rangle$, then

$$\|v\| = \|\langle 3, -4 \rangle\| = \sqrt{3^2 + 4^2} = 5.$$

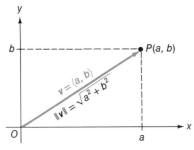

Figure 14–4 Magnitude of a vector. $\|\langle a, b \rangle\| = \sqrt{a^2 + b^2}$.

Unit Vector

A vector of length 1 is called a *unit vector*. It is simple to find a unit vector in the direction of a nonzero vector v. We simply divide v by its magnitude. If $v = \langle 3, -4 \rangle$, for example, then, as we saw above, $\|v\| = 5$. A unit vector in the same direction as v is

$$u = \frac{v}{\|v\|} = \frac{\langle 3, -4 \rangle}{5} = \left\langle \frac{3}{5}, -\frac{4}{5} \right\rangle.$$

14-1 Vectors in the xy-Plane

PROPERTIES OF VECTORS

Several properties of vectors follow from the definitions. Most of these are analogous to important properties of the real number system and have similar names. All these properties can be proved by using the components of the vectors.

THEOREM 14-1

Let u, v, and w be vectors and m and n scalars. Then

Algebraic Properties of Vectors
- (a) $u + v = v + u$ (Commutative law)
- (b) $(u + v) + w = u + (v + w)$ (Associative law)
- (c) $m(u + v) = mu + mv$ (A type of distributive law)
- (d) $(m + n)u = mu + nu$ (A type of distributive law)
- (e) $(mn)u = m(nu)$ (A type of associative law)
- (f) $0 + u = u + 0 = u$ for every vector u
- (g) $0 \cdot u = 0$
- (h) $u + (-u) = 0$

Properties (f), (g), and (h) show that the zero vector $\mathbf{0}$ has properties similar to the number zero. Observe that the "zero" in properties (f) and (h) is the zero vector, while the "zero" on the left-hand side of (g) is the scalar zero.

Because of Property (h), we define subtraction of vectors by

$$u - v = u + (-v).$$

It follows from (h) that

$$u - u = \mathbf{0},$$

a rule that is similar to a rule for subtraction of numbers.

EXAMPLE 4 Prove part (a) of Theorem 14-1: Vector addition is commutative.

Solution Let $u = \langle a, b \rangle$ and $v = \langle c, d \rangle$. Then

$$u + v = \langle a + c, b + d \rangle \quad \text{and} \quad v + u = \langle c + a, d + b \rangle.$$

Since a, b, c, and d are real numbers, then

$$a + c = c + a \quad \text{and} \quad b + d = d + b.$$

Therefore,

$$u + v = \langle a + c, b + d \rangle = \langle c + a, d + b \rangle = v + u. \quad \square$$

THE UNIT VECTORS i AND j

The $\langle a, b \rangle$ notation for vectors is convenient for establishing the properties of vectors. It has the added advantage that it reminds us of the close relation between a vector $\langle a, b \rangle$ and the associated point $P(a, b)$ in the xy-plane. This notation becomes awkward, however, when we apply the principles of the calculus to vector-valued functions. We shall find that the following alternative notation is much more convenient in such settings.

i, j Unit Vectors

We define

$$i = \langle 1, 0 \rangle \quad \text{and} \quad j = \langle 0, 1 \rangle.$$

These two unit vectors are in the directions of the positive coordinate axes. (See Fig. 14–5a.)

Observe that the vector $\langle a, b \rangle$ can now be written as

$$\langle a, b \rangle = \langle a, 0 \rangle + \langle 0, b \rangle$$
$$= a \langle 1, 0 \rangle + b \langle 0, 1 \rangle = ai + bj.$$

(See Fig. 14–5b.) For example,

$$\langle 3, -7 \rangle = 3i - 7j,$$
$$\langle 4, 2 \rangle = 4i + 2j,$$

and so on.

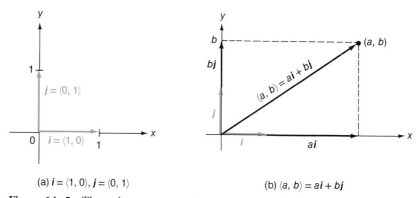

(a) $i = \langle 1, 0 \rangle$, $j = \langle 0, 1 \rangle$ (b) $\langle a, b \rangle = ai + bj$

Figure 14–5 The unit vectors i and j.

The results of Theorem 14–1 ensure that we can operate with vectors (in the i, j notation) as we operate with algebraic expressions. We simply combine terms algebraically, treating i and j much as we treat variables such as x and y.

EXAMPLE 5   (a) $5(2i - 3j) = 10i - 15j.$
             (b) $(2i - 3j) - (3i + 7j) = 2i - 3j - 3i - 7j = -i - 10j.$
             (c) $4(2i - 3j) + 2(3i + 7j) = 8i - 12j + 6i + 14j = 14i + 2j.$ □

DIRECTION ANGLES

The smallest positive angle θ from the positive x-axis to the nonzero position vector v is called the *direction angle of v*. (See Fig. 14–6a.)

Suppose that $u = ai + bj$ is a unit vector (length 1) with direction angle θ. It follows from elementary trigonometry that

$$\cos \theta = a \quad \text{and} \quad \sin \theta = b,$$

14-1 Vectors in the xy-Plane

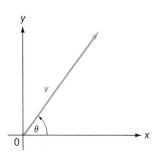

Figure 14–6. (a) θ = the direction angle of **v**   (b) $u = \cos\theta i + \sin\theta j$ is a unit vector in the same direction as **v**

so that

$$u = \cos\theta i + \sin\theta j.$$

(See Fig. 14–6b.) This observation can help us to find the direction angle of any nonzero vector *v*. We first divide *v* by its magnitude, obtaining a unit vector *u* in the same direction. The *x*-component of *u* is $\cos\theta$ and the *y*-component is $\sin\theta$.

A Vector in Terms of its Magnitude and Direction Angle

A similar argument can be used to express any vector in terms of its magnitude and direction angle. If θ is the direction angle of *v*, it can be shown that

$$v = \|v\|(\cos\theta i + \sin\theta j)$$

EXAMPLE 6   Find the direction angle of $v = -3i + 3j$.

Solution $\|v\| = \sqrt{3^2 + 3^2} = 3\sqrt{2}$. Let

$$u = \frac{v}{\|v\|} = \frac{-3i + 3j}{3\sqrt{2}} = -\frac{1}{\sqrt{2}}i + \frac{1}{\sqrt{2}}j.$$

Let θ be the common direction angle of *u* and *v*. Since *u* is a unit vector, then

$$u = \cos\theta i + \sin\theta j = -\frac{1}{\sqrt{2}}i + \frac{1}{\sqrt{2}}j.$$

Therefore,

$$\cos\theta = -\frac{1}{\sqrt{2}} \quad \text{and} \quad \sin\theta = \frac{1}{\sqrt{2}}.$$

The direction angle is

$$\theta = \frac{3\pi}{4}. \quad \square$$

Exercises 14-1

1-6. Write the vector $\overrightarrow{PQ}$ in the $\langle a, b \rangle$ notation and the $ai + bj$ notation. Sketch the vector $\overrightarrow{PQ}$ in the xy-plane.

- **1** $P(2, -3)$, $Q(5, 4)$
- **2** $P(-1, 0)$, $Q(3, -2)$
- **3** $P(6, 4)$, $Q(-3, 1)$
- **4** $P(-7, 2)$, $Q(4, 2)$
- **5** $P(1, 2)$, $Q(-2, 1)$
- **6** $P(5, 2)$, $Q(5, 2)$

7-10. Find the point Q.

- **7** $P(2, 3)$, $\overrightarrow{PQ} = \langle -4, 1 \rangle$
- **8** $P(-10, 2)$, $\overrightarrow{PQ} = \langle 5, -4 \rangle$
- **9** $P(-4, 6)$, $\overrightarrow{PQ} = 4i + 7j$
- **10** $P(4, 3)$, $\overrightarrow{QP} = 2i + 3j$

11-14. Calculate $2v$, $v + w$, $v - w$, $3v + 2w$, and $v - 4w$.

- **11** $v = \langle 2, -3 \rangle$, $w = \langle -2, 4 \rangle$
- **12** $v = \langle -4, 5 \rangle$, $w = \langle -5, 2 \rangle$
- **13** $v = -2i + 4j$, $w = -3i - j$
- **14** $v = i - 4j$, $w = 3i - 2j$

15-20. Find the magnitude and the direction angle of v.

- **15** $v = \langle 1, -1 \rangle$
- **16** $v = \langle -1, \sqrt{3} \rangle$
- **17** $v = 4i$
- **18** $v = 3i + 3j$
- **19** $v = 2\sqrt{3}i - 2j$
- **20** $v = -3\sqrt{3}i - 3j$

21-24. (a) Find a unit vector having the same direction as v.
(b) Find a unit vector that has the opposite direction.

- **21** $v = \langle 0, -5 \rangle$
- **22** $v = \langle 5, -12 \rangle$
- **23** $v = -2i + 3j$
- **24** $v = 4i + 9j$

25(b)-(h) Prove each of the parts (b)-(h) of Theorem 14-1.

14-2 Three-Dimensional Space

Coordinate Axes

It is simple to extend the two-dimensional rectangular coordinate system—the xy-system—to a three-dimensional system that represents points in space. We construct a third axis, the z-axis, mutually perpendicular to the x- and y-axes and with the same origin. It is customary to orient the axes so that the y-axis is horizontal and the z-axis vertical, with the x-axis pointing straight out of the picture toward the reader. (See Fig. 14-7a for a perspective drawing.)

An axis system such as the one in Figure 14-7 is called a *right-handed system*. If we label the thumb, index finger, and middle finger of the right hand with the letters x, y, and z and hold these three fingers at right angles to each other, they will point in the approximate directions of the axes.

Coordinate Planes

The coordinate axes define three *coordinate planes*—the xy-plane, the yz-plane, and the xz-plane. (See Fig. 14-7b.) It might be instructive to think of the coordinate

14-2 Three-Dimensional Space

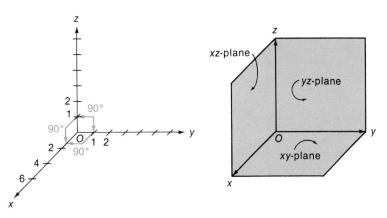

Figure 14-7.   (a) The xyz-coordinate system   (b) The coordinate planes

axes as meeting at a corner of a room. Then the xy-plane is the plane of the floor, and the xz- and yz- planes are the two adjoining walls.

Coordinates of a Point

The point with coordinates (a, b, c) is located $|c|$ units above or below the point $(a, b, 0)$ in the xy-plane. If $c > 0$, it is above $(a, b, 0)$, if $c < 0$, it is below. (See Fig. 14–8.) Observe that a point in the xy-plane can be written either (a, b) or $(a, b, 0)$.

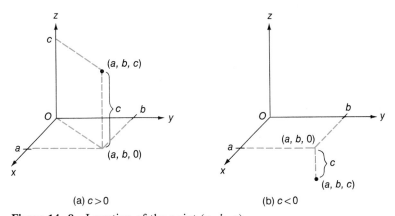

(a) $c > 0$ (b) $c < 0$

Figure 14-8 Location of the point (a, b, c).

EXAMPLE 1   Several points, including $P(-2, 1, 3)$, $Q(1, 2, 3)$, and $R(-1, -2, -3)$ are plotted in Figure 14–9. The guidelines shown in that figure are helpful in locating the points in a perspective drawing. ☐

The coordinate planes separate three-dimensional space into *octants*. The octant with x, y, and z all positive is called Octant I. There is no standard numbering systems for the other seven octants. A small part of Octant I is pictured in Figure 14–10.

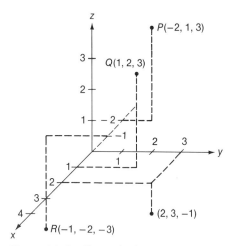

Figure 14-9 Example 1.

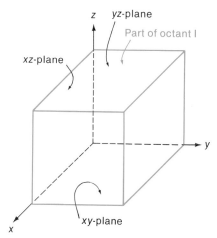

Figure 14-10 Octant I consists of all points $P(x, y, z)$ with $x > 0$, $y > 0$, $z > 0$.

THE DISTANCE FORMULA

Theorem 14-2 extends the distance formula to three-dimensional space.

THEOREM 14-2 Let $P_1(x_1, y_1, z_1)$ and $P_2(x_2, y_2, z_2)$ be points in space. The distance between P_1 and P_2 is

$$|P_1P_2| = \sqrt{(x_2 - x_1)^2 + (y_2 - y_1)^2 + (z_2 - z_1)^2}.$$

Proof We draw a rectangular box in three-dimensional space with P_1 and P_2 at opposite corners, as in Figure 14-11a. Two of the other corners are $Q(x_1, y_2, z_1)$ and $R(x_2, y_2, z_1)$.

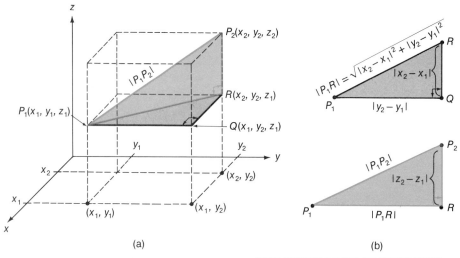

Figure 14-11 The distance formula. $|P_1P_2| = \sqrt{(x_2 - x_1)^2 + (y_2 - y_1)^2 + (z_2 - z_1)^2}$.

14-2 Three-Dimensional Space

Since P_1 and Q are on a line parallel to the y-axis, then

$$|P_1Q| = |y_2 - y_1|.$$

Similarly,

$$|QR| = |x_2 - x_1| \quad \text{and} \quad |RP_2| = |z_2 - z_1|.$$

Since triangle P_1QR is a right triangle, it follows from the Pythagorean Theorem that

$$|P_1R|^2 = |QR|^2 + |P_1Q|^2.$$

(See Fig. 14–11b.) Similarly, since P_1RP_2 is a right triangle, then

$$|P_1P_2|^2 = |P_1R|^2 + |RP_2|^2.$$

It follows that

$$\begin{aligned}|P_1P_2|^2 &= |P_1R|^2 + |RP_2|^2 \\ &= |QR|^2 + |P_1Q|^2 + |RP_2|^2 \\ &= |x_2 - x_1|^2 + |y_2 - y_1|^2 + |z_2 - z_1|^2 \\ &= (x_2 - x_1)^2 + (y_2 - y_1)^2 + (z_2 - z_1)^2.\end{aligned}$$

On taking square roots, we get the desired formula. ∎

EXAMPLE 2 The distance between $P_1(2, -1, 3)$ and $P_2(3, 4, -5)$ is

$$\begin{aligned}|P_1P_2| &= \sqrt{(3-2)^2 + (4+1)^2 + (-5-3)^2} \\ &= \sqrt{1 + 25 + 64} = \sqrt{90}.\end{aligned}$$ □

GRAPHS OF EQUATIONS

The *graph* of an equation in x, y, and z is the set of all points $P(x, y, z)$ with coordinates that satisfy the equation. For example, the point $P_0(5, -2, -5)$ is on the graph of the equation

$$(x - 2)^2 + (y + 1)^2 + (z + 3)^2 = 14.$$

If we substitute 5 for x, -2 for y and -5 for z, the equation becomes an identity.

THE SPHERE $(x - a)^2 + (y - b)^2 + (z - c)^2 = r^2$

The equation of a sphere can be derived by the distance formula.

THEOREM 14-3 The sphere with center $C(a, b, c)$ and radius $r > 0$ is the graph of the equation

Equation of a Sphere

$$(x - a)^2 + (y - b)^2 + (z - c)^2 = r^2.$$

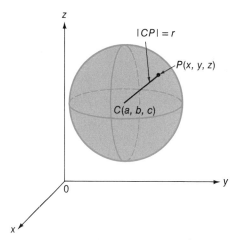

Figure 14-12 The sphere $(x - a)^2 + (y - b)^2 + (z - c)^2 = r^2$.

Proof Let $P(x, y, z)$ be on the sphere. (See Fig. 14-12.) Then P is exactly r units from C, so that $|CP| = r$. By the distance formula

$$\sqrt{(x - a)^2 + (y - b)^2 + (z - c)^2} = r$$
$$(x - a)^2 + (y - b)^2 + (z - c)^2 = r^2.$$

Since all the above steps can be reversed, then $P(x, y, z)$ is on the sphere if and only if the coordinates of P satisfy the equation. ∎

EXAMPLE 3 The sphere with radius $r = 2$ and center $C(2, 1, 1)$ has equation

$$(x - 2)^2 + (y - 1)^2 + (z - 1)^2 = 2^2,$$

which can be written

$$x^2 - 4x + y^2 - 2y + z^2 - 2z + 2 = 0. \ \square$$

EXAMPLE 4 Describe the graphs of the following equations:
(a) $x^2 - 2x + y^2 + 6y + z^2 - 8z = -10$.
(b) $x^2 + 4x + y^2 - 2y + z^2 - 4z = -13$.

Solution (a) We complete the squares on the terms involving x, y, and z:

$$(x^2 - 2x \quad) + (y^2 + 6y \quad) + (z^2 - 8z \quad) = -10$$
$$(x^2 - 2x + 1) + (y^2 + 6y + 9) + (z^2 - 8z + 16) = -10 + 1 + 9 + 16 = 16$$
$$(x - 1)^2 + (y + 3)^2 + (z - 4)^2 \qquad\qquad = 4^2.$$

The graph is the sphere with center $(1, -3, 4)$ and radius $r = 4$.

(b) We complete the square:

$$(x^2 + 4x \quad) + (y^2 - 2y \quad) + (z^2 - 4z \quad) = -13$$
$$(x^2 + 4x + 4) + (y^2 - 2y + 1) + (z^2 - 4z + 4) = -13 + 4 + 1 + 4 = -4$$
$$(x + 2)^2 + (y - 1)^2 + (z - 2)^2 \qquad\qquad = -4.$$

14-2 Three-Dimensional Space

Since the left-hand side is nonnegative and the right-hand side is negative for any choice of x, y, and z, there is no graph. □

THE MIDPOINT FORMULA

Our final result is a formula for the midpoint between two points P_1 and P_2 in three-dimensional space.

THEOREM 14-4 The midpoint of the line segment connecting the points $P_1(x_1, y_1, z_1)$ and $P_2(x_2, y_2, z_2)$ is

$$M\left(\frac{x_1 + x_2}{2}, \frac{y_1 + y_2}{2}, \frac{z_1 + z_2}{2}\right).$$

The proof, which can be based on Figure 14-13, is left for you. (See Exercise 24.) A different type of proof is given in Example 1 and Exercise 1 of Section 14-5.

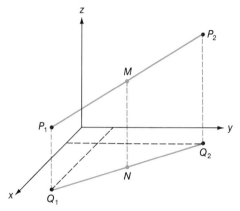

Figure 14-13 $M\left(\dfrac{x_1 + x_2}{2}, \dfrac{y_1 + y_2}{2}, \dfrac{z_1 + z_2}{2}\right)$ is the midpoint between $P_1(x_1, y_1, z_1)$ and $P_2(x_2, y_2, z_2)$.

EXAMPLE 5 Find the equation of the sphere that has the points $P_1(2, -1, 3)$ and $P_2(4, 5, 7)$ at the opposite ends of a diameter.

Solution Since the center of the sphere C must be the midpoint of the diameter connecting P_1 and P_2, then C is the point

$$C\left(\frac{2 + 4}{2}, \frac{-1 + 5}{2}, \frac{3 + 7}{2}\right) = C(3, 2, 5).$$

(See Fig. 14-14.)

The radius is the distance from C to P_1:

$$r = |CP_1| = \sqrt{(2 - 3)^2 + (-1 - 2)^2 + (3 - 5)^2} = \sqrt{14}.$$

The equation of the sphere is

$$(x - 3)^2 + (y - 2)^2 + (z - 5)^2 = 14. \quad □$$

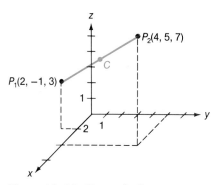

Figure 14–14 Example 5.

Exercises 14–2

1–4. Plot the points P and Q. Calculate the distance between P and Q and find the midpoint of the line segment that connects P and Q.

- **1** $P(4, -1, -2), Q(-4, -2, 2)$
- **2** $P(7, -1, 2)\ Q(9, 3, -2)$
- **3** $P(-3, 8, 1), Q(-6, 2, 3)$
- **4** $P(5, 4, -3), Q(-4, 2, 3)$

5–6. Plot the points P, Q, and R. Prove that P, Q, R are the vertices of a right triangle. Find its area.

- **5** $P(4, -3, 2), Q(6, -2, 1), R(7, -6, 5)$
- **6** $P(2, 1, 0), Q(3, 2, 1), R(1, 0, 2)$

7–8. Points A and B are two corners of a rectangular box with sides parallel to the coordinate planes. Find the coordinates of the other six corners.

- **7** $A(-1, 2, 3), B(2, 5, -2)$
- **8** $A(4, -3, 7), B(-1, 2, 1)$

9–10. Find an equation of the sphere with center C and radius r.

- **9** $C(0, 2, -3), r = 5$
- **10** $C(5, 7, -3), r = 4$

11–16. Identify the graph of the equation. If it is a sphere, find its center and radius.

- **11** $x^2 + y^2 + z^2 - 2x + 6y + 4z - 2 = 0$
- **12** $x^2 + y^2 + z^2 + 6y - 8z = 0$
- **13** $3x^2 + 3y^2 + 3z^2 - 15x + 9y - 18z - \frac{9}{2} = 0$
- **14** $x^2 + y^2 + z^2 - 2x + 4y - 2z + 6 = 0$
- **15** $x^2 + y^2 + z^2 - 2x + 4z - 4 = 0$
- **16** $x^2 + y^2 + z^2 + 4x + 6y - 8z + 33 = 0$

17-20. Find the equation of the sphere that satisfies the given conditions.

•17 Center at $(1, -4, 5)$, tangent to the xy-plane.

18 Center at $(1, -4, 5)$, tangent to the yz-plane.

•19 $P(2, 1, 3)$ is the center, $Q(5, 2, -1)$ is on the sphere.

20 $P(1, 7, -2)$ and $Q(-1, 3, 2)$ are the endpoints of a diameter.

21-22. Find an equation satisfied by the set of all points equidistant from P and Q.

•21 $P(2, -5, 1)$, $Q(0, -1, 3)$ 22 $P(3, 7, 3)$, $Q(3, 1, -5)$

•23 Find the equation of a sphere that is tangent to the three coordinate planes and passes through the point $P(4, 5, 5)$. (Two answers are possible.) (*Note:* $P(4, 5, 5)$ is a point on the surface of the sphere. It is not the center.)

24 Prove Theorem 14-4. (*Hint:* See Fig. 14-13.)

14-3 Vectors in Three-Dimensional Space. Elementary Applications

Three-Dimensional Vectors

Our previous work with vectors can be extended to vectors in three-dimensional space. A *three-dimensional vector* v is an ordered triple of real numbers a, b, c, written

$$v = \langle a, b, c \rangle,$$

which represents the change as we move from a point $P_0(x_0, y_0, z_0)$ to a point $P(x_0 + a, y_0 + b, z_0 + c)$. (See Fig. 14-15a.) As with two-dimensional vectors, the numbers a, b, c are called the *components* of v.

Equality

Two vectors are *equal* if and only if their corresponding components are equal. That is,

$$\langle a_1, a_2, a_3 \rangle = \langle b_1, b_2, b_3 \rangle$$

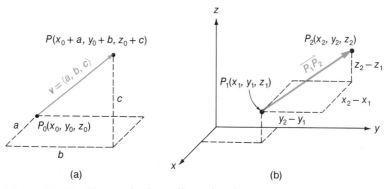

Figure 14-15 Vectors in three-dimensional space.

if and only if
$$a_1 = b_1, a_2 = b_2, a_3 = b_3.$$

The zero vector is $\mathbf{0} = \langle 0, 0, 0 \rangle$.

Geometric Vector

A nonzero vector can be represented by an arrow, called a *geometric vector*, drawn in three-dimensional space. If $P_0(x_0, y_0, z_0)$ is any fixed point, then an arrow representing $v = \langle a, b, c \rangle$ can be drawn from P_0 to the point $P(x_0 + a, y_0 + b, z_0 + c)$. This geometric vector is represented by the symbol $\overrightarrow{P_0 P}$.

The geometric vector corresponding to the change from $P_1(x_1, y_1, z_1)$ to $P_2(x_2, y_2, z_2)$ is
$$\overrightarrow{P_1 P_2} = \langle x_2 - x_1, y_2 - y_1, z_2 - z_1 \rangle.$$

(See Fig. 14–15b.) The geometric vector that corresponds to the zero vector $\langle 0, 0, 0 \rangle$ is drawn as a single point.

The *magnitude* (or *length*) of the vector $v = \langle a, b, c \rangle$ is defined to be

Magnitude

$$\|v\| = \|\langle a, b, c \rangle\| = \sqrt{a^2 + b^2 + c^2}.$$

The zero vector has zero magnitude. If $v \neq \mathbf{0}$, then v has positive magnitude equal to the length of any arrow that represents v.

ADDITION

The sum of $v_1 = \langle a_1, b_1, c_1 \rangle$ and $v_2 = \langle a_2, b_2, c_2 \rangle$ is defined in terms of its components as
$$v_1 + v_2 = \langle a_1 + a_2, b_1 + b_2, c_1 + c_2 \rangle.$$

For example, if $v_1 = \langle 2, 1, -3 \rangle$ and $v_2 = \langle -5, 3, 7 \rangle$, then $v_1 + v_2 = \langle -3, 4, 4 \rangle$.

This type of sum has the same geometrical interpretation as the sum of vectors in the plane. (See Fig. 14–16.) We first draw v_1 from O to $P_1(a_1, b_1, c_1)$. We then draw v_2

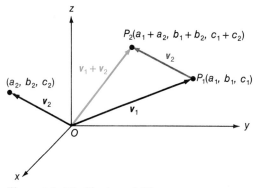

Figure 14–16 Vector addition.

$$\langle a_1, b_1, c_1 \rangle + \langle a_2, b_2, c_2 \rangle = \langle a_1 + a_2, b_1 + b_2, c_1 + c_2 \rangle.$$

14–3 Vectors in Three-Dimensional Space. Elementary Applications

from $P_1(a_1, b_1, c_1)$ to $P_2(a_1 + a_2, b_1 + b_2, c_1 + c_2)$. Then $v_1 + v_2$ is the resultant vector from O to P_2.

SCALAR MULTIPLICATION

If $v = \langle a, b, c \rangle$ and m is a scalar, we define

$$mv = \langle ma, mb, mc \rangle.$$

For example, if $v = \langle 3, 7, -4 \rangle$, then

$$-2v = -2\langle 3, 7, -4 \rangle = \langle -6, -14, 8 \rangle.$$

Properties like those for plane vectors hold for scalar multiplication. In particular,

(1) mv has the same direction as v if $m > 0$.
(2) mv has the direction opposite v if $m < 0$.
(3) $0v = 0$.
(4) The magnitude of mv equals the absolute value of m times the magnitude of v:

$$\|mv\| = |m| \cdot \|v\|.$$

THE i, j, k UNIT VECTORS

Vectors of length 1 are called *unit vectors*. The three unit vectors

$$i = \langle 1, 0, 0 \rangle, \quad j = \langle 0, 1, 0 \rangle, \quad k = \langle 0, 0, 1 \rangle$$

are convenient for representing other vectors. It follows, as in Section 14–1, that the vector $v = \langle a, b, c \rangle$ can be written as

$$v = ai + bj + ck.$$

EXAMPLE 1 Let $u = 2i - 3j + k$, $v = 3i + j - 2k$. Then

$$\begin{aligned} 5u - 2v &= 5(2i - 3j + k) - 2(3i + j - 2k) \\ &= 10i - 15j + 5k - 6i - 2j + 4k \\ &= 4i - 17j + 9k. \end{aligned} \ \square$$

PROPERTIES OF VECTORS

All the properties listed in Theorem 14–1 (which you should review at this time) hold for vectors in space. Thus, vector addition is commutative and associative, scalar multiplication distributes over vector addition, and so on. All these properties can be established by expressing the vectors by their components, as in Example 4 of Section 14–1.

DIRECTION ANGLES

The direction of a nonzero vector v in the xy-plane is determined by a single angle θ, called the *direction angle*. This is the smallest positive angle from the positive x-axis to the vector. (See Fig. 14–17a.) If v is a unit vector in the xy plane, then

$$v = \cos\theta i + \sin\theta j.$$

If v is not a unit vector, then

$$v = \|v\|(\cos\theta i + \sin\theta j).$$

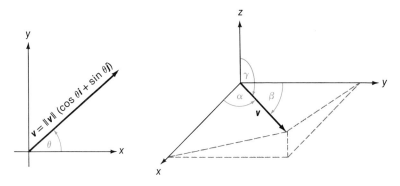

(a) The direction angle of a vector in the xy-plane
$v = \|v\|(\cos\theta i + \sin\theta j)$

(b) The three direction angles for a vector in space
$v = \|v\|(\cos\alpha i + \cos\beta j + \cos\gamma k)$

Figure 14–17 Direction angles.

The situation is more complicated in three-dimensional space. We need three angles to specify the direction of a vector. (See Fig. 14–17b.) Let $v = ai + bj + ck$ be the nonzero vector from the origin to the point $P(a, b, c)$. We define

Direction Angles
α, β, γ

α = smallest positive angle from the positive x-axis to v, $\quad 0 \leq \alpha \leq \pi$.
β = smallest positive angle from the positive y-axis to v, $\quad 0 \leq \beta \leq \pi$.
γ = smallest positive angle from the positive z-axis to v, $\quad 0 \leq \gamma \leq \pi$.

These angles are measured in the planes determined by v and the respective coordinate axes. (See Fig. 14–17b and 14–18a.) They are called the *direction angles* of v.

We now show that a vector $v = ai + bj + ck$ can be expressed by the cosines of its direction angles. We choose the viewing angle so that the triangle determined by the points $(0, 0, 0)$, (a, b, c) and $(a, 0, 0)$, shown in Figure 14–18a, appears as in Figure 14–18b. Since $\cos\alpha = a/\|v\|$, then

$$a = \|v\|\cos\alpha.$$

Similar work with the other two triangles shown in Figure 14–18a establishes that

$$b = \|v\|\cos\beta \quad \text{and} \quad c = \|v\|\cos\gamma.$$

It follows that

A Vector in Terms of its Magnitude and Direction Angles

$$v = ai + bj + ck,$$
$$= \|v\|(\cos\alpha i + \cos\beta j + \cos\gamma k),$$

where α, β, and γ are the direction angles of v.

14–3 Vectors in Three-Dimensional Space. Elementary Applications

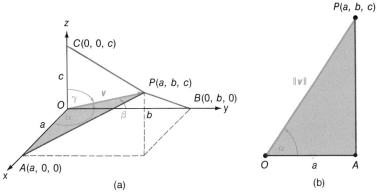

Figure 14–18 Direction angles.

$$a = \|v\| \cos \alpha, \quad b = \|v\| \cos \beta, \quad c = \|v\| \cos \gamma.$$

The numbers $\cos \alpha$, $\cos \beta$, $\cos \gamma$ are called *direction cosines* of v. It is standard always to state direction angles (or cosines) in the order α, β, γ. Thus, if we say that $\pi/2$, $\pi/2$, 0 are direction angles for v, we mean that $\alpha = \pi/2$, $\beta = \pi/2$, and $\gamma = 0$. Since $v = \|v\| (\cos \alpha i + \cos \beta j + \cos \gamma k)$, then

$$\|v\| = \|v\| \sqrt{\cos^2\alpha + \cos^2\beta + \cos^2\gamma}$$

so that

Fundamental Relationship for Direction Cosines

$$\cos^2\alpha + \cos^2\beta + \cos^2\gamma = 1.$$

This relation is the fundamental one relating direction cosines.

Two worthwhile results can be obtained from this work. First, if v has direction angles α, β, γ, then a unit vector in the same direction as v is

$$u = \cos \alpha i + \cos \beta j + \cos \gamma k.$$

Second, not every set of angles α, β, γ can be direction angles for a vector. If $\cos^2\alpha + \cos^2\beta + \cos^2\gamma \neq 1$, the angles do not determine the direction of a vector.

EXAMPLE 2 Find a vector of length 3 units with direction angles $\pi/4$, $2\pi/3$, and $\pi/3$.

Solution We want $\alpha = \pi/4$, $\beta = 2\pi/3$, and $\gamma = \pi/3$. Since

$$\cos^2 \alpha + \cos^2 \beta + \cos^2 \gamma = \left(\frac{1}{\sqrt{2}}\right)^2 + \left(-\frac{1}{2}\right)^2 + \left(\frac{1}{2}\right)^2 = 1$$

the angles are direction angles. The vector is

$$v = \|v\| (\cos \alpha i + \cos \beta j + \cos \gamma k)$$
$$= 3 \left(\frac{1}{\sqrt{2}} i - \frac{1}{2} j + \frac{1}{2} k\right) = \frac{3}{\sqrt{2}} i - \frac{3}{2} j + \frac{3}{2} k. \square$$

ELEMENTARY APPLICATIONS TO PHYSICS (OPTIONAL)

Thus far we have hinted that vectors might be useful for solving certain problems. Here we show how they can be used to solve two elementary problems from physics. Our approach is descriptive rather than rigorous. For simplicity we consider both problems in the *xy*-plane. Similar methods work for the equivalent problems in three-dimensional space.

EXAMPLE 3 (*Direction and Speed*) A plane flies at an air speed of 200 mph heading northwest as indicated on a compass. The plane is subject to a crosswind from the northeast of 50 mph. Find the actual direction in which the plane is flying (relative to the ground) and the equivalent ground speed.

Solution If the air were still, the plane would fly 200 miles northwest in 1 hr. In actual practice, however, due to the movement of the air, the plane travels in a different direction with a different velocity relative to the ground.

The situation can be represented by a vector diagram. We let the vector *p* represent the speed and direction of the plane and *w* the speed and direction of the wind. (See Fig. 14–19a.) The magnitudes and directions of these vectors equal the speeds and

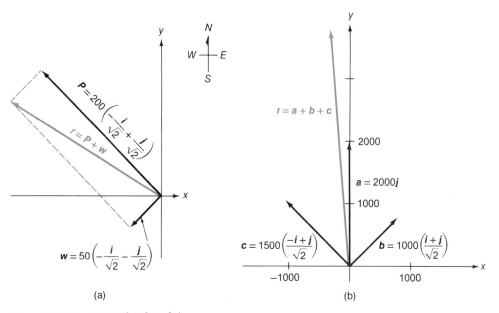

Figure 14–19 Examples 3 and 4.

the directions of the motions. Thus,

$$p = 200\left(\frac{-i}{\sqrt{2}} + \frac{j}{\sqrt{2}}\right) = -100\sqrt{2}\,i + 100\sqrt{2}\,j$$

and

$$w = 50\left(\frac{-i}{\sqrt{2}} - \frac{j}{\sqrt{2}}\right) = -25\sqrt{2}\,i - 25\sqrt{2}\,j.$$

14-3 Vectors in Three-Dimensional Space. Elementary Applications

The resultant vector $r = p + w$ represents the motion of the plane relative to the ground. We calculate

$$r = p + w$$
$$= -125\sqrt{2}\,i + 75\sqrt{2}\,j \approx -176.777\,i + 106.066\,j.$$

The speed of the plane relative to the ground is

$$\|r\| \approx \sqrt{(176.777)^2 + (106.066)^2}$$
$$\approx \sqrt{42500.104} \approx 206.2.$$

To find the direction, we let θ be the direction angle of r. Then

$$\cos\theta \approx \frac{-176.777}{206.2} \approx -0.857308$$

and

$$\sin\theta \approx \frac{106.066}{206.2} \approx 0.514384$$

so that

$$\theta \approx 2.6008 \text{ radians}$$
$$\approx 149°.$$

Relative to the ground, the speed equals approximately 206.2 mph and the direction is approximately 59° west of north. $\square$

EXAMPLE 4 (*Force Diagrams*) A large piece of machinery must be dragged across the floor of a manufacturing plant. Three winches have been attached to it. Winch *A* exerts a force of 2000 lb, Winch *B* a force of 1000 lb, and Winch *C* a force of 1500 lb.

The directions in which the forces are applied depend on the exact location of the piece of machinery. At a given instant, Winch *A* exerts a force toward the north, Winch *B* toward the northeast, and Winch *C* toward the northwest.

Find the magnitude of the resultant force and the direction in which it is applied at that given instant.

Solution Problems of this type are best handled by a diagram in which vectors represent forces. The direction of a vector equals the direction in which the corresponding force is applied, and the magnitude of the vector equals the magnitude of the force. We let a, b, and c be the vectors that describe the forces. (See Fig. 14–19b.) Then

$$a = 2000j$$
$$b = 1000\left(\frac{i+j}{\sqrt{2}}\right) = 500\sqrt{2}\,i + 500\sqrt{2}\,j$$

and

$$c = 1500\left(\frac{-i+j}{\sqrt{2}}\right) = -750\sqrt{2}\,i + 750\sqrt{2}\,j.$$

The resultant vector $r = a + b + c$ describes the resultant force. This vector is

$$r = -250\sqrt{2}\,i + (2000 + 1250\sqrt{2})\,j$$
$$\approx -353.55i + 3767.77j.$$

The magnitude of r is
$$\|r\| \approx \sqrt{(353.55)^2 + (3767.77)^2} \approx 3784.32.$$
The direction angle of r is θ, where
$$\cos \theta \approx \frac{-353.55}{3784.32} \approx -0.0934$$
and
$$\sin \theta \approx \frac{3767.77}{3784.32} \approx 0.9956.$$
Then $\theta \approx 1.664$ radians $\approx 95°20'$.

The resultant force is approximately equal to 3784.32 lb and is directed approximately 5°20′ west of north. □

Exercises 14–3

1–4. Find $\|v\|$, $2v - 3w$, and a unit vector in the opposite direction to v.

- **1** $v = \langle 0, 4, -3 \rangle$, $w = \langle -1, 3, 2 \rangle$
- **2** $v = \langle 7, -2, 1 \rangle$, $w = \langle 1, -8, 4 \rangle$
- **3** $v = 3i - 5j + k$, $w = 2i + j - 3k$
- **4** $v = -i + 2j + 3k$, $w = i - 4k$

5–6. Find the coordinates of point Q for the given conditions.

- **5** $P(2, -1, 3)$, $\overrightarrow{PQ} = 4i + 3j - k$
- **6** $P(-3, 7, 4)$, $\overrightarrow{QP} = 2i - j + k$

7–8. Find the vector that satisfies the given conditions (provided they are consistent).

- **7** Magnitude is 4, direction angles are $\alpha = \pi/4$, $\beta = \pi/3$, $\gamma = \pi/3$.
- **8** Magnitude is $\sqrt{2}$, direction angles are $\alpha = \pi/6$, $\beta = 0$, $\gamma = 3\pi/4$.

9–12. Calculate the direction cosines and the approximate values of the direction angles for the given vectors.

- **9** $3i - 6j + 2k$
- **10** $-6i + 9j + 2k$
- **11** $-i + 4j + 8k$
- **12** $4i - 7j - 4k$

13–15. Find all three-dimensional unit vectors that satisfy the conditions.

- **13** $\cos \alpha = \frac{3}{7}$, $\cos \beta = \frac{-6}{7}$
- **14** $\cos \alpha = \frac{-1}{3}$, $\cos \beta = \frac{2}{3}$
- **15** All three direction angles are equal.

- **16** A pilot must fly due north with a speed of 600 mph relative to the ground. The wind blows from the northeast with a velocity of 50 mph. In what direction must he head and at what air speed to keep on schedule?

17(a)–(h) Prove each of parts (a)–(h) of Theorem 14–1 for vectors in three-dimensional space.

14-4 The Inner Product

Thus far we have said nothing about the product of two vectors. One type of product is the *inner product* (also called the *dot product*, or the *scalar product*). (In Section 14-7 we define a second type of product.)

DEFINITION Let $v_1 = a_1 i + b_1 j + c_1 k$ and $v_2 = a_2 i + b_2 j + c_2 k$ be vectors. The *inner product* of v_1 and v_2, denoted by $v_1 \cdot v_2$, is the scalar defined by

Inner Product
$$v_1 \cdot v_2 = a_1 a_2 + b_1 b_2 + c_1 c_2.$$

Thus, the inner product is the sum of the products of the corresponding components of the two vectors. This same definition holds for vectors in the *xy*-plane. We simply take the *k*-components of the vectors to be zero.

EXAMPLE 1 Let $u = 2i - 3j + k$, $v = 2i - j - 7k$, and $w = i + j - k$. Then
$$u \cdot v = 2 \cdot 2 + (-3)(-1) + 1 \cdot (-7) = 0,$$
$$u \cdot w = 2 \cdot 1 - 3 \cdot 1 + 1 \cdot (-1) = -2,$$
and
$$v \cdot w = 2 \cdot 1 - 1 \cdot 1 - 7 \cdot (-1) = 8. \square$$

Observe that if $v = ai + bj + ck$, then

$$\|v\| = \sqrt{a^2 + b^2 + c^2} = \sqrt{v \cdot v}.$$

Thus, the inner product can be used to calculate the magnitude of a vector.

INNER PRODUCTS AND ANGLES

Angle Between Two Vectors

Let u and v be nonzero vectors that originate at a common point P. (See Fig. 14-20a.) The smallest angle θ formed by the two vectors is called the *angle between* u *and* v. (If u and v have the same direction, we have $\theta = 0$. If they are in opposite directions, then $\theta = \pi$. In the degenerate case where u or v is the zero vector, we define the angle between them to be $\theta = 0$.)

(a) θ is the angle between u and v:
$u \cdot v = \|u\| \|v\| \cos \theta$

(b) If u and v are orthogonal, then
$u \cdot v = \|u\| \cdot \|v\| \cos \frac{\pi}{2} = 0$

Figure 14-20.

Theorem 14–5 shows that the inner product of u and v can be used to calculate the angle between the two vectors. The proof is postponed until later in the section.

THEOREM 14–5

Dot Product and the Angle Between Two Vectors

Let θ be the angle between u and v. Then

$$u \cdot v = \|u\| \cdot \|v\| \cos \theta.$$

EXAMPLE 2 Let $u = 2i + j - k$ and $v = i - 3j + 3k$. Find θ, the angle between u and v.

Solution We calculate

$$\|u\| = \sqrt{4 + 1 + 1} = \sqrt{6},$$
$$\|v\| = \sqrt{1 + 9 + 9} = \sqrt{19},$$

and

$$u \cdot v = 2 - 3 - 3 = -4.$$

Then

$$-4 = u \cdot v = \|u\| \cdot \|v\| \cos \theta = \sqrt{6}\sqrt{19} \cos \theta,$$

so that

$$\cos \theta = \frac{-4}{\sqrt{6}\sqrt{19}} \approx -0.374634.$$

Therefore

$$\theta \approx 1.9548 \text{ radians} \approx 112°. \quad \square$$

Theorem 14–6 is based on Theorem 14–5 and gives an important application of the inner product. It enables us to decide immediately whether two vectors are perpendicular.

THEOREM 14–6 Let u and v be nonzero vectors. Then u and v are perpendicular if and only if $u \cdot v = 0$.

Proof Recall that $u \cdot v = \|u\| \cdot \|v\| \cos \theta$, where θ is the angle between u and v. Since u and v are nonzero vectors, their magnitudes are nonzero. Thus,

$$u \cdot v = 0 \quad \text{if and only if } \cos \theta = 0.$$

Dot Product and Perpendicularity

Since $0 \leq \theta \leq \pi$, then $\cos \theta = 0$ if and only if $\theta = \pi/2$. Thus, $u \cdot v = 0$ if and only if the angle between the two vectors is a right angle. (See Fig. 14–20b.) ∎

For example, since their scalar product is zero, the vectors u and v in Example 1 are perpendicular.

14–4 The Inner Product

ORTHOGONAL VECTORS

It follows from Theorem 14–6 that $u \cdot v = 0$ only in case u and v are perpendicular or at least one of u and v is the zero vector. Since it is awkward to keep track of these three cases, we make the following definition, which extends to vectors in higher dimensions.

DEFINITION The vectors u and v are said to be *orthogonal* provided $u \cdot v = 0$.

Observe that if u and v are nonzero orthogonal vectors, then they are perpendicular.

PROPERTIES OF INNER PRODUCTS

The inner product obeys several algebraic laws, many of which are given in Theorem 14–7. Several of these are similar to properties that hold for numbers, and they are given the same names. All these properties can be proved by expressing the vectors by their components. The proof of the first property is given in Example 3. The proofs of the others are left for you to work. (See Exercise 15.)

THEOREM 14–7 Let u, v, and w be vectors, and m be a scalar. Then

(a) $u \cdot v = v \cdot u$ (Commutative property)
(b) $u \cdot (v + w) = u \cdot v + u \cdot w$ (Distributive property)
(c) $(mu) \cdot v = m(u \cdot v) = u \cdot (mv)$ (A type of associative property)
(d) $0 \cdot u = 0$
(e) $\|v\| = \sqrt{v \cdot v}$.

EXAMPLE 3 Prove part (a) of Theorem 14–7: $u \cdot v = v \cdot u$ for any two vectors u and v.

Solution Let $u = ai + bj + ck$ and $v = di + ej + fk$. Then

$$u \cdot v = ad + be + cf = da + eb + fc = v \cdot u. \ \square$$

PROOF OF THEOREM 14–5

The proof of Theorem 14–5 depends on the law of cosines (which you should review before starting the proof).

THEOREM 14–5 Let θ be the angle between the vectors u and v. Then

$$u \cdot v = \|u\| \|v\| \cos \theta.$$

Proof The result is trivial if u or v is the zero vector or if $\theta = 0$ or π. Thus, we restrict ourselves to the case in which u and v are nonzero vectors that are not parallel.

Let $u = ai + bj + ck$ and $v = di + ej + fk$. We draw u and v with the origin as a common initial point as in Figure 14–21a. Observe that the vector from the terminal

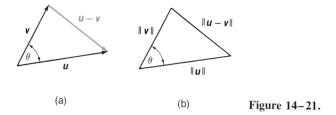

(a)   (b)   **Figure 14–21.**

point of v to the terminal point of u is $u - v$, the vector we add to v to get u. It is the vector

$$u - v = (a - d)i + (b - e)j + (c - f)k.$$

The three vectors u, v, and $u - v$ form a triangle, with sides of length $\|u\|$, $\|v\|$, and $\|u - v\|$. (See Fig. 14–21b.) It follows from the law of cosines that

$$\|u - v\|^2 = \|u\|^2 + \|v\|^2 - 2\|u\| \cdot \|v\| \cos \theta.$$

Rewriting this equation in terms of the components of the vectors, we get

$$(a - d)^2 + (b - e)^2 + (c - f)^2$$
$$= (a^2 + b^2 + c^2) + (d^2 + e^2 + f^2) - 2\|u\| \cdot \|v\| \cos \theta.$$

If we expand the terms on the left and simplify, we get

$$-2ad - 2be - 2cf = -2\|u\| \|v\| \cos \theta.$$

It follows that

$$ad + be + cf = \|u\| \cdot \|v\| \cos \theta$$
$$u \cdot v = \|u\| \cdot \|v\| \cos \theta. \blacksquare$$

PROJECTIONS AND COMPONENTS

Let a and b be nonzero vectors that originate at a common point P. We drop a perpendicular line from the terminal point of a to the vector b (or its extension) as in Figures 14–22a, b. This line intersects b (or its extension) at a point Q.

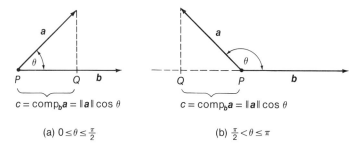

(a) $0 \leq \theta \leq \frac{\pi}{2}$ (b) $\frac{\pi}{2} < \theta \leq \pi$

Figure 14–22 The component of a in the direction of b.

$$\operatorname{comp}_b a = \frac{a \cdot b}{\|b\|}$$

14-4 The Inner Product

Component of a Vector in Direction of Another

The directed distance from P to Q is called the *component of a in the direction of b*, written

$$\text{comp}_b \, a.$$

This directed distance is positive if $\overrightarrow{PQ}$ is in the same direction as b (as in Fig. 14–22a) and negative if $\overrightarrow{PQ}$ is in the opposite direction (as in Fig. 14–22b). In other words, the directed distance is positive if $0 \leq \theta < \pi/2$ and negative if $\pi/2 < \theta \leq \pi$.

Let c be the component of a in the direction of b. Observe in Figure 14–22 that $\cos \theta = c/\|a\|$, so that $c = \|a\| \cos \theta$. It follows that

$$c = \frac{\|a\| \, \|b\| \cos \theta}{\|b\|} = \frac{a \cdot b}{\|b\|}$$

Formula for Component

$$\text{comp}_b \, a = \frac{a \cdot b}{\|b\|}.$$

EXAMPLE 4 Let $a = 2i - j + 2k$ and $b = 3i + 4j - 12k$. Then

$$\text{comp}_b \, a = \frac{a \cdot b}{\|b\|} = \frac{6 - 4 - 24}{\sqrt{169}} = -\frac{22}{13}$$

and

$$\text{comp}_a \, b = \frac{a \cdot b}{\|a\|} = -\frac{22}{3}. \quad \square$$

Projection of One Vector on Another

We now take our development one step further. Recall that Q is the projection of the terminal end of the vector a on b (or its extension). (See Figs. 14–22, 14–23.) The vector $\overrightarrow{PQ}$ is called the *projection of a on b*, written

$$\overrightarrow{PQ} = \text{proj}_b \, a.$$

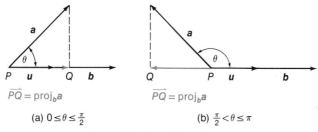

$\overrightarrow{PQ} = \text{proj}_b a$

(a) $0 \leq \theta \leq \frac{\pi}{2}$

$\overrightarrow{PQ} = \text{proj}_b a$

(b) $\frac{\pi}{2} < \theta \leq \pi$

Figure 14–23 The projection of a and b.

$$\text{proj}_b a = \frac{a \cdot b}{b \cdot b} b$$

It is easy to calculate this projection. Let u be the unit vector in the direction of b,

$$u = \frac{b}{\|b\|}.$$

Observe in Figure 14–23 that

$$\overrightarrow{PQ} = (\text{comp}_b\, a)u.$$

Therefore,

$$\overrightarrow{PQ} = \frac{a \cdot b}{\|b\|} \cdot \frac{b}{\|b\|} = \frac{a \cdot b}{\|b\| \cdot \|b\|} b = \frac{a \cdot b}{b \cdot b} b$$

Formula for Projection

$$\text{proj}_b\, a = \frac{a \cdot b}{b \cdot b} b.$$

EXAMPLE 5 Let $a = 2i - j + 2k$ and $b = 3i + 4j - 12k$. Then

$$\text{proj}_b\, a = \frac{a \cdot b}{b \cdot b} b$$

$$= \frac{6 - 4 - 24}{9 + 16 + 144}(3i + 4j - 12k)$$

$$= -\frac{22}{169}(3i + 4j - 12k)$$

$$= -\frac{66}{169}i - \frac{88}{169}j + \frac{264}{169}k$$

and

$$\text{proj}_a\, b = \frac{a \cdot b}{a \cdot a} a$$

$$= \frac{6 - 4 - 24}{4 + 1 + 4}(2i - j + 2k)$$

$$= -\frac{22}{9}(2i - j + 2k)$$

$$= -\frac{44}{9}i + \frac{22}{9}j - \frac{44}{9}k. \quad \square$$

Exercises 14–4

1–6. (a) Calculate $a \cdot b$.
 (b) Let θ be the angle between a and b. Calculate $\cos \theta$ and θ.
 (c) Calculate $\text{comp}_b\, a$, $\text{comp}_a\, b$, and $\text{proj}_b\, a$.

•1 $a = i - 4j + 8k$, $b = 5i - 12k$ 2 $a = -6i + 8k$, $b = 4i + 7j - 4k$

•3 $a = 8i + j - 4k$, $b = 2i - 4j - 4k$ 4 $a = 6i - 2j + 3k$, $b = 6i + 9j - 2k$

•5 $a = i + 2j - 4k$, $b = 3i - 6k$ 6 $a = -4j + 6k$, $b = 2i + 3j$

14–5 Application to Geometry

7–10. Find a unit vector orthogonal to both a and b.

• 7 $a = -i + 3j + 2k, b = 2j + 3k$

  8 $a = 2i - 2j + 3k, b = -2i + 3j - 4k$

• 9 $a = 3i + 5j - k, b = 2i + 3j + k$

  10 $a = -2i + k, b = i + 2j - 3k$.

•11 Determine all values of x such that vectors $3xi + j - 4xk$ and $xi + 4j + 2k$ are orthogonal.

  12 Use vector methods to show that the points $P(2, -3, 1)$, $Q(-5, 1, 7)$, and $R(6, 1, 3)$ are the vertices of a right triangle.

•13 Find the angle between the diagonal of a cube and one of its edges.

  14 Prove that $|a \cdot b| \leq \|a\| \|b\|$. When does equality hold?

  15(b)–(e) Prove each of the parts (b)–(e) of Theorem 14–7.

  16 (*The Triangle Inequality*) **(a)** Prove that
  $$\|u + v\|^2 = \|u\|^2 + \|v\|^2 + 2u \cdot v.$$

  (b) Use (a) and Exercise 14 to prove that

Triangle Inequality
  $$\|u + v\| \leq \|u\| + \|v\|.$$

  This result is known as the *triangle inequality for vectors*.

  (c) Interpret the triangle inequality in terms of the sides of a triangle.

14–5 Application to Geometry

We now consider several isolated problems from analytic geometry that demonstrate the power of vector methods. All these problems can be solved without vectors—some we have already considered—but the arguments are comparatively awkward. Using vectors allows us to get rather elegant solutions with arguments that show the essential nature of the problems.

For simplicity, we consider only problems in the xy-plane. In most cases, similar arguments can be used to solve the comparable problems in three-dimensional space.

EXAMPLE 1 (*Midpoint Formula*) Let $P_1(x_1, y_1)$ and $P_2(x_2, y_2)$ be points in the xy-plane. Find the coordinates of the midpoint of the line segment connecting P_1 and P_2.

Solution Let $M(x, y)$ be the midpoint. (See Fig. 14–24.) Then the vector $\overrightarrow{P_1M}$ points in the same direction as $\overrightarrow{P_1P_2}$ and is half as long. Therefore,

Alternate Proof of Midpoint Formulas

$$\overrightarrow{P_1M} = \frac{1}{2} \overrightarrow{P_1P_2}$$

$$(x - x_1)i + (y - y_1)j = \frac{1}{2}[(x_2 - x_1)i + (y_2 - y_1)j].$$

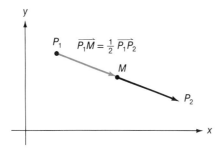

Figure 14-24 Example 1. The midpoint formula.

Since equal vectors have equal corresponding components, then

$$x - x_1 = \frac{1}{2}(x_2 - x_1) \quad \text{and} \quad y - y_1 = \frac{1}{2}(y_2 - y_1)$$

so that

$$x = x_1 + \frac{1}{2}(x_2 - x_1) = \frac{x_1 + x_2}{2}$$

and

$$y = y_1 + \frac{1}{2}(y_2 - y_1) = \frac{y_1 + y_2}{2}.$$

The midpoint is

$$M\left(\frac{x_1 + x_2}{2}, \frac{y_1 + y_2}{2}\right). \quad \square$$

EXAMPLE 2 (*Diagonals of a Parallelogram*) Prove that the diagonals of a parallelogram intersect at their midpoints.

Solution   Let A, B, C, D be the successive vertices of the parallelogram. (See Fig. 14-25.) Let M be the midpoint of diagonal AC and N the midpoint of diagonal BD. The problem will be solved if we prove that $M = N$, for then the diagonals must intersect at the common midpoint.

We shall show that $\overrightarrow{AM} = \overrightarrow{AN}$. Observe in Figure 14-25a that

$$\overrightarrow{AM} = \tfrac{1}{2}\overrightarrow{AC} = \tfrac{1}{2}(\overrightarrow{AB} + \overrightarrow{BC}) = \tfrac{1}{2}(\overrightarrow{AB} + \overrightarrow{AD}).$$

Also (see Fig. 14-25b)

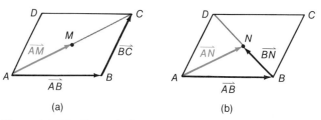

Figure 14-25 Example 2.

14–5 Application to Geometry

$$\vec{AN} = \vec{AB} + \vec{BN} = \vec{AB} + \tfrac{1}{2}\vec{BD} = (\tfrac{1}{2}\vec{AB} + \tfrac{1}{2}\vec{AB}) + \tfrac{1}{2}\vec{BD}$$
$$= \tfrac{1}{2}\vec{AB} + \tfrac{1}{2}(\vec{AB} + \vec{BD}) = \tfrac{1}{2}(\vec{AB} + \vec{AD}) = \vec{AM}$$

Since $\vec{AN} = \vec{AM}$, then $M = N$ and the diagonals intersect at their midpoints. □

Our next problem involves the distance from a point P to a line $\mathscr{L}$. This distance, of course, is measured along a line perpendicular to $\mathscr{L}$. (See Fig. 14–26.) Our work is expedited by the use of a nonzero vector n that is perpendicular to the line. Such a vector is called a *normal vector* to the line $\mathscr{L}$.

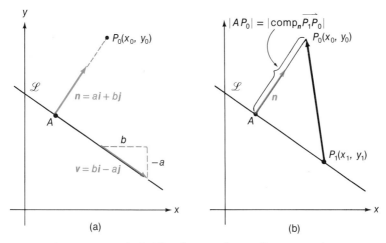

Figure 14–26 Example 3. The distance from a line to a point

$$|\mathscr{L}P_0| = |AP_0| = \frac{|ax_0 + by_0 + c|}{\sqrt{a^2 + b^2}}.$$

EXAMPLE 3 (*Distance From a Line to a Point*) Show that the distance from the line $ax + by + c = 0$ to the point $P_0(x_0, y_0)$ is

$$d = \frac{|ax_0 + by_0 + c|}{\sqrt{a^2 + b^2}}.$$

Solution Our first problem is to find a vector normal to a line. Since the slope of the given line is $m = -a/b$, then a change of b units in the x-direction and a change of $-a$ units in the y-direction from one point on the line results in a second point that also is on the line. Thus, the vector $v = bi - aj$ is parallel to the line.

Observe that the vector $n = ai + bj$ is orthogonal to v. This follows from the fact that

$$n \cdot v = ab - ab = 0.$$

Thus n is normal to the line $ax + by + c = 0$. (See Fig. 14–26a.)

Let A be the point on the line closest to P_0. The distance from the line to the point equals the distance from A to P_0. (See Fig. 14–26b.)

Let $P_1(x_1, y_1)$ be an arbitrary point on the line. Observe in Figure 14–26b that the distance from A to P_0 equals

$$|\text{comp}_n \, \overrightarrow{P_1P_0}|.$$

We now calculate this quantity. Observe that

$$\overrightarrow{P_1P_0} = (x_0 - x_1)i + (y_0 - y_1)j$$

and

$$n = ai + bj.$$

Then

$$\text{Distance from } \mathscr{L} \text{ to } P_0 = |AP_0|$$

$$= |\text{comp}_n \, \overrightarrow{P_1P_0}| = \frac{|\overrightarrow{P_1P_0} \cdot n|}{\|n\|}$$

$$= \frac{|a(x_0 - x_1) + b(y_0 - y_1)|}{\|n\|}$$

$$= \frac{|(ax_0 + by_0) - (ax_1 + by_1)|}{\|n\|}.$$

Since $P_1(x_1, y_1)$ is on the line $ax + by + c = 0$, then $ax_1 + by_1 = -c$. Therefore, the distance from the line to the point is

$$|AP_0| = \frac{|(ax_0 + by_0) - (-c)|}{\|n\|} = \frac{|ax_0 + by_0 + c|}{\sqrt{a^2 + b^2}}. \quad \square$$

You should compare the argument in Example 3 with the argument outlined in Exercise 51 of Section 1–5.

In our final example we derive parametric equations of a curve defined by certain geometric conditions.

EXAMPLE 4 (*Involute of a Circle*) A circle of radius a has its center at the origin. A string is wound around this circle in a clockwise direction until the free end comes to the point $(a, 0)$. The string is then kept taut as it is unwound. Find parametric equations for the curve (called the *involute* of the circle) that is traced out by the free end of the string. (See Fig. 14–27a.)

Involute of a Circle

Solution Let $P(x, y)$ be a typical point on the curve. (See Fig. 14–27a.) Let Q be the point at which the string from P to the circle touches the circle. Let θ be the direction angle of $\overrightarrow{OQ}$.

Since the circle has radius a, then

$$\overrightarrow{OQ} = a(\cos\theta \, i + \sin\theta \, j)$$

so that Q is the point $Q(a\cos\theta, a\sin\theta)$.

Observe in Figure 14–27b that $\theta - \pi/2$ is the direction angle of vector $\overrightarrow{QP}$. Also, the length of $\overrightarrow{QP}$ equals $a\theta$, the length of the portion of the string that has been un-

Exercises 14-5

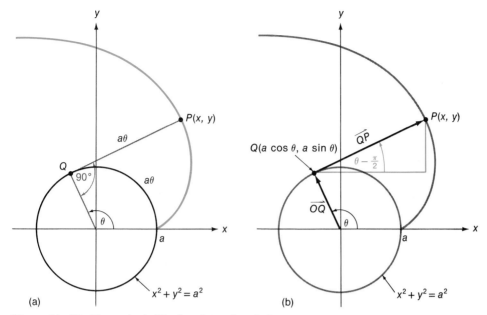

Figure 14-27 Example 4. The involute of a circle.

wound. It follows that

$$\vec{QP} = a\theta\left[\left(\cos\left(\theta - \frac{\pi}{2}\right)i + \sin\left(\theta - \frac{\pi}{2}\right)j\right)\right] = a\theta(\sin\theta i - \cos\theta j).$$

We now write the vector OP as

$$\begin{aligned}\vec{OP} &= xi + yj = \vec{OQ} + \vec{QP} \\ &= a(\cos\theta i + \sin\theta j) + a\theta(\sin\theta i - \cos\theta j) \\ &= a(\cos\theta + \theta\sin\theta)i + a(\sin\theta - \theta\cos\theta)j.\end{aligned}$$

It follows that

$$x = a(\cos\theta + \theta\sin\theta),$$
$$y = a(\sin\theta - \theta\cos\theta). \quad \square$$

Exercises 14-5

Use vector methods for all exercises.

1–2. Derive a formula for the point $Q(x, y, z)$ that satisfies the stated conditions.

1 Q is the midpoint of the line segment that connects $P_1(x_1, y_1, z_1)$ and $P_2(x_2, y_2, z_2)$.

•**2** Q is on the line segment connecting $P_1(x_1, y_1, z_1)$ with $P_2(x_2, y_2, z_2)$ and is two-thirds of the distance from P_1 to P_2.

3-4. Draw a diagram similar to Figure 14–26. Then find the distance from point P to the line through points Q and R.

•3 $P(3, -10, 6)$, $Q(5, -7, 4)$, $R(7, -8, 2)$

4 $P(2, 7, -3)$, $Q(0, -3, 1)$, $R(2, 1, -3)$

5 Prove that the line segment joining the midpoints of two sides of a triangle is parallel to the third side and half as long.

6 A geometrical figure is drawn by connecting the successive midpoints of the sides of a quadrilateral with line segments. Prove that the figure is a parallelogram.

7 Let A, B, C, D be points in space such that $\overrightarrow{OA} - \overrightarrow{OB} = \overrightarrow{OD} - \overrightarrow{OC}$. Prove that figure $ABCD$ is a parallelogram.

•8 Find the vertices of the triangle that has $A(-1, 2, 3)$, $B(2, 1, 4)$, and $C(1, 4, -1)$ as the midpoints of its sides.

9 Prove that a parallelogram is a rhombus if and only if its diagonals are perpendicular.

14–6 Equations of Lines and Planes

It is simple to derive parametric equations of a line in three-dimensional space. Observe that a line is completely determined by a point P_0 on it and a vector parallel to it. The point locates the line in space and the vector determines the direction. (See Fig. 14–28.)

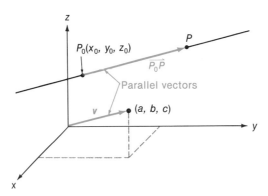

Figure 14–28 Parametric equations of a line: $x = x_0 + at$, $y = y_0 + bt$, $z = z_0 + ct$.

THEOREM 14–8   Let $v = ai + bj + ck$ be a nonzero vector; let $P_0(x_0, y_0, z_0)$ be a point. The line through P_0 that is parallel to v has parametric equations

Parametric Equations of a Line

$$\begin{cases} x = x_0 + at, \\ y = y_0 + bt, \\ z = z_0 + ct. \end{cases}$$

14-6 Equations of Lines and Planes

Proof The point $P(x, y, z)$ is on the line if and only if the vector $\overrightarrow{P_0P}$ is parallel to v. (See Fig. 14–28.) This is the case if and only if there exists a scalar t such that

$$\overrightarrow{P_0P} = tv.$$

Since $\overrightarrow{P_0P} = (x - x_0)i + (y - y_0)j + (z - z_0)k$, then $P(x, y, z)$ is on the line if and only if

$$(x - x_0)i + (y - y_0)j + (z - z_0)k = t(ai + bj + ck).$$

On equating corresponding components of these vectors, we get

$$x - x_0 = at, \qquad y - y_0 = bt, \qquad z - z_0 = ct$$

so that

$$x = x_0 + at, \qquad y = y_0 + bt, \qquad z = z_0 + ct. \blacksquare$$

EXAMPLE 1 The line through $P_0(2, 1, 5)$ that is parallel to the vector $2i - 3j + 4k$ has parametric equations

$$x = 2 + 2t, \qquad y = 1 - 3t, \qquad z = 5 + 4t. \;\square$$

SYMMETRIC FORM OF EQUATIONS OF A LINE

If the numbers a, b, c are nonzero, the parametric equations

$$x = x_0 + at, \qquad y = y_0 + bt, \qquad z = z_0 + ct$$

can be solved for t, with the result

Symmetric Equations of a Line

$$\frac{x - x_0}{a} = \frac{y - y_0}{b} = \frac{z - z_0}{c}.$$

These equations are called the *symmetric form* of the equations of a line. For example, the equations of the line in Example 1 can be rewritten in symmetric form as

$$\frac{x - 2}{2} = \frac{y - 1}{-3} = \frac{z - 5}{4}.$$

In the event that one of the numbers a, b, or c is zero, the symmetric form is slightly different. For example, if $c = 0$, the symmetric form is

$$\frac{x - x_0}{a} = \frac{y - y_0}{b}, \qquad z = z_0$$

similar forms hold if $a = 0$ or $b = 0$.

Remark The equations of a line are not unique. A given line has many different sets of parametric and symmetric equations that do not resemble each other closely.

Note as an example that the point $(-2, \frac{3}{2}, \frac{11}{2})$ is on the line

$$x = -1 + 2t, \qquad y = 2 + t, \qquad z = 3 - 5t \qquad \text{(using } t = -\tfrac{1}{2}\text{)}$$

and that the vector $v = -4i - 2j + 10k$ is parallel to this line. Thus, the line also has the parametric form

$$x = -2 - 4t, \qquad y = \tfrac{3}{2} - 2t, \qquad z = \tfrac{11}{2} + 10t.$$

The corresponding symmetric forms are

$$\frac{x+1}{2} = \frac{y-2}{1} = \frac{z-3}{-5} \qquad \text{and} \qquad \frac{x+2}{-4} = \frac{y-\frac{3}{2}}{-2} = \frac{z-\frac{11}{2}}{10}$$

which are not immediately recognizable as describing the same line.

EQUATION OF A PLANE

A *plane* in three-dimensional space is the union of all lines through a fixed point $P_0(x_0, y_0, z_0)$ that are perpendicular to a fixed nonzero vector n. The vector n is called a *normal vector* to the plane. (See Fig. 14–29a.)

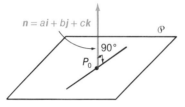
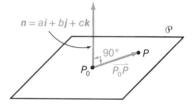

(a) The plane $\mathcal{P}$ is the union of all lines through P_0 that are orthogonal to n

(b) $P(x, y, z)$ is in the plane $\mathcal{P}$ if and only if $\overrightarrow{P_0P}$ is orthogonal to n

Figure 14–29 The equation of a plane: $a(x - x_0) + b(y - y_0) + c(z - z_0) = 0$.

It follows that we should be able to determine an equation for a plane if we know a point P_0 that is on it and the normal vector n.

THEOREM 14–9 The equation of the plane through $P_0(x_0, y_0, z_0)$ with normal vector $n = ai + bj + ck$ is

Equation of a Plane

$$a(x - x_0) + b(y - y_0) + c(z - z_0) = 0.$$

Proof Let $P(x, y, z)$ be a point. This point is on the plane if and only if the vector $\overrightarrow{P_0P}$ is orthogonal to n. (See Fig. 14–29b.) This is the case if and only if

$$n \cdot \overrightarrow{P_0P} = 0.$$

Since $n = ai + bj + ck$, and

$$\overrightarrow{P_0P} = (x - x_0)i + (y - y_0)j + (z - z_0)k,$$

14-6 Equations of Lines and Planes

then $P(x, y, z)$ is on the plane if and only if

$$(a\mathbf{i} + b\mathbf{j} + c\mathbf{k}) \cdot [(x - x_0)\mathbf{i} + (y - y_0)\mathbf{j} + (z - z_0)\mathbf{k}] = 0$$
$$a(x - x_0) + b(y - y_0) + c(z - z_0) = 0. \blacksquare$$

EXAMPLE 2 The plane through $P_0(1, -1, -2)$ with normal vector $\mathbf{n} = 2\mathbf{i} - 3\mathbf{j} + 4\mathbf{k}$ has the equation

$$2(x - 1) - 3(y + 1) + 4(z + 2) = 0$$
$$2x - 3y + 4z = -3. \quad \square$$

EXAMPLE 3 The equation $3x - 4y + 5z = 7$ is the equation of a plane. Observe that the point $(1, -1, 0)$ is on the plane. Observe also that the equation can be rewritten as

$$3(x - 1) - 4(y + 1) + 5(z - 0) = 0$$

so that a normal vector is

$$\mathbf{n} = 3\mathbf{i} - 4\mathbf{j} + 5\mathbf{k}. \quad \square$$

The equation

General Equation of a Plane

$$ax + by + cz = d \quad (a, b, c \text{ not all zero})$$

is called the *general equation* of a plane. The procedure used in Example 3 can always be applied to rewrite the equation in the form

$$a(x - x_0) + b(y - y_0) + c(z - z_0) = 0$$

where (x_0, y_0, z_0) is on the plane. The vector $\mathbf{n} = a\mathbf{i} + b\mathbf{j} + c\mathbf{k}$ is normal to this plane.

TRACES OF GRAPHS

If the equation $F(x, y, z)$ is simple enough, then its graph is a surface in three-dimensional space. The intersection of this surface with a plane parallel to a coordinate plane is called its *trace* in that plane. (See Fig. 14-30.) These traces can be very helpful in sketching the graphs.

Trace of a Graph

EXAMPLE 4 Sketch the graphs of (a) $3x + 2y + 4z = 12$; (b) $2x + 3y = 6$.

Solution (a) $3x + 2y + 4z = 12$. The trace in the xy-plane is found by setting $z = 0$. It is the line

$$\begin{cases} 3x + 2y + 4z = 12 \\ z = 0. \end{cases}$$

These equations reduce to $3x + 2y = 12$, $z = 0$.
Similarly, the trace in the yz-plane (found by setting $x = 0$) is

$$2y + 4z = 12, \quad x = 0$$

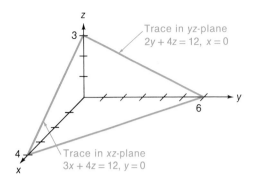
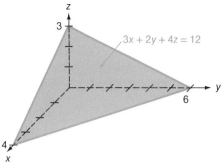

(a) Traces of the plane $3x + 2y + 4z = 12$ in the coordinate planes

(b) The plane $3x + 2y + 4z = 12$

Figure 14–30 Example 4. Parts (c) and (d) are on page 773.

and the trace in the xz-plane (found by setting $y = 0$) is

$$3x + 4z = 12, \qquad y = 0.$$

These traces are shown in Figure 14–30a. The corresponding portion of the plane is shown in Figure 14–30b.

(b) $2x + 3y = 6$. The trace in the xy-plane (found by setting $z = 0$) is

$$2x + 3y = 6, \qquad z = 0.$$

The trace in the yz-plane (found by setting $x = 0$) is

$$y = 2, \qquad x = 0.$$

The trace in the xz-plane (found by setting $y = 0$) is

$$x = 3, \qquad y = 0.$$

These traces are shown in Figure 14–30c. A portion of the plane is shown in Figure 14–30d. Observe that the plane is parallel to the z-axis. □

EXAMPLE 5 Find the point at which the line

$$\frac{x - 3}{2} = \frac{y + 4}{-1} = \frac{z + 1}{3}$$

pierces the plane $2x + 3y - 7z = 11$.

Solution We first rewrite the equations of the line in parametric form:

$$x = 3 + 2t, \qquad y = -4 - t, \qquad z = -1 + 3t.$$

Let $P_0(x_0, y_0, z_0)$ be the point at which the line pierces the plane. (See Fig. 14–31a.) Then there exists a number t_0 such that

$$x_0 = 3 + 2t_0, \; y_0 = -4 - t_0, \qquad \text{and} \qquad z_0 = -1 + 3t_0.$$

14-6 Equations of Lines and Planes

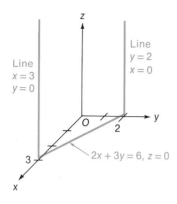

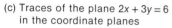

(c) Traces of the plane $2x + 3y = 6$ in the coordinate planes

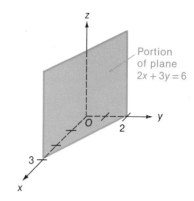

(d) The plane $2x + 3y = 6$

Figure 14-30 (c) and (d).

Since P_0 is on the plane, then

$$2x_0 + 3y_0 - 7z_0 = 11$$
$$2(3 + 2t_0) + 3(-4 - t_0) - 7(-1 + 3t_0) = 11$$
$$-20t_0 = 10$$
$$t_0 = -\tfrac{1}{2}.$$

Then

$$x_0 = 3 + 2(-\tfrac{1}{2}) = 2$$
$$y_0 = -4 - (-\tfrac{1}{2}) = -\tfrac{7}{2}$$
$$z_0 = -1 + 3(-\tfrac{1}{2}) = -\tfrac{5}{2}.$$

the point is $(2, -\tfrac{7}{2}, -\tfrac{5}{2})$. ☐

Example 6 illustrates a method for finding the distance from a point to a plane. Exercise 25 generalizes this method.

EXAMPLE 6 Find the distance from the Point $P_0(2, -1, 3)$ to the plane $5x + 2y + 3z - 7 = 0$.

Solution The plane, the normal vector $n = 5i + 2j + 3k$, and the point $P_0(2, -1, 3)$ are shown in Figure 14-31b. (As with some of our other drawings, the viewing angle has been chosen so as to make the relationships as clear as possible.)

Distance from a Point to a Plane

Our first task is to find a point on the plane. If we substitute $x = 1$, $y = 1$ in the equation of the plane, we can calculate $z = 0$. Thus $P_1(1, 1, 0)$ is on the plane.

Let A be the point on the plane closest to P_0. (See Fig. 14-31b.) Then $|AP_0|$ is the distance from P_0 to the plane. Observe that $|AP_0|$ equals the absolute value of the component of $\overrightarrow{P_1P_0}$ on n. Therefore,

$$\text{Distance} = |AP_0| = |\text{comp}_n \overrightarrow{P_1P_0}| = \frac{|\overrightarrow{P_1P_0} \cdot n|}{\|n\|}.$$

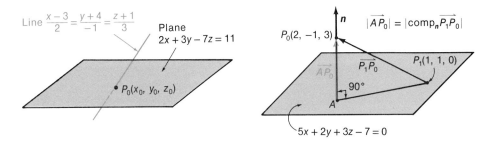

(a) Example 5   (b) Example 6

Figure 14–31 The viewing angles were chosen to clarify the relationships.

We now calculate these quantities. Since

$$\overrightarrow{P_1P_0} = (2-1)i + (-1-1)j + 3k = i - 2j + 3k,$$

then

$$\overrightarrow{P_1P_0} \cdot n = (i - 2j + 3k) \cdot (5i + 2j + 3k) = 5 - 4 + 9 = 10.$$

Therefore,

$$\text{Distance} = \frac{|\overrightarrow{P_1P_0} \cdot n|}{\|n\|} = \frac{10}{\sqrt{25 + 4 + 9}} = \frac{10}{\sqrt{38}}.$$

The point is $10/\sqrt{38}$ units from the plane. □

Exercises 14–6

1–6. Find an equation for the plane that satisfies the given conditions.

•1 Through the origin and perpendicular to the line through $P(3, -3, 4)$ and $Q(6, 6, 0)$.

•2 Through $P(3, 1, -2)$ with normal vector $i - 2j - 2k$.

•3 Through $P(4, -5, 2)$ and parallel to the yz-plane.

•4 Through $P(2, -1, 3)$ and parallel to $3x - 4y + 6z = 5$.

•5 Through $P(1, -3, 2)$ with normal vector $\overrightarrow{OP}$.

•6 Through $P(5, 2, -3)$ and perpendicular to the planes $2x - y + 4z = 7$ and $3y - z = 8$.

7–11. Find a set of parametric equations and a set of symmetric equations for the line that satisfies the given conditions

•7 Through $A(3, -2, 4)$ and $B(-1, 0, 1)$.

8 Through $P(-1, 2, -4)$ and $Q(1, -3, 2)$.

Exercises 14-6

- **9** Through $P(6, -2, 4)$ and parallel to the line $(x + 2)/3 = y/2 = (z - 4)/(-1)$.

- **10** Through $Q(-2, 1, 5)$ and normal to the plane $3x + 7y - 6z + 2 = 0$.

- **11** Through $R(4, -1, 3)$ and perpendicular to the plane $x - 4y + 3z + 10 = 0$.

12–14. Determine whether the line is on the plane, parallel to the plane, or neither. If the line intersects the plane at a single point, determine the point.

- **12** $\dfrac{x-1}{2} = \dfrac{y+1}{3} = \dfrac{z-2}{4};\ 2x - 2y + z = 4$

- **13** $\dfrac{x-2}{2} = \dfrac{y+2}{3} = \dfrac{z-3}{3};\ 3x + 2y - 4z + 10 = 0$

- **14** $x = 1 - t,\ y = 2 + 3t,\ z = 1 + 2t;\ x - 3y + 5z + 4 = 0$

15–18. Determine whether the given lines are (1) identical, (2) parallel, (3) orthogonal, (4) intersect. If they intersect at a single point, determine the point.

- **15** $x = 2 + t,\ y = 1 - 3t,\ z = -1 + t;\ x = 1 + s,\ y = -s,\ z = 2 - s$

- **16** $x = -2t,\ y = -4 - t,\ z = 2 + t;\ x = 3 + t,\ y = 3t,\ z = 6 + 5t$.

- **17** $x - 2 = \dfrac{y+7}{2} = \dfrac{z+3}{5};\ x - 1 = \dfrac{6-y}{3} = z - 4$

- **18** $\dfrac{x+1}{3} = \dfrac{y-3}{-1} = \dfrac{z+5}{2};\ \dfrac{x}{-2} = \dfrac{y-4}{2} = \dfrac{z-1}{4}$

- **19** Find the equation of the plane through $(2, 1, 3)$ that contains the lines

$$\dfrac{x-2}{1} = \dfrac{y-1}{2} = \dfrac{z-3}{2} \quad \text{and} \quad \dfrac{x-2}{5} = \dfrac{y-1}{3} = \dfrac{z-3}{-1}.$$

(*Hint:* First find a vector that is orthogonal to both lines.)

20 Explain how the symmetric form defines a line as the intersection of three planes.

21–22. Find the distance from the point P to the given plane. (See Example 6 and Exercise 25.)

- **21** $P(2, 2, 3),\ 2x + 3y + 5z = 8$ **22** $P(2, 3, 4),\ x + y + z = 1$

- **23** Find the smallest positive angle formed by the lines $x = 4 + t,\ y = 1 - t,\ z = 1 - 4t$ and $x = 1 + 2t,\ y = 1 + t,\ z = 7 - 2t$.

- **24** Find the minimum distance from $P_0(1, 5, 7)$ to the line $x = 2 + t,\ y = 1 + 2t,\ z = -3 + 2t$.

25 Prove that the distance from the plane $ax + by + cz + d = 0$ to the point $P_0(x_0, y_0, z_0)$ is

Distance from Plane to a Point

$$d = \frac{|ax_0 + by_0 + cz_0 + d|}{\sqrt{a^2 + b^2 + c^2}}.$$

(*Hint*: Modify the argument in Example 6.)

14–7 The Vector Product

A brief discussion of 2×2 and 3×3 determinants can be found in the Appendix to this chapter. You are urged to read that Appendix before proceeding further.

In this section we define a new product of vectors, one that yields a vector as the product of two other vectors.

DEFINITION Let $v_1 = a_1 i + b_1 j + c_1 k$ and $v_2 = a_2 i + b_2 j + c_2 k$. The *vector product* (*cross product*) of v_1 and v_2, denoted by $v_1 \times v_2$, is defined to be

Cross Product

$$v_1 \times v_2 = i(b_1 c_2 - b_2 c_1) - j(a_1 c_2 - a_2 c_1) + k(a_1 b_2 - a_2 b_1)$$

$$= i \begin{vmatrix} b_1 & c_1 \\ b_2 & c_2 \end{vmatrix} - j \begin{vmatrix} a_1 & c_1 \\ a_2 & c_2 \end{vmatrix} + k \begin{vmatrix} a_1 & b_1 \\ a_2 & b_2 \end{vmatrix}.$$

There is a simple device for remembering this formula. Suppose that it makes sense to write a determinant with vectors in the first row instead of numbers. In particular, suppose that we write the determinant

$$\begin{vmatrix} i & j & k \\ a_1 & b_1 & c_1 \\ a_2 & b_2 & c_2 \end{vmatrix}$$

If this determinant is expanded about the first row, we get

Determinant Formula for $v_1 \times v_2$

$$\begin{vmatrix} i & j & k \\ a_1 & b_1 & c_1 \\ a_2 & b_2 & c_2 \end{vmatrix} = i \begin{vmatrix} b_1 & c_1 \\ b_2 & c_2 \end{vmatrix} - j \begin{vmatrix} a_1 & c_1 \\ a_2 & c_2 \end{vmatrix} + k \begin{vmatrix} a_1 & b_1 \\ a_2 & b_2 \end{vmatrix} = v_1 \times v_2.$$

Although, strictly speaking, this work is not legal, it does give us a simple way of remembering the formula for the cross product. In the remainder of the book we shall write the cross product of two vectors as a determinant, keeping in mind that this is only a mnemonic device.

EXAMPLE 1 Let $a = 2i + 3j - k$, $b = 4i - j + 2k$. Then

$$a \times b = \begin{vmatrix} i & j & k \\ 2 & 3 & -1 \\ 4 & -1 & 2 \end{vmatrix} = i \begin{vmatrix} 3 & -1 \\ -1 & 2 \end{vmatrix} - j \begin{vmatrix} 2 & -1 \\ 4 & 2 \end{vmatrix} + k \begin{vmatrix} 2 & 3 \\ 4 & -1 \end{vmatrix}$$

$$= 5i - 8j - 14k$$

14–7 The Vector Product

and

$$b \times a = \begin{vmatrix} i & j & k \\ 4 & -1 & 2 \\ 2 & 3 & -1 \end{vmatrix} = i \begin{vmatrix} -1 & 2 \\ 3 & -1 \end{vmatrix} - j \begin{vmatrix} 4 & 2 \\ 2 & -1 \end{vmatrix} + k \begin{vmatrix} 4 & -1 \\ 2 & 3 \end{vmatrix}$$
$$= -5i + 8j + 14k. \ \square$$

Example 1 illustrates one of the basic properties of the vector product—it is *anticommutative*. That is,

Anticommutative Property

$$a \times b = -(b \times a)$$

for every pair of vectors a, b.

Theorem 14–10 states one of the most important properties of the cross product of two vectors—the cross product is orthogonal to each of the original vectors. As we shall see, many problems can be easily solved if we can find a vector orthogonal to two given vectors. The cross product gives us a simple way of finding such a vector.

THEOREM 14–10 Let v_1 and v_2 be vectors. The vector $v_1 \times v_2$ is orthogonal to v_1 and v_2. (See Fig. 14–32.)

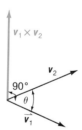

Figure 14–32 The cross product.

Proof Let $v_1 = a_1 i + b_1 j + c_1 k$ and $v_2 = a_2 i + b_2 j + c_2 k$. Then

$$v_1 \times v_2 = \begin{vmatrix} i & j & k \\ a_1 & b_1 & c_1 \\ a_2 & b_2 & c_2 \end{vmatrix} = i \begin{vmatrix} b_1 & c_1 \\ b_2 & c_2 \end{vmatrix} - j \begin{vmatrix} a_1 & c_1 \\ a_2 & c_2 \end{vmatrix} + k \begin{vmatrix} a_1 & b_1 \\ a_2 & b_2 \end{vmatrix}.$$

The dot product $v_1 \cdot (v_1 \times v_2)$ equals

$$v_1 \cdot (v_1 \times v_2) = (a_1 i + b_1 j + c_1 k) \cdot \left(i \begin{vmatrix} b_1 & c_1 \\ b_2 & c_2 \end{vmatrix} - j \begin{vmatrix} a_1 & c_1 \\ a_2 & c_2 \end{vmatrix} + k \begin{vmatrix} a_1 & b_1 \\ a_2 & b_2 \end{vmatrix} \right)$$
$$= a_1 \begin{vmatrix} b_1 & c_1 \\ b_2 & c_2 \end{vmatrix} - b_1 \begin{vmatrix} a_1 & c_1 \\ a_2 & c_2 \end{vmatrix} + c_1 \begin{vmatrix} a_1 & b_1 \\ a_2 & b_2 \end{vmatrix},$$

which equals the expanded value of the determinant

$$\begin{vmatrix} a_1 & b_1 & c_1 \\ a_1 & b_1 & c_1 \\ a_2 & b_2 & c_2 \end{vmatrix}.$$

Since this determinant has two identical rows, its value is zero (Property 3 of the Appendix to this chapter). Since

$$v_1 \cdot (v_1 \times v_2) = \begin{vmatrix} a_1 & b_1 & c_1 \\ a_1 & b_1 & c_1 \\ a_2 & b_2 & c_2 \end{vmatrix} = 0,$$

then the vector $v_1 \times v_2$ is orthogonal to v_1.

A similar argument shows that $v_1 \times v_2$ is orthogonal to v_2. (See Exercise 13.) ∎

EXAMPLE 2   Let $a = 2i + 3j - k$, $b = 4i - j + 2k$. Verify that $a \times b$ is orthogonal to both a and b.

Solution   We saw in Example 1 that $a \times b = 5i - 8j - 14k$. Then

$$(a \times b) \cdot a = 5 \cdot 2 - 8 \cdot 3 + 14 \cdot 1 = 0$$

and

$$(a \times b) \cdot b = 5 \cdot 4 + 8 \cdot 1 - 14 \cdot 2 = 0.$$

Therefore, $a \times b$ is orthogonal to both a and b. □

EXAMPLE 3   Find the equation of the plane that contains the points $O(0, 0, 0)$, $A(1, 1, -1)$, and $B(1, 2, 1)$.

Solution   First we need to find a vector normal to the plane. Since $\overrightarrow{OA}$ and $\overrightarrow{OB}$ lie in the plane, then by Theorem 14–10 their cross product is normal to the plane.

Let

$$n = \overrightarrow{OA} \times \overrightarrow{OB} = \begin{vmatrix} i & j & k \\ 1 & 1 & -1 \\ 1 & 2 & 1 \end{vmatrix}$$

$$= i \begin{vmatrix} 1 & -1 \\ 2 & 1 \end{vmatrix} - j \begin{vmatrix} 1 & -1 \\ 1 & 1 \end{vmatrix} + k \begin{vmatrix} 1 & 1 \\ 1 & 2 \end{vmatrix} = 3i - 2j + k.$$

Since n is normal to the plane and $O(0, 0, 0)$ is on it, then the equation of the plane is

$$3(x - 0) - 2(y - 0) + 1(z - 0) = 0$$
$$3x - 2y + z = 0.$$

(See Fig. 14–33.) □

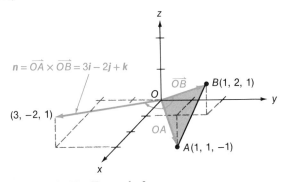

Figure 14–33 Example 3.

14–7 The Vector Product

PROPERTIES OF THE CROSS PRODUCT

The following properties of the cross product can be established by expanding the determinants that are involved. The details are left for you (Exercise 17).

THEOREM 14–11

Algebraic Properties of Cross Product

(a) $i \times j = k, j \times k = i, k \times i = j$

$j \times i = -k, k \times j = -i, i \times k = -j.$

(b) If a and b are parallel or if either is the zero vector, then $a \times b = 0$.

(c) $a \times b = -b \times a$.   (Anticommutative)

(d) In general, $a \times (b \times c) \neq (a \times b) \times c$.   (The cross product is not associative)

(e) $a \times (b \times c) = (a \cdot c)b - (a \cdot b)c$.

(f) $\begin{cases} a \times (b + c) = (a \times b) + (a \times c), \\ (a + b) \times c = (a \times c) + (b \times c). \end{cases}$ (Distributive laws)

(g) $(ma) \times b = a \times (mb) = m(a \times b)$.   (A type of associative property)

(h) $\|a \times b\|^2 = (\|a\| \|b\|)^2 - (a \cdot b)^2$.

THE MAGNITUDE OF $a \times b$

Thus far we have not concerned ourselves with the magnitude of $a \times b$. We know that this vector is normal to the plane determined by a and b but not how long it is. We now establish that its length is

$$\|a \times b\| = \|a\| \|b\| \sin \theta$$

where θ is the angle between a and b. (See Fig. 14–32.) The proof depends on property (h) of Theorem 14–11.

COROLLARY TO THEOREM 14–11

Magnitude of Cross Product

Let θ be the angle between a and b. Then

$$\|a \times b\| = \|a\| \|b\| \sin \theta.$$

Proof   By property (h) of Theorem 14–11,

$$\|a \times b\|^2 = (\|a\| \|b\|)^2 - (a \cdot b)^2.$$

Since $a \cdot b = \|a\| \|b\| \cos \theta$, then

$$\|a \times b\|^2 = (\|a\| \|b\|)^2 - (\|a\| \|b\| \cos \theta)^2$$
$$= (\|a\| \|b\|)^2 (1 - \cos^2 \theta) = \|a\|^2 \|b\|^2 \sin^2 \theta.$$

Therefore,

$$\|a \times b\| = \|a\| \|b\| \sin \theta. \blacksquare$$

Remark   It can be shown that the vectors $a, b,$ and $a \times b$ form a right-handed system. (See Fig. 14–32.) The proof is omitted.

AREA OF A PARALLELOGRAM

The value of $\|a \times b\|$ has an interesting geometrical interpretation. Let a and b be nonzero vectors that emanate from the same point P. Draw a parallelogram with adjacent sides a and b, as in Figure 14-34. We shall show that the area of the parallelogram is equal to the magnitude of $a \times b$.

Let h be the height of the parallelogram. (See Fig. 14-34b.) Then $\sin \theta = h/\|b\|$, where θ is the angle between a and b. It follows that

$$\text{Area of parallelogram} = (\text{base}) \cdot (\text{height}) = \|a\|h$$
$$= \|a\|\,\|b\|\sin\theta = \|a \times b\|.$$

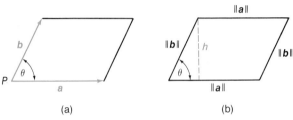

Figure 14-34 Area of the parallelogram $= \|a \times b\|$.

Exercises 14-7

1-4. Calculate $a \times b$ and $(b \times a) \times a$.

- **1** $a = 2i - 2j - k$, $b = i + j + k$ **2** $a = 4i - 8j + k$, $b = 2i + j - 2k$
- **3** $a = 2i + 2j$, $b = 3i - 4j + 2k$ **4** $a = -i + 2j - k$, $b = 3j - 4k$

5-8. Find the equation of the plane that contains the three points.

- **5** $O(0, 0, 0)$, $A(2, -1, 1)$, $B(1, 1, 2)$
- **6** $A(2, -1, 1)$, $B(-3, 2, 0)$, $C(4, -5, 3)$
- **7** $A(-1, 0, 2)$, $B(2, 1, -1)$, $C(1, -2, 2)$
- **8** $A(-2, 1, -4)$, $B(-3, 2, 1)$, $C(-5, 3, 2)$

9-12. Find the area of the triangle that has the points from Exercises 5-8 as its vertices.

- **9**   Exercise 5   **10**   Exercise 6
- **11**   Exercise 7   **12**   Exercise 8

13   Prove that $v_1 \times v_2$ is orthogonal to v_2. (This completes the proof of Theorem 14-10.)

14 Use the vector product to work Exercise 19 of Section 14–6.

•15 Find the equations of the line that is perpendicular to the two lines
$$\frac{x-1}{2} = \frac{y+1}{3} = \frac{z-2}{-1} \quad \text{and} \quad \frac{x-3}{3} = \frac{y-2}{1} = \frac{z-1}{2}$$
and passes through their point of intersection.

•16 Find the line in the plane $2x + 3y - 4z = 1$ that is perpendicular to the line $x = -1 + 2t$, $y = 2 - t$, $z = t$.

17(a)–(h) Prove parts (a)–(h) of Theorem 14–11.

18 (*Volume of a Parallelepiped*) Let a, b, and c be distinct nonzero vectors that originate at the origin. Show that $|a \cdot (b \times c)|$ is the volume of the parallelepiped that has the three vectors as adjacent sides.

Explain why this result implies that
$$|a \cdot (b \times c)| = |b \cdot (a \times c)| = |c \cdot (a \times b)|.$$

19–20. Find the volume of the parallelepiped determined by the three vectors. (See Exercise 18).

•19 $a = -2i + j - 4k$, $b = -3i + 2j + k$, $c = -5i + 3j + 2k$

20 $a = i - j + 3k$, $b = 2i - 3j + 2k$, $c = 3i - 4j + k$

21 Let $v_1 = a_1 i + b_1 j + c_1 k$, $v_2 = a_2 i + b_2 j + c_2 k$, and $v_3 = a_3 i + b_3 j + c_3 k$. Prove that
$$v_1 \cdot (v_2 \times v_3) = \begin{vmatrix} a_1 & b_1 & c_1 \\ a_2 & b_2 & c_2 \\ a_3 & b_3 & c_3 \end{vmatrix}.$$

14–8 Some Problems Involving Vectors

We now consider several geometrical and algebraic problems that involve vectors. In most cases only the general outlines of the solutions are given. The computational details are left for you to work out.

As you will have noted by this time, perspective drawings leave much to be desired. It is very difficult to determine relationships between three-dimensional vectors, for example, unless the viewing angle is chosen so as to emphasize these relations. Unfortunately, this approach loses the relationships that the vectors have with the coordinate system.

Even with a carefully chosen viewing angle, many problems exist because of the built-in distortion of the drawing. For this reason you are advised to construct geometrical models in space to illustrate the relations. These models do not have to be elaborate—nor even very accurate. A table top or the side of a book can represent a plane, and colored pencils can represent vectors.

LINES OF INTERSECTIONS OF PLANES

Two nonparallel planes intersect in a line. Parametric equations of this line can be obtained quite easily from the equations of the two planes.

Our first problem is to find a point on the line. This can be done by arbitrarily choosing a value of x, y, or z, substituting this value in the equations of the two planes and then solving the resulting equations simultaneously.

We need to find, in addition to a point on the line, a vector parallel to the line. Such a vector can be found if we note that the line of intersection of the planes is perpendicular to normal vectors to the planes. (See Fig. 14–35.) If n_1 and n_2 are these normal vectors, then $v = n_1 \times n_2$ is parallel to the line.

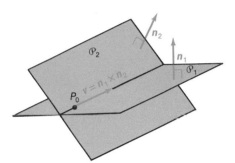

Figure 14–35 Example 1.

EXAMPLE 1 Find the line of intersection of the planes $2x - 3y + 5z = 3$ and $4x + 2y - 7z = 7$.

Solution To find a point on the line of intersection, we choose a convenient value of one of the variables, substitute this value in the equations of the planes, and solve the resulting equations simultaneously. If we choose $x = 1$, then

$$2 - 3y + 5z = 3 \quad \text{and} \quad 4 + 2y - 7z = 7,$$

so that

$$\begin{cases} -3y + 5z = 1 \\ 2y - 7z = 3. \end{cases}$$

The solution of this system is $y = -2$, $z = -1$. The point is $P_0(1, -2, -1)$.

The normal vectors to the given planes are $n_1 = 2i - 3j + 5k$ and $n_2 = 4i + 2j - 7k$. Let

$$v = n_1 \times n_2 = \begin{vmatrix} i & j & k \\ 2 & -3 & 5 \\ 4 & 2 & -7 \end{vmatrix} = i \begin{vmatrix} -3 & 5 \\ 2 & -7 \end{vmatrix} - j \begin{vmatrix} 2 & 5 \\ 4 & -7 \end{vmatrix} + k \begin{vmatrix} 2 & -3 \\ 4 & 2 \end{vmatrix}$$

$$= 11i + 34j + 16k.$$

Since v is perpendicular to both normal vectors, it is parallel to their line of intersection. Parametric equations of the line are

$$x = 1 + 11t, \qquad y = -2 + 34t, \qquad z = -1 + 16t. \quad \square$$

14-8 Some Problems Involving Vectors

ANGLE BETWEEN TWO PLANES

Angle Between Planes Let $\mathcal{P}_1$ and $\mathcal{P}_2$ be intersecting planes. The angle α, $0 \leq \alpha \leq \pi/2$ is defined to be the *angle between* $\mathcal{P}_1$ *and* $\mathcal{P}_2$ if there exist normal vectors n_1 and n_2 such that α is the angle between n_1 and n_2. (See Fig. 14-36.)

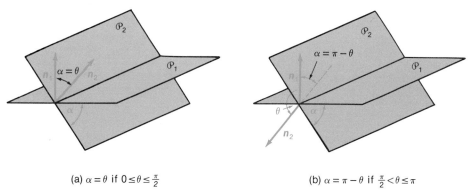

(a) $\alpha = \theta$ if $0 \leq \theta \leq \frac{\pi}{2}$ (b) $\alpha = \pi - \theta$ if $\frac{\pi}{2} < \theta \leq \pi$

Figure 14-36 The angle between two planes.

A certain amount of care must be exercised when we calculate the angle between $\mathcal{P}_1$ and $\mathcal{P}_2$ from given normal vectors n_1 and n_2. Let θ be the angle between n_1 and n_2. If $0 \leq \theta \leq \pi/2$, then $\alpha = \theta$. (See Fig. 14-36a.) If $\pi/2 < \theta \leq \pi$, however, then $\alpha = \pi - \theta$. In this case, α is the angle between the normal vectors n_1 and $-n_2$. In either case,

$$\cos \alpha = |\cos \theta| = \frac{|n_1 \cdot n_2|}{\|n_1\| \, \|n_2\|}.$$

EXAMPLE 2 Find the angle between the planes $2x - 3y + 5z = 3$ and $4x + 2y - 7z = 7$.

Solution Two normal vectors are

$$n_1 = 2i - 3j + 5k \quad \text{and} \quad n_2 = 4i + 2j - 7k.$$

The angle between the normal vectors is θ where

$$\cos \theta = \frac{n_1 \cdot n_2}{\|n_1\| \cdot \|n_2\|} = \frac{2 \cdot 4 - 3 \cdot 2 - 5 \cdot 7}{\sqrt{4 + 9 + 25} \, \sqrt{16 + 4 + 49}} = \frac{-33}{\sqrt{38} \, \sqrt{69}}.$$

The angle between the planes is α, where

$$\cos \alpha = |\cos \theta| = \frac{33}{\sqrt{38} \, \sqrt{69}} \approx 0.64446.$$

Then

$$\alpha \approx \cos^{-1}(0.64446) \approx 0.87048 \text{ radian}$$

$$\approx 49°52.5'. \quad \square$$

PROJECTION OF A VECTOR ON A PLANE

It is occasionally convenient to project a vector v onto a plane $\mathcal{P}$. The situation is pictured in Figure 14–37. The vector

$$p = \overrightarrow{PP_1}$$

is the projection of $v = \overrightarrow{PP_0}$ on the plane $\mathcal{P}$. (Point P is on the plane, angle PP_1P_0 is a right angle.)

Let n be a vector normal to the plane $\mathcal{P}$ and let n_0 be the projection of v on n. Observe in Figure 14–37 that $v = p + n_0$, so that

$$p = v - n_0 = v - \text{proj}_n v.$$

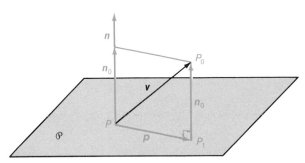

Figure 14–37 The projection of a vector on a plane.

Thus

Formula for Projection of Vector on a Plane

$$\text{Projection of } v \text{ on } \mathcal{P} = v - \text{proj}_n v.$$

EXAMPLE 3 Find the projection of $v = 2i + j + 3k$ on the plane $5x + 2y - 3z = 7$.

Solution A vector normal to the plane is $n = 5i + 2j - 3k$. The projection of v on n is

$$\text{proj}_n v = \frac{v \cdot n}{n \cdot n} n = \frac{10 + 2 - 9}{25 + 4 + 9}(5i + 2j - 3k)$$

$$= \frac{3}{38}(5i + 2j - 3k) = \frac{15i}{38} + \frac{6j}{38} - \frac{9k}{38}.$$

The projection of v on $\mathcal{P}$ is

$$p = v - \text{proj}_n v = (2i + j + 3k) - \left(\frac{15i}{38} + \frac{6j}{38} - \frac{9k}{38}\right)$$

$$= \frac{61i}{38} + \frac{32j}{38} + \frac{123k}{38}. \quad \square$$

Projection of a Point on a Plane

The point P_1 in Figure 14–37 is called the *projection of point P_0 on the plane $\mathcal{P}$*. The formula for the projection of v on $\mathcal{P}$ can be used to derive a comparable formula for the projection of the point P_0 on $\mathcal{P}$. (See Exercise 17.)

14–8 Some Problems Involving Vectors

ORTHOGONAL UNIT VECTORS

Let I, J, K be orthogonal unit vectors that form a right-handed system. That is, $K = I \times J$. We shall show that the simple formulas for the dot product and the cross product hold for vectors written in terms of I, J, K as well as for those written in terms of i, j, k. In symbols, if $v_1 = a_1 I + b_1 J + c_1 K$ and $v_2 = a_2 I + b_2 J + c_2 K$, then

$$v_1 \cdot v_2 = a_1 a_2 + b_1 b_2 + c_1 c_2$$

and

$$v_1 \times v_2 = \begin{vmatrix} I & J & K \\ a_1 & b_1 & c_1 \\ a_2 & b_2 & c_2 \end{vmatrix}.$$

We first observe that since the vectors have unit length, then

$$I \cdot I = J \cdot J = K \cdot K = 1.$$

Also, since they are mutually orthogonal, then

$$I \cdot J = I \cdot K = J \cdot K = 0.$$

These observations simplify our later work considerably.

The dot product of two vectors

$$v_1 = a_1 I + b_1 J + c_1 K \quad \text{and} \quad v_2 = a_2 I + b_2 J + c_2 K$$

can now be calculated by Theorem 14–7. If we expand

$$v_1 \cdot v_2 = (a_1 I + b_1 J + c_1 K) \cdot (a_2 I + b_2 J + c_2 K)$$

by the distributive property, use property (c) of Theorem 14–7, and simplify, we get

$$v_1 \cdot v_2 = (a_1 a_2)(I \cdot I) + b_1 b_2 (J \cdot J) + c_1 c_2 (K \cdot K)$$
$$= a_1 a_2 + b_1 b_2 + c_1 c_2,$$

which is equivalent to our original formula for the dot product.

The cross product can be handled similarly. Observe first that by property (e) of Theorem 14–11,

$$J \times K = J \times (I \times J) = (J \cdot J)I - (J \cdot I)J$$
$$= 1 \cdot I - 0 \cdot J = I.$$

Similarly, $K \times I = J$ and $I \times J = K$.

It follows from Theorems 14–10 and 14–11 that

$$I \times I = J \times J = K \times K = 0$$

and

$$J \times I = -K, \quad K \times J = -I, \quad I \times K = -J.$$

If we now expand the expression

$$v_1 \times v_2 = (a_1 I + b_1 J + c_1 K) \times (a_2 I + b_2 J + c_2 K),$$

using the distributive law and property (g) of Theorem 14–11, and simplify, we get

$$v_1 \times v_2 = I(b_1 c_2 - c_1 b_2) - J(a_1 c_2 - a_2 c_1) + K(a_1 b_2 - a_2 b_1)$$

which can be written in determinant form as

$$\begin{vmatrix} I & J & K \\ a_1 & b_1 & c_1 \\ a_2 & b_2 & c_2 \end{vmatrix}.$$

These formulas allow us to calculate the dot and cross products by simple formulas without expressing the vectors I, J, K in terms of the unit vectors i, j, k.

EXAMPLE 4   Let

$$I = \frac{1}{\sqrt{2}} i - \frac{1}{\sqrt{2}} j, \quad J = \frac{1}{2} i + \frac{1}{2} j + \frac{1}{\sqrt{2}} k, \quad K = -\frac{1}{2} i - \frac{1}{2} j + \frac{1}{\sqrt{2}} k.$$

Then I, J, K are mutually orthogonal unit vectors that form a right-handed system. Let

$$v_1 = 2I + 3J - K \quad \text{and} \quad v_2 = I + J - 2K.$$

Then

$$v_1 \cdot v_2 = 2 \cdot 1 + 3 \cdot 1 - 1(-2) = 7$$

and

$$v_1 \times v_2 = \begin{vmatrix} I & J & K \\ 2 & 3 & -1 \\ 1 & 1 & -2 \end{vmatrix} = -5I + 3J - K. \quad \square$$

Exercises 14–8

1–4. Find symmetric or parametric equations for the line of intersection of the planes.

●1 $2x - y + 3z + 10 = 0, \ x - 2y + 4z + 20 = 0$

2 $x + 2z = 6, \ 2x + 3y + z = 3$

●3 $x + 2y - 2z = 3, \ 2x + y - 2z = 0$

4 $2x + y - 5z + 1 = 0, \ 3x - 3y + 3z - 10 = 0$

5–8. Find the angle of intersection of the planes in Exercises 1–4.

●5   Exercise 1   6   Exercise 2

●7   Exercise 3   8   Exercise 4

Exercises 14–8

9–10. Show that the lines intersect. Then find the angle between them.

● 9 $x = 5 - 3t,\ y = -2 + t,\ z = 1 + 9t;\ x = 6 + 2t,\ y = -9 - 4t,\ z = 8 - t$

10   $x = 7 - 2t,\ y = 4 + 3t,\ z = 5t;\ x = -1 + t,\ y = 5 + 4t,\ z = 13 + t$

11–12. Find the projection of $\overrightarrow{AB}$ on the plane.

● 11 $A(1, 1, 3),\ B(3, 2, 3),\ x + y - 2z + 4 = 0$

12   $A(0, 1, 2),\ B(-2, 1, 3),\ 3x - y - z + 3 = 0$

● 13   Find the point of intersection of the plane $x + 2y - z = 18$ with the line through $(2, -1, 0)$ that is normal to the plane.

● 14   Find parametric equations of the line through $(2, 4, -5)$ that lies in the plane $3x + 2y - z = 19$ and is orthogonal to the line $x = 4 + 2t,\ y = 9 - 4t,\ z = 11 + t$.

15   Calculate $v_1 \cdot v_2$ and $v_1 \times v_2$ for the vectors v_1 and v_2 in Example 4, after expressing v_1 and v_2 in terms of i, j, and k.

16   Let I, J, and K be mutually orthogonal unit vectors with $K = I \times J$. Let $v_1 = a_1 I + b_1 J + c_1 K$ and $v_2 = a_2 I + b_2 J + c_2 K$. Work through the details of the arguments in the text that establish the following properties.
  (a) $I \cdot I = J \cdot J = K \cdot K = 1$ and $I \cdot J = J \cdot K = K \cdot I = 0$
  (b) $J \times K = I$ and $K \times I = J$
  (c) $v_1 \cdot v_2 = a_1 a_2 + b_1 b_2 + c_1 c_2$
  (d) $v_1 \times v_2 = \begin{vmatrix} I & J & K \\ a_1 & b_1 & c_1 \\ a_2 & b_2 & c_2 \end{vmatrix}$.

17   (a) Let $P_0(x_0, y_0, z_0)$ be a point. Let $P_1(x_1, y_1, z_1)$ be the projection of P_0 on the plane $\mathscr{P}$ defined by $ax + by + cz + d = 0$. Prove that
$$x_1 = x_0 - aD, \quad y_1 = y_0 - bD, \quad z_1 = z_0 - cD$$
where
$$D = \frac{ax_0 + by_0 + cz_0 + d}{a^2 + b^2 + c^2}$$

(*Hint:* Let P be a point on the plane $\mathscr{P}$. The vector $\overrightarrow{PP_1}$ is the projection of $\overrightarrow{PP_0}$ on the plane $\mathscr{P}$.)
  ● (b) Use the result of (a) to find the projection of $(5, 5, 4)$ on the plane $2x + 2y + z = 6$. Illustrate with a sketch.

18   (*Distance between Two Lines*) (a) Show that the shortest distance between two nonintersecting, nonparallel lines in space is measured along a line that is mutually perpendicular to the given lines.
  (b) Let d be the shortest distance between the nonintersecting lines $\mathscr{L}_1$ and $\mathscr{L}_2$. Let n be a nonzero vector that is orthogonal to both $\mathscr{L}_1$ and $\mathscr{L}_2$. Let P_1 and P_2 be points on $\mathscr{L}_1$ and $\mathscr{L}_2$, respectively. Prove that
$$d = |\text{comp}_n\ \overrightarrow{P_1 P_2}| = \|\text{proj}_n\ \overrightarrow{P_1 P_2}\|.$$

19–21. Show that the lines are not parallel and do not intersect. Find the minimum distance between them. (See Exercise 18.)

•19 $\dfrac{x-2}{-3} = \dfrac{y-2}{2} = \dfrac{z+1}{3}; \dfrac{x-4}{-1} = \dfrac{y}{4} = \dfrac{z-5}{-5}$

20 $x = 3 + t, y = 2 - 2t, z = t; x = 4 - v, y = 3 + v, z = -2 + 3v$

•21 $x = 2 + t, y = 1 - 3t, z = -1 + t; x = 1 + t, y = -t, z = 2 - 3t$

22 Prove that the medians of a triangle intersect at a point two-thirds of the distance from each vertex to the midpoint of the opposite side. (*Hint:* Let A, B, C be the vertices of the triangle; let A', B', C' be the midpoints of the sides opposite A, B, C. Let Q be the point two-thirds of the distance from A to A'. Let P be an arbitrary point in the plane. Prove that

$$\overrightarrow{PQ} = \dfrac{\overrightarrow{PA} + \overrightarrow{PB} + \overrightarrow{PC}}{3}.$$

Explain why this result implies that the medians intersect at Q.

14–9 Cylinders and Space Curves

Surfaces in three-dimensional space are defined by equations of form $z = f(x, y)$ or $F(x, y, z) = 0$; and space curves are defined by parametric equations $x = f(t), y = g(t), z = h(t)$. A familiarity with the graphs of such equations is needed before we can work with such surfaces and curves.

CYLINDERS

Cylinder

Let $\mathscr{C}$ be a curve in a plane and $\mathscr{L}$ a line that is not parallel to that plane. The union of all lines parallel to $\mathscr{L}$ that intersect $\mathscr{C}$ is called a *cylinder*. (See Fig. 14–38.) We can construct the cylinder by moving a line parallel to $\mathscr{L}$ around the curve $\mathscr{C}$.

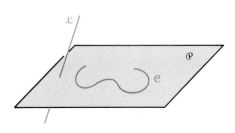

(a) $\mathscr{C}$ is a curve in the plane $\mathscr{P}$; $\mathscr{L}$ is the generating line for the cylinder

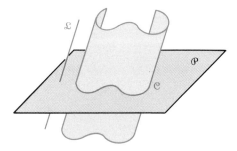

(b) The cylinder is traced out by moving the line $\mathscr{L}$ around the curve $\mathscr{C}$

Figure 14–38 A cylinder.

14-9 Cylinders and Space Curves

Generating Line
Ruling
Right Cylinder

The line $\mathscr{L}$ is called a *generating line* for the cylinder. The lines through $\mathscr{C}$ parallel to $\mathscr{L}$ are called *rulings* of the cylinder.* The curve $\mathscr{C}$ is called the *trace* of the cylinder in the original plane. If $\mathscr{L}$ is perpendicular to the original plane, the cylinder is a *right cylinder*.

We will show that the graph in three-dimensional space of an equation in two variables is a right cylinder. Consider the graph of the equation $F(x, y) = 0$ in three-dimensional space. The trace of this graph in the xy-plane is the curve $\mathscr{C}$ that is defined by the equations

$$\begin{cases} F(x, y) = 0 \\ z = 0. \end{cases}$$

(See Fig. 14–39.) A point is on the curve $\mathscr{C}$ if and only if its x- and y-coordinates satisfy the original equation and its z-coordinate is zero.

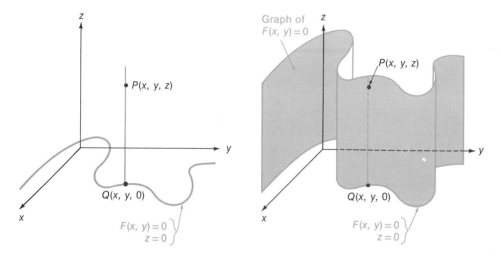

(a) The trace in the xy-plane

(b) The graph of $F(x, y) = 0$

Figure 14–39 The graph of $F(x, y) = 0$ is a cylinder with generating line parallel to the z axis.

Consider an arbitrary point $P(x, y, z)$ in space. It is on the graph of the equation $F(x, y) = 0$ if and only if its x- and y-coordinates satisfy the equation. Thus, it is on the graph if and only if a corresponding point $Q(x, y, 0)$ exists on the curve $\mathscr{C}$. (See Fig. 14–39.) In other words, the graph of $F(x, y) = 0$ in three-dimensional space consists of all points directly above (or on or below) the curve $\mathscr{C}$ in the xy-plane. It follows that the graph of the equation is a cylinder. The curve $\mathscr{C}$ is its trace in the xy-plane and the generating line is parallel to the z-axis. (Note that z is the missing variable in the equation.)

Similarly, the graph of $F(x, z) = 0$ is a right cylinder with generating line parallel to the y-axis, and the graph of $F(y, z) = 0$ is a right cylinder with generating line parallel to the x-axis.

---

* The concept of a ruling is more general than we have indicated. Any straight line that lies on a surface is called a *ruling* of that surface.

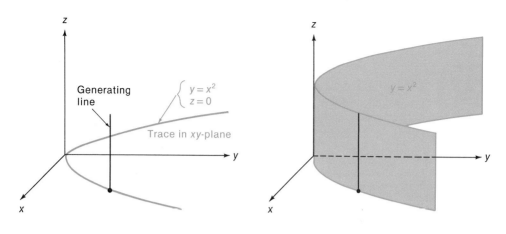

(a) Trace in the xy-plane   (b) The graph is a cylinder

Figure 14–40 Example 1. Graph of $y = x^2$.

EXAMPLE 1   The graph of $y = x^2$ is a right cylinder. The trace in the xy-plane is the parabola $y = x^2$, $z = 0$. The generating line is parallel to the z-axis (Fig. 14–40). ☐

EXAMPLE 2   The graph of $x^2 + z^2 = 1$ is a right cylinder. The unit circle $x^2 + z^2 = 1$, $y = 0$, is the trace in the xz-plane. The generating line is parallel to the y-axis (Fig. 14–41). ☐

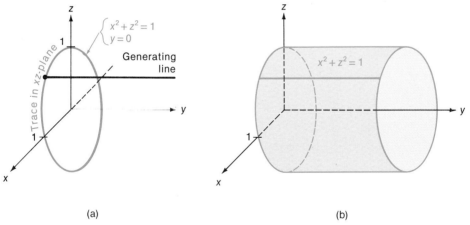

(a)   (b)

Figure 14–41 Example 2. Graph of $x^2 + z^2 = 1$.

SPACE CURVES

The graph of the parametric equations

$$x = f(t), \quad y = g(t), \quad z = h(t),$$

Space Curve   where f, g, and h are continuous, is called a *space curve*. If f, g, and h also have continuous first derivatives, which are not simultaneously zero, the graph is called a
Smooth Curve   *smooth curve*.

A typical space curve, such as the one in Figure 14–42, looks as if it had been

14-9 Cylinders and Space Curves

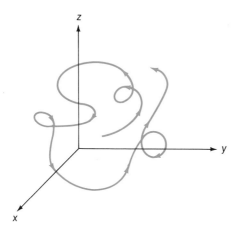

Figure 14-42 A space curve.

traced out by a fly buzzing through the air. As with plane curves, it will be convenient to think of this curve as traced out by a moving particle. The point $(f(t), g(t), h(t))$ represents the location of the particle at time t.

Process for Drawing a Space Curve on a Cylinder

Without some type of guidelines it may be difficult to make an intelligible drawing of a space curve. We will show that each space curve lies on a cylinder. If the cylinder has a simple enough shape, it may be possible to draw the curve on the cylinder in such a way that its properties are apparent.

Our first step is to project the graph of $x = f(t)$, $y = g(t)$, $z = h(t)$ onto the xy-plane. (See Fig. 14-43a.) Each point $P(f(t), g(t), h(t))$ on the curve is projected onto a corresponding point $Q(f(t), g(t), 0)$. Thus, the projection of the original space curve is a curve $\mathcal{C}$ in the xy-plane.

Since each point $P(f(t), g(t), h(t))$ is on a vertical line through $\mathcal{C}$, then the original space curve is on the right cylinder obtained by moving a line parallel to the z-axis

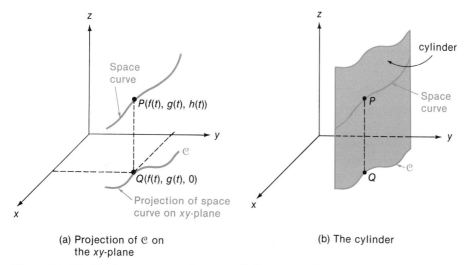

(a) Projection of $\mathcal{C}$ on the xy-plane

(b) The cylinder

Figure 14-43 Each space curve is on a cylinder perpendicular to the xy plane.

around 𝒞. (See Fig. 14–43b.) This cylinder can help to outline the basic shape of the space curve.

EXAMPLE 3   Sketch the graph of the space curve defined by

$$x = t, \quad y = t^2, \quad z = \frac{t}{4} \quad \text{for } t \geq 0.$$

Solution   The projection of the curve on the *xy*-plane is defined by

$$x = t, \quad y = t^2, \quad z = 0 \quad (t \geq 0)$$

Thus, the projection is the portion of the parabola shown in Figure 14–44a. The space curve lies on the parabolic cylinder obtained by moving a line parallel to the *z*-axis around this projection. (See Fig. 14–44b.) Observe that the particle that traces

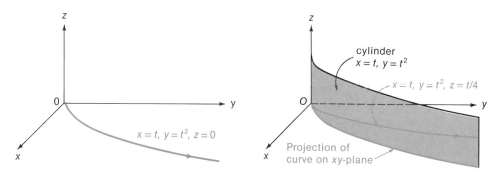

(a) Projection of the space curve on the *xy*-plane

(b) The cylinder and the space curve

Figure 14–44   Example 3. Graph of $x = t$, $y = t^2$, $z = t/4$, $t \geq 0$.

out the space curve is at the origin when $t = 0$ and rises as it moves around the cylinder. ☐

EXAMPLE 4   (*The Helix*) Sketch the graph of the space curve defined by

Helix $x = a \cos t, \quad y = a \sin t, \quad z = bt$

where $a, b > 0$.

Solution   The projection on the *xy*-plane is the graph of

$$x = a \cos t, \quad y = a \sin t, \quad z = 0.$$

Since

$$x^2 + y^2 = a^2(\cos^2 t + \sin^2 t) = a^2,$$

the projection is a circle with center at the origin and radius a. It follows that the curve lies on the right circular cylinder $x^2 + y^2 = a^2$. (See Fig. 14–45.)

At time $t = 0$ the particle that traces out the curve is at the point $(a, 0, 0)$. As t increases, it moves up and around the cylinder tracing out a space curve called a *helix*. ☐

14-10 Level Curves. Quadric Surfaces

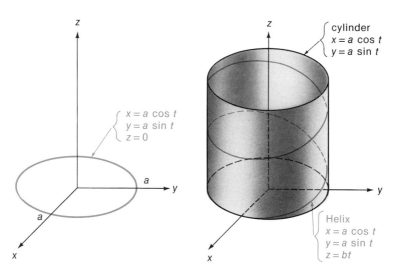

(a) The projection on the xy-plane   (b) The helix

Figure 14-45 Example 4. The helix $x = a \cos t$, $y = a \sin t$, $z = bt$.

Exercises 14-9

1-8. Find the axis that is parallel to the generating line of the cylinder. Sketch the graph.

- 1. $x^2 + y^2 = 4$
- 2. $4x^2 + 9y^2 = 36$
- 3. $z^2 - y = 0$
- 4. $y^2 - x^2 = 9$
- 5. $x^2 = 2z$
- 6. $x + y^2 = 0$
- 7. $4z^2 - 16y^2 = -16$
- 8. $x^2 + y^2 - 2x - 4y + 4 = 0$

9-14. Sketch the graph of the space curve. Show a cylinder that has the curve on its surface.

- 9. $x = \cos t$, $y = \sin t$, $z = t$, $t \geq 0$
- 10. $x = \sec t$, $y = \tan t$, $z = -1$, $-\pi/2 < t < \pi/2$
- 11. $x = t^2 + 1$, $y = t$, $z = 2$
- 12. $x = \cos 2t$, $y = 2$, $z = \sin t$, $0 \leq t \leq \pi$
- 13. $x = t$, $y = t^2$, $z = 3t$, $t \geq 0$
- 14. $x = 4 \cos t$, $y = 9 \sin t$, $z = t$, $t \geq 0$

14-10 Level Curves. Quadric Surfaces

After we examine some general methods for graphing surfaces in three-dimensional space, we make a small catalogue of some special surfaces. In particular, we consider *quadric surfaces*—the graphs of second-degree equations in x, y, and z. These surfaces are the three-dimensional counterparts of the conic sections.

TRACES

Many surfaces can be sketched easily after drawing the traces in the coordinate planes or planes parallel to the coordinate planes.

Trace in Plane Parallel to a Coordinate Plane

If a surface is the graph of the equation $F(x, y, z) = 0$, then its trace in the horizontal plane $z = k$ is defined by the system of equations

$$F(x, y, z) = 0, \quad z = k,$$

which is equivalent to the system

$$F(x, y, k) = 0, \quad z = k.$$

Similarly, the traces in the planes $x = k$ and $y = k$ are defined by the systems of equations

$$F(k, y, z) = 0, \quad x = k$$

and

$$F(x, k, z) = 0, \quad y = k,$$

respectively.

If a surface does not intersect a plane, then it has no trace in that plane. Consequently, the system of equations will have no real solution.

LEVEL CURVES

Perspective drawings of surfaces, such as the one in Figure 14–46a, picture the basic shapes but do not give quantitative relations. There is an alternative method of graphing surfaces that can be used for quantitative relations; unfortunately we lose the advantages of a perspective drawing.

Level (Contour) Curves

Level curves (*contour curves*) are the projections of traces in planes parallel to the *xy*-plane down onto the *xy*-plane. This is illustrated in Figures 14–46a, b. In (a) we see the trace in the plane $z = k$. In (a) and (b) we see that trace projected down on the *xy*-plane. In Figure 14–46c we show several level curves. Such a drawing is called a *contour drawing* of the surface.

The level curve obtained by projecting the trace in the plane $z = k$ onto the *xy*-plane is denoted by $\mathscr{C}_k$. Thus, if the original surface is the graph of $z = f(x, y)$, then

$$\mathscr{C}_k = \{(x, y, 0) \mid f(x, y) = k\}.$$

A point $P(x, y, 0)$ is on $\mathscr{C}_k$ if and only if there is a corresponding point $Q(x, y, k)$ on the surface.

Note that level curves are drawn in the *xy*-plane with no reference to the *z*-axis. Thus, we use the standard orientation of axes with the *x*-axis horizontal and the *y*-axis vertical.

14-10 Level Curves. Quadric Surfaces

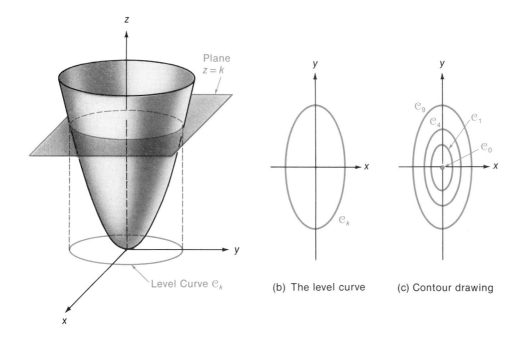

(a) The level curve $\mathcal{C}_k$ is the projection of the trace in the plane $z = k$ on the xy-plane

(b) The level curve

(c) Contour drawing

Figure 14-46 Level Curves

A standard topographic map is shown in Figure 14–47a and a perspective drawing of the same region in (b). Level curves are used to represent the heights of points above sea level.

QUADRIC SURFACES

The *quadric surfaces* are the graphs of second-degree equations in x, y, and z. These surfaces are the generalizations of the conic sections.

In this section we make a brief catalogue of quadric surfaces, and give one example of each type. Each example will have a particular orientation to the axis system. An interchange of the variables in the defining equation would produce a similar surface with a different orientation. For example, our first quadric surface is the graph of $z = x^2/a^2 + y^2/b^2$, a "cuplike" surface that opens upward. (See Fig. 14–48.) The graphs of $x = y^2/a^2 + z^2/b^2$ and $y = x^2/a^2 + z^2/b^2$ are similar but open along other axes. Similar results hold if the variables are interchanged in the equations for the other quadric surfaces.

We consider only one standard form of the equation of each quadric surface. Translations of axes could reduce other equations to these standard forms. The equation

$$z - C = \frac{(x - A)^2}{a^2} + \frac{(y - B)^2}{b^2},$$

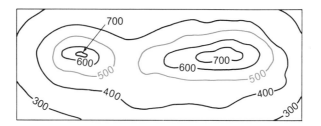

(a) Topographic map

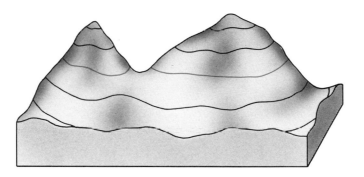

(b) Perspective drawing

Figure 14–47.

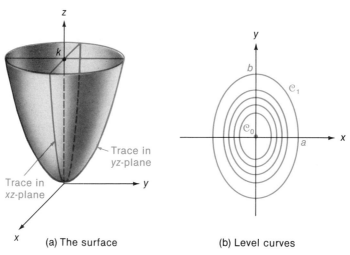

(a) The surface

(b) Level curves

Figure 14–48 The elliptical paraboloid $z = \dfrac{x^2}{a^2} + \dfrac{y^2}{b^2}$.

14-10 Level Curves. Quadric Surfaces

for example, could be reduced to the equation $z' = x'^2/a^2 + y'^2/b^2$ by a translation of axes.

In all our examples the constants a, b, and c are positive.

THE ELLIPTICAL PARABOLOID

The graph of the equation

$$z = \frac{x^2}{a^2} + \frac{y^2}{b^2}$$

can be sketched from its traces (see Fig. 14–48a):

(1) *The trace in the yz-plane* is the graph of $z = y^2/b^2$, $x = 0$. This curve is a parabola that opens upward.

(2) *The trace in the xz-plane* is the graph of $z = x^2/a^2$, $y = 0$. This curve also is a parabola that opens upward.

(3) *The trace in the plane $z = k$* is the graph of

$$\frac{x^2}{a^2} + \frac{y^2}{b^2} = k, \; z = k.$$

If $k = 0$ (the cutting plane is the xy-plane), the trace is a single point. If $k > 0$ (the plane is above the xy-plane), the trace is an ellipse (provided $a \neq b$). If $k < 0$ (the plane is below the xy-plane), there is no trace—the surface does not extend below the xy-plane.

The basic shape of the surface is determined by the parabolas of intersection with the xz- and yz-planes. The cross sections parallel to the xy-plane are ellipses. This surface is called an *elliptical paraboloid*.

Several level curves for the surface are shown in Figure 14–48b. These curves are similar ellipses.

THE ELLIPSOID

The graph of

$$\frac{x^2}{a^2} + \frac{y^2}{b^2} + \frac{z^2}{c^2} = 1$$

is shown in Figure 14–49. This graph can be determined from its traces in planes parallel to the coordinate planes.

The trace in the plane $z = k$ (parallel to the xy-plane) is the graph of

$$\frac{x^2}{a^2} + \frac{y^2}{b^2} = \frac{c^2 - k^2}{c^2}, \quad z = k.$$

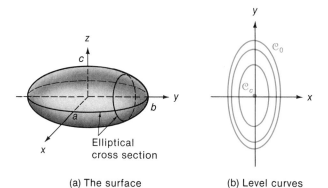

(a) The surface   (b) Level curves

Figure 14-49 The ellipsoid $\dfrac{x^2}{a^2} + \dfrac{y^2}{b^2} + \dfrac{z^2}{c^2} = 1$.

This trace is an ellipse if $|k| < c$ and a single point if $|k| = c$. There is no curve of intersection if $|k| > c$. Thus, the surface is contained between the planes $z = -c$ and $z = c$. It follows similarly that the surface is contained in the region defined by

$$-a \leq x \leq a, \quad -b \leq y \leq b, \quad -c \leq z \leq c.$$

The cross section in any plane parallel to a coordinate plane is an ellipse, a single point, or the empty set. The graph is called an *ellipsoid*.

THE ELLIPTICAL CONE

The graph of

$$\frac{x^2}{a^2} + \frac{y^2}{b^2} - \frac{z^2}{c^2} = 0$$

is the cone pictured in Figure 14-50.

The traces in the xz- and yz-planes are the lines

$$\frac{z}{c} = \pm \frac{x}{a} \quad \text{and} \quad \frac{z}{c} = \pm \frac{y}{b},$$

respectively. The trace in the plane $z = k$ (parallel to the xy-plane) is the ellipse

$$\frac{x^2}{a^2} + \frac{y^2}{b^2} = \frac{k^2}{c^2}, \quad z = k.$$

In the special case where $k = 0$ we get the single point $(0, 0, 0)$, a degenerate ellipse.

14-10 Level Curves. Quadric Surfaces

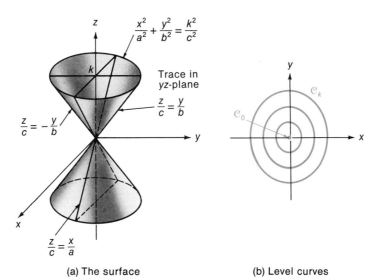

(a) The surface (b) Level curves

Figure 14–50 The elliptical cone $\dfrac{x^2}{a^2} + \dfrac{y^2}{b^2} - \dfrac{z^2}{c^2} = 0.$

THE HYPERBOLOID OF ONE SHEET

The graph of

$$\frac{x^2}{a^2} - \frac{y^2}{b^2} + \frac{z^2}{c^2} = 1$$

is pictured in Figure 14–51.

The traces in the xy- and yz-planes are the hyperbolas

$$\frac{x^2}{a^2} - \frac{y^2}{b^2} = 1, \quad z = 0 \quad \text{(opens along } x\text{-axis)}$$

and

$$-\frac{y^2}{b^2} + \frac{z^2}{c^2} = 1, \quad x = 0 \quad \text{(opens along } z\text{-axis),}$$

respectively. The trace in the plane $y = k$ (parallel to the xz-plane) is the ellipse

$$\frac{x^2}{a^2} + \frac{z^2}{c^2} = \frac{b^2 + k^2}{b^2}.$$

The surface is an *elliptical hyperboloid of one sheet*.

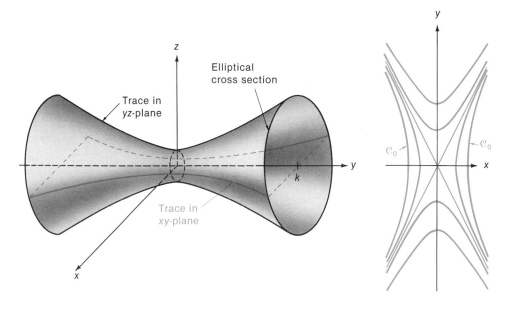

(a) The surface   (b) Level curves

Figure 14–51 The elliptical hyperboloid of one sheet $\frac{x^2}{a^2} - \frac{y^2}{b^2} + \frac{z^2}{c^2} = 1$.

THE HYPERBOLOID OF TWO SHEETS

The graph of

$$-\frac{x^2}{a^2} + \frac{y^2}{b^2} - \frac{z^2}{c^2} = 1$$

is shown in Figure 14–52.

The traces in the xy- and yz-planes are the hyperbolas

$$-\frac{x^2}{a^2} + \frac{y^2}{b^2} = 1, \quad z = 0 \quad \text{(opens along } y\text{-axis)}$$

and

$$\frac{y^2}{b^2} - \frac{z^2}{c^2} = 1, \quad x = 0 \quad \text{(opens along } y\text{-axis)},$$

respectively. The trace in the plane $y = k$ (parallel to xz-plane) is

$$\frac{x^2}{a^2} + \frac{z^2}{c^2} = \frac{k^2 - b^2}{b^2}, \quad y = k.$$

14-10 Level Curves. Quadric Surfaces

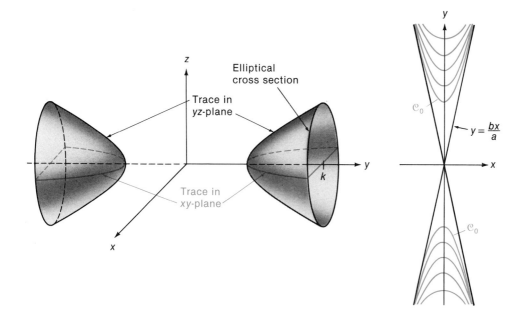

(a) The surface (b) Level curves

Figure 14-52 The elliptical hyperboloid of two sheets $-\dfrac{x^2}{a^2} + \dfrac{y^2}{b^2} - \dfrac{z^2}{c^2} = 1$.

This trace is an ellipse if $|k| > b$ and a single point if $|k| = b$. There is no curve of intersection if $|k| < b$.

The graph is an *elliptical hyperboloid of two sheets*.

THE HYPERBOLIC PARABOLOID

The graph of

$$z = -x^2/a^2 + y^2/b^2$$

is pictured in Figure 14–53. This surface will furnish us with counterexamples for some reasonable, but false, conjectures in Chapter 16.

The traces in the *xz*- and *yz*-planes are the parabolas

$$z = -\frac{x^2}{a^2}, \quad y = 0 \quad \text{(opens downward)}$$

and

$$z = \frac{y^2}{b^2}, \quad x = 0 \quad \text{(opens upward)},$$

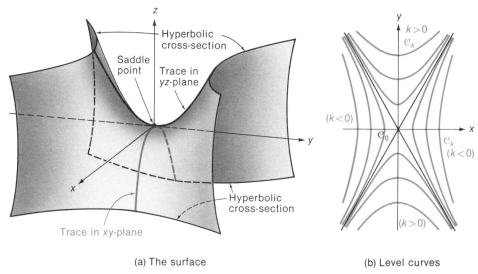

Figure 14–53 The hyperbolic paraboloid $z = -\dfrac{x^2}{a^2} + \dfrac{y^2}{b^2}$.

respectively. Traces in planes parallel to these coordinate planes also are parabolas. The trace in the plane $z = k$ (parallel to the xy-plane) is

$$-\frac{x^2}{a^2} + \frac{y^2}{b^2} = k, \qquad z = k.$$

This trace is:

(1) A hyperbola that opens along a line parallel to the y-axis if $k > 0$.
(2) A pair of intersecting lines if $k = 0$.
(3) A hyperbola that opens along a line parallel to the x-axis if $k < 0$.

Saddle Surface

The graph shown in Figure 14–53 is called a *hyperbolic paraboloid*. It is an example of a *saddle surface*. The origin is a *saddle point*. Observe that there are paths along the surface on which there is a local minimum at the origin and others on which there is a local maximum there.

Exercises 14–10

1–16. Find the traces in planes parallel to coordinate planes. Identify the surface by type. Make a perspective drawing that shows several of the traces and cross sections. Make a contour drawing that shows several level curves.

●1 $4x^2 + z^2 = y$ 2 $25x^2 + 100y^2 + 4z^2 = 100$

●3 $x^2 + y^2 - z^2 = 4$ 4 $25y^2 - x^2 - 4z^2 = 100$

●5 $4x^2 + 16y^2 - z^2 = 0$ 6 $x^2 + y^2 = z$

●7 $x^2 + 4y^2 + 9z^2 = 36$ 8 $25y^2 + 9z^2 = x$

- 9 $x^2 - 36z^2 = 4y^2$
- 10 $x^2 + 9y^2 - z^2 + 9 = 0$
- 11 $36x^2 - 4y^2 + 9z^2 = 0$
- 12 $4x = y^2 - z^2$
- 13 $x^2 - y^2 - z^2 = 9$
- 14 $9x^2 - y^2 + 4z^2 = 36$
- 15 $y = 4z^2 - 16x^2$
- 16 $x^2 - 64y^2 - 4z^2 + 64 = 0$

17 Show that the elliptical level curves $\mathscr{C}_h$, of the paraboloid $z = x^2/a^2 + y^2/b^2$ have the same eccentricity.

18 Show that the hyperbolic level curves $\mathscr{C}_h$ of the hyperboloid $x^2/a^2 - y^2/b^2 + z^2/c^2 = 1$ all have the same asymptotes. How are these asymptotes related to the surface?

Review Problems

1 Define the following concepts: (a) vector; (b) component of a vector; (c) magnitude of a vector.

- 2 Find a unit vector in the same direction as $2i - 3j + 4k$.

3–6. Use components to prove the following properties of vectors:

3 $m(u - v) = mu - mv$

4 $u + (v + w) = (u + v) + w$

5 $u \cdot (v + w) = u \cdot v + u \cdot w$

6 $u \times (v + w) = (u \times v) + (u \times w)$

7 What is meant by the component of a in the direction of b? the projection of a on b? Illustrate these concepts with a sketch. State the formulas for $\text{comp}_b a$ and $\text{proj}_b a$.

- 8 Calculate $\text{comp}_b a$ and $\text{proj}_b a$ for $a = 2i - 3j$ and $b = i + j + k$.

9 (a) Let v be a vector in the xy-plane. What is meant by the *direction angle* for v? Let u be a unit vector in the same direction as v. Express u in terms of its direction angle.

(b) Let v be a vector in three-dimensional space. What are the direction angles of v? Explain why three direction angles are needed in this situation when only one is needed in the xy-plane. Let u be a unit vector in the same direction as v. Express u in terms of its direction angles.

10 State and derive the formula for distance in three-dimensional space.

11 State and derive the equation of a sphere in three-dimensional space.

12–13. Identify and sketch the graph.

- 12 $x^2 + y^2 + z^2 - 8x - 10z + 5 = 0$

- 13 $x^2 + y^2 + z^2 - 4x + 6y + 2z + 14 = 0$

- 14 Find the equation of the sphere that has the points $P(2, 1, 5)$ and $Q(4, -1, 3)$ at the opposite ends of a diameter.

●15 Find the equations of all spheres with radius $r = 6$ that are tangent to both the xy-plane and the yz-plane and that pass through the point $(4, 5, 2)$.

●16 Find the equation satisfied by the set of all points equidistant from $P(2, 1, 5)$ and $Q(4, -1, 3)$. Identify the graph of the equation.

●17 Find the equation satisfied by all points $P(x, y, z)$ that are equidistant from the point $(0, 1, 0)$ and the plane $y = -1$. Identify the graph.

●18 A motorboat can travel 8 mph in still water. The owner wants to travel across a straight river 1 mile wide to reach a point directly opposite on the other side. The river flows downstream at the rate of 4 mph. In what direction should the boat be headed so as to go directly across the river? Find the speed of the boat with respect to the point of departure.

●19 Let $a = 3i + j - 2k$ and $b = i - j + 3k$.

(a) Find the cosine of the angle between a and b and the approximate value of the angle.

(b) Find a unit vector orthogonal to both a and b. Use two methods—one involving the dot product, the other the cross product.

●20 Find the angle between the diagonal of a cube and the diagonal of one of its faces.

21 State and derive the equation of the plane through $P_0(x_0, y_0, z_0)$ with normal vector $n = ai + bj + ck$.

22 State and derive the parametric equations of the line through $P_0(x_0, y_0, z_0)$ that is parallel to the nonzero vector $v = ai + bj + ck$.

●23 Find the equation of the plane that contains the three points $A(-1, 2, 4)$, $B(2, 0, 6)$, and $C(-3, 1, 1)$.

●24 At what point does the line $(x - 1)/2 = -(y + 1) = z/3$ intersect the plane $3x + 2y - z = 5$?

●25 Find the equation of the plane that contains the two lines

$$\frac{x-2}{3} = \frac{y+1}{4} = \frac{z-3}{-1} \quad \text{and} \quad \frac{x-2}{4} = \frac{y+1}{-2} = \frac{z-3}{3}.$$

●26 Find the equations of the line in the plane $2x - 3y + 4z = -9$ that intersects and is perpendicular to the line $(x - 5)/4 = y - 2 = z + 1$.

27 Show that the planes $3x - 5y + z + 12 = 0$ and $7x + y - 16z + 11 = 0$ are perpendicular.

●28 Use vectors to find the distance from the point $P_0(2, -1, 4)$ to the plane $3x - 4y + 12z = 6$.

●29 Find the shortest distance from the sphere $x^2 + y^2 + z^2 - 4x + 2y + 12z + 37 = 0$ to the plane $3x + 4y - 12z + 4 = 0$.

●30 Use vectors to find the minimum distance between the lines $x = 1 - 2t$, $y = t$, $z = 1 + t$ and $x = 3 + t$, $y = 1 + 4t$, $z = -1 - t$.

31–32. Sketch the cylinder.

31 $4x^2 + 25y^2 = 100$

32 $y = z^2 + 1$

33–34. Sketch the space curve. Show a cylinder that has the curve on its surface.

33 $x = 2 \cos t$, $y = 2 \sin t$, $z = t/2$, $t \geq 0$

34 $x = t$, $y = 2t$, $z = t^2$

35–40. (a) Identify and sketch the quadric surface.

(b) Draw several level curves for the surface.

35 $z = x^2 + y^2$

36 $x^2 + y^2 - z^2 = 0$

37 $9x^2 + 4y^2 - 36z^2 = 36$

38 $9x^2 + 4y^2 - 36z^2 = -36$

39 $x^2 + 4y^2 + 9z^2 = 36$

40 $z = x^2 - y^2$

●41 Find the point on the line segment connecting $P_1(x_1, y_1, z_1)$ and $P_2(x_2, y_2, z_2)$ that is three-eighths of the distance from P_1 to P_2.

●42 Find parametric equations of the line of intersection of the planes $2x + 4y - z = 4$ and $x - 5y + 3z + 5 = 0$.

43 How is the graph of $x = 3 + 5 \cos t$, $y = -1 + 2 \cos t$, $z = 4 - \cos t$ related to the line

$$\frac{x - 3}{5} = \frac{y + 1}{2} = \frac{z - 4}{-1}?$$

●44 The point $(1, -2, 5)$ is equidistant from two parallel planes, one of which is $3x + 4y - 3z + 18 = 0$. Find the equation of the other plane.

Appendix. Second-Order and Third-Order Determinants

SECOND-ORDER DETERMINANTS

Second-order determinants are defined by the rule

Second-Order Determinant

$$\begin{vmatrix} a_1 & a_2 \\ b_1 & b_2 \end{vmatrix} = a_1 b_2 - a_2 b_1.$$

For example,

$$\begin{vmatrix} 3 & 1 \\ 7 & 2 \end{vmatrix} = 3 \cdot 2 - 1 \cdot 7 = -1$$

$$\begin{vmatrix} 1 & 0 \\ 3 & 4 \end{vmatrix} = 1 \cdot 4 - 0 \cdot 3 = 4.$$

An application of second-order determinants is furnished by the following method for solving systems of equations. If we solve the system of equations

$$\begin{cases} a_1 x + b_1 y = c_1 \\ a_2 x + b_2 y = c_2 \end{cases}$$

by elementary algebra, we find that

$$x = \frac{c_1 b_2 - c_2 b_1}{a_1 b_2 - a_2 b_1} \quad \text{and} \quad y = \frac{a_1 c_2 - c_1 a_2}{a_1 b_2 - a_2 b_1}$$

provided $a_1 b_2 - a_2 b_1 \neq 0$. If we write these numbers as determinants, we have

Cramer's Rule

$$x = \frac{\begin{vmatrix} c_1 & b_1 \\ c_2 & b_2 \end{vmatrix}}{\begin{vmatrix} a_1 & b_1 \\ a_2 & b_2 \end{vmatrix}} \quad \text{and} \quad y = \frac{\begin{vmatrix} a_1 & c_1 \\ a_2 & c_2 \end{vmatrix}}{\begin{vmatrix} a_1 & b_1 \\ a_2 & b_2 \end{vmatrix}} \quad \text{provided} \quad \begin{vmatrix} a_1 & b_1 \\ a_2 & b_2 \end{vmatrix} \neq 0.$$

This result, known as *Cramer's rule*, can be extended to cover a system of n linear equations in n unknowns for any positive integer n. (See Exercises 15–16 for the case $n = 3$.)

EXAMPLE 1 The solution of the system of equations

$$\begin{cases} 3x - 2y = 4 \\ x + 2y = 7 \end{cases}$$

is

$$x = \frac{\begin{vmatrix} 4 & -2 \\ 7 & 2 \end{vmatrix}}{\begin{vmatrix} 3 & -2 \\ 1 & 2 \end{vmatrix}} = \frac{4 \cdot 2 - (-2)7}{3 \cdot 2 - (-2)1} = \frac{22}{8},$$

$$y = \frac{\begin{vmatrix} 3 & 4 \\ 1 & 7 \end{vmatrix}}{\begin{vmatrix} 3 & -2 \\ 1 & 2 \end{vmatrix}} = \frac{3 \cdot 7 - 4 \cdot 1}{3 \cdot 2 - (-2)1} = \frac{17}{8}. \quad \square$$

Appendix. Second-Order and Third-Order Determinants

THIRD-ORDER DETERMINANTS

Third-order determinants are defined in terms of second-order determinants, fourth-order determinants in terms of third-order, and so on. The definition of a third-order determinant involves the expansion of the determinant by the elements of the first row.

DEFINITION

Third-Order Determinant

$$\begin{vmatrix} a_1 & a_2 & a_3 \\ b_1 & b_2 & b_3 \\ c_1 & c_2 & c_3 \end{vmatrix} = a_1 \begin{vmatrix} b_2 & b_3 \\ c_2 & c_3 \end{vmatrix} - a_2 \begin{vmatrix} b_1 & b_3 \\ c_1 & c_3 \end{vmatrix} + a_3 \begin{vmatrix} b_1 & b_2 \\ c_1 & c_2 \end{vmatrix}$$

$$= a_1(b_2 c_3 - c_2 b_3) - a_2(b_1 c_3 - c_1 b_3) + a_3(b_1 c_2 - c_1 b_2).$$

In expanding a third-order determinant, we multiply each element of the first row by the determinant obtained by deleting the row and column that contains that element. For example, the first element, a_1, is multiplied by the determinant

$$\begin{vmatrix} b_2 & b_3 \\ c_2 & c_3 \end{vmatrix},$$

which is obtained by deleting the row and column containing a_1.

Observe that the signs alternate in the expression

$$a_1 \begin{vmatrix} b_2 & b_3 \\ c_2 & c_3 \end{vmatrix} - a_2 \begin{vmatrix} b_1 & b_3 \\ c_1 & c_3 \end{vmatrix} + a_3 \begin{vmatrix} b_1 & b_2 \\ c_1 & c_2 \end{vmatrix}.$$

The middle product is subtracted from the sum of the other two.

EXAMPLE 2

$$\begin{vmatrix} 2 & -2 & 5 \\ 3 & -1 & 2 \\ -4 & 3 & -2 \end{vmatrix} = 2 \begin{vmatrix} -1 & 2 \\ 3 & -2 \end{vmatrix} - (-2) \begin{vmatrix} 3 & 2 \\ -4 & -2 \end{vmatrix} + 5 \begin{vmatrix} 3 & -1 \\ -4 & 3 \end{vmatrix}$$

$$= 2[(-1)(-2) - 2 \cdot 3] + 2[3(-2) - 2(-4)] + 5[3 \cdot 3 - (-4)(-1)]$$

$$= 2[2 - 6] + 2[-6 + 8] + 5[9 - 4] = 21. \ \square$$

PROPERTIES OF DETERMINANTS

Determinants have several special properties that simplify their calculations. We briefly discuss three of the most important of these properties. In our discussion, the word *line* will mean either a row (horizontal) or a column (vertical) of a determinant.

Property 1 If a line of a determinant consists entirely of zeros, then the value of the determinant is zero. For example,

$$\begin{vmatrix} 5 & -1 & 2 \\ 0 & 0 & 0 \\ 4 & 7 & 3 \end{vmatrix} = 0 \quad \text{and} \quad \begin{vmatrix} 2 & 1 & 0 \\ 3 & 7 & 0 \\ 2 & 5 & 0 \end{vmatrix} = 0.$$

(Each determinant has a line of zeros.)

Property 2 If two parallel lines are interchanged in a determinant, then its sign is changed. For example,

$$\begin{vmatrix} 3 & -1 & 7 \\ 4 & 0 & 2 \\ 5 & 3 & 1 \end{vmatrix} = - \begin{vmatrix} 4 & 0 & 2 \\ 3 & -1 & 7 \\ 5 & 3 & 1 \end{vmatrix}.$$

(The second determinant is obtained from the first by interchanging the first two rows.)

Property 3 If a determinant has two identical parallel lines, then its value is zero. For example,

$$\begin{vmatrix} -8 & 7 & 4 \\ 5 & 3 & 7 \\ -8 & 7 & 4 \end{vmatrix} = 0.$$

(The first and third rows are identical.)

This property follows from Property 2 since the sign of the determinant must be changed when the identical lines are interchanged yet the value is unchanged. If a is the value of the original determinant, then the determinant obtained by interchanging the identical lines has value $a = -a$. Thus $2a = 0$ and $a = 0$.

Many other properties of determinants are proved in a course in matrices and determinants. The properties mentioned above suffice for our purposes.

Exercises

1–8. Evaluate the determinants.

●1 $\begin{vmatrix} 7 & -1 \\ 2 & 3 \end{vmatrix}$

2 $\begin{vmatrix} 4 & 2 \\ 5 & 8 \end{vmatrix}$

●3 $\begin{vmatrix} 0 & 1 & 2 \\ -1 & 3 & 7 \\ 1 & 2 & -1 \end{vmatrix}$

4 $\begin{vmatrix} 4 & 1 & 2 \\ -1 & 3 & 0 \\ 2 & -4 & 1 \end{vmatrix}$

●5 $\begin{vmatrix} 1 & -1 & 1 \\ 2 & 3 & 4 \\ -1 & 0 & 2 \end{vmatrix}$

6 $\begin{vmatrix} 2 & 1 & 5 \\ -1 & 2 & -2 \\ 4 & -3 & 7 \end{vmatrix}$

●7 $\begin{vmatrix} x & x-1 & 2 \\ 3 & -1 & x \\ 4 & -2 & x+1 \end{vmatrix}$

8 $\begin{vmatrix} x & y & z \\ 1 & -2 & 0 \\ 3 & 1 & 7 \end{vmatrix}$

9–11. Prove the properties for second-order determinants by expanding the determinants.

9 Property 1   10 Property 2   11 Property 3

Exercises

12 There is an alternative method of evaluating a third-order determinant

$$\begin{vmatrix} a_1 & a_2 & a_3 \\ b_1 & b_2 & b_3 \\ c_1 & c_2 & c_3 \end{vmatrix}$$

which, unfortunately, cannot be extended to higher-order determinants. We write the determinant in the following array, in which the first two columns are repeated after the third column:

Alternative Expansion of Third-Order Determinant

The arrows indicate the products. (The first arrow that points downward, for example, represents the product $a_1 b_2 c_3$.) A product is given a positive sign if the corresponding arrow points downward and a negative sign if it points upward. Thus, the above array represents the sum

$$a_1 b_2 c_3 + a_2 b_3 c_1 + a_3 b_1 c_2 - c_1 b_2 a_3 - c_2 b_3 a_1 - c_3 b_1 a_2.$$

(a) Show that this sum is the value of the determinant.
(b) Use this method to evaluate the determinant in Problem 5.

13–14. (*Cramer's Rule*) Solve the systems of equations by Cramer's rule.

• **13** $\begin{cases} x - 2y = 4 \\ 3x + y = 5 \end{cases}$

14 $\begin{cases} 5x - 8y = 2 \\ 3x + y = -7 \end{cases}$

15–16. (*Cramer's Rule*) It can be proved that Cramer's rule holds for n equations in n unknowns. For the case of three equations in three unknowns, it has the following form: The system

$$\begin{cases} a_1 x + b_1 y + c_1 z = d_1 \\ a_2 x + b_2 y + c_2 z = d_2 \\ a_3 x + b_3 y + c_3 z = d_3 \end{cases}$$

has the unique solution

Cramer's Rule for Third-Order Determinants

$$x = \frac{\begin{vmatrix} d_1 & b_1 & c_1 \\ d_2 & b_2 & c_2 \\ d_3 & b_3 & c_3 \end{vmatrix}}{\begin{vmatrix} a_1 & b_1 & c_1 \\ a_2 & b_2 & c_2 \\ a_3 & b_3 & c_3 \end{vmatrix}}, \quad y = \frac{\begin{vmatrix} a_1 & d_1 & c_1 \\ a_2 & d_2 & c_2 \\ a_3 & d_3 & c_3 \end{vmatrix}}{\begin{vmatrix} a_1 & b_1 & c_1 \\ a_2 & b_2 & c_2 \\ a_3 & b_3 & c_3 \end{vmatrix}}, \quad z = \frac{\begin{vmatrix} a_1 & b_1 & d_1 \\ a_2 & b_2 & d_2 \\ a_3 & b_3 & d_3 \end{vmatrix}}{\begin{vmatrix} a_1 & b_1 & c_1 \\ a_2 & b_2 & c_2 \\ a_3 & b_3 & c_3 \end{vmatrix}}$$

provided the denominator of these fractions is nonzero.

Use Cramer's rule to solve the given systems of equations.

• **15** $\begin{cases} 2x - y + z = 8 \\ x + 3y - 8z = 2 \\ 3x + y + 4z = 2 \end{cases}$

16 $\begin{cases} x + 4y + 5z = 9 \\ -x + 2y - z = 7 \\ 2x - y + 3z = 6 \end{cases}$

15 The Calculus of Vector-Valued Functions

15–1 Limits, Continuity, Derivatives and Integrals

Position Vector

A *position vector* is a vector $v = ai + bj + ck$ from the origin to a point (a, b, c). The use of position vectors allows us to consider space curves (defined parametrically) as the graphs of vector-valued functions.

DEFINITION

Vector-Valued Function

A *vector-valued function* is a function r that assigns to each real number t in its domain a vector $r(t)$.

As with real functions, unless we specify otherwise, the domain of a vector-valued function is the set of all t for which the given law of correspondence is well defined.

EXAMPLE 1 Two vector-valued functions are

(1) $\qquad r(t) = 3ti + t^2j + e^t k$

and

(2) $\qquad r(t) = \dfrac{3}{t}i + \dfrac{1}{t-1}j + 2k.$

The domain of the first function is the set of all real numbers. The domain of the second is the set of all real numbers different from 0 and 1. □

Observe that every vector-valued function is of form

$$r(t) = f(t)i + g(t)j + h(t)k.$$

15-1 Limits, Continuity, Derivatives, and Integrals

DEFINITION

Graph of Vector-Valued Function

The *graph* of a vector-valued function $r(t) = f(t)i + g(t)j + h(t)k$ consists of all points $(f(t), g(t), h(t))$ such that t is in the domain of r.

Observe that the graph of $r(t) = f(t)i + g(t)j + h(t)k$ is identical to the graph of the parametric equations

$$x = f(t), \quad y = g(t), \quad z = h(t).$$

It is the path traced out by the terminal point of the position vector $r(t)$ as t varies over the domain of r. (See Fig. 15–1a.)

Remark The use of vector-valued functions allows us to work with parametric equations $x = f(t), y = g(t), z = h(t)$ as if they were a single entity. We shall see that this approach is very powerful. It enables us to solve with little difficulty certain problems whose solution would be awkward by any other method.

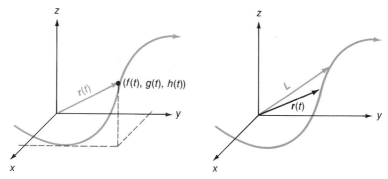

Figure 15–1(a) Graph of the function $r(t) = f(t)i + g(t)j + h(t)k$.

Figure 15–1(b) $\lim_{t \to a} r(t) = L$.

LIMITS

The formal definition of the limit of a vector-valued function is analogous to the definition for a real-valued function.

DEFINITION

Limit of Vector-Valued Function

Let $r(t)$ be defined for all t in an open interval that contains a except possibly at a itself. We say that

$$\lim_{t \to a} r(t) = L$$

provided that for each $\epsilon > 0$ there exists $\delta > 0$ such that

$$\|r(t) - L\| < \epsilon \quad \text{whenever} \quad 0 < |t - a| < \delta.$$

In other words, the vector $r(t) - L$ can be made as close to the zero vector as we wish by requiring t to be sufficiently close to a. Consequently, the position vector $r(t)$ approaches the position vector L as a limiting position as $t \to a$. (See Fig. 15–1b.)

Fortunately, we do not need to repeat the basic work with limits that we developed for real functions (Chapter 2). Theorem 15–1 relates the limit of a vector-valued function $r(t)$ to the limits of the component functions $f(t)$, $g(t)$, $h(t)$. You will find the proof outlined in Exercise 37.

THEOREM 15–1

Limit of a Vector Function in Terms of its Component Functions

Let $r(t) = f(t)i + g(t)j + h(t)k$. Suppose that

$$\lim_{t \to a} f(t) = l, \quad \lim_{t \to a} g(t) = m, \quad \text{and} \quad \lim_{t \to a} h(t) = n.$$

Then

$$\lim_{t \to a} r(t) = li + mj + nk$$
$$= \left[\lim_{t \to a} f(t)\right] i + \left[\lim_{t \to a} g(t)\right] j + \left[\lim_{t \to a} h(t)\right] k.$$

EXAMPLE 2 Let $r(t) = (2t^2 - 3)i + (\cos \pi t)j + e^{t-1}k$. Then

$$\lim_{t \to 1} r(t) = \left[\lim_{t \to 1} (2t^2 - 3)\right] i + \left[\lim_{t \to 1} (\cos \pi t)\right] j + \left[\lim_{t \to 1} e^{t-1}\right] k$$
$$= (-1)i + (-1)j + e^0 k = -i - j + k. \quad \square$$

Properties similar to L1, L2, and so on follow easily from Theorem 15–1. (These are stated for reference purposes in Exercises 23–28.)

CONTINUITY

The vector-valued function $r(t)$ is said to be *continuous* at $t = a$ provided

$$\lim_{t \to a} r(t) = r(a).$$

For example, the function

$$r(t) = (2t^2 - 3)i + (\cos \pi t)j + e^{t-1}k$$

of Example 2 is continuous at $t = 1$. The limit of $r(t)$ as $t \to 1$ equals $-i - j + k$, the value of $r(1)$.

It is easy to show that if the component functions $f(t)$, $g(t)$, $h(t)$ are continuous at $t = a$, then the vector-valued function

$$r(t) = f(t)i + g(t)j + h(t)k$$

is continuous at $t = a$. (See Exercise 29.) Conversely, if $r(t)$ is continuous at $t = a$, then it can be proved that its component functions are also continuous at $t = a$.

Continuity at an Endpoint of the Domain

The definition can be extended to cover endpoints of the domain of definition, as in Section 2–6. If r is defined on $[a, b]$, but is not defined for $t < a$, we say that r is continuous at a, provided

$$\lim_{t \to a^+} r(t) = r(a)$$

A similar definition holds if r is not defined for $t > b$.

15-1 Limits, Continuity, Derivatives, and Integrals

DERIVATIVES

The derivative of a vector function is defined by a difference quotient analogous to the one for the derivative of a real function.

DEFINITION   The *derivative* of the vector-valued function $r(t)$ is

Derivative Defined as a Limit
$$r'(t) = \lim_{\Delta t \to 0} \frac{r(t + \Delta t) - r(t)}{\Delta t}$$

provided the limit exists.

Observe that the derivative defines a new vector function $r'(t)$ at each point where it exists.

Theorem 15-2 shows that the derivative of a vector function can be calculated from the derivatives of the component functions.

THEOREM 15-2   Let $r(t) = f(t)i + g(t)j + h(t)k$, where f, g, and h are differentiable functions. Then

Computation of Derivatives
$$r'(t) = f'(t)i + g'(t)j + h'(t)k.$$

Proof

$$r'(t) = \lim_{\Delta t \to 0} \frac{r(t + \Delta t) - r(t)}{\Delta t}$$

$$= \lim_{\Delta t \to 0} \frac{[f(t + \Delta t)i + g(t + \Delta t)j + h(t + \Delta t)k] - [f(t)i + g(t)j + h(t)k]}{\Delta t}$$

$$= \lim_{\Delta t \to 0} \left[\frac{f(t + \Delta t) - f(t)}{\Delta t} i + \frac{g(t + \Delta t) - g(t)}{\Delta t} j + \frac{h(t + \Delta t) - h(t)}{\Delta t} k \right]$$

$$= \left[\lim_{\Delta t \to 0} \frac{f(t + \Delta t) - f(t)}{\Delta t} \right] i + \left[\lim_{\Delta t \to 0} \frac{g(t + \Delta t) - g(t)}{\Delta t} \right] j$$

$$+ \left[\lim_{\Delta t \to 0} \frac{h(t + \Delta t) - h(t)}{\Delta t} \right] k \quad \text{(by Theorem 15-1)}$$

$$= f'(t)i + g'(t)j + h'(t)k. \blacksquare$$

The converse of this theorem (that is, the differentiability of a vector function implies the differentiability of its component functions) is also true.

Higher-order derivatives can also be calculated by the component functions. That is

$$r''(t) = f''(t)i + g''(t)j + h''(t)k,$$
$$r'''(t) = f'''(t)i + g'''(t)j + h'''(t)k,$$

and so on.

EXAMPLE 3   (a) Sketch the graph of the vector-valued function

$$r(t) = i + t^2 j + 2tk.$$

(b) Draw the position vector $r(1)$ on the graph in (a). Also draw the derivative vectors $r'(1)$ and $r''(1)$ originating at the point at which $t = 1$.

Solution (a) The graph is the space curve defined parametrically by $x = 1$, $y = t^2$, $z = 2t$. This curve is a parabola on the plane $x = 1$. (See Fig. 15-2.)

(b) The first two derivatives of r are

$$r'(t) = 2tj + 2k, \qquad r''(t) = 2j.$$

When $t = 1$, we have

$$r(1) = i + j + 2k, \qquad r'(1) = 2j + 2k, \qquad r''(1) = 2j.$$

These vectors are shown in Figure 15-2. □

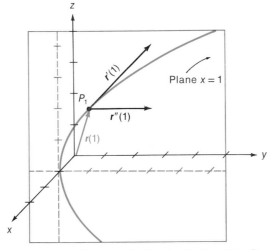

Figure 15-2 Example 3. Graph of $r(t) = i + t^2j + 2tk$.

DERIVATIVE FORMULAS

Most of the derivative formulas can be extended to vector-valued functions. Observe that the product formula D6 can be extended to cover three types of products: (1) the product of a scalar function and a vector-valued function; (2) the scalar product of two vector-valued functions; (3) the vector product of two vector-valued functions. The quotient formula (D7) can be extended in only one way: the quotient of a vector function and a scalar function.

THEOREM 15-3 Let $r(t)$ and $s(t)$ be differentiable vector-valued functions; $a(t)$ and $u(t)$, differentiable scalar functions; m, a constant; and C, a constant vector. Then

D1 $D_t(C) = 0.$

D2 $D_t[r(t) + s(t)] = r'(t) + s'(t).$

D4 $D_t(mr(t)) = mr'(t).$

15–1 Limits, Continuity, Derivatives, and Integrals

D6 $\begin{cases} D_t[a(t)r(t)] = a(t)r'(t) + a'(t)r(t); \\ D_t[r(t) \cdot s(t)] = r(t) \cdot s'(t) + r'(t) \cdot s(t); \\ D_t[r(t) \times s(t)] = r(t) \times s'(t) + r'(t) \times s(t). \end{cases}$

D7 $\quad D_t\left(\dfrac{r(t)}{a(t)}\right) = \dfrac{a(t)r'(t) - a'(t)r(t)}{[a(t)]^2} \quad$ provided $a(t) \neq 0$.

D8 $\quad D_t r(u) = r'(u) \cdot D_t u \quad$ (the Chain Rule).

Each of the formulas in Theorem 15–3 can be proved by using Theorem 15–2 and the equivalent formula for scalar functions. In Example 4, we show that the first part of D6 holds. The other proofs are left for you to work out (Exercises 30–36).

EXAMPLE 4 Let $a(t)$ be a differentiable scalar function, $r(t)$ a differentiable vector function. Show that

$$D_t[a(t)r(t)] = a(t)r'(t) + a'(t)r(t).$$

Solution Let
$$r(t) = f(t)i + g(t)j + h(t)k.$$
Then
$$a(t)r(t) = a(t)f(t)i + a(t)g(t)j + a(t)h(t)k.$$
Therefore,
$$D_t[a(t)r(t)] = D_t[a(t)f(t)]i + D_t[a(t)g(t)]j + D_t[a(t)h(t)]k$$
$$= [a(t)f'(t) + a'(t)f(t)]i + [a(t)g'(t) + a'(t)g(t)]j$$
$$+ [a(t)h'(t) + a'(t)h(t)]k$$
$$= a(t)[f'(t)i + g'(t)j + h'(t)k] + a'(t)[f(t)i + g(t)j + h(t)k]$$
$$= a(t)r'(t) + a'(t)r(t). \quad \Box$$

INTEGRALS

Since the integral has an inverse relation to the derivative, it follows that antiderivatives of vector-valued functions can be obtained from the antiderivatives of the component functions.

Let $r(t) = f(t)i + g(t)j + h(t)k$, where f, g, and h are continuous functions. The *general antiderivative* of $r(t)$ is

Indefinite Integral
$$\int r(t) \, dt = \int f(t) \, dt \, i + \int g(t) \, dt \, j + \int h(t) \, dt \, k + C,$$

where the constant vector C is the constant of integration.

The *definite integral* of r(t) from $t = a$ to $t = b$ is defined to be

$$\int_a^b r(t)\, dt = \int_a^b f(t)\, dt\, i + \int_a^b g(t)\, dt\, j + \int_a^b h(t)\, dt\, k.$$

EXAMPLE 5   Let $r(t) = 2ti + \cos 3t\, j + k$.

(a) $\int r(t)\, dt = t^2 i + \dfrac{\sin 3t}{3} j + tk + C$.

(b) $\int_0^{\pi/2} r(t)\, dt = \left[t^2\right]_0^{\pi/2} i + \left[\dfrac{\sin 3t}{3}\right]_0^{\pi/2} j + \left[t\right]_0^{\pi/2} k$

$= \left[\dfrac{\pi^2}{4} - 0\right] i + \left[\dfrac{\sin(3\pi/2) - \sin 0}{3}\right] j + \left[\dfrac{\pi}{2} - 0\right] k$

$= \dfrac{\pi^2}{4} i - \dfrac{1}{3} j + \dfrac{\pi}{2} k.$ □

Exercises 15–1

1–4.   Evaluate the limits if they exist.

● 1 $\lim\limits_{t \to 1} \left(t^2 i + \dfrac{j}{\sqrt{t-1}} + e^t k \right)$ 2 $\lim\limits_{t \to 0} (e^t i + te^t j + (t^2 + 4)k)$

● 3 $\lim\limits_{t \to 1} (\ln |t|\, i + \sin \pi t\, j + t^2 k)$ 4 $\lim\limits_{t \to 0} \left(\dfrac{\sin t}{t} i + \dfrac{\tan 3t}{t} j + \sin tk \right)$

5–10.   (a)   Find the domain of r and tell where r is continuous.
(b)   Calculate $r'(t)$ and $r''(t)$.

● 5 $r(t) = \dfrac{i}{t} + t^2 j + t^3 k$

6   $r(t) = (t^2 + 2t)i + 3 \sin tj - 4 \cos tk$

● 7 $r(t) = \ln(t - 1)i - e^t k$

8   $r(t) = \sqrt{t + 2}\, i + tj + \dfrac{1}{\sqrt{3 - t}} k$

● 9 $r(t) = t \sin ti + tj + t \cos tk$

10   $r(t) = \tan ti + e^{-t}j + \sec tk$

11–14.   Sketch the graph of the position vector $r(t)$ in the xy-plane. Show $r(t_0)$, $r'(t_0)$, and $r''(t_0)$ at the indicated value of t_0. Draw $r(t_0)$ as a position vector. Let $r'(t_0)$ and $r''(t_0)$ originate at the appropriate point on the graph.

● 11 $r(t) = \sqrt{t}\, i + (t - 4)j,\ t_0 = 4$ 12 $r(t) = -2t^4 i + t^2 j,\ t_0 = 1$

● 13 $r(t) = 3 \sin ti + 2 \cos tj,\ t_0 = \dfrac{\pi}{4}$ 14 $r(t) = \cos ti + \sec tj,\ t_0 = 0$

Exercises 15–1

15–16. Use Formula D6 to calculate $D_t(u \cdot v)$ and $D_t(u \times v)$.

• **15** $u = t^2 i + (2t + 1)j + \sqrt{t}\,k$, $v = e^t(i + k) + j$

16 $u = t^2 i + t^3 j + tk$, $v = (1 + t)i + 2tj + 9t^2 k$

17–18. Calculate

• **17** $\displaystyle\int_0^1 \left(e^{-t} i + \frac{tj}{\sqrt{t^2 + 1}} + te^{t^2} k \right) dt$

18 $\displaystyle\int_4^9 \left(\frac{i}{1 - t} + \sqrt{t}\,j + tk \right) dt$

19–22. Find the function $r(t)$ that satisfies the given conditions.

• **19** $r'(t) = 3t^2 i + (4t + 1)j + 2k$, $r(0) = 2i - j + 3k$

20 $r'(t) = 6t^2 i + 2tj - \sqrt{t}\,k$, $r(1) = -i + 2j - k$

• **21** $r''(t) = e^{-t} i + 6t^2 j + \cos 2t\,k$, $r'(0) = i + j$, and $r(0) = i + j - k$

22 $r''(t) = t^{-2} i + 6tj + e^t k$, $r'(1) = 2i + j + 2ek$, $r(1) = i + 3j + ek$

23–28. Prove the given properties of limits. (*Hint:* Use Theorem 15–1 and the equivalent properties of real-valued functions.)

Limit Formulas

23 L1: $\displaystyle\lim_{t \to a} (u + v) = \lim_{t \to a} u + \lim_{t \to a} v$

24 L2: $\displaystyle\lim_{t \to a} f(t)u(t) = \lim_{t \to a} f(t) \cdot \lim_{t \to a} u(t)$

25 L2: $\displaystyle\lim_{t \to a} u(t) \cdot v(t) = \lim_{t \to a} u(t) \cdot \lim_{t \to a} v(t)$

26 L2: $\displaystyle\lim_{t \to a} u(t) \times v(t) = \lim_{t \to a} u(t) \times \lim_{t \to a} v(t)$

27 L3: $\displaystyle\lim_{t \to a} \frac{u(t)}{f(t)} = \frac{\lim_{t \to a} u(t)}{\lim_{t \to a} f(t)}$ provided $\displaystyle\lim_{t \to a} f(t) \neq 0$

28 L8: $\displaystyle\lim_{t \to a} u(g(t)) = u\left(\lim_{t \to a} g(t)\right)$ provided u is continuous at $L = \displaystyle\lim_{t \to a} g(t)$.

29 Let f, g, h be continuous at $t = a$. Let $r(t) = f(t)i + g(t)j + h(t)k$. Prove that r is continuous at $t = a$. (*Hint:* Use Theorem 15–1.)

30–36. Prove the derivative formulas as stated in this section.

30 D1   **31** D2

32 D4   **33** D6 (Dot product)

34 D6 (Cross product)   **35** D7

36 D8

37 Prove Theorem 15–1. *Hint:* First find positive numbers $\delta_1, \delta_2, \delta_3$ such that $|f(t) - l| < \epsilon/\sqrt{3}$ if $0 < |t - a| < \delta_1$; $|g(t) - m| < \epsilon/\sqrt{3}$ if $0 < |t - a| < \delta_2$; and $|h(t) - n| < \epsilon/\sqrt{3}$ if $0 < |t - a| < \delta_3$.

38 Let $r(t)$ be differentiable at $t = a$. Modify the proof of Theorem 2–6 to prove that $r(t)$ is continuous at $t = a$.

15-2 Velocity and Acceleration Vectors. Arc Length

Let $r(t)$ be a twice differentiable vector-valued function:

$$r(t) = f(t)\mathbf{i} + g(t)\mathbf{j} + h(t)\mathbf{k}.$$

The graph of r is a space curve, defined parametrically by

$$x = f(t), \quad y = g(t), \quad z = h(t),$$

that is traced out by a moving particle. At a fixed time t the position vector $r(t)$ points from the origin to the point $P_t(f(t), g(t), h(t))$—the location of the particle at this time. (See Fig. 15-3a.)

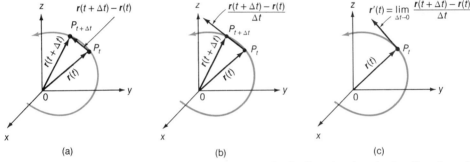

(a) (b) (c)

Figure 15-3 The vector $r'(t)$ is tangent to the graph of $r(t)$ and points in the direction of increasing arc length.

The derivative vector $r'(t)$ has the following properties:

Properties of $r'(t)$

(1) It is tangent to the graph of r at P_t and points in the direction in which the curve is being traced out (see Fig. 15-3c);
(2) Its magnitude $\|r'(t)\|$ equals the rate of change of the arc length function $s(t)$. That is, $\|r'(t)\| = s'(t)$.

To explain the first of these results, recall that for a fixed value of t,

$$r'(t) = \lim_{\Delta t \to 0} \frac{r(t + \Delta t) - r(t)}{\Delta t}.$$

The vector $r(t + \Delta t) - r(t)$, the numerator of the difference quotient, points from the point $P_t(f(t), g(t), h(t))$ on the curve to the point $P_{t+\Delta t}(f(t + \Delta t), g(t + \Delta t), h(t + \Delta t))$, which also is on the curve. (See Fig. 15-3a.)

Let Δt be a small number. If Δt is positive, then the vector

$$\frac{r(t + \Delta t) - r(t)}{\Delta t}$$

points in the same direction as $r(t + \Delta t) - r(t)$. Thus, it originates at P_t and passes through $P_{t+\Delta t}$. (See Fig. 15-3b.) If $\Delta t < 0$, this vector points in the direction opposite $r(t + \Delta t) - r(t)$. In either case, it points approximately in the direction in which the curve is being traced out as the tracing particle passes through P_t.

15-2 Velocity and Acceleration Vectors. Arc Length

As $\Delta t \to 0$, the point $P_{t+\Delta t}$ approaches the point P_t as a limiting position. It can be proved that the vector

$$\frac{r(t + \Delta t) - r(t)}{\Delta t}$$

approaches a limiting position along a line tangent to the curve at P_t. For this reason, we define the *tangent vector* relative to the graph at P_t to be

Tangent Vector
$$r'(t) = \lim_{\Delta t \to 0} \frac{r(t + \Delta t) - r(t)}{\Delta t}.$$

Thus $r'(t)$ is tangent to the graph of r at P_t and points in the direction in which the curves is traced out. (See Fig. 15–3c.)

ARC LENGTH

We now show that the magnitude of the derivative vector equals the rate of change of the arc length of the graph of r.

Recall from Section 6–4 that the arc length of a plane curve defined parametrically by $x = f(t)$, $y = g(t)$, $a \leq t \leq b$, is

$$L_a^b = \int_a^b \sqrt{f'(t)^2 + g'(t)^2}\, dt.$$

This formula has a natural extension to curves in three-dimensional space. If $x = f(t)$, $y = g(t)$, $z = h(t)$, then

Arc Length for Space Curves
$$L_a^b = \int_a^b \sqrt{f'(t)^2 + g'(t)^2 + h'(t)^2}\, dt.$$

It is convenient to fix a point t_0 and let $s(t)$ denote the length of the arc obtained by tracing the curve from time t_0 to time t:

$$s(t) = \int_{t_0}^t \sqrt{f'(t)^2 + g'(t)^2 + h'(t)^2}\, dt.$$

(See Fig. 15–4.) Observe that $s(t) > 0$ if $t > t_0$, and $s(t) < 0$ if $t < t_0$.

The derivative of s, obtained by applying the Fundamental Theorem of Calculus (Theorem 5–3) is

$$s'(t) = D_t \left(\int_{t_0}^t \sqrt{f'(t)^2 + g'(t)^2 + h'(t)^2}\, dt \right)$$
$$= \sqrt{f'(t)^2 + g'(t)^2 + h'(t)^2}.$$

This derivative is called the *speed* of the particle that traces out the curve. It equals the rate at which the arc length function changes with respect to time. It is the number we read from the speedometer as our car travels around a curve.

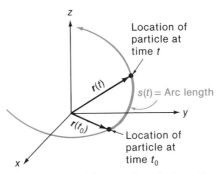

Figure 15-4 $s(t)$ = arc length from time t_0 to time t

$$= \int_{t_0}^{t} \|v(t)\| \, dt$$

$$= \int_{t_0}^{t} \sqrt{f'(t)^2 + g'(t)^2 + h'(t)^2} \, dt.$$

Recall that our original vector-valued function is

$$r(t) = f(t)i + g(t)j + h(t)k$$

and that its derivative is

$$r'(t) = f'(t)i + g'(t)j + h'(t)k.$$

Then

$$\|r'(t)\| = \sqrt{f'(t)^2 + g'(t)^2 + h'(t)^2} = s'(t).$$

Derivative Vector and Speed
The magnitude of the derivative vector equals the speed of the particle that traces out the curve.

VELOCITY AND ACCELERATION VECTORS

Velocity Vector
The vector $r'(t)$ is called the *velocity vector* for the graph of $r(t)$. It is denoted by

$$v(t) = r'(t).$$

As we have seen, the velocity vector $v(t)$ is tangent to the graph of $r(t)$ and points in the direction in which the curve is traced out. Its magnitude $\|v(t)\|$ equals the speed of the particle that traces the curve.

Acceleration Vector
The derivative of the velocity vector is called the *acceleration vector*, denoted by $a(t)$:

$$a(t) = v'(t) = r''(t).$$

Remark
As we shall see, the acceleration vector is related to the force that causes the particle to change its speed and direction as it traces out the curve. If the force is known, then

15-2 Velocity and Acceleration Vectors. Arc Length

the acceleration vector can be determined. The acceleration vector can be integrated to determine the velocity vector; and this in turn can be integrated to determine the position vector.

EXAMPLE 1   Let $r(t) = \cos t\,i + \sin t\,j + \dfrac{t}{2}k$. Calculate (a) $v(t)$; (b) $a(t)$; (c) $s'(t)$; (d) the arc length from $t = 0$ to $t = 2\pi$.

Solution
(a) $v(t) = r'(t) = -\sin t\,i + \cos t\,j + \tfrac{1}{2}k$.
(b) $a(t) = v'(t) = -\cos t\,i - \sin t\,j$.
(c) $s'(t) = \|v(t)\| = \sqrt{\sin^2 t + \cos^2 t + \tfrac{1}{4}} = \sqrt{5}/2$.
(d) The arc length from $t = 0$ to $t = 2\pi$ is

$$s(2\pi) = \int_0^{2\pi} s'(t)\,dt = \int_0^{2\pi} \|r'(t)\|\,dt = \int_0^{2\pi} \frac{\sqrt{5}}{2}\,dt = \frac{\sqrt{5}}{2}\Big[t\Big]_0^{2\pi} = \pi\sqrt{5}. \quad\square$$

The graph of the function

$$r(t) = \cos t\,i + \sin t\,j + \frac{t}{2}k,$$

considered in Example 1, is the helix shown in Figure 15–5. The velocity and accel-

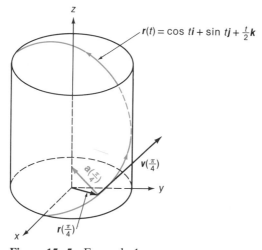

Figure 15–5   Example 1.

eration vectors are shown at the point where $t = \pi/4$. Observe that the velocity vector is tangent to the curve and points in the direction of increasing arc length, and that the acceleration vector points towards the z-axis.

It is standard to draw the velocity and acceleration vectors at the appropriate point on the curve rather than at the origin as with the position vector $r(t)$. We shall show (Section 15–6) that the acceleration vector always points towards the concave side of the curve.

EXAMPLE 2   A particle moves around the circle

$$x^2 + y^2 = a^2 \quad (a > 0)$$

with its location at time t given by the position vector

$$r(t) = a \cos \omega t i + a \sin \omega t j,$$

where $\omega > 0$. (The number ω is called the *angular velocity*. It equals the rate of change of ωt, the direction angle of $r(t)$, with time.)
(a) Show that the speed of the particle is a constant.
(b) Show that the acceleration vector is orthogonal to the velocity vector at each time t.
(c) What effect does doubling the angular velocity have on the speed and acceleration?

Solution   The circle is shown in Figure 15–6.

(a) The velocity vector is

$$v(t) = r'(t) = -a\omega \sin \omega t i + a\omega \cos \omega t j.$$

The speed is

$$s'(t) = \|v(t)\| = \sqrt{(-a\omega \sin \omega t)^2 + (a\omega \cos \omega t)^2}$$
$$= a\omega \sqrt{\sin^2 \omega t + \cos^2 \omega t} = a\omega.$$

Thus, $s'(t)$ is a constant.

(b) The acceleration vector is

$$a(t) = v'(t) = -a\omega^2 \cos \omega t i - a\omega^2 \sin \omega t j.$$

Thus,

$$a(t) \cdot v(t) = a^2 \omega^3 (\sin \omega t \cos \omega t - \cos \omega t \sin \omega t) = 0.$$

The acceleration vector is orthogonal to the velocity vector.

(c) If we double the angular velocity, changing from ω_0 to $2\omega_0$, then $s'(t)$ changes from $a\omega_0$ to $2a\omega_0$ and $a(t)$ changes from

$$-a\omega_0^2 (\cos \omega_0 t i + \sin \omega_0 t j) \quad \text{to} \quad -4a\omega_0^2 (\cos 2\omega_0 t i + \sin 2\omega_0 t j).$$

Thus, the speed is doubled and the magnitude of the acceleration vector is quadrupled. □

The results of Example 2 are verified by our common experience. When we drive around a circular track at 40 mph, we feel as if we are being continually pushed towards the outside of the track. This effect is a reaction to the force that corresponds to the acceleration, which is directed towards the center of the circle. (See Fig. 15–6.) We feel a much greater force if we drive around the same track at 80 mph. Our work in Example 2 shows that the acceleration is four times stronger at 80 mph than at 40 mph.

15-2 Velocity and Acceleration Vectors. Arc Length

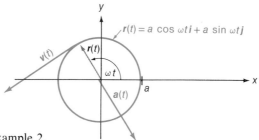

Figure 15-6 Example 2.

The following example shows that the acceleration vector is orthogonal to the velocity vector whenever a particle moves with a constant speed. Thus, the result of Example 2(b) is a consequence of the speed's constancy—and not the circular track, as we might imagine. A more sophisticated argument for this same result is outlined in Exercise 15.

EXAMPLE 3   A particle moves around a track with a constant speed. Show that the acceleration vector is always orthogonal to the velocity vector.

Solution   Let the position at time t be given by the vector

$$r(t) = f(t)i + g(t)j + h(t)k.$$

Then

$$v(t) = f'(t)i + g'(t)j + h'(t)k.$$

If the constant speed is c, then

$$\|v(t)\| = \sqrt{f'(t)^2 + g'(t)^2 + h'(t)^2} = c,$$

so that

$$f'(t)^2 + g'(t)^2 + h'(t)^2 = c^2.$$

If we differentiate this expression with respect to t, we get

$$2f'(t)f''(t) + 2g'(t)g''(t) + 2h'(t)h''(t) = 0$$

so that

$$f'(t)f''(t) + g'(t)g''(t) + h'(t)h''(t) = 0.$$

This last equation can be written in vector form as

$$[f'(t)i + g'(t)j + h'(t)k] \cdot [f''(t)i + g''(t)j + h''(t)k] = 0.$$

Thus,

$$v(t) \cdot a(t) = 0.$$

Therefore, $v(t)$ and $a(t)$ are orthogonal. □

NEWTON'S SECOND LAW OF MOTION

Newton's second law of motion can be formulated in terms of vectors. If a force F causes a particle of mass m to move with an acceleration a, then

$$F = ma$$

(*force equals mass times acceleration*).

In particular, this law implies that the force is applied in the same direction as the acceleration, and that the magnitude of the force is proportional to the magnitude of the acceleration vector:

$$\|F\| = m\|a\|,$$

where m is the mass.

For example, we saw in Example 2 that if the speed of a particle is doubled as it moves around a circular track, then the magnitude of the acceleration is quadrupled. It follows that the force causing the acceleration must also be quadrupled.

In the following section we use the vector form of Newton's second law to generalize some of our previous results concerning moving objects.

Exercises 15–2

1–6. Calculate the velocity, acceleration, and speed of the particle that traces out the graph. Find the arc length over the indicated interval.

- •1 $r(t) = t^2 i + \dfrac{t^3 j}{3} + 2tk$, $[1, 4]$

- 2 $r(t) = ti + \dfrac{t^2 j}{\sqrt{2}} + \dfrac{t^3 k}{3}$, $[1, 4]$

- •3 $r(t) = \dfrac{2t^3 i}{3} + t^2 j + tk$, $[0, 3]$

- 4 $r(t) = \sin 3t i + 2t^{3/2} j + \cos 3t k$, $[0, 3]$

- •5 $r(t) = \cos 2t i + \sin 2t j + \ln(\cos 2t)k$, $\left[0, \dfrac{\pi}{6}\right]$

- 6 $r(t) = e^t \sin t \, i + e^t \cos t \, j + e^t k$, $[0, 1]$

7   Sketch the graph of the function in Exercise 5 for $0 \leq t < \pi/4$. Show the vectors $r(\pi/8)$, $v(\pi/8)$ and $a(\pi/8)$.

Serpentine   8   (*The Serpentine*) Let $a, b > 0$. the graph of

$$r(t) = a \cot t \, i + b \sin 2t \, j$$

is called a *serpentine*.
  (a)   Sketch the graph of the serpentine with $a = 2$, $b = 4$.
•(b)   Find an equation in rectangular coordinates for the serpentine in (a).

9–12. Find parametric equations for the line tangent to the graph of $r(t)$ at the indicated point.

• 9 $r(t) = 4ti + t^2j + (1 - 2t^2)k$; $P(-4, 1, -1)$

10 $r(t) = (1 - 2t)i + (t^2 - 2t)j + (t^3 + t)k$; $P(3, 3, -2)$

• 11 $r(t) = e^{2t}i + \cos 3t j + (t^2 + 2)k$; $P(1, 1, 2)$

12 $r(t) = \ln(t - 1)i + (2t - 1)j + (t^2 - t)k$; $P(0, 3, 2)$

• 13 At what point or points on the graph of $r(t) = (3t + 1)i + 2t^2j + (6t - 1)k$ is the tangent vector orthogonal to $2i - j + k$?

14 Show that the line tangent to the graph of $r(t) = 2ti - tj + \sin^3 t k$ is never orthogonal to $i - 3j - k$.

15 Construct an argument based on the following steps to prove that the velocity vector is orthogonal to the acceleration vector whenever a particle moves with a constant speed: (1) Observe that

$$v(t) \cdot v(t) = C$$

where C is a constant. (2) Differentiate the above equation with respect to t, using D6, to show that

$$2v(t) \cdot a(t) = 0.$$

16 A force acting on a particle is always directed at a right angle to the direction of motion. Show that the speed of the particle is constant.

17 A point moves along a curve in the xy-plane in such a way that its velocity vector is always orthogonal to its position vector. Show that the curve is all or part of a circle with center at the origin. (*Hint:* Let $d(t)$ be the length of the position vector at time t. Write $r(t) = d(t)(\cos \theta(t)i + \sin \theta(t)j)$.

15–3 Velocity and Acceleration Problems

We now reconsider three problems related to Newton's second law

$$F = ma$$

where F is the force acting on a particle, m is its mass, and a is the acceleration caused by the force. Two of these problems involve motion and one an object in equilibrium. Vector methods illuminate the underlying theory.

THE PROJECTILE PROBLEM. I: NO AIR RESISTANCE

We now generalize our previous work on falling bodies to consider a projectile fired at an angle to the earth. In our first discussion, we assume that the air resistance is negligible.

EXAMPLE 1 A projectile is fired from the earth with an initial speed of v_0 ft/sec and at an initial angle of elevation of α radians (rad). Find the equations of motion, provided there is no retarding force due to air resistance.

Solution We set up our axis system in the plane of the projectile with the origin at the point from which it is fired. The x-axis points in the direction in which the horizontal component of velocity increases and the y-axis points upward. (See Fig. 15–7a.) Let

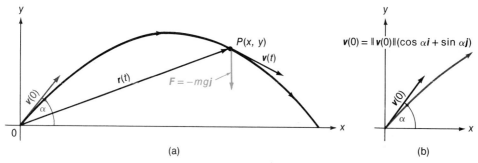

Figure 15–7 Example 1. The projectile problem (no air resistance).

$P(x, y)$ be the location of the projectile at time t. We want to express x and y as functions of t:

$$x = f(t), \quad y = g(t).$$

Let $r(t) = xi + yj = f(t)i + g(t)j$ be the position vector from the origin to $P(x, y)$. The velocity and acceleration vectors at time t are

$$v(t) = r'(t) \quad \text{and} \quad a(t) = v'(t) = r''(t).$$

The only force acting on the projectile is gravity. This force acts downward and has a numerical value of mg, where m is the mass and $g \approx 32$. Thus, this force is

$$F = -mgj.$$

By Newton's second law of motion,

$$F = ma = -mgj,$$

so that the acceleration vector is

$$a(t) = -gj.$$

If we integrate this expression, we get

$$v(t) = \int a(t)\, dt = -gtj + C_1,$$

where C_1 is a constant vector.

To evaluate the vector C_1, we observe that

$$v(0) = -g \cdot 0 \cdot j + C_1 = 0 + C_1 = C_1.$$

Thus,

$$v(t) = -gtj + v(0).$$

15-3 Velocity and Acceleration Problems

Since the original position vector is an antiderivative of the velocity vector, then

$$r(t) = \int v(t)\, dt = -\frac{gt^2}{2}j + v(0)t + C_2,$$

where C_2 is a constant.

To evaluate C_2 we recall that the initial position vector is $r(0) = 0$. Thus,

$$r(0) = \frac{-g \cdot 0^2}{2}j + v(0) \cdot 0 + C_2 = C_2 = 0,$$

so that

$$r(t) = -\frac{gt^2}{2}j + v(0)t.$$

The formula for $r(t)$ can now be expressed in terms of the initial speed v_0 and the initial angle of elevation α.

Observe that v_0 equals the length and that α is the direction angle of the initial vector $v(0)$. (See Fig. 15–7b.) It follows that

$$v(0) = v_0(\cos \alpha\, i + \sin \alpha\, j).$$

Therefore,

$$r(t) = -\frac{gt^2}{2}j + v(0)t$$

$$= -\frac{gt^2}{2}j + v_0(\cos \alpha\, i + \sin \alpha\, j)t$$

$$= (v_0 t \cos \alpha)i + \left(-\frac{gt^2}{2} + v_0 t \sin \alpha\right)j.$$

Since $r(t) = xi + yj$, then the parametric equations for x and y are

$$x = v_0 t \cos \alpha \quad \text{and} \quad y = -\frac{gt^2}{2} + v_0 t \sin \alpha.$$

where $v_0 = \|v(0)\|$. $\square$

The parametric equations

Parametric Equations of Motion

$$x = \|v(0)\| \cos \alpha\, t, \quad y = -\frac{gt^2}{2} + \|v(0)\| \sin \alpha\, t$$

can be used to determine the properties of the motion of the projectile in Example 1. First, if we solve these equations for y in terms of x, we get

$$y = -ax^2 + bx$$

where $a = -g/(2\|v(0)\|^2 \cos^2 \alpha)$ and $b = \tan \alpha$. Thus, the projectile follows a parabolic path.

If we think of the parametric equations as describing the motion of the shadows of the projectile on the coordinate axes, we see that the shadow on the x-axis moves with a uniform speed of $\|v(0)\| \cos \alpha$. The shadow on the y-axis, however, moves as if it had been fired upward with an initial speed of $\|v(0)\| \sin \alpha$ and were subject to the attraction of gravity. The motion of the projectile is completely described by the motion of these two shadows.

By working with $D_t y$ we can find the maximum altitude of the projectile above the earth. By finding when it strikes the ground we can determine the *impact velocity*, the *impact speed* and the *range* (the total horizontal distance traveled from the moment it was fired until the moment of impact).

THE PROJECTILE PROBLEM. II: AIR RESISTANCE (OPTIONAL)

The method used in Example 1 can be modified to solve the corresponding problem that involves a retarding force due to air resistance. We work through a particular example of this type of problem—instead of a general derivation, as in Example 1.

Recall that the numerical coefficient of the retarding force depends on the units of measurement. In the centimeter-gram-second system, this number is much larger than in the foot-pound-second system.

EXAMPLE 2   A projectile of mass 1000 grams (g) is fired from ground level at an angle of elevation of 45° and with an initial speed of 10,000 cm/sec. It is subject to gravity and to a retarding force, due to air resistance, numerically equal to ten times its velocity.
(a) Find the equations of motion.
(b) Use Newton's method to find the speed of the projectile on its impact with the earth.

Solution (a) We follow the same procedure in setting up the axis system as in Example 1. (See Fig. 15–8a.) We let

$$r(t) = xi + yj = f(t)i + g(t)j$$

be the position vector from the origin to $P(x, y)$ at time t.

The projectile is subject to two forces: (1) the force $-mgj = -981,000j$ due to gravity; (2) the force $-10v$ due to air resistance. The resultant force is

$$F = -10v - mgj = -10v - 981,000j.$$

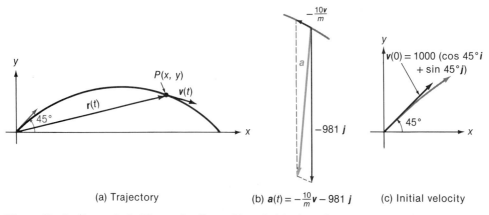

(a) Trajectory   (b) $a(t) = -\frac{10}{m}v - 981\,j$ (c) Initial velocity

Figure 15–8 Example 2. The projectile problem (with air resistance).

15-3 Velocity and Acceleration Problems

By Newton's second law of motion,

$$F = ma = -10v - 981,000j,$$

so that the acceleration is

$$a(t) = -\frac{10}{m}v - \frac{981,000}{m}j = -\frac{1}{100}v - 981j.$$

(See Figure 15-8b.) This leads to the differential equation

$$\frac{dv}{dt} = -\frac{1}{100}v - 981j.$$

We can solve this differential equation by breaking it into component functions. Since $r(t) = f(t)i + g(t)j$, then

$$v(t) = f'(t)i + g'(t)j$$

and

$$\frac{dv}{dt} = a(t) = f''(t)i + g''(t)j.$$

The differential equation can be written as

$$f''(t)i + g''(t)j = -\frac{1}{100}v - 981j$$

$$= -\frac{1}{100}[f'(t)i + g'(t)j] - 981j$$

$$= -\frac{1}{100}f'(t)i + \left[-\frac{1}{100}g'(t) - 981\right]j.$$

If we equate the corresponding component functions, we get two differential equations,

$$f''(t) = -\frac{1}{100}f'(t)$$

$$g''(t) = -\frac{1}{100}g'(t) - 981 = -\frac{1}{100}[g'(t) + 98,100].$$

The first of these differential equations can be simplified by the substitution $u = f'(t)$, the second by the substitution $u = g'(t) + 98,100$. Both of them reduce to the equation

$$\frac{du}{dt} = -\frac{u}{100}.$$

Because of the involved computation in the remainder of this example, we outline the steps of the solution, leaving the details for you.

(1) If we solve the above differential equation by separation of variables, we get $u = Ce^{-t/100}$. Thus, the component functions are of form

$$f'(t) = C_1 e^{-t/100} \quad \text{and} \quad g'(t) = -98,100 + C_2 e^{-t/100},$$

so that
$$v(t) = f'(t)i + g'(t)j$$
$$= C_1 e^{-t/100} i + [-98,100 + C_2 e^{-t/100}]j.$$

(2) To find C_1 and C_2, we observe that
$$v(0) = 10,000 (\cos 45° \, i + \sin 45° \, j) = 5000\sqrt{2} \, (i + j).$$

(See Fig. 15–8c.) If we substitute 0 for t in the expression for $v(t)$ obtained in step (1), we get
$$v(0) = C_1 i + [-98,100 + C_2]j.$$

Thus,
$$C_1 = 5000\sqrt{2} \quad \text{and} \quad C_2 = 98,100 + 5000\sqrt{2},$$

so that
$$f'(t) = C_1 e^{-t/100} = 5000\sqrt{2} e^{-t/100}$$

and
$$g'(t) = -98,100 + C_2 e^{-t/100}$$
$$= -98,100 + (98,100 + 5000\sqrt{2})e^{-t/100}.$$

(3) If we integrate the expressions obtained in step (2) for $f'(t)$ and $g'(t)$, we get
$$f(t) = -500,000\sqrt{2} \, e^{-t/100} + C_3$$
$$g(t) = -98,100t - (9,810,000 + 500,000\sqrt{2})e^{-t/100} + C_4.$$

(4) To evaluate C_3 and C_4, we use the fact that the original position vector is
$$r(0) = 0i + 0j = f(0)i + g(0)j,$$

so that
$$f(0) = 0 \quad \text{and} \quad g(0) = 0.$$

It follows that
$$C_3 = 500,000\sqrt{2} \quad \text{and} \quad C_4 = 9,810,000 + 500,000\sqrt{2}.$$

Therefore,
$$x = f(t) = 500,000 \sqrt{2}(1 - e^{-t/100})$$

and
$$y = g(t) = -98,100t + (9,810,000 + 500,000\sqrt{2})(1 - e^{-t/100}).$$

(b) To find the instant at which the projectile strikes the ground, we must solve the equation
$$y = g(t) = 0$$
$$-98100t + (9,810,000 + 500,000\sqrt{2})(1 - e^{-t/100}) = 0.$$

It follows from Newton's method that the solution is
$$t \approx 14.085 \text{ sec}.$$

15–3 Velocity and Acceleration Problems

The velocity vector is

$$v(t) = f'(t)\mathbf{i} + g'(t)\mathbf{j}$$
$$= 5000\sqrt{2}\, e^{-t/100}\mathbf{i} + [-98{,}100 + (98{,}100 + 5000\sqrt{2})e^{-t/100}]\mathbf{j}.$$

On impact the velocity is

$$v \approx v(14.085) \approx 6142.068\mathbf{i} - 6746.350\mathbf{j}.$$

The speed on impact is approximately equal to

$$\|v(14.085)\| \approx \sqrt{(6142.068)^2 + (6746.350)^2} \approx 9123.499 \text{ cm/sec.} \quad \square$$

THE HANGING-CABLE PROBLEM

The hanging-cable problem was considered in Section 9–8. We now justify some of the assumptions made in that section. Newton's law of motion plays a key role in the argument even though there is no motion.

EXAMPLE 3 A cable of mass m per unit length hangs from points A and B with the lowest point of the sag at point $D\,(0, c)$ on the y-axis. Derive the basic relations used in Section 9–8 to set up the differential equation satisfied by the points on the cable.

Solution Let $P(x, y)$ be a point on the cable. We consider only the portion of the cable from D to P. Let $s = s(x)$ be the length of this part of the cable. Its mass is ms. (See Fig. 15–9a.) We work the remainder of the problem under the assumption that $x > 0$. A similar argument can be used if $x < 0$.

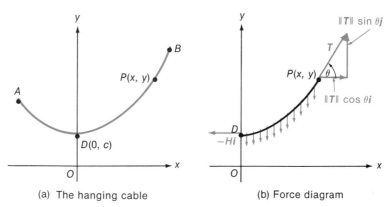

(a) The hanging cable (b) Force diagram

Figure 15–9 Example 3. The hanging cable problem.

The following forces act on this part of the cable:

(1) A gravitational force equal to

$$-ms g \mathbf{j}$$

which acts downward.

(2) A horizontal tension force at D. If H is the numerical value of this force, then the force is

$$-H\mathbf{i}.$$

(3) A tangential tension force T at P. If the direction angle of this force vector is θ and its numerical value is T, then

$$\mathbf{T} = T(\cos\theta \mathbf{i} + \sin\theta \mathbf{j}).$$

Since the cable is in equilibrium, there is no motion. It follows from Newton's second law of motion that the resultant of the forces must be zero. Therefore,

$$-mg\mathbf{j} - H\mathbf{i} + T(\cos\theta \mathbf{i} + \sin\theta \mathbf{j}) = \mathbf{0}$$
$$(-H + T\cos\theta)\mathbf{i} + (-mg + T\sin\theta)\mathbf{j} = \mathbf{0}$$

so that

$$T\cos\theta = H \quad \text{and} \quad T\sin\theta = mg.$$

We can now finish the derivation as in Section 9–8. The final form of the formula is

$$y = \frac{H}{mg}\cosh\frac{mgx}{H} + c$$

where c is the y-coordinate of the point D. □

Exercises 15–3

•1 A projectile is fired from a horizontal plane with an initial speed of 1200 ft/sec and an angle of elevation of 30°. Assume that there is no retarding force due to air resistance. Find the velocity at time t, the maximum altitude, the total time in flight and the range.

2 Work through the details of Example 1 using the data in Exercise 1.

3 Modify the argument in Example 1 to cover the situation in which the initial height of the projectile is y_0. Does the projectile still move along a parabolic path?

•4 As part of a promotional stunt, a motorcycle carrying a dummy is driven horizontally off of a 100-ft-high vertical cliff at a speed of 200 ft/sec. When and where will it strike the ground? (Assume that there is no air resistance.)

•5 A projectile fired from a flat plane at an initial speed of 1000 ft/sec has a range of 25,000 ft. What was its angle of elevation? (Assume that there is no air resistance.)

•6 Mr. W. Tell plans to shoot an apple from his son's head with an arrow. He is 25 ft from his son and will release the arrow at the same height as the apple. He knows that the initial velocity of the arrow will be 40 ft/sec. What must the angle of elevation be for him to hit the apple? (Assume that there is no air resistance.)

7 A projectile is fired into the air at an angle of elevation of α radians. Show that if there is no retarding force due to air resistance, then doubling the initial speed will cause both the maximum height and the range to be quadrupled.

8 Work through Example 2, filling in all the details.

•9 Work through the steps of Example 2 on the assumption that the initial angle of elevation is 30°, the initial speed is 15,000 cm/sec, and the retarding force due to air resistance is equal to ten times the velocity.

10 An object in equilibrium is subject to forces $F_1 = x_1 i + y_1 j + z_1 k$, $F_2 = x_2 i + y_2 j + z_2 k$, and $F_3 = x_3 i + y_3 j + z_3 k$.
  (a) Use Newton's second law of motion to explain why
  $$F_1 + F_2 + F_3 = 0.$$
  (b) Explain why $x_1 + x_2 + x_3 = y_1 + y_2 + y_3 = z_1 + z_2 + z_3 = 0$.

11 A projectile is fired at an angle α to the horizontal with an initial velocity of $\|v(0)\|(\cos \alpha i + \sin \alpha j)$ from a point on a hillside that is inclined at an angle θ to the horizontal, where $0 < \theta < \pi/2$. Assume there is no air resistance.
  (a) Show that the maximum (oblique) range is obtained when $\alpha = \theta/2 + \pi/4$ if the projectile is fired up the hill.
  (b) Find the maximum (oblique) range when the projectile is fired down the hill.

15–4 Curvature. I: Plane Curves

A smooth space curve $\mathscr{C}$, defined as the graph of
$$r(t) = f(t)i + g(t)j + h(t)k,$$
can be considered as if it were traced out by a moving point P. At time t the point is located at $(f(t), g(t), h(t))$.

The location of P can be specified in other ways. We can fix a point P_0 on $\mathscr{C}$ (obtained at time t_0) and measure the arc length s from P_0 to P:
$$s = \int_{t_0}^{t} \sqrt{f'(t)^2 + g'(t)^2 + h'(t)^2}\, dt.$$

The location of the moving point is then determined by the value of s.

There are advantages to locating P by arc length. In particular, it enables us to study intrinsic properties of the curve $\mathscr{C}$ without regard to how the curve is traced out with respect to time. We now consider one of those intrinsic properties—*curvature*, beginning with the curvature of plane curves. Later (Section 15–5) we develop some vector properties related to curvature and extend the concept to space curves.

Let a smooth curve in the plane be the graph of the position vector
$$r(t) = f(t)i + g(t)j.$$
The derivative
$$v(t) = f'(t)i + g'(t)j$$
is a tangent vector to the curve at $(f(t), g(t))$ that points in the direction of increasing arc length.

Let $\phi(t)$ be the direction angle of $v(t)$. (See Fig. 15–10.) The derivative $d\phi/ds$ gives a measure of the rate at which the direction angle changes as the arc length s increases. (Here we relax the restriction that $0 \leq \phi < 2\pi$ and assume that ϕ varies continuously with s.)

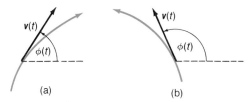

Figure 15–10 $\phi = \phi(t)$ is the direction angle of $v(t)$.

Several properties of the graph can be described in terms of the derivative $d\phi/ds$. If $d\phi/ds > 0$, then the curve bends to the left. (The direction angle increases as the curve is traced out.) If $d\phi/ds < 0$, it bends to the right. (See Figs. 15–11a, b.) If $|d\phi/ds|$ is large, then the curve bends very sharply. If $|d\phi/ds|$ is close to zero, it bends very slightly. (See Figs. 15–11c, d.) In the special case in which $d\phi/ds = 0$ over an interval, there is no bending. The graph is a line.

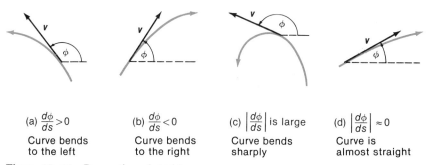

(a) $\frac{d\phi}{ds} > 0$ Curve bends to the left

(b) $\frac{d\phi}{ds} < 0$ Curve bends to the right

(c) $\left|\frac{d\phi}{ds}\right|$ is large Curve bends sharply

(d) $\left|\frac{d\phi}{ds}\right| \approx 0$ Curve is almost straight

Figure 15–11 Properties of $d\phi/ds$.

DEFINITION

Curvature

The number
$$K = \left|\frac{d\phi}{ds}\right|$$

is called the *curvature* of the graph of r at the point $(f(t), g(t))$.

As indicated above, the curvature measures the rate of bending of a curve at a point. If $K = 0$, the curve approximates its tangent line in the vicinity of the point. The greater the value of K, the greater the bend away from the tangent line.

This definition does not give us a technique for calculating the curvature. We now develop some methods for the calculation.

THE GRAPH OF $y = f(x)$

Let f be twice differentiable. Recall that the graph of the function f can be obtained from the vector function $r(t) = ti + f(t)j$. (This is obtained by letting $x = t$, so that $y = f(x) = f(t)$. The graph of these parametric equations is the same as the graph of the vector function $r(t) = ti + f(t)j$.)

15-4 Curvature. I: Plane Curves

The velocity vector for this vector function is

$$v(t) = i + f'(t)j.$$

(See Fig. 15-12.)

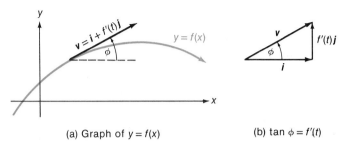

(a) Graph of $y = f(x)$ (b) $\tan \phi = f'(t)$

Figure 15-12 The graph of $y = f(x)$ is the graph of $r(t) = ti + f(t)j$.

The direction angle for the velocity vector is $\phi(t)$. Observe in Figure 15-12 that ϕ is the angle of inclination of the tangent line to the graph. Thus, $\tan \phi(t) = f'(t)$. Taking the derivative with respect to t, we obtain

$$\sec^2 \phi \cdot \frac{d\phi}{dt} = f''(t)$$

$$\frac{d\phi}{dt} = \frac{f''(t)}{\sec^2 \phi} = \frac{f''(t)}{\tan^2 \phi + 1} = \frac{f''(t)}{f'(t)^2 + 1} \quad \text{(by T6)}.$$

Since

$$\frac{ds}{dt} = \|v(t)\| = \sqrt{f'(t)^2 + 1},$$

then, by the Chain Rule,

$$\frac{d\phi}{ds} = \frac{d\phi/dt}{ds/dt} = \frac{\frac{f''(t)}{f'(t)^2 + 1}}{\sqrt{f'(t)^2 + 1}} = \frac{f''(t)}{[f'(t)^2 + 1]^{3/2}}.$$

$$K = \left|\frac{d\phi}{ds}\right| = \frac{|f''(t)|}{[f'(t)^2 + 1]^{3/2}}.$$

Recall that (in this derivation) we have $x = t$ and $y = f(t) = f(x)$. Thus

Curvature Formula for Functions of x

$$K = \frac{|y''|}{[(y')^2 + 1]^{3/2}}.$$

EXAMPLE 1 Let $y = x^2$. Calculate the curvature of the graph at $(1, 1)$.

Solution Since $y = x^2$, then $y' = 2x$ and $y'' = 2$. Then

$$K(x) = \frac{|y''|}{[(y')^2 + 1]^{3/2}} = \frac{2}{(4x^2 + 1)^{3/2}}.$$

When $x = 1$,

$$K = \frac{2}{5^{3/2}}.$$

EXAMPLE 2 Find the point on the graph of $y = x^2$ where the curvature is the greatest. Is there a point where it is zero?

Solution We saw in Example 1 that the curvature at the point (x, x^2) is

$$K(x) = \frac{2}{(4x^2 + 1)^{3/2}}.$$

The maximum value of K is obtained at the point where $x = 0$, the vertex of the parabola. This maximum value is

$$K(0) = 2.$$

The further x is from the origin, the more closely the curvature approaches zero. There is no value of x for which $K(x) = 0$. □

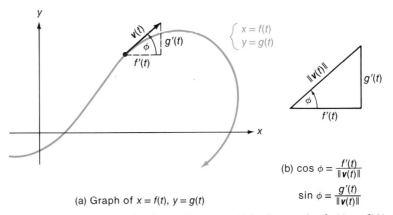

(a) Graph of $x = f(t)$, $y = g(t)$

(b) $\cos \phi = \dfrac{f'(t)}{\|v(t)\|}$

$\sin \phi = \dfrac{g'(t)}{\|v(t)\|}$

Figure 15–13 The graph of $x = f(t)$, $y = g(t)$ is the graph of $r(t) = f(t)i + g(t)j$.

CURVATURE OF PLANE CURVES DEFINED PARAMETRICALLY

Let $\mathscr{C}$ be a smooth curve defined parametrically by $x = f(t)$, $y = g(t)$, where f and g are twice differentiable. The vector function with the same graph is

$$r(t) = f(t)i + g(t)j.$$

The velocity vector is

$$v(t) = f'(t)i + g'(t)j$$

and the speed is

$$\frac{ds}{dt} = s'(t) = \sqrt{f'(t)^2 + g'(t)^2} = \|v(t)\|.$$

15-4 Curvature. I: Plane Curves

Observe that $\phi(t)$, the direction angle for $v(t)$, satisfies the conditions

$$\cos \phi = \frac{f'(t)}{\|v(t)\|} \quad \text{and} \quad \sin \phi = \frac{g'(t)}{\|v(t)\|}$$

(see Fig. 15-13), so that

$$\tan \phi = \frac{\sin \phi}{\cos \phi} = \frac{g'(t)}{f'(t)}.$$

The derivative with respect to t is

$$\sec^2 \phi \cdot \frac{d\phi}{dt} = \frac{d}{dt}\left(\frac{g'(t)}{f'(t)}\right) = \frac{f'(t)g''(t) - g'(t)f''(t)}{f'(t)^2}.$$

If we divide both sides of the above equation by $\sec^2 \phi$, substitute

$$\sec^2 \phi = \tan^2 \phi + 1 = \frac{g'(t)^2 + f'(t)^2}{f'(t)^2},$$

and simplify, we get

$$\frac{d\phi}{dt} = \frac{f'(t)g''(t) - g'(t)f''(t)}{f'(t)^2 + g'(t)^2}.$$

By the Chain Rule,

$$\frac{d\phi}{ds} = \frac{d\phi/dt}{ds/dt} = \frac{\dfrac{f'(t)g''(t) - g'(t)f''(t)}{f'(t)^2 + g'(t)^2}}{\sqrt{f'(t)^2 + g'(t)^2}}$$

$$= \frac{f'(t)g''(t) - g'(t)f''(t)}{[f'(t)^2 + g'(t)^2]^{3/2}}.$$

Taking absolute values, we get the curvature formula

Curvature for Graph of Parametric Equations

$$K(t) = \left|\frac{d\phi}{ds}\right| = \frac{|f'(t)g''(t) - g'(t)f''(t)|}{[f'(t)^2 + g'(t)^2]^{3/2}}.$$

EXAMPLE 3 Show that a circle with radius $a > 0$ has constant curvature $K = 1/a$.

Solution Assume that the circle has center at the origin so that it is the graph of the parametric equations

$$x = f(t) = a \cos t, \quad y = g(t) = a \sin t.$$

Then

$$f'(t) = -a \sin t, \quad g'(t) = a \cos t,$$
$$f''(t) = -a \cos t, \quad g''(t) = -a \sin t.$$

The curvature at $(f(t), g(t))$ is

$$K(t) = \frac{|f'(t)g''(t) - g'(t)f''(t)|}{[f'(t)^2 + g'(t)^2]^{3/2}}$$

$$= \frac{|a^2 \sin^2 t + a^2 \cos^2 t|}{[a^2 \sin^2 t + a^2 \cos^2 t]^{3/2}} = \frac{a^2}{a^3} = \frac{1}{a}. \quad \square$$

THE CIRCLE OF CURVATURE

Recall that the tangent line to the plane curve $\mathscr{C}$ at the point (a, b) is the line that best approximates the curve in the vicinity of the point. We now show how to obtain the circle that best approximates $\mathscr{C}$ in the vicinity of (a, b) provided $\mathscr{C}$ has positive curvature at the point.

Circle of Curvature

The *circle of curvature* (*osculating circle*) to the curve $\mathscr{C}$ at the point (a, b) is defined as the circle with the following properties (see Fig. 15–14):

Properties of Circle of Curvature

(1) The circle passes through (a, b) and has the same tangent line at that point as the original curve $\mathscr{C}$. This condition implies that the center of the circle is on a line that is normal to the curve $\mathscr{C}$ at (a, b).
(2) The center of the circle is on the concave side of the graph.
(3) The circle has the same curvature at (a, b) as the original curve $\mathscr{C}$.

Let ρ be the radius of the circle of curvature that is tangent to the curve $\mathscr{C}$ at the point (a, b). Let K be the curvature of $\mathscr{C}$ at (a, b). It follows from condition (3) and Example 3 that

$$\rho = \frac{1}{K}.$$

Radius of Curvature

Thus, the radius of the circle of curvature equals the reciprocal of the curvature of $\mathscr{C}$ at the point (a, b). The number ρ is called the *radius of curvature*.

In particular cases, it may be easy to find the center of the circle of curvature, but the general procedure, based on our current development, is awkward. As we

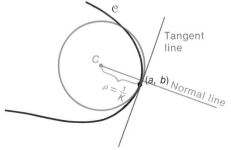

Figure 15–14 The circle of curvature.

develop some additional topics related to curvature, one consequence is a relatively simple method for finding the center of the circle of curvature.

EXAMPLE 4 Find the equation of the circle of curvature to the graph of $y = x^2$ at the point $(0, 0)$.

Solution As we saw in Example 2, the curvature of the parabola at $(0, 0)$ is $K = 2$. Thus, the radius of the circle of curvature is

$$\rho = \frac{1}{K} = \frac{1}{2}.$$

In finding the center of the circle of curvature, we observe that the y-axis is normal to the parabola at $(0, 0)$ and that the concave side is above the curve. Since the radius is $\rho = \frac{1}{2}$, the center must be at the point $(0, \frac{1}{2})$. The equation of the circle of curvature is

$$x^2 + (y - \tfrac{1}{2})^2 = \tfrac{1}{4}.$$

(See Fig. 15–15.) □

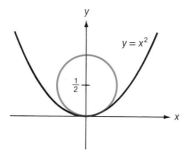

Figure 15–15 Example 4.

Exercises 15–4

1–6. Find the curvature of the graph at the given point.

- **1** $y = x^2 - 2x + 3$, $P(1, 2)$
- **2** $y = xe^x$, $P(1, e)$
- **3** $y = \sin x$, $P\left(\dfrac{\pi}{4}, \dfrac{\sqrt{2}}{2}\right)$
- **4** $y = \ln(\cos x)$, $P(0, 0)$
- **5** $y = e^x - x$, $P(0, 1)$
- **6** $y = \ln x$, $P(1, 0)$

7–10. Find the points at which the curvature is a maximum.

- **7** $y = \ln(\sin x)$ on $(0, \pi)$
- **8** $y = \ln x$
- **9** $y = \sin x$ on $[0, \pi]$
- **10** $y = e^x$

11–14. Find the curvature of the graph of the parametric equations at the given point.

- **11** $x = t - t^2$, $y = 1 - t^3$, $P(-2, 2)$
- **12** $x = t^2 - 2t$, $y = 3t$, $P(-1, 3)$

•13 $x = \sin 2t$, $y = \cos t$, $P(0, -1)$

14 $x = t - \sin t$, $y = 1 - \cos t$, $P\left(\dfrac{\pi}{2} - 1, 1\right)$

•15 Find the curvature at the local maximum of $f(x) = x^3 - x^2 + 1$. Sketch the graph showing the circle of curvature.

•16 Find all points at which the curvature of the graph of $y = \cos x$ is a maximum or a minimum. Sketch the graph showing the circle of curvature at a point at which the curvature is a maximum. (*Hint:* First find the points on the interval $[-\pi/2, \pi/2]$.)

17–20. Find the equation of the circle of curvature that is tangent to the curve at the given point. Sketch the graph and the circle of curvature.

•17 $y = e^x$, $P(0, 1)$ •18 $y = xe^x$, $P(0, 0)$

•19 $y = x^2$, $P(1, 1)$ •20 $y = 2\cos x$, $P(0, 2)$

21 Explain why the curvature of the graph of $y = f(x)$ must be zero at a point of inflection if the curvature is defined there.

22 Let $\mathscr{C}$ be the circle of curvature that is tangent to a parabola at its vertex. Show that the focus of the parabola is midway between the center of the circle $\mathscr{C}$ and the vertex of the parabola. (*Hint:* First perform any necessary translations and rotations of axes so that the parabola has the equation $x^2 = 4py$, $p > 0$.)

23 Let $a, b > 0$. The upper branch of the hyperbola $y^2/a^2 - x^2/b^2 = 1$ is the graph of $y = f(x) = a\sqrt{x^2 + b^2}/b$.
   (a) Show that the maximum curvature of this branch of the hyperbola occurs at the vertex.
   (b) Show that the circle of curvature that is tangent to this curve at its vertex has its center at the point $(0, (a^2 + b^2)/a)$. Illustrate this result with a drawing of the hyperbola $y^2/3^2 - x^2/4^2 = 1$.

24 (*Curvature in Polar Coordinates*) Let $r = f(\theta)$ define a polar curve. Use parametric equations to derive the curvature formula

Curvature in Polar Coordinates

$$K(\theta) = \dfrac{|f(\theta)^2 - f(\theta)f''(\theta) + 2f'(\theta)^2|}{[f(\theta)^2 + f'(\theta)^2]^{3/2}}$$

•25 Use the formula in Exercise 24 to find the curvature of the spiral $r = e^\theta$.

26 (*Center of Curvature*) Let $P_0(x_0, y_0)$ be a point on the graph of $y = f(x)$ where the curvature is not zero. Let $C(\alpha, \beta)$ be the center of the circle of curvature that is tangent to the graph at P_0. Show that

Center of Curvature

$$\alpha = x_0 - \dfrac{f'(x_0)(1 + f'(x_0)^2)}{f''(x_0)}, \quad \beta = y_0 + \dfrac{1 + f'(x_0)^2}{f''(x_0)}.$$

The point $C(\alpha, \beta)$ is called the *center of curvature* for P_0.

15-5 Curvature. II: Unit Tangent and Normal Vectors. Space Curves

As we shall see in this and the following section, certain problems involving plane curves can best be solved by using two auxiliary vectors—a unit tangent vector $T(t)$ and a unit normal vector $N(t)$. It is standard to choose these vectors so that T points in the direction of increasing arc length and N points towards the concave side of the curve. (See Fig. 15–16.)

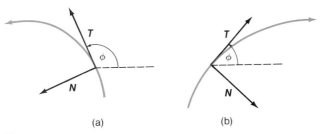

Figure 15–16 The unit tangent $T = T(\phi)$ and principal unit normal $N = N(\phi)$. The vector N points towards the concave side of the curve.

The unit tangent vector $T = T(t)$ is easily calculated. Since it points in the same direction as $v(t)$, then

Unit Tangent Vector T
$$T(t) = \frac{v(t)}{\|v(t)\|} = \frac{v(t)}{ds/dt}.$$

The normal vector N will be defined later.

The vector T can be considered a function of time t or a function of arc length s. As we shall see, dT/ds, the derivative of T with respect to arc length, is important for much of our work.

THE VECTOR dT/ds

Our first task is to study the properties of the vector dT/ds. Observe first, that by the Chain Rule,

$$\frac{dT}{ds} = \frac{dT}{d\phi} \frac{d\phi}{ds},$$

where ϕ is the direction angle of T. (See Fig. 15–16.) Since $K = |d\phi/ds|$ is the curvature, then

$$\left\|\frac{dT}{ds}\right\| = \left\|\frac{dT}{d\phi}\right\| \cdot \left|\frac{d\phi}{ds}\right| = K \left\|\frac{dT}{d\phi}\right\|.$$

We now turn our attention to $dT/d\phi$. Since ϕ is the direction angle for the unit tangent vector T, then

$$T = \cos \phi i + \sin \phi j,$$

so that

$$\frac{dT}{d\phi} = -\sin \phi i + \cos \phi j$$

$$= \cos\left(\phi + \frac{\pi}{2}\right) i + \sin\left(\phi + \frac{\pi}{2}\right) j$$

(by T8 and T9). It follows that $dT/d\phi$ is a unit vector with direction angle $\phi + \pi/2$. Thus, it can be obtained from T by a counterclockwise rotation of 90°. (See Fig. 15–17.)

Since $dT/d\phi$ is a unit vector, then

$$\left\|\frac{dT}{ds}\right\| = K \quad \left\|\frac{dT}{d\phi}\right\| = K$$

that is, *the magnitude of dT/ds equals the curvature.*

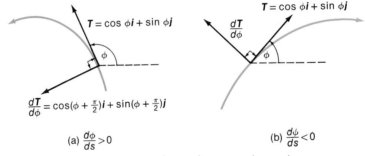

(a) $\frac{d\phi}{ds} > 0$ (b) $\frac{d\psi}{ds} < 0$

Figure 15–17 $dT/d\phi = \cos\left(\phi + \frac{\pi}{2}\right) i + \sin\left(\phi + \frac{\pi}{2}\right) j$. The vector $dT/d\phi$ is obtained from T by a 90° counterclockwise rotation.

THE DIRECTION OF dT/ds

Our next result follows from the remarks about $d\phi/ds$ made at the beginning of Section 15–4. Recall that a curve bends to the left if $d\phi/ds > 0$, and to the right if $d\phi/ds < 0$. We consider these two cases separately.

Case 1: $d\phi/ds > 0$. In this case, the concave side of the curve is to the left of the curve. Since $dT/d\phi$ also points to the left, then $dT/d\phi$ points towards the concave side of the curve. (See Fig. 15–18a.) Since

$$\frac{dT}{ds} = \frac{dT}{d\phi} \frac{d\phi}{ds},$$

then dT/ds also points towards the concave side of the curve.

15-5 Curvature. II: Unit Tangent and Normal Vectors. Space Curves

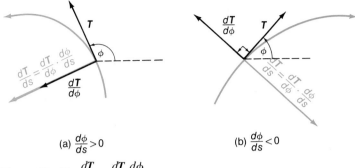

(a) $\dfrac{d\phi}{ds} > 0$ (b) $\dfrac{d\phi}{ds} < 0$

Figure 15-18 $\dfrac{dT}{ds} = \dfrac{dT}{d\phi} \dfrac{d\phi}{ds}.$

Case 2: $d\phi/ds < 0$. In this case the curve bends to the right. Since $dT/d\phi$ points to the left, then it points away from the concave side of the curve. (See Fig. 15–18b.) When we multiply $dT/d\phi$ by the negative number $d\phi/ds$, we reverse the direction of the vector. Thus,

$$\frac{dT}{ds} = \frac{dT}{d\phi}\frac{d\phi}{ds}$$

points towards the concave side of the curve.

In either of the above cases, dT/ds is a normal vector that points towards the concave side of the curve.

The unit normal vector N, mentioned at the beginning of this section, is defined to be

Principal Unit Normal Vector N

$$N = \frac{dT/ds}{\|dT/ds\|} = \frac{1}{K}\frac{dT}{ds}.$$

This vector is called the *principal unit normal vector*. Since it is in the same direction as the normal vector dT/ds, it points towards the concave side of the curve. (See Fig. 15–19.)

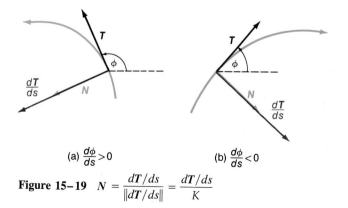

(a) $\dfrac{d\phi}{ds} > 0$ (b) $\dfrac{d\phi}{ds} < 0$

Figure 15-19 $N = \dfrac{dT/ds}{\|dT/ds\|} = \dfrac{dT/ds}{K}$

Remark It can be seen that dT/ds is the key to our development. The magnitude of this vector equals K, the curvature. If we divide dT/ds by its magnitude, we get N, the unit normal vector that points towards the concave side of the curve. Furthermore, since T is a function of t, the vector dT/ds can be calculated by the Chain Rule relation:

$$\frac{dT}{ds} = \frac{dT/dt}{ds/dt} = \frac{dT/dt}{\|v(t)\|}.$$

EXAMPLE 1 Let $r(t) = 4\cos 3t\,i + 4\sin 3t\,j$. Calculate T, dT/ds, K, and N as functions of t.

Solution Observe first that

$$v(t) = -12\sin 3t\,i + 12\cos 3t\,j$$

and

$$\frac{ds}{dt} = \|v(t)\| = 12.$$

Then

(a) $$T = T(t) = \frac{v(t)}{\|v(t)\|} = -\sin 3t\,i + \cos 3t\,j.$$

(b) $$\frac{dT}{ds} = \frac{dT/dt}{ds/dt} = \frac{-3\cos 3t\,i - 3\sin 3t\,j}{12} = \frac{-\cos 3t\,i - \sin 3t\,j}{4}.$$

(c) The curvature is

$$K = \left\|\frac{dT}{ds}\right\| = \left\|\frac{-\cos 3t\,i - \sin 3t\,j}{4}\right\| = \frac{1}{4}.$$

(d) The principal unit normal vector is

$$N = \frac{dT/ds}{\|dT/ds\|} = \frac{\dfrac{-\cos 3t\,i - \sin 3t\,j}{4}}{\dfrac{1}{4}}$$

$$= -\cos 3t\,i - \sin 3t\,j. \ \square$$

THE CIRCLE OF CURVATURE

There is a relatively simple way of finding the center of the circle of curvature. Let $\mathscr{C}$ be the circle of curvature that is tangent to a curve at (a, b). Recall that the center of $\mathscr{C}$ is on a line normal to the curve at (a, b) and is located on the concave side of the curve at a distance $\rho = 1/K$ from (a, b). It follows that the vector

$$\rho N = \frac{1}{K} N$$

points from (a, b) to the center of the circle of curvature. (See Fig. 15–20.)

15-5 Curvature. II: Unit Tangent and Normal Vectors. Space Curves

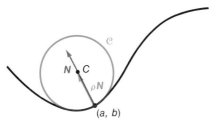

Figure 15–20 The circle of curvature. The vector ρN points from the point of tangency to the center of the circle of curvature.

Example 2 shows how this result can be used to find the center of the circle of curvature.

EXAMPLE 2   Let

$$r(t) = (2t + 2)i + 2 \ln (\cos t)j, \quad -\frac{\pi}{2} < t < \frac{\pi}{2}.$$

(a) Calculate T, N, and the curvature K.
(b) Find the equation of the circle of curvature that is tangent to the graph at $(2, 0)$.

Solution (a) Since $r(t) = (2t + 2)i + 2 \ln (\cos t)j$, then

$$v(t) = 2i + \frac{2}{\cos t}(-\sin t)j = 2i - 2 \tan t j$$

and

$$\frac{ds}{dt} = \|v(t)\| = 2\sqrt{1 + \tan^2 t} = 2\sqrt{\sec^2 t} = 2 \sec t.$$

The unit tangent vector $T(t)$ is

$$T(t) = \frac{v(t)}{ds/dt} = \frac{2i - 2 \tan t j}{2 \sec t} = \cos t i - \sin t j.$$

The derivative of T with respect to time is

$$\frac{dT}{dt} = \frac{d}{dt}(\cos t i - \sin t j) = -\sin t i - \cos t j.$$

The derivative of T with respect to arc length is

$$\frac{dT}{ds} = \frac{dT/dt}{ds/dt} = \frac{-\sin t i - \cos t j}{2 \sec t}.$$

The curvature is

$$K(t) = \left\| \frac{dT}{ds} \right\| = \frac{1}{2 \sec t}\sqrt{\sin^2 t + \cos^2 t} = \frac{1}{2 \sec t}.$$

The principal unit normal vector is

$$N(t) = \frac{dT/ds}{K} = \frac{(-\sin t\,i - \cos t\,j)/(2 \sec t)}{1/(2 \sec t)} = -\sin t\,i - \cos t\,j.$$

(b) We now find the equation of the circle of curvature that is tangent to the graph at $(2, 0)$, the point that corresponds to $t = 0$. Let $C(x_0, y_0)$ be the center of this circle. Since the vector $\rho(0)N(0)$ points from $(2, 0)$ to (x, y), then

$$\rho(0)N(0) = (x_0 - 2)i + (y_0 - 0)j$$
$$2 \sec 0 \cdot (-\sin 0\,i - \cos 0\,j) = (x_0 - 2)i + (y_0 - 0)j$$
$$-2j = (x_0 - 2)i + y_0 j$$

It follows that

$$x_0 = 2 \quad \text{and} \quad y_0 = -2.$$

The center of the circle of curvature is $(2, -2)$.

Since the radius of the circle of curvature is $\rho(0) = 2 \sec 0 = 2$, its equation is

$$(x - 2)^2 + (y + 2)^2 = 2^2. \quad \square$$

CURVATURE OF SPACE CURVES

There is no direct analogue for the angle ϕ when we consider curves in space. Thus we define the vectors T and N and the curvature K by the vector relations we developed for plane curves.

To be specific, let the space curve be the graph of

$$r(t) = f(t)i + g(t)j + h(t)k.$$

We define

T, K, N for Space Curves

$$T(t) = \frac{v(t)}{\|v(t)\|} = \frac{v(t)}{ds/dt}$$

$$K(t) = \left\|\frac{dT}{ds}\right\|$$

and

$$N(t) = \frac{1}{K(t)}\frac{dT}{ds} = \frac{dT/ds}{\|dT/ds\|}.$$

As in the case of plane curves, we calculate dT/ds by the Chain Rule Relation:

$$\frac{dT}{ds} = \frac{dT/dt}{ds/dt} = \frac{dT/dt}{\|v(t)\|}.$$

It can be proved that N is a unit vector that is orthogonal to T. Thus, it is a vector normal to the curve.

EXAMPLE 3 Let $r(t) = \cos t\,i + \sin t\,j + t\,k$. (The graph is a helix.) Find T, N, and the curvature K at the point $(1, 0, 0)$.

Solution $v(t) = -\sin t\,i + \cos t\,j + k$, so that

$$\frac{ds}{dt} = \|v(t)\| = \sqrt{\sin^2 t + \cos^2 t + 1} = \sqrt{2}.$$

The unit tangent vector is

$$T(t) = \frac{v(t)}{ds/dt} = \frac{-\sin t\,i + \cos t\,j + k}{\sqrt{2}}.$$

The derivative of T with respect to t is

$$\frac{dT}{dt} = \frac{-\cos t\,i - \sin t\,j}{\sqrt{2}}.$$

The derivative of T with respect to s is

$$\frac{dT}{ds} = \frac{dT/dt}{ds/dt} = \frac{-\cos t\,i - \sin t\,j}{2}.$$

The curvature is

$$K(t) = \left\|\frac{dT}{ds}\right\| = \frac{1}{2}.$$

The principal unit normal vector is

$$N(t) = \frac{dT/ds}{K(t)} = -\cos t\,i - \sin t\,j.$$

The point $(1, 0, 0)$ is obtained when $t = 0$. At this point we have

$$T = \frac{j + k}{\sqrt{2}}, \qquad N = -i, \qquad \text{and} \qquad K = \frac{1}{2}. \quad \square$$

Remark Our extension of the curvature formulas to space curves is not quite as arbitrary as it first appears. Since T is a unit tangent vector, then any change in T must correspond to a change in direction, not a change in length. Thus, there should be some type of relationship between dT/ds and the bending of the curve. It is, perhaps, surprising that for a plane curve the relationship is so simple—that $K = \|dT/ds\|$.

There is an alternative approach to the curvature problem that avoids the direction angle ϕ altogether. In that approach we simply define the curvature of a plane curve to be $\|dT/ds\|$, the magnitude of dT/ds. If that is done, then the same definition holds for space curves and plane curves.

Exercises 15–5

1–6. Calculate T, N, and K as functions of t.

•1 $r(t) = t^2 i + (1 + t)j$

2 $r = \ln(\sec t)i + tj$, $-\dfrac{\pi}{2} < t < \dfrac{\pi}{2}$

•3 $r(t) = 4ti + t^2 j + 2t^2 k$

4 $r = 3ti + t^2 j + tk$

•5 $r(t) = \sin 2ti + \cos 2tj - 3tk$

6 $r = e^t[\cos t\,i + \sin t\,j + 2k]$

7–10. Use the methods of this section to find the equation of the circle of curvature at the indicated point. Draw a figure showing the graph and the circle of curvature.

• **7** $y = \ln x$, $P(1, 0)$

• **8** $y = e^{2x}$, $P(0, 1)$

• **9** $y = x^2$, $P(-1, 1)$

• **10** $y = \sin x$, $P\left(\dfrac{\pi}{4}, \dfrac{1}{\sqrt{2}}\right)$

---

11 The graphs of $y = f(x)$ and $r(t) = ti + f(t)j$ are the same. Calculate T and dT/ds. Use $K = \|dT/ds\|$ to derive the formula

$$K = \frac{|f''(x)|}{[1 + f'(x)^2]^{3/2}}.$$

12 (a) The graph of the parametric equations $x = f(t)$, $y = g(t)$ is the same as the graph of $r(t) = f(t)i + g(t)j$. Follow the general instructions of Exercise 11 and derive the formula

$$K = \frac{|f'(t)g''(t) - g'(t)f''(t)|}{[f'(t)^2 + g'(t)^2]^{3/2}}.$$

(b) Explain how the process used in (a) can be generalized to derive a formula for the curvature of a space curve. Can you guess why we do not derive such a formula?

15–6 Tangential and Normal Components of Acceleration

Let the vector $r(t)$ describe the position of a particle at time t. The unit tangent and unit normal vectors $T(t)$ and $N(t)$ can be used to get information about the motion of the particle.

Observe first that since

$$T = \frac{v(t)}{ds/dt}$$

then

$$v(t) = \frac{ds}{dt} T.$$

If we differentiate this expression, we get

$$a(t) = v'(t) = \frac{d}{dt}\left(\frac{ds}{dt} T\right) = \frac{d^2s}{dt^2} T + \frac{ds}{dt} \frac{dT}{dt}.$$

Recall that (by the Chain Rule),

$$\frac{dT}{dt} = \frac{ds}{dt} \cdot \frac{dT}{ds}$$

15-6 Tangential and Normal Components of Acceleration

and that

$$\frac{dT}{ds} = KN.$$

Therefore,

$$a(t) = \frac{d^2s}{dt^2} T + \frac{ds}{dt} \cdot \frac{ds}{dt} \cdot \frac{dT}{ds}$$

Acceleration Vector in Terms of T and N

$$a(t) = \frac{d^2s}{dt^2} T + \left(\frac{ds}{dt}\right)^2 KN.$$

This equation expresses the acceleration vector in terms of T and N. In particular, observe that $a(t)$ is in the plane determined by T and N. (See Fig. 15–21.)

Tangential Component

The number d^2s/dt^2 is called the *tangential component of acceleration* or the *scalar acceleration*. It measures the rate at which the speed of the particle changes as the curve is traced out. It is proportional to the rate of movement of the speedometer needle of an automobile.

Normal Component

The number $(ds/dt)^2 K$ is called the *normal component of acceleration*. This component is always nonnegative. It follows that the acceleration vector must always point in the general direction of the concave side of the curve if ds/dt and K are both nonzero. (See Fig. 15–21.)

Several results about constant speeds—some of which were obtained earlier—are simple consequences of the above expression for $a(t)$:

(1) *If a particle moves with a constant speed, then its acceleration vector is orthogonal to its velocity vector and points towards the concave side of the curve* (unless $a(t) = 0$). This follows from the above remarks and the fact that if $ds/dt = c$, then $d^2s/dt^2 = 0$. Thus,

$$a(t) = 0 \cdot T + K c^2 N = Kc^2 N.$$

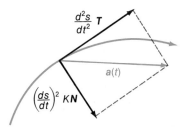

Figure 15–21 The tangential and normal components of acceleration.

$$a(t) = \frac{d^2s}{dt^2} T + \left(\frac{ds}{dt}\right)^2 KN.$$

(2) *If a particle traces out a curve at a constant speed $ds/dt = c$, and then retraces the same curve at the constant speed of $2c$ (twice the original speed), then the magnitude of the acceleration vector changes from Kc^2 to $4Kc^2$. Thus, the force that causes the new acceleration must be four times as strong as the original force.*

(3) *If the entire curve is traced out at a constant speed, then the magnitude of the acceleration vector at a given point is directly proportional to the value of the curvature K at that point. Thus, the force that causes the acceleration has a magnitude that is proportional to the curvature. This force is applied in the direction of the normal vector N.*

CURVATURE

The formulas for velocity and acceleration that we derived above can be used to obtain a very simple formula for the curvature. We write

$$v = \frac{ds}{dt} T$$

and

$$a = \frac{d^2s}{dt^2} T + \left(\frac{ds}{dt}\right)^2 KN.$$

It follows from Theorem 14–11(f, g) that the cross product of these vectors is

$$v \times a = \left(\frac{ds}{dt} T\right) \times \left[\frac{d^2s}{dt^2} T + \left(\frac{ds}{dt}\right)^2 KN\right]$$

$$= \frac{ds}{dt} \cdot \frac{d^2s}{dt^2} (T \times T) + \left(\frac{ds}{dt}\right)^3 K(T \times N)$$

$$= 0 + \left(\frac{ds}{dt}\right)^3 K(T \times N) = \left(\frac{ds}{dt}\right)^3 K(T \times N).$$

Recall that T and N are orthogonal unit vectors. It follows from the corollary to Theorem 14–11 that

$$\|T \times N\| = \|T\| \cdot \|N\| \sin \frac{\pi}{2} = 1.$$

Therefore,

$$\|v \times a\| = \left(\frac{ds}{dt}\right)^3 K \|T \times N\| = \left(\frac{ds}{dt}\right)^3 K.$$

If we divide both sides of this equation by $(ds/dt)^3$, we get

Curvature Formula

$$K = \frac{\|v \times a\|}{(ds/dt)^3} = \frac{\|v \times a\|}{\|v\|^3}.$$

15–6 Tangential and Normal Components of Acceleration

EXAMPLE 1 Let $r(t) = \cos t\,i + \sin t\,j + t\,k$. Find the curvature K at $(1, 0, 0)$. (See Example 3, Section 15–5.)

Solution
$$v(t) = -\sin t\,i + \cos t\,j + k$$
$$a(t) = -\cos t\,i - \sin t\,j.$$

When $t = 0$, we have $v = j + k$ and $a = -i$. Then

$$v \times a = \begin{vmatrix} i & j & k \\ 0 & 1 & 1 \\ -1 & 0 & 0 \end{vmatrix} = -j + k.$$

Therefore,

$$K = \frac{\|v \times a\|}{\|v\|^3} = \frac{\|-j + k\|}{\|j + k\|^3} = \frac{\sqrt{2}}{(\sqrt{2})^3} = \frac{1}{2}. \quad \square$$

The formula

$$K = \frac{\|v \times a\|}{\|v\|^3}$$

also can be used to derive the formulas for curvature used in Section 15–4.

EXAMPLE 2 Derive the formula for the curvature of the graph of $y = f(x)$.

Solution The corresponding vector function is

$$r(t) = t\,i + f(t)\,j.$$

Then

$$v(t) = i + f'(t)\,j$$

and

$$a(t) = f''(t)\,j.$$

The cross product is

$$v \times a = \begin{vmatrix} i & j & k \\ 1 & f'(t) & 0 \\ 0 & f''(t) & 0 \end{vmatrix} = i \cdot 0 - j \cdot 0 + k \cdot f''(t) = f''(t)\,k.$$

Then

$$K(t) = \frac{\|v \times a\|}{\|v\|^3} = \frac{\|f''(t)\,k\|}{\|i + f'(t)\,j\|^3} = \frac{|f''(t)|}{[\sqrt{1 + f'(t)^2}]^3}.$$

Since $x = t$, then

$$K(x) = \frac{|f''(x)|}{[1 + f'(x)^2]^{3/2}}. \quad \square$$

FORMULAS FOR THE COMPONENTS OF ACCELERATION

An alternative method for calculating the tangential and normal components of acceleration is possible without actually calculating T and N. This alternative is useful, for example, when we are interested in the relative sizes of the equivalent forces but not in the exact directions in which they are applied.

The tangential component of acceleration d^2s/dt^2 can, of course, be calculated by differentiating the speed $\|v(t)\| = ds/dt$. An alternative method is based on the fact that

$$v = \frac{ds}{dt} T \quad \text{and} \quad a = \frac{d^2s}{dt^2} T + \left(\frac{ds}{dt}\right)^2 KN.$$

Then

$$v \cdot a = \left(\frac{ds}{dt} T\right) \cdot \left[\frac{d^2s}{dt^2} T + \left(\frac{ds}{dt}\right)^2 KN\right]$$

$$= \frac{ds}{dt} \cdot \frac{d^2s}{dt^2} + 0 = \frac{ds}{dt} \cdot \frac{d^2s}{dt^2}.$$

(See Exercise 16 of Section 14–8.) If we divide both sides of this equation by $ds/dt = \|v\|$, we get

Tangential Component of Acceleration

$$\frac{d^2s}{dt^2} = \frac{v \cdot a}{ds/dt} = \frac{v \cdot a}{\|v\|}.$$

It is also easy to calculate the normal component of acceleration. This component is

$$K\left(\frac{ds}{dt}\right)^2 = \frac{\|v \times a\|}{\|v\|^3}\left(\frac{ds}{dt}\right)^2 = \frac{\|v \times a\|}{\|v\|^3}\|v\|^2,$$

which reduces to

Normal Component of Acceleration

$$K\left(\frac{ds}{dt}\right)^2 = \frac{\|v \times a\|}{\|v\|}.$$

EXAMPLE 3 Let $r(t) = \cos t\, i + \sin t\, j + t k$. Find the tangential and normal components of acceleration at the point at which $t = 0$.

Solution As we saw in Example 1, when $t = 0$,

$$v = j + k, \quad a = -i, \quad \text{and} \quad v \times a = -j + k.$$

The tangential component of acceleration is

$$\frac{d^2s}{dt^2} = \frac{v \cdot a}{\|v\|} = \frac{(j + k) \cdot (-i)}{\sqrt{2}} = 0.$$

The normal component is

$$K\left(\frac{ds}{dt}\right)^2 = \frac{\|v \times a\|}{\|v\|} = \frac{\sqrt{2}}{\sqrt{2}} = 1.$$

The acceleration vector is

$$a(t) = 0 \cdot T + 1 \cdot N = N. \quad \square$$

Exercises 15-6

1–8. Use the methods of this section to calculate the curvature K and the tangential and normal components of acceleration.

- **1** $r(t) = 4ti + t^2j + 2t^2k$
- 2 $r(t) = 2t^2i + t^4j + 4tk$
- **3** $r(t) = (t^2 - 1)i + (2t + 3)j + (t^2 - 4t)k$
- 4 $r(t) = (t^2 - 4t)i + 4tj + 2t^2k$
- **5** $r(t) = t \cos ti + t \sin tj$
- 6 $r(t) = \cosh ti + \sinh tj$
- **7** $r(t) = e^t \cos ti + e^t \sin tj + e^tk$
- 8 $r(t) = e^ti + e^{-t}j$
- 9 Use the methods of this section to derive the formula

$$K(t) = \frac{|f'(t)g''(t) - g'(t)f''(t)|}{[f'(t)^2 + g'(t)^2]^{3/2}}$$

for the curvature of the graph of $x = f(t)$, $y = g(t)$.

10 Suppose that the acceleration vector is always perpendicular to the direction of the motion. Show that the speed must be constant.

11 Suppose that a particle moves with an acceleration $a(t) = 0$ at each instant t. Show that it moves in a straight line with a constant velocity.

12 (a) A particle moves along a curve in the xy-plane. Denote the tangential and normal components of acceleration by a_T and a_N, so that $a(t) = a_TT + a_NN$. Similarly, let the components of i and j be a_i and a_j, so that $a(t) = a_ii + a_jj$. Prove that

$$\|a\|^2 = a_T^2 + a_N^2 = a_i^2 + a_j^2.$$

(b) Use the result of (a) to show that $a_N = \sqrt{\|a\|^2 - a_T^2}$. Since $\|a\|$ and a_T can be calculated easily, this gives an alternative method for calculating a_N.

(c) Use the result of (b) to find the tangential and normal components of acceleration for the curve in Exercise 5.

(d) Use the result of (b) to find the tangential and normal components of acceleration for the curve in Exercise 8.

Review Problems

1. (a) Define "$\lim_{t \to a} r(t) = L$."

   (b) What is meant by saying the function $r(t)$ is *continuous* at $t = a$?

   (c) Define the derivative $r'(t)$.

2. Prove that if $r(t) = f(t)i + g(t)j + h(t)k$, where f, g, and h are differentiable, then
$$r'(t) = f'(t)i + g'(t)j + h'(t)k.$$

3. Sketch the graph of $r(t) = e^t i + (1 - 2e^{2t})j$. Draw the vectors $r(0)$, $r'(0)$, and $r''(0)$ at the appropriate places on the graph.

4–5. (a) Calculate the arc length of the curves.

   (b) Find parametric equations of the line tangent to the curve at the given point.

• 4 $r(t) = e^t i + e^t j + k$, $0 \le t \le \ln 8$; $P(2, 2, 1)$

• 5 $r(t) = t \cos t \, i + t \sin t \, j + tk$, $0 \le t \le 2$; $P\left(0, \dfrac{\pi}{2}, \dfrac{\pi}{2}\right)$

6. Prove that if $u(t) \to L$ and $v(t) \to M$ as $t \to a$, then
$$u(t) \times v(t) \to L \times M \text{ as } t \to a.$$

7. Prove that if $u(t)$ and $v(t)$ are differentiable, then
$$D_t[u(t) \times v(t)] = u(t) \times v'(t) + u'(t) \times v(t)$$

8. Explain why $r'(t)$ is called a *velocity vector* and $r''(t)$ an *acceleration vector*. What properties does $r'(t)$ have that make this name appropriate?

• 9 Let $r(t) = \cos 2t \, i + \sin 2t \, j + tk$. Calculate $v(t)$ and $a(t)$. Calculate the arc length from $t = 0$ to $t = 2\pi$.

10. A particle moves along a path with a constant speed. Prove that the acceleration vector is orthogonal to the velocity vector.

• 11 Find parametric equations of the tangent line to the graph of $r(t) = 4ti + t^2 j + 2t^2 k$ that is parallel to the vector $2i + 3j + 6k$.

12. (a) State Newton's second law of motion in vector form.

   (b) An object in equilibrium is subject to forces $F = ai + bj$ and $G = ci + dj$. Explain why $a = -c$ and $b = -d$.

• 13 A projectile is fired from a horizontal plane at an angle of elevation of 60° and with an initial speed of 1500 ft/sec. Assume that there is no air resistance. Use Newton's second law of motion to derive the equations of motion. (Use $g = 32$.)

Review Problems

•14 Work Problem 13 for an initial speed of 4900 cm/sec. (Use $g = 980$.)

•15 Calculate $\int_0^\pi [\sin t\, i + \cos^3 t\, j + t\, k]\, dt$.

•16 Find $r(t)$ if $r'(t) = 4t^3 i + e^t j + \sin t\, k$ and $r(0) = 2i - j + 2k$.

17 Draw a figure, similar to Figure 15–11, that illustrates the bending of a curve in terms of the sign and the size of $d\phi/ds$.

18 (a) Define the *curvature* of a plane curve.

(b) Define the *curvature* of a space curve.

19 Derive the curvature formula

$$K = \frac{|f''(x)|}{[1 + f'(x)^2]^{3/2}}.$$

20–21. Calculate the curvature.

•20 $y = x^3 - x$

•21 $x = e^t$, $y = e^{-t}$

•22 Find the equation of the circle of curvature that is tangent to the graph of $y = x^3 - x$ at $P(1, 0)$.

•23 Find the points at which the curvature of the graph of $y = x^3$ is a maximum.

24–25. Calculate the unit tangent vector T, the principal unit normal vector N, and the curvature K as functions of t.

•24 $r(t) = e^{2t} \cos t\, i + e^{2t} \sin t\, j$

•25 $r(t) = 10t i + 3t^2 j + 4t^2 k$

•26 Find the equation of the circle of curvature tangent to the graph in Problem 24 at the point $P(1, 0)$.

•27 Find the curvature $K(t)$ and the tangential and normal components of acceleration for the function $r(t) = t \cos t\, i + t \sin t\, j + 2k$.

28 (a) Derive the formula

$$K(t) = \frac{\|v(t) \times a(t)\|}{\|v(t)\|^3}.$$

(b) Use this formula to calculate the curvature of the graph in Problem 24.

29 Use the formula in Problem 28 to derive the formula
$$K = \frac{|f''(x)|}{[1 + f'(x)^2]^{3/2}}.$$

30 A particle moves in such a way that its acceleration vector is always a multiple of the velocity vector. Prove that the particle moves along a straight line. (*Hint:* Use the formula $a = \dfrac{d^2s}{dt^2} T + \left(\dfrac{ds}{dt}\right)^2 KN$.)

16 Partial Differentiation

16–1 Limits and Continuity

Thus far we have worked with functions of one real variable and with real vector-valued functions. Now we shall extend the derivative concept to real functions of two or more variables.

DEFINITION

Function of Two Variables

A *real function of two variables* is a function that has a set of ordered pairs of real numbers for its domain and a set of real numbers for its range. If f is the function and (x, y) is in the domain of f, the corresponding functional value can be denoted by $f(x, y)$.

EXAMPLE 1 The volume of a right circular cone with radius $x > 0$ and height $y > 0$ is

$$V = f(x, y) = \frac{\pi}{3} x^2 y.$$

The domain of f is the set of all ordered pairs (x, y) such that $x > 0$ and $y > 0$. The range is the set of all positive real numbers. □

Convention for Domain

If the functional relation is defined by an equation, such as

$$f(x, y) = x + \frac{1}{y} + \frac{3}{x(y - 1)},$$

then unless it has been specified otherwise, the domain of the function is the set of all ordered pairs (x, y) such that x, y and $f(x, y)$ are real. For example, the domain of the function

$$f(x, y) = x + \frac{1}{y} + \frac{3}{x(y - 1)}$$

is

$$\{(x, y) \mid x \neq 0, y \neq 0, \text{ and } y \neq 1\}.$$

Remark In proper parlance, $f(x, y)$ is the value that the function f assigns to (x, y). We shall occasionally abuse the language and write $f(x, y)$ for the function instead of f. This will be done primarily to avoid confusion when we have several different functions of different variables and to make clear which variables are independent and which are dependent on them. In a similar spirit, we write

$$z = f(x, y)$$

to indicate that x and y are independent variables and that z is a dependent variable that depends on x and y.

GRAPHS

The *graph* of a function $f(x, y)$ is the set of all points $P(x, y, z)$ such that (x, y) is in the domain of f and $z = f(x, y)$. (See Fig. 16–1a.)

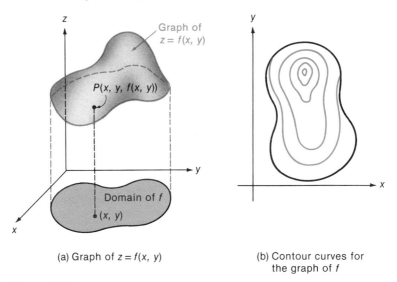

(a) Graph of $z = f(x, y)$

(b) Contour curves for the graph of f

Figure 16–1.

The graph of a function $f(x, y)$ is a surface in three-dimensional space. We can represent a graph by a perspective drawing (as in Fig. 16–1a) or by contour curves (as in Fig. 16–1b).

The graphs of several functions of two variables are shown in Section 14–10. The graph of

$$f(x, y) = \frac{x^2}{a^2} + \frac{y^2}{b^2},$$

for example, is the elliptical paraboloid in Figure 14–48. The graph of

$$f(x, y) = y^2 - x^2$$

is the hyperbolic paraboloid in Figure 14–53.

LIMITS

It may happen that the functional values $f(x, y)$ can be made arbitrarily close to a number L by requiring that (x, y) be sufficiently close to a point (a, b). If this is the

16–1 Limits and Continuity

case, we write

$$\lim_{(x,y)\to(a,b)} f(x, y) = L.$$

In other words, L is the limit of $f(x, y)$ as $(x, y) \to (a, b)$ if we can make

$$|f(x, y) - L| \quad \text{(the distance from } f(x, y) \text{ to } L\text{)}$$

as close to zero as we want by requiring that

$$\sqrt{(x - a)^2 + (y - b)^2} \quad \text{(the distance from } (x, y) \text{ to } (a, b)\text{)}$$

be sufficiently small.

The formal definition follows.

DEFINITION

Limit of a Function of Two Variables

Let $f(x, y)$ be defined everywhere on a circular disk that has its center at (a, b) except possibly at (a, b) itself. We say that

$$\lim_{(x,y)\to(a,b)} f(x, y) = L$$

provided that for each $\epsilon > 0$, there exists $\delta > 0$ such that

$$|f(x, y) - L| < \epsilon \quad \text{wherever } 0 < \sqrt{(x - a)^2 + (y - b)^2} < \delta.$$

The limit concept can be explained by an alternative approach. Let $\delta > 0$. The δ-*neighborhood of* (a, b) consists of all points inside the circular disk with center at (a, b) and radius δ. (See Fig. 16–2a.) Every such disk is called a *neighborhood*.

Neighborhood

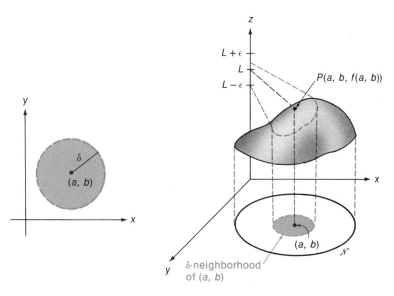

(a) δ-neighborhood of (a, b)

(b) $\lim_{(x, y)\to(a, b)} f(x, y) = L$ provided, for each $\epsilon > 0$ there exists $\delta > 0$ such that $|f(x, y) - L| < \epsilon$ whenever (x, y) is a point in the δ-neighborhood of (a, b)

Figure 16–2 $\lim_{(x,y)\to(a,b)} f(x, y) = L.$

The definition of *limit* can now be restated in terms of the neighborhood concept: Let $f(x, y)$ be defined on some neighborhood $\mathcal{N}$ of (a, b) except possibly at (a, b) itself. We say that

$$\lim_{(x,y)\to(a,b)} f(x, y) = L$$

provided that for each $\epsilon > 0$, there exists a δ-neighborhood of (a, b) such that $|f(x, y) - L| < \epsilon$ wherever (x, y) is a point in the δ-neighborhood different from (a, b).

Example 2 illustrates how the definition can be used to calculate limits. You should observe that the computations are involved, even with such a simple function.

EXAMPLE 2 Let $f(x, y) = x^2 y$. Use the definition to prove that

$$\lim_{(x,y)\to(2,3)} f(x, y) = \lim_{(x,y)\to(2,3)} x^2 y = 12.$$

Solution Let $\epsilon > 0$. We must show that

$$|f(x, y) - 12| = |x^2 y - 12| < \epsilon$$

wherever $0 < \sqrt{(x - 2)^2 + (y - 3)^2} < \delta$ for some δ that depends on ϵ.

We first modify the expression on the left as follows:

$$\begin{aligned}|x^2 y - 12| &= |(x^2 y - 4y) + (4y - 12)| \\ &\le |x^2 y - 4y| + |4y - 12| \quad \text{(by the Triangle Inequality)} \\ &= |y| \cdot |x - 2| \cdot |x + 2| + 4|y - 3|.\end{aligned}$$

(These steps enable us to isolate the terms $|x - 2|$ and $|y - 3|$ so that we can work with them individually.)

Next, we restrict ourselves to the neighborhood of $(2, 3)$ with radius 1—points inside the circle with radius 1 and center at $(2, 3)$. If (x, y) is in this circle, then x is within one unit of 2 and y is within one unit of 3. Thus, $1 < x < 3$ and $2 < y < 4$. It follows that

$$|x + 2| = x + 2 < 5 \quad \text{and} \quad |y| = y < 4.$$

Next, we let δ be the smaller of 1 and $\epsilon/40$.* If (x, y) is in the δ-neighborhood of $(2, 3)$, then in particular, (x, y) is in the 1-neighborhood, so the above inequalities hold.

Let (x, y) be in the δ-neighborhood of (a, b) and different from (a, b). Then

$$|x - 2| \le \sqrt{(x - 2)^2 + (y - 3)^2} < \delta \le \frac{\epsilon}{40}$$

and

$$|y - 3| \le \sqrt{(x - 2)^2 + (y - 3)^2} < \delta \le \frac{\epsilon}{40}.$$

---

* The number $\epsilon/40$ happens to work for this problem. Other numbers could have been used.

16-1 Limits and Continuity

It follows that

$$|f(x, y) - 12| = |x^2y - 12| \leq |y| \cdot |x - 2| \cdot |x + 2| + 4|y - 3|$$
$$< 4 \cdot \frac{\epsilon}{40} \cdot 5 + 4 \cdot \frac{\epsilon}{40} = \frac{\epsilon}{2} + \frac{\epsilon}{10} < \epsilon.$$

Therefore,

$$\lim_{(x,y) \to (2,3)} f(x, y) = 12. \quad \square$$

In actual practice, most limits are calculated by using the following limit theorem, which generalizes Theorem 2–1. The proof is omitted.

THEOREM 16–1 (*The Limit Theorem*) Let

Limit Theorem for Functions of Two Variables

$$\lim_{(x,y) \to (a,b)} f(x, y) = L \quad \text{and} \quad \lim_{(x,y) \to (a,b)} g(x, y) = M.$$

Let C be a constant. Then

L1 $\quad \lim\limits_{(x,y) \to (a,b)} [f(x, y) \pm g(x, y)] = L \pm M$

$\qquad = \lim\limits_{(x,y) \to (a,b)} f(x, y) \pm \lim\limits_{(x,y) \to (a,b)} g(x, y).$

The limit of a sum is the sum of the limits.

L2 $\quad \lim\limits_{(x,y) \to (a,b)} [f(x, y) \cdot g(x, y)] = LM$

$\qquad = \lim\limits_{(x,y) \to (a,b)} f(x, y) \cdot \lim\limits_{(x,y) \to (a,b)} g(x, y).$

The limit of a product is the product of the limits.

L3 $\quad \lim\limits_{(x,y) \to (a,b)} \dfrac{f(x, y)}{g(x, y)} = \dfrac{L}{M}$

$\qquad = \dfrac{\lim\limits_{(x,y) \to (a,b)} f(x, y)}{\lim\limits_{(x,y) \to (a,b)} g(x, y)} \quad \text{provided } M \neq 0.$

The limit of a quotient is the quotient of the limits provided the limit of the denominator is not zero.

L4 $\quad \lim\limits_{(x,y) \to (a,b)} C = C.$

L5 $\quad \lim\limits_{(x,y) \to (a,b)} x = a \quad \text{and} \quad \lim\limits_{(x,y) \to (a,b)} y = b.$

L6 $\quad \lim\limits_{(x,y) \to (a,b)} \sqrt[n]{f(x, y)} = \sqrt[n]{L}$ provided $L > 0$ if n is even.

L7 $\quad \lim\limits_{(x,y) \to (a,b)} Cf(x, y) = CL = C \cdot \lim\limits_{(x,y) \to (a,b)} f(x, y).$

EXAMPLE 3 Use Theorem 16–1 to show that $\lim\limits_{(x,y) \to (2,3)} x^2 y = 12.$

Solution $\lim\limits_{(x,y) \to (2,3)} x^2 y = \lim\limits_{(x,y) \to (2,3)} x \cdot \lim\limits_{(x,y) \to (2,3)} x \cdot \lim\limits_{(x,y) \to (2,3)} y$

$\qquad = 2 \cdot 2 \cdot 3 = 12 \quad$ (by L2 and L5). $\square$

CONTINUITY

The definition of continuity is similar to the definition for a function of one variable.

DEFINITION Let f be a function of two variables. The function f is *continuous at* (a, b) provided

$$\lim_{(x,y) \to (a,b)} f(x, y) = f(a, b).$$

The function f is *continuous on the set* S provided it is continuous at each point of S.

It may be instructive to think of the graph of $z = f(x, y)$ as a surface that hangs in the air. (See Figs. 16–1, 2, 4.) If the surface is not torn, then the function is continuous. If it is torn or has holes in it, then the function is discontinuous at the corresponding points.

EXAMPLE 4 The graph of

$$f(x, y) = \begin{cases} 1 & \text{if } x^2 + y^2 \leq 1 \\ 0 & \text{if } x^2 + y^2 > 1 \end{cases}$$

is shown in Figure 16–3. The function is discontinuous at each point of the unit circle $x^2 + y^2 = 1$. It is continuous everywhere else. □

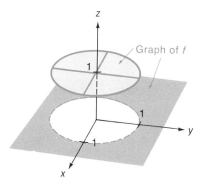

Figure 16–3 Example 4.

Combinations of Continuous Functions

The results of Theorem 16–1 can be applied to show that the sum of two continuous functions is continuous, the product of two continuous functions is continuous, and so on. In particular, this implies that polynomial functions of two variables are continuous everywhere and that rational functions of two variables are continuous except at those points where the denominators are zero.

The composition of two functions is continuous. (See Exercise 24 for a more precise statement.) Thus, for example, the function $f(x, y) = e^{\sqrt{x^2+y^2}}$ is continuous for all real (x, y), and the function $g(x, y) = \ln \cos (x^2 + y^2)$ is continuous wherever $\cos (x^2 + y^2) > 0$.

LIMITS ALONG PATHS

It is more difficult to determine limits and continuity for functions of two variables than for functions of a single variable. Recall that a function $g(x)$ has the limit L as

16–1 Limits and Continuity

$x \to a$ provided $g(x) \to L$ when x approaches a from either the right or the left. For $f(x, y)$ to have L as a limit as $(x, y) \to (a, b)$, it is necessary that the functional values $f(x, y)$ approach L as (x, y) approaches (a, b) along any path $\mathscr{C}$. (See Fig. 16–4.) This

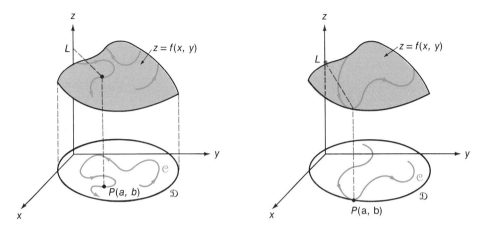

(a) The limit at an interior point   (b) The limit at a boundary point

Figure 16–4 If $\lim_{(x,y)\to(a,b)} f(x, y) = L$, then $f(x, y) \to L$ as $(x, y) \to (a, b)$ along any path $\mathscr{C}$ in the domain of f.

path $\mathscr{C}$ may be a line, it may spiral in towards (a, b), or it may be quite irregular.

The above statement is often used in a negative way—to establish that a particular function f has no limit as (x, y) approaches (a, b). It is sufficient to find two different paths through (a, b) along which f has different limits as $(x, y) \to (a, b)$.

EXAMPLE 5 Let $f(x, y) = \dfrac{x^2 - y^2}{x^2 + y^2}$. Show that $f(x, y)$ has no limit as $(x, y) \to (0, 0)$.

Solution We first consider the path defined by $x = 0$ (the y-axis). Along this path

$$f(x, y) = \frac{0^2 - y^2}{0^2 + y^2} = -1 \qquad (y \neq 0)$$

so that

$$\lim_{\substack{(x,y)\to(0,0)\\ \text{along } y\text{-axis}}} f(x,y) = \lim_{y\to 0}(-1) = -1.$$

Similarly, along the path $y = 0$ (the x-axis), we have

$$\lim_{\substack{(x,y)\to(0,0)\\ \text{along } x\text{-axis}}} f(x,y) = \lim_{\substack{(x,y)\to(0,0)\\ \text{along } x\text{-axis}}} \frac{x^2 - 0^2}{x^2 + 0^2} = \lim_{x\to 0} 1 = 1.$$

Since f has different limits along these paths, then it has no well-defined limit as $(x, y) \to (0, 0)$. □

EXAMPLE 6 Let $f(x, y) = \dfrac{x^2 + 2xy + y^2}{x^2 + y^2}$. Show that f has no limit as $(x, y) \to (0, 0)$.

Solution The approach used in Example 5 will not work with this function, since $f(x, y) \to 1$ as $(x, y) \to (0, 0)$ along either the x- or the y-axis.

Let $(x, y) \to (0, 0)$ along the line $y = mx$. Then

$$\lim_{\substack{(x,y) \to (0,0) \\ \text{along line} \\ y=mx}} \frac{x^2 + 2xy + y^2}{x^2 + y^2} = \lim_{\substack{(x,y) \to (0,0) \\ \text{along line} \\ y=mx}} \frac{x^2 + 2mx^2 + m^2x^2}{x^2 + m^2x^2}$$

$$= \lim_{x \to 0} \frac{1 + 2m + m^2}{1 + m^2} = \frac{1 + 2m + m^2}{1 + m^2}.$$

This limit depends on m. For example, it is equal to $\frac{9}{5}$ when $m = 2$ and is equal to 2 when $m = 1$.

Since there are different limits along different paths through the origin, then f has no well-defined limit as $(x, y) \to (0, 0)$. □

Remark In Examples 5 and 6 we worked with linear paths. This approach can be used with some functions to show that a limit does not exist; it is not enough, however, to prove that a limit does exist. In other words, even if $f(x, y) \to L$ as $(x, y) \to (a, b)$ along every line through (a, b), the function may fail to have a limit as $(x, y) \to (a, b)$. For this function to have the limit L, it is necessary that $f(x, y) \to L$ as $(x, y) \to (a, b)$ along every possible path, not just linear paths.

EXAMPLE 7 Let $f(x, y) = \dfrac{xy^2}{x^2 + y^4}$. Show that f has no limit as $(x, y) \to (0, 0)$.

Solution It is easy to see that the limit is zero as $(x, y) \to (0, 0)$ along the y-axis or along any line $y = mx$. If we let (x, y) approach $(0, 0)$ along the parabola $x = y^2$, however, we obtain

$$\lim_{\substack{(x,y) \to (0,0) \\ \text{along parabola} \\ x=y^2}} \frac{xy^2}{x^2 + y^4} = \lim_{y \to 0} \frac{y^2 \cdot y^2}{y^4 + y^4} = \lim_{y \to 0} \frac{1}{2} = \frac{1}{2}.$$

Since different limits exist along different curves, there is no limit as $(x, y) \to (0, 0)$. □

BOUNDARY POINTS OF THE DOMAIN

The limit concept can be extended to boundary points of the domain $\mathcal{D}$ of f. A typical *boundary point* is shown in Figure 6–4b. The point $P(a, b)$ has the property that every disk with center at P contains at least one point in $\mathcal{D}$ and one not in $\mathcal{D}$. When working with boundary points we restrict ourselves to points in the domain $\mathcal{D}$ as we calculate limits. The formal definition follows.

DEFINITION Let $P(a, b)$ be a boundary point of $\mathcal{D}$, the domain of $f(x, y)$. Suppose that every circular disk about P contains an infinite number of points in $\mathcal{D}$. Then

$$\lim_{(x,y) \to (a,b)} f(x, y) = L$$

Exercises 16–1

provided: If $\epsilon > 0$ there exists $\delta > 0$ such that

$$|f(x, y) - L| < \epsilon \quad \text{whenever} \quad 0 < \sqrt{(x - a)^2 + (y - b)^2} < \delta$$

and $(x, y) \in \mathcal{D}$.

The limit at a boundary point $P(a, b)$ can also be calculated by considering curves through P. It is only necessary that, except for the point P itself, the curves lie in the domain of f. (See Fig. 16–4b.)

EXTENSION TO THREE OR MORE VARIABLES

All the concepts and results of this section can be extended to functions of three or more variables.

A function f of three variables has a set of ordered triples (x, y, z) as its domain. We write $f(x, y, z)$ for the value of f at (x, y, z). Similarly, a function of four variables has a set of ordered quadruples (w, x, y, z) as its domain. Similar definitions hold for any number of variables.

The limit concept for a function of three variables can be stated as follows.

DEFINITION

Limit of Function of Three Variables

Let $f(x, y, z)$ be defined at each point of some sphere with center at (a, b, c) except possibly at the point (a, b, c) itself. We say that

$$\lim_{(x,y,z) \to (a,b,c)} f(x, y, z) = L$$

provided that for each $\epsilon > 0$ there exists $\delta > 0$ such that

$$|f(x, y, z) - L| < \epsilon \quad \text{wherever} \quad 0 < \sqrt{(x - a)^2 + (y - b)^2 + (z - c)^2} < \delta.$$

Similar definitions hold for functions of n variables. All of these definitions can be extended to boundary points of the domain.

The limit theorem (Theorem 16–1) can be extended to functions of any number of variables. You should feel free to use that theorem for functions of three, four, or even n variables.

The definition of continuity and all the results concerning continuity can be extended to functions of three or more variables. For example, if f is a function of x, y, z, we say that f is *continuous at* (a, b, c) provided

$$\lim_{(x,y,z) \to (a,b,c)} f(x, y, z) = f(a, b, c).$$

Exercises 16–1

1–4. Determine the domain and range of the function. Sketch the graph of the domain in the xy-plane.

• 1 $f(x, y) = 4xy \ln (x + y)$

2 $f(x, y) = \dfrac{\cos (xy)}{x^2 - y^2}$

• 3 $f(x, y) = \sqrt{3x - y + 3}$

4 $f(x, y) = \sqrt{1 - \dfrac{y}{x}}$

5–12. If the function has a limit, then use the Limit Theorem to calculate it. If it does not have a limit, then explain how you know that fact. (*Hint for 12:* Use the parabola $y = x^2$.)

- 5. $\displaystyle\lim_{(x,y)\to(0,0)} \frac{x^2 y - 3x^2 + 4}{x^2 - 4}$

- 6. $\displaystyle\lim_{(x,y)\to(2,3)} \frac{2x + y^2}{3x - y}$

- 7. $\displaystyle\lim_{(x,y)\to(0,0)} \frac{5x^2 + 4y^2}{x^2 + y^2}$

- 8. $\displaystyle\lim_{(x,y)\to(0,0)} \frac{2x + y^2}{3x - y}$

- 9. $\displaystyle\lim_{(x,y)\to(0,0)} \frac{x^2 + y^2}{\sqrt{x^2 + 1}}$

- 10. $\displaystyle\lim_{(x,y)\to(0,0)} \frac{xy}{x^2 + y^2}$

- 11. $\displaystyle\lim_{(x,y)\to(0,0)} \frac{2x^3 + xy^2 - 2x^2 y - y^3}{2x^2 + y^2}$

- 12. $\displaystyle\lim_{(x,y)\to(0,0)} \frac{x^4 y}{x^7 + y^3}$

13–18. Discuss the continuity of f.

- 13. $f(x, y, z) = \dfrac{x}{x^2 + y^2 - z^2}$

- 14. $f(x, y) = \sqrt{\dfrac{x - y}{2x + y}}$

- 15. $f(x, y) = \ln(xy + 2)$

- 16. $f(x, y, z) = \dfrac{xy + xz + yz}{xyz}$

- 17. $f(x, y) = \dfrac{x + y}{x^3 + y^3}$

- 18. $f(x, y) = \cos(x + y) + e^{-xy}$

19–20. Find $\delta > 0$ such that $|f(x, y) - 0| < 1/10{,}000$ wherever $0 < \sqrt{(x - 0)^2 + (y - 0)^2} < \delta$. (*Hint for 20:* Write $|xy| = |x| \cdot |y|$. Make both $|x|$ and $|y| < 1/100$.

- 19. $f(x, y) = 3x + 2y$

- 20. $f(x, y) = xy$

21–22. Use the definition to calculate the limits. (*Hint for 22:* Write $|xy| = |x| \cdot |y|$. Make both $|x|$ and $|y| < \sqrt{\epsilon}$.

- 21. $\displaystyle\lim_{(x,y)\to(2,1)} (3x + 2y - 1)$

- 22. $\displaystyle\lim_{(x,y)\to(0,0)} xy$

23. Restate Theorem 16–1 (the Limit Theorem) for functions of three variables x, y, and z.

24. (*Composition of Continuous Functions*) Let $g(x, y)$ be continuous at (x_0, y_0). Let $f(z)$ be continuous at $L = g(x_0, y_0)$. Let $h(x, y) = f(g(x, y))$. Prove that h is continuous at (x_0, y_0). (*Hint:* Modify the proof of Theorem 2–9.)

25. The wheel and two legs of a wheelbarrow form an equilateral triangle with each side 3 ft long. The wheelbarrow is rolled onto a hill that has topography defined by $z = f(x, y)$, where f is a continuous function. Let the wheel be located at the point $P_0(x_0, y_0, z_0)$. Prove that it is possible to set the legs down so that they will be the same height above the xy-plane. (*Hint:* Move the legs around P_0 in a 360° rotation. For each angle θ, let $g_1(\theta)$ be the vertical distance from the xy-plane to the point on the hill directly under the first leg, and let $g_2(\theta)$ be the vertical distance to the point directly under the other leg. Let $g(\theta) = g_1(\theta) - g_2(\theta)$. Show that g is continuous and that it has a maximum value greater than or equal to zero and a minimum value less than or equal to zero. Apply the Intermediate Value Theorem.)

16–2 Partial Derivatives

Let $z = f(x, y)$. In certain problems we need to know the rate at which z changes with respect to a change in x or y. These rates are called the *partial derivatives* of z with respect to x or y, denoted by $\partial z/\partial x = \partial f/\partial x$ and $\partial z/\partial y = \partial f/\partial y$. The formal definition follows.

DEFINITION

Partial Derivatives

Let $z = f(x, y)$. The partial derivatives of z with respect to x and y are

$$\frac{\partial z}{\partial x} = \frac{\partial f}{\partial x} = \lim_{\Delta x \to 0} \frac{f(x + \Delta x, y) - f(x, y)}{\Delta x}$$

and

$$\frac{\partial z}{\partial y} = \frac{\partial f}{\partial y} = \lim_{\Delta y \to 0} \frac{f(x, y + \Delta y) - f(x, y)}{\Delta y},$$

respectively, provided the limits exist.

The partial derivative of z with respect to x is calculated by holding y fixed and letting x vary. It measures the rate of change in z along a line parallel to the x-axis. Similarly, the partial derivative with respect to y is calculated by holding x fixed. It measures the rate of change along a line parallel to the y-axis.

If $z = f(x, y)$ has a simple enough form, then $\partial z/\partial x$ can be calculated from the derivative formulas for functions of one variable by treating y as a constant and z as a function of x. Similarly, $\partial z/\partial y$ can be calculated by treating x as a constant and z as a function of y.

EXAMPLE 1 Let $z = f(x, y) = x^2 e^{3y} + x - y^2$.

(a) $\quad \dfrac{\partial z}{\partial x} = \dfrac{\partial}{\partial x}(x^2 e^{3y} + x - y^2) = 2xe^{3y} + 1.$

(b) $\quad \dfrac{\partial z}{\partial y} = \dfrac{\partial}{\partial y}(x^2 e^{3y} + x - y^2) = 3x^2 e^{3y} - 2y.$ □

NOTATION

We use the symbol

$$\left(\frac{\partial z}{\partial x}\right)_{(x_0, y_0)}$$

to indicate the value of $\partial z/\partial x$ at $P_0(x_0, y_0)$. At the point $P_0(2, 1)$, for example, the function in Example 1 has the partial derivatives

$$\left(\frac{\partial z}{\partial x}\right)_{(2,1)} = (2xe^{3y} + 1)_{(2,1)} = 4e^3 + 1$$

$$\left(\frac{\partial z}{\partial y}\right)_{(2,1)} = (3x^2 e^{3y} - 2y)_{(2,1)} = 12e^3 - 2.$$

In certain cases it is convenient to use functional notation for the partial derivatives. We denote the partial derivatives by

Alternative Symbols for Partial Derivatives

$$f_x = \frac{\partial f}{\partial x} \quad \text{and} \quad f_y = \frac{\partial f}{\partial y}.$$

Then $f_x(x_0, y_0)$ and $f_y(x_0, y_0)$ are the values of the partial derivatives at (x_0, y_0).

EXAMPLE 2 Let $f(x, y) = 3xy + 5x + 7y^{-1}$. Then

$$f_x(x, y) = \frac{\partial f}{\partial x} = 3y + 5$$

$$f_y(x, y) = \frac{\partial f}{\partial y} = 3x - 7y^{-2}$$

At $P_0(2, -1)$, we have

$$f_x(2, -1) = \left(\frac{\partial f}{\partial x}\right)_{(2,-1)} = 3(-1) + 5 = 2$$

$$f_y(2, -1) = \left(\frac{\partial f}{\partial y}\right)_{(2,-1)} = 3 \cdot 2 - 7(-1)^{-2} = -1. \quad \square$$

FUNCTIONS OF SEVERAL VARIABLES

The concepts of this section can be extended to functions of any number of variables. For example, if f is a function of w, x, y, z, then

$$\frac{\partial f}{\partial w} = f_w(w, x, y, z) = \lim_{\Delta w \to 0} \frac{f(w + \Delta w, x, y, z) - f(w, x, y, z)}{\Delta w}$$

provided the limit exists.

If

$$f(w, x, y, z) = e^{wx} + \sin y + 3wx^2yz,$$

for example, then

$$\frac{\partial f}{\partial w} = xe^{wx} + 3x^2yz$$

$$\frac{\partial f}{\partial x} = we^{wx} + 6wxyz$$

$$\frac{\partial f}{\partial y} = \cos y + 3wx^2z$$

and

$$\frac{\partial f}{\partial z} = 3wx^2y$$

GEOMETRICAL INTERPRETATION

The graph of $z = f(x, y)$ is a surface in three-dimensional space. Let (a, b) be a point in the domain of f at which both partial derivatives exist. We shall show that these

16-2 Partial Derivatives

partial derivatives are the slopes of the tangent lines to the curves obtained by intersecting the surface with the vertical planes $x = a$ and $y = b$. (See Fig. 16–5.)

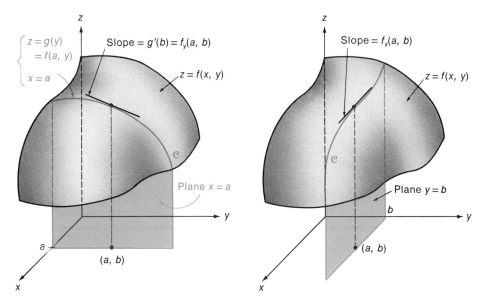

(a) $f_y(a, b)$ is the slope of the tangent line to the trace of the graph of f in the plane $x = a$ at the point (a, b)

(b) $f_x(a, b)$ is the slope of the tangent line to the trace of the graph of f in the plane $y = b$ at the point (a, b)

Figure 16–5 Geometrical interpretation of the partial derivative.

Let $\mathscr{C}$ be the trace of the graph of f in the plane $x = a$. Then $\mathscr{C}$ is the graph of

$$z = f(a, y), \quad x = a.$$

(See Fig. 16–5a.) If we let $g(y) = f(a, y)$, then $\mathscr{C}$ is the graph of $z = g(y)$, $x = a$. The slope of the tangent line to $\mathscr{C}$ at $(a, b, f(a, b))$ is $g'(b)$. Since g is obtained from f by holding x fixed and letting y vary, then

$$g'(b) = f_y(a, b) = \left(\frac{\partial f}{\partial y}\right)_{(a,b)}.$$

Partial Derivative as Slope of a Tangent Line

Thus, the slope of the line tangent to $\mathscr{C}$ at $(a, b, f(a, b))$ equals the partial derivative $f_y(a, b)$.

A similar argument can be used to establish that the slope of the line tangent to the curve

$$z = f(x, b), \quad y = b$$

at $(a, b, f(a, b))$ is $f_x(a, b) = (\partial f/\partial x)_{(a,b)}$. (See Fig. 16–5b.)

EXAMPLE 3 Find parametric equations of the tangent line to the trace of the sphere $x^2 + y^2 + z^2 = 169$ in the plane $x = 3$ at the point $P_0(3, 4, 12)$.

Partial Differentiation

Solution The upper hemisphere, on which the point P_0 is located, is the graph of the equation

$$z = f(x, y) = \sqrt{169 - x^2 - y^2}.$$

(See Fig. 16–6.)

The slope of the trace in the plane $x = 3$ at the point $P_0(3, 4, 12)$ equals

$$\left(\frac{\partial z}{\partial y}\right)_{(3,4)} = \left(\frac{1}{2}(169 - x^2 - y^2)^{-1/2}(-2y)\right)_{(3,4)}$$

$$= \left(\frac{-y}{\sqrt{169 - x^2 - y^2}}\right)_{(3,4)} = -\frac{4}{12} = -\frac{1}{3}.$$

Consequently, a change of -1 unit in the z-direction corresponds to a change of 3 units in the y-direction. It follows that the vector $3\mathbf{j} - \mathbf{k}$ is tangent to the trace at P_0. (See Fig. 16–6.) Thus, the parametric equations of the tangent line are

$$x = 3, \quad y = 4 + 3t, \quad z = 12 - t. \ \square$$

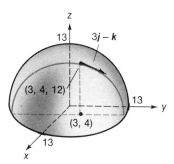

Figure 16–6 Example 3.

Exercises 16–2

1–10. Calculate the partial derivatives.

- **1** $f(x, y) = 3x^3y^2 + x^2y + 5 - 3y$
- **2** $f(x, y) = (2x^2 + y^3)^{3/2}$
- **3** $z = \ln(\sqrt{y^2 - x^2})$
- **4** $z = \sin(x^2 + y^2)$
- **5** $z = \ln x + \dfrac{1}{xy}$
- **6** $z = e^x \cos\left(\dfrac{x}{y}\right)$
- **7** $f(x, y, z) = \cos x - xe^z + ye^{-x}$
- **8** $f(w, x, y, z) = e^{2z}x^2 + e^x\sqrt{y^2 - w} + z^2 \cos y$
- **9** $f(r, s, t) = \tan(sr) - t^3s^{1/2} + \sin^{-1} r$
- **10** $f(x, y, \theta) = x^2 + y^2 - 2xy \cos \theta$

11–12. Find the equations of the tangent line to the trace of the graph in the given plane at the given point.

- **11** $f(x, y) = x^2 + 2y^2 - 16$; $P_0(1, 2, -7)$; Plane $x = 1$
- **12** $f(x, y) = \sqrt{x^2 + y^2}$; $P_0(4, 3, 5)$; Plane $y = 3$

16-3 Higher-Order Partial Derivatives

13 The equation $z = xy$ can be solved for any one variable as a function of the other two. Verify that

$$\frac{\partial x}{\partial y} \cdot \frac{\partial y}{\partial z} \cdot \frac{\partial z}{\partial x} = -1.$$

14 A metal plate is placed over the xy-plane and a heat source is put at the origin. At a given time the temperature at the point (x, y) is

$$T(x, y) = 600(x^2 + y^2 + 1)^{-1}.$$

Find the rate at which T changes with respect to distance at $(3, 4)$ in the following directions:
- •(a) The direction parallel to the positive x-axis.
- (b) The direction parallel to the positive y-axis.

•**15** The general form of *Boyle's law* states that if P, V, and T are the pressure, volume, and temperature of an ideal gas, then

Boyle's Law
$$PV = kT$$

where k is a constant of proportionality.

A container holds 100 ft³ of gas at a pressure of 800 lb/ft² when the temperature is 80°. Find the rate at which pressure changes with temperature at the instant the temperature is 90°. (Assume that the volume is unchanged.)

16-3 Higher-Order Partial Derivatives

Let f be a function of x and y. The partial derivatives $\partial f/\partial x$ and $\partial f/\partial y$ are themselves functions of x and y. There are four partial derivatives of these functions, the *second partial derivatives* of f:

Second Partial Derivatives

$$\frac{\partial^2 f}{\partial x^2} = \frac{\partial}{\partial x}\left(\frac{\partial f}{\partial x}\right) \quad \text{(second partial with respect to } x\text{)}$$

$$\left.\begin{array}{l}\dfrac{\partial^2 f}{\partial x \partial y} = \dfrac{\partial}{\partial x}\left(\dfrac{\partial f}{\partial y}\right) \\[2mm] \dfrac{\partial^2 f}{\partial y \partial x} = \dfrac{\partial}{\partial y}\left(\dfrac{\partial f}{\partial x}\right)\end{array}\right\} \quad \text{(mixed partial derivatives)}$$

$$\frac{\partial^2 f}{\partial y^2} = \frac{\partial}{\partial y}\left(\frac{\partial f}{\partial y}\right) \quad \text{(second partial with respect to } y\text{)}.$$

EXAMPLE 1 Calculate the first and second partial derivatives of

$$z = f(x, y) = \frac{x}{y} + ye^x.$$

Solution The first partial derivatives are

$$\frac{\partial z}{\partial x} = \frac{\partial f}{\partial x} = \frac{1}{y} + ye^x, \qquad \frac{\partial z}{\partial y} = \frac{\partial f}{\partial y} = -\frac{x}{y^2} + e^x.$$

The second partial derivatives are

$$\frac{\partial^2 z}{\partial x^2} = \frac{\partial}{\partial x}\left(\frac{\partial z}{\partial x}\right) = \frac{\partial}{\partial x}\left(\frac{1}{y} + ye^x\right) = ye^x,$$

$$\frac{\partial^2 z}{\partial y \partial x} = \frac{\partial}{\partial y}\left(\frac{\partial z}{\partial x}\right) = \frac{\partial}{\partial y}\left(\frac{1}{y} + ye^x\right) = -\frac{1}{y^2} + e^x,$$

$$\frac{\partial^2 z}{\partial x \partial y} = \frac{\partial}{\partial x}\left(\frac{\partial z}{\partial y}\right) = \frac{\partial}{\partial x}\left(-\frac{x}{y^2} + e^x\right) = -\frac{1}{y^2} + e^x,$$

$$\frac{\partial^2 z}{\partial y^2} = \frac{\partial}{\partial y}\left(\frac{\partial z}{\partial y}\right) = \frac{\partial}{\partial y}\left(-\frac{x}{y^2} + e^x\right) = \frac{2x}{y^3}. \quad \square$$

Remark Observe in Example 1 that the two mixed partial derivatives are equal. We shall see later in this section that if f is sufficiently well behaved, then the mixed partial derivatives must be equal.

If we use functional notation, the second partial derivatives are denoted by

$$f_{xx} = \frac{\partial^2 f}{\partial x^2}, \quad f_{xy} = \frac{\partial^2 f}{\partial y \partial x}, \quad f_{yx} = \frac{\partial^2 f}{\partial x \partial y}, \quad f_{yy} = \frac{\partial^2 f}{\partial y^2}.$$

Observe that the order of the symbols is reversed for the mixed partials. That is,

$$f_{yx} = \frac{\partial^2 f}{\partial x \partial y}.$$

The first symbol stands for $(f_y)_x$, indicating that the partial derivative is first calculated with respect to y, then x. The second symbol stands for

$$\frac{\partial}{\partial x}\left(\frac{\partial f}{\partial y}\right)$$

indicating the same order for calculating the partial derivatives.

Higher partial derivatives are defined analogously. For example, if

$$z = f(x, y) = \frac{x}{y} + ye^x,$$

the function of Example 1, then

$$\frac{\partial^4 z}{\partial x \partial y^2 \partial x} = \frac{\partial}{\partial x}\left(\frac{\partial}{\partial y}\left(\frac{\partial}{\partial y}\left(\frac{\partial z}{\partial x}\right)\right)\right)$$

$$= \frac{\partial}{\partial x}\left(\frac{\partial}{\partial y}\left(\frac{\partial}{\partial y}\left(\frac{1}{y} + ye^x\right)\right)\right) = \frac{\partial}{\partial x}\left(\frac{\partial}{\partial y}\left(-\frac{1}{y^2} + e^x\right)\right)$$

$$= \frac{\partial}{\partial x}\left(\frac{2}{y^3}\right) = 0.$$

16–3 Higher-Order Partial Derivatives

Partial Derivatives of Functions of Several Variables

There is a natural extension of these concepts to functions of three or more variables. For example, if

$$z = f(u, v, x, y) = uv + \frac{x}{y} + 2xu,$$

then

$$\frac{\partial z}{\partial u} = f_u(u, v, x, y) = v + 2x,$$

$$\frac{\partial z}{\partial v} = f_v(u, v, x, y) = u,$$

$$\frac{\partial z}{\partial x} = f_x(u, v, x, y) = \frac{1}{y} + 2u,$$

$$\frac{\partial z}{\partial y} = f_y(u, v, x, y) = -\frac{x}{y^2},$$

$$\frac{\partial^2 z}{\partial u \partial v} = \frac{\partial}{\partial u}\left(\frac{\partial z}{\partial v}\right) = \frac{\partial}{\partial u}(u) = 1,$$

and so on.

EQUALITY OF MIXED PARTIAL DERIVATIVES

That the symbols are reversed in the different notations for the mixed partial derivatives causes almost no difficulty, since in most cases the mixed partials are equal (as in Example 1). It is required only that f be a reasonably well-behaved function of x and y. The exact conditions are given in Theorem 16–2. (The proof is beyond the scope of this book.)

THEOREM 16–2 Let $f = f(x, y)$. Suppose that f, $\partial f/\partial x$, $\partial f/\partial y$, $\partial^2 f/\partial x \partial y$, and $\partial^2 f/\partial y \partial x$ are continuous on some neighborhood $\mathcal{N}$. Then

$$\frac{\partial^2 f}{\partial x \partial y} = \frac{\partial^2 f}{\partial y \partial x}$$

at each point of $\mathcal{N}$.

In particular, if f is a rational function of x and y, then it can be proved that the conditions of Theorem 16–2 are met at each point (x, y) for which the partial derivatives are defined. Thus, the mixed partial derivatives are equal at each point at which they are defined.

The result of Theorem 16–2 can be extended to mixed partial derivatives of any order and to functions of any number of variables. For example, if $f = f(x, y)$ is continuous, and if all the necessary partial derivatives exist and are continuous on some neighborhood $\mathcal{N}$, then

$$\frac{\partial^7 f}{\partial x \partial y^3 \partial x^2 \partial y} = \frac{\partial^7 f}{\partial x^3 \partial y^4}.$$

Remark You may think that some of the conditions in Theorem 16–2 are redundant. In light

Discontinuous Function May Have Partial Derivatives

of Theorem 2–6, it might appear that if $\partial f/\partial x$ and $\partial f/\partial y$ exist at a point, then f must be continuous there. This is not the case, as you can see from the example

$$f(x, y) = \begin{cases} 1 & \text{if } x = 0 \quad \text{or} \quad y = 0 \\ 0 & \text{if } x \neq 0 \quad \text{and} \quad y \neq 0. \end{cases}$$

Both $\partial f/\partial x$ and $\partial f/\partial y$ exist at $(0, 0)$, but f is discontinuous there.

Exercises 16–3

1–6. Calculate the second partial derivatives of the functions. Verify in each exercise that the mixed partial derivatives are equal.

•1 $f(x, y) = 3x^3y^2 + x^2y + 5 - 3y$

2 $f(x, y) = (2x^2 + y^3)^{3/2}$

•3 $z = \ln(\sqrt{y^2 - x^2})$

4 $z = \sin(x^2 + y)$

•5 $z = \ln x - \dfrac{1}{xy}$

6 $z = \cos\left(\dfrac{x}{y}\right)$

7–10. Let $w = x^2/(y + z)$. Calculate the partial derivative.

•7 w_{xx}

8 w_{xy}

•9 w_{xxy}

10 w_{xxyz}

11 Let $w = f(x, y)$ be continuous and have continuous first, second, and third partial derivatives on a neighborhood $\mathcal{N}$. List all the third partial derivatives of w. Which of these are equal?

12 Let $w = f(x, y, z)$ be continuous and have continuous first and second partial derivatives at all points in three-dimensional space. List all the second partial derivatives of w. Which of these are equal?

13 A function $w = f(x, y)$ is said to be *harmonic* if

Harmonic Function

$$\dfrac{\partial^2 f}{\partial x^2} + \dfrac{\partial^2 f}{\partial y^2} = 0.$$

Verify that the following functions are harmonic.

(a) $f(x, y) = \tan^{-1}(y/x)$

(b) $f(x, y) = \ln(x^2 + y^2)$

14 *Laplace's equation* is

Laplace's Equation

$$\dfrac{\partial^2 w}{\partial x^2} + \dfrac{\partial^2 w}{\partial y^2} + \dfrac{\partial^2 w}{\partial z^2} = 0.$$

Verify that the following functions satisfy Laplace's equation.

(a) $w = 2x^2 + 3y^2 - 5z^2$

(b) $w = e^{3x-4y} \sin 5z$

15 Let $z = f(x, y)$. Suppose that f and all its partial derivatives exist and are continuous on a neighborhood $\mathcal{N}$.

16–4 Increments and Approximations

- (a) Use Theorem 16–2 to prove that at each point of $\mathcal{N}$,

$$\frac{\partial^3 w}{\partial x \partial y \partial x} = \frac{\partial^3 w}{\partial x^2 \partial y}.$$

(b) Prove that $\dfrac{\partial^4 w}{\partial x^2 \partial y \partial x} = \dfrac{\partial^4 w}{\partial x^3 \partial y}$ at each point of $\mathcal{N}$.

The following exercise shows that the mixed partial derivatives f_{xy} and f_{yx} may not be equal unless the hypothesis of Theorem 16–2 is satisfied.

16. Let $f(x, y) = \begin{cases} xy \dfrac{x^2 - y^2}{x^2 + y^2} & \text{if } (x, y) \neq (0, 0) \\ 0 & \text{if } (x, y) = (0, 0). \end{cases}$

   (a) Show that $f_x(0, y) = -y$ for all y and $f_y(x, 0) = x$ for all x.

   (b) Show that $f_{xy}(0, 0) = -1$ and $f_{yx}(0, 0) = 1$. (*Hint:* Set up difference quotients for the second partial derivatives.)

   - (c) What hypothesis of Theorem 16–2 is not satisfied by this function?

16–4 Increments and Approximations

Let $y = f(x)$ be a differentiable function of one variable. We established in Section 3–3 that if $\Delta y = f(x_0 + \Delta x) - f(x_0)$, then

$$\Delta y = f'(x_0) \Delta x + \epsilon \, \Delta x,$$

where $\epsilon \to 0$ as $\Delta x \to 0$. Since $\epsilon \, \Delta x$ is small whenever Δx is small, it follows that

$$\Delta y \approx f'(x_0) \, \Delta x$$

whenever Δx is small.

These results can be generalized to functions of two or more variables.

INCREMENTS

Let $w = f(x, y)$ be a function of two variables. Let Δx and Δy be independent increments of x and y. We define the corresponding increment Δw (or Δf) by

Increment of Function of Two Variables

$$\Delta w = \Delta f = f(x + \Delta x, y + \Delta y) - f(x, y).$$

Thus, Δw measures the change in w when we move from the point (x, y) in the domain of f to the point $(x + \Delta x, y + \Delta y)$. (See Fig. 16–7.)

EXAMPLE 1   Let $w = f(x, y) = x^2 + \cos xy + xe^y$. Calculate the change in w when we move from $(1, 2)$ to $(1.01, 1.97)$.

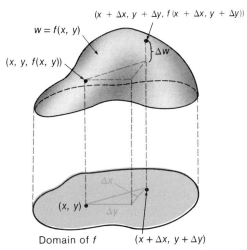

Figure 16–7 Increments: Δw is the change in w that corresponds to a change from (x, y) to $(x + \Delta x, y + \Delta y)$.

Solution In this case, we have $x = 1$, $y = 2$, $\Delta x = 0.01$, and $\Delta y = -0.03$. Then

$$\begin{aligned}\Delta w &= f(x + \Delta x, y + \Delta y) - f(x, y) \\ &= f(1.01, 1.97) - f(1, 2) \\ &= (1.01)^2 + \cos[(1.01)(1.97)] + (1.01)e^{1.97} - (1^2 + \cos 2 + 1 \cdot e^2) \\ &\approx -0.117185. \quad \square\end{aligned}$$

THE APPROXIMATION FORMULA

Our main result is an approximation formula like the one that was proved in Section 3–3 for functions of one variable. We shall establish that under suitable continuity conditions, if $w = f(x, y)$, then

Approximation Formula

$$\Delta w = f_x(x_0, y_0) \, \Delta x + f_y(x_0, y_0) \, \Delta y + \epsilon_1 \, \Delta x + \epsilon_2 \, \Delta y$$

where $\epsilon_1, \epsilon_2 \to 0$ as $\Delta x, \Delta y \to 0$. The conditions on the function are given in the following theorem.

THEOREM 16–3 Let $w = f(x, y)$ be a function of x and y. Let $\mathcal{N}$ be a neighborhood of $P_0(x_0, y_0)$ on which f is defined. Suppose that $\partial f/\partial x$ and $\partial f/\partial y$ exist on $\mathcal{N}$ and are continuous at $P_0(x_0, y_0)$. Let Δx and Δy be increments of x and y that are small enough so that the point $Q(x_0 + \Delta x, y_0 + \Delta y)$ is in $\mathcal{N}$. Then there exist ϵ_1 and ϵ_2 (which depend on Δx and Δy) such that

(1) $\Delta w = f_x(x_0, y_0) \, \Delta x + f_y(x_0, y_0) \, \Delta y + \epsilon_1 \, \Delta x + \epsilon_2 \, \Delta y$, and

(2) ϵ_1 and ϵ_2 both approach 0 as $(\Delta x, \Delta y) \to (0, 0)$.

16–4 Increments and Approximations

Proof Let R be the point $R(x_0 + \Delta x, y_0)$. We shall prove the result in two stages. These are equivalent to moving from P_0 to Q by moving first from P_0 to R, then from R to Q. (See Fig. 16–8a.)

Between P_0 and R the values of y equal y_0. Thus, the functional values of $w = f(x, y)$ depend only on x. To emphasize this fact, we let $g(x) = f(x, y_0)$. Then $w = g(x) = f(x, y_0)$ if (x, y_0) is on the line segment connecting P_0 and R. Observe that if (x, y_0) is on this line segment, then

$$g'(x) = f_x(x, y_0).$$

It follows from the Mean Value Theorem (Theorem 4–3) that there exists a number X (which depends on Δx) between x_0 and $x_0 + \Delta x$ such that

$$g(x_0 + \Delta x) - g(x_0) = g'(X)\Delta x.$$

This result can be rewritten as

$$f(x_0 + \Delta x, y_0) - f(x_0, y_0) = f_x(X, y_0) \Delta x.$$

A similar argument can now be made by restricting ourselves to the line segment between R and Q. (The details of this part of the proof are left for you to work in Exercise 16.) It follows that there must exist a number Y (which depends on Δx and Δy) between y_0 and $y_0 + \Delta y$ such that

$$f(x_0 + \Delta x, y_0 + \Delta y) - f(x_0 + \Delta x, y_0) = f_y(x_0 + \Delta x, Y) \Delta y.$$

We now return to our original problem. Observe that

$$\begin{aligned}\Delta w &= f(x_0 + \Delta x, y_0 + \Delta y) - f(x_0, y_0) \\ &= f(x_0 + \Delta x, y_0 + \Delta y) - f(x_0 + \Delta x, y_0) + f(x_0 + \Delta x, y_0) - f(x_0, y_0) \\ &= f_y(x_0 + \Delta x, Y) \Delta y + f_x(X, y_0) \Delta x.\end{aligned}$$

Because (X, y_0) is between P_0 and R and $(x_0 + \Delta x, Y)$ is between R and Q, it follows that

$$(X, y_0) \to (x_0, y_0) \quad \text{and} \quad (x_0 + \Delta x, Y) \to (x_0, y_0)$$

as $(\Delta x, \Delta y) \to (0, 0)$. (See Fig. 16–8b.) Since f_x and f_y are continuous at $P_0(x_0, y_0)$, then

$$\lim_{(\Delta x, \Delta y) \to (0,0)} f_x(X, y_0) = f_x(x_0, y_0)$$

and

$$\lim_{(\Delta x, \Delta y) \to (0,0)} f_y(x_0 + \Delta x, Y) = f_y(x_0, y_0).$$

Let $\epsilon_1 = f_x(X, y_0) - f_x(x_0, y_0)$ and $\epsilon_2 = f_y(x_0 + \Delta x, Y) - f_y(x_0, y_0)$. It follows from the limits in the preceding paragraph that ϵ_1 and $\epsilon_2 \to 0$ as $(\Delta x, \Delta y) \to 0$.

Finally, we can write

$$\begin{aligned}\Delta w &= f_x(X, y_0) \Delta x + f_y(x_0 + \Delta x, Y) \Delta y \\ &= [f_x(x_0, y_0) + \epsilon_1] \Delta x + [f_y(x_0, y_0) + \epsilon_2] \Delta y \\ &= f_x(x_0, y_0) \Delta x + f_y(x_0, y_0) \Delta y + \epsilon_1 \Delta x + \epsilon_2 \Delta y\end{aligned}$$

where $\epsilon_1, \epsilon_2 \to 0$ as $(\Delta x, \Delta y) \to (0, 0)$. ■

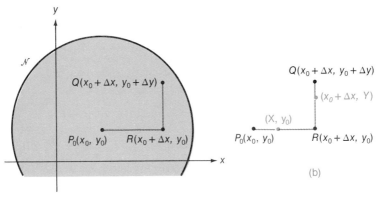

Figure 16-8.

Remark If Δx and Δy are both small numbers, then ϵ_1 and ϵ_2 must also be small. It follows that the products $\epsilon_1 \Delta x$ and $\epsilon_2 \Delta y$ are so small as to be negligible. Thus, we get the following corollary to Theorem 16–3.

COROLLARY Let $w = f(x, y)$ satisfy the hypothesis of Theorem 16–3. Then

Weak Form of the Approximation Formula

$$\Delta w \approx f_x(x_0, y_0) \, \Delta x + f_y(x_0, y_0) \, \Delta y$$

if Δx and Δy are small.

Examples 2 and 3 illustrate the use of this formula.

EXAMPLE 2 Use the above approximation formula to approximate the change in
$$w = f(x, y) = x^2 + \cos xy + xe^y$$
when we move from (1, 2) to (1.01, 1.97). (See Example 1.)

Solution
$$\frac{\partial w}{\partial x} = 2x - y \sin(xy) + e^y,$$

$$\frac{\partial w}{\partial y} = -x \sin(xy) + xe^y.$$

We must approximate Δw when $x_0 = 1$, $y_0 = 2$, $\Delta x = 0.01$, and $\Delta y = -0.03$. By the formula

$$\Delta w \approx f_x(x_0, y_0) \, \Delta x + f_y(x_0, y_0) \, \Delta y$$
$$= (2x_0 - y_0 \sin(x_0 y_0) + e^{y_0}) \, \Delta x + (-x_0 \sin(x_0 y_0) + x_0 e^{y_0}) \, \Delta y$$
$$= [2 \cdot 1 - 2 \sin 2 + e^2] \, (0.01) + [-1 \cdot \sin 2 + 1 \cdot e^2] \, (-0.03)$$
$$\approx (7.570461)(0.01) + (6.479759)(-0.03) \approx -0.118688.$$

Observe that this result differs in the third decimal place from the more accurate approximation obtained in Example 1. ∎

16–4 Increments and Approximations

EXAMPLE 3 A solid right circular cylinder, made of metal, has radius $r = 5$ in. and height $h = 14$ in. When the cylinder is heated, the radius increases to 5.2 in. and the height to 14.6 in. Find the approximate change in (a) the volume; (b) the total surface area.

Solution (a) Since the formula for the volume is

$$V = \pi r^2 h,$$

then

$$\Delta V \approx \frac{\partial V}{\partial r} \Delta r + \frac{\partial V}{\partial h} \Delta h = 2\pi r h\, \Delta r + \pi r^2\, \Delta h.$$

When $r = 5$, $\Delta r = 0.2$, $h = 14$, and $\Delta h = 0.6$, we get

$$\Delta V \approx 2\pi \cdot 5 \cdot 14(0.2) + \pi(5)^2(0.6) = 43\pi \approx 135.09 \text{ in}^3.$$

(b) The formula for the total surface area is

$$S = 2\pi r^2 + 2\pi r h.$$

Then

$$\Delta S \approx \frac{\partial S}{\partial r} \Delta r + \frac{\partial S}{\partial h} \Delta h$$
$$= (4\pi r + 2\pi h)\, \Delta r + 2\pi r\, \Delta h.$$

When we evaluate this expression at the particular values of r, Δr, h, and Δh, we get

$$\Delta S \approx (4\pi \cdot 5 + 2\pi \cdot 14)(0.2) + (2\pi \cdot 5)(0.6)$$
$$= 15.6\pi \approx 49.01 \text{ in}^2. \quad \square$$

DIFFERENTIABLE FUNCTIONS

If a function of one variable has a derivative at a point, then the function must be continuous at that point. The situation is different when we deal with functions of two or more variables. As we saw in the remark at the end of Section 16–3, a function of two variables may even be discontinuous at a point where both partial derivatives exist. We now consider a concept that is closely related to the concept of differentiability for a function of one variable.

DEFINITION

Differentiable Function of Two Variables

The function $w = f(x, y)$ is said to be *differentiable* at (x_0, y_0) provided that there exist ϵ_1 and ϵ_2, which depend on Δx and Δy, such that

$$\Delta w = f(x_0 + \Delta x, y_0 + \Delta y) - f(x_0, y_0)$$
$$= f_x(x_0, y_0)\, \Delta x + f_y(x_0, y_0)\, \Delta y + \epsilon_1 \Delta x + \epsilon_2 \Delta y$$

where

$$\epsilon_1, \epsilon_2 \to 0 \text{ as } (\Delta x, \Delta y) \to (0, 0).$$

The function f is said to be *differentiable* if there exists some neighborhood $\mathcal{N}$ such that f is differentiable at each point of $\mathcal{N}$. Any results that are stated for "differentiable functions" hold on such neighborhoods.

It turns out that differentiable functions of two variables have properties similar to those of differentiable functions of one variable. One of the most important of these properties is given in the following theorem.

THEOREM 16-4 (*See Theorem 2-6*) Let f be differentiable at (x_0, y_0). Then f is continuous at (x_0, y_0).

Proof Let (x, y) be in the domain of f. Let $\Delta x = x - x_0$, $\Delta y = y - y_0$, so that $(x, y) \to (x_0, y_0)$ if and only if $(\Delta x, \Delta y) \to (0, 0)$. It follows from the definition that

$$f(x, y) = f(x_0 + \Delta x, y_0 + \Delta y)$$
$$= f(x_0, y_0) + f_x(x_0, y_0) \Delta x + f_y(x_0, y_0) \Delta y + \epsilon_1 \Delta x + \epsilon_2 \Delta y$$

where $\epsilon_1, \epsilon_2 \to 0$ as $(\Delta x, \Delta y) \to (0, 0)$. Then

$$\lim_{(x,y) \to (x_0, y_0)} f(x, y)$$
$$= \lim_{(\Delta x, \Delta y) \to (0,0)} [f(x_0, y_0) + f_x(x_0, y_0) \Delta x + f_y(x_0, y_0) \Delta y + \epsilon_1 \Delta x + \epsilon_2 \Delta y]$$
$$= f(x_0, y_0) + 0 + 0 + 0 + 0 = f(x_0, y_0).$$

Thus, f is continuous at (x_0, y_0). ∎

Remark Differentiability is clearly stronger than the requirement that partial derivatives exist. Observe that Theorem 16-3 states conditions under which f is known to be differentiable.

THE DIFFERENTIAL

Let $z = f(x, y)$ where f is differentiable. The *differentials* dx and dy are defined to be the increments Δx and Δy:

$$dx = \Delta x \quad \text{and} \quad dy = \Delta y.$$

The differential of z is defined to be

Differential of Function of Two Variables

$$dz = \frac{\partial f}{\partial x} dx + \frac{\partial x}{\partial y} dy.$$

By these definitions, the approximation formula can be written

$$\Delta z \approx f_x(x_0, y_0) \, dx + f_y(x_0, y_0) \, dy$$

or

$$\Delta z \approx dz.$$

EXAMPLE 4 If $z = 3x^2 y + \sin(x + y)$, then

$$dz = \frac{\partial z}{\partial x} dx + \frac{\partial z}{\partial y} dy$$
$$= [6xy + \cos(x + y)] \, dx + [3x^2 + \cos(x + y)] \, dy. \square$$

FUNCTIONS OF THREE OR MORE VARIABLES

The results of this section can be extended to functions of three or more variables. If $w = f(x, y, z)$, then w is said to be *differentiable* if Δw can be written in the form

Differentiable Function of Three Variables

$$\Delta w = f_x(x_0, y_0, z_0) \Delta x + f_y(x_0, y_0, z_0) \Delta y + f_z(x_0, y_0, z_0) \Delta z + \epsilon_1 \Delta x + \epsilon_2 \Delta y + \epsilon_3 \Delta z$$

where $\epsilon_1, \epsilon_2, \epsilon_3 \to 0$ as $\Delta x, \Delta y, \Delta z \to 0$. The corresponding approximation formula is

$$\Delta w \approx \frac{\partial f}{\partial x} \Delta x + \frac{\partial f}{\partial y} \Delta y + \frac{\partial f}{\partial z} \Delta z$$

when $\Delta x, \Delta y$, and Δz are small. These formulas are valid provided continuity conditions similar to those of Theorem 16-3 hold for the function f.

Differentials of functions of several variables are defined in a similar manner to differentials of functions of two variables. For example, if w is a differentiable function of x, y, z, then dx, dy, and dz are independent variables. The differential dw is defined to be

Differential of Function of Three Variables

$$dw = \frac{\partial w}{\partial x} dx + \frac{\partial w}{\partial y} dy + \frac{\partial w}{\partial z} dz.$$

Approximation problems involving several variables can be worked by using either increments or differentials, as with the two-variable case.

Exercises 16–4

1–2. Find the exact value of Δw that corresponds to the indicated change in the independent variables. Compare this result with the one obtained from the approximation formula.

• 1 $w = \sqrt{2x + y}$; (x, y) changes from $(2, 0)$ to $(1.98, 0.01)$.

2   $w = x^2 y z^3$; (x, y, z) changes from $(1, 1, 2)$ to $(1.01, 0.98, 2.01)$.

• 3   (*Error Term*) Calculate the exact value of $\epsilon_1 \Delta x + \epsilon_2 \Delta y$ in terms of $x_0, y_0, \Delta x, \Delta y$ for the function $f(x, y) = xy^2$.

4   (*Error Term*) Calculate the exact value of $\epsilon_1 \Delta x + \epsilon_2 \Delta y + \epsilon_3 \Delta z$ in terms of $x_0, y_0, z_0, \Delta x, \Delta y, \Delta z$ for $f(x, y, z) = xyz$.

5–8.   Write the approximation formula (corollary to Theorem 16–3) in the form

$$f(x_0 + \Delta x, y_0 + \Delta y) \approx f(x_0, y_0) + f_x(x_0, y_0) \Delta x + f_y(x_0, y_0) \Delta y.$$

Use this formula to approximate the following numbers (*Hint for 5:* Let $f(x, y) = \sqrt{x^2 + y^2}$, $x_0 = 3$, $y_0 = 4$.)

- **5** $\sqrt{(3.03)^2 + (3.99)^2}$
- **6** $(3.01)^4 \sqrt[3]{7.98}$
- **7** $\sin(0.01) - \sin(-0.02)$
- **8** $e^{0.01} \cos(0.02)$

9–12. Calculate the differential.

- **9** $z = 3x^4 - 3xy^2 - y^3$
- **10** $z = x \ln y + e^x$
- **11** $u = x \cos z - \tan(xy) + z^3 e^{-y}$
- **12** $w = x^3 \ln(z^2 + y)$

- **13** Two sides and an included angle of a triangle are measured to be $a = 5$, $b = 8$, and $C = 60°$. The actual values are $a = 5.1$, $b = 8.01$, and $C = 60°30'$. Find the approximate error in the computed value of the area. [Area = $\tfrac{1}{2}ab \sin C$.]

- **14** A right circular cone of radius $r = 6$ cm and altitude $h = 8$ cm is heated, whereupon the radius is changed to 6.2 cm and the altitude to 8.5 cm. Find the approximate change in (a) the volume; (b) the lateral surface area.

- **15** An open-topped box has length 3 m, width 5 m, and height 7 m at a certain temperature. When the temperature is lowered, each dimension contracts by 1 percent. Find the change in the surface area.

---

16 Complete the proof of Theorem 16–3. Define h as a function of y between $R(x_0 + \Delta x, y_0)$ and $Q(x_0 + \Delta x, y_0 + \Delta y)$ by

$$h(y) = f(x_0 + \Delta x, y).$$

Use this function to show that there exists Y between y_0 and $y_0 + \Delta y$ such that

$$f(x_0 + \Delta x, y_0 + \Delta y) - f(x_0 + \Delta x, y_0) = f_y(x_0 + \Delta x, Y) \Delta y.$$

16–5 The Chain Rule

In this section we generalize the Chain Rule formula D8 to functions of two or more variables.

Let $z = F(x, y)$ be a differentiable function of x and y. Let $x = f(t)$, $y = g(t)$ be differentiable functions of t. Then z also is a differentiable function of t:

$$z = F(f(t), g(t)) = \phi(t).$$

We shall show that the derivative $dz/dt = \phi'(t)$ can be calculated by a formula analogous to D8.

There is a simple geometrical interpretation for the situation just described. Let S be the surface that is the graph of $z = F(x, y)$ and let $\mathscr{C}$ be the curve in the xy-plane defined by $x = f(t)$, $y = g(t)$. Then $dz/dt = \phi'(t)$ represents the rate of change of z with respect to t as the point $P(x, y)$ moves around the curve $\mathscr{C}$. (See Fig. 16–9.) For example, if $z = F(x, y)$ is the temperature at (x, y), then dz/dt is the rate of change of the temperature with time, as the point $P(x, y)$ moves around the curve $\mathscr{C}$.

16-5 The Chain Rule

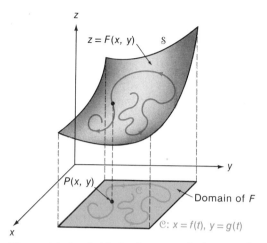

Figure 16-9 dz/dt is the rate of change of z with respect to time as $P(x, y)$ moves around the curve $\mathscr{C}$ defined by $x = f(t)$, $y = g(t)$.

THEOREM 16-5 Let $z = F(x, y)$ be a differentiable function of x and y; let $x = f(t), y = g(t)$ be differentiable functions of t. Then

Chain Rule
Formula
$$\frac{dz}{dt} = \frac{\partial F}{\partial x} \cdot \frac{dx}{dt} + \frac{\partial F}{\partial y} \cdot \frac{dy}{dt}.$$

Proof Let Δt be an increment of t; let

$$\Delta x = f(t + \Delta t) - f(t) \quad \text{and} \quad \Delta y = g(t + \Delta t) - g(t)$$

be the corresponding increments of x and y. Observe that Δx and Δy are not independent variables. They depend on t and Δt.

The increment of z that corresponds to Δt is

$$\Delta z = F(f(t + \Delta t), g(t + \Delta t)) - F(f(t), g(t))$$
$$= F(x + \Delta x, y + \Delta y) - F(x, y).$$

Since F is differentiable, we can write

$$\Delta z = \frac{\partial F}{\partial x} \Delta x + \frac{\partial F}{\partial y} \Delta y + \epsilon_1 \Delta x + \epsilon_2 \Delta y$$

where $\epsilon_1, \epsilon_2 \to 0$ as $(\Delta x, \Delta y) \to (0, 0)$. We first divide this equation by Δt and then take the limit as $\Delta t \to 0$, obtaining

$$\frac{\Delta z}{\Delta t} = \frac{\partial F}{\partial x} \cdot \frac{\Delta x}{\Delta t} + \frac{\partial F}{\partial y} \cdot \frac{\Delta y}{\Delta t} + \epsilon_1 \frac{\Delta x}{\Delta t} + \epsilon_2 \frac{\Delta y}{\Delta t},$$

$$\frac{dz}{dt} = \lim_{\Delta t \to 0} \frac{\Delta z}{\Delta t} = \frac{\partial F}{\partial x} \cdot \left(\lim_{\Delta t \to 0} \frac{\Delta x}{\Delta t}\right) + \frac{\partial F}{\partial y} \cdot \left(\lim_{\Delta t \to 0} \frac{\Delta y}{\Delta t}\right)$$

$$+ \left(\lim_{\Delta t \to 0} \epsilon_1\right) \cdot \left(\lim_{\Delta t \to 0} \frac{\Delta x}{\Delta t}\right) + \left(\lim_{\Delta t \to 0} \epsilon_2\right) \cdot \left(\lim_{\Delta t \to 0} \frac{\Delta y}{\Delta t}\right).$$

Since f and g are differentiable functions of t, then they are continuous. Therefore

$$\lim_{\Delta t \to 0} \Delta x = \lim_{\Delta t \to 0} [f(t + \Delta t) - f(t)] = 0 \quad \text{and} \quad \lim_{\Delta t \to 0} \frac{\Delta x}{\Delta t} = \frac{dx}{dt}.$$

Similarly,

$$\lim_{\Delta t \to 0} \Delta y = 0 \quad \text{and} \quad \lim_{\Delta t \to 0} \frac{\Delta y}{\Delta t} = \frac{dy}{dt}.$$

Since $\Delta x, \Delta y \to 0$ as $\Delta t \to 0$, then $\epsilon_1, \epsilon_2 \to 0$ as $\Delta t \to 0$. It follows that

$$\frac{dz}{dt} = \frac{\partial F}{\partial x} \cdot \frac{dx}{dt} + \frac{\partial F}{\partial y} \cdot \frac{dy}{dt} + 0 \cdot \frac{dx}{dt} + 0 \cdot \frac{dy}{dt}$$

$$= \frac{\partial F}{\partial x} \cdot \frac{dx}{dt} + \frac{\partial F}{\partial y} \cdot \frac{dy}{dt}. \blacksquare$$

EXAMPLE 1 Let $z = x^2 - 2xy^2 + y^3$, where $x = \sin t$ and $y = e^{-t}$. Then

$$\frac{dz}{dt} = \frac{\partial z}{\partial x} \cdot \frac{dx}{dt} + \frac{\partial z}{\partial y} \cdot \frac{dy}{dt}$$

$$= (2x - 2y^2) \cos t + (-4xy + 3y^2)(-e^{-t})$$

$$= 2(x - y^2) \cos t + (4x - 3y) ye^{-t}.$$

If we replace x and y by $\sin t$ and e^{-t}, we get

$$\frac{dz}{dt} = 2(\sin t - e^{-2t}) \cos t + (4 \sin t - 3e^{-t}) e^{-2t}. \ \square$$

A similar Chain Rule formula holds for functions of three or more variables. For example, if w is a differentiable function of x, y, and z, each of which is a differentiable function of t, then

Chain Rule for
Function of
Three Variables

$$\frac{dw}{dt} = \frac{\partial w}{\partial x} \cdot \frac{dx}{dt} + \frac{\partial w}{\partial y} \cdot \frac{dy}{dt} + \frac{\partial w}{\partial z} \cdot \frac{dz}{dt}.$$

(See Exercise 20.)

PARTIAL DERIVATIVES

An analogous Chain Rule formula holds for partial derivatives. Suppose that $u = F(x, y)$ is a differentiable function of x and y, each of which is a differentiable function of s and t, say $x = f(s, t)$, $y = g(s, t)$. Then $u = F(f(s, t), g(s, t))$ is a function of s and t. An argument similar to the proof of Theorem 16–5 can be used to establish that

Chain Rule for
Partial Derivatives

$$\frac{\partial u}{\partial s} = \frac{\partial u}{\partial x} \cdot \frac{\partial x}{\partial s} + \frac{\partial u}{\partial y} \cdot \frac{\partial y}{\partial s},$$

$$\frac{\partial u}{\partial t} = \frac{\partial u}{\partial x} \cdot \frac{\partial x}{\partial t} + \frac{\partial u}{\partial y} \cdot \frac{\partial y}{\partial t}.$$

(See Exercise 21.)

16–5 The Chain Rule

EXAMPLE 2 Let $u = x^2 + 3xy + 5y^2$ where $x = 5s^2$ and $y = 2s - 3t^2$. Then

$$\frac{\partial u}{\partial s} = \frac{\partial u}{\partial x} \cdot \frac{\partial x}{\partial s} + \frac{\partial u}{\partial y} \cdot \frac{\partial y}{\partial s}$$

$$= (2x + 3y) \cdot 10s + (3x + 10y) \cdot 2$$

and

$$\frac{\partial u}{\partial t} = \frac{\partial u}{\partial x} \cdot \frac{\partial x}{\partial t} + \frac{\partial u}{\partial y} \cdot \frac{\partial y}{\partial t}$$

$$= (2x + 3y) \cdot 0 + (3x + 10y) \cdot (-6t). \quad \square$$

Similar Chain Rule formulas hold for differentiable functions of any number of variables. For example, if u is a differentiable function of $w, x, y,$ and z, each of which is a differentiable function of $r, s,$ and t, then

$$\frac{\partial u}{\partial r} = \frac{\partial u}{\partial w} \cdot \frac{\partial w}{\partial r} + \frac{\partial u}{\partial x} \cdot \frac{\partial x}{\partial r} + \frac{\partial u}{\partial y} \cdot \frac{\partial y}{\partial r} + \frac{\partial u}{\partial z} \cdot \frac{\partial z}{\partial r},$$

and so on.

FIRST- AND SECOND-LEVEL VARIABLES

Thus far we have considered only the simplest types of Chain Rule problems. The problems that we consider in the remainder of this section are more complicated and can cause trouble if they are not handled carefully.

It may help you to think of the variables in different levels. If $u = F(x, y, z)$, then $x, y,$ and z are *first-level variables for u*. If $x, y,$ and z can be expressed in terms of s and t, then s and t are *second-level variables for u*. To apply the Chain Rule for the calculation of $\partial u/\partial s$, for example, we first calculate the partial derivative of u with respect to each of the first-level variables, then multiply each of these partial derivatives by the corresponding partial derivative of the first-level variable with respect to s, the second-level variable under consideration.

Our problems with the Chain Rule begin when some of the first-level variables also are second-level. For example, if $u = F(x, y)$, where y is a function of x and t, then we also can write

$$u = G(x, t).$$

In this case the symbol $\partial u/\partial x$ is ambiguous. It could mean either $\partial F/\partial x$ or $\partial G/\partial x$. Furthermore, if we calculate $\partial G/\partial x$, then the Chain Rule formula involves x in two different roles—as a first- and a second-level variable.

In a situation like the one just described, it is usually best to rename the second-level variable. We let x denote the first-level variable and $s = x$ the second-level variable. At the conclusion of the problem, we substitute x for s.

EXAMPLE 3 Let $u = F(x, y) = x^2 + 2y^3$, where $y = x^2 + t^2$. Then u also is a function of x and t (second-level variables), say $u = G(x, t)$. Use the Chain Rule to calculate $\partial G/\partial x$ and $\partial G/\partial t$.

Solution To distinguish between the first- and second-level variables, we let s denote the second-level variable x. Then
$$u = F(x, y) = x^2 + 2y^3,$$
where $x = s$ and $y = s^2 + t^2$.

(a) $\dfrac{\partial G}{\partial s} = \dfrac{\partial F}{\partial x} \cdot \dfrac{\partial x}{\partial s} + \dfrac{\partial F}{\partial y} \cdot \dfrac{\partial y}{\partial s}$

$= 2x \cdot 1 + 6y^2 \cdot 2s = 2x + 12sy^2.$

Thus, since $s = x$,
$$\dfrac{\partial G}{\partial x} = 2x + 12xy^2.$$

(b) $\dfrac{\partial G}{\partial t} = \dfrac{\partial F}{\partial x} \cdot \dfrac{\partial x}{\partial t} + \dfrac{\partial F}{\partial y} \cdot \dfrac{\partial y}{\partial t}$

$= 2x \cdot 0 + 6y^2 \cdot 2t = 12ty^2.$ ☐

IMPLICIT DIFFERENTIATION

Suppose that F is a differentiable function of x and y and that the equation
$$F(x, y) = 0$$
defines y as a differentiable function of x. We shall show that the Chain Rule can be used to calculate dy/dx.

THEOREM 16–6

Theorem for Implicit Differentiation

Let F be a differentiable function of x and y. Suppose the equation
$$F(x, y) = 0$$
defines y as a differentiable function of x. Then
$$\dfrac{dy}{dx} = -\dfrac{\partial F/\partial x}{\partial F/\partial y} \quad \text{provided } \dfrac{\partial F}{\partial y} \neq 0.$$

Proof Let $u = F(x, y)$. If the equation $F(x, y) = 0$ defines $y = g(x)$, then
$$F(x, g(x)) = 0 \quad \text{for all } x.$$

Thus, $u = 0$ whenever (x, y) is on the graph of g.

To distinguish between x as a first- and a second-level variable, we call the second-level variable s. Then
$$u = F(x, y) = 0 \quad \text{when } x = s \text{ and } y = g(s).$$

By the Chain Rule
$$\dfrac{du}{ds} = \dfrac{\partial F}{\partial x} \cdot \dfrac{dx}{ds} + \dfrac{\partial F}{\partial y} \cdot \dfrac{dy}{ds} = \dfrac{d}{ds}(0) = 0$$

$$\dfrac{\partial F}{\partial x} \cdot 1 + \dfrac{\partial F}{\partial y} \cdot \dfrac{dy}{dx} = 0$$

16-5 The Chain Rule

$$\frac{dy}{dx} = -\frac{\partial F/\partial x}{\partial F/\partial y}$$

provided $\partial F/\partial y \neq 0$. ∎

EXAMPLE 4 Use Theorem 16-6 to calculate dy/dx if

$$x^2 y - 3y^3 = 7.$$

Solution The equation can be written in the form $F(x, y) = 0$, where

$$F(x, y) = x^2 y - 3y^3 - 7.$$

Then

$$\frac{dy}{dx} = -\frac{\partial F/\partial x}{\partial F/\partial y} = -\frac{2xy}{x^2 - 9y^2}. \quad \square$$

DIFFERENTIALS

Differential Formulas for Functions of Several Variables

The Chain Rule formula is closely related to the formula for differentials. As we shall see, the differential formulas that were developed for functions of one variable can be extended to hold for functions of any number of variables. These more general formulas are proved by use of the Chain Rule.

For example, if z is a differential function of x and y, it can be shown that

$$d(z^n) = n z^{n-1} \, dz,$$
$$d(\cos z) = -\sin z \, dz,$$
$$d(e^z) = e^z \, dz,$$

and so on. Example 5 illustrates how these formulas can be derived.

EXAMPLE 5 Let $z = f(x, y)$ be a differentiable function. Show that $d(\cos z) = -\sin z \, dz$.

Solution Let $u = \cos z$, where $z = f(x, y)$. Then z is a first-level variable for u, while x and y are second-level variables. By the Chain Rule,

$$\frac{\partial u}{\partial x} = -\sin z \cdot \frac{\partial z}{\partial x} \quad \text{and} \quad \frac{\partial u}{\partial y} = -\sin z \cdot \frac{\partial z}{\partial y}.$$

Therefore,

$$d(\cos z) = du = \frac{\partial u}{\partial x} dx + \frac{\partial u}{\partial y} dy$$

$$= \left(-\sin z \frac{\partial z}{\partial x}\right) dx + \left(-\sin z \frac{\partial z}{\partial y}\right) dy$$

$$= -\sin z \left(\frac{\partial z}{\partial x} dx + \frac{\partial z}{\partial y} dy\right) = -\sin z \, dz. \quad \square$$

EXAMPLE 6   Let $z = 3x^2y - 2y^3$. Use the formula in Example 5 to calculate the differential of $\cos z$.

Solution
$$d(\cos z) = -\sin z \, dz$$
$$= -\sin z[6xy \, dx + (3x^2 - 6y^2) \, dy]$$
$$= -\sin (3x^2y - 2y^3)[6xy \, dx + (3x^2 - 6y^2) \, dy]. \quad \square$$

Exercises 16-5

1-4.   Use the Chain Rule to calculate dz/dt.

- **1** $z = x^4 + x^2y - y^2; x = e^t, y = -\cos t$

2    $z = y \ln x + xy + \tan y; x = \dfrac{t}{t+1}, y = t^3 - t$

- **3** $z = x \sin y + w^2; x = \dfrac{1}{t}, y = \sin^{-1} t, w = \sqrt{1 - t^2}$

4    $z = xy + yw + wx; x = 2t^2 + 1, y = 1 - t^2, w = \ln t$

5-8.   Use the Chain Rule to calculate $\partial w/\partial s$ and $\partial w/\partial t$.

- **5** $w = 6x + 3y \sqrt{xy} - y^3; x = e^t + s^2, y = \cos s$

6    $w = \sqrt{x^2 + y^2} + x \cos z; x = e^{\cos t} - s, y = \ln t, z = \dfrac{s}{1-t}$

- **7** $w = x^2y^2 - z \ln x + xyz; x = \ln s, y = r^2 + s^2, z = s \sin t$

8    $w = x^2 \sin 2y; x = e^{2s+t}, y = \tan^{-1}(r/s)$

9-10.   Use the Chain Rule to calculate $\partial w/\partial x$, $\partial w/\partial y$, and $\partial w/\partial z$.

- **9** $w = \ln r; r = x^2 - y \ln z$.   10 $w = f(v), v = e^x + 3y + \sin z$

11-14.   Use implicit differentiation based on the Chain Rule to calculate dy/dx.

- **11** $3x^2 + 2xy^2 - y^3 = 1$ 12 $x^{1/3} + y^{1/3} = 2$
- **13** $x^y y^x = 1$ 14 $\ln(x^2 + y^2) = 2 \tan x$

15-16.   Calculate dz/dx by two methods.

- **15** $z = x^2 + 3y, y = e^x$ 16 $z = x^2 - 3\sqrt{y}, y = \cos^2 x + 1$

- **17**   As a solid right circular cylinder is heated, its radius increases at the rate of $\frac{1}{10}$ cm/min and its height at the rate of $\frac{1}{5}$ cm/min.
    (a)   Find the rate of increase of the volume with respect to time at the instant at which the radius is 10 cm and the height is 50 cm.
    (b)   Find the rate of increase of the total surface area with respect to time at the instant at which the radius is 10 cm and the height is 50 cm.

- **18**   A quantity of an ideal gas, which is enclosed in a flexible container, has a volume of 100 ft³ at the instant at which its pressure is 10 lb/ft² and its temperature is 100°. The temperature of the gas increases at the rate of 2°/sec. As the gas is heated the container expands so that the pressure increases at the rate of $\frac{1}{10}$ lb/ft²/sec (pound per square foot per second). At what rate is the volume changing with respect to time at the instant when the temperature is 200°? (See Exercise 15 of Section 16–2 for a related problem.)

19 (a) Let f be a differentiable function of u. Prove that $w = f(ax + by)$ satisfies the differential equation

$$b \frac{\partial w}{\partial x} = a \frac{\partial w}{\partial y}.$$

(*Hint:* Let $u = ax + by$. Use the Chain Rule.)

●(b) Use the result of (a) to find three different solutions of the differential equation

$$2 \frac{\partial w}{\partial x} = 3 \frac{\partial w}{\partial y}.$$

20 Let w be a differentiable function of x, y, and z, each of which is a differentiable function of t. Modify the proof of Theorem 16–5 to prove the Chain Rule formula

$$\frac{dw}{dt} = \frac{\partial w}{\partial x} \cdot \frac{dx}{dt} + \frac{\partial w}{\partial y} \cdot \frac{dy}{dt} + \frac{\partial w}{\partial z} \cdot \frac{dz}{dt}.$$

21 Let w be a differentiable function of x and y, each of which is a differentiable function of s and t. Modify the proof of Theorem 16–5 to prove the Chain Rule formula

$$\frac{\partial w}{\partial s} = \frac{\partial w}{\partial x} \cdot \frac{\partial x}{\partial s} + \frac{\partial w}{\partial y} \cdot \frac{\partial y}{\partial s}.$$

22 Let z be a differentiable function of w, x, y.
  (a) Prove that $d(e^z) = e^z \, dz$.
  ●(b) Use (a) to calculate $d(e^z)$ for $z = x^2 y^3 + xw$.

23 Let z be a differentiable function of w, x, and y.
  (a) Prove that $d(z^n) = nz^{n-1} \, dz$ if n is a positive integer.
  ●(b) Use (a) to calculate $d(z^4)$ for $z = x^2 y^3 + xw$.

16–6 The Directional Derivative

Let $f(x, y)$ be differentiable at (x_0, y_0). The partial derivatives $f_x(x_0, y_0)$ and $f_y(x_0, y_0)$ measure the rates of change of f along lines parallel to the x- and y-axes. We now extend this concept to find the rate of change of f at (x_0, y_0) in an arbitrary direction. To be more precise, let

$$u = \cos \alpha \mathbf{i} + \sin \alpha \mathbf{j}$$

be a unit vector in the xy-plane. We shall find the rate of change of f at (x_0, y_0) in the direction defined by $\mathbf{u}$.

For each real number t, we let $P_t(x_t, y_t)$ be the terminal point of the vector $t\mathbf{u}$, drawn with its initial point at P_0. That is,

$$\overrightarrow{P_0P_t} = t\mathbf{u}.$$

(See Fig. 16–10.) Then

$$(x_t - x_0)\mathbf{i} + (y_t - y_0)\mathbf{j} = t(\cos \alpha \, \mathbf{i} + \sin \alpha \, \mathbf{j})$$

so that

$$x_t = x_0 + t \cos \alpha \quad \text{and} \quad y_t = y_0 + t \sin \alpha.$$

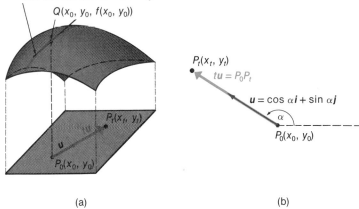

(a) (b)

Figure 16–10 $x_t = x_0 + t \cos \alpha$, $y_t = y_0 + t \sin \alpha$.

DEFINITION

Directional Derivative

Let f be differentiable at (x_0, y_0), let $\mathbf{u}$ be a unit vector. For each real number t, let $P_t(x_t, y_t)$ be the point $P_t(x_0 + t \cos \alpha, y_0 + t \sin \alpha)$. The *directional derivative* of f in the direction of $\mathbf{u}$ at $P_0(x_0, y_0)$, denoted by $D_\mathbf{u}f(x_0, y_0)$, is defined to be

$$D_\mathbf{u}f(x_0, y_0) = \lim_{t \to 0} \frac{f(x_t, y_t) - f(x_0, y_0)}{t}$$

provided the limit exists.

The directional derivative $D_\mathbf{u}f(x_0, y_0)$ measures the *rate of change* of f in the direction of the vector $\mathbf{u}$. This type of derivative is a direct extension of the derivative concept considered in Chapter 3. It is the limit of a difference quotient. The numerator of the difference quotient is the difference in the functional values between two points $P_0(x_0, y_0)$ and $P_t(x_t, y_t)$. The denominator equals the directed distance between P_0 and P_t. Theorem 16–7 gives a simple formula for the calculation of the directional derivative.

THEOREM 16–7 Let f be a differentiable function of x and y, let (x_0, y_0) be in the domain of f, and let $\mathbf{u} = \cos \alpha \mathbf{i} + \sin \alpha \mathbf{j}$ be a unit vector. The directional derivative of f in the direction of $\mathbf{u}$ at $P_0(x_0, y_0)$ is

$$D_\mathbf{u}f(x_0, y_0) = f_x(x_0, y_0) \cos \alpha + f_y(x_0, y_0) \sin \alpha$$

16–6 The Directional Derivative

Proof By definition

$$D_u f(x_0, y_0) = \lim_{t \to 0} \frac{f(x_t, y_t) - f(x_0, y_0)}{t}$$

$$= \lim_{t \to 0} \frac{f(x_0 + t \cos \alpha, y_0 + t \sin \alpha) - f(x_0, y_0)}{t}.$$

It follows from Theorem 16–3 (using $\Delta x = t \cos \alpha$ and $\Delta y = t \sin \alpha$) that

$$f(x_t, y_t) - f(x_0, y_0) = f(x_0 + t \cos \alpha, y_0 + t \sin \alpha) - f(x_0, y_0)$$
$$= f_x(x_0, y_0)t \cos \alpha + f_y(x_0, y_0)t \sin \alpha + \epsilon_1 t \cos \alpha + \epsilon_2 t \sin \alpha$$

where $\epsilon_1, \epsilon_2 \to 0$ as $t \to 0$. If we substitute this expression in the limit of the difference quotient for the directional derivative, we get

$$D_u f(x_0, y_0)$$
$$= \lim_{t \to 0} \frac{f_x(x_0, y_0)t \cos \alpha + f_y(x_0, y_0)t \sin \alpha + \epsilon_1 t \cos \alpha + \epsilon_2 t \sin \alpha}{t}$$
$$= \lim_{t \to 0} [f_x(x_0, y_0)\cos \alpha + f_y(x_0, y_0)\sin \alpha + \epsilon_1 \cos \alpha + \epsilon_2 \sin \alpha]$$
$$= f_x(x_0, y_0) \cos \alpha + f_y(x_0, y_0) \sin \alpha + 0 \cdot \cos \alpha + 0 \cdot \sin \alpha$$
$$= f_x(x_0, y_0) \cos \alpha + f_y(x_0, y_0) \sin \alpha. \blacksquare$$

The directional derivative can be calculated in the direction of any nonzero two-dimensional vector. If $v \neq 0$, we let $u = v/\|v\|$, so that u is a unit vector in the same direction as v. We define

$$D_v f(x_0, y_0) = D_u f(x_0, y_0).$$

Thus, if the nonzero vector v has direction angle α, it follows from Theorem 16–7 that

$$D_v f(x_0, y_0) = f_x(x_0, y_0) \cos \alpha + f_y(x_0, y_0) \sin \alpha.$$

EXAMPLE 1 Let $f(x, y) = x^2 y + 2x - 3y^2$. Calculate the directional derivative of f at $P_0(1, 2)$ in the direction of the vector $v = 3i - 4j$.

Solution Let α be the direction angle of v. Since $\|v\| = 5$, then $\cos \alpha = \frac{3}{5}$ and $\sin \alpha = -\frac{4}{5}$. Therefore,

$$(D_v f)_{(1,2)} = \left(\frac{\partial f}{\partial x}\right)_{(1,2)} \cos \alpha + \left(\frac{\partial f}{\partial y}\right)_{(1,2)} \sin \alpha$$
$$= (2xy + 2)_{(1,2)} \left(\frac{3}{5}\right) + (x^2 - 6y)_{(1,2)} \left(-\frac{4}{5}\right)$$
$$= (2 \cdot 1 \cdot 2 + 2) \left(\frac{3}{5}\right) + (1^2 - 6 \cdot 2) \left(-\frac{4}{5}\right) = \frac{62}{5}. \square$$

Remark It is convenient to drop the subscripts in the expression for the directional derivative.

The directional derivative of f at (x, y) in the direction of $v = ai + bj$ is

$$D_v f = \frac{\partial f}{\partial x} \cos \alpha + \frac{\partial f}{\partial y} \sin \alpha$$

where α is the direction angle of v.

EXAMPLE 2 Let f be a differentiable function of x and y. Show that

$$D_i f = \frac{\partial f}{\partial x} \quad \text{and} \quad D_j f = \frac{\partial f}{\partial y}.$$

Solution (a) The direction angle of i is $\alpha = 0$. Therefore,

$$D_i f = \frac{\partial f}{\partial x} \cos 0 + \frac{\partial f}{\partial y} \sin 0 = \frac{\partial f}{\partial x}.$$

(b) The direction angle of j is $\alpha = \pi/2$. Therefore,

$$D_j f = \frac{\partial f}{\partial x} \cos \frac{\pi}{2} + \frac{\partial f}{\partial y} \sin \frac{\pi}{2} = \frac{\partial f}{\partial y}. \quad \square$$

THE GRADIENT

Let f be a differentiable function of x and y. The *gradient* of f, denoted by ∇f (read "del f") is the vector

Gradient

$$\nabla f = \frac{\partial f}{\partial x} i + \frac{\partial f}{\partial y} j.$$

The gradient vector can be evaluated at any point at which f is differentiable. The value at $P_0(x_0, y_0)$ is

$$(\nabla f)_{(x_0, y_0)} = f_x(x_0, y_0) i + f_y(x_0, y_0) j.$$

EXAMPLE 3 Let $f(x, y) = x^2 y + 2x - 3y^2$.

(a) $\nabla f = \frac{\partial f}{\partial x} i + \frac{\partial f}{\partial y} j = (2xy + 2)i + (x^2 - 6y)j.$

(b) At the point $P_0(1, 2)$, the value of ∇f is

$$(\nabla f)_{(1,2)} = (2xy + 2)_{(1,2)} i + (x^2 - 6y)_{(1,2)} j = 6i - 11j. \quad \square$$

We will show that the gradient simplifies the calculation of the directional derivative. Let $u = \cos \alpha i + \sin \alpha j$ be a unit vector. Then

$$D_u f = \frac{\partial f}{\partial x} \cos \alpha + \frac{\partial f}{\partial y} \sin \alpha$$

$$= \left(\frac{\partial f}{\partial x} i + \frac{\partial f}{\partial y} j \right) \cdot (\cos \alpha i + \sin \alpha j)$$

$$D_u f = \nabla f \cdot u.$$

16-6 The Directional Derivative

Directional Derivative Computed from the Gradient

> The directional derivative of f in the direction of the unit vector $\mathbf{u}$ equals the inner product of the gradient of f with $\mathbf{u}$.

EXAMPLE 4 (See Examples 1, 3.) Let $f(x, y) = x^2y + 2x - 3y^2$. Let $\mathbf{v} = 3\mathbf{i} - 4\mathbf{j}$.
(a) Calculate $D_\mathbf{v} f$.
(b) Evaluate $D_\mathbf{v} f$ at the point $P_0(1, 2)$.

Solution The vector

$$\mathbf{u} = \frac{\mathbf{v}}{\|\mathbf{v}\|} = \frac{3}{5}\mathbf{i} - \frac{4}{5}\mathbf{j}$$

is a unit vector in the same direction as $\mathbf{v}$. The gradient of f is

$$\nabla f = (2xy + 2)\mathbf{i} + (x^2 - 6y)\mathbf{j}.$$

(a) The directional derivative is

$$D_\mathbf{v} f = \nabla f \cdot \mathbf{u} = [(2xy + 2)\mathbf{i} + (x^2 - 6y)\mathbf{j}] \cdot \left[\frac{3}{5}\mathbf{i} - \frac{4}{5}\mathbf{j}\right]$$

$$= \frac{3(2xy + 2)}{5} - \frac{4(x^2 - 6y)}{5}.$$

(b) At $P_0(1, 2)$,

$$(D_\mathbf{v} f)_{(1,2)} = \left(\frac{3(2xy + 2)}{5} - \frac{4(x^2 - 6y)}{5}\right)_{(1,2)} = \frac{3 \cdot 6}{5} - \frac{4 \cdot (-11)}{5} = \frac{62}{5}. \;\square$$

Remark The equation $D_\mathbf{u} f = \nabla f \cdot \mathbf{u}$ expresses the directional derivative in terms of two component functions. The first component is the gradient vector ∇f, which depends only on the function f. The second is the unit vector $\mathbf{u}$, which depends only on the direction. (We shall exploit this relation in Section 16–7 to get additional information about directional derivatives and gradients.)

FUNCTIONS OF THREE VARIABLES

All our results can be extended to functions of three variables. Let f be a differentiable function of x, y, and z, let $P_0(x_0, y_0, z_0)$ be in the domain of f, and let

$$\mathbf{u} = \cos \alpha \mathbf{i} + \cos \beta \mathbf{j} + \cos \gamma \mathbf{k}$$

be a unit vector. We define the directional derivative $D_\mathbf{u} f(x_0, y_0, z_0)$ by a limit similar to the one used for functions of two variables.

For each real number t, let $P_t(x_t, y_t, z_t)$ be the point at the terminal end of the vector $t\mathbf{u}$, drawn with its initial point at $P_0(x_0, y_0, z_0)$. (See Fig. 16–11.) Then

$$\overrightarrow{P_0P_t} = t(\cos \alpha \mathbf{i} + \cos \beta \mathbf{j} + \cos \gamma \mathbf{k}).$$

It follows as in the two-dimensional case that

$$x_t = x_0 + t \cos \alpha, \qquad y_t = y_0 + t \cos \beta, \qquad z_t = z_0 + t \cos \gamma.$$

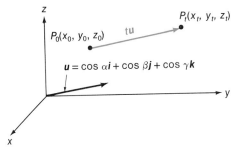

Figure 16–11 $x_t = x_0 + t \cos \alpha$, $y_t = y_0 + t \cos \beta$, $z_t = z_0 + t \cos \gamma$.

The directional derivative of f at $P_0(x_0, y_0, z_0)$ is defined to be

$$D_u f(x_0, y_0, z_0) = \lim_{t \to 0} \frac{f(x_t, y_t, z_t) - f(x_0, y_0, z_0)}{t}$$

$$= \lim_{t \to 0} \frac{f(x_0 + t \cos \alpha, y_0 + t \cos \beta, z_0 + t \cos \gamma) - f(x_0, y_0, z_0)}{t}.$$

It follows from an argument similar to the proof of Theorem 16–7 that

$$D_u f(x_0, y_0, z_0)$$
$$= f_x(x_0, y_0, z_0) \cos \alpha + f_y(x_0, y_0, z_0) \cos \beta + f_z(x_0, y_0, z_0) \cos \gamma.$$

If we drop the subscripts and consider the directional derivative at a general point (x, y, z), we get

Formula for Directional Derivative for Function of Three Variables

$$D_u f = \frac{\partial f}{\partial x} \cos \alpha + \frac{\partial f}{\partial y} \cos \beta + \frac{\partial f}{\partial z} \cos \gamma.$$

EXAMPLE 5 The directional derivative of

$$f(x, y, z) = x^2 + 2xy + y^2 z^2$$

in the direction of the unit vector

$$u = \frac{1}{\sqrt{2}} i - \frac{1}{2} j + \frac{1}{2} k = \cos \alpha i + \cos \beta j + \cos \gamma k$$

is

$$D_u f = \frac{\partial f}{\partial x} \cos \alpha + \frac{\partial f}{\partial y} \cos \beta + \frac{\partial f}{\partial z} \cos \gamma$$

$$= (2x + 2y) \cdot \frac{1}{\sqrt{2}} + (2x + 2yz^2)\left(-\frac{1}{2}\right) + (2y^2 z)\left(\frac{1}{2}\right)$$

$$= (\sqrt{2} - 1)x + \sqrt{2} y - yz^2 + y^2 z. \quad \square$$

Exercises 16-6

The *gradient* of $f(x, y, z)$ is defined to be

Gradient of Function of Three Variables

$$\nabla f = \frac{\partial f}{\partial x}i + \frac{\partial f}{\partial y}j + \frac{\partial f}{\partial z}k.$$

It follows as in the two-dimensional case that

$$D_u f = \nabla f \cdot u$$

provided u is a unit vector.

EXAMPLE 6 Let $f(x, y, z) = \cos x + \sin y + e^z$. Let $v = 4i - 3j + 12k$. Calculate $D_v f(0, 0, 0)$.

Solution Since $\|v\| = \sqrt{4^2 + 3^2 + 12^2} = 13$, then

$$u = \frac{4}{13}i - \frac{3}{13}j + \frac{12}{13}k$$

is a unit vector in the same direction as v.
 The gradient of f is

$$\nabla f = \frac{\partial f}{\partial x}i + \frac{\partial f}{\partial y}j + \frac{\partial f}{\partial z}k$$
$$= -\sin x i + \cos y j + e^z k.$$

The directional derivative of f in the direction of v is

$$D_v f = \nabla f \cdot u = (-\sin x i + \cos y j + e^z k) \cdot \left(\frac{4}{13}i - \frac{3}{13}j + \frac{12}{13}k\right)$$
$$= \frac{-4 \sin x - 3 \cos y + 12 e^z}{13}.$$

At the point $P_0(0, 0, 0)$,

$$D_v f(0, 0, 0) = \frac{-4 \sin 0 - 3 \cos 0 + 12 e^0}{13} = \frac{0 - 3 + 12}{13} = \frac{9}{13}. \quad \square$$

Exercises 16-6

1-6. (a) Calculate the directional derivative of f at P_0 in the direction of the given vector v.
 (b) Calculate the gradient ∇f.

- 1 $f(x, y) = 2x^2 + 4y^2$; $P_0(1, 1)$, $v = i - j$
- 2 $f(x, y) = xe^y - 2e^x$; $P_0(0, 0)$, $v = 3i + 4j$
- 3 $f(x, y) = \sin^{-1}(x/y)$; $P_0(2, -4)$, $v = -i + 3j$
- 4 $f(x, y) = x \ln y + x^2 y$; $P_0(2, 1)$, direction angle of v is $\theta = 2\pi/3$.
- 5 $f(x, y, z) = xy + yz + xz$; $P_0(-1, 1, 1)$, $v = i + 2j + 2k$

6. $f(x, y, z) = ye^{x^2-z^2}$; $P_0(1, 2, 1)$, $v = i + j$
7. Let f be a differentiable function of x and y.
   (a) Show that $D_u f = -\partial f/\partial x$ in the direction of the negative x-axis.
   (b) Show that $D_u f = -\partial f/\partial y$ in the direction of the negative y-axis.
•8. A heated metal plate is placed over the y-axis. The temperature at the point (x, y) is $T(x, y) = 600(x^2 + y^2 + 1)^{-1}$.
   (a) Find the rate of change of the temperature with distance at the point $(3, 4)$ in the direction of the vector $2i - 3j$.
   (b) In what direction is the directional derivative at $(3, 4)$ equal to zero?
9. Modify the proof of Theorem 16–7 to prove the comparable formula for functions of three variables.

16–7 Properties of the Gradient

As we shall see, the gradient of a function f has interesting properties. In our development, we work with differentiable functions of two or three variables.

Let f be a function of three variables. Recall that the gradient of f is

$$\nabla f = \frac{\partial f}{\partial x} i + \frac{\partial f}{\partial y} j + \frac{\partial f}{\partial z} k$$

and that the directional derivative of f at $P_0(x_0, y_0, z_0)$ in the direction of the unit vector u is

$$f_u(x_0, y_0, z_0) = (D_u f)_{(x_0, y_0, z_0)} = (u) \cdot (\nabla f)_{(x_0, y_0, z_0)}$$

Because this notation is awkward, in this and the following section we use the symbols

$$(D_u f)_0 \quad \text{and} \quad (\nabla f)_0$$

for the directional derivative and gradient, respectively, at $P_0(x_0, y_0, z_0)$. Similarly, we write

$$\left(\frac{\partial f}{\partial x}\right)_0 \text{ for } \left(\frac{\partial f}{\partial x}\right)_{(x_0, y_0, z_0)}, \quad \left(\frac{\partial f}{\partial y}\right)_0 \text{ for } \left(\frac{\partial f}{\partial y}\right)_{(x_0, y_0, z_0)},$$

and so on.

In many problems we need to know the direction in which f has its maximum rate of change at P_0. This is equivalent to finding the particular unit vector u for which $(D_u f)_0$ is a maximum.

Let θ be the angle between $(\nabla f)_0$ and the unit vector u. (See Fig. 16–12a.) Then

$$(D_u f)_0 = (\nabla f)_0 \cdot u = \|(\nabla f)_0\| \|u\| \cos \theta = \|(\nabla f)_0\| \cos \theta$$

(by Theorem 14–7). Since $\|(\nabla f)_0\|$ is a constant, then the directional derivative varies as $\cos \theta$ varies. It has its maximum value when $\cos \theta = 1$—that is, when u is in the same direction as $(\nabla f)_0$. (See Fig. 16–12b.) It has its minimum value when $\cos \theta = -1$—that is, when u is in the opposite direction.

16-7 Properties of the Gradient

(a) $D_u f(x_0, y_0, z_0) = \|(\nabla f)_0\| \cdot \|u\| \cos \theta$.

(b) The maximum value of $D_u f(x_0,y_0,z_0)$ is equal to $\|(\nabla f)_0\|$. This maximum is obtained when $\theta = 0$.

Figure 16–12 The directional derivative and the gradient (u is a unit vector).

It follows from this discussion that the maximum value of the directional derivative is

$$\|(\nabla f)_0\| \cdot 1 = \|(\nabla f)_0\|$$

which is obtained when $\theta = 0$. The minimum value is

$$\|(\nabla f)_0\| \cdot (-1) = -\|(\nabla f)_0\|$$

which is obtained when $\theta = \pi$.

Thus, we have two important properties of the gradient vector of f at P_0.

Basic Properties of Gradient

(1) $(\nabla f)_0$ points in the direction in which f has its maximum possible directional derivative at P_0.

(2) The maximum possible value of the directional derivative of f at P_0 is equal to the magnitude of the gradient vector $(\nabla f)_0$.

EXAMPLE 1 Let $f(x, y, z) = \sin(xy) + e^z$. Find the direction in which f has its maximum directional derivative at $P_0(1, 0, 0)$. What is this maximum value?

Solution
$$\nabla f = \frac{\partial f}{\partial x} i + \frac{\partial f}{\partial y} j + \frac{\partial f}{\partial z} k$$
$$= y \cos(xy) i + x \cos(xy) j + e^z k.$$

At $P_0(1, 0, 0)$,

$$(\nabla f)_0 = 0 \cdot i + 1 \cdot j + 1 \cdot k = j + k.$$

The function f has its maximum directional derivative in the direction of the vector $(\nabla f)_0 = j + k$. The maximum value of the directional derivative is

$$\|(\nabla f)_0\| = \|j + k\| = \sqrt{2}. \square$$

EXAMPLE 2 A flat metal plate is placed over an axis system that has a heat source at the origin. The heat is conducted outward from the origin in such a way that at a certain time the temperature at point $P(x, y)$ is

$$T(x, y) = \frac{100}{x^2 + y^2 + 1}.$$

In what direction is there the greatest change in temperature at $P_0(4, 3)$? What is the maximum rate of change in temperature at this point?

Solution The gradient vector is

$$\nabla T = \frac{\partial T}{\partial x} i + \frac{\partial T}{\partial y} j = \frac{-200x}{(x^2 + y^2 + 1)^2} i - \frac{200y}{(x^2 + y^2 + 1)^2} j$$

$$= \frac{200}{(x^2 + y^2 + 1)^2} (-xi - yj).$$

At $P_0(4, 3)$,

$$(\nabla T)_0 = \frac{200}{(4^2 + 3^2 + 1)^2} (-4i - 3j) = \frac{50}{169} (-4i - 3j).$$

The maximum change in temperature occurs in the direction of the gradient $(\nabla T)_0$—that is, in the direction of the vector $-4i - 3j$. Observe that this vector points towards the heat source, indicating that the temperature increases most rapidly in that direction.

The maximum rate of change of T at $(4, 3)$ equals the magnitude of the gradient vector $(\nabla T)_0$. This value is

$$\|(\nabla T)_0\| = \left\|\frac{50}{169}(-4i - 3j)\right\| = \frac{50}{169}\sqrt{4^2 + 3^2}$$

$$= \frac{250}{169} \approx 1.47928.$$

Thus, the maximum increase in temperature at $P_0(4, 3)$ is approximately 1.48 degrees per unit. This occurs in the direction of the vector $-4i - 3j$. (See Fig. 16–13.) □

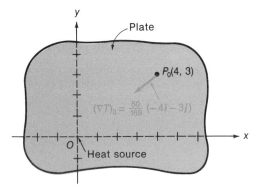

Figure 16–13 Example 2.

GRADIENTS AND CONTOUR CURVES

Let $z = F(x, y)$ be a differentiable function of x and y. There is a close connection between the gradient vector $(\nabla F)_0$ at $P_0(x_0, y_0)$ and the contour curve $\mathscr{C}_h$ that passes through that point. We shall show that the gradient vector $(\nabla F)_0$ is a normal vector to $\mathscr{C}_h$, in other words, that this gradient vector is orthogonal to the vector tangent to $\mathscr{C}_h$ at $P_0(x_0, y_0)$. (See Fig. 16–14.) We shall later generalize this result to find vectors normal to surfaces in three-dimensional space (Section 16–8).

16-7 Properties of the Gradient

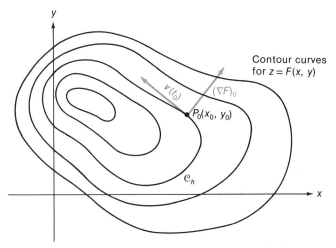

Figure 16-14 The gradient at P_0 is normal to the contour curve at P_0.

THEOREM 16-8

Gradient of Function of Two Variables is Normal to Contour Curve

Let $z = F(x, y)$ be a differentiable function of x and y; let $P_0(x_0, y_0)$ be in the domain of F. Suppose that $\mathscr{C}_h$, the contour curve through (x_0, y_0), can be described parametrically by

$$x = f(t), \qquad y = g(t),$$

where f and g are differentiable. Then $(\nabla F)_0$ is normal to $\mathscr{C}_h$ at P_0.

Proof Observe that $\mathscr{C}_h$ is the graph of the vector-valued function

$$r(t) = f(t)i + g(t)j.$$

If $x_0 = f(t_0)$ and $y_0 = g(t_0)$, then the velocity vector

$$v(t_0) = f'(t_0)i + g'(t_0)j$$

is tangent to $\mathscr{C}_h$ at $P_0(x_0, y_0)$. (See Fig. 16-14.)

Along the curve $\mathscr{C}_h$, we have $z = F(x, y) = h$ (a constant), where $x = f(t)$, $y = g(t)$. By the Chain Rule,

$$\frac{dz}{dt} = \frac{\partial F}{\partial x} \cdot \frac{dx}{dt} + \frac{\partial F}{\partial y} \cdot \frac{dy}{dt} = \frac{d}{dt}(h) = 0$$

$$\frac{\partial F}{\partial x} f'(t) + \frac{\partial F}{\partial y} g'(t) = 0.$$

At the point $P_0(x_0, y_0)$, where $t = t_0$, we have

$$\left(\frac{\partial F}{\partial x}\right)_0 f'(t_0) + \left(\frac{\partial F}{\partial y}\right)_0 g'(t_0) = 0.$$

This equation can be written in vector form as

$$\left[\left(\frac{\partial F}{\partial x}\right)_0 i + \left(\frac{\partial F}{\partial y}\right)_0 j\right] \cdot [f'(t_0)i + g'(t_0)j] = 0$$

$$(\nabla F)_0 \cdot v(t_0) = 0,$$

It follows that the gradient $(\nabla F)_0$ is orthogonal to $v(t_0)$. Since the gradient vector is orthogonal to the tangent vector $v(t_0)$, then it is normal to $\mathscr{C}_h$. ■

EXAMPLE 3   Let $z = F(x, y) = \sin(xy)$. Let $\mathscr{C}_h$ be the level curve through $P_0(\pi, \tfrac{1}{4})$. Find a vector normal to $\mathscr{C}_h$ at P_0.

Solution   Observe first that at $P_0(\pi, \tfrac{1}{4})$ the value of z is $\sin(\pi/4) = 1/\sqrt{2}$. Thus, $h = 1/\sqrt{2}$. The gradient of F is

$$\nabla F = \frac{\partial F}{\partial x}i + \frac{\partial F}{\partial y}j = y\cos(xy)i + x\cos(xy)j.$$

At $P_0(\pi, \tfrac{1}{4})$,

$$(\nabla F)_0 = \frac{1}{4}\cos\frac{\pi}{4}i + \pi\cos\frac{\pi}{4}j = \frac{1}{4}\cdot\frac{\sqrt{2}}{2}i + \frac{\pi\sqrt{2}}{2}j$$
$$= \frac{\sqrt{2}\,(i + 4\pi j)}{8}.$$

It follows from Theorem 16–8 that the vector

$$(\nabla F)_0 = \frac{\sqrt{2}\,(i + 4\pi j)}{8}$$

is orthogonal to the level curve $\mathscr{C}_{1/\sqrt{2}}$ at $P_0(\pi, \tfrac{1}{4})$. Since any vector parallel to $(\nabla F)_0$ is orthogonal to the contour curve, then

$$n = i + 4\pi j$$

also is orthogonal to it. □

Exercises 16–7

1–4.   Find the direction in which f has the greatest rate of increase at P_0. Find the value of this maximum rate of increase.

•1 $f(x, y) = x^3 + y^3 + x^2 y$; $P_0(-3, 1)$

2   $f(x, y) = \sqrt{x^2 + y^2}$; $P_0(2, 1)$

•3 $f(x, y, z) = \dfrac{xyz}{1 + x^2 + y^2}$; $P_0(1, 1, -1)$

4   $f(x, y, z) = x^2 \cos z + y^2 - 2xy \sin z$; $P_0(2, -1, 0)$

5   A highway engineer studies a contour map of a mountain. Explain how he can find the direction of greatest increase of elevation at a certain point $P_0(x_0, y_0)$ on the map.

6   Let $w = f(x, y, z)$ be a differentiable function. Explain why the directional derivative of f at $P_0(x_0, y_0, z_0)$ is always zero in any direction perpendicular to the gradient $(\nabla F)_{(x_0, y_0, z_0)}$.

Exercises 16–7

- **7** The temperature distribution of a flat plate is given by $T(x, y) = 300(2x^2 + 3y^2 + 1)^{-2}$. Find the direction in which the temperature decreases the most rapidly at the point $(4, -1)$. What is the maximum rate of decrease at this point?

- **8** The temperature at point (x, y, z) in a room is given by
$$T(x, y, z) = \frac{100}{\sqrt{2x^2 + 2y^2 + z^2 + 1}}.$$
  (a) Find the direction in which the rate of change of the temperature is a maximum at $(4, 3, 7)$.
  (b) Find the maximum rate of change of the temperature in any direction at $(4, 3, 7)$.

- **9** The electrical potential of a point in three-dimensional space is given by
$$V = \frac{1}{\sqrt{x^2 + y^2 + z^2 + 1}}.$$
  (a) Find the direction in which the electrical potential has the greatest rate of change at $P_0(4, 2, -2)$.
  (b) Find the greatest rate of change of V at $P_0(4, 2, -2)$.

- **10** The directional derivative of $z = f(x, y)$ at $(2, 3)$ is -2 in the direction of $(6, 6)$ and 5 in the direction of $(-1, -1)$.
  (a) Find the directional derivative at $(2, 3)$ in the direction of $(3, 7)$.
  (b) Find the maximum possible value for the directional derivative at $(2, 3)$. In what direction is this value taken?

11–12. Find a vector normal to the level curve of the graph of f that passes through the point $P_0(x_0, y_0)$.

- **11** $f(x, y) = x^2 + y^2 + 2xy$, $P_0(1, 5)$ **12** $f(x, y) = \ln \sqrt{x^2 + y^2}$, $P_0(2, -3)$

13–14. Find a unit vector at P_0 tangent to the level curves in Exercises 11 and 12. (*Hint:* If T is a unit tangent vector and n is a normal vector to a curve in the xy-plane, then T is orthogonal to both n and k.)

- **13** Exercise 11.   **14** Exercise 12

- **15** The graph of $z = f(x, y)$ passes through the point $(0, 1, 2)$. At every point $P(x, y)$ in the xy-plane the vector $6x\mathbf{i} + 10y\mathbf{j}$ is normal to the contour curve through P. Find one such function f.

- **16** (*The Laplacian*) The operator ∇^2 is defined symbolically by
$$\nabla^2 = \nabla \cdot \nabla = \frac{\partial^2}{\partial x^2} + \frac{\partial^2}{\partial y^2} + \frac{\partial^2}{\partial z^2}.$$
If we operate on a function f with ∇^2, we get

Laplacian
$$\nabla^2 f = \frac{\partial^2 f}{\partial x^2} + \frac{\partial^2 f}{\partial y^2} + \frac{\partial^2 f}{\partial z^2},$$

called the *Laplacian* of f.

- (a) Calculate the Laplacian of $f(x, y, z) = e^{xyz}$
- (b) Calculate the Laplacian of $g(x, y, z) = x^2 + y^3 + z^4$
- (c) Explain the relation between the Laplacian and Laplace's equation (Exercise 14 of Section 16–3).

16–8 Normal Vectors and Tangent Planes

Tangent Line to Surface

Let S be a surface in three-dimensional space. Let $P_0(x_0, y_0, z_0)$ be a point on S and let $\mathcal{C}$ be a curve through P_0 that lies on S and is smooth enough to have a tangent line $\mathcal{L}$ at P_0. The line $\mathcal{L}$ is called a *tangent line* to S at P_0. (See Fig. 16–15a.)

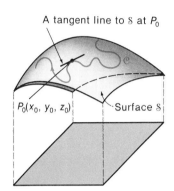
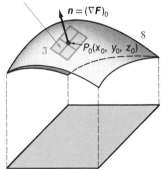

(a) A tangent line to the surface S

(b) The normal vector and tangent plane to S at P_0. [All of the tangent lines to S at P_0 lie on the plane $\mathcal{T}$.]

Figure 16–15.

Normal Vector to Surface

Let n be a nonzero vector. The vector n is said to be *normal* to the surface S at P_0 if n is orthogonal to every line that is tangent to S at P_0.

We consider the case where the surface S is the graph of the equation $F(x, y, z) = 0$, where F is a differentiable function of x, y, and z, and where at least one of the partial derivatives of F is nonzero at P_0. Let the curve $\mathcal{C}$ be defined parametrically by $x = f(t)$, $y = g(t)$, $z = h(t)$; and suppose that $P_0(x_0, y_0, z_0)$ is obtained when $t = t_0$. That is,

$$x_0 = f(t_0), \qquad y_0 = g(t_0) \qquad \text{and} \quad z_0 = h(t_0).$$

Suppose also that $f'(t_0)$, $g'(t_0)$, and $h'(t_0)$ exist and are not all zero. We shall show that the vector tangent to $\mathcal{C}$ at P_0 is orthogonal to $(\nabla F)_0$. Since $\mathcal{C}$ is an arbitrary curve on S, it will follow that

$$n = (\nabla F)_0$$

is normal to S at P_0.

16–8 Normal Vectors and Tangent Planes

To establish this result we let $w = F(x, y, z)$, so that the surface S is the graph of the equation $w = 0$. Along the curve $\mathscr{C}$ we have

$$w = F(f(t), g(t), h(t)) = 0,$$

so that w remains constant as the curve is traced out. By the Chain Rule,

$$\frac{dw}{dt} = \frac{\partial F}{\partial x} \cdot \frac{dx}{dt} + \frac{\partial F}{\partial y} \cdot \frac{dy}{dt} + \frac{\partial F}{\partial z} \cdot \frac{dz}{dt} = 0.$$

If we evaluate this derivative at $t = t_0$, we get

$$\left(\frac{\partial F}{\partial x}\right)_0 f'(t_0) + \left(\frac{\partial F}{\partial y}\right)_0 g'(t_0) + \left(\frac{\partial F}{\partial z}\right)_0 h'(t_0) = 0,$$

which can be written in vector form as

$$\left[\left(\frac{\partial F}{\partial x}\right)_0 i + \left(\frac{\partial F}{\partial y}\right)_0 j + \left(\frac{\partial F}{\partial z}\right)_0 k\right] \cdot [f'(t_0)i + g'(t_0)j + h'(t_0)k] = 0$$

or

$$(\nabla F)_0 \cdot v(t_0) = 0$$

where $v(t) = f'(t_0)i + g'(t_0)j + h'(t_0)k$.

Since $v(t_0)$ is tangent to $\mathscr{C}$ at P_0, then $(\nabla F)_0$ is a normal vector to S at P_0. The above discussion establishes the following results.

THEOREM 16–9

Gradient is Normal to a Surface

Let $F(x, y, z)$ be a differentiable function of x, y, and z. Let $P_0(x_0, y_0, z_0)$ be a point on the graph of $F(x, y, z) = 0$, where ∇F exists and is nonzero. Then $(\nabla F)_0$ is a normal vector at P_0 to the surface defined by $F(x, y, z) = 0$.

TANGENT PLANES

Let n be a normal vector to the surface S at P_0. The plane through P_0 that has n as a normal vector is called the *tangent plane* to S at P_0.

Let S be the graph of $F(x, y, z) = 0$, where F is differentiable and where at least one of the partial derivatives of F is nonzero. It follows from Theorem 16–9 that the plane tangent to S at P_0 has the equation

Equation of Tangent Plane

$$\left(\frac{\partial F}{\partial x}\right)_0 (x - x_0) + \left(\frac{\partial F}{\partial y}\right)_0 (y - y_0) + \left(\frac{\partial F}{\partial z}\right)_0 (z - z_0) = 0.$$

EXAMPLE 1 Find the equation of the tangent plane to the sphere

$$x^2 + y^2 + z^2 = 169$$

at the point $P_0(3, -4, 12)$.

Solution The defining equation is $F(x, y, z) = 0$, where

$$F(x, y, z) = x^2 + y^2 + z^2 - 169.$$

The gradient of F is

$$\nabla F = \frac{\partial F}{\partial x} i + \frac{\partial F}{\partial y} j + \frac{\partial F}{\partial z} k$$

$$= 2xi + 2yj + 2zk.$$

At $P_0(3, -4, 12)$ the gradient is

$$(\nabla F)_0 = 6i - 8j + 24k.$$

The equation of the tangent plane at $P_0(3, -4, 12)$ is

$$6(x - 3) - 8(y + 4) + 24(z - 12) = 0,$$

which reduces to

$$3x - 4y + 12z = 169. \quad \square$$

Remark   Many students confuse the result of Theorem 16–9 with the result of Theorem 16–8. Given a surface defined by $z = f(x, y)$, they assume that $(\nabla f)_0$ is normal to the tangent plane at $P_0(x_0, y_0, z_0)$. It is not. From Theorem 16–8, $(\nabla f)_0$ is orthogonal to the contour curve through (x_0, y_0)—not to the graph of f.

The graphs of equations of form

$$F(x, y, z) = C$$

Equivalue (Level) Surface   are called *equivalue surfaces*, or *level surfaces*. These surfaces play the same role for functions of three variables that contour curves play for functions of two variables.

It follows by an argument similar to the proof of Theorem 16–9 that the gradient vector $(\nabla F)_0$ is orthogonal to the equivalue surface $F(x, y, z) = C$ at $P_0(x_0, y_0, z_0)$, provided F is differentiable at P_0. (See Exercise 15.)

Certain equivalue surfaces have been given special names. If $F(x, y, z)$ measures the electrical potential at (x, y, z), then the surface

$$F(x, y, z) = C$$

is called an *equipotential surface*. If $F(x, y, z)$ is the temperature at (x, y, z), then an equivalue surface is called an *isothermal surface*.

THE TANGENT PLANE TO THE GRAPH OF $z = f(x, y)$

Let f be a differentiable function of x and y. The equation of the plane tangent to the graph of $z = f(x, y)$ at $P_0(x_0, y_0, z_0)$ has a particularly simple form:

$$z - z_0 = \left(\frac{\partial f}{\partial x}\right)_0 (x - x_0) + \left(\frac{\partial f}{\partial y}\right)_0 (y - y_0).$$

To establish this result we define $F(x, y, z)$ by

$$F(x, y, z) = f(x, y) - z.$$

16-8 Normal Vectors and Tangent Planes

Then the graph of $z = f(x, y)$ is the equivalue surface

$$F(x, y, z) = 0.$$

The gradient of F at $P_0(x_0, y_0, z_0)$ is

$$(\nabla F)_0 = \left(\frac{\partial F}{\partial x}\right)_0 i + \left(\frac{\partial F}{\partial y}\right)_0 j + \left(\frac{\partial F}{\partial z}\right)_0 k$$

$$= \left(\frac{\partial f}{\partial x}\right)_0 i + \left(\frac{\partial f}{\partial y}\right)_0 j - k.$$

Since this vector is normal to the plane tangent at P_0, the equation of the tangent plane is

$$\left(\frac{\partial f}{\partial x}\right)_0 (x - x_0) + \left(\frac{\partial f}{\partial y}\right)_0 (y - y_0) - (z - z_0) = 0,$$

which can be written

$$z - z_0 = \left(\frac{\partial f}{\partial x}\right)_0 (x - x_0) + \left(\frac{\partial f}{\partial y}\right)_0 (y - y_0).$$

EXAMPLE 2   Find the equation of the plane tangent to the hyperbolic paraboloid $z = f(x, y) = x^2 - y^2$ at $P_0(2, 1, 3)$.

Solution   Since $\partial f/\partial x = 2x$ and $\partial f/\partial y = -2y$, then

$$\left(\frac{\partial f}{\partial x}\right)_{(2,1,3)} = 4 \quad \text{and} \quad \left(\frac{\partial f}{\partial y}\right)_{(2,1,3)} = -2.$$

The equation of the tangent plane is

$$z - z_0 = \left(\frac{\partial f}{\partial x}\right)_{(2,1,3)} (x - x_0) + \left(\frac{\partial f}{\partial y}\right)_{(2,1,3)} (y - y_0)$$
$$z - 3 = 4(x - 2) - 2(y - 1)$$
$$4x - 2y - z = 3. \square$$

DIFFERENTIALS AND TANGENT PLANES

Let $z = f(x, y)$ be a differentiable function. Recall that the differential of z at (x_0, y_0) is defined to be

$$dz = \left(\frac{\partial f}{\partial x}\right)_0 dx + \left(\frac{\partial f}{\partial y}\right)_0 dy$$

where $dx = \Delta x$ and $dy = \Delta y$ are independent increments. If we let $x_1 = x_0 + dx$, $y_1 = y_0 + dy$, and $z_1 = z_0 + dz$, then

$$dz = z_1 - z_0 = \left(\frac{\partial f}{\partial x}\right)_0 (x_1 - x_0) + \left(\frac{\partial f}{\partial y}\right)_0 (y_1 - y_0)$$

$$= \left[\left(\frac{\partial f}{\partial x}\right)_0 x_1 + \left(\frac{\partial f}{\partial y}\right)_0 y_1\right] - \left[\left(\frac{\partial f}{\partial x}\right)_0 x_0 + \left(\frac{\partial f}{\partial y}\right)_0 y_0\right].$$

Therefore, dz is the change in z from the point (x_0, y_0, z_0) to the point (x_1, y_1, z_1) on the tangent plane to the graph of f.

Recall that the increment Δz equals

$$\Delta z = f(x_0 + dx, y_0 + dy) - f(x_0, y_0).$$

This increment equals the change in the z-coordinates from the point $(x_0, y_0, f(x_0, y_0))$ to the point $(x_1, y_1, f(x_1, y_1))$ on the graph of f. The approximation formula

$$\Delta z \approx dz$$

states that the exact change in z-coordinates between two points on the graph of f is approximately equal to the change in z-coordinates between the corresponding points on a tangent plane. (See Fig. 16–16.)

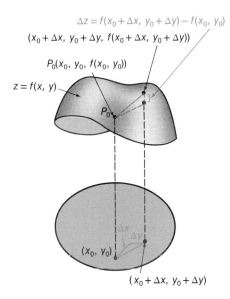
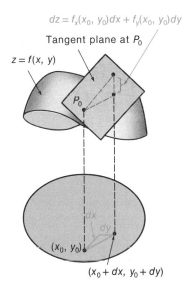

(a) The increment Δz is the change in z that corresponds to a change from (x_0, y_0) to $(x_0 + \Delta x, y_0 + \Delta y)$ as measured on the surface $z = f(x, y)$

(b) The differential dz is the change in z that corresponds to a change from (x_0, y_0) to $(x_0 + dx, y_0 + dy)$ as measured on the tangent plane to the surface at (x_0, y_0)

Figure 16–16 Increments and differentials.

Exercises 16–8

1–6. Find a normal vector and the equation of the tangent plane at the indicated point.

- 1 $z = xy + x \cos y$, $P(2, 0, 2)$
- 2 $x^3 + y^3 + zx = 4$, $P(1, 2, -5)$
- 3 $x^3 + y^3 - \sqrt{z} = 6$, $P(2, -1, 1)$
- 4 $xyz - 2yz + x^3 = 4$, $P(0, 2, -1)$
- 5 $y = \ln(2x/z)$, $P(2, 0, 4)$
- 6 $y = e^{z \cos x}$, $P(\pi, e, -1)$

16-9 Local Maxima and Minima

7-8. At what points is the tangent plane to the surface horizontal?

● **7** $z = 2x + xy - x^2 - y^2$ **8** $xy = e^{yz} + z^2$

● **9** Let $\mathcal{L}$ be the normal line to the graph of $z = 3x^2y + x^3 - y^3$ at the point $P_0(-1, 2, -3)$. At what point does $\mathcal{L}$ intersect the xy-plane?

10 (a) Show that the equation of the tangent plane of the graph of the quadric surface

$$\pm \frac{x^2}{a^2} \pm \frac{y^2}{b^2} \pm \frac{z^2}{c^2} = 1$$

at $P_0(x_0, y_0, z_0)$ is

$$\pm \frac{x_0 x}{a^2} \pm \frac{y_0 y}{b^2} \pm \frac{z_0 z}{c^2} = 1,$$

where the signs are the same as in the original equation.

● (b) Use the result of (a) to calculate the equation of the plane tangent to the graph of $x^2/2 - y^2/3 + z^2/4 = 1$ at the point $(-2, 3, 2\sqrt{2})$.

● **11** Let $P_0(x_0, y_0, z_0)$ be the point on the graph of $z = 4x^2 + 2xy + y^2 + 3$ where the normal vector is orthogonal to both $a = 5i + j - 6k$ and $b = 23i + 17j + 22k$. Find the equation of the tangent plane to the surface at P_0.

Tangent Surfaces

12 Two surfaces are said to be *tangent* at a point of intersection if they have the same tangent plane at that point.

Show that the sphere $x^2 - 4x + y^2 - 6y + z^2 - 6z + 13 = 0$ is tangent to the ellipsoid $x^2 + 2y^2 + 2z^2 = 5$ at the point $(1, 1, 1)$.

Orthogonal Surfaces

13 Two surfaces are said to be *orthogonal* at a point of intersection if their normal vectors are orthogonal there. Show that the sphere $x^2 + y^2 + z^2 = 4$ and the cone $x^2 + y^2 = z^2$ are orthogonal at each point of intersection.

● **14** Find a tangent plane to the cone $z^2 = x^2 + 4y^2$ that is parallel to the plane $3x - 8y - 5z = 7$.

15 Modify the proof of Theorem 16-9 to prove that $(\nabla F)_0$ is normal to the equivalue surface $F(x, y, z) = C$ at the point $P_0(x_0, y_0, z_0)$.

16-9 Local Maxima and Minima

We now extend the results of Sections 4-1 through 4-5 to solve maximum-minimum problems involving functions of two variables. Before we do so, however, we introduce some concepts that simplify the statements of our results.

INTERIOR AND BOUNDARY POINTS

Let S be a set in the xy-plane.

DEFINITION

Interior Point

The point $P(x, y)$ is called an *interior point of* S if there exists a neighborhood $\mathcal{N}$ of P that is completely contained in S.

In other words, $P(x, y)$ is an interior point of S if and only if there exists $\delta > 0$ such that the circular disk with center at P and radius δ lies completely within S. (See Fig. 16–17a.)

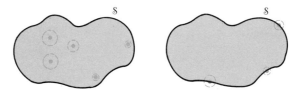

(a) Interior points of S (b) Boundary points of S

Figure 16–17 Interior and boundary points.

DEFINITION

Boundary Point

The point $P(x, y)$ is called a *boundary point of* S provided every neighborhood of P contains a point in S and a point not in S. (See Fig. 16–17b.)

Observe that the interior points of S must be in S but the boundary points may be in S or not in S. Observe also that every point of S is either an interior point or a boundary point. (See Exercise 17.)

REGIONS

Up to this point we have used the word *region* to indicate some portion of the xy-plane. We now make a more restrictive definition. From now on, the word will have the following meaning.

DEFINITION

A *region* in the xy-plane is a set $\mathcal{R}$ that has the following properties:

(1) Every point of $\mathcal{R}$ is an interior point. (The region contains no boundary points.)
(2) Any two points P and Q in $\mathcal{R}$ can be connected by a curve $\mathcal{C}$ that lies completely in $\mathcal{R}$. (See Fig. 16–18a.)

Closed Region
Bounded Region

A set S that is the union of a region $\mathcal{R}$ and its boundary is called a *closed region*. (See Fig. 16–18b.) A region or a closed region is said to be *bounded* if it is contained

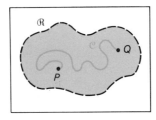 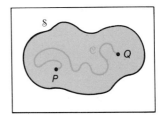

(a) A bounded region $\mathcal{R}$ (b) A bounded closed region S

Figure 16–18 Regions and closed regions.

within some rectangle in the xy-plane. For example, the region $\mathcal{R}$ and the closed region S in Figure 16–18 are bounded.

16-9 Local Maxima and Minima

EXAMPLE 1 Let $\mathcal{R} = \{(x, y) | x^2 + y^2 < 1\}$. Let $\mathcal{S} = \{(x, y) | x^2 + y^2 \leq 1\}$.

(a) Then $\mathcal{R}$ is a bounded region. Every point of $\mathcal{R}$ is an interior point; any two points of $\mathcal{R}$ can be connected by a curve that lies in $\mathcal{R}$; and $\mathcal{R}$ is completely contained in the rectangle with vertices $(-1, 1)$, $(1, 1)$, $(1, -1)$, and $(-1, -1)$.

(b) $\mathcal{S}$ is a bounded, closed region. It is the union of $\mathcal{R}$ and its boundary. □

Remark We shall extend the results of Chapter 4 to cover maximum-minimum problems involving functions of two variables. In Chapter 4 we worked primarily with functions defined over open and closed intervals. In our present setting, bounded regions will correspond to finite open intervals, bounded, closed regions to finite closed intervals, interior points of regions to interior points of intervals, and boundary points to endpoints of intervals.

MAXIMA AND MINIMA

Let $f(x, y)$ be defined on a set $\mathcal{S}$. The number M is called the *maximum* of f on $\mathcal{S}$ provided

Maximum on a Set

(1) $f(x, y) \leq M$ for every $(x, y) \in \mathcal{S}$.

(2) $f(x_0, y_0) = M$ for at least one point $(x_0, y_0) \in \mathcal{S}$.

Minimum on a Set

The minimum of f on $\mathcal{S}$ is defined similarly. If m is the minimum, then $f(x, y) \geq m$ for all $(x, y) \in \mathcal{S}$, while $f(x_1, y_1) = m$ for at least one point $(x_1, y_1) \in \mathcal{S}$.

Not every function defined on a set $\mathcal{S}$ has a maximum or a minimum. The following fundamental theorem states a result similar to Theorem 2-7 that guarantees that certain functions have extreme values over certain sets. The proof is omitted.

THEOREM 16-10 Let $f(x, y)$ be continuous on the bounded, closed region $\mathcal{S}$. Then f has a maximum and a minimum on $\mathcal{S}$.

Observe that an extreme value of a function f on a closed region $\mathcal{S}$ must occur either at a boundary point or at an interior point of $\mathcal{S}$. (There are no other types of points in $\mathcal{S}$.) We handle these two cases separately.

DEFINITION

Local Maximum, Minimum

Let f be a continuous function of x and y. We say that f has a *local maximum* at $P_0(x_0, y_0)$ if there exists a neighborhood $\mathcal{N}$ of P_0, on which f is defined, such that the maximum of f on $\mathcal{N}$ is at P_0. (See Fig. 16-19.)

We say that f has a local minimum at P_0 if there exists a neighborhood $\mathcal{N}$ of P_0 on which f is defined such that the minimum of f on $\mathcal{N}$ is at P_0. (See Fig. 16-19.)

If a function f has an extreme value at an interior point of set $\mathcal{S}$, then it must have a local extremum there. Thus, we turn out attention to methods for finding local extrema.

Suppose that $f(x, y)$ has a local extremum at (a, b) and that $\partial f/\partial x$ and $\partial f/\partial y$ exist at this point. It is easy to see that both of these partial derivatives must be zero.

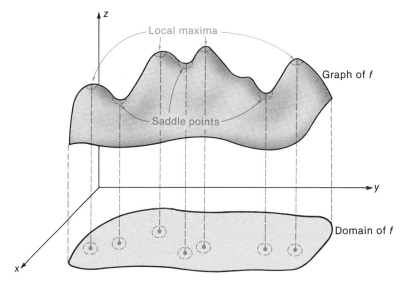

Figure 16-19 Local extrema.

To establish this fact, we let $\mathscr{C}$ be the trace of the graph of f in the vertical plane $x = a$. Then $\mathscr{C}$ is the graph of

$$g(y) = f(a, y).$$

Since f has a local extremum at (a, b), then g has a local extremum at b. (See Fig. 16-20.) It follows from Theorem 4-1 that

$$f_y(a, b) = g'(b) = 0.$$

A similar argument can be used to prove that

$$f_x(a, b) = 0.$$

These remarks prove the following theorem.

THEOREM 16-11 If $f(x, y)$ has a partial derivative at a local extremum, then that partial derivative must equal zero at that point.

It follows from Theorem 16-11 and the remarks that precede it that local extrema of $f(x, y)$ on S can be found only at the following types of interior points:

Interior Critical Points

(1) Points at which $\partial f/\partial x = 0$ and $\partial f/\partial y = 0$.
(2) Points at which one of the partial derivatives is zero and the other fails to exist.
(3) Points at which both partial derivatives fail to exist.

Points that satisfy any of the above conditions are called *interior critical points* of S.

EXAMPLE 2 Let $f(x, y) = x^2 - 6x + y^2 + 4y + 5$. Find all local extrema.

Solution Since f is a polynomial function, then both partial derivatives exist everywhere. To find the interior critical points, we set the partial derivatives equal to zero and solve

16–9 Local Maxima and Minima

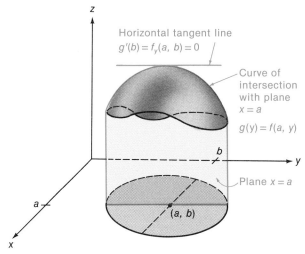

Figure 16–20 The partial derivatives are equal to zero at a local extremum if they exist.

for x and y. Since

$$\frac{\partial f}{\partial x} = 2x - 6 \quad \text{and} \quad \frac{\partial f}{\partial y} = 2y + 4,$$

then we must solve the system

$$2x - 6 = 0, \quad 2y + 4 = 0.$$

There is only one interior critical point, $(3, -2)$.

In this particular example, we can show that f has a local minimum at the critical point. If we complete the square on the x- and the y-terms, we get

$$\begin{aligned} f(x, y) &= x^2 - 6x + 9 + y^2 + 4y + 4 - 8 \\ &= (x - 3)^2 + (y + 2)^2 - 8. \end{aligned}$$

Therefore, $f(3, -2) = -8$, and $f(x, y) > -8$ if $(x, y) \neq (3, -2)$. The function has a local minimum of -8 at $(3, -2)$. □

A function does not necessarily have a local extremum at an interior critical point. It may have a saddle point there. Recall that the graph of f has a saddle point at $P_0(x_0, y_0, f(x_0, y_0))$ if f has a local maximum at P_0 along one curve and a local minimum at P_0 along another.

The function $f(x, y) = y^2 - x^2$ furnishes an example of these remarks. (See Fig. 14–53.) The graph has a saddle point at the origin, the only point at which $\partial f/\partial x$ and $\partial f/\partial y$ are both zero.

THE SECOND-DERIVATIVE TEST

It may be difficult to tell whether a function $f(x, y)$ has a local extremum at a critical point. Theorem 16–12 states a second-derivative test comparable to the test in Theorem 4–10 that can often be used to solve this problem. The proof is omitted.

THEOREM 16–12

Second Derivative Test

(*The Second-Derivative Test*) Let f be continuous and have continuous first and second partial derivatives on a neighborhood of the interior critical point $P_0(x_0, y_0)$. Let

$$\Delta = \Delta(x_0, y_0) = f_{xx}(x_0, y_0)f_{yy}(x_0, y_0) - f_{xy}(x_0, y_0)^2$$

Then:

(1) f has a local minimum at (x_0, y_0) provided

$$\Delta > 0 \quad \text{and} \quad f_{xx}(x_0, y_0) > 0.$$

(2) f has a local maximum at (x_0, y_0) provided

$$\Delta > 0 \quad \text{and} \quad f_{xx}(x_0, y_0) < 0.$$

(3) f has a saddle point at (x_0, y_0) provided

$$\Delta < 0.$$

(4) No conclusion can be drawn if $\Delta = 0$.

EXAMPLE 3 (See Example 2.) Let

$$f(x, y) = x^2 - 6x + y^2 + 4y + 5.$$

Use Theorem 16–12 to show that f has a local minimum at $P_0(3, -2)$.

Solution We saw in Example 2 that

$$\frac{\partial f}{\partial x} = 2x - 6 \quad \text{and} \quad \frac{\partial f}{\partial y} = 2y + 4,$$

and $(3, -2)$ is the only critical point.

The second partial derivatives are

$$\frac{\partial^2 f}{\partial x^2} = 2, \quad \frac{\partial^2 f}{\partial x \partial y} = 0, \quad \text{and} \quad \frac{\partial^2 f}{\partial y^2} = 2.$$

Then

$$\Delta = \left(\frac{\partial^2 f}{\partial x^2}\right)_{(3,-2)} \cdot \left(\frac{\partial^2 f}{\partial y^2}\right)_{(3,-2)} - \left(\frac{\partial^2 f}{\partial x \partial y}\right)_{(3,-2)}^2 = 2 \cdot 2 - 0^2 = 4 > 0.$$

Since $(\partial^2 f/\partial x^2) > 0$ and $\Delta > 0$, then f has a local minimum at $(3, -2)$. □

EXAMPLE 4 Find all local maxima, minima, and saddle points for the graph of

$$f(x, y) = 2x^2 + 2xy + 14x - 2y^2 + 22y - 8.$$

Solution The partial derivatives exist everywhere. They are

$$\frac{\partial f}{\partial x} = 4x + 2y + 14 \quad \text{and} \quad \frac{\partial f}{\partial y} = 2x - 4y + 22.$$

If we set $\partial f/\partial x$ and $\partial f/\partial y = 0$, we get the system of equations

$$\begin{cases} 4x + 2y = -14 \\ 2x - 4y = -22. \end{cases}$$

16–9 Local Maxima and Minima

The only solution is $x = -5$, $y = 3$.

The only critical point is $P_0(-5, 3)$. We apply the second-derivative test. Since

$$\frac{\partial^2 f}{\partial x^2} = 4, \qquad \frac{\partial^2 f}{\partial x \partial y} = 2, \qquad \frac{\partial^2 f}{\partial y^2} = -4,$$

then

$$\Delta = \left(\frac{\partial^2 f}{\partial x^2}\right)_{(-5,3)} \left(\frac{\partial^2 f}{\partial y^2}\right)_{(-5,3)} - \left(\frac{\partial^2 f}{\partial x \partial y}\right)^2_{(-5,3)}$$
$$= 4 \cdot (-4) - 2^2 < 0.$$

The graph has a saddle point at $(-5, 3)$. There are no local extrema. ☐

Example 5 furnishes an instance in which the second-derivative test cannot be applied.

EXAMPLE 5 Show that the function $f(x, y) = (y - x^2)(y - 2x^2)$ has a saddle point at the origin and has no local extrema.

Solution $\dfrac{\partial f}{\partial x} = -6xy + 8x^3$ and $\dfrac{\partial f}{\partial y} = 2y - 3x^2$. If we set these partial derivatives equal to zero, we get the system of equations

$$\begin{cases} -6xy + 8x^3 = 0 \\ 2y - 3x^2 = 0, \end{cases}$$

which has the unique solution $x = 0$, $y = 0$. (The solution can be obtained by substituting $y = 3x^2/2$ from the second equation in the first equation.)

The only critical point is the origin $(0, 0)$. If we try to apply the second-derivative test, we find that $\Delta = 0$ at the origin. Thus, some other approach must be used.

Observe that $f(x, y) = 0$ at $(0, 0)$. The level curve through $(0, 0)$ is the union of the graphs of

$$y = x^2 \quad \text{and} \quad y = 2x^2.$$

(See Fig. 16–21a.) These graphs separate the xy-plane into four regions. (See Fig. 16–21b.) We examine the behavior of f on each region.

The region above the two parabolas is defined by

$$y > 2x^2.$$

Since $2x^2 > x^2$, then $y - 2x^2 > 0$ and $y - x^2 > 0$ on this region. Thus

$$f(x, y) = (y - x^2)(y - 2x^2) > 0$$

if (x, y) is in this region. (See Fig. 16–21b.)

It follows from a similar argument that $f(x, y) > 0$ if (x, y) is in the region below the two parabolas.

The points in the regions between the two parabolas satisfy the conditions $y > x^2$ and $y < 2x^2$. Then

$$f(x, y) = (y - x^2)(y - 2x^2) < 0$$

if (x, y) is in either of these regions.

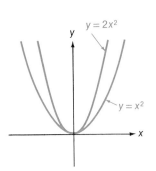
(a) Level curves through the origin

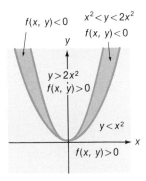
(b) The level curves separate the xy-plane into four regions

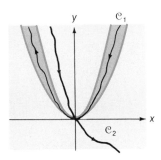
(c) The function f has a local maximum at the origin along the curve $\mathcal{C}_1$, a local minimum at the origin along $\mathcal{C}_2$

Figure 16–21 Example 5.

Let $\mathcal{C}_1$ be a curve through the origin that lies in the regions between the parabolas. (See Fig. 16–21c.) Then f has a local maximum at the origin along $\mathcal{C}_1$. Let $\mathcal{C}_2$ be a curve that passes through the origin from the region above the parabolas to the region below them. (See Fig. 16–21c.) Then f has a local minimum at the origin along $\mathcal{C}_2$.

Since f has a local maximum at the origin along one curve and a local minimum there along another curve, then f has a saddle point at the origin. □

Exercises 16–9

1–8. **(a)** Find all interior critical points.
(b) Find all local extrema and saddle points.

- **1** $f(x, y) = x^2 + 9y^2$
- **2** $f(x, y) = 4x^2 - 9y^2$
- **3** $f(x, y) = x^2 - y^2 + 6x - xy + 2y$
- **4** $f(x, y) = 2x^2 + 2y^2 + xy - 5x$
- **5** $f(x, y) = 2x^3 + x^2y - y^3$
- **6** $f(x, y) = x^3 + 1$
- **7** $f(x, y) = y(e^x - 1)$
- **8** $f(x, y) = xy \sin x$

9 Let f be a differentiable function of x and y. Explain why $\nabla f(x_0, y_0) = \mathbf{0}$ at a local extremum.

10 Show that closed rectangular box of volume V_0 has minimum surface area if it is a cube.

•**11** What is the maximum capacity of an open-topped rectangular box **(a)** for which the sum of the edges is 108 ft? **(b)** for which the surface area is 108 ft²?

12 Show that a triangle of fixed perimeter p has maximum area if it is equilateral. (*Hint:* Use Heron's formula: Area = $\sqrt{s(s-a)(s-b)(s-c)}$, where a, b, c are the sides and $s = (a + b + c)/2$.

Exercises 16–9

●13 A strip of aluminum 24 in. wide is to be bent to form a drain gutter, the cross section of which is to be an isosceles trapezoid (with one base missing). Find the width across the bottom and the measure of the base angles that give the maximum area of the cross section.

●14 Find the shortest distance from the origin to the paraboloid $z = 1 - x^2 - y^2$.

15 A rectangular bin, with open top and open bottom, is to be constructed of sheet metal. (The floor of a room will serve as the bottom of the bin.) Show that there is no theoretical maximum value for the volume if the surface area is to be 40 ft².

16 Use curves through the origin to show that the function $f(x, y) = (x - 2y^2)(x + y^2)$ has a saddle point there.

17 Let S be a set of points in the xy-plane.
   (a) Show that every interior point of S must be in S.
   (b) By examples, show that a boundary point of S may be in S or not in S.
   (c) Prove that every point of S is either a boundary point or an interior point.

Least Squares Method

18 (*Least Squares*) Points plotted from experimental data frequently lie very close to a line. In such a case it may be useful to know the equation of the line that "best fits" the points. There are several ways to approach this problem, leading to different lines of best fit. In this exercise we develop a method, due to Gauss, called the *method of least squares*.

Let $P_1(x_1, y_1), P_2(x_2, y_2), \cdots, P_n(x_n, y_n)$ be points in the xy-plane. We want to find the line $y = mx + b$ such that the sum of the squares of the vertical distances from the points to the line is a minimum. The resulting line is called the *line of regression* for the data. (The vertical distance from $P_k(x_k, y_k)$ to the line is $d_k = (mx_k + b) - y_k$. We want to find m and b such that $d_1^2 + d_2^2 + \cdots + d_n^2$ is a minimum.)

Show that the equation of the line of regression is $y = m_0 x + b_0$ where

Line of Regression

$$m_0 = \frac{n \sum_{k=1}^{n} x_k y_k - \left(\sum_{k=1}^{n} x_k\right)\left(\sum_{k=1}^{n} y_k\right)}{n \sum_{k=1}^{n} x_k^2 - \left(\sum_{k=1}^{n} x_k\right)^2},$$

$$b_0 = \frac{\left(\sum_{k=1}^{n} x_k^2\right)\left(\sum_{k=1}^{n} y_k\right) - \left(\sum_{k=1}^{n} x_k\right)\left(\sum_{k=1}^{n} x_k y_k\right)}{n \sum_{k=1}^{n} x_k^2 - \left(\sum_{k=1}^{n} x_k\right)^2}.$$

(*Hint:* Write $d_1^2 + d_2^2 + \cdots + d_n^2$ as a function of m and b.)

19–21. Use the result of Exercise 18 to find the line of regression for the set of points. Plot the points and the line.

●19 (3, 2), (4, 3), (5, 4), (6, 4), (7, 5)

●20 (−1, 6), (0, 2), (1, 2), (2, 3), (3, 5)

●21 (1, 4.2), (2, 7.3), (3, 10.5), (4, 13.7), (2, 7.4), (5, 16.8), (4, 13.6)

16-10 Maxima and Minima

Theorem 16–11 leads us to the following method for finding the extreme values of a continuous function f defined on a closed region S.

Boundary Critical Points

(1) Locate all interior critical points of f on S.
(2) Locate all points on the boundary of S at which f could have a maximum or a minimum. These points are called *boundary critical points*.
(3) The maximum (minimum) of f on S equals the maximum (minimum) value of f at the points found in steps 1 and 2.

In Section 16–9 we worked with interior critical points. We now turn our attention to boundary critical points. We consider only the case for which the boundary is a curve $\mathscr{C}$ that can be described by simple equations.

If the boundary can be described by the parametric equations

$$x = g(t), \quad y = h(t), \quad a \leq t \leq b,$$

then the values of f on the boundary are given by

$$\phi(t) = f(g(t), h(t)).$$

Thus, the boundary critical points can be found by the methods of Chapter 4.

EXAMPLE 1 Find the maximum and minimum of $f(x, y) = x^2 - 2x + y^2 - 2y + 5$ on the closed region $S = \{(x, y) \mid x^2 + y^2 \leq 4\}$.

Solution The interior of S is the set of all points inside the circle $x^2 + y^2 = 4$. The boundary is that circle. Since

$$\frac{\partial f}{\partial x} = 2x - 2 \quad \text{and} \quad \frac{\partial f}{\partial y} = 2y - 2,$$

the only interior critical point is $(1, 1)$.

Observe that the boundary can be described by the parametric equations

$$x = 2\cos t, \, y = 2\sin t, \quad 0 \leq t \leq 2\pi.$$

The values of f on the boundary are given by

$$\phi(t) = f(2\cos t, 2\sin t), \quad 0 \leq t \leq 2\pi.$$

By the Chain Rule,

$$\phi'(t) = \frac{\partial f}{\partial x} \cdot \frac{dx}{dt} + \frac{\partial f}{\partial y} \cdot \frac{dy}{dt}$$
$$= (2x - 2) \cdot (-2\sin t) + (2y - 2)(2\cos t)$$
$$= (4\cos t - 2)(-2\sin t) + (4\sin t - 2)(2\cos t)$$
$$= 4(\sin t - \cos t).$$

Observe that $\phi'(t) = 0$ only when $t = \pi/4$ or $5\pi/4$. Thus, there are two boundary critical points:

$$(\sqrt{2}, \sqrt{2}) \quad \left(\text{taken when } t = \frac{\pi}{4}\right),$$

16-10 Maxima and Minima

and

$$(-\sqrt{2}, -\sqrt{2}) \quad \left(\text{taken when } t = \frac{5\pi}{4}\right).$$

We have found one interior critical point and two boundary critical points. Also, we must consider the "endpoint" of the boundary where $t = 0$ and $t = 2\pi$, that is, $(2, 0)$. The values of f at these points are

$$f(1, 1) = 1 - 2 + 1 - 2 + 5 = 3$$
$$f(\sqrt{2}, \sqrt{2}) = 2 - 2\sqrt{2} + 2 - 2\sqrt{2} + 5 = 9 - 4\sqrt{2} \approx 3.343$$
$$f(-\sqrt{2}, -\sqrt{2}) = 2 + 2\sqrt{2} + 2 + 2\sqrt{2} + 5 = 9 + 4\sqrt{2} \approx 14.657.$$
$$f(2, 0) = 4 - 4 + 0 + 0 + 5 = 5.$$

The maximum of f on S is $9 + 4\sqrt{2} \approx 14.657$, taken at the boundary critical point $(-\sqrt{2}, -\sqrt{2})$. The minimum is 3, taken at the interior critical point $(1, 1)$. ☐

EXAMPLE 2 Find the maximum and the minimum of $f(x, y) = x^2 - y^2 + 4$ on the closed triangular region with vertices $V_1(-2, 1)$, $V_2(2, 1)$, and $V_3(1, -2)$.

Solution The closed region is pictured in Figure 16–22a. It is not convenient to express the boundary by parametric equations. Instead we observe that between V_1 and V_2 the boundary is the graph of $y = 1$, $-2 \le x \le 2$; between V_3 and V_2 it is the graph of $y = 3x - 5$, $1 \le x \le 2$; between V_1 and V_3 it is the graph of $y = -x - 1$, $-2 \le x \le 1$. (See Fig. 16–22a.)

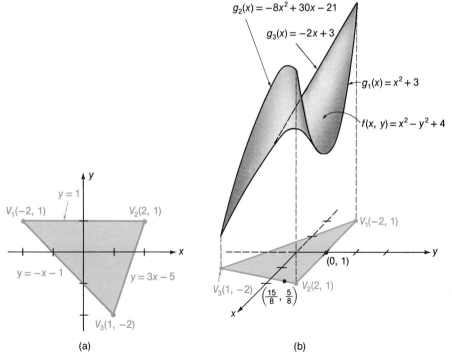

(a) (b)

Figure 16–22 Example 2.

It is easy to see that the function f has a saddle point at $(0, 0)$, the only interior critical point. (See Fig. 16–22b.) Thus, the extrema are located on the boundary.

Along each portion of the boundary the functional values of f can be expressed in terms of x:

(1) Between V_1 and V_2,

$$\begin{aligned} f(x, y) &= f(x, 1) \\ &= g_1(x) = x^2 - 1 + 4 \\ &= x^2 + 3, \qquad -2 \leq x \leq 2. \end{aligned}$$

(2) Between V_3 and V_2,

$$\begin{aligned} f(x, y) &= f(x, 3x - 5) \\ &= g_2(x) = x^2 - (3x - 5)^2 + 4 \\ &= -8x^2 + 30x - 21, \qquad 1 \leq x \leq 2. \end{aligned}$$

(3) Between V_1 and V_3,

$$\begin{aligned} f(x, y) &= f(x, -x - 1) \\ &= g_3(x) = x^2 - (-x - 1)^2 + 4 \\ &= -2x + 3, \qquad -2 \leq x \leq 1. \end{aligned}$$

We consider these portions of the boundary separately.

Between V_1 and V_2, we have

$$g_1(x) = x^2 + 3, \qquad -2 \leq x \leq 2.$$

Then

$$g_1'(x) = 2x.$$

The only critical point interior to the interval $[-2, 2]$ is at $x = 0$. The corresponding value of y is $y = 1$. Thus, $(0, 1)$ is a boundary critical point between V_1 and V_2.

Since x is restricted to the range $-2 \leq x \leq 2$, we must also consider the behavior of the function at the endpoints $x = -2$ and $x = 2$. This gives us the additional boundary critical points $V_1(-2, 1)$ and $V_2(2, 1)$.

Between V_3 and V_2, the functional values are given by

$$g_2(x) = -8x^2 + 30x - 21, \qquad 1 \leq x \leq 2.$$

The only critical point interior to the interval is at $x = \frac{15}{8}$, $y = \frac{5}{8}$. Thus, $(\frac{15}{8}, \frac{5}{8})$ is a boundary critical point between V_3 and V_2. We must also consider the points $V_2(2, 1)$ and $V_3(1, -2)$, which correspond to the endpoints of the interval for x.

Between V_1 and V_3, the functional values are

$$g_3(x) = -2x + 3, \qquad -2 \leq x \leq 1.$$

There are no critical points between -2 and 1. We have only the endpoints to consider—the vertices $V_1(-2, 1)$ and $V_3(1, -2)$ of the triangle.

It follows that we have five distinct boundary points that must be considered: the critical points $(0, 1)$, $(\frac{15}{8}, \frac{5}{8})$, and the vertices $(-2, 1)$, $(2, 1)$, $(1, -2)$. We calculate the values of f at these points:

$$f(0, 1) = 0 - 1 + 4 = 3$$

$$f\left(\frac{15}{8}, \frac{5}{8}\right) = \frac{15^2}{8^2} - \frac{5^2}{8^2} + 4 = \frac{200}{64} + 4 = \frac{57}{8} = 7.125$$

16–10 Maxima and Minima

$$f(-2, 1) = 4 - 1 + 4 = 7$$
$$f(2, 1) = 4 - 1 + 4 = 7$$
$$f(1, -2) = 1 - 4 + 4 = 1.$$

The maximum value of f is $\frac{57}{8}$, taken at the boundary critical point $(\frac{15}{8}, \frac{5}{8})$. The minimum value is 1, taken at the vertex $(1, -2)$. (See Fig. 16–22b.) □

As you can see after working through Example 2, calculating boundary critical points can be laborious—even when the boundary is a simple figure.

LINEAR FUNCTIONS ON POLYGONS

There is one special case in which extreme values can be found easily. If f is a linear function—that is, a function of form $f(x, y) = ax + by + c$—and S is a closed polygon in the xy-plane, then the extreme values of f must be taken at the vertices of the polygon. This result is stated formally in Theorem 16–13. The proof has as its basis the fact that f has no interior critical points and no boundary critical points between the vertices. The details of the proof are left for you. (See Exercise 11.)

THEOREM 16–13   Let $f(x, y) = ax + by + c$. Let S be a closed polygon with vertices at (x_1, y_1), $(x_2, y_2), \ldots, (x_n, y_n)$. Then the extreme values of f on S are taken at the vertices. The maximum (minimum) of f on S is the maximum (minimum) of $f(x_1, y_1)$, $f(x_2, y_2), \ldots, f(x_n, y_n)$.

Theorem 16–13 and its generalization to "polygons" in n-dimensional space are basic to the theory of *linear programming*—a branch of mathematics devoted to maximizing linear functions in many variables that are subject to linear constraints.

EXAMPLE 3   Use Theorem 16–13 to find the maximum and the minimum of

$$f(x, y) = 3x - 2y + 7$$

on the polygon obtained by connecting successive pairs of the points $(0, 3)$, $(1, 1)$, $(2, 1)$, $(2, -1)$, $(1, -3)$, $(-1, -2)$, and $(0, 3)$ with line segments.

Solution   The extreme values are taken at the vertices. We calculate

$$f(0, 3) = 1, \quad f(1, 1) = 8, \quad f(2, 1) = 11$$
$$f(2, -1) = 15, \quad f(1, -3) = 16, \quad f(-1, -2) = 8.$$

The maximum of f on the polygon is $f(1, -3) = 16$. The minimum is $f(0, 3) = 1$. □

EXTENSION TO THREE-DIMENSION SPACE

The methods of this section can be extended to cover functions of three or more variables. The concepts of an interior point, a boundary point, a region, and a closed region have simple extensions to three-dimensional space.

If a function $F(x, y, z)$ is continuous on a bounded, closed region in space, then it has a maximum and a minimum on that region. If $F(x, y, z)$ has an interior local extremum at a point, then $\partial F/\partial x$, $\partial F/\partial y$, and $\partial F/\partial z$ are all zero there provided that they exist.

The point $P_0(x_0, y_0, z_0)$ is an *interior point* of the set S if there is a solid sphere with center at P_0 which is completely contained in S. It is a boundary point if every solid sphere with center at P_0 contains a point in S and a point not in S. The set S is a *solid region* provided every point of S is an interior point and every pair of points can be connected by a smooth curve contained in S. The union of a solid region and its boundary is called a *closed solid region*. A solid region is *bounded* if it is completely contained within a rectangular box in three-dimensional space.

Exercises 16–10

- **1** Find the maximum and minimum of $f(x, y) = x^2 - y^2 + 1$ on the closed region bounded by the unit circle $x^2 + y^2 = 1$.

- **2** Find the maximum and minimum of $f(x, y) = x^2 - 6x + y^2 - 8y + 7$ on the closed region bounded by the unit circle $x^2 + y^2 = 1$.

- **3** Find the maximum and minimum of $f(x, y) = 24x^2 + 16x + 40y^2$ on the circular disk bounded by $x^2 + y^2 = 4$.

- **4** Find the maximum and minimum of $f(x, y) = 2xy - y^2 - x^2 + 10$ on the closed parallelogram with vertices $A(0, 0)$, $B(5, 3)$, $C(3, 5)$, $D(-2, 2)$.

- **5** Find the maximum and minimum of $f(x, y) = 3x^2 - 2xy + y^2 - 8x$ on the closed rectangle bounded by the lines $x = -2$, $x = 2$, $y = -1$, $y = 1$.

- **6** Find the maximum and minimum of $f(x, y) = x^2 - 2xy + 2y + 3y^2$ on the closed region defined by $x \leq 1$, $-x - 4 \leq y \leq x + 4$.

- **7** The temperature distribution of a flat metal plate is given by $T(x, y) = 300(2x^2 + 3y^2 + 1)^{-2}$. Find the maximum and minimum temperatures on the closed triangle bounded by the lines $x = 0$, $y = -3$, $x + y = 3$.

- **8** An electronics company manufactures two types of calculators, Type X and Type Y. The profit of the company is proportional to

$$P(x, y) = 2x + 3y - 100$$

where x and y are the number of production runs of Types X and Y. Various factors influence the number of production runs—availability of parts, contractual commitments, need to manufacture other items, and so on. These factors cause the production runs to be limited as follows:

$$25 \leq x \leq 40, \quad 20 \leq y \leq 50, \quad x + y \leq 80.$$

Find the values of x and y that maximize the profit.

9–10. Use Theorem 16–13 to calculate the extreme values of f on the closed polygon with the given vertices.

- **9** $f(x, y) = 2x + 3y + 9$; $(-1, -1)$, $(0, 1)$, $(1, 0)$, $(0, -2)$

- **10** $f(x, y) = -x + 2y - 3$; $(1, 2)$, $(-1, 6)$, $(-4, 3)$, $(-5, -2)$, $(-3, -3)$, $(0, -2)$

- **11** Prove Theorem 16–13.

16–11 Extremal Problems with Constraints. Lagrange Multipliers

Many problems reduce to finding extreme values of a function, subject to certain restrictions. Such restrictions arise, for example, when we calculate the extreme values on the boundary of a closed region.

In this section we consider general methods of finding extreme values of a function $f(x, y)$ subject to restrictions that can be written as equations. That is, the restrictions are of form

$$\begin{cases} g_1(x, y) = 0 \\ g_2(x, y) = 0 \\ \cdots \cdots \\ g_n(x, y) = 0 \end{cases}$$

Constraints (side conditions)

where $g_1, g_2, \ldots, g_n$ are functions. Such restrictions are called *constraints*, or *side conditions*.

LAGRANGE'S METHOD

An important method for solving extremal problems subject to constraints can be credited to J. L. Lagrange (1736–1813). For simplicity, we consider a function of two variables subject to a single constraint.

THEOREM 16–14 (*Lagrange's Method*) Let f and g be differentiable functions of x and y. The local extrema of f, subject to the constraint $g(x, y) = 0$, are found among the points $P(x, y)$ for which there exists a real number λ such that

Lagrange's Method

$$\begin{cases} \dfrac{\partial f}{\partial x} - \lambda \dfrac{\partial g}{\partial x} = 0 \\ \dfrac{\partial f}{\partial y} - \lambda \dfrac{\partial g}{\partial y} = 0 \\ g(x, y) = 0 \end{cases}$$

Lagrange Multiplier

The number λ is called the *Lagrange multiplier* for the problem.

Proof We define a new function of x, y, and z by

$$F(x, y, z) = f(x, y) - zg(x, y).$$

Observe that $F(x, y, z) = f(x, y)$ if $g(x, y) = 0$. Thus, the extreme values of $F(x, y, z)$, subject to the constraint $g(x, y) = 0$, are the same as the extreme values of $f(x, y)$, subject to the same constraint.

We now investigate the extreme values of $F(x, y, z)$. Observe that F is differentiable since f and g are. Thus, a local extremum of F must exist at a point where

$$\dfrac{\partial F}{\partial x} = 0, \quad \dfrac{\partial F}{\partial y} = 0, \quad \text{and} \quad \dfrac{\partial F}{\partial z} = 0,$$

that is, at a point where

(1)
$$\begin{cases} \dfrac{\partial F}{\partial x} = \dfrac{\partial f}{\partial x} - z\dfrac{\partial g}{\partial x} = 0 \\ \dfrac{\partial F}{\partial y} = \dfrac{\partial f}{\partial y} - z\dfrac{\partial g}{\partial y} = 0 \\ \dfrac{\partial F}{\partial z} = 0 - g(x, y) = 0 \end{cases}$$

The last equation tells us that the constraint $g(x, y) = 0$ must hold at the local extrema of F. Therefore, the extreme values of F exist at points in three-dimensional space where the constraint $g(x, y) = 0$ is in effect. It follows from the remarks at the beginning of this proof that f has the same extreme values.

Suppose now that f has a local extremum at (x_0, y_0) subject to the constraint $g(x, y) = 0$. Then there is a corresponding point (x_0, y_0, λ_0) in three-dimensional space at which F has the same local extremum. It follows from equations (1) that

$$\begin{cases} \left(\dfrac{\partial f}{\partial x}\right)_0 - \lambda_0 \left(\dfrac{\partial g}{\partial x}\right)_0 = 0 \\ \left(\dfrac{\partial f}{\partial y}\right)_0 - \lambda_0 \left(\dfrac{\partial g}{\partial y}\right)_0 = 0 \\ \qquad g(x_0, y_0) = 0. \quad \blacksquare \end{cases}$$

EXAMPLE 1 Use Lagrange's method to find the extreme values of

$$f(x, y) = x^2 - y^2$$

subject to the constraint $g(x, y) = x^2 + y^2 - 1 = 0$.

Solution We are looking for the extreme values of f on the unit circle $x^2 + y^2 = 1$. By Theorem 16–14, these exist at points where there is a number λ such that

$$\begin{cases} \dfrac{\partial f}{\partial x} - \lambda \dfrac{\partial g}{\partial x} = 2x - \lambda \cdot 2x = 0 \\ \dfrac{\partial f}{\partial y} - \lambda \dfrac{\partial g}{\partial y} = -2y - \lambda \cdot 2y = 0 \\ \qquad g(x, y) = x^2 + y^2 - 1 = 0. \end{cases}$$

Thus, we must solve the system of equations

$$\begin{cases} 2x - 2x\lambda = 0 \\ -2y - 2y\lambda = 0 \\ x^2 + y^2 = 1, \end{cases}$$

which is equivalent to the system

$$\begin{cases} x(1 - \lambda) = 0 \\ y(1 + \lambda) = 0 \\ x^2 + y^2 = 1. \end{cases}$$

Observe from the last equation that x and y cannot both be zero. If $x \neq 0$, then from the first equation $\lambda = 1$, so that $y = 0$ (from the second equation). If we substi-

16–11 Extremal Problems with Constraints. Lagrange Multipliers

tute $y = 0$ in the third equation, we get $x = \pm 1$. Thus, if $x \neq 0$, there are two critical points $(-1, 0)$ and $(1, 0)$ corresponding to $\lambda = 1$. Similarly, if $y \neq 0$, there are two critical points $(0, -1)$ and $(0, 1)$ corresponding to $\lambda = -1$.

It follows from Theorem 16–14 that the extreme values of f, subject to the side condition, can be found at these critical points. We calculate

$$f(-1, 0) = 1 - 0 = 1$$
$$f(1, 0) = 1 - 0 = 1$$
$$f(0, -1) = 0 - 1 = -1$$
$$f(0, 1) = 0 - 1 = -1.$$

The maximum value of f on the circle $x^2 + y^2 = 1$ is $f(-1, 0) = 1$. The minimum value is $f(0, 1) = -1$. □

Lagrange's method generalizes to functions of any number of variables subject to any number of constraints. Theorem 16–15 states the conditions for a function of three variables subject to two constraints.

THEOREM 16–15

Lagrange's Method for a Function of Three Variables with Two Constraints

Let $f(x, y, z)$, $g_1(x, y, z)$, and $g_2(x, y, z)$ be differentiable functions. The local extrema of f, subject to the constraints

$$g_1(x, y, z) = 0 \quad \text{and} \quad g_2(x, y, z) = 0$$

are found among the points at which there exist real numbers λ_1 and λ_2 such that

$$\begin{cases} \dfrac{\partial f}{\partial x} - \lambda_1 \dfrac{\partial g_1}{\partial x} - \lambda_2 \dfrac{\partial g_2}{\partial x} = 0 \\[6pt] \dfrac{\partial f}{\partial y} - \lambda_1 \dfrac{\partial g_1}{\partial y} - \lambda_2 \dfrac{\partial g_2}{\partial y} = 0 \\[6pt] \dfrac{\partial f}{\partial z} - \lambda_1 \dfrac{\partial g_1}{\partial z} - \lambda_2 \dfrac{\partial g_2}{\partial z} = 0 \\[6pt] g_1(x, y, z) = 0 \\[3pt] g_2(x, y, z) = 0 \end{cases}$$

EXAMPLE 2 Find the minimum distance from the origin to the line of intersection of the two planes

$$x + y + z = 8 \quad \text{and} \quad 2x - y + 3z = 28.$$

Solution Let (x, y, z) be the point on the line that is the minimum distance from the origin. Our problem reduces to finding this point—that is, finding the point (x, y, z) for which $\sqrt{x^2 + y^2 + z^2}$ is a minimum subject to the side conditions $x + y + z = 8$, $2x - y + 3z = 28$.

Observe that $\sqrt{x^2 + y^2 + z^2}$ has its minimum when $x^2 + y^2 + z^2$ has its minimum. We shall find it convenient to work with $x^2 + y^2 + z^2$ instead of the distance function $\sqrt{x^2 + y^2 + z^2}$. Thus, we wish to minimize the function

$$f(x, y, z) = x^2 + y^2 + z^2$$

subject to the constraints

$$g_1(x, y, z) = x + y + z - 8 = 0$$
$$g_2(x, y, z) = 2x - y + 3z - 28 = 0.$$

We use Lagrange's method. At the critical point there must exist constants λ_1, λ_2 such that

$$\begin{cases} 2x - \lambda_1 - 2\lambda_2 = 0 & \text{(partial derivative with respect to } x\text{)} \\ 2y - \lambda_1 + \lambda_2 = 0 & \text{(partial derivative with respect to } y\text{)} \\ 2z - \lambda_1 - 3\lambda_2 = 0 & \text{(partial derivative with respect to } z\text{)} \\ \left.\begin{array}{l} x + y + z = 8 \\ 2x - y + 3z = 28 \end{array}\right\} & \text{(constraints).} \end{cases}$$

This system of five equations in five unknowns can be solved by the methods of elementary algebra yielding the unique solution

$$x = 4, \quad y = -2, \quad z = 6, \quad \lambda_1 = 0, \quad \lambda_2 = 4.$$

Since it is obvious from the geometrical conditions that a point must exist where the distance is a minimum, then it must be $(4, -2, 6)$. The minimum distance is

$$\sqrt{4^2 + (-2)^2 + 6^2} = \sqrt{56}. \quad \square$$

Remark Several problems are inherent in Lagrange's method. The most obvious problem is that there is no convenient way to tell whether the function f has a maximum or a minimum (or neither) at a point found by the method. As a rule, additional information must be obtained before we can solve this problem.

The Lagrange multipliers λ_1, λ_2, play no actual role in the solution of an extremal problem. They are auxiliary numbers that are used at an intermediate stage. These numbers do have important uses in certain theoretical arguments. One such application follows.

MARGINAL UTILITY (OPTIONAL)

Utility It is assumed in economic theory that each consumer spends his income so as to yield the greatest benefit to him. This benefit, which is unmeasurable in any practical way, is called the *utility* derived from the income.

For simplicity we assume that the income is spent for two commodities, X and Y. Let

$x = $ number of items of X that are bought,
$y = $ number of items of Y that are bought.

Then the utility can be represented by a function $u(x, y)$, which we assume to be differentiable.

The partial derivatives

$$\frac{\partial u}{\partial x} \quad \text{and} \quad \frac{\partial u}{\partial y}$$

Marginal Utility are called the *marginal utilities of x and y*. They measure the rate at which the utility function changes subject to changes in the number of units bought.

Let p and q be the unit prices of X and Y, respectively, and let the total income be I. Then

$$px + qy = I.$$

The problem economists consider is the relation between x and y that maximizes the utility function u, subject to the constraint $px + qy = I$.

It follows from Lagrange's method that $u(x, y)$ will be maximized, subject to the constraint $px + qy = I$, when

$$\begin{cases} \dfrac{\partial u}{\partial x} - \lambda p = 0, \\ \dfrac{\partial u}{\partial y} - \lambda q = 0, \\ px + qy = I, \end{cases}$$

for some constant λ. The first two equations imply that

$$\frac{\partial u}{\partial x} = \lambda p \quad \text{and} \quad \frac{\partial u}{\partial y} = \lambda q.$$

Thus, for maximum utility, the marginal utilities must be proportional to the prices. The Lagrange multiplier is the constant of proportionality.

The relation examined above is used to justify several assumptions about demand functions—functions that describe the number of items that can be sold at a given price.

Exercises 16–11

1–8. Use Lagrange's method to find the extreme values of the function subject to the constraints.

- **1** $f(x, y) = x^2 + y^2 - 1$, $xy = 1$
- **2** $f(x, y) = xy^2$, $\dfrac{x^2}{4} + \dfrac{y^2}{9} = 1$
- **3** $f(x, y) = 4x^2 + 2xy + y^2 + y$, $x + 2y = 1$
- **4** $f(x, y) = y^2 - x^2$, $x^2 + y^2 = 1$
- **5** $f(x, y, z) = x^2 + y^2 + z^2$, $x + y + 2z = 6$
- **6** $f(x, y, z) = 2x + y + z$, $x^2 + y^2 + z^2 = 16$
- **7** $f(x, y, z) = 3x + 2y - 6z$, $4z^2 + y^2 = x$
- **8** $f(x, y, z) = xyz$, $\dfrac{x^2}{4} + \dfrac{y^2}{9} + \dfrac{z^2}{12} = 1$

- **9** Use Lagrange's method to find the minimum distance from the origin to the plane $3x + 4y + 2z = 6$.

- **10** Find the point on the sphere $x^2 + y^2 + z^2 = 2$ that is closest to the point $(3, 1, 2)$.

- **11** The temperature at the point $P(x, y, z)$ on the sphere $x^2 + y^2 + z^2 = 1$ is given by $T(x, y, z) = 100/(x^2y^2z^2 + 1)$. Find the point at which the temperature is a maximum.

●12 An open-topped rectangular box is to be constructed at a cost that cannot exceed $30. The material for the sides costs $1 per square foot and the material for the bottom costs $0.50 per square foot. Find the maximum possible capacity. (Assume that the thickness of the sides and bottom is negligible.)

●13 A cylindrical container is to be constructed at a cost that cannot exceed $36. The sheet metal for the side costs $1.25 per square foot. The metal for the top and bottom costs $1.50 per square foot. Find the dimensions for the maximum volume.

●14 Find the maximum volume of a closed rectangular box inscribed in the ellipsoid

$$\frac{x^2}{a^2} + \frac{y^2}{b^2} + \frac{z^2}{c^2} = 1.$$

Review Problems

1 Define the following terms

(a) $\lim_{(x,y) \to (a,b)} f(x, y) = L$.

(b) The function $f(x, y)$ is *continuous* at (a, b).

(c) The function $f(x, y)$ is *differentiable* at (a, b).

2–5. (a) Calculate the limit as (x, y) approaches the given point or explain why it does not exist.

(b) Discuss the continuity of the function.

●2 $f(x, y) = \dfrac{2x^2 - y^2}{x^2 + y^2}$, $(0, 0)$

●3 $f(x, y) = \dfrac{x + y - 1}{2x + 3y + 1}$, $(0, 0)$

●4 $f(x, y) = \dfrac{x + y - 3}{2x + y}$, $(1, 2)$

●5 $f(x, y) = \dfrac{x + y - 3}{2x + y - 4}$, $(1, 2)$

●6 The function $f(x, y) = (x^2 - y^2)/(x^2 + y^2)$ is defined everywhere except at $(0, 0)$. Is it possible to define $f(0, 0)$ in such a way that f will be continuous at $(0, 0)$?

7 Let $f(x, y, z)$ and $g(x, y, z)$ be continuous at (a, b, c). Let $h(x, y, z) = f(x, y, z) + Cg(x, y, z)$, where C is a constant. Use the Limit Theorem to prove that h is continuous.

8–11. Calculate all first and second partial derivatives.

●8 $f(x, y) = x^2 + 2xy + y^3$

●9 $f(x, y) = \ln x + \ln (x + y)$

●10 $f(x, y) = xe^{xy}$

●11 $f(x, y) = 3xy + \tan^{-1} y$

●12 Find the equations of the line tangent to the curve of intersection of the paraboloid $z = x^2 + 4y^2$ and the plane $2x - 3y + z = 4$ at $(1, 1, 5)$. (*Hint:* First find vectors normal to the surfaces.)

13–14. Verify that the functions satisfy Laplace's equation

$$\frac{\partial^2 f}{\partial x^2} + \frac{\partial^2 f}{\partial y^2} + \frac{\partial^2 f}{\partial z^2} = 0.$$

13 $f(x, y, z) = \dfrac{1}{\sqrt{x^2 + y^2 + z^2}}$

14 $f(x, y, z) = xe^y \sin z$

15 Let $w = f(x, y)$ be a differentiable function of x and y. State the approximation formula for Δw in terms of $\partial f/\partial x$ and $\partial f/\partial y$. Restate the formula using differentials.

•16 Use the approximation formula to calculate the approximate value of

$$e^{0.1}(\sin(0.05) + \cos^2(0.03)).$$

•17 A rectangular metal slab 3.02 ft long, 4.57 ft wide, and 1.12 ft thick is heated, causing the dimensions to change to 3.06 ft, 4.63 ft, and 1.14 ft, respectively. Use differentials to approximate the change in **(a)** the surface area; **(b)** the volume.

•18 **(a)** Give an example of a function $f(x, y)$ that is discontinuous at a point (a, b) even though it has partial derivatives at that point.

(b) Prove that if f is differentiable at (a, b), then f is continuous there.

19–20. Use the Chain Rule to calculate $\partial w/\partial s$.

•19 $w = x^2 + y^2 + z^2,\ x = \sin s + \cos t,\ y = se^t,\ z = s^2 + t^2$

•20 $w = x^2 + e^y \ln z,\ x = 3s + t - u,\ y = s^2 - u^2,\ z = tue^s$

21 Prove the Chain Rule formula for the case in which w is a differentiable function of x, y, and z, each of which is a differentiable function of t.

•22 The equations $w = x^2 \cos y + y^2,\ y = \ln x + t$ define w as a function of x and t, say $w = g(x, t)$. Use the Chain Rule to calculate $\partial g/\partial x$.

23 Let $f(x, y)$ be differentiable at (x_0, y_0).

(a) What is meant by the *directional derivative* $D_u f(x_0, y_0)$, where u is a unit vector?

(b) Prove that $D_u f(x_0, y_0) = (\nabla f)_{(x_0, y_0)} \cdot u$.

24 Prove that $D_u f(x_0, y_0)$ has its maximum value when u is in the direction of $(\nabla f)_{(x_0, y_0)}$ and that this maximum value is equal to $\|(\nabla f)_{(x_0, y_0)}\|$.

•25 Calculate ∇f and $D_v f$.

(a) $f(x, y) = 2x^2 - 3xy + y^3,\ v = 3i + 4j$

(b) $f(x, y, z) = \dfrac{x + y}{y + z},\ v = -2i + j - 2k$

•26 The temperature of a flat metal plate is given by
$$T(x, y) = \frac{81}{x^2 + y^2 + 2}.$$
Find the direction in which the temperature has the greatest rate of change at the point $(-3, 4)$. What is the greatest rate of change of T at $(-3, 4)$?

27 What is meant by a tangent line to a surface? by a tangent plane to a surface?

28 Let $z = f(x, y)$ be a differentiable function of x and y. Derive the formula
$$z - z_0 = f_x(x_0, y_0)(x - x_0) + f_y(x_0, y_0)(y - y_0)$$
for the plane tangent to the graph at $P_0(x_0, y_0, z_0)$.

29–30. Find a normal vector and the equation of the tangent plane to the surface at the given point.

•29 $xyz = 2$, $P_0(1, -1, -2)$

•30 $\dfrac{x^2}{16} + \dfrac{y^2}{4} - \dfrac{z^2}{4} = 1$, $P_0(4, -4, 4)$

•31 Find a tangent plane to the graph of $z^2 = x^2 + y^2 + 6$ that is parallel to the plane $3x - y - 4z = 1$. (Two answers are possible.)

32 (a) Let w be a differentiable function of x, y, and z. Prove that
$$d(w^n) = nw^{n-1}\, dw.$$

•(b) Use the formula in (a) to calculate
$$d((3x^2 - 2\cos y + \ln z)^5).$$

33 (a) What is meant by an *interior point* of a set S? by *boundary point*? by a *region*? by a *closed region*?

(b) Let f be a continuous function of x and y. What is meant by the *maximum* of f on a set S? by the *minimum* of f on S? by a *local maximum*? by an *interior critical point*? by a *boundary critical point*?

34 State the second-derivative test for local extrema.

35–38. Find all local extrema and saddle points.

•35 $f(x, y) = x^3 + y^3 - 9x^2 - 9y$

•36 $f(x, y) = 2xy + \dfrac{1}{x} + \dfrac{1}{2y}$

•37 $f(x, y) = x^3 - 6xy + y^2$

•38 $f(x, y) = x^2 - x^3 + y^3 - y^2$

•39 Let T be the triangle with vertices $(2, 1)$, $(3, 2)$, and $(4, -3)$. Find the point

Review Problems

$P_0(x_0, y_0)$ such that the sum of the squares of the distances from the point to the three vertices of T is a minimum.

- **40** Find the points on the parabola $2y = x^2 - 3$ that are closest to the origin.

- **41** Find the maximum and minimum values of $f(x, y) = 8y^2 + 4x^2 + 4x$ on the circular disk $x^2 + y^2 \leq 1$.

42–43. Use Lagrange's method to find the extreme values of f subject to the given constraint.

- **42** $f(x, y) = xy, \ x^2 + y^2 = 2$

- **43** $f(x, y, z) = x^2 + y^2 + z^2; \ z = 3x^2 + 2y^2 - 1$

- **44** Find the point on the cone $z^2 = x^2 + y^2$ closest to (3, 4, 10).

17 Multiple Integration

17-1 The Double Integral

In generalizing the results of Chapter 5 to integrals of two and three variables, our work will involve two distinct types of integrals. These are denoted (in the two-variable case) by

$$\iint_{\mathcal{R}} F(x, y)\, dA \quad \text{and} \quad \int_a^b \int_{g_1(x)}^{g_2(x)} F(x, y)\, dy\, dx.$$

The first type of integral is called the *double integral over the region* $\mathcal{R}$. It corresponds to the definite integral for a function of one variable. The second is called the *iterated integral*. It can be used to evaluate double integrals by the use of antiderivatives.

Before proceeding further, you should review the development of the definite integral in Chapter 5. The work of this section roughly parallels that development.

BASIC CLOSED REGIONS

Two special type of bounded closed regions, illustrated in Figure 17-1, are fundamental in our work. (The concept of a *bounded closed region* was discussed in Sec-

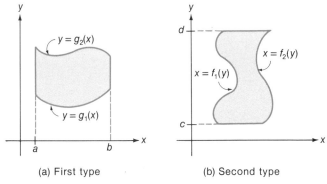

(a) First type (b) Second type

Figure 17-1 "Basic" closed regions are finite unions of the types of closed regions pictured in this figure.

17-1 The Double Integral

tion 16-9.) The closed region in Figure 17-1a has lower and upper boundaries that are graphs of continuous functions of x for $a \leq x \leq b$. The closed region in Figure 17-1b has left- and right-hand boundaries that are graphs of continuous functions of y for $c \leq y \leq d$. These closed regions can, of course, end at points instead of straight lines.

Basic Closed Region

In our work we restrict ourselves to bounded closed regions that are finite unions of nonoverlapping closed regions of the types shown in Figure 17-1. Such closed regions are called *basic closed regions*. A typical basic closed region, along with its decomposition into closed regions of the type described above, is shown in Figure 17-2.

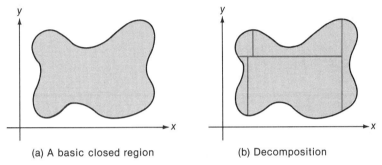

(a) A basic closed region   (b) Decomposition

Figure 17-2 Basic closed regions can be decomposed into closed regions of the type pictured in Figure 17-1.

INNER PARTITIONS

Let $\mathcal{R}$ be a basic closed region with projections $[a, b]$ and $[c, d]$ on the x- and y-axes, respectively. Observe that $\mathcal{R}$ is completely contained within the rectangle with vertices (a, c), (a, d), (b, c), and (b, d). (See Fig. 17-3a.) We partition the closed intervals $[a, b]$ and $[c, d]$ into subintervals. Observe that this induces a partition of the rectangle that contains $\mathcal{R}$ into small rectangular closed regions. (See Fig. 17-3b.) We retain those small rectangles that are completely contained in $\mathcal{R}$ and ignore the others. The resulting set of rectangles is called an *inner partition* of $\mathcal{R}$. (See Fig. 17-3c.)

Inner Partition

Let $\mathcal{P}$ denote an inner partition of $\mathcal{R}$. The individual closed rectangles that make up the partition will be denoted by $R_1, R_2, \ldots, R_n$, and their areas by $\Delta A_1, \Delta A_2, \ldots, \Delta A_n$, respectively. The *norm of* $\mathcal{P}$, denoted by $\|\mathcal{P}\|$, is the length of the longest diagonal of a small rectangle in $\mathcal{P}$.

DOUBLE INTEGRALS

Let $F(x, y)$ be defined on the basic closed region $\mathcal{R}$. Let $R_1, R_2, \ldots, R_n$ be the closed rectangles that form an inner partition $\mathcal{P}$ of $\mathcal{R}$, and let $\Delta A_1, \Delta A_2, \ldots, \Delta A_n$ be the areas of these rectangles.

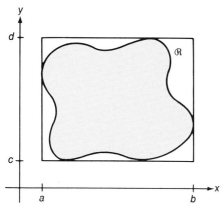

(a) The basic closed region $\mathcal{R}$ is contained in the rectangle with vertices (a, c), (b, c), (a, d) and (b, d)

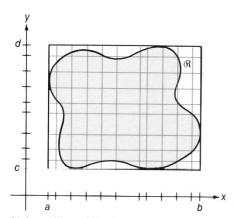

(b) A partition of the intervals $[a, b]$ and $[c, d]$ induces a partition of the rectangle containing $\mathcal{R}$.

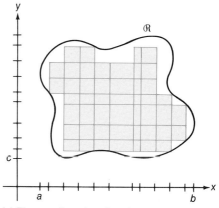

(c) The small rectangles that are contained in $\mathcal{R}$ form an inner partition of $\mathcal{R}$.

Figure 17–3 Inner partition.

Let $(\xi_1, \eta_1), (\xi_2, \eta_2), \ldots, (\xi_n, \eta_n)$ be points in $R_1, R_2, \ldots, R_n$, respectively. (See Fig. 17–4.) The sum

$$S_n = \sum_{k=1}^{n} F(\xi_k, \eta_k) \, \Delta A_k$$

Approximating Sum

is called an *approximating sum based on the function F and the inner partition* $\mathcal{P}$.

It must be emphasized that the value of S_n depends on the function F, the inner partition $\mathcal{P}$, and the particular points $(\xi_1, \eta_1), (\xi_2, \eta_2), \ldots, (\xi_n, \eta_n)$ in the subregions of the partition $\mathcal{P}$. Thus, there are many different approximating sums based on a particular inner partition $\mathcal{P}$.

17-1 The Double Integral

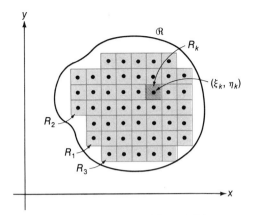

Figure 17-4 An inner partition of $\mathcal{R}$ into rectangles $R_1, R_2, \ldots, R_n$. The area of the rectangle R_k is ΔA_k.

DEFINITION

Limit of Approximating Sums

Let F be defined on the basic closed region $\mathcal{R}$. For each inner partition $\mathcal{P}$ of $\mathcal{R}$ into n rectangles, we indicate the associated approximating sum by S_n. We say that

$$\lim_{\|\mathcal{P}\| \to 0} S_n = L$$

provided that for each $\epsilon > 0$ there exists $\delta > 0$ such that

$$|S_n - L| < \epsilon \quad \text{whenever } \|\mathcal{P}\| < \delta.$$

It is proved in more advanced courses that if F is continuous on the basic closed region $\mathcal{R}$, then

$$\lim_{\|\mathcal{P}\| \to 0} S_n$$

Double Integral

exists and is equal to a real number L. This number is called the *double integral* of F over $\mathcal{R}$, written

$$\iint_{\mathcal{R}} F(x, y)\, dA = L = \lim_{\|\mathcal{P}\| \to 0} S_n$$

$$= \lim_{\|\mathcal{P}\| \to 0} \sum_{k=1}^{n} F(\xi_k, \eta_k)\, \Delta A_k.$$

Observe that the double integral of F over $\mathcal{R}$ is defined by a process that is much like the one used to define the definite integral (Chapter 5). The essential difference is that small rectangles have replaced small intervals in the partitions.

It is practically impossible to calculate a double integral from the definition. We can, however, use an approximating sum S_n to approximate the integral.

EXAMPLE 1 Let $\mathcal{R}$ be the closed rectangle with vertices (1, 1), (1, 3), (4, 1), and (4, 3). Let $\mathcal{P}$ be the inner partition that divides $\mathcal{R}$ into six squares of length 1 unit on a side. Let

(ξ_k, η_k) be the midpoint of the kth square. Use the approximating sum S_6 to approximate

$$\iint_{\mathcal{R}} (\cos x + e^y)\, dA.$$

Solution We number the squares as in Figure 17–5. Then $(\xi_1, \eta_1) = (1.5, 1.5)$, $(\xi_2, \eta_2) = (2.5, 1.5)$, $(\xi_3, \eta_3) = (3.5, 1.5)$, and so on.

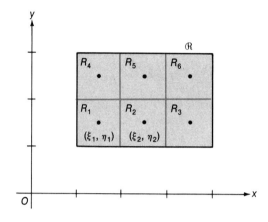

Figure 17–5 Example 1.

Since each ΔA_k equals 1, then

$$\iint_{\mathcal{R}} F(x, y)\, dA = \iint_{\mathcal{R}} (\cos x + e^y)\, dA$$

$$\approx S_6 = f(1.5, 1.5)\,\Delta A_1 + f(2.5, 1.5)\,\Delta A_2 + f(3.5, 1.5)\,\Delta A_3$$
$$+ f(1.5, 2.5)\,\Delta A_4 + f(2.5, 2.5)\,\Delta A_5 + f(3.5, 2.5)\,\Delta A_6$$

$$= (\cos 1.5 + e^{1.5})\cdot 1 + (\cos 2.5 + e^{1.5})\cdot 1$$
$$+ (\cos 3.5 + e^{1.5})\cdot 1 + (\cos 1.5 + e^{2.5})\cdot 1$$
$$+ (\cos 2.5 + e^{2.5})\cdot 1 + (\cos 3.5 + e^{2.5})\cdot 1$$

$$\approx 46.6588$$

It can be shown by the methods of Section 17–2 that the exact value of this integral is

$$\iint_{\mathcal{R}} (\cos x + e^y)\, dA = 2(\sin 4 - \sin 1) + 3(e^3 - e) \approx 48.9052$$

Thus, the approximation S_6 differs from the exact value by more than two units. (See Exercise 12 of Section 17–2.) □

VOLUME

Let $F(x, y)$ be continuous and nonnegative over the basic closed region $\mathcal{R}$. Let R_1, $R_2, \ldots, R_n$ be the subregions of $\mathcal{R}$ defined by an inner partition $\mathcal{P}$. Let (ξ_1, η_1), $(\xi_2, \eta_2), \ldots, (\xi_n, \eta_n)$ be points in $R_1, R_2, \ldots, R_n$, respectively.

17–1 The Double Integral

The term $F(\xi_k, \eta_k) \Delta A_k$ in the approximating sum S_n is equal to the volume of a rectangular prism with base R_k and height $F(\xi_k, \eta_k)$—the distance from the xy-plane to the surface $z = F(x, y)$ at the point (ξ_k, η_k). (See Fig. 17–6a.)

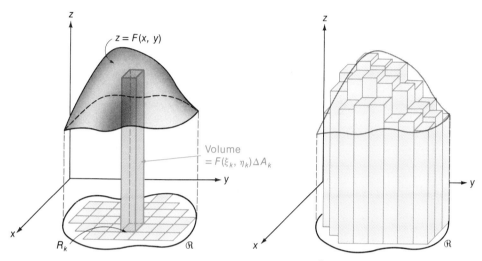

(a) $F(\xi_k, \eta_k) \Delta A_k =$ volume of the rectangular solid defined over R_k.

(b) $S_n = \sum_{k=1}^{n} F(\xi_k, \eta_k) \Delta A_k =$ the sum of the volumes of the rectangular solids defined over the rectangles in the inner partition of $\mathcal{R}$

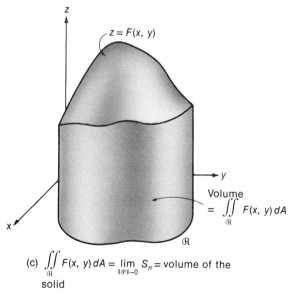

(c) $\iint_{\mathcal{R}} F(x, y) \, dA = \lim_{\|\mathcal{P}\| \to 0} S_n =$ volume of the solid

Figure 17–6 Volume of solid bounded above by graph of $z = F(x, y)$ is $V = \iint_{\mathcal{R}} F(x, y) \, dA$.

The approximating sum

$$S_n = \sum_{k=1}^{n} F(\xi_k, \eta_k) \Delta A_k$$

equals the sum of the volumes of all such prisms. (See Fig. 17–6b.)

It is intuitively obvious that S_n, the sum of the volumes of the rectangular solids pictured in Figure 17–6b, is an approximation of the volume of the solid that has $\mathcal{R}$ as its base and the graph of F as its upper boundary. (See Fig. 17–6c.) Furthermore, this approximation should improve as the norm of the partition approaches zero. For this reason, we define the concept of volume in terms of a double integral.

DEFINITION Let F be a continuous, nonnegative function of x and y over the basic closed region $\mathcal{R}$. Let S be the solid that has $\mathcal{R}$ as its base and the graph of F as its upper surface. The *volume* of S is defined to be

Volume

$$\text{Volume of } S = \lim_{\|\mathcal{P}\| \to 0} \sum_{k=1}^{n} F(\xi_k, \eta_k) \Delta A_k = \iint_{\mathcal{R}} F(x, y) \, dA.$$

EXAMPLE 2 If $F(x, y) = e^{1-xy}$ and $\mathcal{R}$ is the closed triangular region with vertices $(0, 0)$, $(0, 1)$, and $(1, 0)$, then the volume of the solid with base $\mathcal{R}$ and the graph of F as its upper surface is

$$\iint_{\mathcal{R}} e^{1-xy} \, dA.$$

(See Fig. 17–7.) □

AREA

Let $\mathcal{P}$ be an inner partition of the basic closed region $\mathcal{R}$. Let $\Delta A_1, \Delta A_2, \ldots, \Delta A_n$ be the areas of the rectangles in $\mathcal{P}$. The sum

$$S_n = \sum_{k=1}^{n} \Delta A_k$$

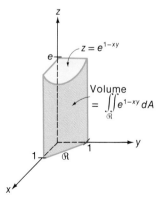

Figure 17–7 Example 2.

17-1 The Double Integral

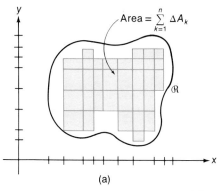
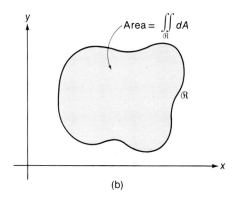

Figure 17-8 $A = \lim_{\|\mathcal{P}\| \to 0} \sum_{k=1}^{n} \Delta A_k = \iint_{\mathcal{R}} dA.$

is an approximation to the area of $\mathcal{R}$. (See Fig. 17–8a.) It is clear that this approximation gets better as $\|\mathcal{P}\| \to 0$. Thus, the area can be written as

Area

$$\text{Area} = \lim_{\|\mathcal{P}\| \to 0} \sum_{k=1}^{n} \Delta A_k = \iint_{\mathcal{R}} dA.$$

This integral generalizes the process used in Chapters 5 and 6 to calculate areas. It can be proved that the two methods give the same result when they can both be applied.

PROPERTIES OF THE DOUBLE INTEGRAL

Several properties of double integrals are listed in Theorem 17–1. All these properties hold for the approximating sums used to define the integrals. The theorem follows from the fact that the properties are preserved by taking limits.

THEOREM 17-1 Let F and G be continuous on the basic closed region $\mathcal{R}$. Let C be a constant. Then:

(a) $\quad \iint_{\mathcal{R}} [F(x, y) \pm G(x, y)]\, dA = \iint_{\mathcal{R}} F(x, y)\, dA \pm \iint_{\mathcal{R}} G(x, y)\, dA.$

(b) $\quad \iint_{\mathcal{R}} C \cdot F(x, y)\, dA = C \iint_{\mathcal{R}} F(x, y)\, dA.$

(c) If $F(x, y) \leq G(x, y)$ for all $(x, y) \in \mathcal{R}$, then

$$\iint_{\mathcal{R}} F(x, y)\, dA \leq \iint_{\mathcal{R}} G(x, y)\, dA.$$

Exercises 17–1

1 Let $\mathcal{R}$ be the closed region bounded above by the graph of $y = \pi \sin(x/2)$, $0 \leq x \leq \pi$, below by the graph of $y = \sin x$, $0 \leq x \leq \pi$, and to the right by the graph of $x = \pi + \sin 2y$, $0 \leq y \leq \pi$.

(a) Sketch the region $\mathcal{R}$.

(b) Show that $\mathcal{R}$ is a basic closed region by decomposing it into subregions of the type shown in Figure 17–1.

•(c) What is the smallest number of these special closed subregions that must be used in (b)? (*Hint:* It is easy to decompose $\mathcal{R}$ into three of these closed subregions. Is it possible to decompose it into two of them?)

2 Let $\mathcal{R}$ be the closed region described in Exercise 1. Partition the interval $[0, \pi + 1]$ on the x-axis into eight equal subintervals and the interval $[0, \pi]$ on the y-axis into eight equal subintervals.

(a) Draw a figure resembling Figure 17–3c, showing the corresponding inner partition of $\mathcal{R}$.

•(b) What is the value of the norm of the partition described in (a)?

3 Let $\mathcal{R}$ be the closed region described in Exercise 1. Express as double integrals (a) the area of $\mathcal{R}$; •(b) the volume of the solid that has $\mathcal{R}$ as its base and the graph of $z = 2x^2 + y^2$ as its upper boundary.

4–5. Let $\mathcal{R}$ be the closed rectangle with vertices $(-1, 0)$, $(-1, 2)$, $(1, 0)$, and $(1, 2)$. Let $\mathcal{P}$ be the inner partition that divides $\mathcal{R}$ into four squares of length 1 unit on a side. Let (ξ_k, η_k) be the midpoint of the kth square. Use the approximating sum S_4 to approximate the integral.

•**4** $\iint\limits_{\mathcal{R}} (x - y)^2 \, dA$

•**5** $\iint\limits_{\mathcal{R}} (x^2 + y^2) \, dA$

6–7. The double integrals in Exercises 4 and 5 are equal to the volumes of solids. Draw figures showing the solids.

6 Exercise 4

7 Exercise 5

8–9. Draw a figure like Figure 17–6b that shows the prisms that represent the approximating sums S_4 in Exercises 4 and 5.

8 Exercise 4

9 Exercise 5

•**10** Work Exercise 4 for the partition of $\mathcal{R}$ into 16 equal squares. Use the approximating sum S_{16}.

17–2 The Iterated Integral

It is almost impossible to calculate the double integral $\iint_{\mathcal{R}} F(x, y) \, dA$ from the definition. The related concept of the *iterated integral* can be used to evaluate double integrals by means of antiderivatives. We will use double integrals to set up the problems, then change to iterated integrals to calculate the values.

17-2 The Iterated Integral

PARTIAL INTEGRALS

Let F be a continuous function of x and y. The *partial integral* with respect to y, denoted by

$$\int F(x, y)\, dy,$$

is calculated by holding x fixed and integrating with respect to y. The partial integral $\int F(x, y)\, dx$ is defined by a similar process.

The process of calculating the partial integral of $F(x, y)\, dy$ reverses the process of calculating the partial derivative with respect to y. For example, since

$$\frac{\partial}{\partial y}(x^2 y^2 + 2x + y) = 2x^2 y + 1.$$

then

$$\int (2x^2 y + 1)\, dy = x^2 y^2 + y + C(x).$$

The function C is an arbitrary function of x that corresponds to the constant of integration for integrals of one variable. In most of our examples, we can use our integration formulas to calculate $\int F(x, y)\, dy$ provided we treat x as a constant.

EXAMPLE 1 (a) $\int (2x - 3x^2 y + y^3)\, dx = x^2 - x^3 y + xy^3 + C_1(y),$

where $C_1(y)$ is an arbitrary function of y.

(b) $\int (2x - 3x^2 y + y^3)\, dy = 2xy - \dfrac{3x^2 y^2}{2} + \dfrac{y^4}{4} + C_2(x).$ □

The integral

$$\int_{g_1(x)}^{g_2(x)} F(x, y)\, dy$$

corresponds to a definite integral. We first calculate the partial integral and then substitute $g_1(x)$ and $g_2(x)$ for y in the lower and upper limits of integration. As with definite integration, it is not necessary to add an arbitrary function of x to the antiderivative. It would only cancel later. The integral

$$\int_{h_1(y)}^{h_2(y)} F(x, y)\, dx$$

is defined by a similar process.

EXAMPLE 2 It follows from Example 1 that

$$\int_4^{2x} (2x - 3x^2 y + y^3)\, dy = \left[2xy - \frac{3x^2 y^2}{2} + \frac{y^4}{4} \right]_{y=4}^{y=2x}$$

$$= \left[2x(2x) - \frac{3x^2(2x)^2}{2} + \frac{(2x)^4}{4} \right] - \left[2x(4) - \frac{3x^2(4)^2}{2} + \frac{4^4}{4} \right]$$

$$= -2x^4 + 28x^2 - 8x - 64. \ \square$$

Observe, as in Example 2, that

$$\int_{g_1(x)}^{g_2(x)} F(x, y)\, dy$$

is a function of x, say

$$G(x) = \int_{g_1(x)}^{g_2(x)} F(x, y)\, dy.$$

If G is continuous for $a \leq x \leq b$, it can be integrated with respect to x over that interval, yielding

$$\int_a^b G(x)\, dx = \int_a^b \left(\int_{g_1(x)}^{g_2(x)} F(x, y)\, dy \right) dx.$$

This last integral is called an *iterated double integral*. It is usually written without the parentheses:

Iterated Integral

$$\int_a^b \int_{g_1(x)}^{g_2(x)} F(x, y)\, dy\, dx = \int_a^b \left(\int_{g_1(x)}^{g_2(x)} F(x, y)\, dy \right) dx$$

To evaluate this iterated integral, we rewrite it as

$$\int_a^b \left(\int_{g_1(x)}^{g_2(x)} F(x, y)\, dy \right) dx$$

and work out the integrals.

EXAMPLE 3

$$\int_0^1 \int_4^{2x} (2x - 3x^2 y + y^3)\, dy\, dx$$

$$= \int_0^1 \left(\int_4^{2x} (2x - 3x^2 y + y^3)\, dy \right) dx$$

$$= \int_0^1 (-2x^4 + 28x^2 - 8x - 64)\, dx \qquad \text{(by Example 2)}$$

$$= \left[\frac{-2x^5}{5} + \frac{28x^3}{3} - \frac{8x^2}{2} - 64x \right]_0^1 = \frac{-2}{5} + \frac{28}{3} - 4 - 64$$

$$= \frac{-886}{15} \approx -59.06667. \; \square$$

A similar type of iterated integral results when we first integrate with respect to x, then with respect to y. If $h_1(y)$ and $h_2(y)$ are continuous functions of y, then

$$\int_c^d \int_{h_1(y)}^{h_2(y)} F(x, y)\, dx\, dy = \int_c^d \left(\int_{h_1(y)}^{h_2(y)} F(x, y)\, dx \right) dy.$$

17–2 The Iterated Integral

EXAMPLE 4

$$\int_0^1 \int_0^y y^2 e^{xy} \, dx \, dy = \int_0^1 \left(\int_0^y y^2 e^{xy} \, dx \right) dy = \int_0^1 \left(\int_0^y y e^{xy} (y \, dx) \right) dy$$

$$= \int_0^1 \left[y e^{xy} \right]_{x=0}^{x=y} dy = \int_0^1 [y e^{y^2} - y e^0] \, dy$$

$$= \int_0^1 (y e^{y^2} - y) \, dy$$

$$= \left[\frac{e^{y^2}}{2} - \frac{y^2}{2} \right]_0^1 = \left[\frac{e}{2} - \frac{1}{2} \right] - \left[\frac{e^0}{2} - 0 \right]$$

$$= \frac{e}{2} - 1. \quad \square$$

DOUBLE AND ITERATED INTEGRALS

Let $\mathcal{R}$ be the closed region in Figure 17–9. The lower boundary of $\mathcal{R}$ is the graph of $y = g_1(x)$, and the upper boundary is the graph of $y = g_2(x)$, $a \leq x \leq b$, where g_1 and g_2 are continuous. Let $F(x, y)$ be continuous on $\mathcal{R}$. It can be shown that

$$\iint_\mathcal{R} F(x, y) \, dA = \int_a^b \int_{g_1(x)}^{g_2(x)} F(x, y) \, dy \, dx.$$

Thus the double integral can be evaluated by an appropriate iterated double integral.

We shall not prove this result in general. Instead we give a plausibility argument, which is based on volume and holds provided $F(x, y)$ is nonnegative on $\mathcal{R}$. We shall show that the volume of the solid with base $\mathcal{R}$ and the graph of F as its upper boundary surface is

$$\int_a^b \int_{g_1(x)}^{g_2(x)} F(x, y) \, dy \, dx.$$

Since this volume is known to equal $\iint_\mathcal{R} F(x, y) \, dA$, then

$$\iint_\mathcal{R} F(x, y) \, dA = \int_a^b \int_{g_1(x)}^{g_2(x)} F(x, y) \, dy \, dx.$$

Let F be nonnegative over $\mathcal{R}$ and let S be the solid that has $\mathcal{R}$ for its base and the graph of F for its upper boundary. (See Fig. 17–9b.) For each x, $a \leq x \leq b$, let $A(x)$ be the area of the cross section of S that is perpendicular to the x-axis at the point x. We know from Section 6–2 that

$$\text{Volume of } S = \int_a^b A(x) \, dx.$$

The cross section at $x = x_0$ is pictured in Figures 17–9b, c. This cross section extends across the region from $(x_0, g_1(x_0))$ to $(x_0, g_2(x_0))$. Furthermore, the height of the

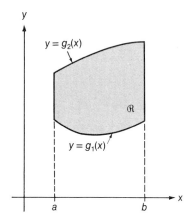

(a) The region $\mathcal{R}$

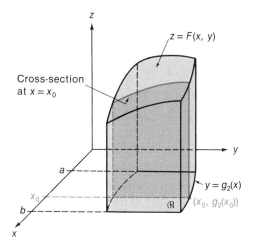

(b) The solid over $\mathcal{R}$ that is bounded above by the graph of F

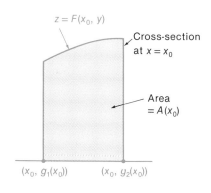

(c) Cross-section of the solid at $x = x_0$
$$\text{Area} = A(x_0) = \int_{g_1(x_0)}^{g_2(x_0)} F(x_0, y)\, dy$$

Figure 17–9 $\displaystyle\iint_{\mathcal{R}} F(x, y)\, dA = \int_a^b \int_{g_1(x)}^{g_2(x)} F(x, y)\, dy\, dx.$

upper boundary at (x_0, y) is $F(x_0, y)$. The area of the cross section is

$$A(x_0) = \int_{g_1(x_0)}^{g_2(x_0)} F(x_0, y)\, dy.$$

Since this holds for every x_0, $a \le x_0 \le b$, then

$$A(x) = \int_{g_1(x)}^{g_2(x)} F(x, y)\, dy, \qquad a \le x \le b.$$

17–2 The Iterated Integral

It follows from our work in Section 6–2 that

$$\iint_{\mathcal{R}} F(x, y)\, dA = \text{volume of } S = \int_a^b A(x)\, dx$$

$$= \int_a^b \left(\int_{g_1(x)}^{g_2(x)} F(x, y)\, dy \right) dx = \int_a^b \int_{g_1(x)}^{g_2(x)} F(x, y)\, dy\, dx.$$

EXAMPLE 5 Let $\mathcal{R}$ be the closed triangular region with vertices $(-1, 0)$, $(3, 1)$, and $(3, 3)$. Find the volume of the solid that has $\mathcal{R}$ for its base and the graph of $z = x^2 + 2y$ for its upper boundary.

Solution The lines $g_1(x) = (x + 1)/4$, $g_2(x) = 3(x + 1)/4$, and $x = 3$ are the boundaries of $\mathcal{R}$. (See Fig. 17–10.) The volume of the solid is

$$\text{Volume} = \iint_{\mathcal{R}} (x^2 + 2y)\, dA$$

$$= \int_{-1}^{3} \int_{(x+1)/4}^{3(x+1)/4} (x^2 + 2y)\, dy\, dx$$

$$= \int_{-1}^{3} \left[x^2 y + y^2 \right]_{y=(x+1)/4}^{y=3(x+1)/4} dx$$

$$= \int_{-1}^{3} \left[\frac{x^3}{2} + x^2 + x + \frac{1}{2} \right] dx$$

$$= \left[\frac{x^4}{8} + \frac{x^3}{3} + \frac{x^2}{2} + \frac{x}{2} \right]_{-1}^{3} = \frac{76}{3}. \quad \square$$

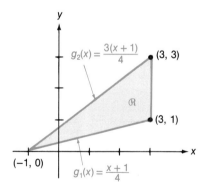

Figure 17–10 Example 5.

A similar integral formula holds if a basic closed region $\mathcal{R}$ has left and right boundary curves defined by $x = h_1(y)$ and $x = h_2(y)$, respectively, for $c \leq y \leq d$. In this case,

$$\iint_{\mathcal{R}} F(x, y)\, dA = \int_c^d \int_{h_1(y)}^{h_2(y)} F(x, y)\, dx\, dy.$$

(See Fig. 17–11.)

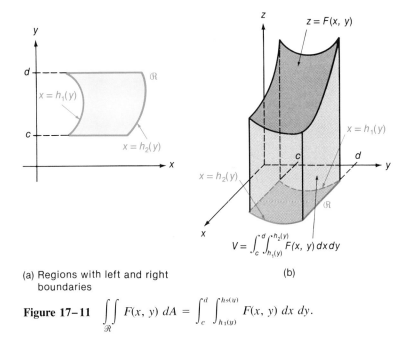

(a) Regions with left and right boundaries

(b)

Figure 17–11 $\iint_{\mathcal{R}} F(x, y)\, dA = \int_c^d \int_{h_1(y)}^{h_2(y)} F(x, y)\, dx\, dy.$

SCHEMATIC REPRESENTATION OF DOUBLE INTEGRATION BY ITERATED INTEGRALS

When we evaluate a double integral by an iterated integral, the first integration is always from a boundary curve to a boundary curve of the closed region $\mathcal{R}$. The second integration is from an extreme point on the projection of $\mathcal{R}$ on an axis to the other extreme point. This is represented schematically in Figure 17–12 for the integral

$$\int_a^b \int_{g_1(x)}^{g_2(x)} F(x, y)\, dy\, dx$$

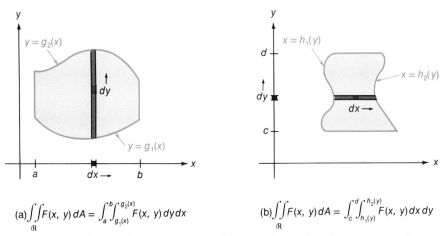

(a) $\iint_{\mathcal{R}} F(x, y)\, dA = \int_a^b \int_{g_1(x)}^{g_2(x)} F(x, y)\, dy\, dx$ (b) $\iint_{\mathcal{R}} F(x, y)\, dA = \int_c^d \int_{h_1(y)}^{h_2(y)} F(x, y)\, dx\, dy$

Figure 17–12 Schematic representation of iterated integrals using elements of area.

17–2 The Iterated Integral

The thin vertical strip in Figure 17–12a is used to represent the first integration, which is from the curve $y = g_1(x)$ to the curve $y = g_2(x)$. A typical element of area is shown as if it moved across the strip. The second integration is from a to b on the x-axis. This is represented by the element dx on that axis. Similar remarks hold for the representation of the integral

$$\int_c^d \int_{h_1(y)}^{h_2(y)} F(x, y) \, dx \, dy$$

in Figure 17–12b.

ORDER OF INTEGRATION

In some cases it may be necessary to reverse the order of integration for an iterated integral. This can be done if all the boundary curves are simple enough. We determine the boundary curves of $\mathcal{R}$ from the limits of integration of the original iterated integral. Then we graph the region. Finally, we determine the equations of the boundary that correspond to the new order of integration.

EXAMPLE 6 Change the integral

$$\int_0^2 \int_{x^2}^{2x} (2x - y) \, dy \, dx$$

to an equivalent integral with a different order of integration. Show, by carrying out the integrations, that the final values of the integrals are the same.

Solution The upper and lower boundaries of $\mathcal{R}$ are the graphs of $y = 2x$ and $y = x^2$, respectively, $0 \le x \le 2$. (See Fig. 17–13a.)

Observe that the left- and right-hand boundaries of $\mathcal{R}$ are the graphs of $x = y/2$ and $x = \sqrt{y}$, respectively, $0 \le y \le 4$. (See Fig. 17–13b.) Therefore,

$$\int_0^2 \int_{x^2}^{2x} (2x - y) \, dy \, dx = \iint_{\mathcal{R}} (2x - y) \, dA = \int_0^4 \int_{y/2}^{\sqrt{y}} (2x - y) \, dx \, dy.$$

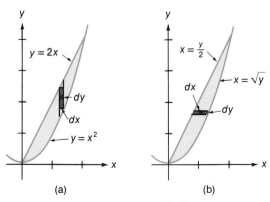

Figure 17–13 Example 6. $\int_0^2 \int_{x^2}^{2x} (2x - y) \, dy \, dx = \int_0^4 \int_{y/2}^{y} (2x - y) \, dx \, dy.$

We show, as a check on our work, the main steps in the evaluation of the two integrals:

$$\int_0^2 \int_{x^2}^{2x} (2x - y)\, dy\, dx = \int_0^2 \left[2xy - \frac{y^2}{2}\right]_{y=x^2}^{y=2x} dx$$

$$= \int_0^2 \left(\frac{x^4}{2} - 2x^3 + 2x^2\right) dx$$

$$= \left[\frac{x^5}{10} - \frac{x^4}{2} + \frac{2x^3}{3}\right]_0^2$$

$$= \frac{32}{10} - 8 + \frac{16}{3} = \frac{8}{15}$$

$$\int_0^4 \int_{y/2}^{\sqrt{y}} (2x - y)\, dx\, dy = \int_0^4 \left[x^2 - xy\right]_{x=y/2}^{x=\sqrt{y}} dy$$

$$= \int_0^4 \left(\frac{y^2}{4} - y^{3/2} + y\right) dy = \left[\frac{y^3}{12} - \frac{2y^{5/2}}{5} + \frac{y^2}{2}\right]_0^4$$

$$= \frac{64}{12} - \frac{64}{5} + 8 = \frac{8}{15}. \quad \square$$

Example 7 shows how reversing the order of integration can help us to evaluate certain iterated integrals.

EXAMPLE 7   The integral

$$\int_0^1 \int_{y^2}^1 e^{x^2} y\, dx\, dy$$

cannot be evaluated by the standard formulas. When we reverse the order of integration, however, we get

$$\int_0^1 \int_0^{\sqrt{x}} e^{x^2} y\, dy\, dx = \int_0^1 \left[\frac{e^{x^2} y^2}{2}\right]_{y=0}^{y=\sqrt{x}} dx$$

$$= \frac{1}{2} \int_0^1 e^{x^2} x\, dx$$

$$= \frac{1}{4} \left[e^{x^2}\right]_0^1 = \frac{e - 1}{4}. \quad \square$$

Exercises 17–2

1–6.   Evaluate the iterated integrals.

- 1 $\int_0^1 \int_0^{1-x} (2x + 1)\, dy\, dx$

2   $\int_0^4 \int_0^x x\, dy\, dx$

- 3 $\int_0^2 \int_{-y}^{y^2} xy\, dx\, dy$

4   $\int_2^3 \int_0^{\sqrt{y}} \frac{x}{y^2 - 3}\, dx\, dy$

- 5 $\int_1^2 \int_0^{2y} y^2 e^{xy}\, dx\, dy$

6   $\int_0^1 \int_{2x}^{3x} e^{x+y}\, dy\, dx$

Exercises 17–2

7–11. Let $\mathcal{R}$ be the closed region enclosed by the graphs of the given equations. Let $\mathcal{S}$ be the solid that has $\mathcal{R}$ as its base and the graph of $z = F(x, y)$ as its upper boundary surface. **(a)** Sketch the region $\mathcal{R}$. **(b)** Express the area of $\mathcal{R}$ and the volume of $\mathcal{S}$ as double integrals. **(c)** Use iterated integrals to calculate the area of $\mathcal{R}$ and the volume of $\mathcal{S}$.

- **7** $\mathcal{R}: x^2 + y^2 = 4,\ x \geq 0,\ y \geq 0;\ z = x + y$

   8 $\mathcal{R}: y = x,\ y = x^2;\ z = x$

- **9** $\mathcal{R}: y = 3x,\ y = 3,\ x = 3;\ z = x^2$

  10 $\mathcal{R}$: The closed rectangle with vertices $(0, 0)$, $(0, 2)$, $(4, 0)$, $(4, 2)$; $z = xy$

- **11** $\mathcal{R}: x = 0,\ y = 0,\ x = 2 - 2y;\ z = 4 - x^2 - 4y^2$

  12   Use an iterated integral to calculate $\int_{\mathcal{R}}\int (\cos x + e^y)\, dA$ for the region $\mathcal{R}$ in Example 1 of Section 17–1.

13–18. Change the iterated integrals to equivalent integrals with a different order of integration. Carry out the integration to show that the final values are the same. (*Hint:* Sketch the region $\mathcal{R}$ defined by the limits of integration.)

- **13** $\displaystyle\int_0^2 \int_0^{x^2} xy\, dy\, dx$ **14** $\displaystyle\int_0^1 \int_y^{\sqrt{y}} y\, dx\, dy$

- **15** $\displaystyle\int_0^2 \int_0^{\sqrt{4-y^2}} x\, dx\, dy$ **16** $\displaystyle\int_0^3 \int_{x^2}^{3x} 2x^2 y\, dy\, dx$

- **17** $\displaystyle\int_0^2 \int_1^{e^x} dy\, dx$ **18** $\displaystyle\int_0^1 \int_{\sqrt{y}}^1 dx\, dy$

- **19**   Evaluate the integral after changing to a different order of integration.

$$\int_0^{\sqrt{\pi}} \int_y^{\sqrt{\pi}} \sin(x^2)\, dx\, dy$$

---

20   Let f and g be continuous functions. Show that

$$\int_a^b \int_c^d f(x)g(y)\, dy\, dx = \left(\int_a^b f(x)\, dx\right)\left(\int_c^d g(y)\, dy\right).$$

21   Let f and g be continuous functions. Extend the result of Exercise 20 to improper integrals. Show that

$$\int_a^\infty \int_c^\infty f(x)g(y)\, dy\, dx = \left(\int_a^\infty f(x)\, dx\right)\left(\int_c^\infty g(y)\, dy\right)$$

provided the integrals on the right converge. [Remark: We define the integral on the left by

$$\int_a^\infty \int_c^\infty f(x)g(y)\, dy\, dx = \int_a^\infty \left(\int_c^\infty f(x)g(y)\, dy\right) dx$$

provided the integral on the right-hand side of this expression converges.]

22 Use the results of Exercises 20 and 21 to evaluate

● (a) $\int_0^2 \int_1^3 e^x e^y \, dy \, dx$ (b) $\int_1^\infty \int_1^\infty e^{-x-y} \, dx \, dy$

17–3 Volume and Mass

Let S be the solid in three-dimensional space that has the graphs of $z = F(x, y)$ and $z = G(x, y)$ as its upper and lower boundaries, where F and G are continuous functions. Let the projection of S on the xy-plane be the basic closed region $\mathcal{R}$. Then the volume of S is defined by the integral,

Volume

$$\text{Volume of } S = \iint_\mathcal{R} [F(x, y) - G(x, y)] \, dA.$$

This integral formula can be justified by the following argument. Let $\mathcal{P}$ be an inner partition of $\mathcal{R}$ into rectangles $R_1, R_2, \ldots, R_n$ of areas $\Delta A_1, \Delta A_2, \ldots, \Delta A_n$, respectively. Let (ξ_k, η_k) be a point in $\Delta \mathcal{R}_k$. Then

$$[F(\xi_k, \eta_k) - G(\xi_k, \eta_k)] \Delta A_k$$

is the volume of a rectangular solid above R_k that extends from the lower boundary surface of S to the upper boundary surface. (See Fig. 17–14a.)

The sum

$$S_n = \sum_{k=1}^n [F(\xi_k, \eta_k) - G(\xi_k, \eta_k)] \Delta A_k,$$

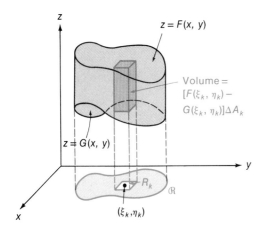
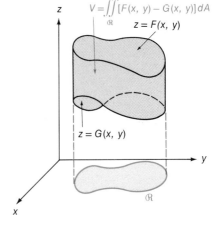

(a) Volume of prism $= [F(\xi_k, \eta_k) - G(\xi_k, \eta_k)] \Delta A_k$ (b) Volume of solid $= \iint_\mathcal{R} [F(x, y) - G(x, y)] \, dA$

Figure 17–14 Volume of a solid.

17–3 Volume and Mass

the sum of the volumes of all such rectangular solids, is an approximation to the number that we normally call the volume of S. We define the volume of S to be the limit of all such sums as $\|\mathcal{P}\| \to 0$:

$$\text{Volume} = \lim_{\|\mathcal{P}\| \to 0} \sum_{k=1}^{n} [F(\xi_k, \eta_k) - G(\xi_k, \eta_k)] \Delta A_k$$

$$\text{Volume} = \iint_{\mathcal{R}} [F(x, y) - G(x, y)] \, dA.$$

Observe that this definition coincides with the definition in Section 17–1 when the lower surface of S is the xy-plane. In that case, $G(x, y) = 0$ for all (x, y) in $\mathcal{R}$ so that

$$\text{Volume} = \iint_{\mathcal{R}} [F(x, y) - 0] \, dA = \iint_{\mathcal{R}} F(x, y) \, dA.$$

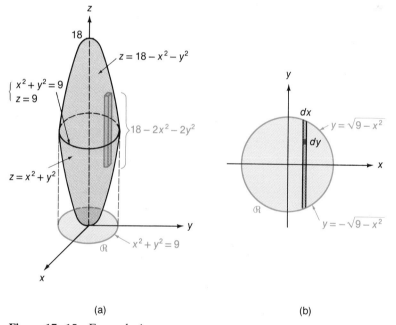

(a)   (b)

Figure 17–15 Example 1.

EXAMPLE 1 Let S be the solid bounded above by the paraboloid $z = 18 - x^2 - y^2$ and below by the paraboloid $z = x^2 + y^2$. Find the volume of S.

Solution The paraboloids intersect when

$$x^2 + y^2 = 18 - x^2 - y^2$$
$$x^2 + y^2 = 9.$$

Thus, the curve of intersection is the circle

$$x^2 + y^2 = 9, \quad z = 9.$$

See (Fig. 17–15a.) It follows that the projection of S on the xy-plane is the closed circular disk $\mathcal{R}$ bounded by the circle

$$x^2 + y^2 = 9, \quad z = 0.$$

The volume of the solid is

$$V = \iint_{\mathcal{R}} [(18 - x^2 - y^2) - (x^2 + y^2)] \, dA = \iint_{\mathcal{R}} (18 - 2x^2 - 2y^2) \, dA.$$

To express the volume as an iterated integral, we note that the boundaries of $\mathcal{R}$ are the graphs of $y = \sqrt{9 - x^2}$ and $y = -\sqrt{9 - x^2}$, $-3 \leq x \leq 3$. (See Fig. 17–15b.) Thus,

$$V = \iint_{\mathcal{R}} (18 - 2x^2 - 2y^2) \, dA = \int_{-3}^{3} \int_{-\sqrt{9-x^2}}^{\sqrt{9-x^2}} (18 - 2x^2 - 2y^2) \, dy \, dx.$$

In this particular example, a considerable saving in labor can be made by using the symmetry of S about the xz- and yz-planes. Since the total volume of S is four times the volume of the portion in Octant I, then

$$V = 4 \int_{0}^{3} \int_{0}^{\sqrt{9-x^2}} (18 - 2x^2 - 2y^2) \, dy \, dx$$

$$= 8 \int_{0}^{3} \int_{0}^{\sqrt{9-x^2}} (9 - x^2 - y^2) \, dy \, dx$$

$$= 8 \int_{0}^{3} \left[9y - x^2 y - \frac{y^3}{3} \right]_{0}^{\sqrt{9-x^2}} dx = \frac{2}{3} \cdot 8 \int_{0}^{3} (9 - x^2)^{3/2} \, dx$$

$$= \frac{16}{3} \left[\frac{x}{4} (9 - x^2)^{3/2} + \frac{27x}{8} \sqrt{9 - x^2} + \frac{243}{8} \sin^{-1} \frac{x}{3} \right]_{0}^{3} \quad \text{(by I83)}$$

$$= \frac{16}{3} \left[0 + 0 + \frac{243}{8} \sin^{-1} 1 \right] - \frac{16}{3} [0 + 0 + 0] = \frac{16}{3} \cdot \frac{243}{8} \cdot \frac{\pi}{2}$$

$$= 81\pi \text{ cu units.} \quad \square$$

Thus far we have worked with the projections of the solids on the xy-plane. In certain examples it may be more convenient to use projections on the xz- or yz-planes.

EXAMPLE 2 Let S be the solid in Octant I bounded by the coordinate planes and the paraboloid $y = 16 - x^2 - z^2$. Calculate the volume of S.

Solution Let $\mathcal{R}$ be the projection of S on the xz-plane. Then $\mathcal{R}$ is the closed region bounded by the x-axis, the z-axis and the quarter circle

$$x^2 + z^2 = 16, \quad x \geq 0, \quad z \geq 0.$$

(See Figs. 17–16a, b.)
The volume of S is

$$V = \iint_{\mathcal{R}} (16 - x^2 - z^2) \, dA = \int_{0}^{4} \int_{0}^{\sqrt{16-x^2}} (16 - x^2 - z^2) \, dz \, dx$$

$$= \int_{0}^{4} \left[16z - x^2 z - \frac{z^3}{3} \right]_{z=0}^{z=\sqrt{16-x^2}} dx = \frac{2}{3} \int_{0}^{4} (16 - x^2)^{3/2} \, dx$$

17-3 Volume and Mass

$$= \frac{2}{3} \left[\frac{x}{4} (16 - x^2)^{3/2} + \frac{48x}{8} \sqrt{16 - x^2} + \frac{3 \cdot 256}{8} \sin^{-1} \frac{x}{4} \right]_0^4 \quad \text{(by I83)}$$

$$= \frac{2}{3} [0 + 0 + 96 \sin^{-1} 1] - \frac{2}{3} [0 + 0 + 96 \sin^{-1} 0]$$

$$= \frac{2}{3} \cdot 96 \cdot \frac{\pi}{2} = 32\pi. \quad \square$$

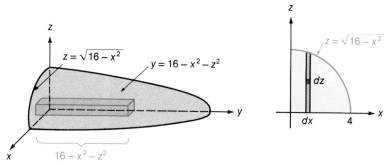

Figure 17-16 Example 2.

PLANE LAMINAS

Recall that a plane lamina is a thin slab that covers a closed region in the xy-plane. We have already worked with laminas of constant density (mass per unit of surface area) in Section 6-8. We now extend our results to laminas of variable density.

Let $\mathcal{R}$ be a basic closed region in the xy-plane that is covered by a lamina of variable density. Suppose that the density at each point $(x, y) \in \mathcal{R}$ is given by $\delta(x, y)$, where δ is continuous on $\mathcal{R}$. Then the total mass of the lamina is given by the double integral

Mass of Lamina

$$\text{Mass} = \iint_{\mathcal{R}} \delta(x, y) \, dA.$$

In relating the above integral formula to the mass, suppose that $\mathcal{P}$ is an inner partition of $\mathcal{R}$ into rectangles $R_1, R_2, \ldots, R_n$ of areas $\Delta A_1, \Delta A_2, \ldots, \Delta A_n$, respectively. Let (ξ_k, η_k) be a point in R_k. Then $\delta(\xi_k, \eta_k) \Delta A_k$ is approximately equal to the mass of the portion of the lamina that is directly above R_k. It follows that

$$S_n = \sum_{k=1}^{n} \delta(\xi_k, \eta_k) \Delta A_k$$

is an approximation to the total mass of the lamina.

Since δ is continuous, this approximation improves as the norm of the partition approaches zero. Thus,

$$\text{Mass of lamina} = \lim_{\|\mathcal{P}\| \to 0} \sum_{k=1}^{n} \delta(\xi_k, \eta_k) \Delta A_k$$

$$= \iint_{\mathcal{R}} \delta(x, y) \, dA.$$

EXAMPLE 3   A lamina covers the region $\mathcal{R}$ in the xy-plane that is bounded above by the graph of $y = x$ and below by the graph of $y = x^2$. The density of the lamina at the point (x, y) is

$$\delta(x, y) = \sqrt{xy}.$$

Find the mass.

Solution   The region $\mathcal{R}$ is pictured in Figure 17–17. The mass of the lamina is

$$\text{Mass} = \iint_{\mathcal{R}} \delta(x, y) \, dA = \int_0^1 \int_{x^2}^{x} \sqrt{xy} \, dy \, dx$$

$$= \int_0^1 \left[x^{1/2} \frac{y^{3/2}}{3/2} \right]_{y=x^2}^{y=x} dx = \frac{2}{3} \int_0^1 [x^2 - x^{7/2}] \, dx$$

$$= \frac{2}{3} \left[\frac{x^3}{3} - \frac{x^{9/2}}{9/2} \right]_0^1 = \frac{2}{3} \left[\frac{1}{3} - \frac{2}{9} \right] = \frac{2}{27}. \quad \square$$

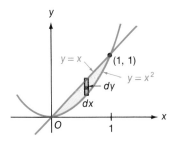

Figure 17–17   Example 3.

Exercises 17–3

1–7.   Find the volume of the solid bounded by the graphs of the equations.

• 1   The planes $x = 0$, $y = 2$, $y = x$, $z = 0$, and $z = 2 + x$

• 2   The planes $x = 0$, $y = 0$, $z = 0$, and $3x + 2y + 6z = 6$

• 3   The solid in Octant I bounded by the cylinder $x^2 + y^2 = 9$, the xy-plane and the plane $z = 2x$

• 4   The paraboloid $z = x^2 + y^2$ and the plane $z = 4$

• 5   The cylinder $x = 4y^2$ and the planes $z = 2$, $z = 0$, and $x = 8$

17–4 Moments and Centroids

- **6** The planes $x = 0$, $y = 0$, $x + y + z = 3$, and $z = 3x + 5y - 9$
- **7** The cylinders $x^2 + y^2 = 9$ and $y^2 + z^2 = 9$
- **8** (a) Interpret the integral

$$\int_{-2}^{0} \int_{0}^{x+2} \frac{y - x + 2}{\sqrt{2}} \, dy \, dx$$

as the volume of a solid. Describe the solid.
(b) Describe a plane lamina with variable density that has the integral in (a) equal to its mass. What is the density function?

9–14. A lamina covers the region $\mathcal{R}$ that is bounded by the graphs of the given equations. Its density at (x, y) is given by $\delta(x, y)$. Find the area and the mass of the lamina.

- **9** $\mathcal{R}$: $y = 0$, $y = 2$, $x = 0$, and $x = 3$; $\delta(x, y) = ky$, where k is a constant
- **10** $\mathcal{R}$: $y = \sqrt{x}$, $y = 0$, and $x = 4$; $\delta(x, y) = x + y$
- **11** $\mathcal{R}$: $y = \sqrt[3]{x}$, $x = 8$, $y = 0$; $\delta(x, y) = y^2$
- **12** $\mathcal{R}$: $y = 2 - 3x^2$, $3x + 2y = 1$; $\delta(x, y) = k$, a constant.
- **13** $\mathcal{R}$: The circle $x^2 + y^2 = 1$; $\delta(x, y) = |x|$
- **14** $\mathcal{R}$: The triangle with vertices $(0, 0)$, $(1, 0)$ and $(1, 1)$; $\delta(x, y) = \dfrac{y}{x + 1}$

17–4 Moments and Centroids

The work of Section 6–8 can be extended to cover moments and centroids of laminas of variable density. As we shall see, the new development is much simpler than the original one and includes the previous results as special cases. (See Exercise 9.) (In other words, these topics can be handled better by using double integrals than definite integrals.)

Let the basic closed region $\mathcal{R}$ in the xy-plane be covered by a lamina. Let the density of the lamina at the point $(x, y) \in \mathcal{R}$ be given by $\delta(x, y)$, where δ is continuous on $\mathcal{R}$. Let $\mathcal{P}$ be an inner partition of $\mathcal{R}$ into closed rectangles $R_1, R_2, \ldots, R_n$ of areas $\Delta A_1, \Delta A_2, \ldots, \Delta A_n$. Let (ξ_k, η_k) be a point in R_k.

The mass of the part of the lamina above R_k is approximately equal to $\delta(\xi_k, \eta_k) \Delta A_k$. Since this small portion of the lamina is approximately ξ_k units from the y-axis, then its moment about the y-axis is approximately equal to

$$\underbrace{\xi_k}_{\text{distance}} \cdot \underbrace{\delta(\xi_k, \eta_k) \Delta A_k}_{\text{mass}}$$

(See Fig. 17–18a.)

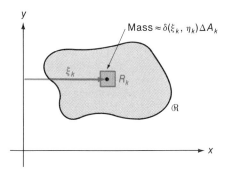
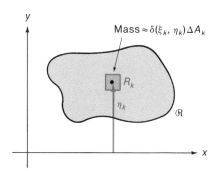

(a) First moment of rectangle about
y-axis $\approx \xi_k \, \delta(\xi_k, \eta_k) \Delta A_k$
Second moment of rectangle about
y-axis $\approx \xi_k^2 \, \delta(\xi_k, \eta_k) \Delta A_k$

(b) First moment of rectangle about
x-axis $\approx \eta_k \, \delta(\xi_k, \eta_k) \Delta A_k$
Second moment of rectangle about
x-axis $\approx \eta_k^2 \, \delta(\xi_k, \eta_k) \Delta A_k$

Figure 17–18 Moments about the coordinate axes:

$$M_y = \iint_{\mathcal{R}} x \delta(x, y) \, dA \qquad M_x = \iint_{\mathcal{R}} y \delta(x, y) \, dA$$

$$I_y = \iint_{\mathcal{R}} x^2 \delta(x, y) \, dA \qquad I_x = \iint_{\mathcal{R}} y^2 \delta(x, y) \, dA$$

The sum $S_n = \sum_{k=1}^{n} \xi_k \, \delta(\xi_k, \eta_k) \, \Delta A_k$ is an approximation to the moment of the entire lamina about the y-axis. If we calculate the limit of this sum as the norm of the partition approaches zero, we get the exact value of the moment:

$$M_y = \lim_{\|\mathcal{P}\| \to 0} \sum_{k=1}^{n} \xi_k \, \delta(\xi_k, \eta_k) \, \Delta A_k = \iint_{\mathcal{R}} x \, \delta(x, y) \, dA.$$

A similar argument can be used to show that the moment about the x-axis is

$$M_x = \lim_{\|\mathcal{P}\| \to 0} \sum_{k=1}^{n} \eta_k \, \delta(\xi_k, \eta_k) \, \Delta A_k = \iint_{\mathcal{R}} y \, \delta(x, y) \, dA.$$

(See Fig. 17–18b.)

The integral formulas for the mass and the moments about the coordinate axes can be remembered most easily by using elements of area. A typical element of area is pictured in Figure 17–19. Its area is $dA = dx \, dy$. Its mass is $\delta(x, y) \, dA$, its moment about the x-axis is $y \, \delta(x, y) \, dA$, and its moment about the y-axis is $x \, \delta(x, y) \, dA$. The corresponding integrals give the area, mass, and moments of the lamina:

Integrals for Area, Mass, Moments

$$\text{Area} = \iint_{\mathcal{R}} dA \qquad \text{Mass} = \iint_{\mathcal{R}} \delta(x, y) \, dA$$

$$M_x = \iint_{\mathcal{R}} y \, \delta(x, y) \, dA \qquad M_y = \iint_{\mathcal{R}} x \, \delta(x, y) \, dA.$$

17–4 Moments and Centroids

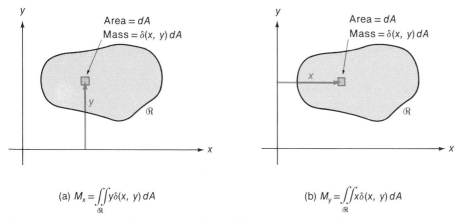

Figure 17–19 Schematic representation of the moment formulas.

The *centroid* (*center of mass*) of a lamina is the point $(\bar{x}, \bar{y})$ where

Coordinates of Centroid

$$\bar{x} = \frac{M_y}{\text{mass}} \quad \text{and} \quad \bar{y} = \frac{M_x}{\text{mass}}.$$

It is the point at which the mass could be concentrated without changing the moments about the coordinate axes.

EXAMPLE 1 (See Example 3 of Section 17–3.) A lamina covers the region $\mathcal{R}$ bounded above by the graph of $y = x$ and below by the graph of $y = x^2$. The density of the lamina at the point (x, y) is $\delta(x, y) = \sqrt{xy}$. Find its centroid.

Solution The region $\mathcal{R}$ is pictured in Figure 17–17. We saw in Example 3 of Section 17–3 that

$$\text{Mass} = \iint_{\mathcal{R}} \delta(x, y)\, dA = \int_0^1 \int_{x^2}^x \sqrt{xy}\, dy\, dx = \frac{2}{27}.$$

The moment about the x-axis is

$$M_x = \iint_{\mathcal{R}} y\, \delta(x, y)\, dA = \int_0^1 \int_{x^2}^x y\sqrt{xy}\, dy\, dx$$

$$= \int_0^1 \int_{x^2}^x x^{1/2} y^{3/2}\, dy\, dx = \int_0^1 \left[\frac{2x^{1/2} y^{5/2}}{5}\right]_{y=x^2}^{y=x} dx$$

$$= \frac{2}{5} \int_0^1 (x^3 - x^{11/2})\, dx = \frac{2}{5}\left[\frac{x^4}{4} - \frac{2x^{13/2}}{13}\right]_0^1$$

$$= \frac{2}{5}\left[\frac{1}{4} - \frac{2}{13}\right] = \frac{1}{26}.$$

Similarly,

$$M_y = \iint_{\mathcal{R}} x\, \delta(x, y)\, dA = \int_0^1 \int_{x^2}^x x\sqrt{xy}\, dy\, dx = \frac{1}{22}.$$

Then

$$\bar{x} = \frac{M_y}{\text{mass}} = \frac{1/22}{2/27} = \frac{27}{44} \approx 0.61364,$$

$$\bar{y} = \frac{M_x}{\text{mass}} = \frac{1/26}{2/27} = \frac{27}{52} \approx 0.51923.$$

The centroid is $(\frac{27}{44}, \frac{27}{52})$. □

MOMENTS AND CENTROIDS OF PLANE REGIONS

The moments of the plane region $\mathcal{R}$ are defined as if the density function were the constant function $\delta(x, y) = 1$:

Moments of Plane Region

$$M_x = \iint_{\mathcal{R}} y \, dA \quad \text{and} \quad M_y = \iint_{\mathcal{R}} x \, dA.$$

The *centroid* of the plane region $\mathcal{R}$ is $(\bar{x}, \bar{y})$, where

$$\bar{x} = \frac{M_y}{\text{area}} \quad \text{and} \quad \bar{y} = \frac{M_x}{\text{area}}.$$

EXAMPLE 2 Let $\mathcal{R}$ be the plane region bounded above by the circle $x^2 + y^2 = 2$ and below by the line $y = 1$. Find the centroid of $\mathcal{R}$.

Solution The region $\mathcal{R}$ is pictured in Figure 17–20. The area of $\mathcal{R}$ is

$$A = \iint_{\mathcal{R}} dA = \int_{-1}^{1} \int_{1}^{\sqrt{2-x^2}} dy \, dx$$

$$= \int_{-1}^{1} [\sqrt{2 - x^2} - 1] \, dx$$

$$= \left[\frac{x}{2} \sqrt{2 - x^2} + \sin^{-1} \frac{x}{\sqrt{2}} - x \right]_{-1}^{1} = \frac{\pi}{2} - 1 \quad \text{(by I72).}$$

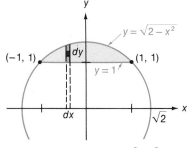

Figure 17–20 Example 2.

17-4 Moments and Centroids

The moment about the x-axis is

$$M_x = \iint_{\mathcal{R}} y \, dA = \int_{-1}^{1} \int_{1}^{\sqrt{2-x^2}} y \, dy \, dx = \int_{-1}^{1} \left[\frac{y^2}{2}\right]_{y=1}^{y=\sqrt{2-x^2}} dx$$

$$= \frac{1}{2} \int_{-1}^{1} (1-x^2) \, dx = \frac{1}{2} \left[x - \frac{x^3}{3}\right]_{-1}^{1} = \frac{2}{3}.$$

The moment about the y-axis is

$$M_y = \iint_{\mathcal{R}} x \, dA = \int_{-1}^{1} \int_{1}^{\sqrt{2-x^2}} x \, dy \, dx,$$

which reduces to

$$M_y = 0.$$

The centroid is $(\bar{x}, \bar{y})$, where

$$\bar{x} = \frac{M_y}{\text{area}} = 0,$$

$$\bar{y} = \frac{M_x}{\text{area}} = \frac{2/3}{(\pi/2) - 1} = \frac{4}{3(\pi - 2)} \approx 1.16796. \ \square$$

MOMENT OF INERTIA

Moment of Inertia of Particle

The *moment of inertia* (or *second moment*) of a particle about a line is defined to be

$$d^2 m,$$

where d is the distance from the particle to the line and m is the mass of the particle.

This definition can be extended to define moments of inertia of laminas about coordinate axes. Let a lamina of continuous density $\delta(x, y)$ cover the basic closed region $\mathcal{R}$. Partition the lamina into closed rectangles $R_1, R_2, \ldots, R_n$, of areas $\Delta A_1, \Delta A_2, \ldots, \Delta A_n$, by the inner partition $\mathcal{P}$. Let $(\xi_1, \eta_1), (\xi_2, \eta_2), \ldots, (\xi_n, \eta_n)$ be the midpoints of $R_1, R_2, \ldots, R_n$.

If R_k is small, then the moment of inertia about the y-axis of the part of the lamina above R_k is approximately equal to

$$\underbrace{\xi_k^2}_{\substack{\text{square of} \\ \text{distance} \\ \text{from } y\text{-axis}}} \cdot \underbrace{\delta(\xi_k, \eta_k) \, \Delta A_k}_{\text{mass}},$$

(See Fig. 17–18a) Thus, the sum

$$S_n = \sum_{k=1}^{n} \xi_k^2 \, \delta(\xi_k, \eta_k) \, \Delta A_k$$

is approximately equal to the sum of the moments of inertia of $R_1, R_2, \ldots, R_n$ about the y-axis.

We define the *moment of inertia (second moment) of the lamina* about the y-axis to be the number I_y that is the limit of the sum S_n as the norm of the partition approaches zero:

I_y = moment of inertia about y-axis

$$= \lim_{\|\mathcal{P}\|\to 0} \sum_{k=1}^{n} \xi_k^2 \, \delta(\xi_k, \eta_k) \, \Delta A_k.$$

Integral for Moment of Inertia About y-axis

$$I_y = \iint_{\mathcal{R}} x^2 \, \delta(x, y) \, dA.$$

The *moment of inertia about the x-axis* is defined similarly (see Fig. 17–18b):

I_x = moment of inertia about the x-axis

$$= \lim_{\|\mathcal{P}\|\to 0} \sum_{k=1}^{n} \eta_k^2 \, \delta(\xi_k, \eta_k) \, \Delta A_k.$$

$$I_x = \iint_{\mathcal{R}} y^2 \, \delta(x, y) \, dA.$$

EXAMPLE 3   A lamina of constant density $\delta(x, y) = 1$ covers the closed rectangular region with vertices $(0, 0)$, $(4, 0)$, $(0, 3)$, and $(4, 3)$. Find I_x and I_y.

Solution   The region is pictured in Figure 17–21. The moments of inertia are

$$I_x = \iint_{\mathcal{R}} y^2 \, \delta(x, y) \, dA = \int_0^4 \int_0^3 y^2 \, dy \, dx$$

$$= \int_0^4 \left[\frac{y^3}{3}\right]_0^3 dx = 9 \int_0^4 dx = 36.$$

$$I_y = \iint_{\mathcal{R}} x^2 \, \delta(x, y) \, dA = \int_0^4 \int_0^3 x^2 \, dy \, dx$$

$$= \int_0^4 \left[x^2 y\right]_{y=0}^{y=3} dx = 3 \int_0^4 x^2 \, dx = 3 \left[\frac{x^3}{3}\right]_0^4 = 64. \quad \square$$

APPLICATION. KINETIC ENERGY

A close relation exists between the moment of inertia and the kinetic energy of a spinning object. (The *kinetic energy* is the energy possessed by a moving object because of its motion. It equals the work that must be done to stop the motion.) If a particle of mass m moves with a speed of v units per second, then its kinetic energy is defined to be

Kinetic Energy of Moving Object

Kinetic energy $= \frac{1}{2}mv^2$.

17-4 Moments and Centroids

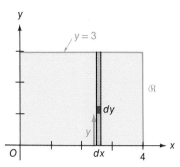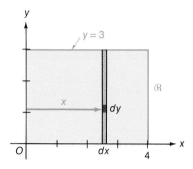

Figure 17-21 Example 3.

This formula can easily be applied to a particle that revolves about an axis in a circular orbit. If r is the radius and ω the angular velocity, then the speed is $v = \omega r$ and

$$\text{Kinetic energy} = \tfrac{1}{2}v^2 m = \tfrac{1}{2}\omega^2 r^2 m.$$

(See Fig. 17-22a.) Since $I = r^2 m$ is the moment of inertia of the particle, then

Kinetic Energy of Spinning Particle

$$\text{Kinetic energy} = \tfrac{1}{2}\omega^2 I = \tfrac{1}{2}\omega^2 r^2 m.$$

Let a lamina of continuous density $\delta(x, y)$ cover the basic closed region $\mathcal{R}$ in the xy-plane. Now revolve this lamina about the y-axis in three-dimensional space in a circular orbit. (See Fig. 17-22b.)

Let R_k be a typical rectangle in an inner partition of $\mathcal{R}$. Let (ξ_k, η_k) be the midpoint of R_k. It follows from the formula for kinetic energy, $\tfrac{1}{2}\omega^2 r^2 m$, that the kinetic energy of the portion of the lamina above R_k is approximately equal to

$$\tfrac{1}{2}\omega^2 \xi_k^2 \, \delta(\xi_k, \eta_k) \, \Delta A_k,$$

where ω is the angular velocity of the lamina.

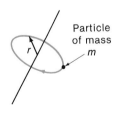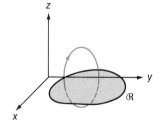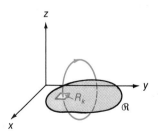

(a) Kinetic energy of particle $= \tfrac{1}{2}m\omega^2 r^2$ where m is the mass and ω is the angular velocity

(b) Lamina in xy-plane is revolved about the y-axis

(c) Kinetic energy of the portion of the lamina over R_k is approximately equal to $\tfrac{1}{2}\omega^2 \xi_k^2 \delta(\xi_k, \eta_k)\Delta A_k$

Figure 17-22 Kinetic energy of lamina revolved about y-axis is $\dfrac{1}{2}\omega^2 \iint_{\mathcal{R}} x^2 \delta(x, y)\, dA = \dfrac{1}{2}\omega^2 I_y$.

The total kinetic energy of the lamina is approximately equal to

$$S_n = \sum_{k=1}^{n} \tfrac{1}{2}\omega^2 \xi_k^2 \, \delta(\xi_k, \eta_k) \, \Delta A_k.$$

Since δ is continuous, this approximation improves as the norm of the partition approaches zero. Thus

Kinetic Energy of Spinning Lamina

$$\text{Kinetic energy} = \lim_{\|\mathcal{P}\| \to 0} \sum_{k=1}^{n} \tfrac{1}{2}\omega^2 \xi_k^2 \, \delta(\xi_k, \eta_k) \, \Delta A_k$$

$$= \tfrac{1}{2}\omega^2 \iint_{\mathcal{R}} x^2 \, \delta(x, y) \, dA = \tfrac{1}{2}\omega^2 I_y.$$

A similar result holds if the lamina is revolved about any line. In particular, if it is revolved about the x-axis with an angular velocity of ω, then

$$\text{Kinetic energy} = \tfrac{1}{2}\omega^2 I_x.$$

RADIUS OF GYRATION

Radius of Gyration

Imagine that a lamina, such as the one discussed above, is revolved about the y-axis with an angular velocity of ω revolutions per second. In certain theoretical considerations it is convenient to imagine that the mass of the lamina is concentrated at a single point that spins about the y-axis with the same angular velocity. The distance that this particle must be located from the y-axis so as to have the same kinetic energy as the original lamina is called the *radius of gyration of the lamina about the y-axis*, denoted by R_y.

It is simple to calculate R_y. Since the kinetic energies of the two systems must be the same, then

$$\underbrace{\frac{\omega^2}{2} R_y^2 \cdot \text{mass}}_{\substack{\text{kinetic energy of} \\ \text{``mass'' particle}}} = \underbrace{\frac{\omega^2}{2} I_y}_{\substack{\text{kinetic energy} \\ \text{of lamina}}}$$

so that

$$R_y^2 = \frac{I_y}{\text{mass}},$$

Radius of Gyration About y-axis

$$R_y = \sqrt{\frac{I_y}{\text{mass}}}.$$

Exercises 17–4

The radius of gyration about the x-axis is calculated similarly:

Radius of Gyration About x-axis
$$R_x = \sqrt{\frac{I_x}{\text{mass}}}.$$

EXAMPLE 4 Find R_x and R_y for the lamina in Example 3.

Solution The lamina is the rectangle shown in Figure 17–21. We saw in Example 3 that $I_x = 36$ and $I_y = 64$. Since the density function is $\delta(x, y) = 1$, the mass is

$$\text{Mass} = \delta \cdot \text{area} = 1 \cdot 12 = 12.$$

The radii of gyration are

$$R_x = \sqrt{\frac{I_x}{\text{mass}}} = \sqrt{\frac{36}{12}} = \sqrt{3},$$

$$R_y = \sqrt{\frac{I_y}{\text{mass}}} = \sqrt{\frac{64}{12}} = \frac{4}{\sqrt{3}}. \quad \square$$

Exercises 17–4

1–4. A lamina of density $\delta(x, y)$ covers the region $\mathcal{R}$, which is bounded by the graphs of the given equations. Find the centroid of the lamina.

- **1** $\mathcal{R}$: $y = 0$, $y = x + 1$, $x = 0$, and $x = 3$; $\delta(x, y) = kx$
- **2** $\mathcal{R}$: $y = 1 - x^2$, $x + y + 1 = 0$; $\delta(x, y) = x + 1$
- **3** $\mathcal{R}$: $y = \sqrt{x}$, $y = 0$, and $x = 1$; $\delta(x, y) = x + y$
- **4** $\mathcal{R}$: $\sqrt{x} + \sqrt{y} = 1$, $x = 0$, and $y = 0$; $\delta(x, y) = xy$

5–8. Find the centroids of the plane regions $\mathcal{R}$ in Exercises 1–4.

- **5** Exercise 1
- **6** Exercise 2
- **7** Exercise 3
- **8** Exercise 4

The following exercise shows that the formulas for moments developed in this section are generalizations of the formulas of Section 6–8.

9 A lamina of constant density δ is bounded above by the graph of $y = f(x)$ and below by the graph of $y = g(x)$, for $a \leq x \leq b$.

(a) Express the moment about the y-axis as a double integral. Show that this integral is numerically equal to

$$\delta \int_a^b x[f(x) - g(x)]\, dx.$$

(b) Express the moment about the x-axis as a double integral. Show that this integral is numerically equal to

$$\frac{\delta}{2} \int_a^b [f(x)^2 - g(x)^2]\, dx.$$

10 (See Exercise 8 of Section 17–3.) Describe a plane region $\mathcal{R}$ and a line $\mathcal{L}$ such that the moment of $\mathcal{R}$ about $\mathcal{L}$ equals

$$\int_{-2}^{0} \int_{0}^{x+2} \frac{y - x + 2}{\sqrt{2}} \, dy \, dx.$$

(See Exercise 7 of Section 6–7 for the concept of a directed distance from a line to a point.)

11–14. A lamina of density $\delta(x, y)$ covers the region $\mathcal{R}$ bounded by the graphs of the equations. Find the moment of inertia about the indicated axis.

- **11** $\mathcal{R}$: $x = y^2$ and $x = 2y - y^2$; $\delta(x, y) = y + 1$; x-axis
- **12** $\mathcal{R}$: $y = e^x$, $y = 0$, $x = 0$, and $x = 1$; $\delta(x, y) = 1$; x-axis
- **13** $\mathcal{R}$: $y = \sqrt{x}$, $y = 0$, and $x = 4$; $\delta(x, y) = 1$, y-axis
- **14** $\mathcal{R}$: $x = y - y^2$ and $x + y = 0$; $\delta(x, y) = x + y$, x-axis

15–18. Calculate the radius of gyration about the indicated axis for the laminas in Exercises 11–14.

- **15** Exercise 11
- **16** Exercise 12
- **17** Exercise 13
- **18** Exercise 14

17–5 Double Integrals in Polar Coordinates

When the boundary curves of a basic closed region $\mathcal{R}$ are expressed in polar coordinates, it is necessary to use a different type of inner partition to define the double integral over $\mathcal{R}$. In this section we concern ourselves with basic closed regions that have the graphs of $r = g_1(\theta)$ and $r = g_2(\theta)$ as "inner" and "outer" boundaries, respectively, where θ varies from α to β. It is assumed, furthermore, that g_1 and g_2 are continuous for $\alpha \leq \theta \leq \beta$. A typical closed region of this type is pictured in Figure 17–23a.

We partition $\mathcal{R}$ by a system of rays from the origin and circular arcs with centers at the origin. (See Fig. 17–23b.) These rays and arcs partition a region that contains $\mathcal{R}$ into closed subregions of a very simple type. We use those closed subregions that are completely contained in $\mathcal{R}$ and ignore the others. The resulting set of closed subregions is called an *inner partition* $\mathcal{P}$ of $\mathcal{R}$. (See Fig. 17–23c.) The *norm* of the partition equals the length of the largest "diagonal" between opposite vertices of a closed subregion.

Let $R_1, R_2, \ldots, R_n$ be the closed subregions of an inner partition of $\mathcal{R}$. Let (r_k, θ_k) be the midpoint of R_k—the point midway between the boundary arcs and midway between the boundary rays that form R_k. (See Fig. 17–23d.) Let $\Delta\theta_k$ be the angle between the rays that bound R_k; and let Δr_k be the distance between the con-

17-5 Double Integrals in Polar Coordinates

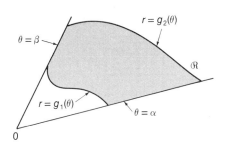

(a) The closed region $\mathcal{R}$

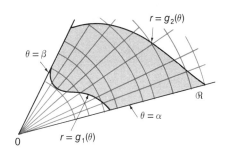

(b) A partition of a closed region that contains $\mathcal{R}$ formed by arcs and rays

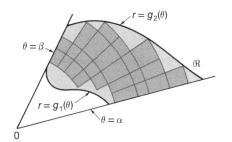

(c) An inner partition of $\mathcal{R}$

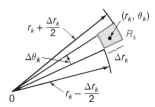

(d) Area of $R_k = r_k \Delta r_k \Delta \theta_k$

Figure 17-23 Double integral over a polar region:

$$\iint_{\mathcal{R}} F(r, \theta) \, dA = \lim_{\|\mathcal{P}\| \to 0} \sum_{k=1}^{n} F(r_k, \theta_k) r_k \Delta r_k \Delta \theta_k = \int_{\alpha}^{\beta} \int_{g_1(\theta)}^{g_2(\theta)} F(r, \theta) \, r \, dr \, d\theta.$$

centric arcs that bound R_k. (See Fig. 17-23d.) Then the inner boundary of R_k is the arc

$$r = r_k - \frac{\Delta r_k}{2}$$

and the outer boundary is the arc

$$r = r_k + \frac{\Delta r_k}{2}.$$

It follows from the formula for the area of a sector of a circle that the area of R_k is

$$\Delta A_k = \frac{\Delta \theta_k}{2} \left(r_k + \frac{\Delta r_k}{2} \right)^2 - \frac{\Delta \theta_k}{2} \left(r_k - \frac{\Delta r_k}{2} \right)^2$$

$$= r_k \, \Delta r_k \, \Delta \theta_k.$$

Let $F(r, \theta)$ be a continuous function defined on $\mathcal{R}$. Then

$$S_n = \sum_{k=1}^{n} F(r_k, \theta_k) \, \Delta A_k = \sum_{k=1}^{n} F(r_k, \theta_k) r_k \, \Delta r_k \, \Delta \theta_k$$

is an approximating sum for the double integral

$$\iint_{\mathcal{R}} F(r, \theta) \, dA.$$

It follows that

$$\iint_{\mathcal{R}} F(r, \theta) \, dA = \lim_{\|\mathcal{P}\| \to 0} \sum_{k=1}^{n} F(r_k, \theta_k) r_k \, \Delta r_k \, \Delta \theta_k.$$

It can be proved that this double integral can be evaluated by an iterated integral. The approximating sums are of form

$$S_n = \sum_{k=1}^{n} F(r_k, \theta_k) r_k \, \Delta r_k \, \Delta \theta_k,$$

where r varies from $g_1(\theta)$ to $g_2(\theta)$ and θ from α to β. The equivalent integrals are

Double and Iterated Integrals in Polar Coordinates

$$\iint_{\mathcal{R}} F(r, \theta) \, dA = \int_{\alpha}^{\beta} \int_{g_1(\theta)}^{g_2(\theta)} F(r, \theta) r \, dr \, d\theta.$$

Observe that the first integration for the iterated integral is with respect to r. This is because the boundary curves of $\mathcal{R}$ are usually given with r as a function of θ. We can think of this integration as occurring outward from the inner boundary to the outer boundary along a fixed ray. (See Fig. 17–24a.) The second integration is with respect to θ. We let θ vary from α, the smaller of the extreme angles that form the boundary of $\mathcal{R}$, to β, the larger of the angles. (See Fig. 17–24b.) The other order of integration is possible, but is not as common.

EXAMPLE 1   A closed region $\mathcal{R}$ has the graph of $r = 2 \cos \theta$ for its outer boundary and $r = 1$ for its inner boundary. Calculate its area.

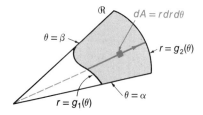

(a) The first integration is from the inner boundary curve to the outer boundary curve along a ray

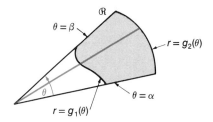

(b) The second integration is from $\theta = \alpha$ to $\theta = \beta$

Figure 17–24   Schematic representation of double integration in polar coordinates:

$$\iint_{\mathcal{R}} F(r, \theta) \, dA = \int_{\alpha}^{\beta} \int_{g_1(\theta)}^{g_2(\theta)} F(r, \theta) \, r \, dr \, d\theta.$$

17–5 Double Integrals in Polar Coordinates

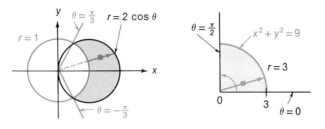

(a) Example 1   (b) Example 2   **Figure 17–25**

Solution The region is shown in Figure 17–25a. The inner boundary is $r = 1$ and the outer boundary is $r = 2\cos\theta$, $-\pi/3 \le \theta \le \pi/3$.

The area of $\mathcal{R}$ is

$$A = \iint_{\mathcal{R}} dA$$

$$= \int_{-\pi/3}^{\pi/3} \int_{1}^{2\cos\theta} r \, dr \, d\theta = \int_{-\pi/3}^{\pi/3} \left[\frac{r^2}{2}\right]_{r=1}^{r=2\cos\theta} d\theta$$

$$= \int_{-\pi/3}^{\pi/3} \left[2\cos^2\theta - \frac{1}{2}\right] d\theta = \left[\theta + \frac{\sin 2\theta}{2} - \frac{\theta}{2}\right]_{-\pi/3}^{\pi/3} \quad \text{(by I107)}$$

$$= \left[\frac{\pi}{6} + \frac{\sin(2\pi/3)}{2}\right] - \left[-\frac{\pi}{6} + \frac{\sin(-2\pi/3)}{2}\right]$$

$$= \frac{\pi}{3} + \frac{\sqrt{3}}{2} \approx 1.9132. \quad \square$$

CONVERSION FROM RECTANGULAR TO POLAR COORDINATES

Many problems originally posed in rectangular coordinates can best be solved by converting all terms to polar coordinates. In particular, this may be the case if an integral is defined over a region $\mathcal{R}$ that is a sector of a circle with center at the origin. It is necessary, in carrying out the conversion, to convert all quantities to polar coordinates, including the boundary curves and the function being integrated. We use the conversion formulas

Conversion Equations

$$x = r\cos\theta, \quad y = r\sin\theta, \quad dA = r \, dr \, d\theta, \quad \text{and} \quad x^2 + y^2 = r^2.$$

Observe that in rectangular coordinates we use

$$dA = dx \, dy \quad \text{or} \quad dA = dy \, dx$$

while in polar coordinates we use

$$dA = r \, dr \, d\theta.$$

EXAMPLE 2   Convert the integral

$$\int_0^3 \int_0^{\sqrt{9-x^2}} (x^2 + y^2) \, dy \, dx$$

to an equivalent integral in polar coordinates. Evaluate the new integral.

Solution   We first convert the integral to a double integral, then to an iterated integral in polar coordinates.

Let $\mathcal{R}$ be the region over which the original integral is defined. Then $\mathcal{R}$ is bounded above by the semicircle $y = \sqrt{9 - x^2}$ and below by the x-axis, $0 \le x \le 3$. (See Fig. 17–25b.)

In polar coordinates the boundaries of $\mathcal{R}$ are the circle $r = 3$ and the rays $\theta = 0$ and $\theta = \pi/2$.

In converting the integral to an equivalent integral in polar coordinates, we recall that $r^2 = x^2 + y^2$. Then

$$\int_0^3 \int_0^{\sqrt{9-x^2}} (x^2 + y^2) \, dy \, dx = \iint_{\mathcal{R}} (x^2 + y^2) \, dA = \iint_{\mathcal{R}} r^2 \, dA$$

$$= \iint_{\mathcal{R}} r^2 r \, dr \, d\theta = \int_0^{\pi/2} \int_0^3 r^3 \, dr \, d\theta$$

$$= \int_0^{\pi/2} \left[\frac{r^4}{4}\right]_0^3 d\theta = \int_0^{\pi/2} \frac{81}{4} \, d\theta$$

$$= \frac{81}{4} \cdot \frac{\pi}{2} = \frac{81\pi}{8}. \quad \square$$

MOMENT FORMULAS

Let $\delta(r, \theta)$ be the density function for a lamina. The integral formulas

$$\text{Mass} = \iint_{\mathcal{R}} \delta \, dA, \quad M_x = \iint_{\mathcal{R}} y\delta \, dA, \quad M_y = \iint_{\mathcal{R}} x\delta \, dA,$$

$$I_x = \iint_{\mathcal{R}} y^2 \delta \, dA, \quad \text{and} \quad I_y = \iint_{\mathcal{R}} x^2 \delta \, dA$$

hold for regions with polar boundaries. It is only necessary to convert all terms to polar coordinate form.

One word of warning. The coordinates of the centroid

$$\bar{x} = \frac{M_y}{\text{mass}}, \quad \bar{y} = \frac{M_x}{\text{mass}}$$

are in rectangular coordinates. This is the case even when all other terms are in polar coordinates.

EXAMPLE 3   A lamina of density $\delta(r, \theta) = r$ covers the closed region $\mathcal{R}$ between the first-quadrant portions of the circles $r = 2$ and $r = 3$. Find the centroid.

17-5 Double Integrals in Polar Coordinates

Solution The region is shown in Figure 17-26. The mass is

$$\text{Mass} = \iint_{\mathcal{R}} \delta(r, \theta) \, dA = \int_0^{\pi/2} \int_2^3 r \cdot r \, dr \, d\theta$$

$$= \frac{1}{3} \int_0^{\pi/2} \left[r^3 \right]_{r=2}^{r=3} d\theta = \frac{1}{3} \int_0^{\pi/2} 19 \, d\theta = \frac{19\pi}{6}$$

The moments about the coordinate axes are

$$M_x = \iint_{\mathcal{R}} y\delta(r, \theta) \, dA = \int_0^{\pi/2} \int_2^3 r \sin\theta \cdot r \cdot r \, dr \, d\theta$$

$$= \frac{1}{4} \int_0^{\pi/2} \left[r^4 \sin\theta \right]_{r=2}^{r=3} d\theta = \frac{1}{4} \int_0^{\pi/2} 65 \sin\theta \, d\theta$$

$$= \frac{65}{4} \left[-\cos\theta \right]_0^{\pi/2} = \frac{65}{4} \left[-\cos\frac{\pi}{2} + \cos 0 \right] = \frac{65}{4}.$$

$$M_y = \iint_{\mathcal{R}} x\delta(r, \theta) \, dA = \int_0^{\pi/2} \int_2^3 r \cos\theta \cdot r \cdot r \, dr \, d\theta$$

$$= \frac{1}{4} \int_0^{\pi/2} \left[r^4 \cos\theta \right]_{r=2}^{r=3} d\theta = \frac{1}{4} \int_0^{\pi/2} 65 \cos\theta \, d\theta = \frac{65}{4} \left[\sin\theta \right]_0^{\pi/2}$$

$$= \frac{65}{4} \left[\sin\frac{\pi}{2} - \sin 0 \right] = \frac{65}{4}.$$

The coordinates (in rectangular coordinates) of the centroid are

$$\bar{x} = \frac{M_y}{\text{mass}} = \frac{65/4}{19\pi/6} = \frac{195}{38\pi} \approx 1.6334,$$

$$\bar{y} = \frac{M_x}{\text{mass}} = \frac{65/4}{19\pi/6} = \frac{195}{38\pi} \approx 1.6334. \ \square$$

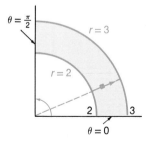

Figure 17-26 Example 3.

NORMAL PROBABILITY DENSITY FUNCTION (OPTIONAL)

Under the proper conditions, the results of this chapter can be extended to improper integrals. The following discussion illustrates, without proof, how this can be done.

It was stated in Section 10-6 that the normal probability density function

$$f(x) = \frac{1}{\sqrt{2\pi}\sigma} e^{-(x-\mu)^2/2\sigma^2}$$

Area Under the Normal Curve is 1

satisfies the condition

$$\int_{-\infty}^{\infty} f(x)\, dx = 1.$$

We shall show that this result can be obtained by changing a related double integral to an equivalent integral in polar coordinates. For simplicity, we consider only the case for which $\sigma = 1$ and $\mu = 0$. We must show that

$$\int_{-\infty}^{\infty} \frac{1}{\sqrt{2\pi}} e^{-x^2/2}\, dx = 1.$$

Observe first that

$$\int_0^{\infty} e^{-x^2/2}\, dx = \int_0^{\infty} e^{-y^2/2}\, dy,$$

so that

$$\left(\int_0^{\infty} e^{-x^2/2}\, dx\right)^2 = \int_0^{\infty} e^{-x^2/2}\, dx \int_0^{\infty} e^{-y^2/2}\, dy$$

$$= \int_0^{\infty} \int_0^{\infty} e^{-x^2/2} e^{-y^2/2}\, dx\, dy$$

(See Exercises 20 and 21 of Section 17–2.)

$$= \int_0^{\infty} \int_0^{\infty} e^{-(x^2+y^2)/2}\, dx\, dy$$

$$= \iint_{\mathcal{R}} e^{-(x^2+y^2)/2}\, dA,$$

where $\mathcal{R}$ is the infinite closed region that consists of Quadrant I and its boundary lines. (See Fig. 17–27.)

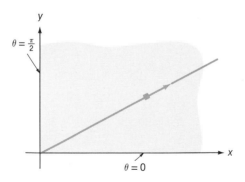

Figure 17–27

We now convert this integral to an equivalent integral in polar coordinates, observing that r varies from 0 to ∞ and θ from 0 to $\pi/2$:

$$\left(\int_0^{\infty} e^{-x^2/2}\, dx\right)^2 = \iint_{\mathcal{R}} e^{-(x^2+y^2)/2}\, dA = \int_0^{\pi/2} \int_0^{\infty} e^{-r^2/2} r\, dr\, d\theta$$

$$= \int_0^{\pi/2} \lim_{t \to \infty} \left[-e^{-r^2/2} \right]_0^t d\theta = \int_0^{\pi/2} [-0 + 1] \, d\theta = \frac{\pi}{2}.$$

Therefore,

$$\int_0^\infty e^{-x^2/2} \, dx = \sqrt{\frac{\pi}{2}} = \frac{\sqrt{2\pi}}{2}$$

so that

$$\int_0^\infty \frac{1}{\sqrt{2\pi}} e^{-x^2/2} \, dx = \frac{1}{2}.$$

Since the function $e^{-x^2/2}/\sqrt{2\pi}$ is symmetric about the y-axis, then

$$\int_{-\infty}^\infty \frac{1}{\sqrt{2\pi}} e^{-x^2/2} \, dx = 2 \int_0^\infty \frac{1}{\sqrt{2\pi}} e^{-x^2/2} \, dx = 2 \cdot \frac{1}{2} = 1.$$

Exercises 17–5

1–4. Evaluate the iterated integral.

- 1 $\displaystyle\int_0^{\pi/2} \int_{2\sin\theta}^2 r \, dr \, d\theta$ 2 $\displaystyle\int_{\pi/2}^{\pi} \int_0^{1-\sin\theta} r \, dr \, d\theta$

- 3 $\displaystyle\int_{\pi/4}^{\pi/2} \int_{\pi/2}^r \csc^2\theta \, d\theta \, dr$ 4 $\displaystyle\int_0^{\pi/4} \int_{\sqrt{2}}^{2\cos\theta} r \, dr \, d\theta$

5–8. Use an iterated integral to calculate the area of the closed region $\mathscr{R}$ bounded by the graphs of the given equations.

- 5   Inside $r = 1 + \cos\theta$ and outside $r = 1$
- 6   Inside $r = \cos\theta$ and outside $r = 1 - \cos\theta$
- 7 $r = 2\cos 3\theta$ • 8 $r = 1 + \sin\theta$

9–12. A lamina of density $\delta(r, \theta)$ covers the closed region bounded by the given graph. Find its centroid.

- 9 $r = 1 + \cos\theta$; $\delta(r, \theta) = 1$ • 10 $r = 2\cos\theta$, $\dfrac{-\pi}{2} \leq \theta \leq \dfrac{\pi}{2}$, $\delta(r, \theta) = r$

- 11 $r = \cos 2\theta$, $-\dfrac{\pi}{4} \leq \theta \leq \dfrac{\pi}{4}$; $\delta(r, \theta) = 1$

- 12 $r = 1 + \cos\theta$, $\delta(r, \theta) = r$

13–16. Convert the integral to an equivalent integral in polar coordinates. Evaluate the new integral.

- 13 $\displaystyle\int_{-4}^4 \int_{-\sqrt{16-x^2}}^{\sqrt{16-x^2}} dy \, dx$ • 14 $\displaystyle\int_0^3 \int_{-\sqrt{18-y^2}}^{-y} \sin(x^2 + y^2) \, dx \, dy$

• 15 $\int_0^1 \int_y^{\sqrt{2-y^2}} x^2 \, dx \, dy$

• 16 $\int_0^2 \int_0^{\sqrt{4-x^2}} e^{-x^2-y^2} \, dy \, dx$

17–18. Convert the integral to an equivalent integral in rectangular coordinates. Evaluate the new integral and the original integral.

• 17 $\int_{\pi/4}^{\pi/2} \int_0^{2\csc\theta} r^3 \, dr \, d\theta$

• 18 $\int_{-\pi/2}^{0} \int_0^3 r^3 \sin\theta \cos\theta \, dr \, d\theta$

19 Let $\mathcal{R}$ be the region bounded by the graph of $r = f(\theta)$, and the rays $\theta = \alpha$ and $\theta = \beta$. Show that the iterated integral for the area of $\mathcal{R}$ reduces to the integral

$$\tfrac{1}{2} \int_\alpha^\beta f(\theta)^2 \, d\theta$$

that was used in Chapter 13.

17–6 Surface Area

Let $f(x, y)$ have continuous partial derivatives at each point of a basic closed region $\mathcal{R}$. We will show how to calculate the surface area of the portion of the graph of f that is defined over $\mathcal{R}$. Our work is a generalization of the work on arc length in Chapter 6.

Let $\mathcal{P}$ be an inner partition of $\mathcal{R}$ into rectangles $R_1, R_2, \ldots, R_n$. Let (ξ_k, η_k) be a point in R_k. Let $\mathcal{T}_k$ be the tangent plane to the graph of f at $(\xi_k, \eta_k, f(\xi_k, \eta_k))$. (See Fig. 17–28a.) We project the region R_k vertically on to the tangent plane $\mathcal{T}_k$, forming a parallelogram $\mathcal{P}_k$. (See Fig. 17–28b.) The area of this parallelogram is an approximation to the area of the portion of the graph of f that is defined over R_k.

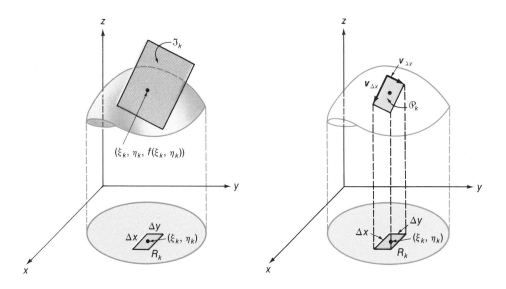

Figure 17–28

17–6 Surface Area

Let Δx and Δy be the lengths of the sides of the rectangle R_k. Let $v_{\Delta x}$ and $v_{\Delta y}$ be the vectors that form the corresponding sides of the parallelogram $\mathcal{P}_k$. It follows from our work on partial derivatives that

$$v_{\Delta x} = \Delta x \mathbf{i} + f_x(\xi_k, \eta_k)\, \Delta x \mathbf{k}$$
$$v_{\Delta y} = \Delta y \mathbf{j} + f_y(\xi_k, \eta_k)\, \Delta y \mathbf{k}.$$

Recall from Section 14–7 that the area of a parallelogram with sides formed by the vectors a and b is $\|a \times b\|$. Therefore, the area of the parallelogram $\mathcal{P}_k$ is $\|v_{\Delta x} \times v_{\Delta y}\|$.

The cross product of $v_{\Delta x}$ and $v_{\Delta y}$ is

$$v_{\Delta x} \times v_{\Delta y} = \begin{vmatrix} \mathbf{i} & \mathbf{j} & \mathbf{k} \\ \Delta x & 0 & f_x(\xi_k, \eta_k)\, \Delta x \\ 0 & \Delta y & f_y(\xi_k, \eta_k)\, \Delta y \end{vmatrix}$$
$$= \Delta x\, \Delta y\, [-f_x(\xi_k, \eta_k)\mathbf{i} - f_y(\xi_k, \eta_k)\mathbf{j} + \mathbf{k}].$$

Therefore,

$$\text{Area of } \mathcal{P}_k = \|v_{\Delta x} \times v_{\Delta y}\|$$
$$= \sqrt{f_x(\xi_k, \eta_k)^2 + f_y(\xi_k, \eta_k)^2 + 1}\ \Delta x\, \Delta y.$$

The number that we would consider to be the total surface area of the graph of f over $\mathcal{R}$ is approximately equal to the sum of the areas of the parallelograms $\mathcal{P}_1, \mathcal{P}_2, \ldots, \mathcal{P}_k$:

$$S_n = \sum_{k=1}^{n} \sqrt{f_x(\xi_k, \eta_k)^2 + f_y(\xi_k, \eta_k)^2 + 1}\ \Delta x\, \Delta y.$$

We define the surface area to be the integral that is the limit of this sum as the norm of the partition approaches zero.

DEFINITION

Surface Area

Let f have continuous partial derivatives over the basic closed region $\mathcal{R}$. The *surface area* of the portion of the graph of f that is defined over $\mathcal{R}$ is

$$S = \iint_\mathcal{R} \sqrt{\left(\frac{\partial f}{\partial x}\right)^2 + \left(\frac{\partial f}{\partial y}\right)^2 + 1}\ dA.$$

EXAMPLE 1 Find the surface area of the portion of the plane $3x + 2y + 6z = 12$ that is inside the cylinder $x^2 + y^2 = 1$.

Solution The plane is the graph of

$$z = f(x, y) = \frac{12 - 3x - 2y}{6} = 2 - \frac{x}{2} - \frac{y}{3}.$$

Therefore,

$$\frac{\partial f}{\partial x} = -\frac{1}{2} \quad \text{and} \quad \frac{\partial f}{\partial y} = -\frac{1}{3}.$$

The projection of the cylinder on the xy-plane is the closed region $\mathcal{R}$ that is enclosed by the circle $x^2 + y^2 = 1$. (See Fig. 17–29.) The surface area is

$$S = \iint_{\mathcal{R}} \sqrt{\left(\frac{\partial f}{\partial x}\right)^2 + \left(\frac{\partial f}{\partial y}\right)^2 + 1} \, dA = \int_{-1}^{1} \int_{-\sqrt{1-x^2}}^{\sqrt{1-x^2}} \sqrt{\frac{1}{4} + \frac{1}{9} + 1} \, dy \, dx$$

$$= \frac{7}{6} \int_{-1}^{1} \left[y \right]_{-\sqrt{1-x^2}}^{\sqrt{1-x^2}} dx = 2 \cdot \frac{7}{6} \int_{-1}^{1} \sqrt{1 - x^2} \, dx$$

$$= \frac{2 \cdot 7}{6 \cdot 2} \left[x\sqrt{1 - x^2} + \sin^{-1} x \right]_{-1}^{1} \quad \text{(by I72)}$$

$$= \frac{7}{6} [0 + \sin^{-1} 1 - 0 - \sin^{-1}(-1)] = \frac{7\pi}{6}. \quad \square$$

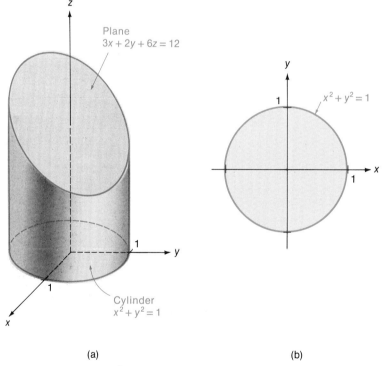

(a)　　　　　　　　　　(b)

Figure 17–29 Example 1.

It is difficult to calculate many of the integrals for surface area. Some can be simplified by being converted to equivalent integrals in polar coordinates. A certain amount of care must be exercised in this conversion because the surface area integral is defined in terms of rectangular coordinates. We must first set up the integral in rectangular coordinates and then convert all terms to the equivalent terms in polar coordinates.

EXAMPLE 2　Find the area of the portion of the paraboloid $z = x^2 + y^2$ that is contained within the sphere $x^2 + y^2 + z^2 = 6z$.

17-6 Surface Area

Solution The intersection of the paraboloid and the sphere is the graph of the system of equations

$$\begin{cases} x^2 + y^2 + z^2 = 6z \\ x^2 + y^2 = z. \end{cases}$$

This system reduces to

$$\begin{cases} z + z^2 = 6z \\ x^2 + y^2 = z. \end{cases}$$

The first equation implies that $z = 0$ or $z = 5$. If $z = 0$, we get the single point $(0, 0, 0)$. If $z = 5$, we get the circle

$$x^2 + y^2 = 5, \quad z = 5.$$

(See Fig. 17–30a.) Thus, the projection on the *xy*-plane of the portion of the paraboloid is the region $\mathcal{R}$ bounded by the circle $x^2 + y^2 = 5$.

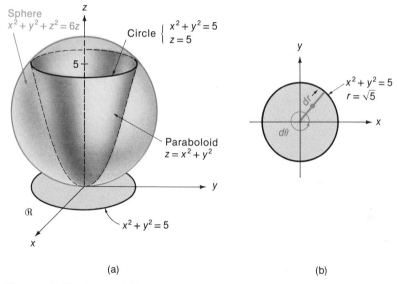

(a)   (b)

Figure 17–30 Example 2.

The area of this portion of the paraboloid is

$$S = \iint\limits_{\mathcal{R}} \sqrt{\left(\frac{\partial z}{\partial x}\right)^2 + \left(\frac{\partial z}{\partial y}\right)^2 + 1}\, dA = \iint\limits_{\mathcal{R}} \sqrt{4x^2 + 4y^2 + 1}\, dA.$$

We can now convert this integral to an equivalent integral in polar coordinates. Recall that $x = r \cos \theta$, $y = r \sin \theta$, and $dA = r\, dr\, d\theta$. The region $\mathcal{R}$ is swept out as r varies from 0 to $\sqrt{5}$ and θ from 0 to 2π. (See Fig. 17–30b.) Therefore,

$$S = \int_0^{2\pi} \int_0^{\sqrt{5}} \sqrt{4r^2 \cos^2 \theta + 4r^2 \sin^2 \theta + 1}\, r\, dr\, d\theta$$

$$= \int_0^{2\pi} \int_0^{\sqrt{5}} \sqrt{4r^2 + 1}\, r\, dr\, d\theta = \frac{1}{8} \int_0^{2\pi} \int_0^{\sqrt{5}} (4r^2 + 1)^{1/2} (8r\, dr)\, d\theta$$

$$= \frac{1}{8} \cdot \frac{2}{3} \int_0^{2\pi} \left[(4r^2 + 1)^{3/2} \right]_0^{\sqrt{5}} d\theta = \frac{1}{12} \int_0^{2\pi} [21^{3/2} - 1] \, d\theta$$

$$= \frac{1}{12} [21\sqrt{21} - 1] \cdot 2\pi = \frac{(21\sqrt{21} - 1)\pi}{6}. \quad \square$$

EXAMPLE 3 Derive the formula for the area of a sphere of radius a.

Solution We shall calculate the area of the upper hemisphere. If we set up our axis system so that the center of the sphere is at the origin, then the equation of the sphere is $x^2 + y^2 + z^2 = a^2$ and the equation of the upper hemisphere is

$$z = \sqrt{a^2 - x^2 - y^2}.$$

Observe in Figure 17–31 that the projection of the upper hemisphere on the xy-plane is the closed region $\mathcal{R}$ bounded by the circle $x^2 + y^2 = a^2$.

The surface area of the hemisphere is

$$S = \iint_{\mathcal{R}} \sqrt{\left(\frac{\partial z}{\partial x}\right)^2 + \left(\frac{\partial z}{\partial y}\right)^2 + 1} \, dA$$

$$= \iint_{\mathcal{R}} \sqrt{\left(\frac{-x}{\sqrt{a^2 - x^2 - y^2}}\right)^2 + \left(\frac{-y}{\sqrt{a^2 - x^2 - y^2}}\right)^2 + 1} \, dA$$

$$= \iint_{\mathcal{R}} \frac{a}{\sqrt{a^2 - x^2 - y^2}} \, dA.$$

If we convert all terms to polar coordinates, the integral is changed to the equivalent integral:

$$S = \iint_{\mathcal{R}} \frac{a}{\sqrt{a^2 - r^2}} r \, dr \, d\theta = a \int_0^{2\pi} \int_0^a (a^2 - r^2)^{-1/2} r \, dr \, d\theta$$

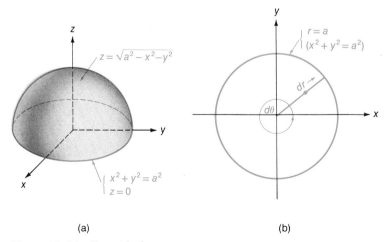

(a)   (b)

Figure 17–31 Example 3.

$$= -\frac{a}{2} \int_0^{2\pi} \int_0^a (a^2 - r^2)^{-1/2}(-2r\,dr)\,d\theta$$

$$= -\frac{a}{2} \int_0^{2\pi} 2\left[\sqrt{a^2 - r^2}\right]_0^a d\theta = -\frac{a}{2} \int_0^{2\pi}(-2a)\,d\theta$$

$$= a^2 \int_0^{2\pi} d\theta = 2\pi a^2.$$

The total area of the sphere is twice the area of the upper hemisphere:

Surface area of sphere $= 4\pi a^2$. □

Exercises 17–6

1–4. **(a)** Set up an integral in rectangular coordinates for the area of the surface.

(b) Calculate the area of the surface. (It may be necessary to convert the integral to an equivalent integral in polar coordinates.)

• **1** The portion of the plane $z = ax + by$ that is interior to the cylinder $x^2 + y^2 = c^2$.

• **2** The portion of the hemisphere $x^2 + y^2 + z^2 = a^2$, $z \geq 0$, that is interior to the cylinder $x^2 + y^2 = c^2$, where $a > c > 0$.

• **3** The portion of the hyperbolic paraboloid $z = y^2 - x^2$ that is interior to the "cylinder" enclosed by the vertical planes $y = 0$, $x = y$ and the graph of $x^2 + y^2 = 4$, $x \geq 0$, $y \geq 0$.

• **4** The portion of the sphere $x^2 + y^2 + z^2 = 8$ that is interior to the first-octant portion of the cone $z^2 = x^2 + y^2$.

• **5** Calculate the area of the portion of the paraboloid $z = x^2 + y^2$ that is between the planes $z = 2$ and $z = 6$.

6 Use an integral to derive the formula for the lateral area of a right circular cylinder of radius a and height h. (*Hint:* Let the cylinder be the graph of $x^2 + z^2 = a^2$, $0 \leq y \leq h$.)

7 Use an integral to derive the formula for the lateral area of a right circular cone of radius a and height h.

8 (*Surfaces of Revolution*) Let f be a differentiable function of y for $a \leq y \leq b$, where $a \geq 0$. Revolve the graph of $z = f(y)$ (located in the yz-plane) about the z-axis, obtaining a surface of revolution. (See Fig. 17–32.)

(a) Show that the surface of revolution is the graph of $z = f(\sqrt{x^2 + y^2})$, $a \leq \sqrt{x^2 + y^2} \leq b$. (*Hint:* It may be convenient to work with r, the distance from $(0, 0)$ to (x, y).)

(b) Show that in polar coordinates the area of the surface of revolution is

$$S = \int_0^{2\pi} \int_a^b \sqrt{f'(r)^2 + 1}\, r\, dr\, d\theta.$$

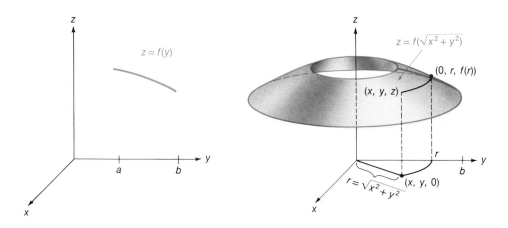

(a) The graph of $z = f(y)$, $x = 0$
(b) The surface of revolution

Figure 17–32 Exercise 8.

9 **(a)** Use the formula in Exercise 8b to work Example 3.

(b) Use the formula in Exercise 8b to work Exercise 7.

•10 A sphere of radius $a > 0$ is intercepted by two parallel planes that are b units apart. Prove that the surface area of the section of the sphere between the planes is independent of the location of the planes, depending only on a and b. (*Hint:* Assume that the axes have been rotated and translated until a convenient orientation and position have been obtained relative to the sphere and the planes.)

17–7 Triple Integration

The results of the previous sections can be extended to *triple integrals* defined over solids in three-dimensional space. We consider special types of solids in our development. Each of these is a solid figure that includes all its boundary points, is finite, and has the property that every pair of points in its interior can be connected by a curve that lies in the interior. Furthermore, we assume that the solid is contained between a lower boundary surface which is the graph of the continuous function $z = f_1(x, y)$, and an upper boundary surface which is the graph of the continuous function $z = f_2(x, y)$, and that the projection of the solid on each coordinate plane is a basic closed region. (See Fig. 17–33a.) Unless we specify otherwise, in this and the following section the word *solid* means a solid figure with the properties just mentioned.

INNER PARTITIONS

Let $\mathcal{R}$ be the projection of the solid $\mathcal{S}$ on the xy-plane, let $[c, d]$ be the projection of $\mathcal{S}$ on the z-axis. We choose an inner partition of $\mathcal{R}$ and a partition of $[c, d]$ into subintervals. These partitions induce a partition of $\mathcal{S}$ into small rectangular blocks. (See Figs. 17–33b, c.) Some of these blocks are completely contained in $\mathcal{S}$; others are

17–7 Triple Integration

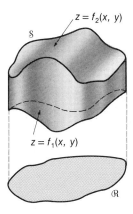

(a) The basic closed region $\mathcal{R}$ is the projection of the solid S on the xy-plane

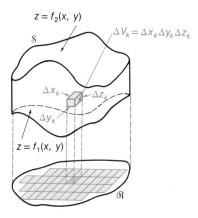

(b) A typical block in an inner partition of S: $\Delta V_k = \Delta x_k \Delta y_k \Delta z_k$

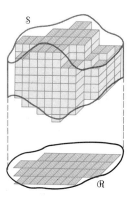

(c) An inner partition of S

Figure 17–33 Inner partitions of solids.

partly or completely exterior to S. We use those blocks that are completely in S and ignore the others. The set of blocks remaining in S is called an *inner partition of* S. If $\mathcal{P}$ is an inner partition of S, then the length of the longest diagonal of a block in $\mathcal{P}$ is the *norm* of the partition, denoted by $\|\mathcal{P}\|$.

THE TRIPLE INTEGRAL

Let $S_1, S_2, \ldots, S_n$ be the blocks in an inner partition $\mathcal{P}$ of the solid S. Let ΔV_1, $\Delta V_2, \ldots, \Delta V_n$, respectively, be the volumes of these blocks. Let (ξ_1, η_1, ζ_1), $(\xi_2, \eta_2, \zeta_2), \ldots, (\xi_n, \eta_n, \zeta_n)$, respectively, be points in these blocks. Let the function $F(x, y, z)$ be defined on S. The sum

$$\sum_{k=1}^{n} F(\xi_k, \eta_k, \zeta_k)\, \Delta V_k$$

is called an *approximating sum* based on the function F and the partition $\mathcal{P}$.

If F is continuous on S, this approximating sum can be made arbitrarily close to a limit L, provided that the norm of the corresponding partition is sufficiently close to zero. The proof of this statement is analogous to the proof of the same property for a function of a single variable. The limit L is called the *triple integral of F over S*:

Triple Integral
$$\iiint_S F(x, y, z)\, dV = L = \lim_{\|\mathcal{P}\| \to 0} \sum_{k=1}^{n} F(\xi_k, \eta_k, \zeta_k)\, \Delta V_k.$$

APPLICATIONS

The triple integral $\iint_S \int F(x, y, z)\, dv$ has several interpretations according to the function F. Three typical interpretations follow:

(1) *Volume* Let $F(x, y, z) = 1$ for all $(x, y, z) \in S$. Since ΔV_k is the volume of the kth block in the partition $\mathcal{P}$, then the approximating sum

$$\sum_{k=1}^{n} \Delta V_k$$

is an approximation of the volume of S. The exact value of the volume is the limit of this sum as $\|\mathcal{P}\| \to 0$:

Volume
$$V = \iiint_S dV$$

(2) *Mass* If $\delta(x, y, z)$ is the *density* (mass per unit of volume) of S at (x, y, z), then

$$\delta(\xi_k, \eta_k, \zeta_k)\, \Delta V_k$$

is approximately equal to the mass of the kth block S_k. It follows that

$$\sum_{k=1}^{n} \delta(\xi_k, \eta_k, \zeta_k)\, \Delta V_k$$

is approximately equal to the mass of S. The exact value of the mass is

Mass
$$\text{Mass} = \iiint_S \delta(x, y, z)\, dV.$$

17-7 Triple Integration

(3) *Heat* If $T(x, y, z)$ is the temperature at the point (x, y, z) then

$$T(\xi_k, \eta_k, \zeta_k) \Delta V_k$$

is approximately equal to the total amount of heat stored in the kth block of S. It follows that

$$\sum_{k=1}^{n} T(\xi_k, \eta_k, \zeta_k) \Delta V_k$$

is approximately equal to the total heat stored in the solid S. The exact value of the heat stored in S is

Heat

$$\text{Heat} = \iiint_S T(x, y, z)\, dV.$$

ITERATED INTEGRALS

The triple integrals that we consider can be evaluated by iterated triple integrals. The *iterated triple integral*

$$\int_a^b \int_{g_1(x)}^{g_2(x)} \int_{f_1(x,y)}^{f_2(x,y)} F(x, y, z)\, dz\, dy\, dx$$

is evaluated as

Iterated Triple Integral

$$\int_a^b \int_{g_1(x)}^{g_2(x)} \int_{f_1(x,y)}^{f_2(x,y)} F(x, y, z)\, dz\, dy\, dx$$
$$= \int_a^b \left[\int_{g_1(x)}^{g_2(x)} \left[\int_{f_1(x,y)}^{f_2(x,y)} F(x, y, z)\, dz \right] dy \right] dx.$$

The first integration is with respect to z. After $F(x, y, z)$ has been integrated with respect to z, we substitute $z = f_1(x, y)$ and $z = f_2(x, y)$ for the upper and lower limits of integration, obtaining a function of x and y. We then proceed as with an iterated integral in x and y.

Iterated integrals with other orders of integration are defined similarly.

EXAMPLE 1

$$\int_0^1 \int_{x^2}^x \int_{x-y}^{x+y} xyz\, dz\, dy\, dx$$

$$= \int_0^1 \left\{ \int_{x^2}^x \left[\int_{x-y}^{x+y} xyz\, dz \right] dy \right\} dx = \int_0^1 \left\{ \int_{x^2}^x \left[\frac{xyz^2}{2} \right]_{z=x-y}^{z=x+y} dy \right\} dx$$

$$= \int_0^1 \left\{ \int_{x^2}^x \left[\frac{xy}{2}(x+y)^2 - \frac{xy}{2}(x-y)^2 \right] dy \right\} dx$$

$$= \int_0^1 \left[\int_{x^2}^x 2x^2 y^2\, dy \right] dx = \int_0^1 \left[\frac{2x^2 y^3}{3} \right]_{y=x^2}^{y=x} = \frac{2}{3} \int_0^1 (x^5 - x^8)\, dx$$

$$= \frac{2}{3} \left[\frac{x^6}{6} - \frac{x^9}{9} \right]_0^1 = \frac{2}{3} \left[\frac{1}{6} - \frac{1}{9} \right] = \frac{1}{27}. \quad \square$$

Multiple Integration

Triple Integral Evaluated as an Iterated Integral

We now show how to evaluate a triple integral $\iiint_S F(x, y, z)\, dV$ by an iterated integral. Let $z = f_1(x, y)$ be the lower and $z = f_2(x, y)$ the upper boundary surface of the solid S. Let $\mathcal{R}$ be the projection of S on the xy-plane. (See Fig. 17–34a.) Let $\mathcal{R}$ be bounded by the graphs of $y = g_1(x)$ and $y = g_2(x)$, $a \leq x \leq b$, as pictured in Figure 17–34b. If f_1 and f_2 are continuous on $\mathcal{R}$ and g_1 and g_2 are continuous for $a \leq x \leq b$, then the triple integral of F over S is evaluated as

$$\iiint_S F(x, y, z)\, dV = \iint_\mathcal{R} \left[\int_{f_1(x,y)}^{f_2(x,y)} F(x, y, z)\, dz \right] dA$$

$$= \int_a^b \int_{g_1(x)}^{g_2(x)} \int_{f_1(x,y)}^{f_2(x,y)} F(x, y, z)\, dz\, dy\, dx.$$

In setting up this iterated integral, we note that:

Order of Integration

(1) The first integration is from the lower surface of S to the upper surface and is performed along a line parallel to the z-axis (Fig. 17–34a).

(2) The second integration is from the lower boundary curve of $\mathcal{R}$ to the upper boundary curve and is performed along a line parallel to the y-axis (Fig. 17–34b).

(3) The third integration is from the left-hand endpoint to the right-hand endpoint of the interval $[a, b]$—the projection of $\mathcal{R}$ on the x-axis—and is performed along the x-axis.

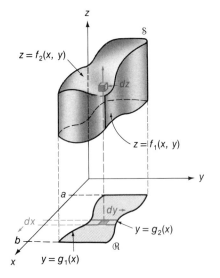

(a) The first integration is from a boundary surface to a boundary surface

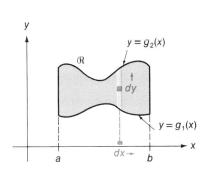

(b) The second and third integrations are over the projection of the solid on a plane

Figure 17–34 Schematic representation of triple integration:
$$\iiint_S F(x, y, z)\, dV = \int_a^b \int_{g_1(x)}^{g_2(x)} \int_{f_1(x,y)}^{f_2(x,y)} F(x, y, z)\, dz\, dy\, dx.$$

17–7 Triple Integration

Similar results hold if R has left and right boundaries defined by continuous functions of x.

EXAMPLE 2   A solid S is bounded by the xz- and yz-planes and the planes $z = 2 - 2x - y$ and $z = 4x + 2y - 4$. Find its volume.

Solution   The upper and lower boundary planes intersect when

$$4x + 2y - 4 = 2 - 2x - y$$
$$6x + 3y = 6$$
$$2x + y = 2.$$

(See Fig. 17–35a.) Thus, the projection on the xy-plane is the closed region $\mathcal{R}$ bounded by the coordinate axes and the line $2x + y = 2$.

The volume of S is

$$V = \iiint_S dV = \int_0^1 \int_0^{2-2x} \int_{4x+2y-4}^{2-2x-y} dz\, dy\, dx$$

$$= \int_0^1 \int_0^{2-2x} \left[z \right]_{z=4x+2y-4}^{z=2-2x-y} dy\, dx$$

$$= \int_0^1 \int_0^{2-2x} (6 - 6x - 3y)\, dy\, dx$$

$$= \int_0^1 \left[6y - 6xy - \frac{3y^2}{2} \right]_{y=0}^{y=2-2x} dx$$

$$= 6 \int_0^1 (1 - 2x + x^2)\, dx$$

$$= 6 \left[x - x^2 + \frac{x^3}{3} \right]_0^1 = 6 \cdot \frac{1}{3} = 2. \quad \square$$

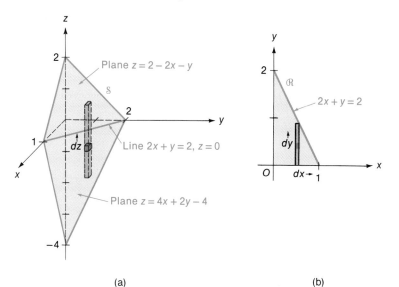

(a) (b)

Figure 17–35   Example 2.

EXAMPLE 3 The solid S of Example 2 has density $\delta(x, y, z) = x + y$ at the point (x, y, z). Find its mass.

Solution
$$\text{Mass} = \iiint_S \delta(x, y, z) \, dV$$

$$= \int_0^1 \int_0^{2-2x} \int_{4x+2y-4}^{2-2x-y} (x + y) \, dz \, dy \, dx$$

$$= \int_0^1 \int_0^{2-2x} \left[xz + yz \right]_{z=4x+2y-4}^{z=2-2x-y} dy \, dx$$

$$= \int_0^1 \int_0^{2-2x} [-6x^2 - 9xy - 3y^2 + 6x + 6y] \, dy \, dx$$

$$= \int_0^1 \left[-6x^2 y - \frac{9xy^2}{2} - y^3 + 6xy + 3y^2 \right]_{y=0}^{y=2-2x} dx$$

$$= \int_0^1 (2x^3 - 6x + 4) \, dx = \left[\frac{x^4}{2} - 3x^2 + 4x \right]_0^1$$

$$= \left[\frac{1}{2} - 3 + 4 \right] = \frac{3}{2}. \ \square$$

Projections on xz and xy-planes

In this introductory section we have worked with solids that can be projected onto a basic closed region $\mathcal{R}$ on the xy-plane. Similar results hold for solids that can be projected onto basic closed regions on other coordinate planes.

EXAMPLE 4 Find the volume of the solid in Octant I bounded by the coordinate planes and the paraboloid $y = 16 - x^2 - z^2$.

Solution The solid S is pictured in Figure 17–16(a). Its projection on the xz-plane is the quarter circle bounded by the x- and z-axes and the graph of $x^2 + z^2 = 16$, $0 \le x \le 4$. The volume of S is

$$V = \iiint_S dV = \int_0^4 \int_0^{\sqrt{16-x^2}} \int_0^{16-x^2-z^2} dy \, dz \, dx,$$

which can be evaluated as $V = 32\pi$. $\square$

Exercises 17–7

1–4. Evaluate the integrals

• 1 $\displaystyle\int_0^2 \int_0^x \int_{y^2}^{x-y} x \, dz \, dy \, dx$

• 2 $\displaystyle\int_0^2 \int_{y^2}^{3y} \int_0^x dz \, dx \, dy$

• 3 $\displaystyle\int_0^1 \int_{x^2}^x \int_0^{xy} dz \, dy \, dx$

• 4 $\displaystyle\int_0^4 \int_0^{4-y} \int_0^{4-x-y} z^8 \, dz \, dx \, dy$

17-8 Moments and Centroids

5–10. A solid S is bounded by the graphs of the equations. Use triple integration to calculate its volume.

- 5 $2x + 5y + z = 20$, $x = 0$, $y = 0$, $z = 0$
- 6 $z = 4 - x^2$, $x + y = 2$, the coordinate planes
- 7 $y = 2 - z^2$, $y = z^2$, $x + z = 4$, and $x = 0$
- 8 $z = 4y^2$, $z = 2$, $x = 2$, and $x = 0$
- 9 $x^2 + z^2 = 4$, $y^2 + z^2 = 4$
- 10 $z = x^2 + y^2$, $y + z = 2$

11–12. Change to an equivalent iterated integral with the order of integration given by $dV = dy\, dx\, dz$.

- 11 $\displaystyle\int_0^3 \int_0^{3-x} \int_0^{6-2x-2y} (x + 2y)\, dz\, dy\, dx$

12 $\displaystyle\int_0^3 \int_0^{(12-4x)/3} \int_0^{9-x^2} dz\, dy\, dx$

17-8 Moments and Centroids

Triple integration has applications related to moments and centroids. We assume throughout this section that S is a solid in three-dimensional space with a continuous density function (mass per unit of volume) equal to $\delta(x, y, z)$ at the point (x, y, z). As discussed in Section 17–7, the mass of S is

$$\text{Mass} = \iiint_S \delta(x, y, z)\, dV.$$

In our earlier work we considered moments of laminas in the xy-plane about lines in that plane. When we generalize these results to three-dimensional space, we consider *moments about planes*. If a particle of mass m is located at the point (x, y, z), then its *moment about the xy-plane* is the product of its directed distance from that plane and its mass:

Moment about xy-plane $= zm$.

(See Fig. 17–36a.) The moments about the xz- and yz-planes are defined similarly.

Suppose now that $\mathscr{P}$ is an inner partition of the solid S into rectangular blocks $S_1, S_2, \ldots, S_n$. Let (ξ_k, η_k, ζ_k) be a point on S_k and let ΔV_k be the volume of S_k. The mass of S_k is approximately equal to

$$\delta(\xi_k, \eta_k, \zeta_k)\, \Delta V_k,$$

and its directed distance from the xy-plane is approximately equal to ζ_k. It follows

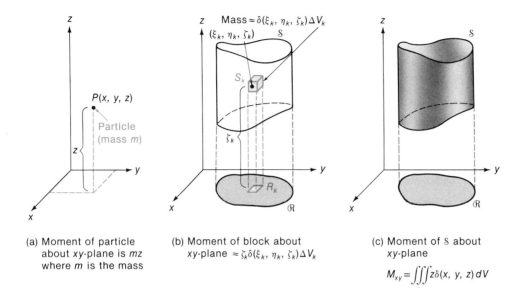

Figure 17–36.

(a) Moment of particle about xy-plane is mz where m is the mass

(b) Moment of block about xy-plane $\approx \zeta_k \delta(\xi_k, \eta_k, \zeta_k) \Delta V_k$

(c) Moment of S about xy-plane
$$M_{xy} = \iiint_S z\delta(x, y, z)\, dV$$

that the moment of S_k about the xy-plane is approximately equal to

$$\zeta_k\, \delta(\xi_k, \eta_k, \zeta_k)\, \Delta V_k.$$

(See Fig. 17–36b.)

The sum

$$\sum_{k=1}^{n} \zeta_k\, \delta(\xi_k, \eta_k, \zeta_k)\, \Delta V_k$$

is approximately equal to the sum of the moments of the blocks about the xy-plane. The *moment of* S *about the* xy-*plane* is defined to be the limit of this sum as the norms of all possible inner partitions approach zero. This moment, denoted by M_{xy}, is

Moment About
xy-Plane

$$M_{xy} = \lim_{\|\mathcal{P}\| \to 0} \sum_{k=1}^{n} \zeta_k\, \delta(\xi_k, \eta_k, \zeta_k)\, \Delta V_k$$

$$= \iiint_{\mathcal{S}} z\, \delta(x, y, z)\, dV.$$

The moments of S about the xz- and yz-planes are defined by similar processes. The integral formulas for the three moments are

Moment
Formulas

$$M_{yz} = \iiint_S x\, \delta(x, y, z)\, dV \quad \text{(moment about } yz\text{-plane)}$$

$$M_{xz} = \iiint_S y\, \delta(x, y, z)\, dV \quad \text{(moment about } xz\text{-plane)}$$

$$M_{xy} = \iiint_S z\, \delta(x, y, z)\, dV \quad \text{(moment about } xy\text{-plane)}$$

17–8 Moments and Centroids

The *centroid* (*center of mass*) of $\mathcal{S}$ is the point $(\bar{x}, \bar{y}, \bar{z})$ where

Coordinates of Centroid of Solid

$$\bar{x} = \frac{M_{yz}}{\text{mass}}, \quad \bar{y} = \frac{M_{xz}}{\text{mass}}, \quad \bar{z} = \frac{M_{xy}}{\text{mass}}.$$

The centroid is the point where the total mass of $\mathcal{S}$ could be concentrated without changing the moments about the coordinate planes.

If no density function is given for $\mathcal{S}$, then we assume that it has the constant density function $\delta(x, y, z) = 1$. In that case the mass is numerically equal to the volume and the formulas for the moments and centroid reduce to

$$M_{yz} = \iiint_\mathcal{S} x \, dV, \quad M_{xz} = \iiint_\mathcal{S} y \, dV, \quad M_{xy} = \iiint_\mathcal{S} z \, dV$$

$$\bar{x} = \frac{M_{yz}}{\text{volume}}, \quad \bar{y} = \frac{M_{xz}}{\text{volume}}, \quad \bar{z} = \frac{M_{xy}}{\text{volume}}.$$

EXAMPLE 1 A solid is bounded by the three coordinate planes and the plane $6x + 3y + 4z = 12$. Find its centroid.

Solution The solid is pictured in Figure 17–37. Its projection on the xy-plane is the closed triangle $\mathcal{R}$ bounded by the x- and y-axes and the line $2x + y = 4$.
The volume of the solid is

$$\iiint_\mathcal{S} dV = \int_0^2 \int_0^{4-2x} \int_0^{(12-6x-3y)/4} dz \, dy \, dx$$

$$= \frac{1}{4} \int_0^2 \int_0^{4-2x} (12 - 6x - 3y) \, dy \, dx$$

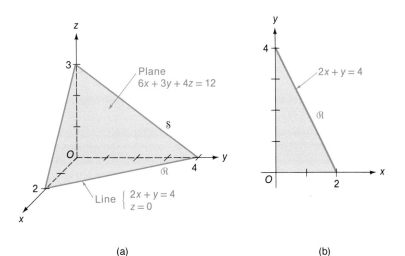

(a)   (b)

Figure 17–37 Example 1.

$$= \frac{1}{4} \int_0^2 \left[12y - 6xy - \frac{3y^2}{2} \right]_0^{4-2x} dx$$

$$= \frac{6}{4} \int_0^2 (x-2)^2 \, dx = \frac{3}{2} \left[\frac{(x-2)^3}{3} \right]_0^2 = 4.$$

The moments about the coordinate planes are

$$M_{yz} = \iiint_S x \, dV$$

$$= \int_0^2 \int_0^{4-2x} \int_0^{(12-6x-3y)/4} x \, dz \, dy \, dx = 2,$$

$$M_{xz} = \iiint_S y \, dV$$

$$= \int_0^2 \int_0^{4-2x} \int_0^{(12-6x-3y)/4} y \, dz \, dy \, dx = 4,$$

and

$$M_{xy} = \iiint_S z \, dV$$

$$= \int_0^2 \int_0^{4-2x} \int_0^{(12-6x-3y)/4} z \, dz \, dy \, dx = 3.$$

The coordinates of the centroid are

$$\bar{x} = \frac{M_{yz}}{V} = \frac{2}{4} = \frac{1}{2},$$

$$\bar{y} = \frac{M_{xz}}{V} = \frac{4}{4} = 1,$$

$$\bar{z} = \frac{M_{xy}}{V} = \frac{3}{4}. \quad \square$$

MOMENTS OF INERTIA

If a particle of mass m is located at the point $P(x, y, z)$, then its distance from the z-axis is $\sqrt{x^2 + y^2}$. The product of the mass and the square of this distance is called the *moment of inertia* of the particle about the z-axis.

Moment of inertia about z-axis = $\underbrace{(x^2 + y^2)}_{\substack{\text{square of} \\ \text{distance} \\ \text{from } z\text{-axis}}} \underbrace{m}_{\text{mass}}.$

(See Fig. 17–38a.) Moments of inertia about the x- and y-axes are defined similarly.

If a small rectangular solid of volume ΔV_k has its center at (ξ_k, η_k, ζ_k), then its mo-

17–8 Moments and Centroids

ment of inertia about the z-axis is approximately equal to

$$\underbrace{(\xi_k{}^2 + \eta_k{}^2)}_{\text{square of distance from z-axis}} \underbrace{\delta(\xi_k, \eta_k, \zeta_k) \, \Delta V_k}_{\text{mass}}$$

where $\delta(\xi_k, \eta_k, \zeta_k)$ is the density at its center. The sum of the moments of inertia about the z-axis of the blocks in an inner partition of a solid S is approximately equal to

$$\sum_{k=1}^{n} (\xi_k{}^2 + \eta_k{}^2) \, \delta(\xi_k, \eta_k, \zeta_k) \, \Delta V_k.$$

(See Fig. 17–38b.)

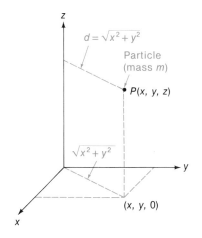 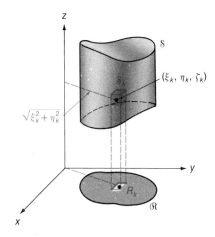

(a) Moment of inertia of particle about z-axis
$= d^2 m = (x^2 + y^2) m$

(b) Moment of inertia of block about z-axis is approximately equal to
$\underbrace{(\xi_k^2 + \eta_k^2)}_{\text{Square of distance}} \underbrace{\delta(\xi_k, \eta_k, \zeta_k) \Delta V_k}_{\text{Mass}}$

Figure 17–38 $I_z = \iiint\limits_S (x^2 + y^2) \, \delta(x, y, z) \, dV.$

We define the moment of inertia of S about the z-axis to be the limit of this sum as the norms of all possible partitions approach zero:

$$I_z = \lim_{\|\mathcal{P}\| \to 0} \sum_{k=1}^{n} (\xi_k{}^2 + \eta_k{}^2) \, \delta(\xi_k, \eta_k, \zeta_k) \, \Delta V_k$$

$$= \iiint\limits_S (x^2 + y^2) \, \delta(x, y, z) \, dV.$$

Multiple Integration

Similar integrals can be used to calculate the moments of inertia about the x- and the y-axes:

Moments of Inertia Formula

$$I_x = \iiint_S (y^2 + z^2)\, \delta(x, y, z)\, dV \quad \text{(moment of inertia about } x\text{-axis)},$$

$$I_y = \iiint_S (x^2 + z^2)\, \delta(x, y, z)\, dV \quad \text{(moment of inertia about } y\text{-axis)},$$

$$I_z = \iiint_S (x^2 + y^2)\, \delta(x, y, z)\, dV \quad \text{(moment of inertia about } z\text{-axis)}.$$

As discussed in Section 17–4, the moment of inertia I_z is proportional to the numerical value of the kinetic energy of S as it spins around the z-axis in a circular orbit with an angular velocity of ω revolutions per second:

Kinetic Energy

$$\text{Kinetic energy} = \frac{\omega^2}{2} I_z.$$

EXAMPLE 2 A solid cylinder S of constant density $\delta(x, y, z) = 1$ is bounded by the graph of $x^2 + y^2 = 1$, the xy-plane, and the plane $z = 4$.
(a) Find the moment of inertia about the z-axis.
(b) Assume that the cylinder spins about the z-axis at a rate of 15 rev/sec. Find the numerical value of its kinetic energy.

Solution (a) The cylinder S is pictured in Figure 17–39a. The projection of S on the xy-plane is the circular disk bounded by the circle $x^2 + y^2 = 1$. (See Fig. 17–39b.)

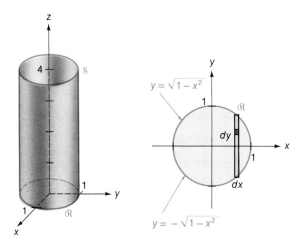

Figure 17–39 Example 2.

The moment of inertia about the z-axis is

$$I_z = \iiint_S (x^2 + y^2)\,\delta(x, y, z)\,dV$$

$$= \int_{-1}^{1} \int_{-\sqrt{1-x^2}}^{\sqrt{1-x^2}} \int_{0}^{4} (x^2 + y^2)\,dz\,dy\,dx$$

$$= 2\pi.$$

(b) If the cylinder spins about the z-axis at a rate of 15 rev/sec, then its angular velocity is

$$\omega = 15 \cdot 2\pi = 30\pi \text{ rad/sec}.$$

The kinetic energy is numerically equal to

$$\text{KE} = \frac{1}{2}\omega^2 I_z = \frac{1}{2}(30\pi)^2 \cdot 2\pi = 900\pi^3. \quad \square$$

RADIUS OF GYRATION

The *radius of gyration* of a solid S with respect to an axis is the distance from the axis at which the mass of S could be concentrated without changing the moment of inertia about that axis. It follows from an argument similar to the one in Section 17–4 that the radii of gyration of S with respect to the coordinate axes are

Radius of Gyration

$$R_x = \sqrt{\frac{I_x}{\text{mass}}}, \quad R_y = \sqrt{\frac{I_y}{\text{mass}}}, \quad R_z = \sqrt{\frac{I_z}{\text{mass}}}.$$

EXAMPLE 3 Find the radius of gyration of the cylinder in Example 2 with respect to the z-axis.

Solution We saw in Example 2 that $I_z = 2\pi$. Since the density is $\delta(x, y, z) = 1$, then the mass is numerically equal to the volume:

$$\text{Mass} = \text{volume} = \pi \cdot 1^2 \cdot 4 = 4\pi.$$

The radius of gyration about the z-axis is

$$R_z = \sqrt{\frac{I_z}{\text{mass}}} = \sqrt{\frac{2\pi}{4\pi}} = \frac{1}{\sqrt{2}}.$$

If the entire mass of the cylinder were to be concentrated at a point $1/\sqrt{2}$ units from the z-axis, the particle would have the same moment of inertia about the z-axis as the original cylinder. $\square$

Exercises 17–8

1–6. A solid S is bounded by the graphs of the given equations. The density of S at the point (x, y, z) is $\delta(x, y, z)$. Find the centroid of S.

• 1 $x^2 + y^2 = 9$, $z = 0$, and $z = 4$; $\delta(x, y, z) = z$

• 2 $z = 4y^2$, $z = 2$, $x = 2$, and $x = 0$; $\delta(x, y, z) = 1$

- 3 $2x + 5y + z = 10$, the coordinate planes; $\delta(x, y, z) = y$
- 4 $z = 4 - x^2$, $x + y = 2$, the coordinate planes; $\delta(x, y, z) = x$
- 5 $x^2 + y^2 = 4$, $y^2 + z^2 = 4$; $x \geq 0$, $y \geq 0$, $z \geq 0$; $\delta(x, y, z) = 1$
- 6 $y = 4 - x^2$, $x - y + 2 = 0$, $z = y$, $x = 0$, $z = 0$; $\delta(x, y, z) = 1$

7–10. A solid of constant density $\delta(x, y, z) = 1$ is bounded by the graphs of the given equations. Set up the integral for the moment of inertia I_z.

- 7 $x^2 + y^2 + z^2 = 4$
- 8 $z = x^2 + y^2$, $z = 9$
- 9 $4x^2 + y^2 = 16$, $z = 0$, and $z = 7$
- 10 $z = 36 - 4x^2 - 9y^2$, $z = 0$

- 11 The sphere in Exercise 7 spins about the z-axis with an angular velocity of $\omega = 20\pi$ rad/sec.
    (a) Find the radius of gyration about the z-axis.
    (b) Find the numerical value of its kinetic energy.

17–9 Cylindrical and Spherical Coordinate Systems

Two alternative coordinate systems for three-dimensional space are of interest. As we shall see, certain types of triple integrals can best be handled by using these coordinate systems instead of the rectangular coordinate system.

THE CYLINDRICAL COORDINATE SYSTEM

The cylindrical coordinate system is a hybrid that combines elements of a polar system and a rectangular system. Let P be a point in space; let Q be its projection on the xy-plane. Let (r, θ) be the polar coordinates of Q (considered as a point in the xy plane), and let z be the z-coordinate of P (measured in rectangular coordinates). The *cylindrical coordinates* of P are (r, θ, z). (See Fig. 17–40a.)

Interpretation of Cylindrical Coordinates

The cylindrical coordinates of a point $P_0(r_0, \theta_0, z_0)$ can be interpreted as follows: (1) The coordinate r_0 is the directed distance from the origin to the projection of P_0 on the xy-plane. Thus P_0 is on a cylinder of radius $|r_0|$ that is symmetric about the z-axis. (See Fig. 17–40b.) (2) The coordinate θ_0 is the angle from the positive x-axis to a vertical plane through P_0 and the z-axis. (3) The coordinate z_0 is the directed distance from the xy-plane to a horizontal plane through P_0. Thus, the cylindrical coordinates locate P_0 as the point of intersection of a cylinder, a vertical plane through the z-axis, and a plane parallel to the xy-plane. (See Fig. 17–40b.)

Several special surfaces have very simple equations in cylindrical coordinates. In particular, right circular cylinders that are symmetric about the z-axis, vertical planes through the z-axis, and horizontal planes have simple equations.

EXAMPLE 1 (a) The graph of the equation $r = 2$ is a right circular cylinder with radius 2, center at the origin, and generating line parallel to the z-axis. (See Fig. 17–41a.) This cylinder also is the graph of the equation $r = -2$.

17-9 Cylindrical and Spherical Coordinate Systems

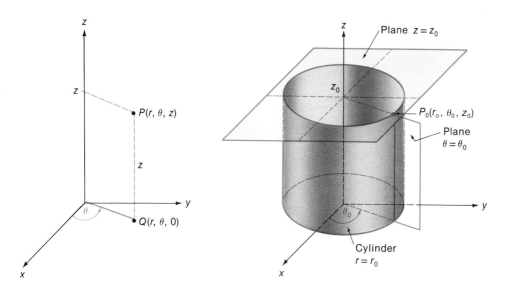

(a) Cylindrical coordinates

(b) $P(r_0, \theta_0, z_0)$ is the point of intersection of the cylinder $r = r_0$, the plane $\theta = \theta_0$ and the the plane $z = z_0$

Figure 17-40.

(b) The graph of the equation $z = 5$ is a plane parallel to the xy-plane. (See Fig. 17-41b.) This equation has the same graph in either rectangular or cylindrical coordinates.

(c) The graph of the equation $\theta = \pi/3$ is a plane perpendicular to the xy-plane. (See Fig. 17-41c.)

(d) The graph of the equation $r = f(\theta)$ is a right cylinder with generating line parallel to the z-axis. (The graph of $r = 1 + \sin \theta$ is shown in Fig. 17-41d.)

(e) The graph of the equation $z = r$ $(r \geq 0)$ is the cone generated by revolving a 45° line in the yz-plane about the z-axis. (See Fig. 17-41e.) □

CONVERSION FROM RECTANGULAR TO CYLINDRICAL COORDINATES

To convert from rectangular to cylindrical coordinates, we simply convert x and y to polar coordinates, using the relations

Conversion Equations
$$x = r \cos \theta, \quad y = r \sin \theta, \quad x^2 + y^2 = r^2.$$

The z-coordinate, of course, is unchanged. These equations also can be used to convert from cylindrical to rectangular coordinates.

EXAMPLE 2 Convert the equation $z = r$ $(r \geq 0)$ to rectangular coordinates.

Solution Since $z = r = \sqrt{x^2 + y^2}$, then
$$z^2 = x^2 + y^2.$$

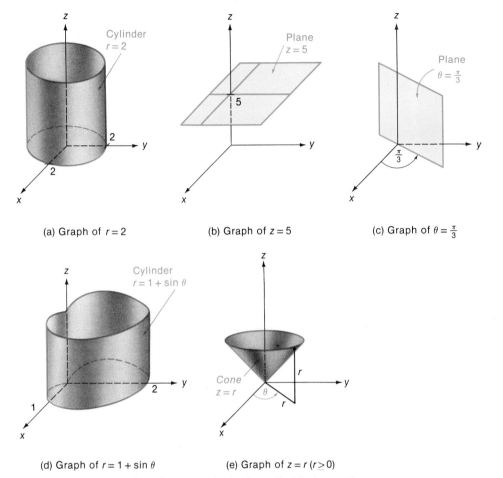

(a) Graph of $r = 2$
(b) Graph of $z = 5$
(c) Graph of $\theta = \frac{\pi}{3}$
(d) Graph of $r = 1 + \sin\theta$
(e) Graph of $z = r \ (r \geq 0)$

Figure 17-41 Example 1. Some graphs in the cylindrical coordinate system.

Since $z \geq 0$, the graph is the top half of the cone obtained by revolving the 45° line about the z-axis. (See Fig. 17-41e.) □

EXAMPLE 3 Convert the equation of the sphere

$$x^2 + y^2 + z^2 = 4 \quad (rectangular\ coordinates)$$

to cylindrical coordinates.

Solution Since $r^2 = x^2 + y^2$, then

$$x^2 + y^2 = 4 - z^2$$
$$r^2 = 4 - z^2. \ \square$$

EXAMPLE 4 Convert the equation $z = r^2$ to rectangular coordinates. Describe the graph.

17-9 Cylindrical and Spherical Coordinate Systems

Solution Since $r^2 = x^2 + y^2$, then

$$z = r^2,$$
$$z = x^2 + y^2.$$

The graph is a paraboloid that opens upward. Each cross section parallel to the xy-plane is a circle. □

THE SPHERICAL COORDINATE SYSTEM

Cylindrical coordinates are useful for problems that involve symmetry about the z-axis. A different system, the *spherical coordinate system,* is frequently used for problems that involve symmetry about the origin.

Let P be a point in space; let Q be its projection on the xy-plane. (See Fig. 17–42a.) Let θ be the angle from the positive x-axis to the line segment OQ, ϕ the smallest nonnegative angle from the positive z-axis to the line segment OP, and $\rho = |OP|$, the distance from O to P. The numbers ρ, ϕ, θ are called the *spherical coordinates* of P. We indicate this fact by the writing $P(\rho, \theta, \phi)$. (See Fig. 17–42a.)

To locate a point in spherical coordinates, we first determine the ray from the origin fixed by the angles θ and ϕ. If $\rho > 0$, we move a distance of ρ units along this ray from the origin to find $P(\rho, \phi, \theta)$. (See Fig. 17–42a.) If $\rho = 0$, the point is the origin. In this case, ϕ and θ are arbitrary.

Interpretation of Spherical Coordinates

The spherical coordinates of a point $P_0(\rho_0, \phi_0, \theta_0)$ can be interpreted as follows: (1) Since ρ_0 is the directed distance from the origin to P_0, then P_0 is on a sphere with radius ρ_0. (See Fig. 17–42b.) (2) The coordinate θ_0 determines a vertical half-plane through P_0 and the z-axis. (3) The coordinate ϕ_0 is the angle of a cone that is symmetric about the z-axis. (Here we extend our definition of a cone somewhat. The xy-plane is such a cone, with $\phi = \pi/2$.) Thus, the spherical coordinates determine $P_0(\rho_0, \phi_0, \theta_0)$ as the point of intersection of a sphere, a vertical half-plane through z-axis, and a cone. (See Fig. 17–42b.)

It is relatively straightforward to convert from rectangular to spherical coordinates. Let P have rectangular coordinates (x, y, z) and spherical coordinates (ρ, ϕ, θ). Let Q, the projection of P on the xy-plane, have cylindrical coordinates $(r, \theta, 0)$. (See Fig. 17–43a.) Then

$$x = r \cos \theta \quad \text{and} \quad y = r \sin \theta.$$

Observe also that $\cos \phi = z/\rho$ and $\sin \phi = r/\rho$, so that

$$z = \rho \cos \phi \quad \text{and} \quad r = \rho \sin \phi.$$

(See Fig. 17–43b.) If we substitute this value of r in the expressions for x and y, we get the conversion equations

Conversion Equations

$$x = \rho \sin \phi \cos \theta, \quad y = \rho \sin \phi \sin \theta, \quad z = \rho \cos \phi.$$

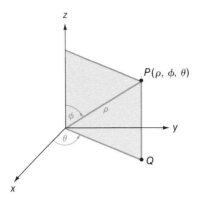

(a) $P(\rho, \phi, \theta)$, $\rho > 0$

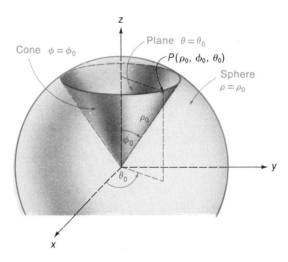

(b) $P_0(\rho_0, \phi_0, \theta_0)$ is the point of intersection of the sphere $\rho = \rho_0$, the half-plane $\theta = \theta_0$, and the cone $\phi = \phi_0$

Figure 17–42 The spherical coordinate system.

EXAMPLE 5 Convert the equation of the sphere

$$x^2 + y^2 + z^2 = a^2$$

to an equivalent equation in spherical coordinates.

Solution Using the conversion equations, we get

$$x^2 + y^2 + z^2 = a^2$$
$$(\rho \sin \phi \cos \theta)^2 + (\rho \sin \phi \sin \theta)^2 + (\rho \cos \phi)^2 = a^2$$
$$\rho^2 \sin^2 \phi (\cos^2 \theta + \sin^2 \theta) + \rho^2 \cos^2 \phi = a^2$$

17-9 Cylindrical and Spherical Coordinate Systems

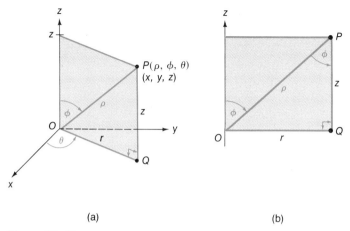

(a) (b)

Figure 17–43.

$$\rho^2(\sin^2\phi \cdot 1 + \cos^2\phi) = a^2$$
$$\rho^2 = a^2$$
$$\rho = a.$$

See Fig. 17–44a. □

EXAMPLE 6 (a) The graph of $\phi = \pi/6$ consists of all points $P(\rho, \phi, \theta)$ with $\phi = \pi/6$. The graph is the cone shown in Figure 17–44b.

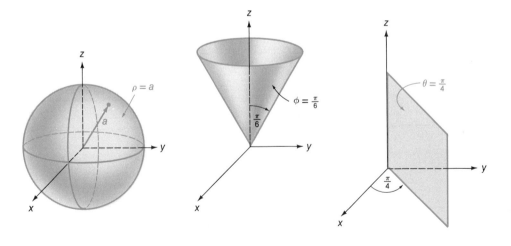

(a) Graph of $\rho = a$ (b) Graph of $\phi = \frac{\pi}{6}$ (c) Graph of $\theta = \frac{\pi}{4}$

Figure 17–44 Examples 5 and 6. Some graphs in the spherical coordinate system.

(b) The graph of $\theta = \pi/4$ is the vertical half-plane that passes through the z-axis and the line $\theta = \pi/4$ in the xy-plane. (See Fig. 17–44c.) □

EXAMPLE 7   Describe the graph of the system of equations

$$\begin{cases} \rho = 2 \\ \phi = \dfrac{\pi}{4}. \end{cases}$$

Solution   The graph of $\rho = 2$ is a sphere of radius 2 with center at the origin. The graph of $\phi = \pi/4$ is a cone generated by revolving the line $z = y$ (in the yz-plane) about the z-axis. (See Fig. 17–45a.) The points that satisfy the system of equations $\rho = 2$, $\phi = \pi/4$ are the points where the cone intersects the sphere $\rho = 2$. The graph of the system is a circle of radius $2/\sqrt{2} = \sqrt{2}$ that is parallel to the xy-plane and is above the xy-plane. (See Fig. 17–45b.) □

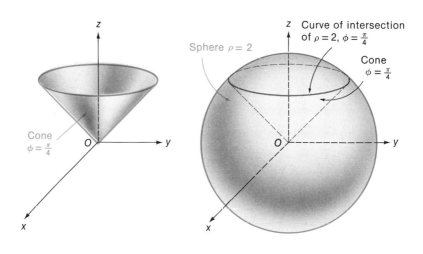

(a)   (b)

Figure 17–45 Example 7.

Exercises 17–9

1–4. Sketch the graph. (Use cylindrical coordinates.)

• 1 $\theta = \dfrac{\pi}{4}, \; 2 \leq r \leq 4$ 2 $r = 2$

• 3 $r = 3, \; \dfrac{\pi}{4} \leq \theta \leq \dfrac{3\pi}{4}, \; 1 \leq z \leq 5$ 4 $z = 4$

5–8. Convert the equation from rectangular to cylindrical coordinates. Sketch the graph.

- 5. $x^2 + y^2 + z = 4$
- 6. $x^2 + y^2 + z^2 = 36$
- 7. $x + 2y = 3$
- 8. $x^2 + y^2 - 2x = 0$

9–12. Convert the equation from cylindrical to rectangular coordinates. Sketch the graph.

- 9. $r = -4$
- 10. $z + r = 1$
- 11. $r\cos\theta + r\sin\theta + 2z = 5$
- 12. $r = -z^2$

13–16. Sketch the graph. (Use spherical coordinates.)

- 13. $\phi = \dfrac{\pi}{3}$
- 14. $\rho = 1,\ 0 \le \phi \le \dfrac{\pi}{3}$
- 15. $\rho = 2,\ 0 \le \theta \le \dfrac{\pi}{2},\ 0 \le \phi \le \dfrac{\pi}{2}$
- 16. $\theta = \dfrac{\pi}{4},\ 0 \le \phi \le \dfrac{\pi}{4},\ 0 \le \rho \le 5$

17–20. Convert the equation to spherical coordinates. Sketch the graph.

- 17. $x^2 + y^2 = 9$
- 18. $x = -y$
- 19. $x^2 + y^2 + z^2 = -4z + 13$
- 20. $r = \cos 2\theta$ (cylindrical coordinates)

21–24. Convert the equation from spherical to rectangular coordinates. Sketch the graph.

- 21. $\rho = 6$
- 22. $\theta = \dfrac{\pi}{2}$
- 23. $\phi = \dfrac{3\pi}{4}$
- 24. $\rho = 1,\ \phi = \dfrac{\pi}{2}$

17–10 Triple Integrals in Cylindrical and Spherical Coordinates

CYLINDRICAL COORDINATES

Let S be a solid in three-dimensional space. Let $\mathcal{R}$ be the projection of S on the xy-plane. We partition $\mathcal{R}$ into closed subregions formed by concentric circles with centers at the origin and rays from the origin, as with polar coordinates. These subregions are the projections of cylinders through S. (See Fig. 17–46a.) Next, we slice these cylinders with horizontal planes, forming a set of blocks. (See Figs. 17–46a, b.) Finally, we discard all the blocks that are not completely contained in S. The resulting set of blocks forms an inner partition $\mathcal{P}$ of S. The *norm* of the partition, denoted by $\|\mathcal{P}\|$, equals the length of the longest diagonal of a block in the partition.

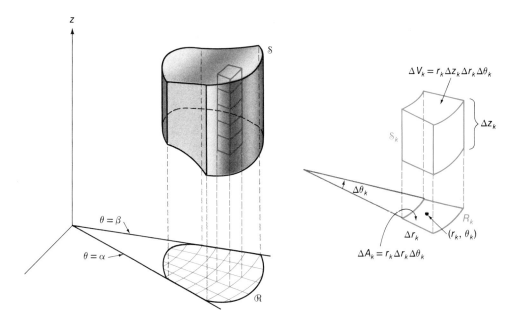

(a) Part of an inner partition of S

(b) The volume of the kth block
$\Delta V_k = r_k \Delta z_k \Delta r_k \Delta \theta_k$

Figure 17–46 Cylindrical coordinates. Volume of $S = \iiint\limits_{S} dV = \iiint\limits_{S} r\, dz\, dr\, d\theta$.

The volume of a typical block S_k in the inner partition of S is

$$\Delta V_k = r_k\, \Delta r_k\, \Delta z_k\, \Delta \theta_k,$$

where (r_k, θ_k, z_k) is the center of the kth block. (See Fig. 17–46b.) The total volume of S is approximately equal to

$$\sum_{k=1}^{n} r_k\, \Delta z_k\, \Delta r_k\, \Delta \theta_k.$$

The exact value of the volume is the limit of this sum as $\|\mathcal{P}\| \to 0$:

$$\text{Volume} = \lim_{\|\mathcal{P}\| \to 0} \sum_{k=1}^{n} r_k\, \Delta z_k\, \Delta r_k\, \Delta \theta_k = \iiint\limits_{S} r\, dz\, dr\, d\theta.$$

A similar integration formula holds for any continuous function $F(r, \theta, z)$ defined on S. The *triple integral* of F over S can be written

Triple Integral in Cylindrical Coordinates
$$\iiint\limits_{S} F(r, \theta, z)\, dV = \iiint\limits_{S} F(r, \theta, z) r\, dz\, dr\, d\theta.$$

All the previously established integral formulas can be extended to integrals in cylindrical coordinates. For example, if $\delta(r, \theta, z)$ is the density of S at the point

17-10 Triple Integrals in Cylindrical and Spherical Coordinates

(r, θ, z), then the mass of S and the moments of S about the coordinate planes are given by the integrals:

$$\text{Mass} = \iiint_S \delta(r, \theta, z)\, dV = \iiint_S \delta(r, \theta, z) r\, dz\, dr\, d\theta,$$

$$M_{xy} = \iiint_S z\, \delta(r, \theta, z)\, dV = \iiint_S z\, \delta(r, \theta, z) r\, dz\, dr\, d\theta,$$

$$M_{yz} = \iiint_S \underbrace{r \cos \theta}_{x}\, \delta(r, \theta, z)\, dV = \iiint_S r \cos \theta \cdot \delta(r, \theta, z) \cdot r\, dz\, dr\, d\theta,$$

$$M_{xz} = \iiint_S \underbrace{r \sin \theta}_{y}\, \delta(r, \theta, z)\, dV = \iiint_S r \sin \theta \cdot \delta(r, \theta, z) \cdot r\, dz\, dr\, d\theta.$$

These formulas can be used to calculate the centroid of S. The coordinates of the centroid are, of course, in rectangular coordinates.

ITERATED INTEGRALS IN CYLINDRICAL COORDINATES

Let S be a solid with lower boundary surface $z = g_1(r, \theta)$ and upper boundary surface $z = g_2(r, \theta)$. (See Fig. 17-47a.) Let $\mathcal{R}$ be the projection of S on the xy-plane. Suppose that the inner boundary of $\mathcal{R}$ is the graph of $r = f_1(\theta)$ and the outer bound-

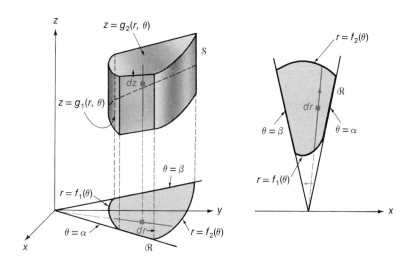

(a) The first integration is from the lower surface to the upper surface

(b) The second and third integrations are over the projection on the xy-plane

Figure 17-47 Schematic representation of triple integration in cylindrical coordinates:

$$\iiint_S F(r, \theta, z)\, dV = \int_\alpha^\beta \int_{f_1(\theta)}^{f_2(\theta)} \int_{g_1(r,\theta)}^{g_2(r,\theta)} F(r, \theta, z)\, r\, dr\, d\theta$$

ary the graph of $r = f_2(\theta)$, for $\alpha \le \theta \le \beta$. (See Figs. 17–47a, b.) Let $F(r, \theta, z)$ be continuous on S. Let g_1 and g_2 be continuous on $\mathcal{R}$; let f_1 and f_2 be continuous for $\alpha \le \theta \le \beta$. It can be proved that

$$\iiint_S F(r, \theta, z)\, dV = \iint_\mathcal{R} \left[\int_{g_1(r,\theta)}^{g_2(r,\theta)} F(r, \theta, z)\, dz \right] dA$$

$$= \int_\alpha^\beta \int_{f_1(\theta)}^{f_2(\theta)} \int_{g_1(r,\theta)}^{g_2(r,\theta)} F(r, \theta, z)\, r\, dz\, dr\, d\theta.$$

Remark It is convenient to set up integrals in cylindrical coordinates so that the order of integration is given in the integral by $dz\, dr\, d\theta$. If this is done, the following observations can help to determine the limits of integration.

(1) The first integration, which is with respect to z, is along a line parallel to the z-axis from the lower boundary of S to the upper boundary (*surface to surface*). (See Fig. 17–47a.)

(2) The second integration, which is with respect to r, is defined over the region $\mathcal{R}$—the projection of S on the xy-plane. This integration is along a ray through the origin from the inner boundary of $\mathcal{R}$ to the outer boundary (*curve to curve*). (See Fig. 17–47b.)

(3) The third integration, which is with respect to θ, is from the lower value of θ that intersects the region $\mathcal{R}$ to the upper value. (See Fig. 17–47b.)

EXAMPLE 1 Let S be the solid bounded below by the cone $z = r$ ($r \ge 0$) and above by the hemisphere $z = \sqrt{8 - x^2 - y^2}$. Find the volume of S.

Solution The equation of the hemisphere can be written as $z = \sqrt{8 - r^2}$. The cone and hemisphere intersect when

$$z = r = \sqrt{8 - r^2}$$
$$r^2 = 8 - r^2$$
$$2r^2 = 8$$
$$r = 2.$$

(See Fig. 17–48.) Thus, the projection of the solid on the xy-plane is the circular disk bounded by the circle $r = 2$.

Observe that z varies from r to $\sqrt{8 - r^2}$, r from 0 to 2, and θ from 0 to 2π. Therefore:

$$V = \int_0^{2\pi} \int_0^2 \int_r^{\sqrt{8-r^2}} r\, dz\, dr\, d\theta = \int_0^{2\pi} \int_0^2 \left[rz \right]_{z=r}^{z=\sqrt{8-r^2}} dr\, d\theta$$

$$= \int_0^{2\pi} \int_0^2 [(8 - r^2)^{1/2} r - r^2]\, dr\, d\theta = -\frac{1}{3} \int_0^{2\pi} \left[(8 - r^2)^{3/2} + r^3 \right]_0^2 d\theta$$

$$= -\frac{1}{3} [16 - 8\sqrt{8}] \int_0^{2\pi} d\theta$$

$$= \frac{8\sqrt{8} - 16}{3} \cdot 2\pi. \ \square$$

17–10 Triple Integrals in Cylindrical and Spherical Coordinates

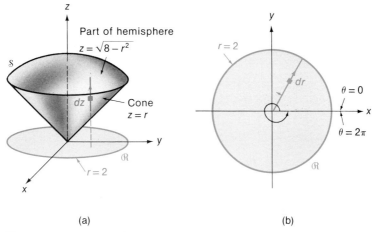

(a)　　　　　　　　　　　　　　　　(b)

Figure 17–48 Example 1.

EXAMPLE 2　Find the centroid of the solid in Octant I that is bounded below by the cone $z = r$ ($r \geq 0$) and above by the plane $z = 3$.

Solution　The solid is shown in Figure 17–49. Its projection on the xy-plane is bounded by the quarter circle $r = 3$ and the rays $\theta = 0$, $\theta = \pi/2$. The volume is

$$V = \int_0^{\pi/2} \int_0^3 \int_r^3 r \, dz \, dr \, d\theta = \frac{9\pi}{4}.$$

(The details of the computation are left for you to work out.) The moments about the coordinate planes are

$$M_{yz} = \iiint_S x \, dV = \iiint_S r \cos \theta \, dV$$

$$= \int_0^{\pi/2} \int_0^3 \int_r^3 r \cos \theta \cdot r \, dz \, dr \, d\theta = \frac{27}{4},$$

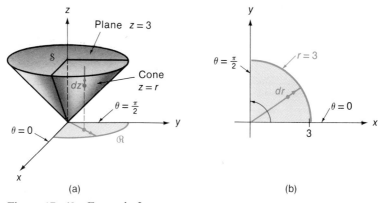

(a)　　　　　　　　　　　　　　　　(b)

Figure 17–49 Example 2.

$$M_{xz} = \iiint_S y\, dV = \iiint_S r\sin\theta\, dV$$

$$= \int_0^{\pi/2}\int_0^3\int_r^3 r\sin\theta \cdot r\, dz\, dr\, d\theta = \frac{27}{4},$$

$$M_{xy} = \iiint_S z\, dV = \int_0^{\pi/2}\int_0^3\int_r^3 zr\, dz\, dr\, d\theta = \frac{81\pi}{16}.$$

The coordinates of the centroid (in rectangular coordinates) are

$$\bar{x} = \frac{M_{yz}}{V} = \frac{27/4}{9\pi/4} = \frac{3}{\pi},$$

$$\bar{y} = \frac{M_{xz}}{V} = \frac{27/4}{9\pi/4} = \frac{3}{\pi},$$

$$\bar{z} = \frac{M_{xy}}{V} = \frac{81\pi/16}{9\pi/4} = \frac{9}{4}. \quad \square$$

EXAMPLE 3 Find the volume of the solid interior to the cylinder $x^2 + y^2 = 1$ that is bounded above by the paraboloid $z = 4 - x^2 - y^2$ and below by the xy-plane.

Solution The solid is shown in Figure 17–50. In cylindrical coordinates the upper boundary is the graph of $z = 4 - r^2$ and the lower boundary is the graph of $z = 0$. The projection of the solid on the xy-plane is bounded by the circle $r = 1$.

The volume equals

$$V = \iiint_S r\, dz\, dr\, d\theta = \int_0^{2\pi}\int_0^1\int_0^{4-r^2} r\, dz\, dr\, d\theta$$

$$= \int_0^{2\pi}\int_0^1 r\left[z\right]_0^{4-r^2} dr\, d\theta = \int_0^{2\pi}\int_0^1 (4r - r^3)\, dr\, d\theta$$

$$= \int_0^{2\pi}\left[2r^2 - \frac{r^4}{4}\right]_0^1 d\theta = \frac{7\pi}{2}. \quad \square$$

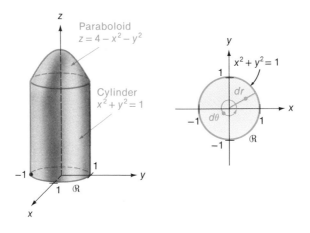

Figure 17–50 Example 3.

17–10 Triple Integrals in Cylindrical and Spherical Coordinates

SPHERICAL COORDINATES

Let $F(\rho, \phi, \theta)$ be continuous on a solid $\mathcal{S}$. We decompose $\mathcal{S}$ into small blocks of the type shown in Figure 17–51. The volume of a typical block in such a partition is

$$\Delta V_k \approx \rho_k^2 \sin \phi_k \, \Delta\rho_k \, \Delta\phi_k \, \Delta\theta_k,$$

where $(\rho_k, \phi_k, \theta_k)$ is the midpoint of the kth block.

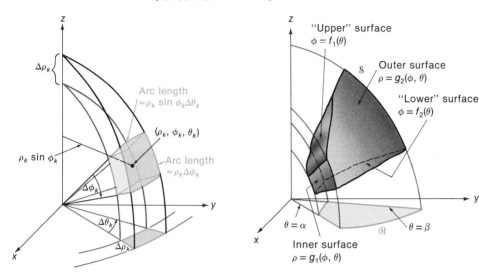

(a) A typical block in an inner partition of a solid $\mathcal{S}$: $V_k \approx \rho_k^2 \sin \phi_k \, \Delta\rho_k \Delta\phi_k \Delta\theta_k$

(b) A solid $\mathcal{S}$: $V = \iiint_\mathcal{S} \rho^2 \sin \phi \, d\rho \, d\phi \, d\theta$

Figure 17–51.

The triple integral of F over $\mathcal{S}$ is approximately equal to the sum

$$\sum_{k=1}^{n} \underbrace{F(\rho_k, \phi_k, \theta_k)}_{\substack{\text{value of } F \text{ at} \\ \text{midpoint of } k\text{th} \\ \text{block}}} \underbrace{\rho_k^2 \sin \phi_k \, \Delta\rho_k \, \Delta\phi_k \, \Delta\theta_k}_{\substack{\text{approximate} \\ \text{volume of } k\text{th block}}}.$$

If we take the limit of this expression as each dimension of each block approaches zero, we get the exact value of the integral:

Triple Integral in Spherical Coordinates

$$\iiint_\mathcal{S} F(\rho, \phi, \theta) \, dV = \iiint_\mathcal{S} F(\rho, \phi, \theta) \, \rho^2 \sin \phi \, d\rho \, d\phi \, d\theta.$$

ITERATED INTEGRALS IN SPHERICAL COORDINATES

A typical solid $\mathcal{S}$ is pictured in Figure 17–51b. This solid has its "inner" boundary surface defined by $\rho = g_1(\phi, \theta)$ and its "outer" boundary surface by $\rho = g_2(\phi, \theta)$.

Multiple Integration

For each value of θ, the angle ϕ varies from $\phi = f_1(\theta)$ to $\phi = f_2(\theta)$. Finally, θ varies from α to β.

To express a triple integral over S as an iterated integral, we note that ρ varies from $g_1(\phi, \theta)$ to $g_2(\phi, \theta)$, ϕ varies from $f_1(\theta)$ to $f_2(\theta)$, and θ from α to β. It can be proved that

Triple and Iterated Integrals in Spherical Coordinates

$$\iiint_S F(\rho, \phi, \theta)\, dV = \int_\alpha^\beta \int_{f_1(\theta)}^{f_2(\theta)} \int_{g_1(\phi,\theta)}^{g_2(\phi,\theta)} F(\rho, \phi, \theta)\, \rho^2 \sin\phi\, d\rho\, d\phi\, d\theta.$$

Remark We perform the integrations for the integral

$$\int_\alpha^\beta \int_{f_1(\theta)}^{f_2(\theta)} \int_{g_1(\phi,\theta)}^{g_2(\phi,\theta)} F(\rho, \phi, \theta)\, \rho^2 \sin\phi\, d\rho\, d\phi\, d\theta,$$

as follows:

(1) The first integration is with respect to ρ. We integrate along a ray through the origin from the inner boundary of S to the outer boundary. (See Fig. 17–52a.)

(2) The second integration is with respect to ϕ. In carrying out this integration, we temporarily fix the value of θ and let ϕ vary from $f_1(\theta)$ to $f_2(\theta)$. Here $f_1(\theta)$ is the smallest value of ϕ that intersects the solid and $f_2(\theta)$ is the largest value of ϕ. (See Fig. 17–52b.)

(3) The third integration is with respect to θ. We let θ vary from α, the smallest value of θ that intersects the solid, to β, the largest value. (See Fig. 17–52c.)

EXAMPLE 4   A solid sphere bounded by the graph of $\rho = 4$ has density $\delta(\rho, \phi, \theta) = \sqrt{\rho}$ at the point (ρ, ϕ, θ). A wedge bounded by the planes $\theta = \pi/4$ and $\theta = \pi/2$ is cut from the sphere, and the rest of the sphere is discarded. Find the volume and the mass of the wedge.

Solution   The wedge cut from the sphere is pictured in Figure 17–35a. Observe that ρ varies from 0 to 4, ϕ from 0 to π, and θ from $\pi/4$ to $\pi/2$. Then

$$\text{Volume} = \iiint_S dV = \int_{\pi/4}^{\pi/2} \int_0^\pi \int_0^4 \underbrace{\rho^2 \sin\phi\, d\rho\, d\phi\, d\theta}_{dV} = \frac{32\pi}{3}$$

and

$$\text{Mass} = \iiint_S \delta(\rho, \phi, \theta)\, dV = \int_{\pi/4}^{\pi/2} \int_0^\pi \int_0^4 \underbrace{\sqrt{\rho}}_{\delta} \cdot \underbrace{\rho^2 \sin\phi\, d\rho\, d\phi\, d\theta}_{dV}$$

$$= \frac{128\pi}{7}. \qquad \square$$

EXAMPLE 5   Calculate the volume and the z-coordinate of the centroid of the hemisphere of radius 3 that is symmetric about the z-axis and bounded below by the xy-plane.

17–10 Triple Integrals in Cylindrical and Spherical Coordinates

Solution The hemisphere is pictured in Figure 17–53b. Observe that ρ varies from 0 to 3, ϕ from 0 to $\pi/2$, and θ from 0 to 2π.

The volume is

$$V = \iiint_S dV = \int_0^{2\pi} \int_0^{\pi/2} \int_0^3 \underbrace{\rho^2 \sin\phi \, d\rho \, d\phi \, d\theta}_{dV} = 18\pi.$$

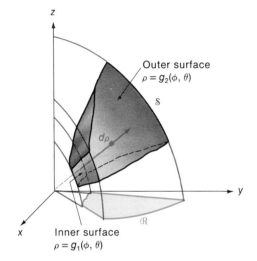

(a) The first integration is from the inner surface to the outer surface along a ray

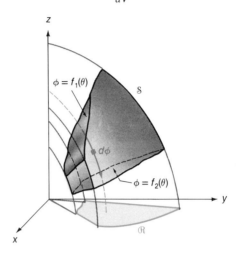

(b) The second integration is from the upper surface to the lower surface with respect to ϕ

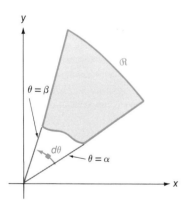

(c) The third integration is over the angle from α to β in the xy-plane

Figure 17–52 Schematic representation of triple integration in spherical coordinates:

$$\iiint_S F(\rho, \phi, \theta) \, dV = \int_\alpha^\beta \int_{f_1(\theta)}^{f_2(\theta)} \int_{g_1(\phi,\theta)}^{g_2(\phi,\theta)} F(\rho, \phi, \theta) \rho^2 \sin\phi \, d\rho \, d\phi \, d\theta.$$

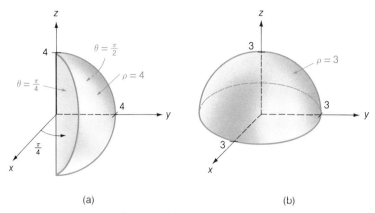

(a)   (b)

Figure 17–53 Examples 4 and 5.

The moment about the xy-plane is

$$M_{xy} = \iiint_S z\, dV$$

$$= \iiint_S \underbrace{\rho \cos \phi}_{z} \cdot \underbrace{\rho^2 \sin \phi\, d\rho\, d\phi\, d\theta}_{dV}$$

$$= \int_0^{2\pi} \int_0^{\pi/2} \int_0^3 \rho^3 \cos \phi \sin \phi\, d\rho\, d\phi\, d\theta = \frac{81\pi}{4}.$$

The z-coordinate of the centroid is

$$\bar{z} = \frac{M_{xy}}{V} = \frac{81\pi/4}{18\pi} = \frac{9}{8}.$$

Since the hemisphere is symmetric about the xy- and yz-planes, then its centroid is on the z-axis. Thus, the centroid is the point $(0, 0, \frac{9}{8})$ (in rectangular coordinates). □

EXAMPLE 6   A solid is bounded by the upper nappe of the cone $\phi = \pi/4$ and by the spheres $\rho = 5$ and $\rho = 8$. Calculate its volume.

Solution   The solid is pictured in Figure 17–54. Observe that ρ varies from 5 to 8, ϕ from 0 to $\pi/4$, and θ from 0 to 2π. The volume is

$$V = \iiint_S \rho^2 \sin \phi\, d\rho\, d\phi\, d\theta = \int_0^{2\pi} \int_0^{\pi/4} \int_5^8 \rho^2 \sin \phi\, d\rho\, d\phi\, d\theta$$

$$= \int_0^{2\pi} \int_0^{\pi/4} \sin \phi \left[\frac{\rho^3}{3}\right]_5^8 d\phi\, d\theta = 129 \int_0^{2\pi} \left[-\cos \phi\right]_0^{\pi/4} d\theta$$

$$= 129 \left(1 - \frac{\sqrt{2}}{2}\right) \int_0^{2\pi} d\theta = 129 \left(1 - \frac{\sqrt{2}}{2}\right) 2\pi \approx 237.4. \quad \square$$

Exercises 17-10

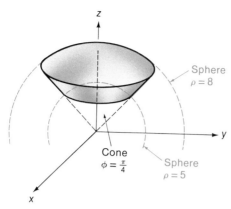

Figure 17-54 Example 6.

Exercises 17-10

1-4. Evaluate the iterated integral.

● 1 $\displaystyle\int_0^{\pi/2}\int_0^4\int_0^r r^2\,dz\,dr\,d\theta$

2   $\displaystyle\int_0^{\pi}\int_0^{\theta}\int_0^{r\theta} r\,dz\,dr\,d\theta$

● 3 $\displaystyle\int_{\pi/4}^{\pi/2}\int_0^{\pi}\int_0^4 \rho^2 \sin\phi\,d\rho\,d\phi\,d\theta$

4   $\displaystyle\int_0^{2\pi}\int_0^{\pi/2}\int_0^3 \rho^3 \cos\phi \sin\phi\,d\rho\,d\phi\,d\theta$

5-10. Use an integral in cylindrical coordinates to calculate the volume of the solid described.

● 5   Bounded by the graphs of $x^2 + y^2 + z = 4$, $z = 0$.

● 6   Interior to the sphere $x^2 + y^2 + z^2 = 36$ and the cylinder $x^2 + y^2 = 9$.

● 7   Bounded below by the paraboloid $z = x^2 + y^2$ and above by the plane $z = 2y$.

● 8   Bounded above by the sphere $x^2 + y^2 + z^2 = 32$ and below by the paraboloid $4z = x^2 + y^2$.

● 9   Bounded by the graphs of $z = x^2 + y^2$ and $2z = x^2 + y^2 + 1$.

●10   In Octant I. Bounded by the cylinder $(x - 1)^2 + y^2 = 1$, the paraboloid $z = x^2 + y^2$, and the xy-plane.

11-12. Use an integral in cylindrical coordinates to find the centroid of the solid described.

●11   Bounded by the graphs of $z = 5 - x^2 - y^2$, $x^2 + y^2 = 1$ and $z = 1$; $\delta(x, y, z) = z$.

●12   Interior to the cylinder $x^2 + y^2 - 2y = 0$ and the sphere $x^2 + y^2 + z^2 = 4$ and above the xy-plane, $\delta(x, y, z) = z$.

13–14. Use an integral in cylindrical coordinates to find the moment of inertia of the given solid about the z-axis.

•13 Bounded above by the sphere $x^2 + y^2 + z^2 = 16$ and below by the paraboloid $6z = x^2 + y^2$; $\delta(x, y, z) = z$.

•14 Bounded below by the upper nappe of the cone $z^2 = x^2 + y^2$, bounded above by the plane $z = 4$, $\delta(r, \theta, z) = r$.

•15 A solid S of constant density $\delta(\rho, \phi, \theta) = 1$ is bounded above by the sphere $\rho = 4$ and below by the cone $\phi = \pi/6$. Use spherical coordinates to find (a) the volume of S; (b) the centroid of S.

•16 A solid S is bounded below by the cone $\phi = \pi/4$ and above by the plane $z = 3$ (rectangular coordinates).
    (a) Use spherical coordinates to find the volume of S.
    (b) The density of S is given by $\delta(\rho, \phi, \theta) = \rho$. Use spherical coordinates to find the mass of S.

•17 A solid of constant density $\delta = 1$ is bounded above by the sphere $\rho = 2$ and below by the cone $\phi = \pi/3$.
    (a) Find the moment of inertia I_z.
    (b) Find the radius of gyration R_z.

•18 A hemisphere is bounded by the graph of $\rho = 1$, $0 \leq \phi \leq \pi/2$, and the xy-plane. Its density is given by $\delta(\rho, \phi, \theta) = \rho$.
    (a) Find moment of inertia I_z.
    (b) Find the radius of gyration R_z.

19 Convert the integral

$$\int_{\pi/2}^{\pi} \int_0^2 \int_0^{\sqrt{4-r^2}} r^2 \cos\theta \, dz \, dr \, d\theta$$

to an equivalent integral in (a) rectangular coordinates; (b) spherical coordinates.

Review Problems

1 Define the following concepts in two-dimensional space (rectangular coordinates):
    (a) "Basic closed region."
    (b) "Inner partition of a basic closed region."
    (c) "Norm of an inner partition."

2 Define the double integral as a limit of a sum.

3–6. Evaluate the integrals.

•3 $\int_1^2 \int_0^{x-1} y^5 \, dy \, dx$

•4 $\int_0^1 \int_0^{2x} \sin(\pi x^2) \, dy \, dx$

Review Problems

- **5** $\displaystyle\int_0^1 \int_{x^2}^{\sqrt{x}} (x^2 + 2xy)\, dy\, dx$

- **6** $\displaystyle\int_0^3 \int_{y^2}^{3y} x\, dx\, dy$

7–10. Change the integrals in Problems 3 to 6 to equivalent integrals with a different order of integration.

- **7** Problem 3
- **8** Problem 4
- **9** Problem 5
- **10** Problem 6

- **11** Let S be the solid bounded by the graphs of $z = (x - 1)^2 + (y - 1)^2$ and $2x + 2y + z = 6$.

    (a) Express the volume of S as an iterated double integral. Calculate its value.

    (b) Express the volume of S as an iterated triple integral.

12 Let $\mathcal{R}$ be the plane region bounded by the graphs of the continuous functions $y = f(x)$ and $y = g(x)$, $a \le x \le b$. Suppose that $f(x) \ge g(x)$ for $a \le x \le b$. Express the moments M_x and M_y as iterated integrals. Show that these integrals reduce to

$$M_y = \int_a^b x\, [f(x) - g(x)]\, dx \quad \text{and} \quad M_x = \frac{1}{2}\int_a^b [f(x)^2 - g(x)^2]\, dx.$$

- **13** A lamina of density $\delta(x, y) = y$ covers the region $\mathcal{R}$ bounded by the graphs of $y = x^3$ and $y = 4x$, $0 \le x \le 2$. Use iterated integrals to calculate (a) the area of $\mathcal{R}$; (b) the mass of the lamina; (c) the centroid of the lamina; (d) the moment of inertia I_y of the lamina; (e) the radius of gyration of the lamina about the y-axis.

14–15. Change to equivalent integrals in polar coordinates. Evaluate the new integrals.

- **14** $\displaystyle\int_0^2 \int_0^{\sqrt{4-x^2}} e^{x^2} e^{y^2}\, dy\, dx$

- **15** $\displaystyle\int_{-1}^0 \int_{-y}^{\sqrt{2-y^2}} x\, dx\, dy$

16–17. Use an iterated integral in polar coordinates to calculate the area of the closed region.

- **16** The region bounded by the cardioid $r = 1 + \cos\theta$.

- **17** The region that is inside the circle $r = 3\cos\theta$ and outside the cardioid $r = 1 + \cos\theta$.

- **18** Calculate the rectangular coordinates of the centroid of the region enclosed by the cardioid $r = 1 + \cos\theta$.

19–20. (a) Write an integral for the area of the surface.

(b) Evaluate the integral. (*Hint:* It may be convenient to use polar coordinates.)

- **19** The portion of the paraboloid $z = 4 - x^2 - y^2$ that is above the xy-plane.

- **20** The portion of the sphere $x^2 + y^2 + z^2 = 9$ that is outside the paraboloid $z = 9 - x^2 - y^2$.

21 (a) State the formula for the surface area of a hemisphere of radius r.

(b) Derive the formula in (a).

22–23. Evaluate the integral.

- **22** $\displaystyle\int_0^2 \int_0^{2-x} \int_0^{4-2x-2y} \sqrt{z}\ dz\ dy\ dx$

- **23** $\displaystyle\int_{\pi/2}^{\pi} \int_{\pi/4}^{\pi/2} \int_2^4 \rho^2 \sin\phi\ d\rho\ d\phi\ d\theta$

- **24** Change the integral in Problem 22 to an equivalent iterated integral with the order of integration given by $dV = dx\ dy\ dz$.

- **25** A solid in Octant I is bounded by the coordinate planes and the cylinders $z = 9 - x^2$ and $y = 9 - x^2$. Calculate (a) the volume; (b) the coordinates of the centroid.

- **26** Write an integral for the moment of inertia of the ellipsoid $x^2 + 4y^2 + 4z^2 = 4$ about the x-axis.

27–32. Sketch the graphs.

27 $r = 2$ (cylindrical coordinates)   **28** $z = 2$ (cylindrical coordinates)

29 $\theta = -\pi/4$ (cylindrical coordinates)   **30** $\rho = 2$ (spherical coordinates)

31 $\phi = \pi/6$ (spherical coordinates)

32 $\theta = \pi/4$, $\rho = 2$ (spherical coordinates)

- **33** Let S be the solid bounded by the graphs of $z = 0$, $x^2 + y^2 = 16$ and $z = 32 - x^2 - y^2$ that is inside the cylinder and the paraboloid.

  (a) Set up an integral in cylindrical coordinates for the volume of S.

  (b) Calculate the volume.

- **34** Let S be the solid bounded above by the sphere $\rho = 6$ and below by the cone $\phi = \pi/4$. Let $\mathcal{T}$ be the portion of S in Octant I.

  (a) Set up an integral in spherical coordinates for the volume of $\mathcal{T}$.

  (b) Evaluate the integral in (a).

35 A solid cone with vertex at the center of a sphere is cut out of the sphere leaving a hole. After a suitable translation and rotation of axes, the volume of the remaining part of the sphere can be described by the integral

$$\int_0^{2\pi} \int_{\pi/6}^{\pi} \int_0^5 \rho^2 \sin\phi \, d\rho \, d\phi \, d\theta.$$

(a) Describe the cone and the sphere.

•(b) Calculate the volume of the part of the sphere that remains and the volume of the part that was cut out.

36 (a) Set up an integral in spherical coordinates for the volume of the solid in Octant I bounded by the planes $x = \sqrt{3}y$, $y = \sqrt{3}x$ and the sphere $\rho = 6$.
•(b) Evaluate the integral in (a).

18 Introduction to Vector Analysis

18–1 Vector Fields

Vector-valued functions can be defined at points in three-dimensional space as well as at points on the t-axis. For example, the gravitational attraction in the earth's gravitational field is applied (theoretically) at every point in space. If we locate the origin of the xyz-coordinate system at the center of the earth, then the gravitational attraction of a particle at $P(x, y, z)$ is represented by a vector at P that points towards the origin. (See Fig. 18-1).

Another example of a vector-valued function is furnished by the gradient. If $f(x, y, z)$ is differentiable on its domain, then the function F defined by

$$F(x, y, z) = \nabla f(x, y, z)$$
$$= f_x(x, y, z)\mathbf{i} + f_y(x, y, z)\mathbf{j} + f_z(x, y, z)\mathbf{k}$$

is a vector-valued function defined on the domain of f.

Observe that the domain of a vector-valued function can be an interval on an axis (as in our work in Chapter 15), a subset of the xy-coordinate plane, or a subset of

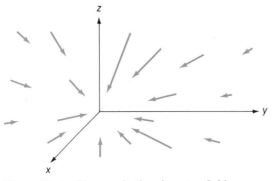

Figure 18–1 The gravitational vector field.

18–1 Vector Fields

Vector Field

three-dimensional space. If a vector-valued function is defined over a region $\mathcal{R}$ (open or closed) in two- or three-dimensional space, then the function is called a *vector field* over $\mathcal{R}$.

A vector field F is said to be *continuous* on the region $\mathcal{R}$ if each of its component functions is continuous on $\mathcal{R}$. Most of the vector fields we work with are continuous on their domains.

EXAMPLE 1 (*Gravitational Fields*) Newton's law of gravitational attraction states that the gravitational attraction of a particle of mass m towards the center of the earth is numerically equal to

$$\frac{gm}{d^2},$$

where g is the gravitational constant and d is the distance from the center of the earth.

If the center of the earth is located at the origin of the xyz-coordinate system and the particle is at $P(x, y, z)$, then the attraction is numerically equal to

$$\frac{gm}{\|\overrightarrow{OP}\|^2} = \frac{gm}{x^2 + y^2 + z^2}.$$

Since the attraction is towards the origin, the effect of gravity on the particle is given by the vector

$$F(x, y, z) = \frac{gm}{\|\overrightarrow{OP}\|^2} u,$$

where

$$u = -\frac{\overrightarrow{OP}}{\|\overrightarrow{OP}\|}$$

is a unit vector that points from P towards O. Thus

Formula for Gravitational Vector Field

$$F(x, y, z) = \frac{gm}{\|\overrightarrow{OP}\|^2} \left(\frac{-\overrightarrow{OP}}{\|\overrightarrow{OP}\|} \right) = \frac{-gm(x\mathbf{i} + y\mathbf{j} + z\mathbf{k})}{(x^2 + y^2 + z^2)^{3/2}}$$

A few of the vectors in this gravitational field are pictured in Figure 18-1. Observe the lessening effect of gravity as distance from the earth increases. □

VELOCITY FIELDS

Velocity Field

Imagine that particles move through a fixed region. At a given instant, a particle is in motion at each point P of the region. The velocity of the particle at P defines a vector at P. The resulting set of vectors forms a vector field called a *velocity field*. A complete description of a velocity field takes into account how the vectors at a given point change with time. We will avoid this complication. We consider only velocity fields that are independent of time. The velocity vector at P depends only on P and not on time.

EXAMPLE 2   A right circular cylinder of radius $a = 2$ ft is full of water. The water is stirred with a constant motion, causing the molecules to spin around the axis of the cylinder in circular orbits with a constant positive angular velocity of ω rad/sec. Describe the motion of the surface of the water by a vector field.

Solution   We superimpose the xy-coordinate system on the surface of the water so that the surface is the circular disk $x^2 + y^2 \leq 4$. (See Fig. 18-2a.)

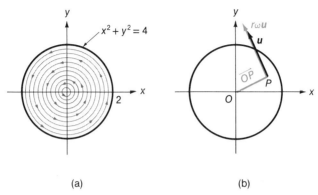

(a)   (b)

Figure 18–2   Example 2.

Let $P(x, y)$ be a point in this disk. The molecule of water at P is $r = \|\overrightarrow{OP}\|$ units from the origin and has an angular velocity of ω rad/sec. Thus, the molecule is moving counterclockwise around the disk with a speed of $r\omega$ ft/sec. Since the motion is in a direction perpendicular to $\overrightarrow{OP}$, the velocity of the particle is

$$r\omega u,$$

where u is a unit vector orthogonal to $\overrightarrow{OP}$ that is obtained by a 90° counterclockwise rotation from $\overrightarrow{OP}$. (See Fig. 18-2b.) Since $\overrightarrow{OP} = xi + yj$, then

$$u = \frac{-yi + xj}{\sqrt{x^2 + y^2}}$$

Thus, the motion of the surface is described by the vector field

$$v(x, y, z) = r\omega u = \sqrt{x^2 + y^2}\, \omega \, \frac{(-yi + xj)}{\sqrt{x^2 + y^2}}$$

$$= -y\omega i + x\omega j. \;\square$$

EXAMPLE 3   As water flows through a circular pipe, the various molecules flow at different speeds—with those near the walls of the pipe flowing most slowly. A given pipe has radius 1 unit. Assume that a molecule of water located r units from the axis of the pipe moves parallel to the axis of the pipe with a speed proportional to $1 - r$ units/sec. Describe the motion of the water by a vector field.

Solution   We set up the axis system in three-dimensional space so that the y-axis is the axis of the pipe and the water flows in the positive direction of that axis. Then the interior of the pipe is the interior of the cylinder $x^2 + z^2 = 1$. (See Fig. 18-3a.)

18–1 Vector Fields

If $P(x, y, z)$ is a point interior to the pipe, then the distance from P to the y-axis is

$$r = \sqrt{x^2 + z^2}.$$

If m is the constant of proportionality, then the speed of the particle at P is

$$m(1 - r) = m(1 - \sqrt{x^2 + z^2}).$$

Since the water flows in the positive direction of the y-axis, the velocity at $P(x, y, z)$ is

$$v(x, y, z) = m(1 - \sqrt{x^2 + z^2})j.$$

Several vectors in this vector field are shown at points of a typical cross section in Figure 18-3b. □

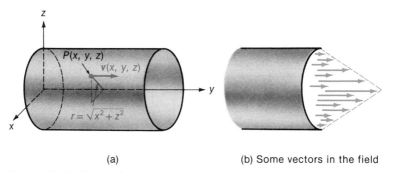

(a)   (b) Some vectors in the field

Figure 18–3 Example 3.

GRADIENT FIELDS

Many of the vector fields studied in physics are defined as gradients; that is,

$$F(x, y, z) = \nabla f(x, y, z) = f_x(x, y, z)i + f_y(x, y, z)j + f_z(x, y, z)k$$

Gradient (Conservative) Field

for some differentiable function f. Such a vector field is called a *gradient field*, or a *conservative field*. The function f is called a *potential function* for the field.

EXAMPLE 4 Show that the gravitational field

$$F(x, y, z) = \frac{-gm}{(x^2 + y^2 + z^2)^{3/2}} (xi + yj + zk)$$

is conservative. Find the potential function.

Solution We first show that a potential function f (if one exists) must have a certain form, and then show that every function of that form is a potential function.
Part 1 If f is a potential function for the field, then

$$\frac{\partial f}{\partial x} = -\frac{gmx}{(x^2 + y^2 + z^2)^{3/2}}, \quad \frac{\partial f}{\partial y} = -\frac{gmy}{(x^2 + y^2 + z^2)^{3/2}},$$

and

$$\frac{\partial f}{\partial z} = -\frac{gmz}{(x^2 + y^2 + z^2)^{3/2}}.$$

We reconstruct the function f from the first equation by partial integration:

$$f(x, y, z) = \int \frac{-gmx \, dx}{(x^2 + y^2 + z^2)^{3/2}} + K(y, z)$$

$$= \frac{gm}{(x^2 + y^2 + z^2)^{1/2}} + K(y, z)$$

where $K(y, z)$ corresponds to the constant of integration.

To find $K(y, z)$, we note that

$$\frac{\partial f}{\partial y} = \frac{-gmy}{(x^2 + y^2 + z^2)^{3/2}} + \frac{\partial K}{\partial y} = \frac{-gmy}{(x^2 + y^2 + z^2)^{3/2}}$$

so that $\partial K/\partial y = 0$. Similarly, $\partial K/\partial z = 0$. Since these partial derivatives are zero, then $K(y, z)$ is independent of both y and z. Thus, $K(y, z)$ is a constant function, say $K(y, z) = C$. It follows that

$$f(x, y, z) = \frac{gm}{(x^2 + y^2 + z^2)^{1/2}} + C.$$

Part 2 It is simple to show that if $f(x, y, z)$ is of the form found in Part 1, then

$$\nabla f - F(x, y, z).$$

Thus, F is a gradient function. $\square$

INVERSE-SQUARE FIELDS

The gravitational field is a special case of an *inverse-square field*—one of form

$$F(x, y, z) = \frac{cr}{\|r\|^3},$$

where $r = xi + yj + zk$.

Many of the important vector fields of physics, such as gravitational fields, magnetic fields, and electrical fields, are inverse-square fields. The result of Example 4 can be generalized to show that any inverse-square field is conservative and that a potential function is of form

$$f(x, y, z) = \frac{-c}{\sqrt{x^2 + y^2 + z^2}}$$

The term "inverse-square field" is used because the field is of form

$$F(x, y, z) = \frac{cu}{\|r\|^2},$$

where $u = r/\|r\|$ is a unit vector in the direction of r.

Exercises 18–1

1–2. Find a region in the xy-plane on which the vector field is defined.

•1 $F(x, y) = x^2 i + \dfrac{1}{\sqrt{y}} j$

2 $F(x, y) = \sqrt{xy} \, i + e^x j$

3–6. Calculate the gradient field $F = \nabla f$ for the potential function f.

● 3 $f(x, y) = x^2 + y^2$ 4 $f(x, y) = \tan^{-1}(x/y)$

● 5 $f(x, y, z) = xy + yz + xz$ 6 $f(x, y, z) = \sqrt{x^2 + y^2 + z^2}$

7–12. Each of the vector fields is conservative. Find a potential function.

● 7 $F(x, y) = (2x + y)i + (2y + x)j$ 8 $F(x, y) = y^2 i + 2xy j$

● 9 $F(x, y) = y \cos(xy) i + x \cos(xy) j$ 10 $F(x, y) = \dfrac{yi + xj}{1 + x^2 y^2}$

● 11 $F(x, y, z) = yz i + xz j + xy k$ 12 $F(x, y, z) = 2xi + 3y^2 j + 4z^3 k$

13–14. Describe the motion of the two-dimensional velocity field.

● 13 $F(x, y) = 2j$, $-1 \le x \le 1$

14 $F(x, y) = \dfrac{xi + yj}{\sqrt{x^2 + y^2}}$, $(x, y) \ne (0, 0)$

15 Review Examples 2 and 3. Then desribe the motion of a fluid that flows through the cylinder $x^2 + y^2 = 1$ with the velocity field

$$F(x, y, z) = -yi + xj + (1 - \sqrt{x^2 + y^2})k.$$

16–17. Let f, F, and G be continuous functions of x, y, z. Use component functions to establish the given results.

16 fF is continuous on its domain.

17 $F + G$ is continuous on its domain.

18 Modify the argument of Example 4 to show that an inverse-square field is conservative.

18–2 Line Integrals. I: Definition and Properties

In this section we take a temporary break from the subject of vector fields to consider yet another extension of the integral concept. We first defined definite integrals over intervals; later we considered double integrals over regions of the plane and triple integrals over solids. We now consider integrals that are defined along *curves* (sometimes called *paths*) in two or three dimensions.

We have already seen that it is natural to use functions of a parameter t to describe curves. It is now a simple matter to partition an interval on the t-axis for purposes of defining a Riemann sum, whose limit can be considered to be a definite integral. Integrals of this type have important applications in physics and engineering, as we will see in Section 18–4.

Let $\mathscr{C}$ be a smooth curve in the xy-plane defined parametrically by

$$x = f(t), \quad y = g(t), \quad a \le t \le b.$$

Let F be a function of x and y that is continuous at each point of $\mathscr{C}$.

Introduction to Vector Analysis

We begin by partitioning the interval $[a, b]$ on the t-axis into n subintervals by the partition P defined by:

$$a = t_0 < t_1 < t_2 < \cdots < t_n = b.$$

We also choose points $\xi_1, \xi_2, \ldots, \xi_n$ with ξ_k on the kth subinterval in the partition.

The partition of $[a, b]$ on the t-axis induces a partition of the curve $\mathscr{C}$. Corresponding to each point t_k on the t-axis there is a point $P_k(f(t_k), g(t_k))$ on $\mathscr{C}$. Corresponding to ξ_k is a point $(f(\xi_k), g(\xi_k))$ on $\mathscr{C}$ between P_{k-1} and P_k. (See Fig. 18-4.)

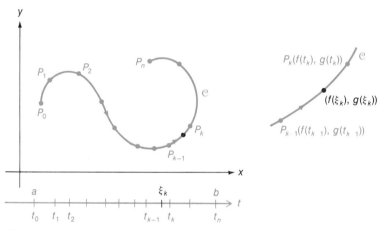

Figure 18-4.

Let Δs_k be the length of the arc connecting P_{k-1} and P_k. Form the Riemann sum

$$\sum_{k=1}^{n} F(f(\xi_k), g(\xi_k))\, \Delta s_k$$

Line Integral The limit of this sum as the norm of the partition approaches 0 is called the *line integral* of F over $\mathscr{C}$, denoted by

$$\int_{\mathscr{C}} F(x, y)\, ds.$$

That is,

$$\int_{\mathscr{C}} F(x, y)\, ds = \lim_{\|\mathscr{P}\| \to 0} \sum_{k=1}^{n} F(f(\xi_k), g(\xi_k))\, \Delta s_k.$$

(Obviously, the name *curve integral* would be more descriptive, but we use the traditional name.)

To evaluate a line integral $\int_{\mathscr{C}} F(x, y)\, ds$, we change it to a definite integral with respect to t. This can be accomplished by the substitutions

$$x = f(t), \quad y = g(t), \quad ds = \sqrt{f'(t)^2 + g'(t)^2}\, dt.$$

18–2 Line Integrals. I: Definition and Properties

Then

Evaluation of Line Integral

$$\int_{\mathscr{C}} F(x, y) \, ds = \int_a^b F(f(t), g(t)) \sqrt{f'(t)^2 + g'(t)^2} \, dt.$$

There are two obvious elementary applications. First, if $F(x, y) = 1$ for all $(x, y) \in \mathscr{C}$, then the line integral reduces to the integral for arc length. That is,

Arc Length

$$\text{Arc length of } \mathscr{C} = \int_{\mathscr{C}} ds = \int_a^b \sqrt{f'(t)^2 + g'(t)^2} \, dt.$$

Next, suppose that a thin wire is bent into the shape of $\mathscr{C}$. If $\delta(x, y)$ is the density (mass per unit of arc length) at (x, y), then

$$\delta(f(\xi_k), g(\xi_k)) \, \Delta s_k$$

is the approximate mass of the arc connecting P_{k-1} and P_k. It follows that the total mass of the wire is

$$\int_a^b \delta(f(t), g(t)) \sqrt{f'(t)^2 + g'(t)^2} \, dt,$$

so that

Mass of Wire

$$\text{Mass} = \int_{\mathscr{C}} \delta(x, y) \, ds.$$

EXAMPLE 1 A thin wire is bent around the quarter circle

$$x = \cos t, \quad y = \sin t, \quad 0 \le t \le \frac{\pi}{2}.$$

The density of the wire at the point (x, y) is $\delta(x, y) = x^2 y$. Calculate the mass of the wire.

Solution The mass is $M = \int_{\mathscr{C}} \delta(x, y) \, dx = \int_{\mathscr{C}} x^2 y \, ds$.
Along the curve we have

$$x = f(t) = \cos t, \quad y = g(t) = \sin t,$$

and

$$ds = \sqrt{f'(t)^2 + g'(t)^2} \, dt = \sqrt{\sin^2 t + \cos^2 t} \, dt = dt.$$

Then

$$M = \int_0^{\pi/2} \cos^2 t \sin t \, dt = -\left[\frac{\cos^3 t}{3}\right]_0^{\pi/2}$$

$$= -\frac{\cos^3 (\pi/2)}{3} + \frac{\cos^3 0}{3} = \frac{1}{3}. \; \square$$

BASIC PROPERTIES

Since line integrals can be written as definite integrals, then some of their properties can be established from corresponding properties of definite integrals. In particular, if F and G are continuous on $\mathcal{C}$ and if k is a constant, then

Properties of Line Integrals

$$\int_{\mathcal{C}} [F(x, y) \pm G(x, y)] \, ds = \int_{\mathcal{C}} F(x, y) \, ds \pm \int_{\mathcal{C}} G(x, y) \, ds,$$

$$\int_{\mathcal{C}} kF(x, y) \, ds = k \int_{\mathcal{C}} F(x, y) \, ds.$$

If the orientation of a curve is reversed, then the new line integral is the negative of the original one. More precisely, suppose that $\mathcal{C}_1$ is a particular curve that is traced out from A to B and that $\mathcal{C}_2$ consists of the same set of points traced out from B to A. (See Fig. 18–5.) It can be proved that

$$\int_{\mathcal{C}_1} F(x, y) \, ds = - \int_{\mathcal{C}_2} F(x, y) \, ds.$$

This property corresponds to the property

$$\int_a^b f(x) \, dx = - \int_b^a f(x) \, dx$$

for definite integrals.

The value of a line integral depends on the integrand $F(x, y)$ and on the curve $\mathcal{C}$, not on the particular parametrization that is used for $\mathcal{C}$. In other words, suppose that $\mathcal{C}$ is defined by

$$x = f(t), \qquad y = g(t), \qquad a \leq t \leq b$$

and also by

$$x = p(s), \qquad y = q(s), \qquad c \leq s \leq d.$$

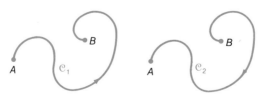

Figure 18–5

$$\int_{\mathcal{C}_1} F(x, y) \, ds = - \int_{\mathcal{C}_2} F(x, y) \, ds.$$

18–2 Line Integrals. I: Definition and Properties

It can be proved that

$$\int_b^a F(f(t), g(t)) \sqrt{f'(t)^2 + g'(t)^2}\, dt$$
$$= \int_c^d F(p(s), q(s)) \sqrt{p'(s)^2 + q'(s)^2}\, ds.$$

Thus, we get the same value for the line integral with the two parametrizations.

DEPENDENCE ON PATH

In light of the Fundamental Theorem of Calculus, you are apt to think that the value of a line integral may depend only on the endpoints of the curve $\mathscr{C}$ and not on $\mathscr{C}$ itself. Example 2 shows that this is not the case. We integrate the same function as in Example 1 over a different curve with the same endpoints, $(1, 0)$ and $(0, 1)$, getting a different value. Thus the value of a line integral depends on the particular curve $\mathscr{C}$. (We will see in the next section, however, that under certain conditions a line integral can be evaluated at the endpoints of the curve $\mathscr{C}$.)

EXAMPLE 2 Let $\mathscr{C}$ be the line segment connecting $(1, 0)$ (initial point) and $(0, 1)$ (terminal point). Calculate

$$\int_{\mathscr{C}} x^2 y\, ds.$$

Solution Parametric equations for $\mathscr{C}$ are $x = 1 - t$, $y = t$, $0 \le t \le 1$. Then

$$\int_{\mathscr{C}} x^2 y\, ds = \int_0^1 (1 - t)^2 t \sqrt{1^2 + 1^2}\, dt = \sqrt{2} \int_0^1 (t - 2t^2 + t^3)\, dt$$
$$= \sqrt{2} \left[\frac{t^2}{2} - \frac{2t^3}{3} + \frac{t^4}{4}\right]_0^1 = \frac{\sqrt{2}}{12}. \square$$

ALTERNATIVE FORM

If the curve $\mathscr{C}$ is the graph of $y = g(x)$, $a \le x \le b$, then we can use x as the parameter for a line integral. To emphasize this, we temporarily write

$$x = f(t) = t \quad \text{and} \quad y = g(t).$$

Then

$$ds = \sqrt{f'(t)^2 + g'(t)^2}\, dt = \sqrt{1 + g'(x)^2}\, dx,$$

so that

(1) $$\int_{\mathscr{C}} F(x, y)\, ds = \int_a^b F(x, g(x)) \sqrt{1 + g'(x)^2}\, dx.$$

Line Integral with x as a Parameter

Integrals of the type shown in (1) are usually written

$$\int_{\mathscr{C}} P(x, y)\, dx$$

where
$$P(x, y) = f(x, y) \sqrt{1 + g'(x)^2}.$$

To evaluate such an integral, we replace y by $g(x)$ and integrate over the interval $[a, b]$. That is,

$$\int_{\mathscr{C}} P(x, y) \, dx = \int_a^b P(x, g(x)) \, dx.$$

The integral $\int_{\mathscr{C}} Q(x, y) \, dy$ is defined similarly. If the smooth curve $\mathscr{C}$ is the graph of $x = f(y)$, $c \leq y \leq d$, and $Q(x, y)$ is continuous at each point of $\mathscr{C}$, then

$$\int_{\mathscr{C}} Q(x, y) \, dy = \int_c^d Q(f(y), y) \, dy.$$

The integral $\int_{\mathscr{C}} P(x, y) \, dx + Q(x, y) \, dy$ is defined to be

Alternative Form of Line Integral

$$\int_{\mathscr{C}} P(x, y) \, dx + Q(x, y) \, dy = \int_{\mathscr{C}} P(x, y) \, dx + \int_{\mathscr{C}} Q(x, y) \, dy.$$

provided both integrals on the right exist.

If the curve $\mathscr{C}$ is defined parametrically, we can evaluate

$$\int_{\mathscr{C}} P(x, y) \, dx + Q(x, y) \, dy$$

by an alternative method. We express x, y, dx, and dy in terms of the parameter and integrate. The first part of Example 3 illustrates the method.

EXAMPLE 3 Let $\mathscr{C}$ be the line $y = 2x - 1$, $0 \leq x \leq 1$. Evaluate

$$\int_{\mathscr{C}} (x + y) \, dx + 3x^2 y \, dy$$

by two different methods.

Solution **Method 1** Write the curve parametrically as
$$x = t, \quad y = 2t - 1, \quad 0 \leq t \leq 1.$$
Then $dx = dt$ and $dy = 2dt$. The integral is

$$\int_{\mathscr{C}} (x + y) \, dx + 3x^2 y \, dy = \int_0^1 (t + 2t - 1) \, dt + \int_0^1 3t^2(2t - 1) \cdot 2dt$$

$$= \int_0^1 (12t^3 - 6t^2 + 3t - 1) \, dt$$

$$= \left[3t^4 - 2t^3 + \frac{3t^2}{2} - t \right]_0^1 = \frac{3}{2}.$$

Method 2 Write

$$\int_{\mathcal{C}} (x+y)\,dx + 3x^2 y\,dy = \int_{\mathcal{C}} (x+y)\,dx + \int_{\mathcal{C}} 3x^2 y\,dy.$$

For the first integral we use the fact that $y = 2x - 1$, $0 \le x \le 1$. For the second we write $x = (y+1)/2$, $-1 \le y \le 1$. Then

$$\int_{\mathcal{C}} (x+y)\,dx + \int_{\mathcal{C}} 3x^2 y\,dy$$

$$= \int_0^1 [x + (2x-1)]\,dx + \int_{-1}^1 3\left(\frac{y+1}{2}\right)^2 y\,dy$$

$$= \int_0^1 (3x - 1)\,dx + \frac{3}{4}\int_{-1}^1 (y^3 + 2y^2 + y)\,dy$$

$$= \left[\frac{3x^2}{2} - x\right]_0^1 + \frac{3}{4}\left[\frac{y^4}{4} + \frac{2y^3}{3} + \frac{y^2}{2}\right]_{-1}^1$$

$$= \left[\frac{3}{2} - 1\right] + \frac{3}{4}\left[\frac{1}{4} + \frac{2}{3} + \frac{1}{2}\right] - \frac{3}{4}\left[\frac{1}{4} - \frac{2}{3} + \frac{1}{2}\right] = \frac{3}{2}. \quad \square$$

Remark   Do not confuse the integrals

$$\int_{\mathcal{C}} F(x,y)\,dx \quad \text{and} \quad \int_{g_1(y)}^{g_2(y)} F(x,y)\,dx.$$

The first is a line integral; the second, a partial integral. The two integrals are conceptually different and generally have different values.

SPACE CURVES

The concepts of this section can be extended to smooth curves in three-dimensional space. If the smooth curve $\mathcal{C}$ is the graph of

$$x = f(t), \qquad y = g(t), \qquad z = h(t), \qquad a \le t \le b,$$

and $F(x, y, z)$ is continuous on $\mathcal{C}$, then

Line Integrals Defined Over Space Curves

$$\int_{\mathcal{C}} F(x, y, z)\,ds = \int_a^b F(f(t), g(t), h(t))\sqrt{f'(t)^2 + g'(t)^2 + h'(t)^2}\,dt.$$

If $P(x, y, z)$, $Q(x, y, z)$, and $R(x, y, z)$ are continuous on $\mathcal{C}$, then

Evaluation of Line Integral

$$\int_{\mathcal{C}} P(x, y, z)\,dx + Q(x, y, z)\,dy + R(x, y, z)\,dz$$

is defined to be

$$\int_a^b [P(f(t), g(t), h(t))f'(t) + Q(f(t), g(t), h(t))g'(t) + R(f(t), g(t), h(t))h'(t)]\,dt.$$

This line integral can also be evaluated by the second method used in Example 3 provided each variable can be expressed as a function of the other variables.

Exercises 18–2

1–12. Evaluate the line integral over $\mathscr{C}$.

- **1** $\int_{\mathscr{C}} (\sin x + \cos y) \, ds$; $\mathscr{C}$ is the line segment from $(0, 0)$ to $(\pi, 2\pi)$.

- **2** $\int_{\mathscr{C}} xe^y \, ds$; $\mathscr{C}$ is the line segment from $(1, 1)$ to $(-1, 2)$.

- **3** $\int_{\mathscr{C}} (2x + 3xy) \, ds$; $\mathscr{C}$ is the quarter-circle with center at the origin traced out from $(1, 0)$ to $(0, 1)$.

- **4** $\int_{\mathscr{C}} (2x + 3xy) \, ds$; $\mathscr{C}$ is the quarter-circle with center at the origin traced out from $(2, 0)$ to $(0, -2)$.

- **5** $\int_{\mathscr{C}} x^5 e^{xy} \, ds$; $\mathscr{C}$ is the portion of the hyperbola $y = 1/x$ from $(1, 1)$ to $(2, \tfrac{1}{2})$.

- **6** $\int_{\mathscr{C}} x \, ds$; $\mathscr{C}$ is the portion of the parabola $y = x^2$ from $(1, 1)$ to $(0, 0)$.

- **7** $\int_{\mathscr{C}} (2x + 9z) \, ds$; $\mathscr{C}$ is the portion of the twisted cubic $\mathbf{r} = t\mathbf{i} + t^2\mathbf{j} + t^3\mathbf{k}$ with $0 \le t \le 1$.

- **8** $\int_{\mathscr{C}} (x^2 + y^2 + z^2) \, ds$; $\mathscr{C}$ is the portion of the helix $\mathbf{r} = 4\cos t\,\mathbf{i} + 4\sin t\,\mathbf{j} + t\mathbf{k}$ with $0 \le t \le 2\pi$.

- **9** $\int_{\mathscr{C}} (x + 2y) \, dx + (x - 2y) \, dy$; $\mathscr{C}$ is the line segment from $(1, 1)$ to $(3, -1)$.

- **10** $\int_{\mathscr{C}} y \, dx + x \, dy$; $\mathscr{C}$ is the graph of $y = x^2$ for $-1 \le x \le 1$.

- **11** $\int_{\mathscr{C}} (x + y + z) \, dx + x \, dy - yz \, dz$; $\mathscr{C}$ is the line segment from $(2, 1, 0)$ to $(1, 2, -1)$.

- **12** $\int_{\mathscr{C}} xy \, dx + 3y^2 \, dy + z \, dz$; $\mathscr{C}$ is the portion of the helix $\mathbf{r} = \cos t\,\mathbf{i} + \sin t\,\mathbf{j} + t\mathbf{k}$ with $0 \le t \le \pi/4$.

13 (*Different Parametrizations*) Evaluate $\int_{\mathscr{C}} (2x + y) \, ds$, where $\mathscr{C}$ is the semicircle $y = \sqrt{1 - x^2}$ from $(-1, 0)$ to $(1, 0)$, using the following parametrizations.

Compare the two answers.
(a) $x = t$, $y = \sqrt{1 - t^2}$, $-1 \le t \le 1$
(b) $x = -\cos t$, $y = \sin t$, $0 \le t \le \pi$

•14 A wire is bent in the shape of the portion of the hyperbola $y = 1/x$ from $(\frac{1}{2}, 2)$ to $(2, \frac{1}{2})$. The density (mass per unit of length) at (x, y) is $\delta(x, y) = x^5/21$. Calculate the mass of the wire.

15 A wire is bent in the shape of the circle $x^2 + y^2 = 1$. The density at (x, y) is $\delta(x, y) = x^2 y^2/100$. Calculate the mass of the wire.

•16 (*Dependence of Path*) Evaluate

$$\int_{\mathcal{C}} (x^2 + y) \, dx + (3x + 2y^2) \, dy$$

on $\mathcal{C}_1$ and on $\mathcal{C}_2$, where $\mathcal{C}_1$ is the x-axis from $(-1, 0)$ to $(1, 0)$ and $\mathcal{C}_2$ is the semicircle $y = \sqrt{1 - x^2}$ from $(-1, 0)$ to $(1, 0)$. Compare the answers.

17 Let f be a continuous function of x for $a \le x \le b$. Let $\mathcal{C}$ be the portion of the x-axis from a to b where $a < b$. The line integral and definite integral

$$\int_{\mathcal{C}} f(x) \, dx \quad \text{and} \quad \int_a^b f(x) \, dx$$

Line Integrals and Definite Integrals

are conceptually different.
(a) Show that these two integrals have the same value. (*Hint:* Find a parameter for $\mathcal{C}$.)
(b) Does a similar result hold if $\mathcal{C}$ is a line segment parallel to the x-axis, say the segment from $(a, 1)$ to $(b, 1)$? Explain.
• (c) Does a similar result hold on a smooth curve $y = g(x)$ for $a \le x \le b$? Explain.

18–3 Line Integrals. II: Independence of Path

Under certain conditions the value of a line integral

$$\int_{\mathcal{C}} P(x, y) \, dx + Q(x, y) \, dy$$

Exact Differential

is independent of the particular curve $\mathcal{C}$, depending only on the endpoints of $\mathcal{C}$. This is the case if $P(x, y) \, dx + Q(x, y) \, dy$ is the *exact differential* of a function $G(x, y)$; that is,

$$dG = P(x, y) \, dx + Q(x, y) \, dy$$

on some region $\mathcal{R}$ that contains $\mathcal{C}$. The result is stated in the following theorem. The proof is omitted.

THEOREM 18–1 Let $G(x, y)$ be differentiable on a region $\mathcal{R}$. Let

$$dG = P(x, y)\, dx + Q(x, y)\, dy$$

where P and Q are continuous on $\mathcal{R}$. If $\mathcal{C}$ is a smooth curve in $\mathcal{R}$ from $P_1(x_1, y_1)$ to $P_2(x_2, y_2)$, then

Line Integral Evaluated at Endpoints of Curve

$$\int_{\mathcal{C}} P(x, y)\, dx + Q(x, y)\, dy = \int_{\mathcal{C}} dG = G(x_2, y_2) - G(x_1, y_1).$$

This theorem, which can be generalized to line integrals along space curves, is a direct extension of the fundamental Theorem of Calculus. It states that the line integral of dG can be calculated by evaluating G at the endpoints of the curve.

EXACT DIFFERENTIALS

Not every expression $P(x, y)\, dx + Q(x, y)\, dy$ is an exact differential. Observe that *if* the expression is the differential of a function G, then

$$\frac{\partial G}{\partial x} = P \quad \text{and} \quad \frac{\partial G}{\partial y} = Q.$$

If in addition, G has continuous partial derivatives, then we must have

$$\frac{\partial P}{\partial y} = \frac{\partial^2 G}{\partial y \partial x} = \frac{\partial^2 G}{\partial x \partial y} = \frac{\partial Q}{\partial x}.$$

These conditions also are sufficient to prove that

$$P(x, y)\, dx + Q(x, y)\, dy$$

Conditions for Exact Differential

is an exact differential. That is, if

$$\frac{\partial P}{\partial y} = \frac{\partial Q}{\partial x}$$

on some region $\mathcal{R}$ where P and Q are continuous, then it can be proved that there exists a differentiable function G such that

$$dG = P(x, y)\, dx + Q(x, y)\, dy$$

at each point of $\mathcal{R}$. The function G can be calculated from P and Q by partial integration.

EXAMPLE 1 Let $\mathcal{C}$ be a smooth curve from $P_1(2, 1)$ to $P_2(3, -5)$. Use Theorem 18–1 to calculate

$$\int_{\mathcal{C}} (3x^2 + 2y)\, dx + (2x + e^y)\, dy.$$

Solution Let $P(x, y) = 3x^2 + 2y$ and $Q(x, y) = 2x + e^y$. Since

$$\frac{\partial P}{\partial y} = 2 = \frac{\partial Q}{\partial x}$$

at every point in the xy-plane, then $P(x, y)\, dx + Q(x, y)\, dy$ is an exact differential dG. To find G, we note that

$$\frac{\partial G}{\partial x} = P(x, y) = 3x^2 + 2y,$$

so that

$$G(x, y) = \int (3x^2 + 2y)\, dx = x^3 + 2xy + K(y),$$

where $K(y)$ corresponds to the constant of integration. To find $K(y)$, we note that

$$\frac{\partial G}{\partial y} = Q(x, y) = 2x + e^y$$
$$2x + K'(y) = 2x + e^y$$
$$K'(y) = e^y$$
$$K(y) = e^y + C.$$

The function is

$$G(x, y) = x^3 + 2xy + K(y) = x^3 + 2xy + e^y + C.$$

Since we need a particular function, we choose $C = 0$. Then

$$G(x, y) = x^3 + 2xy + e^y.$$

It follows from Theorem 18–1 that

$$\int_\mathscr{C} (3x^2 + 2y)\, dx + (2x + e^y)\, dy$$
$$= G(3, -5) - G(2, 1)$$
$$= [3^3 + 2 \cdot 3 \cdot (-5) + e^{-5}] - [2^3 + 2 \cdot 2 \cdot 1 + e^1]$$
$$= e^{-5} - e - 15. \quad \square$$

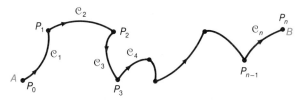

Figure 18–6 A piecewise smooth curve.

PIECEWISE SMOOTH CURVES

A curve $\mathscr{C}$ is said to be *piecewise smooth* if it is the union of a finite number of smooth arcs. In other words, a curve with endpoints A and B is piecewise smooth if there exist points $A = P_0, P_1, P_2, \ldots, P_n = B$ on $\mathscr{C}$ such that the portion of $\mathscr{C}$ between P_{k-1} and P_k is a smooth curve, $k = 1, 2, \ldots, n$. (See Fig. 18–6.)

Let $\mathscr{C}$ be a piecewise smooth curve that is the union of the smooth arcs $\mathscr{C}_1$,

$\mathscr{C}_2, \ldots, \mathscr{C}_n$. Let $P(x, y)$ and $Q(x, y)$ be continuous on $\mathscr{C}$. The line integral

$$\int_{\mathscr{C}} P(x, y)\, dx + Q(x, y)\, dy$$

is defined to be the sum

$$\int_{\mathscr{C}_1} P(x, y)\, dx + Q(x, y)\, dy + \int_{\mathscr{C}_2} P(x, y)\, dx + Q(x, y)\, dy$$

$$\cdots + \int_{\mathscr{C}_n} P(x, y)\, dx + Q(x, y)\, dy.$$

There is a simple corollary to Theorem 18–1 relating to line integrals defined over piecewise smooth curves.

COROLLARY TO THEOREM 18–1 Let $P(x, y)\, dx + Q(x, y)\, dy$ be the differential of $G(x, y)$ on a region $\mathscr{R}$ where P and Q are continuous. Let $\mathscr{C}$ be a piecewise smooth curve in $\mathscr{R}$ from $A(a_1, a_2)$ to $B(b_1, b_2)$. Then

$$\int_{\mathscr{C}} P(x, y)\, dx + Q(x, y)\, dy = G(b_1, b_2) - G(a_1, a_2).$$

Proof There exist successive points

$$A = P_0, P_1, P_2, \ldots, P_n = B$$

on $\mathscr{C}$ such that the portion of $\mathscr{C}$ between $P_{k-1}(x_{k-1}, y_{k-1})$ and $P_k(x_k, y_k)$ is a smooth curve in $\mathscr{R}$. Let $\mathscr{C}_k$ be the portion of the curve between P_{k-1} and P_k. Then

$$\int_{\mathscr{C}} P\, dx + Q\, dy = \int_{\mathscr{C}_1} P\, dx + Q\, dy + \int_{\mathscr{C}_2} P\, dx + Q\, dy + \cdots + \int_{\mathscr{C}_n} P\, dx + Q\, dy$$
$$= [G(x_1, y_1) - G(x_0, y_0)] + [G(x_2, y_2) - G(x_1, y_1)]$$
$$+ \cdots + [G(x_n, y_n) - G(x_{n-1}, y_{n-1})]$$
$$= G(x_n, y_n) - G(x_0, y_0) = G(b_1, b_2) - G(a_1, a_2). \blacksquare$$

LINE INTEGRALS OVER CLOSED CURVES

If an exact differential is integrated over a piecewise smooth closed curve, the value of the integral must be zero. This result, which follows immediately from the corollary to Theorem 18–1, is proved in Theorem 18–2.

THEOREM 18–2 Let $P(x, y)\, dx + Q(x, y)\, dy$ be an exact differential on a region $\mathscr{R}$ where P and Q are continuous. Let $\mathscr{C}$ be a piecewise smooth closed curve in $\mathscr{R}$. Then

$$\int_{\mathscr{C}} P(x, y)\, dx + Q(x, y)\, dy = 0.$$

Proof Let $P\, dx + Q\, dy$ be the differential of $G(x, y)$ on $\mathscr{R}$. Let (x_0, y_0) be a point on $\mathscr{C}$. We can assume that $\mathscr{C}$ is traced out from (x_0, y_0) to (x_0, y_0). It follows from the corollary

18-3 Line Integrals. II: Independence of Path

to Theorem 18-1 that

$$\int_{\mathcal{C}} P(x, y) \, dx + Q(x, y) \, dy = G(x_0, y_0) - G(x_0, y_0) = 0. \quad \blacksquare$$

EXAMPLE 2 Let $G(x, y) = \tan^{-1}(x/y)$, so that

$$dG = \frac{y}{x^2 + y^2} \, dx - \frac{x}{x^2 + y^2} \, dy.$$

Let $\mathcal{C}$ be the closed unit circle, traced out counterclockwise from $(1, 0)$ to $(1, 0)$. Show that

$$\int_{\mathcal{C}} \frac{y}{x^2 + y^2} \, dx - \frac{x}{x^2 + y^2} \, dy = -2\pi.$$

Explain why this does not contradict Theorem 18-2.

Solution   **Part 1** Parametric equations for $\mathcal{C}$ are

$$x = \cos t, \quad y = \sin t, \quad 0 \le t \le 2\pi.$$

Then $dx = -\sin t \, dt$, $dy = \cos t \, dt$, and

$$\int_{\mathcal{C}} \frac{y}{x^2 + y^2} \, dx - \frac{x}{x^2 + y^2} \, dy$$

$$= \int_0^{2\pi} \left[\frac{\sin t}{\cos^2 t + \sin^2 t} (-\sin t) - \frac{\cos t}{\cos^2 t + \sin^2 t} \cdot \cos t \right] dt$$

$$= \int_0^{2\pi} \left(-\frac{\sin^2 t + \cos^2 t}{\cos^2 t + \sin^2 t} \right) dt = -\int_0^{2\pi} dt = -2\pi.$$

Part 2 It appears from Part 1 that we have contradicted Theorem 18-2. What happened?

The answer to this question concerns the region $\mathcal{R}$. Observe that $G(x, y) = \tan^{-1}(x/y)$ is not defined when $y = 0$. This function is differentiable on the region above the x-axis and on the region below the x-axis but not at any point on the x-axis. Thus, there is no region containing $\mathcal{C}$ on which G is differentiable. $\square$

SPACE CURVES

All of our results can be extended to space curves. Let $G(x, y, z)$ be differentiable on a solid region $\mathcal{D}$ and let

$$dG = P(x, y, z) \, dx + Q(x, y, z) \, dy + R(x, y, z) \, dz$$

where P, Q, and R are continuous on $\mathcal{D}$.

(1) If $\mathcal{C}$ is a piecewise smooth curve from $P_1(x_1, y_1, z_1)$ to $P_2(x_2, y_2, z_2)$ that lies in $\mathcal{D}$, then

$$\int_{\mathcal{C}} P(x, y, z) \, dx + Q(x, y, z) \, dy + R(x, y, z) \, dz$$

$$= \int_{\mathcal{C}} dG = G(x_2, y_2, z_2) - G(x_1, y_1, z_1).$$

(2) If $\mathscr{C}$ is a piecewise smooth closed curve that lies in $\mathscr{D}$, then

$$\int_{\mathscr{C}} P(x, y, z)\, dz + Q(x, y, z)\, dy + R(x, y, z)\, dz = 0.$$

VECTOR FORM

Let F be the vector field

$$F(x, y, z) = P(x, y, z)\mathbf{i} + Q(x, y, z)\mathbf{j} + R(x, y, z)\mathbf{k},$$

where P, Q, and R are continuous on a solid region $\mathscr{D}$. Let $\mathscr{C}$ be a piecewise smooth curve in $\mathscr{D}$. The line integral

$$\int_{\mathscr{C}} P(x, y, z)\, dx + Q(x, y, z)\, dy + R(x, y, z)\, dz$$

can be written in the compact form

Vector Form of Line Integral

$$\int_{\mathscr{C}} F(x, y, z) \cdot d\mathbf{r}$$

where

$$d\mathbf{r} = dx\,\mathbf{i} + dy\,\mathbf{j} + dz\,\mathbf{k}.$$

If $\mathscr{C}$ is the graph of

$$r(t) = f(t)\mathbf{i} + g(t)\mathbf{j} + h(t)\mathbf{k}, \quad a \le t \le b,$$

then we evaluate this line integral by writing

$$\begin{aligned} d\mathbf{r} &= dx\,\mathbf{i} + dy\,\mathbf{j} + dz\,\mathbf{k} \\ &= f'(t)\,dt\,\mathbf{i} + g'(t)\,dt\,\mathbf{j} + h'(t)\,dt\,\mathbf{k}. \end{aligned}$$

Then

$$\begin{aligned} \int_{\mathscr{C}} F(x, y, z) \cdot d\mathbf{r} &= \int_{\mathscr{C}} P(x, y, z)\, dx + Q(x, y, z)\, dy + R(x, y, z)\, dz \\ &= \int_{a}^{b} [P(f(t), g(t), h(t))f'(t) + Q(f(t), g(t), h(t))g'(t) \\ &\quad + R(f(t), g(t), h(t))h'(t)]\, dt. \end{aligned}$$

EXAMPLE 3 Calculate $\int_{\mathscr{C}} F \cdot d\mathbf{r}$, where

$$F(x, y, z) = 2x\mathbf{i} + y\mathbf{j} + z\mathbf{k}$$

and $\mathscr{C}$ is the graph of

$$r = t^2\mathbf{i} + t\mathbf{j} + t^3\mathbf{k}, \quad -1 \le t \le 1.$$

Solution Note that

$$\begin{aligned} d\mathbf{r} &= dx\,\mathbf{i} + dy\,\mathbf{j} + dz\,\mathbf{k} \\ &= 2t\,dt\,\mathbf{i} + dt\,\mathbf{j} + 3t^2\,dt\,\mathbf{k}. \end{aligned}$$

Then

$$\int_{\mathcal{C}} F(x, y, z) \cdot dr = \int_{\mathcal{C}} 2x \, dx + y \, dy + z \, dz$$

$$= \int_{-1}^{1} (2t^2 \cdot 2t + t + t^3 \cdot 3t^2) \, dt$$

$$= \int_{-1}^{1} (4t^3 + t + 3t^5) \, dt$$

$$= \left[t^4 + \frac{t^2}{2} + \frac{t^6}{2} \right]_{-1}^{1} = 0. \quad \square$$

GRADIENT FIELDS

Let f have continuous partial derivatives. Let

$$\nabla f(x, y, z) = \frac{\partial f}{\partial x} i + \frac{\partial f}{\partial y} j + \frac{\partial f}{\partial z} k$$

be a gradient field. Let $\mathcal{C}$ be a piecewise smooth curve in the domain of f from $P_1(x_1, y_1, z_1)$ to $P_2(x_2, y_2, z_2)$.

Observe that

$$\nabla f \cdot dr = \left(\frac{\partial f}{\partial x} i + \frac{\partial f}{\partial y} j + \frac{\partial f}{\partial z} k \right) \cdot (dx \, i + dy \, j + dz \, k)$$

$$= \frac{\partial f}{\partial x} dx + \frac{\partial f}{\partial y} dy + \frac{\partial f}{\partial z} dz = df.$$

It follows from the extension of Theorem 18–1 to space curves that

Line Integral Defined by Gradient Field

$$\int_{\mathcal{C}} \nabla f \cdot dr = \int_{\mathcal{C}} df = f(x_2, y_2, z_2) - f(x_1, y_1, z_1).$$

Exercises 18–3

1–6. Show that the integrand is an exact differential. Use Theorem 18–1 to evaluate the line integral.

• 1 $\int_{\mathcal{C}} 2xy \, dx + x^2 \, dy$; $\mathcal{C}$ is a smooth curve from $(2, 4)$ to $(-1, -1)$.

2 $\int_{\mathcal{C}} (3x^2 + 10xy) \, dx + (5x^2 + 3y^2) \, dy$; $\mathcal{C}$ is a smooth curve from $(0, 0)$ to $(1, 1)$.

• 3 $\int_{\mathcal{C}} (y \cos x + \cos y) \, dx + (\sin x - x \sin y) \, dy$; $\mathcal{C}$ is a smooth curve from $(\pi/2, 0)$ to $(\pi, \pi/2)$.

4 $\int_{\mathcal{C}} \cos(x + y) \, dx + \cos(x + y) \, dy$; $\mathcal{C}$ is a smooth curve from $(0, \pi)$ to $(\pi, 0)$.

●5 $\int_{\mathcal{C}} (2x + y \cos xy) \, dx + x \cos xy \, dy$; $\mathcal{C}$ is the graph of $x = \sin t$, $y = \cos t$, $0 \leq t \leq \pi/2$.

6 $\int_{\mathcal{C}} ye^{xy} \, dx + xe^{xy} \, dy$; $\mathcal{C}$ is the graph of $x = 3t$, $y = t$, $0 \leq t \leq 1$.

7–8. Each integrand is an exact differential. Use the extension of Theorem 18–1 to functions in three-dimensional space to evaluate the integral.

●7 $\int_{\mathcal{C}} (y + z) \, dx + (x + z) \, dy + (x + y) \, dz$; $\mathcal{C}$ is the helix $r = \cos t\,i + \sin t\,j + t\,k$, $0 \leq t \leq \pi$.

8 $\int_{\mathcal{C}} yz \, dx + xz \, dy + xy \, dz$; $\mathcal{C}$ is the twisted cubic $x = t$, $y = t^2$, $z = t^3$, $-1 \leq t \leq 1$.

9–10. Evaluate $\int_{\mathcal{C}} (\nabla f) \cdot d\mathbf{r}$.

● 9 $f(x, y) = x^2 y$; $\mathcal{C}$ is a smooth curve from $(0, 0)$ to $(1, 1)$.

10 $f(x, y, z) = x^2 y + xyz + z^3$; $\mathcal{C}$ is the twisted cubic $r(t) = t\,i + t^2 j + t^3 k$, $0 \leq t \leq 1$.

●11 Let $\mathcal{C}$ be the union of the top half of the unit circle from $(1, 0)$ to $(-1, 0)$ and the x-axis from $(-1, 0)$ to $(1, 0)$.

(a) Evaluate $\int_{\mathcal{C}} ye^{xy} \, dx + xe^{xy} \, dy$ after expressing the two integrals involved in terms of parameters.

(b) Use the corollary to Theorem 18–1 to evaluate the integral in (a).

●12 Evaluate $\int_{\mathcal{C}} (x^2 + y) \, dx + (3x + 2y^2) \, dy$ where $\mathcal{C}$ is the x-axis from $(0, 0)$ to $(1, 0)$ and the line $x = 1$ from $(1, 0)$ to $(1, 1)$. Explain why Theorem 18–1 cannot be used for this exercise.

●13 Evaluate $\int_{\mathcal{C}} (x^2 + y) \, dx + (2x + y^2) \, dy$ along the unit circle traced out counterclockwise from $(1, 0)$ to $(1, 0)$. Explain why your answer does not contradict the corollary to Theorem 18–1.

The following exercise shows that the value of a line integral around a simple closed curve is independent of the initial point on the curve.

14 Let A and B be points on a piecewise smooth, simple closed curve. Let $\mathcal{C}_A$ represent the curve with A as its initial and terminal point; let $\mathcal{C}_B$ represent the curve with B as its initial and terminal point. Let P and Q be continuous on $\mathcal{C}_A$. (The same orientation is used for both curves.) Show that

$$\int_{\mathcal{C}_A} P(x, y) \, dx + Q(x, y) \, dy = \int_{\mathcal{C}_B} P(x, y) \, dx + Q(x, y) \, dy.$$

(*Hint:* Let $\mathcal{C}_1$ be the portion of the curve from A to B and $\mathcal{C}_2$ the remaining portion. Express the integrals in terms of $\mathcal{C}_1$ and $\mathcal{C}_2$.)

Line Integral on Closed Curve Independent of Starting Point

18–4 Applications

Line integrals occur naturally in many applications to physics. For example, let $F(x, y)$ be a two-dimensional force field that describes the magnetic attraction of a particle in the xy-plane toward a magnet placed at the origin. Let $\mathscr{C}$ be a simple smooth curve in the xy-plane. We will show how line integrals can be used to define the work done in moving the particle along $\mathscr{C}$.

Note that if the particle is moved towards the magnet, then the work is done by the magnetic field. If it is moved away from the magnet, then the work must be done by some external force. To avoid considering two cases, we will define the work done by the vector field in moving from point A to point B along the curve. If the final answer is positive, then the main part of the work was done by the vector field. If the final answer is negative, the main part was done by an external force.

If a particle is moved along a straight line from A to B in a constant force field F, then the work is defined to be the product of the effective force in the direction of $\overrightarrow{AB}$ with the magnitude of $\overrightarrow{AB}$. That is,

Work (Motion Along a Line)

$$\text{Work} = (\text{comp}_{\overrightarrow{AB}} F) \, \|\overrightarrow{AB}\| = \frac{F \cdot \overrightarrow{AB}}{\|\overrightarrow{AB}\|} \|\overrightarrow{AB}\|$$

$$= F \cdot \overrightarrow{AB}.$$

(See Fig. 18–7a.)

Now consider the same problem for a particle moving along the curve $\mathscr{C}$. Suppose $\mathscr{C}$ is the graph of

$$x = f(t), \quad y = g(t), \quad a \le t \le b.$$

At a particular point $P_k(x_k, y_k)$ on $\mathscr{C}$ we move in the direction of the velocity vector $v(t_k)$. Thus, if we move a distance Δs_k, the work done is approximately

(1) $\quad F(x_k, y_k) \cdot \dfrac{v(t_k)}{\|v(t_k)\|} \Delta s_k = F(f(t_k), g(t_k)) \cdot \dfrac{v(t_k)}{\|v(t_k)\|} \Delta s_k.$

(See Fig. 18–7b.)

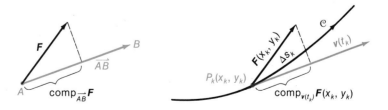

(a) Work $= (\text{comp}_{\overrightarrow{AB}} F) \|\overrightarrow{AB}\|$
$= F \cdot \overrightarrow{AB}$

(b) Work $\approx F(x_k, y_k) \cdot \dfrac{v(t_k)}{\|v(t_k)\|} \Delta s_k$

Figure 18–7

$$\text{Work} = \int_{\mathscr{C}} F \cdot \frac{v}{\|v\|} \, ds = \int_{\mathscr{C}} F \cdot dr.$$

The work done by the force field in moving along $\mathscr{C}$ is defined to be the limit of the sum of terms of form (1). That is,

$$\text{Work} = \lim_{\|\mathscr{P}\|\to 0} \sum_{k=1}^{n} F(f(t_k), g(t_k)) \cdot \frac{v(t_k)}{\|v(t_k)\|} \Delta s_k$$

$$= \int_a^b F(f(t), g(t)) \cdot \frac{v(t)}{\|v(t)\|} ds.$$

Since $ds = \|v(t)\| \, dt$ and $v(t) \, dt = r'(t) \, dt = dr$, this integral reduces to

Work (Motion Along a Curve)

$$\text{Work} = \int_a^b F \cdot v \, dt = \int_{\mathscr{C}} F \cdot dr.$$

EXAMPLE 1 A magnet placed at the origin of the xy-plane exerts an inverse-square force field on each particle in the plane. Let the force of the attraction be given by

$$F(x, y) = \frac{-K(xi + yj)}{(x^2 + y^2)^{3/2}},$$

where K is a positive constant. Let $\mathscr{C}$ be a smooth curve from $A(3, 4)$ to $B(1, 1)$. How much work is done by the force in moving a particle along $\mathscr{C}$ from A to B?

Solution The work is

$$W = \int_{\mathscr{C}} F \cdot dr = \int_{\mathscr{C}} \frac{-K(xi + yj)}{(x^2 + y^2)^{3/2}} \cdot (dx\, i + dy\, j)$$

$$= \int_{\mathscr{C}} \frac{-Kx\, dx}{(x^2 + y^2)^{3/2}} + \frac{-Ky\, dy}{(x^2 + y^2)^{3/2}}.$$

Let

$$P = \frac{-Kx}{(x^2 + y^2)^{3/2}}, \qquad Q = \frac{-Ky}{(x^2 + y^2)^{3/2}}.$$

Then

$$\frac{\partial P}{\partial y} = \frac{3Kxy}{(x^2 + y^2)^{5/2}} = \frac{\partial Q}{\partial x}$$

so that

$$\frac{-Kx\, dx}{(x^2 + y^2)^{3/2}} + \frac{-Ky\, dy}{(x^2 + y^2)^{3/2}}$$

is an exact differential. It follows from partial integration that it is the differential of the potential function for the field

$$f(x, y) = \frac{K}{\sqrt{x^2 + y^2}}.$$

18-4 Applications

It follows from Theorem 18–1 that the line integral for work can be evaluated from the endpoints of the curve $\mathscr{C}$:

$$W = \int_{\mathscr{C}} \frac{-Kx\,dx}{(x^2+y^2)^{3/2}} + \frac{-Ky\,dy}{(x^2+y^2)^{3/2}}$$

$$= f(1,1) - f(3,4) = \frac{K}{\sqrt{2}} - \frac{K}{\sqrt{25}} = K\left(\frac{1}{\sqrt{2}} - \frac{1}{5}\right). \quad \square$$

CONSERVATIVE FIELDS

The result of Example 1 is typical of conservative vector fields. Let $F(x, y)$ be a conservative vector field with potential function $f(x, y)$. [In physics, the potential function is usually written $-f(x, y)$.] Since $F = \nabla f$ is a gradient field, then $F \cdot dr = \nabla f \cdot dr$ is the differential of f. Thus, the work done by the force in moving a particle along a smooth curve from $P_1(x_1, y_1)$ to $P_2(x_2, y_2)$ is

$$W = \int_{\mathscr{C}} F \cdot dr = f(x_2, y_2) - f(x_1, y_1).$$

Potential Energy The *potential energy* of a particle at (x, y) is defined to be $f(x, y)$. Consequently, the work done by the force in moving the particle from P_1 to P_2 equals the difference in the potential energy at the two points.

SPACE CURVES

For simplicity, we have worked with plane curves. Analogous results hold for particles moved along space curves. If $\mathscr{C}$ is a smooth space curve in the domain of a vector field F, then the work done in moving a particle along $\mathscr{C}$ equals

$$\int_{\mathscr{C}} F \cdot dr$$

where $dr = dx\,i + dy\,j + dz\,k$.

If the field is conservative with potential function f, then the work equals the change in potential energy at the endpoints of the curve.

FLUID FLOW

A stream of constant depth flows across a portion of a plane, which we assume to be the xy-plane. Let $\mathscr{C}$ be a smooth curve in the plane. (This curve is fixed; it does not move with the flow of the stream.) Let $F(x, y)$ be the continuous velocity vector field that describes the flow of water at the point (x, y). [We assume that the velocity of water does not vary with depth, so $F(x, y)$ describes the velocity at each particle over (x, y).] We shall show how to describe the flow of water across $\mathscr{C}$ by a line integral.

One technical problem must be decided before we begin the process. It will be necessary to find a unit normal vector to $\mathscr{C}$ at each point of $\mathscr{C}$. Obviously there are two such unit normals. Which one do we choose?

To be definite, we consider that the xy-plane is part of three-dimensional space.

Let T be a unit vector tangent to $\mathscr{C}$ at a given point that points in the direction of increasing arc length. If $\mathscr{C}$ is the graph of

$$r(t) = f(t)i + g(t)j,$$

then

$$T(t) = \frac{f'(t)i + g'(t)j}{\sqrt{f'(t)^2 + g'(t)^2}} = \frac{v(t)}{\|v(t)\|}$$

We choose

Outer Unit Normal

$$n = T \times k.$$

Then n is a unit normal to $\mathscr{C}$. The direction of n is such that if $\mathscr{C}$ is the boundary of a closed region that is traced out in a counterclockwise direction, then n points away from the region. For this reason, n is called an *outer normal* to $\mathscr{C}$. (See Fig. 18–8c.)

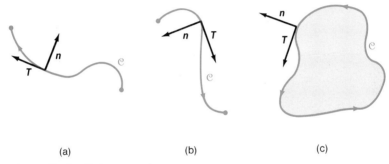

(a)   (b)   (c)

Figure 18–8 Outer normal.

Let Δs_k be the length of a short arc of $\mathscr{C}$ that contains the point $P_k(x_k, y_k)$. Let $n_k(x_k, y_k)$ be the outer normal to $\mathscr{C}$ at P_k. Since F is a continuous vector field, the velocity of the stream at each point of the arc is approximately equal to $F(x_k, y_k)$.

Let δ be the depth of the stream. The quantity of water that crosses the arc in a unit of time is approximately equal to

$$\delta \cdot (\text{comp}_n F) \cdot \Delta s_k = \frac{\delta F(x_k, y_k) \cdot n_k(x_k, y_k) \, \Delta s_k}{\|n_k(x_k, y_k)\|}$$

$$= \delta F(x_k, y_k) \cdot n_k(x_k, y_k) \, \Delta s_k,$$

since n_k is a unit vector. (See Fig. 18–9.)

If $\mathscr{C}$ is decomposed into n such arcs, then the total flow across $\mathscr{C}$ is approximately

$$\delta \sum_{k=1}^{n} F(x_k, y_k) \cdot n_k(x_k, y_k) \, \Delta s_k.$$

If we take the limit of this expression as each Δs_k approaches 0, we obtain the integral

$$\delta \int_{\mathscr{C}} F(x, y) \cdot n(x, y) \, ds,$$

which equals the total flow across $\mathscr{C}$.

18-4 Applications

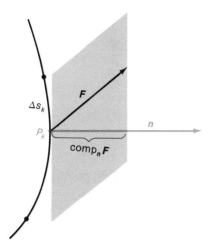

Figure 18-9 The flow across the arc over a unit of time is approximately equal to the area of the parallelogram.

The line integral

$$\int_{\mathcal{C}} F(x, y) \cdot n(x, y) \, ds$$

Flux   is called the *flux* of **F** across $\mathcal{C}$. We use this terminology even when **F** does not represent fluid flow.

EXAMPLE 2   A stream that is 1 m deep flows in a plane between parallel banks 2 m apart. If we consider that the stream flows in the xy-plane, we may take the banks to be the lines $y = 1$ and $y = -1$. The velocity of the stream at $P(x, y)$ is given by

$$F(x, y) = (1 - |y|)i, \quad -1 \leq y \leq 1.$$

(See Fig. 18-10a.)

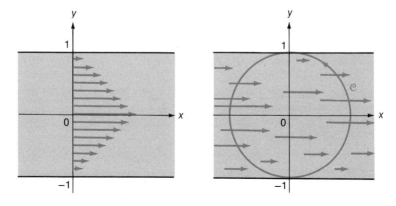

(a) Some vectors in the field   (b) The total flow across $\mathcal{C}$ is zero

Figure 18-10 Example 2.

(a) Let $\mathscr{C}$ be the unit circle, defined parametrically by

$$x = \cos t\mathbf{i}, \quad y = \sin t\mathbf{j}, \quad 0 \leq t \leq 2\pi.$$

Calculate what the flow across $\mathscr{C}$ is over a unit of time.
(b) Interpret the answer in (a) in terms of the physical situation.

Solution (a) The flow across $\mathscr{C}$ equals the flux integral

$$\text{Flux} = \int_{\mathscr{C}} \mathbf{F}(x, y) \cdot \mathbf{n}(x, y) \, ds.$$

Since the curve $\mathscr{C}$ is the graph of the position vector

$$\mathbf{r}(t) = \cos t\mathbf{i} + \sin t\mathbf{j}, \quad 0 \leq t \leq 2\pi,$$

then

$$\mathbf{v}(t) = -\sin t\mathbf{i} + \cos t\mathbf{j}.$$

Since $\mathbf{v}$ is a unit vector, then

$$\mathbf{T} = \mathbf{v} = -\sin t\mathbf{i} + \cos t\mathbf{j}.$$

The outer unit normal vector is defined in terms of t by

$$\mathbf{n} = \mathbf{T} \times \mathbf{k} = \begin{vmatrix} \mathbf{i} & \mathbf{j} & \mathbf{k} \\ -\sin t & \cos t & 0 \\ 0 & 0 & 1 \end{vmatrix} = \cos t\mathbf{i} + \sin t\mathbf{j}.$$

Furthermore,

$$ds = \|\mathbf{v}\| \, dt = \sqrt{\sin^2 t + \cos^2 t} \, dt = dt.$$

Therefore, the flux is

$$\text{Flux} = \int_{\mathscr{C}} \mathbf{F} \cdot \mathbf{n} \, ds$$

$$= \int_0^{2\pi} (1 - |\sin t|)\mathbf{i} \cdot (\cos t\mathbf{i} + \sin t\mathbf{j}) \, dt$$

$$= \int_0^{2\pi} (1 - |\sin t|) \cos t \, dt.$$

To calculate this integral, we rewrite it as the sum of two integrals.

$$\text{Flux} = \int_0^{\pi} (1 - |\sin t|) \cos t \, dt + \int_{\pi}^{2\pi} (1 - |\sin t|) \cos t \, dt$$

$$= \int_0^{\pi} (1 - \sin t) \cos t \, dt + \int_{\pi}^{2\pi} (1 + \sin t) \cos t \, dt$$

$$= -\left[\frac{(1 - \sin t)^2}{2}\right]_0^{\pi} + \left[\frac{(1 + \sin t)^2}{2}\right]_{\pi}^{2\pi}$$

$$= -\left[\frac{(1 - \sin \pi)^2}{2} - \frac{(1 - \sin 0)^2}{2}\right]$$

$$\quad + \left[\frac{(1 + \sin 2\pi)^2}{2} - \frac{(1 + \sin \pi)^2}{2}\right]$$

$$= -\left[\frac{1}{2} - \frac{1}{2}\right] + \left[\frac{1}{2} - \frac{1}{2}\right] = 0.$$

Exercises 18-4

(b) If we examine Figure 18–10b, we see that the water is entering the disk bounded by $\mathscr{C}$ when it passes across the semicircle in Quadrants II and III and is leaving the disk when it passes across the semicircle in Quadrants I and IV. The flux integral gives the net flow across the disk. Since the flux is zero, the same quantity of water leaves the disk as enters it over a unit period of time. □

Interpretation of Flux Let the smooth closed curve $\mathscr{C}$ be the boundary of a closed region $\mathscr{R}$. Let $F(x, y)$ be a vector field defined over $\mathscr{R}$. We say that F has a *source* in $\mathscr{R}$ if the flux of F over $\mathscr{C}$ is positive and a *sink* in $\mathscr{R}$ if the flux is negative. These terms have a natural interpretation in terms of fluid flow. The field F has a source in $\mathscr{R}$ if more liquid flows out of $\mathscr{R}$ than flows in, a sink if more flows in than out.

Exercises 18-4

•1 A particle is moved counterclockwise around the circle $x^2 + y^2 = 1$ from $(1, 0)$ to $(-1, 0)$ in the force field $F(x, y) = -xi - yj$. Calculate the work.

•2 A particle is moved along a line from $(3, 2, 1)$ to $(1, 1, 0)$ in the force field

$$F(x, y, z) = -\left(\frac{xi + yj + zk}{\sqrt{x^2 + y^2 + z^2}}\right).$$

Calculate the work.

•3 A particle is moved along the twisted cubic $r = ti + t^2j + t^3k$, $1 \le t \le 2$, in the force field

$$F(x, y, z) = \frac{-xi - yj}{\sqrt{x^2 + y^2}}.$$

Calculate the work.

•4 Assume that the acceleration due to gravity in a small region near the surface of the earth is $-gk$, where g is the gravitational constant.

A sliding board of a swimming pool is shaped like a helix. The helix has radius 5 ft and height 8 ft and makes one complete revolution. How much work is done by gravity when a 64-lb child slides down the board? (Mass $= \frac{64}{32} = 2$ slugs. Take $g = 32$; use Newton's second law of motion.)

5 Let $F(x, y, z) = f(x)i + g(y)j + h(z)k$ be a force field, where f, g, and h are continuous in a solid region $\mathscr{R}$. Let P_1 and P_2 be points in $\mathscr{R}$. Explain why the work done by the field in moving a particle along a smooth curve in $\mathscr{R}$ from P_1 to P_2 is independent of the particular curve.

6 (a) Let F be a force field defined on a solid region $\mathscr{R}$. Let $\mathscr{C}$ be a smooth curve from P_1 to P_2 that lies in $\mathscr{R}$. Suppose that as we move around $\mathscr{C}$, the direction of movement at each point is orthogonal to the force field. Explain why the work done by the field in moving from P_1 to P_2 along $\mathscr{C}$ must be zero.

•(b) Let $\mathscr{C}$ be a smooth curve from P_1 to P_2 in the xy-plane. Assume that the gravitational field in a region that contains $\mathscr{C}$ is $F(x, y, z) = -gk$, where g is the gravitational constant. Use the result of (a) to calculate the work done by gravity in moving from P_1 to P_2 around $\mathscr{C}$.

7-10. The velocity field for a stream in the xy-plane is described by $F(x, y)$. The surface density (mass per unit of surface area) of the stream is 1. Describe the motion of the stream. Calculate the mass of the liquid that crosses $\mathcal{C}$ in one unit of time.

• 7 $F(x, y) = 2i + j$; $\mathcal{C}$ is the quarter circle $x^2 + y^2 = 1$ from $(1, 0)$ to $(0, 1)$.

8 $F(x, y) = (1 - x^2)j$, $-1 \leq x \leq 1$; $\mathcal{C}$ is the line segment $y = x$, $-1 \leq x \leq 1$.

• 9 $F(x, y) = \dfrac{xi + yj}{x^2 + y^2}$, $(x, y) \neq (0, 0)$; $\mathcal{C}$ is the circle $x^2 + y^2 = 1$ traced out counterclockwise from $(1, 0)$ to $(1, 0)$.

10 $F(x, y) = (1 - x^2)j$, $-1 \leq x \leq 1$; $\mathcal{C}$ is the unit circle $x^2 + y^2 = 1$.

11 Let $\mathcal{C}$ be a circle with center at the origin, that is traced out counterclockwise. Show that the flow across $\mathcal{C}$ for the velocity field of Exercise 9 is the same as the flow across the unit circle. Use this result to locate the source of the stream.

18–5 Green's Theorem

Theorem 18–1 is a generalization of the Fundamental Theorem of Calculus. It states that under certain conditions on the integrand, a line integral can be calculated by evaluating a related function at the endpoints of the curve.

A second generalization of the Fundamental Theorem, which we now consider, shows that under certain conditions a double integral over a basic closed region can be calculated by evaluating a line integral along the boundary of the region.

This result is not as surprising as we might imagine. We showed in Example 2 of Section 18–4 that the flow of a stream across a basic closed region $\mathcal{R}$ can be calculated from a flux integral around the boundary of $\mathcal{R}$. It would seem reasonable that the flow across $\mathcal{R}$ could be calculated by a double integral over $\mathcal{R}$. Thus, there should be some sort of relation between a double integral over $\mathcal{R}$ and a line integral along the boundary of $\mathcal{R}$.

We will find it convenient to have a standard orientation for our curves. For this purpose, we assume that the simple closed curve $\mathcal{C}$ is oriented so that the region enclosed by $\mathcal{C}$ is always on our left as we trace around $\mathcal{C}$. If a curve has this orientation, we indicate a line integral by

$$\oint_{\mathcal{C}} P(x, y)\, dx + Q(x, y)\, dy \quad \text{or} \quad \oint_{\mathcal{C}} P(x, y)\, dx + Q(x, y)\, dy.$$

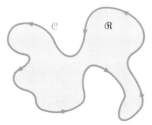

Figure 18–11 Standard orientation for a simple closed curve.

18-5 Green's Theorem

(See Fig. 18-11.) If a curve has the opposite orientation, we write the line integral as

$$\oint_{\mathscr{C}} P(x, y)\, dx + Q(x, y)\, dy.$$

Observe that

$$\oint_{\mathscr{C}} P(x, y)\, dx + Q(x, y)\, dy = -\oint_{\mathscr{C}} P(x, y)\, dx + Q(x, y)\, dy.$$

The relation between a double integral over a region $\mathscr{R}$ and a line integral around the boundary of $\mathscr{R}$ was discovered by the mathematician-physicist George Green (English, 1793–1841). We give a general statement of Green's theorem. Our proof, however, holds for special types of curves only.

THEOREM 18-3 (*Green's Theorem*) Let $\mathscr{C}$ be a piecewise smooth, simple closed curve that is the boundary of a region $\mathscr{R}$. Let $P(x, y)$ and $Q(x, y)$ have continuous partial derivatives in a region that contains $\mathscr{C}$ and $\mathscr{R}$. Then

Green's Theorem
$$\oint_{\mathscr{C}} P(x, y)\, dx + Q(x, y)\, dy = \iint_{\mathscr{R}} \left(-\frac{\partial P}{\partial y} + \frac{\partial Q}{\partial x}\right) dA.$$

Proof We prove Green's theorem for regions of the type shown in Figure 18-12. The boundary curve $\mathscr{C}$ is the union of the graphs of the continuous functions $y = f_1(x)$ and $y = f_2(x)$ for $a \leq x \leq b$. We also can write $\mathscr{C}$ as the union of the graphs of the continuous functions $x = g_1(y)$ and $x = g_2(y)$ for $c \leq y \leq d$. In the remainder of the proof, the numbers a, b, c, d and the points A, B, C, and D refer to Figure 18-12.

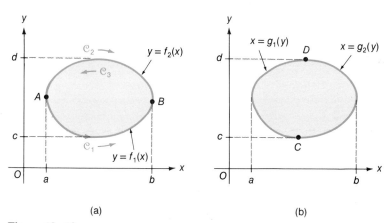

Figure 18-12.

Recall from Exercise 14 of Section 18-3 that the value of a line integral over a piecewise smooth, simple closed curve is independent of the point at which the curve starts. Thus, we can evaluate $\oint_{\mathscr{C}} P(x, y)\, dx + Q(x, y)\, dy$ by tracing $\mathscr{C}$ from A to B to A or from C to D to C.

We integrate the terms $P(x, y)\, dx$ and $Q(x, y)\, dy$ separately around $\mathscr{C}$.

Let $\mathscr{C}_1$ be the graph of $y = f_1(x)$, $a \leq x \leq b$, and $\mathscr{C}_2$ the graph of $y = f_2(x)$, $a \leq x \leq b$. It will be convenient to let $\mathscr{C}_3$ be the graph of $y = f_2(x)$, traced out from B to A.

Then

$$\oint_{\mathscr{C}} P(x, y)\, dx = \int_{\mathscr{C}_1} P(x, y)\, dx + \int_{\mathscr{C}_3} P(x, y)\, dx$$

$$= \int_{\mathscr{C}_1} P(x, y)\, dx - \int_{\mathscr{C}_2} P(x, y)\, dx$$

$$= \int_a^b P(x, f_1(x))\, dx - \int_a^b P(x, f_2(x))\, dx$$

$$= \int_a^b [P(x, f_1(x)) - P(x, f_2(x))]\, dx.$$

On the other hand, since $P(x, y)$ is an antiderivative with respect to y of $\partial P/\partial y$, then

$$\iint_{\mathscr{R}} \left(\frac{\partial P}{\partial y}\right) dA = \int_a^b \int_{f_1(x)}^{f_2(x)} \left(\frac{\partial P}{\partial y}\right) dy\, dx$$

$$= \int_a^b \left[P(x, y)\right]_{f_1(x)}^{f_2(x)} dx = \int_a^b [P(x, f_2(x)) - P(x, f_1(x))]\, dx.$$

Comparing this result with the value of the line integral, we find that

$$\oint_{\mathscr{C}} P(x, y)\, dx = -\iint_{\mathscr{R}} \left(\frac{\partial P}{\partial y}\right) dA.$$

A similar argument, based on tracing $\mathscr{C}$ from C to D to C, can be used to show that

$$\oint_{\mathscr{C}} Q(x, y)\, dy = \iint_{\mathscr{R}} \left(\frac{\partial Q}{\partial x}\right) dA.$$

(See Exercise 19.) On combining the two integrals, we get

$$\oint_{\mathscr{C}} P(x, y)\, dx + Q(x, y)\, dy = \oint_{\mathscr{C}} P(x, y)\, dx + \int_{\mathscr{C}} Q(x, y)\, dy$$

$$= \iint_{\mathscr{R}} \left(-\frac{\partial P}{\partial y} + \frac{\partial Q}{\partial x}\right) dA. \blacksquare$$

EXAMPLE 1   Calculate $\int_{\mathscr{C}} (xy - y^2 + e^{x^2})\, dx + (xy + \sin y^3)\, dy$ along the curve $\mathscr{C}$ that consists of the portion of the parabola $y = x^2$ from $(-1, 1)$ to $(1, 1)$ and the line $y = 1$ from $(1, 1)$ to $(-1, 1)$.

Solution   The curve $\mathscr{C}$ is shown in Figure 18–13. The line integral is

$$\oint_{\mathscr{C}} P(x, y)\, dx + Q(x, y)\, dy,$$

where $P(x, y) = xy - y^2 + e^{x^2}$ and $Q(x, y) = xy + \sin y^3$. By Green's theorem,

$$\oint_{\mathscr{C}} (xy - y^2 + e^{x^2})\, dx + (xy + \sin y^3)\, dy$$

$$= \iint_{\mathscr{R}} \left(-\frac{\partial P}{\partial y} + \frac{\partial Q}{\partial x}\right) dA$$

18–5 Green's Theorem

$$= \iint_{\mathcal{R}} [(-x + 2y) + y] \, dA$$

$$= \int_{-1}^{1} \int_{x^2}^{1} (-x + 3y) \, dy \, dx = \int_{-1}^{1} \left[-xy + \frac{3y^2}{2} \right]_{x^2}^{1} dx$$

$$= \int_{-1}^{1} \left(-\frac{3x^4}{2} + x^3 - x + \frac{3}{2} \right) dx = \left[-\frac{3x^5}{10} + \frac{x^4}{4} - \frac{x^2}{2} + \frac{3}{2}x \right]_{-1}^{1} = \frac{12}{5}. \quad \square$$

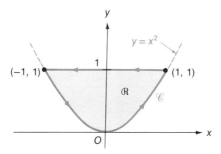

Figure 18–13 Example 1.

Green's theorem has a number of important applications. We give, as an example, a new proof of Theorem 18–2 for the special case for which $P(x, y) \, dx + Q(x, y) \, dy$ is an exact differential on a region that contains both the simple closed curve $\mathcal{C}$ and the region $\mathcal{R}$ that it encloses.

EXAMPLE 2   Let $G(x, y)$ be a differentiable function of x and y that has continuous second partial derivatives on a region that contains both the piecewise smooth, simple closed curve $\mathcal{C}$ and the region $\mathcal{R}$ that has $\mathcal{C}$ for its boundary. Let $dG = P(x, y) \, dx + Q(x, y) \, dy$. Use Green's theorem to prove that

$$\oint_{\mathcal{C}} P(x, y) \, dx + Q(x, y) \, dy = 0.$$

Solution   Since $dG = P(x, y) \, dx + Q(x, y) \, dy$, then

$$\frac{\partial G}{\partial x} = P(x, y) \quad \text{and} \quad \frac{\partial G}{\partial y} = Q(x, y).$$

Since G has continuous second partial derivatives, the mixed partial derivatives are equal. Therefore,

$$\frac{\partial P}{\partial y} = \frac{\partial^2 G}{\partial y \partial x} = \frac{\partial^2 G}{\partial x \partial y} = \frac{\partial Q}{\partial x}.$$

By Green's Theorem

$$\oint_{\mathcal{C}} P(x, y) \, dx + Q(x, y) \, dy = \iint_{\mathcal{R}} \left(-\frac{\partial P}{\partial y} + \frac{\partial Q}{\partial x} \right) dA$$

$$= \iint_{\mathcal{R}} 0 \cdot dA = 0. \quad \square$$

EXAMPLE 3 Let $\mathcal{R}$ be a region that has a piecewise smooth simple closed curve $\mathcal{C}$ for its boundary. Show that

Area
$$\text{Area of } \mathcal{R} = \frac{1}{2} \oint_{\mathcal{C}} (-y\, dx + x\, dy).$$

Solution Let $P(x, y) = -y$, $Q(x, y) = x$. It follows from Green's theorem that

$$\oint_{\mathcal{C}} (-y)\, dx + x\, dy = \iint_{\mathcal{R}} \left[-\frac{\partial P}{\partial y} + \frac{\partial Q}{\partial x} \right] dA$$

$$= \iint_{\mathcal{R}} [-(-1) + 1]\, dA$$

$$= 2 \iint_{\mathcal{R}} dA = 2A.$$

Therefore
$$A = \frac{1}{2} \oint_{\mathcal{C}} (-y)\, dx + x\, dy. \quad \square$$

EXAMPLE 4 Use the result of Example 3 to calculate the area of the ellipse

$$\frac{x^2}{2^2} + \frac{y^2}{3^2} = 1.$$

Solution Parametric equations for the ellipse are

$$x = 2 \cos t, \quad y = 3 \sin t, \quad 0 \le t \le 2\pi.$$

Then

$$A = \frac{1}{2} \oint_{\mathcal{C}} (-y)\, dx + x\, dy$$

$$= -\frac{1}{2} \oint_{\mathcal{C}} y\, dx + \frac{1}{2} \oint_{\mathcal{C}} x\, dy$$

$$= -\frac{1}{2} \int_0^{2\pi} 3 \sin t (-2 \sin t)\, dt + \frac{1}{2} \int_0^{2\pi} 2 \cos t \cdot 3 \cos t\, dt$$

$$= \frac{1}{2} \int_0^{2\pi} (6 \sin^2 t + 6 \cos^2 t)\, dt = 3 \int_0^{2\pi} dt = 6\pi. \quad \square$$

Green's theorem can be extended to other types of curves and regions. In particular, it holds over rectangles and figures that can be built out of rectangles. It holds also for curves such as those pictured in Figure 18–14.

Consider the curve in Figure 18–14a. Let $\mathcal{C}$ be the total boundary of the figure, where for $\mathcal{C}$ to be a closed curve, the portion marked $\mathcal{C}_3$ and $\mathcal{C}_4$ is traced out twice, once in each direction. Then

$$\oint_{\mathcal{C}} P\, dx + Q\, dy = \int_{\mathcal{C}_1} P\, dx + Q\, dy + \int_{\mathcal{C}_3} P\, dx + Q\, dy$$

$$+ \int_{\mathcal{C}_2} P\, dx + Q\, dy + \int_{\mathcal{C}_4} P\, dx + Q\, dy.$$

18–5 Green's Theorem

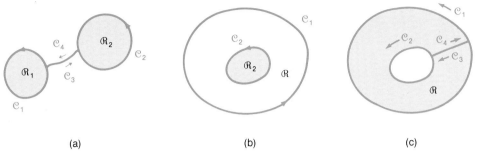

Figure 18–14 Some extensions of Green's theorem.

Since the second and fourth of these integrals cancel each other, then

$$\int_{\mathscr{C}} P\,dx + Q\,dy = \int_{\mathscr{C}_1} P\,dx + Q\,dy + \int_{\mathscr{C}_2} P\,dx + Q\,dy$$

$$= \iint_{\mathscr{R}_1} \left(-\frac{\partial P}{\partial y} + \frac{\partial Q}{\partial x}\right) dA + \iint_{\mathscr{R}_2} \left(-\frac{\partial P}{\partial y} + \frac{\partial Q}{\partial x}\right) dA$$

$$= \iint_{\mathscr{R}_1 \cup \mathscr{R}_2} \left(-\frac{\partial P}{\partial y} + \frac{\partial Q}{\partial x}\right) dA$$

provided P and Q have continuous partial derivatives on a region that contains $\mathscr{C}$, $\mathscr{R}_1$, and $\mathscr{R}_2$.

A similar result holds for the curves in Figure 18–14b. Let $\mathscr{R}$ be the region between the two curves, $\mathscr{R}_1$ the total region enclosed by $\mathscr{C}_1$, and $\mathscr{R}_2$ the total region enclosed by $\mathscr{C}_2$. Then $\mathscr{R}_1 = \mathscr{R}_2 \cup \mathscr{R}$. We connect $\mathscr{C}_1$ and $\mathscr{C}_2$ by the line marked $\mathscr{C}_3$ and $\mathscr{C}_4$ shown in Figure 18–14c and let $\mathscr{C}$ be the union of $\mathscr{C}_1$, $\mathscr{C}_3$, $\mathscr{C}_2$, and $\mathscr{C}_4$. If we consider that the added line cuts the region $\mathscr{R}$, it follows as in the preceeding argument that

$$\oint_{\mathscr{C}} P\,dx + Q\,dy = \oint_{\mathscr{C}_1} P\,dx + Q\,dy + \oint_{\mathscr{C}_2} P\,dx + Q\,dy$$

$$= \oint_{\mathscr{C}_1} P\,dx + Q\,dy - \oint_{\mathscr{C}_2} P\,dx + Q\,dy$$

$$= \iint_{\mathscr{R}} \left(-\frac{\partial P}{\partial y} + \frac{\partial Q}{\partial x}\right) dA$$

provided P and Q have continuous partial derivatives on a region that contains $\mathscr{C}_1$, $\mathscr{C}_2$, and $\mathscr{R}$.

EXAMPLE 5 Let $\mathscr{C}_1$ and $\mathscr{C}_2$ be nonintersecting, piecewise smooth, simple closed curves such that the origin is in the region enclosed by $\mathscr{C}_2$ and $\mathscr{C}_2$ is in the region enclosed by $\mathscr{C}_1$. (See Figure 18–15.) Show that

$$\oint_{\mathscr{C}_1} \frac{-y}{x^2 + y^2}\,dx + \frac{x}{x^2 + y^2}\,dy = \oint_{\mathscr{C}_2} \frac{-y}{x^2 + y^2}\,dx + \frac{x}{x^2 + y^2}\,dy.$$

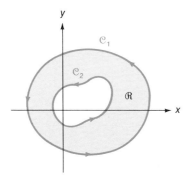

Figure 18-15 Example 5.

Solution Let $\mathcal{R}$ be the region enclosed by the two curves. It follows from the remarks preceding this example that

$$\oint_{\mathcal{C}_1}\left(\frac{-y}{x^2+y^2}dx+\frac{x}{x^2+y^2}dy\right)-\oint_{\mathcal{C}_2}\left(\frac{-y}{x^2+y^2}dx+\frac{x}{x^2+y^2}dy\right)$$

$$=\iint_{\mathcal{R}}\left(-\frac{\partial P}{\partial y}+\frac{\partial Q}{\partial x}\right)dA$$

$$=\iint_{\mathcal{R}}\left(\frac{x^2-y^2}{(x^2+y^2)^2}+\frac{-x^2+y^2}{(x^2+y^2)^2}\right)dA$$

$$=\iint_{\mathcal{R}} 0\cdot dA = 0.$$

Therefore,

$$\oint_{\mathcal{C}_1}\frac{-y}{x^2+y^2}dx+\frac{x}{x^2+y^2}dy=\oint_{\mathcal{C}_2}\frac{-y}{x^2+y^2}dx+\frac{x}{x^2+y^2}dy. \quad\square$$

It is simple to extend the result of Example 5 to any two piecewise smooth, simple closed curves that enclose the origin. It follows from Example 2 of Section 18–3 that if $\mathcal{C}$ is the unit circle, then

$$\oint_{\mathcal{C}}\frac{-y}{x^2+y^2}dx+\frac{x}{x^2+y^2}dy=2\pi.$$

It then follows from Example 5 that if $\mathcal{C}$ is *any* piecewise smooth, simple closed curve that encloses the origin, then

$$\oint_{\mathcal{C}}\frac{-y}{x^2+y^2}dx+\frac{x}{x^2+y^2}dy=2\pi.$$

Exercises 18-5

1–6. Use an equivalent double integral to evaluate the line integral. It may be convenient to convert the double integral to polar coordinates.

•1 $\oint_{\mathcal{C}}(x^2y+e^x)\,dx+(\ln y-3x^2)\,dy$; $\mathcal{C}$ is the rectangle with vertices $(1, 1)$, $(1, 5)$, $(3, 5)$, and $(3, 1)$.

Exercises 18–5

- **2** $\oint_{\mathcal{C}} (xy - \sin^{-1} x)\, dx + (e^y + x^2 y)\, dy$; $\mathcal{C}$ consists of the portion of the parabola $y = x^2$ from $(-1, 1)$ to $(1, 1)$ and the portion of the parabola $y = 2 - x^2$ from $(1, 1)$ to $(-1, 1)$.

- **3** $\oint_{\mathcal{C}} (e^x + e^y)\, dx + (\tan^{-1} x + \tan^{-1} y)\, dy$; $\mathcal{C}$ is the triangle with vertices $(0, 0)$, $(0, 3)$, and $(3, 3)$.

- **4** $\oint_{\mathcal{C}} (\sec x + \sin y)\, dx + (\sin x + x \cos y)\, dy$; $\mathcal{C}$ is the square with vertices $(-1, -1)$, $(1, -1)$, $(1, 1)$, and $(-1, 1)$.

- **5** $\oint_{\mathcal{C}} (xy + x^3)\, dx + (x^2 + y)\, dy$; $\mathcal{C}$ is the circle $x^2 + y^2 = 4$.

- **6** $\oint_{\mathcal{C}} (x^2 + y^2)\, dx + (x^3 + y^2)\, dy$; $\mathcal{C}$ is the circle $x^2 + y^2 = 1$.

7–10. Use the formula $A = \tfrac{1}{2} \oint_{\mathcal{C}} (-y)\, dx + x\, dy$ to calculate the area of the region bounded by the graphs of the equations.

- **7** $y = x^2$, $y = 2 - x$
- **8** $x^2 + y^2 = 1$
- **9** $x = \sin t$, $y = -\cos t \sin t$, $0 \le t \le \pi$
- **10** $x = \cos^3 t$, $y = \sin^3 t$, $0 \le t \le 2\pi$

11 Let $a, b > 0$. Use a line integral to prove that the area of the ellipse $x^2/a^2 + y^2/b^2 = 1$ is $A = ab\pi$.

12 Let A be the area of the region $\mathcal{R}$ bounded by the piecewise smooth, simple closed curve $\mathcal{C}$. Modify the argument in Example 3 to show that

$$A = \oint_{\mathcal{C}} (-y)\, dx = \oint_{\mathcal{C}} x\, dy.$$

13–16. Use one of the formulas in Exercise 12 to calculate the areas of the regions in Exercises 7–10.

13 Exercise 7
14 Exercise 8
15 Exercise 9
16 Exercise 10

17 Let $\mathcal{R}$ be the region bounded by the piecewise smooth, simple closed curve $\mathcal{C}$. Prove that the moments of $\mathcal{R}$ about the x- and y-axes are

$$M_x = -\tfrac{1}{2} \oint_{\mathcal{C}} y^2\, dx \quad \text{and} \quad M_y = \tfrac{1}{2} \int_{\mathcal{C}} x^2\, dy.$$

- **18** Use the formulas in Exercise 17 to calculate the centroids of the following regions.
    - (a) The region in Exercise 9.
    - (b) The region in Exercise 10.

19 Complete the proof of Theorem 18–3 by proving that

$$\oint_C Q(x, y)\, dy = \iint_R \left(\frac{\partial Q}{\partial x}\right) dA.$$

20 Let C be a piecewise smooth, simple closed curve that encloses the origin. Use the result of Example 5 to show that

$$\int_C \frac{-y}{x^2 + y^2}\, dx + \frac{x}{x^2 + y^2}\, dy = 2\pi.$$

Explain why it is necessary that this curve enclose the origin. (Or is it?)

18–6 Surface Integrals

In this section we generalize the concept of a line integral to an integral defined on a surface. Later in the chapter we generalize Green's Theorem to surface integrals.

Let $f(x, y)$ have continuous partial derivatives over the basic closed region R in the xy-plane. Let S be the graph of f over R. In Section 17–6 we defined the surface area of S by the integral

$$A = \iint_R \sqrt{1 + f_x^2 + f_y^2}\, dA.$$

This integral can be generalized by the following process. Let $g(x, y, z)$ be continuous at each point of S. Let P be an inner partition of R into small rectangles R_1, $R_2, \ldots, R_n$. Let (ξ_k, η_k) be a point in R_k. Let ΔA_k be the area of R_k. Corresponding to the rectangle R_k there is a parallelogram P_k of area

$$\Delta S_k = \sqrt{f_x(\xi_k, \eta_k)^2 + f_y(\xi_k, \eta_k)^2 + 1}\, \Delta x_k\, \Delta y_k$$

tangent to S at $(\xi_k, \eta_k, f(\xi_k, \eta_k))$. (See Fig. 17–28.) The sum

$$\sum_{k=1}^n g(\xi_k, \eta_k, f(\xi_k, \eta_k))\, \Delta S_k$$

$$= \sum_{k=1}^n g(\xi_k, \eta_k, f(\xi_k, \eta_k)) \sqrt{f_x(\xi_k, \eta_k)^2 + f_y(\xi_k, \eta_k)^2 + 1}\, \Delta A_k$$

is an approximating sum for an integral. The limit of this sum, as the norm of the partition approaches 0, is called the *surface integral of g over S*, written

Surface Integral
$$\iint_S g(x, y, z)\, dS.$$

To evaluate this surface integral, we write it as an equivalent double integral over R:

Evaluation of Surface Integral
$$\iint_S g(x, y, z)\, dS = \iint_R g(x, y, f(x, y)) \sqrt{f_x^2 + f_y^2 + 1}\, dA.$$

18–6 Surface Integrals

Similar definitions hold for surface integrals when the surface S is the graph of $x = f(y, z)$ or the graph of $y = f(x, z)$. In these cases we project the surface S onto a basic closed region $\mathcal{R}$ in the yz- or xz-plane and express the surface integral as a double integral over $\mathcal{R}$.

Two simple applications are of special interest:

Surface Area

(1) If $g(x, y, z) = 1$ for all (x, y, z) on S, then the surface integral of g over S is

$$\iint_S dS = \iint_\mathcal{R} \sqrt{f_x^2 + f_y^2 + 1}\, dA,$$

which according to our work in Section 17–6, equals the *surface area* of S.

Mass

(2) Let S be covered by a thin metal sheet. Let the surface density (mass per unit of surface area) of the sheet at $P(x, y, z)$ be $\delta(x, y, z)$. Then the total *mass* of the sheet is

$$\iint_S \delta(x, y, z)\, dS.$$

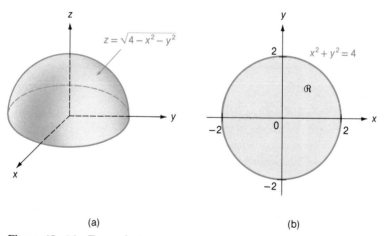

(a)   (b)

Figure 18–16 Example 1.

EXAMPLE 1   The hemisphere

$$f(x, y) = \sqrt{4 - x^2 - y^2}, \quad 0 \le x^2 + y^2 \le 4,$$

is covered by a thin sheet of metal. The surface density at (x, y, z) is $\delta(x, y, z) = z/10$. Calculate the mass of the sheet.

Solution   The projection of the hemisphere on the xy-plane is the disk

$$\mathcal{R} = \{(x, y)\,|\, x^2 + y^2 \le 4\}.$$

(See Fig. 18–16.) The mass of the sheet is

$$M = \iint_S \delta(x, y, z) \, dS = \iint_S \frac{z}{10} \, dS = \iint_{\mathcal{R}} \frac{z}{10} \sqrt{1 + f_x^2 + f_y^2} \, dA.$$

Since the surface is the graph of $x^2 + y^2 + z^2 = 4$, implicit differentiation yields

$$f_x = -\frac{x}{z} \quad \text{and} \quad f_y = -\frac{y}{z},$$

so that

$$\sqrt{1 + f_x^2 + f_y^2} = \sqrt{1 + \frac{x^2}{z^2} + \frac{y^2}{z^2}} = \frac{2}{z}.$$

Inserting this expression in the integral for M gives

$$M = \iint_{\mathcal{R}} \frac{z}{10} \cdot \frac{2}{z} \, dA = \frac{1}{5} \iint_{\mathcal{R}} dA$$
$$= \frac{1}{5} \cdot 4\pi = \frac{4\pi}{5},$$

since the double integral is equal to the area of $\mathcal{R}$. □

APPLICATION TO FLUID FLOW

Let $f(x, y)$ have continuous partial derivatives over a basic closed region $\mathcal{R}$ in the xy-plane. Let S be the graph of f over $\mathcal{R}$. Imagine that S has been submerged in a fluid in which the velocity at (x, y, z) is given by the vector field $F(x, y, z)$. (We assume that S does not interfere with the flow of the fluid. Think of S as an imaginary boundary or as an open-weave net woven with very fine threads.) We will extend the techniques of Section 18–4 to calculate the total fluid flow across S during a unit of time.

It follows from the conditions on f that there are nonzero normal vectors to S at each point of S. At each point $P(x, y, z)$ of S we choose a unit normal vector $n(x, y, z)$ with direction angles α, β, γ, where $0 \leq \gamma \leq 90°$. (See Fig. 18–17a.) Such a normal is called an *upper normal* to S.

Upper Normal

Let (ξ_k, η_k) be a point in a small rectangle R_k of an inner partition of $\mathcal{R}$. Let $\mathcal{P}_k$ be the corresponding parallelogram tangent to the surface S at $P_k(\xi_k, \eta_k, f(\xi_k, \eta_k))$. Let n_k be the upper unit normal to S at P_k and let F_k be the velocity vector for the fluid flow at P_k. (See Fig. 18–17b.)

The total amount of the fluid that crosses the parallelogram $\mathcal{P}_k$ over a unit of time is approximated by the prism shown in Figure 18–17c. The volume of this prism equals

$$(\text{comp}_{n_k} F_k) \, \Delta S_k = \frac{F_k \cdot n_k}{\|n_k\|} \Delta S_k = F_k \cdot n_k \, \Delta S_k,$$

where ΔS_k is the area of $\mathcal{P}_k$. Note that this number may be positive or negative depending on the directions of F_k and n_k. The net amount of fluid that crosses S over a

18-6 Surface Integrals

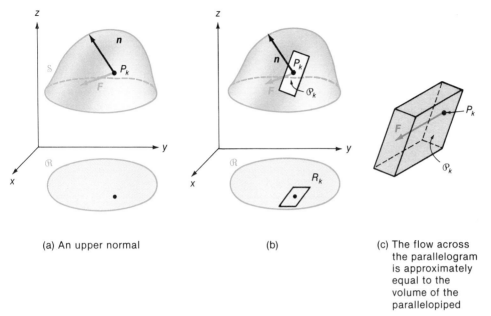

(a) An upper normal

(b)

(c) The flow across the parallelogram is approximately equal to the volume of the parallelopiped

Figure 18-17 Fluid flow.

unit of time is approximately

$$\sum_{k=1}^{n} F_k \cdot n_k \, \Delta S.$$

The exact amount of fluid that crosses S is given by the corresponding surface integral

$$\iint_S F(x, y, z) \cdot n(x, y, z) \, dS.$$

This integral is called the *flux* of F across S. That is,

Flux
$$\text{Flux} = \iint_S F(x, y, z) \cdot n(x, y, z) \, dS.$$

Procedure to Find Upper Unit Normal

Before going ahead, we show how to find the upper unit normal $n(x, y, z)$. We first write the equation of the surface S as

$$G(x, y, z) = 0$$

where $G(x, y, z) = z - f(x, y)$. Then one vector normal to S at (x, y, z) is

$$\nabla G = -\frac{\partial f}{\partial x} i - \frac{\partial f}{\partial y} j + k.$$

Since the k-component of ∇G is positive, ∇G is an upper normal. We get n by dividing

∇G by its magnitude:

$$n = \frac{\nabla G}{\|\nabla G\|} = \frac{-(\partial f/\partial x)\mathbf{i} - (\partial f/\partial y)\mathbf{j} + \mathbf{k}}{\sqrt{(\partial f/\partial x)^2 + (\partial f/\partial y)^2 + 1}}.$$

EXAMPLE 2   The hemisphere $z = f(x, y) = \sqrt{4 - x^2 - y^2}$ marks a boundary surface S in a stream with fluid flow described by the vector field

$$\mathbf{F}(x, y, z) = -y\mathbf{i} + x\mathbf{j} + 2\mathbf{k}.$$

(The stream flows in the direction of the z-axis, rotating circularly. Calculate the flux of $\mathbf{F}$ across S.)

Solution   The upper unit normal is

$$n = \frac{-(\partial f/\partial x)\mathbf{i} - (\partial f/\partial y)\mathbf{j} + \mathbf{k}}{\sqrt{(\partial f/\partial x)^2 + (\partial f/\partial y)^2 + 1}}.$$

Since the surface is the graph of $x^2 + y^2 + z^2 = 4$, as in Example 1 we find that $\partial f/\partial x = -x/z$ and $\partial f/\partial y = -y/z$. Then

$$n = \frac{\frac{x}{z}\mathbf{i} + \frac{y}{z}\mathbf{j} + \mathbf{k}}{\sqrt{\frac{x^2}{z^2} + \frac{y^2}{z^2} + 1}} = \frac{\frac{x}{z}\mathbf{i} + \frac{y}{z}\mathbf{j} + \mathbf{k}}{2/z}$$

$$= \frac{x}{2}\mathbf{i} + \frac{y}{2}\mathbf{j} + \frac{z}{2}\mathbf{k}.$$

The flux of $\mathbf{F}$ across S is

$$\text{Flux} = \iint_S \mathbf{F}(x, y, z) \cdot \mathbf{n}(x, y, z)\, dS$$

$$= \iint_S (-y\mathbf{i} + x\mathbf{j} + 2\mathbf{k}) \cdot \left(\frac{x}{2}\mathbf{i} + \frac{y}{2}\mathbf{j} + \frac{z}{2}\mathbf{k}\right) dS$$

$$= \iint_S \left(-\frac{yx}{2} + \frac{xy}{2} + z\right) dS = \iint_S z\, dS.$$

It follows from the same technique used in Example 1 that this integral can be written as

$$\text{Flux} = \iint_R z\sqrt{f_x^2 + f_y^2 + 1}\, dA$$

$$= \iint_R z\sqrt{\frac{x^2}{z^2} + \frac{y^2}{z^2} + 1}\, dA = \iint_R 2\, dA$$

$$= 2 \cdot 4\pi = 8\pi.$$

(See Fig. 18–16.) The total fluid flow across S over one unit of time is 8π cubic units. □

Exercises 18–6

SURFACE INTEGRALS ON CLOSED SURFACES

Outer Normal

Our results can be extended to certain closed surfaces in three-dimensional space—the boundaries of solids such as spheres or ellipsoids. In this case we choose an *outer normal* **n** rather than an *upper normal*. (An outer normal is one that points away from the solid enclosed by the surface. See Fig. 18–18a.)

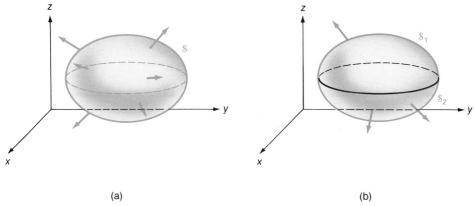

(a) (b)

Figure 18–18 Outer normals to a closed surface.

If S is a sufficiently simple closed surface, such as the boundary of an ellipsoid, we can split S into an upper surface S_1 and a lower surface S_2. (See Fig. 18–18b.) On the upper surface we use an upper normal for the outer normal; on the lower surface we use a *lower normal*. Then the integral for flux is

$$\iint_S \mathbf{F} \cdot \mathbf{n}\, dS = \iint_{S_1} \mathbf{F} \cdot \mathbf{n}\, dS + \iint_{S_2} \mathbf{F} \cdot \mathbf{n}\, dS.$$

Source Sink

If $\mathbf{F}$ represents fluid flow and S is a closed surface, then the flux measures the flow out of S over a unit period of time. We make the same definitions of source and sink as with line integrals. If the flux is positive, we say that $\mathbf{F}$ has a *source* in S. (More fluid leaves S than enters.) If the flux is negative, $\mathbf{F}$ has a *sink* in S. (More fluid enters S than leaves.)

Exercises 18–6

1–6. Evaluate the surface integral.

●1 $\iint_S (x^2 + y)\, dS$; S is the portion of the plane $z = 2x - y$ within the vertical planes $x = 1$, $y = 0$, $x - 2y + 1 = 0$.

- **2** $\iint_S (3x - y - 2z)\, dS$; S is the portion of the plane $x = y + z$ within the planes $y = 0$, $y = 2$, $z = 0$, and $z = 2$.

- **3** $\iint_S dS$; S is the graph of $z = x^2 + y^2$, $0 \le x^2 + y^2 \le 2$.

- **4** $\iint_S (2y^2 + z)\, dS$; S is the graph of $z = x^2 - y^2$, $0 \le x^2 + y^2 \le 1$.

- **5** $\iint_S (x + y)\, dS$; S is the portion of the half-cylinder $z = \sqrt{4 - x^2}$ with $0 \le x \le \sqrt{3}$, $0 \le y \le 1$.

- **6** $\iint_S (x^2 - y^2 + z^2 + 1)\, dS$; S is the portion of the half-cone $z = \sqrt{y^2 - x^2}$ with $0 \le x \le \sqrt{2}$, $2 \le y \le 2\sqrt{2}$.

7–10. Calculate the mass of S if the surface density at (x, y, z) is $\delta(x, y, z)$.

- **7** $\delta(x, y, z) = x^2 + y^2$; S is the hemisphere $z = -\sqrt{1 - x^2 - y^2}$.
- **8** $\delta(x, y, z) = \sqrt{z}$; S is portion of the cone $z = \sqrt{x^2 + y^2}$ with $x^2 + y^2 \le 16$.
- **9** $\delta(x, y, z) = \sqrt{z}$; S is the portion of the paraboloid $z = x^2 + y^2$ with $z \le 4$.
- **10** $\delta(x, y, z) = z$; S is the portion of the hyperboloid of two sheets $-x^2 - y^2 + z^2 = 1$, with $z \ge 0$, that is interior to the cylinder $x^2 + y^2 = 4$.

11–14. Use an upper normal n to S to calculate the flux of F over S.

- **11** $F(x, y, z) = -yi + xj$; S is the portion of the plane $f(x, y) = 8x - 4y + 5$ that is above the triangle with vertices $(0, 0, 0)$, $(0, 1, 0)$, $(1, 0, 0)$.

- **12** $F(x, y, z) = (9 - x^2)j$, $-3 \le x \le 3$; S is the Octant I portion of the plane $2x + 3y + 6z = 6$.

- **13** $F(x, y, z) = i + j + k$; S is the hemisphere $z = \sqrt{4 - x^2 - y^2}$.

- **14** $F(x, y, z) = yi - xj + 2k$; S is the portion of the half-cylinder $z = \sqrt{1 - y^2}$, with $0 \le x \le 5$.

15–16. Calculate the flux $\iint_S F \cdot n\, dS$ over the sphere $x^2 + y^2 + z^2 = 1$. Is there a source or a sink in S? (The vector n is an outer normal to S. It is an upper normal on the upper hemisphere and a lower normal on the lower hemisphere.)

- **15** $F(x, y, z) = \dfrac{xi + yj}{x^2 + y^2}$

- **16** $F(x, y, z) = yi + xj + 3k$

18–7 Divergence and Curl

17–18. Let F and G be continuous over the surface S that is the graph of $z = f(x, y)$, $(x, y) \in \mathcal{R}$. Let c be a constant. Express the surface integrals as double integrals over $\mathcal{R}$ to verify the given results.

17 $\iint_S [F(x, y, z) + G(x, y, z)]\, dS = \iint_S F(x, y, z)\, dS + \iint_S G(x, y, z)\, dS$

18 $\iint_S cF(x, y, z)\, dS = c \iint_S F(x, y, z)\, dS$

18–7 Divergence and Curl

We now discuss two operations on a vector field F that are roughly comparable to derivatives of F. One operation yields a scalar function, the other a vector field.

DIVERGENCE

Let
$$F(x, y, z) = f(x, y, z)\mathbf{i} + g(x, y, z)\mathbf{j} + h(x, y, z)\mathbf{k}$$

be a vector field. The *divergence* of F, denoted by $\operatorname{div} F$ or $\nabla \cdot F$ is defined to be

Divergence
$\nabla \cdot F$

$$\operatorname{div} F = \frac{\partial f}{\partial x} + \frac{\partial g}{\partial y} + \frac{\partial h}{\partial z}.$$

The divergence of F is defined at each point at which the three partial derivatives exist.

The symbol $\nabla \cdot F$ is a mnemonic device for remembering the divergence formula. If we symbolically calculate the dot product of the operator ∇ with F, we get

$$\nabla \cdot F = \left(\frac{\partial}{\partial x}\mathbf{i} + \frac{\partial}{\partial y}\mathbf{j} + \frac{\partial}{\partial z}\mathbf{k}\right) \cdot (f\mathbf{i} + g\mathbf{j} + h\mathbf{k})$$

$$= \frac{\partial f}{\partial x} + \frac{\partial g}{\partial y} + \frac{\partial h}{\partial z} = \operatorname{div} F.$$

Note that "multiplication" of an operator such as $\partial/\partial x$ by a function means that we calculate the appropriate partial derivative.

EXAMPLE 1 Let $F(x, y, z) = 3xz\mathbf{i} + y^2\mathbf{j} + xy\mathbf{k}$. Then

$$\operatorname{div} F = \nabla \cdot F = \left(\frac{\partial}{\partial x}\mathbf{i} + \frac{\partial}{\partial y}\mathbf{j} + \frac{\partial}{\partial z}\mathbf{k}\right) \cdot (3xz\mathbf{i} + y^2\mathbf{j} + xy\mathbf{k})$$

$$= \frac{\partial}{\partial x}(3xz) + \frac{\partial}{\partial y}(y^2) + \frac{\partial}{\partial z}(xy)$$

$$= 3z + 2y. \quad \square$$

Example 2 establishes a property of divergence comparable to the product rule for derivatives. Note that divergence acts as the derivative of the vector field and the gradient as the derivative of the scalar function.

EXAMPLE 2   Let $u(x, y, z)$ be a differentiable scalar function and $F(x, y, z)$ a vector field. Show that

$$\text{div } (uF) = u \text{ div } F + F \cdot \nabla u.$$

Solution   Write F in terms of its component functions:

$$F(x, y, z) = f(x, y, z)i + g(x, y, z)j + h(x, y, z)k.$$

Then

$$uF = ufi + ugj + uhk$$

and

$$\begin{aligned}
\text{div } (uF) &= \frac{\partial}{\partial x}(uf) + \frac{\partial}{\partial y}(ug) + \frac{\partial}{\partial z}(uh) \\
&= \left(u\frac{\partial f}{\partial x} + f\frac{\partial u}{\partial x}\right) + \left(u\frac{\partial g}{\partial y} + g\frac{\partial u}{\partial y}\right) + \left(u\frac{\partial h}{\partial z} + h\frac{\partial u}{\partial z}\right) \\
&= u\left(\frac{\partial f}{\partial x} + \frac{\partial g}{\partial y} + \frac{\partial h}{\partial z}\right) + \left(f\frac{\partial u}{\partial x} + g\frac{\partial u}{\partial y} + h\frac{\partial u}{\partial z}\right) \\
&= u \text{ div } F + (fi + gj + hk) \cdot \left(\frac{\partial u}{\partial x}i + \frac{\partial u}{\partial y}j + \frac{\partial u}{\partial z}k\right) \\
&= u \text{ div } F + F \cdot \nabla u. \quad \square
\end{aligned}$$

We shall show in Section 18–8 that divergence has a natural interpretation in terms of fluid flow. Let $F(x, y, z)$ be the velocity field for a fluid. We shall establish that div $F(x, y, z)$ is proportional to the rate of change of the mass of the fluid at (x, y, z) over a unit of time. That is, there is no change in mass at a point if div $F = 0$; there is a *source* if div $F > 0$ and a *sink* if div $F < 0$.

CURL

The other generalization of the derivative concept is the *curl* of a vector field F. Let

$$F(x, y, z) = f(x, y, z)i + g(x, y, z)j + h(x, y, z)k.$$

The *curl* of F, denoted by curl F, is the vector

Curl
$\nabla \times F$

$$\text{curl } F = \left(\frac{\partial h}{\partial y} - \frac{\partial g}{\partial z}\right)i - \left(\frac{\partial h}{\partial x} - \frac{\partial f}{\partial z}\right)j + \left(\frac{\partial g}{\partial x} - \frac{\partial f}{\partial y}\right)k.$$

The curl of F is defined at each point at which the partial derivatives exist.

The formula for curl F can be remembered symbolically as $\nabla \times F$. This is based on

18–7 Divergence and Curl

Calculation of Curl

$$\nabla \times F = \begin{vmatrix} i & j & k \\ \dfrac{\partial}{\partial x} & \dfrac{\partial}{\partial y} & \dfrac{\partial}{\partial z} \\ f & g & h \end{vmatrix} = \left(\dfrac{\partial h}{\partial y} - \dfrac{\partial g}{\partial z}\right) i - \left(\dfrac{\partial h}{\partial x} - \dfrac{\partial f}{\partial z}\right) j + \left(\dfrac{\partial g}{\partial x} - \dfrac{\partial f}{\partial y}\right) k.$$

EXAMPLE 3 Let $F(x, y, z) = -yi + xj + k$. Then

$$\text{curl } F = \nabla \times F = \begin{vmatrix} i & j & k \\ \dfrac{\partial}{\partial x} & \dfrac{\partial}{\partial y} & \dfrac{\partial}{\partial z} \\ -y & x & 1 \end{vmatrix}$$

$$= \left(\dfrac{\partial(1)}{\partial y} - \dfrac{\partial x}{\partial z}\right) i - \left(\dfrac{\partial(1)}{\partial x} + \dfrac{\partial y}{\partial z}\right) j + \left(\dfrac{\partial x}{\partial x} + \dfrac{\partial y}{\partial y}\right) k$$

$$= 2k. \ \square$$

EXAMPLE 4 Let $F(x, y, z) = f(x, y, z)i + g(x, y, z)j + h(x, y, z)k$. Let f, g, and h have continuous second partial derivatives on an open solid region S. Show that at each point of S

$$\text{div (curl } F) = 0.$$

Solution

$$\text{div (curl } F) = \dfrac{\partial}{\partial x}\left(\dfrac{\partial h}{\partial y} - \dfrac{\partial g}{\partial z}\right) - \dfrac{\partial}{\partial y}\left(\dfrac{\partial h}{\partial x} - \dfrac{\partial f}{\partial z}\right) + \dfrac{\partial}{\partial z}\left(\dfrac{\partial g}{\partial x} - \dfrac{\partial f}{\partial y}\right)$$

$$= \dfrac{\partial^2 h}{\partial x \partial y} - \dfrac{\partial^2 g}{\partial x \partial z} - \dfrac{\partial^2 h}{\partial y \partial x} + \dfrac{\partial^2 f}{\partial y \partial z} + \dfrac{\partial^2 g}{\partial z \partial x} - \dfrac{\partial^2 f}{\partial z \partial y}$$

$$= \dfrac{\partial^2 h}{\partial x \partial y} - \dfrac{\partial^2 g}{\partial x \partial z} - \dfrac{\partial^2 h}{\partial x \partial y} + \dfrac{\partial^2 f}{\partial y \partial z} + \dfrac{\partial^2 g}{\partial x \partial z} - \dfrac{\partial^2 f}{\partial y \partial z}$$

$$= 0.$$

(Recall that the mixed partial derivatives must be equal on S.) $\square$

Paddle-Wheel Interpretation of Curl

We shall show in Section 18–9 that the curl of a vector field also can be interpreted in terms of fluid flow. Water flowing along a stream, for example, may have a combination of movements. If there is an eddy (small whirlpool) in the stream, molecules of water tend to rotate laterally as they flow along the stream. The curl gives a measure of the rotation. If we place a very small paddle wheel in the stream, we find that it rotates fastest when the axis of the wheel points in the direction of curl F (where F is the velocity vector at the point). The maximum speed of rotation is proportional to $\|\text{curl } F\|$. (See Fig. 18–19.)

This interpretation can be illustrated by the vector field of Example 3:

$$F(x, y, z) = -yi + xj + k.$$

If this vector field represents fluid flow, then motion at $P(x, y, z)$ is a combination of a rotation determined by $-yi + xj$ and a flow in the direction of the z-axis determined by the vector k. That the curl of F is constantly equal to $2k$ means that each particle

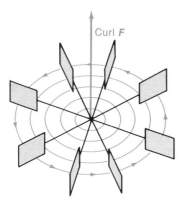

Figure 18-19 The paddle wheel rotates fastest when its axis is parallel to curl F.

tends to rotate in a plane perpendicular to the z-axis and that the angular velocity is constant.

INDEPENDENCE OF COORDINATE SYSTEMS

Physicists often define vector fields without reference to a coordinate system. For example, it makes sense to think of the velocity of a particle at point P as a function of the location of P, not of the coordinates of P with respect to a particular axis system. The velocity would be written $v(P)$, where v is the velocity vector at P.

Since the divergence and curl describe physical properties of a vector field, their values should be independent of the particular coordinate system used to define them. That is, if $F(x, y, z)$ is a vector field and $G(u, v, w)$ is the equivalent vector field obtained after a rotation and translation of axis, we should expect the divergence and curl of F at a given point to be the same as the divergence and curl of G at the same point. In fact, it can be proved that they are the same. The gradient and curl are invariant with respect to rotation and translation of axes.

Exercises 18-7

1-6. Calculate **(a)** div F and **(b)** curl F.

1 $F(x, y, z) = xyi + 3zj$

2 $F(x, y, z) = \cos x i + \sin x j + y k$

•3 $F(x, y, z) = x^2 i + y^2 j + z^2 k$

4 $F(x, y, z) = -yi + xj + zk$

•5 $F(x, y, z) = \dfrac{xi + yj + zk}{\sqrt{x^2 + y^2 + z^2}}$

6 $F(x, y, z) = \nabla(xyz)$

7-10. Verify the derivative properties for differentiable functions u, F, and G.

7 div $(F + G) =$ div $F +$ div G

8 curl $(F + G) =$ curl $F +$ curl G

9. curl $(uG) = u$ curl $G + \nabla u \times G$

10. div $(F \times G) = ($curl $F) \cdot G - ($curl $G) \cdot F$

11–12. Let f have continuous second derivatives.

11. Show that curl $(\nabla f) = \mathbf{0}$.

12. Show that div $(\nabla f) = \dfrac{\partial^2 f}{\partial x^2} + \dfrac{\partial^2 f}{\partial y^2} + \dfrac{\partial^2 f}{\partial z^2}$.

13. Let F be an inverse-square field. Show that curl $F = \mathbf{0}$ at every point in the domain of F. Interpret this result in terms of fluid flow.

14. Let F be an inverse-square field. Show that div $F = 0$ at each point in the domain of F. Interpret this result in terms of fluid flow.

15. Let $r = xi + yj + zk$; let $r = \|r\|$.
    (a) Show that div $(r\,r) = 4r$.
    (b) Show that div $(r^2\,r) = 5r^2$.
    (c) It can be shown that
    $$\text{div } (r^3\,r) = 6r^3 \quad \text{and} \quad \text{div } (r^4\,r) = 7r^4.$$
    Make a conjecture about div $(r^n\,r)$. Can you prove this conjecture?

16. (*Green's Theorem*) Let $F(x, y)$ be a differentiable vector field. Let $\mathscr{C}$ be a piecewise smooth, simple closed curve in the domain of F that is the boundary of a region $\mathscr{R}$ that also lies in the domain of F. Show that Green's theorem implies that

Divergence Theorem in the Plane

$$\iint_{\mathscr{R}} \text{div } F \, dA = \oint_{\mathscr{C}} F \cdot n \, ds,$$

where n is a unit outer normal to $\mathscr{C}$ at (x, y, z). This form of Green's theorem is called the *divergence theorem in the plane*. [*Hint:* Write $F(x, y) = P(x, y)i + Q(x, y)j$; let $x = f(t), y = g(t)$ be parametric equations of $\mathscr{C}$. Express the line integral in terms of t, then show that this integral equals $\int_{\mathscr{C}} (-Q\, dx + P\, dy)$. Apply Green's theorem.]

17. (*Green's Theorem*) Let F, $\mathscr{C}$, and $\mathscr{R}$ be as described in Exercise 16. Show that Green's theorem implies that

Stokes' Theorem in the Plane

$$\iint_{\mathscr{R}} (\text{curl } F) \cdot k \, dA = \oint_{\mathscr{C}} F \cdot dr.$$

This form of Green's theorem is called *Stokes' theorem in the plane*. [*Hint:* Write $F(x, y) = P(x, y)i + Q(x, y)j$.]

18–8 The Divergence Theorem

In this section we discuss an important generalization of Green's theorem to functions defined over solid domains in three-dimensional space. Green's theorem states that the double integral of a certain function over a basic closed region $\mathscr{R}$ is equal to the line integral of a related function around the boundary of $\mathcal{R}$. The generalization, the *divergence* theorem, states that the triple integral of a certain function over a solid region equals the surface integral of a related function over the boundary surface of the solid.

The key to the generalization is found in Exercise 16 of Section 18–7. We showed that under certain conditions Green's theorem implies that

$$\iint_{\mathcal{R}} \operatorname{div} \mathbf{F} \, dA = \oint_{\mathscr{C}} \mathbf{F} \cdot \mathbf{n} \, ds,$$

where $\mathbf{n}$ is an outer normal to $\mathscr{C}$, the boundary of $\mathcal{R}$.

The divergence theorem extends this result to vector fields defined over closed solid regions in three-dimensional space. If $\mathcal{D}$ is a solid region and S is its boundary surface, then according to the divergence theorem, under certain conditions on $\mathbf{F}$,

$$\iiint_{\mathcal{D}} \operatorname{div} \mathbf{F} \, dV = \iint_{S} \mathbf{F} \cdot \mathbf{n} \, dS,$$

where $\mathbf{n}(x, y, z)$ is a unit outer normal to S at (x, y, z).

We shall prove the divergence theorem for closed solid domains of a special type, namely solids such as spheres and cubes that have closed basic regions for their projections in the three coordinate planes. We also assume that the solids have continuous upper and lower surfaces, continuous front and back surfaces, and continuous left and right surfaces. (See Fig. 18–20.) Such a solid will be called a *basic closed solid*.

Basic Closed Solid

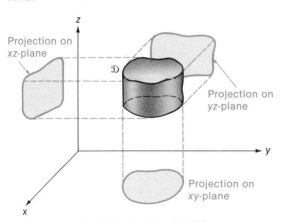

Figure 18–20 A basic closed solid.

THEOREM 18–4 (*The Divergence Theorem*) Let $\mathcal{D}$ be a basic closed solid with boundary surface S. Let

Divergence Theorem

$$F(x, y, z) = P(x, y, z)\mathbf{i} + Q(x, y, z)\mathbf{j} + R(x, y, z)\mathbf{k}$$

be a vector field defined on $\mathcal{D}$. Let $\partial P/\partial x$, $\partial Q/\partial y$, and $\partial R/\partial z$ be continuous on $\mathcal{D}$. Let $\mathbf{n}(x, y, z)$ be a unit outer normal to S at (x, y, z). Then

$$\iiint_{\mathcal{D}} \operatorname{div} \mathbf{F} \, dV = \iint_{S} \mathbf{F} \cdot \mathbf{n} \, dS.$$

18–8 The Divergence Theorem

Proof Write n in terms of its direction cosines as

$$n = \cos \alpha \, i + \cos \beta \, j + \cos \gamma \, k$$

where α, β, γ are functions of $x, y,$ and z. We need to show that

$$\iiint_\mathcal{D} \left(\frac{\partial P}{\partial x} + \frac{\partial Q}{\partial y} + \frac{\partial R}{\partial z}\right) dV = \iint_S (P \cos \alpha + Q \cos \beta + R \cos \gamma) \, dS.$$

We shall establish that

$$\iiint_\mathcal{D} \frac{\partial R}{\partial z} \, dV = \iint_S R \cos \gamma \, dS.$$

Similar arguments can be used to show that

$$\iiint_\mathcal{D} \frac{\partial P}{\partial x} \, dV = \iint_S P \cos \alpha \, dS \quad \text{and} \quad \iiint_\mathcal{D} \frac{\partial Q}{\partial y} \, dV = \iint_S Q \cos \beta \, dS.$$

Let $\mathcal{R}$ be the projection of $\mathcal{D}$ on the xy-plane. Let S_1 be the upper surface of $\mathcal{D}$; S_2 be the lower surface and S_3 the lateral surface. (See Fig. 18–21.) Then

$$\iint_S R \cos \gamma \, dS = \iint_{S_1} R \cos \gamma \, dS + \iint_{S_2} R \cos \gamma \, dS + \iint_{S_3} R \cos \gamma \, dS.$$

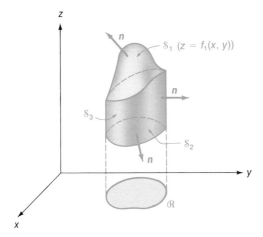

Figure 18–21.

Let S_1 be the graph of $z = f_1(x, y)$. Observe that the unit normal to S_1 is an upper normal. Thus, it is of form

$$n = \cos \alpha \, i + \cos \beta \, j + \cos \gamma \, k,$$

where $\cos \gamma \geq 0$. In calculating n, observe that one normal to S_1 is given by the gradient of the function $z - f_1(x, y)$:

$$-\frac{\partial f_1}{\partial x} i - \frac{\partial f_1}{\partial y} j + k.$$

Since the component of **k** is positive, this vector is an upper normal. If we divide it by its magnitude, we get

$$n = \frac{\dfrac{\partial f_1}{\partial x} i - \dfrac{\partial f_1}{\partial y} j + k}{\sqrt{\left(\dfrac{\partial f_1}{\partial x}\right)^2 + \left(\dfrac{\partial f_1}{\partial y}\right)^2 + 1}}.$$

Thus, the **k**-component of **n** is

$$\cos \gamma = \frac{1}{\sqrt{(\partial f_1/\partial x)^2 + (\partial f_1/\partial y)^2 + 1}}.$$

Let S_2 be the graph of $z = f_2(x, y)$. Since **n** is a lower normal to S_2, it follows that

$$n = \frac{\dfrac{\partial f_2}{\partial x} i + \dfrac{\partial f_2}{\partial y} j - k}{\sqrt{\left(\dfrac{\partial f_2}{\partial x}\right)^2 + \left(\dfrac{\partial f_2}{\partial y}\right)^2 + 1}}.$$

Thus, on S_2 we have

$$\cos \gamma = -\frac{1}{\sqrt{\left(\dfrac{\partial f_2}{\partial x}\right)^2 + \left(\dfrac{\partial f_2}{\partial y}\right)^2 + 1}}.$$

Finally, note that each normal to S_3 is orthogonal to **k**. Thus,

$$\cos \gamma = 0$$

at each point of S_3.

The surface integral is

$$\iint_S R(x, y, z) \cos \gamma \, dS$$

$$= \iint_{S_1} R(x, y, z) \cos \gamma \, dS$$

$$+ \iint_{S_2} R(x, y, z) \cos \gamma \, dS + \iint_{S_3} R(x, y, z) \cos \gamma \, dS$$

$$= \iint_{S_1} R(x, y, z) \frac{dS}{\sqrt{\left(\dfrac{\partial f_1}{\partial x}\right)^2 + \left(\dfrac{\partial f_1}{\partial y}\right)^2 + 1}}$$

$$+ \iint_{S_2} R(x, y, z) \frac{-dS}{\sqrt{\left(\dfrac{\partial f_2}{\partial x}\right)^2 + \left(\dfrac{\partial f_2}{\partial y}\right)^2 + 1}} + \iint_{S_3} R(x, y, z) \cdot 0 \cdot dS$$

$$= \iint_{\mathcal{R}} R(x, y, f_1(x, y)) \frac{1}{\sqrt{\left(\dfrac{\partial f_1}{\partial x}\right)^2 + \left(\dfrac{\partial f_1}{\partial y}\right)^2 + 1}} \sqrt{\left(\dfrac{\partial f_1}{\partial x}\right)^2 + \left(\dfrac{\partial f_1}{\partial y}\right)^2 + 1} \, dA$$

18–8 The Divergence Theorem

$$+ \iint_{\mathcal{R}} R(x, y, f_2(x, y)) \frac{-1}{\sqrt{\left(\frac{\partial f_2}{\partial x}\right)^2 + \left(\frac{\partial f_2}{\partial y}\right)^2 + 1}} \sqrt{\left(\frac{\partial f_2}{\partial x}\right)^2 + \left(\frac{\partial f_2}{\partial y}\right)^2 + 1}\, dA$$

$$= \iint_{\mathcal{R}} [R(x, y, f_1(x, y)) - R(x, y, f_2(x, y))]\, dA.$$

We now calculate the triple integral as an iterated integral:

$$\iiint_{\mathcal{D}} \frac{\partial R}{\partial z}\, dV = \iint_{\mathcal{R}} \left[\int_{z=f_2(x,y)}^{z=f_1(x,y)} \frac{\partial R}{\partial z}\, dz\right] dA$$

$$= \iint_{\mathcal{R}} \left[R(x, y, z)\right]_{z=f_2(x,y)}^{z=f_1(x,y)} dA$$

$$= \iint_{\mathcal{R}} [R(x, y, f_1(x, y)) - R(x, y, f_2(x, y))]\, dA.$$

Thus,

$$\iint_S R(x, y, z) \cos \gamma\, dS = \iiint_{\mathcal{D}} \frac{\partial R}{\partial z}\, dV.$$

Similar arguments can be used to show that

$$\iiint_{\mathcal{D}} \frac{\partial P}{\partial x}\, dV = \iint_S P \cos \alpha\, dS \quad \text{and} \quad \iiint_{\mathcal{D}} \frac{\partial Q}{\partial y}\, dV = \iint_S Q \cos \beta\, dS.$$

The details are omitted. ■

The divergence theorem holds for solids that are more general than the ones considered in our statement of the theorem. The proof for these solids can be found in a textbook on advanced calculus.

EXAMPLE 1 Verify the divergence theorem for

$$\mathbf{F}(x, y, z) = x\mathbf{i} + y\mathbf{j} + z\mathbf{k}$$

on the unit sphere defined by $x^2 + y^2 + z^2 \leq 1$.

Solution We must verify that

$$\iiint_{\mathcal{D}} \text{div } \mathbf{F}\, dV = \iint_S \mathbf{F} \cdot \mathbf{n}\, dS.$$

The triple integral is

$$\iiint_{\mathcal{D}} \left(\frac{\partial x}{\partial x} + \frac{\partial y}{\partial y} + \frac{\partial z}{\partial z}\right) dV = \iiint_{\mathcal{D}} 3\, dV$$

$$= 3 \cdot \frac{4\pi}{3} = 4\pi.$$

(Recall that the volume of a sphere is $4\pi r^3/3$.)

The outer normal to the boundary surface of the sphere at (x, y, z) is $\mathbf{n} = x\mathbf{i} + y\mathbf{j} + z\mathbf{k}$. Then

$$\iint_S \mathbf{F} \cdot \mathbf{n}\, dS = \iint_S (x\mathbf{i} + y\mathbf{j} + z\mathbf{k}) \cdot (x\mathbf{i} + y\mathbf{j} + z\mathbf{k})\, dS$$

$$= \iint_S (x^2 + y^2 + z^2)\, dS$$

$$= \iint_S dS \quad \text{(since } x^2 + y^2 + z^2 = 1 \text{ on } S\text{)}$$

$$= 4\pi \quad \text{(the surface area of } S\text{)}. \quad \square$$

EXAMPLE 2 Let $\mathbf{F}(x, y, z) = x\mathbf{i} + xy\mathbf{j} + xyz\mathbf{k}$. Use the divergence theorem to calculate

$$\iint_S \mathbf{F} \cdot \mathbf{n}\, dS$$

on the surface of the cube bounded by the planes

$$x = 0, \quad x = 1, \quad y = 0, \quad y = 1, \quad z = 0, \quad z = 1.$$

Solution
$$\iint_S \mathbf{F} \cdot \mathbf{n}\, dS = \iiint_\mathcal{D} \operatorname{div} \mathbf{F}\, dV = \iiint_\mathcal{D} (1 + x + xy)\, dV$$

$$= \int_0^1 \int_0^1 \int_0^1 (1 + x + xy)\, dz\, dy\, dx$$

$$= \int_0^1 \int_0^1 \left[(1 + x + xy)z\right]_{z=0}^{z=1} dy\, dx$$

$$= \int_0^1 \int_0^1 (1 + x + xy)\, dy\, dx = \int_0^1 \left[y + xy + \frac{xy^2}{2}\right]_{y=0}^{y=1} dx$$

$$= \int_0^1 \left(1 + x + \frac{x}{2}\right) dx = \left[x + \frac{3x^2}{4}\right]_0^1 = \frac{7}{4}.$$

Note that this surface integral could have been evaluated by surface integrals on the six faces of the cube. (See Fig. 18–22.) $\square$

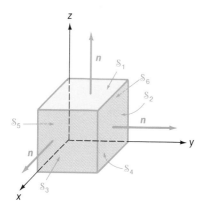

Figure 18–22 Example 2.

18–8 The Divergence Theorem

SPECIAL FORM OF THE DIVERGENCE THEOREM

Let $F(x, y, z)$ be a gradient field,

$$F(x, y, z) = \nabla f(x, y, z),$$

where $\partial^2 f/\partial x^2$, $\partial^2 f/\partial y^2$, and $\partial^2 f/\partial z^2$ exist on the closed domain $\mathcal{D}$. Then the divergence theorem has the special form

$$\iint_S D_n f(x, y, z) \, dS = \iiint_\mathcal{D} \left(\frac{\partial^2 f}{\partial x^2} + \frac{\partial^2 f}{\partial y^2} + \frac{\partial^2 f}{\partial z^2} \right) dV,$$

where n is a unit outer normal to S, the surface of $\mathcal{D}$, at (x, y, z), and $D_n f$ is the directional derivative. To show that this is the case, recall that

$$D_n f(x, y, z) = (\nabla f) \cdot n = F(x, y, z) \cdot n$$

so that

$$\iint_S D_n f(x, y, z) \, dS = \iint_S F \cdot n \, dS.$$

It follows from the divergence theorem that

$$\iint_S D_n f(x, y, z) \, dS = \iiint_\mathcal{D} \text{div} \, (\nabla f) \, dV$$

$$= \iiint_\mathcal{D} \text{div} \left(\frac{\partial f}{\partial x} i + \frac{\partial f}{\partial y} j + \frac{\partial f}{\partial z} k \right) dV$$

$$= \iiint_\mathcal{D} \left(\frac{\partial^2 f}{\partial x^2} + \frac{\partial^2 f}{\partial y^2} + \frac{\partial^2 f}{\partial z^2} \right) dV.$$

Laplacian ∇^2

The *Laplacian operator* ∇^2 is defined by

$$\nabla^2 f = \frac{\partial^2 f}{\partial x^2} + \frac{\partial^2 f}{\partial y^2} + \frac{\partial^2 f}{\partial z^2}.$$

Using this notation, we can write the special form of the divergence theorem as

$$\iint_S D_n f(x, y, z) \, dS = \iiint_\mathcal{D} \nabla^2 f \, dV$$

or

$$\iint_S \nabla f \cdot n \, dS = \iiint_\mathcal{D} \nabla^2 f \, dV.$$

Harmonic Function A function $f(x, y, z)$ is *harmonic* if

$$\nabla^2 f = \frac{\partial^2 f}{\partial x^2} + \frac{\partial^2 f}{\partial y^2} + \frac{\partial^2 f}{\partial z^2} = 0.$$

EXAMPLE 3 Let f be harmonic on a closed solid domain $\mathscr{D}$. Let $n(x, y, z)$ be a unit outer normal to S, the boundary surface of $\mathscr{D}$. Show that

$$\iint_S D_n f(x, y, z) \, dS = 0.$$

Solution It follows from the special form of the divergence theorem that

$$\iint_S D_n f(x, y, z) \, dS = \iint_S \nabla f \cdot n \, dS$$

$$= \iiint_{\mathscr{D}} \left(\frac{\partial^2 f}{\partial x^2} + \frac{\partial^2 f}{\partial y^2} + \frac{\partial^2 f}{\partial z^2} \right) dV$$

$$= \iiint_{\mathscr{D}} 0 \cdot dV = 0. \quad \square$$

PHYSICAL INTERPRETATION OF DIVERGENCE

Let $F(x, y, z)$ be a continuous velocity field for an incompressible fluid which has a continuous divergence function. For simplicity, assume that the density (mass per unit of volume) of the fluid is 1. Let $\mathscr{D}$ be a small sphere with center $P_0(x_0, y_0, z_0)$. If n is a unit outer normal to S, the boundary surface of $\mathscr{D}$, then the flux,

$$\iint_S F \cdot n \, dS$$

represents the total mass of the fluid that flows across S over a unit of time.

If $\mathscr{D}$ is very small, then the value of div F does not vary much over $\mathscr{D}$. Thus,

$$\text{div } F \approx (\text{div } F)_0,$$

the value of div F at P_0. It follows that

$$\iint_S F \cdot n \, dS = \iiint_{\mathscr{D}} \text{div } F \, dV \approx \iiint_{\mathscr{D}} (\text{div } F)_0 \, dV$$

$$= (\text{div } F)_0 \iiint_{\mathscr{D}} dV = (\text{div } F)_0 V,$$

where V is the volume of $\mathscr{D}$. If we solve this expression for $(\text{div } F)_0$, we get

$$(\text{div } F)_0 \approx \frac{\int_S \int F \cdot n \, dS}{V}.$$

In the limiting case, we get equality:

Divergence as a Limit
$$(\text{div } F)_0 = \lim_{V \to 0} \frac{\int_S \int F \cdot n \, dS}{V}.$$

That is, the divergence of F at P_0 equals the limit of the flux per unit of volume.

It follows from this relation that the divergence of F at P_0 equals the rate at which the fluid gains or loses mass at P_0. If $(\operatorname{div} F)_0 > 0$, there is a *source* at P_0; if $(\operatorname{div} F)_0 < 0$, a *sink* at P_0. If $(\operatorname{div} F)_0 = 0$, there is neither a source nor a sink at P_0.

Let $\mathcal{D}$ be an open solid region in the domain of

$$F(x, y, z) = P(x, y, z)i + Q(x, y, z)j + R(x, y, z)k.$$

If there are no sources and no sinks in $\mathcal{D}$, then

Equation of Continuity

$$\operatorname{div} F = \frac{\partial P}{\partial x} + \frac{\partial Q}{\partial y} + \frac{\partial R}{\partial z} = 0$$

at each point of $\mathcal{D}$. This equation is known as the *equation of continuity* for incompressible fluids.

Exercises 18–8

1–2. Verify the divergence theorem for F on the solid region $\mathcal{D}$ by calculating $\iiint_{\mathcal{D}} \operatorname{div} F \, dV$ and $\iint_{S} F \cdot n \, dS$.

●1 $F(x, y, z) = xyi + yzj$; $\mathcal{D}$ is the cube bounded by the planes $x = 0$, $x = 1$, $y = 0$, $y = 1$, $z = 0$, $z = 1$.

●2 $F(x, y, z) = yi + zj + xk$; $\mathcal{D}$ is the hemisphere $0 \leq z \leq \sqrt{1 - x^2 - y^2}$.

3–6. Use the divergence theorem to calculate the integral $\int_S \int F \cdot n \, dS$.

●3 $F(x, y, z) = x^2yzi + xy^2zj + xyz^2k$; $\mathcal{D}$ is the rectangular box bounded by the planes $x = 0$, $x = a$, $y = 0$, $y = b$, $z = 0$, $z = c$, where a, b, c are positive.

●4 $F(x, y, z) = x^2i + y^2j + z^2k$; $\mathcal{D}$ is the sphere $x^2 + y^2 + z^2 \leq 1$.

●5 $F(x, y, z) = x^2i + y^2j + z^2k$; $\mathcal{D}$ is the solid bounded by the paraboloid $z = 4 - x^2 - y^2$ and the xy-plane.

●6 $F(x, y, z) = 2xyi - y^2j + z^2k$; $\mathcal{D}$ is the semi-ellipsoid defined by $0 \leq z \leq c\sqrt{1 - x^2/a^2 - y^2/b^2}$ where a, b, c are positive.

7–8. Find all sinks and sources for F.

●7 $F(x, y, z) = xi + yj + zk$ ●8 $F(x, y, z) = x^2i + y^2j + z^2k$

9 Show that an inverse-square field has no sinks and no sources in its domain.

10 Show that the sinks and sources of a conservative field are determined by the sign of the Laplacian of the potential function.

11 Use the divergence theorem to show that $\int_S \int (x + y + z) \, dS = 0$ on the unit sphere $x^2 + y^2 + z^2 = 1$.

12 Let $f(y, z)$, $g(x, z)$, $h(x, y)$ be continuous on the solid sphere S defined by $0 \le x^2 + y^2 + z^2 \le 1$. Use the divergence theorem to show that

$$\iint_S [xf(y, z) + yg(x, z) + zh(x, y)] \, dS = 0.$$

---

13 Let $\mathscr{D}$ and S be as described in the divergence theorem. Show that V, the volume of $\mathscr{D}$, is equal to the surface integral

$$V = \frac{1}{3} \iint_S (x\mathbf{i} + y\mathbf{j} + z\mathbf{k}) \cdot \mathbf{n} \, dS.$$

14 Let S be a sphere of radius a. Use Exercise 13 to prove that the volume of S is $a/3$ times the surface area of S. Thus, if the volume is $4\pi a^3/3$, it follows that the surface area is $4\pi a^2$. If the surface area is $4\pi a^2$, it follows that the volume is $4\pi a^3/3$.

15 Let the component functions of $\mathbf{F}$ have continuous second partial derivatives on the closed solid $\mathscr{D}$. Let $\mathscr{D}$ and its boundary surface satisfy the conditions of the divergence theorem. Use the divergence theorem to show that

$$\iint_S \text{curl } \mathbf{F} \cdot \mathbf{n} \, dS = 0.$$

18–9 Stokes' Theorem

Stokes' theorem, our second generalization of Green's theorem, states that under suitable conditions, a surface integral can be calculated by a corresponding line integral around the boundary of the surface.

Before we state Stokes' theorem, we must introduce a new concept. A surface S is said to be *oriented* provided there is a vector $\mathbf{n}(x, y, z)$ at each point of S not on the boundary curve of S such that:

Oriented Surface

(1) $\mathbf{n}(x, y, z)$ is a unit normal to S at (x, y, z).
(2) $\mathbf{n}$ is a continuous function of (x, y, z) on S.

We restrict ourselves to surfaces of a special type. Each surface must be oriented, have finite area, and have a single piecewise smooth, simple closed curve for its boundary curve.

If S is such a surface and $\mathscr{C}$ is its boundary curve, then the orientation of S allows us to give a standard orientation to $\mathscr{C}$, called the *orientation induced* by the orientation of S.

Induced Orientation of a Curve

The induced orientation of $\mathscr{C}$ can be described as follows. Set the thumb, index finger, and middle finger of the right hand at right angles to each other. If the thumb lies along the surface S and the index finger points in the direction of $\mathbf{n}$, then the middle finger points in the direction of the induced orientation of $\mathscr{C}$. (See Fig. 18–23.)

The Stokes generalization of Green's theorem is named after G. G. Stokes (English mathematical physcist, 1819–1903), who first publicly proposed it. This theorem gen-

18–9 Stokes' Theorem

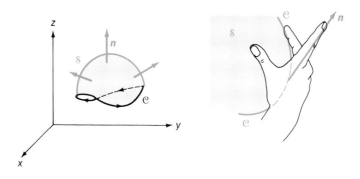

(a) The surface S and boundary curve $\mathcal{C}$

(b) The induced orientation of $\mathcal{C}$

Figure 18–23 Orientable surface.

eralizes the statement of Green's theorem given in Exercise 17 of Section 18–7. The proof is omitted.

THEOREM 18–5

Stokes' Theorem

(*Stokes' Theorem*) Let S be an oriented surface with finite area and unit normal $n(x, y, z)$ at (x, y, z). Let the boundary of S be a piecewise smooth, simple closed curve $\mathcal{C}$ with orientation induced by the orientation of S. Let F be a vector field defined on S, and assume that the component functions of F have continuous partial derivatives at each point of S not on $\mathcal{C}$. Then

$$\iint_S (\operatorname{curl} F) \cdot n \, dS = \oint_\mathcal{C} F \cdot dr,$$

where the symbol $\oint$ indicates the line integral in the positive direction around $\mathcal{C}$.

EXAMPLE 1 Let S be the portion of the plane $z = f(x, y) = -x + y + 3$ that is bounded by the vertical planes $x = 0$, $x = 2$, $y = 0$, and $y = 2$. Let $F(x, y, z) = x i + xy j + xyz k$. Verify Stokes' Theorem for F on S.

Solution The surface S is shown in Figure 18–24. If we use an upper normal to orient S, then the boundary curve $\mathcal{C}$ consists of the four line segments $\mathcal{C}_1$, $\mathcal{C}_2$, $\mathcal{C}_3$, and $\mathcal{C}_4$ shown in the figure.

Note that the upper unit normal to S (obtained from the gradient of f) is

$$n = \frac{i - j + k}{\sqrt{3}}$$

and that

$$\operatorname{curl} F = \begin{vmatrix} i & j & k \\ \dfrac{\partial}{\partial x} & \dfrac{\partial}{\partial y} & \dfrac{\partial}{\partial z} \\ x & xy & xyz \end{vmatrix}$$

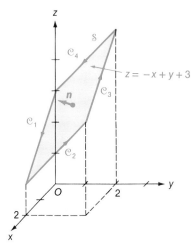

Figure 18-24 Example 1.

$$= \left[\frac{\partial}{\partial y}(xyz) - \frac{\partial}{\partial z}(xy)\right]i - \left[\frac{\partial}{\partial x}(xyz) - \frac{\partial}{\partial z}(x)\right]j$$
$$+ \left[\frac{\partial}{\partial x}(xy) - \frac{\partial}{\partial y}(x)\right]k$$
$$= xzi - yzj + yk.$$

Then

$$\iint_S \text{curl } \mathbf{F} \cdot \mathbf{n} \, dS = \iint_S (xzi - yzj + yk) \cdot \left(\frac{i - j + k}{\sqrt{3}}\right) dS$$

$$= \frac{1}{\sqrt{3}} \iint_S (xz + yz + y) \, dS = \frac{1}{\sqrt{3}} \iint_S [(x + y)z + y] \, dS$$

$$= \frac{1}{\sqrt{3}} \int_0^2 \int_0^2 [(x + y)(-x + y + 3) + y] \sqrt{\left(\frac{\partial f}{\partial x}\right)^2 + \left(\frac{\partial f}{\partial y}\right)^2 + 1} \, dy \, dx$$

$$= \frac{1}{\sqrt{3}} \int_0^2 \int_0^2 (-x^2 + y^2 + 3x + 4y) \sqrt{3} \, dy \, dx$$

$$= \int_0^2 \left[-x^2 y + \frac{y^3}{3} + 3xy + 2y^2\right]_{y=0}^{y=2} dx$$

$$= \int_0^2 \left[-2x^2 + 6x + \frac{32}{3}\right] dx = \left[-\frac{2x^3}{3} + 3x^2 + \frac{32x}{3}\right]_0^2$$

$$= -\frac{16}{3} + 12 + \frac{64}{3} = 28.$$

The corresponding line integral around the boundary $\mathscr{C}$ is

$$\oint_{\mathscr{C}} \mathbf{F} \cdot d\mathbf{r} = \int_{\mathscr{C}_1} \mathbf{F} \cdot d\mathbf{r} + \int_{\mathscr{C}_2} \mathbf{F} \cdot d\mathbf{r} + \int_{\mathscr{C}_3} \mathbf{F} \cdot d\mathbf{r} + \int_{\mathscr{C}_4} \mathbf{F} \cdot d\mathbf{r}.$$

18-9 Stokes' Theorem

The curve $\mathcal{C}_1$ is the graph of
$$x = t, \quad y = 0, \quad z = 3 - t, \quad 0 \le t \le 2.$$

Thus, $dx = dt$, $dy = 0$, $dz = -dt$. The corresponding integral is

$$\int_{\mathcal{C}_1} \mathbf{F} \cdot d\mathbf{r} = \int_{\mathcal{C}_1} (x\mathbf{i} + xy\mathbf{j} + xyz\mathbf{k}) \cdot (dx\,\mathbf{i} + dy\,\mathbf{j} + dz\,\mathbf{k})$$

$$= \int_{\mathcal{C}_1} x\,dx + xy\,dy + xyz\,dz$$

$$= \int_0^2 [t + t \cdot 0 + t \cdot 0 \cdot (3 - t)(-1)]\,dt$$

$$= \int_0^2 t\,dt = \left[\frac{t^2}{2}\right]_0^2 = 2.$$

Similar computations show that

$$\int_{\mathcal{C}_2} \mathbf{F} \cdot d\mathbf{r} = \frac{40}{3}, \quad \int_{\mathcal{C}_3} \mathbf{F} \cdot d\mathbf{r} = \frac{38}{3}, \quad \int_{\mathcal{C}_4} \mathbf{F} \cdot d\mathbf{r} = 0.$$

Then

$$\oint_{\mathcal{C}} \mathbf{F} \cdot d\mathbf{r} = 2 + \frac{40}{3} + \frac{38}{3} + 0 = 28.$$

Thus,

$$\iint_S \text{curl } \mathbf{F} \cdot \mathbf{n}\,dS = \int_{\mathcal{C}} \mathbf{F} \cdot d\mathbf{r}. \quad \square$$

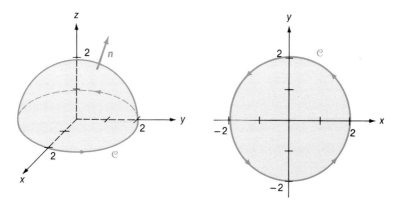

(a) The surface and boundary curve   (b) The projection in the xy-plane

Figure 18-25 Example 2.

EXAMPLE 2 Let S be the graph of $z = \sqrt{4 - x^2 - y^2}$. Let $\mathbf{n}$ be the unit upper normal to S at (x, y, z). Let $\mathbf{F}(x, y, z) = x^2\mathbf{i} + z^2\mathbf{j} + y\mathbf{k}$. Use Stokes' Theorem to calculate

$$\iint_S \text{curl } \mathbf{F} \cdot \mathbf{n}\,dS.$$

Solution The boundary $\mathscr{C}$ of S is the circle $x^2 + y^2 = 4$ in the xy-plane. (See Fig. 18–25.) The orientation of S induces the usual counterclockwise orientation of $\mathscr{C}$. Thus, $\mathscr{C}$ is the graph of

$$x = 2\cos t, \quad y = 2\sin t, \quad z = 0, \quad 0 \le t \le 2\pi,$$

so that

$$dx = -2\sin t\, dt, \quad dy = 2\cos t\, dt, \quad dz = 0 \cdot dt$$

on $\mathscr{C}$.

It follows from Stokes' Theorem that

$$\iint_S \operatorname{curl} \mathbf{F} \cdot \mathbf{n}\, dS = \oint_{\mathscr{C}} \mathbf{F} \cdot d\mathbf{r}$$

$$= \oint_{\mathscr{C}} (x^2\mathbf{i} + z^2\mathbf{j} + y\mathbf{k}) \cdot (dx\,\mathbf{i} + dy\,\mathbf{j} + dz\,\mathbf{k})$$

$$= \oint_{\mathscr{C}} x^2\, dx + z^2\, dy + y\, dz$$

$$= \int_0^{2\pi} [4\cos^2 t(-2\sin t) + 0^2 \cdot 2\cos t + 2\sin t \cdot 0]\, dt$$

$$= 8\int_0^{2\pi} \cos^2 t(-\sin t\, dt) = 8\left[\frac{\cos^3 t}{3}\right]_0^{2\pi} = 8\left[\frac{1}{3} - \frac{1}{3}\right] = 0. \quad \square$$

Let S_1 and S_2 be oriented surfaces that have the same boundary curve $\mathscr{C}$. If S_1, S_2, and $\mathbf{F}$ satisfy the conditions of Stokes' theorem, it follows that

$$\iint_{S_1} \operatorname{curl} \mathbf{F} \cdot \mathbf{n}_1\, dS = \oint_{\mathscr{C}} \mathbf{F} \cdot d\mathbf{r} = \iint_{S_2} \operatorname{curl} \mathbf{F} \cdot \mathbf{n}_2\, dS,$$

where $\mathbf{n}_1$ is a unit normal to S_1 and $\mathbf{n}_2$ a unit normal to S_2. Since the two surface integrals are the same, then either one can be calculated to evaluate the other.

EXAMPLE 3 Let S_1 be the portion of the paraboloid $z = 4 - x^2 - y^2$ that is on or above the xy-plane. Let $\mathbf{n}_1(x, y, z)$ be an upper normal to S_1 at (x, y, z). Use Example 2 to calculate

$$\iint_{S_1} \operatorname{curl} \mathbf{F} \cdot \mathbf{n}_1\, dS,$$

where $\mathbf{F}(x, y, z) = x^2\mathbf{i} + z^2\mathbf{j} + y\mathbf{k}$.

Solution The surface S_1 has the same boundary curve $\mathscr{C}$ as the surface S of Example 2. Thus,

$$\iint_{S_1} \operatorname{curl} \mathbf{F} \cdot \mathbf{n}_1\, dS = \oint_{\mathscr{C}} \mathbf{F} \cdot d\mathbf{r} = \iint_S \operatorname{curl} \mathbf{F} \cdot \mathbf{n}\, dS = 0. \quad \square$$

PHYSICAL INTERPRETATION OF THE CURL

Let S be a circular disk with center $P_0(x_0, y_0, z_0)$ in the domain of a velocity field $\mathbf{F}(x, y, z)$ that describes the velocity of a fluid. Let $\mathbf{n}$ be a unit normal to S; let $\mathscr{C}$ be the boundary of S with the orientation induced by $\mathbf{n}$.

18-9 Stokes' Theorem

The integral

Circulation
$$\oint_{\mathcal{C}} \mathbf{F} \cdot d\mathbf{r}$$

is called the *circulation of* $\mathbf{F}$ *around* $\mathcal{C}$. It measures the tendency of the fluid to circulate around $\mathcal{C}$ in the direction in which $\mathcal{C}$ is oriented.

To make the analogy more concrete, suppose that a paddle wheel is placed in the fluid with its center at P_0 and its paddles on $\mathcal{C}$. (See Fig. 18–26.) The speed with which the wheel rotates is proportional to $|\oint_{\mathcal{C}} \mathbf{F} \cdot d\mathbf{r}|$. If $\oint_{\mathcal{C}} \mathbf{F} \cdot d\mathbf{r} > 0$, the wheel rotates in the positive direction of $\mathcal{C}$. If $\oint_{\mathcal{C}} \mathbf{F} \cdot d\mathbf{r} < 0$, it rotates in the opposite direction.

It follows from Stokes' theorem that

$$\oint_{\mathcal{C}} \mathbf{F} \cdot d\mathbf{r} = \iint_{S} \text{curl } \mathbf{F} \cdot \mathbf{n} \, dS.$$

Let ρ be the radius of S. If ρ is small, and curl $\mathbf{F}$ is continuous at P_0, then the value of curl $\mathbf{F}$ does not vary appreciably from (curl $\mathbf{F}$)$_0$, the value at P_0. Thus,

$$\int_{\mathcal{C}} \mathbf{F} \cdot d\mathbf{r} = \iint_{S} \text{curl } \mathbf{F} \cdot \mathbf{n} \, dS$$

$$\approx \iint_{S} (\text{curl } \mathbf{F})_0 \cdot \mathbf{n} \, dS$$

$$= (\text{curl } \mathbf{F})_0 \cdot \mathbf{n} \iint_{S} dS = (\text{curl } \mathbf{F})_0 \cdot \mathbf{n} \cdot (\pi \rho^2).$$

This approximation improves as $\rho \to 0$. Thus,

Curl as a Limit
$$(\text{curl } \mathbf{F})_0 \cdot \mathbf{n} = \lim_{\rho \to 0} \frac{1}{\pi \rho^2} \oint_{\mathcal{C}} \mathbf{F} \cdot d\mathbf{r}.$$

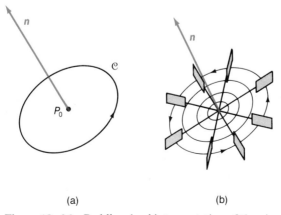

(a)   (b)

Figure 18–26 Paddle wheel interpretation of the circulation.

| Rotation of Vector Field | The expression $(\text{curl } F)_0 \cdot n$ is called the *rotation of F at P_0 about n*. It measures the tendency of the fluid at P_0 to rotate about n in a plane normal to n.

If we now let n vary over all possible unit vectors, we see that $(\text{curl } F)_0 \cdot n$ has its maximum value when n is in the direction of $(\text{curl } F)_0$. Thus, the maximum rotation of a particle at P_0 is in a plane normal to $(\text{curl } F)_0$. This maximum rotation equals $\|(\text{curl } F)_0\|$.

Returning to our earlier analogy, we find that if a paddle wheel of very small diameter is placed at P_0, it rotates fastest when the axis points in the direction of $(\text{curl } F)_0$.

Exercises 18-9

1-4. Verify Stokes' theorem on the given surfaces.

●1 $F(x, y, z) = x^2 i + y^2 j + z^2 k$ on the hemisphere $z = -\sqrt{1 - x^2 - y^2}$.

●2 $F(x, y, z) = y^2 i + x^2 j + z^2 k$ on the hemisphere $z = \sqrt{1 - x^2 - y^2}$.

●3 $F(x, y, z) = xyi + yzj + xzk$ on the triangle with vertices $(0, 0, 0)$, $(1, 0, 0)$ and $(0, 2, 1)$. Let n be an upper normal.

●4 $F(x, y, z) = y^2 i + z^2 j + x^2 k$ on the part of the plane $z = 3x - 5y + 2$ that is inside the cylinder $x^2 + y^2 = 1$. Let n be an upper normal.

●5 Use Stokes' theorem to evaluate $\oint_{\mathscr{C}} xz\, dx + yz\, dy + z^2\, dz$ on the unit circle $x^2 + y^2 = 1$, $z = 0$. Use an upper normal to orient the surface.

●6 Use Stokes' theorem to evaluate $\oint_{\mathscr{C}} xy\, dx + 3yz\, dy + x\, dz$ on the boundary of the rectangle with vertices $(0, 0, -5)$, $(2, 0, 1)$, $(2, 3, 7)$, $(0, 3, 1)$. Use an upper normal to orient the surface.

●7 Use Stokes' theorem to evaluate $\int_S \int \text{curl } F \cdot n\, dS$, where $F(x, y, z) = xyi + yzj + xzk$ and S is the portion of the paraboloid $z = 1 - x^2 - y^2$ that is on or above the xy-plane. Use an upper normal to orient the surface.

●8 Use Stokes' theorem to evaluate $\int_S \int \text{curl } F \cdot n\, dS$, where $F(x, y, z) = yi + zj + xk$ and S is the portion of the plane $z = x + y$ that is interior to the cylinder $x^2 + y^2 = 1$. Use an upper normal to orient the surface.

9 Let $\mathscr{C}$ be a simple closed curve that is the boundary of a surface S. Assume that $\mathscr{C}$ and S satisfy the conditions of Stokes' theorem. Let $r = xi + yj + zk$.
  (a) Use Stokes' theorem to show that $\int_{\mathscr{C}} r \cdot dr = 0$.
  (b) Assume that the origin is not on S. Use Stokes' theorem to show that
  $$\oint_{\mathscr{C}} \frac{r \cdot dr}{\|r\|} = 0.$$

10 Let the surface S and its boundary curve $\mathscr{C}$ satisfy the conditions of Stokes' theorem. Let f have continuous partial derivatives on S. Use Stokes' theorem to show that
$$\oint_{\mathscr{C}} \nabla f \cdot dr = 0.$$ |
|---|---|

Review Problems

Möbius Strip

11 Let the surface S and its boundary curve $\mathcal{C}$ satisfy the conditions of Stokes' theorem. Let F be a vector field such that curl $F = n$, where n is the vector used to orient S. Use Stokes' theorem to interpret the integral $\oint_{\mathcal{C}} F \cdot dr$.

12 (*Nonorientable Surface*) A *Möbius strip* is constructed by giving a strip of paper a half-twist and gluing the ends together. (See Fig. 18–27.) Show that if we start with a unit normal at one point of a Möbius strip and move it around the strip, we eventually get a normal at the same point that points in the opposite direction from the original normal. Explain why this shows that a Möbius strip cannot be oriented.

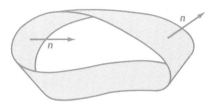

Figure 18–27 Exercise 12. A Möbius strip.

Review Problems

1 Define the concept and illustrate it with an example.

(a) Vector field (b) Gradient field

(c) Conservative field (d) Inverse-square field

(e) Potential function

2 What relations hold among the concepts in Problem 1?

3–6. Let F be a velocity field. Describe the motion in space.

• **3** $F(x, y, z) = i + j$

4 $F(x, y, z) = -xi - yj - zk$

• **5** $F(x, y, z) = \dfrac{yi - xj}{\sqrt{x^2 + y^2}}$

6 $F(x, y, z) = -yi + xj + 3k$

7 What is meant by a smooth curve? a piecewise smooth curve?

8 Define each line integral over the smooth curve $\mathcal{C}$ that is the graph of $x = f(t)$, $y = g(t)$, $z = h(t)$, $a \leq t \leq b$:

(a) $\displaystyle\int_{\mathcal{C}} F(x, y, z)\, ds$ (b) $\displaystyle\int_{\mathcal{C}} P(x, y, z)\, dx + Q(x, y, z)\, dy + R(x, y, z)\, dz$

9–12. Evaluate the line integral.

• **9** $\displaystyle\int_{\mathcal{C}} (x + y^2)\, ds$; $\mathcal{C}$ is the line segment from $(0, 0, 0)$ to $(2, 1, 3)$.

•10 $\int_{\mathcal{C}} (x^2 - y^2)\, ds$; $\mathcal{C}$ is the quarter-circle from $(1, 0)$ to $(0, -1)$

•11 $\int_{\mathcal{C}} xy\, dx + (x^2 - y^2)\, dy + x^2 z\, dz$; $\mathcal{C}$ is the graph of $r = ti + t^3 j + t^2 k$, $-1 \leq t \leq 1$.

•12 $\int_{\mathcal{C}} (x + y)\, dx + yz\, dy + xz\, dz$; $\mathcal{C}$ is the graph of $x = \cos t$, $y = \sin t$, $z = \tan t$, $0 \leq t \leq \pi/4$.

•13 A wire, bent in the shape of the helix $x = \cos t$, $y = \sin t$, $z = t/4$, $0 \leq t \leq 4\pi$, has density $\delta(x, y, z) = z/10$. Find the mass of the wire.

•14 The force field $F(x, y, z) = xi + yj + zk$ causes a particle to move along the graph of $r = ti + t^2 j + t^3 k$ from $(1, 1, 1)$ to $(2, 4, 8)$. Calculate the work done by the field.

•15 The force field $F(x, y, z) = -k$ causes a particle to move along the helix defined by $x = \cos t$, $y = \sin t$, $z = t/2\pi$ from $(1, 0, 1)$ to $(1, 0, 0)$. Calculate the work done by the field.

16–17. Calculate the flux of F over the plane curve $\mathcal{C}$. Is there a source or a sink in the region enclosed by $\mathcal{C}$?

•16 $F(x, y) = \dfrac{-(xi + yj)}{(x^2 + y^2)^{3/2}}$; $\mathcal{C}$ is the circle $x^2 + y^2 = 1$.

•17 $F(x, y) = xi + yj$; $\mathcal{C}$ is the square with vertices $(1, 1)$, $(-1, 1)$, $(-1, -1)$, and $(1, -1)$.

18 Let $\mathcal{C}$ be a piecewise smooth, simple curve from point A to point B in the xy-plane. Under what conditions is the line integral $\int_{\mathcal{C}} P(x, y)\, dx + Q(x, y)\, dy$ independent of the curve $\mathcal{C}$ and dependent only on the points A and B?

19–20. Calculate the line integral along a piecewise smooth curve $\mathcal{C}$ from A to B.

•19 $\int_{\mathcal{C}} x^3\, dx + 2y\, dy + \dfrac{1}{z}\, dz$; $A(0, 0, 1)$, $B(0, 1, 2)$. Is it essential that $\mathcal{C}$ not intersect the xy-plane? Explain.

•20 $\int_{\mathcal{C}} yz\, dx + xz\, dy + xy\, dz$; $A(1, 2, 3)$, $B(-1, 7, 4)$.

21 Let $\mathcal{C}$ be a smooth curve in the xy-plane. What does it mean to say that n is a *unit outer normal* to $\mathcal{C}$ at (x, y)?

Review Problems

22 State Green's theorem. Give alternative statements related to the divergence theorem and Stokes' theorem.

•**23** Use Green's theorem to calculate $\oint_{\mathcal{C}} y^2\, dx + x^2\, dy$, where $\mathcal{C}$ is the square with vertices $(0, 0)$, $(1, 0)$, $(1, 1)$, and $(0, 1)$.

•**24** Use Green's theorem to prove that $\int_{\mathcal{C}} x^2\, dx + y^2\, dy = 0$ over any piecewise smooth, simple closed curve in the xy-plane.

25 Let $F(x, y)$ have continuous second partial derivatives everywhere in the xy-plane. Prove that $\int_{\mathcal{C}} \operatorname{curl} F \cdot n\, ds = 0$ over any smooth simple closed curve in the xy-plane.

26 Use the area formula obtained from Green's theorem to show that the area of a circle of radius a is πa^2.

27 (a) Explain the geometrical setting for the definition of a surface integral $\int_S \int f(x, y, z)\, dS$.

(b) Exactly how is the integral in (a) defined?

(c) Give two elementary applications of surface integrals.

28–29. Calculate the surface integral.

•**28** $\int_S \int z\, dS$; S is the hemisphere $z = \sqrt{1 - x^2 - y^2}$.

•**29** $\int_S \int 2\, dS$; S is the part of the cone $z = \sqrt{x^2 + y^2}$ interior to the cylinder $x^2 + y^2 = 4$.

•**30** Assume that the part of the paraboloid $z = x^2 + y^2$ with $z \leq 1$ is covered by a thin sheet of density $\delta(x, y, z) = x^2 y^2 z / 100$. Write an iterated double integral for the mass of the sheet.

•**31** Calculate the flux of $F(x, y, z) = xi + yj$ over the sphere $x^2 + y^2 + z^2 = 1$.

32 (1) Define the concept. (2) Give a physical interpretation in terms of fluid flow. (3) Explain the alternative notation.

(a) *Divergence* of a vector field F; $\nabla \cdot F$.

(b) *Curl* of a vector field F; $\nabla \times F$.

33–34. Calculate (a) the divergence and (b) the curl of F.

•**33** $F(x, y, z) = yi - xj + 2k$ •**34** $F(x, y, z) = xzi + xyj + yzk$.

35 Let f have continuous second partial derivatives. Show that

(a) $\operatorname{div}(\nabla f) = \nabla^2 f$;   (b) $\operatorname{curl}(\nabla f) = 0$.

•**36** Find all sinks and sources for $F(x, y, z) = x^2 i - y^2 j + k$.

37 Show that the divergence of an inverse-square field is zero at each point different from the origin.

38 State the divergence theorem. How is this theorem related to Green's theorem?

●39 Use the divergence theorem to calculate $\int_S \int F \cdot n \, dS$, where $F(x, y, z) = xyi + yzj + xzk$, the lower surface of S is the graph of $z = x^2 + y^2$, $0 \leq x^2 + y^2 \leq 4$, and the upper surface is the plane $z = 4$.

40 Let S satisfy the conditions of the divergence theorem. Let c be a constant vector. Use the divergence theorem to show that

$$\int\int_S c \cdot n \, dS = 0.$$

41 **(a)** What is meant by an *oriented surface?*

(b) Can every surface be oriented?

(c) Let the simple closed curve $\mathcal{C}$ be the boundary curve of an oriented surface. What is the *induced orientation* of $\mathcal{C}$?

42 State Stokes' theorem.

●43 Use Stokes' theorem to calculate $\oint_\mathcal{C} x \, dx + xy \, dy + xyz \, dz$, where $\mathcal{C}$ is the intersection of the plane $z = x + y + 1$ with the cylinder $x^2 + y^2 = 1$. Use an upper normal to orient the surface.

●44 Use Stokes' theorem to calculate $\int_S \int \text{curl} \, F \cdot n \, dS$, where S is the portion of the paraboloid $z = x^2 + y^2$ with $z \leq 4$, $F(x, y, z) = xyi + yzj + xzk$, and n is a lower normal to the surface.

19 Introduction to Differential Equations

19–1 Review of Basic Concepts. Separation of Variables

The first part of this section partially duplicates the material in Section 8-4.

Differential Equation A *differential equation* is an equation that relates a function with one or more of its derivatives. For example, if y denotes a function of x, then

$$y' + 3y = x + 1$$
$$y'' + yy' = 0$$
$$y''' = x^2 + 4x - 1$$

are differential equations.

Order The order of the highest-order derivative of the unknown function that appears in a differential equation is called the *order* of the differential equation. Thus, the first of the above differential equations has order 1; the second, order 2; and the third, order 3.

Solution By a *solution* of a differential equation in y we mean any function $f(x)$ that, when substituted for y, makes the equation an identity for all values of x on some interval. For example, the function $f(x) = x/3 + 2/9$ is a solution of the differential equation

$$y' + 3y = x + 1.$$

If we substitute $y = x/3 + 2/9$, $y' = \frac{1}{3}$, the equation reduces to the identity

$$x + 1 = x + 1$$

In most cases, solving a first-order differential equation leads to one integration; solving a second-order differential equation, to two integrations; and so on. In each of these integrations, an arbitrary constant results. Thus, we expect to have one arbitrary constant in the solution of a first-order differential equation, two arbitrary constants in the solution of a second-order equation, and so on.

Most of the nth-order differential equations that are considered in elementary work

General Solution

Particular Solution

have solutions that involve n arbitrary constants. Such solutions are called the *general solutions* of the differential equations. Solutions obtained by assigning particular values to the arbitrary constants are called *particular solutions*.

EXAMPLE 1 The general solution of the differential equation

$$y' + 3y = x + 1$$

is

$$g(x) = Ce^{-3x} + \frac{x}{3} + \frac{2}{9}.$$

(The order of the equation equals 1 and the solution has one arbitrary constant.) The solution

$$f(x) = \frac{x}{3} + \frac{2}{9}$$

is the particular solution obtained by choosing $C = 0$ in the general solution. □

Remark This definition of *general solution* is sufficient for our purposes. A more precise definition is given in advanced courses. You can consult almost any textbook on differential equations for a more thorough treatment of the subject.

In most cases every particular solution can be obtained from the general solution by assigning values to the arbitrary constants. There are a few exceptions to this rule. Occasionally solutions exist that are not obtainable from the general solution. These are called *singular solutions*.

Singular Solution

EXAMPLE 2 It can be verified by substitution that the general solution of the differential equation

$$y' = y(y + 1)$$

is

$$y = \frac{Ce^x}{1 - Ce^x},$$

where C is a constant. The particular solution $y = -1$ is singular. It cannot be obtained from the general solution by assigning a value to C. □

Initial Conditions

Many differential equations must be solved subject to *initial conditions*. These are conditions that the solution function and possibly some of its derivatives must satisfy at a particular point. We might, for example, want to solve the differential equation

$$y'' + 4y = 0$$

subject to the initial conditions that $y = 2$ and $y' = 3$ when $x = 0$.

Boundary Conditions

Other differential equations must be solved subject to *boundary conditions*. These are conditions that the function must satisfy at one or more points in its domain. We

19-1 Basic Concepts. Separation of Variables

might, for example, want to solve
$$y'' + 4y = 0$$
subject to the boundary conditions that $y = 1$ when $x = 0$ and $y = 2$ when $x = \pi/4$.

If a differential equation must be solved subject to initial or boundary conditions, we usually try to find the general solution, then determine the values of the constants that make the solution satisfy the auxiliary conditions.

EXAMPLE 3 We can verify that
$$f(x) = C_1 \cos 2x + C_2 \sin 2x$$
is the general solution of the differential equation
$$y'' + 4y = 0.$$
Find the particular solutions that satisfy the following conditions.

(a) $y = 2$ and $y' = 3$ when $x = 0$.
(b) $y = 1$ when $x = 0$, $y = 2$ when $x = \pi/4$.

Solution (a) Since the general solution is
$$f(x) = C_1 \cos 2x + C_2 \sin 2x,$$
then
$$f'(x) = -2C_1 \sin 2x + 2C_2 \cos 2x.$$
We need $f(0) = 2$ and $f'(0) = 3$. If we substitute 0 in the expressions for f and f', we get the system of equations
$$\begin{cases} C_1 \cos 0 + C_2 \sin 0 = 2 \\ -2C_1 \sin 0 + 2C_2 \cos 0 = 3, \end{cases}$$
which, since $\cos 0 = 1$ and $\sin 0 = 0$, has the solution
$$C_1 = 2, \quad C_2 = \tfrac{3}{2}.$$
The particular solution is
$$f(x) = 2 \cos 2x + \tfrac{3}{2} \sin 2x.$$

(b) We need $f(0) = 1$ and $f\left(\dfrac{\pi}{4}\right) = 2$. Substituting 0 and $\pi/4$ in the expression for f, we get the system of equations
$$\begin{cases} C_1 \cos 0 + C_2 \sin 0 = 1 \\ C_1 \cos \dfrac{\pi}{2} + C_2 \sin \dfrac{\pi}{2} = 2, \end{cases}$$
which has the solution
$$C_1 = 1 \quad \text{and} \quad C_2 = 2.$$
The particular solution is
$$f(x) = \cos 2x + 2 \sin 2x. \quad \square$$

SEPARATION OF VARIABLES

The method of separation of variables has already been discussed (Section 8-4). It involves writing a differential equation in the form

$$M(x)\,dx = N(y)\,dy$$

and integrating both sides of the equation. This method for solving differential equations is one of the easiest to describe. Unfortunately, it may require horrendous computations.

EXAMPLE 4 Solve the differential equation

$$\frac{dy}{dx} = y(y+1).$$

Solution We separate the variables, writing the differential equation in the form

$$\frac{dy}{y(y+1)} = dx.$$

The expression on the left can be rewritten by the use of partial fractions (Section 11-4) as

$$\frac{dy}{y(y+1)} = \left(\frac{1}{y} - \frac{1}{y+1}\right) dy.$$

Therefore

$$\int \frac{dy}{y(y+1)} = \int dx$$

$$\int \left(\frac{1}{y} - \frac{1}{y+1}\right) dy = \int dx$$

$$\ln|y| - \ln|y+1| = x + C_1$$

$$\ln\left|\frac{y}{y+1}\right| = x + C_1.$$

If we now raise e to the powers given by the two sides of this equation, we get

$$e^{\ln|y/(y+1)|} = e^{x+C_1} = e^{C_1}\,e^x$$

$$\left|\frac{y}{y+1}\right| = e^{C_1}\,e^x$$

$$\frac{y}{y+1} = Ce^x$$

where $C = \pm e^{C_1}$. If we solve this last equation for y, we get

$$y = \frac{Ce^x}{1 - Ce^x}. \quad \square$$

19–1 Basic Concepts. Separation of Variables

HOMOGENEOUS FIRST-ORDER DIFFERENTIAL EQUATIONS

A first-order differential equation is said to be *homogeneous* if it can be written in the form

$$\frac{dy}{dx} = F\left(\frac{y}{x}\right).$$

Homogeneous first-order differential equations can be transformed into equivalent differential equations with separated variables by making the substitution

Substitutions for Homogeneous Equations

$$v = \frac{y}{x}.$$

Observe that since $y = vx$, we must also make the substitution

$$\frac{dy}{dx} = v + x\frac{dv}{dx}.$$

These substitutions change the equation $dy/dx = F(y/x)$ to

$$v + x\frac{dv}{dx} = F(v).$$

We separate the variables, solve for v, then use the substitution $y = vx$.

EXAMPLE 5 Solve the differential equation

$$x^2 y' = y^2 + xy.$$

Solution Since the differential equation can be rewritten as

$$\frac{dy}{dx} = \frac{y^2 + xy}{x^2} = \left(\frac{y}{x}\right)^2 + \frac{y}{x},$$

it is homogeneous. We make the substitution

$$v = \frac{y}{x}, \quad \frac{dy}{dx} = v + x\frac{dv}{dx},$$

obtaining

$$v + x\frac{dv}{dx} = v^2 + v$$

$$x\frac{dv}{dx} = v^2$$

$$\frac{dv}{v^2} = \frac{dx}{x}$$

$$\int \frac{dv}{v^2} = \int \frac{dx}{x}$$

$$-\frac{1}{v} = \ln|x| + C$$

$$v = \frac{-1}{\ln|x| + C}.$$

Since $v = y/x$, then

$$\frac{y}{x} = \frac{-1}{\ln |x| + C}$$

$$y = \frac{-x}{\ln |x| + C}. \quad \square$$

EXISTENCE THEOREMS

Frequently we need to know that a differential equation has a solution that satisfies certain initial conditions. For that reason much of the advanced theory of differential equations is concerned with existence theorems. A typical theorem states that any differential equation of a certain form has a solution that satisfies certain types of conditions.

The following two existence theorems illustrate the results that are established in more advanced courses. They guarantee the existence of solutions for most of the differential equations that we shall consider. The proofs are omitted.

THEOREM 19–1 Let $a_1(x)$ and $a_2(x)$ be continuous for all x. Let x_0 and y_0 be real numbers. The differential equation

$$y' + a_1(x)y = a_2(x)$$

has a unique solution $y = f(x)$ such that $f(x_0) = y_0$.

In other words, if we choose any point (x_0, y_0) in the plane, there is a solution of the differential equation that passes through that point. (See Fig. 19–1.) Furthermore, there is only one such solution. (This is the "uniqueness" part of the theorem.)

THEOREM 19–2 Let $a_1(x)$, $a_2(x)$, and $a_3(x)$ be continuous for all x. Let x_0, y_0, and m be real numbers. The differential equation

$$y'' + a_1(x) y' + a_2(x) y = a_3(x)$$

has a unique solution $y = f(x)$ such that $f(x_0) = y_0$ and $f'(x_0) = m$.

In other words, if we choose any point (x_0, y_0) and any nonvertical line through the point (m is the slope of the line), there is one and only one solution of the differential equation that passes through the point and is tangent to the given line at the point. (See Fig. 19–2.)

Remark These existence theorems are not stated in their full generality. The results hold on any open interval on which $a_1(x)$, $a_2(x)$, and $a_3(x)$ are continuous. In that case x_0 must be in the open interval and the uniqueness of the solution applies only to functions defined on that interval.

Exercises 19–1

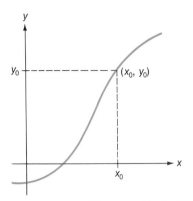

Figure 19–1 *Theorem 19–1.* The differential equation $y' + a_1(x)y = a_2(x)$ has a unique solution $y = f(x)$ such that $f(x_0) = y_0$ provided $a_1(x)$ and $a_2(x)$ are continuous for all x.

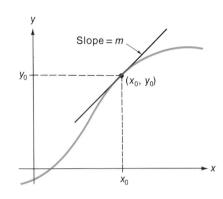

Figure 19–2 *Theorem 19–2.* The differential equation $y'' + a_1(x)y' + a_2(x)y = a_3(x)$ has a unique solution $y = f(x)$ such that $f(x_0) = y_0$ and $f'(x_0) = m$ provided $a_1(x)$, $a_2(x)$, $a_3(x)$ are continuous for all x.

Exercises 19–1

1–4. Find the general solution. (*Hint for 9 and 10:* Substitute $u = y'$.)

- 1 $y' = \dfrac{y}{x}$
- 2 $(x^2 - 1)\, y' = y$
- 3 $y' = 4x^3 y^2$
- 4 $y' = e^{x-y}$
- 5 $(x - 1)\, y' = y(y + 2)$
- 6 $y^2 y' = (y^3 + 1)\, x$
- 7 $xy + y' \ln y = 0$
- 8 $y'' = 2x + e^x$
- 9 $y'' = 2y'$ (See hint.)
- 10 $y'' = 2x \sqrt{y'}$ (See hint.)
- 11 $x + 2y = xy'$
- 12 $(x - y)\, dy + (x + y)\, dx = 0$
- 13 $y' = \dfrac{y}{x} + \cot\left(\dfrac{y}{x}\right)$
- 14 $y' = e^{y/x} + \dfrac{y}{x}$

15–18. Find the particular solution that satisfies the initial or boundary condition.

- 15   Differential equation in Exercise 2, $f(4) = 1$
- 16   Differential equation in Exercise 3, $f(0) = 2$
- 17   Differential equation in Exercise 9, $f(0) = -1, f'(0) = 3$
- 18   Differential equation in Exercise 10, $f(0) = -2, f'(0) = 1$

19   Based on the form of the general solution given in the answer to Exercise 3, verify that $f(x) = 0$ is a singular solution of the differential equation in Exercise 3.

- 20   Based on the form of the general solution given in the answer to Exercise 5, verify that $f(x) = -2$ is a singular solution of the differential equation in Exercise 5. Is the solution $f(x) = 0$ a singular solution?

21 Is the solution $y = x$ a singular solution of the differential equation in Exercise 4?

22 Show that there is no real solution of the differential equation in Exercise 10 that satisfies the initial conditions $f(0) = 1$, $f'(0) = -1$. Explain why this does not contradict Theorem 19–2.

23 (*Homogeneous Equations*) **(a)** Show that the substitutions $x = r \cos \theta$, $y = r \sin \theta$ enable us to separate the variables in the homogeneous equation $y' = F(y/x)$.
•**(b)** Use this method to solve the differential equation $y' = x^2/y^2$.

•**24** The tangent line to a curve has slope $m(x) = -1 + 2y/x$ at the point (x, y), provided $x \neq 0$. Find the equation of the curve provided it passes through $(-1, 3)$.

Family of Curves

25–26. (*Family of Curves*) A differential equation usually defines a *family of curves*. These are the curves obtained by assigning particular values to the arbitrary constants. As a rule, these curves have a common geometrical property that can be determined from the original differential equation.

Describe the families of curves defined by the given differential equations.

•**25** $y' = \dfrac{2y}{x}$ **26** $y' = -\dfrac{x}{2y}$

Orthogonal Trajectories

27 (*Orthogonal Trajectories*) A family of curves is said to form a set of *orthogonal trajectories* for a second family of curves if each curve in the first family is orthogonal to each curve in the second family.

Show that the two families of curves in Exercises 25 and 26 are orthogonal trajectories of each other.

19–2 Exact Differential Equations

A differential equation of form

$$M(x, y) \, dx + N(x, y) \, dy = 0,$$

Exact Differential Equation

where M and N are continuous functions, is said to be *exact* if its left-hand side is the exact differential of a function $u = u(x, y)$. If that is the case, the equation can be written as

$$du = 0,$$

which has the solution

$$u(x, y) = C.$$

Test for Exactness

There is a simple test for exactness provided M and N have continuous partial derivatives. If there is a function u such that

$$du = M(x, y) \, dx + N(x, y) \, dy,$$

then

$$M = \dfrac{\partial u}{\partial x} \quad \text{and} \quad N = \dfrac{\partial u}{\partial y}.$$

19-2 Exact Differential Equations

If M and N have continuous partial derivatives, then the mixed partial derivatives are equal:

$$\frac{\partial M}{\partial y} = \frac{\partial^2 u}{\partial y\, \partial x} = \frac{\partial^2 u}{\partial x\, \partial y} = \frac{\partial N}{\partial x}$$

so that

$$\frac{\partial M}{\partial y} = \frac{\partial N}{\partial x}.$$

This last condition also is sufficient for exactness. If $\partial M/\partial y = \partial N/\partial x$, we can use partial integration* to determine a function u such that

$$du = M(x, y)\, dx + N(x, y)\, dy.$$

The process is illustrated in Example 1.

EXAMPLE 1 Solve the differential equation

$$\cos y\, dx + (-x \sin y + \cos y)\, dy = 0.$$

Solution In this example, $M(x, y) = \cos y$ and $N(x, y) = -x \sin y + \cos y$. Since

$$\frac{\partial M}{\partial y} = -\sin y \quad \text{and} \quad \frac{\partial N}{\partial x} = -\sin y,$$

then the differential equation is exact. Therefore, there exists a function $u(x, y)$ such that

$$\frac{\partial u}{\partial x} = M = \cos y \quad \text{and} \quad \frac{\partial u}{\partial y} = N = -x \sin y + \cos y.$$

We can determine the function u by partial integration. Since $\partial u/\partial x = \cos y$, then

$$u = \int \cos y\, dx + K(y) = x \cos y + K(y),$$

where K is a function of y that corresponds to a constant of integration.† (See Section 17–2.) To determine the function $K(y)$, recall that $\partial u/\partial y = N$. Then

$$\frac{\partial u}{\partial y} = \frac{\partial}{\partial y}(x \cos y + K(y)) = N = -x \sin y + \cos y,$$

$$-x \sin y + K'(y) = -x \sin y + \cos y$$
$$K'(y) = \cos y,$$
$$K(y) = \int \cos y\, dy + C_1 = \sin y + C_1.$$

Therefore,

$$u = x \cos y + K(y)$$
$$= x \cos y + \sin y + C_1.$$

---

*See Section 17–2.
†The partial integral should not be confused with the line integral. (Section 18–2.)

Since the original differential equation is exact, it is of form
$$du = 0.$$
Its solution is
$$u(x, y) = C_2$$
$$x \cos y + \sin y + C_1 = C_2.$$

It follows that y, the solution of the original differential equation, satisfies the equation
$$x \cos y + \sin y = C,$$
where $C = C_2 - C_1$. □

INTEGRATING FACTORS

It is proved in advanced courses that every differential equation of form
$$M(x, y)\,dx + N(x, y)\,dy = 0,$$
where M and N are continuous, can be transformed into an exact differential equation by multiplying the left-hand side by a function $\mu = \mu(x, y)$, which depends on both $M(x, y)$ and $N(x, y)$. This function μ is called an *integrating factor* for the original differential equation.

Integrating Factor

EXAMPLE 2 Show that $\mu = x$ is an integrating factor for the differential equation
$$\left(2y + \frac{3x}{y}\right) dx + \left(x - \frac{x^2}{y^2}\right) dy = 0.$$
Solve the differential equation.

Solution You should verify that the differential equation is not exact.
If we multiply both sides of the differential equation by $\mu = x$, we get
$$\underbrace{\left(2xy + \frac{3x^2}{y}\right)}_{M} dx + \underbrace{\left(x^2 - \frac{x^3}{y^2}\right)}_{N} dy = 0.$$

Then
$$\frac{\partial M}{\partial y} = 2x - \frac{3x^2}{y^2} = \frac{\partial N}{\partial x},$$
so that the new differential equation is exact. Its solution is
$$u(x, y) = C,$$
where
$$\frac{\partial u}{\partial x} = M = 2xy + \frac{3x^2}{y} \quad \text{and} \quad \frac{\partial u}{\partial y} = N = x^2 - \frac{x^3}{y^2}.$$

19–2 Exact Differential Equations

Since

$$\frac{\partial u}{\partial x} = 2xy + \frac{3x^2}{y},$$

then

$$u = \int \left(2xy + \frac{3x^2}{y}\right) dx + K(y)$$

$$= x^2 y + \frac{x^3}{y} + K(y).$$

To find $K(y)$, we use the fact that

$$\frac{\partial u}{\partial y} = N = x^2 - \frac{x^3}{y^2}.$$

Then

$$x^2 - \frac{x^3}{y^2} + K'(y) = x^2 - \frac{x^3}{y^2}$$

$$K'(y) = 0,$$

$$K(y) = C_1.$$

Therefore

$$u(x, y) = x^2 y + \frac{x^3}{y} + C_1.$$

The solution of the differential equation is

$$u(x, y) = C_2$$

$$x^2 y + \frac{x^3}{y} + C_1 = C_2$$

$$x^2 y + \frac{x^3}{y} = C.$$

If we wish, we can solve this equation for y by the quadratic formula, obtaining

$$y = \frac{C \pm \sqrt{C^2 - 4x^5}}{2x^2}. \quad \square$$

In general, it is not a simple matter to determine an integrating factor for a non-exact differential equation. The best we can do is to show that a specific integrating factor will work under certain conditions on M and N.

In Example 3, we show that if

$$\frac{\partial M/\partial y - \partial N/\partial x}{N}$$

is a function of x only, say

$$\frac{\partial M/\partial y - \partial N/\partial x}{N} = f(x),$$

then $\mu = e^{\int f(x)dx}$ is an integrating factor. This result is mentioned to indicate the difficulties that arise when we try to find integrating factors. It is not intended as a practical method for finding many of these factors.

EXAMPLE 3 Let $M(x, y)$ and $N(x, y)$ have continuous partial derivatives. Show that $\mu = e^{\int f(x)dx}$ is an integrating factor for

$$M\,dx + N\,dy = 0$$

if and only if

$$\frac{\partial M/\partial y - \partial N/\partial x}{N} = f(x).$$

Solution Suppose that $\mu = e^{\int f(x)dx}$ is an integrating factor. Then the differential equation

$$e^{\int f(x)dx} M\,dx + e^{\int f(x)dx} N\,dy = 0$$

is exact. Therefore,

$$\frac{\partial}{\partial y}(e^{\int f(x)dx} M) = \frac{\partial}{\partial x}(e^{\int f(x)dx} N)$$

$$e^{\int f(x)dx}\frac{\partial M}{\partial y} = e^{\int f(x)dx}\frac{\partial N}{\partial x} + N\frac{\partial}{\partial x}(e^{\int f(x)dx})$$

$$e^{\int f(x)dx}\frac{\partial M}{\partial y} = e^{\int f(x)dx}\frac{\partial N}{\partial x} + Ne^{\int f(x)dx}f(x).$$

If we cancel $e^{\int f(x)dx}$ from both sides of this last equation, we get

$$\frac{\partial M}{\partial y} = \frac{\partial N}{\partial x} + N \cdot f(x),$$

$$f(x) = \frac{\partial M/\partial y - \partial N/\partial x}{N}.$$

Since all the above steps can be reversed, then $e^{\int f(x)dx}$ is an integrating factor if and only if

$$f(x) = \frac{\partial M/\partial y - \partial N/\partial x}{N}. \quad \square$$

EXAMPLE 4 Use the result of Example 3 to find an integrating factor for the differential equation

$$(2y - 3xy^3 + 12x^2)\,dx + (x - 3x^2y^2)\,dy = 0.$$

Solution Let $M = 2y - 3xy^3 + 12x^2$, $N = x - 3x^2y^2$. Then

$$\frac{\partial M/\partial y - \partial N/\partial x}{N} = \frac{(2 - 9xy^2) - (1 - 6xy^2)}{x - 3x^2y^2}$$

$$= \frac{1 - 3xy^2}{x - 3x^2y^2} = \frac{1}{x} = f(x).$$

The integrating factor is

$$\mu = e^{\int f(x)dx} = e^{\int dx/x} = e^{\ln |x|} = |x|. \quad \square$$

Exercises 19-2

1-8. Test the differential equation for exactness. If an equation is exact, use the methods of this section to solve it.

- **1** $(x^2 + 4y) \, dx + (4x - 3y) \, dy = 0$
- **2** $2e^y \, dx + (xe^y + 2x) \, dy = 0$
- **3** $3xy^2 \, dx + (3x^2y - y) \, dy = 0$
- **4** $e^y \, dx + (xe^y + \cos y) \, dy = 0$
- **5** $(2y - 6x^2y) \, dx + (2x + 4y^2 - 2x^3) \, dy = 0$
- **6** $(y^3 e^x + 4xy + 1) \, dx + (3y^2 e^x + 2x^2) \, dy = 0$
- **7** $y^2 \cos x \, dx + (2y \sin x - \sin y) \, dy = 0$
- **8** $(y + 1) \, dx + (x + \sin y + e^y) \, dy = 0$

9-14. Solve the differential equation by using an integrating factor. If an integrating factor is not given, then use the result of Example 3 or Exercise 15 to find one.

- **9** $(2y^2 + 3xy) \, dx + (3xy + 2x^2) \, dy = 0, \ \mu = xy$
- **10** $(xy + 1) \, dx + x^2 \, dy = 0, \ \mu = e^{xy}$
- **11** $(x^2 + 2x + y) \, dx + dy = 0$
- **12** $y \, dx + (y^3 - x) \, dy = 0$
- **13** $(3x + y^2 - 2y) \, dx + x(y - 1) \, dy = 0$
- **14** $(3x^2 - x - 2y) \, dx + 2x \, dy = 0$
- **15** Show that $\mu = e^{\int f(y) \, dy}$ is an integrating factor for the differential equation $M \, dx + N \, dy = 0$ if and only if

Special Formula for Integrating Factor

$$\frac{\partial N/\partial x - \partial M/\partial y}{M} = f(y).$$

- **16** (a) Solve the differential equation $xy' + y = e^x$.
  (b) Find a particular solution $y = f(x)$ such that $f(1) = 2$ and $f(-1) = -5$. Find another particular solution $y = g(x)$ such that $g(1) = 2$ and $g(-1) = 0$. Explain why these functions do not contradict the uniqueness part of Theorem 19-1.

19-3 First-Order Linear Differential Equations

A differential equation of form

$$y^{(n)} + p_{n-1}(x)y^{(n-1)} + p_{n-2}(x)y^{(n-2)} + \cdots + p_0(x)y = q(x),$$

where the functions $p_0, p_1, p_2, \ldots, p_{n-1}, q$ are continuous on some open interval, is called a *linear differential equation of order n*. We shall derive a formula for the solution of the first-order linear differential equation—the equation

Linear Differential Equation

$$y' + p_0(x)y = q(x).$$

THEOREM 19–3

Formula for First-Order Linear Differential Equation

The general solution of the linear differential equation

$$y' + p(x)y = q(x)$$

is

$$y = e^{-\int p(x)dx} \left\{ \int e^{\int p(x)dx} q(x)\, dx + C \right\}$$

where C is a constant. This formula holds on any open interval on which p and q are continuous.

Proof We multiply each side of the differential equation by $e^{\int p(x)dx}$, getting

$$e^{\int p(x)dx} y' + e^{\int p(x)dx} p(x)y = e^{\int p(x)dx} q(x).$$

We now make the substitution $u = e^{\int p(x)dx} y$. Then

$$\frac{du}{dx} = D_x \left(e^{\int p(x)dx} y \right) = e^{\int p(x)dx} y' + e^{\int p(x)dx} p(x)y,$$

which is equal to the left-hand side of the above differential equation. Consequently, the differential equation reduces to

$$\frac{du}{dx} = e^{\int p(x)dx} q(x).$$

We now separate the variables and integrate, obtaining

$$du = e^{\int p(x)dx} q(x)\, dx,$$

$$u = \int du = \int e^{\int p(x)dx} q(x)\, dx + C$$

Since $u = e^{\int p(x)dx} y$, then

$$e^{\int p(x)dx} y = \int e^{\int p(x)dx} q(x)\, dx + C$$

$$y = e^{-\int p(x)dx} \left\{ \int e^{\int p(x)dx} q(x)\, dx + C \right\}. \blacksquare$$

EXAMPLE 1 Solve the differential equation

$$y' - \cos x \cdot y = \cos x.$$

Solution The differential equation is first-order linear with

$$p(x) = -\cos x \quad \text{and} \quad q(x) = \cos x.$$

Then

$$\int p(x)\, dx = \int (-\cos x)\, dx = -\sin x.$$

(We drop the constant of integration. It would cancel out later.)

19–3 First-Order Linear Differential Equations

By the formula,

$$y = e^{-\int p(x)dx} \left\{ \int e^{\int p(x)dx} q(x)\, dx + C \right\}$$

$$= e^{\sin x} \left\{ \int e^{-\sin x} \cos x\, dx + C \right\}.$$

This integral can be evaluated by making the substitution $v = -\sin x$, $dv = -\cos x\, dx$. The solution is

$$y = -e^{\sin x} \{e^{-\sin x} - C\} = -1 + Ce^{\sin x}. \quad \square$$

EXAMPLE 2 Solve the differential equation

$$\frac{dy}{dx} + \frac{1}{x} y = 3x^2 + 5.$$

Solution The differential equation is first-order linear with $p(x) = 1/x$, $q(x) = 3x^2 + 5$. Then

$$\int p(x)dx = \int \frac{1}{x}\, dx = \ln |x|.$$

It follows that

$$e^{-\int p(x)dx} = e^{-\ln|x|} = e^{\ln 1/|x|} = \frac{1}{|x|}$$

and

$$e^{\int p(x)dx} = e^{\ln|x|} = |x|.$$

The general solution is

$$y = e^{-\int p(x)dx} \left\{ \int e^{\int p(x)dx} q(x)\, dx + C \right\}$$

$$= \frac{1}{|x|} \left\{ \int |x|\, (3x^2 + 5)\, dx + C \right\}.$$

We now consider two cases. (1) If $x > 0$, then $|x| = x$. The solution is

$$y = \frac{1}{x} \left\{ \int x(3x^2 + 5)\, dx + C \right\} = \frac{1}{x} \left\{ \int (3x^3 + 5x)\, dx + C \right\}$$

$$= \frac{1}{x} \left\{ \frac{3x^4}{4} + \frac{5x^2}{2} + C \right\} = \frac{3x^3}{4} + \frac{5x}{2} + \frac{C}{x}.$$

(2) If $x < 0$, then $|x| = -x$. We leave it for you to show that the solution has the same form. $\square$

EXAMPLE 3 Solve $y' + e^x y = 0$.

Solution The differential equation can be solved by the formula or by separation of variables. We use the formula with

$$p(x) = e^x, \qquad q(x) = 0.$$

Then
$$e^{-\int p(x)dx} = e^{-e^x} \quad \text{and} \quad e^{\int p(x)dx} = e^{e^x}.$$

The solution is
$$y = e^{-e^x}\left\{\int e^{e^x} \cdot 0\, dx + C\right\} = Ce^{-e^x}. \quad \square$$

Remark As you may have guessed, much of the work in an elementary course on differential equations involves classifying the equations by type and studying methods of solving each type. You should approach the classification systematically. A tentative checklist might contain the following questions:

(1) Can the variables be separated?
(2) Is the equation first-order homogeneous?
(3) Is it first-order linear?
(4) Is it exact?

Observe that any affirmative answer leads to a method of solution.

This checklist should be extended as we work through the remaining sections. Exercises 9–16 are a collection of differential equations to test your skill.

APPLICATION. LEARNING CURVES

Psychologists have made numerous theoretical studies of learning. For simplicity we consider a case in which a simple skill is learned by a completely unskilled person. When he is first exposed to the process, learning is rapid. As his proficiency increases, the rate of learning lowers until eventually there is no learning at all.

Assume that the learner practices the skill continuously so that his proficiency is a function of time, say $y = p(t)$. Then $y' = p'(t)$ is the rate at which the learning occurs at time t. If m is the maximum level of proficiency of which the learner is capable, then
$$m - y = m - p(t)$$

is a measure of his capacity to learn. It is reasonable to assume that the rate of learning is proportional to the capacity to learn. That is,
$$y' = k(m - y).$$

This differential equation can be rewritten as the linear differential equation
$$y' + ky = km$$

or the variables can be separated. The solution is
$$y = m + Ce^{-kt}.$$

To evaluate the constant C, we recall that $y = 0$ when $t = 0$. (The worker was completely unskilled at the beginning of the learning.) Then
$$0 = m + Ce^{-0} = m + C,$$

so that
$$C = -m.$$
The solution is

Learning Curve Formula
$$y = m(1 - e^{-kt}),$$

where m is the particular worker's maximum level of proficiency and k is a constant that depends on the individual rate of learning.

The graph of the function
$$y = m(1 - e^{-kt})$$
is shown in Figure 19-3. The graph is an example of a *learning curve*.

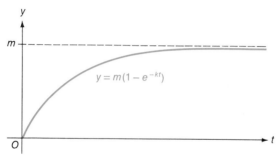

Figure 19-3 The learning curve.

Exercises 19-3

1-8. Solve the first-order linear differential equations.

- 1. $\sin x \cdot y' + \cos x \cdot y - 2 = 0$
- 2. $x^{-2} y' + 3y = 2$
- 3. $y' + \sin x \cdot y = 2 \sin x$
- 4. $y' + y = e^{2x}$
- 5. $(x + 1) y' + 3y = e^x (x + 1)^{-2}$
- 6. $xy' - 2y = x^3 e^{-2x}$
- 7. $(x^2 + 1)y' + 2xy = x - 1$
- 8. $y' + y \sec x = \tan x$

9-16. (a) Classify the differential equations by type (variables separable, homogeneous, exact, first-order linear).
(b) Find the general solution.

- 9. $\sin x \cos y \, dx + \cos x \sin y \, dy = 0$
- 10. $x^2 \, dy = (3x \, e^{-x^2} - 2xy) \, dx$
- 11. $(4x^4 - 4x^3 y) y' = x^4$
- 12. $(1 + y^2) \, dx = (1 + x^2) \, dy$
- 13. $(y - 2xy) \, dx + (x + xy) \, dy = 0$
- 14. $e^{x+y} \, dy = e^{x-y} \, dx$
- 15. $x \, dy = (x^2 e^{-x} + y) \, dx$
- 16. $(y \, e^{2x} + 2y^2) \, dy + (y^2 e^{2x} + 2x) \, dx = 0$

17 (*Learning Curves*) A foreman must train the workers on an assembly line in a new skill. All workers are completely untrained at the beginning, so that the equation

$$y = m(1 - e^{-kt})$$

describes the proficiency at time t. The foreman measures the number of times per hour each worker can perform the necessary operations. These results are listed for the end of the first and second days of training. Calculate m, the maximum number of times each worker will be able to perform the operations when fully trained.

(a) Worker A: 50 times per hour after one day, 75 times per hour after two days.

●(b) Worker B: 40 times per hour after one day, 75 times per hour after two days.

(c) Worker C: 60 times per hour after one day, 70 times per hour after two days.

Bernouilli Equation

18 (*The Bernoulli Equation*) The differential equation

$$y' + p(x)\, y = q(x)\, y^n$$

is called the *Bernoulli equation* (after *Jacques Bernoulli*, 1654–1705.) Note that if $n = 0$, the Bernoulli equation is first-order linear, and if $n = 1$, the variables can be separated. Show that if $n \neq 1$, the substitution $v = y^{1-n}$ allows us to reduce the Bernoulli equation to the first-order linear equation

$$v' + (1 - n)\, p(x)\, v = (1 - n)\, q(x).$$

19–22. Use the results of Exercise 18 to solve the following Bernoulli equations.

●**19** $y' + xy = \dfrac{x}{y^3}$

20 $xy' + y = x^2 y^2$

●**21** $(x^2 - 1)\, y' + 4xy = \sqrt{y}$

22 $y' + \tan x \cdot y = \dfrac{\sin x}{\sqrt{y}}$

Alternative Method of Solution of First-Order Linear Differential Equation

23 First-order linear differential equations can be solved by an alternative method. Let $u(x)$ be a nonzero particular solution of the differential equation

$$y' + p(x)y = 0.$$

Let $v(x)$ be a particular solution of the equation

$$y' + p(x)y = q(x).$$

(a) Prove that $y = Cu(x) + v(x)$ is the general solution of

$$y' + p(x)y = q(x).$$

●(b) Use the result of (a) to solve the differential equation

$$y' + xy = (x + 1)\, e^x.$$

(*Hint:* A particular solution is $v(x) = e^x$.)

19–4 Homogeneous Second-Order Linear Differential Equations with Constant Coefficients. I: Real Solutions of the Characteristic Equation

The linear differential equation

$$y^{(n)} + p_{n-1}(x)y^{(n-1)} + \cdots + p_1(x)y^{(1)} + p_0(x)y = q(x)$$

is said to be *homogeneous* if $q(x)$ is identically zero and *nonhomogeneous* if $q(x)$ is not identically zero. As we shall see later, the solutions of a nonhomogeneous linear differential equation can be generated from the solutions of the related homogeneous equation obtained by replacing $q(x)$ by 0. [A special case of this result is discussed in Exercise 23 of the preceding section.]

Note that in this section the word *homogeneous* has a different meaning from its meaning in Section 19–1.

In the next few sections we work primarily with second-order linear differential equations with *constant coefficients*. These equations can be written in the form

$$ay'' + by' + cy = q(x) \qquad (a \neq 0).$$

DIFFERENTIAL OPERATORS

If we work exclusively with functions of x, the subscript is not needed on the symbol D_x for the derivative. Thus, we define

$$D(f(x)) = f'(x), \qquad D^2 f(x) = f''(x),$$

Differential Operator

and so on. The symbols $D, D^2, \ldots$ represent mathematical entities called *differential operators*.

A differential operator of form

$$aD + b$$

is defined by the rule

$$(aD + b)f(x) = aD(f(x)) + bf(x) = af'(x) + bf(x),$$

an operator of form

$$aD^2 + bD + c$$

by the rule

$$(aD^2 + bD + c)f(x) = aD^2(f(x)) + bD(f(x)) + cf(x)$$
$$= af''(x) + bf'(x) + cf(x),$$

and so on. Observe that a differential operator acts on a function $f(x)$ to produce a new function of x provided f has enough derivatives.

EXAMPLE 1 What effect does the differential operator $D^2 - 3D + 2$ have on the function

(a) $f(x) = x^2 - 1$? (b) $f(x) = e^{3x}$?

Solution (a)
$$(D^2 - 3D + 2)(x^2 - 1)$$
$$= D^2(x^2 - 1) - 3D(x^2 - 1) + 2(x^2 - 1)$$
$$= 2 - 3(2x) + 2(x^2 - 1) = 2x^2 - 6x.$$

The differential operator $D^2 - 3D + 2$ acts on $x^2 - 1$ to produce $2x^2 - 6x$.

(b)
$$(D^2 - 3D + 2)(e^{3x}) = D^2(e^{3x}) - 3D(e^{3x}) + 2e^{3x}$$
$$= 9e^{3x} - 3(3e^{3x}) + 2e^{3x}$$
$$= 2e^{3x}.$$

The operator $D^2 - 3D + 2$ acts on e^{3x} to product $2e^{3x}$. □

If L and M are differential operators, we define the sum $L + M$ and the product LM by

Sum and Product
$$(L + M)(f(x)) = L(f(x)) + M(f(x)),$$
$$(LM)(f(x)) = L(M(f(x))).$$

In the expression for LM, we first operate on $f(x)$ with M, then on $M(f(x))$ with L.

Equality Two differential operators are *equal* provided they can be applied to the same functions and have the same effect on these functions. A differential operator is *linear* if it is of form

$$a_n(x)D^n + a_{n-1}(x)D^{n-1} + \cdots + a_1(x)D + a_0(x).$$

EXAMPLE 2 Show that $(D - 2)(D - 3) = D^2 - 5D + 6$.

Solution Let y be a twice differentiable function of x. Then
$$(D - 2)(D - 3)y = (D - 2)[(D - 3)y]$$
$$= (D - 2)(y' - 3y)$$
$$= D(y' - 3y) - 2(y' - 3y)$$
$$= y'' - 3y' - 2y' + 6y$$
$$= y'' - 5y' + 6y$$
$$= (D^2 - 5D + 6)y.$$

Since the operators $(D - 2)(D - 3)$ and $D^2 - 5D + 6$ have the same effect, then they are equal. It follows that $(D - 2)(D - 3)$ is a linear operator. □

FACTORABLE DIFFERENTIABLE OPERATORS

The method used in Example 2 can be generalized to show that a differential operator

$$a_n D^n + a_{n-1} D^{n-1} + \cdots + a_1 D + a_0,$$

where $a_0, a_1, \ldots, a_{n-1}, a_n$ are real numbers, can be factored as if it were a polynomial in D.

EXAMPLE 3
(a) $D^2 - 5D + 6 = (D - 2)(D - 3) = (D - 3)(D - 2)$.
(b) $D^3 + 2D^2 + D = D(D + 1)(D + 1)$.
(c) $D^4 - 1 = (D^2 - 1)(D^2 + 1) = (D - 1)(D + 1)(D^2 + 1)$. □

19–4 Linear Differential Equations with Constant Coefficients. I.

This type of factorization only holds if the coefficients $a_0, a_1, a_2, \ldots, a_n$ are real numbers. A factorization may be possible if the coefficients are nonconstant functions of x, but it can not be determined by using rules for the factorization of polynomials.

EXAMPLE 4 Show that
$$(D - x)(D + x) = D^2 + 1 - x^2 \neq (D + x)(D - x).$$

Solution Let y be a twice differentiable function. Then
$$\begin{aligned}(D - x)(D + x)y &= (D - x)(y' + xy) \\ &= D(y' + xy) - x(y' + xy) \\ &= y'' + [xy' + y] - xy' - x^2 y \\ &= y'' + y - x^2 y \\ &= (D^2 + 1 - x^2)y.\end{aligned}$$

Since $(D - x)(D + x)y = (D^2 + 1 - x^2)y$ for every twice differentiable function y, then
$$(D - x)(D + x) = D^2 + 1 - x^2.$$

A similar argument shows that
$$(D + x)(D - x) = D^2 - 1 - x^2.$$

Since $D^2 + 1 - x^2 \neq D^2 - 1 - x^2$, then
$$(D - x)(D + x) \neq (D + x)(D - x). \quad \square$$

HOMOGENEOUS SECOND-ORDER LINEAR DIFFERENTIAL EQUATIONS WITH CONSTANT COEFFICIENTS

Suppose the linear operator $D^2 + aD + b$ can be factored as
$$D^2 + aD + b = (D - m_1)(D - m_2),$$
where m_1 and m_2 are real numbers. Then the homogeneous second-order linear differential equation $y'' + ay' + by = 0$ can be written as
$$(D - m_1)(D - m_2)y = 0.$$

The following theorem gives the solution of this differential equation in a closed form. The theorem is proved after Example 6.

THEOREM 19–4 Let m_1 and m_2 be real numbers.

General Solution (a) If $m_1 \neq m_2$, then the general solution of the differential equation
$$(D - m_1)(D - m_2)y = 0$$
is
$$y = C_1 e^{m_1 x} + C_2 e^{m_2 x}.$$

(b) ($m_1 = m_2$) The general solution of the differential equation

$$(D - m_1)(D - m_1)y = 0$$

is

$$y = (C_1 x + C_2)e^{m_1 x}.$$

EXAMPLE 5 Solve $y'' - 6y' + 5y = 0$.

Solution The differential equation can be written as

$$(D^2 - 6D + 5)y = 0$$
$$(D - 5)(D - 1)y = 0.$$

It follows from Theorem 19–4 that the solution of the differential equation is

$$y = C_1 e^{5x} + C_2 e^x. \quad \square$$

THE CHARACTERISTIC EQUATION

The numbers m_1 and m_2 of Theorem 19-4 for the differential equation

$$(D^2 + aD + b)y = 0$$

can be obtained from the related polynomial equation

$$m^2 + am + b = 0,$$

Characteristic Equation

called the *characteristic equation* of the differential equation. We get the characteristic equation by replacing the nth-order derivative of y with the nth power of m.

For example, the characteristic equation of the differential equation considered in Example 5 is

$$m^2 - 6m + 5 = 0$$

which factors as

$$(m - 5)(m - 1) = 0.$$

Since its solutions are $m_1 = 5$ and $m_2 = 1$, the solution of the original differential equation is

$$y = C_1 e^{5x} + C_2 e^x.$$

EXAMPLE 6 Solve $y'' + 4y' + 4y = 0$.

Solution The characteristic equation is

$$m^2 + 4m + 4 = 0$$
$$(m + 2)(m + 2) = 0.$$

The only solution of the characteristic equation is $m_1 = -2$. It follows from Theorem

19-4 Linear Differential Equations with Constant Coefficients. I.

19-4 that the general solution of the differential equation is

$$y = (C_1 x + C_2) e^{m_1 x} = C_1 x e^{-2x} + C_2 e^{-2x}. \quad \Box$$

We are now ready to prove Theorem 19-4.

Proof of Theorem 19-4 We write the differential equation in the form

$$(D - m_1)(D - m_2) y = 0$$

and make the substitution $(D - m_2) y = u$, obtaining the first-order linear differential equation

$$(D - m_1) u = 0.$$

It follows from our work in Section 19-3 that this equation has the solution

$$u = C_0 e^{m_1 x}.$$

Since $(D - m_2) y = u$, then

$$y' - m_2 y = C_0 e^{m_1 x}.$$

This differential equation also is first-order linear. It has the solution

$$y = e^{\int m_2 dx} \left\{ \int e^{-\int m_2 dx} C_0 e^{m_1 x} \, dx + C_2 \right\}$$

$$= e^{m_2 x} \left\{ C_0 \int e^{(m_1 - m_2) x} \, dx + C_2 \right\}.$$

At this point we distinguish the two cases.

(a) If $m_1 \neq m_2$, the solution of the differential equation is

$$y = e^{m_2 x} \left\{ \frac{C_0}{m_1 - m_2} e^{(m_1 - m_2) x} + C_2 \right\}$$

$$= \frac{C_0}{m_1 - m_2} e^{m_1 x} + C_2 e^{m_2 x}$$

$$= C_1 e^{m_1 x} + C_2 e^{m_2 x}$$

where $C_1 = C_0 / (m_1 - m_2)$.

(b) If $m_1 = m_2$, the solution is

$$y = e^{m_1 x} \{ C_0 \int e^0 dx + C_2 \} = e^{m_1 x} \{ C_0 x + C_2 \}$$
$$= (C_1 x + C_2) e^{m_1 x}$$

where $C_1 = C_0$. ■

LINEAR INDEPENDENCE

Up to this point, we have been rather careless about the arbitrary constants in the general solution. We would say that

$$y = C_1 e^x + C_2 e^{-x}$$

is the general solution of

$$y'' - y = 0,$$

but would not say that

$$y = C_1 e^x + C_2 \left(\frac{e^x}{2}\right)$$

is the general solution, even though both solutions involve two constants that can be assigned arbitrary values. The concept of *linear independence* can help us to clear some of the ambiguity.

Linear Independence and Dependence

Two functions $u_1(x)$ and $u_2(x)$ are said to be *linearly independent* if neither is a constant multiple of the other. They are *linearly dependent* if they are not linearly independent. For example, the functions e^x and e^{-x} are linearly independent, while the functions e^x and $e^x/2$ are linearly dependent.

For

$$y = C_1 u_1(x) + C_2 u_2(x)$$

to be the general solution of a second-order homogeneous linear differential equation, it is necessary and sufficient that $u_1(x)$ and $u_2(x)$ are linearly independent solutions. If this is not the case, then the solution could be rewritten with a single arbitrary constant. For example, since e^x and $e^x/2$ are linearly dependent, we can write

$$C_1 e^x + C_2 \left(\frac{e^x}{2}\right) = \left(C_1 + \frac{C_2}{2}\right) e^x = C_3 e^x.$$

On the other hand, e^x and e^{-x} are linearly independent. There is no way to write

$$C_1 e^x + C_2 e^{-x}$$

with a single arbitrary constant.

Exercises 19-4

1 What effect does the differential operator $D^2 + 4D + 4$ have on the function f?
   - (a) $f(x) = 2x^3 - 3x - 2$
   - (b) $f(x) = e^{-2x}$
   - (c) $f(x) = \cos 2x$

2 Write the operator in the form $D^2 + a(x)D + b(x)$.
   - (a) $(D - 2)(D + 5)$
   - (b) $D(D - 1)$
   - (c) $(D + 1)(D + x)$
   - (d) $(D + 2x)(D + e^x)$

3-10 (a) Use the methods of this section to find the general solution.
   (b) Find the particular solution that satisfies the given initial or boundary condition.

- 3 $\begin{cases} y'' + 5y' + 6y = 0 \\ f(0) = 1, f'(0) = -1 \end{cases}$

4 $\begin{cases} y'' + 3y' = 9y' - 8y \\ f(0) = -2, f(\ln 2) = 4 \end{cases}$

- 5 $\begin{cases} (D^2 - 6D + 9)y = 0 \\ f(0) = 2, f'(0) = 3 \end{cases}$

6 $\begin{cases} (3D^2 - 3D + 1)y = y'' \\ f(0) = 1, f(\ln 4) = 0 \end{cases}$

19–5 Linear Differential Equations with Constant Coefficients. II.

- **7** $\begin{cases} y'' = 16y \\ f(0) = 0, f'(0) = 0 \end{cases}$

8 $\begin{cases} 2y'' + 6y' + 3y = 0 \\ f(0) = 0, f(2) = 0 \end{cases}$

- **9** $\begin{cases} y'' - y' = 0 \\ f(1) = 3, f(1 + \ln 2) = 4 \end{cases}$

10 $\begin{cases} \dfrac{y''}{2} + y' + y = \dfrac{y}{2} \\ f(1) = -\dfrac{1}{e}, f(2) = \dfrac{1}{e^2} \end{cases}$

11 Work through the steps of the proof of Theorem 19–4 with the following examples.
- **(a)** $y'' - 5y' + 6y = 0$
- **(b)** $y'' - 6y' + 9y = 0$

- **12** Use the method illustrated in the proof of Theorem 19–4 to solve the differential equation $(D - 2)\left(D + \dfrac{1}{x}\right) y = 0$.

13 **(a)** Let $u_1(x)$ and $u_2(x)$ be particular solutions of the homogeneous equation $y'' + a(x)y' + b(x)y = 0$. Show that
$$y = C_1 u_1(x) + C_2 u_2(x)$$
is a solution of the differential equation.

(b) Must the solution $C_1 u_1(x) + C_2 u_2(x)$ be the general solution? Either prove that the answer is *yes* or find an example to show that it may be *no*. (*Hint:* Consider whether $u_1(x)$ and $u_2(x)$ are linearly independent.)

- **(c)** Use the method outlined in part (a) to find the general solution of
$$x^2 y'' + (x - x^2)y' - (x + 1)y = 0.$$

(*Hint:* Show that $u_1(x) = 1/x$ and $u_2(x) = e^x(x - 1)/x$ are linearly independent solutions.)

19–5 Homogeneous Second-Order Linear Differential Equations with Constant Coefficients. II: Imaginary Solutions of the Characteristic Equation

We have just considered the case in which the characteristic equation of $(D^2 + aD + b)y = 0$ has real solutions. We now extend our results to allow imaginary solutions of the characteristic equation.

In more advanced courses it is established that the results of the calculus can be extended to cover complex-valued functions of complex-valued variables. It can be shown that the solution
$$y = C_1 e^{m_1 x} + C_2 e^{m_2 x}$$
of the differential equation $(D - m_1)(D - m_2) = 0$ also is obtained if m_1 and m_2 are imaginary numbers. Our immediate problem is to make some sense out of such a solution and relate it to real functions.

Introduction to Differential Equations

EULER'S IDENTITY

Our starting point is the theorem devised by *Leonhard Euler* (Swiss, 1707–1783), which relates the complex-valued function e^{ix} to the real-valued sine and cosine functions.

THEOREM 19–5 (*Euler's Identity*)

$$e^{ix} = \cos x + i \sin x$$

Euler's identity has several interesting consequences. First, if we let $x = \pi$, we obtain

$$e^{i\pi} = \cos \pi + i \sin \pi.$$

Since $\cos \pi = -1$ and $\sin \pi = 0$, this reduces to

$$e^{i\pi} = -1,$$

a simple equation that relates four of the most important mathematical constants—e, i, π, and 1.

A second result (obtained by replacing x with $-x$ in Euler's identity) is

$$e^{-ix} = \cos(-x) + i \sin(-x),$$

which reduces to

$$e^{-ix} = \cos x - i \sin x.$$

Euler's identity can be explained by infinite series. It can be established that all our work on infinite series, with suitable modifications, can be extended to cover complex numbers. In particular, the Maclaurin series

$$e^x = 1 + \frac{x}{1!} + \frac{x^2}{2!} + \frac{x^3}{3!} + \frac{x^4}{4!} + \frac{x^5}{5!} + \frac{x^6}{6!} + \cdots$$

is valid if x is real or complex. If we substitute ix for x and recall that $i^2 = -1$, $i^3 = -i$, $i^4 = 1$, we get

$$e^{ix} = 1 + \frac{(ix)}{1!} + \frac{(ix)^2}{2!} + \frac{(ix)^3}{3!} + \frac{(ix)^4}{4!} + \frac{(ix)^5}{5!} + \frac{(ix)^6}{6!} + \cdots$$

$$= 1 + \frac{ix}{1!} - \frac{x^2}{2!} - \frac{ix^3}{3!} + \frac{x^4}{4!} + \frac{ix^5}{5!} - \frac{x^6}{6!} - \cdots$$

$$= \left(1 - \frac{x^2}{2!} + \frac{x^4}{4!} - \frac{x^6}{6!} + \cdots\right) + i\left(\frac{x}{1!} - \frac{x^3}{3!} + \frac{x^5}{5!} - \cdots\right)$$

$$= \cos x + i \sin x.$$

19–5 Linear Differential Equations with Constant Coefficients. II.

This argument outlines the proof of Theorem 19–5. (The main task of justifying the steps in the argument is beyond the scope of this book.)

APPLICATION TO DIFFERENTIAL EQUATIONS

Let the differential equation

$$(D^2 + aD + b)y = 0$$

have real coefficients and the characteristic equation

$$m^2 + am + b = 0$$

have imaginary solutions m_1 and m_2. It is proved in algebra that m_1 and m_2 are complex conjugates of each other. In other words, there exist real numbers α and β such that $m_1 = \alpha + \beta i$ and $m_2 = \alpha - \beta i$.

We shall show that the general solution of the above differential equation is

$$\boxed{y = e^{\alpha x} (C_1 \cos \beta x + C_2 \sin \beta x)}$$

To establish this result, we write the solution of the differential equation as

$$y = K_1 e^{m_1 x} + K_2 e^{m_2 x}$$
$$= K_1 e^{(\alpha + \beta i)x} + K_2 e^{(\alpha - \beta i)x}$$
$$= e^{\alpha x}[K_1 e^{\beta i x} + K_2 e^{-\beta i x}].$$

By Euler's identity,

$$e^{\beta i x} = \cos \beta x + i \sin \beta x$$

and

$$e^{-\beta i x} = \cos \beta x - i \sin \beta x.$$

If we substitute these values in the general solution of the differential equation, we get

$$y = e^{\alpha x}[K_1(\cos \beta x + i \sin \beta x) + K_2(\cos \beta x - i \sin \beta x)]$$
$$= e^{\alpha x}[(K_1 + K_2) \cos \beta x + (K_1 - K_2) i \sin \beta x]$$
$$= e^{\alpha x}[C_1 \cos \beta x + C_2 \sin \beta x],$$

where $C_1 = K_1 + K_2$ and $C_2 = (K_1 - K_2) i$.

Assuming the work with complex exponents is legal, this argument establishes the following theorem.

THEOREM 19–6 Let the characteristic equation of the differential equation

$$y'' + ay' + by = 0$$

have imaginary solutions

$$m_1 = \alpha + \beta i \text{ and } m_2 = \alpha - \beta i.$$

The general solution of the differential equation is

General Solution
$$y = K_1 e^{m_1 x} + K_2 e^{m_2 x} = e^{\alpha x}[C_1 \cos \beta x + C_2 \sin \beta x].$$

Note In the special case in which the solutions of the characteristic equation are the pure imaginary numbers $m_1 = \beta i$ and $m_2 = -\beta i$, the solution of the differential equation reduces to

$$y = C_1 \cos \beta x + C_2 \sin \beta x.$$

EXAMPLE 1 Solve $y'' + 9y = 0$.

Solution The characteristic equation is

$$m^2 + 9 = 0.$$

The solutions of this equation are $m = \pm 3i$. The general solution of the differential equation is

$$y = C_1 \cos 3x + C_2 \sin 3x. \quad \square$$

EXAMPLE 2 Solve $4y'' - 16y' + 17y = 0$.

Solution The characteristic equation is

$$4m^2 - 16m + 17 = 0,$$

which has the solutions

$$m = \frac{16 \pm \sqrt{16^2 - 4 \cdot 4 \cdot 17}}{8} = \frac{16 \pm \sqrt{-16}}{8},$$

$$= \frac{16 \pm 4i}{8} = 2 \pm \frac{i}{2}.$$

The solution of the differential equation is

$$y = e^{\alpha x}[C_1 \cos \beta x + C_2 \sin \beta x]$$
$$= e^{2x}\left[C_1 \cos \frac{x}{2} + C_2 \sin \frac{x}{2}\right]. \quad \square$$

Exercises 19-5

1-8. Solve the differential equation.

- •1 $y'' + 16y = 0$
- 2 $y'' + 5y = 0$
- •3 $y'' - 2y' + 2y = 0$
- 4 $2y'' - y' - 3y = 0$
- •5 $y'' - 6y' + 10y = 0$
- 6 $y'' - 6y' + 13y = 0$
- •7 $y'' - 4y' + 7y = 0$
- 8 $y'' - 10y' + 38y = 0$

9-12. Find the particular solution of the differential equation in the given exercise that satisfies the initial or boundary condition.

Exercises 19–5

- •9 Exercise 1: $f(0) = 1$, $f(4\pi/3) = 1$
- 10 Exercise 2: $f(0) = 2$, $f'(0) = 1$
- •11 Exercise 3: $f(0) = 2$, $f(\pi/4) = e^{\pi/4}$
- 12 Exercise 4: $f(0) = 1$, $f'(0) = 4$

13 (*The Pendulum*) A simple pendulum consists of a weight that swings along a circular arc at the end of a wire. (See Fig. 19–4.) Let $\theta = \theta(t)$ be the angle of displace-

Figure 19–4 Exercise 13. A simple pendulum.

ment of the wire from the vertical at time t. It can be proved that θ satisfies the differential equation

$$\frac{d^2\theta}{dt^2} + \frac{g}{l}\sin\theta = 0,$$

where g is the gravitational constant and l the length of the wire.

The above differential equation is not easily solved. Observe, however, that if θ is a small angle, then it follows from the series

$$\sin\theta = \theta - \frac{\theta^3}{3!} + \frac{\theta^5}{5!} - \cdots$$

that $\sin\theta \approx \theta$. Consequently, the differential equation

$$\frac{d^2\theta}{dt^2} + \frac{g}{l}\theta = 0$$

approximates the motion of the pendulum when θ is small.

- •(a) Find the particular solution of

$$\frac{d^2\theta}{dt^2} + \frac{g}{l}\theta = 0$$

that satisfies the initial conditions $\theta = \theta_0 \neq 0$ and $d\theta/dt = 0$ when $t = 0$.
- (b) Find the period of the solution in (a). Show that the period depends only on g and l, not on the initial displacement θ_0.

14 De Moivre's theorem in trigonometry establishes that

$$(\cos x + i \sin x)^n = \cos nx + i \sin nx$$

for each positive integer n. Use de Moivre's theorem and Euler's identity to show that

$$(e^{ix})^n = e^{inx}$$

for every positive integer n.

- •15 Let C_1 and C_2 be arbitrary real numbers. Show that we can find complex numbers K_1 and K_2 such that

$$C_1 = K_1 + K_2 \quad \text{and} \quad C_2 = (K_1 - K_2)i.$$

It follows that if

$$e^{\alpha x}[(K_1 + K_2) \cos \beta x + (K_1 - K_2)i \sin \beta x]$$

is a solution of a differential equation for every pair of complex numbers K_1 and K_2, then we can obtain real solutions of form

$$e^{\alpha x}[C_1 \cos \beta x + C_2 \sin \beta x],$$

where C_1 and C_2 are arbitrary real numbers.

19–6 Nonhomogeneous Second-Order Linear Differential Equations with Constant Coefficients

It is a simple matter to solve a nonhomogeneous differential equation

$$y'' + ay' + by = f(x)$$

provided we can find one particular solution $y_p(x)$. We shall show that if $y_h(x)$ is the general solution of the related homogeneous equation

$$y'' + ay' + by = 0,$$

then

General Solution of Nonhomogeneous Equation

$$y(x) = y_h(x) + y_p(x)$$

is the general solution of the original differential equation. That is, the general solution is obtained by adding the general solution of the related homogeneous differential equation to any particular solution of the nonhomogeneous equation.

THEOREM 19–7 Let $y_p(x)$ be a particular solution of the differential equation

$$y'' + ay' + by = f(x),$$

and let $y_h(x)$ be the general solution of the related homogeneous equation

$$y'' + ay' + by = 0.$$

Then the general solution of the original nonhomogeneous equation is

$$y(x) = y_h(x) + y_p(x).$$

Proof It follows from Theorems 19–4 and 19–6 that there exist linearly independent functions $y_1(x)$ and $y_2(x)$ such that the general solution of the homogeneous equation

$$y'' + ay' + by = 0$$

19-6 Nonhomogeneous Second-Order Linear Differential Equations

is

$$y_h(x) = C_1 y_1(x) + C_2 y_2(x).$$

More specifically, $y_h(x)$ is one of the functions $C_1 e^{m_1 x} + C_2 e^{m_2 x}$, $(C_1 x + C_2)e^{mx}$, or $C_1 e^{\alpha x} \cos \beta x + C_2 e^{\alpha x} \sin \beta x$. Furthermore, it can be proved that every solution of the homogeneous equation can be obtained from the general solution. (There are no singular solutions.)

Observe first that if we let $y(x) = y_h(x) + y_p(x)$, then

$$\begin{aligned} y''(x) + ay'(x) + by(x) &= [y_h'' + y_p''] + a[y_h' + y_p'] + b[y_h + y_p] \\ &= [y_h'' + ay_h' + by_h] + [y_p'' + ay_p' + by_p] \\ &= 0 + f(x) = f(x), \end{aligned}$$

so that $y(x)$ is a solution of the nonhomogeneous equation.

We now show that every solution is of the type described above. Let $y(x)$ be a solution of the nonhomogeneous equation

$$y'' + ay' + by = f(x).$$

Let $Y(x) = y(x) - y_p(x)$. Then

$$\begin{aligned} Y''(x) + aY'(x) + bY(x) &= [y'' - y_p''] + a[y' - y_p'] + b[y - y_p] \\ &= [y'' + ay' + by] - [y_p'' + ay_p' + by_p] \\ &= f(x) - f(x) = 0 \end{aligned}$$

for all x. Thus, $Y(x)$ is a solution of the homogeneous differential equation

$$y'' + ay' + by = 0.$$

Since this equation has no singular solutions, then there exist constants C_1 and C_2 such that

$$\begin{aligned} Y(x) &= C_1 y_1(x) + C_2 y_2(x), \\ y(x) - y_p(x) &= C_1 y_1(x) + C_2 y_2(x) \\ y(x) &= C_1 y_1(x) + C_2 y_2(x) + y_p(x). \end{aligned}$$

Therefore

$$y(x) = y_h(x) + y_p(x)$$

for the proper choice of constants C_1 and C_2. It follows that

$$y(x) = y_h(x) + y_p(x)$$

is the general solution. ■

EXAMPLE 1 We can verify by substitution that

$$y_p = \frac{4x}{3}$$

is a particular solution of

$$(D^2 + 9)y = 12x.$$

We saw in Example 1 of Section 19–5 that

$$y_h = C_1 \cos 3x + C_2 \sin 3x$$

is the general solution of the related homogeneous equation. It follows from Theorem 19–7 that the general solution of the nonhomogeneous differential equation is

$$y = y_h + y_p = C_1 \cos 3x + C_2 \sin 3x + \frac{4x}{3}. \quad \square$$

UNDETERMINED COEFFICIENTS

The procedure for solving the nonhomogeneous second-order linear differential equation

$$y'' + ay' + by = f(x)$$

requires that we find a particular solution y_p. If $f(x)$ is a sufficiently simple function, we may be able to guess the form of this solution. For example, if $f(x)$ is a second-degree polynomial, we should expect that y_p also would be a polynomial—probably of second degree. In that case, we should assume that

$$y_p = Ax^2 + Bx + C.$$

We then substitute this function in the differential equation and find the specific values of A, B, and C that give a solution.

For obvious reasons this technique is called the *method of undetermined coefficients*.

EXAMPLE 2 Solve the differential equation

$$y'' - 5y' + 6y = x^2 - 3.$$

Solution (a) The related homogeneous equation is

$$y'' - 5y' + 6y = 0,$$

which has the characteristic equation

Example of Solution by Undetermined Coefficients

$$m^2 - 5m + 6 = 0$$
$$(m - 2)(m - 3) = 0.$$

The general solution of the homogeneous equation is

$$y_h = C_1 e^{2x} + C_2 e^{3x}.$$

(b) Since the right-hand side of the original differential equation is a second-degree polynomial, we should expect that a particular solution might be of form

$$y_p = Ax^2 + Bx + C.$$

If we substitute this function in the differential equation, we get

$$y_p'' - 5y_p' + 6y_p = x^2 - 3$$
$$(2A) - 5(2Ax + B) + 6(Ax^2 + Bx + C) = x^2 - 3$$
$$6Ax^2 + (-10A + 6B)x + (2A - 5B + 6C) = x^2 - 3.$$

19-6 Nonhomogeneous Second-Order Linear Differential Equations

If we equate the corresponding coefficients, we get the system of equations

$$\begin{cases} 6A = 1 \\ -10A + 6B = 0 \\ 2A - 5B + 6C = -3, \end{cases}$$

which has the solution

$$A = \frac{1}{6}, \quad B = \frac{5}{18}, \quad C = \frac{-35}{108}.$$

The particular solution is

$$y_p = \frac{1}{6}x^2 + \frac{5}{18}x - \frac{35}{108}.$$

The general solution is

$$y = y_h + y_p = C_1 e^{2x} + C_2 e^{3x} + \frac{1}{6}x^2 + \frac{5}{18}x - \frac{35}{108}. \quad \square$$

To apply the method of undetermined coefficients, we must guess the form of the terms in the solution. We start with the terms of $f(x)$—the function on the right-hand side of the differential equation. We must also include all terms that can be obtained from the individual terms of $f(x)$ by calculating derivatives of all orders. This explains why we must include the terms Bx and C in Example 2 even though $f(x)$ does not have corresponding terms.

Suppose, for example, that

$$f(x) = xe^x + \cos x + x^2.$$

The term xe^x causes us to consider the terms xe^x and e^x. The term $\cos x$ causes us to consider the terms $\cos x$ and $\sin x$, the term x^2 causes us to consider the terms x^2, x, and 1. Consequently, we try to find a particular solution of form

First Modification
$$y_p = Axe^x + Be^x + C \cos x + D \sin x + Ex^2 + Fx + G.$$

There is one major exception to the above rule. Let $f_1(x)$ be a term of $f(x)$. If $f_1(x)$ or one of its derivatives is a solution of the related homogeneous equation $y'' + ay' + by = 0$, then we must replace $f_1(x)$ and its derivatives by $xf_1(x)$ and its derivatives. To solve the differential equation

$$y'' + y = x \cos x,$$

for example, we should normally start with $f_1(x) = x \cos x$ and list all its derivatives. Two of the terms of the derivatives are $\cos x$ and $\sin x$, which are solutions of the homogeneous equation. Thus we must start over with the function $x^2 \cos x$ and list all the terms in its derivatives, obtaining the functions

$$x^2 \cos x, \quad x^2 \sin x, \quad x \cos x, \quad x \sin x, \quad \cos x, \quad \text{and} \sin x.$$

Since $\cos x$ and $\sin x$ are part of the solution of the homogeneous equation, they can be deleted from the list. Thus we must consider a particular solution of form

$$y(x) = Ax^2 \cos x + Bx^2 \sin x + Cx \cos x + Dx \sin x.$$

EXAMPLE 3 Solve $y'' + y = x \cos x$.

Solution As indicated above, we try to find a solution of form

$$y_p(x) = Ax^2 \cos x + Bx^2 \sin x + Cx \cos x + Dx \sin x.$$

We calculate

$$y_p'(x) = [-Ax^2 + (2B - C)x + D] \sin x$$
$$+ [Bx^2 + (2A + D)x + C] \cos x$$

and

$$y_p''(x) = [-Ax^2 + (4B - C)x + (2A + 2D)] \cos x$$
$$+ [-Bx^2 - (4A + D)x + (2B - 2C)] \sin x.$$

If we substitute these expressions in the differential equation

$$y_p''(x) + y_p(x) = x \cos x$$

and simplify, we obtain

$$4Bx \cos x - 4Ax \sin x + (2A + 2D) \cos x + (2B - 2C) \sin x = x \cos x.$$

Consequently, we must have

$$4B = 1, \quad -4A = 0$$
$$2A + 2D = 0, \quad 2B - 2C = 0,$$

so that

$$A = 0, \quad B = \tfrac{1}{4}, \quad C = \tfrac{1}{4}, \quad D = 0.$$

The particular solution is

$$y_p(x) = \tfrac{1}{4}x^2 \sin x + \tfrac{1}{4}x \cos x.$$

The general solution is

$$y(x) = C_1 \cos x + C_2 \sin x + \tfrac{1}{4}x^2 \sin x + \tfrac{1}{4}x \cos x. \quad \square$$

Second Modification A similar modification must be made in the form of the particular solution if a term $f_1(x)$ and a term in one of its derivatives are solutions of the homogeneous equation or if two terms in derivatives of different orders are solutions. In that case, we must replace the term $f_1(x)$ and its derivatives by $x^2(f_1(x))$ and its derivatives in our list of terms for $y_p(x)$.

EXAMPLE 4 Solve $y'' - 4y' + 4y = e^{2x} + x + 1$.

Solution The general solution of the related homogeneous equation

$$y'' - 4y' + 4y = 0$$

is

$$y_h = C_1 e^{2x} + C_2 x e^{2x}.$$

Since the term e^{2x} of $f(x) = e^{2x} + x + 1$ is a solution of the homogeneous equation, we consider a term xe^{2x} for the particular solution. Since this term (along with one of its derivatives) is also a solution of the homogeneous equation, we must consider a term x^2e^{2x} for the particular solution. When we calculate the derivatives of x^2e^{2x}, we get terms x^2e^{2x}, xe^{2x} and e^{2x}. The only one of these terms that is not a solution of the homogeneous equation is x^2e^{2x}. Thus the only term needed in the particular solution because of the term e^{2x} in $f(x)$ is Ax^2e^{2x}.

The terms x and 1 lead us to consider first-degree polynomials for the particular solution. Since no polynomial of first degree has a derivative that satisfies the homogeneous equation, we must consider terms Bx and C for the particular solution.

It follows that we should consider a particular solution of form

$$y_p(x) = Ax^2e^{2x} + Bx + C.$$

If we calculate the derivatives, we get

$$y_p'(x) = 2Ax^2e^{2x} + 2Axe^{2x} + B$$
$$y_p''(x) = 4Ax^2e^{2x} + 8Axe^{2x} + 2Ae^{2x}.$$

If we substitute these expressions in the differential equation

$$y'' - 4y' + 4y = e^{2x} + x + 1$$

and simplify, we get

$$2Ae^{2x} + 4Bx + (4C - 4B) = e^{2x} + x + 1.$$

Thus

$$2A = 1, \quad 4B = 1, \quad 4C - 4B = 1$$

so that

$$A = \tfrac{1}{2}, \quad B = \tfrac{1}{4}, \quad C = \tfrac{1}{2}.$$

The particular solution of $y'' - 4y' + 4y = e^{2x} + x + 1$ is

$$y_p(x) = \tfrac{1}{2}x^2e^{2x} + \tfrac{1}{4}x + \tfrac{1}{2}.$$

The general solution is

$$y(x) = y_h(x) + y_p(x) = C_1e^{2x} + C_2xe^{2x} + \tfrac{1}{2}x^2e^{2x} + \tfrac{1}{4}x + \tfrac{1}{2}. \quad \square$$

Remark It is not uncommon to make a mistake when we try to determine the form of $y_p(x)$. If a mistake is made, then the resulting system of equations for $A, B, C, \ldots$ will be inconsistent. For example, we might get a system of equations such as

$$3A = 0, \quad 4B + C = 1, \quad A - C = 2, \quad C = 4,$$

which has no solution. Do not despair if this occurs. Unless there is a computational error, this is a sign that an incorrect form was chosen for $y_p(x)$. Study the differential equation and try to find the correct form.

Exercises 19–6

Use the method of undetermined coefficients to find a particular solution of the differential equation. Then find the general solution.

• 1 $y'' + y' - 6y = e^{-2x}$ 2 $y'' - 9y = 3x^2 + 9x$

- 3 $y'' + 4y' + 4y = \cos x + 8$
- 5 $y'' + y = \sin x + 2x$
- 7 $y'' + 4y' + 4y = 4e^{-2x}$
- 9 $2y'' - y' = 2e^{x/2}$

4 $y'' - 6y' + 9y = 5 \cos x$

6 $y'' + 2y' + 2y = 6e^x + \cos 2x$

8 $y'' - 5y' + 4y = 3e^x + 8x^2$

10 $y'' - 3y' + 2y = 3x^2 e^x$

19-7 Variation of Parameters

It should be apparent from the work in Section 19–6 that the method of undetermined coefficients can be applied only in special cases to find a particular solution of a second-order linear differential equation with constant coefficients. An alternative method can always be applied—the method of *variation of parameters*. Unfortunately, this alternative method usually requires us to integrate some complicated functions.

To develop this method, we suppose that $u_1(x)$ and $u_2(x)$ are linearly independent solutions of the homogeneous differential equation

$$y'' + ay' + by = 0.$$

Suppose that the nonhomogeneous equation

$$y'' + ay' + by = f(x)$$

has a particular solution of form

$$y_p = v_1(x)u_1(x) + v_2(x)u_2(x),$$

where v_1 and v_2 are unknown functions of x. [We get the expression for y_p by replacing C_1 and C_2 by $v_1(x)$ and $v_2(x)$ in the general solution of the homogeneous equation. Hence the name "variation of parameters."]

Thus far we have only one condition that $v_1(x)$ and $v_2(x)$ must satisfy. We now add the additional restriction that

$$v_1'(x)\, u_1(x) + v_2'(x)\, u_2(x) = 0.$$

This restriction will simplify the calculations that must be made in the derivation.

The first derivative of y_p is

$$\begin{aligned} y_p' &= v_1 u_1' + v_1' u_1 + v_2 u_2' + v_2' u_2 \\ &= v_1 u_1' + v_2 u_2' \end{aligned}$$

(since we assumed that $v_1' u_1 + v_2' u_2 = 0$). The second derivative is

$$y_p'' = v_1 u_1'' + v_2 u_2'' + v_1' u_1' + v_2' u_2'.$$

If we substitute these expressions in the differential equation, we get

$$y_p'' + ay_p' + by_p = f(x),$$

$$[v_1 u_1'' + v_2 u_2'' + v_1' u_1' + v_2' u_2'] + a[v_1 u_1' + v_2 u_2'] + b[v_1 u_1 + v_2 u_2] = f(x),$$

19–7 Variation of Parameters

$$v_1[u_1'' + au_1' + bu_1]$$
$$+ v_2[u_2'' + au_2' + bu_2] + [v_1'u_1' + v_2'u_2'] = f(x).$$

Since u_1 and u_2 are solutions of the homogeneous equation, it follows that the first two bracketed terms equal zero. Thus, the equation reduces to

$$v_1'u_1' + v_2'u_2' = f(x).$$

Recall that we also have the restriction $v_1'u_1 + v_2'u_2 = 0$. Thus, v_1' and v_2' satisfy the system of equations

$$\begin{cases} v_1'u_1 + v_2'u_2 = 0 \\ v_1'u_1' + v_2'u_2' = f(x). \end{cases}$$

This system can be solved algebraically for v_1' and v_2', yielding

$$v_1' = \frac{-u_2 f(x)}{u_1 u_2' - u_1' u_2}, \qquad v_2' = \frac{u_1 f(x)}{u_1 u_2' - u_1' u_2}.$$

Finally, we integrate to get v_1 and v_2.

Variation of Parameters Formulas

$$v_1 = \int \frac{-u_2 f(x)\, dx}{u_1 u_2' - u_1' u_2}, \qquad v_2 = \int \frac{u_1 f(x)\, dx}{u_1 u_2' - u_1' u_2}.$$

EXAMPLE 1 Use variation of parameters to solve

$$y'' + y = \cos x.$$

Solution The general solution of the related homogeneous equation is

$$y_h = C_1 \cos x + C_2 \sin x.$$

Since $\cos x$ and $\sin x$ are linearly independent, we take

$$u_1 = \cos x \quad \text{and} \quad u_2 = \sin x.$$

Then

$$v_1 = \int \frac{-u_2 f(x)\, dx}{u_1 u_2' - u_1' u_2} = \int \frac{-\sin x \cdot \cos x\, dx}{\cos^2 x - (-\sin^2 x)}$$

$$= \int (-\sin x) \cos x\, dx = -\frac{\sin^2 x}{2}$$

$$v_2 = \int \frac{u_1 f(x)\, dx}{u_1 u_2' - u_1' u_2} = \int \frac{\cos x \cdot \cos x\, dx}{\cos^2 x - (-\sin^2 x)}$$

$$= \int \cos^2 x\, dx$$

$$= \frac{x}{2} + \frac{\sin 2x}{4} = \frac{x}{2} + \frac{\sin x \cos x}{2}.$$

(Since we need only one solution v_1 and one solution v_2, we omit the constants of integration.)

The particular solution is

$$y_p = v_1 \cos x + v_2 \sin x$$
$$= -\frac{\sin^2 x}{2} \cos x + \left(\frac{x}{2} + \frac{\sin x \cos x}{2}\right) \sin x$$
$$= \frac{x}{2} \sin x.$$

The general solution is

$$y = y_h + y_p = C_1 \cos x + C_2 \sin x + \frac{x}{2} \sin x. \quad \square$$

EXAMPLE 2 Solve $y'' - 2y' + y = x^2 e^x$.

Solution The general solution of the related homogeneous equation $y'' - 2y' + y = 0$ is

$$y_h = C_1 e^x + C_2 x e^x.$$

Thus, we take

$$u_1(x) = e^x \quad \text{and} \quad u_2(x) = xe^x.$$

Then

$$v_1 = \int \frac{-u_2 f(x)\,dx}{u_1 u_2' - u_1' u_2} = -\int \frac{x^3 e^{2x} dx}{e^{2x}} = -\int x^3 dx = -\frac{x^4}{4}$$

and

$$v_2 = \int \frac{u_1 f(x)\,dx}{u_1 u_2' - u_1' u_2} = \int \frac{x^2 e^{2x} dx}{e^{2x}} = \int x^2 dx = \frac{x^3}{3}.$$

A particular solution of the nonhomogeneous differential equation is

$$y_p = u_1 v_1 + u_2 v_2 = e^x \left(-\frac{x^4}{4}\right) + xe^x \left(\frac{x^3}{3}\right) = \frac{x^4 e^x}{12}.$$

The general solution is

$$y = y_h + y_p = C_1 e^x + C_2 x e^x + \frac{x^4 e^x}{12}. \quad \square$$

Exercises 19-7

1-6. Use variation of parameters to solve the differential equation.

- 1 $y'' + 9y = \tan 3x$
- 2 $y'' - 6y' + 9y = x^2 e^{3x}$
- 3 $y'' - y' = e^x \sin x$
- 4 $y'' - y = e^x \sin x$
- 5 $y'' + y = \csc x$
- 6 $y'' - 4y' + 4y = e^{2x} \ln x$

7 Work through the steps of the derivation of the variation of parameters formula with the following differential equations.
(a) The equation in Exercise 2
(b) The equation in Exercise 5

8 Show that the method of variation of parameters can be used to solve differential equations of form

$$y'' + a(x)y' + b(x)y = f(x).$$

• **9** It follows from Exercise 13(c) of Section 19–4 that $u_1 = 1/x$ and $u_2 = e^x(x-1)/x$ are linearly independent solutions of the differential equation

$$x^2 y'' + (x - x^2)y' - (x+1)y = 0.$$

Use this result and Exercise 8 to solve the differential equation

$$x^2 y'' + (x - x^2)y' - (x+1)y = x^2 e^x.$$

(*Hint:* First divide the equation by x^2.)

10 A second variation-of-parameters method can be used to find a second particular solution of a homogeneous second-order linear differential equation provided one particular solution can be found. Let $u_1(x)$ be a particular solution of

$$a(x)y'' + b(x)y' + c(x)y = 0.$$

Let $u_2(x) = v(x)u_1(x)$. Show that u_2 is a solution of the differential equation if and only if

Variation of Parameters
$$v''(x) + \left[\frac{2u_1'(x)}{u_1(x)} + \frac{b(x)}{a(x)}\right] v'(x) = 0.$$

• **11** The differential equation $x^2 y'' + 3xy' + y = 0$ has a solution of form $y = x^k$. Find it. Then use Exercise 10 to find the general solution.

19–8 Series Solutions

If a function $f(x)$ has a power series expansion

$$f(x) = a_0 + a_1 x + a_2 x^2 + a_3 x^3 + \cdots + a_n x^n + \cdots,$$

then it can be differentiated term by term with the result

$$f'(x) = a_1 + 2a_2 x + 3a_3 x^2 + 4a_4 x^3 + \cdots + (n+1)a_{n+1} x^n + \cdots,$$

$$f''(x) = 2a_2 + 6a_3 x + 12a_4 x^2 + 20a_5 x^3 + \cdots + (n+2)(n+1) a_{n+2} x^n + \cdots,$$

and so on. The series for the derivatives have the same radii of convergence as the original series. (See Theorem 12–20.)

In some cases, the best way to solve a differential equation is by a power series. We assume that the solution can be written as a power series, substitute it in the differential equation, simplify the resulting expression, and determine the values of the coefficients $a_0, a_1, a_2, \ldots$. We must then find the interval of convergence of the series. The resulting power series represents a solution of the differential equation on its interval of convergence.

As our first illustration, we solve the differential equation $y' = 2y$. (This particular example was chosen because of its simplicity. It can be solved by other methods with less work.)

EXAMPLE 1 Use the power series method to solve $y' = 2y$.

Solution Assume that the solution is

$$y = a_0 + a_1 x + a_2 x^2 + a_3 x^3 + \cdots + a_n x^n + \cdots,$$

where $a_0, a_1, a_2, \ldots$ are to be determined. Then

$$y' = a_1 + 2a_2 x + 3a_3 x^2 + \cdots + (n+1) a_{n+1} x^n + \cdots.$$

If we substitute the series for y and y' in the differential equation $y' = 2y$, we get

$$a_1 + 2a_2 x + 3a_3 x^2 + \cdots + (n+1) a_{n+1} x^n + \cdots$$
$$= 2a_0 + 2a_1 x + 2a_2 x^2 + \cdots + 2a_n x^n + \cdots.$$

Equating coefficients of like powers of x, we get

$$a_1 = 2a_0, \quad 2a_2 = 2a_1, \quad 3a_3 = 2a_2, \quad \ldots, \quad (n+1)a_{n+1} = 2a_n, \quad \ldots.$$

It follows that

$$a_1 = 2a_0 = \frac{2a_0}{1!},$$

$$a_2 = a_1 = 2a_0 = \frac{2^2 a_0}{2!},$$

$$a_3 = \frac{2}{3} a_2 = \frac{2}{3} \cdot \frac{2^2 a_0}{2!} = \frac{2^3 a_0}{3!},$$

and so on. In general, we find that

$$a_n = \frac{2^n a_0}{n!} \quad (n = 1, 2, 3, \ldots).$$

The solution is

$$y = a_0 + a_1 x + a_2 x^2 + a_3 x^3 + \cdots + a_n x^n + \cdots$$

$$= a_0 + \frac{2a_0}{1!} x + \frac{2^2 a_0}{2!} x^2 + \frac{2^3 a_0}{3!} x^3 + \cdots + \frac{2^n a_0}{n!} x^n + \cdots$$

$$= a_0 \left(1 + \frac{(2x)}{1!} + \frac{(2x)^2}{2!} + \frac{(2x)^3}{3!} + \cdots + \frac{(2x)^n}{n!} + \cdots \right)$$

$$= a_0 e^{2x}.$$

Note that the series converges for all x. ☐

Our second example is more interesting. It involves a differential equation that cannot be solved by the other methods developed in this chapter.

EXAMPLE 2 Find a series solution of $xy'' + y' + xy = 0$.

Solution Let $y = a_0 + a_1 x + a_2 x^2 + a_3 x^3 + \cdots$. Then

$$y' = a_1 + 2a_2 x + 3a_3 x^2 + 4a_4 x^3 + \cdots,$$
$$y'' = 2a_2 + 2 \cdot 3a_3 x + 3 \cdot 4 a_4 x^2 + 4 \cdot 5 a_5 x^3 + \cdots.$$

19-8 Series Solutions

For convenience we multiply the series for y'' and y by x before substituting in the differential equation:

$$xy'' = 2a_2x + 2\cdot 3a_3x^2 + 3\cdot 4a_4x^3 + 4\cdot 5a_5x^4 + \cdots,$$
$$xy = a_0x + a_1x^2 + a_2x^3 + a_3x^4 + \cdots.$$

If we substitute these series in the differential equation

$$xy'' + y' + xy = 0,$$

we get

$$\begin{aligned}&2a_2x + 2\cdot 3a_3x^2 + 3\cdot 4a_4x^3 + 4\cdot 5a_5x^4 + \cdots\\&+\ a_1 + 2a_2x + 3a_3x^2 + 4a_4x^3 + 5a_5x^4 + \cdots\\&+\ a_0x + a_1x^2 + a_2x^3 + a_3x^4 + \cdots = 0\end{aligned}$$

When like powers of x are added, this expression reduces to

$$a_1 + (2^2 a_2 + a_0)x + (3^2 a_3 + a_1)x^2 + (4^2 a_4 + a_2)x^3 + \cdots = 0.$$

If we now set each of these coefficients equal to zero, we find that

$$a_1 = 0,$$

$$a_2 = -\frac{a_0}{2^2} = -\frac{a_0}{2^2(1!)^2},$$

$$a_3 = -\frac{a_1}{3^2} = 0,$$

$$a_4 = -\frac{a_2}{4^2} = -\frac{1}{4^2}\left(-\frac{a_0}{2^2}\right) = \frac{a_0}{2^2\cdot 4^2} = \frac{a_0}{(2\cdot 1)^2(2\cdot 2)^2} = \frac{a_0}{2^4(2!)^2}.$$

The next few terms are

$$a_5 = 0, \quad a_6 = -\frac{a_0}{2^6(3!)^2}, \quad a_7 = 0, \quad a_8 = \frac{a_0}{2^8(4!)^2}.$$

In general, we find that

$$a_n = \begin{cases} 0 & \text{if } n \text{ is odd} \\ \dfrac{(-1)^{n/2}a_0}{2^n((n/2)!)^2} & \text{if } n \text{ is even.} \end{cases}$$

The solution of the differential equation is

$$\begin{aligned}y &= a_0 + a_1x + a_2x^2 + a_3x^3 + a_4x^4 + \cdots\\&= a_0\left(1 - \frac{x^2}{2^2(1!)^2} + \frac{x^4}{2^4(2!)^2} - \frac{x^6}{2^6(3!)^2} + \frac{x^8}{2^8(4!)^2} - \cdots\right).\end{aligned}$$

It can be proved that this series converges for all x. □

The series

Bessel Function

$$1 - \frac{x^2}{2^2(1!)^2} + \frac{x^4}{2^4(2!)^2} - \frac{x^6}{2^6(3!)^2} + \cdots$$

obtained in Example 2 defines a new function, denoted by $J_0(x)$, called a *Bessel function*. The graph of $J_0(x)$ is similar to the graph obtained in damped oscillations. (See Fig. 19–5.) The zero subscript in $J_0(x)$ is used to distinguish this function from several other Bessel functions.

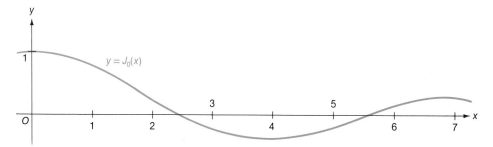

Figure 19–5 Example 2. The Bessel function:

$$J_0(x) = 1 - \frac{x^2}{2^2(1!)^2} + \frac{x^4}{2^4(2!)^2} - \frac{x^6}{2^6(3!)^2} + \cdots.$$

Note that our series solution of $xy'' + y' + xy = 0$ involves only one arbitrary constant, while the general solution must involve two arbitrary constants. It follows that we have not found the general solution. The general solution is of the form

$$y = C_1 J_0(x) + C_2 Y_0(x)$$

where $Y_0(x)$, another Bessel function, also is a particular solution of the differential equation. The function $Y_0(x)$, however, cannot be expressed as a Maclaurin series.

Exercises 19–8

1–6. Find a series solution of the differential equation

- •1 $y' + y = 0$
- 2 $y'' - y = 0$
- •3 $y'' + xy' - y = 0$
- 4 $xy'' + xy' + 2y = 0$
- •5 $x^2 y'' + xy' + (x^2 - 1)y = 0$
- 6 $xy'' + xy' + (x - 2)y = 0$ (first five terms only)

•7 Find a particular solution of the differential equation in Exercise 3 that satisfies the initial conditions $f(0) = 2$, $f'(0) = 2$.

8 (a) Find the general solution of $xy' + 2y = 0$.
   (b) Find a series solution of $xy' + 2y = \cos(x^2)$.
   •(c) Find the general solution of the differential equation in (b).

•9 Use series to solve the differential equation

$$y'' - 2xy' - 2y = 0.$$

Write the general solution in the form $C_1 u_1 + C_2 u_2$, where u_1 and u_2 are written as series.

19-9 Approximate Solutions

10 The methods of this section depend on the fact that if $\sum_{n=0}^{\infty} a_n x^n$ and $\sum_{n=0}^{\infty} b_n x^n$ converge to the same function on their common interval of convergence, then

$$a_n = b_n \quad (n = 0, 1, 2, \ldots).$$

Prove that the result is valid provided each series has a nonzero radius of convergence. (*Hint:* Use Theorem 12-20. Evaluate the derivatives at $x = 0$.)

19-9 Approximate Solutions

As scientists and engineers try to solve basic problems in their fields, they find that more and more of the problems reduce to differential equations that cannot be solved by elementary methods. For this reason, many methods have been developed that give approximate solutions.

In a typical example we are given a differential equation of form

$$y' = f(x, y)$$

and a boundary condition that the solution $y = \phi(x)$ must satisfy—say $\phi(a) = y_0$. We are required to find the value of the solution function ϕ at some point b.

One approximation method was developed by Euler. We have chosen it because it can be easily understood and because its application is simple. Unfortunately, it usually produces approximations that are inaccurate compared with results from other methods. Euler's method can be modified to give better results, but we do not consider the modifications.

EULER'S METHOD

Let a solution $y = \phi(x)$ of the differential equation

$$y' = f(x, y)$$

satisfy the boundary condition $\phi(a) = y_0$. Let b be a number different from a. We want to approximate $\phi(b)$. For purposes of this discussion, we assume that $b > a$. The argument is similar if $b < a$.

We subdivide the interval $[a, b]$ into n equal subintervals of length $\Delta x = (b - a)/n$ by the points

$$a = x_0 < x_1 < x_2 < \cdots < x_n = b,$$

where

$$x_0 = a, \quad x_1 = a + \Delta x, \quad x_2 = a + 2\Delta x, \quad \cdots, \quad x_n = a + n\,\Delta x = b.$$

For simplicity of notation, we let $y_k = \phi(x_k)$, $k = 0, 1, 2, \ldots, n$.

Since the solution of the differential equation that satisfies the initial condition is $y = \phi(x)$, then the differential equation can be written as

$$\phi'(x) = f(x, \phi(x)).$$

Then

$$\int_{x_{k-1}}^{x_k} f(x, \phi(x))\, dx = \int_{x_{k-1}}^{x_k} \phi'(x)\, dx = \Big[\phi(x)\Big]_{x_{k-1}}^{x_k} = \phi(x_k) - \phi(x_{k-1}) = y_k - y_{k-1},$$

so that

$$y_k = y_{k-1} + \int_{x_{k-1}}^{x_k} f(x, \phi(x))\, dx.$$

If ϕ and f do not vary much over the interval $[x_{k-1}, x_k]$, then

$$f(x, \phi(x)) \approx f(x_{k-1}, \phi(x_{k-1})) = f(x_{k-1}, y_{k-1})$$

when x is between x_{k-1} and x_k. In that case,

$$\begin{aligned} y_k &= y_{k-1} + \int_{x_{k-1}}^{x_k} f(x, \phi(x))\, dx \\ &\approx y_{k-1} + \int_{x_{k-1}}^{x_k} f(x_{k-1}, y_{k-1})\, dx \\ &= y_{k-1} + f(x_{k-1}, y_{k-1})(x_k - x_{k-1}), \end{aligned}$$

so that

Formula for Euler's Method

$$y_k \approx y_{k-1} + f(x_{k-1}, y_{k-1})\, \Delta x.$$

This approximation formula is illustrated in Figure 19–6. In (a), we start at the point (x_0, y_0) on the graph of ϕ. The tangent line to the graph of ϕ at (x_0, y_0) has slope $\phi'(x_0) = f(x_0, y_0)$. The value $y_1 = \phi(x_0 + \Delta x)$ is approximately equal to $y_0 + f(x_0, y_0)\, \Delta x$, the value obtained by moving along the tangent line instead of the graph of ϕ. Observe that there is a small error in this approximation. The result of calculating several values of y in succession is shown in Figure 19–6b. Observe that the errors tend to accumulate, causing a large total error.

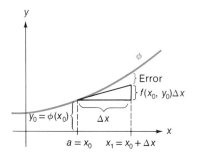

(a) The first approximation

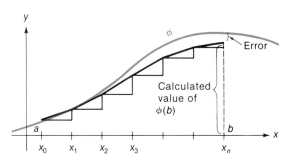

(b) The nth approximation

Figure 19–6 Euler's method for the approximate value of a solution of a differential equation $y' = f(x, y)$.

19-9 Approximate Solutions

To apply Euler's method, we start with the known value of y_0 (given in the boundary condition) and compute $y_1, y_2, \ldots, y_n$ in succession.

EXAMPLE 1   Let $y = \phi(x)$ be the solution of the differential equation

$$y' = x^2 + y$$

that satisfies the boundary condition $\phi(0) = 1$.
(a) Use Euler's method with $\Delta x = 0.1$ to calculate the approximate value of $\phi(1)$.
(b) Compare the result of (a) with the exact value of $\phi(1)$.

Solution (a) Since $\Delta x = 0.1$, then the points $x_0 = 0, x_1, x_2, \ldots, x_{10}$ partition $[0, 1]$ into ten equal subintervals of length 0.1. Therefore,

$$x_k = k\,\Delta x = k(0.1)$$

and

$$y_k = \phi(x_k) = \phi(k\,\Delta x) \quad (k = 1, 2, 3, \ldots, 10).$$

Since $f(x, y) = x^2 + y$ and $\Delta x = 0.1$, then the approximation formula

$$y_k \approx y_{k-1} + f(x_{k-1}, y_{k-1})\,\Delta x$$

reduces to

$$y_k \approx y_{k-1} + (x_{k-1}^2 + y_{k-1})\,\Delta x$$
$$\approx (1 + \Delta x)y_{k-1} + (k-1)^2\,\Delta x^3$$

$$\boxed{y_k \approx (1.1)y_{k-1} + (k-1)^2\,(0.1)^3.}$$

Since $y_0 = 1$, then

$$y_1 \approx (1.1)\cdot 1 + 0^2(0.1)^3 = 1.1 \qquad (k = 1),$$
$$y_2 \approx (1.1)(1.1) + 1^2\cdot(0.1)^3 = 1.211 \qquad (k = 2),$$
$$y_3 \approx (1.1)(1.211) + 2^2\,(0.1)^3 = 1.3361 \qquad (k = 3).$$

If we continue in this fashion, we find that

$$\phi(1) = y_{10} \approx 2.9406.$$

(b) The general solution of $y' = x^2 + y$ (obtained by the methods of Section 19–3) is

$$y = Ce^x - x^2 - 2x - 2.$$

The particular solution ϕ that satisfies the boundary condition $\phi(0) = 1$ is

$$\phi(x) = 3e^x - x^2 - 2x - 2.$$

The approximate value of $\phi(1)$ is

$$\phi(1) = 3e - 1^2 - 2 - 2 = 3e - 5 \approx 3.154845.$$

The error in the approximation obtained in (a) is

$$\text{Error} \approx 0.214245,$$

which is about 7 percent of the true value. □

Exercises 19–9

1–6 (a) Let $\phi(x)$ be the solution that satisfies the given initial condition. Use Euler's method with the given value of Δx to find the approximate value of $\phi(b)$.

c(b) (*For students with programmable calculators or access to a computer*) Use Euler's method with $\Delta x = 0.01$.

● 1 $y' = y/x$; $\phi(1) = 2$, $b = 2$, $\Delta x = 0.2$

● 2 $(x + 3y)\, dx + x\, dy = 0$; $\phi(1) = 0$, $b = 2$, $\Delta x = 0.1$

● 3 $y' = xy + x$; $\phi(0) = 2$, $b = 1$, $\Delta x = 0.2$

● 4 $(x + 1)y' = x - 2y$, $\phi(0) = 1$, $b = 0.5$, $\Delta x = 0.1$

● 5 $y' = x^2 + \cos(xy)$, $\phi(0) = 0$, $b = 1$, $\Delta x = 0.1$

● 6 $y' = \sqrt{x} + \ln y$, $\phi(1) = 1$, $b = 3$, $\Delta x = 0.2$

7–10. Find the exact solutions of the differential equations in Exercises 1–4. Calculate the approximate percentage error in the answer obtained by Euler's method with the given value of Δx.

● 7 Exercise 1

● 8 Exercise 2

● 9 Exercise 3

● 10 Exercise 4

11 Make a flowchart for the steps in Euler's method.

Review Problems

1 (a) What is meant by a *solution* of a first-order differential equation? a *particular solution*? the *general solution*? a *singular solution*?

(b) Illustrate these concepts with examples.

2 Classify the differential equations in Problems 3–30 according to the method of solution that seems most likely to succeed (that is, separable variables, homogeneous, exact, linear, and so on).

3–30. Solve the differential equations.

● 3 $y' = xy + 2x$

● 4 $\left(y - \dfrac{xy}{\sqrt{x^2 + 1}}\right) dx + (x - \sqrt{x^2 + 1})\, dy = 0$

● 5 $e^x\, dy + (ye^x + x^3)\, dx = 0$

● 6 $y\, y' = \sqrt{x} + x$

● 7 $(4x^2 + 4y)\, dx + (4x - 2y)\, dy = 0$

● 8 $xy' = 6xe^{2x} + y(2x - 1)$

Review Problems

- 9 $y' - y = 3x^2 e^x$
- 10 $\sin\left(\dfrac{y}{x}\right) + \dfrac{y}{x} = y'$
- 11 $(y^2 + xy)\, dx - x^2\, dy = 0$
- 12 $y' - y + x = 5$
- 13 $x \cot y - y' \csc x = 0$
- 14 $2(y - 4x^2)\, dx + x\, dy = 0$
- 15 $(x + 1)y\, dx = (x^2 - 4)\, dy$
- 16 $y'' - 4y' + 5y = 0$
- 17 $x^2 y' + 2x^2 = y^2$
- 18 $(x^2 + y e^{xy})\, dx + (\sin y + x e^{xy})\, dy = 0$
- 19 $6y'' = y' + y$
- 20 $y'' - 3y' = 10y + 2$
- 21 $(x e^{xy} + 2x^2) y' + (y e^{xy} + 4xy + 3) = 0$
- 22 $xy' + y - x^2 \ln x = 0$
- 23 $y' + y = e^{3x} + e^{-2x} + 1$
- 24 $\sqrt{9 - x^2}\, y' + \sec^2 y = 0$
- 25 $4y'' + y' = e^x + x + 1$
- 26 $x^2\, dy = (y^2 + xy + x^2)\, dx$
- 27 $x e^y y' = e^y + 1$
- 28 $y' \cos x + y \sin x = \cos^3 x \sin^2 x$
- 29 $y'' + 49y = e^{x+1}$
- 30 $(2xy^3 + e^x)\, dx + (3x^2 y^2 + e^{-y})\, dy = 0$

31 (a) Prove that $M = e^{\int f(x)\, dx}$ is an integrating factor for the equation
$$M(x, y)\, dx + N(x, y)\, dy = 0$$
if and only if
$$f(x) = \frac{\partial M/\partial y - \partial N/\partial x}{N}.$$

- (b) Use the result of (a) to solve the differential equation
$$(x + \sqrt{y})\, dx + \frac{x}{\sqrt{y}}\, dy = 0.$$

32 Derive the formula for the general solution of the first-order linear differential equation $y' + p(x) y = q(x)$.

33 (a) What is meant by a *second-order linear differential equation*? a *homogeneous* second-order linear differential equation?

(b) Let $u_1(x)$ and $u_2(x)$ be solutions of a homogeneous second-order linear differential equation. Prove that
$$y = C_1 u_1(x) + C_2 u_2(x)$$
is a solution for any constants C_1 and C_2. Under what conditions is this the general solution?

34 (a) Show that $(D - 2)(D - 3) = D^2 - 5D + 6$.

(b) Show that $(D - x)(D + x) \neq D^2 - x^2$.

•(c) What operator of form $D^2 + a(x) D + b(x)$ is equal to $(D - x)(D + x)$?

35–46 (a) Solve the differential equations. Use the method of undetermined coefficients for 41–46.

(b) Find the particular solution $f(x)$ of the differential equations that satisfies the conditions $f(0) = 1$, $f'(0) = -2$.

•**35** $(D^2 - 3D + 2)y = 0$

•**36** $(D^2 + 6D + 9)y = 0$

•**37** $(D^2 + 3)y = 0$

•**38** $(D^2 + 5D + 6)y = 0$

•**39** $y'' + 4y' + 5y = 0$

•**40** $y'' + 16y = 0$

•**41** $y'' + 2y' - 3y = e^x$

•**42** $8y'' - y' = 13 \sin x + 2x$

•**43** $y'' - 6y' + 16y = -2 \cos x + 5 \sin x$ •**44** $y'' + 4y = \sin 2x + \sinh 2x$

•**45** $y'' - 4y' + 4y = e^{2x}$

•**46** $y'' + y = x \sin x + \cos x$

47–49. Use variation of parameters to solve the differential equation.

47 Problem 41

48 Problem 45

49 Problem 46

50–51. Find series solutions of these differential equations.

•**50** $y' - 4y = 0$

•**51** $x^2 y'' + y' - y = 0$
(First five terms only)

•**52** Let $\phi(x)$ be the solution of the differential equation

$$y' = x + e^{-y}$$

that satisfies the condition $\phi(0) = 1$. Use Euler's approximation method with $\Delta x = 0.1$ to find the approximate value of $\phi(1)$.

APPENDIX A

Mathematical Induction

Many theorems in mathematics are concerned with propositions about positive integers. Let $\mathcal{P}_n$ represent a proposition about the positive integer n. Then $\mathcal{P}_1$ is the equivalent proposition about the integer 1, $\mathcal{P}_2$ is the proposition about the integer 2, and so on.

Suppose, for example, that $\mathcal{P}_n$ is the proposition that

$$(ab)^n = a^n b^n$$

for every positive integer n. Then $\mathcal{P}_1$ is the proposition that

$$(ab)^1 = a^1 b^1,$$

$\mathcal{P}_2$ is the proposition that

$$(ab)^2 = a^2 b^2,$$

and so on.

Many theorems about positive integers can be proved by a powerful method known as *mathematical induction*. This method is based on the following axiom:

AXIOM OF MATHEMATICAL INDUCTION

Let $\mathcal{P}_n$ be a proposition about the positive integer n. Suppose that

(1) $\mathcal{P}_1$ is true, and
(2) $\mathcal{P}_{k+1}$ is true whenever $\mathcal{P}_k$ is true.

Then $\mathcal{P}_n$ is true for every positive integer n.

The axiom of mathematical induction is illustrated in the following diagram:

$$\circled{\mathcal{P}_1} \to \mathcal{P}_2 \to \mathcal{P}_3 \to \mathcal{P}_4 \to \mathcal{P}_5 \cdots$$

A proposition $\mathcal{P}_k$ is circled whenever it is known to be true. Statement (1) of the axiom establishes that $\mathcal{P}_1$ is true, which gives us a starting point. Statement (2) and the truth of $\mathcal{P}_1$ establish that $\mathcal{P}_2$ is true, Statement (2) and the truth of $\mathcal{P}_2$ establish that $\mathcal{P}_3$ is true, and so on. Since this line of reasoning can be continued forever, it is apparent that $\mathcal{P}_n$ is true for every positive integer n.

To prove a theorem by mathematical induction, we must prove that conditions (1) and (2) of the axiom are satisfied for the particular proposition with which we are concerned. Essentially this means that we must prove the following auxiliary theorems.

AUXILIARY THEOREM 1 (*Fundamental step*) Prove that $\mathcal{P}_1$ is true.

AUXILIARY THEOREM 2 (*Inductive Step*) Prove that $\mathcal{P}_{k+1}$ is true whenever $\mathcal{P}_k$ is true.

It must be emphasized that these two auxiliary theorems are independent of each other and have distinct proofs.

In most cases it is simple to prove that $\mathcal{P}_1$ is true (the fundamental step). To prove the inductive step, we usually assume that the proposition is true for the integer k and then prove that by this assumption, it must be true for the integer $k + 1$.

EXAMPLE 1 Prove that $2^n \geq n + 1$ for every positive integer n.

Solution Let $\mathcal{P}_n$ be the proposition that
$$2^n \geq n + 1.$$

Fundamental Step Prove that $\mathcal{P}_1$ is true.
Proof Since $2^1 = 1 + 1$, then $\mathcal{P}_1$ is true.

Inductive Step Prove that $\mathcal{P}_{k+1}$ is true whenever $\mathcal{P}_k$ is true.
Proof Assume that $\mathcal{P}_k$ is true. Then
$$2^k \geq k + 1.$$
It follows that
$$2^{k+1} = 2 \cdot 2^k \geq 2(k + 1) = 2k + 2 \geq k + 2,$$
so that
$$2^{k+1} \geq (k + 1) + 1.$$
Therefore, $\mathcal{P}_{k+1}$ is true.

Since the fundamental and inductive steps have both been proved, it follows that $\mathcal{P}_n$ is true for every positive integer n. Therefore,
$$2^n \geq n + 1 \quad \text{for every positive integer } n. \quad \square$$

EXAMPLE 2 Let a be a real number. Define positive integral powers of a by
$$a^1 = a$$
and
$$a^{k+1} = a^k \cdot a \quad \text{if } k \geq 1.$$
Prove that a^n is defined for every positive integer n.

Solution Let $\mathcal{P}_n$ be the proposition
$$\mathcal{P}_n: a^n \text{ is defined for the integer } n.$$

Fundamental Step Since $a^1 = a$, then a^1 is defined. Therefore, $\mathcal{P}_1$ is true.

Inductive Step Assume that $\mathcal{P}_k$ is true. Then a^k is defined. Since $a^{k+1} = a^k \cdot a$, it follows that a^{k+1} is defined. Therefore $\mathcal{P}_{k+1}$ is true whenever $\mathcal{P}_k$ is true.

It follows from the axiom of mathematical induction that $\mathcal{P}_n$ is true for every positive integer n. Therefore, a^n is defined for every positive integer n. □

EXAMPLE 3   Let a and b be real numbers. Prove that

$$(ab)^n = a^n b^n$$

for every positive integer n.

Solution   Let $\mathcal{P}_n$ be the proposition

$$\mathcal{P}_n: (ab)^n = a^n b^n.$$

Fundamental Step   Since $(ab)^1 = (ab) = a^1 b^1$, then $\mathcal{P}_1$ is true.

Inductive Step   Assume that $(ab)^k = a^k b^k$. Then

$$\begin{aligned}(ab)^{k+1} &= (ab)^k(ab) = a^k b^k \cdot a \cdot b \\ &= (a^k \cdot a)(b^k \cdot b) = a^{k+1} b^{k+1}.\end{aligned}$$

Therefore, $\mathcal{P}_{k+1}$ is true whenever $\mathcal{P}_k$ is true.

It follows from the above two steps and the axiom of mathematical induction that $\mathcal{P}_n$ is true for every positive integer n. Therefore,

$$(ab)^n = a^n b^n$$

for every positive integer n. □

EXAMPLE 4   Let $\lim_{x \to a} f(x) = L$. Use L2 to prove that

$$\lim_{x \to a} [f(x)]^n = L^n$$

for every positive integer n.

Solution   Let $\mathcal{P}_n$ be the proposition that

$$\lim_{x \to a} [f(x)]^n = L^n.$$

It follows from the hypothesis that

$$\lim_{x \to a} [f(x)]^1 = \lim_{x \to a} f(x) = L = L^1,$$

so that $\mathcal{P}_1$ is true.

Suppose that $\lim_{x \to a} [f(x)]^k = L^k$. Then

$$\lim_{x \to a} [f(x)]^{k+1} = \lim_{x \to a} [f(x)^k \cdot f(x)]$$

$$= \lim_{x \to a} [f(x)]^k \cdot \lim_{x \to a} f(x) \quad \text{(by L2)}$$

$$= L^k \cdot L = L^{k+1}.$$

Therefore, $\mathcal{P}_{k+1}$ is true whenever $\mathcal{P}_k$ is true.

It follows from the axiom of mathematical induction that $\mathcal{P}_n$ is true for every positive integer n. Therefore,

$$\lim_{x \to a} [f(x)]^n = L^n$$

for every positive integer n. □

EXAMPLE 5 Prove that $\sum_{j=1}^{n} j = \frac{n(n+1)}{2}$.

Solution Let $\mathcal{P}_n$ be the proposition that

$$\sum_{j=1}^{n} j = \frac{n(n+1)}{2}.$$

Since

$$\sum_{j=1}^{1} j = 1 \quad \text{and} \quad \frac{1(1+1)}{2} = 1,$$

then $\mathcal{P}_1$ is true.
  Suppose that

$$\sum_{j=1}^{k} j = \frac{k(k+1)}{2}.$$

Then

$$\sum_{j=1}^{k+1} j = \sum_{j=1}^{k} j + (k+1) = \frac{k(k+1)}{2} + (k+1)$$

$$= \frac{k(k+1)}{2} + \frac{2(k+1)}{2} = \frac{(k+1)(k+2)}{2}$$

$$= \frac{(k+1)(k+1+1)}{2}.$$

Thus, $\mathcal{P}_{k+1}$ is true whenever $\mathcal{P}_k$ is true.
  It follows from the axiom of mathematical induction that $\mathcal{P}_n$ is true for every positive integer n. Therefore,

$$\sum_{j=1}^{n} j = \frac{n(n+1)}{2}$$

for every positive integer n. □

Exercises. Appendix A

1 Each of the following propositions about the positive integer n may be true or false. Write out a careful statement of the two auxiliary theorems that would be needed to prove the proposition by mathematical induction. Then decide to your own satisfaction whether the proposition is true or false. If it is false, decide which of the auxiliary theorems is not true. The first auxiliary theorem is the "fundamental step"; the second is the "inductive step."

Exercises. Appendix A

(a) $\mathcal{P}_n: \sum_{k=1}^{n} k^2 = \dfrac{n(n+1)(2n+1)}{6} + (n-1)(n-2).$

(b) $\mathcal{P}_n: \sum_{k=1}^{n} k^2 = \dfrac{n(n+1)(2n+1)}{6} + 3.$

(c) $\dfrac{1}{1 \cdot 2} + \dfrac{1}{2 \cdot 3} + \dfrac{1}{3 \cdot 4} + \cdots + \dfrac{1}{n(n+1)} = \dfrac{n}{n+1}.$

Prove the propositions in Exercises 2–10 by mathematical induction. The symbols m and n represent positive integers.

2 $(1.5)^{n+2} \geq n + 2$

3 $D_x(x^n) = nx^{n-1}$ (Use the Product Rule D6.)

4 $D_x(u^n) = nu^{n-1} \cdot D_x u$ (Use the Product Rule D6.)

5 $D_x(u_1 u_2 u_3 \cdots u_n) = (u_1' u_2 u_3 \cdots u_n) + (u_1 u_2' u_3 \cdots u_n)$
   $\qquad + (u_1 u_2 u_3' \cdots u_n) + \cdots + (u_1 u_2 u_3 \cdots u_n').$

6 Prove that $a^m \cdot a^n = a^{m+n}$. (*Hint:* Hold m fixed and run induction on n.)

7 Prove that $a^n/a^m = a^{n-m}$, $a \neq 0$. (*Hint:* Hold m fixed and run induction on n.)

8 Prove that $(a^m)^n = a^{mn}$. (*Hint:* Hold m fixed and run induction on n.)

9 Prove that $\sum_{j=1}^{n} j^2 = \dfrac{n(n+1)(2n+1)}{6}.$

10 Prove that $\sum_{j=1}^{n} j^3 = \dfrac{n^2(n+1)^2}{4}.$

APPENDIX B.
The Trigonometric Functions

Radian Measure

The *radian* system for measuring angles is defined by the following process. Mark off an arc of a circle equal to the radius of the circle. The corresponding central angle is defined to have a measure of *one radian*. (See Fig. B–1a.)

A similar process is used to construct an angle of α radians where $\alpha > 0$. We mark off an arc of length αr, where r is the radius of the circle. The corresponding central angle has a measure of α radians. (See Fig. B–1b.)

It is simple to convert from the degree-minute-second system for measuring angles to the radian system. Observe that one complete revolution around a circle corresponds to an angle of 360° in the degree-minute-second system and to 2π radians in the radian system. (See Fig. B-1c.) Thus,

$$2\pi \text{ radians } = 360°.$$

It follows that

Radian and Degree Measure

$$1 \text{ radian} = \frac{360°}{2\pi} \approx 57°17'44.8''$$

$$1° = \frac{2\pi}{360} \approx 0.0174532925 \text{ radian}$$

The "standard" angles of 0°, 30°, 45°, 60°, 90°, and 180° have the following radian measures:

Radian Measure of Standard Angles

$$0° = 0 \text{ radian} \qquad 45° = \frac{\pi}{4} \text{ radians} \qquad 90° = \frac{\pi}{2} \text{ radians}$$

$$30° = \frac{\pi}{6} \text{ radians} \qquad 60° = \frac{\pi}{3} \text{ radians} \qquad 180° = \pi \text{ radians}$$

Appendix B

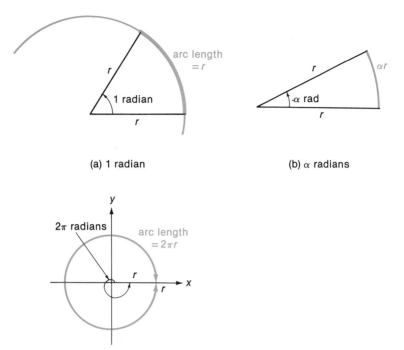

(a) 1 radian

(b) α radians

(c) 2π radians = 360°

Figure B-1 Radian measure.

Angles greater than 2π radians can be measured in these systems. For example, an angle of $495° = 2\pi + 3\pi/4$ radian corresponds to one and three-eighths of a complete revolution. (See Fig. B–2.)

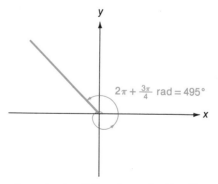

Figure B-2 $495° = 2\pi + 3\pi/4$ radians.

Arc Length and Area of Sectors

There is a direct relation between the length of an arc of a circle and the radian measure of the corresponding angle. Since a central angle of one radian corresponds to an arc with length equal to the radius r, it follows that a central angle of α radians corresponds to an arc with length αr. If s denotes the arc length, then

Arc Length
$$s = \alpha r.$$

(See Fig. B–3a.)

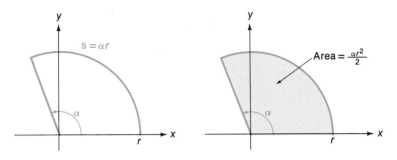

(a) Arc length $= \alpha r$ (b) Area $= \alpha r^2/2$

Figure B–3 Area and arc length of a sector.

The area of a sector can be calculated by a similar argument. Recall that the area of a circle of radius r is πr^2. Since this is the area of a sector with a central angle of 2π radians, then

$$\text{Area of sector of } 2\pi \text{ radians} = \pi r^2.$$

If we take proportional parts, we find that

$$\text{Area of sector of 1 radian} = \frac{\pi r^2}{2\pi} = \frac{r^2}{2}.$$

Similarly, if the central angle measures α radians, then

Area of Sector
$$\text{Area of sector of } \alpha \text{ radians} = \frac{\alpha r^2}{2}.$$

(See Fig. B–3b.)

EXAMPLE 1   A circle has radius $r = 3$ cm. A sector has a central angle of $60°$ ($\pi/3$ radians).
(a) The length of the corresponding arc is

$$s = \alpha r = \frac{\pi}{3} \cdot 3 = \pi \approx 3.14159 \text{ cm}.$$

(b) The area of the sector is

$$A = \frac{\alpha r^2}{2} = \frac{\pi}{3} \cdot \frac{3^2}{2} \approx 4.712 \text{ cm}^2. \quad \square$$

Appendix B

With suitable interpretation, these results can be extended to negative angles and angles greater than 2π. For example, if a wheel of diameter 18 in. is revolved twice, then a point on the edge of the wheel travels a total distance

$$s = \alpha r = 4\pi \cdot 9 = 36\pi \approx 113.1 \text{ in.}$$

The Trigonometric Functions

An angle α is said to be in *standard position* if its vertex is at the origin in the xy-plane and its initial side is on the positive x-axis. If $\alpha \geq 0$, the terminal side can be determined by a counterclockwise rotation of α radians (or the equivalent angle in degrees). If $\alpha < 0$, the terminal side is determined by a clockwise rotation of $|\alpha|$ radians. (See Fig. B–4.)

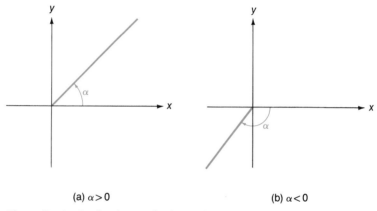

(a) $\alpha > 0$ (b) $\alpha < 0$

Figure B–4 Angles in standard position.

The *trigonometric functions* are defined by using central angles of the unit circle $x^2 + y^2 = 1$. For each angle α in standard position, we let $P_\alpha(x_\alpha, y_\alpha)$ be the point at which the terminal side of angle α intersects the unit circle $x^2 + y^2 = 1$. (See Fig. B–5a.)

The *cosine* and *sine* of α are defined to be the coordinates of $P_\alpha(x_\alpha, y_\alpha)$:

Sine and Cosine

$$\cos \alpha = x_\alpha, \quad \sin \alpha = y_\alpha.$$

The *tangent, cotangent, secant,* and *cosecant* of α are defined in terms of the sine and cosine of α.

T1 $\quad \tan \alpha = \dfrac{\sin \alpha}{\cos \alpha} \quad$ (provided $\cos \alpha \neq 0$)

T2 $\quad \cot \alpha = \dfrac{\cos \alpha}{\sin \alpha} \quad$ (provided $\sin \alpha \neq 0$)

T3	$\sec \alpha = \dfrac{1}{\cos \alpha}$	(provided $\cos \alpha \neq 0$)
T4	$\csc \alpha = \dfrac{1}{\sin \alpha}$	(provided $\sin \alpha \neq 0$).

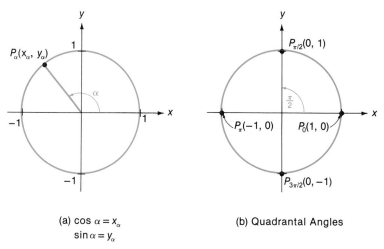

(a) $\cos \alpha = x_\alpha$
$\sin \alpha = y_\alpha$

(b) Quadrantal Angles

Figure B–5.

EXAMPLE 2 The angle $\alpha = \pi/2$ intersects the unit circle at the point $(0,1)$. (See Fig. B–5b.) Thus,

$$\cos \frac{\pi}{2} = 0, \qquad \sin \frac{\pi}{2} = 1;$$

$$\cot \frac{\pi}{2} = \frac{\cos \dfrac{\pi}{2}}{\sin \dfrac{\pi}{2}} = 0, \qquad \csc \frac{\pi}{2} = \frac{1}{\sin \dfrac{\pi}{2}} = 1,$$

while $\tan \pi/2$ and $\sec \pi/2$ are not defined.
Similarly,

$$\cos \pi = -1 \quad \text{and} \quad \sin \pi = 0,$$
$$\cos \frac{3\pi}{2} = 0 \quad \text{and} \quad \sin \frac{3\pi}{2} = -1,$$

and so on. (See Fig. B–5b.) □

The trigonometric functions of the "standard" angles $\theta = 0$, $\pi/6$, $\pi/4$, $\pi/3$, and $\pi/2$ are given in Table B–1.

Appendix B

Table B-1 Trigonometric Functions of Standard Angles

Angle, in Radians	Angle, in Degrees	sin	cos	tan	cot	sec	csc
0	0°	0	1	0	—	1	—
$\frac{\pi}{6}$	30°	$\frac{1}{2}$	$\frac{\sqrt{3}}{2}$	$\frac{1}{\sqrt{3}}$	$\sqrt{3}$	$\frac{2}{\sqrt{3}}$	2
$\frac{\pi}{4}$	45°	$\frac{1}{\sqrt{2}}$	$\frac{1}{\sqrt{2}}$	1	1	$\sqrt{2}$	$\sqrt{2}$
$\frac{\pi}{3}$	60°	$\frac{\sqrt{3}}{2}$	$\frac{1}{2}$	$\sqrt{3}$	$\frac{1}{\sqrt{3}}$	2	$\frac{2}{\sqrt{3}}$
$\frac{\pi}{2}$	90°	1	0	—	0	—	1

Graphs

The trigonometric functions can be considered to be functions of a real variable x. For each real number x we define sin x and cos x to be the functions of the corresponding angle measured in radians. [For example, sin π = sin (π radians) = 0.] The tangent, cotangent, secant, and cosecant functions are defined in terms of the sine and cosine functions as in T1–T4. The graphs of these functions are shown in Figure B–6.

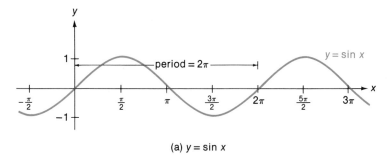

(a) $y = \sin x$

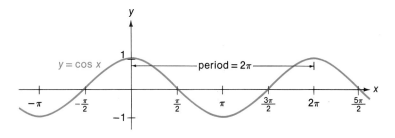

(b) $y = \cos x$

Figure B-6 The trigonometric functions. (Parts c through f on p. A.12.)

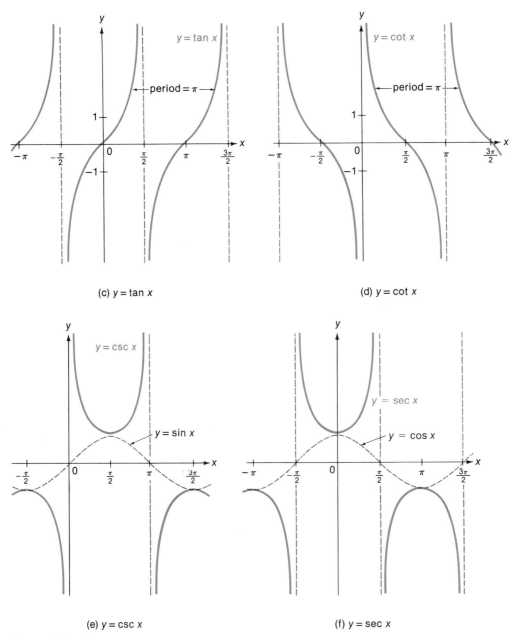

(c) $y = \tan x$

(d) $y = \cot x$

(e) $y = \csc x$

(f) $y = \sec x$

Figure B–8c–f.

Periodic Function

The trigonometric functions are *periodic*. If f is one of these functions, there exists a positive number p such that

$$f(x + p) = f(x)$$

whenever $f(x)$ is defined. The smallest such number p is called the *period* of f. The

Appendix B

sine, cosine, secant, and cosecant functions have period 2π, and the tangent and cotangent functions have period π.

The similarity between the graphs of the sine and cosine functions is not coincidental. We shall show in Example 4 that

$$\cos\left(x - \frac{\pi}{2}\right) = \sin x.$$

Thus, the graph of the cosine function can be obtained by shifting the graph of the sine function $\pi/2$ units to the left. This shift corresponds to a translation of axes. (See Fig. B–7.)

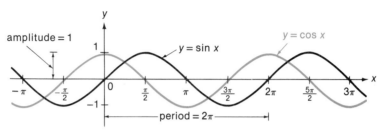

Figure B–7 $\cos(x - \pi/2) = \sin x$. The graph of the cosine function is a translation of the graph of the sine function.

The values taken by the sine and cosine functions range between -1 and $+1$. (This follows from the fact that $(\sin x, \cos x)$ is a point on the unit circle.) One-half of this range of values is called the *amplitude* of the functions. Thus

Amplitude   Amplitude of sine and cosine functions = 1.

(See Figs. B–6a, b.)

The graph of $y = a \sin bx$, where a and b are positive, can be obtained from the graph of $y = \sin x$ by a horizontal "compression" by a factor of b and a vertical "stretch" by a factor of a. (See

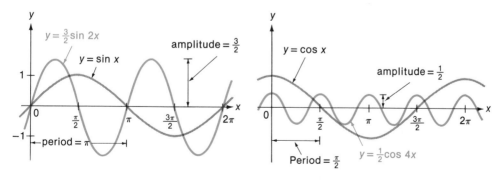

(a) Graph of $y = \frac{3}{2}\sin 2x$ (b) Graph of $y = \frac{1}{2}\cos 4x$

Figure B–8.

Fig. B–8a.) The period of $y = a \sin bx$ is

$$p = \frac{2\pi}{b}$$

and the amplitude is a.

The graph of $y = a \cos bx$ can be obtained in a similar way from the graph of the cosine function. The period is

$$\text{Period} = \frac{2\pi}{b}$$

and the amplitude equals a. (See Fig. B–8b.)

Computation of Trigonometric Functions

Table III (at the end of this book) gives the approximate values of the sine, cosine, tangent, and cotangent for first quadrant angles measured in radians. For example,

$$\sin 1.32 \approx 0.96872.$$

Trigonometric functions of angles in other quadrants can be calculated by the use of *reference angles*.

Let α be an angle in standard position. Let β be the smallest angle that can be drawn from the x-axis to the terminal side of α. (See Fig. B–9.) The angle β is called the *reference angle that corresponds to* α. Observe that $0 \leq \beta \leq \pi/2$ for every choice of α, so that β is a first-quadrant angle.

Reference Angle

We now draw the reference angle β in standard position in the xy-plane. Let $P_\alpha(\cos \alpha, \sin \alpha)$ and $P_\beta(\cos \beta, \sin \beta)$ be the points where the terminal sides of α and β intersect the unit circle. (See Fig. B–9.) It is obvious from the symmetry of the angles that the coordinates of P_α and P_β are equal in absolute value. Since the coordinates of P_β are nonnegative, then

$$\cos \alpha = x_\alpha = \pm x_\beta = \pm \cos \beta$$

and

$$\sin \alpha = y_\alpha = \pm y_\beta = \pm \sin \beta.$$

It follows that

$$\tan \alpha = \pm \tan \beta,$$
$$\cot \alpha = \pm \cot \beta,$$

and so on, the sign being determined by the quadrant.

EXAMPLE 3   Use Table III to calculate the approximate value of (a) $\cos 2.91$; (b) $\tan 4.21$.

Solution   We use the approximation $\pi \approx 3.14$.

Appendix B

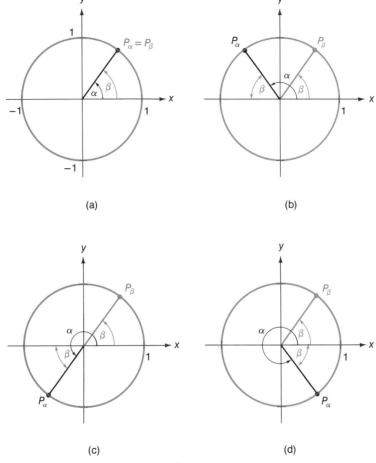

Figure B–9 Reference angles: $\begin{cases} \cos \alpha = \pm \cos \beta \\ \sin \alpha = \pm \sin \beta. \end{cases}$

(a) 2.91 is a Quadrant II angle. The reference angle is
$$\beta = \pi - 2.91 \approx 3.14 - 2.91 = 0.23.$$
Since the cosine is negative in Quadrant II, then
$$\cos 2.91 = -\cos \beta = -\cos 0.23 \approx -0.97367.$$

(b) 4.21 is a Quadrant III angle. The reference angle is
$$\beta = 4.21 - \pi \approx 4.21 - 3.14 = 1.07.$$
Since the tangent is positive in Quadrant III, then
$$\tan 4.21 = \tan \beta = \tan 1.07 \approx 1.8270. \quad \square$$

Identities

Numerous identities relate the trigonometric functions. Some of the most important are developed in the remainder of this Appendix.

Pythagorean Identities

The most basic of the identities is the *Pythagorean Identity*.

Pythagorean Identity

T5 $\cos^2 \alpha + \sin^2 \alpha = 1.$

This identity follows from the fact that $P_\alpha(\cos \alpha, \sin \alpha)$ is on the unit circle $x^2 + y^2 = 1$. Other forms of the Pythagorean Identity are obtained by dividing T5 by $\cos^2 \alpha$ and $\sin^2 \alpha$, respectively:

T6 $1 + \tan^2 \alpha = \sec^2 \alpha.$
T7 $\cot^2 \alpha + 1 = \csc^2 \alpha.$

Sum and Difference Identities

The sum and difference identities express the sum and difference of two angles in terms of the original angles.

Sum and Difference Identities

T8 $\sin(\alpha \pm \beta) = \sin \alpha \cos \beta \pm \cos \alpha \sin \beta.$
T9 $\cos(\alpha \pm \beta) = \cos \alpha \cos \beta \mp \sin \alpha \sin \beta.$

EXAMPLE 4   Show that $\cos\left(x - \dfrac{\pi}{2}\right) = \sin x$. Interpret this result geometrically.

Solution   By T9,

$$\cos\left(x - \frac{\pi}{2}\right) = \cos x \cos \frac{\pi}{2} + \sin x \sin \frac{\pi}{2}$$
$$= \cos x \cdot 0 + \sin x \cdot 1 = \sin x.$$

As mentioned earlier in this section, this relation shows that the graph of $y = \cos x$ can be obtained by shifting the graph of $y = \sin x$ a distance of $\pi/2$ units to the left. (See Fig. B–7.) □

Appendix B

In the special cases where the first angle is 0 or $\pi/2$, the sum and difference identities reduce to

T10 $\sin(-\alpha) = -\sin\alpha, \quad \cos(-\alpha) = \cos\alpha,$

T11 $\sin\left(\dfrac{\pi}{2} - \alpha\right) = \cos\alpha, \quad \cos\left(\dfrac{\pi}{2} - \alpha\right) = \sin\alpha.$

Double-Angle and Half-Angle Identities

If we take $\alpha = \beta$, identities T8 and T9 reduce to

$$\sin 2\alpha = \sin\alpha\cos\alpha + \cos\alpha\sin\alpha = 2\sin\alpha\cos\alpha$$

and

$$\cos 2\alpha = \cos\alpha\cos\alpha - \sin\alpha\sin\alpha = \cos^2\alpha - \sin^2\alpha.$$

This last identity can be modified by using the Pythagorean Identity T5:

$$\cos 2\alpha = \cos^2\alpha - (1 - \cos^2\alpha) = 2\cos^2\alpha - 1$$
$$= 2(1 - \sin^2\alpha) - 1 = 1 - 2\sin^2\alpha.$$

Thus we have the following *double-angle identities*:

Double-Angle Identities

T12 $\sin 2\alpha = 2\sin\alpha\cos\alpha.$
T13 $\cos 2\alpha = \cos^2\alpha - \sin^2\alpha.$
T14 $\cos 2\alpha = 2\cos^2\alpha - 1.$
T15 $\cos 2\alpha = 1 - 2\sin^2\alpha.$

If we solve T14 for $\cos\alpha$, we obtain

$$2\cos^2\alpha = 1 + \cos 2\alpha$$
$$\cos\alpha = \pm\sqrt{\dfrac{1 + \cos 2\alpha}{2}}.$$

Similarly, it follows from T15 that

$$\sin\alpha = \pm\sqrt{\dfrac{1 - \cos 2\alpha}{2}}.$$

Since $\tan\alpha = \sin\alpha/\cos\alpha$, then

$$\tan\alpha = \pm\sqrt{\dfrac{1 - \cos 2\alpha}{1 + \cos 2\alpha}} = \dfrac{1 - \cos 2\alpha}{\sin 2\alpha} = \dfrac{\sin 2\alpha}{1 + \cos 2\alpha}.$$

If we now substitute $\theta/2$ for α, we obtain the *half-angle identities*:

Half-Angle Identities

> T16 $\quad \sin \dfrac{\theta}{2} = \pm \sqrt{\dfrac{1 - \cos \theta}{2}}.$
>
> T17 $\quad \cos \dfrac{\theta}{2} = \pm \sqrt{\dfrac{1 + \cos \theta}{2}}.$
>
> T18 $\quad \tan \dfrac{\theta}{2} = \pm \sqrt{\dfrac{1 - \cos \theta}{1 + \cos \theta}} = \dfrac{1 - \cos \theta}{\sin \theta} = \dfrac{\sin \theta}{1 + \cos \theta}.$

The sign before the radical is determined by the quadrant.

EXAMPLE 5   Let $\cos \theta = \tfrac{119}{169}$, where $3\pi/2 < \theta < 2\pi$. Calculate $\sin(\theta/2)$ and $\cos(\theta/2)$.

Solution   Since $3\pi/4 < \theta/2 < \pi$, then $\theta/2$ is a second-quadrant angle. Therefore, $\sin(\theta/2) > 0$ and $\cos(\theta/2) < 0$. It follows from T16 and T17 that

$$\sin \dfrac{\theta}{2} = \sqrt{\dfrac{1 - \cos \theta}{2}} = \sqrt{\dfrac{1 - \tfrac{119}{169}}{2}} = \sqrt{\dfrac{25}{169}} = \dfrac{5}{13}$$

$$\cos \dfrac{\theta}{2} = -\sqrt{\dfrac{1 + \cos \theta}{2}} = -\sqrt{\dfrac{1 + \tfrac{119}{169}}{2}} = -\sqrt{\dfrac{144}{169}} = -\dfrac{12}{13}. \quad \square$$

Product Identities

The two identities listed as T8 are

$$\sin(\alpha + \beta) = \sin \alpha \cos \beta + \cos \alpha \sin \beta,$$
$$\sin(\alpha - \beta) = \sin \alpha \cos \beta - \cos \alpha \sin \beta.$$

If we add these two equations and divide by 2, we obtain the *product* identity

$$\sin \alpha \cos \beta = \dfrac{1}{2}[\sin(\alpha + \beta) + \sin(\alpha - \beta)].$$

A similar identity can be obtained by subtracting one of the identities for T8 from the other or by adding or subtracting the two identities for T9. The complete list of product identities is

Product Identities

> T19 $\quad \sin \alpha \cos \beta = \tfrac{1}{2}[\sin(\alpha + \beta) + \sin(\alpha - \beta)].$
> T20 $\quad \cos \alpha \sin \beta = \tfrac{1}{2}[\sin(\alpha + \beta) - \sin(\alpha - \beta)].$
> T21 $\quad \cos \alpha \cos \beta = \tfrac{1}{2}[\cos(\alpha + \beta) + \cos(\alpha - \beta)].$
> T22 $\quad \sin \alpha \sin \beta = \tfrac{1}{2}[\cos(\alpha - \beta) - \cos(\alpha + \beta)].$

Law of Cosines and Law of Sines

The laws of cosines and sines allow us to apply the trigonometric functions to problems involving oblique triangles.

The standard labeling system for triangles is shown in Figure B–10. The sides op-

Appendix B

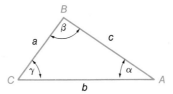

Figure B–10 Standard labeling of triangles.

posite vertices A, B, and C are a, b, and c. The angles with these points as vertices are α, β, and γ.

The *law of cosines* is a generalization of the Pythagorean Theorem and can be proved from that theorem. (See Exercise 11.)

The square of any side of a triangle equals the sum of the squares of the other two sides minus twice the product of the other two sides and the cosine of the angle between them.

Thus,

Law of Cosines
$$a^2 = b^2 + c^2 - 2bc \cos \alpha;$$
$$b^2 = a^2 + c^2 - 2ac \cos \beta;$$
$$c^2 = a^2 + b^2 - 2ab \cos \gamma.$$

The *law of sines* relates the sines of the angles α, β, and γ to the lengths of the sides a, b, and c:

Law of Sines
$$\frac{\sin \alpha}{a} = \frac{\sin \beta}{b} = \frac{\sin \gamma}{c}$$

(See Exercise 12.)

Exercises. Appendix B

- **1** Convert from radian to degree measure or from degree to radian measure.
  - (a) 3°
  - (b) 210°
  - (c) 330°
  - (d) −135°
  - (e) 2 rad
  - (f) $\frac{5\pi}{6}$ rad
  - (g) $\frac{7\pi}{3}$ rad
  - (h) $-\frac{2\pi}{3}$ rad

- **2** A circle has diameter 7 in. A sector has central angle 135°. Calculate (a) the area of the sector; (b) the total perimeter of the sector.

- **3** Draw the appropriate reference angle. Calculate the required trigonometric function. For (a)–(d) use Table B–1 of this Appendix. For (e)–(h) use Table III (at the end of the book) and the approximation $\pi \approx 3.14$.
  - (a) $\tan\left(\frac{3\pi}{4}\right)$
  - (b) $\cos \pi$
  - (c) $\sec 240°$
  - (d) $\sin\left(\frac{7\pi}{6}\right)$

(e) sin 6.53 (f) cos (−2.15)
(g) tan (−4.13) (h) cos (−1.02)

●4 Calculate the period and amplitude of the function. Sketch the graph.

(a) $f(x) = 2 \cos x$ (b) $f(x) = -\frac{1}{3} \sin x$

(c) $y = 3 \sin 4x$ (d) $y = -4 \cos \left(\frac{x}{2}\right)$

●5 Calculate the period of the function. Sketch the graph.

(a) $y = \tan 2x$ (b) $y = \sec \left(\frac{x}{3}\right)$

(c) $y = \cot 3x$ (d) $y = \tan(-x)$

●6 Let P_α be in Quadrant II with $\sin \alpha = \frac{3}{5}$ and P_β be in Quadrant III with $\tan \beta = \frac{5}{12}$. Calculate the following.

(a) $\sin \beta$ (b) $\cos \beta$
(c) $\sin(\alpha + \beta)$ (d) $\cos(\alpha + \beta)$
(e) $\sin(\alpha - \beta)$ (f) $\cos(\alpha - \beta)$
(g) $\sin 2\alpha$ (h) $\cos 2\beta$
(i) $\sin \frac{\beta}{2}$ (j) $\cos \frac{\beta}{2}$

7 Use T8 and T9 to derive the following identities.

(a) $\tan(\alpha \pm \beta) = \dfrac{\tan \alpha \pm \tan \beta}{1 \mp \tan \alpha \tan \beta}$ (b) $\tan 2\alpha = \dfrac{2 \tan \alpha}{1 - \tan^2 \alpha}$

8 (a) Prove T17. (b) Prove T18.

9 Use the identities of this appendix to prove the following identities:
(a) $\cos^4 x - \sin^4 x = \cos 2x$
(b) $1 - \sin^2 2\theta = (1 - 2 \sin^2 \theta)^2$
(c) $\dfrac{1 - \sin \theta}{\cos \theta} = \dfrac{\cos \theta}{1 + \sin \theta}$
(d) $\cos 3\phi = 4 \cos^3 \phi - 3 \cos \phi$
(e) $\sin\left(\dfrac{3\pi}{2} - \phi\right) = -\cos \phi = \sin\left(\dfrac{3\pi}{2} + \phi\right)$
(f) $\cos\left(\dfrac{\pi}{3} - \phi\right) - \sin\left(\dfrac{\pi}{6} - \phi\right) = \sqrt{3} \sin \phi$

10 Many of the trigonometric identities follow from T8 and T9. In this exercise we outline the proofs of several of these identities. The identity for $\cos(\alpha - \beta)$ is basic to the rest of the work.

(a) Prove that $\cos(\alpha - \beta) = \cos \alpha \cos \beta + \sin \alpha \sin \beta$. [*Hint:* Prove that the arc of the unit circle joining P_β ($\cos \beta$, $\sin \beta$) and P_α ($\cos \alpha$, $\sin \alpha$) equals the arc joining $P_{\alpha-\beta}$ ($\cos(\alpha - \beta)$, $\sin(\alpha - \beta)$) and $P_0(1, 0)$. Conclude from this that $|P_\alpha P_\beta| = |P_{\alpha-\beta} P_0|$. Simplify the resulting expression to prove the identity.]

(b) By choosing special values in (a), prove that

$$\cos(-\beta) = \cos\beta \quad \text{and} \quad \cos\left(\frac{\pi}{2} - \beta\right) = \sin\beta.$$

(c) Use the second identity in (b) to prove that

$$\sin\left(\frac{\pi}{2} - \alpha\right) = \cos\alpha.$$

(*Hint:* Let $\beta = \frac{\pi}{2} - \alpha$.)

(d) Use the identity in (a) and the fact that

$$\sin(-\alpha) = \sin\left(\frac{\pi}{2} - \left(\frac{\pi}{2} + \alpha\right)\right) = \cos\left(\frac{\pi}{2} + \alpha\right) = \cos\left(\alpha - \left(-\frac{\pi}{2}\right)\right)$$

to prove that

$$\sin(-\alpha) = -\sin\alpha.$$

(e) Prove that $\cos(\alpha + \beta) = \cos\alpha\cos\beta - \sin\alpha\sin\beta$.

(f) Use the results of (a), (b) and (e) to prove that

$$\sin(\alpha - \beta) = \sin\alpha\cos\beta - \cos\alpha\sin\beta.$$

(g) Prove that $\sin(\alpha + \beta) = \sin\alpha\cos\beta + \cos\alpha\sin\beta$.

11 Make an argument based on Figure B–11 to prove the following special case of the law of cosines:

$$a^2 = b^2 + c^2 - 2bc\cos\alpha.$$

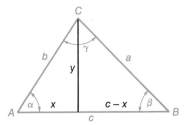

Figure B–11 Problems 11 and 12.

(*Hint:* Use the Pythagorean Theorem on each of the right triangles. By combining the results, show that $c^2 - 2cx = a^2 - b^2$.)

12 Make an argument based on Figure B–11 to prove the special case of the law of sines:

$$\frac{\sin\alpha}{a} = \frac{\sin\beta}{b}.$$

(*Hint:* Express y in terms of $\sin\alpha$ and in terms of $\sin\beta$.)

•**13** A triangle with standard labeling has sides $a = 12$, $b = 17$, $c = 15$. Calculate the magnitude of the angles α, β, γ.

•**14** A triangle with standard labeling has $\alpha = 17°$, $b = 4$, and $c = 20$. Calculate the length of side a and the magnitude of the angles β and γ.

TABLES

The numerical tables are intended for emergency use only. Answers based on these tables may differ slightly from the answers in the back of the book, which were obtained by use of a scientific calculator.

Table I Four-place Natural Logarithms **1.00–5.49** (ln x = $\log_e x$)

N	.00	.01	.02	.03	.04	.05	.06	.07	.08	.09
1.0	0.0000	0.0100	0.0198	0.0296	0.0392	0.0488	0.0583	0.0677	0.0770	0.0862
1.1	0.0953	0.1044	0.1133	0.1222	0.1310	0.1398	0.1484	0.1570	0.1655	0.1740
1.2	0.1823	0.1906	0.1989	0.2070	0.2151	0.2231	0.2311	0.2390	0.2469	0.2546
1.3	0.2624	0.2700	0.2776	0.2852	0.2927	0.3001	0.3075	0.3148	0.3221	0.3293
1.4	0.3365	0.3436	0.3507	0.3577	0.3646	0.3716	0.3784	0.3853	0.3920	0.3988
1.5	0.4055	0.4121	0.4187	0.4253	0.4318	0.4383	0.4447	0.4511	0.4574	0.4637
1.6	0.4700	0.4762	0.4824	0.4886	0.4947	0.5008	0.5068	0.5128	0.5188	0.5247
1.7	0.5306	0.5365	0.5423	0.5481	0.5539	0.5596	0.5653	0.5710	0.5766	0.5822
1.8	0.5878	0.5933	0.5988	0.6043	0.6098	0.6152	0.6206	0.6259	0.6313	0.6366
1.9	0.6419	0.6471	0.6523	0.6575	0.6627	0.6678	0.6729	0.6780	0.6831	0.6881
2.0	0.6931	0.6981	0.7031	0.7080	0.7129	0.7178	0.7227	0.7275	0.7324	0.7372
2.1	0.7419	0.7467	0.7514	0.7561	0.7608	0.7655	0.7701	0.7747	0.7793	0.7839
2.2	0.7885	0.7930	0.7975	0.8020	0.8065	0.8109	0.8154	0.8198	0.8242	0.8286
2.3	0.8329	0.8372	0.8416	0.8459	0.8502	0.8544	0.8587	0.8629	0.8671	0.8713
2.4	0.8755	0.8796	0.8838	0.8879	0.8920	0.8961	0.9002	0.9042	0.9083	0.9123
2.5	0.9163	0.9203	0.9243	0.9282	0.9322	0.9361	0.9400	0.9439	0.9478	0.9517
2.6	0.9555	0.9594	0.9632	0.9670	0.9708	0.9746	0.9783	0.9821	0.9858	0.9895
2.7	0.9933	0.9969	1.0006	1.0043	1.0080	1.0116	1.0152	1.0188	1.0225	1.0260
2.8	1.0296	1.0332	1.0367	1.0403	1.0438	1.0473	1.0508	1.0543	1.0578	1.0613
2.9	1.0647	1.0682	1.0716	1.0750	1.0784	1.0818	1.0852	1.0886	1.0919	1.0953
3.0	1.0986	1.1019	1.1053	1.1086	1.1119	1.1151	1.1184	1.1217	1.1249	1.1282
3.1	1.1314	1.1346	1.1378	1.1410	1.1442	1.1474	1.1506	1.1537	1.1569	1.1600
3.2	1.1632	1.1663	1.1694	1.1725	1.1756	1.1787	1.1817	1.1848	1.1878	1.1909
3.3	1.1939	1.1969	1.2000	1.2030	1.2060	1.2090	1.2119	1.2149	1.2179	1.2208
3.4	1.2238	1.2267	1.2296	1.2326	1.2355	1.2384	1.2413	1.2442	1.2470	1.2499
3.5	1.2528	1.2556	1.2585	1.2613	1.2641	1.2669	1.2698	1.2726	1.2754	1.2782
3.6	1.2809	1.2837	1.2865	1.2892	1.2920	1.2947	1.2975	1.3002	1.3029	1.3056
3.7	1.3083	1.3110	1.3137	1.3164	1.3191	1.3218	1.3244	1.3271	1.3297	1.3324
3.8	1.3350	1.3376	1.3403	1.3429	1.3455	1.3481	1.3507	1.3533	1.3558	1.3584
3.9	1.3610	1.3635	1.3661	1.3686	1.3712	1.3737	1.3762	1.3788	1.3813	1.3838
4.0	1.3863	1.3888	1.3913	1.3938	1.3962	1.3987	1.4012	1.4036	1.4061	1.4085
4.1	1.4110	1.4134	1.4159	1.4183	1.4207	1.4231	1.4255	1.4279	1.4303	1.4327
4.2	1.4351	1.4375	1.4398	1.4422	1.4446	1.4469	1.4493	1.4516	1.4540	1.4563
4.3	1.4586	1.4609	1.4633	1.4656	1.4679	1.4702	1.4725	1.4748	1.4770	1.4793
4.4	1.4816	1.4839	1.4861	1.4884	1.4907	1.4929	1.4951	1.4974	1.4996	1.5019
4.5	1.5041	1.5063	1.5085	1.5107	1.5129	1.5151	1.5173	1.5195	1.5217	1.5239
4.6	1.5261	1.5282	1.5304	1.5326	1.5347	1.5369	1.5390	1.5412	1.5433	1.5454
4.7	1.5476	1.5497	1.5518	1.5539	1.5560	1.5581	1.5602	1.5623	1.5644	1.5665
4.8	1.5686	1.5707	1.5728	1.5748	1.5769	1.5790	1.5810	1.5831	1.5851	1.5872
4.9	1.5892	1.5913	1.5933	1.5953	1.5974	1.5994	1.6014	1.6034	1.6054	1.6074
5.0	1.6094	1.6114	1.6134	1.6154	1.6174	1.6194	1.6214	1.6233	1.6253	1.6273
5.1	1.6292	1.6312	1.6332	1.6351	1.6371	1.6390	1.6409	1.6429	1.6448	1.6467
5.2	1.6487	1.6506	1.6525	1.6544	1.6563	1.6582	1.6601	1.6620	1.6639	1.6658
5.3	1.6677	1.6696	1.6715	1.6734	1.6752	1.6771	1.6790	1.6808	1.6827	1.6845
5.4	1.6864	1.6882	1.6901	1.6919	1.6938	1.6956	1.6974	1.6993	1.7011	1.7029

ln .1 = .6974−3   ln .01 = .3948−5   ln .001 = .0922−7

(continued)

Table I (*Cont.*) Four-place Natural Logarithms **5.50–9.99**

N	.00	.01	.02	.03	.04	.05	.06	.07	.08	.09
5.5	1.7047	1.7066	1.7084	1.7102	1.7120	1.7138	1.7156	1.7174	1.7192	1.7210
5.6	1.7228	1.7246	1.7263	1.7281	1.7299	1.7317	1.7334	1.7352	1.7370	1.7387
5.7	1.7405	1.7422	1.7440	1.7457	1.7475	1.7492	1.7509	1.7527	1.7544	1.7561
5.8	1.7579	1.7596	1.7613	1.7630	1.7647	1.7664	1.7681	1.7699	1.7716	1.7733
5.9	1.7750	1.7766	1.7783	1.7800	1.7817	1.7834	1.7851	1.7867	1.7884	1.7901
6.0	1.7918	1.7934	1.7951	1.7967	1.7984	1.8001	1.8017	1.8034	1.8050	1.8066
6.1	1.8083	1.8099	1.8116	1.8132	1.8148	1.8165	1.8181	1.8197	1.8213	1.8229
6.2	1.8245	1.8262	1.8278	1.8294	1.8310	1.8326	1.8342	1.8358	1.8374	1.8390
6.3	1.8405	1.8421	1.8437	1.8453	1.8469	1.8485	1.8500	1.8516	1.8532	1.8547
6.4	1.8563	1.8579	1.8594	1.8610	1.8625	1.8641	1.8656	1.8672	1.8687	1.8703
6.5	1.8718	1.8733	1.8749	1.8764	1.8779	1.8795	1.8810	1.8825	1.8840	1.8856
6.6	1.8871	1.8886	1.8901	1.8916	1.8931	1.8946	1.8961	1.8976	1.8991	1.9006
6.7	1.9021	1.9036	1.9051	1.9066	1.9081	1.9095	1.9110	1.9125	1.9140	1.9155
6.8	1.9169	1.9184	1.9199	1.9213	1.9228	1.9242	1.9257	1.9272	1.9286	1.9301
6.9	1.9315	1.9330	1.9344	1.9359	1.9373	1.9387	1.9402	1.9416	1.9430	1.9445
7.0	1.9459	1.9473	1.9488	1.9502	1.9516	1.9530	1.9544	1.9559	1.9573	1.9587
7.1	1.9601	1.9615	1.9629	1.9643	1.9657	1.9671	1.9685	1.9699	1.9713	1.9727
7.2	1.9741	1.9755	1.9769	1.9782	1.9796	1.9810	1.9824	1.9838	1.9851	1.9865
7.3	1.9879	1.9892	1.9906	1.9920	1.9933	1.9947	1.9961	1.9974	1.9988	2.0001
7.4	2.0015	2.0028	2.0042	2.0055	2.0069	2.0082	2.0096	2.0109	2.0122	2.0136
7.5	2.0149	2.0162	2.0176	2.0189	2.0202	2.0215	2.0229	2.0242	2.0255	2.0268
7.6	2.0281	2.0295	2.0308	2.0321	2.0334	2.0347	2.0360	2.0373	2.0386	2.0399
7.7	2.0412	2.0425	2.0438	2.0451	2.0464	2.0477	2.0490	2.0503	2.0516	2.0528
7.8	2.0541	2.0554	2.0567	2.0580	2.0592	2.0605	2.0618	2.0631	2.0643	2.0656
7.9	2.0669	2.0681	2.0694	2.0707	2.0719	2.0732	2.0744	2.0757	2.0769	2.0782
8.0	2.0794	2.0807	2.0819	2.0832	2.0844	2.0857	2.0869	2.0882	2.0894	2.0906
8.1	2.0919	2.0931	2.0943	2.0956	2.0968	2.0980	2.0992	2.1005	2.1017	2.1029
8.2	2.1041	2.1054	2.1066	2.1078	2.1090	2.1102	2.1114	2.1126	2.1138	2.1150
8.3	2.1163	2.1175	2.1187	2.1199	2.1211	2.1223	2.1235	2.1247	2.1258	2.1270
8.4	2.1282	2.1294	2.1306	2.1318	2.1330	2.1342	2.1353	2.1365	2.1377	2.1389
8.5	2.1401	2.1412	2.1424	2.1436	2.1448	2.1459	2.1471	2.1483	2.1494	2.1506
8.6	2.1518	2.1529	2.1541	2.1552	2.1564	2.1576	2.1587	2.1599	2.1610	2.1622
8.7	2.1633	2.1645	2.1656	2.1668	2.1679	2.1691	2.1702	2.1713	2.1725	2.1736
8.8	2.1748	2.1759	2.1770	2.1782	2.1793	2.1804	2.1815	2.1827	2.1838	2.1849
8.9	2.1861	2.1872	2.1883	2.1894	2.1905	2.1917	2.1928	2.1939	2.1950	2.1961
9.0	2.1972	2.1983	2.1994	2.2006	2.2017	2.2028	2.2039	2.2050	2.2061	2.2072
9.1	2.2083	2.2094	2.2105	2.2116	2.2127	2.2138	2.2148	2.2159	2.2170	2.2181
9.2	2.2192	2.2203	2.2214	2.2225	2.2235	2.2246	2.2257	2.2268	2.2279	2.2289
9.3	2.2300	2.2311	2.2322	2.2332	2.2343	2.2354	2.2364	2.2375	2.2386	2.2396
9.4	2.2407	2.2418	2.2428	2.2439	2.2450	2.2460	2.2471	2.2481	2.2492	2.2502
9.5	2.2513	2.2523	2.2534	2.2544	2.2555	2.2565	2.2576	2.2586	2.2597	2.2607
9.6	2.2618	2.2628	2.2638	2.2649	2.2659	2.2670	2.2680	2.2690	2.2701	2.2711
9.7	2.2721	2.2732	2.2742	2.2752	2.2762	2.2773	2.2783	2.2793	2.2803	2.2814
9.8	2.2824	2.2834	2.2844	2.2854	2.2865	2.2875	2.2885	2.2895	2.2905	2.2915
9.9	2.2925	2.2935	2.2946	2.2956	2.2966	2.2976	2.2986	2.2996	2.3006	2.3016

$\ln .0001 = .7897 - 10$ $\ln .00001 = .4871 - 12$ $\ln .000\,001 = .1845 - 14$

Table I (*Cont.*)   Four-place Natural Logarithms **10.0–54.9**

N	.0	.1	.2	.3	.4	.5	.6	.7	.8	.9
10	2.3026	2.3125	2.3224	2.3321	2.3418	2.3514	2.3609	2.3702	2.3795	2.3888
11	2.3979	2.4069	2.4159	2.4248	2.4336	2.4423	2.4510	2.4596	2.4681	2.4765
12	2.4849	2.4932	2.5014	2.5096	2.5177	2.5257	2.5337	2.5416	2.5494	2.5572
13	2.5649	2.5726	2.5802	2.5878	2.5953	2.6027	2.6101	2.6174	2.6247	2.6319
14	2.6391	2.6462	2.6532	2.6603	2.6672	2.6741	2.6810	2.6878	2.6946	2.7014
15	2.7081	2.7147	2.7213	2.7279	2.7344	2.7408	2.7473	2.7537	2.7600	2.7663
16	2.7726	2.7788	2.7850	2.7912	2.7973	2.8034	2.8094	2.8154	2.8214	2.8273
17	2.8332	2.8391	2.8449	2.8507	2.8565	2.8622	2.8679	2.8736	2.8792	2.8848
18	2.8904	2.8959	2.9014	2.9069	2.9124	2.9178	2.9232	2.9285	2.9339	2.9392
19	2.9444	2.9497	2.9549	2.9601	2.9653	2.9704	2.9755	2.9806	2.9857	2.9907
20	2.9957	3.0007	3.0057	3.0106	3.0155	3.0204	3.0253	3.0301	3.0350	3.0397
21	3.0445	3.0493	3.0540	3.0587	3.0634	3.0681	3.0727	3.0773	3.0819	3.0865
22	3.0910	3.0956	3.1001	3.1046	3.1091	3.1135	3.1179	3.1224	3.1268	3.1311
23	3.1355	3.1398	3.1442	3.1485	3.1527	3.1570	3.1612	3.1655	3.1697	3.1739
24	3.1781	3.1822	3.1864	3.1905	3.1946	3.1987	3.2027	3.2068	3.2108	3.2149
25	3.2189	3.2229	3.2268	3.2308	3.2347	3.2387	3.2426	3.2465	3.2504	3.2542
26	3.2581	3.2619	3.2658	3.2696	3.2734	3.2771	3.2809	3.2847	3.2884	3.2921
27	3.2958	3.2995	3.3032	3.3069	3.3105	3.3142	3.3178	3.3214	3.3250	3.3286
28	3.3322	3.3358	3.3393	3.3429	3.3464	3.3499	3.3534	3.3569	3.3604	3.3638
29	3.3673	3.3707	3.3742	3.3776	3.3810	3.3844	3.3878	3.3911	3.3945	3.3979
30	3.4012	3.4045	3.4078	3.4111	3.4144	3.4177	3.4210	3.4243	3.4275	3.4308
31	3.4340	3.4372	3.4404	3.4436	3.4468	3.4500	3.4532	3.4563	3.4595	3.4626
32	3.4657	3.4689	3.4720	3.4751	3.4782	3.4812	3.4843	3.4874	3.4904	3.4935
33	3.4965	3.4995	3.5025	3.5056	3.5086	3.5115	3.5145	3.5175	3.5205	3.5234
34	3.5264	3.5293	3.5322	3.5351	3.5381	3.5410	3.5439	3.5467	3.5496	3.5525
35	3.5553	3.5582	3.5610	3.5639	3.5667	3.5695	3.5723	3.5752	3.5779	3.5807
36	3.5835	3.5863	3.5891	3.5918	3.5946	3.5973	3.6000	3.6028	3.6055	3.6082
37	3.6109	3.6136	3.6163	3.6190	3.6217	3.6243	3.6270	3.6297	3.6323	3.6350
38	3.6376	3.6402	3.6428	3.6454	3.6481	3.6507	3.6533	3.6558	3.6584	3.6610
39	3.6636	3.6661	3.6687	3.6712	3.6738	3.6763	3.6788	3.6814	3.6839	3.6864
40	3.6889	3.6914	3.6939	3.6964	3.6988	3.7013	3.7038	3.7062	3.7087	3.7111
41	3.7136	3.7160	3.7184	3.7209	3.7233	3.7257	3.7281	3.7305	3.7329	3.7353
42	3.7377	3.7400	3.7424	3.7448	3.7471	3.7495	3.7519	3.7542	3.7565	3.7589
43	3.7612	3.7635	3.7658	3.7682	3.7705	3.7728	3.7751	3.7773	3.7796	3.7819
44	3.7842	3.7865	3.7887	3.7910	3.7932	3.7955	3.7977	3.8000	3.8022	3.8044
45	3.8067	3.8089	3.8111	3.8133	3.8155	3.8177	3.8199	3.8221	3.8243	3.8265
46	3.8286	3.8308	3.8330	3.8351	3.8373	3.8395	3.8416	3.8437	3.8459	3.8480
47	3.8501	3.8523	3.8544	3.8565	3.8586	3.8607	3.8628	3.8649	3.8670	3.8691
48	3.8712	3.8733	3.8754	3.8774	3.8795	3.8816	3.8836	3.8857	3.8877	3.8898
49	3.8918	3.8939	3.8959	3.8979	3.9000	3.9020	3.9040	3.9060	3.9080	3.9100
50	3.9120	3.9140	3.9160	3.9180	3.9200	3.9220	3.9240	3.9259	3.9279	3.9299
51	3.9318	3.9338	3.9357	3.9377	3.9396	3.9416	3.9435	3.9455	3.9474	3.9493
52	3.9512	3.9532	3.9551	3.9570	3.9589	3.9608	3.9627	3.9646	3.9665	3.9684
53	3.9703	3.9722	3.9741	3.9759	3.9778	3.9797	3.9815	3.9834	3.9853	3.9871
54	3.9890	3.9908	3.9927	3.9945	3.9964	3.9982	4.0000	4.0019	4.0037	4.0055

$\ln 100 = 4.6052 \qquad \ln 1000 = 6.9078 \qquad \ln 10{,}000 = 9.2103$

(*continued*)

Table I (*Cont.*) Four-place Natural Logarithms 55.0–99.9

N	.0	.1	.2	.3	.4	.5	.6	.7	.8	.9
55	4.0073	4.0091	4.0110	4.0128	4.0146	4.0164	4.0182	4.0200	4.0218	4.0236
56	4.0254	4.0271	4.0289	4.0307	4.0325	4.0342	4.0360	4.0378	4.0395	4.0413
57	4.0431	4.0448	4.0466	4.0483	4.0500	4.0518	4.0535	4.0553	4.0570	4.0587
58	4.0604	4.0622	4.0639	4.0656	4.0673	4.0690	4.0707	4.0724	4.0741	4.0758
59	4.0775	4.0792	4.0809	4.0826	4.0843	4.0860	4.0877	4.0893	4.0910	4.0927
60	4.0943	4.0960	4.0977	4.0993	4.1010	4.1026	4.1043	4.1059	4.1076	4.1092
61	4.1109	4.1125	4.1141	4.1158	4.1174	4.1190	4.1207	4.1223	4.1239	4.1255
62	4.1271	4.1287	4.1304	4.1320	4.1336	4.1352	4.1368	4.1384	4.1400	4.1415
63	4.1431	4.1447	4.1463	4.1479	4.1495	4.1510	4.1526	4.1542	4.1558	4.1573
64	4.1589	4.1604	4.1620	4.1636	4.1651	4.1667	4.1682	4.1698	4.1713	4.1728
65	4.1744	4.1759	4.1775	4.1790	4.1805	4.1821	4.1836	4.1851	4.1866	4.1881
66	4.1897	4.1912	4.1927	4.1942	4.1957	4.1972	4.1987	4.2002	4.2017	4.2032
67	4.2047	4.2062	4.2077	4.2092	4.2106	4.2121	4.2136	4.2151	4.2166	4.2180
68	4.2195	4.2210	4.2224	4.2239	4.2254	4.2268	4.2283	4.2297	4.2312	4.2327
69	4.2341	4.2356	4.2370	4.2384	4.2399	4.2413	4.2428	4.2442	4.2456	4.2471
70	4.2485	4.2499	4.2513	4.2528	4.2542	4.2556	4.2570	4.2584	4.2599	4.2613
71	4.2627	4.2641	4.2655	4.2669	4.2683	4.2697	4.2711	4.2725	4.2739	4.2753
72	4.2767	4.2781	4.2794	4.2808	4.2822	4.2836	4.2850	4.2863	4.2877	4.2891
73	4.2905	4.2918	4.2932	4.2946	4.2959	4.2973	4.2986	4.3000	4.3014	4.3027
74	4.3041	4.3054	4.3068	4.3081	4.3095	4.3108	4.3121	4.3135	4.3148	4.3162
75	4.3175	4.3188	4.3202	4.3215	4.3228	4.3241	4.3255	4.3268	4.3281	4.3294
76	4.3307	4.3320	4.3334	4.3347	4.3360	4.3373	4.3386	4.3399	4.3412	4.3425
77	4.3438	4.3451	4.3464	4.3477	4.3490	4.3503	4.3516	4.3529	4.3541	4.3554
78	4.3567	4.3580	4.3593	4.3605	4.3618	4.3631	4.3644	4.3656	4.3669	4.3682
79	4.3694	4.3707	4.3720	4.3732	4.3745	4.3758	4.3770	4.3783	4.3795	4.3808
80	4.3820	4.3833	4.3845	4.3858	4.3870	4.3883	4.3895	4.3907	4.3920	4.3932
81	4.3944	4.3957	4.3969	4.3981	4.3994	4.4006	4.4018	4.4031	4.4043	4.4055
82	4.4067	4.4079	4.4092	4.4104	4.4116	4.4128	4.4140	4.4152	4.4164	4.4176
83	4.4188	4.4200	4.4212	4.4224	4.4236	4.4248	4.4260	4.4272	4.4284	4.4296
84	4.4308	4.4320	4.4332	4.4344	4.4356	4.4368	4.4379	4.4391	4.4403	4.4415
85	4.4427	4.4438	4.4450	4.4462	4.4473	4.4485	4.4497	4.4509	4.4520	4.4532
86	4.4543	4.4555	4.4567	4.4578	4.4590	4.4601	4.4613	4.4625	4.4636	4.4648
87	4.4659	4.4671	4.4682	4.4694	4.4705	4.4716	4.4728	4.4739	4.4751	4.4762
88	4.4773	4.4785	4.4796	4.4807	4.4819	4.4830	4.4841	4.4853	4.4864	4.4875
89	4.4886	4.4898	4.4909	4.4920	4.4931	4.4942	4.4954	4.4965	4.4976	4.4987
90	4.4998	4.5009	4.5020	4.5031	4.5042	4.5053	4.5065	4.5076	4.5087	4.5098
91	4.5109	4.5120	4.5131	4.5142	4.5152	4.5163	4.5174	4.5185	4.5196	4.5207
92	4.5218	4.5229	4.5240	4.5250	4.5261	4.5272	4.5283	4.5294	4.5304	4.5315
93	4.5326	4.5337	4.5347	4.5358	4.5369	4.5380	4.5390	4.5401	4.5412	4.5422
94	4.5433	4.5444	4.5454	4.5465	4.5475	4.5486	4.5497	4.5507	4.5518	4.5528
95	4.5539	4.5549	4.5560	4.5570	4.5581	4.5591	4.5602	4.5612	4.5623	4.5633
96	4.5643	4.5654	4.5664	4.5675	4.5685	4.5695	4.5706	4.5716	4.5726	4.5737
97	4.5747	4.5757	4.5768	4.5778	4.5788	4.5799	4.5809	4.5819	4.5829	4.5839
98	4.5850	4.5860	4.5870	4.5880	4.5890	4.5901	4.5911	4.5921	4.5931	4.5941
99	4.5951	4.5961	4.5971	4.5981	4.5992	4.6002	4.6012	4.6022	4.6032	4.6042

$\ln 100{,}000 = 11.5129 \qquad \ln 1{,}000{,}000 = 13.8155 \qquad \ln 10{,}000{,}000 = 16.1181$

Table II Exponential and Hyperbolic Functions

x	e^x	e^{-x}	sinh x	cosh x	tanh x	x
0.0	1.0000	1.0000	.00000	1.0000	.00000	**0.0**
0.1	1.1052	.90484	.10017	1.0050	.09967	0.1
0.2	1.2214	.81873	.20134	1.0201	.19738	0.2
0.3	1.3499	.74082	.30452	1.0453	.29131	0.3
0.4	1.4918	.67032	.41075	1.0811	.37995	0.4
0.5	1.6487	.60653	.52110	1.1276	.46212	0.5
0.6	1.8221	.54881	.63665	1.1855	.53705	0.6
0.7	2.0138	.49659	.75858	1.2552	.60437	0.7
0.8	2.2255	.44933	.88811	1.3374	.66404	0.8
0.9	2.4596	.40657	1.0265	1.4331	.71630	0.9
1.0	2.7183	.36788	1.1752	1.5431	.76159	**1.0**
1.1	3.0042	.33287	1.3356	1.6685	.80050	1.1
1.2	3.3201	.30119	1.5095	1.8107	.83365	1.2
1.3	3.6693	.27253	1.6984	1.9709	.86172	1.3
1.4	4.0552	.24660	1.9043	2.1509	.88535	1.4
1.5	4.4817	.22313	2.1293	2.3524	.90515	1.5
1.6	4.9530	.20190	2.3756	2.5775	.92167	1.6
1.7	5.4739	.18268	2.6456	2.8283	.93541	1.7
1.8	6.0496	.16530	2.9422	3.1075	.94681	1.8
1.9	6.6859	.14957	3.2682	3.4177	.95624	1.9
2.0	7.3891	.13534	3.6269	3.7622	.96403	**2.0**
2.1	8.1662	.12246	4.0219	4.1443	.97045	2.1
2.2	9.0250	.11080	4.4571	4.5679	.97574	2.2
2.3	9.9742	.10026	4.9370	5.0372	.98010	2.3
2.4	11.023	.09072	5.4662	5.5569	.98367	2.4
2.5	12.182	.08208	6.0502	6.1323	.98661	2.5
2.6	13.464	.07427	6.6947	6.7690	.98903	2.6
2.7	14.880	.06721	7.4063	7.4735	.99101	2.7
2.8	16.445	.06081	8.1919	8.2527	.99263	2.8
2.9	18.174	.05502	9.0596	9.1146	.99396	2.9

(*continued*)

Table II (Cont.)

x	e^x	e^{-x}	sinh x	cosh x	tanh x	x
3.0	20.086	.04979	10.018	10.068	.99505	**3.0**
3.1	22.198	.04505	11.076	11.122	.99595	3.1
3.2	24.533	.04076	12.246	12.287	.99668	3.2
3.3	27.113	.03688	13.538	13.575	.99728	3.3
3.4	29.964	.03337	14.965	14.999	.99777	3.4
3.5	33.115	.03020	16.543	16.573	.99818	3.5
3.6	36.598	.02732	18.285	18.313	.99851	3.6
3.7	40.447	.02472	20.211	20.236	.99878	3.7
3.8	44.701	.02237	22.339	22.362	.99900	3.8
3.9	49.402	.02024	24.691	24.711	.99918	3.9
4.0	54.598	.01832	27.290	27.308	.99933	**4.0**
4.1	60.340	.01657	30.162	30.178	.99945	4.1
4.2	66.686	.01500	33.336	33.351	.99955	4.2
4.3	73.700	.01357	36.843	36.857	.99963	4.3
4.4	81.451	.01228	40.719	40.732	.99970	4.4
4.5	90.017	.01111	45.003	45.014	.99975	4.5
4.6	99.484	.01005	49.737	49.747	.99980	4.6
4.7	109.95	.00910	54.969	54.978	.99983	4.7
4.8	121.51	.00823	60.751	60.759	.99986	4.8
4.9	134.29	.00745	67.141	67.149	.99989	4.9
5.0	148.41	.00674	74.203	74.210	.99991	**5.0**
5.1	164.02	.00610	82.008	82.014	.99993	5.1
5.2	181.27	.00552	90.633	90.639	.99994	5.2
5.3	200.34	.00499	100.17	100.17	.99995	5.3
5.4	221.41	.00452	110.70	110.71	.99996	5.4
5.5	244.69	.00409	122.34	122.35	.99997	5.5
5.6	270.43	.00370	135.21	135.22	.99997	5.6
5.7	298.87	.00335	149.43	149.44	.99998	5.7
5.8	330.30	.00303	165.15	165.15	.99998	5.8
5.9	365.04	.00274	182.52	182.52	.99998	5.9
6.0	403.43	.00248	201.71	201.72	.99999	**6.0**
6.1	445.86	.00224	222.93	222.93	.99999	6.1
6.2	492.75	.00203	246.37	246.38	.99999	6.2
6.3	544.57	.00184	272.29	272.29	.99999	6.3
6.4	601.85	.00166	300.92	300.92	.99999	6.4
6.5	665.14	.00150	332.57	332.57	1.0000	6.5

Tables

Table III Trigonometric Functions For Angles in Radians

Rad.	Sin	Tan	Cot	Cos	Rad.	Sin	Tan	Cot	Cos
0.00	.00000	.00000	–	1.00000	**0.40**	.38942	.42279	2.3652	.92106
.01	.01000	.01000	99.997	0.99995	.41	.39861	.43463	2.3008	.91712
.02	.02000	.02000	49.993	.99980	.42	.40776	.44657	2.2393	.91309
.03	.03000	.03001	33.323	.99955	.43	.41687	.45862	2.1804	.90897
.04	.03999	.04002	24.987	.99920	.44	.42594	.47078	2.1241	.90475
.05	.04998	.05004	19.983	.99875	.45	.43497	.48306	2.0702	.90045
.06	.05996	.06007	16.647	.99820	.46	.44395	.49545	2.0184	.89605
.07	.06994	.07011	14.262	.99755	.47	.45289	.50797	1.9686	.89157
.08	.07991	.08017	12.473	.99680	.48	.46178	.52061	1.9208	.88699
.09	.08988	.09024	11.081	.99595	.49	.47063	.53339	1.8748	.88233
0.10	.09983	.10033	9.9666	.99500	**0.50**	.47943	.54630	1.8305	.87758
.11	.10978	.11045	9.0542	.99396	.51	.48818	.55936	1.7878	.87274
.12	.11971	.12058	8.2933	.99281	.52	.49688	.57256	1.7465	.86782
.13	.12963	.13074	7.6489	.99156	.53	.50553	.58592	1.7067	.86281
.14	.13954	.14092	7.0961	.99022	.54	.51414	.59943	1.6683	.85771
.15	.14944	.15114	6.6166	.98877	.55	.52269	.61311	1.6310	.85252
.16	.15932	.16138	6.1966	.98723	.56	.53119	.62695	1.5950	.84726
.17	.16918	.17166	5.8256	.98558	.57	.53963	.64097	1.5601	.84190
.18	.17903	.18197	5.4954	.98384	.58	.54802	.65517	1.5263	.83646
.19	.18886	.19232	5.1997	.98200	.59	.55636	.66956	1.4935	.83094
0.20	.19867	.20271	4.9332	.98007	**0.60**	.56464	.68414	1.4617	.82534
.21	.20846	.21314	4.6917	.97803	.61	.57287	.69892	1.4308	.81965
.22	.21823	.22362	4.4719	.97590	.62	.58104	.71391	1.4007	.81388
.23	.22798	.23414	4.2709	.97367	.63	.58914	.72911	1.3715	.80803
.24	.23770	.24472	4.0864	.97134	.64	.59720	.74454	1.3431	.80210
.25	.24740	.25534	3.9163	.96891	.65	.60519	.76020	1.3154	.79608
.26	.25708	.26602	3.7591	.96639	.66	.61312	.77610	1.2885	.78999
.27	.26673	.27676	3.6133	.96377	.67	.62099	.79225	1.2622	.78382
.28	.27636	.28755	3.4776	.96106	.68	.62879	.80866	1.2366	.77757
.29	.28595	.29841	3.3511	.95824	.69	.63654	.82534	1.2116	.77125
0.30	.29552	.30934	3.2327	.95534	**0.70**	.64422	.84229	1.1872	.76484
.31	.30506	.32033	3.1218	.95233	.71	.65183	.85953	1.1634	.75836
.32	.31457	.33139	3.0176	.94924	.72	.65938	.87707	1.1402	.75181
.33	.32404	.34252	2.9195	.94604	.73	.66687	.89492	1.1174	.74517
.34	.33349	.35374	2.8270	.94275	.74	.67429	.91309	1.0952	.73847
.35	.34290	.36503	2.7395	.93937	.75	.68164	.93160	1.0734	.73169
.36	.35227	.37640	2.6567	.93590	.76	.68892	.95045	1.0521	.72484
.37	.36162	.38786	2.5782	.93233	.77	.69614	.96967	1.0313	.71791
.38	.37092	.39941	2.5037	.92866	.78	.70328	.98926	1.0109	.71091
.39	.38019	.41105	2.4328	.92491	.79	.71035	1.0092	.99084	.70385
0.40	.38942	.42279	2.3652	.92106	**0.80**	.71736	1.0296	.97121	.69671

(*continued*)

Table III (*Cont.*) Trigonometric Functions for Angles in Radians

Rad.	Sin	Tan	Cot	Cos	Rad.	Sin	Tan	Cot	Cos
0.80	.71736	1.0296	.97121	.69671	**1.20**	.93204	2.5722	.38878	.36236
.81	.72429	1.0505	.95197	.68950	1.21	.93562	2.6503	.37731	.35302
.82	.73115	1.0717	.93309	.68222	1.22	.93910	2.7328	.36593	.34365
.83	.73793	1.0934	.91455	.67488	1.23	.94249	2.8198	.35463	.33424
.84	.74464	1.1156	.89635	.66746	1.24	.94578	2.9119	.34341	.32480
.85	.75128	1.1383	.87848	.65998	1.25	.94898	3.0096	.33227	.31532
.86	.75784	1.1616	.86091	.65244	1.26	.95209	3.1133	.32121	.30582
.87	.76433	1.1853	.84365	.64483	1.27	.95510	3.2236	.31021	.29628
.88	.77074	1.2097	.82668	.63715	1.28	.95802	3.3413	.29928	.28672
.89	.77707	1.2346	.80998	.62941	1.29	.96084	3.4672	.28842	.27712
0.90	.78333	1.2602	.79355	.62161	**1.30**	.96356	3.6021	.27762	.26750
.91	.78950	1.2864	.77738	.61375	1.31	.96618	3.7471	.26687	.25785
.92	.79560	1.3133	.76146	.60582	1.32	.96872	3.9033	.25619	.24818
.93	.80162	1.3409	.74578	.59783	1.33	.97115	4.0723	.24556	.23848
.94	.80756	1.3692	.73034	.58979	1.34	.97438	4.2556	.23498	.22875
.95	.81342	1.3984	.71511	.58168	1.35	.97572	4.4552	.22446	.21901
.96	.81919	1.4284	.70010	.57352	1.36	.97786	4.6734	.21398	.20924
.97	.82489	1.4592	.68531	.56530	1.37	.97991	4.9131	.20354	.19945
.98	.83050	1.4910	.67071	.55702	1.38	.98185	5.1774	.19315	.18964
.99	.83603	1.5237	.65631	.54869	1.39	.98370	5.4707	.18279	.17981
1.00	.84147	1.5574	.64209	.54030	**1.40**	.98545	5.7979	.17248	.16997
1.01	.84683	1.5922	.62806	.53186	1.41	.98710	6.1654	.16220	.16010
1.02	.85221	1.6281	.61420	.52337	1.42	.98865	6.5811	.15195	.15023
1.03	.85730	1.6652	.60051	.51482	1.43	.99010	7.0555	.14173	.14033
1.04	.86240	1.7036	.58699	.50622	1.44	.99146	7.6018	.13155	.13042
1.05	.86742	1.7433	.57362	.49757	1.45	.99271	8.2381	.12139	.12050
1.06	.87236	1.7844	.56040	.48887	1.46	.99387	8.9886	.11125	.11057
1.07	.87720	1.8270	.54734	.48012	1.47	.99492	9.8874	.10114	.10063
1.08	.88196	1.8712	.53441	.47133	1.48	.99588	10.983	.09105	.09067
1.09	.88663	1.9171	.52162	.46249	1.49	.99674	12.350	.08097	.08071
1.10	.89121	1.9648	.50897	.45360	**1.50**	.99749	14.101	.07091	.07074
1.11	.89570	2.0143	.49644	.44466	1.51	.99815	16.428	.06087	.06076
1.12	.90010	2.0660	.48404	.43568	1.52	.99871	19.670	.05084	.05077
1.13	.90441	2.1198	.47175	.42666	1.53	.99917	24.498	.04082	.04079
1.14	.90863	2.1759	.45959	.41759	1.54	.99953	32.461	.03081	.03079
1.15	.91276	2.2345	.44753	.40849	1.55	.99978	48.078	.02080	.02079
1.16	.91680	2.2958	.43558	.39934	1.56	.99994	92.621	.01080	.01080
1.17	.92075	2.3600	.42373	.39015	1.57	1.00000	1255.8	.00080	.00080
1.18	.92461	2.4273	.41199	.38092	1.58	.99996	− 108.65	− .00920	− .00920
1.19	.92837	2.4979	.40034	.37166	1.59	.99982	− 52.067	− .01921	− .01920
1.20	.93204	2.5722	.38878	.36236	**1.60**	.99957	− 34.233	− .02921	− .02920

Tables

Table IV Radians to Degrees, Minutes and Seconds

	Radians	Tenths	Hundredths	Thousandths	Ten-thousandths
1	57°17′44.″8	5°43′46.″5	0°34′22.″6	0° 3′26.″3	0°0′20.″6
2	114°35′29.″6	11°27′33.″0	1° 8′45.″3	0° 6′52.″5	0°0′41.″3
3	171°53′14.″4	17°11′19.″4	1°43′07.″9	0°10′18.″8	0°1′01.″9
4	229°10′59.″2	22°55′05.″9	2°17′30.″6	0°13′45.″1	0°1′22.″5
5	286°28′44.″0	28°38′52.″4	2°51′53.″2	0°17′11.″3	0°1′43.″1
6	343°46′28.″8	34°22′38.″9	3°26′15.″9	0°20′37.″6	0°2′03.″8
7	401° 4′13.″6	40° 6′25.″4	4° 0′38.″5	0°24′03.″9	0°2′24.″4
8	458°21′58.″4	45°50′11.″8	4°35′01.″2	0°27′30.″1	0°2′45.″0
9	515°39′43.″3	51°33′58.″3	5° 9′23.″8	0°30′56.″4	0°3′05.″6

$1° \approx 0.0174532925$ rad.
$1′ \approx 0.0002908882$ rad.
$1″ \approx 0.0000048481$ rad.

Table V Indefinite Integrals (Continued from endpapers)

Forms Involving $ax + b$

I32 $\quad \int x(ax + b)^m \, dx = \dfrac{(ax + b)^{m+2}}{a^2(m + 2)} - \dfrac{b(ax + b)^{m+1}}{a^2(m + 1)}, \qquad (m \neq -1, -2)$

I33 $\quad \int \dfrac{x \, dx}{ax + b} = \dfrac{x}{a} - \dfrac{b}{a^2} \ln |ax + b|$

I34 $\quad \int \dfrac{x \, dx}{(ax + b)^2} = \dfrac{b}{a^2(ax + b)} + \dfrac{1}{a^2} \ln |ax + b|$

I35 $\quad \int \dfrac{x \, dx}{(ax + b)^3} = \dfrac{b}{2a^2(ax + b)^2} - \dfrac{1}{a^2(ax + b)}$

I36 $\quad \int x^2(ax + b)^m \, dx = \dfrac{1}{a^3} \left[\dfrac{(ax + b)^{m+3}}{m + 3} - \dfrac{2b(ax + b)^{m+2}}{m + 2} + \dfrac{b^2(ax + b)^{m+1}}{m + 1} \right] \qquad (m \neq -1, -2, -3)$

I37 $\quad \int \dfrac{x^2 \, dx}{ax + b} = \dfrac{1}{a^3} \left[\dfrac{1}{2} (ax + b)^2 - 2b(ax + b) + b^2 \ln |ax + b| \right]$

I38 $\quad \int \dfrac{x^2 \, dx}{(ax + b)^2} = \dfrac{1}{a^3} \left[(ax + b) - \dfrac{b^2}{ax + b} - 2b \ln |ax + b| \right]$

I39 $\quad \int \dfrac{x^2 \, dx}{(ax + b)^3} = \dfrac{1}{a^3} \left[\ln |ax + b| + \dfrac{2b}{ax + b} - \dfrac{b^2}{2(ax + b)^2} \right]$

I40 $\quad \int x^m(ax + b)^n \, dx = \dfrac{1}{a(m + n + 1)} \left[x^m(ax + b)^{n+1} - mb \int x^{m-1}(ax + b)^n \, dx \right]$

$\qquad \qquad \qquad \qquad = \dfrac{1}{m + n + 1} \left[x^{m+1}(ax + b)^n + nb \int x^m(ax + b)^{n-1} \, dx \right]$

$\qquad \qquad \qquad \qquad \qquad \qquad \qquad \qquad (m > 0, \; m + n + 1 \neq 0)$

(continued)

Table V (*Cont.*) Indefinite Integrals (Continued)

I41 $\quad \int x^m(ax+b)^n \, dx = \dfrac{1}{a^{m+1}} \int u^n(u-b)^m \, du \qquad\qquad (u = ax+b)$

I42 $\quad \int \dfrac{x^m \, dx}{(ax+b)^n} = \dfrac{1}{a^{m+1}} \int \dfrac{(u-b)^m \, du}{u^n} \qquad\qquad (u = ax+b)$

I43 $\quad \int \dfrac{dx}{x^m(ax+b)^n} = -\dfrac{1}{b^{m+n-1}} \int \dfrac{(v-a)^{m+n-2} \, dv}{v^n} \qquad\qquad \left(v = \dfrac{ax+b}{x}\right)$

Forms Involving $\sqrt{ax+b}$, Where $ax+b > 0$

I44 $\quad \int x\sqrt{ax+b} \, dx = \dfrac{2(3ax-2b)}{15a^2}\sqrt{(ax+b)^3}$

I45 $\quad \int x^2\sqrt{ax+b} \, dx = \dfrac{2(15a^2x^2 - 12abx + 8b^2)}{105a^3}\sqrt{(ax+b)^3}$

I46 $\quad \int x^m\sqrt{ax+b} \, dx = \dfrac{2}{a(2m+3)}\left[x^m\sqrt{(ax+b)^3} - mb \int x^{m-1}\sqrt{ax+b} \, dx\right]$

I47 $\quad \int \dfrac{\sqrt{ax+b} \, dx}{x} = 2\sqrt{ax+b} + \sqrt{b} \ln \left|\dfrac{\sqrt{ax+b} - \sqrt{b}}{\sqrt{ax+b} + \sqrt{b}}\right| \qquad\qquad (b > 0)$

I48 $\quad \int \dfrac{\sqrt{ax+b} \, dx}{x} = 2\sqrt{ax+b} - 2\sqrt{-b}\, \tan^{-1}\sqrt{\dfrac{ax+b}{-b}} \qquad\qquad (b < 0)$

I49 $\quad \int \dfrac{\sqrt{ax+b}}{x^2} \, dx = -\dfrac{\sqrt{ax+b}}{x} + \dfrac{a}{2}\int \dfrac{dx}{x\sqrt{ax+b}}$

I50 $\quad \int \dfrac{\sqrt{ax+b} \, dx}{x^m} = -\dfrac{1}{(m-1)b}\left[\dfrac{\sqrt{(ax+b)^3}}{x^{m-1}} + \dfrac{(2m-5)a}{2}\int \dfrac{\sqrt{ax+b} \, dx}{x^{m-1}}\right] \qquad (m \neq 1)$

I51 $\quad \int \dfrac{dx}{\sqrt{ax+b}} = \dfrac{2\sqrt{ax+b}}{a}$

I52 $\quad \int \dfrac{x \, dx}{\sqrt{ax+b}} = \dfrac{2(ax-2b)\sqrt{ax+b}}{3a^2}$

I53 $\quad \int \dfrac{x^m \, dx}{\sqrt{ax+b}} = \dfrac{2}{a(2m+1)}\left[x^m\sqrt{ax+b} - mb\int \dfrac{x^{m-1} \, dx}{\sqrt{ax+b}}\right] \qquad \left(m \neq -\dfrac{1}{2}\right)$

I54 $\quad \int \dfrac{dx}{x\sqrt{ax+b}} = \dfrac{1}{\sqrt{b}} \ln \left|\dfrac{\sqrt{ax+b} - \sqrt{b}}{\sqrt{ax+b} + \sqrt{b}}\right| \qquad\qquad (b > 0)$

I55 $\quad \int \dfrac{dx}{x\sqrt{ax+b}} = \dfrac{2}{\sqrt{-b}}\tan^{-1}\sqrt{\dfrac{ax+b}{-b}} \qquad\qquad (b < 0)$

I56 $\quad \int \dfrac{dx}{x^m\sqrt{ax+b}} = -\dfrac{\sqrt{ax+b}}{(m-1)bx^{m-1}} - \dfrac{(2m-3)a}{(2m-2)b}\int \dfrac{dx}{x^{m-1}\sqrt{ax+b}} \qquad (m \neq 1)$

Forms Involving $\sqrt{a^2 + x^2}$ or $\sqrt{(a^2 + x^2)^3}$

I57 $\quad \int \sqrt{a^2 + x^2} \, dx = \dfrac{x}{2}\sqrt{a^2 + x^2} + \dfrac{a^2}{2}\ln(x + \sqrt{a^2 + x^2})$

I58 $\quad \int x\sqrt{a^2 + x^2} \, dx = \tfrac{1}{3}\sqrt{(a^2 + x^2)^3}$

Tables

Table V (*Cont.*) Indefinite Integrals (Continued)

I59 $\quad \int x^2 \sqrt{a^2 + x^2}\, dx = \dfrac{x}{4}\sqrt{(a^2+x^2)^3} - \dfrac{a^2 x}{8}\sqrt{a^2+x^2} - \dfrac{a^4}{8}\ln(x + \sqrt{a^2+x^2})$

I60 $\quad \int x^3 \sqrt{a^2 + x^2}\, dx = (\tfrac{1}{5}x^2 - \tfrac{2}{15}a^2)\sqrt{(a^2+x^2)^3}$

I61 $\quad \int \dfrac{\sqrt{a^2+x^2}\, dx}{x} = \sqrt{a^2+x^2} - a\ln\left|\dfrac{a+\sqrt{a^2+x^2}}{x}\right|$

I62 $\quad \int \dfrac{\sqrt{a^2+x^2}\, dx}{x^2} = -\dfrac{\sqrt{a^2+x^2}}{x} + \ln(x + \sqrt{a^2+x^2})$

I63 $\quad \int \dfrac{\sqrt{a^2+x^2}\, dx}{x^3} = -\dfrac{\sqrt{a^2+x^2}}{2x^2} - \dfrac{1}{2a}\ln\left|\dfrac{a+\sqrt{a^2+x^2}}{x}\right|$

I64 $\quad \int \dfrac{dx}{\sqrt{a^2+x^2}} = \ln(x + \sqrt{a^2+x^2})$

I65 $\quad \int \dfrac{x\, dx}{\sqrt{a^2+x^2}} = \sqrt{a^2+x^2}$

I66 $\quad \int \dfrac{x^2\, dx}{\sqrt{a^2+x^2}} = \dfrac{x}{2}\sqrt{a^2+x^2} - \dfrac{a^2}{2}\ln(x + \sqrt{a^2+x^2})$

I67 $\quad \int \dfrac{x^3\, dx}{\sqrt{a^2+x^2}} = \dfrac{1}{3}\sqrt{(a^2+x^2)^3} - a^2\sqrt{a^2+x^2}$

I68 $\quad \int \dfrac{dx}{x\sqrt{a^2+x^2}} = -\dfrac{1}{a}\ln\left|\dfrac{a+\sqrt{a^2+x^2}}{x}\right|$

I69 $\quad \int \dfrac{dx}{x^2\sqrt{a^2+x^2}} = -\dfrac{\sqrt{a^2+x^2}}{a^2 x}$

I70 $\quad \int \dfrac{dx}{x^3\sqrt{a^2+x^2}} = -\dfrac{\sqrt{a^2+x^2}}{2a^2 x^2} + \dfrac{1}{2a^3}\ln\left|\dfrac{a+\sqrt{a^2+x^2}}{x}\right|$

I71 $\quad \int \sqrt{(a^2+x^2)^3}\, dx = \dfrac{x}{4}\sqrt{(a^2+x^2)^3} + \dfrac{3a^2 x}{8}\sqrt{a^2+x^2} + \dfrac{3a^4}{8}\ln(x + \sqrt{a^2+x^2})$

Forms Involving $\sqrt{a^2 - x^2}$ or $\sqrt{(a^2-x^2)^3}$, Where $a^2 > x^2$, $a > 0$

I72 $\quad \int \sqrt{a^2-x^2}\, dx = \dfrac{1}{2}\left(x\sqrt{a^2-x^2} + a^2 \sin^{-1}\dfrac{x}{a}\right)$

I73 $\quad \int x^2 \sqrt{a^2-x^2}\, dx = -\dfrac{x}{4}\sqrt{(a^2-x^2)^3} + \dfrac{a^2}{8}\left(x\sqrt{a^2-x^2} + a^2\sin^{-1}\dfrac{x}{a}\right)$

I74 $\quad \int x^3 \sqrt{a^2-x^2}\, dx = (-\tfrac{1}{5}x^2 - \tfrac{2}{15}a^2)\sqrt{(a^2-x^2)^3}$

I75 $\quad \int \dfrac{\sqrt{a^2-x^2}\, dx}{x} = \sqrt{a^2-x^2} - a\ln\left|\dfrac{a+\sqrt{a^2-x^2}}{x}\right|$

I76 $\quad \int \dfrac{\sqrt{a^2-x^2}\, dx}{x^2} = -\dfrac{\sqrt{a^2-x^2}}{x} - \sin^{-1}\dfrac{x}{a}$

I77 $\quad \int \dfrac{\sqrt{a^2-x^2}\, dx}{x^3} = -\dfrac{\sqrt{a^2-x^2}}{2x^2} + \dfrac{1}{2a}\ln\left|\dfrac{a+\sqrt{a^2-x^2}}{x}\right|$

(*continued*)

Table V (*Cont.*) Indefinite Integrals (Continued)

I78 $\quad \displaystyle\int \frac{x^2\,dx}{\sqrt{a^2-x^2}} = -\frac{x}{2}\sqrt{a^2-x^2} + \frac{a^2}{2}\sin^{-1}\frac{x}{a}$

I79 $\quad \displaystyle\int \frac{x^3\,dx}{\sqrt{a^2-x^2}} = \frac{1}{3}\sqrt{(a^2-x^2)^3} - a^2\sqrt{a^2-x^2}$

I80 $\quad \displaystyle\int \frac{dx}{x\sqrt{a^2-x^2}} = -\frac{1}{a}\ln\left|\frac{a+\sqrt{a^2-x^2}}{x}\right|$

I81 $\quad \displaystyle\int \frac{dx}{x^2\sqrt{a^2-x^2}} = -\frac{\sqrt{a^2-x^2}}{a^2 x}$

I82 $\quad \displaystyle\int \frac{dx}{x^3\sqrt{a^2-x^2}} = -\frac{\sqrt{a^2-x^2}}{2a^2 x^2} - \frac{1}{2a^3}\ln\left|\frac{a+\sqrt{a^2-x^2}}{x}\right|$

I83 $\quad \displaystyle\int \sqrt{(a^2-x^2)^3}\,dx = \frac{1}{4}x\sqrt{(a^2-x^2)^3} + \frac{3a^2 x}{8}\sqrt{a^2-x^2} + \frac{3a^4}{8}\sin^{-1}\frac{x}{a}$

I84 $\quad \displaystyle\int x^2\sqrt{(a^2-x^2)^3}\,dx = -\frac{1}{6}x\sqrt{(a^2-x^2)^5} + \frac{a^2 x}{24}\sqrt{(a^2-x^2)^3} + \frac{a^4 x}{16}\sqrt{a^2-x^2} + \frac{a^6}{16}\sin^{-1}\frac{x}{a}$

Forms Involving $\sqrt{x^2-a^2}$ or $\sqrt{(x^2-a^2)^3}$, Where $a^2 < x^2$, $a > 0$

I85 $\quad \displaystyle\int \sqrt{x^2-a^2}\,dx = \frac{x}{2}\sqrt{x^2-a^2} - \frac{a^2}{2}\ln\left|x+\sqrt{x^2-a^2}\right|$

I86 $\quad \displaystyle\int x^2\sqrt{x^2-a^2}\,dx = \frac{x}{4}\sqrt{(x^2-a^2)^3} + \frac{a^2 x}{8}\sqrt{x^2-a^2} - \frac{a^4}{8}\ln\left|x+\sqrt{x^2-a^2}\right|$

I87 $\quad \displaystyle\int x^3\sqrt{x^2-a^2}\,dx = \frac{1}{5}\sqrt{(x^2-a^2)^5} + \frac{a^2}{3}\sqrt{(x^2-a^2)^3}$

I88 $\quad \displaystyle\int \frac{\sqrt{x^2-a^2}\,dx}{x} = \sqrt{x^2-a^2} - a\cos^{-1}\frac{a}{x}$

I89 $\quad \displaystyle\int \frac{\sqrt{x^2-a^2}\,dx}{x^2} = -\frac{1}{x}\sqrt{x^2-a^2} + \ln\left|x+\sqrt{x^2-a^2}\right|$

I90 $\quad \displaystyle\int \frac{\sqrt{x^2-a^2}\,dx}{x^3} = -\frac{\sqrt{x^2-a^2}}{2x^2} + \frac{1}{2a}\cos^{-1}\frac{a}{x}$

I91 $\quad \displaystyle\int \frac{dx}{\sqrt{x^2-a^2}} = \ln\left|x+\sqrt{x^2-a^2}\right|$

I92 $\quad \displaystyle\int \frac{x^2\,dx}{\sqrt{x^2-a^2}} = \frac{x}{2}\sqrt{x^2-a^2} + \frac{a^2}{2}\ln\left|x+\sqrt{x^2-a^2}\right|$

I93 $\quad \displaystyle\int \frac{x^3\,dx}{\sqrt{x^2-a^2}} = \frac{1}{3}\sqrt{(x^2-a^2)^3} + a^2\sqrt{x^2-a^2}$

I94 $\quad \displaystyle\int \frac{dx}{x\sqrt{x^2-a^2}} = \frac{1}{a}\cos^{-1}\frac{a}{x}$

I95 $\quad \displaystyle\int \frac{dx}{x^2\sqrt{x^2-a^2}} = \frac{\sqrt{x^2-a^2}}{a^2 x}$

Tables

Table V (Cont.) Indefinite Integrals (Continued)

I96 $\quad \displaystyle\int \frac{dx}{x^3\sqrt{x^2-a^2}} = \frac{\sqrt{x^2-a^2}}{2a^2x^2} + \frac{1}{2a^3}\cos^{-1}\frac{a}{x}$

Forms Involving sin x

I97 $\quad \displaystyle\int \sin^2 x\, dx = \frac{x}{2} - \frac{\sin 2x}{4}$

I98 $\quad \displaystyle\int \sin^3 x\, dx = \frac{\cos^3 x}{3} - \cos x$

I99 $\quad \displaystyle\int \sin^4 x\, dx = \frac{3x}{8} - \frac{\sin 2x}{4} + \frac{\sin 4x}{32}$

I100 $\quad \displaystyle\int \sin^{2m} x\, dx = -\frac{\sin^{2m-1} x \cos x}{2m} + \frac{2m-1}{2m}\int \sin^{2(m-1)} x\, dx$

I101 $\quad \displaystyle\int \sin^{2m+1} x\, dx = \int (1-\cos^2 x)^m \sin x\, dx$

I102 $\quad \displaystyle\int x \sin x\, dx = \sin x - x\cos x$

I103 $\quad \displaystyle\int x^m \sin x\, dx = -x^m \cos x + mx^{m-1}\sin x - m(m-1)\int x^{m-2}\sin x\, dx$

I104 $\quad \displaystyle\int x \sin^2 x\, dx = \frac{x^2}{4} - \frac{x\sin 2x}{4} - \frac{\cos 2x}{8}$

I105 $\quad \displaystyle\int x^2 \sin^2 x\, dx = \frac{x^3}{6} - \left(\frac{x^2}{4} - \frac{1}{8}\right)\sin 2x - \frac{x\cos 2x}{4}$

I106 $\quad \displaystyle\int \frac{x\, dx}{\sin^2 x} = -x\cot x + \ln|\sin x|$

Forms Involving cos x

I107 $\quad \displaystyle\int \cos^2 x\, dx = \frac{x}{2} + \frac{\sin 2x}{4}$

I108 $\quad \displaystyle\int \cos^3 x\, dx = \sin x - \frac{\sin^3 x}{3}$

I109 $\quad \displaystyle\int \cos^4 x\, dx = \frac{3x}{8} + \frac{\sin 2x}{4} + \frac{\sin 4x}{32}$

I110 $\quad \displaystyle\int \cos^{2m} x\, dx = \frac{1}{2m}\cos^{2m-1} x \sin x + \frac{2m-1}{2m}\int \cos^{2(m-1)} x\, dx$

I111 $\quad \displaystyle\int \cos^{2m+1} x\, dx = \int (1-\sin^2 x)^m \cos x\, dx.$

I112 $\quad \displaystyle\int \frac{dx}{\cos^m x} = \int \sec^m x\, dx = \frac{\sin x}{(m-1)\cos^{m-1} x} + \frac{m-2}{m-1}\int \frac{dx}{\cos^{m-2} x} \qquad (m \neq 1)$

I113 $\quad \displaystyle\int x \cos x\, dx = \cos x + x\sin x$

(continued)

Table V (Cont.) Indefinite Integrals (Continued)

I114 $\quad \int x^m \cos x \, dx = x^m \sin x + m x^{m-1} \cos x - m(m-1) \int x^{m-2} \cos x \, dx$

I115 $\quad \int x \cos^2 x \, dx = \dfrac{x^2}{4} + \dfrac{x \sin 2x}{4} + \dfrac{\cos 2x}{8}$

I116 $\quad \int x^2 \cos^2 x \, dx = \dfrac{x^3}{6} + \left(\dfrac{x^2}{4} - \dfrac{1}{8} \right) \sin 2x + \dfrac{x \cos 2x}{4}$

I117 $\quad \int \dfrac{x \, dx}{\cos^2 x} = x \tan x + \ln |\cos x|$

Forms Involving tan x, cot x, sec x, csc x

I118 $\quad \int \tan x \, dx = \ln |\sec x|$

I119 $\quad \int \tan^2 x \, dx = \tan x - x$

I120 $\quad \int \tan^m x \, dx = \dfrac{\tan^{m-1} x}{m - 1} - \int \tan^{m-2} x \, dx \qquad\qquad (m \neq 1)$

I121 $\quad \int \cot x \, dx = \ln |\sin x|$

I122 $\quad \int \cot^2 x \, dx = - \cot x - x$

I123 $\quad \int \cot^m x \, dx = - \dfrac{\cot^{m-1} x}{m - 1} - \int \cot^{m-2} x \, dx \qquad\qquad (m \neq 1)$

I124 $\quad \int \sec x \, dx = \ln |\sec x + \tan x| = \ln \left| \tan \left(\dfrac{\pi}{4} + \dfrac{x}{2} \right) \right|$

I125 $\quad \int \sec^m x \, dx = \dfrac{\sin x}{(m-1) \cos^{m-1} x} + \dfrac{m-2}{m-1} \int \sec^{m-2} x \, dx \qquad (m \neq 1)$

I126 $\quad \int \csc x \, dx = \ln |\csc x - \cot x| = \ln \left| \tan \dfrac{x}{2} \right|$

I127 $\quad \int \csc^m x \, dx = - \dfrac{\cos x}{(m-1) \sin^{m-1} x} + \dfrac{m-2}{m-1} \int \csc^{m-2} x \, dx \qquad (m \neq 1)$

Exponential Forms

For integrals involving a^x, substitute $a^x = e^{(\ln a)x}$ and use the forms that follow.

I128 $\quad \int x e^{ax} \, dx = \dfrac{e^{ax}}{a^2} (ax - 1)$

I129 $\quad \int x^m e^{ax} \, dx = \dfrac{1}{a} x^m e^{ax} - \dfrac{m}{a} \int x^{m-1} e^{ax} \, dx$

I130 $\quad \int \dfrac{e^{ax} \, dx}{x} = \ln |x| + ax + \dfrac{(ax)^2}{2 \cdot 2!} + \dfrac{(ax)^3}{3 \cdot 3!} + \cdots \qquad\qquad (0 < x^2 < \infty)$

I131 $\quad \int \dfrac{e^{ax} \, dx}{x^m} = - \dfrac{e^{ax}}{(m-1) x^{m-1}} + \dfrac{a}{m-1} \int \dfrac{e^{ax} \, dx}{x^{m-1}} \qquad\qquad (m \neq 1)$

Tables

Table V (*Cont.*) Indefinite Integrals (Continued)

I132 $\quad \displaystyle\int \frac{e^{ax}\, dx}{b + ce^{ax}} = \frac{1}{ac} \ln |b + ce^{ax}|$

I133 $\quad \displaystyle\int \frac{dx}{b + ce^{ax}} = \frac{1}{ab} \ln \left| \frac{e^{ax}}{b + ce^{ax}} \right|$

I134 $\quad \displaystyle\int \frac{dx}{ae^{cx} + be^{-cx}} = \frac{1}{c\sqrt{ab}} \tan^{-1}\left(e^{cx} \sqrt{\frac{a}{b}} \right) \qquad (ab > 0)$

I135 $\quad \displaystyle\int e^{ax} \sin bx\, dx = \frac{e^{ax}}{a^2 + b^2}(a \sin bx - b \cos bx)$

I136 $\quad \displaystyle\int e^{ax} \sin^m bx\, dx = \frac{e^{ax}(a \sin bx - mb \cos bx)\sin^{m-1} bx}{a^2 + m^2 b^2} + \frac{m(m-1)b^2}{a^2 + m^2 b^2}\int e^{ax} \sin^{m-2} bx\, dx$

I137 $\quad \displaystyle\int e^{ax} \cos bx\, dx = \frac{e^{ax}}{a^2 + b^2}(a \cos bx + b \sin bx)$

I138 $\quad \displaystyle\int e^{ax} \cos^m bx\, dx = \frac{e^{ax}(a \cos bx + mb \sin bx)\cos^{m-1} bx}{a^2 + m^2 b^2} + \frac{m(m-1)b^2}{a^2 + m^2 b^2}\int e^{ax} \cos^{m-2} bx\, dx$

I139 $\quad \displaystyle\int e^{ax} \ln bx\, dx = \frac{1}{a} e^{ax} \ln bx - \frac{1}{a} \int \frac{e^{ax}\, dx}{x} \qquad (bx > 0)$

Logarithmic Forms

In these forms, $x > 0$.

I140 $\quad \displaystyle\int (\ln x)^m\, dx = x(\ln x)^m - m \int (\ln x)^{m-1}\, dx$

I141 $\quad \displaystyle\int \frac{dx}{(\ln x)^m} = -\frac{x}{(m-1)(\ln x)^{m-1}} + \frac{1}{m-1} \int \frac{dx}{(\ln x)^{m-1}}$

I142 $\quad \displaystyle\int x^m \ln x\, dx = x^{m+1}\left[\frac{\ln x}{m+1} - \frac{1}{(m+1)^2} \right] \qquad (m \neq -1)$

I143 $\quad \displaystyle\int x^m (\ln x)^n\, dx = \frac{x^{m+1}}{m+1}(\ln x)^n - \frac{n}{m+1} \int x^m (\ln x)^{n-1}\, dx \qquad (m, n \neq -1)$

I144 $\quad \displaystyle\int \frac{x^m\, dx}{(\ln x)^n} = -\frac{x^{m+1}}{(n-1)(\ln x)^{n-1}} + \frac{m+1}{n-1} \int \frac{x^m\, dx}{(\ln x)^{n-1}} \qquad (n \neq 1)$

Forms Involving Inverse Trigonometric Functions

I145 $\quad \displaystyle\int \sin^{-1} x\, dx = x \sin^{-1} x + \sqrt{1 - x^2}$

I146 $\quad \displaystyle\int (\sin^{-1} x)^2\, dx = x(\sin^{-1} x)^2 - 2x + 2\sqrt{1 - x^2} \sin^{-1} x$

I147 $\quad \displaystyle\int x \sin^{-1} x\, dx = \tfrac{1}{4}[(2x^2 - 1)\sin^{-1} x + x\sqrt{1 - x^2}]$

I148 $\quad \displaystyle\int x^m \sin^{-1} x\, dx = \frac{x^{m+1}}{m+1} \sin^{-1} x - \frac{1}{m+1} \int \frac{x^{m+1}\, dx}{\sqrt{1 - x^2}} \qquad (m \neq -1)$

I149 $\quad \displaystyle\int \cos^{-1} x\, dx = x \cos^{-1} x - \sqrt{1 - x^2}$

(continued)

Table V (Cont.) Indefinite Integrals (Continued)

I150 $\int (\cos^{-1} x)^2 \, dx = x(\cos^{-1} x)^2 - 2x - 2\sqrt{1-x^2} \cos^{-1} x$

I151 $\int x \cos^{-1} x \, dx = \frac{1}{4}[(2x^2 - 1) \cos^{-1} x - x\sqrt{1-x^2}]$

I152 $\int x^m \cos^{-1} x \, dx = \frac{x^{m+1}}{m+1} \cos^{-1} x + \frac{1}{m+1} \int \frac{x^{m+1} \, dx}{\sqrt{1-x^2}}$ $\quad (m \neq -1)$

I153 $\int \tan^{-1} x \, dx = x \tan^{-1} x - \ln \sqrt{1+x^2}$

I154 $\int x^m \tan^{-1} x \, dx = \frac{x^{m+1} \tan^{-1} x}{m+1} - \frac{1}{m+1} \int \frac{x^{m+1} \, dx}{1+x^2}$ $\quad (m \neq -1)$

I155 $\int \cot^{-1} x \, dx = x \cot^{-1} x + \ln \sqrt{1+x^2}$

I156 $\int x^m \cot^{-1} x \, dx = \frac{x^{m+1}}{m+1} \cot^{-1} x + \frac{1}{m+1} \int \frac{x^{m+1} \, dx}{1+x^2}$ $\quad (m \neq -1)$

I157 $\int \sec^{-1} x \, dx = x \sec^{-1} x - \ln |x + \sqrt{x^2-1}|$

I158 $\int x^m \sec^{-1} x \, dx = \frac{x^{m+1}}{m+1} \sec^{-1} x - \frac{1}{m+1} \int \frac{x^m \, dx}{\sqrt{x^2-1}}$ $\quad (m \neq -1)$

I159 $\int \csc^{-1} x \, dx = x \csc^{-1} x + \ln |x + \sqrt{x^2-1}|$

I160 $\int x^m \csc^{-1} x \, dx = \frac{x^{m+1}}{m+1} \csc^{-1} x + \frac{1}{m+1} \int \frac{x^m \, dx}{\sqrt{x^2-1}}$ $\quad (m \neq -1)$

Table VI Formulas From Algebra

Special Products and Factors. (Also see the Binomial Theorem)

$(x \pm y)^2 = x^2 \pm 2xy + y^2$

$(x \pm y)^3 = x^3 \pm 3x^2y + 3xy^2 \pm y^3$

$(x + y + z + \cdots)^2 = x^2 + y^2 + z^2 + \cdots + 2xy + 2xz + \cdots + 2yz + \cdots$

$x^3 + y^3 = (x + y)(x^2 - xy + y^2)$

$x^3 - y^3 = (x - y)(x^2 + xy + y^2)$

$x^n - y^n = (x - y)(x^{n-1} + x^{n-2}y + x^{n-3}y^2 + \cdots + y^{n-1})$

$x^n - y^n = (x + y)(x^{n-1} - x^{n-2}y + x^{n-3}y^2 - \cdots - y^{n-1})$ $\quad (n \text{ even})$

$x^n + y^n = (x + y)(x^{n-1} - x^{n-2}y + x^{n-3}y^2 - \cdots + y^{n-1})$ $\quad (n \text{ odd})$

Tables

Table VI (*Cont.*) Formulas From Algebra

Radicals.

If $r^q = a$, then r is a qth root of a. The principal qth root of a, denoted by $\sqrt[q]{a}$ or $a^{1/q}$ satisfies the conditions

$$\sqrt[q]{a} = a^{1/q} \text{ is } \begin{cases} \text{nonnegative if } a \geq 0 \\ \text{negative if } a < 0 \text{ and } q \text{ is odd} \\ \text{imaginary if } a < 0 \text{ and } q \text{ is even.} \end{cases}$$

Then

$$\sqrt[q]{a^q} = \begin{cases} a \text{ if } a \geq 0 \\ a \text{ if } a < 0 \text{ and } q \text{ is odd} \\ |a| = -a \text{ if } a < 0 \text{ and } q \text{ is even.} \end{cases}$$

In particular, $\sqrt{a^2} = |a|$.

Radicals obey the following rules unless $a < 0$, $b < 0$ and q is even

$$\sqrt[q]{a} \sqrt[q]{b} = a^{1/q} b^{1/q} = (ab)^{1/q} = \sqrt[q]{ab}$$

$$\frac{\sqrt[q]{a}}{\sqrt[q]{b}} = \frac{a^{1/q}}{b^{1/q}} = \left(\frac{a}{b}\right)^{1/q} = \sqrt[q]{\frac{a}{b}}$$

Exponents.

If n is a positive integer, then $b^n = \underbrace{b \cdot b \cdot b \cdots b}_{n \text{ factors}}$

$$b^0 = 1, \quad b^{-n} = \frac{1}{b^n} \quad (b \neq 0),$$

$$b^{m/n} = \sqrt[n]{b^m} = (\sqrt[n]{b})^m$$

Exponents obey the following rules

$$b^m b^n = b^{m+n} \qquad (ab)^n = a^n b^n$$

$$\frac{b^m}{b^n} = b^{m-n} \qquad \left(\frac{a}{b}\right)^n = \frac{a^n}{b^n}$$

Logarithms.

Let $b > 0$, $b \neq 1$. Let M and N be positive. Then

$$\log_b N = x \text{ if, and only if, } N = b^x.$$

Logarithms obey the following rules

$$\log_b (MN) = \log_b M + \log_b N \qquad \log_b(M^k) = k \log_b M$$

$$\log_b \frac{M}{N} = \log_b M - \log_b N \qquad \log_b b = 1, \quad \log_b 1 = 0, \quad b^{\log_b M} = M$$

(continued)

Table VI (Cont.) Formulas From Algebra

Quadratic Equations.

The roots of
$$ax^2 + bx + c = 0 \quad (a \neq 0)$$
are
$$x = \frac{-b \pm \sqrt{b^2 - 4ac}}{2a}.$$

If a, b, c are real and
$b^2 - 4ac > 0$, the roots are real and unequal
$b^2 - 4ac = 0$, the roots are real and equal
$b^2 - 4ac < 0$, the roots are imaginary and unequal

Factorials.

$0! = 1$,
$n! = 1 \cdot 2 \cdot 3 \cdots n = n \cdot (n-1)!$ if n is a positive integer.
Stirling's Approximation: $n! \approx \sqrt{2\pi n}\, n^n e^{-n}$

Sums.

$$\sum_{k=1}^{n} k = 1 + 2 + 3 + \cdots + n = \frac{n(n+1)}{2}$$

$$\sum_{k=1}^{n} k^2 = 1^2 + 2^2 + 3^2 + \cdots + n^2 = \frac{n(n+1)(2n+1)}{6}$$

$$\sum_{k=1}^{n} k^3 = 1^3 + 2^3 + 3^3 + \cdots + n^3 = \frac{n^2(n+1)^2}{4} = \left(\sum_{k=1}^{n} k\right)^2$$

The sum of the first n terms of an *arithmetic progression* is
$$a + (a+d) + (a+2d) + (a+3d) + \cdots + (a + [n-1]d) = \frac{n(2a + [n-1]d)}{2}$$

The sum of the first n terms of a *geometric progression* is
$$a + ar + ar^2 + ar^3 + \cdots + ar^{n-1} = \frac{a(1 - r^n)}{1 - r} \quad (r \neq 1)$$

Binomial Theorem.

Let n be a positive integer, k an integer in the range $0 \leq k \leq n$. The *binomial coefficient* $\binom{n}{k}$ is defined by
$$\binom{n}{k} = \frac{n!}{k!\,(n-k)!} = \frac{n(n-1)(n-2)\cdots(n-k+1)}{k!}.$$

The *binomial theorem* states that
$$(x+y)^n = \binom{n}{0}x^n + \binom{n}{1}x^{n-1}y + \binom{n}{2}x^{n-2}y^2 + \cdots + \binom{n}{k}x^{n-k}y^k + \cdots + \binom{n}{n}y^n$$

Tables

Table VII Formulas From Geometry

Rectangle.

Area = ab
Perimeter = $2a + 2b$

Triangle.

Area = $\frac{1}{2} bh$
Perimeter = $a + b + c$
$A + B + C = 180°$
$h = c \sin A = a \sin C$
Heron's Formula: Area = $\sqrt{s(s - a)(s - b)(s - c)}$ where
$$s = \frac{a + b + c}{2}$$

Right Triangle.

$A + B = 90°$
Pythagorean Theorem: $a^2 + b^2 = c^2$

Parallelogram.

Area = bh
$\alpha + \beta = 180°$

Trapezoid.

Area = $\dfrac{b_1 + b_2}{2} h$

Circle.

r = radius, d = diameter
Area = $\pi r^2 = \dfrac{\pi d^2}{4}$
Circumference = $2\pi r = \pi d$

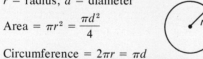

Cube.

a = length of side
Volume = a^3 Diagonal of Face = $a\sqrt{2}$
Surface Area = $6a^2$ Diagonal of Cube = $a\sqrt{3}$

Cylinder or Prism.

Volume = (Area of base) · (height)
Lateral Area = (Perimeter of base) · (height)

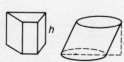

Right Circular Cylinder.

r = radius, h = height
Volume = $\pi r^2 h$
Lateral Area = $2\pi r h$

(continued)

Table VII (*Cont.*)   Formulas From Geometry

Pyramid or Cone.

Volume = ⅓ (area of base) · (height)

Right Circular Cone.

r = radius, h = height
Volume = $\frac{1}{3}\pi r^2 h$
Lateral Area = $\pi r \sqrt{r^2 + h^2}$

Sphere.

r = radius
Volume = $\frac{4}{3}\pi r^3$
Surface Area = $4\pi r^2$

Answers

Most of the numerical answers were calculated on a Texas Instrument SR-52 Programmable Calculator, a few on a home computer. The values may differ slightly from answers calculated from the tables or on other calculators.

Chapter 1

Exercises 1-1, p. 5–6

1 $>$ 3 $=$ 5 $>$ 7 $>$ 9 (3,5) open, finite   11 $(-1000,83000.1)$ open, finite
13 $\{x \mid 0.0012 \leq x < 0.0013\}$ half-open, finite   15 $\{x \mid x \leq -2\}$ infinite   17 $[4,\infty)$ 19 $(-\infty,-3)$
21 $(-\infty,9]$ 23 $(-\infty,-24]$ 25 $(0,\infty)$ 27 $(1,2)$ 29 No solution   31 $-1 < x < 2$
33 $-1 \leq x \leq 3$ 35 $(2,3)$ 37 No solution

Exercises 1-2, p. 12–13

1 $(-3/2,2)$ 3 $[0,1]$ 5 $(-5,1)$ 7 $(-\infty,-4) \cup (1,\infty)$ 9 No solution   11 $(-3,0) \cup (3,\infty)$
13 $(-2,2)$ 15 $(-5,-7/2)$ 17 $(-\infty,2) \cup (3,\infty)$ 19 $(-\infty,1) \cup [2,\infty)$ 21 $[-2/3,4)$
23 $(-\infty,0) \cup (3,\infty)$ 25 $(-\infty,-1) \cup (2/3,8/7)$

Exercises 1-3, p. 18–19

1 $-4 \leq x \leq 1$ 3 $x \leq -7$ or $x \geq 9$ 5 $x \neq 5/2$ 7 $x < -3$ or $x > 27$ 9 $-2 \leq x \leq 8$
11 $1 < x < 5$ 13 $x < -4$ or $x > -1$ 15 $3 - \epsilon < x < 3 + \epsilon$ 17 $x < -2$ or $x > 4$ 19 $x < -1$
21 $-2 < x < 2$

Exercises 1-4, p. 26–28

1 5   3 7   5 Right   7 Isosceles, right   9 Right   11 Right   13 (3,5)   15 $(-2,-2)$
17 $(3,-8), (3,12)$ 18 $(1,2),(1,6); (-7,2),(-7,6); (-5,4),(-1,4)$ 19 $x^2 + y^2 - 2x - 4y - 11 = 0$
21 $x^2 + y^2 - 8x + 4y + 11 = 0$ 23 $x^2 + y^2 - 6x + 10y + 25 = 0$ 25 $x^2 + y^2 - 2x - 6y - 7 = 0$
27 $(-1,2), 2$ 29 $(-3,-4), 5$ 31 Point $(-5,5)$ 33 $(2,-1/2), 2$ 35 $\emptyset$ 37(a) Inside; (b) On;
(c) Outside; (d) On   38 $(x - 7)^2 + (y - 7)^2 = 20$ 42 Set $|PA| = |PB|$ and simplify

Exercises 1-5, p. 37–39

1(a) $\Delta x = 0.05, \Delta y = 0.02$; (b) $\Delta x = -0.05, \Delta y = -0.13$; (c) $\Delta x = -0.98, \Delta y = 0.1$; (d) $\Delta x = 0.05, \Delta y = -0.01$
2(a) (1.01,2.98); (c) $(-1.02,3.03)$ 3 3/2   5 $-6/7$ 7 2   11 No   13 Yes   15 Yes
17 $2x - y = 7$ 19 $x = -1$ 21 $2x + y = 3$ 23 $5x + 2y = 10$ 25 $y = 4$ 27 $-3/8, 3$
29 4/3, 2   31 $y = 2x, x + 2y = 0$ 33 $2x + 3y + 7 = 0, 3x - 2y + 4 = 0$ 35 $(3, \pm 4)$
37 No intersection   39 Show that opposite sides have equal slopes; (3,1), (2,-1), (-3,3), (-4,1)
40 $x - 4y + 1 = 0$ 41 $x^2 + y^2 - 6x + 2y = 15$ 43 (1,1) or (2,8)   44 $(6,-4)$ or $(7,-4)$
45(a) $(5,2), 2\sqrt{5}$; (b) $(8/5,21/5), 4$ 47 $x^2 + y^2 + 2x + 2y - 23 = 0$ 49 $x^2 + y^2 - 4x - 6y - 12 = 0$
52 $4/\sqrt{13}$ 54 $m_2 = 3$ or $m_2 = -1/3$

Exercises 1-6, p. 45–47

1 $x \leq 1$ 3 $y \geq 4$ 5 All z 7 $x \neq \pm 3$ 9 All z 11 $x < 0$ or $x \geq 1$
13 $5 < x < 7$ 15 $x > 0$ 17(a) 0; (b) 5; (c) $(a + h)^2 + 2(a + h) - 3$; (d) $a^2 + h^2 + 2(a + h) - 6$
19(a) Not defined; (b) -3, (c) $\dfrac{1 + a + h}{1 - a - h}$; (d) $\dfrac{1 + a}{1 - a} + \dfrac{1 + h}{1 - h}$ 21(a) 1; (b) 6; (c) $(a + h)^2 + 2(a + h) - 2$;

A-43

(d) $(a^2 + h^2) + 2(a + h) - 4$ **25** $y = (x - 1)/x, x \neq 0; x = 1/(1 - y), y \neq 1$ **27** y is not a function of x; $x = y^2 + 1$, all y. **29** 0 **31** $x + a$ **33** $-1/ax$ **52(c)** $[f(x) + g(x)]/2 + |f(x) - g(x)|/2$

Exercises 1-7, p. 52-53

1(a) 0.542853; **(b)** 0.6542898; **(c)** 1.557408; **(d)** 1.716526 **3** $4x^2 + 28x + 50$, all x; $2x^2 + 9$, all x **5** $x^2/(3 - x^2), x \neq 0, \pm\sqrt{3}$; $(3x - 1)^2, x \neq 1/3$ **7** and **9** $f(g(x)) = g(f(x)) = x$, all x **11** 7, all x; 4, all x **13** $3x^2 + 1$; $9x + 4$; $9x^2 + 6x + 1$ **15** $x^2 - 1$ **17** Two possible answers are $g(x) = x - 1$ and $g(x) = 1 - x$ **19** $h(x) = f(g(x)), f(x) = 3x^2 - 2x - 5, g(x) = x + 1$ **21** $h(x) = f(g(x)), f(x) = \sqrt{x} - 1/x, g(x) = (x - 1)/(x + 1)$

Exercises 1-8, p. 60

5 Yes **7** Yes **11** $(x - 9)/4$, all x **13** $\sqrt{(x + 1)/3}, x \geq -1$ **15** $1/x, x \neq 0$ **17** $\sqrt[3]{x + 1/2}$, all x **19** $(x - 2)^3/27$, all x

Review Problems (Chapter 1), p. 61-64

5 $x < -3/5$ or $x > 7/2$ **6** $-2 < x < 2$ **7** $x < 1/3$ or $x > 2$ **8** $x > 1$ **9** $1/3 < x < 3$ **10** $-3 < x < 3$ **11** $x < 1$ **12** No solution **15** Circle, $C(2,2), r = 2$ **16** Point $(3, -4)$ **17** No graph **18** Line, slope $m = -3/2$ **19** Horizontal line segments one unit long **20** Horizontal line segments one-half unit long **21** $x^2 + y^2 - 6x + 4y = 19$ **22** Inside **23** $C(1,1), r = 1$ or $C(5,5), r = 5$ **24** $2x - y = 7$ and $2x - y + 23 = 0$ **25** $(3,1)$ and $(-27,41)$ **26** $3x + 4y = 25$ and $4x - 3y = 25$ **29(a)** $x + 3y + 1 = 0$; **(b)** $4x + y = 9$ **30** $3x + 2y = 46$ **31(a)** $\frac{4x^2 + 1}{x^2}$; **(b)** $(x^2 + 4)/x^2$; **(c)** 26; **(d)** $2x + a$ **33(b)** $x < -\sqrt{3}$ or $x > \sqrt{3}$; **(c)** $|x| \geq \sqrt{3}$ **34(a)** $\sqrt{\sqrt{16 - x^2} - 2}, \sqrt{18 - x}$; **(b)** $\{x \mid -2\sqrt{3} \leq x \leq 2\sqrt{3}\}, \{x \mid 2 \leq x \leq 18\}$ **36** $(x^2 + 1)/2, x > 0$ **37(c)** Volume of rectangular box with sides $x, x, x + 1$ **39** $\{x \mid x < -1 \text{ or } x \geq 0\}$ **41** $2(x_2 - x_1)x + 2(y_2 - y_1)y = x_2^2 - x_1^2 + y_2^2 - y_1^2$ **42** 2 **43** $\sqrt{13} - 3/2$

Chapter 2

Exercises 2-1, p. 72-73

1 5 **3** 3 **5** 1/12 **7** 1/6 **9** $-1/4$ **11** No limit **13** 2 **15** 1 **17** 1 **19** No limit **21** $m = 4$ **23** $4x + y + 4 = 0, x - 4y + 18 = 0$ **25** $8x - y = 7, x + 8y = 74$ **27** $5x + y = 0, x - 5y + 26 = 0$

Exercises 2-2, p. 79-81

1(a) 3; **(b)** $y = 3x + 5, x + 3y = 15$ **3(a)** $2x - 1$; **(b)** $x - y = 3, x + y + 1 = 0$ **5(a)** $2x$; **(b)** $4x - y = 6, x + 4y = 10$ **7(a)** $9x^2 - 4$; **(b)** $5x - y + 1 = 0, x + 5y + 21 = 0$ **9(a)** $1/2\sqrt{x + 1}$; **(b)** $x - 4y + 5 = 0, 4x + y - 14$ **11(a)** $1/2\sqrt{3}$ **(b)** $t + 2y - 4, 2t - y = 3$ **13(a)** $-2/z^3$; **(b)** $z + 4y = 3, 16z - 4y = 31$ **17** $2x - 1$ **19** $4x$ **21** $x - 1$ **23** $3 + 8t$ **25** $-14v$ **27** $-1/2$ **29** $1/4$ **31** $\pm 2/3$ **33** None **35(c)** $D_x(x^n) = nx^{n-1}$ if n is a positive integer **36** $2x + y = 1, 6x + y + 7 = 0$

Exercises 2-3, p. 87-88

1(a) 3; **(b)** $v(1) = 3$, speed $= 3, a(1) = 0$; **(c)** $v(t) \neq 0$; **(d)** Anytime **3(a)** -6; **(b)** $v(1) = -6$, speed $= 6, a(1) = 6$; **(c)** $t = 2$; **(d)** $t = 1$ **5(a)** 7, **(b)** $v(1) = 6$, speed $= 6, a(1) = 2$; **(c)** $t = -2$; **(d)** $t = 3/2$ **7(a)** $1/2$; **(b)** $v(1) = 0$, speed $= 0, a(1) = 2$; **(c)** $t = \pm 1$; **(d)** $t = \pm\sqrt{2}$ **9** $s = -16t^2 + 96t, v = -32t + 96$, 3 sec., 6 sec. **11** $s = -16t^2 + 48t, v = -32t + 48$, 3/2 sec., 3 sec. **13** $s = -16t^2 + 100, v = -32t$, in the air **15** $s = -16t^2 + 36, v = -32t$, on the ground **17** 784, $t = 12, -224$ **19** 144, $t = 5, -96$ **21** 484 ft

Exercises 2-4, p. 95

1 One number is $\delta = 0.01/3$ **3** One number is $\delta = 0.002$ **5** Take $0 < \delta \leq 4\epsilon$ **7** Take $0 < \delta \leq \epsilon/4$ **9** Take $0 < \delta \leq \epsilon$ **11** 1, -1, no limit as $x \to 0$ **12** 5, 4, no limit as $x \to 2$ **13** $-6, -6$, yes **14** 1, 0, no

Exercises 2-5, p. 102-104

1 -6 **3** 1 **5** $1/3$ **7** $-1/9$ **9** -4 **11** $1/4$ **13** 0 **15** $1/2$ **17** Possible **19** Not possible **26(a)** Use L2; **(b)** Let $f(x) = 1$ if $x > 0, f(x) = -1$ if $x < 0$

Exercises 2-6, p. 111-112

1 None **3** $x = 4$ (infinite) **5** Integers (jump) **7** None **9** None **11** $x \leq 0, x = 1$ (infinite at $x = 0, 1$, not defined for $x \leq 0$) **13** $x \leq 1, x = 2$ (infinite at $x = 1, 2$, not defined for $x \leq 1$) **15** $x = -1$

Answers

(removable), define $f(-1) = -2$ to make continuous   **17** $x = -2$ (removable), define $f(-2) = 12$ to make continuous   **19** Possible   **21** Possible   **23** No   **25** Yes   **27** Yes

Exercises 2-7, p. 119–120

1 7   **3** 8   **5** 3   **7** Continuous everywhere   **9** Discontinuous if $-2 < x < 2$ **11** Discontinuous if $-1 < x < 1$ **13** $m = 0$, $M = 1$ **15** $m = 0$, no max.   **17** $M = 3$, no min.   **18** $m = -1$, no max.   **19** $M = 1$, no min.   **21** $f(1) = -2 < 0 < f(2) = 6$ **23** $f(0) = -12 < 0 < f(1) = 4$ **26** Apply the Intermediate Value Theorem

Exercises 2-8, p. 131–132

1(a) ∞; (b) $x = \pm 1$ **3**(a) ∞; (b) $x = 2$ **5**(a) $-\infty$; (b) $x = -2$ **7**(a) ∞; (b) $x = 1, -2$ **9**(a) 0; (b) x-axis   **11**(a) $-\infty$; (b) None   **13**(a) $4/3$; (b) $y = 4/3$ **15**(a) 0; (b) x-axis   **17**(a) 2; (b) $y = 2$ **19**(a) 0; (b) x-axis   **21** Possible   **23** Not possible   **25** Possible   **27** If $M < 0$, there exists $\delta > 0$ such that $f(x) < M$ whenever $0 < |x - c| < \delta$ **29** If $M < 0$, there exists $\delta > 0$ such that $f(x) < M$ whenever $c - \delta < x < c$ **31** If $M > 0$, there exists $N < 0$ such that $f(x) > M$ whenever $x < N$

Review Problems (Chapter 2), p. 132–134

7(a) 0; (b) $y = 1$; (c) $x = -3$ **9**(a) ∞; (b) $y = 1$; (c) $x = 2$ **10**(a) -1; (b) $y = \pm 1$; (c) none   **11**(a) 0; (b) $y = 0$; (c) $x = 0, \pm 1$ **13** $4x - y = 3$ **15** $x - 2y = 1$ **18**(a) $-1/2(x - 4)^{3/2}$; (b) $2/(x + 1)^2$ **19**(a) 256 ft (b) 7 sec, 128 ft/sec   **28**(b) 0

Chapter 3

Exercises 3-1, p. 144–145

1 0   **3** 3   **5** $6t - 2$ **7** $2ax + b$ **9** $-12z^2 - 2z + 1$ **11** $18t - 12$ **13** $24x^2$ **15** 1 ($t \neq 0$)   **17** $2x - y = 1$, $x + 2y = 3$ **19** $43x - y = 63$, $x + 43y = 991$ **21** $3/4$ **23** $-1, 4$ **25** $-1, 2$ **27** $a = 2$, $b = -1$, $c = 1$ **28**(a) $D_r A = 2\pi r$; (b) $D_r V = 4\pi r^2$ **29** At $x = -1, 7/3$ **30** 160 ft/sec

Exercises 3-2, p. 149–150

1 $3 - x^{-2}$ **3** $-3/z^4 - 4/z^3 + 4/z^2$ **5** $45(2x - 3)(x^2 - 3x)^{44}$ **7** $5x^4 - 12x^3 - 3x^2 + 12x - 2$ **9** $2(t^2 - 2t + 1)^4(t^2 - 1)^2(8t^3 - 11t^2 - 2t + 5) = 2(t - 1)^{12}(t + 1)^2(8t + 5)$ **11** $-20(2y^2 - 3y + 1)^{-21}(4y - 3)$ **13** $(-2t^3 + 6t^2 + 4)/(t^3 + 2t)^2$ **15** $(3z^2 - z^4)/(1 - z^2)^2$ **17** $9(x - 1)^2/(x + 2)^4$ **19** $(x^4 - 3x^2 + 6)/x^4$ **21** $x(x + 2)/(x + 1)^2$ **23** $20x - y = 25$, $x + 20y = -99$ **25** $x = 0, 3, 3/2$ **27** $x = -3, 1/2, 4$ **29** $x = 0, \pm 3$ **31** $(1/2, 4)$ **32**(b) $(5x^2 - 1)^3(4x + 2)^2(7x^2 + 1)^4[(5x^2 - 1)(4x + 2) \cdot 70x + (5x^2 - 1) \cdot 12 \cdot (7x^2 + 1) + 40x(4x + 2)(7x^2 + 1)]$

Exercises 3-3, p. 156–158

1 0   **3** $-\Delta x$ **5** -0.02 **7** -0.001 **9** 0.768   **11** 1.04   **13** 81.216   **15** $2x \, dx$ **17** $(2x + 3) \, dx$ **19** $(1 - s^{-2}) \, ds$ **21** $-(x - 1)^{-2} \, dx$ **23** $15/4 = 3.75$ **25** $1/8 = 0.125$ **27** 8.120601, 8.12, error = 0.000601   **28** 0.2   **29** $0.12\pi \approx 0.37699$ **30** $2.4\pi \approx 7.539822$ **33**(a) $2\Delta r/r$; (b) Decreases

Exercises 3-4, p. 163–164

1 $8x(x^2 + 4)(u - 1)$ **3** $2(v + 1/v^2)(2x + 1/x^2)$ **5** $6(2v - 1)(x - 1)$ **7** $-(u^2 + 2u - 1)(u + 1)^{-2}(x^3 + 1)x^{-2}$ **9** $-36u(3u^2 - 1)^2(x - 1)^{-2}$ **11** $2x \cdot f'(x^2 - 4)$ **13** $-\dfrac{4x}{(x^2 - 2)^2} f'\left(\dfrac{x^2}{x^2 - 2}\right)$ **15** $-(3x^2 + 1)/ZAG(x^3 + x - 1)$ **17** $-2/ZAG(2x + 3)$ **19** $\left[4\left(\dfrac{x^2 - 1}{x}\right)^3 - 6\left(\dfrac{x^2 - 1}{x}\right) + 1\right]\left(1 + \dfrac{1}{x^2}\right)$ **21** $-18/x^4 + 10x^5/3$ **23** $(3u^2 + 1)/3$ **25** $3/4u$ **27** $dy/dx = -4(10u + 7)(4 + 3v^{-2} - 4v^{-3})(2 + x)^{-2}$ **29** $f'(w)g'(s)h'(x) = \dfrac{dy}{dw} \cdot \dfrac{dw}{ds} \cdot \dfrac{ds}{dx}$

Exercises 3-5, p. 170–171

1 $f(x) = (x + 2)/(x - 1)$, $x \neq 1$ **3** $f(x) = 3x$ or $f(x) = -x$, x real   **5** $f(x) = (x^3 + 2x - 1)/(x^2 - 4)$, $x \neq \pm 2$ **7** $(1 - y)/x$ **9** $(2x - y)/(x - 2y)$ **11** $x/16y$ **13** $-\sqrt[3]{y/x}$ **15** $(3x^2 - 6x - y^3 + 3)/3y^2(x - 1)$ **17** $3x^{1/2}/2 + 4x^{1/3}$ **19** $r^{-2/3} + r^{-1/2} + r^{-3/2}$ **21** $(3t^2 - 5t + 2)^{-2/3}(6t - 5)/3$ **23** $-(r + 1)^{-1/2}(r - 1)^{-3/2}$ **25** $-16/z^2 \sqrt{16 - z^2}$ **27** $3x + 4y = 25$ **29** $4x - 3y = -10$ **31** $13x - 6y = 7$ **33** $x = 1$ **35** $x = -5$ **37** $x = \pm 1$ **39** 9.95   **41** 7.9988   **43** 3.0011   **45**(b) $D_x(|u(x)|) = u(x)u'(x)/|u(x)|$

Exercises 3-6, p. 175–176

1. 0 3. $-3/2$ 5. 3/4 ft/sec 6. $1/4\pi \approx 0.0796$ ft/sec 7. 5 ft/sec 8. 36 ft²/min 9. 10 ft/sec
10. $98/13 \approx 7.5385$ ft/sec 11. $1/8\pi \approx 0.0398$ ft/min 12. 11.8 mph 13. -2 ft/sec
14(a). $-512/1250 \approx -0.4096$ lb/sec; (b) $-51200/68921 \approx -0.7429$ lb/sec 15. -5.5 in³/min
16. $125\pi \approx 393$ in³/hr

Exercises 3-7, p. 178

1. $x = \pm 2$ 3. $x = -2$ 5. $x = 0$ 7. $x = \pm 10$ 9. $x = 0$

Exercises 3-8, p. 181–182

1. $6x - 4, 6, 0$ 3. $4y^{-2} - 4y^{-3}, -8y^{-3} + 12y^{-4}, 24y^{-4} - 48y^{-5}$ 5. $2y + y^{-2}, 2 - 2y^{-3}, 6y^{-4}$
7. $-(2 - 3x)^{-2/3}, -2(2 - 3x)^{-5/3}, -10(2 - 3x)^{-8/3}$ 9. $-x/y, -4/y^3$ 11. $-\sqrt{y}/\sqrt{x}, 1/2x^{3/2}$
13. $(y - x^2)/(y^2 - x), -4xy/(y^2 - x)^3$ 15(a). $gt - 1, g$; (b) $a(1/g) = g$ 17(a). $1/2\sqrt{t} - 1/2t^{3/2}$,
$-1/4t^{3/2} + 3/4t^{5/2}$; (b) $a(1) = 1/2$ 19(a). $1 - 4t^{-2}, 8t^{-3}$; (b) $a(-2) = -1$ 21(a). $1 - 2t - 3t^2, -2 - 6t$;
(b) $a(1/3) = -4, a(-1) = 4$ 23(b). $f^{(n)}(x) = n!(1 - x)^{-(n+1)}$
24(b). $f^{(n)}(x) = \dfrac{(-1)^{n+1} \cdot 1 \cdot 3 \cdot 5 \cdots (2n - 3) \cdot x^{-(2n-1)/2}}{2^n}$ if $n \geq 2$

Exercises 3-9, p. 185–186

1. $f^{-1}(x) = (x - 1)/3$ 3(c). $g(-1) = 2, g'(-1) = 1/3$ (use Theorem 3-10) 9. $x = 7y - 10$

Review Problems (Chapter 3), p. 186–187

3. $-3x^{-4} - 6x^{-3} - 5x^{-2}$ 4. $(x^3 - 3x^2 + 2x - 1)(5x^4 + 6x^2 - 1) + (x^5 + 2x^3 - x + 2)(3x^2 - 6x + 2) =$
$8x^7 - 21x^6 + 24x^5 - 35x^4 + 12x^3 + 9x^2 - 16x + 5$ 5. $(3x^2 + 6x + 2)/(x + 1)^2$
6. $3(2u - 1)(x - 1)/\sqrt{u^2 - u}$ 7. $-(2xy + y^2 + 6x)/(x^2 + 2xy - 3y^2)$
8. $\dfrac{1}{2\sqrt{x + \sqrt{2x + \sqrt{3x}}}}\left[1 + \dfrac{1}{2\sqrt{2x + \sqrt{3x}}}\left(2 + \dfrac{3}{2\sqrt{3x}}\right)\right]$ 9(a). $x = \pm\sqrt{3}$; (b) $x = \pm 1$ 10. $(2,16)$
11. 0.48 12(a). 2.9167, 2.0417 14(a). $(-3x^{-4} - 6x^{-3} - 5x^{-2})dx$ (b) $(3x^2 + 6x + 2)dx/(x + 1)^2$
15. 0.04 16. $\left[21\left(\dfrac{5-x}{x^2}\right)^2 + \dfrac{1}{2}\sqrt{\dfrac{x^2}{5-x}} + \left(\dfrac{x^2}{5-x}\right)^2\right]\dfrac{x-10}{x^3}$
17. $2x + y = 10$ 18. $-x^2/y^2, -2x/y^5$ 19(a). $x = 0$; (b) $x = 1$ 20. 9/10 ft/sec 21. $\sqrt{2}$ ft/min
22. 9 units per sec 23. $g(x) = (x^2 + 1)/2, x > 0$

Chapter 4

Exercises 4-1, p. 198–199

1. $-1/2$ 3. $0, \pm 1$ 5. $1, -2, -1/2$ 7. $3, -3, -2/3$ 9. 32 at $x = 2$, -4 at $x = -1$
11. 29 at $x = 2$, $-1/4$ at $x = 1/2$ 13. 0 at $x = 1$, -3 at $x = 0$ 15. 8 at $x = 2$, -1 at $x = 1$
17. 49/5 at $x = 7$, 8 at $x = 4$ 19. 10/3 at $x = 3$, 2 at $x = 1$ 21. 10 at $x = 1$, 1 at $x = 0$ 23. 64 at $x = 2$,
-3 at $x = 1$ 25. 9 27. 1 29. Possible 31. Not possible. (Why?) 33. Not possible. (Why?)

Exercises 4-2, p. 204–205

1. 1/2 3. $\sqrt{2}$ 5. 2 7. Not continuous on [0,1] 9. Not differentiable at $x = 0$ 15(a). No; (b) Yes

Exercises 4-3, p. 209–210

1. Max $(-2,5)$, min $(0,1)$ 3. Max $(0,0)$, min $(\pm 1, -1)$ 5. Max $(-2, -16/15)$, $(1,38/15)$, min $(-1, -38/15)$,
$(2,16/15)$ 7. Max $(-3/2, 9^7/2^8)$, min $(\pm 3, 0)$ 9. Max $(1,1)$, min $(0,0)$ 11. Min $(2, -3)$ 13. Max $(1, 1/3)$
15. Max $(0,1)$, min $(\pm 1, 0)$ 17. Max $(0, -1)$ 18. Max $(-3, -9/2)$, min $(3, 9/2)$ 19. -8 21. 405 ft

Exercises 4-4, p. 216–218

1. 1/2 3. $2'' \times 1.5''$ 4. $3\sqrt{2}'' \times 3\sqrt{2}''$ 5. $(1/2, 1/\sqrt{2})$ 6. 1 7. $2' \times 2' \times 1'$ 8. 432 in³
9. $r = 6\sqrt{2}''$, $h = 12''$ 10(a). Radius $= r\sqrt{2/3}$, height $= 2r/\sqrt{3}$; (b) Radius $= r/\sqrt{2}$, height $= r\sqrt{2}$
11(a). Length $= p/3$, height $= p/6$; (b) Length $=$ height $= p/4$ 12. $5\sqrt{5} \approx 11.18$ ft
13. Length $=$ diameter $= 24/(\pi + 4)$, height $= 12/(\pi + 4)$ 14. $8'' \times 12''$, 54 in² 15. $3''$ on each side
16. $100m \times 200/\pi m$ 17. 100 yds $\times$ 150 yds 18. 4 mi from B' 19. width $= 2r/\sqrt{3}$, depth $= 2r\sqrt{6}/3$

Answers   A-47

20 1 hr, 16 min   **21** Go diagonally across the river to a point 600 feet downstream, then along land for 1000 ft. Minimum cost: $8000   **22** 200   **24** $6\sqrt[3]{I_1}/(\sqrt[3]{I_1} + \sqrt[3]{I_2})$

Exercises 4-5, p. 224–225

1(a) Increasing on $[-1/2, \infty)$, decreasing on $(-\infty, -1/2]$; **(b)** concave upward everywhere; **(c)** Local min at $x = -1/2$ **3(a)** Increasing on $(-\infty, 2]$ and $[4, \infty)$, decreasing on $[2, 4]$; **(b)** Concave upward on $[3, \infty)$, downward on $(-\infty, 3]$; **(c)** Local max at $x = 2$, local min at $x = 4$, point of inflection at $x = 3$
5(a) Increasing on $(-\infty, -1]$ and $[1, \infty)$, decreasing on $[-1, 1]$; **(b)** Concave upward on $[0, \infty)$, downward on $(-\infty, 0]$; **(c)** Local max at $x = -1$, local min at $x = 1$, point of inflection $(0, 5)$ **7(a)** Increasing on $(-\infty, 0]$, decreasing on $[0, \infty)$; **(b)** Concave upward on $(-\infty, -1]$ and $[1, \infty)$, downward on $[-1, 1]$; **(c)** Local max at $x = 0$, inflection points at $x = \pm 1$ **9(a)** Increasing on $(-\infty, 5]$, decreasing on $[5, \infty)$; **(b)** Concave downward everywhere; **(c)** Local max at $x = 5$ **11(a)** Decreasing on $(-\infty, 0)$ and $(0, \infty)$; **(b)** Concave upward on $(0, \infty)$, downward on $(-\infty, 0)$; **(c)** None   **13(a)** Increasing on $[2, \infty)$, decreasing on $(-\infty, 0)$ and $(0, 2]$; **(b)** Concave upward on $(-\infty, -\sqrt[3]{16}]$ and $(0, \infty)$, downward on $[-\sqrt[3]{16}, 0)$; **(c)** Local min at $x = 2$, inflection point at $x = -\sqrt[3]{16}$
15 Possible   **17** Possible

Exercises 4-6, p. 227–228

1 Local min at $x = 1/5$ **3** Local max at $x = 8$ **5** Local max at $x = 0$, local min at $x = 2$, inflection point at $x = 1$ **7** Local max at $x = 0$, local min at $x = 4$, inflection point at $x = 3$ **9** Local max at $x = 4$, local min at $x = 0$, inflection point at $x = 2$ **11** Local max at $x = 1$, local min at $x = -1$, inflection points at $x = 0$, $\pm\sqrt{3}$ **13** Local max at $x = 0$, local min at $x = \pm 2$, inflection points at $x = \pm 2/\sqrt{3}$ **15** Local max at $x = 3/4$, local min at $x = 1/2$ and $x = 1$, inflection points at $x = (9 \pm \sqrt{3})/12$ **17** Local min at $x = -1$ and $x = 2$, local max at $x = 1$, inflection points at $x = 1 \pm 2/\sqrt{10}$ **19** $9'' \times 18''$ **21** $(5^{2/3} + 6^{2/3})^{3/2} \approx 15.535$ ft

Exercises 4-7, p. 233–234

1 $x = 6$ **3** $x = 12$ **5(a)** 2660 at $x = 6$ and $x = 15$; **(b)** Increasing for $3 \le x \le 6$ and $12 \le x \le 15$, decreasing for $6 \le x \le 12$ **7** $x = 40$ **9** $x = 448/60 \approx \$7.47$, 226 units   **11** $520 or $530 per month (assuming that the rent can only be increased in $10 increments)   **13** 20000 per shipment (use optimal lot size formula)   **14** $x = y = 400$ **15(a)** 3300 tuners at $70 each; **(b)** Theoretical values: $x = 32.4$, $d = \$69$, $d + 2 = \$71.00$. Practical values (assuming x must be an integer): $x = 32$, $d = \$69.67$, $d + 2 = \$71.67$. Note that part of the tax must be absorbed by the manufacturer to maintain reasonable sales.

Exercises 4-8, p. 240–242 (*Numerical answers depend on the accuracy of the calculator*)

1 1.41421356   **3** 1.25992105   **5** 1.16403514 (use $x_1 = 1$)   **7** 0.564099733   **9** -0.2679491924
10(a) 0.1984372145   **11** Show that $x_{n+1} = -2x_n$. Thus x_{n+1} is twice as far from $c = 0$ as x_n.
13(b) $x_{n+1} = [(k - 1)x_n + a/x_n^{k-1}]/k$

Exercises 4-9, p. 248–249

1(b) $x^4 - 3x^2/2 + 2x + 1/2$ **3(b)** $x^3/3 + 2x^{3/2}/3 + x$ **5(b)** $-1/x - 2\sqrt{x} + 5$
7(b) $2x^{5/2}/5 + 2x^{3/2}/3 - x^2/2 - x + 73/30$ **9(b)** $-1/x - 1/x^2 + 1/3x^3 + 11/3$
11(b) $3x^{5/3}/5 + 3x^{1/3} - 8/5$ **13** $-(2x + 1)^{-2}/4 + C$ **15** $-2\sqrt{1 - x} + C$ **17** $-(1 - x^2)^{3/2}/3 + C$
19 $(2x^4 + 4x^2 + 1)^{11}/88 + C$ **21** $(2x + x^{-1})^3/3 + C$ **23** $-3\sqrt{1 - x^2} + C$ **25** $(1 + 3\sqrt{x})^4/6 + C$

Exercises 4-10, p. 253–254

1 $x^3 - 3x^2 + x + 5$ **3** $2x^{3/2}/3 - 3x + 7$ **5** $x^3 - x^2/2 + 2x + 11/2$ **7** $t^3/3 + t^2 + 2$
9 $t^4/4 + 2\sqrt{t} - 58$ **11** $s(t) = t^3/6 - t^2/2 + t + 3$ **13** $s(t) = t^4/12 - t^2 - 2t - 1/3$
15 $s = -16t^2 + 96t$; 144 ft   **16** 14 ft/sec^2; 56 ft/sec $\approx$ 38 mph   **17** 352 ft   **19** 48 ft/sec
20 $C(100) = 2890.01$ **21** $-x^3 + x^2 + 118x - 36$ **22** 70 ft/sec

Review Problems (Chapter 4), p. 254–256

3 Max of 27 at $x = -2$, min of 0 at $x = \pm 1$, critical points at 0, ± 1 **4** Max of $\sqrt[3]{2}$ at $x = 1$, min of 0 at $x = -1$, critical points at $x = 0, -1$ **5** Max of 2 at $x = 2$, min of 0 at $x = 0$, critical point at $x = 0$ **7** Min at $x = -1/3$, no point of inflection   **8** Local max at $x = 0$, local min at $x = 2$, point of inflection at $x = 1$
9 Local max at $x = 0$, local min at $x = \pm 1$, points of inflection at $x = \pm 1/\sqrt{7}$ **10** Local min at $x = 1/4$, points of inflection at $x = 1/2$ and $x = 1$ **11** Local max at $x = 0$, local min at $x = 4/3$, point of inflection at $x = 1$ **12** Local max at $x = -27$, local min at $x = 27$, point of inflection at $x = 0$ **13** Increasing if $x \ge -1/3$, decreasing if $x \le -1/3$, concave upward everywhere   **14** Increasing if $x \le 0$ or $x \ge 2$, decreasing if $0 \le x \le 2$, concave upward if $x \ge 1$, downward if $x \le 1$ **15** Increasing if $-1 \le x \le 0$ or $x \ge 1$, decreasing if

$x \leq -1$ or $0 \leq x \leq 1$, concave upward if $|x| \geq 1/\sqrt{7}$, downward if $|x| \leq 1/\sqrt{7}$ **16** Decreasing if $x \leq 1/4$, increasing if $x \geq 1/4$, concave upward if $x \leq 1/2$ or $x \geq 1$, downward if $1/2 \leq x \leq 1$ **17** Increasing if $x \leq 0$ or $x \geq 4/3$, decreasing if $0 \leq x \leq 4/3$, concave downward if $x \leq 1$, upward if $x \geq 1$ **18** Increasing if $x \leq -27$ or $x \geq 27$, decreasing if $-27 \leq x \leq 27$, concave downward if $x \leq 0$, upward if $x \geq 0$ **19** $x = y = 5$
20 $6' \times 6' \times 3'$ **21** $40\pi/(4 + \pi) \approx 17.6$ ft from one end **22** $125' \times 250'$ **23** 2:12 p.m.
24 $V = 729\pi/8 \approx 286.28$ in^3 **25(a)** 10; **(b)** 25 **26** 200 **27** 5000 **28** 162 or 163 **29** Apply the Intermediate Value Theorem; $x_0 \approx 2.33005874$ **31** Modify the proof of Theorem 4-5 using $a = -1$, $b = 2$.
35 $x^3/3 + 3x^2/2 - x + C$ **36** $-1/x - 1/2x^2 + C$ **37** $2x^{1/2} + 2x^{3/2}/3 + x + C$ **38** $5x^4/4 - x^{-1} + C$
39 $y = x^3 - x^2/2 + x + 2$ **40** $y = -(x-1)^{-1} + 1$ **41** 120 ft/sec
42 $x^4/4 - 2x^3/3 - 3x^2/2 + 6x + 11/12$ **43** 440 ft

Chapter 5

Exercises 5-1, p. 265–266

1 $\sum_{k=1}^{5} k$ **3** $\sum_{k=1}^{4} k(k+1)^{k+2}$ **5** $\sum_{k=1}^{n} (2k)^3$ **7** $\sum_{k=m}^{n} k^3$ **9** 92 **11** 440 **13** 329
15 $n(n-1)(n+1)/3$ **17** $(2n^3 + 5n^2 + 3n)/2 - 110$ **19** 1/2 **21** 1/3 **23** 4 **25** $A = 6$
26 $A = 8/3$ **27** $A = 1/6$

Exercises 5-2, p. 273–274

1 8/3 **3** -6 **8** $\int_a^b x^n\, dx = (b^{n+1} - a^{n+1})/(n+1)$ provided $n \neq -1$ **9** 27/2 **11** $(\pi^3 + 1)/3$
15 30 **16** 2π **17** Let $\Delta x = 1/n$, $x_k = k/n$. The sum is $\sum_{k=1}^{n} f(x_k)\Delta x$. Express the limit as an integral.
18 Let $\Delta x = 2/n$, $x_k = 1 + 2k/n$. The sum is $\sum_{k=1}^{n} f(x_k)\Delta x$. Express the limit as an integral.

Exercises 5-3, p. 281–282

1 6 **3** 0 **5** $(x^3+1)/3$ **7** $(x^2-400)/2$ **9** $7x^3/3$ **11** $[(x+1)^4 - x^4]/4$ **13** If $a = b$, then both are equal to zero. If $b < a$, then $\int_a^b x\, dx = -\int_b^a x\, dx$. Use Exercise 6, Section 5-2. **14** Hint: If $f(x) \leq x$ for $0 \leq x \leq 1$, then $\int_{0.5}^{1} f(x)\,dx \leq \int_{0.5}^{1} x\, dx$. Evaluate the integral on the right. **15** Since $1/4 \leq 1/x \leq 1$, then $\int_1^4 dx/4 \leq \int_1^4 dx/x \leq \int_1^4 dx$. Evaluate the integrals on the left and right. **16(a)** If $b \leq a \leq c$, then $\int_b^c f(x)\,dx = \int_b^a f(x)\,dx + \int_a^c f(x)\,dx$ (Property 1). Solve for $\int_a^c f(x)\,dx$ and use the fact that $\int_b^a f(x)\,dx = -\int_a^b f(x)\,dx$. **17** 1 **19** 2 **20** $-\sqrt{3}$ **21** $m = 0$, $M = 16$, $x_1 = 0$, $x_2 = 4$, $K = 13/3$, $\xi = \sqrt{13/3}$

Exercises 5-4, p. 287–288

1 25 **3** 703/3 **5** 1 **7** 633/5 **9** 23/24 **11** 33/2 **13** 7/6 **15** $\sqrt{1+x^2}$ **17** $-1/(x^2+1)$
19 0 **21** $2/\sqrt{1-t^2} + C$ **23** $G(x) = x^3/3$ **24(a)** $1/c\sqrt{c^2-1}$; **(b)** $1/\sqrt{x^2+1}$ **25** The proofs can be patterned after the proof in the text for the case where $x \to c$.

Exercises 5-5, p. 293–294

1 $x^4 - x^3 + x^2 + x + C$ **3** $2x^3/3 - 5x^2/2 + 2x + C$ **5** $8x^{3/2}/3 + 8x^{1/2} + C$
7 $2(5x-1)^{3/2}/15 + C$ **9** $3(x^2+1)^{2/3}/4 + C$ **11** $x^3/3 + x + C$ **13** 2 **15** 7/200
17 $2(\sqrt{2}-1)$ **19** 1/3 **21** 1/6 **23** 98/3 **25** 2/5 **27** 422/5 **28** $[(2t+1)^{3/2} + 11]/3$
31 $v = t^3/3 - t^2/2 + 5$, $s = t^4/12 - t^3/6 + 5t$ **32(a)** Concave upward if $x > 0$, downward if $x < 0$; **(b)** Yes, $f(x) = 2x + 3/x - 3$ **33** If $x > 0$, then $A(x) = \int_0^x f(t)\,dt = \sqrt[3]{x^2+4} - \sqrt[3]{4}$. Apply the Fundamental Theorem of Calculus to show that $f(x) = 2x/3(x^2+4)^{2/3}$

Answers

A-49

Exercises 5-6, p. 302–303

1(a) 1.10156; (b) 1.09866   2(a) 0.87953; (b) 0.88138   3(a) 1.10632; (b) 1.10714   4(a) 1.62897; (b) 1.61085   5(a) 1.81025; (b) 1.81009   6(a) 3.67370; (b) 3.65347   7(a) $1/75 \approx 0.01333$ (use $M = 2$); (b) $4/9375 \approx 0.00043$ (use $N = 24$)   8(a) 0.16667 (use $M = 2$); (b) 0.03333 (use $N = 24$)   10 $V = 4$

Review Problems (Chapter 5), p. 304–305

1 $\sum_{t=1}^{k-1}\left(\frac{t}{t+1}\right)^{t+1}$ 2 1   4(a) 26/3; (b) 9   9 $-2x^{-1/2} - x^{-1} + C$ 10 5/6
11 $2x^{5/2}/5 + 3x^2/2 + 2x^{3/2} + x + C$ 12 $(3^{14} - 2^{14})/7$ 13 $-1/12(x^2 - 4x + 3)^6 + C$
14 $(8 + 7x^3)^{1/3}/7 + C$ 15 8/5   16 129/7   17 $x^5/5 - x + C$ 18 $(x^4 - 1)^{13}/52 + C$ 19 32/3
20 9/2   21 $[(2x - 4)^{3/2} - 2]/3$ 22(b) 0.69702; (c) 1.61724   23(b) 0.69325; (c) 1.61037   24(a) $-1/x$; (b) $2/x$ 25 Between 6/11 and 3/5 (apply Property 2 of Section 5-3)

Chapter 6

Exercises 6-1, p. 314–315

1 32/3   3 253/12   5 1/12   7 9/2   9 9/2   11 9/2   13 $4^{2/3}$ 15(a) $\lim_{n \to \infty} \sum_{k=1}^{n} (8 - 2\xi_k^2) \Delta x_k$
16 Two regions. Total area $= 148/3$ 17 Two regions. Total area $= 71/6$ 18 The area of one region would be added to the negative of the area of the other, yielding 0.

Exercises 6-2, p. 324–325

1(a) 18π; (b) 18π 2 $512\pi/15$ 3(a) $256\pi/5$; (b) 8π; (c) $512\pi/15$ 4(a) 8π; (b) $128\pi/5$; (c) $256\pi/15$; (d) $40\pi/3$ 5(a) $768\pi/7$; (b) $96\pi/5$; (c) $144\pi/5$; (d) $576\pi/7$ 6 $2\pi/3$ 7(a) $2\pi/15$; (b) $\pi/6$; (c) $5\pi/6$; (d) $\pi/5$ 8 $56\pi/15$ 10 36π 11 $128/3$ 12 54 ft³   13 $8\sqrt{3}$ cu units   14 4π cu units

Exercises 6-3, p. 332–333

1(a) $128\pi/3$; (b) $128\pi/3$ 2 4π 3 $8\pi/3$ 4 $128\pi/3$ 5 $416\pi/3$ 6(a) 16π; (b) $512\pi/3$ 7 $8\pi/3$
8 18π 13 $(64 - 15^{3/2})\pi/6 \approx 3.092$ 14 $4\pi \cdot 15^{3/2}/3 \approx 243.347$ 15(a) $2\pi \sum_{k=1}^{n} (4 - \xi_k)(8 - 2\xi_k^2)\Delta x_k$;
(b) $2\pi \sum_{k=1}^{n} 2\xi_k \sqrt{16 - \xi_k^2} \Delta x_k$ 16(b) $\pi \int_0^2 [5^2 - (y^2 + 1)^2]dy = 544\pi/15$

Exercises 6-4, p. 341–342

1 $\int_{-1}^{2} \sqrt{1 + 4x^2}\,dx$ 3 $\int_{1}^{3} \sqrt{64x^6 + 16x^3 + 2}\,dx$ 5 $\int_{1}^{9} \sqrt{4t^2 + t^{-4}}\,dt$ 7 $8^{3/2} - 5^{3/2}$ 9 74   11 9
13 31/6   15 $2[5^{3/2} - 1]$ 17 12   19 21/2   25 6.13236   26 3.16799   27(a) If $t \geq 0$,
$s(t) = \int_{1}^{t} \sqrt{4u^2 + 9u^4}\,du = [(4 + 9t^2)^{3/2} - 13^{3/2}]/27$, if $t < 0$, $s(t) = [16 - 13^{3/2} - (4 + 9t^2)^{3/2}]/27$;
(b) $s'(t) = \sqrt{4t^2 + 9t^4}$; (c) $(0.2)\sqrt{13} \approx 0.72$

Exercises 6-5, p. 347–348

1 675 ft-lbs   2 24 in.-lbs   3 15 in.-lbs   4 600000π ft-lbs   5 $(62.5)10^4\pi/9$ ft-lbs
6 $96448(62.5)\pi/675 \approx 8930.37\pi$ ft-lbs   7(a) $(62.5)(2500\pi)$ ft-lbs; (b) $62.5(17500\pi)/3$ ft-lbs   8 675 ft-lbs

Exercises 6-6, p. 354–356

1 1500   2 10500   3 1000   4 375,000   5 1250π 6 9000   7 1031.25   9(b) 3200.96
10 18000   11 7800

Exercises 6-7, p. 362–363

1 19/14   3 $(-29/22, 49/22)$ 5 $(23/22, 37/22)$ 8(a) $\sum_{k=1}^{n} (ax_k + by_k + c)m_k / \sqrt{a^2 + b^2}$

Exercises 6-8, p. 370–371

1 $M_x = \dfrac{\rho}{2}\int_0^4 x\,dx = 4\rho$, $M_y = \rho\int_0^4 x^{3/2}\,dx = 64\rho/5$, $(12/5, 3/4)$ 2 $(8/5, 16/7)$ 3 $M_x = 256\rho/15$, $M_y = 64\rho/3$, $(2, 8/5)$ 4 $(-1/2, -3/5)$ 5 $M_x = 512\rho/15$, $M_y = -20\rho/3$, $(-1/2, 64/25)$ 6 $(4, 0)$ 7 $M_x = 1/35$, $M_y = 1/20$, $(3/5, 12/35)$ 8 $(3/2, 12/5)$ 9 $M_x = 32/3$, $M_y = 416/15$, $(13/5, 1)$ 10 $M_x = -9/4$, $M_y = 54/5$, $(12/5, -1/2)$ 11 $M_x = 245/3$, $M_y = 5/3$, $(2/(15\pi + 6), 98/(15\pi + 6))$ 12 $M_x = 111/5$, $M_y = 15/4$, $(5/12, 37/15)$ 14(b) $M_{\mathcal{L}} = 2652\rho/175$ 16 $(3/(2\pi + 2), 1/(2\pi + 2))$ 17 $(-1/2(2\pi - 1), 29/6(2\pi - 1))$

Exercises 6-9, p. 377–378

1 Centroid is 4/3 ft below surface, $F = 1000$ ft-lbs   2 $F \approx 3927$ ft-lbs   3 Centroid is 3 ft below surface, $F = 18{,}000$ ft-lbs   4 $\pi R^2 H/3$ 5 $40\pi^2$ cu units   6 $40\pi^2$ sq units   7(a) 72π cu units; (b) $48\pi\sqrt{2}$ cu units   8(a) $2\pi^2 Rr^2$; (b) $4\pi^2 Rr$

Review Problems (Chapter 6), p. 378–380

3 $1/12$ 4 $1/3$ 5 Two regions, total area $71/6$ 6 $16/3$ 7 $32/\sqrt{3}$ 8 $81/8$ 9 π 10 $640\pi/3$ 11 16π 12 6π 13 $140\pi/3$ 14 $16\pi/15$ 15 $29\pi/15$ 18 $2(64^{3/2} - 1)/27 = 1022/27$ 19 $(25^{3/2} - 16^{3/2})/9 = 61/9$ 20 $19/2$ 21 $(1300^{3/2} - 13^{3/2})/27 = 481\sqrt{13}$ 22 15 in-lbs   23 9000π ft-lbs   24 $(62.5)(11\pi)/12 \approx 180$ ft-lbs   25 $500/3$ lbs   26 $3(62.5)/2$ lbs   28 $(0, 8/3\pi)$ 29 $(1/2, 8/5)$ 30 $(10/13, 123/65)$ 31 $(1/3, 0)$ 32 $M_0 = 181$, $\bar{x} = 181/34 \approx 5.3235$ 34 $22\pi/3$

Chapter 7

Exercises 7-1, p. 393–395

1 $y = -1$, $x = 5$ 3 $y = 0$, $x = 3$ 5 $y = -1$, $x = \pm 3$ 7 $x = 3$, $y = x + 8$ 9 $x = 1$, $x = 2$, $y = 5x + 22$ 11 $y = 3x - 5$ as $x \to \infty$, $y = -3x + 5$ as $x \to -\infty$ 13 $x = \pm 2$ 15 $x = \pm 10$, $y = \pm 10$ 17 $x = -2, 6$ 18 $x = 2$, $y = 2$ 19 $x = -2$ 20 $x = \pm 4$, $y = \pm 4$ 21 Origin, line $y = x$ 23 Origin, x-axis, y-axis, line $y = x$ 25 Origin   27 Line $y = x$ 29 x-projection $= \{x \,|\, x \neq 4\}$, y-projection $= \{y \,|\, y \neq 0\}$ 31 x-projection $= \{x \,|\, x \neq 3\}$, y-projection $= \{y \,|\, y \neq 1\}$ 33 x-projection $= \{x \,|\, -7 \leq x \leq 5\}$, y-projection $= \{y \,|\, -3 \leq y \leq 9\}$ 35 x-projection $= \{x \,|\, x \geq 2\}$, y-projection $= \{y \,|\, y \text{ is real}\}$

Exercises 7-2, p. 403–405

1 $x^2 = 4y$ 3 $y^2 = -16x$ 5 $(x + 2)^2 = -8(y - 1)$ 7 $(y - 2)^2 = -12(x - 2)$ 9 $x = 2y^2$, $x^2 = 4y$ 11 $V(0, 0)$, $F(0, -1/4)$, $y = 1/4$ 13 $V(-1, 5)$, $F(-1, 19/4)$, $y = 21/4$ 15 $V(-1, -4)$, $F(-1, -15/4)$, $y = -17/4$ 17 $V(-1/4, -1/2)$, $F(0, -1/2)$, $x = -1/2$ 19 $V(1, -1)$, $F(1, -11/12)$, $y = -13/12$ 21 $V(-11/4, 1/4)$, $F(-45/16, 1/4)$, $x = -43/16$ 25 $9x^2 - 24xy + 16y^2 - 400x - 300y + 2500 = 0$ 26 Hint for (a): Explain why the parabola never crosses the x-axis   27(a) $2\pi p^3$; (b) $(0, 3p/5)$ 28 $x^2 = -60y$

Exercises 7-3, p. 412–414

1 $9x^2 + 25y^2 = 900$ 3 $x^2 + 9(y - 5)^2 = 9$ 5 $13(x - 2)^2 + 9y^2 = 117$ 7 $V(0, \pm 3)$, $F(0, \pm\sqrt{5})$ 9 $V(\pm 6, 0)$, $F(\pm 4\sqrt{2}, 0)$ 11 $V(0, \pm 6)$, $F(0, \pm 4\sqrt{2})$ 13 $V(1, 2 \pm 3)$, $F(1, 2 \pm \sqrt{5})$ 15 $V(2 \pm 3, -1)$, $F(2 \pm \sqrt{6}, -1)$ 17 $V(1, 5 \pm 3\sqrt{3})$, $F(1, 5 \pm 3\sqrt{3}/2)$ 19 $V(-1 \pm 4/\sqrt{3}, 2)$, $F(-1 \pm 4\sqrt{2/15}, 2)$ 21 $x^2/(92.9)^2 + y^2/(92.89)^2 = 1$ 22 The argument is similar to the proof of Theorem 7-2

Exercises 7-4, p. 423–424

1 $16x^2 - 9y^2 = 144$ 3 $x^2 - 4y^2 = 16$ 5 $(x + 2)^2 - 12(y - 2)^2 = 16$ 7 $V(\pm 3, 0)$, $F(\pm\sqrt{13}, 0)$ 9 $V(\pm 6, 0)$, $F(\pm\sqrt{40}, 0)$ 11 $V(0, \pm 2)$, $F(0, \pm\sqrt{40})$ 13 $V(2 \pm 4, 1)$, $F(2 \pm \sqrt{24}, 1)$ 15 $V(6, 2 \pm 2)$, $F(6, 2 \pm \sqrt{40})$ 17 $V(-1, -5 \pm \sqrt{3})$, $F(-1, -5 \pm \sqrt{57}/4)$ 19 $V(-2 \pm 5/3, 1)$, $F(-2 \pm \sqrt{70}/3, 1)$ 21(a) $x'y' = 1$ (b) One answer (using $x' = x - 3$, $y' = y - 4$) is $(x' + 1)(y' + 1) = 1$ 22 $4x^2 - y^2 = c$, where $c \neq 0$ 23(a) $x^2 - y^2 = a^2$ 25(a) Portion of hyperbola $y^2 - x^2 = 1$ in Quadrant II; (b) Branch of hyperbola $4x^2 - 9y^2 = 36$ in Quadrants I and IV   28 $(-6.03, -0.77)$ approx.

Exercises 7-5, p. 433–434

1 $e = 13/5$, directrices $x = \pm 25/13$ 3 $e = \sqrt{35}/5$, directrices $x = 1 \pm 25/\sqrt{35}$ 5 $e = 1$, directrix $y = 55/12$ 7 $8x^2 - y^2 = A$ or $8y^2 - x^2 = A$, $A > 0$ 9 $3x^2 + 4y^2 = A$ or $4x^2 + 3y^2 = A$, $A > 0$ 12 $b = a\sqrt{3}$, c is arbitrary

Answers

Exercises 7-6, p. 441–442

1 $4x'^2 + y'^2 = 16$ 2 $4x'^2 + 9y'^2 = 36$ 3 $4x'^2 - 9y'^2 + 36 = 0$ 4 $4x'^2 + 9y'^2 = 36$
5 $2x'^2 - y'^2 = 4$ 6 $y''^2 = -4x''$ 7 $4x''^2 + y''^2 = 4$ 8 $8y''^2 = x''$ 9 $2x''^2 = y''$
10 $4y''^2 - x''^2 = 8$ 11 Ellipse (Ex. 1)   12 Ellipse (Ex. 2)   13 Hyperbola (Ex. 3)   14 Ellipse (Ex. 4)   15 Hyperbola (Ex. 5)   16 Parabola (Ex. 6)   17 Ellipse (Ex. 7)   18 Parabola (Ex. 8)
19 Parabola (Ex. 9)   20 Hyperbola (Ex. 10)   21 Pair of parallel lines (degenerate parabola)   23 Circle (degenerate ellipse)

Review Problems (Chapter 7), p. 442–444

3 Vertical asymptote $x = 2$, oblique $y = 3x + 2$, local min at $x = 2 + \sqrt{11/3}$, local max at $x = 2 - \sqrt{11/3}$
4 Vertical asymptotes $x = \pm 1$, horizontal asymptote $y = 1$, symmetric about y-axis, local min at $x = 0$.
5 Asymptotes $x = \pm 3$, $y = 0$, symmetric about origin, no local extrema.   6 Asymptotes $x = \pm 1$, $y = 0$, local min at $x = 4 - \sqrt{15}$, local max at $x = 4 + \sqrt{15}$.   7 Asymptotes $x = \pm 3$, $y = 1$, symmetric about y-axis.
8 No asymptotes, vertical tangent lines (from one side) at $x = \pm\sqrt{8}$, no critical points, symmetric about origin, graph defined for $x = 0$ and $|x| \geq \sqrt{8}$, function is increasing for $x \leq -\sqrt{8}$ and $x \geq \sqrt{8}$.
13 $(x - 3)^2 = 8(y - 3)$ 14 $25(x - 2)^2 + 16(y + 1)^2 = 400$ 15 $(y - 1)^2 - 4(x - 2)^2 = 16$
16 $27x^2 - 24xy + 20y^2 - 66x - 88y = 121$ 17 Parabola, $V(8,0)$, $F(65/8,0)$ 18 Ellipse, $V(3, -1 \pm 4)$, $F(3, -1 \pm \sqrt{12})$ 19 Hyperbola, $V(-4,1 \pm 1)$, $F(-4,1 \pm \sqrt{17})$ 20 Parabola, $V(-1,2)$, $F(-1,15/8)$
21 Ellipse, $V(-5 \pm 3, -2)$, $F(-5 \pm \sqrt{8}, -2)$ 22 Hyperbola, $V(3 \pm 2, -4)$, $F(3 \pm \sqrt{29}, -4)$ 23 Pair of lines $2(y + 1) = \pm 3(x - 1)$ 24 No graph   25 $e = 1$, $x = 63/8$ 26 $e = \sqrt{3}/2$, $y = -1 \pm 8/\sqrt{3}$
27 $e = \sqrt{17}$, $y = 1 \pm 1/\sqrt{17}$ 28 $e = 1$, $y = 17/8$ 29 $e = \sqrt{8}/3$, $x = -5 \pm 9/\sqrt{8}$ 30 $e = \sqrt{29}/2$, $x = 3 \pm 4/\sqrt{29}$ 31 and 32 Not conics   33 $8y'^2 = x' + 2$, $8y''^2 = x''$ 34 $(y' - 1)^2 = x' + 2$, $y''^2 = x''$
35 $x'^2 - y'^2 + 16 = 0$ 36 $y'^2 = -4\sqrt{2}(x' - 2)$, $y''^2 = -4\sqrt{2}x''$ 37 $y'^2 - 4x'^2 = 4$
38 $9x'^2 + 25y'^2 = 225$ 40 $D = 0$ 41 $D = 0$ 42 $D = 1$ 43 $D = -562500$ 45 $y = -3x^2 + 9x$

Chapter 8

Exercises 8-1, p. 452–454

1(a) 4; (c) -5 2 3^{7x} 4 $\log_8 13$ 6(a) 1.89167 (approx.); (c) 0.40302 (approx.)   7(b) 2.48127
13 $2x/(x^2 - 1)$ 15 $-1/x$ 17 $x(1 + 2\ln x)$ 19 $3(\ln x)^2/x$ 21 $-2/(x^2 - 1)$
23 $x/(x^2 + 1)\sqrt{\ln(x^2 + 1)}$ 25 $(x + 2\ln x)(2 + 4/x)$ 27 $-y/x(y + 1)$ 29 $-(x\ln y + y)y/x(x + y\ln x)$
31 1.61008   32 Local min at $x = 1$, concave upward for $x > 0$

Exercises 8-2, p. 459–461

1 $y = \ln(3x)$ 3 $y = e^{2x+4}$ 5(a) $e^{\sqrt{2}\ln 4} \approx 7.103$ (c) $e^{(\pi-1)\ln 3} \approx 10.515$ 7 $6e^{2x}$ 9 $5(e^x + e^{-x})^4(e^x - e^{-x})$
11 $4e^{2x}/(e^{2x} + 1)^2$ 13 $e^x(\ln x + 1/x)$ 15 $2e^x/(1 - e^{2x})$ 17 $2e^x/(e^x + 2)$ 19 Local min at $x = 1$, concave downward if $x < 0$, upward if $x > 0$, no point of inflection   21 Local max at $x = e$, point of inflection at $x = e^{3/2}$ 24 1, 2   26 $(7 \pm \sqrt{41})/2$ 28(a) $50! \approx (3.03634)10^{64}$

Exercises 8-3, p. 469–470

1 $2\ln|x| + C$ 3 $9x - 24\ln|x| - 16x^{-1} + C$ 5 $(\ln x)^4/4 + C$ 7 $2x^2 + 7x + 8\ln|x - 1| + C$
9 $e^{3x}/3 + C$ 11 $(0.5)e^{x^2-2x+2} + C$ 13 $-1/(e^x + 1) + C$ 15 $2e^{\sqrt{x}} + C$ 17 $\ln 2$
19 $2\ln 3 - 2\ln 2$ 21 $\ln(e^2 + 1) - \ln(e + 1)$ 23 $5^x \cdot \ln 5$ 25 $2x/(x^2 + 1)\ln 3$ 27 $5^x/\ln 5 + C$
29 $-4^{1/x}/\ln 4 + C$ 31 $(1 + x)^x[\ln(1 + x) + x/(1 + x)]$ 33 $u\left[\dfrac{17(10x + 3)}{5x^2 + 3x + 2} - \dfrac{20}{4x + 3}\right]$
34 $x^{x^2}[x + 2x\ln x]$ 36 8   37(a) $\pi\ln 9$; (b) $112\pi/3$ 38 $\bar{x} = 13/3$, $\bar{y} = \ln 9/8 \approx 0.27465$
39 $\bar{x} = 25/(20 - 4\ln 6) \approx 1.94811$, $\bar{y} = 25/(30 - 6\ln 6) \approx 1.29874$ 40 $\sqrt{3}(e^2 - 1)/8$

Exercises 8-4, p. 474–475

1 $(y + 1)^2 = (x + 1)^2 + C$ or $y = -1 \pm \sqrt{x^2 + 2x + C_1}$ 3 $y = 2 + Ce^{(x+1)^2/2}$
5 $y^2 + 2\ln|y| = 2x^2 + C$ 7 $y = Ce^{4x}$ 9 $y = C_1 e^{2x} + C_2$ 11 $(y + 1)^2 = (x + 1)^2 + 8$
13 $y = 2 - 6e^{x+x^2/2}$ 15 $y^2 + 2\ln|y| = 2x^2 - 31$ 17 $f(x) = -4e^{4x}$ 19 $y = -2e^{2x} + 3$
20 $y = (33 - e^{-4x})/8$ 21 $-4\ln|x| + 2x^3 + x^2 - 2x + 1$
23 $-4\ln|x| + 2x^3 + x^2 + (4\ln 2 - 20)x - 4\ln 2 + 19$ 25(a) $200e^{-(t\ln 2)/1620} = 200(1/2)^{t/1620}$;
(b) 2567.64 years   26(a) 27.9% (approx); (b) $2\ln(0.5)/\ln(0.6) \approx 2.714$ years   27 846 years (approx)
28 231 weeks (approx)   29 $y = 5 - 3x$ 30(a) $dT/dt = m(T - K)$ 31(a) $75°$; (b) $72.04°$; (c) $64.07°$;
(d) $66.56°$

Exercises 8-5, p. 481–482

1. 5.91% (approx.)   3. 7.84% (approx.)   5. 5.13% (approx.)   7. 8.33% (approx.)   9(a) 9.9 million; (b) 23.5 million   10. Double in 14 years (approx.), triple in 22.5 years (approx.)   11. $26,321.48   12. $2,330 (approx.)   13(a) $126.50   14. Predicted population is 252 million (approx.)   15(c) $P = 20e^{(t\ln 2)/10}/[e^{(t\ln 2)/10} + 3] = 20 \cdot 2^{t/10}/[2^{t/10} + 3]$ 16(a) $P = 10 \cdot 6^{t/3}/(6^{t/3} + 9)$ thousand

Exercises 8-6, p. 488–489

1. $v = 150e^{-3t}$, $s = 50(1 - e^{-3t})$ 3. $v = 200e^{-t}$, $s = 200(1 - e^{-t})$ 5. $v = -160 + 130e^{-0.2t}$, $s = 1650 - 160t - 650e^{-0.2t}$ 6. $v = 65400(e^{-3t/200} - 1)$, $s = 4380000 - 65400t - 4360000e^{-3t/200}$ 7. $v = -400 + 900e^{-2t/25}$, $s = -400t + 11250(1 - e^{-2t/25})$ 8. $v = -9810 + 19810e^{-t/10}$, $s = -9810t + 198100(1 - e^{-t/10})$ 9(a) 2280 ft (approx) in 10.3 sec (approx); (b) -271.3 ft/sec   10. -48 ft/sec (approx)

Review Problems (Chapter 8), p. 489–491

7. Use the fact that $\ln 27 - \ln 8 = \ln 3^3 - \ln 2^3$ 8. $24x/(3x^2 - 2) + 45x^2/(5x^3 + 7)$ 9. $x(1 + 2\ln x)$ 10. $(\ln x + \ln 2x)/x = (\ln(2x^2))/x$ 11. $1/x\ln x$ 12. $e^x(x + 1)$ 13. $2x$ 14. $e^x/(e^x + 1)$ 15. $e^{\sqrt{x}}/2\sqrt{x}$ 16. $y' = -y^2 e^{xy}/(xye^{xy} + 1)$ 17. $y' = (2x - y^2 + e^y)/(2xy - xe^y)$ 18. $y[3/x + 1/(x + 1) - 2x/(x^2 + 3)]/2$ 19. $-e^{-2x}/2 + C$ 20. $\ln|\ln x| + C$ 21. $\ln|e^x - e^{-x}| + C$ 22. $e^{x^2}/2 + C$ 23. $(3e^x + 2)^{21}/63 + C$ 24. $9e^{2x}/2 + 12e^x + 4x + C$ 25. $x^2/2 - x + 6\ln|x + 4| + C$ 26. $(0.5)\ln(x^2 + 2x + 5) + C$ 27. $\pi(1 - e^{-2})/2 \approx 1.3582$ 29(b) $y = x^2 + x + 1$ 30(b) $y = (7e^{2x-2} - 1)/2$ 31(b) $y = 4e^{[(x+1)^2 - 4]/2} - 1$ 32(b) $y = 3x$ 33. $10^3 e^{-0.12} \approx 887$ grams/cm^2 34. 2.31 grams (approx.)   35(a) 155000 (approx.); (b) 556000 (approx.)   36. $298000 (approx.)   37. $151 (approx.)   38(a) $dq/dt = k(A - q)$ where A is the total quantity of sugar and q is the quantity of dextrose; (b) 57 grams (approx.)   40(c) Max height occurs at $t = 25\ln(1.5) \approx 10.1366$, $s(10.1366) \approx 1891$ ft.

Chapter 9

Exercises 9-1, p. 498–500

1. $3/2$ 3. $1/2$ 5. $2/3$ 7. 0   9. $-5\sin 5x$ 11. $2x\cos(x^2 + 5)$ 13. $-3\cos^2 x \sin x$ 15. $-2[\sin x \sin 2x + \cos x \cos 2x]/\sin^3 x$ 17. $-3\sin 2x \sin 3x + 2\cos 2x \cos 3x$ 19. $2x(\cos x^2)e^{\sin x^2}$ 21. $[\sin(2x - 1)]/2 + C$ 23. $-(\cos 2x)/2 - (\cos 4x)/4 + C$ 25. $(\sin^3 3x)/9 + C$ 27. $-2\cos\sqrt{x + 1} + C$ 29. $-1/3$ 31–35 Information given for interval $[0, 2\pi]$. Use the periodicity for other intervals.   31. Local max at $x = \pi/2$, local min at $x = 3\pi/2$, concave downward on $[0, \pi]$, upward on $[\pi, 2\pi]$ 33. No local extrema, concave upward on $[0, \pi/2) \cup [\pi, 3\pi/2)$, downward on $(\pi/2, \pi] \cup (3\pi/2, 2\pi]$, vertical asymptotes at $x = \pi/2, 3\pi/2$.   35. Local min at $x = 3\pi/4$, local max at $x = 7\pi/4$, concave upward on $[\pi/4, 5\pi/4]$, downward on $[0, \pi/4] \cup [5\pi/4, 2\pi]$ 38. 2   40. $\pi^2/2$ 41. $3/\sqrt{2} \approx 2.12$ ft above the table

Exercises 9-2, p. 505–507

1. $2x\sec^2 x^2$ 3. $2\tan x \sec^2 x$ 5. $10\tan^4 2x \sec^2 2x$ 7. $-4\tan 4x \csc^2 4x = -4\sec 4x \csc 4x$ 9. $e^x(\sec x \tan x + \sec x)$ 11. $-\sec^2 x/2(\tan x + 3)^{3/2}$ 13. $-(\cot 2x)/2 + C$ 15. $\ln|\sec 2x^3 + \tan 2x^3|/6 + C$ 17. $\ln|\sec 4x|/4 + C$ 19. $\tan x + 2\ln|\sec x| + C$ 21. $-(\cos^3 2x)/6 + C$ 23. $(\ln 3)/2$ 25. Period = 2π, local min at $x = 0, \pm 2\pi, \pm 4\pi, \ldots$, local max at $x = \pm\pi, \pm 3\pi, \pm 5\pi, \ldots$, no points of inflection, vertical asymptotes at $x = \pm\pi/2, \pm 3\pi/2, \ldots$ 27. $4\pi/3 - \ln(2 + \sqrt{3}) + \ln(2 - \sqrt{3}) \approx 1.5549$ 28. $2\pi(4\pi - 3\sqrt{3})/3 \approx 15.436$ 29. $\cos\theta = (\sqrt{3} - 1)/2$, $\theta \approx 1.196$ rad $\approx 68°31'45''$ 30. $(4^{2/3} + 6^{2/3})^{3/2} \approx 14.047$ ft   31. Height = Diameter = $4\sqrt{2}$ 32. $\pi\csc^2(4\pi/9)/18$ mi/sec ≈ 648 mi/hr   33. $11/60$ rad/sec $\approx 10°30'15''$ per sec   34. $400\pi \approx 1256.6$ ft/sec   35(b) $\ln(2 + \sqrt{3})$

Exercises 9-3, p. 513–515

1. $x = -4\sin(\pi t + \pi/2) = -4\cos\pi t$ 2. $x = 2\sin(-t/\sqrt{3} + \pi/6) = -2\sin(t/\sqrt{3} - \pi/6)$ 3. $x = \sin(2\pi t/3 - \pi/4)$ 5. $x = 3\cos 2t - 5/4 = 3\sin(2t + \pi/2) - 5/4$ 6. $x = -3(\cos 2t)/4 - 3\sqrt{3}(\sin 2t)/4 - 5/4 = -3\sin(2t + \pi/6)/2 - 5/4$ 7. $x = [-3(\sin 2t) - 5]/4$ 8. No solution   9. If we use centimeters and grams, the solution is $x = 100\sin(2\pi t/3 + \pi/2) = 100\cos(2\pi t/3)$, $k = 8000\pi^2/9$ 10(c) Period = $2\pi\sqrt{m/k}$, amplitude $= x_0 - gm/k$ 11. The graph of f has local extrema at $x = 3\pi/4 \pm n\pi$, points of inflection at $x = \pm n\pi$; the graph of g has local extrema at $x = \pi/4 \pm n\pi$, points of inflection at $x = \pi/2 \pm n\pi$ [n represents an integer]   13(a) $x = e^{-3t}[4\cos 4t + 3\sin 4t]$

Exercises 9-4, p. 524–525

1. 1.30   3. 2.64   5. $\sqrt{2}$ 7. $(3\sqrt{3} - 1)/2\sqrt{10}$ 8. $(1 - \sqrt{8})/(1 + \sqrt{8}) = (2\sqrt{8} - 9)/7$ 9. $(\sqrt{1 - 4x^2})/2x$ 11. $-1/(2\sqrt{x}\sqrt{1 - x})$ 13. $(1 - x^2)^{-1/2} - 2(1 - 4x^2)^{-1/2}$

Answers

15 $-4/(\sqrt{1-16x^2}\cos^{-1}4x)$ 17 $-1/[(\tan^{-1}x)^2(1+x^2)]$ 19 $4(\sin^{-1}2x - 1)/\sqrt{1-4x^2}$
21 $\sqrt{1-y^2}(1-\cos^{-1}y)/(1-x)$ 22 $2xy(1+y^2)/[1+(e^y - x^2)(1+y^2)]$ 25 Concave upward for $-1 \le x \le 1$, downward for $1 \le x \le 3$ 26(b) $(-1,0)$ 29 $\sqrt{66} \approx 8.1$ ft from the wall

Exercises 9-5, p. 528–529

1 $(\tan^{-1}(x/3))/3 + C$ 3 $2\sin^{-1}(x/2) + C$ 5 $\sin^{-1}(e^x) + C$ 7 $-\tan^{-1}(\cos x) + C$
9 $\sec^{-1}|3x/2|/2 + C$ 11 $\pi/8$ 13 $\pi/3$ 15 $\pi/12$ 17(a) $\pi^2/4$ (b) $2\pi(\sqrt{2}-1)$ 18 $\pi/6$

Exercises 9-6, p. 532–534

1 $4x^3\cosh x^4$ 3 $4x\,\text{sech}^2(2x^2 + 1)$ 5 $-2\,\text{sech}\,2x\,\text{csch}\,2x$ 7 $2e^{2x}\,\text{csch}\,x^2 \cdot (1 - x\coth x^2)$
9 $\frac{1}{2}\cosh(2x+1) + C$ 11 $\frac{1}{8}\cosh^4(2x) + C$ 13 $-\frac{1}{3}\text{sech}^3 x + C$ 15 $-\frac{1}{4}\coth x + C$ 17 Increasing for all x, concave downward if $x \le 0$, upward if $x \ge 0$, symmetric about origin.   19 Symmetric about the origin, increasing for all x, concave upward if $x \le 0$, downward if $x \ge 0$, horizontal asymptotes: $y = 1$ as $x \to \infty$, $y = -1$ as $x \to -\infty$.   28 $K_1 = C_1 + C_2$, $K_2 = C_1 - C_2$

Exercises 9-7, p. 539–540

1 $4x/\sqrt{1+4x^4}$ 3 $e^{\cosh^{-1}x}/\sqrt{x^2-1}$ 5 $-1/(2x\sqrt{1-x^2}\sqrt{\text{sech}^{-1}x})$ 7 $\coth^{-1}(\sqrt{x}) - \sqrt{x}/(2x-2)$
9 $\frac{1}{3}\tanh^{-1}(3x) + C$ if $|3x| < 1$, $\frac{1}{3}\coth^{-1}(3x) + C$ if $|3x| > 1$, $\frac{1}{6}[\ln|3x+1| - \ln|3x-1|] + C$
11 $\frac{3}{2}\cosh^{-1}(2x/3) + C$ if $x > 3/2$, $-\frac{3}{2}\cosh^{-1}|2x/3| + C$ if $x < -3/2$, $\frac{3}{2}\ln|2x + \sqrt{4x^2 - 9}| + C$
13 $\frac{1}{2}[\coth^{-1}2 - \coth^{-1}(3/2)] = [\ln 3 - \ln 5]/4$ 15 $\ln 5 - \ln 3$

Exercises 9-8, p. 543–544

4 $a \approx 11.3831$, $C \approx 6.0355$, $(0, 17.4186)$

Review Problems (Chapter 9), p. 544–546

1 $-1/(1 + \sin x)$ 2 $2\cos 2x \cdot e^{\sin 2x}$ 3 $\sec x(\sec^2 x + \tan^2 x)$ 4 $-2x\cot x^2$ 5 $3\sec x(\tan x + \sec x)^3$
6 $-\csc^2 x \cdot \ln 5 \cdot 5^{\cot x}$ 7 $-2x^2/\sqrt{1-x^4} + \cos^{-1}(x^2)$ 8 $x/(\sqrt{x^2-1}\sqrt{1-(x^2-1)})$
9 $4\sinh 2x\cosh 2x$ 10 $\cosh^2 x + \sinh^2 x$ 11 $e^x(\text{sech}^2 x + \tanh x)$ 12 $-4\,\text{sech}(4x-1)\tanh(4x-1)$
13 $2x(x^4-1)^{-1/2}$ 14 $2x(4x^2+1)^{-1/2} + \sinh^{-1}(2x)$ 15 $\sec(x+y) - 1$
16 $y/(\sec^2 y - x) = y/(1 - x + x^2 y^2)$ 17 $e - 1$ 18 $\ln 2$ 19 $-\frac{1}{3}\csc(3x) + C$ 20 $\ln(\sqrt{2}+1)$
21 $3x + 4\ln|\sin x| - \cot x + C$ 22 $\frac{1}{2}e^{\tan 2x} + C$ 23 $\frac{1}{6}\tan^{-1}(3x/2) + C$ 24 $\frac{2}{3}\sin^{-1}(3x/2) + C$
25 $\pi/3$ 26 $\pi/6$ 27 $\frac{1}{15}\tan^{-1}(x^3/5) + C$ 28 $\frac{1}{4}\sin^{-1}(x^4/2) + C$ 29 $2\cosh(\sqrt{x}) + C$
30 $\frac{1}{5}\sinh^5 x + C$ 31 $\frac{1}{2}\ln\cosh(x^2) + C$ 32 $\frac{1}{12}[\ln|3x+2| - \ln|3x-2|] + C$ 33 $2\ln(x + \sqrt{x^2+4}) + C$
34 $\ln(x\sqrt{3} + \sqrt{3x^2+2})/\sqrt{3} + C$ 37 0 38 $2/3$ 41 See Fig 6, Appendix B   42–44 See Figures 9-11, 9-18   51 $x = 2\cos 4t + \frac{1}{4}\sin 4t$ 52 $x = \cos 4t + \sqrt{3}\sin 4t$ 53 $x = \frac{1}{2}\cos(2\pi t/3)$ 56(a) 2;
(b) $\pi^2/2$ 57 45 rad/hr $= (1/80)$ rad/sec $\approx 0.716°$ per sec

Chapter 10

Exercises 10-1, p. 552–553

1 1 3 ∞ 5 $1/3$ 7 0 9 -1

Exercises 10-2, p. 556

1 0 3 -2 5 1 7 e 9 e^{-2} 11 e 13 1

Exercises 10-3, p. 562–563

1 2 3 Diverges to ∞ 5 10 6 $23 - \sqrt{2} - \sqrt{3} - \sqrt{4} - \sqrt{5} - \sqrt{6} - \sqrt{7} - \sqrt{8} \approx 7.694$ 7 0
9 e 11 Not defined   13 6 15 Not defined   16 12

Exercises 10-4, p. 568–569

1 ∞ 3 1 5 ∞ 7 $1/\ln 3$ 9 $\ln(e+1) - \ln(e-1) \approx 0.77194$ 11 $5/6$ 13 ∞ 15 $\pi/3$
16 $(9/5, 6/175)$

Exercises 10-5, p. 574–575

1 $8/3$ 3 $2/3$ 5 $2a$ 6(a) -240 ft/sec; (b) -240 ft/sec   7 0.25 8 0.75 9 0.5 10 0.75
11(b) $1/256$ 12(a) $1 - 14(0.6)^{13} + 13(0.6)^{14} \approx 0.9919$; (b) $14(0.6)^{13} - 13(0.6)^{14} \approx 0.0081$ 13 $5/9$

Exercises 10-6, p. 577–578

1(a) 0.3413; (c) 0.383   2(a) 69% (approx.) have higher scores, 31% (approx.) have lower   3(a) 0.6826; (c) 0.0013   4(a) 0.5328 (c) 0.1056   7(a) Local max at $x = 0$, concave upward if $x \leq -1$ or $x \geq 1$, downward if $-1 \leq x \leq 1$, points of inflection at $x = \pm 1$, the x-axis is a horizontal asymptote as $x \to \pm\infty$

Review Problems (Chapter 10), p. 578–580

3 $-1/2$ 4 $1/3$ 5 1   6 1   7 $-1/2$ 8 ∞ 9 1   10 $e^{-1/2}$ 11 $e^{1/2}$ 12 1   13 1
14 $1/e$ 15 e^2 16 e^3 19 $9/4$ 20 ∞ 21 $1/6$ 22 ∞ 23 ∞ 24 $1/8$ 25 ∞
26 $1/(n-1)$ 27 $(3,3/224)$ 28(a) $[4\ln(\sqrt{2}+1)]/\pi \approx 1.122$ (b) $[\sqrt{2} + \ln(\sqrt{2}+1)]/2 \approx 1.148$
29(b) $\int_{-1}^{3} f(x)\,dx = \int_{1}^{3} x^{-2}\,dx = 2/3$; (c) $1/5$ 31(a) 0.8186; (b) 0.0227

Chapter 11

Exercises 11-1, p. 585–586

1 $x\tan x + \ln|\cos x| + C$ 3 $x\sin x + \cos x + C$ 5 $x\ln(1 + x^2) - 2x + 2\tan^{-1}x + C$
7 $[\sin x^2 - x^2\cos x^2]/2 + C$ 9 $\cos x[1 - \ln(\cos x)] + C$ 11 $\cos x(2 - x^2) + 2x\sin x + C$ 13 1
15 $2 - 5/e$ 17 $3(e^{-2\pi} + 1)/13$ 19 1   21 $-4/9$ 27 $[\sec x \tan x + \ln|\sec x + \tan x|]/2 + C$
29 $-x^2\cos x + 2x\sin x + 2\cos x + C$ 31 $e^x[x^4 - 4x^3 + 12x^2 - 24x + 24] + C$ 33(a) $\pi(e - 2)$;
(b) $\pi(e^2 + 1)/2$ 34 $\bar{x} = (e^2 + 1)/(e^2 - 1)$, $\bar{y} = (e^2 + 1)/4$

Exercises 11-2, p. 593–594

1 $(\cos^6 2x)/12 - (\cos^4 2x)/8 + C$ or $(\sin^4 2x)/8 - (\sin^6 2x)/12 + C$ 3 $-\cos x + 2(\cos^3 x)/3 - (\cos^5 x)/5 + C$
5 $-\csc x - \sin x + C$ 7 $\tan x + C$ 9 $\frac{1}{3}\cos 3x - \frac{1}{3}\ln|\csc 3x + \cot 3x| + C$ 11 $(\sec^5 x)/5 + C$
13 $(\sec^5 x)/5 - (\sec^3 x)/3 + C$ 15 $(\tan^4 x)/4 - (\tan^2 x)/2 + \ln|\sec x| + C$
17 $[-\csc x \cot x - \ln|\cot x + \csc x|]/2 + C$ 19 $-(\cos 2x)/2 + C$ 21 $-(\csc^3 x)/3 + C$
23 $(\sin x)/2 - (\sin 5x)/10 + C$ 25 $(\sin 4x)/8 + (\sin 2x)/4 + C$ 27 $1 - \pi/4$ 28 $\ln(1 + \sqrt{2}) - 1/\sqrt{2}$
29 $\pi/64 + 1/48$ 30 $(3\pi - 2)/24$ 31 $\pi/8 + 9\sqrt{3}/128$ 32 $2/15 - 8/45\sqrt{3}$ 34 $(\pi/2, \pi/8)$
35 $(3\pi/4, 0)$ 36(a) $3\pi^2\sqrt{2}$; (b) $4\pi\sqrt{2}$

Exercises 11-3, p. 597–598

1 $-\sqrt{25 - x^2}/x - \sin^{-1}(x/5) + C$ 2 $\sqrt{9 - x^2} - 3\ln|3 + \sqrt{9 - x^2}| + 3\ln|x| + C$
3 $(e^x/2)\sqrt{9 - e^{2x}} + \frac{9}{2}\sin^{-1}(e^x/3) + C$ 4 $\ln|\ln x + \sqrt{9 + (\ln x)^2}| + C$ 5 $(-x\sqrt{1 - x^2} + \sin^{-1}x)/2 + C$
6 $-[\ln|4 + \sqrt{16 - x^2}| - \ln|x|]/4 + C$ 7 $-\sqrt{1 - 9x^2}/x + C$ 8 $(-\ln|\sqrt{25 + x^2} + 5| + \ln|x|)/5 + C$
9 $x/(4\sqrt{4 - 9x^2}) + C$ 10 $x/\sqrt{36 - x^2} - \sin^{-1}(x/6) + C$ 11 $\frac{1}{4}(10x - x^3)\sqrt{4 - x^2} + 6\sin^{-1}(x/2) + C$
12 $\frac{1}{10}\left[\dfrac{5x}{25 - x^2} - \ln\left|\dfrac{5 + x}{\sqrt{25 - x^2}}\right|\right] + C$ 13 $1/4\sqrt{x^2 + 4} - \frac{1}{8}\ln\left|\dfrac{2 + \sqrt{x^2 + 4}}{x}\right| + C$
14 $\frac{1}{256}\left(\ln\left|\dfrac{x}{\sqrt{16 + x^2}}\right| - \dfrac{1}{2}\cdot\dfrac{x^2}{16 + x^2}\right) + C$ 15 $\frac{1}{8}[\tan^{-1}x - x(1 - x^2)/(1 + x^2)^2] + C$
16 $\frac{1}{7}(4 + x^2)^{7/2} - \frac{4}{5}(4 + x^2)^{5/2} + C$ 17 $\frac{1}{5}(x^2 - 6)(9 + x^2)^{3/2} + C$
18 $\frac{1}{16}[\frac{1}{7}(4x^2 - 25)^{7/2} + 5(4x^2 - 25)^{5/2}] + C$ 19 $\ln|e^x + \sqrt{e^{2x} - 1}| + C$
20 $\sqrt{4e^{2x} - 1}/(2e^{2x}) + 2\sec^{-1}(2e^x) + C$ 21 $\frac{1}{4}x^2\sqrt{4 + x^4} + \ln(x^2 + \sqrt{4 + x^4}) + C$
22 $-x^2/(2\sqrt{16 + x^4}) + \frac{1}{8}\ln(x^2 + \sqrt{16 + x^4}) + C$ 23 $1/\sqrt{3} - \pi/6$ 24 $(7\sqrt{17} - \sqrt{5})/24$
25 $\sqrt{10} - \sqrt{37}/2$ 26 $3 - 3\pi/4$ 27 $255/2 - 72\ln 2$ 28 $4^5/5 + 3\cdot 4^3 = 1984/5$
29 Diverges to $-\infty$ 30 $7.5 + 8\ln 2$ 31 $1/180$ 32 $\pi/16$ 33 $\frac{1}{2}[\sqrt{2} + \ln(1 + \sqrt{2})]$ 34 $ab\pi$
35 $(2 + \pi)/4$ 36 $5\pi/3$ 37 $\pi - 2$

Exercises 11-4, p. 604

1 $3\ln|x + 2| - 2\ln|x - 1| + C$ 2 $\ln|x - 1| - \ln|x| + C$ 3 $\frac{1}{3}[\ln|x - 3| - \ln|x + 3|] + C$
4 $2\ln|2x - 1| - \ln|x - 3| + C$ 5 $x^2 - 3x + 2\ln|x - 2| - \ln|x + 3| + C$
6 $x^2/2 - \ln|x| + \frac{3}{2}\ln|x^2 - 4| + C$ 7 $x^2/2 + 4x + 2\ln|x^2 - x| + C$
8 $3\ln|x| + 3\ln|x - 1| - 2\ln|x + 1| + C$ 9 $\ln|x| - 2\tan^{-1}x + C$ 10 $6\ln|x + 2| - \ln(x^2 + 1) + \tan^{-1}x + C$
11 $4\tan^{-1}x - 3/[2(x^2 + 1)] + C$ 12 $x^2 - 2\ln|x| - 3/x - \frac{1}{2}\ln(x^2 + 4) - 3\tan^{-1}(x/2) + C$
13 $2\ln|x| - 3\ln|x - 2| + 5\ln|x + 4| + C$ 14 $\frac{3}{2}\tan^{-1}(x/2) + 2\ln|x - 2| - 2\ln|x + 2| + C$
15 $-\tan^{-1}(x/2) + \frac{1}{2}\ln(x^2 + 1) + C$ 16 $\frac{7}{2}\ln|2x + 1| - \frac{3}{2}\ln(x^2 + 1) + 2\tan^{-1}x + C$
17 $2\ln|x| + 3/x + \frac{3}{2}\ln(x^2 + 1) - \tan^{-1}x + C$

Answers

18 $x^3/3 - 2x + 2/x + \frac{5}{2}\tan^{-1}(x/2) + C$
19 $\sqrt{e^2 + 1}(e - 1)/e + \ln(\sqrt{e^2 + 1} + e) - \ln(\sqrt{e^2 + 1} + 1) + 1 \approx 3.196$

Exercises 11-5, p. 606-607

1 $\ln(x^2 - 2x + 5) + \tan^{-1}[(x - 1)/2] + C$ 2 $\frac{1}{2}\tan^{-1}[(x + 3)/2] + C$ 3 $\sin^{-1}[(\sqrt{2}(x + 1))/3]/\sqrt{2} + C$
4 $-2\sqrt{21 - 4x - x^2} - \sin^{-1}[(x + 2)/5] + C$ 5 $\frac{1}{4}\{2(x + 1)\sqrt{5 - 4x^2 - 8x} + 9\sin^{-1}[(2x + 2)/3]\} + C$
6 $-\frac{1}{9}\sqrt{16 - 18x - 9x^2} - \frac{1}{3}\sin^{-1}[(3x + 3)/5] + C$ 7 $\frac{1}{8}\ln(4x^2 - 8x + 13) + \frac{1}{8}\tan^{-1}[(2x - 2)/3] + C$
8 $\frac{1}{12}[\ln|2x - 9| - \ln|2x - 3|] + C$ 9 $\frac{1}{9}[(x - 2)/\sqrt{5 + 4x - x^2}] + C$ 10 $(7 - 9x)/144\sqrt{7 - 18x - 9x^2} + C$
11 $\frac{1}{5}(x^2 + 6x + 8)^{5/2} + \frac{1}{3}(x^2 + 6x + 8)^{3/2} + C$ 12 $(x - 1)/[8(x^2 - 2x + 5)] + \frac{1}{16}\tan^{-1}[(x - 1)/2] + C$
13 $3\sqrt{x^2 - 4x + 13} + 6\ln|\sqrt{x^2 - 4x + 13} + x - 2| + C$ 14 $\frac{1}{4}\ln|2x - 5| - \frac{1}{8}\ln|4x^2 - 12x + 5| + C$
15 $\frac{1}{8}\tan^{-1}[(x^2 - 3)/4] + C$ 16 $\frac{1}{2}\ln|x^2 + 4x - 5| + \frac{1}{6}\ln|x - 1| - \frac{1}{6}\ln|x + 5| + C$

Exercises 11-6, p. 610-611

1 $2\ln(1 + \sqrt{x}) + C$ 2 $x - \ln(1 + e^x) + C$ 3 $2\ln|\sqrt{x + 1} - 1| + C$ 4 $(x - 2)/\sqrt{2x - 5} + C$
5 $\frac{2}{5}(3 - x)^{5/2} - \frac{10}{3}(3 - x)^{3/2} + C$ 6 $2(x + 1)^{3/2} - 4(x + 1)^{1/2} + C$ 7 $\frac{2}{5}(x^3 + 2)^{3/2} - \frac{4}{3}(x^3 + 2)^{1/2} + C$
8 $\frac{3}{7}(1 - x)^{7/3} - \frac{3}{4}(1 - x)^{4/3} + C$ 9 $2\sqrt{x} - 3\sqrt[3]{x} + 6\sqrt[6]{x} - 6\ln|\sqrt[6]{x} + 1| + C$
10 $\frac{4}{5}x^{5/4} - x + \frac{4}{3}x^{3/4} - 2x^{1/2} + 4x^{1/4} - 4\ln(1 + \sqrt[4]{x}) + C$ 11 $2e^{\sqrt{x}}(\sqrt{x} - 1) + C$
12 $(3x^{2/3} - 6)\sin(x^{1/3}) + 6x^{1/3}\cos(x^{1/3}) + C$ 13 $\tan(x/2) + C$ 14 $-2/[\tan(x/2) + 1] + C$
15 $x - \tan(x/2) + C$ 16 $\sqrt{2}\ln|\sec(x/2)| + x/\sqrt{2} + C$ 17 $\frac{1}{2}[\ln|3\tan(x/2) + 1| - \ln|\tan(x/2) - 1|] + C$
18 $-\ln|\tan(x/2) - 1| + C$

Exercises 11-7, p. 612-614

25 $\frac{1}{2}[\tan^{-1}4 - \tan^{-1}2] \approx 0.10934$ 26 $7 + (\ln 9)/8 \approx 7.27465$ 27 $\ln 2 \approx 0.69315$ 28 $\pi^2/36 \approx 0.27415$
29 1.46268   30 6.20876   31 1.2498   32 1/2   33 0.07410   34 4.2359   35 $3\pi/4 \approx 2.3562$
36 $6 - 2e \approx 0.5634$ 37 $\frac{1}{32}(\frac{9}{25} - \frac{17}{81}) \approx 0.00469$ 38 $\frac{1}{4}[\sqrt{10} + 9\ln(1 + \sqrt{10}) - 9\ln 3] \approx 3.055$
39 $-(\ln 15)/7 \approx -0.38686$ 40 $\frac{1}{2}\ln(2 + \sqrt{5}) \approx 0.72182$

Review Problems (Chapter 11), p. 614-615

1 $\frac{1}{2}[\sin 2x - \sin^3 2x + \frac{3}{5}\sin^5 2x - \frac{1}{7}\sin^7 2x] + C$ 2 $\tan x - x + C$ 3 $\frac{1}{2}[3\ln|x - 5| - \ln|x - 1|] + C$
4 $x\ln 2x - x + C$ 5 $\frac{1}{8}\ln(4x^2 + 1) + C$ 6 $-x + \frac{3}{2}[\ln|x + 3| - \ln|x - 3|] + C$
7 $3\ln|x| + 1/x + 2\ln|x - 1| + 2/(x - 1) + C$ 8 $6\ln|\sec(x/2)| - 4\ln|\tan(x/2) - 3| - 2\ln|\tan(x/2) + 3| + C$
9 $e^{x^2}(x^2 - 1)/2 + C$ 10 $\ln|\ln\sin x| + C$ 11 $2[\sqrt{x + 3} + \ln|\sqrt{x + 3} - 2| - \ln|\sqrt{x + 3} + 2|] + C$
12 $\ln(x^2 - 4x + 8) + \frac{1}{2}\tan^{-1}[(x - 2)/2] + C$ 13 $-\frac{1}{3}\csc^3 3x + C$ 14 $e^x[\cos x + \sin x]/2 + C$
15 $\frac{1}{5}(x^2 - 6)(x^2 + 9)^{3/2} + C$ 16 $\frac{2}{45}(3x + 1)^{5/2} - \frac{8}{27}(3x + 1)^{3/2} + C$
17 $-\ln|\tan^2(x/2) + 4| + 2\ln|\sec(x/2)| - 2\tan^{-1}[\frac{1}{2}\tan(x/2)] + C$
18 $-4\ln|x| + 1/x + \frac{3}{2}\ln|x - 1| + \frac{3}{2}\ln|x + 1| + C$
19 $[\ln|\tan(x/2) + 2 - \sqrt{7}| - \ln|\tan(x/2) + 2 + \sqrt{7}|]/\sqrt{7} + C$
20 $(\sec^7 x)/7 + C$ 21 $(\cos 2x)/4 - (\cos 8x)/16 + C$ 22 $\frac{1}{12}(\ln|2x + 3| - \ln|2x - 3|) + C$
23 $(x + 1)\ln\sqrt{x + 1} - x/2 + C$ 24 $\frac{9}{2}\sin^{-1}[(x - 3)/3] + \frac{1}{2}(x - 3)\sqrt{6x - x^2} + C$
25 $\ln|x| + \ln(x^2 + 1) + \tan^{-1}x + C$ 26 $x/2 - (\sin 2x)/4 + C$ 27 1/9   28 1
31 $((e^2 + 1)/4, (e - 2)/2)$ 32 (1,1/4)   33 $\frac{1}{4}[\sqrt{2} + \ln(1 + \sqrt{2})]$ 34 $\ln(\sqrt{2} + 1)$ 36(b) Let

$f(n) = \int_0^{\pi/2} \sin^n x\, dx$. Show that $f(1) = 1$, $f(3) = 2/3$, $f(5) = (2 \cdot 4)/(3 \cdot 5)$, $f(7) = (2 \cdot 4 \cdot 6)/(3 \cdot 5 \cdot 7)$. It also can be shown that $f(2) = \frac{1}{2} \cdot \frac{\pi}{2}$, $f(4) = \frac{1 \cdot 3}{2 \cdot 4} \cdot \frac{\pi}{2}$, $f(6) = \frac{1 \cdot 3 \cdot 5}{2 \cdot 4 \cdot 6} \cdot \frac{\pi}{2}$. See if you can generalize these formulas and prove them by mathematical induction.

Chapter 12

Exercises 12-1, p. 621-622

1 5/2, 9/4, 17/8, 33/16; $L = 2$ 3 No limit   5 $L = e$ 7 2, 1.75, 1.732143, 1.732051; $L = \sqrt{3}$
9 2.613706, 2.716244, 2.718281, 2.718282; $L = e$ 11(a) 0.5, 0.375, 0.351852, 0.34375; (b) 1/3, 1/3, 1/3, 1/3;
(c) $L = 1/3$ 12(a) 8.389056, 6.9128099, 6.623953, 6.5216101; (b) 6.420728, 6.391210, 6.389489, 6.389194;
(c) $L = e^2 - 1 \approx 6.389056$ 13 1, 3/2, 7/4, 15/8, 31/16; $L = 2$ 15 1, 2, 5/2, 8/3, 65/24   17 1, 2, 3, 4, 5; diverges   19 1/3, 4/9, 13/27, 40/81, 121/243; converges to 1/2   21 8/33   23 5/9   26(a) If $M > 0$, there exists $N > 0$ such that if $n \geq N$, then $a_n \geq M$.

Exercises 12-2, p. 629–630

1   $s_1 = -1/4$, $s_2 = -3/16$, $s_3 = -13/64$, $s_4 = -51/256$, geometric series, $a = -1/4$, $r = -1/4$, $L = -1/5$
3   diverges   5 $\lim_{n\to\infty} a_n = 2/3 \neq 0$, diverges

7   $\sum_{n=1}^{\infty} [1/2^{n-1} + (-\frac{1}{3})^n] = 2 - 1/4 = 7/4$ 9 $\sum_{n=1}^{\infty} \frac{3}{n} = 3\sum_{n=1}^{\infty} \frac{1}{n}$, diverges   11   40 ft   12(b) $s_2 = 2$,
$s_5 \approx 2.708333$; (c) $s_{20} \approx 2.718281828$, correct to 9 decimal places   13(b) $s_2 = 0.5$, $s_5 \approx 0.783333$;
(c) $s_{20} \approx 0.668771$, $s_{100} \approx 0.688172$, s_{100} approximates ln 2 correct to 2 decimal places   15   1 and 0
17   1/2 and 0

Exercises 12-3, p. 637

1(a)   diverges   3(a)   converges; (b) $2.089973 \approx s_5 + 1/\ln 7 \leq \Sigma \leq s_6 + 1/\ln 7 \approx 2.1277$; (c) converges to a number between 2.109124 and 2.110356   5(a)   converges;
(b) $0.53411 \approx s_5 + [\pi/2 - \tan^{-1}(e^6)] \leq \Sigma \leq s_6 + [\pi/2 - \tan^{-1}(e^6)] \approx 0.536597$; (c) converges to a number approximately equal to $s_{50} \approx s_{51} \approx 0.535560665$ 7(a)   Diverges   9   Diverges   11   Converges to a number between 1.199551 and 1.204181

Exercises 12-4, p. 645

1   Converges   3   converges   5   Diverges   7   converges   9   converges   11   Diverges

Exercises 12-5, p. 648–649

1(a)   converges; (b) $0.40921 < L < 0.81746$; (c) $0.53454 < L < 0.67457$ 3(a)   converges;
(b) $-1.83032 < L < -1.28978$; (c) $-1.56629 < L < -1.50672$ 5(a)   converges; (b) $0.005055 < L < 0.30368$;
(c) $0.1216 < L < 0.1987$ 6(a)   converges; (b) $4.336428 < L < 4.880759$; (c) $4.547422 < L < 4.692941$
7   Converges   9   Converges   11   Diverges   13   Diverges   16(a) $n > 4 \cdot 10^{10} - 1$

Exercises 12-6, p. 656

1   Conditional   3   Absolute   5   Divergent   7   Absolute

Exercises 12-7, p. 663–664

1   All x 3   Convergent if $-1/2 \leq x < 1/2$, absolutely convergent if $-1/2 < x < 1/2$ 5   Diverges if $x \neq -3$ 7 $1 \leq R \leq 1.5$ 9(b) $x + 2y = 2$; (c) $c + x - x^2/(2 \cdot 2!) + x^3/(3 \cdot 4!) - x^4/(4 \cdot 6!) + \cdots$
10(c)   2.716667; (d) 1.648698   11(c) $\pi/4 \approx 0.7853253$

Exercises 12-8, p. 672–674

1, 3(a)   The series and intervals of convergence are given in the text.   1(b)   0.4791667; (c) 0.4794256
3(b)   0.520833; (c) 0.5210953   5(a) $\ln(x + 1) = x - x^2/2 + x^3/3 - x^4/4 + \cdots + (-1)^{n+1}x^n/n + \cdots$,
$-1 < x \leq 1$; (b) 0.40104; (c) 0.405465   7   1/6   9   2
11   $C + x - x^3/(3 \cdot 1!) + x^5/(5 \cdot 2!) - x^7/(7 \cdot 3!) + \cdots$ 13 $C + x^2/2! - x^4/4! + 9x^6/6! + \cdots$
23(a) $\sin(ix) = i\sinh x$; (b) $\cos(ix) = \cosh x$

Exercises 12-9, p. 680–682

1(a) $1/2 - \sqrt{3}(x - \pi/3)/2 \cdot 1! - (x - \pi/3)^2/2 \cdot 2! + \sqrt{3}(x - \pi/3)^3/2 \cdot 3! + (x - \pi/3)^4/2 \cdot 4! - \cdots$, converges for all x; (b) $|r_n(1.5)| < (1.5 - \pi/3)^{n+1}/(n + 1)! < (0.46)^{n+1}/(n + 1)!$; (c) $n = 6$
3(a) $(x - 1) - (x - 1)^2/2 + (x - 1)^3/3 - (x - 1)^4/4 + \cdots$, $R = 1$; (b) $|r_n(1.5)| < (0.5)^{n+1}/(n + 1)$; (c) $n = 13$
5(a) $28 + 35(x - 2) + 13(x - 2)^2 + (x - 2)^3$, converges for all x; (b) $r_n(1.5) = 0$ if $n \geq 4$; (c) $n = 4$
7(a) $1 - (x - 1) + (x - 1)^2 - (x - 1)^3 + (x - 1)^4 - \cdots$, $R = 1$; (b) $|r_n(1.5)| < (0.5)^{n+1}$;
(c) $n + 1 > (\ln 5 - 6\ln 10)/\ln(0.5)$, $n \geq 17$ 11(c) $\sqrt[3]{1.5} = f(0.5) \approx 1.144714$ (use $r = 1/3$, $n = 13$)
13(a) $1 + x^2/(2 \cdot 1!) + (1 \cdot 3x^4)/(2^2 \cdot 2!) + (1 \cdot 3 \cdot 5x^6)/(2^3 \cdot 3!) + \cdots$, $R = 1$;
(b) $x + x^3/(2 \cdot 3) + (1 \cdot 3x^5)/(2^2 \cdot 2! \cdot 5) + (1 \cdot 3 \cdot 5x^7)/(2^3 \cdot 3! \cdot 7) + (1 \cdot 3 \cdot 5 \cdot 7x^9)/(2^4 \cdot 4! \cdot 9) + \cdots$ (Evaluate the series at $x = 0$ to show the constant of integration is 0.)

Review Problems (Chapter 12), p. 682–685

5   2   6   Diverge   7   1/2   8   1   9(a) $123/10^4 + 123/10^7 + 123/10^{10} + \cdots = 123/9990 = 41/3330$
13   Convergent geometric series   14   Converges (sum of two convergent series)   15   Diverges (integral test or comparison test)   16   Diverges (Corollary to Theorem 12-2)   17   Converges (integral test or comparison test)
18   Multiple of divergent p-series   19   Converges (compare with p-series)   20   Absolutely convergent (compare

Answers   A-57

with geometric series)   **21** Converges (ratio test)   **22** Converges (ratio test)   **23** Conditionally convergent (alternating series)   **24** Conditionally convergent (alternating series)   **29** Convergent on $[-3,3)$, absolutely convergent on $(-3,3)$ **30** Absolutely convergent everywhere   **31** Absolutely convergent on $(1,3)$ **32** Convergent on $(-3, -2]$, absolutely convergent on $(-3, -2)$

36 $2x - (2x)^2/2 + (2x)^3/3 - \cdots$, $-\frac{1}{2} < x \leq \frac{1}{2}$ **37** $\ln x = \sum_{n=1}^{\infty} (-1)^{n+1}(x-1)^n/n$, $0 < x \leq 2$

38 $\sin x = \frac{1}{\sqrt{2}}[1 + (x - \pi/4)/1! - (x - \pi/4)^2/2! - (x - \pi/4)^3/3! + \cdots]$, all x **39** $\sum_{n=0}^{\infty} (x^{2n}/n!)$

40 $\sum_{n=0}^{\infty} 1/(2^n \cdot n!)$ **41** $C + x - x^2/(2 \cdot 2!) + x^3/(3 \cdot 4!) - x^4/(4 \cdot 6!) + x^5/(5 \cdot 8!) - x^6/(6 \cdot 10!) + \cdots$

42 $C + x - x^5/5 + x^9/9 - x^{13}/13 + x^{17}/17 - \cdots$ **43** $C - x/2! + x^3/(3 \cdot 4!) - x^5/(5 \cdot 6!) + x^7/(7 \cdot 8!) - \cdots$

44 $C + \ln|x| + \sum_{n=1}^{\infty} x^n/(n \cdot n!)$ **45** $-1/2$ **46** $1/3$ **47** 1 **48** 1 **51** $n \geq 8$

Chapter 13

Exercises 13-1, p. 692–694

1 The y-axis from $(0,0)$ to $(0,1)$ and the line $y = 1$ from $(0,1)$ to $(1,1)$.   **5** Part of parabola $x = -2y^2$ with $y \geq 0$ **7** First quadrant part of graph of $y = 1/x^2$ **9** Part of line $y = x + 2$ with $x \geq 1$ **11** Ellipse $x^2/3^2 + y^2/2^2 = 1$ traced out once in clockwise direction   **13** Part of parabola $x = 2y^2 - 1$ with $-1 \leq y \leq 1$ **15** Part of hyperbola $x^2 - y^2 = 1$ in Quadrant IV   **17(b)** $0 \leq t \leq 1$ **18** $x = \sinh^{-1} s = \ln(s + \sqrt{s^2 + 1})$, $y = \sqrt{1 + s^2}$ **21** $x = 15\phi - 10\sin\phi$, $y = 15 - 10\cos\phi$

Exercises 13-2, p. 701–702

1(a) $1, 0$; **(b)** $x - y + 7 = 0$ **3(a)** $2t^2, 4t^2$; **(b)** $2x - y = -1$ **5(a)** $(-2\tan t)/3, (-2\sec^3 t)/9$, $t \neq \pm\pi/2, \pm 3\pi/2, \ldots$; **(b)** $y = -2$ **7(a)** $2\sin t, -2, x \neq \pm\pi/2, \pm 3\pi/2, \ldots$; **(b)** $4x - 4y = 3$ **9** Local max at $(0,1)$ when $t = 0$ **11** Local min at $(3, -16)$ when $t = 2$ **13** Cycloid. Horizontal at $t = \pm\pi, \pm 3\pi, \pm 5\pi, \ldots$, vertical at $t = 0, \pm 2\pi, \pm 4\pi, \ldots$ **15(a)** $24x - y = 24\ln 4 - 8$; **(b)** $12x - y = 12\ln 3 - 3$ **17(a)** $\sin\phi_0/(1 - \cos\phi_0) = \cot(\phi_0/2)$; **(b)** $(\pm 2n\pi, 0)$, n an integer

Exercises 13-3, p. 708–709

1(a) $(1, 2\pi/3), (-1, 5\pi/3), (-1/2, \sqrt{3}/2)$; **(c)** $(\sqrt{2}, 5\pi/4), (-\sqrt{2}, 9\pi/4), (-1, -1)$; **(e)** $(-4, \pi/6), (4, -5\pi/6), (-2\sqrt{3}, -2)$ **2(a)** $(2, \pi), (-2, 0)$; **(c)** $(2, 4\pi/3), (-2, \pi/3)$ **3** $x = 2$ **5** $y = x$ **7** $x^2 + y^2 = 1$ **9** $(x - 1)(y - 1) = 1$ **11** $r\cos\theta = 2$ **13** $r = 6$ **14** Use the quadratic formula to obtain $r = 3/(2 + \sin\theta)$ or $r = 3/(\sin\theta - 2)$ (both equations have the same graph)   **15** $r = 1/(1 - \cos\theta)$ or $r = -1/(1 + \cos\theta)$ **17** $(x^2 + y^2)^3 = (x^2 - y^2)^2$ **19** circle, radius 3   **21** circle, radius 1   **23** Line $y = 3$ **25** Spiral. Note that the line $y = 1$ is a horizontal asymptote as $\theta \to 0$ **27** $(1, \pm\pi/3)$ **28** $(2,0)$, $(2,\pi)$, $(2,\pi/2)$, $(2,3\pi/2)$

Exercises 13-4, p. 717–718

3 Cardioid, symmetric about y-axis   **5** Limaçon, symmetric about y-axis   **7** Limaçon, symmetric about x-axis   **9** Three-leaved rose, symmetric about y-axis, tangent lines at origin: $\theta = 0, \pi/3, 2\pi/3, \pi, 4\pi/3, 5\pi/3$ **11** Lemniscate, symmetric about origin, defined when $0 \leq \theta \leq \pi/2$ and $\pi \leq \theta \leq 3\pi/2$. Graph has one loop in Quadrant I and another in Quadrant III. The coordinate axes are tangent lines at the origin.   **13** Cardioid, symmetric about the x-axis   **15(a)** Replace r by $-r$, θ by $-\theta$. If the new equation is equivalent to the original one, the graph is symmetric about the y-axis; **(b)** Symmetric about y-axis   **16(a)** Replace θ by $\pi + \theta$. If the new equation is equivalent to the original one, the graph is symmetric about the origin; **(b)** Symmetric about origin   **17–20** Find the points where $D_x y$ is zero or does not exist to locate the relative extrema   **17** Cardioid, $(3/2, \pm\pi/3)$ **19** Four-leaved rose. Extreme points are $(-1, \pm\pi/2)$. The extreme points on the loops about the x-axis are at $\theta = \pm\pi/6$ and $\theta = \pm 5\pi/6$

Exercises 13-5, p. 723–724

1 line $x = 2$ **3** lines $y = \pm\sqrt{3}x$ **5** Circle, center $(1, \pi)$, $r = 1$ **7** Parabola $12x + 4y^2 = 9$ **9** Ellipse $8x^2 + 9y^2 + 12x = 36$ **11** Ellipse $4x^2 + 3y^2 + 8y = 16$ **13** $r = 2/(1 - \sin\theta)$ **14** $r = 24/(5 + \cos\theta)$ **15** $r = 5/(2 - 3\sin\theta)$ **17–20** Answers are not unique. They depend on the orientation of the axes.   **17(a)** 0.206; **(b)** $r \approx 34.5/(1 + 0.206\cos\theta)$ **19(a)** 0.017; **(b)** $r \approx 92.9/(1 + 0.017\cos\theta)$

Exercises 13-6, p. 730-731

1 Circle, π 3 Cardioid, 6π 5 9   7 Three leaved rose, $\pi/4$ 9 $\pi/3 + \sqrt{3}/2$ 10 $2 + \pi/4$ 11 π
12 $\sqrt{3} - \pi/3$ 13 2   14 $5\pi/4$ 15 $\sqrt{5}(e - \sqrt{e})$ 17 $\pi\sqrt{4\pi^2 + 1} + \frac{1}{2}\ln(2\pi + \sqrt{4\pi^2 + 1})$
19 $3\pi/4$ 21 $5\pi/6 - 3\sqrt{3}/2$ 22 $25\pi/6 + 3\sqrt{3}/2$ 23 $10\pi/3 + 3\sqrt{3}$ 24 $5\pi/24 - \sqrt{3}/4$
25 $5\pi/4$ 28(d) $4\pi a^2$

Review Problems, (Chapter 13), p. 731-733

1 $x + y = 1, 0 \le x \le 1$ 3 $4x^2 - y^2 = 4, x \ge 1$ 7 $(3t^2 + 14t)/(2t + 1), (6t^2 + 6t + 14)/(2t + 1)^3$
8 $-\cot 2\theta, -\csc^3 2\theta$ 9 $8x - y = 12$ 10 $x + \sqrt{3}y = 3$ 11 Horizontal at $t = 1/2$, vertical at $t = 0$
12 Horizontal at $(3, \pm\pi/3)$, vertical at $(4,0), (1, \pm 2\pi/3)$ 13 $g'(t)/f'(t), [f'(t)g''(t) - f''(t)g'(t)]/[f'(t)]^3$
14 $4x - y = 36$ 15(a) $x^2 + y^2 = x + y$; (b) $x = \sin\theta\cos\theta + \cos^2\theta, y = \sin\theta\cos\theta + \sin^2\theta$ 16(a) $y = x^2$;
(b) $x = \tan\theta, y = \tan^2\theta$ 17(a) $xy = 1$; (b) $x = \sqrt{\cot\theta}, y = \sqrt{\tan\theta}$ if $0 < \theta < \pi/2$; $x = -\sqrt{\cot\theta}, y = -\sqrt{\tan\theta}$
if $\pi < \theta < 3\pi/2$ 18(a) $8x^2 - y^2 + 12x + 4 = 0$; (b) $x = 2\cos\theta/(1 - 3\cos\theta), y = 2\sin\theta/(1 - 3\cos\theta)$
19 $\theta = \pi/3$ 20 $r = 4\cos\theta$ 21 $r = \sin 2\theta$ 22 $r = \cot\theta\csc\theta$ 23 Four-leaved rose
24 Leminiscate   25 Circle   26 Three-leaved rose   27 Limaçon   28 Cardioid   29 Hyperbola,
$x = -2/3$ 30 Parabola, $y = 6$ 31 $\pi/3 + \sqrt{3}/2$ 32 $\sqrt{3} + 8\pi/3$ 33 4   34 Area inside large
loop $= 3(\pi + 1)/2$, area inside small loop $= (\pi - 3)/2$, area between loops $= \pi + 3$ 35 $\pi/3$ 36 8
37 $\sqrt{2}(e^{2\pi} - 1)$ 38 $2\sqrt{2}(1 - e^{-2\pi})$

Chapter 14

Exercises 14-1, p. 742

1 $\langle 3,7 \rangle = 3\mathbf{i} + 7\mathbf{j}$ 3 $\langle -9,-3 \rangle = -9\mathbf{i} - 3\mathbf{j}$ 5 $\langle -3,-1 \rangle = -3\mathbf{i} - \mathbf{j}$ 7 $(-2,4)$ 9 $(0,13)$
11 $\langle 4,-6 \rangle, \langle 0,1 \rangle, \langle 4,-7 \rangle, \langle 2,-1 \rangle, \langle 10,-19 \rangle$ 13 $-4\mathbf{i} + 8\mathbf{j}, -5\mathbf{i} + 3\mathbf{j}, \mathbf{i} + 5\mathbf{j}, -12\mathbf{i} + 10\mathbf{j}, 10\mathbf{i} + 8\mathbf{j}$
15 $\sqrt{2}, 7\pi/4$ 17 4, 0   19 4, $11\pi/6$ 21(a) $\langle 0,-1 \rangle$; (b) $\langle 0,1 \rangle$ 23(a) $-2\mathbf{i}/\sqrt{13} + 3\mathbf{j}/\sqrt{13}$;
(b) $2\mathbf{i}/\sqrt{13} - 3\mathbf{j}/\sqrt{13}$

Exercises 14-2, p. 748-749

1 9, $(0,-3/2,0)$ 3 7, $(-9/2,5,2)$ 7 $(2,2,3), (2,2,-2), (2,5,3), (-1,2,-2), (-1,5,-2), (-1,5,3)$
9 $x^2 + y^2 + z^2 - 4y + 6z = 12$ 11 $C(1,-3,-2), r = 4$ 13 $C(5/2,-3/2,3), r = \sqrt{19}$ 15 $C(1,0,-2)$,
$r = 3$ 17 $x^2 + y^2 + z^2 - 2x + 8y - 10z + 17 = 0$ 19 $x^2 + y^2 + z^2 - 4x - 2y - 6z = 12$
21 $x - 2y - z = 5$ 23 $x^2 + y^2 + z^2 - 6x - 6y - 6z + 18 = 0, x^2 + y^2 + z^2 - 22x - 22y - 22z + 242 = 0$

Exercises 14-3, p. 756

1 5, $\langle 3,-1,-12 \rangle, \frac{1}{5}\langle 0,-4,3 \rangle$ 3 $\sqrt{35}, -13\mathbf{j} + 11\mathbf{k}, (-3\mathbf{i} + 5\mathbf{j} - \mathbf{k})/\sqrt{35}$ 5 $(6,2,2)$ 7 $2\sqrt{2}\mathbf{i} + 2\mathbf{j} + 2\mathbf{k}$
9 $\cos\alpha = 3/7, \cos\beta = -6/7, \cos\gamma = 2/7, \alpha \approx 64.6°, \beta \approx 149°, \gamma \approx 73.4°$ 11 $\cos\alpha = -1/9, \cos\beta = 4/9$,
$\cos\gamma = 8/9, \alpha \approx 96.4°, \beta \approx 63.6°, \gamma \approx 27.3°$ 13 $(3\mathbf{i} - 6\mathbf{j} \pm 2\mathbf{k})/7$ 15 $\pm(\mathbf{i} + \mathbf{j} + \mathbf{k})/\sqrt{3}$ 16 Approx $3°11'$
east of north, approx 636.34 mph

Exercises 14-4, p. 762-763

1(a) -91; (b) $\cos\theta = -91/117, \theta \approx 141.1° \approx 2.462$ rad; (c) $-7, -91/9, -91(5\mathbf{i} - 12\mathbf{k})/169$ 3(a) 28;
(b) $\cos\theta = 14/27, \theta \approx 58.8° \approx 1.03$ rad; (c) $14/3, 28/9, (14\mathbf{i} - 28\mathbf{j} - 28\mathbf{k})/9$ 5(a) 27; (b) $\cos\theta = 9/\sqrt{105}$,
$\theta \approx 28.6° \approx 0.4985$ rad; (c) $9/\sqrt{5}, 27/\sqrt{21}, (9\mathbf{i} - 18\mathbf{k})/5$ 7 $(5\mathbf{i} + 3\mathbf{j} - 2\mathbf{k})/\sqrt{38}$ 9 $(8\mathbf{i} - 5\mathbf{j} - \mathbf{k})/3\sqrt{10}$
11 2/3, 2   13 $\cos\theta = 1/\sqrt{3}, \theta \approx 54°44'8'' \approx 0.9553$ rad

Exercises 14-5, p. 767-768

2 $Q((x_1 + 2x_2)/3, (y_1 + 2y_2)/3, (z_1 + 2z_2)/3)$ 3 $8\sqrt{2}/3$ 8 $(-2,5,-2), (0,-1,8), (4,3,0)$

Exercises 14-6, p. 774-776

1 $3x + 9y - 4z = 0$ 2 $x - 2y - 2z = 5$ 3 $x = 4$ 4 $3x - 4y + 6z = 28$ 5 $x - 3y + 2z = 14$
6 $11x - 2y - 6z = 69$ 7 $x = 3 + 4t, y = -2 - 2t, z = 4 + 3t; (x - 3)/4 = (y + 2)/(-2) = (z - 4)/3$
9 $x = 6 + 3t, y = -2 + 2t, z = 4 - t; (x - 6)/3 = (y + 2)/2 = 4 - z$ 11 $x = 4 + t, y = -1 - 4t$,
$z = 3 + 3t; x - 4 = (y + 1)/(-4) = (z - 3)/3$ 12 $(-1,-4,-2)$ 13 On the plane
14 Parallel to the plane   15 Not orthogonal, not parallel, intersect at $(3,-2,0)$ 17 Orthogonal, intersect at
$(4,-3,7)$ 19 $8x - 11y + 7z = 26$ 21 $17/\sqrt{38}$ 23 $\theta = \pi/4$ 24 6

Answers

A-59

Exercises 14-7, p. 780–781

1 $-i - 3j + 4k$; $-11i - 7j - 8k$ 3 $4i - 4j - 14k$; $-28i + 28j - 16k$ 5 $x + y - z = 0$
7 $3x + 3y + 4z = 5$ 9 $3\sqrt{3}/2$ 11 $\sqrt{34}$ 15 $x - 3 = 2 - y = 1 - z$ 16 $x = 1 + t, y = 1 + 10t, z = 1 + 8t$ 19 5

Exercises 14-8, p. 786–788

1–4 The form of the answer is not unique. 1 $x = 2t, y = 10 - 5t, z = -3t$ 3 $x = -3 + 2t, y = 2t, z = -3 + 3t$ 5 $\alpha = \cos^{-1}(16/(7\sqrt{6})) \approx 0.3677$ rad $\approx 21°4'14''$ 7 $\alpha = \cos^{-1}(8/9) \approx 0.4759$ rad $\approx 27°23'57''$
9 $(2,-1,10), \theta = \cos^{-1}(19/(7\sqrt{39})) \approx 1.1212$ rad $\approx 64°14'17''$ 11 $\frac{1}{2}(3i + j + 2k)$ 13 $(5,5,-3)$
14 $x = 2 + 2t, y = 4 + 5t, z = -5 + 16t$ 17(b) $(1,1,2)$ 19 $34/\sqrt{227}$ 21 $4/\sqrt{30}$

Exercises 14-9, p. 793

1 Circular cylinder parallel to the z-axis 3 Parabolic cylinder parallel to x-axis 5 Parabolic cylinder parallel to the y-axis 7 Hyperbolic cylinder parallel to the x-axis 9 Portion of a helix on the cylinder $x^2 + y^2 = 1$
11 Parabola 2 units above the xy-plane 13 Intersection of plane $z = 3x$ with a portion of the parabolic cylinder $y = x^2$

Exercises 14-10, p. 802–803

1 Elliptical paraboloid, opens along the y-axis 3 Hyperboloid of one sheet, opens along the z-axis
5 Elliptical cone, opens along the z-axis 7 Ellipsoid 9 Elliptical cone, opens along the x-axis
11 Elliptical cone, opens along y-axis 13 Circular hyperboloid of two sheets, opens along x-axis
15 Hyperbolic paraboloid

Review Problems (Chapter 14), p. 803–805

2 $(2i - 3j + 4k)/\sqrt{29}$ 8 $-1/\sqrt{3}, \frac{1}{3}(-i - j - k)$ 12 Sphere, $C(4,0,5), r = 6$ 13 Point $(2,-3,-1)$
14 $x^2 + y^2 + z^2 - 6x - 8z + 22 = 0$ 15 $x^2 + y^2 + z^2 - 12x - 2y - 12z + 37 = 0$;
$x^2 + y^2 + z^2 - 12x - 18y - 12z + 117 = 0$ 16 Plane $x - y - z + 1 = 0$ 17 Paraboloid $4y = x^2 + z^2$
18 Upstream in direction of vector $8(\cos 30°i + \sin 30°j)$, speed: $4\sqrt{3} \approx 6.9$ mph 19(a) $\cos\theta = -4/\sqrt{154}$, $\theta \approx 1.8990$ rad $\approx 108°48'14''$ (b) $(i - 11j - 4k)/\sqrt{138}$ 20 $\theta = \cos^{-1}(2/\sqrt{6}) \approx 0.6155$ rad $\approx 35°15'52''$
23 $8x + 5y - 7z + 26 = 0$ 24 $(9,-5,12)$ 25 $10x - 13y - 22z + 33 = 0$ 26 $x = 1 + t, y = 1 - 2t, z = -2 - 2t$ 28 4 29 4 30 $7/\sqrt{107}$ 41 $((5x_1 + 3x_2)/8, (5y_1 + 3y_2)/8, (5z_1 + 3z_2)/8)$ 42 $x = t, y = 1 - t, z = -2t$ 44 $3x + 4y - 3z + 22 = 0$

Appendix (Chapter 14), p. 808–809

1 23 3 -4 5 17 7 $2x^2 - 5x - 1$ 13 $x = 2, y = -1$ 15 $x = 3, y = -3, z = -1$

Chapter 15

Exercises 15-1, p. 816–817

1 No limit 3 k 5(a) $t \neq 0$; (b) $-t^{-2}i + 2tj + 3t^2k, 2t^{-3}i + 2j + 6tk$ 7(a) $t > 1$; (b) $i/(t-1) - e^t k$, $-i/(t-1)^2 - e^t k$ 9(a) All reals; (b) $(\sin t + t\cos t)i + j + (\cos t - t\sin t)k, (2\cos t - t\sin t)i - (2\sin t + t\cos t)k$
11 $r'(4) = \frac{1}{4}i + j, r''(4) = -\frac{1}{32}i$ 13 $3i/\sqrt{2} - \sqrt{2}j, -3i/\sqrt{2} - \sqrt{2}j$ 15 $e^t(t^2 + 2t + \sqrt{t} + \frac{1}{2}t^{-1/2}) + 2,$
$[(2t + 3)e^t - \frac{1}{2}t^{-1/2}]i - e^t[t^2 + 2t - \sqrt{t} - \frac{1}{2}t^{-1/2}]j + [2t - e^t(2t + 3)]k$ 17 $(1 - e^{-1})i + (\sqrt{2} - 1)j + \frac{1}{2}(e - 1)k$
19 $(t^3 + 2)i + (2t^2 + t - 1)j + (2t + 3)k$ 21 $(e^{-t} + 2t)i + (\frac{1}{2}t^4 + t + 1)j - \frac{1}{4}(\cos 2t + 3)k$

Exercises 15-2, p. 824–825

1 $v = 2ti + t^2j + 2k, a = 2i + 2tj, s'(t) = t^2 + 2, 27$ 3 $v = 2t^2i + 2tj + k, a = 4ti + 2j, s'(t) = 2t^2 + 1, 21$
5 $v = -2\sin 2ti + 2\cos 2tj - 2\tan 2tk, a = -4\cos 2ti - 4\sin 2tj - 4\sec^2 2tk, s'(t) = 2|\sec 2t|, \ln(2 + \sqrt{3})$
8(b) $y = 16x/(x^2 + 4)$ 9 $x = -4 + 4s, y = 1 - 2s, z = -1 + 4s$ 11 $x = 1 + 2s, y = 1, z = 2$
13 $(10,18,17)$

Exercises 15-3, p. 832–833

1 $v = 600\sqrt{3}i + (600 - 32t)j$, max altitude $= 600^2/64 = 5625$ ft, total time $= 1200/32 = 37.5$ sec, range $= (600^2)2\sqrt{3}/32 \approx 38{,}971.14$ ft 4 2.5 sec, 500 ft 5 $\alpha = \frac{1}{2}\sin^{-1}(0.8) \approx 0.92730$ rad $\approx 26°33'54''$
6 $\alpha = \frac{1}{2}\sin^{-1}(0.5) = \pi/12$ rad $= 15°$ 9 $v(0) = 7500\sqrt{3}i + 7500j, C_1 = 7500\sqrt{3}, C_2 = 105,600,$

$f'(t) = 7500\sqrt{3}e^{-t/100}$, $g'(t) = -98100 + 105600e^{-t/100}$,
$r(t) = 750000\sqrt{3}(1 - e^{-t/100})i + [10560000(1 - e^{-t/100}) - 98100t]j$

Exercises 15-4, p. 839–840

1 2 **3** $2/3^{3/2}$ **5** 1 **7** $(\pi/2, 0)$ **9** $(\pi/2, 1)$ **11** $\sqrt{2}/9$ **13** 1/4 **15** 2 **16** Max curvature of 1 at $x = n\pi$ (n an integer) **17** $(x + 2)^2 + (y - 3)^2 = 8$ **18** $(x + 1)^2 + (y - 1)^2 = 2$ **19** $(x + 4)^2 + (y - 7/2)^2 = 125/4$ **20** $x^2 + (y - 3/2)^2 = 1/4$ **25** $1/(e^\theta \sqrt{2})$

Exercises 15-5, p. 847–848

1 $T = (2ti + j)/\sqrt{4t^2 + 1}$, $N = (i - 2tj)/\sqrt{4t^2 + 1}$, $K = 2/(4t^2 + 1)^{3/2}$ **3** $T = (2i + tj + 2tk)/\sqrt{5t^2 + 4}$, $N = (-5ti + 2j + 4k)/\sqrt{25t^2 + 20}$, $K = \sqrt{5}/(5t^2 + 4)^{3/2}$ **5** $T = (2\cos 2t i - 2\sin 2t j - 3k)/\sqrt{13}$, $N = -\sin 2t i - \cos 2t j$, $K = 4/13$ **7** $(x - 3)^2 + (y + 2)^2 = 8$ **8** $(x + 5/2)^2 + (y - 9/4)^2 = 125/16$ **9** $(x - 4)^2 + (y - 7/2)^2 = 125/4$ **10** $(x - (\pi + 6)/4)^2 + (y + \sqrt{2})^2 = 27/4$

Exercises 15-6, p. 853

1 $K = \sqrt{5}/(4 + 5t^2)^{3/2}$, Tan comp $= 10t/\sqrt{4 + 5t^2}$, Nor comp $= 4\sqrt{5}/\sqrt{4 + 5t^2}$
3 $K = \sqrt{6}/(2(2t^2 - 4t + 5)^{3/2})$, Tan comp $= (4t - 4)/\sqrt{2t^2 - 4t + 5}$, Nor comp $= 2\sqrt{6}/\sqrt{2t^2 - 4t + 5}$
5 $K = (t^2 + 2)/(t^2 + 1)^{3/2}$, Tan comp $= t/\sqrt{t^2 + 1}$, Nor comp $= (t^2 + 2)/\sqrt{t^2 + 1}$ **7** $K = \sqrt{2}/(3e^t)$, Tan comp $= e^t\sqrt{3}$, Nor comp $= e^t\sqrt{2}$

Review Problems (Chapter 15), p. 854–856

4(a) $7\sqrt{2}$; (b) $x = 2 + 2s, y = 2 + 2s, z = 1$ **5**(a) $\sqrt{6} + \ln(\sqrt{2} + \sqrt{3})$; (b) $x = -\pi s/2, y = \pi/2 + s, z = \pi/2 + s$ **9** $v = -2\sin 2t i + 2\cos 2t j + k$, $a = -4\cos 2t i - 4\sin 2t j$, $2\pi\sqrt{5}$ **11** $x = 12 + 2s, y = 9 + 3s, z = 18 + 6s$ **13** $x = 750t, y = -16t^2 + 750\sqrt{3}t$ **14** $x = 2450t, y = -490t^2 + 2450\sqrt{3}t$ **15** $2i + \pi^2 k/2$ **16** $(t^4 + 2)i + (e^t - 2)j + (3 - \cos t)k$ **20** $|6x|/(9x^4 - 6x^2 + 2)^{3/2}$ **21** $2(e^{2t} + e^{-2t})^{-3/2}$ **22** $(x + 2/3)^2 + (y - 5/6)^2 = 125/36$ **23** $(\pm(45)^{-1/4}, \pm(45)^{-3/4})$ **24** $T = [(-\sin t + 2\cos t)i + (\cos t + 2\sin t)j]/\sqrt{5}$, $N = [-(\cos t + 2\sin t)\vec{i} + (2\cos t - \sin t)\vec{j}]/\sqrt{5}$, $K = 1/(\sqrt{5}e^{2t})$ **25** $T = (5i + 3tj + 4tk)/(5\sqrt{t^2 + 1})$, $N = (-5ti + 3j + 4k)/(5\sqrt{t^2 + 1})$, $K = 1/[10(t^2 + 1)^{3/2}]$ **26** $x^2 + (y - 2)^2 = 5$ **27** $K = (t^2 + 2)/(t^2 + 1)^{3/2}$, Tan comp $= t/\sqrt{t^2 + 1}$, Nor comp $= (t^2 + 2)/\sqrt{t^2 + 1}$

Chapter 16

Exercises 16-1, p. 865–866

1 $\{(x, y) | x + y > 0\}$; R **3** Points on or below the line $y = 3x + 3$; $f(x, y) \geq 0$ **5** -1 **7** No limit (check limit along line $y = mx$) **9** 0 **11** 0 **13** Continuous on domain **15** Continuous on domain **17** Continuous on domain **19** $\delta = 1/50000$ will work **21** Take $\delta = \epsilon/5$

Exercises 16-2, p. 870–871

1 $f_x = 9x^2y^2 + 2xy$, $f_y = 6x^3y + x^2 - 3$ **3** $x/(x^2 - y^2)$, $y/(y^2 - x^2)$ **5** $(xy - 1)/x^2y$, $-1/xy^2$ **7** $-\sin x - e^z - ye^{-x}$, e^{-x}, $-xe^z$ **9** $s\sec^2(sr) + (1 - r^2)^{-1/2}$, $r\sec^2(sr) - t^3/2\sqrt{s}$, $-3t^2s^{1/2}$ **11** $x = 1, y = 2 + t, z = -7 + 8t$ **14**(a) $-900/169$ **15** 10

Exercises 16-3, p. 874–875

1 $f_{xx} = 18xy^2 + 2y$, $f_{yy} = 6x^3$, $f_{xy} = f_{yx} = 18x^2y + 2x$ **3** $z_{xx} = z_{yy} = -(x^2 + y^2)/(x^2 - y^2)^2$, $z_{xy} = z_{yx} = 2xy/(x^2 - y^2)^2$ **5** $z_{xx} = -(xy + 2)/x^3y$, $z_{yy} = -2/xy^3$, $z_{xy} = z_{yx} = -1/x^2y^2$ **7** $2/(y + z)$ **9** $-2/(y + z)^2$ **15**(a) Hint: Rewrite the term on the left as $\dfrac{\partial}{\partial x}\left(\dfrac{\partial^2 w}{\partial y \partial x}\right)$ **16**(c) Show that $\partial^2 f/\partial y \partial x$ is not continuous at $(0, 0)$ by considering the limit as $(x, y) \to (0, 0)$ along the line $y = mx$.

Exercises 16-4, p. 881–882

1 $\Delta w = \sqrt{3.97} - 2 \approx -0.00751$, $dw = -0.0075$ **3** $x_0\Delta y^2 + 2y_0\Delta x \Delta y + \Delta x \Delta y^2$ **5** 5.01 **7** 0.03 **9** $(12x^3 - 3y^2)dx - (6xy + 3y^2)dy$ **11** $[\cos z - y\sec^2(xy)]dx - [x\sec^2(xy) + z^3e^{-y}]dy - (x\sin z - 3z^2e^{-y})dz$ **13** 0.4553 **14**(a) 12.4π; (b) 5.12π **15** -2.54

Exercises 16-5, p. 888–889

1 $(4x^3 + 2xy)e^t + (x^2 - 2y)\sin t$ **3** $-\sin y/t^2 + (x\cos y - 2tw)/\sqrt{1 - t^2}$
5 $\partial w/\partial s = (12 + 3x^{-1/2}y^{3/2})s - (9x^{1/2}y^{1/2}/2 - 3y^2)\sin s$; $\partial w/\partial t = (12 + 3x^{-1/2}y^{3/2})e^t/2$

Answers

7 $\partial w/\partial s = (2xy^2 - z/x + yz)/s + 2(2x^2y + xz)s + (xy - \ln x)\sin t$; $\partial w/\partial t = s(xy - \ln x)\cos t$ **9** $\partial w/\partial x = 2x/r$; $\partial w/\partial y = -(\ln z)/r$; $\partial w/\partial z = -y/rz$ **11** $(6x + 2y^2)/(3y^2 - 4xy)$ **13** $-y(y + x\ln y)/x(x + y\ln x)$ **15** $2x + 3e^x$ **17(a)** 120π cu. cm/min; **(b)** 18π sq. cm/min **18** 4/9 cu. ft/sec **19(b)** Three solutions are $(3x + 2y)^{13}$, $\sin(3x + 2y)$, e^{3x+2y} **22(b)** $e^{x^2y^3+xw}[(2xy^3 + w)dx + 3x^2y^2 dy + x\,dw]$ **23(b)** $4(x^2y^3 + xw)^3[(2xy^3 + w)dx + 3x^2y^2 dy + x\,dw]$

Exercises 16-6, p. 895-896

1(a) $-2\sqrt{2}$; **(b)** $4\mathbf{i} + 8\mathbf{j}$ **3(a)** $-1/4\sqrt{30}$; **(b)** $-(2\mathbf{i} + \mathbf{j})/4\sqrt{3}$ **5(a)** 2/3; **(b)** $2\mathbf{i}$ **8(a)** $1800/13^{5/2}$; **(b)** $4\mathbf{i} - 3\mathbf{j}$ or $-4\mathbf{i} + 3\mathbf{j}$

Exercises 16-7, p. 900-902

1 $7\mathbf{i} + 4\mathbf{j}$, $3\sqrt{65}$ **3** $-\mathbf{i} - \mathbf{j} + 3\mathbf{k}$, $\sqrt{11/9}$ **7** $8\mathbf{i} - 3\mathbf{j}$, $-100\sqrt{73}/(3 \cdot 36^2)$ **8(a)** $-8\mathbf{i} - 6\mathbf{j} - 7\mathbf{k}$; **(b)** $\sqrt{149}/10$ **9(a)** $-2\mathbf{i} - \mathbf{j} + \mathbf{k}$; **(b)** $2\sqrt{6}/125$ **10(a)** $-35/\sqrt{17}$; **(b)** $5\sqrt{5}$ in direction of $\mathbf{i} - 2\mathbf{j}$ **11** $\mathbf{i} + \mathbf{j}$ **13** $(\mathbf{i} - \mathbf{j})/\sqrt{2}$ **15** One such function is $3x^2 + 5y^2 - 3$; answer is not unique. **16(a)** $e^{xyz}(y^2z^2 + x^2z^2 + x^2y^2)$

Exercises 16-8, p. 906-907

1 $\mathbf{i} + 2\mathbf{j} - \mathbf{k}$, $z = x + 2y$ **3** $24\mathbf{i} + 6\mathbf{j} - \mathbf{k}$, $24x + 6y - z = 41$ **5** $2\mathbf{i} - 4\mathbf{j} - \mathbf{k}$, $2x - 4y - z = 0$ **7** $(4/3, 2/3, 4/3)$ **9** $(26, 29, 0)$ **10(b)** $2x + 2y - \sqrt{2}z + 2 = 0$ **11** $2x - 4y + z + 4 = 0$ **14** $3x - 8y - 5z = 0$

Exercises 16-9, p. 914-915

1 Local min at $(0,0)$ **3** Saddle point at $(-2,2)$ **5(a)** $(0,0)$; **(b)** Saddle point at $(0,0)$ (The second derivative test does not work. Consider paths through the origin composed of parts of the coordinate axes.) **7** Saddle point at $(0,0)$ **11(a)** $9 \times 9 \times 9$; **(b)** $6 \times 6 \times 3$ **13** $8''$, $2\pi/3$ **14** $\sqrt{3}/2$ **19** $y = 0.7x + 0.1$ **20** $y = -0.1x + 3.7$ **21** $y = 3.15x + 1.05$

Exercises 16-10, p. 920

1 2 at $(\pm 1, 0)$, 0 at $(0, \pm 1)$ **2** 18 at $(-3/5, -4/5)$, -2 at $(3/5, 4/5)$ **3** 164 at $1/2, \pm\sqrt{15}/2$, $-8/3$ at $(-1/3, 0)$ **4** 10 on line $y = x$, -6 at $(-2, 2)$ **5** 33 at $(-2, 1)$, $-22/3$ at $(5/3, 1)$ **6** 76 at $(1, \pm 5)$, $-1/2$ at $(-1/2, -1/2)$ **7** 300 at $(0, 0)$, 0.03 at $(6, -3)$ **8** $(30, 50)$ **9** 12 at $(0, 1)$, 3 at $(0, -2)$ **10** 10 at $(-1, 6)$, -7 at $(0, -2)$

Exercises 16-11, p. 925-926

1 Min of 1 at $(\pm 1, \pm 1)$, no max **2** Max of $4\sqrt{3}$ at $(2/\sqrt{3}, \pm\sqrt{6})$, Min of $-4\sqrt{3}$ at $(-2/\sqrt{3}, \pm\sqrt{6})$ **3** Min of 3/4 at $(0, 1/2)$, no max **4** Max of 1 at $(0, \pm 1)$, Min of -1 at $(\pm 1, 0)$ **5** Min of 6 at $(1,1,2)$, no max **6** Max of $4\sqrt{6}$ at $(8/\sqrt{6}, 4/\sqrt{6}, 4/\sqrt{6})$, min of $-4\sqrt{6}$ at $(-8/\sqrt{6}, -4/\sqrt{6}, -4/\sqrt{6})$ **7** Min of $-13/12$ at $(13/36, -1/3, 1/4)$, no max **8** Critical points at eight points given by all possible signs of $(\pm 2/\sqrt{3}, \pm\sqrt{3}, \pm 2)$. Max of 4 at $(2/\sqrt{3}, \sqrt{3}, 2)$ and three other points. Min of -4 at $(-2/\sqrt{3}, \sqrt{3}, 2)$ and three other points. **9** $\sqrt{1044/29} \approx 1.114$ at $(18/29, 24/29, 12/29)$ **10** $(3/\sqrt{7}, 1/\sqrt{7}, 2/\sqrt{7})$ **11** Max of 100 along the circles defined on the sphere by $x = 0$, $y = 0$, $z = 0$ **12** $10\sqrt{5}$, dimensions $2\sqrt{5} \times 2\sqrt{5} \times \sqrt{5}/2$ **13** Rad $= 2/\sqrt{\pi}$, height $= 24/5\sqrt{\pi}$ **14** $8abc/3\sqrt{3}$

Review Problems (Chapter 16), p. 926-929

2(a) No limit; **(b)** Continuous except at $(0,0)$ **3(a)** -1; **(b)** continuous except on line $2x + 3y + 1 = 0$ **4(a)** 0; **(b)** Continuous except on the line $2x + y = 0$ where denominator is zero. **5(a)** No limit; **(b)** Discontinuous at each point of line $2x + y = 4$ **6** No **8** $f_x = 2x + 2y$, $f_y = 2x + 3y^2$, $f_{xx} = f_{xy} = 2$, $f_{yy} = 6y$ **9** $f_x = 1/x + 1/(x + y)$, $f_y = 1/(x + y)$, $f_{xx} = -1/x^2 - 1/(x + y)^2$, $f_{xy} = f_{yy} = -1/(x + y)^2$ **10** $f_x = (xy + 1)e^{xy}$, $f_y = x^2 e^{xy}$, $f_{xx} = (xy^2 + 2y)e^{xy}$, $f_{xy} = (x^2y + 2x)e^{xy}$, $f_{yy} = x^3 e^{xy}$ **11** $f_x = 3y$, $f_y = 3x + 1/(1 + y^2)$, $f_{xx} = 0$, $f_{xy} = 3$, $f_{yy} = -2y/(1 + y^2)^2$ **12** $x = 1 + 5t$, $y = 1 - 4t$, $z = 5 - 22t$ **16** Formula value 1.15 (Calculator value ≈ 1.159) **17(a)** ≈ 1.2556; **(b)** ≈ 0.6837 **18(a)** See the Remark at the end of Section 16-3. **19** $2x\cos s + 2ye^t + 4sz$ **20** $6x + 2se^y \ln z + tue^{s+y}/z$ **22** $2x\cos y - x\sin y + 2y/x$ **25(a)** $(4x - 3y)\mathbf{i} - 3(x - y^2)\mathbf{j}$, $(12y^2 - 9y)/5$; **(b)** $[(y + z)\mathbf{i} + (z - x)\mathbf{j} - (x + y)\mathbf{k}]/(y + z)^2$, $(x - z)/3(y + z)^2$ **26** $3\mathbf{i} - 4\mathbf{j}$; 10/9 **29** $2\mathbf{i} - 2\mathbf{j} - \mathbf{k}$, $2x - 2y - z = 6$ **30** $\mathbf{i} - 4\mathbf{j} - 4\mathbf{k}$, $x - 4y - 4z = 4$ **31** $3x - y - 4z = \pm 6$ at $(3, -1, 4)$ and $(-3, 1, -4)$ **32(b)** $5(3x^2 - 2\cos y + \ln z)^4(6x\,dx + 2\sin y\,dy + dz/z)$ **35** Local max at $(0, -\sqrt{3})$, local min at $(6, \sqrt{3})$, saddle points at $(0, \sqrt{3})$ and $(6, -\sqrt{3})$ **36** Local min at $(1, 1/2)$, no local max **37** Local min at $(6, 18)$, saddle point at $(0, 0)$ **38** Local max at $(2/3, 0)$, local min at $(0, 2/3)$, saddle points at $(0, 0)$, $(2/3, 2/3)$ **39** $(3, 0)$ **40** $(\pm 1, -1)$ **41** Max of 9 at $(1/2, \pm\sqrt{3}/2)$, min of -1 at $(-1/2, 0)$ **42** Max of 1 at $(\pm 1, \pm 1)$, min of -1 at $(\pm 1, \mp 1)$ **43** Min of 11/36 at $(\pm\sqrt{5/18}, 0, -1/6)$, no max **44** $(9/2, 6, 15/2)$

Chapter 17

Exercises 17-1, p. 938

1(c) 2   2(b) $\sqrt{2\pi^2 + 2\pi + 1}/8$ 3(b) $\iint_{\mathcal{R}} (2x^2 + y^2)\,dA$ 4 6   5 6   10 13/2

Exercises 17-2, p. 946–948

1 5/6   3 10/3   5 $(e^8 - e^2 - 6)/4$ 7 $A = \int_0^2 \int_0^{\sqrt{4-x^2}} dy\,dx = \pi$, $V = \int_0^2 \int_0^{\sqrt{4-x^2}} (x+y)\,dy\,dx$
9 $A = \int_3^9 \int_{y/3}^3 dx\,dy = 6$, $V = \int_3^9 \int_{y/3}^3 x^2\,dx\,dy = 34$ 11 $A = 1$, $V = 8/3$ 13 $\int_0^4 \int_{\sqrt{y}}^2 xy\,dx\,dy = 16/3$
15 8/3   17 $e^2 - 3$ 19 1   22(a) $(e^2 - 1)(e^3 - e) = e^5 - 2e^3 + e$

Exercises 17-3, p. 952–953

1 16/3   2 1   3 18   4 8π 5 $64\sqrt{2}/3$ 6 12   7 144   9 6k   10 84/5   11 32/3
12 27k/16   13 4/3   14 $(2\ln 2 - 1)/4$

Exercises 17-4, p. 961–962

1 $M_x = 171k/8$, $M_y = 117k/4$, $(13/6, 19/12)$ 2 $M_x = -243/40$, $M_y = 27/5$, $(4/5, -9/10)$ 3 $M_x = 3/10$, $M_y = 19/42$, $(190/273, 6/13)$ 4 $M_x = M_y = 1/1260$, $(2/9, 2/9)$ 5 $(9/5, 7/5)$ 6 $(1/2, -3/5)$ 7 $(3/5, 3/8)$
8 $(1/5, 1/5)$ 11 $1/6$ 12 $(e^3 - 1)/9$ 13 256/7   14 64/105   15 $1/\sqrt{3}$ 16 $\sqrt{e^2 + e + 1}/3$
17 $4\sqrt{3/7}$ 18 $\sqrt{8/7}$

Exercises 17-5, p. 969–970

1 $\pi/2$ 3 $-(\ln 2)/2$ 5 $2 + \pi/4$ 6 $\sqrt{3} - \pi/3$ 7 π 8 $3\pi/2$ 9 $(5/6, 0)$ 10 $(6/5, 0)$
11 $(128\sqrt{2}/105\,\pi, 0)$ 12 $(21/20, 0)$ 13 16π 14 $\pi[1 - \cos 18]/8 \approx 0.13339$ 15 $(\pi + 2)/8$
16 $\pi(1 - e^{-4})/4$ 17 16/3   18 $-81/8$

Exercises 17-6, p. 975–976

1(b) $\pi c^2 \sqrt{a^2 + b^2 + 1}$ 2(b) $2\pi a(a - \sqrt{a^2 - c^2})$ 3(b) $\pi(17^{3/2} - 1)/48$ 4(b) $(4 - 2\sqrt{2})\pi$ 5 $49\pi/3$
10 $S = 2\pi ab$

Exercises 17-7, p. 982–983

1 2   2 44/5   3 1/24   4 $4^{11}/990$ 5 400/3   6 20/3   7 32/3   8 $8\sqrt{2}/3$ 9 128/3
10 $81\pi/32$ 11 $\int_0^6 \int_0^{(6-z)/2} \int_0^{(6-2x-z)/2} (x+2y)\,dy\,dx\,dz$

Exercises 17-8, p. 989–990

1 $M = 72\pi$, $M_{xy} = 192\pi$, $M_{yz} = M_{xz} = 0$, $\bar{x} = 0$, $\bar{y} = 0$, $\bar{z} = 8/3$ 2 $M = 8\sqrt{2}/3$, $M_{xy} = 16\sqrt{2}/5$, $M_{xz} = 0$, $M_{yz} = 8\sqrt{2}/3$, $\bar{x} = 1$, $\bar{y} = 0$, $\bar{z} = 6/5$ 3 $M = 25/3$, $M_{xy} = 50/3$, $M_{xz} = 20/3$, $M_{yz} = 25/3$, $\bar{x} = 1$, $\bar{y} = 4/5$, $\bar{z} = 2$ 4 $M = 56/15$, $M_{xy} = 608/105$, $M_{xz} = 32/15$, $M_{yz} = 16/5$, $\bar{x} = 6/7$, $\bar{y} = 4/7$, $\bar{z} = 76/49$ 5 $M = 16/3$, $M_{xy} = 3\pi/2$, $M_{yz} = 3\pi/2$, $M_{xz} = 4$, $\bar{x} = 9\pi/32$, $\bar{y} = 3/4$, $\bar{z} = 9\pi/32$ 6 $M = 18/5$, $M_{xy} = 1587/280$, $M_{xz} = 1587/140$, $M_{yz} = 31/24$, $\bar{x} = 155/432$, $\bar{y} = 529/168$, $\bar{z} = 529/336$ 7 $\int_{-2}^2 \int_{-\sqrt{4-x^2}}^{\sqrt{4-x^2}} \int_{-\sqrt{4-x^2-y^2}}^{\sqrt{4-x^2-y^2}} (x^2 + y^2)\,dz\,dy\,dx$
9 $\int_{-2}^2 \int_{-2\sqrt{4-x^2}}^{2\sqrt{4-x^2}} \int_0^7 (x^2 + y^2)\,dz\,dy\,dx$ 11(a) $2\sqrt{10}/5$; (b) $10240\pi^3/3$

Exercises 17-9, p. 996–997

1 Vertical strip on plane $\theta = \pi/4$ 3 One quarter of right circular cylinder of radius 3 and height 4
5 $z = 4 - r^2$, paraboloid   7 $r(\cos\theta + 2\sin\theta) = 3$, vertical plane   9 $x^2 + y^2 = 16$, cylinder
11 $x + y + 2z = 5$, plane   13 Cone   15 First octant portion of sphere of radius 2   17 $\rho\sin\phi = 3$, cylinder   19 $\rho^2 + 4\rho\cos\phi = 13$, sphere   21 $x^2 + y^2 + z^2 = 36$ 23 $x^2 + y^2 = z^2$, $z \leq 0$

Exercises 17-10, p. 1007–1008

1 32π 3 $32\pi/3$ 5 8π 6 $36\pi(8 - 3\sqrt{3})$ 7 $\pi/2$ 8 $32\pi(8\sqrt{2} - 7)/3$ 9 $\pi/4$ 10 $3\pi/4$
11 Mass $= 29\pi/3$, $(0, 0, 365/116)$ 12 Mass $= 5\pi/4$, $(0, 4/5, 128(15\pi - 16)/3^2 \cdot 5^3\pi)$ 13 216π

Answers

14 $4096\pi/15$ 15(a) $64\pi(2 - \sqrt{3})/3$; (b) $(0,0,3/(8 - 4\sqrt{3}))$ 16(a) 9π; (b) $27\pi(2\sqrt{2} - 1)/2$ 17(a) $8\pi/3$; (b) 1   18(a) $2\pi/9$; (b) $2/3$

Review Problems (Chapter 17), p. 1008–1011

3 $1/42$ 4 $2/\pi$ 5 $53/210$ 6 $81/5$ 7 $\int_0^1 \int_{y+1}^2 y^5 \, dx \, dy$ 8 $\int_0^2 \int_{y/2}^1 \sin(\pi x^2) \, dx \, dy$

9 $\int_0^1 \int_{y^2}^{\sqrt{y}} (x^2 + 2xy) \, dx \, dy$ 10 $\int_0^9 \int_{x/3}^{\sqrt{x}} x \, dy \, dx$ 11(a) $\iint_\mathcal{R} (4 - x^2 - y^2) \, dA = 8\pi$;

(b) $\int_{-2}^2 \int_{-\sqrt{4-x^2}}^{\sqrt{4-x^2}} \int_{(x-1)^2+(y-1)^2}^{6-2x-2y} dz \, dy \, dx$ 13(a) 4; (b) $256/21$; (c) $(21/16, 21/5)$; (d) $1024/45$; (e) $2\sqrt{7/15}$

14 $\pi(e^4 - 1)/4$ 15 $2/3$ 16 $3\pi/2$ 17 π 18 $(5/6, 0)$

19 $\int_{-2}^2 \int_{-\sqrt{4-x^2}}^{\sqrt{4-x^2}} \sqrt{1 + 4x^2 + 4y^2} \, dy \, dx = \int_0^{2\pi} \int_0^2 \sqrt{1 + 4r^2} \, r \, dr \, d\theta = \pi(17^{3/2} - 1)/6$

20 $\int_0^{2\pi} \int_{\sqrt{8}}^3 (3/\sqrt{9 - r^2}) \, r \, dr \, d\theta = 6\pi$ 22 $256/105$ 23 $28\pi/3\sqrt{2}$ 24 $\int_0^4 \int_0^{2-z/2} \int_0^{2-y-z/2} \sqrt{z} \, dx \, dy \, dz$

25(a) $648/5$; (b) $(15/16, 27/7, 27/7)$ 26 $\int_{-2}^2 \int_{-\sqrt{4-x^2}/2}^{\sqrt{4-x^2}/2} \int_{-\sqrt{4-x^2-4y^2}/2}^{\sqrt{4-x^2-4y^2}/2} (y^2 + z^2) \, dz \, dy \, dx$

33(a) $\int_0^{2\pi} \int_0^4 \int_0^{32-r^2} r \, dz \, dr \, d\theta$; (b) 384π 34(a) $\int_0^{\pi/2} \int_0^{\pi/4} \int_0^6 \rho^2 \sin\phi \, d\rho \, d\phi \, d\theta$; (b) $18\pi(2 - \sqrt{2})$

35(b) $125\pi(2 + \sqrt{3})/3$, $125\pi(2 - \sqrt{3})/3$ 36(b) 12π

Chapter 18

Exercises 18-1, p. 1016

1 First and second quadrants   3 $2x\mathbf{i} + 2y\mathbf{j}$ 5 $(y + z)\mathbf{i} + (x + z)\mathbf{j} + (y + x)\mathbf{k}$ 7 $x^2 + xy + y^2 + C$
9 $\sin(xy) + C$ 11 $xyz + C$ 13 Particles move with constant velocity parallel to the y-axis in a stream between (and including) the lines $x = -1$, $x = 1$.

Exercises 18-2, p. 1024

1 $2\sqrt{5}$ 3 $7/2$ 5 $e(17^{3/2} - 2^{3/2})/6$ 7 $(14^{3/2} - 1)/6$ 9 0   11 $-11/6$ 14 $17^{3/2}/128$

16 First integral is $\int_{-1}^1 t^2 \, dt = 2/3$, second is $\int_0^\pi (\cos^2 t + \sin t) \sin t \, dt + \int_0^\pi (-3\cos t + 2\sin^2 t) \cos t \, dt = -\pi + 2/3$

17(c) Yes

Exercises 18-3, p. 1031

1 $G(x,y) = x^2y + C$, -17 3 $G(x,y) = x\cos y + y\sin x + C$, $-\pi/2$ 5 $G(x,y) = x^2 + \sin(xy) + C$, 1
7 $G(x,y,z) = xy + yz + xz + C$, $-\pi$ 9 1   11(a),(b) 0   12 4   13 π

Exercises 18-4, p. 1039

1 0   2 $\sqrt{14} - \sqrt{2}$ 3 $\sqrt{2} - 2\sqrt{5}$ 4 512 ft-lbs   6(b) 0   7 Particles move with constant velocity in direction of $2\mathbf{i} + \mathbf{j}$; mass = 3   9 Particles move away from the origin with diminishing velocity, flux = 2π

Exercises 18-5, p. 1046

1 $-392/3$ 2 0   3 $-2e^3 + 3\tan^{-1}3 - \frac{1}{2}\ln 10 - 1$ 4 $4\sin 1$ 5 0   6 $3\pi/4$ 7 $9/2$ 9 $2/3$
10 $3\pi/8$ 18(a) $(3\pi/16, 0)$; (b) $(0,0)$

Exercises 18-6, p. 1053

1 $\int_{-1}^1 \int_0^{(1+x)/2} \sqrt{6}(x^2 + y) \, dy \, dx = 2\sqrt{6}/3$ 2 $12\sqrt{3}$ 3 $13\pi/3$ 4 $\pi(5^{5/2} + 1)/60$ 5 $2 + \pi/3$
6 $[6(\sqrt{3} - 1) + \pi]/3\sqrt{2}$ 7 $4\pi/3$ 8 $128\pi\sqrt{2}/5$ 9 $\pi[132\sqrt{17} - \ln(4 + \sqrt{17})]/32$
10 $26\pi/3$ 11 $\int_0^1 \int_0^{1-x} (8y + 4x) \, dy \, dx = 2$ 12 $45/4$ 13 4π 14 20   15 4π 16 0

Exercises 18-7, p. 1058

1(a) y; (b) $-3\mathbf{i} - x\mathbf{k}$ 3(a) $2x + 2y + 2z$; (b) $\mathbf{O}$ 5(a) $2(x^2 + y^2 + z^2)^{-1/2}$; (b) $\mathbf{O}$

Exercises 18-8, p. 1067

1. The surface integrals have the value 0 except for the integrals on the planes $x = 1$ and $y = 1$, which have value $1/2$ **2.** 0. Both surface integrals have value 0.   **3.** $3a^2b^2c^2/4$ **4.** 0   **5.** $64\pi/3$ **6.** $abc^2\pi/2$ **7.** A source at every point.   **8.** Every point above the plane $x + y + z = 0$ is a source, every point below is a sink.

Exercises 18-9, p. 1074

1. Both integrals are 0   **2.** Both integrals are 0   **3.** $-1/6$ **4.** 12π **5.** 0   **6.** 87   **7.** 0   **8.** π

Review Problems (Chapter 18), p. 1075

3. Particles move with constant speed in direction of $i + j$ **5.** Particles move about the z-axis in clockwise circular orbits   **9.** $4\sqrt{14}/3$ **10.** 0   **11.** $14/15$ **12.** $\sqrt{2} - 5/4$ **13.** $\pi^2\sqrt{17}/20 \approx 2.0347$ **14.** $81/2$ **15.** 1   **16.** -2π **17.** 0   **19.** $1 + \ln 2$ **20.** -34 **23.** 0   **24.** Calculate $\iint_{\mathcal{R}} (-\partial(x^2)/\partial y + \partial(y^2)/\partial x) dA$ **28.** π **29.** $8\pi\sqrt{2}$ **30.** $\frac{1}{100} \int_0^{2\pi} \int_0^1 \cos^2\theta \sin^2\theta r^7 \sqrt{1 + 4r^2}\, dr\, d\theta$ **31.** $8\pi/3$ (use both hemispheres)   **33.**(a) 0; (b) $-2k$ **34.**(a) $x + y + z$; (b) $zi + xj + yk$ **36.** A source at (x,y,z) if $x > y$, a sink if $x < y$ **39.** $64\pi/3$ **43.** 0   **44.** 0

Chapter 19

Exercises 19-1, p. 1085

1. $y = cx$ **3.** $y = 1/(C - x^4)$ **5.** $y = 2C(x - 1)^2/[1 - C(x - 1)^2]$ **7.** $(\ln y)^2 = C - x^2$ **9.** $y = C_1 e^{2x} + C_2$ **11.** $y = Cx^2 - x$ **13.** $y = x\sec^{-1}(cx)$ **15.** $y = \sqrt{5/3}\sqrt{|x - 1|/|x + 1|}$ **17.** $y = (3e^{2x} - 5)/2$ **20.** $f(x) = 0$ is not a singular solution.   **23.**(b) Intermediate solution: $r = C_1(\sin^3\theta - \cos^3\theta)^{-1/3}$; Final solution: $y^3 = x^3 + C$ **24.** $y = 4x^2 + x$ **25.** The x-axis and the set of all parabolas with vertex at the origin that are symmetric about the y-axis.

Exercises 19-2, p. 1091

1. $x^3/3 + 4xy - 3y^2/2 = C$ **3.** $3x^2y^2 - y^2 = C$ **5.** $2xy - 2x^3y + 4y^3/3 = C$ **7.** $y^2 \sin x + \cos y = C$ **9.** $x^2y^3 + x^3y^2 = C$ **11.** Use Example 3, $x^2e^x + ye^x = C$ **13.** $2x^3 + x^2y^2 - 2x^2y = C$ **16.**(a) $y = (e^x + C)/x$; (b) $f(x) = (e^x + 5 - e^{-1})/x$ if $x < 0$, $f(x) = (e^x + 2 - e)/x$ if $x > 0$, $g(x) = (e^x - e^{-1})/x$ if $x < 0$, $g(x) = (e^x + 2 - e)/x$ if $x > 0$

Exercises 19-3, p. 1095

1. $y = (2x + C)/\sin x$ **3.** $y = 2 + Ce^{\cos x}$ **5.** $y = (e^x + C)/(x + 1)^3$ **7.** $y = [(x^2/2) - x + C]/(x^2 + 1)$ **9.** $\cos x \cos y = C$ **10.** $2x^2y + 3e^{-x^2} = C$ **11.** $\ln|2y - x| + x/(2y - x) = C$ **12.** $\tan^{-1} y = \tan^{-1} x + C$ or $y = (C + x)/(1 - Cx)$ **13.** $\ln|x| - 2x + \ln|y| + y = C$ **14.** $2y = \ln(2x + C)$ **15.** $y = x(C - e^{-x})$ **16.** $3y^2e^{2x} + 6x^2 + 4y^3 = C$ **17.**(b) $m = 320$ **19.** $y^4 = 1 + Ce^{-2x^2}$ **21.** $\sqrt{y} = (C + x/2)/(x^2 - 1)$ **23.**(b) $y = Ce^{-x^2/2} + e^x$

Exercises 19-4, p. 1102

1.(a) $8x^3 + 24x^2 - 20$ **2.**(a) $D^2 + 3D - 10$; (c) $D^2 + (x + 1)D + (x + 1)$ **3.**(a) $C_1 e^{-2x} + C_2 e^{-3x}$; (b) $2e^{-2x} - e^{-3x}$ **5.**(a) $(C_1 + C_2 x)e^{3x}$; (b) $(2 - 3x)e^{3x}$ **7.**(a) $C_1 e^{-4x} + C_2 e^{4x}$; (b) 0   **9.**(a) $C_1 + C_2 e^x$; (b) $2 + e^x/e = 2 + e^{x-1}$ **11.**(a) Solution: $y = C_1 e^{2x} + C_2 e^{3x}$ **12.** $y = C_1(2e^{2x} - e^{2x}/x) + C_2/x$ **13.**(c) $y = C_1 u_1 + C_2 u_2 = C_1/x + C_2 e^x(x - 1)/x$

Exercises 19-5, p. 1106

1. $C_1 \cos 4x + C_2 \sin 4x$ **3.** $e^x[C_1 \cos x + C_2 \sin x]$ **5.** $e^{3x}[C_1 \cos x + C_2 \sin x]$ **7.** $e^{2x}[C_1 \cos \sqrt{3}x + C_2 \sin \sqrt{3}x]$ **9.** $\cos 4x - \sqrt{3} \sin 4x$ **11.** $e^x[2\cos x + (\sqrt{2} - 2)\sin x]$ **13.**(a) $\theta = \theta_0 \cos(t\sqrt{g/l})$ **15.** $K_1 = (C_1 - iC_2)/2$, $K_2 = (C_1 + iC_2)/2$

Exercises 19-6, p. 1113

1. $C_1 e^{-3x} + C_2 e^{2x} - e^{-2x}/4$ **3.** $(C_1 + C_2 x)e^{-2x} + \frac{3}{25}\cos x + \frac{4}{25}\sin x + 2$ **5.** $C_1 \cos x + C_2 \sin x - (x\cos x)/2 + 2x$ **7.** $(C_1 + C_2 x + 2x^2)e^{-2x}$ **9.** $C_1 + (C_2 + 2x)e^{x/2}$

Exercises 19-7, p. 1116

1. $C_1 \cos 3x + C_2 \sin 3x - \frac{1}{9}\cos 3x \cdot \ln|\sec 3x + \tan 3x|$ **3.** $C_1 + C_2 e^x - \frac{1}{2}e^x(\sin x + \cos x)$

Answers

A-65

5 $C_1\cos x + C_2\sin x - x\cos x + \sin x \cdot \ln|\sin x|$
9 $C_1/x + C_2e^x(x-1)/x + xe^x - 2e^x + 2e^x/x = C_1'/x + C_2'e^x(x-1)/x + xe^x$ 11 $(C_1 + C_2\ln|x|)/x$

Exercises 19-8, p. 1120

1 $C[1 - (x/1!) + (x^2/2!) - (x^3/3!) + (x^4/4!) - \cdots]$
3 $C_0 + C_1x + (C_0x^2/2!) - (C_0x^4/4!) + (1 \cdot 3C_0x^6/6!) - (1 \cdot 3 \cdot 5C_0x^8/8!) + \cdots$
5 $C[x - x^3/(3^2 - 1) + x^5/[(3^2 - 1)(5^2 - 1)] - x^7/[(3^2 - 1)(5^2 - 1)(7^2 - 1)] + \cdots$
7 $2\left(1 + x + \dfrac{1}{2!}x^2 - \dfrac{1}{4!}x^4 + \dfrac{1 \cdot 3}{6!}x^6 - \dfrac{1 \cdot 3 \cdot 5}{8!}x^8 + \cdots\right)$
8(c) $C/x^2 + 1/2 - x^4/(2 \cdot 3!) + x^8/(2 \cdot 5!) - x^{12}/(2 \cdot 7!) + \cdots$
9 $u_1 = 1 + \dfrac{x^2}{1!} + \dfrac{x^4}{2!} + \dfrac{x^6}{3!} + \dfrac{x^8}{4!} + \cdots = e^{x^2}$
$u_2 \doteq \dfrac{x}{1} + 2x^3/(1 \cdot 3) + 2^2x^5/(1 \cdot 3 \cdot 5) + 2^3x^7/(1 \cdot 3 \cdot 5 \cdot 7) + 2^4x^9/(1 \cdot 3 \cdot 5 \cdot 7 \cdot 9) + \cdots$

Exercises 19-9, p. 1124

1(a) 4; (b) 4   2(a) -0.478328; (b) -0.4696898 3(a) 3.377784; (b) 3.913461   4(a) 0.461538;
(b) 0.513287   5(a) 1.190685; (b) 1.204183   6(a) 5.354819; (b) 5.788181   7 4 (answer exact)
8 $-15/32 = -0.46875$, 2% error   9 3.946164, 14% error   10 0.518519, 11% error

Review Problems (Chapter 19), p. 1124

3 $-2 + Ce^{x^2/2}$ 4 $C/(x - \sqrt{x^2+1}) = C_1(x + \sqrt{x^2+1})$ 5 $(C - x^4)/(4e^x)$
6 $3y^2 = 4x^{3/2} + 3x^2 + C$ 7 $4x^3 + 12xy - 3y^2 = C$ 8 $e^{2x}(3 + C/x)$ 9 $e^x(x^3 + C)$
10 $\csc(y/x) - \cot(y/x) = Cx$ 11 $x/(C - \ln|x|)$ 12 $x + Ce^x - 4$ 13 $\sec y = Ce^{\sin x - x\cos x}$
14 $2x^2 + C/x^2$ 15 $y^4 = C(x-2)^3(x+2)$ 16 $e^{2x}[C_1\cos x + C_2\sin x]$ 17 $(2x + cx^4)/(1 - Cx^3)$
18 $x^3/3 + e^{xy} - \cos y = C$ 19 $C_1e^{-x/3} + C_2e^{x/2}$ 20 $C_1e^{-2x} + C_2e^{5x} - 1/5$ 21 $e^{xy} + 2x^2y + 3x = C$
22 $x^2\ln x/3 - x^2/9 + C/x$ 23 $e^{3x}/4 - e^{-2x} + 1 + Ce^{-x}$ 24 $2y + \sin 2y + 4\sin^{-1}(x/3) = C$
25 $C_1 + C_2e^{-x/4} + e^x/5 + x^2/2 - 3x$ 26 $x\tan(\ln(cx))$ 27 $\ln(cx - 1)$ 28 $\cos x(\sin^3 x + C)/3$
29 $C_1\cos 7x + C_2\sin 7x + e^{x+1}/50$ 30 $x^2y^3 + e^x - e^{-y} = C$ 31(b) $(C - x^{3/2})^2/(9x)$
34(c) $D^2 + (1 - x^2)$ 35(a) $C_1e^x + C_2e^{2x}$; (b) $4e^x - 3e^{2x}$ 36(a) $(C_1 + C_2x)e^{-3x}$; (b) $(1 + x)e^{-3x}$
37(a) $C_1\cos\sqrt{3}x + C_2\sin\sqrt{3}x$; (b) $\cos\sqrt{3}x - 2(\sin\sqrt{3}x)/\sqrt{3}$ 38(a) $C_1e^{-2x} + C_2e^{-3x}$; (b) e^{-2x}
39(a) $e^{-2x}[C_1\cos x + C_2\sin x]$; (b) $e^{-2x}\cos x$ 40(a) $C_1\cos 4x + C_2\sin 4x$; (b) $\cos 4x - (\sin 4x)/2$
41(a) $C_1e^x + C_2e^{-3x} + (xe^x)/4$; (b) $3e^x/16 + 13e^{-3x}/16 + xe^x/4$
42(a) $C_1 + C_2e^{x/8} - (8\sin x)/5 + (\cos x)/5 - x^2 - 16x$;
(b) $-124 + (624e^{x/8})/5 - (8\sin x)/5 + (\cos x)/5 - x^2 - 16x$ 43(a) $e^{3x}[C_1\cos\sqrt{7}x + C_2\sin\sqrt{7}x] + (\sin x)/3$;
(b) $e^{3x}[\cos\sqrt{7}x - 16(\sin\sqrt{7}x)/3\sqrt{7}] + (\sin x)/3$ 44(a) $C_1\cos 2x + C_2\sin 2x - (x\cos 2x)/4 + (\sinh 2x)/8$;
(b) $\cos 2x - \sin 2x - (x\cos 2x)/4 + (\sinh 2x)/8$ 45(a) $(C_1 + C_2x + x^2/2)e^{2x}$; (b) $(1 - 4x + x^2/2)e^{2x}$
46(a) $C_1\cos x + C_2\sin x - (x^2\cos x)/4 + (3x\sin x)/4$; (b) $\cos x - 2\sin x - (x^2\cos x)/4 + (3x\sin x)/4$
50 $a_0[1 + (4x)/1! + (4x)^2/2! + (4x)^3/3! + \cdots]$
51 $a_0[1 + x/1! + x^2/2! - x^3/3! + 5x^4/4! - 55x^5/5! + 19 \cdot 55x^6/6! - \cdots]$ 52 1.827485

Appendix B

Exercises, Appendix B, p. A.19–A.21

1(a) $\pi/60$; (b) $7\pi/6$; (c) $11\pi/6$; (d) $-3\pi/4$; (e) $(360/\pi)°$; (f) $150°$; (g) $420°$; (h) $-120°$ 2(a) $147\pi/8$;
(b) $7(8 + 3\pi)/4$ 3(a) -1; (b) -1; (c) -2; (d) -0.5; (e) 0.24740; (f) -0.54869; (g) -1.5237; (h) 0.52337
4(a) $2\pi, 2$; (b) $2\pi, 1/3$; (c) $\pi/2, 3$; (d) $4\pi, 4$ 5 The periods are given: (a) $\pi/2$; (b) 6π: (c) $\pi/3$; (d) π
6(a) $-5/13$; (b) $-12/13$; (c) $-16/65$; (d) $63/65$; (e) $-56/65$; (f) $33/65$; (g) $-24/25$; (h) $119/169$; (i) $5/\sqrt{26}$;
(j) $-1/\sqrt{26}$ 13 $\alpha = \cos^{-1}(37/51) \approx 43°29'25''$, $\beta \approx 77°9'37''$, $\gamma \approx 59°20'57''$ 14 $a \approx 16.217$, $\beta \approx 4°8'8''$,
$\gamma \approx 158°51'52''$

Index

A

Absolute convergence, 650ff.
  and convergence, 651
  ratio test for, 652
Absolute value, 13ff.
  properties of, 14, 16–17
Acceleration, 83, 180, 249, 820
  normal and tangential components, 849ff.
  units of measurement, 83
Acceleration vector, 820
Algebra
  formulas from, A.38–A.40
  of functions, 48
  topics from, 1–64, 601, 604
Algebraic function, 43
  derivative of, 169
Alternating series, 645ff.
Amplitude of function, A.13
Analytic geometry, 381–444
Angle
  between lines, 39
  between vectors, 757
  radian measure, A.6
  reference, A.14–A.15
Antiderivative, 242ff. (*see also* Integral)
  definition, 243
  general 243, 288
  of vector-valued functions, 815–816
  rules for, 244, 288
Approximate (numerical) integration, 294ff., 612
Approximate solution (differential equation), 1121ff.
Approximating (Riemann) sum, 269, 294–295, 932, 964, 978
Approximation formula, 152ff., 154, 876ff., 878, 881
Approximations, accuracy of, 301
Arc length, 334ff., 729, 819, 1019
  and improper integrals, 561
  as parameter, 693
  curves defined parametrically, 336
  formula for, 335, 339, 729, 819, 1019
  of sector of circle, A.8
  polar curve, 729
  space curve, 819
Arc length function, 339–340, 833ff.
Area, 257ff., 306ff., 937ff.
  and improper integrals, 561, 567
  by approximating sums, 259ff.
  definition, 308
  double integral for, 937
  line integral for, 1044, 1047
  of parallelogram, 780
  of sector of circle, A.8
  polar coordinates, 724ff.
  surface (*see* Surface area)
  under normal curve, 576
  under parabola, 298–299
Arithmetic progression, A.40
Asymptote, 125ff., 382ff.
  horizontal, 128, 382, 384
  oblique, 382, 384, 385
  vertical, 125, 384
    location of, 125, 384
Asymptotic equality, 382
Average (mean), 569
Average value, 569–570

B

Babylonian square root algorithm, 241
Ballistic problem (*see* Projectile problem)
Base (logarithm), 446
Bernoulli equation, 1096
Bernoulli, Jacques, 1096
Bernoulli, Jean, 547
Bessel function, 1120
Binomial series, 681
Binomial Theorem, 681, A.40
Boundary condition, 471, 1080
Boundary point, 864, 908, 920
Boyle's law, 871
Business problems, 215, 228ff.

C

CRC Standard Mathematical Tables, 611
Cable, 541ff., 831–832
Calculus, early development, 78–79, 88–89
Cardioid, 713–714
Catenary, 541
Cauchy, A. L., 88, 650
Cauchy's Theorem, 548
Center
  of curvature, 840
  of ellipse, 408
  of hyperbola, 417
  of mass, 357, 360, 366, 955, 985
Centimeter-gram-second system, 483
  conversion to foot-pound-second, 486
Centroid, 369, 376, 955–956, 985
  of boundary curve, 376
  orbit of, 373
Chain Rule, 158ff., 883ff.
  and differentials, 162–163
  applications of, 162
  for partial derivatives, 884
  proof of, 159, 883
Characteristic equation, 1100
Circle, 23–26
  area and arc length, A.8
  equation, 23
  general form, 24
  in polar coordinates, 719–720
  involute of, 766
  of curvature, 838
  three point, 38
Circulation, 1073
Comparison tests, 638ff., 641
Completeness axiom, 625
Components of vector, 734, 749, 760ff.
Composite function, 47ff.
  derivative of, 158ff.
  limit of, 118–119, 862, 866
Compound interest, 459, 475ff.
  effective rate, 476
  instantaneous compounding, 459

Concavity, 218ff.
  and tangent lines, 223
Conditional covergence, 650ff.
Cone, 798–799
Conic section, 395–396, 430ff., 720ff.
  identification by eccentricity, 426
  in polar coordinates, 720ff.
Conjugate axis, 417
Constraints, 921
Continuity, 104ff., 812, 862, 866, 1013
  and differentiability, 110, 880
  and integrability, 269, 933
  at endpoint of domain, 108–109
  equation of, 1067
  of inverse functions, 183, 188–189
  of vector-valued function, 812
  on closed interval, 109
  on open interval, 108
Continuous function, 104ff., 812, 1013
  and differentiable function, 110, 880
  and extrema, 115
  properties of, 106–107, 110, 113ff.
Contour curve, 794–795
Contour drawing, 794
Convergence
  absolute, 650ff.
  conditional, 650ff.
  interval of, 660–661
  of improper integral, 557, 559, 563
  of sequence, 617, 625
  of series, 619–620, 626ff.
  radius of, 660–661
Cooling, Newton's law, 475
Coordinate axes, 19, 315, 742–743
Coordinate plane, 19, 742–743
Cosecant function, 1136–1138
  derivative of, 500
  integral of, 504
Cosine function, A.9, A.11
  derivative of, 497
  integral of, 497
Cosines, law of, A.19
Cotangent function, A.9, A.12
  derivative of, 500
  integral of, 504
Cramer's rule, 806, 809
Critical points, 197, 910, 916
  boundary, 916
  interior, 910
Cross product, *see* Vector product
Curl, 1056ff.
  interpretation of, 1057, 1072ff.
Curvature, 834ff.
  center of, 840
  circle of, 838, 844–845
  formulas for, 835, 837, 840, 850
  in polar coordinates, 840
  plane curve, 834

  radius of, 838
  space curve, 846–847
Curve, 337, 687ff.
  closed, 687–688
  induced orientation, 1068
  on cylinder, 791
  piecewise smooth, 1027
  simple, 687–688
  smooth, 333, 337, 687–688, 790
  space, 790, 819
  standard orientation, 1040
Curves, family of, 1086
Curve sketching, 220–221, 392
  in parametric equations, 700
  in polar coordinates, 709ff.
Cycloid, 690–691
  area and arc length, 731
Cylinder, 788ff.
  and space curves, 791
  right, 789
Cylindrical coordinates, 990ff.
  and rectangular coordinates, 991
  triple integrals, 997ff.

D

Damping effect, 512
Decreasing function, 182ff., 202ff.
Definite integral, 269ff. (*see also* Integral)
  and substitution principle, 291
  definition, 269, 274
  of vector-valued function, 816
  properties of, 274ff.
Demand function, 229
DeMoivre's Theorem, 1107
Depreciation, 481
Derivative (*see also* the particular functions)
  and continuity, 110
  and polar graphs, 711
  definition, 73, 77, 178–179
  directional, 890ff., 894
  formulas, inside cover
  higher order, 178ff., 698
  interpretation of, 86
  of vector-valued functions, 813–814
  partial, 867ff.
  parametric equations, 696–698
  symbols for, 75, 179
Derivative test
  first, 206
  second, 225, 912
Determinant, 805ff.
  properties, 807–808
  second-order, 805
  third-order, 807
Difference quotient, 77

Differentiable function, 73, 879
  and continuity, 110, 880
Differential, 155ff., 880–881, 887
  and chain rule, 162–163
  and increment, 155, 906
  and substitution principle, 292–293
  and tangent plane, 905–906
  exact, 1025ff., 1086
  formulas for, 156, 887
Differential equation, 249, 251ff., 471ff., 1079ff.
  boundary condition, 471
  exact, 1086ff.
  existence theorems for, 1084–1085
  first-order homogeneous, 1083, 1086
  first-order linear, 1092ff., 1096
  general solution, 1080
  initial condition, 471
  linear, 1091ff.
  order of, 1079
  particular solution, 1080
  second order linear, 1099ff.
    homogeneous, 1099ff., 1103ff., 1117
    nonhomogeneous, 1108ff., 1114ff.
  series solution, 1117ff.
  solution of, 252, 1079–1080
Differential operator, 1097–1099
Differentiation (*see also* Derivative)
  implicit, 165ff.
  logarithmic, 468
Directed distance, 361ff., 363
Direction angle, 740–741, 751–753
Direction cosine, 753
Directional derivative, 890ff.
Directrix
  and focus, 396–428
  of ellipse and hyperbola, 428
  of parabola, 396
Discontinuous function, 94, 105ff.
Discontinuity, 105ff.
  infinite, 106
  jump, 106
  removable, 105
Discriminant of second-degree equation, 439
Discriminant test, 440–441
Disk method, 318ff.
Distance
  between lines, 787
  between points, 20, 744
  line to a point, 39, 765–766
    directed, 361ff., 363
  plane to point, 774, 776
Distance formula, 19–21, 744
Distance (position) function, 81, 249
Distributive Property, 264
Divergence
  of improper integral, 557, 559, 563
  of sequence, 617, 625, 627, 628

Index

of series, 619–620, 626ff.
of vector field, 1055ff.
Divergence theorem, 1059, 1060ff.
  physical interpretation, 1066ff.
Domain of function, 39, 40, 857
  unspecified, 41, 857
Dot product, *see* Inner Product
Double-angle identities, A.17
Double integral, 933ff., 962ff.
  polar coordinates, 962ff.
Dyne, 483

E

e, number, 454
Eccentricity, 424ff., 426
  and similar conics, 425
  identification of conics, 426
  of parabola, 426
Element
  of area, 311, 351, 368
  of volume, 319, 329
Ellipse, 396, 405ff.
  area, 413
  construction, 413
  definition, 405
  properties, 407ff.
  standard equation, 406
Ellipsoid, 797–798
Endpoint of interval, 3
Epicycloid, 693–694
Equipotential surface, 904
Equivalue surface, 904
Euler, L., 1104
Euler's identity, 674, 1104
Euler's method (differential equations) 1121ff.
Exact differential, 1025ff., 1086
  conditions for, 1026, 1086
Exponent
  integral, 445
  properties of, 447, 461
  rational, 445
  real, 456–457, 461
Exponential function, 445 (*see also* Natural exponential function)
  general derivative and integral formulas, 467
Extremal problems
  steps in solution, 211–212
Extremum, 113ff., 192ff., 909ff.
  and continuity, 115
  and derivative, 194–195, 910
  local, 194ff., 909
  steps for finding, 197

F

Factor theorem, 604
Factorial, 460, A.40
Stirling's formula, 460, A.40
Falling body problem, 84ff., 249ff. (*see also* Projectile problem)
  with air resistance, 484ff., 540
Family of terms (partial fractions), 601
Fourier analysis, 592
First derivative test, 206
Fluid flow, 1035ff., 1055ff.
Fluid pressure problems, 348ff.
  and centroid, 371ff.
Flux, 1037ff., 1051, 1053
Focus (foci)
  and directrix, 428
  of ellipse, 406
  of hyperbola, 415
  of parabola, 396
Foot-pound-second system, 486
  conversion to metric system, 486
Force diagram, 755
Fraction, decimal, 619, 645
Function, 39ff.
  algebraic, 43
  Bessel, 1120
  composition of, 47ff., 118–119
  constant (and zero derivative), 203
  continuous, 104ff., 812, 862, 866
  decreasing, 182ff., 188, 202ff.
  demand, 229
  difference of, 48
  differentiable, 73, 879
  discontinuous, 94, 105ff.
  domain, 39, 40, 857
  equality of, 48
  explicit, 165
  exponential, 445, 454
  graph of, 43, 858
  harmonic, 874
  hyperbolic, 529ff.
  identity, 55
  implicit, 165
  increasing, 182ff., 188, 202ff.
  integrable, 270
  inverse, 53ff.
  inverse hyperbolic, 534ff.
  inverse trigonometric, 515ff.
  linear, 919
  natural exponential, 455ff.
  natural logarithm, 448ff.
  of several variables, 857ff.
  one-to-one, 56, 182ff.
  periodic, 493, 1138
  polynomial, 42, 600
  product of, 48
  quotient of, 48
  range, 39, 40
  rational, 43, 599ff.
  root, 184–185
  sum of, 47–48
  transcendental, 43
  trigonometric, 492ff., 1135ff.
  vector-valued, *see* Vector-valued function
Fundamental theorem of calculus, 283ff.
  proof of, 283–284

G

General power rule, *see* Power rule
Generating line for cylinder, 789
Geometric series, 619, 628
  convergence of, 628
Geometric progression, A.40
Geometric vector, 735, 750
Geometry, formulas from, A.41–A.42
Gradient, 892ff., 895
  and contour curves, 898–899
  and directional derivative, 893, 895
  and normal vector, 903
  properties of, 896ff.
Graph
  of equation, 22, 745
  of parametric equations, 337, 687
  of polar equation, 703, 709ff.
  of vector-valued function, 811
Gravitational constant, 483
Greatest integer function, 44–45
Greatest lower bound, 625
Green, George, 1041
Green's theorem, 1041ff., 1059
Growth law
  restricted (inhibited), 479ff., 482
  unrestricted, 478

H

Half-angle identities, A.18
Hanging cable problem, 541ff., 831–832
Harmonic function, 874
Harmonic motion, simple, 507ff., 511ff.
  differential equation for, 509, 511
Harmonic series, 626–627
  alternating, 630
Heat, 979
Helix, 792–793
Heron's formula, 914
Hodgman, C. D., 611
Hooke's law, 344
Hyperbola, 396, 414ff.
  asymptotes of, 417
  conjugate, 433
  construction of, 418
  definition, 414

equilateral, 423
properties, 416ff.
standard equation, 415, 418–419
Hyperbolic function, 529ff.
  derivatives, 530, 532
  graphs, 530
  identities involving, 531, 533
  integrals, 532
  table of, A.27–A.28
Hyperbolic paraboloid, 801–802
Hyperboloid
  one sheet, 799–800
  two sheets, 800–801
Hypocycloid, 693–694

I

IQ (intelligence quotient), 577
Image (function), 40
Implicit differentiation, 165ff., 886–887
  and higher order derivatives, 180–181
  and related rates, 171
  steps in, 165
Improper integral, 557ff., 563ff.
  and concepts defined by integrals, 561, 567
  convergent and divergent, 557, 559, 563
  discontinuous integrand, 557ff., 566
  over infinite intervals, 563ff.
Inclination, angle of, 30
Increasing function, 182ff., 202ff.
  and continuity, 182ff.
  and sign of derivative, 203
Increment, 28, 151, 875
  and differential, 155, 906
Indefinite integral, 278, 288
  and derivative, 286
  of vector-valued function, 815–816
Indeterminate form, 547, 550
  exponential forms, 554
  form 0/0, 547
  form ∞/∞, 550
  form $0 \cdot \infty$, 553
  form $\infty - \infty$, 553
Inequality, 4, 7–19
  equivalent, 4
  in one variable, 4, 7–19
  involving absolute value, 13–19
  involving fractions, 10
  linear, 7
  quadratic, 8
  solution, 4
  solution set, 4
Infinite limit, 120ff.
  at infinity, 128ff.
  rules for, 122, 124
Infinite series, see Series
Inflection, point of, 219
Initial condition, 471, 1080
Initial point of vector, 734
Inner product, 757ff.
  and angle between vectors, 757
  and orthogonal vectors, 759
  properties of, 759
Integrable function, 270
  and continuity, 269
Integral
  and substitution, 290–292
  double, 933ff., 937, 941ff., 962ff.
  general formulas, 288ff., 937
  improper, 557ff.
  iterated, 940ff., 979ff.
  line, 1017ff., (see also Line integral)
  of vector-valued function, 815–6
  partial, 939
  surface, 1048ff., (see also Surface integral)
  triple, 978ff.
  table of, 611–612, inside cover, A.31ff.
Integral function (indefinite integral), 278
Integral test, 631, 634–635
Integrand, 288
Integrating factor, 1088ff., 1091
Integration, also see Techniques of Integration
  by parts, 581ff., 584
  numerical, 294ff., 612
  order of, 945, 980
Intelligence quotient (IQ), 577
Interest, see Compound interest
Interior point, 907, 920
Intermediate Value Theorem, 116
Intersection of sets, 7
Interval, 3
  closed, 3
  finite, 3
  half open, 3, 4
  infinite, 3, 4
  of convergence, 660–661
  open, 3
Invariant, 439
Inverse function, 53ff., 183ff., 188ff.
  calculation of, 57
  continuity of, 183, 187
  derivative of, 183, 188
  graph of, 58ff.
Inverse hyperbolic functions, 534ff.
  and logarithms, 535–536
  derivatives of, 535, 538
  integrals involving, 538–539
Inverse trigonometric functions, 515ff.
  derivatives of, 518ff.
  integrals involving, 526ff.
Involute of circle, 766
Irreducible polynomial, 600
Isothermal surface, 904

K

Kinetic energy, 958ff., 988

L

Lagrange, J. L., 921
Lagrange multiplier, 921, 925
Lagrange's formula, 678
Lagrange's method, 921ff., 923
Lamina, see also Moment
  homogeneous, 363ff.
  mass of, 366, 951
Laplacian, 901, 1065
Laplace's equation, 874
Larsen, H. D., 611
Law of cosines, A.19
Law of sines, A.19
Learning curve, 1094–1095
Least squares, 915
Least upper bound, 625
Leibniz, G. W., 78–79, 662
Less than or equal, 3
Level curve, 794–795
Level surface, 904
l'Hôpital, Marquis G. de, 547
l'Hôpital's rule, 547ff., 553ff., (see also Indeterminate form)
Limaçon, 715–716, 718
Limit, 65ff., 88ff., 120ff., 617, 858ff.
  along paths, 862–863
  at infinity, 126ff.
  definition, 90
  evaluated by series, 671
  formulas for, inside front cover
  infinite, 120ff., 128ff.
  of composite function, 118–119
  of function of several variables, 858ff., 865
  of root function, 98
  of sequence, 617
  of series, 619
  of vector-valued function, 811–812, 817
  one-sided, 67, 93, 94, 99
  positive and negative, 101
  properties of, 96ff.
Limit theorem, 96, 99, 127
  618, 861
  for functions of two variables, 861
  for limits at infinity, 127
  for one-sided limits, 99

Index

for sequences, 618
proof of, 135ff.
Limits of integration, 269
  variable, 277ff.
Line
  angle of intersection, 39
  distance to point, 39, 361ff., 363, 765–766
  equations of, 33–36, 693, 768–769
  general form, 35
  in polar coordinates, 718–719
  intercept form, 38
  normal, 71
  of regression, 915
  parametric equations of, 693, 768
  perpendicular, 32–33
  point-slope form, 33–34
  slope of, 29–33
  slope-intercept form, 35
  symmetric form, 769
  tangent, 68ff.
  vertical, 36
Line integral, 1017ff.
  and definite integral, 1025
  and exact differential, 1025ff.
  definition, 1018
  dependence of path, 1021, 1025
  different parametrizations, 1024
  independence of path, 1025ff.
  over closed curves, 1028, 1032
  over space curves, 1023–1024, 1029ff.
  properties of, 1020ff.
Linear dependence and independence, 1102
Linear inequality, 7
Local maximum, minimum, extremum, see Extremum, local
Logarithm, 446ff., A.23ff. (see also Natural logarithm)
  conversion to different base, 467
  properties of, 447, A.37, A.39
Logarithmic differentiation, 468
LORAN, 424
Lower bound, 625

M

Maclaurin, Colin, 667
Maclaurin series, 666ff.
Major axis, 408
Marginal analysis, 229ff.
Marginal cost, profit, revenue, 229ff.
Mass, 485
  of lamina, 366, 951
  of surface, 1049
  of wire, 1019
  triple integral for, 978

Mathematical induction, A.1ff.
Maximum, 113ff., 192ff., 909ff.
  and continuity, 115
  and derivative, 194–195, 910
  local, 194ff., 909
  steps for finding, 197
Mean (average), 569
  of function, 570
  of normally distributed data, 575
Mean value theorem
  for derivatives, 199ff.
  for integrals, 279–280
Metric system, 483
Midpoint formula, 21–22, 27, 747, 763–764
Minimum, 113ff., 192ff., 909ff.
  and continuity, 115
  and derivative, 194–195, 910
  local, 194ff., 909
  steps for finding, 197
Minor axis, 408
Mobius strip, 1075
Moment, 356ff., 953ff., 983ff.
  about arbitrary line, 361–362
  about axis, 359, 365–366, 369, 954
  about coordinate plane, 983–984
  about origin, 356–357
  and improper integrals, 561, 567
  line integral for, 1047
  of finite system, 356ff., 359
  of lamina, 365ff., 953ff.
  of plane region, 369, 956
  of solid, 983ff.
Moment of inertia, 957ff., 986ff.
Monotone sequence, 624ff.

N

Napier, John, 446
Natural exponential function, 455ff.
  derivative of, 458
  integral of, 464
  properties of, 455, 458
  table of, A.27–A.28
Natural logarithm function, 448ff.
  definition, 448
  derivative of, 448, 462
  properties of, 449ff.
  table of, A-23ff.
Neighborhood, 859
Newton, Isaac, 78–79, 235, 241
Newton's laws
  cooling, 475
  gravitational attraction, 483
  motion, 482
Newton's method for roots, 234ff., 238, 241–242
  accuracy, 238ff., 241–242

Norm of partition, 267, 931, 962, 977
Normal line, 71
Normal probability density function, 575ff., 967–969
  area under graph, table, 576
  standard, 578
Normal vector, 770, 843, 846, 902
  and gradient, 903
  lower, 1053
  outer, 1036, 1053
  upper, 1050–1052
  to curve, 843, 846, 903
  to surface, 902, 1036
Notation
  for derivative, 75, 179
  for partial derivative, 867–868, 871
  for vector, 735, 751
Numerical integration, 294ff.

O

One-sided limit, 67, 93–94, 99
  and limit theorem, 99
  definition, 94
One-to-one function, 56, 182ff.
  and continuity, 182ff.
Optimal lot size formula, 232
Orbit of centroid, 373
Order, 1–6
  properties of, 1–2
Orthogonal surfaces, 907
Orthogonal trajectories, 1086
Orthogonality, 759

P

π (number), 662–663
Pappus, 374
  theorems of, 374, 377
Parabola, 396ff.
  area under, 289–299
  by paper folding, 405
  definition, 396
  reflection property, 400, 404
  standard equation, 396, 399
  uniqueness of shape, 399
Paraboloid
  elliptical, 796–797
  hyperbolic, 801–802
Parallelepiped, volume of, 781
Parallelogram
  area of, 780
  diagonals of, 764–765
Parameter, 336
Parametric equations, 336ff., 686ff.
  and arc length, 339
  and curve sketching, 700

and derivatives, 694ff.
and polar equations, 708
and rectangular equations, 688ff.
and vector-valued functions, 811
Partial derivative, 867ff.
  and tangent line, 869
  equality of mixed, 873, 875
  higher order, 871ff.
  symbols for, 867–868
Partial fractions, 598ff.
  general procedure, 601–602
Partial sums, sequence of, 618, 664
Partition, 266, 270, 931, 977
  inner, 931, 962, 977
Pendulum, 1107
Periodic function, 493, A.12
Plane, 770ff.
  angle between, 783
  coordinate, 19, 742–743
  definition, 770
  equations of, 770–771
  tangent, 903
Plane region, *see also* Region
  centroid of, 369, 956
  moment of, 369, 956
Planimeter, 375
Point-slope form, 33
Polar axis, 702
Polar coordinates, 702ff.
  and rectangular coordinates, 704–705, 965
  double integrals, 962ff.
Polar equation
  and parametric equations, 708
  area bounded by, 724ff., 962ff.
  graph of, 703, 706, 709ff.
  symmetry, 709ff.
Polynomial, 42, 600
  derivative of, 143
  Factor Theorem, 604
  irreducible and reducible, 600
Population growth, 222, 477ff.
  restricted (inhibited) law, 479ff., 482
  unrestricted, 478
Position (distance) function, 81
Position vector, 735, 810
Positive term series, 624
Potential energy, 1035
Potential function, 1015
Power rule for derivatives, 142, 147ff.
  fractional exponents, 167ff.
  general, 147–148
  negative exponents, 148
Power series, 657ff.
  convergence, 658
  derivative and integral, 661
  properties of, 659
Probability, 571ff., 575ff.

Probability density function, 571ff., 575ff.
  normal, 575ff., 578
Product identities, A.18
Product rule for derivatives, 145, 815
Projectile problem, 825ff.
  with air resistance, 828ff.
Projection
  of point on plane, 784, 787
  of vector on another, 761
  of vector on plane, 784
  on axis, 309, 390
  determination of, 391
p-series, 635
Pythagorean identities, A.16
Pythagorean theorem, 20

Q

Quadrant, 19
Quadratic equation, A.39
Quadratic expression (integrals involving), 605ff.
Quadratic inequality, 8
Quadric surface, 794ff.
Quotient rule for derivative, 146, 815

R

Radian measure, 1132
  conversion to degree measure, 1132, A.31
Radioactive decay, 472–473
Radius of convergence, 660–661
Radius of curvature, 638
Radius of gyration, 960–961, 989
Range finding and hyperbola, 421ff.
Range of function, 39–40
Rate of change, 86
Ratio test, 643, 652
  absolute convergence, 652
Rational function, 43, 599ff.
  and partial fractions, 599ff.
  of sines and cosines, 608ff.
Rearrangement of series, 654ff.
Rectangular coordinates, 19, 742–743
  and cylindrical coordinates, 991
  and polar coordinates, 704–705
  and spherical coordinates, 993
Recursion formula, 621
Reducible and irreducible polynomials, 600
Reduction formulas, 583
Reference angle, 1140–1141
Reflection property
  of ellipse, 411, 413
  of parabola, 400, 404

Region, 908–909, 920, 930
  basic, 930
  bounded, 908, 920
  closed, 908, 920
  solid, 920
Related rates, 171ff.
  and implicit differentiation, 171
  steps in solving problems, 171
Relative change, 158
Relative maximum, minimum, extremum, *see* Extremum, local
Remainder terms (Taylor series), 675
Riemann sum, 269, 294–295, 932, 964, 978
Rinehart Mathematical Tables, 611
Rolle, Michel, 199
Rolle's theorem, 199–200
Root function
  derivative of, 168, 184–185
  limit of, 98
Rose, n-leaved, 716–717
Rotation of axes, 434ff.
  and second-degree equations, 437ff.
Rotation of vector field, 1074
Ruling of surface, 789

S

SAT scores, 577
Saddle point, 802
Saddle surface, 802
Sandwich theorem, 100–101, 139–140
Scalar, 737, 751
Scalar product, *see* Inner product
Secant function, A.10, A.12
  derivative of, 500
  integral of, 504
Second-degree equation
  and rotation of axes, 437ff.
Second derivative test, 225
Separation of variables, 471ff., 1082ff.
Sequence, 616ff.
  convergent and divergent, 617
  increasing and decreasing, 624
  limit of, 617
  limit theorem for, 618
  monotone, 624ff.
  of partial sums, 618, 664
  subscript notation, 616
Series, 618ff.
  absolute and conditional convergence, 650ff.
  alternating, 645ff.
  alternating harmonic, 630
  accuracy of approximation, 634–635, 646, 649, 679–680
  binomial, 681
  bounds for limit, 634–635, 646, 649, 679–680

Index

convergence and divergence, 619
  tests for, 622–623, 626, 628, 631, 635–636, 638, 641, 643, 646, 652, 658
  for e, 665
  for π, 662–663
  geometric, 619, 628
  harmonic, 626–627
  Maclaurin, 666ff.
  positive term, 624, 626
  power, 657ff. (*see also* Power series)
  properties of, 623
  p-series, 635
  subseries, 654
  Taylor, 666ff.
  to evaluate limits, 671
Serpentine, 824
Set building notation, 4
Shell method, 325ff.
Side conditions, 921
Sigma notation, 263
Simpson's rule, 298–300
Sine function, A.9, A.11
  derivative of, 496
  integral of, 497
Sines, law of, A.19
Sink, 1039, 1053
Slope of line, 29–33
  geometrical interpretation, 30–31
  of perpendicular lines, 32–33
  properties of, 31–33
Smooth curve, 333, 337
Solid, 976
  basic closed, 1060
Solid of revolution, 318
  and theorems of Pappus, 373ff., 377
  surface area of, 377, 975
  volume of, 318ff., 325ff., 373ff.
Solution
  of differential equation, 252, 1079–1080
    general, 1080
    particular, 1080
    singular, 1080
  of inequality, 4
Solution set of inequality, 4
Source, 1039, 1053
Speed, 83, 754
  and derivative, 820
Sphere, 745
Spherical coordinates, 993ff.
  and rectangular coordinates, 993
  triple integrals, 1003ff.
Spring problems
  oscillating, 507ff., 510ff.
    damping effect, 512, 514
    effect of air resistance, 512
    effect of gravity, 512–514
  work, 344ff.

Square root, 14
  Babylonian algorithm, 241
Standard deviation, 575
Statistics, applications to, 569ff.
Stirling's formula, 460, A.40
Stokes, G. G., 1068
Stokes' theorem, 1059, 1069ff.
  in plane, 1059
Subseries, 653–654
Substitution principle, 290ff.
  and differentials, 292–293
  problems with, 290–291
  with definite integral, 291
Sum and difference identities, A.16
Summation (sigma) notation, 263
Sums from algebra, 263–264, A.40
Surface, oriented, 1068, 1075
Surface, area, 377, 971ff., 1049
  and surface integral, 1049
Surface integral, 1048ff.
Symbols
  for derivative, 75, 179
  for partial derivative, 867–868, 871
  for vector, 735
Symmetry, 386ff., 709ff.
  about line $y = x$, 388
  about origin, 388, 389, 710
  about x-axis, 388, 709–710
  about y-axis, 387, 710, 717
  tests for, 388, 710

T

Tables of integrals, inside covers, A.31ff.
  use of, 611–612
Tangent function, A.9, A.12
  derivative of, 500
  integral of, 503
Tangent line, 688ff.
  and concavity, 223
  at origin (polar coordinates), 712–713
  definition, 70
  point of tangency, 69
  to surface, 902
  vertical, 176ff.
Tangent plane, 903ff.
  and differential, 905–906
Tangent surfaces, 907
Tangent vector, 819, 841, 846
Taylor, Brook, 666
Taylor series, 666ff.
  accuracy of, 679–680
  table of, 670
Taylor's formula, 675ff.
Techniques of integration, 290ff., 581ff.

integration by parts, 581ff., 584
involving quadratic expressions, 605–606
miscellaneous substitutions, 607ff.
partial fractions, 598ff.
powers of trigonometric functions, 586ff.
substitution principle, 290ff.
trigonometric substitution, 594ff.
Term of sequence, 616
Terminal point of vector, 734
Three-dimensional coordinate system, 315
Three-point test for extrema, 210
Trace, 771, 794
Transcendental function, 43
Transitive property of order, 2
Translation of axes, 400ff., 795
Transverse axis, 417
Trapezoidal rule, 295–296, 302–303
Triangle
  medians of, 788
  standard labeling, A.19
Triangle inequality, 14, 19, 763
  second, 19
  vector form, 763
Trichotomy property of order, 2
Trigonometric function, 492ff., A.6ff., A.9ff., *see also* the particular functions
  derivatives of, 496ff.
  graphs of, A.11–A.12
  integrals of, 497, 503–504, 586ff., 592
  of standard angles, A.11
  periods of, A.12–A.13
  substitutions involving, 594ff.
  table of, 1137, A.29–A.30
Trigonometric identities, A.16ff.
  double angle, A.17
  half angle, A.18
  product, A.18
  Pythagorean, A.16
  sum and difference, A.16
Trigonometric substitutions, 594ff.
  table of, 595
Trigonometry
  historical remark, 492
  review of, A.6ff.
Triple integral, 978ff.
  cylindrical coordinates, 997ff.
  spherical coordinates, 1003ff.

U

Undetermined coefficients, 1110ff.
Union of sets, 7
Upper bound, 624
  for sequence, 625
Utility, 924

V

Variation of parameters, 1114ff, 1117
Vector, 734ff., 749ff.
  acceleration, 820
  addition, 736, 750
  components of, 734, 749
  direction angles, 740–741, 751–753
  direction cosines, 753
  equality of, 736, 749
  inner (dot, scalar) product, 757ff.
  magnitude, 737–738, 741, 752
  multiplication by scalar, 737, 751
  normal, 770, 843, 846, 902
  notations for, 735, 740
  orthogonal, 759, 785–786
  properties of, 739, 751, 759
  tangent, 819
  three-dimensional, 749ff.
  two-dimensional, 734ff.
  unit, 738ff.
  unit normal N, 843, 846
  unit tangent T, 841, 846
  vector (cross) product, 776ff.
  velocity, 820
Vector field, 1012ff.
  gradient (conservative), 1015
  gravitational, 1013
  inverse-square, 1016
  velocity, 1013
Vector (cross) product, 776ff.
  magnitude of, 779
  properties of, 777, 779ff.
Vector-valued function, 810ff.
  continuous, 812
  definition, 810
  derivative formulas, 813–814
  derivative of, 813
  graph of, 811
  limit of, 811
Velocity, 82ff., 180, 249, 820
  angular, 822
  average, 81
  units of measurement, 82
  vector, 820
Vertex
  of ellipse, 408
  of hyperbola, 417
  of parabola, 396
Vertical tangent line, 176ff.
  at endpoint of domain, 177
  possible locations for, 178
Volume, 315ff.
  and improper integral, 561, 567
  and theorem of Pappus, 373ff.
  by slicing, 316ff.
  disk method, 318ff, 331
  double integral for, 936, 949
  of parallelepiped, 781
  of solid of revolution, 318ff., 373ff.
  shell method, 325ff., 331
  triple integral for, 978ff.

W

Weierstrass, Karl, 88
Weight and mass, 485
Whispering gallery, 411
Witch of Agnesi, 731
Work, 343ff., 1033ff.
  definition, 344
  line integral for, 1034–1035

Z

Zero vector, 734, 750

Integrals

(A constant of integration should be added to each integral)
This list is continued in Table V.

FUNDAMENTAL FORMS

I1 $\quad \int (u \pm v)\,dx = \int u\,dx \pm \int v\,dx$

I2 $\quad \int k u\,dx = k \int u\,dx$

I3 $\quad \int u^n\,du = \dfrac{u^{n+1}}{n+1} \quad (n \neq -1)$

I4 $\quad \int \dfrac{1}{u}\,du = \ln |u| \quad (u(x) \neq 0)$

I5 $\quad \begin{cases} \int e^u\,du = e^u \\[4pt] \int a^u\,du = \dfrac{1}{\ln a} \cdot a^u \quad (a > 0) \end{cases}$

I6 $\quad \int \ln u\,du = u \ln u - u$

I7 $\quad \int \cos u\,du = \sin u$

I8 $\quad \int \sin u\,du = -\cos u$

I9 $\quad \int \sec^2 u\,du = \tan u$

I10 $\quad \int \csc^2 u\,du = -\cot u$

I11 $\quad \int \sec u \tan u\,du = \sec u$

I12 $\quad \int \csc u \cot u\,du = -\csc u$

I13 $\quad \int \tan u\,du = \ln |\sec u|$

I14 $\quad \int \cot u\,du = \ln |\sin u|$

I15 $\quad \int \sec u\,du = \ln |\sec u + \tan u|$